Fully Solved

3500 MCQs

Useful For

ESE | GATE | PSUs

Electrical Engineering

MADE EASY
Publications

MADE EASY
Publications

MADE EASY Publications

Corporate Office: 44-A/4, Kalu Sarai (Near Hauz Khas Metro Station), New Delhi-110016
E-mail: infomep@madeeasy.in
Contact: 011-45124660, 0-8860378007
Visit us at: www.madeeasypublications.org

3500 Multiple Choice Questions for ESE, GATE, PSUs : Electrical Engineering

First Edition: 2016
Second Edition: 2017
Third Edition: 2018
Reprint : 2019

Reprint : 2020

Published by: MADE EASY Publications, New Delhi-110016

PREFACE

B. Singh (Ex. IES)

It gives me great happiness to introduce the **Revised Edition** on Electrical Engineering containing nearly 3500 MCQs which focuses in-depth understanding of subjects at basic and advanced level which has been segregated topicwise to disseminate all kind of exposure to students in terms of quick learning and deep apt. The topicwise segregation has been done to align with contemporary competitive examination pattern. Attempt has been made to bring out all kind of probable competitive questions for the aspirants preparing for ESE, GATE & PSUs. The content of this book ensures threshold level of learning and wide range of practice questions which is very much essential to boost the exam time confidence level and ultimately to succeed in all prestigious engineer's examinations. It has been ensured from MADE EASY team to have broad coverage of subjects at chapter level.

While preparing this book utmost care has been taken to cover all the chapters and variety of concepts which may be asked in the exams. The solutions and answers provided are upto the closest possible accuracy. The full efforts have been made by MADE EASY Team to provide error free solutions and explanations.

I have true desire to serve student community by way of providing good sources of study and quality guidance. I hope, this book will be proved an important tool to succeed in competitive examinations. Any suggestions from the readers for the improvement of this book are most welcome.

B. Singh (Ex. IES)
Chairman and Managing Director
MADE EASY Group

CONTENTS

UNIT 1

Electromagnetic Theory

1. Vector Analysis

Q.1 If $\vec{P} = x^2y^2\vec{i} + (x-y)\vec{k}$, $\vec{Q} = zx\vec{i}$ and $\phi = xy^2z^3$, then match **List-I** with **List-II** and select the correct answer using the codes given below the lists:

List-I	List-II
A. Div. $\vec{Q}$	1. $y^2z^3\vec{i} + 2yxz^0\vec{j} + 3z^2y^2x\vec{k}$
B. Grad ϕ	2. $-\vec{i} + \vec{k}x^2$
C. Curl $\vec{P}$	3. z

Codes:

	A	B	C
(a)	1	2	3
(b)	2	1	3
(c)	3	1	2
(d)	3	2	1

Q.2 The unit vector $\vec{a}_R$ which points from $z = h$ on the z-axis towards $(r, \phi, 0)$ in cylindrical co-ordinates as shown below is given by

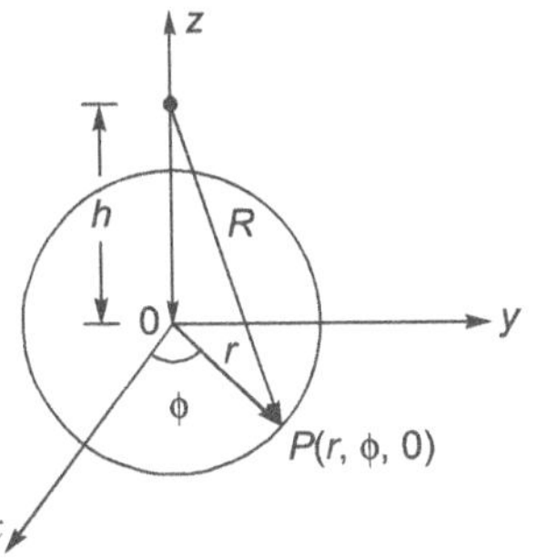

(a) $\dfrac{h\vec{a}_r - r\vec{a}_z}{\sqrt{r^2 + h^2}}$

(b) $\dfrac{r\vec{a}_r - h\vec{a}_z}{\sqrt{r^2 + h^2}}$

(c) $\dfrac{h\vec{a}_\phi - r\vec{a}_z}{\sqrt{r^2 + h^2}}$

(d) $\dfrac{r\vec{a}_z - h\vec{a}_\phi}{\sqrt{r^2 + h^2}}$

Q.3 If the vector V given below is irrotational, then the values of a, b and c will be respectively

$V = (x + 2y + az)\vec{i} + (bx - 3y - z)\vec{j} + (4x + cy + 2z)\vec{k}$

(a) $a = 4$, $b = 2$ and $c = -1$
(b) $a = 2$, $b = -1$and $c = 4$
(c) $a = 4$, $b = -1$ and $c = 2$
(d) $a = 2$, $b = 4$ and $c = -1$

Q.4 Match **List-I (Vector Identities)** with **List-II (Equivalent expression)** and select the correct answer using the codes given below the lists:

List-I

A. $(\vec{A}\times\vec{B})\cdot(\vec{C}\times\vec{D})$

B. $\vec{A}\times(\vec{B}\times\vec{C})$

C. $(\vec{A}\times\vec{B})\times(\vec{C}\times\vec{D})$

List-II

1. $(\vec{A}\cdot\vec{C}\cdot\vec{D})\vec{B} - (\vec{B}\cdot\vec{C}\cdot\vec{D})\vec{A}$

2. $[(\vec{A}\cdot\vec{C})(\vec{B}\cdot\vec{D}) - (\vec{A}\cdot\vec{D})(\vec{B}\cdot\vec{C})]$

3. $(\vec{A}\cdot\vec{C})\vec{B} - (\vec{A}\cdot\vec{B})\vec{C}$

Codes:

	A	B	C
(a)	1	3	2
(b)	3	1	2
(c)	2	1	3
(d)	2	3	1

Q.5 The vector differential operator, Del(∇) in spherical co-ordinate system is given by

(a) $\nabla = \vec{a}_r\dfrac{\partial}{\partial r} + \vec{a}_\theta\dfrac{1}{r}\dfrac{\partial}{\partial\theta} + \vec{a}_\varphi\dfrac{1}{r\sin\theta}\dfrac{\partial}{\partial\varphi}$

(b) $\nabla = \vec{a}_r\dfrac{1}{r}\dfrac{\partial}{\partial r} + \vec{a}_\theta\dfrac{1}{r\sin\theta} + \vec{a}_\varphi\dfrac{\partial}{\partial\varphi}$

(c) $\nabla = \vec{a}_r\dfrac{\partial}{\partial r} + \vec{a}_\theta\dfrac{\partial}{\partial\theta} + \vec{a}_\varphi\dfrac{1}{r\cos\theta}\dfrac{\partial}{\partial\varphi}$

(d) $\nabla = \vec{a}_r\dfrac{1}{r}\dfrac{\partial}{\partial r} + \vec{a}_\theta\dfrac{1}{r}\dfrac{\partial}{\partial\theta} + \vec{a}_\varphi\dfrac{1}{r\cos\theta}\dfrac{\partial}{\partial\varphi}$

Q.6 **Assertion (A):** Divergence of a vector function $\vec{A}$ at each point gives the rate per unit volume at which the physical entity is issuing from that point.

Reason (R): If some physical entity is generated or absorbed within a certain region of the field, then that region is known as source or sink respectively and if there are no sources or sinks in the field, the net outflow of the incompressible physical entity over any part of the region is zero. However, the net outflow is said to be positive, if the total strength of the sources are greater than the total strength of sink and vice-versa.

(a) Both A and R are true and R is a correct explanation of A.
(b) Both A and R are true but R is not a correct explanation of A.
(c) A is true but R is false.
(d) A is false but R is true.

Q.7 Which of the following identity is not true?

(a) $\vec{A}(\vec{B}\cdot\vec{C}) = (\vec{A}\cdot\vec{C})\vec{B} - (\vec{A}\cdot\vec{B})\vec{C}$
(b) $\nabla\cdot(\nabla\times\vec{A}) = 0$
(c) $\nabla\times\nabla\varphi \neq 0$
(d) None of the above

Q.8 The vector $\vec{A}$ directed from $(2, -4, 1)$ to $(0, -2, 0)$ in Cartesian coordinates is given by

(a) $-2\vec{a}_x + 2\vec{a}_y + \vec{a}_z$ (b) $-2\vec{a}_x + 2\vec{a}_y - \vec{a}_z$
(c) $-\vec{a}_x - 2\vec{a}_y + 2\vec{a}_z$ (d) $\vec{a}_x - 2\vec{a}_y - \vec{a}_z$

Q.9 What is the value of $\iint_s \vec{F}\cdot d\vec{s}$, where $\vec{F} = 4xz\vec{i}_1 - y^2\vec{i}_2 + yz\vec{i}_3$?

Here, s is the surface bounded by $x = 0$, $x = 1$, $y = 0$, $y = 1$, $z = 0$, $z = 1$ and $\vec{i}_1, \vec{i}_2, \vec{i}_3$ are unit vectors along x, y and z axes respectively.

(a) 1/2 (b) 5/2
(c) 2 (d) 3/2

Q.10 The vector field given by

$$\vec{A} = yz\vec{a}_x + xz\vec{a}_y + xy\vec{a}_z \text{ is}$$

(a) rotational and solenoidal
(b) rotational but not solenoidal
(c) irrotational and solenoidal
(d) irrotational but not solenoidal

Q.11 If $\vec{A} = \dfrac{\vec{a}_x}{\sqrt{x^2+y^2}}$, then the value of $\nabla\cdot\vec{A}$ at (2, 2, 0) will be

(a) –0.0884 (b) 0.0264
(c) –0.0356 (d) 0.0542

Q.12 If $\vec{r} = x\vec{i} + y\vec{j} + z\vec{k}$, then the value of $\vec{i}\times(\vec{r}\times\vec{i}) + \vec{j}\times(\vec{r}\times\vec{j}) + \vec{k}\times(\vec{r}\times\vec{k})$ is

(a) $\vec{r}$ (b) $2\vec{r}$
(c) $3\vec{r}$ (d) $6\vec{r}$

Q.13 What is the value of constant b so that the vector

$$\vec{V} = (x+3y)\vec{i} + (y-2x)\vec{j} + (x+bz)\vec{k}$$

is solenoidal?

(a) 2 (b) –1
(c) 3 (d) –2

Q.14 Match **List-I** with **List-II** and select the correct answer using the codes given below the lists:

List-I
A. Gauss's divergence theorem
B. Stroke's theorem
C. The divergence
D. The curl

List-II
1. $\nabla\cdot\vec{A}$
2. $\oint_L \vec{A}\cdot d\vec{l} = \iint_s (\nabla\times\vec{A})\cdot d\vec{s}$
3. $\iint_s \vec{A}\cdot d\vec{s} = \iint_s \vec{A}\cdot\vec{e}\,d\vec{s}$ ($\vec{e}$ - An unit vector)
4. $\nabla\times\vec{A}$

Codes:

	A	B	C	D
(a)	3	2	4	1
(b)	2	3	1	4
(c)	3	2	1	4
(d)	2	3	4	1

Q.15 **Assertion (A):** Vector differential operator is a vector quantity and it signifies that certain operations of a differentiation are to be carried out on the scalar function following it.

Reason (R): Vector differential operator posses properties similar to ordinary vectors.

(a) Both A and R are true and R is a correct explanation of A.
(b) Both A and R are true but R is not a correct explanation of A.
(c) A is true but R is false.
(d) A is false but R is true.

Q.16 Consider the following statements:

1. Divergence of a vector function $\vec{A}$ at each point gives the rate per unit volume at which the physical entity is issuing from that point.
2. If a vector function ϕ represents temperature, then grad. ϕ or $\nabla\phi$ will represents rate of change of temperature with distance.
3. The curl of a vector function A gives the measure of the angular velocity at every point of the vector field.

Which of the above statements is/are correct?

(a) 2 and 3 only (b) 1, 2 and 3
(c) 1 and 3 only (d) 2 only

Q.17 **Assertion (A):** The Gauss's divergence theorem permits us to express certain integrals by means of surface integrals.

Reason (R): Gauss's divergence theorem states that "the surface integral of the curl of a vector field taken over any surface s is equal to the line integral of the vector field around the closed periphery (contour) of the surface.

(a) Both A and R are true and R is a correct explanation of A.
(b) Both A and R are true but R is not a correct explanation of A.
(c) A is true but R is false.
(d) A is false but R is true.

Q.18 Match **List-I (Physical quantities)** with **List-II (Dimensions)** and select the correct answer using the codes given below the lists:

List-I	List-II
A. Electric potential	1. $MT^{-2}I^{-1}$
B. Magnetic flux	2. $ML^2T^{-3}I^{-1}$
C. Magnetic field intensity	3. IL^{-1}
D. Magnetic flux density	4. $ML^2T^{-2}I^{-1}$

Codes:

	A	B	C	D
(a)	2	4	3	1
(b)	4	2	3	1
(c)	1	2	1	3
(d)	4	2	1	3

Q.19 Which of the following statements is not true regarding vector algebra?

(a) Dot product of like unit vector is unity.
(b) Dot product of unlike unit vector is zero.
(c) Cross product of two like unit vectors is a third unit vector having positive sign for normal rotation and negative for reverse rotation.
(d) All the above statements are true.

Q.20 A rigid body is rotating with an angular velocity of ω where, $\vec{\omega} = \omega_x\vec{i} + \omega_y\vec{j} + \omega_z\vec{k}$ and v is the line velocity. If $\vec{r}$ is the position vector given by $\vec{r} = x\vec{i} + y\vec{j} + z\vec{k}$, then the value of curl $\vec{v}$ will be equal to

(a) $\frac{1}{2}\omega$ (b) ω

(c) $\frac{1}{3}\omega$ (d) 2ω

Q.21 If $\vec{r} = x\vec{I}_x + y\vec{I}_y + z\vec{I}_z$, then which of the following relation will hold true?

(a) $\nabla\vec{r} = 3$ (b) $\nabla\times\vec{r} = 0$
(c) Both (a) and (b) (d) Neither (a) nor (b)

Q.22 If $\vec{E}$ is any vector field in cartesian co-ordinates system, then

(a) $\nabla.(\nabla\times\vec{E}) = \nabla\times\nabla\times\vec{E} - \nabla^2\vec{E}$

(b) Div. curl $\vec{E} = 0$

(c) $\nabla.(\nabla\times\vec{E}) = \nabla^2\vec{E} - \nabla\times\nabla\times\vec{E}$

(d) Div. curl $\vec{E} \neq 0$

Q.23 If $\vec{c} = \vec{a}\times\vec{b}$ and $\vec{b} = \vec{a}\times\vec{c}$, then

(a) $\vec{b} = 0$ and $\vec{c} = 0$ (b) Only $\vec{b} = 0$
(c) Only $\vec{c} = 0$ (d) $\vec{b} \neq 0$ and $\vec{c} \neq 0$

Q.24 If S is any closed surface enclosing a volume V and $\vec{A} = ax\vec{i} + by\vec{j} + cz\vec{k}$, then the value of $\iint_S \vec{A}\cdot\hat{n}d\vec{s}$ ($\hat{n}$ is a unit vector) will be equal to

(a) $\frac{1}{3}(a+b+c)V$ (b) $(a-b-c)V$

(c) $\frac{1}{2}(a+b+c)V$ (d) $(a+b+c)V$

Q.25 **Assertion (A):** The laplacian operator of a scalar function ϕ can be defined as "Gradient of the divergence of the scalar ϕ".

Reason (R): Laplacian operator may be a "scalar laplacian" or a "vector laplacian" depending upon whether it is operated with a scalar function or a vector, respectively.

(a) Both A and R are true and R is a correct explanation of A.
(b) Both A and R are true but R is not a correct explanation of A.
(c) A is true but R is false.
(d) A is false but R is true.

Q.26 Match **List-I (Terms)** with **List-II (Type)** and select the correct answer using the codes given below the lists:

List-I	List-II
A. Curl $(\vec{F}) = 0$	1. Laplace equation
B. Div $(\vec{F}) = 0$	2. Irrotational
C. Div grad $(\phi) = 0$	3. Solenoidal
D. Div div $(\phi) = 0$	4. Not defined

Codes:

	A	B	C	D
(a)	2	3	1	4
(b)	4	1	3	2
(c)	2	1	3	4
(d)	4	3	1	2

Q.27 Which of the following relations are not correct?
(a) $[B \times C, C \times A, A \times B] = [ABC]^2$
(b) $A \times [B \times (C \times D)] = B \cdot D (A \times C) - B \cdot C (A \times D)$
(c) $(B \times C) \cdot (A \times D) + (C \times A) \cdot (B \times D) + (A \times B) \cdot (C \times D) = 0$
(d) $(A \times B)^2 = A^2 B^2 - (A \cdot B)^2$

Q.28 If $uF = \nabla v$, where u and v are scalar fields and F is a vector field, then F. curl F is equal to
(a) zero
(b) $\frac{\nabla^2 v}{u^2}$
(c) $\frac{(\nabla v \cdot \nabla) v}{u^2}$
(d) not defind

Q.29 Which of the following option is not correct?
(a) A vector field $\vec{A}$ is solenoid, if $\nabla \cdot \vec{A} = 0$
(b) A vector field $\vec{A}$ is irrotational, if $\nabla \times \vec{A} = 0$
(c) A vector field V is harmonics, if $\nabla^2 V \neq 0$
(d) All options are correct

Q.30 Which of the following statements is not true of a phasor?
(a) It may be a scalar or a vector.
(b) It is a time dependent quantity.
(c) It is a complex quantity.
(d) All are true

Q.31 A scalar function, V is given by $V = xyz^2$. The gradient of V is given by
(a) $xz^2 \hat{a}_x + 2xyz \hat{a}_y + xz^2 \hat{a}_z$
(b) $yz^2 \hat{a}_x + xz^2 \hat{a}_y + xyz \hat{a}_z$
(c) $2xyz \hat{a}_z + yz^2 \hat{a}_y + xz^2 \hat{a}_z$
(d) $yz^2 \hat{a}_x + xz^2 \hat{a}_y + 2xyz \hat{a}_z$

Q.32 The scalar potential is given by
$$V = (x^2 - y^2 - z^2) \text{ Volts}$$
The laplacian of V is
(a) 0
(b) –2
(c) 1
(d) –1

2. Electrostatics

Q.33 What is the ratio of electrostatic repulsion and gravitational attraction between two electrons? Given:
Q = Charge on an electron = 1.6×10^{-19}C
G = Gravitation constant = 6.67×10^{-11} Nm²/kg²
m = Mass of an electron = 9.1×10^{-31} kg
(a) 3.23×10^{42}
(b) 4.17×10^{42}
(c) 2.33×10^{42}
(d) 1.47×10^{42}

Q.34 What is the electric field strength at a distance of 200 mm from a charge of 2×10^{-6} Coulomb in vacuum?
(a) 450 kV/m
(b) 236 kV/m
(c) 525 kV/m
(d) 328 kV/m

Q.35 Two point charges of 3×10^{-9} C and -2×10^{-9} C are spaced two meters apart. What is the electric field at a point which is one meter from each of the two point charges?
(a) 11 V/m
(b) 5 V/m
(c) 9 V/m
(d) 6 V/m

Q.36 Five equal point charges of $Q = 20 \times 10^{-9}$ C, are placed at x = 2,3,4,5 and 6 cm. The electric potential at origin will be
(a) 416 V
(b) 128 V
(c) 325 V
(d) 261 V

Q.37 What is the equivalent capacitance of a parallel plate capacitor having two dielectrics, $\epsilon_{r1} = 1.5$ and $\epsilon_{r2} = 3.5$, each comprising one-half the volume as shown in figure below with $A = 2\ m^2$ and $d = 1$ mm?

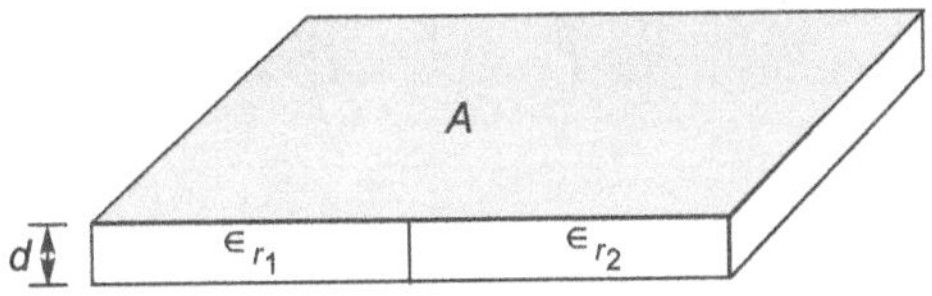

(a) 44.27 nF (b) 24.47 nF
(c) 27.44 nF (d) 42.47 nF

Q.38 Four point charges 1, –2, –3 and 4 μC are located on the x-axis at x = 1, 2, 3 and 4 metre respectively. The energy stored in the field is
(a) –2 Joule (b) 4 Joule
(c) 0 Joule (d) –1 Joule

Q.39 The electric field strength at a point in front of an infinite sheet of charge is
(a) independent of the distance of the point from the sheet
(b) inversely proportional to the distance of the point from the sheet
(c) inversely proportional to the square of distance of the point from the sheet
(d) none of the above

Q.40 **Assertion (A):** If all the points in space which have the same potential are jointed, then equipotential surfaces are obtained.
Reason (R): The field and the equipotential surfaces are orthogonal to each other.
(a) Both A and R are true and R is a correct explanation of A.
(b) Both A and R are true but R is not a correct explanation of A.
(c) A is true but R is false.
(d) A is false but R is true.

Q.41 Consider the following statements associated with various medias encountered in electostatics:
1. An isotropic media is one whose properties are independent of direction
2. A homogeneous media is one whose physical characteristics vary from point-to-point.
3. A non-homogenous media is one whose physical characteristics do not vary from point-to-point.
4. A linear media is a one in which the electric flux density is proportional to the electric field intensity.

Which of the above statements are correct?
(a) 2 and 3 (b) 3 and 4
(c) 1, 2 and 4 (d) 1 and 4

Q.42 **Assertion (A):** Continuity of current equation is derived from law of conservation of charge.
Reason (R): Continuity of current equation states that there can be no accumulation of charge at any point.
(a) Both A and R are true and R is a correct explanation of A.
(b) Both A and R are true but R is not a correct explanation of A.
(c) A is true but R is false.
(d) A is false but R is true.

Q.43 Match **List-I** with **List-II** and select the correct answer using the codes given below the lists:

List-I	List-II
A. Joule's law	1. $\vec{J} = \sigma\vec{E}$
B. Ohm's law	2. $\nabla \cdot \vec{J} = \frac{\partial \rho}{\partial t}$
C. Solenoidal current	3. $W = I^2Rt$
D. Continuity equation	4. $\iint_s \vec{J}d\vec{s} = 0$

Codes:

	A	B	C	D
(a)	4	2	3	1
(b)	3	1	2	4
(c)	3	1	4	2
(d)	4	3	1	2

Q.44 **Assertion (A):** An electric field produces no migration of charge in a dielectric.
Reason (R): The permittivity of a dielectric is always less than the permittivity of vaccum.
(a) Both A and R are true and R is a correct explanation of A.
(b) Both A and R are true but R is not a correct explanation of A.
(c) A is true but R is false.
(d) A is false but R is true.

Q.45 Which one of the following pairs is NOT correctly matched?

(a) Gauss theorem: $\oint_s \vec{D}\cdot d\vec{s} = \oint_s \nabla\cdot\vec{D}\,dV$

(b) Gauss's law: $\oint \vec{D}\cdot d\vec{s} = \int \rho\, dV$

(c) Coulomb's law: $V = -\dfrac{d\phi_m}{dt}$

(d) Stoke's theorem: $\oint_L \vec{E}\cdot d\vec{l} = \iint_s (\nabla\times\vec{E})\cdot d\vec{s}$

Q.46 Match **List-I (Law/Quantity)** with **List-II (Mathematical expression)** and select the correct answer using the codes given below the lists:

List-I	List-II
A. Gauss's law	1. $\nabla\cdot\vec{D} = \rho$
B. Ampere's law	2. $\nabla\times\vec{E} = -\dfrac{\partial\vec{B}}{\partial t}$
C. Faraday's law	3. $s = \vec{E}\times\vec{H}$
D. Poynting vector	4. $\vec{F} = q(\vec{E}\times\vec{v}\vec{B})$
	5. $\nabla\times\vec{H} = \vec{J}_c + \dfrac{\partial\vec{D}}{\partial t}$

Codes:

	A	B	C	D
(a)	1	2	4	3
(b)	3	5	2	1
(c)	1	5	2	3
(d)	3	2	4	1

Q.47 What is the value of total electric flux coming out of a closed surface?

(a) Zero
(b) Equal to volume charge density
(c) Equal to the total charge enclosed by the surface
(d) Equal to the surface charge density

Q.48 Which one of the following is the Poisson's equation for a linear and isotropic but in homogeneous medium?

(a) $\nabla^2 \nabla = -\rho/\epsilon$ (b) $\vec{\nabla}\cdot(\epsilon\,\vec{\nabla}V) = -\rho$

(c) $\vec{\nabla}\cdot\vec{\nabla}(\epsilon\, V) = -\rho$ (d) $\nabla^2 V = -\rho/\epsilon$

Q.49 Two concentric spherical shells carry equal and opposite uniformly distributed charges over their surfaces as shown in figure below.

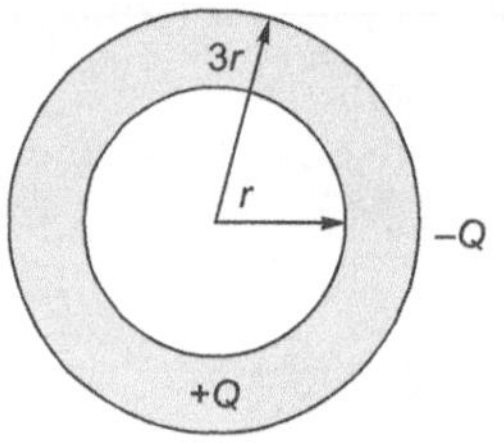

Electric field on the surface of inner shell will be

(a) zero (b) $\dfrac{Q}{4\pi\epsilon_0 r^2}$

(c) $\dfrac{Q}{8\pi\epsilon_0 r^2}$ (d) $\dfrac{Q}{16\pi\epsilon_0 R^2}$

Q.50 Electric field intensity due to line charge of infinite length is

(a) $\dfrac{\rho_L}{2\pi\epsilon r}$ (b) $\dfrac{\rho_L}{4\pi\epsilon r}$

(c) $\dfrac{\rho_L}{\pi\epsilon r}$ (d) $\dfrac{2\rho_L}{\pi\epsilon r}$

Q.51 Consider the following statements associated with equipotential surface:

1. Potential is same everywhere.
2. No current flows on this surface.
3. Work done in moving charge from one point to another is zero.
4. Potential is different everywhere.

Which of the above statements is/are not correct?

(a) 1 and 3 only (b) 3 and 4 only
(c) 4 only (d) 2 and 4 only

Q.52 A finite sheet $0 \le x \le 1$, $0\ y \le 1$ on the $z = 0$ plane has a charge density

$$\rho_s = xy\,(x^2 + y^2 + 25)^{3/2}\ \text{nC/m}^2$$

The total charge on the sheet would be

(a) $\left(\dfrac{(25)^{7/2} - (27)^{7/2} - 2(26)^{7/2}}{25}\right)$ nC

(b) $\left(\dfrac{(26)^{7/2} - 2(25)^{7/2} + (27)^{7/2}}{35}\right)$ nC

(c) $\left(\dfrac{(26)^{7/2} - (27)^{7/2} + 2(25)^{7/2}}{25}\right)$ nC

(d) $\left(\dfrac{(27)^{7/2} + (25)^{7/2} - 2(26)^{7/2}}{35}\right)$ nC

Q.53 **Assertion (A):** Gauss's law is an alternative statement of Coulomb's law.
Reason (R): Proper application of the divergence theorem to Coulomb's law results in Gauss's law.
(a) Both A and R are true and R is a correct explanation of A.
(b) Both A and R are true but R is not a correct explanation of A.
(c) A is true but R is false.
(d) A is false but R is true.

Q.54 If the potential $V = \frac{10}{r^2}\sin\theta\cos\phi$, then the work done in moving a 10 μC charge from point $A(1, 30°, 120°)$ to $B(4, 90°, 60°)$ would be equal to
(a) 16.65 μJ (b) 28.125 μJ
(c) 25.25 μJ (d) 30.625 μJ

Q.55 Three point charges –1 nC, 4 nC and 3 nC are located at (0, 0, 0) , (0,0,1) and (1,0,0,) respectively. To total energy contained in the system would be
(a) 22.25 nJ (b) 15.56 nJ
(c) 12.84 nJ (d) 13.37 nJ

Q.56 A wire of diameter 1 mm and conductivity 5×10^7 S/m has 10^{29} free electrons /m^3 when an electric field of 10 mV/m is applied.
The current density and the drift velocity of the electrons will be respectively given by (take charge on an electrons $e = -1.6 \times 10^{-19}$ C)
(a) 500 kA/m^2 and 3.125×10^{-5} m/s
(b) 300 kA/m^2 and 2.225×10^{-5} m/s
(c) 500 kA/m^2 and 2.225×10^{-5} m/s
(d) 300 kA/m^2 and 3.125×10^{-5} m/s

Q.57 The equivalent capacitance between the terminals *A* and *B* for the capacitor shown below would be (the area occupied by each dielectric is 30 cm^2)

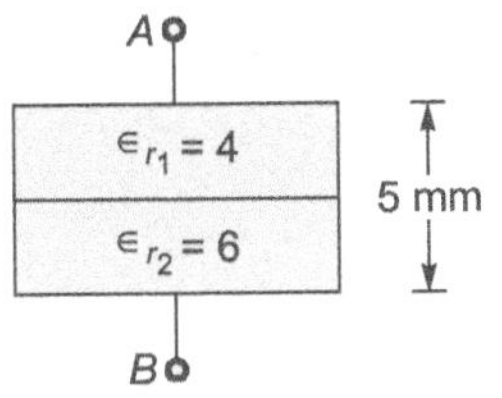

(a) 26.53 pF (b) 28.68 pF
(c) 25.46 pF (d) 22.22 pF

Q.58 The cylindrical coordinates equation
$$\frac{\partial^2\psi}{\partial\rho^2} + \frac{1}{\rho}\frac{\partial\psi}{\partial\rho} + \frac{\partial^2\psi}{\partial z^2} + 10 = 0$$ is called
(a) Laplace's equation
(b) Poisson's equation
(c) Larentz's equation
(d) Maxwell's equation

Q.59 Image theory is applicable to problems involving
(a) electrostatic field only
(b) magnetostatic fields only
(c) both electrostatic and magnetostatic fields
(d) neither electrostatic nor magnetostatic fields

Q.60 The potential field at any point in space containing a dielectric material of relative permittivity 3 is given by
$V = (5x^2y + 3yz^2 + 6xz)$ Volts, where x, y, z are in meters. The magnetude of volume charge density at point (1, 2, 3) will be given by
(a) $48\,\epsilon_0$ (b) $96\,\epsilon_0$
(c) $144\,\epsilon_0$ (d) $192\,\epsilon_0$

Q.61 If $\vec{A}$ and $\vec{J}$ are the vector potential and current density vectors associated with a coil, then $\int_V \vec{A}\cdot\vec{J}\,dV$ has the unit of
(a) flux-linkage (b) power
(c) energy (d) inductance

Q.62 In the infinite plane, y = 6 m, there exists a uniform surface charge density of (1/600π) μC/m^2. The associated electric field strength is
(a) $30\hat{i}$ V/m (b) $30\hat{j}$ V/m
(c) $30\hat{k}$ V/m (d) $60\hat{i}$ V/m

Q.63 Consider the following statements:
1. Electric field intensity at any point is the negative of the potential gradient at that point.
2. Electric field due to a finite line of charge at a point is independent of the length of the line of charge.
3. The direction of the electric field is the direction in which the gradient is greatest.
4. Electric field due to an infinite line of charge at a point is inversely proportional to square of the distance of the point from the infinite line charge.

Which of the above statements is/are correct?
(a) 2, 3 and 4 only (b) 1, 2 and 4 only
(c) 3 only (d) 1 and 3 only

Q.64 A positive charge of magnitude 2 μC is situated in air at the origin of a rectangular co-ordinate system and a second positive charge of magnitude 10 μC is situated on the positive z-axis at a separation of 0.30 meter from the origin as shown below.

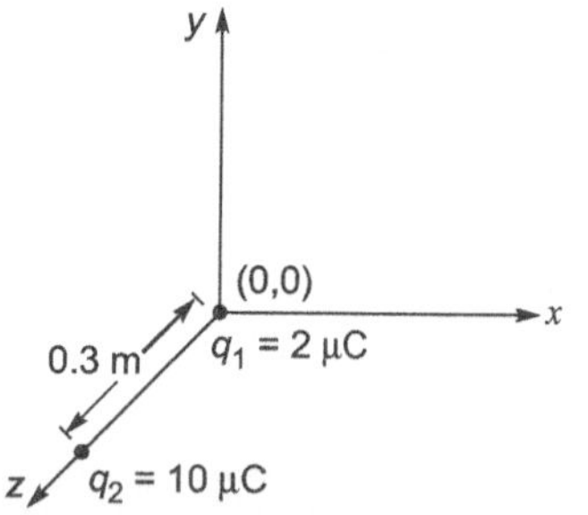

The force on the second charge is
(a) 2 N (b) 4 N
(c) 6 N (d) $\frac{1}{2}$ N

Q.65 What is the flux crossing a 1 m^2 area that is normal to the x-axis at $x = 3$ m?

(Given, $\vec{D} = 10x\vec{a}_x$ C/m^2)
(a) 15 Coulombs (b) 20 Coulombs
(c) 25 Coulombs (d) 30 Coulombs

Q.66 What is the value of electric potential at the centre of a square as shown below whose sides are equal to 1.0 meter and charges Q_1, Q_2, Q_3 and Q_4 are respectively + 0.01 μF, –0.02 μF, +0.03 μF and +0.02 μF?

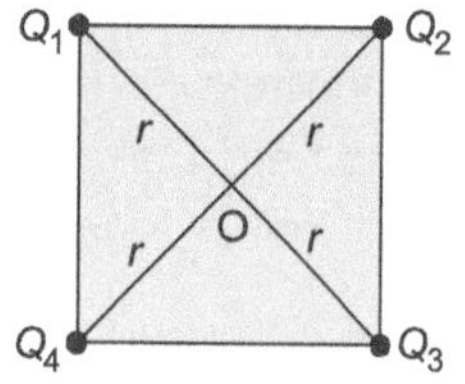

(a) 426.8 V (b) 509.2 V
(c) 625.6 V (d) None of these

Q.67 Conservative properties are applicable only to
(a) electrostatic field
(b) time varying electric field
(c) time varying magnetic field
(d) time varying electric and magnetic field

Q.68 The resultant potential at a distance r from the electric dipole is
(a) inversely proportional to length of the dipole.
(b) directly proportional to dipole moment.
(c) inversely proportional to square of the distance r.
(d) both (b) and (c)

Q.69 **Assertion (A)** Total electric flux out of a closed surface is equal to the net charge within that surface.
Reason (R): Gauss's law states that the total displacement or electric flux through any closed surfaces surrounding the charge is equal to the amount of charge enclosed.
(a) Both A and R are true and R is a correct explanation of A.
(b) Both A and R are true but R is not a correct explanation of A.
(c) A is true but R is false.
(d) A is false but R is true.

3. Magnetostatics

Q.70 The magnitude of H at a radius of 1 meter from a long linear conductor is 1 A/m. The current in the wire is
(a) 2π Ampere (b) $\frac{\pi}{2}$ Ampere
(c) 4π Ampere (d) $\frac{\pi}{4}$ Ampere

Q.71 What is the value of magnetic flux density at the centre of the square current loop shown in figure below?

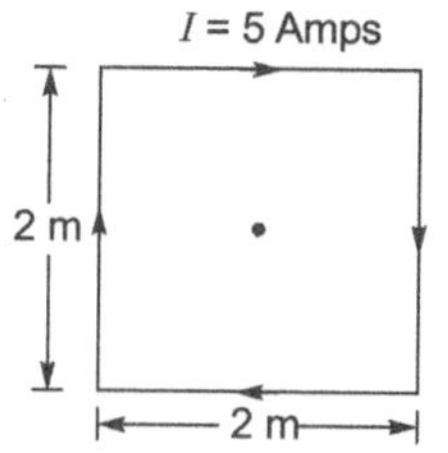

(a) 1.25 mW/m^2 (b) 2.26 μWb/m^2
(c) 2.82 μWb/m^2 (d) 3.43 μWb/m^2

Q.72 A solenoid of 20 cm long and 1 cm diameter has a uniform winding of 1000 turns. If the solenoid is placed in a uniform field of 2 Wb/m^2 flux density and a current of 10 amps. is passed through the solenoid winding, then the maximum torque on the solenoid will be

(a) 10 N-m (b) 20 N-m
(c) 30 N-m (d) 60 N-m

Q.73 What is the value of magnetic flux density at the centre of a current carrying loop when the loop radius is 2 cm, loop current is 1 mA and the loop is placed in air?
(a) 1 μWb (b) 0.1 μWb
(c) 10 μWb (d) 0.01 μWb

Q.74 A magnetic field known to be directed in a cartesian co-ordinate system so that $\vec{H}$ exists in the x-direction where H_x = Constant. The value of curl $\vec{H}$ is
(a) 0 (b) –1
(c) +1 (d) none of these

Q.75 Match **List-I** with **List-II** and select the correct answer using the codes given below the lists:

List-I
A. Faraday's induction law
B. Maxwell equation in point form
C. Electric field strength general equation

List-II
1. $\nabla \times \vec{E}_e = -\dfrac{\partial \vec{B}}{\partial t}$
2. $\oint E_e . d\vec{l} = -\dfrac{d}{dt}\iint_s \vec{B} d\vec{s}$
3. $\vec{E} = -\nabla V - \dfrac{\partial \vec{A}}{\partial t}$
4. $\vec{E} = -\nabla V + \dfrac{\partial \vec{A}}{\partial t}$

Codes:

	A	B	C
(a)	1	2	3
(b)	2	1	4
(c)	1	2	4
(d)	2	1	3

Q.76 The force acting between two parallel wires carrying currents I_1 and I_2 and separated by a distance r is given by
(a) $\dfrac{\mu}{2\pi}\dfrac{I_1 I_2}{r^2}$ (b) $\dfrac{\mu}{4\pi}\dfrac{I_1 I_2}{r}$
(c) $\dfrac{\mu}{2\pi}\dfrac{I_1 I_2}{r}$ (d) $\dfrac{\mu}{4\pi}\dfrac{I_1 I_2}{r^2}$

Q.77 Flux lines are received at an iron-air boundary at an angle of 45° from the normal on the iron side of the boundary. If the iron has a relative permeability of 350, then the angle from the normal with which the flux emerges into the air would be
(a) 0 degree (b) $\tan^{-1}\left(\dfrac{1}{350}\right)$ degree
(c) 45 degree (d) $\tan^{-1}(350)$ degree

Q.78 The unit of magnetic flux is given by
(a) $ML^2 T^{-2} I^{-1}$ (b) $ML^2 T^{-3} I^{-2}$
(c) $ML^2 T^{-1} I^{-1}$ (d) $ML^2 T^{-2} I^{-2}$

Q.79 **Assertion (A):** The potential in case of magnetic field is a vector potential.
Reason (R): The potential in case of electric field is a scalar potential
(a) Both A and R are true and R is a correct explanation of A.
(b) Both A and R are true but R is not a correct explanation of A.
(c) A is true but R is false.
(d) A is false but R is true.

Q.80 A circular conductor of 1 cm radius has an internal magnetic field

$$H = \frac{1}{r}\left(\frac{1}{a^2}\sin ar - \frac{r}{a}\cos ar\right)\vec{i}_\phi \text{ mA/m}$$

where $a = \dfrac{\pi}{2r_0}$, r_0 is the radius of the conductor and $\vec{a}_\phi$ is the unit vector. The total current in the conductor is given by
(a) 0.15 μA (b) 0.50 μA
(c) 0.25 μA (d) 0.75 μA

Q.81 What is the value of mutual inductance between an infinitely long straight conductor along the y-axis and a rectangular single turn coil situated in x-y plane with its corners located at point (a, 0), (a + d, 0), (a, h) and (a + d, h) as shown in figure below?

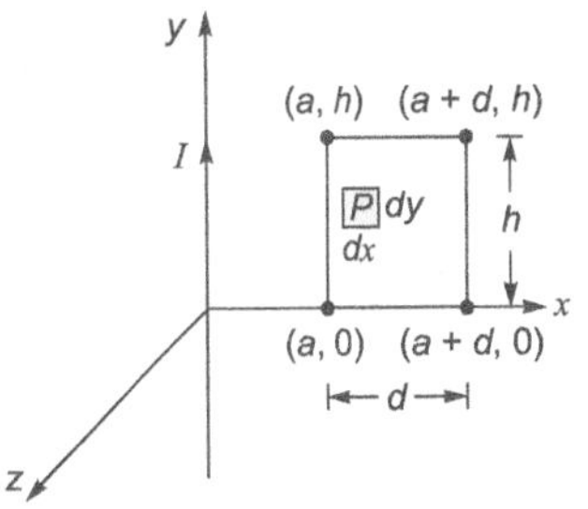

(a) $\frac{\mu h}{4\pi} ln\left(\frac{a+h}{a}\right)$ (b) $\frac{\mu h}{2\pi} ln\left(\frac{a}{a+h}\right)$

(c) $\frac{\mu h}{4\pi} ln\left(\frac{a}{a+h}\right)$ (d) $\frac{\mu h}{2\pi} ln\left(\frac{a+h}{a}\right)$

Q.82 **Assertion (A):** The force acting between two parallel wires carrying currents is directly proportional to the individual currents and inversely proportional to the square of the distance between them.

Reason (R): The force is repulsive if the two currents I_1 and I_2 are in the same directions and attractive if in opposite directions.

(a) Both A and R are true and R is a correct explanation of A.
(b) Both A and R are true but R is not a correct explanation of A.
(c) A is true but R is false.
(d) A is false but R is true.

Q.83 The energy stored in a magnetic field is given by

(a) $\frac{H^2}{2\mu}$ (b) $\frac{\mu H^2}{2}$

(c) $\frac{\mu H}{2}$ (d) none of these

Q.84 Lorentz force law is

(a) $\vec{F} = \vec{v} \times \vec{B}$ (b) $\vec{F} = Q(\vec{E} + \vec{v} \times \vec{B})$

(c) $\vec{F} = Q(\vec{v} \times \vec{B})$ (d) $\vec{F} = Q\vec{E}$

Q.85 The equation $\nabla \times \vec{E} = -\frac{\partial \vec{B}}{\partial t}$ is the generalization of

(a) Faraday's law (b) Biot-Savart's law
(c) Gauss's law (d) Ampere's law

Q.86 The unit of relative permeability is

(a) meter/Henry (b) Henry
(c) Henry/meter (d) it is dimensionless

Q.87 Consider the volume current density distribution in cylindrical co-ordinates as

$$J(r, \phi, z) = 0, 0 < r < a$$

$$= J_0\left(\frac{r}{a}\right)\vec{i}_z, a < r < b$$

$$= 0, b < r < \infty$$

Where a and b are inner and outer radii of the cylinder as shown in figure below.

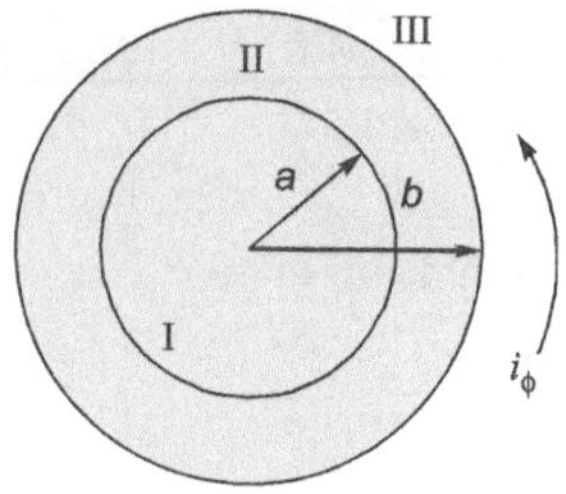

Now, consider the value of magnetic field intensity in various regions I, II and III respectively obtained as:

1. $\vec{H} = 0$ for region-I ($0 < r < a$).
2. $\vec{H} = \frac{J_0}{3ar}(r^2 - a^3)\hat{i}_\phi$ for region-II ($a < r < b$).
3. $\vec{H} = 0$ for region-III ($b < r < \infty$).

Which of the above found values of H are correct for the respective regions?

(a) 1 and 3 only (b) 2 only
(c) 1, 2 and 3 (d) 1 only

Q.88 **Assertion (A):** It is not possible to have isolated magnetic charges.

Reason (R): The magnetic flux lines always close upon themselves.

(a) Both A and R are true and R is a correct explanation of A.
(b) Both A and R are true but R is not a correct explanation of A.
(c) A is true but R is false.
(d) A is false but R is true.

Q.89 Match **List-I** with **List-II** and select the correct answer using the codes given below the lists:

List-I

A. Gauss's law for magnetostatic fields
B. Gauss's law for electrostatic fields
C. Conservativeness of electrostatic fields
D. Ampere's law

List-II

1. $\nabla \cdot V = \rho_v$
2. $\oint B \cdot ds = Q$
3. $\oint B \cdot ds = 0$
4. $\oint_L E \cdot dl = 0$
5. $\oint H \cdot dl = \int_S J \cdot ds$

Codes:

	A	B	C	D
(a)	5	4	1	2
(b)	3	1	4	5
(c)	3	4	1	5
(d)	5	1	4	2

Q.90 The magnetic vector potential is given by

$$A = \frac{-\rho^2}{4} a_z \text{ Wb/m}$$

The total magnetic flux crossing the surface

$\phi = \frac{\pi}{2}, 1 \le \rho \le 2$ m, $0 \le z \le 5$ m is

(a) 3.75 Wb (b) 2.25 Wb
(c) 6.25 Wb (d) 1.75 Wb

Q.91 Which of the following is not a source of magnetostatic fields?
(a) An electric field linearly changing with time
(b) An accelerated charge
(c) A permanent magnet
(d) Both (a) and (b)

Q.92 Consider the following statements associated with the characteristics of static magnetic field:
1. It is solenoidal.
2. Magnetic flux lines are not closed in certain cases.
3. The total number of flux lines entering a given region is equal to the total number of flux lines leaving the region.
4. It is conservative.

Which of the above statements is/are not correct?
(a) 2 only (b) 1 and 4 only
(c) 2 and 4 only (d) 1, 2 and 3 only

Q.93 What are the units for the product of two values whose respective units are Henrys and Amperes?
(a) Ω/s (b) V-s
(c) A^2/s (d) Ω-V

Q.94 What are the possible dimensions for a rectangular coil that has a magnetic flux of 9.5 webers when in a magnetic field of strength 19 Tesla at an angle of 60° from the perpendicular to the plane of the coil?
(a) 360 cm × 30 cm (b) 1000 cm × 100 cm
(c) 80 cm × 70 cm (d) 250 cm × 40 cm

Q.95 What is the change in magnetic flux in a coil of area 5 m^2 as its orientation relative to the perpendicular of a uniform 3.0 T magnetic field changes from 45° to 90°?
(a) –11 Wb (b) 7.5 Wb
(c) 0 Wb (d) 11 Wb

Q.96 If a vector field $\vec{B}$ is solenoidal, which of these is true?

(a) $\oint_L \vec{B} \cdot d\vec{l} = 0$ (b) $\oint_S \vec{B} \cdot d\vec{s} = 0$

(c) $\nabla \times \vec{B} = 0$ (d) $\nabla \cdot \vec{B} = 0$

Q.97 If $\vec{A} = \hat{a}_\rho + \hat{a}_\phi + \hat{a}_z$, the value of $\oint \vec{A} \cdot d\vec{l}$ around the closed circular quadrant shown in the given figure is

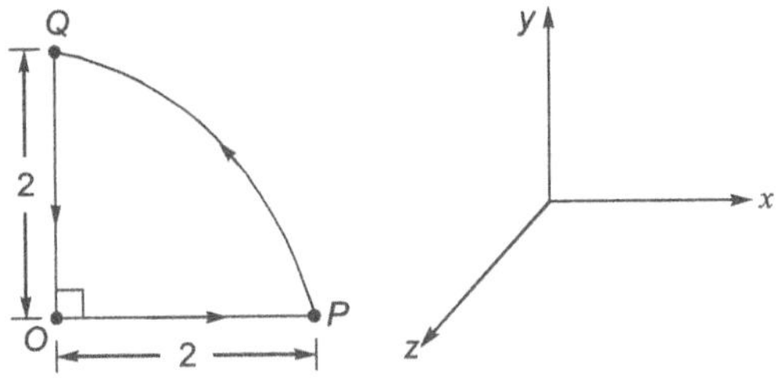

(a) π (b) $\frac{\pi}{2} + 4$

(c) $\pi + 4$ (d) $\frac{\pi}{2} + 2$

Q.98 The Maxwell's equation, $\nabla \times \vec{H} = \vec{J} + \frac{\partial \vec{D}}{\partial t}$ is based on
(a) Gauss's law (b) Ampere's law
(c) Faraday's law (d) Coulomb's law

Q.99 Given a vector field $\vec{A} = 2r\cos\phi\,\hat{a}_r$ in cylindrical coordinates. For the contour as shown below, $\oint \vec{A} \cdot d\vec{l}$ is

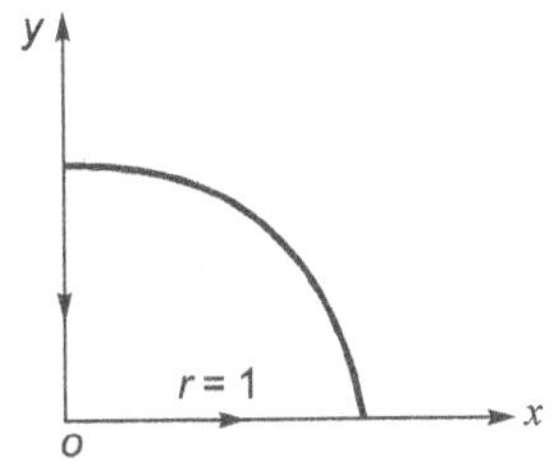

(a) 1 (b) $1-\frac{\pi}{2}$

(c) $1+\frac{\pi}{2}$ (d) −1

Q.100 Match **List-I** with **List-II** and select the correct answer using the codes given below the lists:

List-I	List-II
A. Continuity equation	1. $\nabla\times\vec{H}=\vec{J}+\frac{\partial\vec{D}}{\partial t}$
B. Ampere's law	2. $\vec{J}=\frac{\partial\vec{D}}{\partial t}$
C. Displacement current	3. $\nabla\times\vec{E}=-\frac{\partial\vec{B}}{\partial t}$
D. Faraday's law	4. $\nabla\times\vec{J}=-\frac{\partial\rho_v}{\partial t}$

Codes:

	A	B	C	D
(a)	4	3	2	1
(b)	4	1	2	3
(c)	2	3	4	1
(d)	2	1	4	3

Q.101 In the given diagram, if $\phi = 90°$, and $I_0 = 2$ A, then the total force acting on the coil will be ($i_1(t) = I_0 \cos\omega t$ and $i_2(t) = I_0 \sin(\omega t + \phi)$)

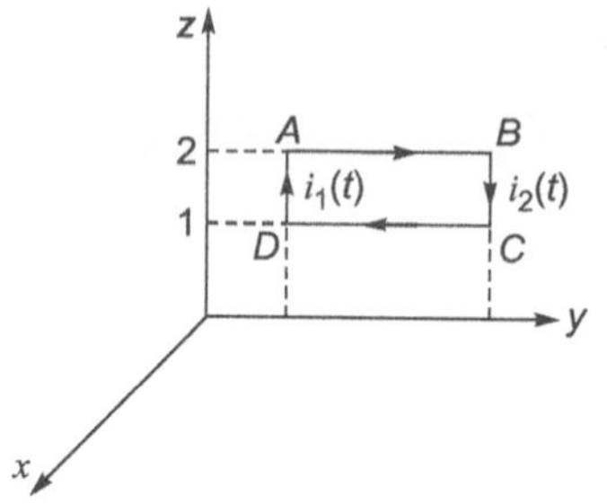

(a) $\frac{\mu_0}{\pi}\cos^2\omega t$ (b) $\frac{\mu_0}{3\pi}\cos^2\omega t$

(c) $\frac{4}{3\pi}\mu_0\cos^2\omega t$ (d) zero

Q.102 What is the value of magnetomotive force generated by a multilayer coil of 100 turns of fine wire carrying a current of 100 mA?

(a) 10 AT (b) 20 AT
(c) 5 AT (d) none of these

Q.103 The unit of $\nabla\times\vec{H}$ is

(a) Ampere (b) Ampere/meter
(c) Ampere/meter^2 (d) Ampere-meter

Q.104 At any surface of discontinuity the following statements in respect of electric and magnetic fields may be stated:

1. The tangential component of $\vec{E}$ is continuous at the surface i.e. it is the same just inside the surface as just outside the surface.
2. The tangential component of $\vec{H}$ is continuous across a surface except at the surface of a perfect conductor.
3. The normal component of $\vec{B}$ is continuous at the surface of discontinuity.
4. The normal component of $\vec{D}$ is continuous if there is no surface charge density ρ_s, otherwise $\vec{D}$ is discontinuous by an amount equal to the surface charge density.

Which of the above statements are correct?

(a) 2, 3 and 4 (b) 1, 2, 3 and 4
(c) 1, 3 and 4 (d) 1, 2 and 3

4. Time Varying EMT Fields

Q.105 If the magnetic field vector $\vec{H}$ has only $\vec{a}_z$ component given by $H_z = 3x\cos\beta + 6y\sin\gamma$ and if the field is invariant with time, what is the expression for current density $\vec{J}$?

(a) $(3\cos\beta)\,\vec{a}_x + (6\sin\gamma)\,\vec{a}_y$
(b) $(6\sin\gamma)\,\vec{a}_x - (3\cos\beta)\,\vec{a}_y$
(c) $(6\cos\beta)\,\vec{a}_x + (3\sin\gamma)\,\vec{a}_y$
(d) $(3\sin\gamma)\,\vec{a}_x - (6\cos\beta)\,\vec{a}_y$

Q.106 The conduction current in a straight copper wire of cross-sectional area A = 1.5×10^{-5} m^2 is I = 2 A. If the constitutive parameters of copper are $\mu = \mu_0 = 4\pi\times10^{-7}$ H/m, $\varepsilon = \varepsilon_0 = \frac{10^{-9}}{36\pi}$ F/m and $\sigma = 5.8\times10^7$ mho/m, then the displacement current I_d in the wire at a frequency of f = 1 GHz would be

(a) 1.9 nA (b) 1.0 nA
(c) 2.8 nA (d) 3.0 nA

Q.107 A rectangular loop of length a = 1 meter and width b = 80 cm is placed in a uniform magnetic field. What is the maximum value of induced emf if the magnetic flux density B = 0.1 Wb/m^2

is constant and the loop rotates about the x-axis with a frequency of 50 Hz?

(a) 32.65 volts (b) 18.54 volts
(c) 28 volts (d) 25.13 volts

Q.108 A straight conducting wire of length 50 cm is moved in a direction at right angles to its length in a region in air permitted by a uniform flux density B of magnitude B = 1 Wb/m^2. If the magnetic flux density $\vec{B}$ is perpendicular to both the direction of the motion and the length of the wire and the magnitude of the velocity of motion v = 10 m/s, the induced voltage in the conductor would be

(a) 2.5 V (b) 10 V
(c) 5 volt (d) 1.5 V

Q.109 A Faraday disc of 10 cm radius rotates at 5000 rpm in a uniform magnetic field of 250 mWb/m^2, the field being normal to the plane of the disc. What is the emf between rpm and axis assuming the diameter of the axis very small?

(a) 0.35 volt (b) 0.65 volt
(c) 0.45 volt (d) 0.55 volt

Q.110 What is the emf developed about the path r = 0.5, z = 0 and at t = 0, if $\vec{B} = 0.1 \sin 377t\, \hat{a}_x$?

(a) 4.7 volt (b) 3.8 volt
(c) 2.2 volt (d) 1.6 volt

Q.111 A square loop of wire 25 cm has a voltmeter (of infinite impedance) connected in series with one side. The plane of the loop is perpendicular to the magnetic field and the frequency is 10 MHz. If the maximum intensity is 1 Amp/m, then the voltage indicated by the meter when the loop is placed in the alternating field would be

(a) 3.6 volt (b) 2.8 volt
(c) 4.9 volt (d) none of these

Q.112 Assertion (A): The total e.m.f. induced in a circuit is equal to the time rate of decrease of the total magnetic flux linking the circuit.

Reason (R): Changing magnetic field will induce on electric field.

(a) Both A and R are true and R is a correct explanation of A.
(b) Both A and R are true but R is not a correct explanation of A.
(c) A is true but R is false.
(d) A is false but R is true.

Q.113 Assertion (A): Motional induction or flux cutting law gives the e.m.f. induced in a moving conductor w.r.t. observer in a magnetic field.

Reason (R): The motional emf equation depends on the velocity of the conductor and its position.

(a) Both A and R are true and R is a correct explanation of A.
(b) Both A and R are true but R is not a correct explanation of A.
(c) A is true but R is false.
(d) A is false but R is true.

Q.114 In a material for which conductivity σ = 5 Siemen/m and ϵ_r = 1, the electric field intensity is $E = 250 \sin 10^{10} t$ V/m. What is the frequency at which the displacement and conduction current densities will be equal?

(a) 60 GHz (b) 90 MHz
(c) 60 MHz (d) 90 GHz

Q.115 Consider the following statements:

1. Electric charge at rest produces an electrostatic field.
2. Electric charge in motion produces a magnetic field.
3. Electric and magnetic fields are fundamentally fields of force which originate from electric charges.
4. Faraday's induction law is given by

$$V = \oint \vec{E}_e \cdot d\vec{l} = \frac{-d}{dt}\iint_s \vec{B} \cdot ds$$

Which of the above statements are correct?

(a) 1, 2, 3 and 4 (b) 1 and 2 only
(c) 1, 2 and 3 only (d) 3 and 4 only

Q.116 A circular coil begins completely outside any magnetic field, and then during the next 10 sec. it moves completely into a uniform 15 T magnetic field. Assuming that the plane of the loop and the direction of motion are both perpendicular to the magnetic field, the magnitude of induced emf in the coil with 25 loops and radius of 7 cm is

(a) 0 volt (b) 0.58 volt
(c) 1.2 volt (d) 0.23 volt

Q.117 A 18 cm conducting rod attached to a U-shaped conductor moves at a speed of 1 cm/s perpendicular to a magnetic field of strength 37.5*T* as shown in figure below.

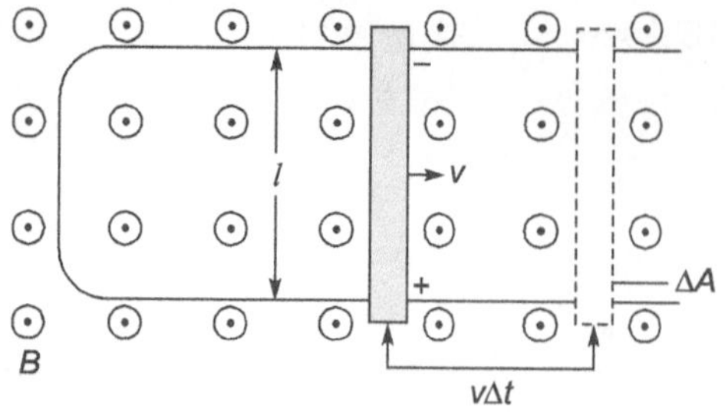

What is the emf induced in the conducting rod?

(a) 30 mV (b) 2.1 V
(c) 70 mV (d) 3.0 V

Q.118 What is the strength of the electric field produced in the conductor by the change in flux in the above problem?

(a) 0.75 V/m (b) 1.5 N/C
(c) 3.8 V/m (d) 0.38 N/C

Q.119 What is the induced emf produced when the radius of one loop or wire in a 12*T* magnetic field changes from 2 cm to 9 cm in 3 seconds? Assume that the plane of the loop remains parallel to the direction of the magnetic field.

(a) 0 V (b) 0.96 V
(c) 0.32 V (d) 2.21 V

Q.120 When a magnetic flux cuts across 200 turns at the rate of 2 Wb/s, the induced voltage is

(a) 400 V (b) 100 V
(c) 600 V (d) 0 V

Q.121 A conducting bar can slide freely over two conducting rails as shown in figure.

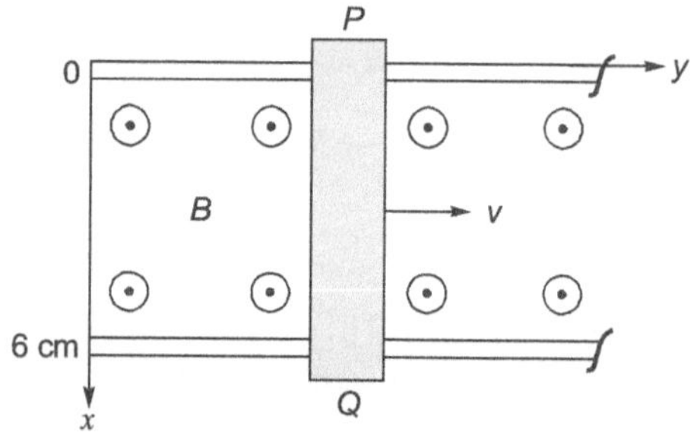

What is the magnitude of maximum induced voltage in the bar if the bar is stationed at $y = 8$ cm and $B = 4\cos 10^6 t\, a_z$ mWb/m^2?

(a) 8.80 V (b) 19.20 V
(c) 23.60 V (d) None of these

Q.122 A parallel plate capacitor with plate area of 5 cm^2 and plate separation of 3 mm has a voltage $50\sin 10^3 t$ applied to its plate. Calculate the displacement current assuming $\varepsilon = 2\,\varepsilon_0$.

(a) $147.4\cos 10^3\, t$ nA
(b) $125.8\cos 10^3\, t\,\mu$A
(c) $166.2\cos 10^3\, t$ nA
(d) $99\cos 10^3\, t\,\mu$A

Q.123 The flux through each turn of a 100-turn coil is $(t^3 - 2t)$ mWb, where t is in seconds. The induced emf at $t = 2$ sec. is

(a) –1 V (b) 4 mV
(c) 0.4 V (d) 1 V

Q.124 A loop is rotating about the y-axis in a magnetic field $B = (B_0 \sin\omega t)\, a_x$ Wb/m^2. The voltage induced in the loop is due to

(a) motional emf
(b) transformer emf
(c) a combination of motional and transformer emf
(d) none of the above

Q.125 The concept of displacement current is a major contribution attributed to

(a) Lenz (b) Maxwell
(c) Faraday (d) Lorentz

Q.126 Which of the following expressions does not represent Maxwell's equations for time-varying fields?

(a) $\nabla \cdot E = -\dfrac{\partial B}{\partial t}$ (b) $\nabla \cdot D = \rho_v$

(c) $\oint B \cdot dS = 0$ (d) $\nabla \cdot J + \dfrac{\partial \rho_v}{\partial t} = 0$

Q.127 An electromagnetic field is said to be conservative when

(a) $\nabla^2 E = \mu E\left(\dfrac{\partial^2 E}{\partial t^2}\right)$

(b) $\nabla^2 H = \mu\epsilon\left(\dfrac{\partial^2 H}{\partial t^2}\right)$

(c) Curl of the field is zero
(d) divergence of the field is zero

Q.128 A loop is inside a uniform magnetic field of $B = 50\, a_x$ mWb/m^2 as shown in figure. If side DC of the loop cuts the flux lines at the frequency of 50 Hz and the loop lies in the yz plane at time $t = 0$, what is the induced emf at $t = 5$ ms?

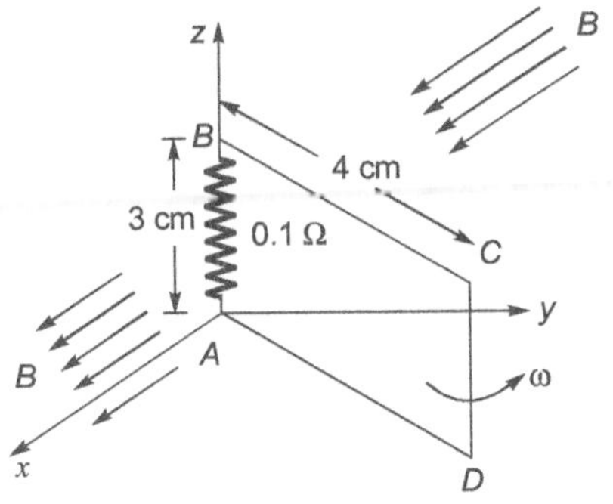

(a) π mV (b) 2π mV
(c) 4π mV (d) 6π mV

Q.129 If $J = \frac{50}{\rho}u_p + \frac{40}{\rho^2 + (0.1)^2}u_z$ A/m^2, the total current crossing the plane $z = 0.2$ in the u_z direction for $\rho < 0.4$ is
(a) 356 A (b) 178 A
(c) 712 A (d) zero

Q.130 The magnetic vector potential is given as $A = \frac{-\rho^2}{4}a_z$ Wb/m. What is the total magnetic flux crossing the surface $\phi = \frac{\pi}{2}$, $1 \le \rho \le 2$ m, $0 \le z \le 5$ m?

(a) $\frac{11}{4}$ Wb (b) $\frac{13}{4}$ Wb

(c) $\frac{15}{4}$ Wb (d) $\frac{17}{4}$ Wb

Q.131 Consider the following statements associated with conductors and dielectrics:
1. When the displacement current is much smaller than the conduction current, the medium corresponds to a good conductor.
2. When the displacement current is very much greater than the conduction current, the medium behaves as a semi-conductor or quasi-conductor.
3. When the displacement current is neither very high nor very low (i.e. midway between the two), the medium behaves as a good dielectric.

Which of the above statements is/are correct?
(a) 1 only (b) 2 and 3 only
(c) 3 only (d) 1 and 2 only

5. Electromagnetic Waves

Q.132 Consider the following statements associated with an electromagnetic wave in free space:
1. At every point in space, the electric field $\vec{E}$ and the magnetic vector field $\vec{H}$ are perpendicular to each other and to the direction of propagation.
2. Velocity of propagation of wave in free space is equal to $\sqrt{\frac{\mu_0}{\varepsilon_0}}$.
3. $\vec{E}$ and $\vec{H}$ oscillate in phase opposition and the ratio of their amplitude is constant being equal to $\sqrt{\frac{\mu_0}{\varepsilon_0}}$.
4. Whatever may be the frequency, the electromagnetic wave travels in space with velocity of light.

Which of the above statements are correct?
(a) 1, 2 and 3 only (b) 1 and 3 only
(c) 1 and 4 only (d) 1, 3 and 4

Q.133 Intrinsic impedance or charcteristic impedance of the non-conducting medium in the free space is
(a) 120π Ohms (b) 80π Ohms
(c) 160π Ohms (d) 60π Ohms

Q.134 If the magnitude of $\vec{E}$ in a plane wave is 1 V/m, what is the magnitude of $\vec{H}$ for a plane wave in free space?
(a) 0.65 mA/meter (b) 2.6 mA/meter
(c) 1.3 mA/meter (d) 3.9 mA/meter

Q.135 What is the velocity of a plane wave in a lossless medium having a relative permittivity of 5 and relative permeability of unity?
(a) 3×10^8 m/s (b) 10^7 m/s
(c) 2×10^8 m/s (d) 1.3×10^8 m/s

Q.136 A plane wave in free space has an average Poynting vector of 1 Wm^{-2}. The average energy density is
(a) 1.0 nJ/m^3 (b) 3.3 nJ/m^3
(c) 2 nJ/m^3 (d) 4.4 nJ/m^3

Q.137 In free space, $\vec{E}$ and $\vec{H}$ are respectively given by

$$\vec{E} = 30\pi e^{j(10^8 t + \beta z)}\vec{a}_x \text{ V/m}$$

and $$\vec{H} = H_0 e^{j(10^8 t + \beta z)}\vec{a}_y \text{ A/m}$$

The value of β and H_0 are respectively

(a) $\frac{1}{3}$ rad/m and $\pm\frac{1}{4}$ A/m

(b) $\frac{1}{2}$ rad/m and $\pm\frac{1}{3}$ A/m

(c) $\frac{1}{3}$ rad/m and $\pm\frac{1}{2}$ A/m

(d) $\frac{1}{2}$ rad/m and $\pm\frac{1}{6}$ A/m

Q.138 In a homogeneous non-conducting region where $\mu_r = 1$, it is given that $\vec{E} = 30\pi \cdot e^{j(\omega t - (4/3)y)}\vec{a}_y$ V/m and $\vec{H} = 1.0 \cdot e^{j(\omega t - (4/3)y)}\vec{a}_x$ A/m

The value of ε_r and ω are

(a) 4 and 3×10^8 rad/s respectively

(b) 16 and 3×10^8 rad/s respectively

(c) 16 and 10^8 rad/s respectively

(d) 4 and 10^8 rad/s respectively

Q.139 In free space it is given that,

$$\vec{E}(z,t) = 10^3 \sin(\omega t - \beta z)\ \vec{a}_y \text{ V/m}$$

$\vec{H}(z,t)$ will be given by

(a) $\vec{H}(z,t) = \frac{125}{3\pi}\sin(\omega t - \beta z)\vec{a}_x$ A/m

(b) $\vec{H}(z,t) = \frac{-100}{2\pi}\sin(\omega t - \beta z)\vec{a}_x$ A/m

(c) $\vec{H}(z,t) = \frac{10}{2\pi}\sin(\omega t - \beta z)\vec{a}_x$ A/m

(d) $\vec{H}(z,t) = \frac{-25}{3\pi}\sin(\omega t - \beta z)\vec{a}_x$ A/m

Q.140 Assertion (A): Uniform plane electromagnetic waves are transverse in nature.

Reason (R): Uniform plane electromagnetic waves have components of $\vec{E}$ and $\vec{H}$ only in direction perpendicular to the direction of propagation and not in the direction of propagation.

(a) Both A and R are true and R is a correct explanation of A.

(b) Both A and R are true but R is not a correct explanation of A.

(c) A is true but R is false.

(d) A is false but R is true.

Q.141 Assertion (A): The term $\sigma/\omega\varepsilon$ is sometimes referred to the loss tangent.

Reason (R): The two vectors $\vec{J}_c$ (conduction current density) and $\vec{J}_d$ (displacement current density) are 180° out of phase and are in the opposite direction in space.

(a) Both A and R are true and R is a correct explanation of A.

(b) Both A and R are true but R is not a correct explanation of A.

(c) A is true but R is false.

(d) A is false but R is true.

Q.142 The propagation constant γ for a material with $\mu_r = 1$, $\varepsilon_r = 8$, $\sigma = 0.25$ pS/m and having the wave frequency of 1.6 MHz would be approximately equal to

(a) $(0.02 + j1.1)$ m^{-1} (b) $(j0.95)$ m^{-1}

(c) $(0.1 + j0.95)$ m^{-1} (d) $j1.1$ m^{-1}

Q.143 What is the skin depth in copper at 10 GHz? Assuming conductivity of copper $\sigma = 5.8 \times 10^7$ mhos/m and permeability equal to that of free space.

(a) 0.55 μm (b) 0.33 nm

(c) 0.44 nm (d) 0.66 μm

Q.144 The velocity of a plane wave in a lossless medium having a relative permittivity of 4 and relative permeability of unity would be

(a) 1.5×10^8 m/s (b) 2.5×10^8 m/s

(c) 1×10^8 m/s (d) 2×10^8 m/s

Q.145 An elliptically polarized wave travelling in the $+z$ direction in air has x and y components

$$E_x = 2\sin(\omega t - \beta z) \text{ V/m};$$
$$E_y = 4\sin(\omega t - \beta z + 75°) \text{ V/m}$$

The average power per unit area conveyed by the wave is

(a) 15.8 mW/m^2 (b) 12 mW/m^2

(c) 26.5 mW/m^2 (d) none of these

Q.146 Consider the following statements:

1. A wave whose phase is constant over a set of planes is called plane wave.
2. A wave whose phase is constant while magnitude is variable is called uniform plane wave.
3. The $\vec{E}$ and $\vec{H}$ fields are non-uniform over plane surfaces.
4. Earth is considered as good conductor in one part of the radio frequency range and as dielectric in other frequency range.

Which of the above statements is/are not correct?

(a) 1 and 4 (b) 2 and 3
(c) 3 and 4 (d) 4 only

Q.147 Assertion (A): Displacement current directly passes through the capacitor.

Reason (R): The current through a capacitive element is called displacement current.

(a) Both A and R are true and R is a correct explanation of A.
(b) Both A and R are true but R is not a correct explanation of A.
(c) A is true but R is false.
(d) A is false but R is true.

Q.148 In a homogeneous, isotropic, linear and stationary media, a plane electromagnetic wave has the following characteristics.

(a) Both $\vec{E}$ and $\vec{H}$ are transverse
(b) $\rho = 0$
(c) Both (a) and (b)
(d) Neither (a) nor (b)

Q.149 Assertion (A): The distance to which a wave penetrates a conductor is known as depth of penetration.

Reason (R): The depth of penetration (δ) is defined as that depth in which the wave has been attenuated to '*e*' percent of its original value.

(a) Both A and R are true and R is a correct explanation of A.
(b) Both A and R are true but R is not a correct explanation of A.
(c) A is true but R is false.
(d) A is false but R is true.

Q.150 The polarization of a dielectric is given by

(a) $\vec{P} = \varepsilon_r \vec{E}$ (b) $\vec{P} = (\varepsilon_r - 1)\vec{E}$
(c) $\vec{P} = \vec{E}\varepsilon_0(\varepsilon_r - 1)$ (d) $\vec{P} = (\varepsilon_r - 1)\varepsilon_0$

Q.151 In a travelling electromagnetic wave, *E* and *H* vector fields are

(a) parallel in space
(b) perpendicular in space
(c) E is in the direction of wave travel
(d) H is in the direction of wave travel

Q.152 A wave is incident normally on a good conductor. If the frequency of a plane electromagnetic wave increases four times, the skin depth, will

(a) remain the same
(b) decrease by a factor of 4
(c) increase by a factor of 2
(d) decrease by a factor of 2

Q.153 The electric field of a travelling wave is given by $100\cos(10^9 t - 4x)$. The velocity and the wavelength are respectively

(a) 2.5×10^8 m/s and $\frac{\pi}{2}$ m
(b) 10^9 m/s and $\pi/4$ m
(c) 3×10^8 m/s and 4 m
(d) 4×10^9 m/s and 2π m

Q.154 In a dielectric-conductor boundary (interface), the tangential component of electric field is

(a) $2E_t$ (b) E_t
(c) infinity (d) zero

Q.155 Which of the following statements is not true of waves in general?

(a) It may be a function of time and space.
(b) For practical reasons, it must be finite in extent.
(c) It may be a function of time only.
(d) It may be sinusoidal or cosinusoidal.

Q.156 When an EM wave is incident on a dielectric, it is

(a) fully transmitted
(b) fully reflected
(c) partially transmitted and partially reflected
(d) none of the above

Q.157 Poynting vector gives

(a) rate of energy flow

(b) intensity of electric field
(c) intensity of magnetic field
(d) direction of polarization

Q.158 Match **List-I** with **List-II** and select the correct answer using the codes given below the lists:

List-I
A. Free space
B. Lossless dielectrics
C. Lossy dielectrics
D. Good conductors

List-II
1. $\sigma \neq 0, \varepsilon = \varepsilon_0\varepsilon_r, \mu = \mu_0\mu_r$
2. $\sigma = 0, \varepsilon = \varepsilon_0, \mu = \mu_0$
3. $\sigma \approx \infty, \varepsilon = \varepsilon_0, \mu = \mu_0\mu_r$
4. $\sigma = 0, \varepsilon = \varepsilon_0 \varepsilon_r, \mu = \mu_0\mu_r$

Codes:

	A	B	C	D
(a)	4	2	3	1
(b)	2	4	1	3
(c)	4	2	1	3
(d)	2	4	3	1

Q.159 Match **List-I (EM phenomena)** with **List-II (Uses)** and select the correct answer using the codes given below the lists:

List-I	List-II
A. Cosmic rays	1. Sterilization
B. Ultraviolet radiation	2. Cancer therapy
C. Infra-red radiation	3. Astronomy
D. Gamma rays	4. Photography

Codes:

	A	B	C	D
(a)	4	3	2	1
(b)	3	4	1	2
(c)	1	2	3	4
(d)	3	1	4	2

Q.160 A lossy dielectric has an intrinsic impedance of $200\angle 30°\ \Omega$ at a particular frequency. If, at that frequency, the plane wave propagating through the dielectric has the magnetic field component

$H = 10^{-\alpha x}\cos\left(\omega t - \frac{1}{2}x\right)a_y$ A/m, the skin depth will be

(a) $\frac{1}{2\sqrt{3}}$ m
(b) $\frac{1}{\sqrt{3}}$ m
(c) $2\sqrt{3}$ m
(d) $\sqrt{3}$ m

Q.161 What is the major factor for determining whether a mediam is free space, lossless dielectric, or good conductor?
(a) Attenuation constant
(b) Constitutive parameters (σ, ε, μ)
(c) Loss tangent
(d) Reflection coefficient

Q.162 Assuming perfect conductors of a transmission line, pure TEM propagation is not possible in
(a) coaxial cable
(b) air-filled cylindrical waveguide
(c) parallel twin-wire line in air
(d) semi-infinite parallel plate wave guides.

Q.163 For an electromagnetic wave incident on a conducting medium, the depth of penetration
(a) is directly proportional to the attenuation constant.
(b) is inversely proportional to the attenuation constant.
(c) has a logrithmic relationship with the attenuation constant.
(d) is independent of the attenuation constant.

Q.164 In a uniform plane wave, the value of $|\vec{E}|/|\vec{H}|$ is

(a) $\sqrt{\frac{\mu}{\varepsilon}}$
(b) $\sqrt{\frac{\varepsilon}{\mu}}$
(c) 1
(d) $\sqrt{\mu\varepsilon}$

Q.165 A plane wave whose electric field is given by $\vec{E} = 100\cos(\omega t - 6\pi x)\,\hat{z}$ passes normally from a material '*A*' having $\varepsilon_r = 4, \mu_r = 1$ and $\sigma = 0$ to a material '*B*' having $\varepsilon_r = 9, \mu_r = 4$ and $\sigma = 0$.
Match **List-I** with **List-II** and select the correct answer using the codes given below the lists:

List-I
A. Intrinsic impedance of medium B
B. Reflection coefficient
C. Transmission coefficient
D. Phase shift constant of medium A

List-II
1. 6π
2. 80π
3. 1/5
4. 6/5

Codes:

	A	B	C	D
(a)	4	1	2	3
(b)	2	3	4	1
(c)	4	3	2	1
(d)	2	1	4	3

Q.166 A plane slab of dielectric having dielectric constant 5, placed normal to a uniform field with a flux density of 2 C/m^2, is uniformally polarized. The polarization of the slab is
(a) 0.4 C/m^2 (b) 1.6 C/m^2
(c) 2.0 C/m^2 (d) 6.4 C/m^2

Q.167 The electric field of a uniform plane wave is given by $E = [10\sin(\omega t - \pi z)\hat{a}_x + 10\cos(\omega t - \pi z)\hat{a}_y]$ V/m.
The polarization of the wave is
(a) circular (b) elliptical
(c) linear (d) undefined

Q.168 In a lossless medium the intrinsic impedance η = 60π and μ_r = 1. What is the value of the dielectric constant ε_r?
(a) 2 (b) 1
(c) 4 (d) 8

Q.169 Match **List-I (Dominant mode of propagation)** with **List-II (Type of transmission structure)** and select the correct answer using the codes given below the lists:

List-I	List-II
A. Coaxial cable	1. TE
B. Rectangular waveguide	2. Quasi TEM
C. Microstrip line	3. Hybrid
D. Coplanar waveguide	4. TEM

Codes:

	A	B	C	D
(a)	1	4	2	3
(b)	4	1	3	2
(c)	1	4	3	2
(d)	4	1	2	3

Q.170 The wavelength of a wave with propagation constant $(0.1\pi + j0.2\pi)$ m^{-1} is
(a) $\frac{2}{\sqrt{0.05}}$ m (b) 10 m
(c) 20 m (d) 30 m

6. Transmission Lines

Q.171 For a reciprocal transmission line, the three transmission parameters are given by A = 4, B = 7 and C = 5. The value of D is equal to
(a) 8.5 (b) 9
(c) –8.5 (d) –9

Q.172 The real part of the propagation constant shows:
(a) reduction in voltage, current values of signal amplitude
(b) reduction of only voltage amplitude
(c) reduction of only current amplitude
(d) variation of phase shift/position of voltage

Q.173 In a transmission line terminated by characteristic impedance, Z_0
(a) the incident current is zero for only applied voltage.
(b) there are large number of maximum and minimum on the line.
(c) the reflection is maximum due to termination.
(d) there is no reflection of the incident wave.

Q.174 A symmetrical two-part network have
(a) $Z_{SC_1} = Z_{SC_2}$
(b) $Z_{OC_1} \neq Z_{OC_2}$
(c) $Z_{SC_1} = Z_{OC_2}$
(d) $Z_{OC_1} \neq Z_{OC_2}$ and $Z_{SC_1} = Z_{SC_2}$

Q.175 A pure resistance, R_L when connected at the load end of a loss-less 100 Ω line produces a VSWR of 2. The R_L would be
(a) 400 Ω (b) 50 Ω or 200 Ω
(c) 50 Ω only (d) 200 Ω only

Q.176 The reflection coefficient of a transmission line with a short-circuited load is
(a) infinite (b) $1\angle 180°$
(c) zero (d) none of these

Q.177 The characteristic impedance Z_0 of a transmission line is given by (where R, L, G, C are the unit length parameters)
(a) $\left[\frac{G + j\omega L}{R + j\omega L}\right]^{1/2}$ (b) $\frac{(R + j\omega L)}{(G + j\omega C)}$
(c) $\left[\frac{R + j\omega L}{G + j\omega C}\right]^{1/2}$ (d) $\left[\frac{G + j\omega C}{R + j\omega L}\right]$

Q.178 The standing wave ratio (VSWR) in terms of reflection coefficient ρ is given by

(a) $\frac{1+\rho}{1-\rho}$ (b) $\frac{1-\rho}{1+\rho}$

(c) $\frac{\rho-1}{\rho+1}$ (d) $\frac{\rho}{1+\rho}$

Q.179 If a transmission line of length less than λ/4 is short circuited, it behaves as:

(a) series resonant circuit

(b) pure capacitive reactance

(c) pure inductive reactance

(d) parallel resonant circuit

Q.180 A line becomes distortionless if

(a) it is terminated into Z_0

(b) $LR = GC$

(c) $LG = CR$

(d) it is properly matched

Q.181 Double stub matching eliminates standing waves on the

(a) load side of right stub

(b) source side of the left stub

(c) both sides of the stub

(d) in between the two stubs

Q.182 If $Z_{SC} = 64\ \Omega$ and $Z_{OC} = 100\ \Omega$, the characteristic impedance will be given by

(a) 60 Ω (b) 170 Ω

(c) 400 Ω (d) 80 Ω

Q.183 Consider a lossless line with characteristic impedance R_0 and VSWR = S. Then, the impedance at the point of voltage maxima and voltage minima are respectively given by

(a) SR_0 and $\frac{R_0}{S}$ (b) $\frac{R_0}{S}$ and SR_0

(c) $\frac{R_0}{S}$ and $\frac{R_0}{S}$ (d) SR_0 and SR_0

Q.184 If K is the reflection coefficient and S is the voltage standing wave ratio, then

(a) $K = \frac{VSWR+1}{VSWR-1}$ (b) $|K| = \frac{VSWR-1}{VSWR+1}$

(c) $K = \frac{VSWR-1}{VSWR+1}$ (d) $|K| = \frac{VSWR+1}{VSWR-1}$

Q.185 In terms of R, L, G and C the propagation constant of a transmission line is given by

(a) $\sqrt{G+j\omega C}$

(b) $\sqrt{\frac{R+j\omega L}{G+j\omega C}}$

(c) $\sqrt{(R+j\omega L)(G+j\omega C)}$

(d) $\sqrt{R+j\omega L}$

Q.186 A line has $Z_0 = 300\angle 0°\ \Omega$ and $Z_L = 150\angle 0°\ \Omega$. Voltage standing wave ratio (S) is given by

(a) 2 (b) 1

(c) 0.5 (d) infinite

Q.187 The characteristic impedance of a distortionless line is

(a) inductive (b) capacitive

(c) complex (d) real

Q.188 A transmission line works as a

(a) LPF (b) HPF

(c) attenuator (d) none of these

Q.189 A transmission line with a characteristic impedance Z_0 is connected to a transmission line with characteristic impedance Z_c. If the systme is being driven by a generator connected to the first line, then the overall transmission coefficient of current will be

(a) $\frac{2Z_0}{Z_0+Z_c}$ (b) $\frac{Z_0}{Z_0+Z_c}$

(c) $\frac{2Z_c}{Z_0+Z_c}$ (d) $\frac{Z_c}{Z_0+Z_c}$

Q.190 Which of the following statements related to a transmission line is/are correct?

1. Transmission line elements are integral parts of the antenna, in some antenna system.
2. A feeder is a particular case of a transmission line.
3. A lossless transmission line doesn't has resistance but, has a non-zero value of leakage conductance.
4. At radio frequency (RF), R and G both are neglected.

(a) 1 and 2 only (b) 1, 2 and 4 only

(c) 2 and 4 only (d) 1, 2, 3 and 4

Q.191 Assertion (A): A finite transmission line terminated in its characteristic impedance Z_0, is equivalent to an infinite line.

Reason (R): the input impedance of an infinite line is the characteristic impedance of the line.

(a) Both A and R are true and R is a correct explanation of A.
(b) Both A and R are true but R is not a correct explanation of A.
(c) A is true but R is false.
(d) A is false but R is true.

Q.192 Consider the following statements:

1. Propagation constant is a dimensionless quantity.
2. When the line is lossless, propagation constant is directly proportional to the frequency.
3. Propagation constant when multiplied with the frequency gives the electrical length of the line.

Which of the above statements is/are true?

(a) 3 only (b) 2 only
(c) 1, 2 and 3 (d) 2 and 3 only

Q.193 Assertion (A): The group velocity is usually more than the phase velocity.

Reason (R): If the transmission line or transmission medium is such that different frequencies travel with different velocities, then the line or the medium is said to be dispersive.

(a) Both A and R are true and R is a correct explanation of A.
(b) Both A and R are true but R is not a correct explanation of A.
(c) A is true but R is false.
(d) A is false but R is true.

Q.194 Assertion (A): A transmission line act as resonant circuit and is used in many applications at high frequencies in antenna design and other ratio circuitory.

Reason (R): An open and short-circuited lines behaves like resonant circuit when length of line is an integral multiple of $\lambda/3$.

(a) Both A and R are true and R is a correct explanation of A.
(b) Both A and R are true but R is not a correct explanation of A.
(c) A is true but R is false.
(d) A is false but R is true.

Q.195 Assertion (A): Sometimes a quarter wave line is called as impedance inverter.

Reason (R): The quarter wavelength line transforms a load impedance Z_R that is smaller than Z_0 into a value Z_S that is larger than Z_0 and vice-versa.

(a) Both A and R are true and R is a correct explanation of A.
(b) Both A and R are true but R is not a correct explanation of A.
(c) A is true but R is false.
(d) A is false but R is true.

Q.196 In an impedance smith chart, a clockwise movement along a constant resistance circle gives rise to

(a) a decrease in the value of reactance
(b) an increase in the value of reactance
(c) no change in the reactance value
(d) no change in the impedance value

Q.197 A transmission line whose characteristic impedance is purely resistive

(a) must be lossless line
(b) must be a distortionless line
(c) may not be a lossless oline
(d) may not be a distortionless line

Q.198 The input impedance of a short circuited quarter wave long transmission line is

(a) purely reactive
(b) purely resistive
(c) dependent on the characteristic impedance of the line
(d) none of the above

Q.199 A rectangular air filled waveguide has cross section of 4 cm × 10 cm. The minimum frequency which can propagate in the waveguide is

(a) 1.5 GHz (b) 2.0 GHz
(c) 2.5 GHz (d) 3.0 GHz

Q.200 Consider the transmission system shown below:

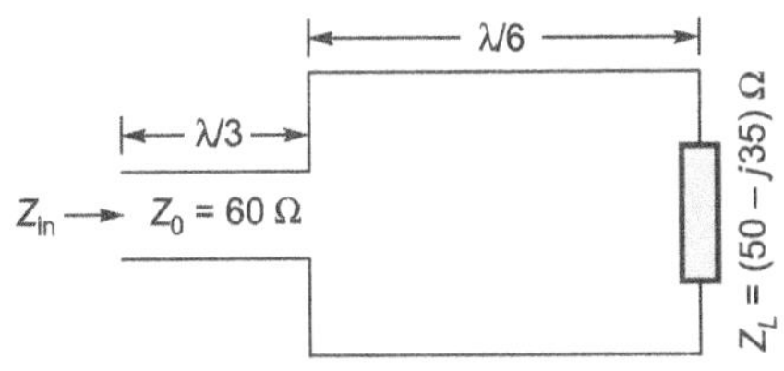

The input impedance Z_{in} is given by
(a) $(42.6 - j22.8)\ \Omega$ (b) $(22.8 - j42.6)\ \Omega$
(c) $(42.6 + j22.8)\ \Omega$ (d) $(22.8 + j42.6)\ \Omega$

Q.201 A loss-less transmission line of characteristic impedance Z_0 and $l < \frac{\lambda}{4}$ is terminated at the load end by a short circuit. Its input impedance Z_s is
(a) $Z_s = -jZ_0 \tan\beta l$ (b) $Z_s = jZ_0 \cot\beta l$
(c) $Z_s = jZ_0 \tan\beta l$ (d) $Z_s = -jZ_0 \cot\beta l$

Q.202 In a twin-wire transmission line in air, the adjacent voltage maxima are at 12.5 cm and 27.5 cm. The operating frequency is
(a) 300 MHz (b) 1 GHz
(c) 2 GHz (d) 6.28 GHz

Q.203 Consider a 300 Ω, quarter-wave long (at 1 GHz) transmission line as shown in the figure. It is connected to a 10 V, 50 Ω source at one end and is left open circuited at the other end. The magnitude of the voltage at the open circuit of the line is

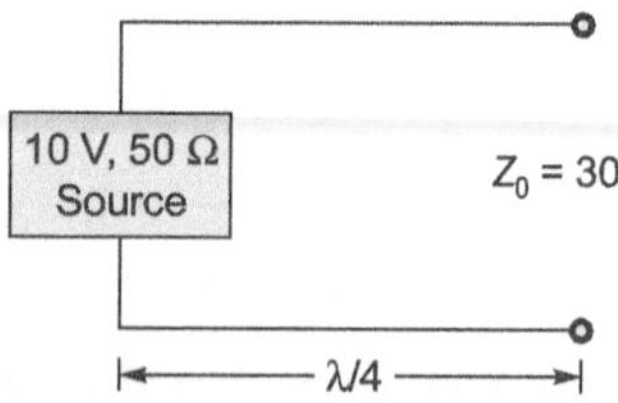

(a) 10 V (b) 5 V
(c) 60 V (d) 60/7 V

Q.204 Match **List-I (Quantity)** with **List-II (Range of values)** and select the correct answer using the codes given below the lists:

List-I	List-II
A. Input impedance	1. –1 to 1
B. Reflection coefficient	2. 1 to ∞
C. VSWR	3. 0 to ∞

Codes:

	A	B	C
(a)	2	3	1
(b)	3	2	1
(c)	3	1	2
(d)	2	1	3

Q.205 A transmission line of characteristic impedance 400 Ω is to be matched to a load of 25 Ω through a quarter wavelength line. The quarter wave line characteristic impedance must be
(a) 40 Ω (b) 100 Ω
(c) 400 Ω (d) 425 Ω

Q.206 Assertion (A): Single stub matching is useful for a fixed frequency only.
Reason (R): It single stub matching, the range of terminating impedances which can be transferred is limited.
(a) Both A and R are true and R is a correct explanation of A.
(b) Both A and R are true but R is not a correct explanation of A.
(c) A is true but R is false.
(d) A is false but R is true.

Q.207 A transmission line has series inductance of 0.56 millihenry and capacitance of 0.1 microfrad per km. If the losses due to conductor resistance and insulation leakage are negligible, then the characteristic impedance of the line woule be
(a) 65 Ω (b) 55 Ω
(c) 35 Ω (d) 75 Ω

■■■■

Answers Electromagnetic Theory

1. (c)	2. (b)	3. (a)	4. (d)	5. (a)	6. (a)	7. (c)	8. (b)
9. (d)	10. (c)	11. (a)	12. (b)	13. (d)	14. (c)	15. (d)	16. (b)
17. (c)	18. (a)	19. (c)	20. (d)	21. (c)	22. (b)	23. (a)	24. (d)
25. (d)	26. (a)	27. (d)	28. (a)	29. (c)	30. (a)	31. (d)	32. (b)
33. (b)	34. (a)	35. (c)	36. (d)	37. (a)	38. (c)	39. (a)	40. (b)
41. (d)	42. (a)	43. (c)	44. (c)	45. (c)	46. (c)	47. (c)	48. (b)
49. (a)	50. (a)	51. (c)	52. (d)	53. (a)	54. (b)	55. (d)	56. (a)
57. (c)	58. (b)	59. (a)	60. (b)	61. (c)	62. (b)	63. (d)	64. (a)
65. (d)	66. (b)	67. (a)	68. (d)	69. (a)	70. (a)	71. (c)	72. (b)
73. (d)	74. (a)	75. (d)	76. (d)	77. (b)	78. (a)	79. (b)	80. (c)
81. (d)	82. (c)	83. (b)	84. (b)	85. (a)	86. (d)	87. (c)	88. (a)
89. (b)	90. (a)	91. (b)	92. (c)	93. (b)	94. (d)	95. (a)	96. (b)
97. (a)	98. (b)	99. (a)	100. (b)	101. (c)	102. (a)	103. (c)	104. (b)
105. (b)	106. (a)	107. (d)	108. (c)	109. (b)	110. (a)	111. (c)	112. (b)
113. (c)	114. (d)	115. (a)	116. (b)	117. (c)	118. (d)	119. (a)	120. (a)
121. (b)	122. (a)	123. (a)	124. (c)	125. (b)	126. (d)	127. (c)	128. (d)
129. (a)	130. (c)	131. (a)	132. (c)	133. (a)	134. (b)	135. (d)	136. (b)
137. (a)	138. (c)	139. (d)	140. (a)	141. (c)	142. (b)	143. (d)	144. (a)
145. (c)	146. (b)	147. (d)	148. (c)	149. (c)	150. (b)	151. (c)	152. (d)
153. (a)	154. (d)	155. (c)	156. (c)	157. (a)	158. (b)	159. (d)	160. (c)
161. (c)	162. (b)	163. (b)	164. (a)	165. (b)	166. (b)	167. (a)	168. (c)
169. (d)	170. (b)	171. (b)	172. (a)	173. (d)	174. (d)	175. (b)	176. (b)
177. (c)	178. (a)	179. (c)	180. (c)	181. (b)	182. (d)	183. (a)	184. (b)
185. (c)	186. (a)	187. (d)	188. (a)	189. (a)	190. (b)	191. (a)	192. (c)
193. (d)	194. (c)	195. (a)	196. (b)	197. (a)	198. (d)	199. (a)	200. (a)
201. (c)	202. (b)	203. (c)	204. (c)	205. (b)	206. (b)	207. (d)	

Explanations

1. (c)

Here, $$\text{Div. } \vec{Q} = \nabla\cdot\vec{Q} = \left(\vec{i}\frac{\partial}{\partial x}+\vec{j}\frac{\partial}{\partial y}+\vec{k}\frac{\partial}{\partial y}\right)\cdot(\vec{i}zx)$$

$$= (\vec{i}\cdot\vec{i})\left(\frac{\partial}{\partial x}\cdot zx\right) = 1\cdot z = z$$

Also, $$\text{Grad}\,\phi = \nabla\phi = \left(\frac{\partial}{\partial x}\vec{i}+\frac{\partial}{\partial y}\vec{j}+\frac{\partial}{\partial z}\vec{k}\right)(xy^2z^3) = (y^2z^3\,\vec{i}+2yxz^3\,\vec{j}+3z^2y^2x\vec{k})$$

And, $$\text{Curl } \vec{P} = \nabla\times\vec{P} = \begin{vmatrix} \vec{i} & 0 & \vec{k} \\ \frac{\partial}{\partial x} & \frac{\partial}{\partial y} & \frac{\partial}{\partial z} \\ x^2y & 0 & (x-y) \end{vmatrix}$$

$$= \vec{i}\left[\frac{\partial}{\partial y}(x-y)-0\right]+\vec{k}\left[0-\frac{\partial}{\partial y}(x^2y)\right]$$

$$= \vec{i}\,[-1]+\vec{k}\,[x^2] = -\vec{i}+x^2\,\vec{k}$$

2. (b)

Let the unit vector be given by $\vec{a}_R$.

Now, $$\vec{R} = \text{Difference of two vectors}$$

$$= r\vec{a}_r - h\vec{a}_z$$

∴ Unit vector, $$\vec{a}_R = \frac{\vec{R}}{|\vec{R}|} = \frac{r\,\vec{a}_r - h\vec{a}_z}{\sqrt{r^2+h^2}}$$

3. (a)

Since the given vector V is irrotational, therefore curl $V = 0$ or , $\nabla\times V = 0$.

Now, $$\nabla\times V = \begin{vmatrix} \vec{i} & \vec{j} & \vec{k} \\ \frac{\partial}{\partial x} & \frac{\partial}{\partial y} & \frac{\partial}{\partial z} \\ (x+2y+az) & (bx-3y-z) & (4x+cy+2z) \end{vmatrix}$$

$$= \left\{\frac{\partial}{\partial y}(4x+cy+2z)-\frac{\partial}{\partial z}(bx-3y-z)\right\}\vec{i}+\left\{\frac{\partial}{\partial z}(x+2y+az)-\frac{\partial}{\partial x}(4x+cy+2z)\right\}\vec{j}$$

$$+\left\{\frac{\partial}{\partial x}(bx-3y-z)-\frac{\partial}{\partial y}(x+2y+az)\right\}\vec{k}$$

$$= (c+1)\vec{i}+(a-4)\vec{j}+(b-2)\vec{k}$$

Since, $\nabla\times V = 0$, therefore, $a = 4$, $b = 2$, and $c = -1$

4. (d)

- $(\vec{A}\times\vec{B})\cdot(\vec{C}\times\vec{D})$ is called "product of four vectors".

- $\vec{A}\times(\vec{B}\times\vec{C})$ is called "vector triple product".
- $(\vec{A}\times\vec{B})\cdot(\vec{C}\times\vec{D})$ is called "vector product of four vectors".

6. (a)

Both assertion and reason are true and reason is the correct explanation of assertion. Reason is the physical Interpretation of divergence.

7. (c)

- $\vec{A}\times(\vec{B}\times\vec{C})$ is called "vector triple product" which is a correct expression.
- $$\nabla = \left(\vec{i}\frac{\partial}{\partial x}+\vec{j}\frac{\partial}{\partial y}+\vec{k}\frac{\partial}{\partial z}\right)$$

and let $\vec{A} = A_x\vec{i} + A_y\vec{j} + A_z\vec{k}$

Then,
$$\nabla\times\vec{A} = \begin{vmatrix} \vec{i} & \vec{j} & \vec{k} \\ \frac{\partial}{\partial x} & \frac{\partial}{\partial y} & \frac{\partial}{\partial z} \\ A_x & A_y & A_z \end{vmatrix}$$

$$\therefore \quad \nabla(\nabla\times\vec{A}) = \left(\vec{i}\frac{\partial}{\partial x}+\vec{j}\frac{\partial}{\partial y}+\vec{k}\frac{\partial}{\partial z}\right)\times\left[\vec{i}\left(\frac{\partial A_z}{\partial y}-\frac{\partial A_y}{\partial z}\right)+\vec{j}\left(\frac{\partial A_x}{\partial z}-\frac{\partial A_z}{\partial x}\right)+\vec{k}\left(\frac{\partial A_y}{\partial x}-\frac{\partial A_x}{\partial y}\right)\right]$$

$$= \frac{\partial}{\partial x}\left(\frac{\partial A_z}{\partial y}-\frac{\partial A_y}{\partial z}\right)+\frac{\partial}{\partial y}\left(\frac{\partial A_z}{\partial x}-\frac{\partial A_x}{\partial z}\right)+\frac{\partial}{\partial z}\left(\frac{\partial A_y}{\partial x}-\frac{\partial A_x}{\partial y}\right)$$

$$= \frac{\partial^2 A_z}{\partial x\partial y}-\frac{\partial^2 A_y}{\partial x\partial z}+\frac{\partial^2 A_z}{\partial y\partial x}-\frac{\partial^2 A_x}{\partial y\partial z}+\frac{\partial^2 A_y}{\partial z\partial x}-\frac{\partial^2 A_x}{\partial z\partial y} = 0$$

- $$\nabla\times\nabla\varphi = \left(\frac{\partial}{\partial x}\vec{i}+\frac{\partial}{\partial y}\vec{j}+\frac{\partial}{\partial z}\vec{k}\right)\times\left(\frac{\partial\varphi}{\partial x}\vec{i}+\frac{\partial\varphi}{\partial y}\vec{j}+\frac{\partial\varphi}{\partial z}\vec{k}\right)$$

$$= \left\{(\vec{i}\times\vec{j})\frac{\partial^2\varphi}{\partial x\partial y}+(\vec{i}\times\vec{k})\frac{\partial^2\varphi}{\partial x\partial z}\right\}+\left\{(\vec{j}\times\vec{i})\frac{\partial^2\varphi}{\partial x\partial y}+(\vec{j}\times\vec{k})\frac{\partial^2\varphi}{\partial y\partial z}\right\}+\left\{(\vec{k}\times\vec{i})\frac{\partial^2\varphi}{\partial z\partial x}+(\vec{k}\times\vec{j})\frac{\partial^2\varphi}{\partial y\partial z}\right\}$$

$$= \vec{k}\frac{\partial^2\varphi}{\partial x\partial y}-\vec{j}\frac{\partial^2\varphi}{\partial x\partial z}-\vec{k}\frac{\partial^2\varphi}{\partial y\partial x}+\vec{i}\frac{\partial^2\varphi}{\partial y\partial z}+\vec{j}\frac{\partial^2\varphi}{\partial z\partial x}-\vec{i}\frac{\partial^2\varphi}{\partial z\partial y} = 0$$

8. (b)

The vector $\vec{A}$ is given as

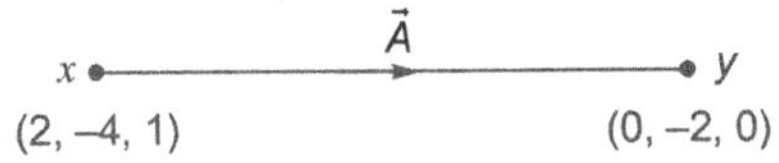

$$\vec{A} = (0-2)\vec{a}_x + [-2-(-4)]\vec{a}_y + (0-1)\vec{a}_z$$
$$= -2\vec{a}_x + 2\vec{a}_y - \vec{a}_z$$

9. (d)

By divergence theorem,

$$\iint_S \vec{F}\cdot d\vec{s} = \iiint_V \nabla\cdot\vec{F}dv$$

$$= \iiint_V \left[\vec{i}_1\frac{\partial}{\partial x}+\vec{i}_2\frac{\partial}{\partial y}+\vec{i}_3\frac{\partial}{\partial z}\right]\times(4xz\vec{i}_1 - y^2\vec{i}_2 + yz\vec{i}_3)dV$$

$$= \iiint_V \left[\frac{\partial}{\partial x}(4xz) - \frac{\partial}{\partial y}(y^2) + \frac{\partial}{\partial z}(yz)\right] dV$$

$$= \iiint_V [4z - 2y + y]\, dV = \iiint_V [4z - y]\, dV$$

Since, the surface s is bounded by $x = 0, 1$; $y = 0, 1$ and $z = 0, 1$ so, putting the limits, we have:

$$\iint_s \vec{F} \cdot d\vec{s} = \int_0^1\int_0^1\int_0^1 (4z - y)\, dx\,dy\,dz = \int_0^1\int_0^1 \left(\frac{4z^2}{2} - yz\right)_0^1 dx\,dy$$

$$= \int_0^1\int_0^1 (2 - y)\,dx\,dy = \int_0^1 \left(2y - \frac{y^2}{2}\right)_0^1 dx = \int_0^1 \frac{3}{2} dx = \frac{3}{2}$$

10. (c)

The vector field $\vec{A}$ will be irrotational, if $\nabla \times \vec{A} = 0$.

Now,
$$\nabla \times \vec{A} = \begin{vmatrix} \vec{a}_x & \vec{a}_y & \vec{a}_z \\ \frac{\partial}{\partial x} & \frac{\partial}{\partial y} & \frac{\partial}{\partial z} \\ yz & xz & xy \end{vmatrix}$$

$$= \left[\frac{\partial}{\partial y}(xy) - \frac{\partial}{\partial z}(xz)\right]\vec{a}_x + \left[\frac{\partial}{\partial x}(xy) - \frac{\partial}{\partial z}(yz)\right]\vec{a}_y + \left[\frac{\partial}{\partial x}(xz) - \frac{\partial}{\partial y}(yz)\right]\vec{a}_z$$

$$= [x - x]\,\vec{a}_x + [y - y]\,\vec{a}_y + [z - z]\,\vec{a}_z = 0$$

Hence, $\vec{A}$ is irrotational.

The vector field $\vec{A}$ will be solenoidal, if $\nabla \cdot \vec{A} = 0$

Here,
$$\nabla \cdot \vec{A} = \left(\vec{a}_x \frac{\partial}{\partial x} + \vec{a}_y \frac{\partial}{\partial y} + \vec{a}_z \frac{\partial}{\partial z}\right) \cdot (yz\vec{a}_x + xz\vec{a}_y + xy\vec{a}_z)$$

$$= \vec{a}_x \cdot \vec{a}_x \frac{\partial}{\partial x}(yz) + \vec{a}_y \cdot \vec{a}_y \frac{\partial}{\partial y}(xz) + \vec{a}_z \cdot \vec{a}_z \frac{\partial}{\partial z}(xy) = 0 + 0 + 0 = 0$$

Hence, $\vec{A}$ is solenoidal.

11. (a)

Given,
$$\vec{A} = \frac{1}{\sqrt{x^2 + y^2}}\vec{a}_x$$

$\therefore$
$$\nabla \cdot \vec{A} = \frac{\partial}{\partial x}(A_x) + \frac{\partial}{\partial y}(A_y) + \frac{\partial}{\partial z}(A_z)$$

$$= \frac{\partial}{\partial x}\left(\frac{1}{\sqrt{x^2 + y^2}}\right) + 0 + 0 = \frac{\partial}{\partial x}(x^2 + y^2)^{-1/2} = -\frac{1}{2}(x^2 + y^2)^{-3/2} \cdot 2x$$

$$= \nabla \cdot \vec{A} = -\frac{x}{\sqrt{(x^2 + y^2)(x^2 + y^2)}}$$

Now,
$$(\nabla \cdot \vec{A})_{2,2,0} = -\frac{2}{\sqrt{(2^2 + 2^2) \cdot (2^2 + 2^2)}} = -\frac{2}{\sqrt{8}.8} = -0.0884$$

12. (b)

Given, $\vec{r} = x\vec{i} + y\vec{j} + z\vec{k}$

$\therefore$ $\vec{r} \times \vec{i} = (x\vec{i} + y\vec{j} + z\vec{k}) \times \vec{i} = -y\vec{k} + z\vec{j}$

Also, $\vec{i} \times (\vec{r} \times \vec{i}) = \vec{i} \times (-y\vec{k} + z\vec{j}) = \vec{j}y + z\vec{k}$

Similarly, $\vec{j} \times (\vec{r} \times \vec{j}) = \vec{i}x + \vec{k}z$

and $\vec{k} \times (\vec{r} \times \vec{k}) = \vec{i}x + \vec{j}y$

Thus, $\vec{i} \times (\vec{r} \times \vec{i}) + \vec{j} \times (\vec{r} \times \vec{j}) + \vec{k} \times (\vec{r} \times \vec{k}) = 2(x\vec{i} + y\vec{j} + z\vec{k}) = 2\vec{r}$

13. (d)

Since vector $\vec{V}$ is solenoidal, therefore

$$\nabla \cdot V = 0$$

$$\therefore \left[\vec{i}\frac{\partial}{\partial x} + \vec{j}\frac{\partial}{\partial y} + \vec{k}\frac{\partial}{\partial z}\right] \cdot \begin{bmatrix} \vec{i}(x+3y) + \vec{j}(y-2x) \\ +\vec{k}(x+bz) \end{bmatrix} = 0$$

or, $[1 + 1 + b] = 0$ or $b = -2$

15. (d)

Vector differential operator ('∇') is not a vector quantity. Hence, assertion is a false statement.

16. (b)

All the given statements are correct.

17. (c)

Reason is a statement of stroke's theorem not that of Gauss's divergence theorem.

18. (a)

- Electric potential, $V = \dfrac{\text{Work done}}{\text{Test charge}} = \dfrac{\text{Force} \times \text{Displacement}}{\text{Current} \times \text{Time}}$

$$\therefore \quad [V] = \frac{[MLT^{-2}][L]}{[I][T]} = [ML^2T^{-3}I^{-1}]$$

- Emf induced, $V = \dfrac{d\phi}{dt}$

  or, $\phi = V.t = \text{Magnetic flux}$

  $\therefore$ $[\phi] = [ML^2T^{-3}I^{-1}][T] = [ML^2T^{-2}I^{-1}]$

- Magnetic field intensity,

$$H = \frac{NI}{l}$$

$$\therefore \quad [H] = \frac{[I]}{[L]} = [IL^{-1}]$$

- Magnetic flux density $B = \dfrac{\phi}{A}$

$$\therefore \quad [B] = \frac{[ML^2T^{-2}I^{-1}]}{[L^2]} = [MT^{-2}I^{-1}]$$

19. **(c)**

Option (c) is not correct because cross product of two unlike vectors is a third unit vector having positive sign for normal rotation and negative for reverse rotation while cross product of two like unit vectors is zero.

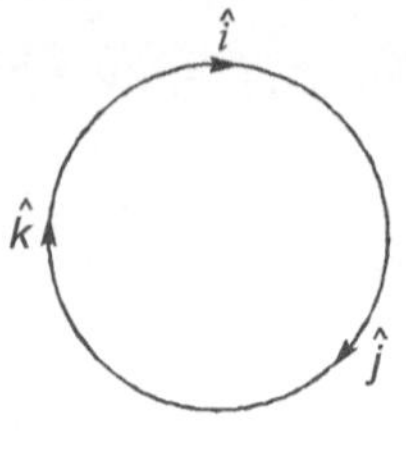

20. **(d)**

Taking the curl, we have:

$$\nabla \times \vec{v} = \nabla \times \vec{\omega} \times \vec{r} \qquad (\text{Since } \vec{v} = \vec{\omega} \times \vec{r})$$

or,
$$\nabla \times \vec{v} = \left(\vec{i}\frac{\partial}{\partial x} + \vec{j}\frac{\partial}{\partial y} + \vec{k}\frac{\partial}{\partial z}\right) \times \begin{vmatrix} \vec{i} & \vec{j} & \vec{k} \\ \omega_x & \omega_y & \omega_z \\ x & y & z \end{vmatrix}$$

$$= \left[\vec{i}\frac{\partial}{\partial x} + \vec{j}\frac{\partial}{\partial y} + \vec{k}\frac{\partial}{\partial z}\right] \times \left[\vec{i}(\omega_y z - \omega_z y) + \vec{j}(\omega_z x - \omega_x z) + \vec{k}(\omega_x y - \omega_y x)\right]$$

$$= (\vec{i} \times \vec{j})\frac{\partial}{\partial x}(\omega_z x - \omega_x z) + (\vec{i} \times \vec{k})\frac{\partial}{\partial x}(\omega_x y - \omega_y x) + (\vec{j} \times \vec{i})\frac{\partial}{\partial y}(\omega_y z - \omega_z y)$$

$$+ (\vec{j} \times \vec{k})\frac{\partial}{\partial y}(\omega_x y - \omega_y x) + (\vec{k} \times \vec{i})\frac{\partial}{\partial z}(\omega_y z - \omega_z y) + (\vec{k} \times \vec{j})\frac{\partial}{\partial z}(\omega_z x - \omega_x z)$$

$$= \vec{k}(\omega_z - 0) - \vec{j}(0 - \omega_y) - \vec{k}(0 - \omega_z) + \vec{i}(\omega_x - 0) + \vec{j}(\omega_y - 0) - \vec{i}(0 - \omega_x)$$

$$= \vec{i}(\omega_x + \omega_x) + \vec{j}(\omega_y + \omega_y) + \vec{k}(\omega_z + \omega_z)$$

$$= 2\omega_x\vec{i} + 2\omega_y\vec{j} + 2\omega_z\vec{k} = 2(\omega_x\vec{i} + \omega_y\vec{j} + \omega_z\vec{k}) = 2\omega$$

$\therefore$
$$\nabla \times \vec{v} = \text{Curl } \vec{v} = 2\omega$$

21. **(c)**

We have:
$$\nabla\vec{r} = \left(\vec{I}_x\frac{\partial}{\partial x} + \vec{I}_y\frac{\partial}{\partial y} + \vec{I}_z\frac{\partial}{\partial z}\right)(x\vec{I}_x + y\vec{I}_y + z\vec{I}_z)$$

$$= (\vec{I}_x \cdot \vec{I}_x)\frac{\partial x}{\partial x} + (\vec{I}_y \cdot \vec{I}_y)\frac{\partial y}{\partial y} + (\vec{I}_z \cdot \vec{I}_z)\frac{\partial z}{\partial z}$$

$$= 1.1 + 1.1 + 1.1 = 1 + 1 + 1 = 3$$

Also,
$$\nabla \times \vec{r} = \begin{vmatrix} \vec{I}_x & \vec{I}_y & \vec{I}_z \\ \frac{\partial}{\partial x} & \frac{\partial}{\partial y} & \frac{\partial}{\partial z} \\ x & y & z \end{vmatrix} = \vec{I}_x\left[\frac{\partial z}{\partial y} - \frac{\partial y}{\partial z}\right] + \vec{I}_y\left[\frac{\partial x}{\partial z} - \frac{\partial z}{\partial x}\right] + \vec{I}_z\left[\frac{\partial y}{\partial x} - \frac{\partial x}{\partial y}\right]$$

$$= \vec{I}_x[0] + \vec{I}_y[0] + \vec{I}_z[0] = 0 + 0 + 0 = 0$$

Thus,
$$\nabla \times \vec{r} = 0$$

Hence, both (a) and (b) will hold true.

22. **(b)**

Let, $\vec{E} = E_x\vec{i} + E_y\vec{j} + E_z\vec{k}$ be any vector field in cartesian co-ordinate system then, we can prove that

$$\nabla\times(\nabla\times\vec{E}) = \nabla\cdot(\nabla\times\vec{E}) - \nabla^2\vec{E} \quad \text{or} \quad \nabla\cdot(\nabla\times\vec{E}) = \nabla\times\nabla\times\vec{E} + \nabla^2\vec{E}$$

Also, Div. Curl $\vec{E} = \nabla\cdot\nabla\times\vec{E}$

$$= \left(\vec{i}\frac{\partial}{\partial x} + \vec{j}\frac{\partial}{\partial y} + \vec{k}\frac{\partial}{\partial z}\right)\cdot\begin{vmatrix} \vec{i} & \vec{j} & \vec{k} \\ \frac{\partial}{\partial x} & \frac{\partial}{\partial y} & \frac{\partial}{\partial z} \\ E_x & E_y & E_z \end{vmatrix}$$

$$= \left(\vec{i}\frac{\partial}{\partial x} + \vec{j}\frac{\partial}{\partial y} + \vec{k}\frac{\partial}{\partial z}\right)\cdot\left[\vec{i}\left(\frac{\partial E_z}{\partial y} - \frac{\partial E_y}{\partial z}\right) + \vec{j}\left(\frac{\partial E_x}{\partial z} - \frac{\partial E_z}{\partial x}\right) + \vec{k}\left(\frac{\partial E_y}{\partial x} - \frac{\partial E_x}{\partial y}\right)\right]$$

$$= \frac{\partial}{\partial x}\left(\frac{\partial E_z}{\partial y} - \frac{\partial E_y}{\partial z}\right) + \frac{\partial}{\partial y}\left(\frac{\partial E_x}{\partial z} - \frac{\partial E_z}{\partial x}\right) + \frac{\partial}{\partial z}\left(\frac{\partial E_y}{\partial x} - \frac{\partial E_x}{\partial y}\right) = 0$$

23. **(a)**

Given, $\vec{b} = \vec{a}\times\vec{c}$ and $\vec{c} = \vec{a}\times\vec{b}$

∴ $\vec{b} = \vec{a}\times(\vec{a}\times\vec{b}) = (\vec{a}\cdot\vec{b})\vec{a} - (\vec{a}\cdot\vec{a})\vec{b} = a^2\vec{b} - a^2\vec{b} = 0$

Also, $\vec{c} = \vec{a}\times\vec{b} = \vec{a}\times(\vec{a}\times\vec{c}) = (\vec{a}\cdot\vec{c})\vec{a} - (\vec{a}\cdot\vec{a})\vec{c} = a^2\vec{c} - a^2\vec{c} = 0$

24. **(d)**

$$\iint_s \vec{A}\cdot\vec{n}\,d\vec{s} = \iiint_s \nabla\cdot\vec{A}\,dV$$

$$= \iiint_V \left[\left(\vec{i}\frac{\partial}{\partial x} + \vec{j}\frac{\partial}{\partial y} + \vec{k}\frac{\partial}{\partial z}\right).(\vec{i}ax + \vec{j}by + \vec{k}cz\right]dV$$

$$= \iiint_V \left[\frac{\partial}{\partial x}(ax) + \frac{\partial}{\partial y}(by) + \frac{\partial}{\partial z}(cz)\right]dV = \iiint_V [(a+b+c)]dV$$

$$= (a+b+c)\iiint_V dV = (a+b+c)V$$

25. **(d)**

Assertion is not true because the laplician operatore (∇^2) of a scalar function ϕ can be defined as "Divergence of the gradient of the scalar ϕ". i.e. $\nabla.\ \nabla\phi$.

27. **(d)**

$$(A\times B)^2 = A^2B^2 - (A\cdot B)^2$$

28. **(a)**

Given, $F = \frac{1}{u}\nabla v$

∴ Curl $F = \nabla\times\left(\frac{1}{u}\nabla v\right)$

or, Curl $F = \nabla\frac{1}{u}\times\nabla v + \frac{1}{u}\nabla\times(\nabla v) = \nabla\frac{1}{u}\times\nabla v$

Hence, $F.\text{Curl}\,F = \frac{1}{u}\nabla v\cdot\left(\nabla\frac{1}{u}\times\nabla v\right) = 0$

29. (c)

A scalar field V is harmonics, if $\nabla^2 V = 0$. Hence, option (c) is not correct.

30. (a)

A phasor is always a vector quantity.

31. (d)

$$\text{Grad } V = \nabla V$$
$$= \left(\frac{\partial V}{\partial x}\right)\hat{a}_x + \left(\frac{\partial V}{\partial y}\right)\hat{a}_y + \left(\frac{\partial V}{\partial z}\right)\hat{a}_z$$
$$= \frac{\partial}{\partial x}(xyz^2)\,\hat{a}_x + \frac{\partial}{\partial y}(xyz^2)\,\hat{a}_y + \frac{\partial}{\partial z}(xyz^2)\,\hat{a}_z$$

or, $$\nabla V = yz^2\hat{a}_x + xz^2\,\hat{a}_y + 2xyz\,\hat{a}_z$$

32. (b)

The laplacian of $V = \nabla^2 V = \frac{\partial^2 V}{\partial x^2} + \frac{\partial^2 V}{\partial y^2} + \frac{\partial^2 V}{\partial z^2}$

$= 2 - 2 - 2 = -2$

33. (b)

By columb's law,

$$\text{electrostatic repulsion} = \frac{1}{4\pi\epsilon_0} \times \frac{Q_1 Q_2}{r^2}$$
$$= \frac{1}{4\pi\epsilon_0} \times \frac{Q^2}{r^2} = 9\times10^9 \times \frac{Q^2}{r^2}$$
$$\left(\because \frac{1}{4\pi\,\epsilon_0} = 9\times10^9\right)$$

Also, by Newton's law of gravitation,

$$\text{gravitational attraction} = G\frac{m_1 m_2}{r^2} = G\,\frac{m^2}{r^2}$$
$$\therefore \frac{\text{Electrostatic repulsion}}{\text{Gravitational attraction}} = \frac{9\times10^9 \times Q^2}{r^2} \times \frac{r^2}{Gm^2}$$
$$= \frac{9\times10^9 \times (1.6\times10^{-19})^2}{6.67\times10^{-11}\times(9.1\times10^{-31})^2}$$
$$= 4.17 \times 10^{42}$$

34. (a)

$$E = \frac{Q}{4\pi\,\epsilon_0\,r^2}\text{ V/m} = \frac{9\times10^9\times2\times10^{-6}}{(0.2)^2}$$
$$= \frac{18}{0.04}\times10^3 \text{ V/m} = 450 \text{ kV/m}$$

35. (c)

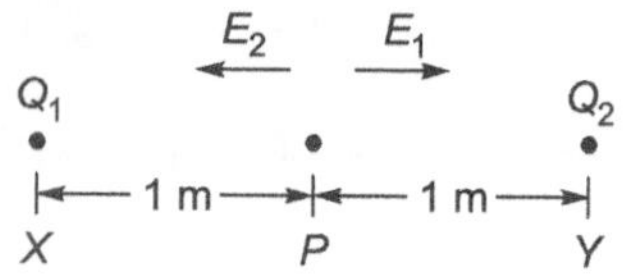

Given, $Q_1 = +3\times10^{-9}$ C and $Q_2 = -2\times10^{-9}$ C

Here, $$E_1 = \frac{Q_1}{4\pi\,\epsilon_0\,r^2} = \frac{9\times10^9\times3\times10^{-9}}{1^2}$$
$$= 27 \text{ V/m} \qquad \text{(Along } PY\text{)}$$

and $$E_2 = \frac{Q_2}{4\pi\,\epsilon_0\,r^2} = \frac{9\times10^9\times(-2\times10^9)}{1^2}$$
$$= -18 \text{ V/m} \qquad \text{(Along } PX\text{)}$$

Hence, electric field intensity at point P is

$$E = E_1 + E_2 = 27 - 18 = 9 \text{ V/m}$$

36. (d)

Electric potential at origin is given as

$$V = \frac{1}{4\pi\,\epsilon_0}\left[\frac{Q_1}{r_1} + \frac{Q_2}{r_2} + \frac{Q_3}{r_3} + \frac{Q_4}{r_4} + \frac{Q_5}{r_5}\right]$$

Since, $Q_1 = Q_2 = ...Q_5$
$= Q = 20\times10^{-9}$ C

Therefore,

$$V = 9\times10^9\times20\times10^{-9}\left[\frac{1}{2}+\frac{1}{3}+\frac{1}{4}+\frac{1}{5}+\frac{1}{6}\right]$$
$$= 180\left[\frac{30+20+15+12+10}{60}\right] \text{ Volts}$$
$$= 180\times\frac{87}{60} \text{ V} = 3\times87 = 261 \text{ Volts}$$

37. (a)

Given, $\epsilon_{r1} = 1.5$, $\epsilon_{r2} = 3.5$,

$$A_1 = A_2 = \frac{2}{2}\text{m}^2 = 1\text{m}^2$$

and $d = 1\times10^{-3}$ m

Here, $$C_1 = \frac{\epsilon_0\epsilon_{r_1} A_1}{d}$$
$$= \frac{8.854\times10^{-12}\times1.5\times(2/2)}{10^{-3}}$$
$$= 13.281\times10^{-9}\text{ F} = 13.281 \text{ nF}$$

and $$C_2 = \frac{\epsilon_0\epsilon_{r_2} A_2}{d} = \frac{8.854\times3.5\times10^{-12}\times1}{10^{-3}}$$
$$= 30.989 \text{ nF}$$

$\therefore \quad C_{equ} = C_1 + C_2$
$= 13.281 + 30.989 = 44.27$ nF

38. (c)

Energy stored $W = \frac{1}{2}VQ$ Joules

$$\therefore \quad W = \frac{1}{2}Q\left[\frac{1}{4\pi\epsilon}\left\{\frac{Q_1}{r_1}+\frac{Q_2}{r_2}+\frac{Q_3}{r_3}+\frac{Q_4}{r_4}\right\}\right]$$
$$= \frac{Q}{8\pi\epsilon}\left[\frac{1}{1}+\frac{-2}{2}+\frac{-3}{3}+\frac{4}{4}\right]$$
$$= \frac{Q}{8\pi\epsilon}[1-1-1+1] = 0 \text{ Joule}$$

39. (a)

The electric field strength at a point in front of an infinite sheet of charge is given by

$$\vec{E} = \frac{\rho_s}{\epsilon_0}\vec{a}_x \text{ Volts/meter}$$

where, ρ_s = charge density and
$\vec{a}_x$ = unit vector normal to the sheet and directed away from the sheet.

Here, $\vec{E}$ is independent of the distance of the point from the sheet.

40. (b)

Both assertion and reason are individually true. However, the correct reason for assertion is that equipotential surfaces are those on which the potential is everywhere the same. Thus, if all the points in space which have the same potential are joined, then equipotential surfaces are obtained.

41. (d)

Statements 2 and 3 are not correct because a homogenous media is one whose physical characteristics do not vary from point-to-point while a non-homogeneous media is one whose physical characteristics vary from point-to-point.

44. (c)

- Assertion is a true statement.
- Reason is a false statement because the permittivity of dielectric is always more than the permittivity of vaccum which makes it a good insulator.
- In a dielectric or insulator the electrons are so tightly bound by their parent nuclei in the equilibrium positions that they can not easily be detached by the applications of electric field. Due to this reason an electric field produces no migration of charge In a dielectric.

45. (c)

Option (c) is Faraday's law given by $V = -\frac{d\phi_m}{dt}$

47. (c)

From Gauss's law, total electric flux from a closed surface is $\psi = \oint \vec{D}\cdot d\vec{s} = Q_{enclosed}$.

48. (b)

For homogeneous medium, $\nabla^2 V = -\rho/\epsilon$ and for non-homogeneous medium ϵ is a variable quantity due to which Poisson's equation becomes $\vec{\nabla}\cdot(\epsilon\vec{\nabla}V) = -\rho$.

51. (c)

Potential is same everywhere on an equipotential surface. Hence, statement 4 is not correct. Statements 1, 2 and 3 are correct.

52. (d)

The total charge on the sheet is

$$Q = \int \rho_s\, dS = \int_0^1\int_0^1 xy(x^2+y^2+25)^{3/2}\,dx\,dy \text{ nC}$$

Since, $x\,dx = \frac{1}{2}d(x^2)$, we now integrate w.r.t. x^2

(or change variables: $x^2 = u$ so that $x\,dx = \frac{du}{2}$)

$$\therefore Q = \frac{1}{2}\int_2^1 y\int_0^1 (x^2+y^2+25)^{3/2}d(x^2)\,dy \text{ nC}$$
$$= \frac{1}{2}\int_0^1 y\cdot\frac{2}{5}(x^2+y^2+25)^{5/2}\Big|_0^1 dy$$
$$= \frac{1}{5}\int_0^1 \frac{1}{2}[(y^2+26)^{5/2}-(y^2+25)^{5/2}]\,d(y^2)$$
$$= \frac{1}{10}\times\frac{2}{7}[(y^2+26)^{7/2}-(y^2+25)^{7/2}]\Big|_0^1$$
$$= \frac{1}{35}[(27)^{7/2}+(25)^{7/2}-2(26)^{7/2}]\text{nC}$$

54. (b)

Required work done,

$$W = -Q\int_A^B E \cdot dl = QV_{AB}$$

$$= Q(V_B - V_A)$$

$$= 10\left(\frac{10}{16}\sin 90° \cos 60° - \frac{10}{1}\sin 30° \cos 120°\right) \times 10^{-6}$$

$$= 10\left(\frac{10}{32} - \frac{-5}{2}\right) \times 10^{-6} = 28.125\ \mu J$$

55. (d)

Total energy contained in the system is:

$$W = W_1 + W_2 + W_3$$

$$= 0 + Q_2V_{21} + Q_3(V_{31} + V_{32})$$

$$= Q_2 \cdot \frac{Q_1}{4\pi \epsilon_0 |(0,0,1) - (0,0,0)|} +$$

$$\frac{Q_3}{4\pi \epsilon_0}\left[\frac{Q_1}{|(1,0,0,) - (0,0,0)|} + \frac{Q_2}{|(1,0,0,) - (0,0,1)|}\right]$$

$$= \frac{1}{4\pi\epsilon_0}\left(Q_1Q_2 + Q_1Q_3 + \frac{Q_2Q_3}{\sqrt{2}}\right)$$

$$= \frac{1}{4\pi \times \dfrac{10^{-9}}{36\pi}}\left(-4 - 3 + \frac{12}{\sqrt{2}}\right) \times 10^{-18}$$

$$= 9\left(\frac{12}{\sqrt{2}} - 7\right) nJ = 13.37\ nJ$$

56. (a)

The charge density of free electron is:

$$\rho_v = ne = (10^{29}) \times (-1.6 \times 10^{-19})$$

$$= -1.6 \times 10^{10}\ C/m^2$$

Given, $E = 10$ mV/m,

$d = 1$ mm,

$\sigma = 5 \times 10^7$ S/m,

$n = 10^{29}/m^3$

Now, current density,

$$J = \sigma E$$

$$= (5 \times 10^7) \times (10 \times 10^{-3})$$

$$= 500\ kA/m^2$$

Also, drift velocity,

$$V_d = \frac{J}{\rho_v} = \frac{\text{Current density}}{\text{Charge density}}$$

$$= \frac{5 \times 10^5}{1.6 \times 10^{10}} = 3.125 \times 10^{-5}\ m/s$$

57. (c)

Given, $d = 5$ mm, $\frac{d}{2} = 2.5$ mm,

$\epsilon_{r_1} = 4$, $\epsilon_{r_2} = 6$,

$$\frac{A}{2} = 15\ cm^2$$

For the given arrangement of dielectrics, D and E are parallel to the dielectric interface. Therefore, we can treat the capacitor as consisting of two capacitors C_1 and C_2 in parallel (same voltage across C_1 and C_2.

Thus, $$C_1 = \frac{\epsilon_0 \epsilon_{r_1} A}{d/2} = \frac{2\epsilon_0 \epsilon_{r_1} A}{d}$$

and $$C_2 = \frac{2\epsilon_0 \epsilon_{r_2} A}{d}$$

The total capacitance/equivalent capacitance between terminals A and B is

$$C_{AB} = \frac{C_1C_2}{C_1 + C_2} = \frac{2\epsilon_0 A}{d}\left(\frac{\epsilon_{r_1}\epsilon_{r_2}}{\epsilon_{r_1} + \epsilon_{r_2}}\right)$$

$$= 2 \times \frac{10^{-9}}{36\pi} \times \frac{30 \times 10^{-4}}{5 \times 10^{-3}} \times \frac{4 \times 6}{10}$$

$$= 25.46\ pF$$

60. (b)

Given, $V = (5x^2y + 3yz^2 + 6xz)$ Volts and $\epsilon_r = 3$

Using Poisson's equation,

$$\nabla^2 V = -\frac{\rho_v}{\epsilon} \text{ or } \rho_v = -\epsilon \nabla^2 V$$

Here, $$\nabla^2 V = \frac{\partial^2 V}{\partial x^2} + \frac{\partial^2 V}{\partial y^2} + \frac{\partial^2 V}{\partial z^2}$$

$$= 10y + 0 + 6y = 16y$$

So, $$\rho_v = -\epsilon\,(10y + 6y)$$

$$\therefore \quad \rho_v|_{at\,(1,2,3)} = -\epsilon_0\epsilon_r(10 \times 2 + 6 \times 2)$$

$$= -\epsilon_0 \times 3 \times 32 = -96\epsilon_0$$

$$= 96\epsilon_0 \text{ (Magnitude only)}$$

61. (c)

On comparing $W_m = \int_V \vec{A} \cdot \vec{J}\, dV$ with electrical energy stored i.e. $W = \frac{1}{2}\int_V \rho_v \cdot V\, dV$ we can find that W_m has the unit of energy.

62. (b)

Given, $y = 6$ m

$$\rho = \frac{1}{600\pi}\ \mu C/m^2$$

The electric field strength due to an infinite charged sheet is given by

$$\vec{E} = \left(\frac{\rho_s}{2\epsilon_0}\right)\hat{a}_n$$

$$\therefore \quad \vec{E} = \left(\frac{10^{-6}}{600\pi \times 2 \times 8.854 \times 10^{-12}}\right)\hat{a}_y$$

$$= 30\hat{a}_y \text{ V/m} = 30\hat{j} \text{ V/m}$$

63. (d)

- Since, $\nabla V \cdot d\vec{s} = -\vec{E}d\vec{s}$

  i.e. $\vec{E} = -\nabla V = -\text{Grad V}$,

  therefore electric field intensity at any point is the negative of the potential gradient at that point. Hence, statement-1 is correct.
- Electric field at a point due to a finite line of charge is given by

$$E = \frac{\rho_L a}{2\pi \in r\sqrt{a^2 + r^2}}$$

Where,

$2a$ = Length of finite lines of charge and

r = Distance of the point on the r-axis from the finite line of charge.

Thus E depends on length of finite lines of charge. Hence, statement-2 is not correct.

- Statement-3 is correct since $\vec{E} = -\text{Grad V}$.
- Electric field at a point due to an infinite line of charge is given by

$$E = \frac{\rho_L}{2\pi \in r}$$

(Where, r = Distance of the point from the infinite line of the charge)

Thus, $E \propto \frac{1}{r}$. Hence, statement -4 is not correct.

64. (a)

$$\vec{F} = \frac{Q_1 Q_2}{4\pi \epsilon_0 r^2}\vec{a}_z \text{ N}$$

$$= \frac{2\times 10^{-6} \times 10 \times 10^{-6}}{4\pi \epsilon_0 \times (0.3)^2}\vec{a}_z \text{ N}$$

$$= \frac{20\times 10^{-12} \times 9 \times 10^{9}}{0.09}\vec{a}_z \text{ N}$$

$$= \frac{20\times 10^{-12} \times 10^{11} \times 9}{9}\vec{a}_z = 2\vec{a}_z \text{ N}$$

$$\therefore \quad |\vec{F}| = 2 \text{ Newton}$$

65. (d)

Since $\vec{D}$ is constant over the area and perpendicular to it, therefore flux $\psi = D{\cdot}A$ Coulomb.

Given, $A = 1\ m^2$,

$$\vec{D} = 10x\vec{a}_x$$

$$\therefore \quad |\vec{D}|_{x=3} = 10 \times 3\vec{a}_x = 30\vec{a}_x$$

or, $|\vec{D}| = 30\ C/m^2$

$$\therefore \quad \psi = 30 \times 1 = 30 \text{ Coulombs}$$

66. (b)

Here, $r = \frac{\sqrt{1^2 + 1^2}}{2} = \frac{\sqrt{2}}{2} = \frac{1}{\sqrt{2}} = 0.707$ m

$\therefore$ Electric potential at point 0 is

$$V = \frac{1}{4\pi \epsilon_0}\left[\frac{Q_1}{r} + \frac{Q_2}{r} + \frac{Q_3}{r} + \frac{Q_4}{r}\right]$$

$$= \frac{9\times 10^9}{0.707}(0.01 - 0.02 + 0.03 + 0.02) \times 10^{-6}$$

$$= \frac{9\times 10^9 \times 0.04 \times 10^{-6}}{0.707} = \frac{9 \times 4 \times 10}{1/\sqrt{2}} = 360\sqrt{2}$$

$= 509.116$ V ≈ 509.2 V

67. (a)

Since $\oint \vec{E}\cdot d\vec{l} = 0$ for a conservative field, therefore it is applicable to only electrostatic field.

68. (d)

Electric potential at a distance r from the electric dipole is

$$V = \frac{Ql\cos\theta}{4\pi \in r^2} = \frac{1}{4\pi \in} \times (Ql) \times \frac{1}{r^2} \times \cos\theta$$

$$\therefore \quad V \propto Ql \propto \text{dipole moment}$$

$$\propto \frac{1}{r^2} \propto \cos\theta \propto l$$

69. (a)

$$\psi = \iint \vec{D} \cdot d\vec{s}$$

or, $$\oint \vec{D} \cdot d\vec{s} = Q_{enc} = \psi$$

70. (a)

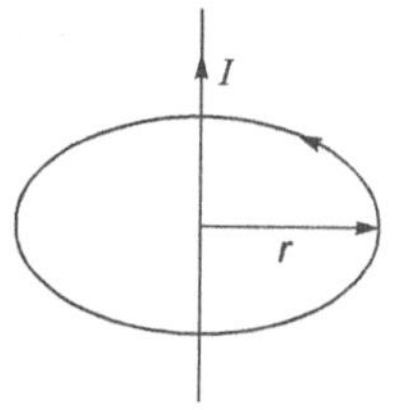

Given, $r = 1$ m, $H = 1$ A/m

Using Ampere's circuit law, we have:

$$\oint \vec{H} \cdot dl = I \text{ or } H \times 2\pi r = I$$

or, $$1 \times 2\pi \times 1 = I \text{ or } I = 2\pi \text{ Ampere}$$

71. (c)

The magnetic flux density at the centre of the square having each side a is given by

$$B = \frac{2\sqrt{2}\mu_0 I}{\pi a}$$

Given, $a = 2$ m, $I = 5$ Amp

$$\therefore \quad B = \frac{2\sqrt{2} \times 4\pi \times 10^{-7} \times 5}{\pi \times 2}$$

$$= 2\sqrt{2} \times 10^{-6}$$

$$= 2.828\ \mu\text{Wb/m}^2$$

72. (b)

Given, $l = 20$ cm $= 0.2$ m,

$r = 0.5$ cm = radius = 0.005 m

$N = 1000$,

$B = 2$ Wb/m^2,

$I = 10$ A

The torque acting on the solenoid is

$$T = \text{NIAB} \sin\theta$$

T will be maximum when

$\theta = 90°$, i.e. $\sin 90° = 1$

$$\therefore \quad T_{max} = 1000 \times 10\,(0.005 \times 0.2) \times 2$$

$$= 20 \text{ N-m}$$

73. (d)

Magnetic flux density at the centre of a current carrying loop is given by

$$B = \frac{\mu I}{2\pi r} = \frac{\mu_0 \mu_r I}{2\pi r}$$

Given, $\mu_r = 1$ (Air), $I = 1$ mA,

$r = 2$ cm $= 0.02$ m

$$\therefore \quad B = \frac{4\pi \times 10^{-7} \times 1 \times 10^{-3}}{2\pi \times 2 \times 10^{-2}}$$

$$= 10^{-8} = 0.01\ \mu\text{Wb}$$

74. (a)

Curl $\vec{H}$ in cartesian coordinate is given by

$$\nabla \times \vec{H} = \begin{vmatrix} \vec{a}_x & \vec{a}_y & \vec{a}_z \\ \frac{\partial}{\partial x} & \frac{\partial}{\partial y} & \frac{\partial}{\partial z} \\ H_x & H_y & H_z \end{vmatrix}$$

$$= \vec{a}_z\left(\frac{\partial H_z}{\partial y} - \frac{\partial H_y}{\partial z}\right) + \vec{a}_y\left(\frac{\partial H_x}{\partial z} - \frac{\partial H_z}{\partial x}\right) + \vec{a}_z\left(\frac{\partial H_y}{\partial x} - \frac{\partial H_x}{\partial y}\right)$$

Since H_x = constant and $H_y = H_z = 0$, therefore

$$\nabla \times \vec{H} = \vec{a}_x(0-0) + \vec{a}_y\left(\frac{\partial}{\partial z}(\text{constant}) - 0\right) + \vec{a}_z\left(0 - \frac{\partial}{\partial y}(\text{constant})\right)$$

$$= 0 + 0 + 0$$

$$\therefore \text{Curl } \vec{H} = \nabla \times \vec{H} = 0$$

77. (b)

Given, $\theta_1 = 45°$, $\mu_1 = 350$ and $\mu_2 = 1$

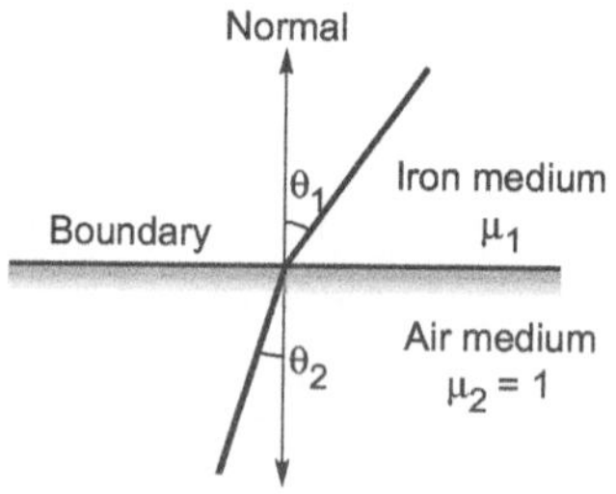

Let medium-1 and medium-2 are respectively iron and air, then

$$\frac{\tan\theta_1}{\tan\theta_2} = \frac{\mu_1}{\mu_2} = \frac{350}{1}$$

or, $$\tan\theta_2 = \frac{1}{350}\tan\theta_1$$

or, $$\theta_2 = \tan^{-1}\left(\frac{1}{350}\tan 45°\right)$$

$$= \tan^{-1}\left(\frac{1}{350}\right)$$

79. (b)

- Both assertion and reason are correct statements.
- The potential in case of magnetic field is a vector potential because source for producing a magnetic field is a current element which has both magnitude and direction.
- The potential in case of electric field i.e. electric potential is a scalar potential because the source for producing an electric field is a charge which is a scaler quantity (having only magnitude).

80. (c)

By Ampere's law,

$$\oint \vec{H} \cdot \vec{dl} = I_{\text{enclosed}} = I$$

Hence, current enclosed $= I = \int_{r=r_0} \vec{H} \cdot \vec{dl}$

$$= \int_0^{2\pi} \frac{1}{r_0}\left[\frac{1}{a^2}\sin ar_0 - \frac{r_0}{a}\cos ar_0\right]\vec{i}_\phi \text{ mA/m}$$

$$= \int_0^{2\pi} \frac{1}{r_0}\left[\frac{4r_0^2}{\pi^2}\sin\frac{\pi r_0}{2r_0} - \frac{r_0 \times 2r_0}{\pi}\cos\left(\frac{\pi}{2r_0}\cdot r_0\right)\right] \times r_0\, d\phi$$

$$= \int_0^{2\pi} \frac{1}{r_0}\left[\frac{4r_0^2}{\pi^2}\sin\frac{\pi}{2} - \frac{2r_0^2}{\pi}\cos\frac{\pi}{2}\right] r_0\, d\phi$$

$$= \int_0^{2\pi} \frac{1}{r_0}\left[\frac{4r_0^2}{\pi^2} - \frac{2r_0^2}{\pi} \times 0\right] r_0\, d\phi = \left[\frac{4r_0^2}{\pi^2}\{\phi\}_0^{2\pi}\right]$$

$$= \frac{4r_0^2}{\pi^2} \cdot 2\pi = \frac{8r_0^2}{\pi} \text{ mA}$$

Given, $r_0 = 1 \text{ cm} = 0.01 \text{ m}$

$$\therefore \quad I = \frac{8}{\pi} \times (0.01)^2 \text{ mA} = 0.25\ \mu\text{A}$$

81. (d)

The mutual inductance is given by

$$M = \frac{\Phi}{I}$$

Flux density at any point P in the x-y plane is

$$B = \frac{\mu I}{2\pi x}$$

and

$$\Phi = \int_{y=0}^{y=h}\int_{x=a}^{x=a+h} \frac{\mu I}{2\pi x} dx\, dy$$

$$= \frac{\mu I}{2\pi}[y]_0^h [ln\, x]_a^{a+h}$$

or,

$$\Phi = \frac{\mu I h}{2\pi} ln\left[\frac{a+h}{a}\right] \text{ Wb}$$

$$\therefore \quad M = \frac{\Phi}{I} = \frac{\mu h}{2\pi} ln\left[\frac{a+h}{a}\right]$$

82. (c)

The force acting between the two parallel wires carrying currents I_1 and I_2 respectively is given by

$$F = \frac{\mu}{4\pi}\frac{I_1 I_2}{r^2} \text{ Newton}$$

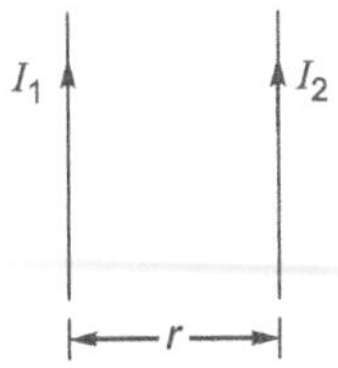

Thus, $F \propto I_1 I_2$

$$\propto \frac{1}{r^2}$$

Hence, assertion is a correct statement.

Reason is a false statement because F will be attractive if the two currents I_1 and I_2 will be in the same directions and repulsive if in opposite directions.

86. (d)

Relative permeability,

$$\mu_r = \frac{\text{Permeability of medium } (\mu)}{\text{Permeability of air}}$$

μ_r is a dimensionless quantity.

87. (c)

- For region-I $(0 < r < a)$, $J(r, \phi, z) = 0$

  Using Ampere's law, we have:

  $$\oint \vec{H} \cdot \vec{dl} = \int \vec{J} \cdot \vec{ds} \int_{r=0}^{a}\int_{\phi=0}^{2\pi} J r\, dr\, d\phi\, \vec{i}_z = 0$$

  or, $H \cdot 2\pi r = 0$ or $H = 0$

- For region-II $(a < r < b)$, $J = J_0\left(\frac{r}{a}\right)\vec{i}_z$

  Again,

  $$\oint \vec{H} \cdot \vec{dl} = \int_{\phi=0}^{2\pi}\int_{r=a}^{r} J_0 \cdot \frac{r}{a} \cdot r\, dr\, d\phi = \frac{J_0}{a} \cdot \frac{r^3}{3}\Bigg|_a^r \phi\Big|_0^{2\pi}$$

  or, $$H \cdot 2\pi r = \frac{2\pi}{3a} J_0 (r^3 - a^3),$$

  or, $$H = \frac{J_0}{3ar}(r^3 - a^3)\vec{i}_\phi$$

- For region-III $(b < r < \infty)$, $J = 0$

  Again,

  $$\oint \vec{H} \cdot \vec{dl} = \int_{\phi=0}^{2\pi}\int_{r=a}^{r=b} J r\, dr\, d\phi\, \vec{i}_z = 0$$

88. (a)

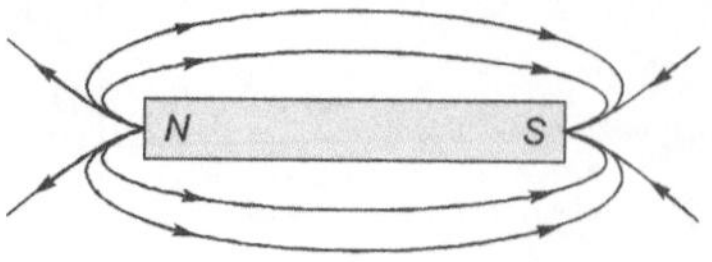

Since, magnetic flux lines always close upon themselves therefore it is not possible to have isolated magnetic charges (or poles).

90. (a)

The total magnetic flux crossing the surface $\phi = \frac{\pi}{2}$, $1 \le \rho \le 2$ m, $0 \le z \le 5$ m is given as:

$$\Psi = \int B \cdot ds = \frac{1}{2}\int_{z=0}^{5}\int_{\rho=1}^{2} \rho d\rho dz = \frac{1}{4}\rho^2\Big|_1^2 (5) = \frac{15}{4}$$

$= 3.75$ Wb

91. (b)

An accelerated charge will produce both time varying electric and magnetic field called electromagnetic field.

92. (c)

- Static magnetic field is non-conservative unlike static electric field. Hence, statement-4 is not correct.
- Magnetic flux lines are always closed. Hence, statement-2 is not correct.
- Statement 1 and 3 are correct.

93. (b)

The Henry, the unit for mutual inductance is equivalent to V-sec /Ampere.
So, $H \times$ Ampere = Volt-sec = V-s

94. (d)

Magnetic flux $\phi_B = BA\cos\theta$

or, $A = \frac{\phi_B}{B\cos\theta} = \frac{9.5}{19\cos 60°} = 1\,\text{m}^2$

Hence, the area of the rectangular coil should be 1 m^2. Only option (d) is matching the answer i.e. 250 cm × 40 cm = 10000 cm^2 = 1 m^2.

95. (a)

Flux $\phi = BA\cos\theta$

At $\theta = 45°$, $\phi_1 = 3 \times 5 \times \cos 45° = 11$ Wb

At $\theta = 90°$, $\phi_2 = 3 \times 5 \times \cos 90° = 0$ Wb

$\therefore$ Change in flux $= \phi_2 - \phi_1 = -11$ Wb

96. (b)

$$\oint_s \vec{B} \cdot d\vec{s} = 0 = \nabla \cdot \vec{B}$$

(Non-existence of monopole)

97. (a)

$$\oint \vec{A} \cdot \vec{dl} = \oint(\hat{a}_\rho + \hat{a}_\phi + \hat{a}_z) \cdot (d\rho\hat{a}_\rho + \rho d\phi\hat{a}_\phi + dz\,\hat{a}_z)$$

For path OP, $\oint \vec{A} \cdot \vec{dl} = \int_0^2 d\rho = 2$

For path PQ, $\oint \vec{A} \cdot \vec{dl} = \int_0^{\pi/2} \rho d\phi = 2 \times \frac{\pi}{2} = \pi$

For path QO, $\oint \vec{A} \cdot \vec{dl} = \int_2^0 d\rho = -2$

$\therefore \quad \oint \vec{A} \cdot \vec{dl} = 2 + \pi - 2 = \pi$

99. (a)

$$\oint \vec{A} \cdot \vec{dl} = \int_0^1 2r\cos\phi T_r \cdot dr T_r + \int_0^{\pi/2} 2r\cos\phi \cdot r d\phi (T_r \cdot T_\phi) + \int_1^0 2r\cos\phi \cdot T_r \cdot dr T_r$$

As, $T_r \cdot T_\phi = 0$ and $\cos\frac{\pi}{2} = 0$, therefore

$\oint \vec{A} \cdot \vec{dl} = 1 + 0 + 0 = 1$

101. (c)

Total resultant force is

$$F = F_{AD} - F_{BC}$$

$$= \frac{\mu_0\, i_1(t) \cdot i_2(t)}{2\pi}\left(1 - \frac{1}{3}\right)$$

$$= \frac{4}{3\pi}\mu_0 \cos^2 \omega t$$

102. (a)

$Mmf = \oint(\vec{H}.\vec{dl})$ = NI Ampere-turn

$= 100 \times 100 \times 10^{-3} = 10$ AT

103. (c)

$\nabla \times \vec{H} = \frac{\partial \vec{D}}{\partial t} + \vec{J}$, where J is current density in A/m^2. Hence, unit of $\nabla \times \vec{H}$ = Ampere/m^2

104. (b)

Maxwell equations can be used to determine the conditions at the boundary surface of two different media.

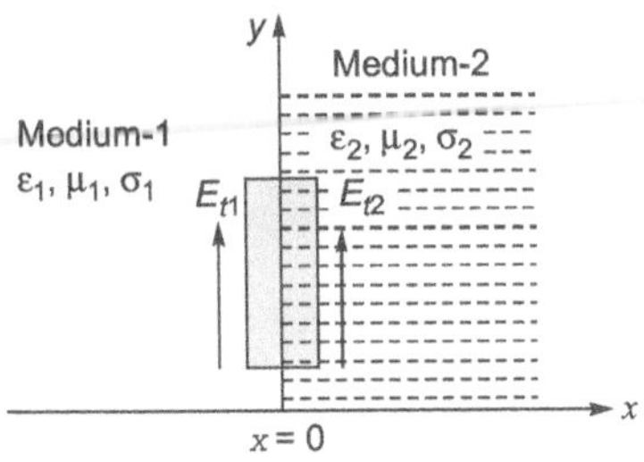

Boundary surface between two media

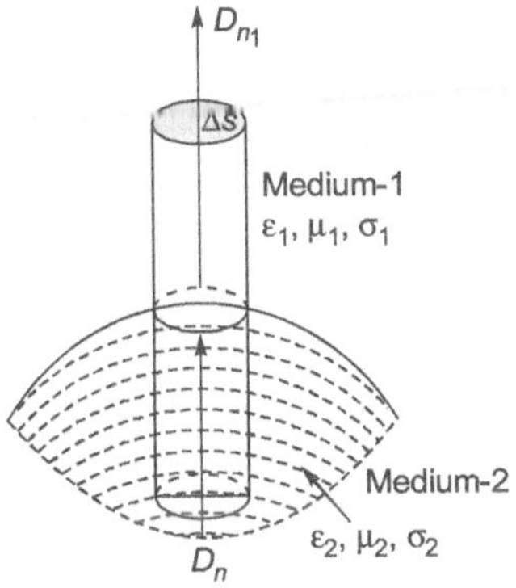

A "pill-box" volume enclosing a portion of boundary surface

Here, $E_{t1} = E_{t2}$;

$H_{t1} = H_{t2}$;

$B_{n1} = B_{n2}$

and $D_{n1} = D_{n2}$

105. (b)

Since, $\nabla \times \vec{H} = \vec{J} + \dfrac{\partial \vec{D}}{\partial t}$ and as the field is invariant, therefore

$$\nabla \times \vec{H} = \vec{J} + 0 = \vec{J}$$

$$\text{or, } \vec{J} = \begin{vmatrix} \vec{a}_x & \vec{a}_y & \vec{a}_z \\ \dfrac{\partial}{\partial x} & \dfrac{\partial}{\partial y} & \dfrac{\partial}{\partial z} \\ 0 & 0 & (3x\cos\beta + 6y\sin\gamma) \end{vmatrix}$$

$$= \vec{a}_x\left[\frac{\partial}{\partial y}\{3x\cos\beta + 6y\sin\gamma\} - 0\right]$$

$$-\vec{a}_y\left[\frac{\partial}{\partial x}\{3x\cos\beta + 6y\sin\gamma\} - 0\right]$$

$$= \vec{a}_x[0 + 6\times 1\times\sin\gamma] - \vec{a}_y[3\times 1\times\cos\beta + 0]$$

$$\text{or, } \vec{J} = (6\sin\gamma)\vec{a}_x - (3\cos\beta)\vec{a}_y$$

106. (a)

Let the copper wire lies along the z-axis of the cylindrical coordinate system. The magnetic field $\vec{H}$ is then directed in azimuthal direction $\vec{a}_\phi$ and electric field $\vec{E}$ is directed in the z-direction.

Given, $I = 2$ A, $\sigma = 5.8 \times 10^7$ mho/m,

$A = 1.5 \times 10^{-5}$ m^2,

$f = 10^9$ Hz

Now, $\vec{J} = \sigma\vec{E}$

$$\text{or, } \vec{E} = \frac{J_z}{\sigma}\vec{a}_z = \frac{I}{\sigma A}\vec{a}_z$$

$$\text{or, } \vec{E} = \frac{2}{5.8\times 10^7 \times 1.5\times 10^{-5}}\vec{a}_z$$

$$= 2.298 \times 10^{-3}\ \vec{a}_z \text{ V/m}$$

Since, $\vec{J}_d = \omega\epsilon\vec{E}$

and $\vec{J}_c = \sigma E$, therefore

$$I_d = \int \vec{J}_d \cdot d\vec{s} = \omega\epsilon|\vec{E}|A$$

$$= 2\pi f\epsilon_0 \times 2.298 \times 10^{-3} \times 1.5 \times 10^{-5}$$

$$(\because \int d\vec{s} = A)$$

$$= 2\pi\times 10^9 \times \frac{1}{36\pi\times 10^9} \times 2.298 \times 10^{-7} \times 1.5$$

$$= \frac{0.3447\times 10^{-7}}{18} = 1.9 \times 10^{-9}\text{ A}$$

or, $I_d = 1.9$ nA

107. (d)

The induced emf in uniform magnetic field is given by

$$V = \omega\,(ab)\,B\sin\omega t$$

$$\text{or, } V = \omega AB\sin\omega t$$

(where, $A = ab$ = Area of rectangular loop)

$$\text{or, } |V| = |2\pi f\, AB\sin\omega t|$$

$$= |2\pi\times 50\times 1\times 80\times 10^{-2}\times 0.1|$$

$$= 25.13 \text{ volts}$$

108. (c)

Induced emf in the conductor is

$$V = \oint(\vec{v}\times\vec{B})\cdot d\vec{l} = \oint(\vec{v}\times\vec{B})\cdot d\vec{l}$$

$$= vBl = 10 \times 1 \times 0.5$$

$$= 5 \text{ volts}$$

109. (b)

$$\omega = \frac{2\pi}{60}N = 2\pi f$$

where, N = Number of revolutions/min

$$\therefore \quad \omega = \frac{2\pi}{60}\times 5000 \text{ rad/sec;}$$

$a = 0.01$ m

$B = 250 \times 10^{-3}$ Wb/m^2

The emf induced is,

$$V = \frac{1}{2}\omega B a^2$$

$$= \frac{1}{2}\times\frac{2\pi}{60}\times 5000\times 250\times 10^{-3}\times(0.10)^2$$

$$= 0.654 \text{ volt} \approx 0.65 \text{ volt}$$

110. (a)

Given, $B = 0.1$ Wb/m^2

$\omega = 377$ rad/s

$r = 0.5$ m

$$\text{Emf developed} = \oint \vec{E}\cdot d\vec{l} = \frac{1}{2}\omega B a^2$$

$$= \frac{1}{2}\times 377\times 0.1\times(0.5)^2$$

$$= 4.71 \text{ volt} \approx 4.7 \text{ volt}$$

111. (c)

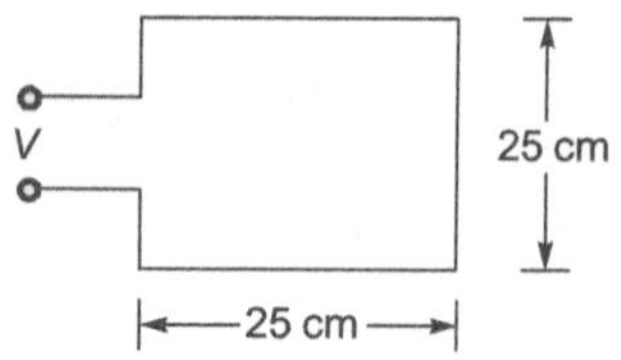

Given, $A = 25 \times 25 \times 10^{-4}$ m^2

$= 0.0625$ m^2

$\omega = 2\pi f = 2 \times 3.14 \times 10 \times 10^6$

$= 6.28 \times 10^7$ rad/sec

$H = A \sin \omega t$

$= 1 \sin (6.28 \times 10^7)t$ Amp/m

The required induced voltage is given by

$$V = \oint \vec{E}\cdot d\vec{l} = -\int \frac{\partial \vec{B}}{\partial t}\cdot d\vec{s}$$

Since, $\vec{B} = \mu_0 \vec{H}$, therefore

$$\frac{\partial \vec{B}}{\partial t} = \mu_0 \frac{\partial \vec{H}}{\partial t}$$

$$= \mu_0 \frac{\partial}{\partial t}(\sin 6.28\times 10^7 t)$$

$$= 4\pi\times 10^{-7}[6.28\times 10^7 \cos 6.28\times 10^7 t]$$

$$= 4\times 3.14\times 6.28 \cos(6.28\times 10^7)t$$

$$= 78.87 \cos(6.28\times 10^7)\, t$$

Hence, the maximum induced voltage is

$$V = -\int \frac{\partial \vec{B}}{\partial t}\cdot d\vec{s}$$

$$= -78.87 \cos(6.28\times 10^7)t \int d\vec{s}$$

$$= -78.87 \cos(6.28\times 10^7)t \times 0.0625$$

or, $|V| = |78.87 \times 0.0625|$

$= 4.9$ volt

112. (b)

Both assertion and reason are individually correct statements.

Since induced emf,

$$e = \frac{-d\psi}{dt} \quad (\psi = \text{total flux}),$$

thus due to negative sign, the total emf induced is equal to the time rate of decrease of the total magnetic flux linking the circuit. (The negative sign supports the Lenz's law).

113. (c)

- Assertion (A) is a correct statement since the motional induction or flux cutting law are given by

$$V = \int(\vec{v}\times\vec{B})\cdot d\vec{l} \text{ volts}$$

or, $V = vBl$ volts

- Motional emf equation depends only on the velocity of the conductor and not its position. Thus, reason (R) is not a correct statement.

114. (d)

Given, $|J_c| = |J_d|$

or, $\left|\frac{J_c}{J_d}\right| = 1$

Given, $E = 250 \sin 10^{10}\, t$ V/m

$= E_0 \sin\omega t$

$\therefore \quad |J_c| = \sigma E_0$

= Conduction current density

and $\vec{J}_d = \frac{\partial \vec{D}}{\partial t}$

= Displacement current density

$$= \epsilon\frac{\partial \vec{E}}{\partial t} = \epsilon_0\epsilon_r \frac{\partial}{\partial t}[E_0 \sin\omega t]$$

$\therefore \quad |\vec{J}_d| = \epsilon_0 \epsilon_r E_0 \omega$

Now, $\quad \left|\dfrac{\vec{J}_c}{\vec{J}_d}\right| = \dfrac{\sigma E_0}{\epsilon_0 \epsilon_r E_0 \omega} = 1$

or, $\quad \dfrac{\sigma}{\epsilon_0 \omega} = 1 \quad (\epsilon_r = 1)$

or, $\quad \omega = \dfrac{\sigma}{\epsilon_0} = \dfrac{5}{(1/36\pi \times 10^9)}$

$= 5 \times 36\,\pi \times 10^9 = 2\pi f$

or, $\quad f = \dfrac{5 \times 36\pi \times 10^9}{2\pi}$

$= 5 \times 18 \times 10^9 \text{ Hz} = 90 \text{ GHz}$

116. (b)

Since the loop begins outside of any magnetic field, therefore initial magnetic flux is zero.
The final flux is given by

$\Phi = B \cdot A = 15 \times (\pi \times (0.07)^2)$

$= 0.23 \text{ Wb}$

Using Faraday's law, emf induced in the loop is given by

$e = -N\dfrac{\Delta\Phi}{\Delta t}$

$= -25 \times \dfrac{0.23}{10} = -0.58 \text{ V}$

$\therefore \quad |e| = 0.58 \text{ volt}$

117. (c)

The motional emf induced when the conductor's velocity, length and the magnetic field are all perpendicular is given by

$e = Blv$

$= 37.5 \times 0.18 \times 0.01$

$= 0.07 \text{ V}$

118. (d)

The electric field produced is given by

$E = vB = 0.01 \times 37.5$

$= 0.375 \text{ N/C} = 0.375 \text{ V/m}$

or, $\quad E \approx 0.38 \text{ N/C or V/m}$

119. (a)

An emf is not induced if the plane of the coil is parallel to the lines of the magnetic field since the magnetic flux through the coil does not change when the coil shrinks.

120. (a)

Induced emf,

$e = N\dfrac{d\Phi}{dt} = 200 \times 2 = 400 \text{ volt}$

121. (b)

The transformer emf is given by

$V_{emf} = -\int \dfrac{\partial B}{\partial t} \cdot dS$

$= \int_{y=0}^{0.08} \int_{x=0}^{0.06} 4 \times 10^{-3} \times 10^6 \sin 10^6 t \, dx\, dy$

$= 19.2 \sin 10^6\, t \text{ V}$

$\therefore \quad |V_{emf}|_{max} = 19.2 \text{ V}$

122. (a)

$D = \epsilon E = \epsilon \dfrac{V}{d}$

$J_d = \dfrac{\partial D}{\partial t} = \dfrac{\epsilon}{d}\dfrac{dV}{dt} \qquad (S \to \text{Area})$

Hence, $\quad I_d = J_d \cdot S = \dfrac{\varepsilon S}{d} \cdot \dfrac{dV}{dt} = C\dfrac{dV}{dt}$

Which is the same as the conduction current given by

$I_c = \dfrac{dQ}{dt} = S\dfrac{d\rho_s}{dt} = S\dfrac{dD}{dt}$

$= \epsilon S \dfrac{dE}{dt} = \dfrac{\epsilon S}{d}\dfrac{dV}{dt} = C\dfrac{dV}{dt}$

$\therefore \quad I_d = 2 \times \dfrac{10^{-9}}{36\pi} \times \dfrac{5 \times 10^{-4}}{3 \times 10^{-3}}$

$\times 10^3 \times 50 \cos 10^3 t$

$= 147.4 \cos 10^3 t \text{ nA}$

123. (a)

Given, $\quad \Phi = (t^3 - 2t) \text{ mWb}$

$\therefore \quad \dfrac{d\Phi}{dt} = (3t^2 - 2) \text{ mWb/sec}$

Emf induced is

$e = -N\dfrac{d\Phi}{dt}$

$= -[100 \times (3t^2 - 2)] \times 10^{-3} \text{ volt}$

$\therefore \quad e|_{t=2s} = -[100 \times (3 \times 2^2 - 2) \times 10^{-3}] \text{ volt}$

$= -(100 \times 10^{-3} \times 10) = -1 \text{ volt}$

127. (c)

$\nabla \times E = 0$ for conservative field

128. (d)

$$V_{emf} = \int (u \times B)\cdot dl$$

Given, $dl = dl_{DC} = dZ_{az}$

$$u = \frac{dl}{dt} = \frac{\rho d\phi}{dt} a_\phi = \rho\omega a_\phi$$

$$\rho = AD = 4 \text{ cm}, \omega = 2\pi f = 100\,\pi$$

$$B = B_0\, a_x = B_0 (\cos\phi\, a_\rho - \sin\phi\, a_\phi)$$

and
$$u \times B = \begin{vmatrix} a_\rho & a_\phi & a_z \\ 0 & \rho\omega & 0 \\ B_0 \cos\phi & -B_0 \sin\phi & 0 \end{vmatrix}$$

$$= -\rho\omega B_0 \cos\phi\, a_z$$

Now, $(u \times B)\cdot dl = -\rho B_0 \omega \cos\phi\, dz$

$$= -0.2\,\pi \cos\phi\, dz$$

$$\therefore \quad V_{emf} = \int_{z=0}^{z=0.03} (-0.2\pi \cos\phi) dz$$

$$= -6\,\pi \cos\phi \text{ mV}$$

Also, $V_{emf} = -6\,\pi \cos(\omega t + \pi/2)$, at $t = 5$ ms

$$\therefore \quad |V_{emf}| = 6\,\pi \text{ mV}$$

129. (a)

$$I = \iint_S J\cdot n\Big|_{z=0.2} ds$$

or,
$$I = \int_0^{2\pi}\int_0^{4} \left(\frac{40}{\rho^2 + (0.1)^2}\right) \rho d\rho d\phi$$

$$= \frac{40}{2} \log[\rho^2 + (0.1)^2]\Big|_0^{0.4} \cdot (2\pi)$$

or, $I = 40\pi \log 17 = 356$ A

130. (c)

$$B = \nabla \times A = \frac{-\partial A_z}{\partial\rho} a_\phi = \frac{\rho}{2} a_\phi$$

$$dS = d\rho \cdot dz \cdot a_\phi$$

$\therefore$ ψ = Total magnetic flux

$$= \frac{1}{2}\int_{z=0}^{5}\int_{\rho=1}^{2} \rho d\rho dz = \frac{15}{4} \text{ Wb}$$

131. (a)

We have: $\left|\frac{\vec{J}_c}{\vec{J}_d}\right| = \frac{\sigma}{\omega \in}$

If $\frac{\sigma}{\omega \in} >> 1$, then $|\vec{J}_d| << |\vec{J}_c|$ so the medium behaves as a good conductor.

If $\frac{\sigma}{\omega \in} = 1$, then $|\vec{J}_d| = |\vec{J}_c|$ so the medium behaves as a semi-conductor.

If $\frac{\sigma}{\omega \in} << 1$, then $|\vec{J}_d| >> |\vec{J}_c|$ so the medium behaves as a good insulator.

Hence, only statement-1 is correct.

132. (c)

- Velocity of propagation of EM-wave in free space is equal to $\frac{1}{\sqrt{\mu_0 \varepsilon_0}}$ = speed of light (3×10^8 m/s). Hence statement-2 is not correct.
- $\vec{E}$ and $\vec{H}$ oscillate in phase and the ratio of their amplitude is constant being equal to $\sqrt{\frac{\mu_0}{\varepsilon_0}}$ = 120π or 377. Hence, statement-3 is not correct.

Thus, statements-1 and 4 are only correct.

133. (a)

Intrinsic impedance or characteristic impedance of the non-conducting medium in free space is given as

$$\eta = \eta_0 = \sqrt{\frac{\mu_0}{\varepsilon_0}} = \sqrt{\frac{4\pi \times 10^{-7}}{\left(\frac{1}{36\pi \times 10^9}\right)}} = 120\,\pi \text{ Ohms}$$

134. (b)

We know that,

$$\eta = \frac{E}{H} \text{ or } H = \frac{E}{\eta}$$

Here, $\eta = 120\,\pi$ ohm in free space

So, $H = \frac{1}{120\pi}$ A/m $= 0.0026$ A/m

= 2.6 mA/meter

135. (d)

Velocity of propagtion is given by

$$v = \frac{1}{\sqrt{\varepsilon\mu}} = \frac{1}{\sqrt{\varepsilon_0\, \varepsilon_r \times \mu_0\, \mu_r}}$$

Given, $\varepsilon_r = 5$, $\mu_r = 1$

$$\therefore \quad v = \frac{1}{\sqrt{5\varepsilon_0 \times \mu_0}} = \frac{3 \times 10^8}{\sqrt{5}} = 1.3 \times 10^8 \text{ m/s}$$

136. (b)

Poynting vector $P = EH$ and total energy density $W = \mu H^2$.

$$\therefore \quad \frac{P}{W} = \frac{EH}{\mu H^2} = \frac{1}{\mu}\frac{E}{H} = \frac{1}{\mu}\sqrt{\frac{\mu}{\varepsilon}}$$

$$= \frac{1}{\sqrt{\mu\varepsilon}} = \text{Velocity}$$

Hence, average energy density

$$= \frac{\text{Poynting vector}}{\text{Average velocity}} = \frac{1\,\text{W/m}^2}{3\times10^8\,\text{m/s}}$$

$$= \frac{1}{3}\times10^{-8}\,\text{W-s/m}^2 = 3.3\,\text{nJ/m}^3$$

137. (a)

Given, $\omega = 10^8$ rad/sec,

$E_0 = 30\pi$ V/m

This is a plane wave.

For plane wave in space,

$$c = \frac{\omega}{\beta} = 3\times10^8\,\text{m/s}$$

or, $$\beta = \frac{\omega}{c} = \text{Wavelength}$$

$$= \frac{10^8}{3\times10^8} = \frac{1}{3}\,\text{rad/m}$$

Also, $$\frac{E_0}{H_0} = 120\pi\,\Omega$$

and $$H_0 = \pm\frac{E_0}{120\pi} = \pm\frac{30\pi}{120\pi} = \pm\frac{1}{4}\,\text{A/m}$$

138. (c)

We have: $$\frac{\omega}{\beta} = c = \frac{1}{\sqrt{\mu\varepsilon}} = \frac{c}{\sqrt{\mu_r\varepsilon_r}} = \frac{3\times10^8}{\sqrt{\mu_r\varepsilon_r}}$$

Given, $\mu_r = 1$; $\beta = \frac{4}{3}$; $E_0 = 30\pi$ V/m

and $H_0 = 1$ A/m

$$\therefore \quad \frac{\omega}{\beta} = \frac{3\times10^8}{\sqrt{\varepsilon_r}}$$

or, $$\omega = \frac{4}{3}\times\frac{3\times10^8}{\sqrt{\varepsilon_r}} = \frac{4\times10^8}{\sqrt{\varepsilon_r}}$$

Now, $$\frac{E_0}{H_0} = \sqrt{\frac{\mu}{\varepsilon}} = \sqrt{\frac{\mu_0\varepsilon_r}{\varepsilon_0\varepsilon_r}} = \sqrt{\frac{\mu_0}{\varepsilon_0}}\sqrt{\frac{\mu_r}{\varepsilon_r}}$$

$$= 120\pi\sqrt{\frac{\mu_r}{\varepsilon_r}} = \frac{120\pi}{\sqrt{\varepsilon_r}}$$

or, $$\frac{30\pi}{1} = \frac{120\pi}{\sqrt{\varepsilon_r}} \text{ or } \varepsilon_r = 16$$

Hence, $$\omega = \frac{3\times10^8}{\sqrt{\mu_r\,\varepsilon_r}}\times\beta = \frac{3\times10^8}{\sqrt{16\times1}}\times\frac{4}{3}$$

$$= 10^8\,\text{rad/s}$$

139. (d)

Looking at the phase $(\omega t - \beta z)$, we can find that direction of propagation is $+z$.

Since $\vec{E}\times\vec{H}$ must also lie in $a + z$ direction, $\vec{H}$ must have the direction $-\vec{a}_x$.

Hence, $$\frac{E_y}{-H_x} = \eta_0 = 120\pi$$

or, $$H_x = \frac{E_y}{-120\pi} = \frac{10^3\sin(\omega t - \beta z)}{-120\pi}$$

or, $$H_x = \frac{-10^3}{120\pi}\sin(\omega t - \beta z)\,\text{A/m}$$

Writing in vector form, we have:

$$\vec{H}(z,t) = \frac{-10^3}{120\pi}\sin(\omega t - \beta)\vec{a}_x\ \text{A/m}$$

$$= \frac{-25}{3\pi}\sin(\omega t - \beta)\vec{a}_x\ \text{A/m}$$

141. (c)

$$\frac{\vec{J}_c}{\vec{J}_d} = \frac{\sigma}{j\omega\varepsilon}$$

which indicates that the two vectors $\vec{J}_c$ and $\vec{J}_d$ are out of phase in time and are in the same direction in the space. Hence, reason is not a correct statement.

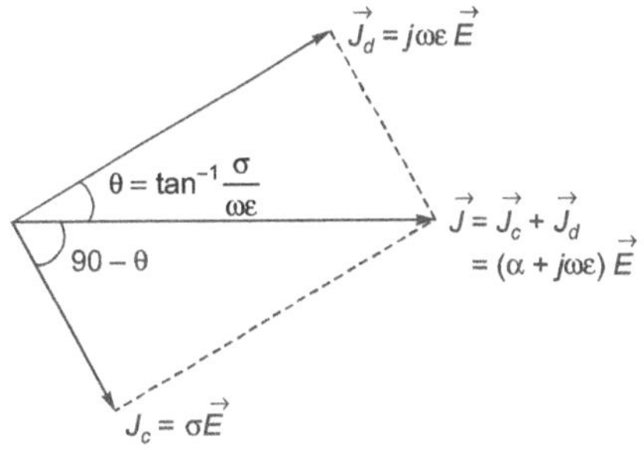

Here, $$\tan\theta = \frac{\sigma}{\omega\varepsilon} = \text{loss tangent}$$

142. (b)

Given, $\sigma = 0.25$ pS/m; $\mu_r = 1$; $\varepsilon_r = 8$;

$f = 1.6\times10^6$ Hz

Now, $$\frac{\sigma}{\omega\varepsilon} = \frac{\sigma}{\omega\varepsilon_0\varepsilon_r}$$

$$= \frac{0.25\times10^{-12}\times36\pi\times10^9}{2\pi\times1.6\times10^6\times8\times1}$$

$$= \frac{4.5\times10^{-9}}{12.8}$$

or, $$\frac{\sigma}{\omega\varepsilon} = 0.3515\times10^{-9} \approx 0$$

or, $$\sigma \approx 0$$

Since $\sigma = 0$, therefore the material will behave as a perfect dielectric ($\alpha = 0$).

$$\therefore \quad \beta = \omega\sqrt{\mu\varepsilon} = 2\pi\sqrt{\mu_0\mu_r\cdot\varepsilon_0\varepsilon_r}$$

$$= 2\pi\times1.6\times10^6\sqrt{4\pi\times10^{-7}\times1\times\frac{1}{36\pi\times10^9}\times8}$$

$$= \frac{2\pi\times1.6\times10^6\times10^{-8}\times2\sqrt{2}}{3}$$

$= 9.471\times10^{-2}$ rad/m

$\therefore$ Propagation constant,

$$\gamma = \alpha + j\beta = 0 + j9.471\times10^{-2}$$

$$\approx j0.95\ \text{m}^{-1}$$

143. (d)

Skin depth,

$$\delta = \frac{1}{\sqrt{\pi f\mu_0\sigma}}$$

$$= \frac{1}{\sqrt{\pi\times10\times10^9\times4\pi\times5.8\times10^7\times10^{-7}}}$$

$$= \frac{1}{3.14\times2\sqrt{10^{10}}\times5.8}\ \text{metre}$$

$$= \frac{10^{-5}}{6.28\times2.40}\ \text{metre} = \frac{10^{-5}}{15.072}\ \text{metre}$$

$= 0.66\times10^{-6}$ m $= 0.66\ \mu$m

144. (a)

For a lossless or perfect medium, velocity of plane wave is given by

$$v_p = \frac{\omega}{\beta} = \frac{\omega}{\omega\sqrt{\mu\varepsilon}} = \frac{1}{\sqrt{\mu\varepsilon}}$$

$$= \frac{1}{\sqrt{\mu_0\,\varepsilon_0\,\mu_r\,\varepsilon_r}} = \frac{c}{\sqrt{\mu_r\,\varepsilon_r}}$$

$$= \frac{3\times10^8}{\sqrt{1\times4}} = \frac{3}{2}\times10^8\ \text{m/s}$$

$= 1.5\times10^8$ m/s

145. (c)

The average power per unit area is equal to the Poynting vector, the magnitude of which is given by

$$|\vec{P}_{av}| = \frac{E_0^2}{2\eta_0}\ \text{W/m}^2 = \frac{E_1^2+E_2^2}{2\eta_0}\ \text{W/m}^2$$

$$= \frac{2^2+4^2}{2\times377}\ \text{W/m}^2 = 26.5\ \text{mW/m}^2$$

146. (b)

- A wave whose phase and magnitude both are constant is called uniform plane wave. Hence, statement-2 is not correct.
- A uniform plane is one in which $\vec{E}$ and $\vec{H}$ lie in a plane and have the same value everywhere in that plane at any fixed instant.

  The $\vec{E}$ and $\vec{H}$ fields are uniform over plane surfaces. Hence, statement-3 is also not correct.

147. (d)

Assertion is not correct because it is not the displacement current which directly passes through a capacitor but it is only apparant current representing the rate at which the flow of charge takes place from one plate of the capacitor to the other plate.

148. (c)

In a homogeneous, isotropic, linear and stationary media, $\nabla\cdot E = \rho = 0$ and both $\vec{E}$ and $\vec{H}$ are transverse.

149. (c)

The depth of penetration (δ) is defined as that depth in which the wave has been attenuated to '$1/e$' percent (37%) of its original value. Hence, reason is not a correct statement.

152. (d)

$$\text{Skin depth } (\delta) = \frac{1}{\alpha} = \frac{1}{\sqrt{\pi f\mu\sigma}}$$

$$\therefore \quad \delta \propto \frac{1}{\sqrt{f}}$$

When the frequency is increased 4-times, then skin depth will become half.

153. (a)

Given, $E = 100 \cos(10^9 t - 4x)$

or, $E = E_0 \cos(\omega t - \beta x)$

Here, $\omega = 10^9$ rad/s and $\beta = 4$

Thus, velocity of propagation is

$$v = \frac{\omega}{\beta} = \frac{10^9}{4} = 2.5 \times 10^8 \text{ m/s}$$

and wavelength,

$$\lambda = \frac{2\pi}{\beta} = \frac{2\pi}{4} = \frac{\pi}{2} \text{m}$$

155. (c)

A wave can't be a function of time only. It is function of both space and time.

157. (a)

The quantity $\vec{E} \times \vec{H}$ is known as Poynting vector, measured in watt/m^2.

$$\vec{P} = \vec{E} \times \vec{H}$$

Poynting vector $\vec{P}$ represents the instantaneous power density vector associated with the EM field at a given point. In other words, it gives rate of energy flow.

160. (c)

Given, $H = 10^{-\alpha x} \cos\left(\omega t - \frac{1}{2}x\right) a_y$ A/m

Here, $\beta = \frac{1}{2}$, $\theta_n = 30°$

Now, $$\alpha = \omega\sqrt{\frac{\mu\varepsilon}{2}\left[\sqrt{1+\left(\frac{\sigma}{\omega\varepsilon}\right)^2} - 1\right]}$$

and $$\beta = \omega\sqrt{\frac{\mu\varepsilon}{2}\left[\sqrt{1+\left(\frac{\sigma}{\omega\varepsilon}\right)^2} + 1\right]}$$

$$\therefore \quad \frac{\alpha}{\beta} = \left[\frac{\sqrt{1+\left(\frac{\sigma}{\omega\varepsilon}\right)^2} - 1}{\sqrt{1+\left(\frac{\sigma}{\omega\varepsilon}\right)^2} + 1}\right]$$

But, $\frac{\sigma}{\omega\varepsilon} = \tan 2\theta_n = \tan 60° = \sqrt{3}$

Hence, $$\frac{\alpha}{\beta} = \left(\frac{2-1}{2+1}\right)^{1/2} = \frac{1}{\sqrt{3}}$$

or, $$\alpha = \frac{\beta}{\sqrt{3}} = \frac{1}{2\sqrt{3}}$$

So, skin depth

$$\delta = \frac{1}{\alpha} = 2\sqrt{3} \text{ m} = 3.464 \text{ m}$$

163. (b)

Depth of penetration,

$$\delta = \frac{1}{\sqrt{\pi f \mu_0 \sigma}} = \frac{1}{\alpha}$$

$$\therefore \quad \delta \propto \frac{1}{\alpha}$$

165. (b)

Intrinsic impedance of medium-B

$$= \sqrt{\frac{\mu}{\varepsilon}} = \sqrt{\frac{\mu_0 \mu_r}{\varepsilon_0 \varepsilon_r}} = 120\pi\sqrt{\frac{\mu_r}{\varepsilon_r}}$$

$$= \frac{120\pi \times 2}{3} = 80\pi$$

Given, $\vec{E} = 100\cos(\omega t - 6\pi x)\,\hat{z}$

$= E_m \cos(\omega t - \beta x)\,\hat{z}$

Here, $\beta = 6\pi$ = Phase shift constant

Reflection coefficient

$$= \left(\frac{\sqrt{\varepsilon_2} - \sqrt{\varepsilon_1}}{\sqrt{\varepsilon_2} + \sqrt{\varepsilon_1}}\right) = \left(\frac{3-2}{3+2}\right) = \frac{1}{5}$$

Transmission coefficient

$$= \frac{2\sqrt{\varepsilon_2}}{\sqrt{\varepsilon_2} + \sqrt{\varepsilon_1}} = \frac{2 \times 3}{3+2} = \frac{6}{5}$$

166. (b)

Polarization, $P = D\left(1 - \frac{1}{\varepsilon_r}\right) = 2\left(1 - \frac{1}{5}\right)$

$= 1.6$ C/m^2

167. (a)

Here, $|\vec{E}_x| = |\vec{E}_y| = 10$

and angle between $\vec{E}_x$ and $\vec{E}_y$ is 90° which means polarization is circular.

168. (c)

$$\eta = \eta_0\sqrt{\frac{\mu_r}{\varepsilon_r}}$$

or, $$60\pi = 120\pi\sqrt{\frac{1}{\varepsilon_r}} \text{ or } \varepsilon_r = 4$$

170. (b)

Given, $\gamma = (0.1\,\pi + j\,0.2\pi)\text{ m}^{-1}$
$= (\alpha + j\beta)$

Here, $\beta = 0.2\,\pi$

∴ Wavelength,

$$\lambda = \frac{2\pi}{\beta} = \frac{2\pi}{0.2\pi} = 10\text{ m}$$

171. (b)

Since the transmission line is reciprocal, therefore,

$AD - BC = 1$

or, $$D = \frac{1+BC}{A} = \frac{1+7\times5}{4} = \frac{36}{4} = 9$$

173. (d)

When a transmission line is terminated by an impedance Z, then reflection coefficient are :

$$\rho_V = \frac{Z-Z_0}{Z+Z_0};\quad \rho_I = \frac{Z_0-Z}{Z+Z_0}$$

Here, Z_0 = Characteristic impedance of transmission line.

Since $Z = Z_0$, therefore reflection co-efficient of voltage (ρ_V) and current (ρ_I) both will be zero. i.e there will be no reflection of incident wave.

175. (b)

VSWR = 2 (Given)

Now, $$\text{VSWR} = \frac{R_0}{R_L} = \frac{100}{R_L} = 2$$

or, $R_L = 50\ \Omega$

Also, $$\text{VSWR} = \frac{R_L}{R_0} = \frac{R_L}{100} = 2$$

or, $R_L = 200\ \Omega$

176. (b)

For a short-circuited load,
reflection coefficient of voltage,

$$\rho_V = -1 = 1\angle180°$$

and reflection coefficient of current,

$$\rho_I = +1 = 1\angle0°$$

Here, only option (b) matches the answer

i.e., $\rho_V = =1\angle180°$

177. (c)

The characteristic impedance (Z_0) of a transmission line is given by

$$Z_0 = \sqrt{\frac{Z}{Y}}$$

Here, $Z = (R + j\omega L)\ \Omega$

and $Y = (G + j\omega C)\ ℧$

∴ $$Z_0 = \sqrt{\frac{R+j\omega L}{G+j\omega C}}$$

180. (c)

A distortionless transmission line is one in which the attenuation constant α is independent of frequency while the phase constant β is linearly dependent on frequency.

Thus, $$\frac{R}{L} = \frac{G}{C} \text{ or } LG = CR$$

181. (b)

Double stub matching eliminates standing waves on the source side of the left stub.

182. (d)

Characteristic impedance is given by

$$Z_C = \sqrt{Z_{OC}\cdot Z_{SC}}$$
$$= \sqrt{100\times64} = 80\ \Omega$$

183. (a)

Given, $Z_0 = R_0$

we have:

$$|Z_{in}|_{max} = \frac{V_{max}}{I_{max}} = SZ_0 = SR_0$$

and $$|Z_{in}|_{min} = \frac{V_{max}}{I_{min}} = \frac{Z_0}{S} = \frac{R_0}{S}$$

184. (b)

Voltage standing wave ratio is given by

$$S = \frac{V_{max}}{V_{min}} = \frac{I_{max}}{I_{min}} = \frac{1+|K|}{1-|K|}$$

or, $\frac{S+1}{S-1} = \frac{2}{2|K|}$

(Using componendo and dividendo)

or, $|K| = \frac{S-1}{S+1} = \frac{VSWR-1}{VSWR+1}$

185. (c)

Propagation constant of a transmission line is given by

$$\gamma = \sqrt{Z \cdot Y} = \sqrt{(R+j\omega L)(G+j\omega C)}$$

186. (a)

Given, $Z_0 = 300\angle 0°\ \Omega$

and $Z_L = 150\angle 0°\ \Omega$

$\therefore$ $S = VSWR = \frac{Z_0}{Z_L} = \frac{300\angle 0°}{150\angle 0°} = 2$

187. (d)

Characteristic impedance,

$$Z_0 = \sqrt{\frac{(R+j\omega L)}{(G+j\omega C)}} = \sqrt{\frac{R(1+j\omega L/R)}{G(1+j\omega C/G)}}$$

or, $Z_0 = \sqrt{\frac{R}{G}}$

(Since $\frac{L}{R} = \frac{C}{G}$ for a distortionless line)

$$= \sqrt{\frac{L}{C}} = R_0 + jX_0$$

where, $R_0 = \sqrt{\frac{R}{G}} = \sqrt{\frac{L}{C}}$ and $X_0 = 0$

Hence, characteristic impedance of a distortionless line is purely real.

189. (a)

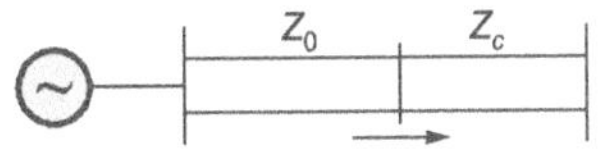

The overall transmission coefficient is

$$\tau_v = \left(\frac{2Z_0}{Z_0 + Z_c}\right)$$

190. (b)

- Statements-1 and 2 are true.
- Statement-3 is not correct because in a lossless transmission line there is neither resistive loss (i.e. $R = 0$) nor any leakage current (i.e. $G = 0$).
- At radio frequency (RF), the inductive reactance is much larger than the resistance and so also the capacitie susceptance in comparison to the shunt conductance. Hence, at radio frequencies R and G both are neglected. Thus, statement-4 is correct.

191. (a)

The input impedance of a finite line terminated in its characteristic impedance (Z_0), is equivalent to an infinite line because the input impedance of an infinite line is the characteristic impedance of the line (Z_0). Hence, both assertion and reason are true and reason is the correct explanation of assertion.

192. (c)

- Propagation constant is dimensionless quantity because it is the ratio of voltages or currents.

$$P = \log_e\left(\frac{I_S}{I_R}\right)$$

or, $$P = \log_e\left(\frac{V_S}{V_R}\right)$$

(where, P = Propagation constant)

Hence, statement-1 is correct.

- For a lossless transmission line,

$$V_P = \frac{\omega}{\beta} = \frac{1}{\sqrt{LC}} = f$$

$\therefore$ $V_P \propto f$

Thus, statement - 2 is also correct.

- Statement-3 is also true.

Thus, all statements are true.

193. (d)

Group velocity is given by

$$V_g = \frac{d\omega}{d\beta} = \left(\frac{\omega_2 - \omega_1}{\beta_2 - \beta_1}\right)$$

Phase velocity or velocity of propagation is given by

$$V_P = \left(\frac{\omega}{\beta}\right) \text{ km/sec}$$

The group velocity is usually less than the phase velocity. Hence, assertion is a false statement.

194. (c)

Assertion is correct because when a transmission line is open or short-circuited it behaves as resonant circuit. However, reason is false because this happens when length of the line is an integral multiple of $\lambda/4$.

We know that,

$$Z_{oc} = -j\cot\beta l$$

and $$Z_{sc} = jZ_0\tan\beta l$$

Thus, when βl = length of line = $\frac{n\lambda}{4}$, then

$$\cot\beta l = 0$$

and $$\tan\beta l = \infty$$

$$\therefore \quad Z_{oc} = 0$$

and $$Z_{sc} = \infty$$

This means a quarter wave short-circuit line represents an infinite impedance at input terminals, just like a parallel resonant (LC) circuit and a $\lambda/4$ open circuit line present zero impedance at input terminals just like a series resonant LC circuit.

198. (d)

Input impedance of a short-circuited quarter wave line depends on the nature of load impedance Z_L.

199. (a)

$$f = \frac{1}{2\sqrt{\mu\varepsilon}}\left[\left(\frac{1}{10\times10^{-2}}\right)^2 + 0\right]^{1/2}$$

$$= 1.5 \text{ GHz}$$

201. (c)

$$Z_c = Z_{01}\left[\frac{Z_{02} + jZ_{01}\tan\beta l}{Z_{01} + jZ_{02}\tan\beta l}\right]$$

Since, $Z_{01} = Z_0$ and $Z_{02} = 0$ (s.c), therefore

$$Z_s = jZ_0\tan\beta l$$

202. (b)

Adjacent volt maximas are at $\lambda/2$ distance.

$$\therefore \quad 27.5 - 12.5 = \frac{\lambda}{2}$$

or, $$\lambda = 30 \text{ cm} = 0.3 \text{ m}$$

Thus, operating frequency

$$f = \frac{C}{\lambda} = \frac{3\times10^8}{0.3} = 1 \text{ GHz}$$

203. (c)

$$\frac{V_L}{V_{in}} = \frac{Z_0}{Z_{in}}$$

or, $$V_L = \frac{10\times300}{50} = 60 \text{ V}$$

205. (b)

$$Z = \sqrt{Z_C Z_L}$$

$$= \sqrt{400\times25} = 100\ \Omega$$

206. (b)

Both assertion and reason are individually true. However, the correct reason is that as the frequency varies, the position of the stub has to be varied due to which single stub matching is useful for a fixed frequency only.

207. (d)

Characteristic impedance,

$$Z_0 = \sqrt{\frac{L}{C}}$$

$$\therefore \quad Z_0 = \sqrt{\frac{0.56\times10^{-3}}{0.1\times10^{-6}}} = \sqrt{0.56\times10^4}$$

$$= 10\sqrt{56} = 74.83 \approx 75\ \Omega$$

■■■■

UNIT 2

Electrical Materials

1. Introduction to Engineering Materials

Q.1 Match **List-I (Engineering materials)** with **List-II (Uses)** and select the correct answer using the codes given below the lists:

List-I
A. Ferrous metals
B. Non-ferrous metals
C. Ceramics
D. Metal- metal composites

List-II
1. Semiconductors making
2. Conductors in HV transmission
3. Alloy making
4. Refractories

Codes:

	A	B	C	D
(a)	3	1	2	4
(b)	3	1	4	2
(c)	1	2	4	3
(d)	2	4	3	1

Q.2 Match **List-I (Properties of engineering materials)** with **List-II (Examples)** and select the correct answer using the codes given below the lists:

List-I	List-II
A. Physical	1. Malleability
B. Mechanical	2. Reluctivity
C. Magnetic	3. Passivity
D. Chemical	4. Density

Codes:

	A	B	C	D
(a)	3	4	2	1
(b)	4	1	3	2
(c)	1	4	3	2
(d)	4	1	2	3

Q.3 Material science deals with
(a) solid materials
(b) molten metals
(c) gases and liquids
(d) solids and vapours

Q.4 Which of the following is/are mechanical property of a material?
(a) Specific heat and thermal fatigue
(b) Passivity and pH-value
(c) Proof resilience and creep
(d) Susceptibility and retentivity

Q.5 A unit cell is
(a) an agglomerated structure.
(b) the basic building block of crystal.
(c) the smallest group of atoms which when regularly repeated forms the crystal.
(d) a cube containing the largest number of atoms.

Q.6 **Assertion (A) :** A unit cell is analogous to a brick used in the building construction.
Reason (R) : A unit cell is defined as the basic structural part in the composition of materials.
(a) Both A and R are true and R is the correct explanation of A.
(b) Both A and R are true but R is not the correct explanation of A.
(c) A is true but R is false.
(d) A is false but R is true.

Q.7 Consider the following statements:
1. The crystal directions of a family must be parallel to one another.
2. Crystal directions and crystal plane are denoted by the Miller indices.
3. In cubic crystals, a crystal plane and a crystal direction normal to it have different indices.
4. The effective number of lattice points in unit cell is highest in case of face centered cubic space lattice.

Which of the statements given above is/are correct?
(a) 2, 3 and 4 (b) 2 and 4
(c) 1, 2 and 3 (d) 4 only

Q.8 The number of Bravais space lattices with two lattice points are

(a) 2 (b) 3
(c) 4 (d) 5

Q.9 The atomic diameter of an FCC crystal (lattice parameter is *a*) is

(a) $\frac{a\sqrt{2}}{2}$ (b) $\frac{a\sqrt{2}}{4}$
(c) $\frac{a\sqrt{3}}{4}$ (d) $\frac{a}{2}$

Q.10 Consider the following statements associated with the atomic structure and chemical bonding:
1. Covalent bond is non-directional.
2. The metallic bond is directional and generally weaker than ionic and covalent bonds.
3. Free electrons are responsible for the high thermal and electrical conductivities of metals.
4. The order of occupation of quantum states by electrons is determined by the Pauli exclusion principle, the Hund's rule, and the minimum energy criterion.
5. The bond energy is related to the enthalpy of atomization of the solid.

Which of the statements given above are correct?
(a) 2, 3 and 5 (b) 1, 2 and 4
(c) 3, 4 and 5 (d) 1, 3 and 5

Q.11 If there are six electrons in the *d* orbital of a transition metal, the number of unpaired electrons are
(a) 6 (b) 5
(c) 4 (d) 0

Q.12 The thermal expansion of materials arises from
(a) strong bonds
(b) thermal vibrations
(c) weak bonds
(d) asymmetry of potential energy curve

Q.13 **Assertion (A) :** Diamond is very hard and brittle.
Reason (R) : Diamond is covalently bonded.
(a) Both A and R are true and R is the correct explanation of A.
(b) Both A and R are true but R is not the correct explanation of A.
(c) A is true but R is false.
(d) A is false but R is true.

Q.14 Hydrogen bonds are stronger than
(a) Vander walls bonds
(b) Metallic bonds
(c) Ionic bonds
(d) Covalent bonds

Q.15 The Miller indices of a plane are proportional to
(a) the reciprocal of numerical parameters of the intercepts
(b) the square of unit cell dimensions
(c) the intercepts of the planes on the coordinate axes
(d) interplaner spacing

Q.16 **Assertion (A):** Nano-mineral, nano-gold and nano-silver nitrate are useful to remove the harmful element arsenic (As) from water.
Reason (R): Nanomaterials are prepared by smaller to larger doing technique.
(a) Both A and R are true and R is the correct explanation of A.
(b) Both A and R are true but R is not the correct explanation of A.
(c) A is true but R is false.
(d) A is false but R is true.

Q.17 A material having different properties in different directions is known as
(a) Isotropic (b) Anisotropic
(c) Crystalline (d) Metal

Q.18 What is the Miller indices (*h*, *k*, *l*) of a plane whose intercepts are *a*, *b*/2 and 3*c* on *x*, *y* and *z* axes respectively in a simple cubic unit cell?
(a) (3, 1, 6) (b) (6, 3, 1)
(c) (1, 3, 6) (d) (3, 6, 1)

Q.19 A metal which possesses body centered tetragonal structure is
(a) Vanadium (b) Tin
(c) Aluminium (d) Nickel

Q.20 Which of the following is not a member of ceramic family?
(a) Bainite (b) Garnet
(c) Ferrites (d) Lime

Q.21 Match **List-I (Engineering materials)** with **List-II (Characteristics)** and select the correct answer using the codes given below the lists:

List-I
A. Alloys
B. Organic Polymers
C. Ceramics
D. Composites

List-II
1. Inert and light
2. Metallic or non-metallic oxides
3. Two or more constituent of dissimilar properties
4. Combination of two or more metals.

Codes:

	A	B	C	D
(a)	1	2	4	3
(b)	4	3	2	1
(c)	4	1	2	3
(d)	1	4	3	2

Q.22 Match **List-I (Non-ferrous metals)** with **List-II (Properties)** and select the correct answer using the codes given below the lists:

List-I	List-II
A. Gold	1. Most ductile
B. Silver	2. Most malleable
C. Chromium	3. Corrosion resistance
D. Nickel	4. Imparts strength and creep resistance

Codes:

	A	B	C	D
(a)	1	2	3	4
(b)	2	1	4	3
(c)	4	2	3	1
(d)	4	3	2	1

Q.23 Which of the following material is used for making helicopter structure?
(a) pure aluminium
(b) stetite
(c) carbon fibre polymeric composite
(d) thoriated nickel alloy

Q.24 Point imperfections, during interaction with each other,
(a) lower their total energy
(b) are not affected at all
(c) are thermodynamically stable
(d) both (b) and (c)

Q.25 The imperfections may be minimized by
(a) thermal energy
(b) making metal in fibre form
(c) surface treatment
(d) all of these

Q.26 The forces involved in the chemical bonding are
(a) Pauli's forces (b) Newton's forces
(c) Plank's forces (d) Coulomb's forces

Q.27 The nature of bond is governed by
(a) geometry of unit cell
(b) electron structure of the atom
(c) orientation of crystal
(d) basis of filling of space lattice

Q.28 A material is most stable when its potential energy is
(a) maximum (b) infinite
(c) minimum (d) zero

Q.29 The ACSR (aluminium conductor steel reinforced) is an example of
(a) agglomerated composite
(b) metal fibre composite
(c) bulk moulding compound
(d) whisker reinforced composite

Q.30 The structure of quartz (SiO_2) is
(a) hexagonal
(b) body centered cubic
(c) face centered cubic
(d) orthorhombic

Q.31 Crystals like diamond and silicon are brittle, because
(a) they contain no dislocations
(b) they are non-crystalline
(c) the stress to move a dislocation is high
(d) they contain very few dislocations

Q.32 The crystal structure of most of the common metals is
(a) hexagonal (b) cubic
(c) orthorhombic (d) linear

Q.33 In a crystal lattice, what are the vacancies created by the absence of certain atoms known as?
(a) Hertz defects (b) Schottky defects
(c) Pauli defects (d) Crystal defects

2. Dielectric Properties of Materials

Q.34 **Assertion (A):** Dielectrics are very poor conductors of heat and electricity.
Reason (R): Dielectrics allow movement of some electrons at abnormally high temperature causing a small flow of current.
(a) Both A and R are true and R is the correct explanation of A.
(b) Both A and R are true but R is not the correct explanation of A.
(c) A is true but R is false.
(d) A is false but R is true.

Q.35 Match **List-I (Form of Dielectric materials)** with **List-II (Purposes)** and select the correct answer using the codes given below the lists:

List-I	List-II
A. Coating materials	1. Glass wool
B. Filling materials	2. Teflon
C. Adhesive materials	3. Elastomers
D. Bonding materials	4. Plastic films

Codes:

	A	B	C	D
(a)	1	3	2	4
(b)	1	2	3	4
(c)	2	3	4	1
(d)	2	1	4	3

Q.36 Dielectric materials
1. have high specific resistance.
2. have positive temperature coefficient of resistive.
3. may belong to the ceramic family.
4. exists only in solid and liquid form but not in gas form.

Which of the above characteristics of dielectric material holds true?
(a) 1 and 3 only (b) 2 and 4 only
(c) 1, 3 and 4 only (d) 2, 3 and 4 only

Q.37 Which of the following dielectric material possess maximum dielectric strength?
(a) Bakelite (b) CO_2
(c) Glass (d) Mica

Q.38 The dielectric strength of a natural rubber is 40 kV/mm at 60 Hz. What is the thickness of insulation on a wire carrying 33 kV to sustain the breakdown?
(a) 0.082 mm (b) 8.2 mm
(c) 0.82 mm (d) 0.0082 mm

Q.39 A good insulating material should have
(a) high dielectric strength but low dielectric constant.
(b) high dielectric strength and dielectric constant.
(c) low dielectric strength and dielectric constant.
(d) low dielectric strength but high dielectric constant.

Q.40 In general, the effect of temperature on relative permittivity of a material is to
(a) reduce it (b) increase it
(c) maintain it (d) all of these

Q.41 Condensers to be used at high frequencies require a material whose loss angle is
(a) low (b) high
(c) uniform (d) unity

Q.42 In which of the following ceramics, the ionic character is maximum?
(a) CaF_2 (b) ZnS
(c) SiO_2 (d) NaC*l*

Q.43 Which of the following is a basic refractory?
(a) Quartz (b) Dolomite
(c) SiO_2 (d) Chromite

Q.44 Which of the following is not a desirable property possessed by a good refractory?
(a) High electrical conductivity
(b) High temperature withstanding capability
(c) Chemically inactiveness
(d) Both (a) and (c)

Q.45 The dielectric strength and dielectric constant of a ceramic should be
(a) high and low respectively
(b) low
(c) low and high respectively
(d) high

Q.46 Which of the following items is not made of a ceramic material?
(a) Ferrite
(b) Insulators
(c) Magnet
(d) Reinforcing bar for cement concrete

Q.47 The relative dielectric constant of a barium titanate crystal, which when inserted in a parallel plate condenser of area 10 mm × 10 mm and distance of separation of 2 mm and giving a capacitance of 10^{-9} F would be equal to

(a) 3002 (b) 1986
(c) 2259 (d) 2042

Q.48 **Assertion (A) :** Electronic polarization results in a dipole moment within the atom.

Reason (R) : Electronic polarization has an inverse relation with the field strength.

(a) Both A and R are true and R is the correct explanation of A.
(b) Both A and R are true but R is not the correct explanation of A.
(c) A is true but R is false.
(d) A is false but R is true.

Q.49 **Assertion (A) :** The intrinsic breakdown of a dielectric material is due to the excitation of electrons into the conduction band across the energy gap under conditions of excessive voltage.

Reason (R) : The excited electrons moving under a high accelerating force can excite more electrons in turn, resulting in an avalanche of conducting electrons and consequent physical breakdown.

(a) Both A and R are true and R is the correct explanation of A.
(b) Both A and R are true but R is not the correct explanation of A.
(c) A is true but R is false.
(d) A is false but R is true.

Q.50 Match **List-I (Types of Polarization)** with **List-II (Relation with temperature)** and select the correct answer using the codes given below the lists:

List-I
A. Ionic
B. Electronic
C. Space charge
D. Orientation

List-II
1. Independent of temperature
2. Proportional to temperature
3. Inversely proportional to temperature

Codes:

	A	B	C	D
(a)	1	2	1	3
(b)	1	1	3	3
(c)	2	3	1	1
(d)	1	1	2	1

Q.51 Which of the following types of polarization occurs in a multi-phase material?

(a) Orientation polarization
(b) Ionic polarization
(c) Electronic polarization
(d) Space charge polarization

Q.52 Consider the following statements associated with different types of electric polarization occurring in electric materials:

1. Orientation polarization is slower than ionic polarization.
2. Ionic polarization is slower than electronic polarization.
3. Space charge polarization is the slowest process.

Which of the statements given above is/are correct?

(a) 1, 2 and 3 (b) 1 and 3 only
(c) 2 only (d) 1 only

Q.53 What is the polarization of a $BaTiO_3$ crystal if the shift of the titanium ion from the body centre is 0.06 Å, the oxygen anions of the side faces shift by 0.06 Å, while the oxygen anions of the top and bottom faces shift by 0.08 Å, all in a direction to that of the titanium ion?

(a) 0.12 C-m^{-2} (b) 0.16 C-m^{-2}
(c) 0.20 C-m^{-2} (d) 0.24 C-m^{-2}

Q.54 Consider the following statements:

1. The charge stored in a capacitor increases if a dielectric material is inserted between the capacitor plates which happens due to the polarization in the dielectric.
2. The dielectric constant of a material is independent of the frequency of the alternating field applied but depends on the temperature.
3. Electrical energy loss is a maximum when the relaxation time of a polarization process matches the period of the alternating field.

Which of the statements given above is/are **not** correct?
(a) 2 and 3 (b) 1 only
(c) 1 and 3 (d) 2 only

Q.55 The unit of relative dielectric constant is
(a) Fm^{-1} (b) CV^{-1}
(c) dimensionless (d) FC^{-1}

Q.56 Among the common dielectric materials, the highest dielectric strength is possessed by
(a) transformer oil (b) mica
(c) PVC (d) polyethylene

Q.57 During melting, the relative dielectric constant
(a) always increases
(b) always decreases
(c) may increase or decrease
(d) none of the above

Q.58 In the polarization versus field strength plot for a ferroelectric crystal, P_s stands for
(a) space charge polarization
(b) saturation polarization
(c) spontaneous polarization
(d) none of the above

Q.59 Expressed in base SI units, the electronic polarizability has the units
(a) $F\text{-}m^2$ (b) $CV^{-1}m^{-2}$
(c) $A^2\ kg^{-1}\ s^4$ (d) $A\ kg^{-1}\ m^{-2}\ s^2$

Q.60 The materials in which properties are independent and dependent on direction are called
(a) anisotropic and isotropic materials respectively.
(b) isotropic and anisotropic materials respectively.
(c) anisotropic and orthotropic materials respectively.
(d) orthotropic and isotropic materials respectively.

Q.61 Which of the following pairs does **not** represent piezoelectric materials?
(a) Lead titanate, Rochelle's salt
(b) Barium titanate, Quartz
(c) Lead zirconate, Sodium Neobate
(d) Potassium dihydrogen phosphate, Rochelle's salt

Q.62 **Assertion (A)** : Above a particular value of temperature, spontaneous polarization is reduced to zero.
Reason (R) : The material starts behaving like a ferroelectric material.
(a) Both A and R are true and R is the correct explanation of A.
(b) Both A and R are true but R is not the correct explanation of A.
(c) A is true but R is false.
(d) A is false but R is true.

Q.63 Clausius Mossotti equation is valid for an internal field constant of
(a) $\frac{1}{2}$ (b) $\frac{1}{4}$
(c) 1 (d) none of these

Q.64 Match **List-I (Types of dielectric materials)** with **List-II (Characteristics)** and select the correct answer using the codes given below the lists:

List-I
A. Antiferroelectric
B. Piezoelectric
C. Ferroelectric

List-II
1. Exhibits property of Hysteresis
2. Electric field gives rise to mechanical strain
3. Zero polarization in absence of electric field

Codes:

	A	B	C
(a)	2	3	1
(b)	3	2	1
(c)	1	2	3
(d)	2	1	3

Q.65 The relative dielectric constant of monoatomic gas is related to radius R of atom and the density N atom/metre^3 by the relation
(a) $\varepsilon_r = 1 + 4\pi R^2 N$ (b) $\varepsilon_r = 1 + 4\pi R^3 N$
(c) $\varepsilon_r = 1 + 4\pi R^2 N^2$ (d) $\varepsilon_r = 1 + 4\pi R^3 N^2$

Q.66 Condition under which spontaneous polarization in a dielectric occurs is
(a) $\frac{N\alpha\gamma^2}{\varepsilon_0} = 1$ (b) $\frac{N\alpha^2\gamma}{\varepsilon_0} = 1$
(c) $\frac{N^2\alpha\gamma}{\varepsilon_0} = 1$ (d) $\frac{N\alpha\gamma}{\varepsilon_0} = 1$

Q.67 Ferrites have
(a) low copper loss
(b) low oddy current loss
(c) low resistivity
(d) high specific gravity compared to iron

Q.68 The polarization *P* in a solid dielectric is related to the electric field *E* and the electric flux density *D* by the relation
(a) $E = \varepsilon_0 D + P$ (b) $D = E + \varepsilon_0 P$
(c) $D = \varepsilon_0 E + P$ (d) $D = \varepsilon_0 (E + P)$

Q.69 The insulator filling the space between the plates of parallel plate capacitor is of area 10^4 mm^2 and thickness 1 mm. Its resistivity is 10^{10} ohm-meter. When a d.c. voltage of 1 kV is applied, the dielectric loss will be
(a) 10^{-3} W (b) 10^{-2} W
(c) 10^2 W (d) 10^3 W

Q.70 Loss in a dielectric may occur due to
(a) polarization (b) conductivity
(c) ionization (d) any of the above

Q.71 Dielectric loss due to ionization is peculiar to
(a) non-ferrous dielectrics
(b) non-metallic dielectrics
(c) non-uniform dielectrics
(d) gaseous dielectrics

Q.72 When subjected to an alternating electrical stresses, an insulating material may be characterized by a complex dielectric constant $(r' - jr'')$. The dielectric losses in such materials when subjected to alternating electric stress will be proportional to
(a) r'' (b) $(r' - jr'')^{1/2}$
(c) r''^2 (d) $(r'^2 - r''^2)$

Q.73 Which of the following statements are true about piezoelectric crystal?
1. Their crystal structure possesses a centre of symmetry.
2. They always exhibits ferroelectricity.
3. They are necessarily poor electrical conductors.
4. They exhibit a relative permittivity less than unity.

Select the correct answer using the codes given in the options.
(a) 1, 2 and 4 (b) 2, 3 and 4
(c) 1 and 3 (d) 1, 2 and 3

Q.74 When a dielectric is subjected to an alternating field, the real part of the complex relative dielectric constant is found to be 2.5. The loss tangent is 0.004. The imaginary part of the relative dielectric constant is
(a) 0.01 (b) 2.496
(c) 25.04 (d) 625.63

3. Magnetic Properties of Materials

Q.75 **Assertion (A)** : Soft magnetic materials are suitable for AC appliances.
Reason (R) : Hard magnetic materials are not used for AC appliances because large hysteresis loss will take place owing to large area of hysteresis loop.
(a) Both A and R are true and R is the correct explanation of A.
(b) Both A and R are true but R is **not** the correct explanation of A.
(c) A is true but R is false.
(d) A is false but R is true.

Q.76 Consider the following statements associated with various magnetic materials:
1. A diamagnetic material obey Curie's law.
2. Susceptibility of a paramagnetic material is directly proportional to temperature.
3. A ferromagnetic material is freely repelled by a magnet.
4. Anti-ferromagnetic materials have small permanent dipole moment.

Which of the statements given above is/are correct?
(a) 1 only (b) 3 and 4
(c) 2 only (d) none of these

Q.77 The relative permeability and the magnetic field intensity of a diamagnetic materials are respectively 1.00038 and 0.31 A/m. If the magnitude of magnetic susceptibility is 4×10^{-6}, then the magnitude of magnetization in the material is approximately equal to
(a) 0.12 mA/m (b) 0.32 mA/m
(c) 0.63 mA/m (d) none of these

Q.78 A solenoid with 2000 turns is 300 mm long and 20 mm in diameter. If the current is 600 mA, then the inductance of the solenoid would be equal to

(a) 2.65 mH (b) 5.26 mH
(c) 4.75 mH (d) 3.45 mH

Q.79 A transformer core is wound with a coil carrying an alternating current at a frequency of 50 Hz. The hysteresis loop has an area of 60000 units, when the axes are drawn in units of 10^{-4} Wb-m^{-2} and 10^2 A-m^{-1}. Assuming the magnetization to be uniform throughout the core volume of 0.01 m^3, the hysteresis loss is
(a) 100 W (b) 200 W
(c) 300 W (d) 400 W

Q.80 Match **List-I (Magnetic materials)** with **List-II (Applications)** and select the correct answer using the codes given below the lists:

List-I
A. Permalloy and Supermalloy
B. Ferrites and garnets
C. Nickel- Zinc ferrites
D. Magnesium-maganese ferrites

List-II
1. Audio and TV transformers
2. Communication equipments
3. Microwave isolators
4. Suitable in high frequency range

Codes:

	A	B	C	D
(a)	2	4	1	3
(b)	1	4	2	3
(c)	4	2	3	1
(d)	4	1	3	2

Q.81 In a paramagnetic material of susceptibility equal to 10^{-3}, the induction in Wb-m^{-2} at an applied field of 100 kA m^{-1} is
($\mu_0 = 1.25 \times 10^{-6}$ H-m^{-1})
(a) 0.125 (b) 0.225
(c) 0.638 (d) none of these

Q.82 The factor that obstruct domain wall motion in Iron is/are
1. dislocation tangles
2. impurity atoms
3. voids
4. non-magnetic inclusions

Choose the correct code from the given options.
(a) 1 and 3 only (b) 2, 3 and 4 only
(c) 2 and 3 only (d) 1, 2, 3 and 4

Q.83 The temperature of the anti-ferromagnetic-to-paramagnetic transition is called
(a) Antiferromagnetic Curie temperature
(b) Curies-Weiss temperature
(c) Neel temperature
(d) Debye temperature

Q.84 **Assertion (A):** The power loss due to hysteresis in a transformer core is independent of the number of times the full loop is traversed per second.
Reason (R): When the induction is large for a small applied field, the loop area is small and the hysteresis loss is reduced.
(a) Both A and R are true and R is the correct explanation of A.
(b) Both A and R are true but R is **not** the correct explanation of A.
(c) A is true but R is false.
(d) A is false but R is true.

Q.85 In a 440 V, 50 Hz transformer, the total iron loss is 2300 W. When the applied voltage is 220 V at 25 Hz, the total iron loss is 750 W. The eddy current loss at the normal voltage and frequency is
(a) 1200 Watt (b) 1600 Watt
(c) 1000 Watt (d) 1800 Watt

Q.86 The transition from the ferromagnetic to the paramagnetic state is named after
(a) Curie (b) Curie-Weiss
(c) Neel (d) Debye

Q.87 The garnet crystal used in a microprocessor has the formula
(a) $YO. Fe_2O_3$ (b) $Y_3. Fe_5O_{12}$
(c) $YO. 6Fe_2O_3$ (d) $Y_2O_3. Fe_2O_3$

Q.88 A suitable material for audio and TV transformers is
(a) ferrite (b) Fe - 4% Si
(c) Fe - 30% Ni (d) very pure Fe

Q.89 Match **List-I (Types of magnetic material)** with **List-II (Characteristics)** and select the correct answer using the codes given below the lists:

List-I
A. Paramagnetic
B. Diamagnetic
C. Ferrimagnetic
D. Ferromagnetic

List-II

1. Possess very high resistivity
2. Attract magnetic lines of force
3. Attract the magnetic lines of force very strongly
4. Repel magnetic lines of force

Codes:

	A	B	C	D
(a)	2	1	4	3
(b)	3	1	4	2
(c)	2	4	1	3
(d)	3	4	1	2

Q.90 **Assertion (A) :** Diamagnetic materials cannot be made magnets.
Reason (R) : In diamagnetic materials, the magnetic moments of equal magnitude are aligned in opposite directions.
(a) Both A and R are true and R is the correct explanation of A.
(b) Both A and R are true but R is **not** the correct explanation of A.
(c) A is true but R is false.
(d) A is false but R is true.

Q.91 Ferrites
1. have high electrical resistivities.
2. have low power loss at all frequencies.
3. are suitable for temporary magnets only.
4. with respect to conductivity behaves like those of semiconductors.
5. have almost rectangular hysteresis loop.

Select the correct code from the given options which holds good for ferrites.
(a) 1, 4 and 5 (b) 2, 4 and 5
(c) 1, 2 and 4 (d) 3 and 4

Q.92 Magnetic susceptibility is
(a) inversely proportional to both temperature and magnetizing field.
(b) inversely proportional to temperature but independent of magnetizing field.
(c) proportional to temperature but inversely proportional to magnetizing field.
(d) proportional to temperature but independent of magnetizing field.

Q.93 Match **List-I (Susceptibility vs Temperature)** with **List-II (Type of magnetic material)** and select the correct answer using the codes given below the lists:

List-I | List-II

A.
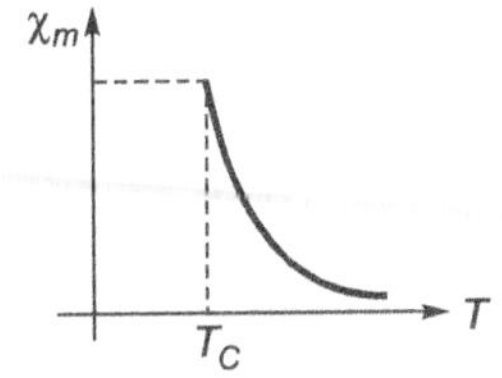

1. Antiferromagnetic

B.
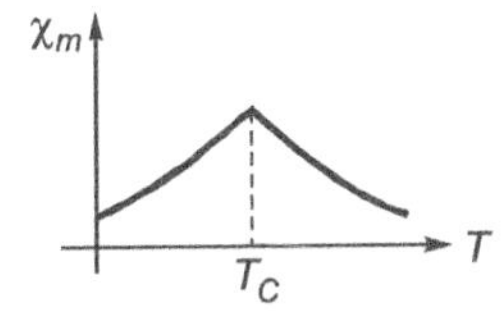

2. Ferromagnetic

C.
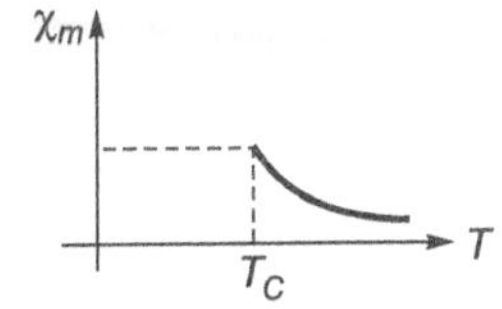

3. Ferrimagnetic

Codes:

	A	B	C
(a)	1	3	2
(b)	1	2	3
(c)	3	1	2
(d)	2	1	3

Q.94 The B-H curve for a magnetic material is shown in figure. Assuming the loop to be roughly a rectangle, the hysteresis loss per cycle for it would be equal to

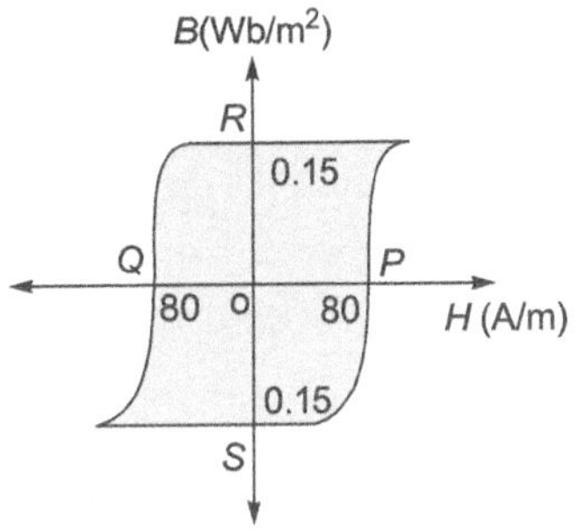

(a) 42 J (b) 24 J
(c) 48 J (d) 21 J

Q.95 The relative permeability of 'Supermalloy' is 200000. It has a magnetization of 6000 A/m. The strength of the magnet thus produced is (Take $\mu_0 = 4\pi \times 10^{-7}$ H/m)
(a) 0.007 Tesla (b) 0.002 Tesla
(c) 0.004 Tesla (d) 0.003 Tesla

Q.96 The value of magnetic moments (in Bohr magneton) in cobalt and nickel are respectively equal to

(a) 3 and 4 (b) 2 and 3
(c) 2 and 4 (d) 3 and 2

Q.97 A coil of wire 250 mm long and having 400 turns carries a current of 15 A. The flux density of the coil in vaccum is
(a) 0.01 Wb/m^2 (b) 0.25 Wb/m^2
(c) 0.03 Wb/m^2 (d) 0.15 Wb/m^2

Q.98 For a compass needle and computer disc drive
(a) soft and hard magnetic materials respectively are required
(b) hard magnetic material is required
(c) hard and soft magnetic materials respectively are required
(d) soft magnetic material is required

Q.99 For high speed of reading and storing the informations in a computer, the use is made of
(a) piezoelectrics
(b) pyroelectrics
(c) ferrites
(d) ferromagnetics above 768°C

Q.100 Magnetic susceptibility of a material is –1. It belongs to the kind of solid known as
(a) ferrite
(b) superconductor, type I
(c) anti-ferromagnetic
(d) garnet

Q.101 Match **List-I** with **List-II** and select the correct answer using the codes given below the lists:

List-I
A. Iron oxide
B. Memory core of computer
C. High hysteresis loss
D. Neel temperature

List-II
1. Ferrite
2. Hard magnetic materials
3. Anti-ferromagnetic materials
4. Paramagnetic

Codes:

	A	B	C	D
(a)	4	1	2	3
(b)	2	3	1	4
(c)	3	4	2	1
(d)	3	1	2	4

Q.102 Magnetostriction is a phenomenon which occurs in
(a) paramagnetic materials
(b) antiferromagnetic materials
(c) ferrimagnetic materials
(d) ferromagnetic materials

Q.103 Bohr magneton is defined as
(a) magnetic moment of nucleus spin
(b) magnetic moment of an electron spin
(c) magnetic moment of an electron orbital motion
(d) none of the above

Q.104 The value of Larmor angular frequency (ω_L) is given by
(a) $\frac{Bm}{2e}$ (b) $\frac{2m}{eB}$
(c) $\frac{eB}{2m}$ (d) $\frac{2em}{B}$

Q.105 At Neel temperature
(a) permeability is minimum
(b) permeability is maximum
(c) susceptibility is minimum
(d) susceptibility is maximum

Q.106 Magnetostriction is the phenomenon
(a) of producing electricity in ferromagnetic materials
(b) of producing magnetism in conductors
(c) of changes in permeability of ferromagnetic material during magnetization
(d) of changes in physical dimensions of ferromagnetic materials during magnetization

Q.107 The converse of magnetostriction effect is
(a) Thompson effect (b) Villari effect
(c) Curie effect (d) Bitter powder effect

Q.108 The magnetic permeability is maximum for
(a) paramagnetic materials
(b) ferromagnetic materials
(c) diamagnetic materials
(d) none of the above

Q.109 Which of the following is **not** matched correctly?
(a) Diamagnetic : Al_2O_3
(b) Anti-ferromagnetic : $Mn F_2$
(c) Paramagnetic : Fe_2O_3
(d) Ferrimagnetic : Graphite

Q.110 The correct sequence of the following materials in the increasing order of magnetic susceptibility is

(a) Diamagnetic - Ferromagnetic - Paramagnetic - Superconductor
(b) Ferromagnetic - Paramagnetic - Diamagnetic - Superconductor
(c) Paramagnetic-Diamagnetic- Superconductor - Ferromagnetic
(d) Superconductor-Diamagnetic - Paramagnetic - Ferromagnetic

Q.111 Consider the following statements:
Magnetic field induced orbital dipole moment in an atom is
1. in a direction opposite to that of the applied field.
2. larger for larger atoms.
3. independent of effective electron mass.
4. 1.3 times the zero-field orbital dipole, moment.

Of these, the correct statements are
(a) 1 and 2 (b) 2 and 3
(c) 3 and 4 (d) 1 and 4

4. Conducting Materials

Q.112 The acceleration (m-s^{-2}) of a free electron in an electric field of 100 V-m^{-1} is
(a) 1.76×10^3 (b) 1.1×10^{13}
(c) 1.76×10^{13} (d) 1.1×10^{32}

Q.113 The SI units of electrical conductivity are
(a) $kg^{-1}m^{-3}s^3A^2$ (b) $kg\ m^3s^{-3}A^2$
(c) ohm-m (d) ohm^{-1} m^{-3}

Q.114 The correct order of increasing resistivity of the following is
(a) nickel, doped silicon, sodium silicate, pure silica.
(b) doped silicon, pure silica, nickel, sodium silicate.
(c) pure silica, doped silicon, sodium silicate, nickel.
(d) nickel, pure silica, doped silicon, sodium silicate.

Q.115 The Fermi level E_F depends on the length L of a linear solid as
(a) $\frac{1}{L^2}$ (b) $\frac{1}{L^3}$
(c) $\frac{1}{L}$ (d) is independent of L

Q.116 The classical equation for the kinetic energy can be used for the free electrons, because, as compared to the velocity of light, their velocity is
(a) more (b) much less
(c) slightly less (d) zero

Q.117 The probability of occupation of an energy level E, when $E - E_F = kT$, is given by
(a) 0.73 (b) 0.63
(c) 0.27 (d) 0.50

Q.118 The magnetization of a superconductor is
(a) 0 (b) $-B$
(c) -1 (d) $-H$

Q.119 **Assertion (A) :** In the superconducting state, the flux lines of a magnetic field are penetrated into the superconductor.
Reason (R) : The maximum current that a superconductor carries at a given temperature below T_C is limited by the magnetic field it produces at the surface of the superconductor.
(a) Both A and R are true and R is the correct explanation of A.
(b) Both A and R are true but R is not the correct explanation of A.
(c) A is true but R is false.
(d) A is false but R is true.

Q.120 A superconducting material exhibits
1. infinite relative permeability
2. zero specific resistance
3. negligible copper loss
4. zero value of magnetic flux density
5. very large negative value of susceptibility

Of these, the correct statements are
(a) 2 and 4 (b) 2, 3, 4 and 5
(c) 1, 2 and 3 (d) 1 and 3

Q.121 Critical temperature of lead in superconducting state is 7.17 K under zero magnetic field. The value of critical field for it is 0.0803 A/m at 0 K. Its critical field at 3 K is
(a) –0.075 A/m (b) 0.066 A/m
(c) –0.026 A/m (d) 0.035 A/m

Q.122 What is the critical current which can pass through a long thin superconducting wire of aluminium of diameter 2 mm if the critical magnetic field for aluminium is 7.9×10^3 A/m?

(a) 38.68 A (b) 26.29 A
(c) 52.20 A (d) 49.64 A

Q.123 The Nb_3Sn is to be used as magnet grade superconductor to make the rail for a wheelless train. Its critical field at 7 K and 14 K are 21 A/m and 10 A/m respectively. Its transition temperature is
(a) 15.76 K (b) 13.49 K
(c) 18.15 K (d) 16.20 K

Q.124 **Assertion (A) :** Meissner effect is helpful in protecting devices from the undesired magnetic fields.
Reason (R) : Meissner effect determines the critical current density in a superconductor.
(a) Both A and R are true and R is the correct explanation of A.
(b) Both A and R are true but R is not the correct explanation of A.
(c) A is true but R is false.
(d) A is false but R is true.

Q.125 An example of liquid conducting material is
(a) mercury (b) irridium
(c) silicon resin (d) boiling sodium

Q.126 Match **List-I** with **List-II** and select the correct answer using the codes given below the lists:
List-I
A. Debye's temperature
B. Matteheissen's rule
C. Weidmann frenz law
D. Meissner's effect
List-II
1. Total resistivity is the sum of different effects
2. Magnetic flux inside a superconductor is zero
3. $\propto$ (isotropic mass)$^{-1.2}$
4. Lorenz number
Codes:

	A	B	C	D
(a)	2	1	4	3
(b)	3	1	4	2
(c)	1	4	3	2
(d)	4	3	2	1

Q.127 A good conductor should have
1. high electrical conductivity
2. low thermal conductivity
3. low melting point
4. good oxidation resistance
Of these, the correct statements are
(a) 1, 2, 3 and 4 (b) 1 and 4
(c) 2, 3 and 4 (d) 1, 3 and 4

Q.128 Presence of phosphorus, sulphur and iron in copper
(a) lower its electrical conductivity
(b) increases its electrical conductivity
(c) lower its thermal conductivity
(d) increases its thermal conductivity

Q.129 The resistivity of materials are influenced by
1. plastic deformation
2. impurities
3. temperature
Select the correct code from the given options.
(a) 3 only (b) 1 and 2
(c) 1 and 3 (d) 1, 2 and 3

Q.130 **Assertion (A) :** In an ideal crystal at 0 Kelvin, as the mean free length tends to infinity, conductivity tends to infinity.
Reason (R) : The average distance travelled by an electron between two successive collisions is called mean free path.
(a) Both A and R are true and R is the correct explanation of A.
(b) Both A and R are true but R is not the correct explanation of A.
(c) A is true but R is false.
(d) A is false but R is true.

Q.131 The carrier concentration in a material of conductivity 0.018/ohm-m is 10^{19} electrons/m^3. A voltage of 0.16 volts is applied across the 0.29 mm thick material. What is the drift velocity of the carriers? Take the standard values of mass of electron and electric charge.
(a) 6.2 m/s (b) 3.8 m/s
(c) 5.5 m/s (d) 4.7 m/s

Q.132 Probability of occupation by an electron at Fermi energy level is
(a) 100% (b) 25%
(c) 50% (d) 0%

Q.133 The resistivity based on impurity and plastic deformation is approximately zero at
(a) low temperature (b) high temperature
(c) 0 Kelvin (d) all temperature

Q.134 Consider the following statements:

1. The potential energy is assumed constant in free electron theory.
2. Energy band theory consider only kinetic energy of free electrons.
3. Drift velocity of an electron is equal to its normal velocity.
4. Conductivity of pure copper will be more than the conductivity of brass.

Which of the statements given above are correct?

(a) 2 and 4 (b) 1 and 4
(c) 1, 2 and 3 (d) 1 and 3

Q.135 Superconductivity of a superconductor is destroyed

(a) at high temperature
(b) at high frequency
(c) both (a) and (b)
(d) neither (a) nor (b)

Q.136 The resistivity component arising due to impurities and defects is commonly known as

(a) residual component of resistivity
(b) thermal component of resistivity
(c) Joule's component of resistivity
(d) none of the above

Q.137 The superconducting materials in the increasing order of critical temperature are

(a) Pb, Ti, Nb_3Sn, Nb-Ti
(b) Ti, Pb, Nb_3Sn, Nb-Ti
(c) Pb, Ti, Nb-Ti, Nb_3Sn
(d) Ti, Pb, Nb-Ti, Nb_3Sn

Q.138 The value of H_c(Critical magnetic field intensity) indicates that the material is

(a) in normal state
(b) in superconducting state
(c) either in normal or superconducting state
(d) neither in normal nor in superconducting state

Q.139 Thermal conductivity of a solid will decrease when

(a) it is placed in a weak magnetic field.
(b) it is a single crystal structure.
(c) it is polycrystalline and placed in a strong magnetic field.
(d) it is polycrystalline and placed in a weak magnetic field.

Q.140 The magnetic lines of force are repelled back completely due to an effect called

(a) BCS effect (b) Silsbee effect
(c) Meissner effect (d) Bednorz effect

Q.141 As compared to the elements, the compounds and alloys can be made superconductors at

(a) higher temperature
(b) lower temperature
(c) the same temperature
(d) much lower temperature

Q.142 With the increase in temperature, the mean free path

(a) decreases
(b) increases
(c) remains unchanged
(d) vanishes

Q.143 Thermal expansion of materials arises from

(a) strong bonds
(b) weak bonds
(c) thermal vibrations
(d) asymmetry of potential energy curve

Q.144 On which of the following factors does the value of critical current density in a superconductor depend?

(a) Temperature
(b) Applied magnetic field
(c) Temperature and applied magnetic field
(d) Silsbee's rule

Q.145 The correct sequence of the following materials in the increasing order of conductivity is

(a) Silicon doped with boron - silver - aluminium - intrinsic silicon
(b) Intrinsic silicon - aluminium - silver - silicon doped with boron
(c) Aluminium - intrinsic silicon - silicon doped with boron - silver
(d) Intrinsic silicon - silicon - silicon doped with boron - aluminium - silver

Q.146 The electrical conductivity of copper reduces with the addition of a small amount of nickel because

(a) the mobility decreases due to creation of new scattering for electrons.

(b) nickel is ferromagnetic.
(c) less free electrons are available for conduction.
(d) the electrons of copper become bound to nickel and are not free to move.

Q.147 The average drift velocity v_d of electrons in a metal is related to electric field E and collision time T as

(a) $v_d = Q_e ET/m_e$ (b) $v_d = m_e Q_e T$
(c) $v_d = m_e Q_e T/2E$ (d) $v_d = Q_e ET/2m_e$

Q.148 Match **List-I (Materials)** with **List-II (Equation/ Rule)** and select the correct answer using the codes given below the lists:

List-I
A. Dielectric material
B. Ferromagnetic material
C. Conductors
D. Superconductors

List-II
1. Debye equation
2. Curie Weiss law
3. Matthiessen's rule
4. Meissner effect

Codes:

	A	B	C	D
(a)	3	4	1	2
(b)	1	4	3	2
(c)	3	2	1	4
(d)	1	2	3	4

Q.149 Match **List-I (Parameter)** with **List-II (Unit)** and select the correct answer using the codes given below the lists:

List-I
A. Boltzmann constant
B. Permeability of free space
C. Permittivity of free space
D. Mobility

List-II
1. Farad/meter
2. cm^2/volt-second
3. Henry/meter
4. cm^2/second
5. electron-volt/Kelvin

Codes:

	A	B	C	D
(a)	5	2	1	4
(b)	1	2	5	4
(c)	5	3	1	2
(d)	1	3	5	2

5. Semiconductor Materials

Q.150 Match **List-I (Type of solids)** with **List-II (Energy gap)** and select the correct answer using the codes given below the lists:

List-I	List-II
A. Conductor	1. $3\ eV > E_g > 1\ eV$
B. Insulator	2. $E_g > 3\ eV$
C. Semiconductor	3. $E_g = 0\ eV$

Codes:

	A	B	C
(a)	1	2	3
(b)	3	1	2
(c)	3	2	1
(d)	1	3	2

Q.151 If the wavelength of the absorption edge of a semiconductor material is 1771 mm, then the band gap energy is around
(Use Plank's constant $h = 6.626 \times 10^{-34}$ J-s)

(a) 0.6 eV (b) 0.5 eV
(c) 0.2 eV (d) 0.7 eV

Q.152 During an electron transition across the energy gap in Si,
1. the momentum of the electron changes.
2. the direction of motion of the electron changes.
3. the potential energy of the electron changes.
4. the kinetic energy of the electron remains constant.

Of these, the correct statements are
(a) 1, 2, 3 and 4 (b) 2, 3 and 4
(c) 1, 2 and 4 (d) 1, 2 and 3

Q.153 The energy gap in divalent metals is
(a) very small (b) fairly small
(c) zero (d) large

Q.154 Pure silicon at 0 K is
(a) an intrinsic semiconductor
(b) an extrinsic semiconductor
(c) a metal
(d) an insulator

Q.155 If the drift velocity of holes under a field gradient of 100 V m^{-1} is 5 ms^{-1}, their mobility (in the same SI units) is
(a) 0.05 (b) 0.50
(c) 50 (d) 500

Q.156 The majority charge carrier in *p*-type Ge are
(a) free electrons (b) ions
(c) holes (d) conduction electrons

Q.157 For silicon doped with *B*,
(a) $n_e >> n_h$ (b) $n_h >> n_e$
(c) $n_h >> n_i$ (d) both b and c

Q.158 **Assertion (A) :** The mobility decreases with increasing temperature.
Reason (R) : Increasing temperature results in decreasing mean free path of the conduction electrons or holes.
(a) Both A and R are true and R is the correct explanation of A.
(b) Both A and R are true but R is not the correct explanation of A.
(c) A is true but R is false.
(d) A is false but R is true.

Q.159 As the concentration of electrons in a semiconductor is changed by changing the impurity level, the conductivity also changes. If n_i is the intrinsic concentration, then the minimum value of conductivity would be equal to (assume the temperature to be constant)
(a) $n_e e\mu_e$ (b) $\frac{n_e e\mu_e}{2}$
(c) $3n_e e\mu_e$ (d) $2n_e e\mu_e$

Q.160 The resistivity of pure silicon at room temperature is 3000 ohm-m. If μ_e = 0.14 $m^2 V^{-1} s^{-1}$ and μ_h = 0.05 $m^2 V^{-1} s^{-1}$, then the intrinsic carrier density is approximately equal to
(a) 1.25×10^{13} m^{-3} (b) 1.1×10^{16} m^{-3}
(c) 0.90×10^{14} m^{-3} (d) none of these

Q.161 **Assertion (A) :** At room temperature, a large fraction of the donor level electrons are excited into the conduction band.
Reason (R) : According to the law of mass action, the product of the number of electrons in the conduction band and the number of holes in the valence band must be constant.
(a) Both A and R are true and R is the correct explanation of A.
(b) Both A and R are true but R is not the correct explanation of A.
(c) A is true but R is false.
(d) A is false but R is true.

Q.162 Match **List-I** with **List-II** and select the correct answer using the codes given below the lists:

List-I
A. *n*-type semiconductor
B. *p*-type semiconductor
C. Zener diode
D. Diffusion

List-II
1. Acceptor
2. Concentration gradient
3. Faster
4. Zener breakdown

Codes:

	A	B	C	D
(a)	3	1	2	4
(b)	1	3	4	2
(c)	3	1	4	2
(d)	1	3	2	4

Q.163 As compared to Si, the electron mobility in GaAs is
(a) slower by about five times
(b) same
(c) faster by about six times
(d) faster by about 200 times

Q.164 Consider the following statements:
1. Photoconductors are devices used for detecting and measuring light energy.
2. GaAs is cheaper and widely used for making inexpensive chips.
3. Metals, insulators and semiconductors have a finite forbidden gap.
4. The ability of a semiconductor or an insulator to conduct electricity depends on thermal excitation and the presence of impurities.
5. Electrons can take the energy values that lie in the forbidden gap.

Which of the statements given above are correct?
(a) 1, 3 and 5 (b) 2, 3 and 4
(c) 1, 2, 3 and 5 (d) 1 and 4

Q.165 Match **List-I (Devices)** with **List-II (Applications)** and select the correct answer using the codes given below the lists:

List-I

A. Photoconductors
B. Galium Arsenide
C. Zener diode
D. Crystal Oscillator

List-II

1. Automatic door opening
2. Satellite communication
3. Voltage regulator
4. Sine wave generator

Codes:

	A	B	C	D
(a)	1	2	3	4
(b)	2	1	3	4
(c)	3	4	1	2
(d)	4	3	2	1

Q.166 GaAs has an energy gap of 1.43 eV. The wavelength of the radiation emitted during an electronic transition in GaAs will be in the
(a) visible range (b) ultraviolet range
(c) infrared region (d) far ultraviolet region

Q.167 The value of energy gap of Si is
(a) 0.08 eV (b) 0.66 eV
(c) 5.4 eV (d) 1.1 eV

Q.168 Match **List-I (Basic Semiconducting devices)** with **List-II (Operating principle)** and select the correct answer using the codes given below the lists:

List-I

A. Thermistors
B. Varistors
C. Photoconductive cells
D. Hall effect generators

List-II

1. Utilize the light illumination effect of semiconductors
2. Utilize the temperature dependency effect of semiconductors
3. Utilize the carrier drift effect in semiconductors
4. Utilize the voltage dependency effect of semiconductors

Codes:

	A	B	C	D
(a)	2	1	4	3
(b)	2	4	1	3
(c)	1	2	3	4
(d)	4	3	2	1

Q.169 A current of density 1000 A/m^2 flows through a *n*-type germanium crystal. This crystal has a resistivity of 0.05 ohm-m and the electron mobility of 0.4 m^2/V-s. The drift velocity of the electron is
(a) 20 m/s (b) 10 m/s
(c) 5 m/s (d) none of these

Q.170 Consider the following statements associated with semiconducting materials:
1. They consume considerable amount of power.
2. They are much smaller is size.
3. They require very high voltage to operate.
4. They operate over a wide range of temperature.

Which of the above statements is/are **not** correct?
(a) 3 only (b) 4 only
(c) 1 and 3 (d) 2 and 4

Q.171 The rate of diffusion process will increase if temperature is
(a) decreased (b) increased
(c) constant (d) sub-zero

Q.172 The unit of ionization potential is the same as that of
(a) electronic charge (b) electron energy
(c) electron affinity (d) radiation's frequency

Q.173 Lattice thermal conductivity is more than electronic thermal conductivity in
(a) dielectric (b) metals
(c) semiconductors (d) all of these

Q.174 A *p*-type germanium rod, 10 mm long and 1 mm in diameter, has a resistance of 100 ohm. If the mobility of hole in germanium is 0.19 m^2/V-s, then the impurity concentration in the rod is around
(a) 3.8×10^{19} /m^3 (b) 2.6×10^{18} / m^3
(c) 5.2×10^{22} /m^3 (d) 4.2×10^{21} /m^3

Q.175 If silicon is doped with 10^{19} donor / m^3 and the intrinsic carrier concentration is 1.5×10^{16} /m^3, then the hole density at room temperature would be equal to
(a) $1.15 \times 10^{12}/m^3$ (b) $3.35 \times 10^{11}/m^3$
(c) 2.25×10^{13} /m^3 (d) none of these

Q.176 The energy gap in diamond is
(a) 5.4 eV (b) 2 - 3 eV
(c) 1.1 eV (d) 0.08 eV

Q.177 If light is falling on one end of a p-type semiconductor which has an infinite recombination lifetime, the excess electron density in the semiconductor will vary ______ with distance, as one moves away from the end on which light is falling.
(a) linearly (b) parabolically
(c) exponentially (d) logarithmically

Q.178 In a semiconductor, electron exist in
(a) the conduction band
(b) the valence band
(c) the forbidden energy band
(d) no band

Q.179 In silicon, arsenic acts as an impurity that ______ electrons, and aluminium acts as an impurity that ______ electrons.
(a) donates, donates (b) accepts, donates
(c) accepts, accepts (d) donates, accepts

Q.180 **Assertion (A)** : Diffusion and mobility are two independent phenomena.
Reason (R) : Diffusion and mobility of charge carriers depends on temperature.
(a) Both A and R are true and R is the correct explanation of A.
(b) Both A and R are true but R is not the correct explanation of A.
(c) A is true but R is false.
(d) A is false but R is true.

Q.181 Even though carbon is in the group IV of the Periodic table, it is **not** used as a semiconductor because it has
(a) high dielectric constant
(b) large energy gap > 5 eV
(c) low temperature coefficient
(d) low thermal conductivity

Q.182 A thermistor is a
(a) thermocouple
(b) thermometer
(c) miniature resistance
(d) heat sensitive explosive

Q.183 An intrinsic semiconductor at a temperature of absolute zero behaves like an insulator because of
(a) non-availability of free electrons
(b) non-recombination of electrons with holes
(c) low drift velocity of free electrons
(d) low (almost zero) electron energy

Q.184 The Hall coefficient of an intrinsic semiconductor is
(a) positive (b) negative
(c) zero (d) infinite

Q.185 Fermi level in a *p*-type semiconductor lies close to
(a) the top of the valence band
(b) the bottom of the valence band
(c) the top of the conduction band
(d) the bottom of the conduction band

Q.186 For which one of the following materials, is the Hall coefficient zero?
(a) Metal
(b) Insulator
(c) Intrinsic semiconductor
(d) Alloy

Q.187 A semiconductor specimen of breadth d, width W and carrying current I is placed in a magnetic field B to develop Hall voltage V_H in a direction perpendicular to I and B. V_H is **not** proportional to
(a) B (b) I
(c) $1/W$ (d) $1/d$

Q.188 Match **List-I (Type of material)** with **List-II (Type of bonding)** and select the correct answer using the codes given below the lists:

List-I	List-II
A. Elemental semiconductor	1. Ionic
B. Hydrogen molecule	2. Covalent
C. Copper	3. Vander Wall
D. AgI	4. Metallic

Codes:

	A	B	C	D
(a)	3	2	1	4
(b)	2	3	1	4
(c)	2	3	4	1
(d)	3	2	4	1

Q.189 In a silicon crystal, the arrangement of atoms repeats periodically. This type of material is classified as
(a) amorphous and non-crystalline
(b) non-crystalline and epitaxial
(c) epitaxial and single crystal
(d) amorphous and single crystal

6. Miscellaneous

Q.190 Which of the following is a low resistivity material?
(a) Manganin (b) Nichrome
(c) Nickel (d) Gold

Q.191 Match **List-I (Alloys)** with **List-II (Applications)** and select the correct answer using the codes given below the lists:

List-I	List-II
A. Fe-C alloy	1. Soldering
B. Pb-Sn alloy	2. Manufacture of semiconductor devices
C. Au-Si alloy	3. Making of steel

Codes:

	A	B	C
(a)	1	2	3
(b)	3	1	2
(c)	2	1	3
(d)	3	2	1

Q.192 The unit of electrical permittivity is
(a) $C\text{-}m^{-2}$ (b) $H\text{-}m^{-1}$
(c) $F\text{-}m^{-1}$ (d) dimensionless

Q.193 **Assertion (A)** : Insulators, also known as dielectrics, are very poor conductors of heat and electricity.
Reason (R) : The solids having an energy gap of 3 eV or more are termed as insulators.
(a) Both A and R are true and R is the correct explanation of A.
(b) Both A and R are true but R is not the correct explanation of A.
(c) A is true but R is false.
(d) A is false but R is true.

Q.194 Match **List-I (Insulating materials)** with **List-II (Applications)** and select the correct answer using the codes given below the lists:

List-I
A. Wax paper-aluminium foil
B. Teflon
C. Glass wool
D. Resins

List-II
1. As filling material
2. As impregnating materials
3. As adhesive materials
4. As coating materials

Codes:

	A	B	C	D
(a)	2	4	1	3
(b)	2	1	4	3
(c)	1	2	3	4
(d)	3	2	4	1

Q.195 Which of the following insulating material is used for high voltage services?
(a) Porcelain (b) Glass tapes
(c) Air (d) N_2

Q.196 Which of the following is a member of ceramic family?
(a) Bainite (b) Garnet
(c) LDFE (d) High speed steel

Q.197 In an imperfection free crystal, the resistivity will be
(a) infinite (b) zero
(c) negative (d) unity

Q.198 Highest electrical resistivity exists in
(a) platinum wire (b) nichrome wire
(c) silver wire (d) kanthal wire

Q.199 Which of the following pairs is not correctly matched?
(a) Low resistivity material : Brass
(b) Commutator segments : Cadmium bronze
(c) High resistivity material : Tantalum
(d) Sealing of glass tubes : Palladium

Q.200 Gold is the most
(a) malleable metal (b) tough metal
(c) ductile metal (d) both (a) and (b)

Q.201 Which material among the following possesses excellent dielectric properties and good reliability for use in making capacitors?

(a) Silicon monoxide (b) Silicon dioxide
(c) Tin oxide (d) Chromium oxide

Q.202 In an integrated circuit, the SiO_2 layer provides
(a) electrical connection to external circuit
(b) physical strength
(c) isolation
(d) conducting path

Q.203 German silver is an alloy of
(a) Copper, Silver and aluminium
(b) Silver, tin and tungsten
(c) Copper, manganese and nickel
(d) Copper, zinc and nickel

Q.204 Dielectric materials are used primarily for
(a) insulation
(b) charge storage
(c) reducing dielectric loss
(d) none of the above

Q.205 In metals, resistivity is composed of two parts: one part is characteristic of the particular substance. The other part is due to
(a) applied voltage
(b) crystal imperfections
(c) applied magnetic field
(d) supplied thermal energy

Q.206 A dielectric material has the real part of the dielectric constant (ε_r') as 4 and its loss tangent is 0.004. What is the complex dielectric constant (ε_r^*) represented by?
(a) $4 + j0.016$ (b) $4 - j0.016$
(c) $4 + j0.001$ (d) $4 - j0.001$

Q.207 Which of the following materials has the highest dielectric strength?
(a) Polystyrene (b) Marble
(c) Cotton (d) Transformer oil

■■■■

Answers Electrical Materials

1.	(b)	2.	(d)	3.	(a)	4.	(c)	5.	(c)	6.	(a)	7.	(b)	8.	(d)
9.	(a)	10.	(c)	11.	(c)	12.	(d)	13.	(a)	14.	(a)	15.	(a)	16.	(b)
17.	(b)	18.	(d)	19.	(b)	20.	(a)	21.	(c)	22.	(a)	23.	(c)	24.	(d)
25.	(d)	26.	(d)	27.	(b)	28.	(c)	29.	(b)	30.	(a)	31.	(c)	32.	(b)
33.	(b)	34.	(b)	35.	(d)	36.	(a)	37.	(d)	38.	(c)	39.	(a)	40.	(d)
41.	(a)	42.	(a)	43.	(b)	44.	(a)	45.	(d)	46.	(d)	47.	(c)	48.	(c)
49.	(a)	50.	(b)	51.	(d)	52.	(a)	53.	(b)	54.	(d)	55.	(c)	56.	(b)
57.	(c)	58.	(c)	59.	(c)	60.	(b)	61.	(c)	62.	(c)	63.	(d)	64.	(b)
65.	(b)	66.	(d)	67.	(b)	68.	(c)	69.	(a)	70.	(d)	71.	(d)	72.	(a)
73.	(b)	74.	(a)	75.	(b)	76.	(d)	77.	(a)	78.	(b)	79.	(c)	80.	(a)
81.	(a)	82.	(d)	83.	(c)	84.	(d)	85.	(b)	86.	(a)	87.	(b)	88.	(a)
89.	(c)	90.	(c)	91.	(a)	92.	(b)	93.	(d)	94.	(c)	95.	(a)	96.	(d)
97.	(c)	98.	(b)	99.	(c)	100.	(b)	101.	(a)	102.	(d)	103.	(b)	104.	(c)
105.	(d)	106.	(d)	107.	(b)	108.	(b)	109.	(d)	110.	(d)	111.	(a)	112.	(c)
113.	(a)	114.	(a)	115.	(d)	116.	(b)	117.	(c)	118.	(d)	119.	(d)	120.	(a)

121. (b)	122. (d)	123. (c)	124. (c)	125. (a)	126. (b)	127. (b)	128. (a)
129. (d)	130. (b)	131. (a)	132. (c)	133. (d)	134. (b)	135. (c)	136. (a)
137. (d)	138. (a)	139. (c)	140. (c)	141. (a)	142. (a)	143. (d)	144. (d)
145. (d)	146. (a)	147. (d)	148. (d)	149. (c)	150. (c)	151. (d)	152. (a)
153. (c)	154. (d)	155. (a)	156. (c)	157. (d)	158. (a)	159. (d)	160. (b)
161. (b)	162. (c)	163. (c)	164. (d)	165. (a)	166. (c)	167. (d)	168. (b)
169. (a)	170. (c)	171. (b)	172. (c)	173. (a)	174. (d)	175. (c)	176. (a)
177. (a)	178. (a)	179. (d)	180. (d)	181. (b)	182. (c)	183. (a)	184. (b)
185. (a)	186. (a)	187. (d)	188. (c)	189. (c)	190. (d)	191. (b)	192. (c)
193. (b)	194. (a)	195. (d)	196. (b)	197. (b)	198. (d)	199. (d)	200. (c)
201. (d)	202. (c)	203. (d)	204. (b)	205. (b)	206. (b)	207. (a)	

Explanations

2. (d)

Table below shows various properties of engineering materials with their examples:

Properties	Examples
Physical	Density, Colour, Shape, Size, Melting point etc.
Mechanical	Ductility, Malleability, Brittleness, Hardness etc.
Magnetic	Hysteresis, Retentivity, Permeability, Reluctivity etc.
Chemical	Atomic number, Passivity, Acidity etc.

3. (a)

The scope of material science is generally restricted to the study of solid materials, and that too only to those which are useful as engineering materials.

4. (c)

- Specific heat and thermal fatigue are thermal properties.
- Passivity and pH-value are chemical properties.
- Proof resilience and creep are mechanical properties.
- Susceptibility and retentivity are magnetic properties.

5. (c)

When many unit cells repeat in a three-dimensional space, a crystal is obtained. The structure of a crystal is same as that of a repeating unit cell.

7. (b)

- The crystal directions of a family are not necessarily parallel to one another. Hence, statement-1 is not correct.
- Statement-2 is correct.
- In cubic crystals, a crystal plane and a crystal direction normal to it have the same indices. Hence, statement-3 is not correct.
- The effective number of lattice points in unit cell is highest i.e. 4 in case of face-centered cubic space lattice. In case of simple cubic it is 1 and in case of BCC it is 2. Hence, statement-4 is correct.

9. (a)

In FCC crystal, diagonal of each face

$$= 4r = \sqrt{2}a$$

(Where, r = radius of each atom)

The atomic diameter of an FCC crystal

$$= 2r = \frac{\sqrt{2}a}{2} = \frac{a}{\sqrt{2}}$$

10. (c)

- The sharing of electrons between neighbouring atoms results in a covalent bond, which is directional. Hence, statement-1 is not correct.
- The metallic bond is non-directional and generally weaker than ionic and covalent bonds. Thus, statement-2 is not correct.

- Free electrons are responsible for the high thermal and electrical conductivities. Hence, statement-3 is correct.
- Statement-4 is correct.
- The magnitude of the energy released, when two atoms come together from a large distance of separation to the equilibrium distance, is called the bond energy. It is related to the enthalpy of atomization of the solid. Hence, statement-5 is correct.

11. (c)

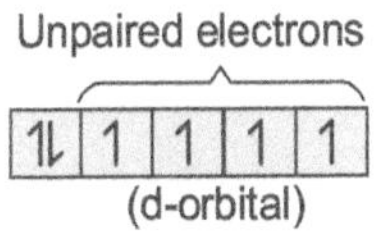

The number of unpaired electrons = 4

12. (d)

The thermal expansion of materials arises from the asymmetry of the potential energy versus distance curve as shown in figure below.

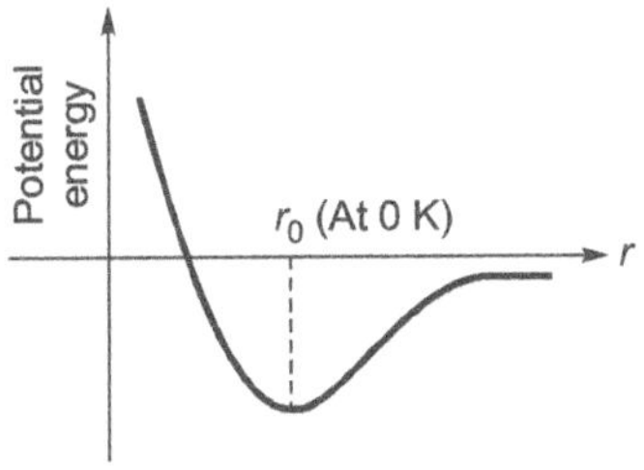

Deep potential wells are more symmetrical about the equilibrium position r_0 than shallow potential wells. So, the thermal expansion at a given temperature tends to be less for strongly bonded materials than for weakly bonded materials.

13. (a)

The mechanical properties of solids like hardness and brittleness etc are dependent on the strength of the bonds as well as the directional nature of bonding. Solids with strong and directional bonds tend to be brittle. Since diamond is covalently bonded which is very strong and directional in nature therefore, diamond is very hard and brittle. Hence, both assertion and reason are true and reason is the correct explanation of assertion.

14. (a)

The bond that is formed between water molecules due to attraction between the positively charged hydrogen end of a molecule and the negatively charged oxygen end of another molecule is called the hydrogen bond.

The strength of various bonds in increasing order are as follows:

Vander walls bonds < Hydrogen bonds < Ionic bonds < Metallic bonds < Covalent bonds.

16. (b)

Nanotechnology can be used for water purification by using nano-mineral, nano-gold and nano-silver nitrate which removes the harmful element arsenic (As) from water. Under nanotechnology, the size of material is reduced to the size of nano-domain. For small sized materials, the surface to volume ratio plays important role and most of the molecules make the material hyperactive by coming on the surface. Consequence of this effect, is to change the 'chemical effectiveness' of material. This is the correct reason for assertion.

Nanomaterials are prepared by two methods:

(i) Larger to smaller doing methods.

(ii) Smaller to larger doing methods.

Hence, both assertion and reason are true but reason is not the correct explanation of assertion.

17. (b)

The nature of identical properties in all the directions is called isotropy while dissimilar properties in all the direction is called anisotorpy.

18. (d)

The intercepts C_1 (along x-axis) = a,

C_2 (along y-axis) = $b/2$,

and C_3 (along z-axis) = $3c$

Therefore, $$p = \frac{c_1}{a} = \frac{a}{a} = 1$$

$$q = \frac{c_2}{b} = \frac{b/2}{b} = \frac{1}{2}$$

and $$r = \frac{c_3}{c} = \frac{3c}{c} = 3$$

$\therefore$ $$h = \frac{1}{p} = \frac{1}{1} = 1$$

$$k = \frac{1}{q} = \frac{1}{1/2} = 2$$

and $l = \frac{1}{r} = \frac{1}{3}$

Hence, $(h, k, l) = \left(1, 2, \frac{1}{3}\right) = \frac{1}{3}(3, 6, 1) = (3, 6, 1)$

(Since Miller indices is always an integer)

20. (a)

Rocks, glasses, fireclay and firebricks, cements and limes are ceramics. Ferrites, garnets, ferroelectrics and ceramic superconductors are the latest developed ceramics.

21. (c)

- An alloy is a combination of two or more metals.
- Organic polymers are relatively inert and light, and generally have a high degree of plasticity.
- Ceramics are generally metallic or non-metallic oxides.
- Composites may be inorganic or organic. They have two or more constituents of dissimilar properties.

22. (a)

Non-ferrous metals do not contain Fe and C as their constituents. Al, Cu, Ag, Au, Ni, Zn, Sn, Cr etc are some examples.

23. (c)

More and more structural parts of helicopters, aircrafts and space going vehicles are built with light weight (but tough) materials such as a aluminium alloys and composites. Pure aluminium can't be used for such purposes.

24. (d)

Point imperfections are imperfect point-like regions in the crystal. These defects are of one or two atomic diameters only. Hence, these are called two-dimensional defects.
Point imperfections of different types interact with each other and in doing so lower the total energy. These are thermodynamically stable.

25. (d)

The imperfections may be minimized by:

- use of thermal energy.
- more careful control on crystallization i.e. making metal in fibre form.
- surface treatment.
- use of whisker form of material.

26. (d)

Bonding is the nature of a material to hold the atoms together. In materials, atoms are arranged systematically. Coulomb's force, both attractive and repulsive, work on them. The net force out of these two opposite natured forces, are responsible for the bonding.

34. (b)

Dielectrics are very poor conductors of heat and electricity because in such materials it is almost impossible to excite the electrons from the valence band to the conduction band by means of an applied field or normal thermal energy. Reason is also true but not the correct explanation of assertion.

35. (d)

Form of Dielectric Materials	Examples
Coating materials	Polyethylene, Teflon
Filling materials	Glass wool, Asbestos wool
Adhesive materials	Plastic films, Resins
Bonding materials	Elastomers, Glass-reinforced-mica

36. (a)

Dielectric materials

- have high specific resistance (ρ).
- have negative temperature coefficient of resistance.
- may belong to the ceramic family such as glass, silica, mica etc.
- exists in all three forms i.e. solid, liquid and gas.

Examples:

(*i*) Solid dielectric : Bakelite, PVC etc.
(*ii*) Liquid dielectric : Mineral oils, transformer oil etc.
(*iii*) Gaseous dielectric : N_2, H_2, Air etc.

37. (d)

Dielectric strength of mica is 100 MV/m which is maximum among all conventional insulating materials.

38. **(c)**

The required thickness of insulation will be

$$= \frac{33\text{ kV}}{40\text{ kV/mm}}$$

$$= \frac{33 \times 10^3\text{ mm}}{40 \times 10^3} = 0.82\text{ mm}$$

39. **(a)**

The dielectric constant ε_r (or relative permittivity) is a measure of polarization of dielectric materials. Dielectric strength is the voltage per unit thickness that can be sustained by an insulating material before its breakdown. A good insulating material or a dielectric material should have high value of dielectric strength but low value of dielectric constant (ε_r).

41. **(a)**

Condensers to be used at high frequencies require a material whose dielectric loss should be small i.e. low value of loss angle since dielectric loss $\propto$ tan δ or δ.

43. **(b)**

Refractories are ceramic materials of specific nature which are capable of withstanding high temperature. The examples of the common or basic refractories are silica, magnesite, forsterite, dolomite, silicon carbide and zircon.

44. **(a)**

A good refractory must have low electrical conductivity.

45. **(d)**

Ceramics have high values of dielectric constant ε_r (8 for mica, 7 for soda-lime glass, and 6 for porcelain). Dielectric strength of ceramic is high.

47. **(c)**

The capacitance of a parallel plate condenser is

$$C = \varepsilon_0 \varepsilon_r \frac{A}{d}$$

or, $$\varepsilon_r = \frac{Cd}{\varepsilon_0 A}$$

Given, $C = 10^{-9}$ F, $d = 2$ mm $= 2 \times 10^{-3}$ m, $A = 100 \times 10^{-6}$ m^2 and $\varepsilon_0 = 8.854 \times 10^{-12}$ F/m

$$\therefore \quad \varepsilon_r = \frac{10^{-9} \times 2 \times 10^{-3}}{8.854 \times 10^{-12} \times 100 \times 10^{-6}}$$

$$= 2259$$

48. **(c)**

Electronic polarization is the result of the displacement of the positively charged nucleus and the (negative) electrons of an atom in opposite directions on application of an electric field. On applying a field, the electron cloud around the nucleus readily shifts towards the positive end of the field. Hence, such a shift results in a dipole moment within the atom, as a certain distance now separates the nucleus and the centre of the electron clouds. Hence, assertion is a true statement.

The extent of the shift is proportional to the field strength. As the dipole moment is the product of charge and the shift distance, it is also proportional to the field strength. Hence, Reason is a false statement.

50. **(b)**

Ionic and electronic polarizations are independent of temperature while space charge and Orientation polarization are inversely proportional to temperature.

51. **(d)**

Space charge polarization occurs due to the accumulation of charges at the electrodes or at the interfaces in a multi-phase material.

52. **(a)**

- Orientation polarization is slower than ionic polarization. Orientation polarization occurs, when the frequency of the applied voltage is in the audio range. Hence, statement-1 is correct.
- Ionic polarization is slower than electronic polarization, as the displacement involved here is that of the much heavier ion, as compared to the electron cloud in the case of electronic polarization. Hence, statement-2 is correct.

- Space charge polarization is the slowest process, as it involves the diffusion of ions over several interatomic distances. Thus, statement-3 is correct.

Hence, all the three statements are correct.

53. (b)

The dipole moment due to the effective number of each type of ion in the unit cell are due to

(*i*) Two O^{2-} on four side faces:
$2 \times 2 \times 1.6 \times 10^{-19} \times 0.06 \times 10^{-10}$ C-m

(*ii*) One O^{2-} on top and bottom:
$1 \times 2 \times 1.6 \times 10^{-19} \times 0.08 \times 10^{-10}$ C-m

(*iii*) One Ti^{4+} at body centre:
$1 \times 4 \times 1.6 \times 10^{-19} \times 0.06 \times 10^{-10}$ C-m

∴ Total dipole moment = 1.02×10^{-29} C-m

Polarization is the sum of the dipole moments per unit volume. Ignoring the effect due to the barium ions, we get

$$P = \frac{1.02 \times 10^{-29}}{4.03 \times 3.98^2 \times 10^{-30}} \text{ C-m}^{-2}$$

$$= 0.16 \text{ C-m}^{-2}$$

54. (d)

Statements-1 and 3 are correct.

Statement-2 is not correct because the dielectric constant of a material is dependent on the frequency of the alternating field applied and the temperature.

55. (c)

Relative dielectric constant is

$$\varepsilon_r = \frac{\varepsilon}{\varepsilon_0}$$

which is a dimensionless quantity.

56. (b)

Dielectric Material	Dielectric Strength
Transformer Oil	10
Mica	100
PVC	2
Polyethylene	4

57. (c)

During melting, the relative dielectric constant may increase or decrease depending on the type of solid.

58. (c)

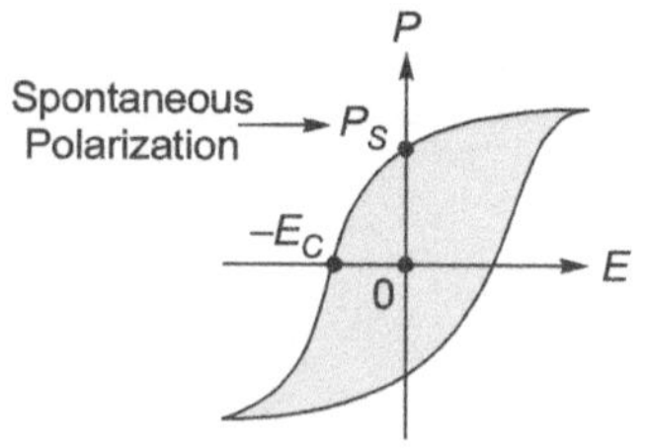

60. (b)

- Isotropic materials are those whose properties are direction independent.
- Anisotropic materials are those whose properties are direction dependent.
- Orthotropic material is a special kind of anisotropic material. A unidirectional fibre composite is an example of such material.

61. (c)

Examples of piezoelectric materials are:

Quartz, Potassium dihydrogen phosphate, Lead titanate, Barium titanate, Rochelle's salt.

Lead zirconate and sodium neobate are examples of anti ferroelectric materials.

62. (c)

Above a particular value of temperature called "Curie temperature", spontaneous polarization in the material is reduced to zero because with the increase in temperature, randomness in the orientation of dipoles increases so the size of domain decreases and at Curie-temperature all the dipoles will be randomly oriented. Hence, assertion is a true statement.

Above curie temperature material starts behaving like a non-ferroelectric material and below Curie temperature material behaves as a ferroelectric material. Hence, reason is a false statement.

63. (d)

Clausius Mossotti equation is valid for Lorentz internal field only and for Lorentz internal field, internal field constant is $\gamma = \frac{1}{3}$.

65. (b)

We know that,

$$\alpha = \frac{(\varepsilon_r - 1)\varepsilon_0}{N} = 4\pi\varepsilon_0 R^3$$

(for monoatomic gas)

$$\therefore \quad \varepsilon_r = 1 + 4\pi R^3 N$$

67. (b)

Ferrites have high resistivity due to which eddy current loss is reduced.

68. (c)

In the presence of dielectric,

$$D = (\varepsilon_0 E + P) = (\varepsilon_0 E + \varepsilon_0 \gamma_e E)$$

69. (a)

$$R = \frac{\rho l}{A}$$

$$\therefore \quad R = \frac{10^{10} \times 10^{-3}}{10^4 \times 10^{-6}} = 10^9\ \Omega$$

$$\text{Dielectric loss} = \frac{V^2}{R} = \frac{(10^3)^2}{10^9} = 10^{-3}\ \text{Watt}$$

72. (a)

$$\text{Dielectric loss} = \frac{r''}{r}$$

i.e. dielectric loss $\propto r''$

74. (a)

$$\varepsilon^* = \varepsilon_r' - j\varepsilon_r''$$

$$\text{Loss tangent} = \frac{\text{Imaginary part of } \varepsilon^*}{\text{Real part of } \varepsilon^*} = \frac{\varepsilon_r''}{\varepsilon_r'}$$

$$\therefore \quad \varepsilon_r'' = 0.004 \times 2.5 = 0.01$$

75. (b)

Soft magnetic materials are suitable for AC appliances because they have a small remanence, a small coercive force and have small area of hysteresis loop and therefore small losses (Hysteresis loss $\propto$ area of hysteresis loop). Hence, assertion is a true statement. Reason is also a true statement but not the correct explanation of assertion.

76. (d)

- A diamagnetic material does not obey Curie's law.
- Paramagnetic material obeys Curie's law i.e. susceptibility of a paramagnetic material is inversely proportional to temperature.

$$\chi_m \propto \frac{1}{T} \text{ or } \chi_m T = \text{Constant}$$

- A ferromagnetic material is strongly attracted by a magnet.
- In antiferromagnetic materials the permanent dipoles tend to align themselves so that alternate dipoles are antiparallel to the applied field. The result is a cancellation of the effects of each dipole and a zero net increase in flux density in the material. Hence, the permanent dipole moment is zero in these materials.

77. (a)

Magnetization,

$$M = \chi_m H$$

Where, χ_m = Magnetic susceptibility which is negative for diamagnetic materials.

So, $\chi_m = -4 \times 10^{-6}$ (Given)

and $\mu_r = 1 + \chi_m$

or, $\chi_m = \mu_r - 1 = 1.00038 - 1 = 0.00038$

Hence, $M = \chi_m H = 0.00038 \times 0.31 = 0.0001178$ A/m $= 0.1178$ mA/m ≈ 0.12 mA/m

78. (b)

Given, $l = 300 \times 10^{-3}$ m; $N = 2000$; $I = 600$ mA;

$$r = \frac{20}{2} \times 10^{-3}\ \text{m} = 10^{-2}\ \text{m}$$

We know that,

$$L = \frac{\mu_0 A N^2}{l}$$

$$\therefore \quad L = \frac{4\pi \times 10^{-7} \times \pi \times (10^{-2})^2 \times (2000)^2}{300 \times 10^{-3}}$$

$$= \frac{4\pi^2 \times 10^{-4}}{3} = 5.264\ \text{mH}$$

$$\approx 5.26\ \text{mH}$$

79. (c)

Energy lost during each cycle

$= \text{Area of loop}$

$= 10^{-4} \times 10^2 \times 60000$

$= 600 \text{ J-m}^{-3}$

Volume of the transformer core = 0.01 m^3

Energy lost in the core in each cycle

$= 600 \times 0.01 = 6 \text{ J}$

$\therefore$ Power loss due to hysteresis

$= 6 \text{ J} \times 50 \text{ Hz} = 300 \text{ W}$

81. (a)

We know that,

$$B = \mu_0(H + \chi_m H) = \mu_0(H + M)$$
$$= \mu_0 H(1 + \chi_m)$$

Given, $\mu_0 = 1.25 \times 10^{-6}$ H/m,

$\chi_m = 10^{-3}$,

$H = 100 \text{ kA/m} = 10^5 \text{ A/m}$

Putting the values in above equation, we have

$$B = 1.25 \times 10^{-6} \times 10^5 (1 + 10^{-3}) \text{ Wb/m}^2$$
$$= 0.125 \text{ Wb/m}^2$$

84. (d)

Each time the hysteresis loop is traversed, energy equal to the area of the loop is dissipated as heat. Thus, power loss due to hysteresis in a transformer core is dependent on the number of times the full loop is traversed per second. Hence, assertion is a false statement. Reason is a true statement.

85. (b)

Hysteresis loss $\propto f$ and eddy current loss $\propto f^2$. Let, W_e and W_h be the eddy current and the hysteresis losses at 25 Hz. Then,

$$W_e + W_h = 750 \text{ W}$$

At 440 V and 50 Hz, we have

$$4\,W_e + 2\,W_h = 2300 \text{ W} \quad \text{or} \quad 2\,W_e = 800 \text{ W}$$

So, eddy current loss at normal voltage and frequency = $4W_e$ = 1600 W.

86. (a)

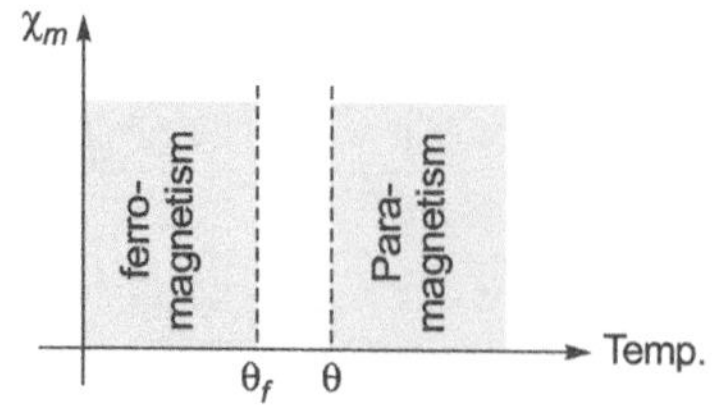

Here, θ_f = Ferromagnetic Curie temperature

87. (b)

Garnets such as $Y_3Fe_5O_{12}$ (yttrium-iron garnet) have a narrow resonance line width and are widely used as microwave isolators and in a microprocessor in the high frequency GHz range.

88. (a)

Ferrite (Nickel-Zinc ferrites) is used for audio and TV transformers. They have a much higher resistivity than alloys. This reduces the eddy current losses to a negligible value.

90. (c)

Diamagnetic materials cannot be made magnets because the permanent dipoles are absent in these materials. Hence, assertion is true but reason is false.

91. (a)

Ferrites have following properties:

- High electrical resistivities.
- Low power loss at high frequencies.
- Suitable for temporary as well as permanent magnetic application due to their spontaneous magnetization.
- Conductivity behaviour like those of semiconductors.
- Hysteresis loop is almost rectangular.

Hence, statements 1, 4 and 5 are correct.

92. (b)

Magnetic susceptibility,

$$\chi_m \propto \frac{1}{T}$$

or, $$\chi_m = \frac{C_c}{T}$$

Where, C_c = Curie constant and

T = Absolute temperature

Thus, χ_m is inversely proportional to temperature but independent of magnetizing field.

93. (d)

Ferromagnetic materials have very high χ_m. Hence, curve-A will be represented by them.

94. (c)

The width of the loop = $(OP + OQ)$

$= 80 + 80 = 160$ A/m

The height of the loop = $(OR + OS)$

$= 0.15 + 0.15 = 0.30$ Wb/m^2

$\therefore$ Area of the loop = $(160 \times 0.3) = 48$ Joule

95. (a)

Given, $\mu_r = 200000$ and $M = 6000$ A/m

$$M = \chi_m H = (\mu_r - 1)\,H$$

or, $$6000 = (200000 - 1)\,H$$

or, $$H = \frac{6000}{199999} \text{ A/m} = 0.03 \text{ A/m}$$

Now, $$B = \mu_0 \mu_r H$$
$$= (4\pi \times 10^{-7}) \times 200000 \times 0.03$$
$$= 7.539 \times 10^{-3} \text{ Wb/m}^2$$
$$= 0.007 \text{ tesla}$$

96. (d)

Cobalt: | ↑↓ | ↑↓ | ↑ | ↑ | ↑ |

Since no. of unpaired electrons = 3, therefore value of magnetic moment = 3 Bohr magneton.

Nickel: | ↑↓ | ↑↓ | ↑↓ | ↑ | ↑ |

Since no. of unpaired electrons = 2, therefore value of magnetic moment = 2 Bohr magneton.

97. (c)

Given, l = 250 mm = 0.25 m,

N = 400 turns

I = 15 A,

$\mu_0 = 4\pi \times 10^{-7}$ H/m

Now, $$H = \frac{NI}{l} = \frac{400 \times 15}{0.25} = 4000 \text{ AT/m}$$

As $B = \mu_0 \mu_r H$

(for coil in vaccum, $\mu_r = 1$),

$\therefore$ magnetic flux density in vaccum is

$$B = \mu_0 H$$
$$= 4\pi \times 10^{-7} \times 24000$$
$$= 0.03 \text{ Wb/m}^2$$

98. (b)

- Since a compass needle has to retain its magnetism, hence it should have permanent magnetism. Thus, a compass needle requires a hard magnetic material.
- For computer disc drive, high grade high energy hard magnetic materials like $Nd_2Fe_{14}B$ made by sintering process is suitable as only a small volume of magnetic material is required. It also reduces the access time.

99. (c)

Ferrites exhibit hysteresis during magnetization. The hysteresis loops of Cu-Mn ferrite, Mg-Mn ferrite and Zn-Mn ferrite are square, and they are used as memory cores in computers.

100. (b)

The diamagnetic susceptibility is negative and very small in magnitude, typical values of the order of -10^{-5}. An important exception, however is a superconductor (type I), which is perfectly diamagnetic with a susceptibility value of -1.

104. (c)

Larmor angular frequency is given as

$$\omega_L = \frac{eB}{2m}$$

105. (d)

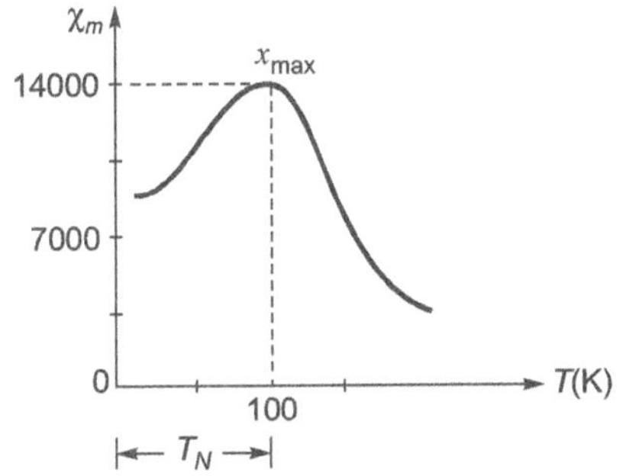

At Neel temperature (T_N), susceptibility (χ_m) is maximum.

106. (d)

Magnetostriction is responsible for humming noise in transformer due to changes in physical dimensions of ferromagnetic materials during magnetization.

109. (d)

Graphite is a diamagnetic solid having susceptibility of around -12×10^{-5}.

112. (c)

The force acting on a free electron is

$$F = eE \quad \text{or} \quad ma = eE$$

or, $$a = \frac{eE}{m} = \frac{1.6 \times 10^{-19} \times 100}{9.1 \times 10^{-31}}$$
$$= 1.76 \times 10^{13} \text{ m/s}^2$$

115. (d)

Since the probability of finding the free electron is same anywhere within the solid, therefore the Fermi level E_F is independent of length L of a linear solid.

116. (b)

The free electrons that are accelerated inside the conducting material towards the positive end of the field do not continue to increase their velocity indefinitely. They collide with obstacles on their way. Depending on the time interval between two successive collisions, the electrons acquire an average incremental of velocity called "drift velocity", all of which they loss during a collision. The drift velocity is the extra velocity that electrons acquire over and above their normal velocity in the absence of a field and this velocity is much less than the velocity of light.

117. (c)

The probability of occupation $P(E)$ of an energy level E by an electron is given by

$$P(E) = \frac{1}{1+e^{(E-E_f)/kT}}$$

when $(E-E_f) = kT$, then

$$P(E) = \frac{1}{1+e^1} = 0.27$$

118. (d)

Magnetization, $M = \chi_m H$

Since superconductor shows perfect diamagnetism,

therefore, $\chi_m = -1$

Hence, $M = -H$

119. (d)

In the superconducting state, the flux lines of a magnetic field are ejected out of the superconductor as shown in figure below. Hence, assertion is a false statement.

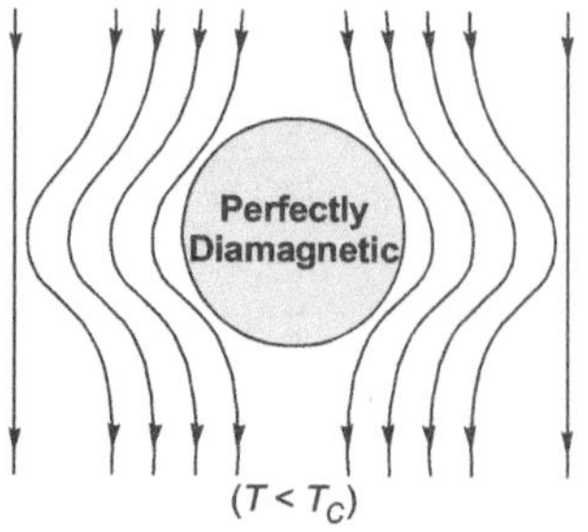

Reason is a true statement because at $T = T_C$, $J_C = 0$. J_C increases as the temperature decreases below T_C in a manner similar to H_C.

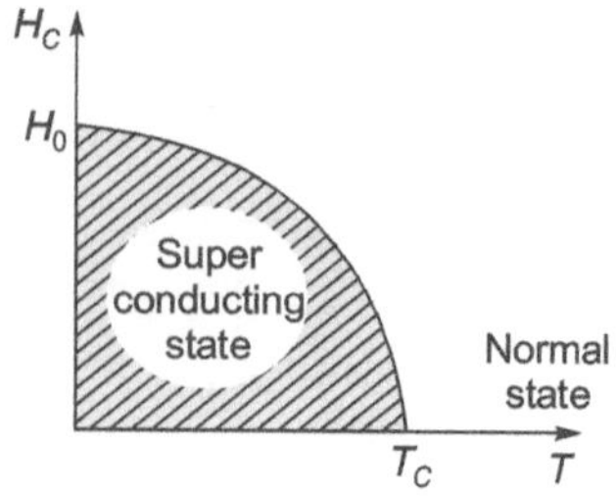

120. (a)

A superconducting material exhibits

- zero relative permeability (μ_r)
- zero specific resistance (ρ)
- zero copper loss (I^2R loss)
- zero value of magnetic flux density (B)
- magnetic susceptibility (χ_m) of –1.

121. (b)

We know that,

$$H_c = H_0\left[1-\left(\frac{T}{T_c}\right)^2\right]$$

Given, $H_0 = 0.0803$ A/m, $T = 3$ K and $T_c = 7.17$ K

$$\therefore \quad H_c = 0.0803\left[1-\left(\frac{3}{7.17}\right)^2\right]$$

$$= 0.06624 \text{ A/m} \approx 0.066 \text{ A/m}$$

122. (d)

The critical current in superconducting state may be determined by silsbee's rule.

According to silsbee's rule,

$$I_c = 2\pi r H_c \qquad \left(r = \frac{d}{2} = 1\text{mm}\right)$$

$$= 2\pi \times (1\times 10^{-3}) \times 7.9 \times 10^3$$

$$= 49.64 \text{ A}$$

123. (c)

Given, $H_c = 71$ A/m at $T_c = 7$ K and $H_c = 10$ A/m at $T_C = 14$ K.

We know that,

$$H_C = H_0\left[1-\left(\frac{T}{T_c}\right)^2\right]$$

thus, $$21 = H_0\left[1-\frac{T^2}{T_c^2}\right] \quad ...(i)$$

and $$10 = H_0\left[1-\frac{14^2}{T_c^2}\right] \quad ...(ii)$$

On dividing equation (*i*) by equation (*ii*), we get

$$\frac{21}{10} = \frac{\left(1-\frac{49}{T_c^2}\right)}{\left(1-\frac{196}{T_c^2}\right)}$$

or, $$\frac{3626}{T_c^2} = 11$$

or, $$T_c^2 = 329.63$$

or, $$T_c = 18.15\text{ K}$$

Hence, transition temperature of Nb_3Sn is

$$T_c = 18.15\text{ K}$$

124. (c)

- When a magnetic field is applied, the superconducting materials repel back all the flux lines. This is the perfectly diamagnetic behaviour of a superconducting material and is known as "Meissner effect". It is helpful in protecting devices from the undesired magnetic fields. Thus, assertion is a true statement.
- The critical current density in a superconductor is determined by silsbee's rule. Thus, reason is a false statement.

126. (b)

- The temperature at which thermal component of resistivity has maximum value is called Debye's temperature. Here, $M^{1/2}\,T_c$ = const. or, $T_c \propto$ (isotropic mass)$^{-1/2}$
- According to Matteheissen's rule, in impure metals and dilute alloys, the total resistivity is the sum of two components namely thermal and residual.
- According to Weidman frenz law, at a particular temperature, the ratio of thermal conductivity to electrical conductivity is a constant which is equal to Lorenz number.
- According to Meissner's effect, the magnetic flux inside a superconductor is zero.

127. (b)

A good conductor should have

- high electrical conductivity i.e. low resistivity
- high thermal conductivity
- high melting point
- good oxidation resistance

128. (a)

Presence of phosphorus, sulphur and iron in copper is undesired as they lower its electrical conductivity.

130. (b)

In an ideal crystal at 0 kelvin, as the mean free path (or mean free length) tends to infinity, conductivity tends to infinity because there are no collisions in an ideal crystal as it is free from imperfections, where as the real crystals have too many imperfections.

Reason is also a true statement but not the correct explanation of assertion.

131. (a)

Given:

$$n = 10^{19}\text{ electrons/m}^3$$
$$\sigma = 0.018\text{ per ohm-m}$$
$$V = 0.16\text{ volt}$$
$$e = 1.6\times10^{-19}\text{ C}$$
$$m = 9.1\times10^{-31}\text{ kg}$$
$$t = 0.29\text{ mm}$$

The electric field,

$$E = \frac{V}{t} = \frac{0.16}{0.29\times10^{-3}}$$
$$= 551.72\text{ V/m}$$

Now, drift velocity,

$$v_d = \frac{eE}{m}t$$

or, $$\frac{v_d}{E} = \frac{et}{m} \quad ...(i)$$

Also, conductivity,

$$\sigma = \frac{nev_d}{E}$$
$$= ne\left(\frac{et}{m}\right) \quad \text{(using equation (i))}$$

or, $$\sigma = \frac{ne^2t}{m}$$

or, $0.018 = \frac{10^{19} \times 1.6 \times 10^{-19} \times v_d}{551.72}$

or, $0.018 = 0.0029\, v_d$

or, $v_d = \frac{0.018}{0.0029} = 6.2$ m/s

132. (c)

$$P(E) = \frac{1}{1+e^{E-E_f/kT}}$$

At fermi energy level,

$$P(E_f) = \frac{1}{1+e^0} = \frac{1}{1+1} = \frac{1}{2} = 0.5$$

∴ Probability of occupation by as electron at Fermi energy level is 50%.

134. (b)

- Statement-1 is correct.
- Energy band theory considers potential energy of free electrons. Hence, statement-2 is false.
- Statement-3 is false because drift velocity is more than the normal velocity of an electron but less than the speed of light.
- Brass is an alloy of copper, therefore its conductivity will be less than pure copper. Hence, statement-4 is correct.

135. (c)

Superconductivity of a superconductor is destroyed at high temperature and frequency (above radio frequency).

136. (a)

Thermal component of resistivity arises due to lattice vibration and residual component of resistivity arises due to impurities and defects present in the material.

137. (d)

Materials	Critical Temperature (K)
Ti	0.53
Pb	7.17
Nb-Ti	10.20
Nb_3 Sn	18.30

138. (a)

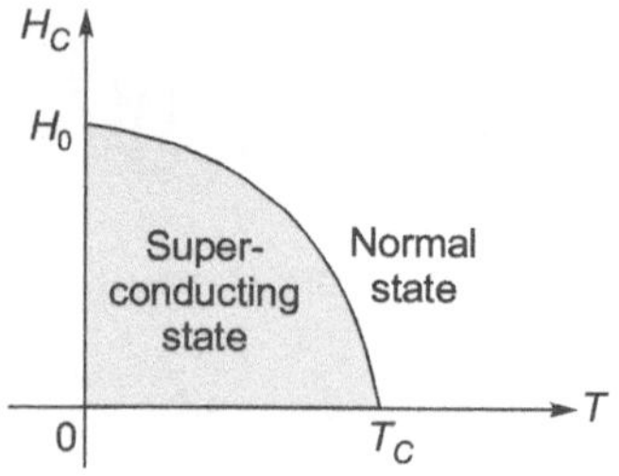

For $H_c < 0$, material will be in normal state.

141. (a)

Superconductors, whose $T_c > 25$ K, are known as high temperature superconductors and are ceramics alloys or compounds i.e. NB_3Al, V_3Ga, Nb-Ti, Nb-Zr etc.

142. (a)

The effect of high temperature is to decrease the mean free path and conductivity of the material.

144. (d)

Superconductivity in a long circular wire of radius r may be destroyed when the current "I" exceeds the value "I_c", which at the surface of the wire would produce the critical magnetic field H_C. Hence, the critical current $I_c = 2\pi r H_c$. This is referred to as silsbee's rule. This rule prevents the use of superconductors as coils for the production of strong magnetic fields.

147. (d)

$$S = uT + \frac{1}{2}aT^2$$

Initial velocity, $u = 0$

∴ $S = \frac{1}{2}aT^2$

Also, $a = \frac{Q_eE}{m}$

∴ $v_d = \frac{S}{T} = \frac{Q_eTE}{2m_e}$

151. (d)

Since, $E_g = \frac{hc}{\lambda}$

therefore after putting the values, we get

$$E_g = \frac{(2.99\times10^{8})\times(6.626\times10^{-34})}{1.771\times10^{-6}}$$

$$= 1.12 \times 10^{-19} \text{ Joules}$$

$$= \frac{1.12\times10^{-19}}{1.6\times10^{-19}} \text{ eV} = 0.7 \text{ eV}$$

153. (c)

Divalent metals such as Mg and Be have overlapping conduction and valence bands. Therefore, they conduct even if the valence band is full. Hence, energy gap in divalent metals is zero.

154. (d)

At 0 K, the conduction band is totally unfilled while the valence band is completely filled (i.e. no. of free electrons in conduction band is nil). Therefore, pure silicon at 0 K behaves as perfect insulator.

155. (a)

Mobility, $\mu = \dfrac{\text{Drift velocity}}{\text{Field gradient}}$

$\therefore \quad \mu = \dfrac{5}{100} = 0.05 \text{ m}^2\text{/V-s}$

156. (c)

In any *p*-type semiconductor, majority carriers are holes and minority carriers are electrons.

157. (d)

When silicon is doped with Boron (*B*), the semiconductor becomes *p*-type because Boron acts as a trivalent impurity.
For *p*-type semiconductor,

$$n_h >> n_i \text{ and } n_h >> n_e$$

159. (d)

The conductivity of a semiconductor is given by

$$\sigma = n_e e\mu_e + n_h e\,\mu_h$$

Replacing n_h in terms of n_i and n_e ($n_i^2 = n_e n_h$),

we have: $\quad \sigma = n_e e\mu_e + \dfrac{n_i^2 e\mu_h}{n_e} \quad \ldots(i)$

(Here, n_i = constant at constant temperature)
For minimum value of σ,

$$\frac{d\sigma}{dn_e} = 0$$

or, $\quad e\mu_e - \dfrac{n_i^2 e\mu_h}{n_e^2} = 0 \text{ or } n_e = n_i\sqrt{\dfrac{\mu_h}{\mu_e}}$

or, $\quad n_i^2 = \dfrac{n_e^2\mu_e}{\mu_h}$

Putting the value of n_i^2 in equation (*i*), we have:

$$\sigma_{min} = 2\,n_e e\mu_e$$

160. (b)

The intrinsic charge carriers in pure silicon are electrons and holes in equal numbers.

$$\therefore \quad n_i = n_e = n_h = \frac{\sigma}{(\mu_e+\mu_h)e}$$

$$= \frac{1}{(0.14+0.05)\times3000\times1.6\times10^{-19}}$$

$$= 1.096 \times 10^{16} \text{ m}^{-3}$$

$$\approx 1.1 \times 10^{16} \text{ m}^{-3}$$

161. (b)

At room temperature, a large fraction of the donor level electrons are excited into the conduction band because as compared to the energy gap, the ionization energy of an impurity atom is very small.
According to the "law of mass action,"

$$np = n_i^2$$

Hence, both assertion and reason are true but reason is not the correct explanation of assertion.

162. (c)

- Since $\mu_n > \mu_p$, therefore *n*-type semiconductor is faster than *p*-type semiconductor.
- A pure semiconductor becomes p-type when acceptor impurity is added to it.
- Zener breakdown phenomenon is associated with Zener diode.
- Diffusion is due to concentration gradient of charge carriers.

163. (c)

$$\mu_n \text{ for Si} = 1300 \text{ cm}^2\text{/V-s}$$

$$\mu_n \text{ for GaAs} = 8500 \text{ cm}^2\text{/V-s}$$

$$= 6.53 * (\mu_n)_{Si}$$

164. (d)

- Statement-1 is correct.
- GaAs is costly and will not replace silicon, which will continue to be used for making inexpensive chips. Hence, statement-2 is not correct.
- Metals have partially filled bands or overlapping bands i.e. zero forbidden gap. However, both insulators and semiconductors have finite forbidden gap. Hence, statement-3 is not correct.
- Statement-4 is correct.
- The net energy gap taking into account all directions of motion is called the forbidden gap. Electrons cannot take the energy values that lie in the forbidden gap. Hence, statements-5 is not correct.

166. (c)

The wavelength of the radiation for a GaAs junction laser is 8700 Å which lies in the infrared region.

167. (d)

For Si, $E_{G(0\,K)} = 1.21$ eV
and $E_{G(300\,K)} = 1.1$ eV

169. (a)

Given, $J = 1000$ A/m^2,
$\rho = 0.05$ ohm-m,
$\mu_e = 0.4$ m^2/ V-s

Let a n-type semiconductor, $n_e >> n_h$, therefore neglecting the conductivity due to minority carriers (holes), conductivity is

$$\sigma = n_e e \mu_e$$

or, $$\frac{1}{0.05} = n_e \times 1.6 \times 10^{-19} \times 0.4$$

or, $$n_e = 3.125 \times 10^{20} \text{ /m}^3$$
= concentration of electrons

We know that, current density,

$$J = n_e \,.\, e.\, v_d$$

or, $$v_d = \frac{J}{en_e}$$

$$= \frac{1000}{1.6 \times 10^{-19} \times 3.125 \times 10^{20}}$$
$$= 20 \text{ m/s}$$

170. (c)

A semiconducting material:

- consume negligible power. For example, the calculator works at an expense of 0.0004 watt only for years.
- is much smaller in size. For example, a semiconductor strain gauge has a gauge length as small as 0.25 mm.
- operate on low voltage i.e. 9 V, 3 V, 1.5 V etc. (ICs in calculators).
- operate over a practically fruitful wide range of temperatures.

Hence, statement-1 and 3 are not correct.

171. (b)

With the increase in temperature, rate of diffusion increases.

174. (d)

The area of cross-section of the rod is

$$a = \frac{\pi d^2}{4} = \frac{\pi(1 \times 10^{-3})^2}{4}$$
$$= 7.85 \times 10^{-7} \text{ m}^2$$

$\therefore$ The resistivity of the rod is

$$\rho = \frac{aR}{l} = \frac{7.85 \times 10^{-7} \times 100}{10 \times 10^{-3}}$$
$$= 7.85 \times 10^{-3}\ \Omega\text{-m}$$

In a p-type semiconductor, the holes are majority carriers as $n_h >> n_e$.
Hence, conductivity,

$$\sigma = n_h e \mu_h$$

$\therefore$ $$\rho = \frac{1}{\sigma} = \frac{1}{n_h e \mu_h}$$

$\therefore$ $$7.85 \times 10^{-3} = \frac{1}{n_h \times 1.6 \times 10^{-19} \times 0.19}$$

or, $$n_h = 4.19 \times 10^{21} \text{ per m}^3$$
$$\approx 4.2 \times 10^{21} \text{ per m}^3$$

Neglecting the intrinsic concentration and assuming that each impurity atom produces one electron in the rod, the impurity concentration N_h is

$$N_h = n_h = 4.2 \times 10^{21} \text{ /m}^3$$

175. (c)

Using 'mass action law'

$$np = n_i^2$$

$$\therefore \quad p = \frac{n_i^2}{n} = \frac{(1.5\times10^{16})^2}{10^{19}}$$
$$= 2.25\times10^{13}/m^3$$

176. (a)

Diamond is an insulator for which the energy gap is more than 3 eV (ie. 5.4 eV).

178. (a)

In a semiconductor, electron always exists in conduction band while hole always exists in valence band. Staying of electron is prohibited in the forbidden energy band.

179. (d)

Arsenic is a pentavalent impurity (valence electrons = 5) which donates electrons while aluminium is a trivalent impurity (valence electrons = 3) which accepts electrons.

180. (d)

Since both diffusion and mobility are statistical thermodynamic phenomena, therefore they are not independent. Hence, assertion is a false statement.

Reason is a correct statement because according to Einstein equation,

$$\frac{D}{\mu} = V_T = \frac{T}{11600}$$

Thus, D and μ both depends on temperature.

181. (b)

A good semiconductor conductor material should have energy gap of less than 2 eV.

183. (a)

At 0 Kelvin, an intrinsic semiconductor behaves like a perfect insulator because of non-availability of free electrons since conduction band is empty while valence band is completely filled.

184. (b)

$$R_H = \frac{1}{e}\frac{p\mu_p^2 - n\mu_n^2}{(p\mu_p + n\mu_n)^2}$$

For intrinsic semiconductor,

$$n = p = ni$$
$$\therefore \quad \mu_p < \mu_n$$

$$R_H = \frac{1}{n_i e}\frac{\mu_p - \mu_n}{\mu_p + \mu_n}$$

$$\therefore \quad \mu_p < \mu_n$$
$$R_H = -ve$$

Large Hall coefficient corresponds to low carrier densities (i.e intrinsic semiconductor) and hence to pure samples.

186. (a)

$$R_H \propto \frac{1}{n}$$

For metals, there is very high concentration of free electrons, therefore hall coefficient is almost zero.

187. (d)

Hall voltage, $V_H = \frac{BI}{neW}$

$$\therefore \quad V_H \propto B$$
$$\propto I \propto \frac{1}{W}$$

190. (d)

Examples of low resistivity materials are:
Au, Ag, Cu, Al, brass, bronze etc.
Examples of high resistivity materials are:
W, Pt, C, Ni, Ta, nichrome, manganin etc.

191. (b)

- The eutectic reaction in the iron-carbon (Fe-C) alloys facilitates the melting of cast irons at a lower temperature which is required for making steels.
- Pb-Sn eutectic alloys are useful as soldering material.
- Gold Silicon (Au-Si) eutectic alloys finds uses in the manufacture of semiconductor devices.

192. (c)

Capacitance, $C = \frac{\varepsilon_0 \varepsilon_r A}{d} = \frac{\varepsilon A}{d}$

or, $\varepsilon_0 = \frac{Cd}{\varepsilon_r A}$

$$\therefore \quad (\varepsilon_0) = \frac{(F)(m)}{(m^2)} = F/m = F\text{ - }m^{-1}$$

193. (b)

Insulators (or dielectrics) are poor conductors of heat and electricity because in such materials it is almost impossible to excite the electrons from the valence band to the conduction band by means of an applied field or normal thermal energy. Reason is also a true statement but is not the correct explanation of assertion.

195. (d)

- Porcelain and air are for low voltage services.
- For high temperature application, glass tape is used.
- N_2 (in glass-filled cables alongwith oil-treated paper insulation) is used for high voltage applications.

196. (b)

Ceramics are generally metallic or nonmetallic oxides. Rocks, glasses, fireclay, firebricks, cements, limes, ferrites, garnets etc are some examples of ceramic.

200. (c)

Gold is the most ductile metal existing in the nature.

206. (b)

$$\tan\delta = \frac{\varepsilon_r''}{\varepsilon_r'}$$

and $\varepsilon_r^* = \varepsilon_r' - j\varepsilon_r''$

$\therefore$ $\varepsilon_r'' = \varepsilon_r' \times \tan\delta$

$= 4 \times 0.004 = 0.016$

Thus, $\varepsilon_r^* = 4 - j0.016$

UNIT 3

Electrical Circuits

1. Basic and Electric Circuits

Q.1 If the power supplied by the 20 V voltage source shown in figure below is 60 W then, the power supplied by the unknown circuit element (×) will be

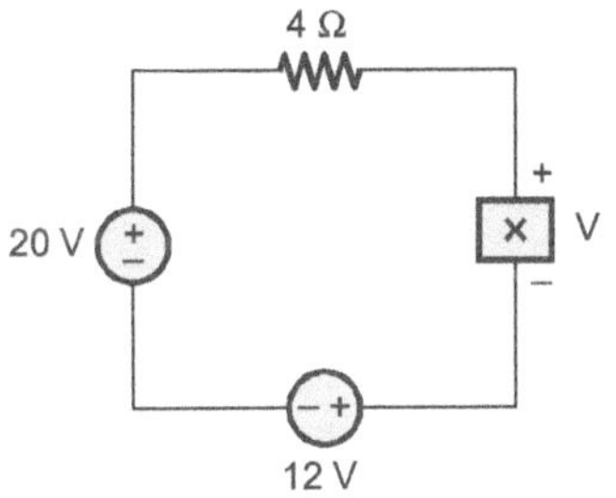

(a) −12 W (b) 3 W
(c) 12 W (d) −3 W

Q.2 In the network of figure shown below, $I_1 = 0.05$ A, $I_2 = 0.1$ A, $R = 50\ \Omega$. The unknown current source has strength of i. If 100 J of energy is delivered to I_2 in 1 min, the magnitude of i will be

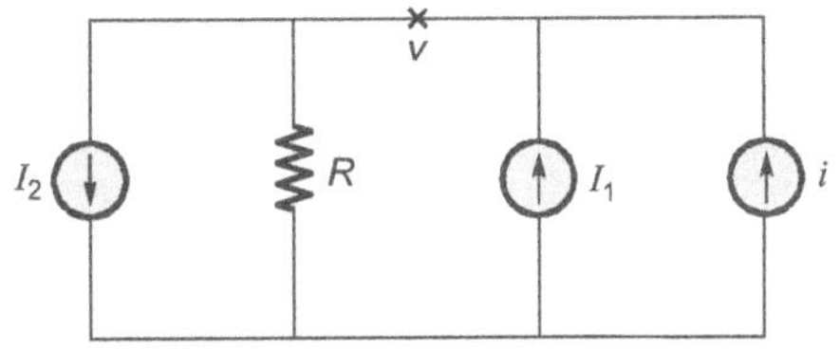

(a) 0.38 A (b) 0.50 A
(c) −0.25 A (d) 0.05 A

Q.3 The energy required to move 120 C charge through 3 V is
(a) 540 J (b) 25 mJ
(c) 360 J (d) 2.78 mJ

Q.4 For a given voltage, four heating coils will produce maximum heat, when connected
(a) all in parallel
(b) all in series
(c) with two parallel pairs in series
(d) one pair in parallel with the other two in series

Q.5 Given figure shows a plot of d.c. voltage applied to a resistor of 5 Ω for 15 seconds. The total energy consumed by the resistor is

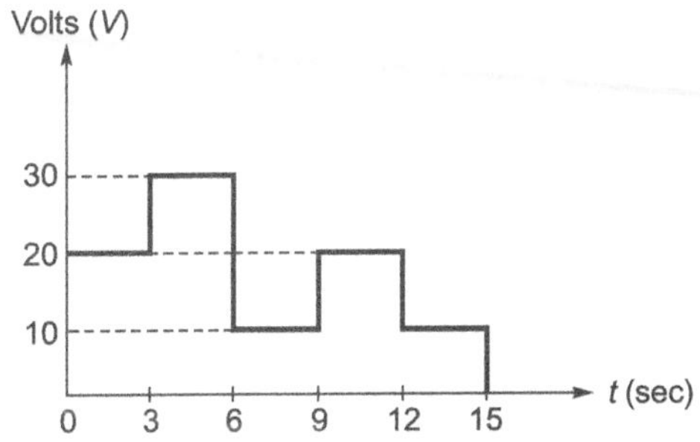

(a) 540 J (b) 625 J
(c) 1220 J (d) 1140 J

Q.6 The charge in a capacitor is given by

$$q = \left(v + \frac{1}{3}v^3\right)$$

If the voltage across this capacitor be $v(t) = \sin t$, the current $i(t)$ through the capacitor is
(a) $(1 + \sin^2 t)\cos t$ (b) $(1 + \sin^2 t)$
(c) $(1 + \cos^2 t)\sin t$ (d) $\sin^2 t \cos t$

Q.7 The unit of reactive power is
(a) Watt (b) Kilo-Watt
(c) Var (d) Volt ampere

Q.8 The unit of electrical energy is
(a) Watt (b) Kilo-Watt
(c) Kilo-watt-hour (d) Joule

Q.9 The given signal is

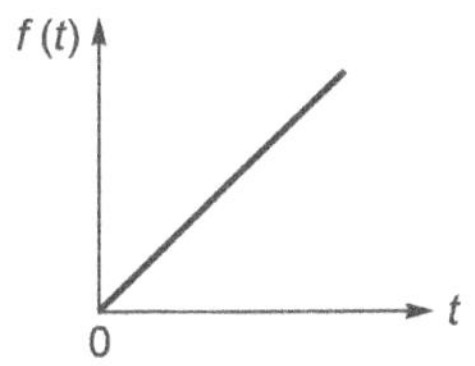

(a) a power signal
(b) an energy signal
(c) an energy and a power signal
(d) neither an energy nor a power signal

Q.10 The two voltage sources connected in parallel as shown below, must satisfy the conditions

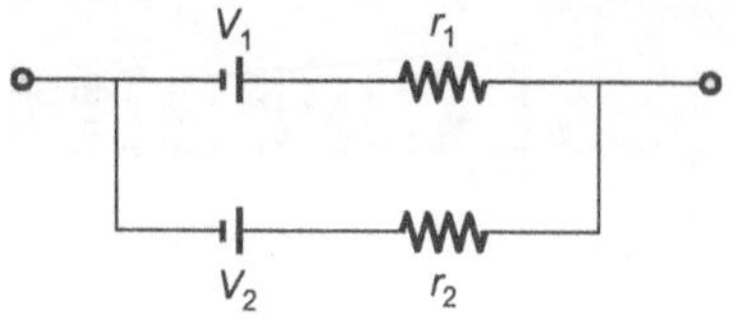

(a) $V_1 = V_2, r_1 = r_2$
(b) $V_1 \neq V_2$ but $r_1 = r_2$
(c) $V_1 \neq V_2, r_1 \neq 0$ or $r_2 \neq 0$
(d) $V_1 = V_2, r_1 \neq r_2$

Q.11 A current of 3 A flows through a resistor of 20 ohms. The energy dissipated in the resistor per minute is

(a) 0.18 W (b) 1.8 W
(c) 180 W (d) 3.6 W

Q.12 A long uniform coil of a inductance *L* henries and associated resistance *R* ohms is physically cut into two exact halves which are then rewound in parallel. The resistance and inductance of the combination are

(a) *R* and *L* (b) $\frac{R}{2}$ and $\frac{L}{2}$
(c) $\frac{R}{4}$ and $\frac{L}{4}$ (d) 2*R* and 2*L*

Q.13 In the circuit shown, the power dissipated in the resistor *R* is 1 W when only source '1' is present and '2' is replaced by short circuit. The power dissipated in the same resistor *R* is 4 W when only source '2' is present and '1' is replaced by a short circuit. When both the sources '1' and '2' are present, the power dissipated in *R* will be

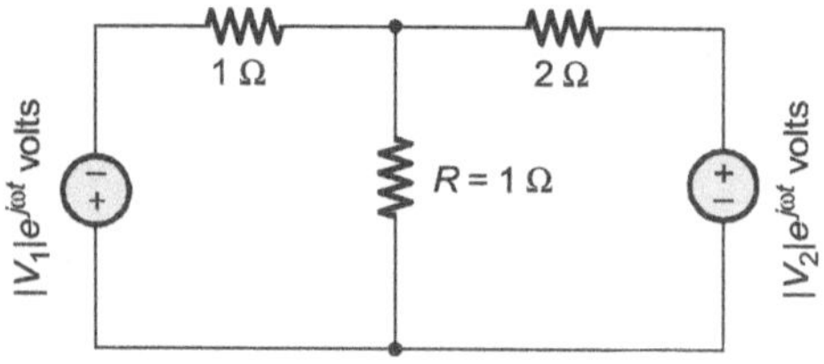

(a) 5 W (b) 4 W
(c) 3 W (d) 1 W

Q.14 If *v*, *w*, *q* stand for voltage, energy and charge, then *v* can be expressed as:

(a) $v = \frac{dq}{dw}$ (b) $dv = \frac{dw}{dq}$
(c) $v = \frac{dw}{dq}$ (d) $dv = \frac{dq}{dw}$

Q.15 In the figure shown below, ϕ = power-factor angle, *W* = watts, *VA* = volt ampere and *VAr* = volt-ampere reactive for an ac circuit. The correct figure is

(a) 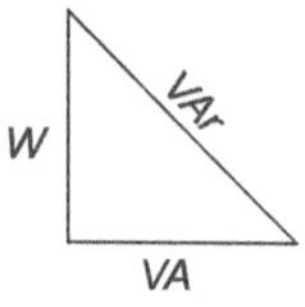

(b) 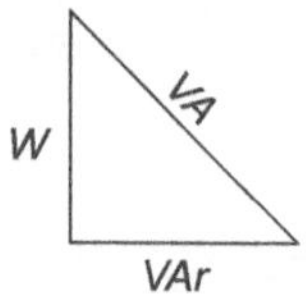

(c) 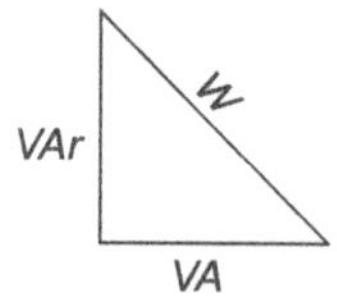

(d) 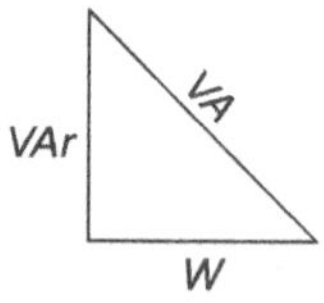

Q.16 A circuit possesses resistance *R* and inductive reactance X_L in series. Its susceptance is given by

(a) $\frac{R}{Z}$ (b) $\frac{X_L}{Z}$
(c) $\frac{R}{Z^2}$ (d) $\frac{X_L}{Z^2}$

Q.17 If $V = a + jb$ and $I = c + jd$, then the power is given by

(a) $bc + ad$ (b) $ac + bd$
(c) $bc - bd$ (d) $ac + ad$

Q.18 A 230 V, 100 W bulb has resistance R_A and a 230 V, 200 W bulb has resistance R_B. Here,

1. $R_A > R_B$ 2. $R_B > R_A$
3. $R_A = 2\,R_B$ 4. $R_B = 2\,R_A$
5. $R_A = 4\,R_B$

From these, the correct answer is

(a) 2 and 4 (b) 1 only
(c) 1 and 3 (d) 1 and 5

Q.19 The voltage phasor of a circuit is $10\angle 15°$ V and the current phasor is $2\angle -45°$ A. The active and the reactive powers in the circuit are

(a) 10 W and 17.32 VAr
(b) $20\sqrt{2}$ W and $10\sqrt{2}$ VAr
(c) 200 W and 60 VAr
(d) 5 W and 8.66 VAr

Q.20 In the circuit shown below, if the power consumed by the 5 Ω resistor is 10 W, then power factor of the circuit will be

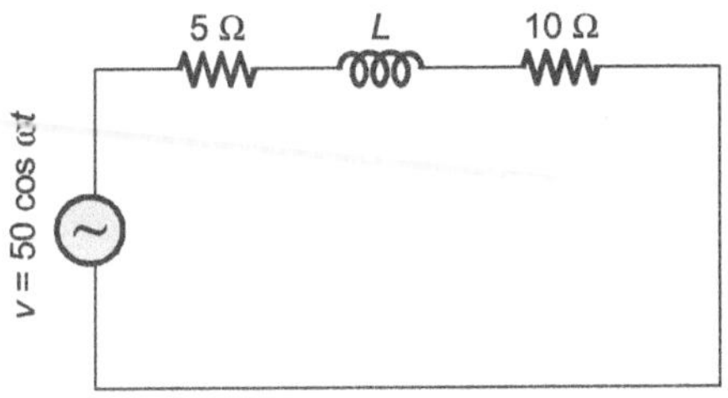

(a) zero (b) 0.5
(c) 0.8 (d) 0.6

Q.21 The minimum requirements for causing flow of current are
(a) a voltage source, a resistor and a switch.
(b) a voltage source, a conductor, an ammeter and a switch.
(c) a voltage source and a conductor.
(d) a power source and a bulb.

Q.22 Ohm's law is applicable to
(a) Carbon resistor (b) Semiconductors
(c) both (a) and (b) (d) neither (a) nor (b)

Q.23 For a fixed supply voltage, the current flowing through a conductor will increase when its
(a) area of cross-section is reduced
(b) length is reduced
(c) length is increased
(d) length is increased and area of cross-section is reduced

Q.24 Two registors R_1 and R_2 give combined resistance of 4.5 Ω when in series and 1 Ω when in parallel. The resistances are
(a) 1 Ω and 3.5 Ω (b) 4 Ω and 0.5 Ω
(c) 2 Ω and 2.5 Ω (d) 1.5 Ω and 3 Ω

Q.25 Which of the following is not equivalent to watts?
(a) Amperes × volts (b) Joules per second
(c) Amperes / volts (d) $(\text{Amperes})^2 \times$ ohm

Q.26 Two heaters, rated at 1000 W, 250 V each are connected in series across a 250 V, 50 Hz ac mains. The total power drawn from the supply would be
(a) 2000 W (b) 1000 W
(c) 250 W (d) 500 W

Q.27 A 100 watt light bulb burns on an average of 10 hours a day for one week. The weekly consumption of energy will be
(a) 7 units (b) 700 units
(c) 70 units (d) 0.7 units

Q.28 **Assertion (A):** The direction of flow of conventional current is taken opposite to that of electrons.
Reason (R): Electrons have negative charge.
(a) Both A and R are true and R is a correct explanation of A.
(b) Both A and R are true but R is not a correct explanation of A.
(c) A is true but R is false.
(d) A is false but R is true.

Q.29 Match **List-I (Materials)** with **List-II (Range of resistivity)** and select the correct answer using the codes given below the lists:
List-I
A. Conducting materials
B. Semiconductor materials
C. Insulating material
List-II
1. 10^0 to 10^2 Ω-m
2. 10^{-8} to 10^{-6} Ω-m
3. 10^{12} to 10^{18} Ω-m
4. 10^{20} to 10^{30} Ω-m

Codes:

	A	B	C
(a)	2	1	3
(b)	2	1	4
(c)	1	2	3
(d)	1	2	4

Q.30 A constant current source supplies a current of 200 mA to a load of 2 kΩ. When the load is changed to 100 Ω, the load current is
(a) 2 mA (b) 200 mA
(c) 4 mA (d) 20 mA

Q.31 A parallel plate capacitor is filled with two dielectrics of ε_1 and ε_2 lengthwise equally. The capacitance of the combination is
(a) $\dfrac{\varepsilon_0 A(\varepsilon_1 + \varepsilon_2)}{2d}$ (b) $\dfrac{2\varepsilon_0 \varepsilon_1 \varepsilon_2 A^2}{d^2}$
(c) $\dfrac{2\varepsilon_0 \varepsilon_1 \varepsilon_2 A}{d}$ (d) $\dfrac{A\varepsilon_1 \varepsilon_2}{d}$

Q.32 **Assertion (A):** Elements like resistances, inductances and capacitances are linear elements.

Reason (R): A circuit or network whose parameters are always constant irrespective of the change in time, voltage, temperature etc. is known as a linear network.

(a) Both A and R are true and R is the correct explanation of A.
(b) Both A and R are true but R is not the correct explanation of A.
(c) A is true but R is false.
(d) A is false but R is true.

Q.33 **Assertion (A):** A network in which the circuit elements like resistance, inductance etc. cannot be physically separated for analysis purposes, is called lumped network.

Reason (R): Most of the electric networks are lumped in nature.

(a) Both A and R are true and R is the correct explanation of A.
(b) Both A and R are true but R is not the correct explanation of A.
(c) A is true but R is false.
(d) A is false but R is true.

2. Basic Laws

Q.34 The equivalent resistance of the circuit shown in figure below across the terminals *a-b* will be

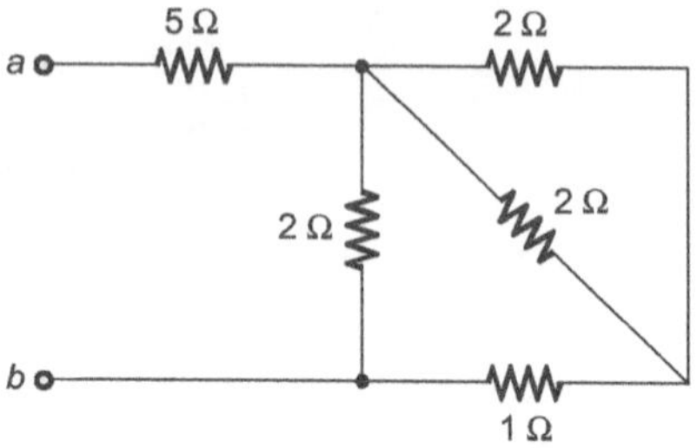

(a) 4.50 Ω (b) 4.25 Ω
(c) 2.50 Ω (d) 6.0 Ω

Q.35 For the circuit shown below, the equivalent resistance across terminals *a-b* will be

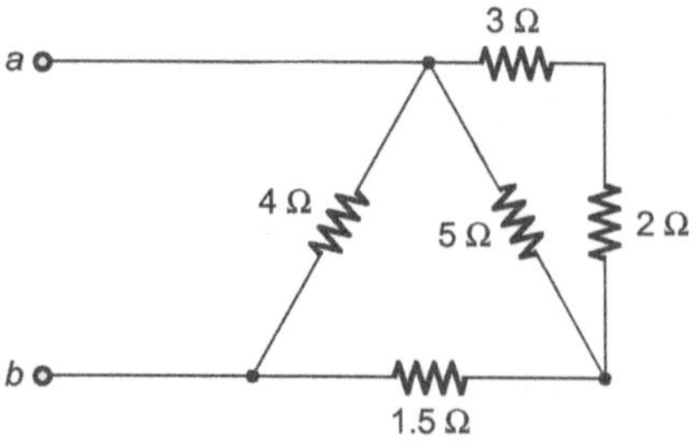

(a) 2 Ω (b) 2.5 Ω
(c) 5 Ω (d) 6.25 Ω

Q.36 What is the value of *m* so that the current *I* in the circuit shown below is maximum?

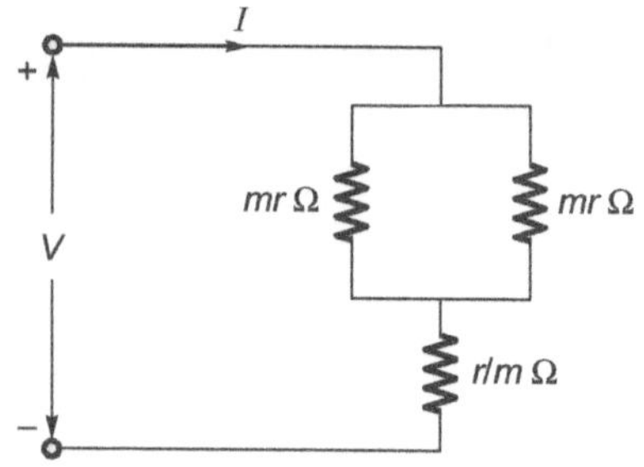

(a) $\sqrt{3}$ (b) $\frac{1}{\sqrt{2}}$
(c) $\frac{1}{\sqrt{3}}$ (d) $\sqrt{2}$

Q.37 The given circuit shown below is converted to an equivalent voltage source *V′* connected in series with an equivalent resistance *R′*.

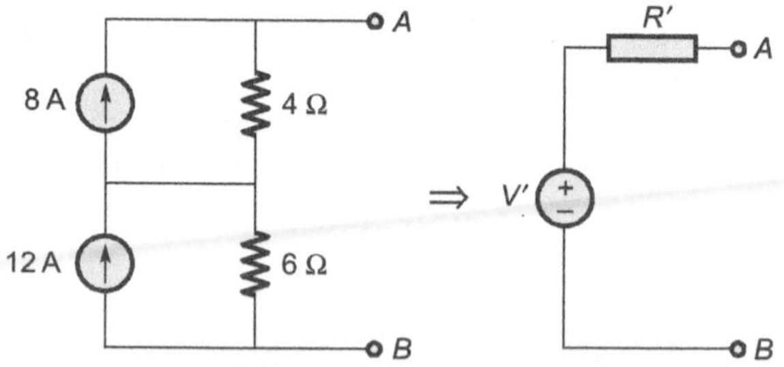

The values of *V′* and *R′* are respectively
(a) 40 volts and 2.4 Ω
(b) 104 volts and 2.4 Ω
(c) 104 volts and 10 Ω
(d) 40 volts and 10 Ω

Q.38 The equivalent current source for the circuit shown below will be represented as

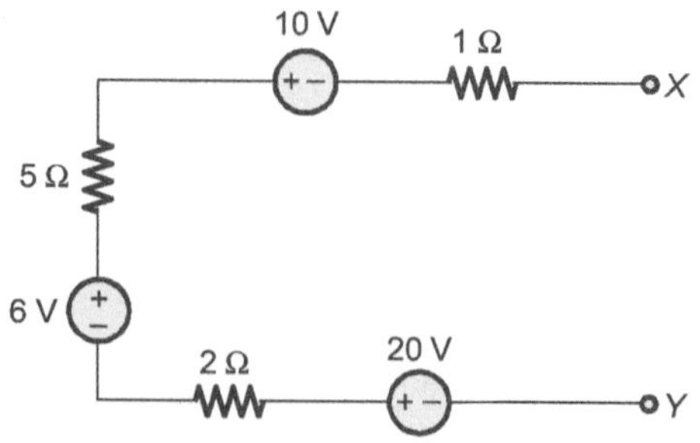

(a) 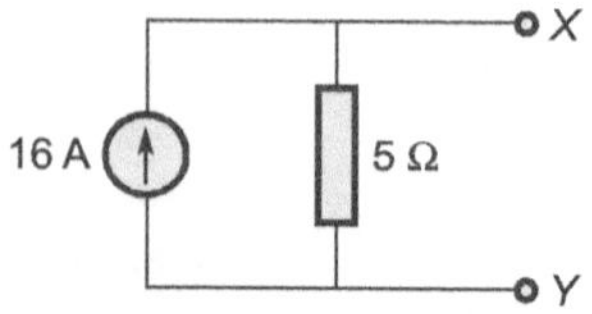

(b) 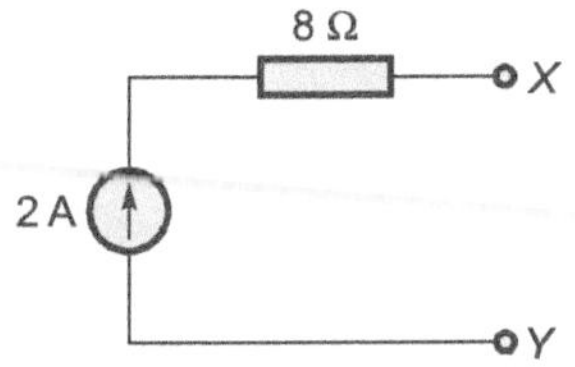

(c) 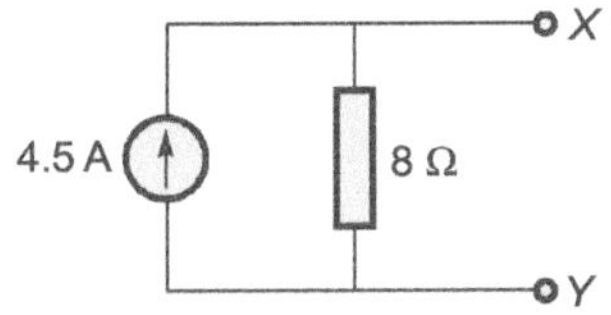

(d) 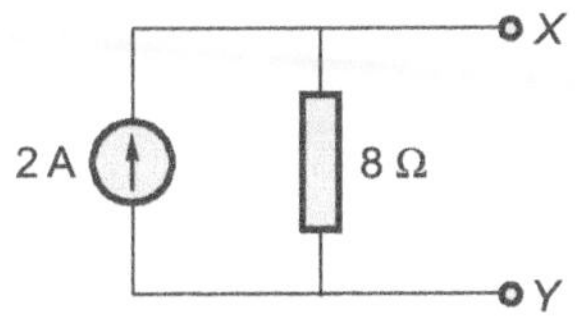

Q.39 Assuming the batteries to be ideal voltage sources in the figure shown below, the value of current i is

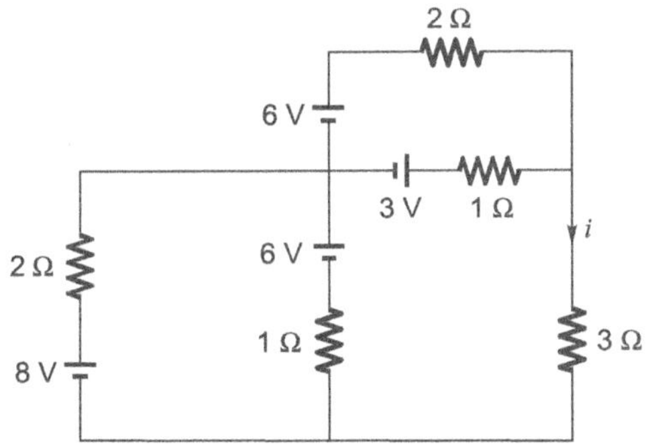

(a) 1.15 A (b) 2.46 A
(c) 3.25 A (d) 8.88 A

Q.40 Two 100 W, 220 V bulbs are required to be connected across a 400 volts supply. The value of resistance to be inserted in the line so that the voltage across the bulbs does not exceed 220 V is

(a) 198 Ω (b) 0 Ω
(c) 125.5 Ω (d) 76.2 Ω

Q.41 The equivalent resistance across the terminal X-Y for the circuit shown below is

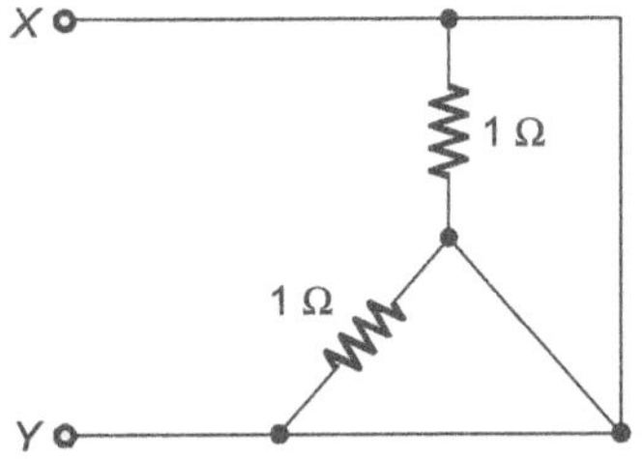

(a) 0 Ω (b) 1 Ω
(c) $\frac{3}{2}$ Ω (d) $\frac{1}{2}$ Ω

Q.42 **Assertion (A) :** KCL states that at any node of a circuit, at any instant of time, the sum of incoming currents is equal to the sum of outgoing currents.

Reason (R): KCL indicates the law of conservation of energy.

(a) Both A and R are true and R is the correct explanation of A.
(b) Both A and R are true but R is not the correct explanation of A.
(c) A is true but R is false.
(d) A is false but R is true.

Q.43 The total power absorbed in the circuit below is

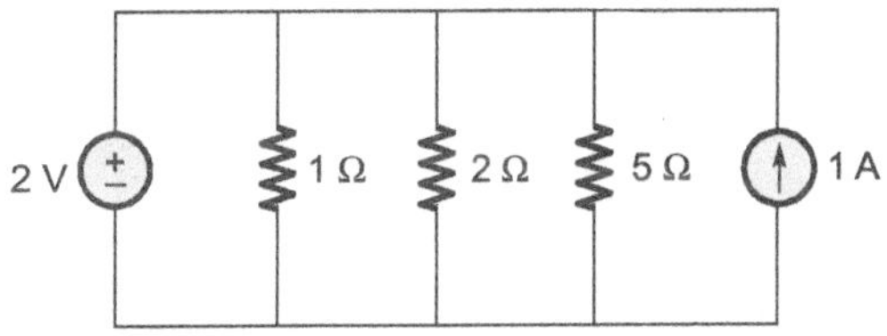

(a) 2.2 W (b) 6.8 W
(c) 5.5 W (d) 12 W

Q.44 The current I in the circuit shown below is

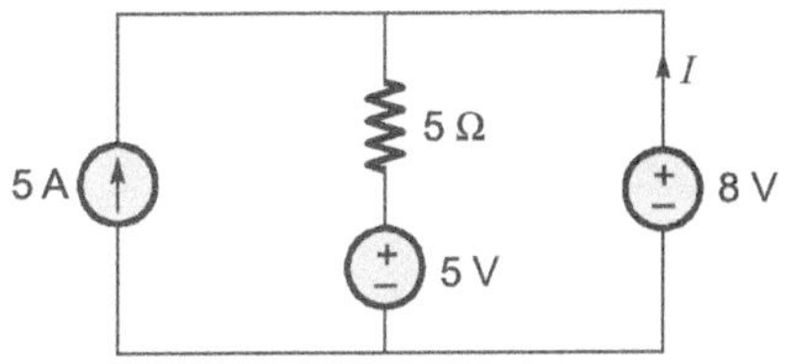

(a) 0 A (b) 4.4 A
(c) 1.5 A (d) –4.4 A

Q.45 For the circuit shown below, match **List-I** with **List-II** and select the correct answer using the codes given below the lists:

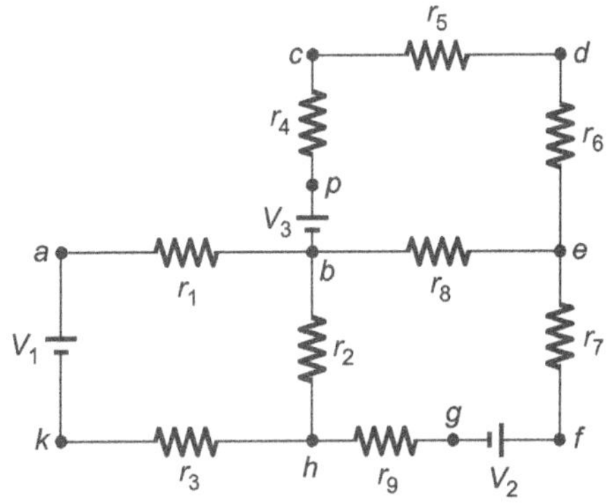

List-I	List-II
A. Number of nodes	1. 3
B. Number of junction points	2. 5
C. Number of branches	3. 10
D. Number of meshes	4. 8
	5. 2

Codes:

	A	B	C	D
(a)	4	5	2	1
(b)	3	4	5	2
(c)	3	1	2	1
(d)	4	2	1	3

Q.46 The value of current I in the circuit shown below for $V = 2$ volt is

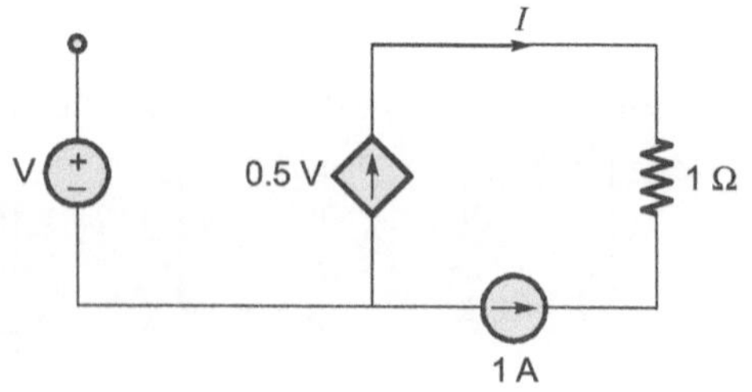

(a) 0 A (b) –0.5 A
(c) 0.5 A (d) 1 A

Q.47 For the circuit shown below, the value of $(V_1 + V_2)$ is

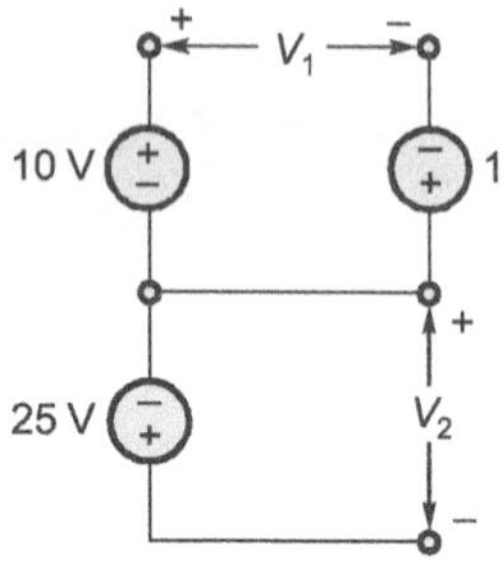

(a) 50 volts (b) –25 volts
(c) 0 volts (d) 25 volts

Q.48 The value of voltage source to be connected across the terminals X and Y so that drop across the 10 Ω resistor is 45 V is

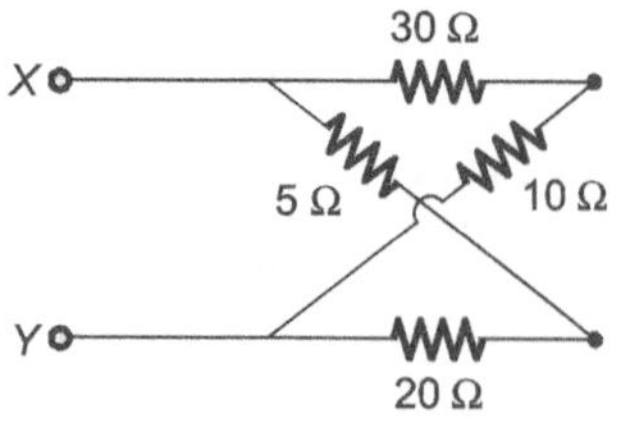

(a) 36 volts (b) 180 volts
(c) 95 volts (d) 120 volts

Q.49 In a voltage divider circuit as shown below with two resistances R_1 and R_2, the voltage drop V_1 is twice the drop V_2 across R_2. The value of R_1 and R_2 will be

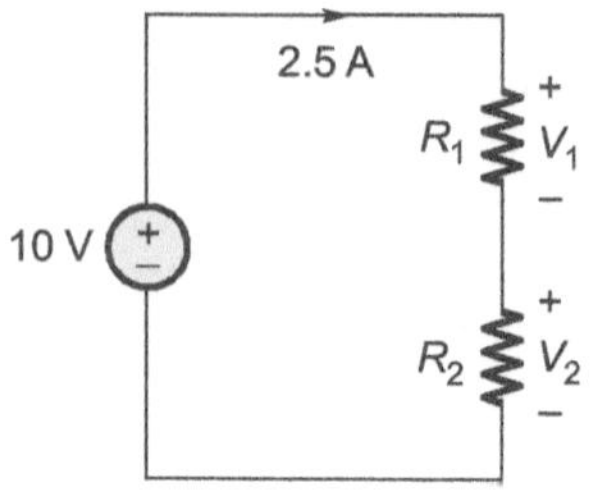

(a) $R_1 = \frac{8}{3}\,\Omega$ and $R_2 = \frac{32}{3}\,\Omega$

(b) $R_1 = \frac{4}{3}\,\Omega$ and $R_2 = \frac{8}{3}\,\Omega$

(c) $R_1 = \frac{4}{3}\,\Omega$ and $R_2 = \frac{32}{3}\,\Omega$

(d) $R_1 = \frac{8}{3}\,\Omega$ and $R_2 = \frac{4}{3}\,\Omega$

Q.50 A delta connected load is shown in figure below. The equivalent star connection has a value of R in Ω is

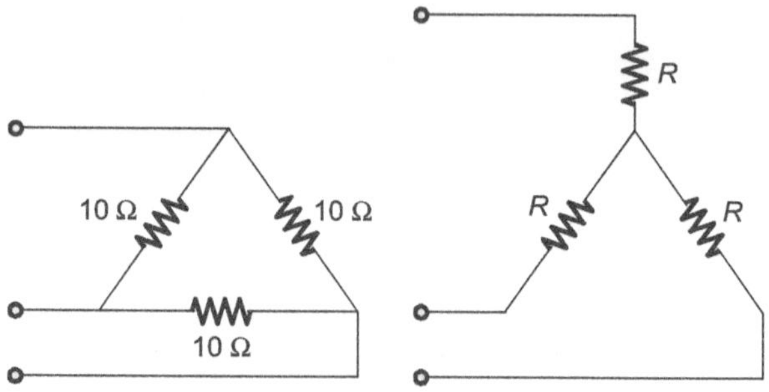

(a) 10 (b) 30
(c) $\frac{10}{3}$ (d) $\frac{20}{3}$

Q.51 The voltage drop across the 2 Ω resistor for the circuit shown below is

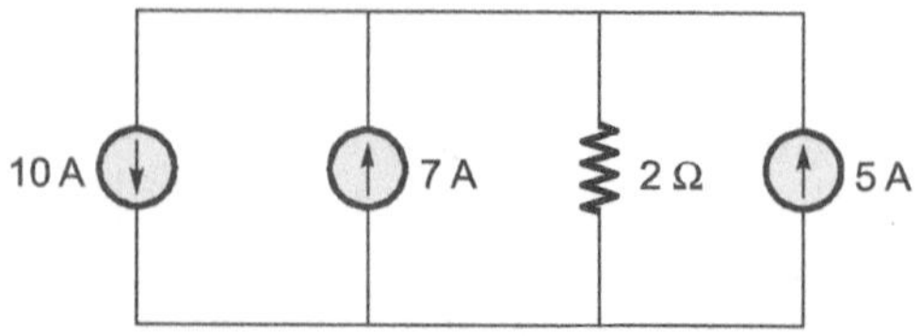

(a) 2 volts (b) 14 volts
(c) 4 volts (d) –20 volts

Q.52 For the circuit shown below, the sum of the unknown currents I_1, I_2, I_3 and I_4 is

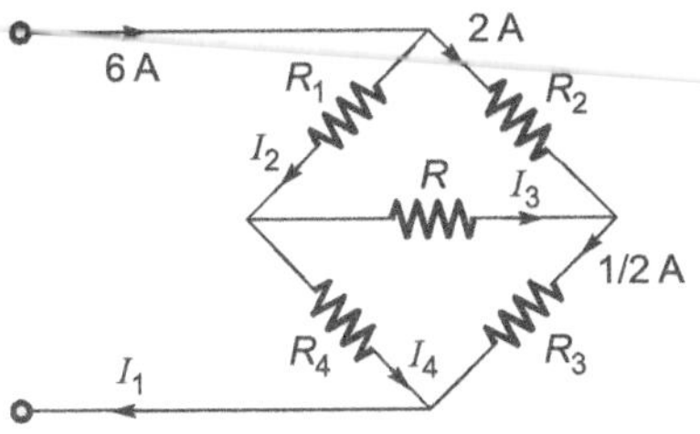

(a) 0 A (b) 14 A
(c) −1.5 A (d) 16.5 A

Q.53 In the delta equivalent of the given star connected circuit $\vec{Z}_{BC}$ is equal to

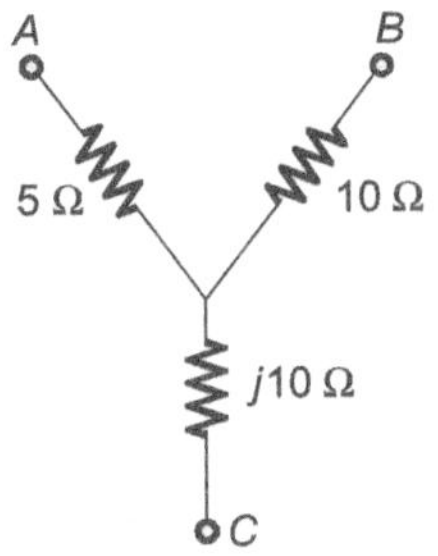

(a) 40 Ω (b) (20 + j10) Ω
(c) $\left(5+j\frac{10}{3}\right)\Omega$ (d) (10 + j30) Ω

Q.54 Four resistances 80 Ω, 50 Ω, 25 Ω and R are connected in parallel. Current through 25 Ω resistance is 4 A. The total current of the supply is 10 A. The value of R will be
(a) 66.66 Ω (b) 36.36 Ω
(c) 40.25 Ω (d) 76.56 Ω

Q.55 For the circuit shown below, the equivalent resistance will be

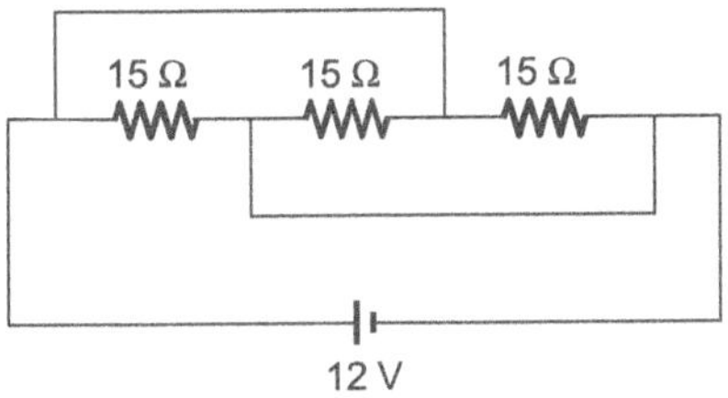

(a) 45 Ω (b) 15 Ω
(c) 5 Ω (d) 7.5 Ω

Q.56 Three resistors of R Ω each are connected to form a triangle. The resistance between any two terminals will be
(a) R Ω (b) $3R$ Ω
(c) $\frac{2}{3}R$ Ω (d) $\frac{3}{2}R$ Ω

Q.57 A network has 10 nodes and 17 branches. The number of different node pair voltage would be
(a) 10 (b) 45
(c) 7 (d) 9

Q.58 Kirchhoff's laws are not applicable to circuits with
(a) passive elements
(b) lumped parameters
(c) distributed parameters
(d) non-linear resistances

Q.59 Consider the following statements:
1. Solution of an electric circuit will give the same result whether the source is treated as a voltage source or a current source.
2. According to Kirchhoff's voltage law, at any junction of an electrical network, the sum of incoming currents is equal to sum of outgoing currents.
3. If a network has 4 nodes and 3 independent loops, then the number of branches in the network will be 5.

Which of the above statements is/are true?
(a) 1 only (b) 1 and 3 only
(c) 2 only (d) None of these

Q.60 Match **List-I (Basic Laws)** with **List-II (Characteristics)** and select the correct answer using the codes given below the lists:

List-I
A. Kirchhoff's voltage law
B. Ohm's law
C. Kirchhoff's current law

List-II
1. For bilateral elements only
2. Conservation of charge
3. Conservation of energy
4. For linear and non-linear elements both

Codes:

	A	B	C
(a)	3	4	2
(b)	2	4	3
(c)	2	1	3
(d)	3	1	2

3. Basic Nodal and Mesh Analysis

Q.61 Of the two methods of loop and node variable analysis
- (a) loop analysis is always preferable.
- (b) node analysis is always preferable.
- (c) loop analysis may be preferable in some situations while node analysis may be preferable in other situations.
- (d) there is nothing to choose between them.

Q.62 Which of the following is not true about the circuit shown below?

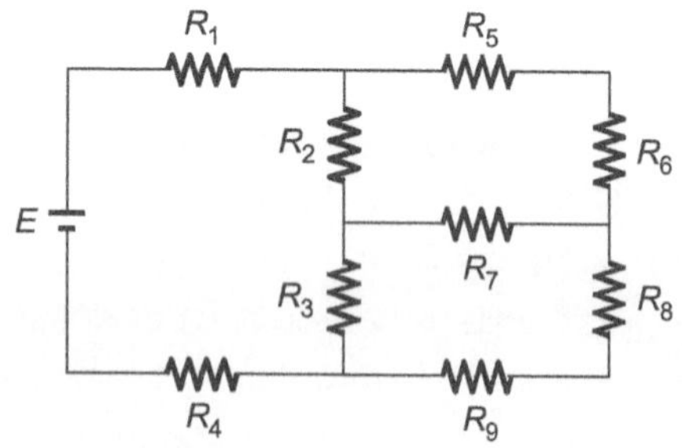

- (a) Number of nodes for the above circuit is 8.
- (b) Number of branches for the above circuit is 6.
- (c) Number of meshes for the above circuit is 3.
- (d) Number of junction points for the above circuit is 6.

Q.63 For the circuit shown below, the current through the 1 Ω resistor is

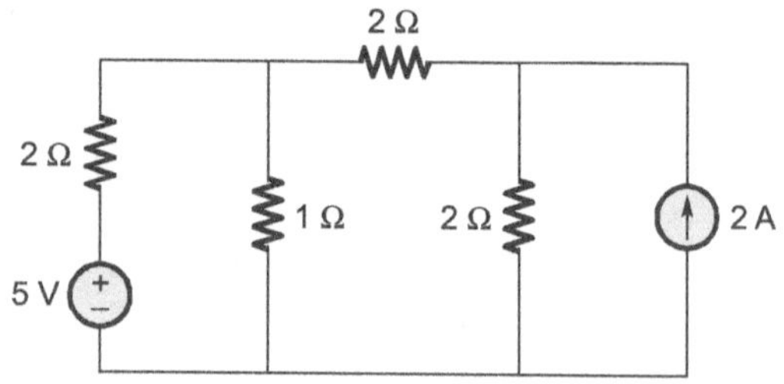

(a) $\frac{1}{2}$ A (b) $\frac{1}{4}$ A

(c) 2 A (d) $\frac{3}{2}$ A

Q.64 An electric circuit with 8 branches and 4 nodes will have
- (a) 5 loop equations (b) 11 loop equations
- (c) 3 loop equations (d) 7 loop equations

Q.65 For the circuit shown below, the current through the 10 V battery is

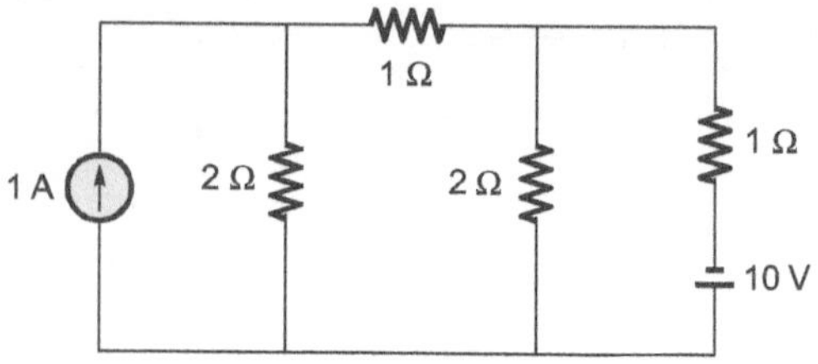

(a) 2.36 A (b) 4.91 A
(c) –2.36 A (d) None of these

Q.66 In the circuit shown in figure below, the value of V_2/V_1 is

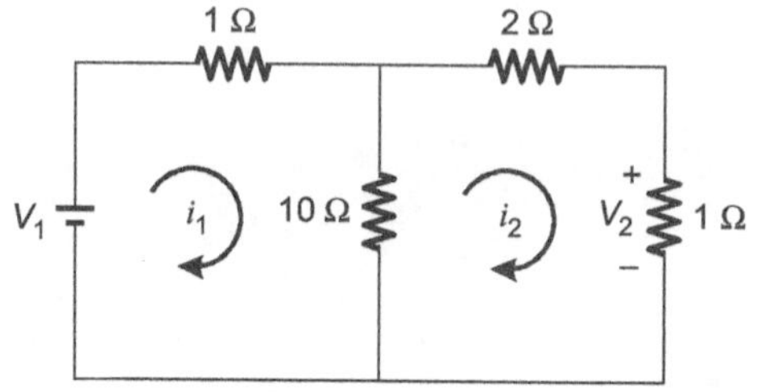

(a) $\frac{-19}{2}$ (b) $\frac{19}{2}$

(c) 1.5 (d) $\frac{2}{19}$

Q.67 The value of V in volts for the circuit shown below is

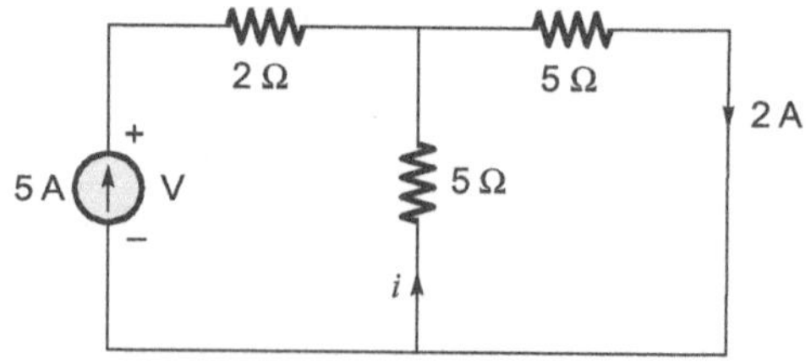

(a) zero (b) 16.65
(c) 22.50 (d) –3.6

Q.68 The number of KVL and KCL equations for the circuit shown below are

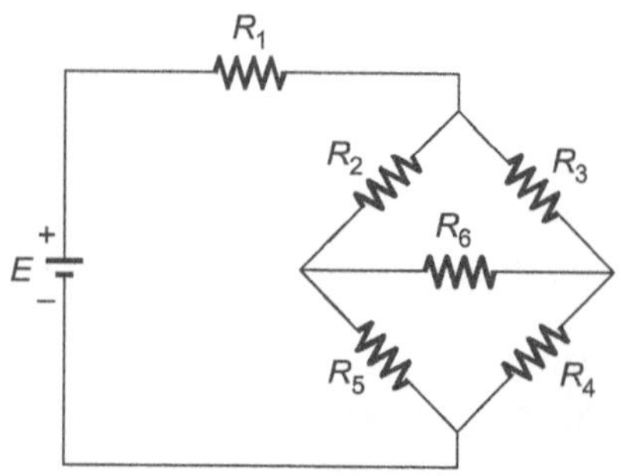

(a) 6 and 5 respectively
(b) 2 and 4 respectively
(c) 6 and 4 respectively
(d) 6 and 2 respectively

Q.69 The current I flowing in the given figure is

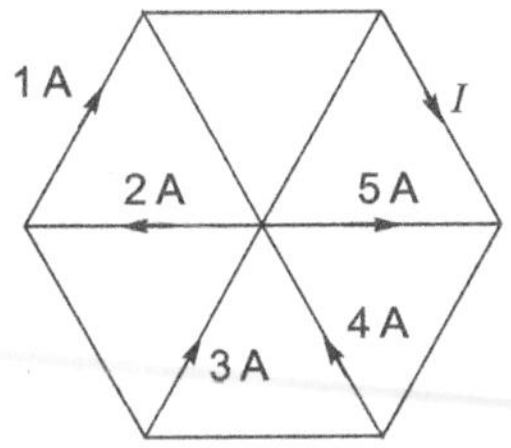

(a) 1 A (b) 2 A
(c) 3 A (d) 4 A

Q.70 The value of dependent source for the circuit shown below is

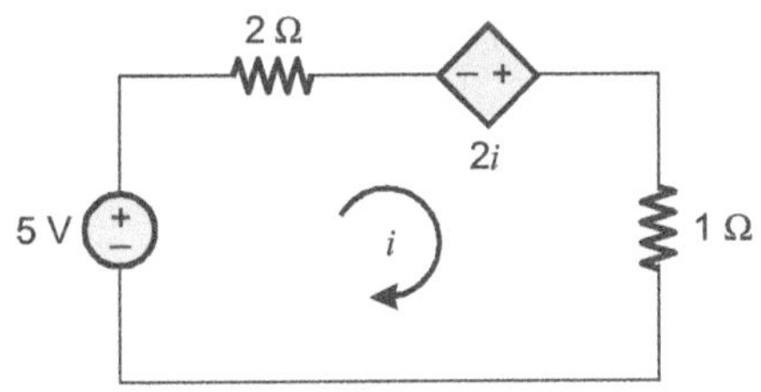

(a) 5 A (b) –10 V
(c) –5 A (d) 10 V

Q.71 For the circuit shown below, the value of current through R_L is

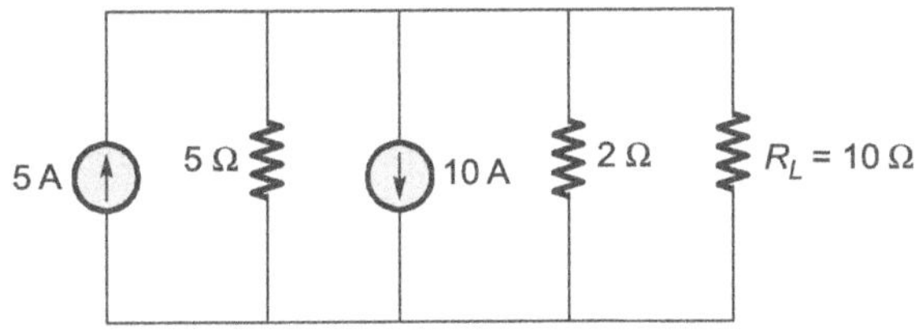

(a) zero A (b) – 1 A
(c) $\frac{-5}{8}$ A (d) ∞

Q.72 In nodal analysis, if there are N nodes in the circuit, then how many equations will be written to solve the network ?
(a) $N-1$ (b) N
(c) $N+1$ (d) $N-2$

Q.73 In the node voltage technique of solving networks, choice of a reference node does not
1. affect the operation of the circuit.
2. change the voltage across any element.
3. affect the pd between any pair of nodes.
4. affect the voltages of various nodes.

Which of the above statements is/are true?
(a) 1 only (b) 2, 3 and 4 only
(c) 2 and 4 only (d) 1 and 3 only

Q.74 Nodal analysis is based on
(a) KCL
(b) KVL
(c) both (a) and (b)
(d) law of conservation of energy

Q.75 In the circuit shown in figure below, what is the value of current i through source when $V = 4$ volts?

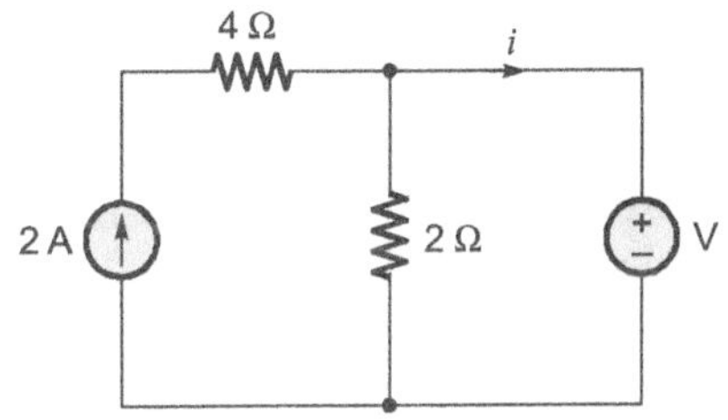

(a) 1 A (b) 2 A
(c) 3 A (d) 0 A

Q.76 In the circuit shown below, the value of I_x is

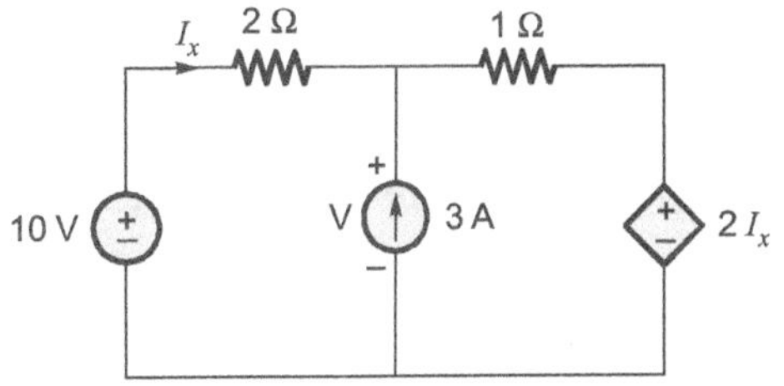

(a) 2 A (b) –0.6 A
(c) 2.6 A (d) 1.4 A

Q.77 The value of V_1 in the circuit shown in the given figure is

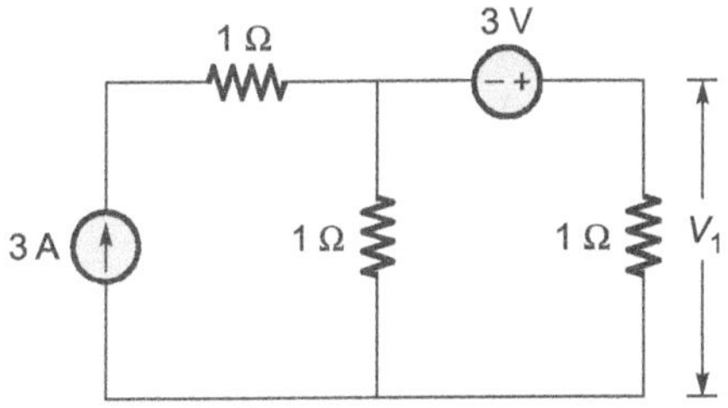

(a) 4 V (b) 3 V
(c) 2 V (d) 1 V

Q.78 Consider the electrical network shown below.

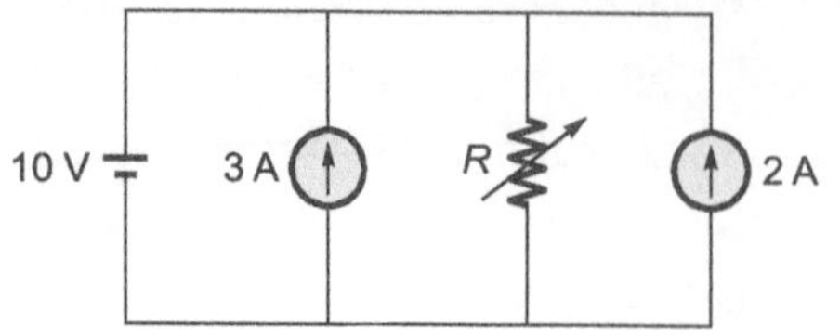

What is the value of R so that current through the resistor R is zero?

(a) 2 Ω (b) 5 Ω
(c) 4 Ω (d) 3 Ω

Q.79 **Assertion (A):** In the method of solution of the network using nodal analysis, the assumption is made that the node potential is always higher than the other voltages appearing in the equation.

Reason (R): Nodal analysis is used to find the branch currents.

(a) Both A and R are true and R is the correct explanation of A.
(b) Both A and R are true but R is not the correct explanation of A.
(c) A is true but R is false.
(d) A is false but R is true.

Q.80 **Assertion (A):** "Mesh analysis", the name being derived from the similarities in appearance between the closed loops of a network and a physical "fence" or "Mesh".

Reason (R): In "Mesh analysis", a distinct current is assumed in the loop and the polarities of drops in each element in the loop is determined by the assumed direction of loop current for that loop. KVL is then applied around each closed loop and by solving these linear simultaneous equations, the branch currents are determined.

(a) Both A and R are true and R is the correct explanation of A.
(b) Both A and R are true but R is not the correct explanation of A.
(c) A is true but R is false.
(d) A is false but R is true.

Q.81 The number of mess equations needed to solve an electrical network is $m = (b - n + 1)$, where b is the number of branches and n is the number of nodes in the electrical network.

Consider the following statements related to mesh and nodal-analysis for analysing an electrical network.

1. If $m < n$, the Mesh method offers less advantages.
2. If $m > n$, i.e. when the number of parallel paths in the network is more, mess method is preferred.

Which of the above statements is/are true?

(a) Neither 1 nor 2 (b) 1 only
(c) 2 only (d) Both 1 and 2

Q.82 The nodal equations of the circuit shown below are respectively given by

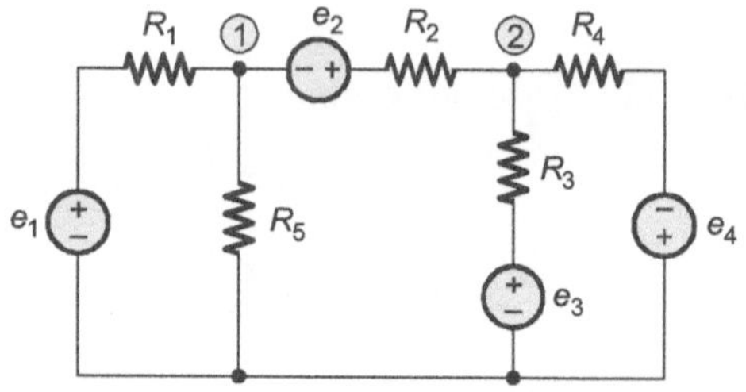

(a) $V_1\left[\frac{1}{R_1}+\frac{1}{R_2}+\frac{1}{R_5}\right]+\frac{V_2}{R_2}=\frac{e_1}{R_1}-\frac{e_2}{R_2}$

$V_2\left[\frac{1}{R_2}+\frac{1}{R_3}+\frac{1}{R_4}\right]+\frac{V_1}{R_2}=\frac{e_2}{R_2}+\frac{e_3}{R_3}+\frac{e_4}{R_4}$

(b) $V_1\left[\frac{1}{R_1}+\frac{1}{R_2}+\frac{1}{R_5}\right]-\frac{V_2}{R_2}=\frac{e_1}{R_1}-\frac{e_2}{R_2}$

$V_2\left[\frac{1}{R_2}+\frac{1}{R_3}+\frac{1}{R_4}\right]-\frac{V_1}{R_2}=\frac{e_2}{R_2}+\frac{e_3}{R_3}+\frac{e_4}{R_4}$

(c) $V_1\left[\frac{1}{R_1}+\frac{1}{R_2}+\frac{1}{R_5}\right]-\frac{V_2}{R_2}=\frac{e_1}{R_1}-\frac{e_2}{R_2}$

$V_2\left[\frac{1}{R_2}+\frac{1}{R_3}+\frac{1}{R_4}\right]-\frac{V_1}{R_2}=\frac{e_2}{R_2}+\frac{e_3}{R_3}-\frac{e_4}{R_4}$

(d) $V_1\left[\frac{1}{R_1}+\frac{1}{R_2}+\frac{1}{R_5}\right]-\frac{V_2}{R_2}=\frac{e_2}{R_2}-\frac{e_1}{R_1}$

$V_2\left[\frac{1}{R_2}+\frac{1}{R_3}+\frac{1}{R_4}\right]+\frac{V_1}{R_2}=\frac{e_2}{R_2}+\frac{e_3}{R_3}+\frac{e_4}{R_4}$

Q.83 The current I in the circuit shown below is

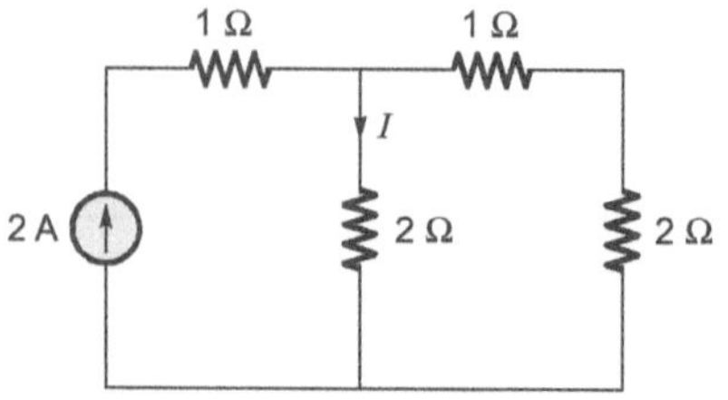

(a) $\frac{4}{5}$ A (b) $\frac{6}{5}$ A

(c) $\frac{2}{5}$ A (d) $\frac{7}{5}$ A

4. Circuit Theorems

Q.84 Superposition theorem is applicable only to network that are

(a) passive (b) time-invariant

(c) linear (d) non-linear

Q.85 In any lumped network with elements in b branches, $\sum_{i=1}^{b} V_i(t)\, i_i(t) = 0$, for all t, holds good according to

(a) Millman's theorem

(b) Tellegen's theorem

(c) Dual-Millman's theorem

(d) Compensation theorem

Q.86 The Thevenin's equivalent resistance to the left of AB in figure given below is given by

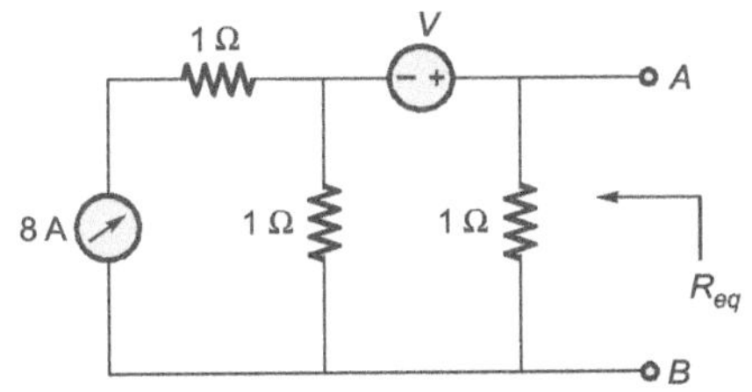

(a) $\frac{3}{2}\,\Omega$ (b) 1 Ω

(c) $\frac{1}{2}\,\Omega$ (d) $\frac{1}{3}\,\Omega$

Q.87 For which of the following, Thevenin's and Norton's equivalents can't be developed?

(a) DC independent sources

(b) AC independent sources

(c) Independent and dependent sources

(d) Independent and dependent sources with all controlling current and voltage contained within the network whose equivalent is being developed.

Q.88 The value of Z_{22} (Ω) for the circuit shown below is

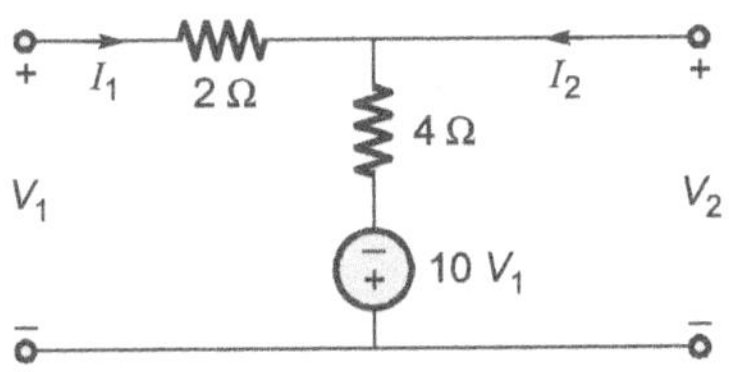

(a) $\frac{11}{4}$ (b) $\frac{4}{9}$

(c) $\frac{9}{4}$ (d) $\frac{4}{11}$

Q.89 In which of the following it is not desired to attain the condition of maximum power transfer?

(a) Electronic circuits

(b) Communicational circuits

(c) Computer circuits

(d) Electric circuits

Q.90 A simple equivalent circuit of the two terminal network shown in figure below is

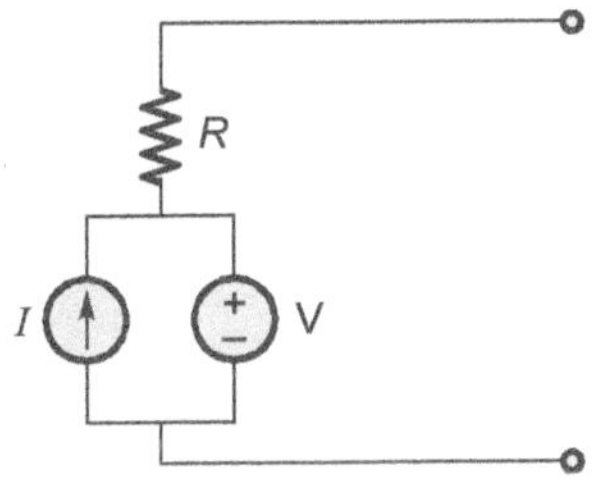

(a) 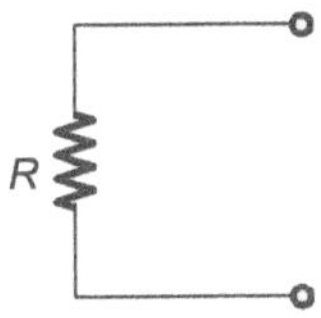

(b) 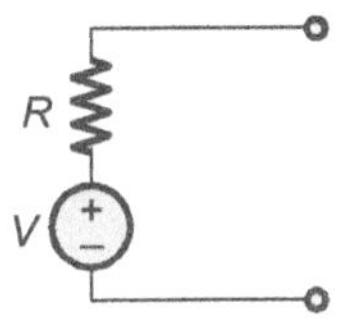

(c) 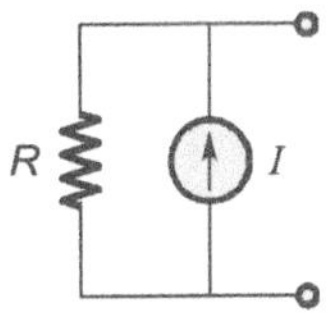

(d) 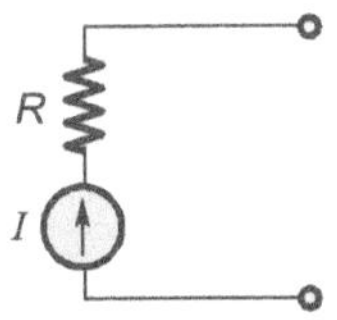

Q.91 For the circuit shown below, a resistance R is to be connected across an active network. It is given that when

(*i*) $R = \infty$, $V = 5$ volts

(*ii*) $R = 0$, current output of the active source is 2.5 A.

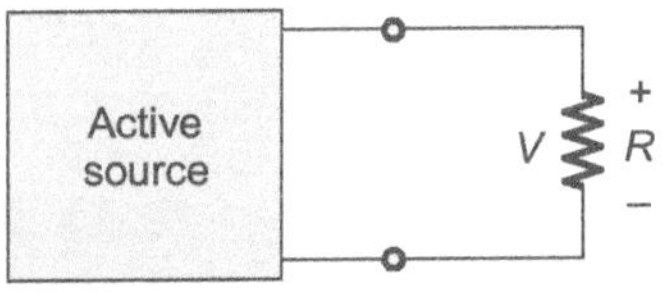

What would be the current through R when the value of R is 5 Ω?

(a) 2.5 A (b) 1.4 A

(c) 0.71 A (d) None of these

Q.92 A generator of internal impedance $|Z_g|$ delivers maximum power to a load impedance Z_p, only if

(a) $Z_p < Z_g$ (b) $Z_p = Z_g$

(c) $Z_p > Z_g$ (d) $Z_g = \frac{Z_p}{2}$

Q.93 In the circuit shown below, the current through the 5 Ω resistor is

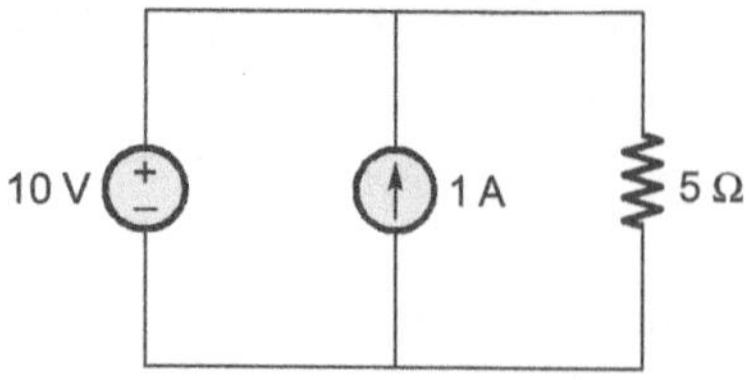

(a) 0 ampere (b) 1 ampere
(c) 3 ampere (d) none of these

Q.94 The current flowing through the resistance of 10 Ω for the circuit shown below is

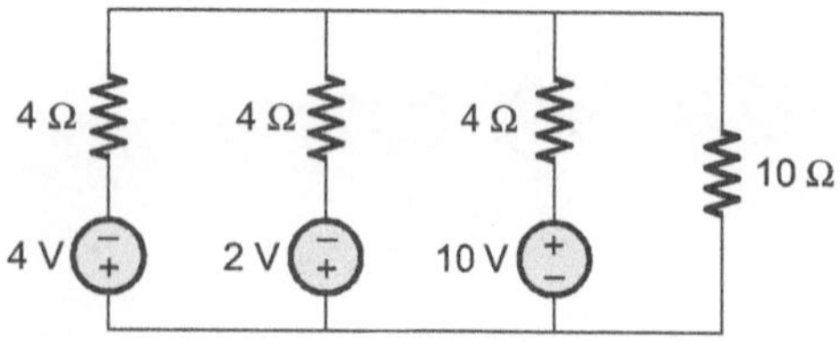

(a) $\frac{2}{17}$ A (b) $\frac{-1}{17}$ A

(c) $\frac{3}{17}$ A (d) $\frac{-4}{17}$ A

Q.95 The Thevenin's equivalent voltage source across the terminals *X*-*Y* shown below is

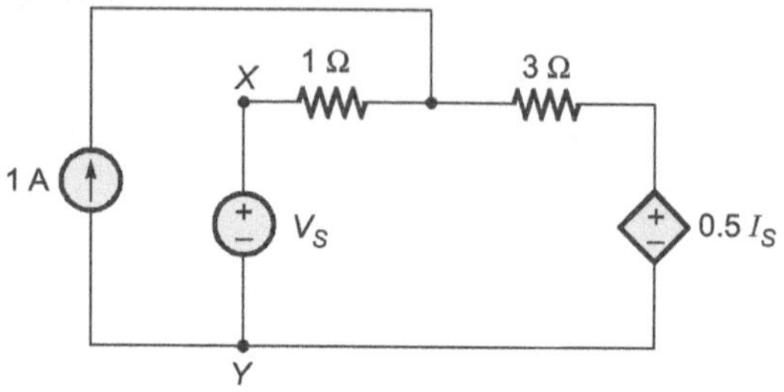

(a) 0 volts (b) 4 volts
(c) –0.5 volts (d) 3 volts

Q.96 The Norton's equivalent resistance across the diode in the given circuit is

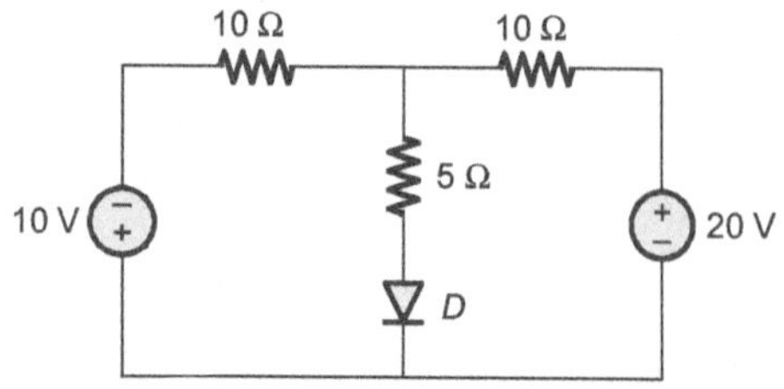

(a) 5 Ω (b) 0 Ω

(c) $\frac{5}{2}$ Ω (d) 10 Ω

Q.97 Assuming the op-amp to be ideal, the input admittance of the op-amp shown below is

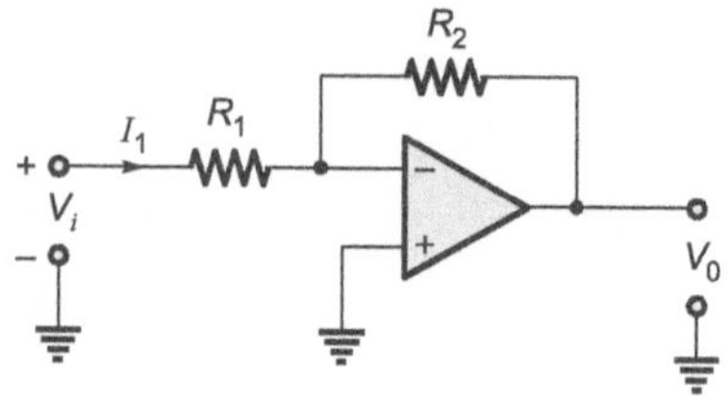

(a) $(R_1 + R_2)$ (b) $\frac{1}{R_2}$

(c) $\frac{1}{R_1}$ (d) $\frac{1}{(R_1 + R_2)}$

Q.98 In the π-circuit shown below, the current thorugh the 2 Ω resistor is

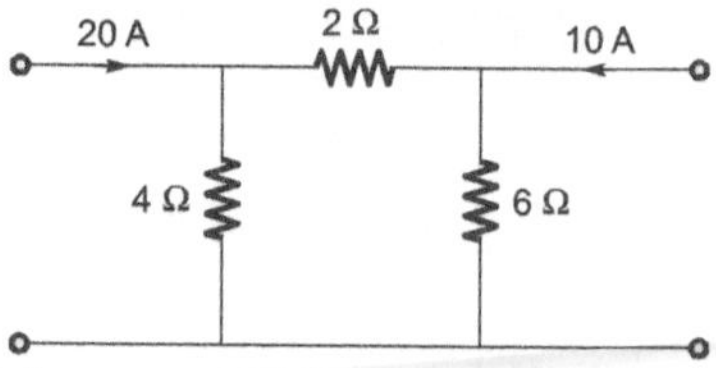

(a) 5 A (b) –6.67 A
(c) 11.67 A (d) 1.67 A

Q.99 Match **List-I (Elements)** with **List-II (Type)** and select the correct answer using the codes given below the lists:

List-I	List-II
A. Vaccum tube	1. Linear
B. Capacitor	2. Passive
C. Transistor	3. Active
D. Thermistor	4. Unilateral

Codes:

	A	B	C	D
(a)	4	3	1	2
(b)	3	1	4	2
(c)	4	2	3	1
(d)	2	1	4	3

Q.100 The equivalent resistance of the given circuit is

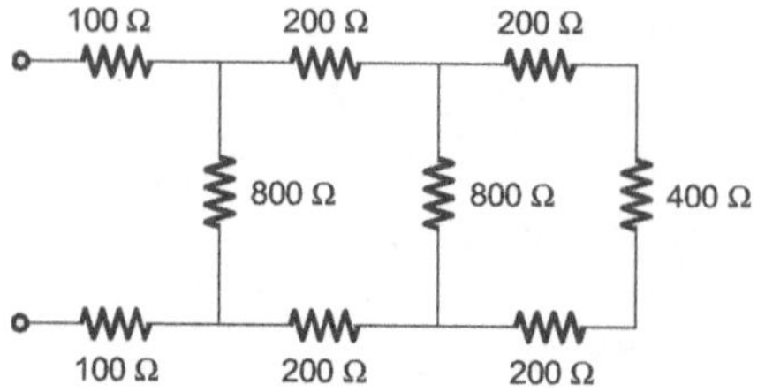

(a) 200 Ω (b) 400 Ω
(c) 600 Ω (d) 1600 Ω

Q.101 The maximum power that a 12 V d.c. source with an internal resistance of 2 Ω can supply to a resistive load is
(a) 12 W (b) 18 W
(c) 36 W (d) 48 W

Q.102 The Norton's equivalent circuit for the circuit shown below across the terminals *X-Y* can be represented as a current source (I_N) connected across a resistance (R_N). The value of I_N and R_N are respectively

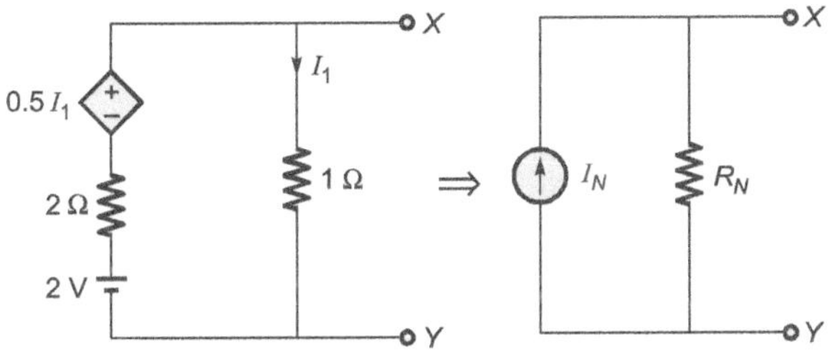

(a) 1.25 A and 0.8 Ω
(b) 1 A and 0.5 Ω
(c) 1 A and 0.8 Ω
(d) 1.25 A and 0.5 Ω

Q.103 To effect maximum power transfer to the load, Z_L (W) for the circuit shown below should be

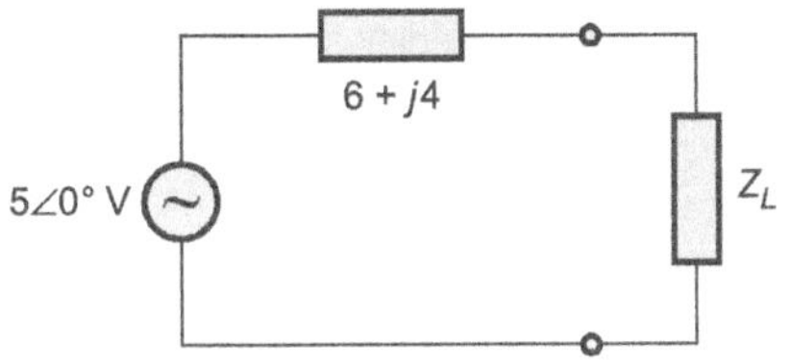

(a) 6 (b) 4
(c) 7.211∠33°.69 (d) 7.211∠–33°.69

Q.104 In order to apply superposition theorem, it is necessary that the network be only
(a) linear and reciprocal
(b) time-invariant and reciprocal
(c) linear
(d) linear and time-invariant

Q.105 A network *N* is to be connected to a load of 500 ohms. If the Thevenin's equivalent voltage and Norton's equivalent current of *N* are 5 volts and 10 mA respectively, the current through the load will be
(a) 2.5 mA (b) 1 mA
(c) 10 mA (d) 5 mA

Q.106 If all the elements in a particular network are linear, then the superposition theorem would hold, when the excitation is
(a) dc only (b) ac only
(c) ac or dc (d) an impulse

Q.107 The maximum value of power transferred to the load Z_L shown in figure below is

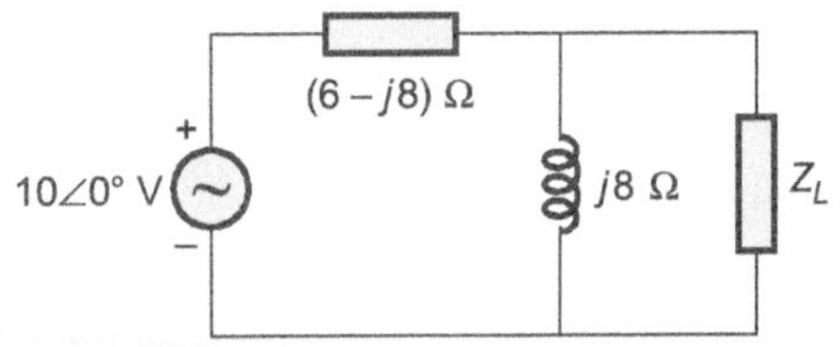

(a) 4.16 W (b) 2.22 W
(c) 1.25 W (d) 6.25 W

Q.108 In the circuit shown below, the current through the inductive reactance is

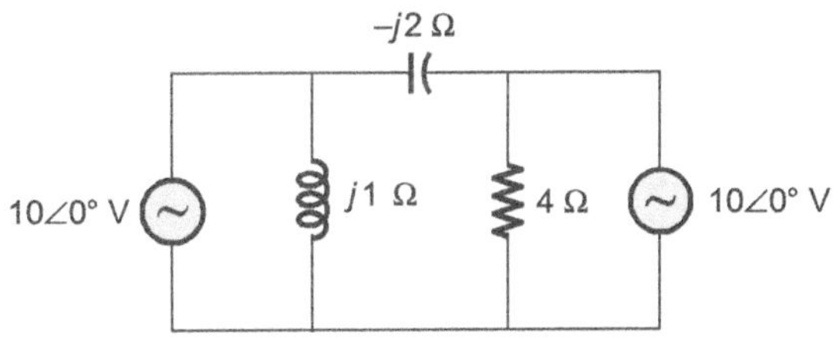

(a) 0 A (b) –j10 A
(c) –j2.25 A (d) j5 A

Q.109 Which of the following theorems is applicable for both linear and non-linear circuits?
(a) Superposition (b) Thevenin's
(c) Norton's (d) None of these

Q.110 While determining R_{Th} of a circuit
1. voltage and current sources should be left as they are.
2. all independent current and voltage sources are short-circuited.
3. all sources should be replaced by their source resistances.
4. all dependent sources remain as they are.

Which of the statements is/are true in relation of R_{Th}?
(a) 2 and 3 only (b) 3 only
(c) 3 and 4 only (d) 4 only

Q.111 For the circuit given in figure below, the Thevenin's voltage and resistance as seen at "*AB*" are respectively

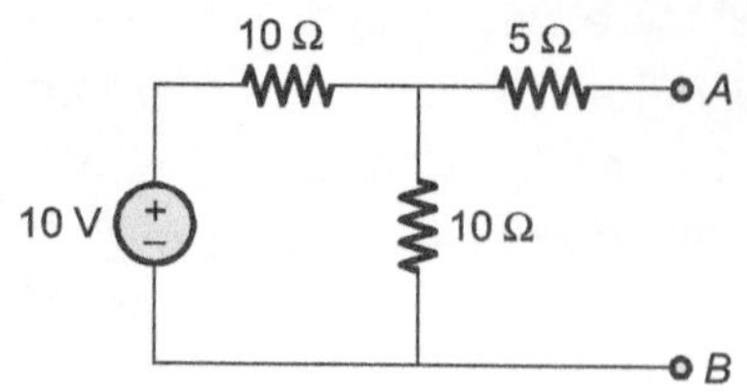

(a) 5 V, 10 Ω (b) 10 V, 10 Ω
(c) 5 V, 5 Ω (d) 5 V, 15 Ω

Q.112 Consider the following statements about Norton's equivalent of a circuit across a given two load terminals:

1. Norton's equivalent resistance is the same as Thevenin's equivalent resistance R_{Th}.
2. Norton's equivalent is the voltage equivalent of the network.
3. The load is connected in parallel to the Norton's equivalent resistance.

Out of the above statements, which statements hold true?

(a) 1, 2 and 3 (b) 1 and 3 only
(c) 1 only (d) 2 and 3 only

Q.113 When the power transferred to the load is maximum, the efficiency is

(a) 25% (b) 75%
(c) 50% (d) 100%

Q.114 "For a linear network containing generators and impedances, the ratio of the voltage to the current produced in other loop is the same as the ratio of voltage and current obtained if the positions of the voltage source and the ammeter measuring the current are interchanged."

The above stated network theorem is known as:

(a) Millman's theorem
(b) Norton's theorem
(c) Tellegen's theorem
(d) Reciprocity theorem

Q.115 Millman's theorem yields equivalent

(a) impedance or resistance
(b) current source
(c) voltage or current source
(d) voltage source

Q.116 Match **List-I (Theorem/Law)** with **List-II (Property)** and select the correct answer using the codes given below the lists:

List-I
A. Norton's theorem
B. Superposition theorem
C. Thevenin's theorem
D. Kirchhoff's current law

List-II
1. Effects of independent sources in a linear circuit are additive.
2. Law of non-accumulation of charge holds good at nodes.
3. Current source with shunt resistor.
4. Voltage source with series resistor.

Codes:

	A	B	C	D
(a)	2	4	1	3
(b)	3	1	4	2
(c)	2	1	4	3
(d)	3	4	1	2

Q.117 The *v-i* characteristic of an element is shown in figure below. The element is

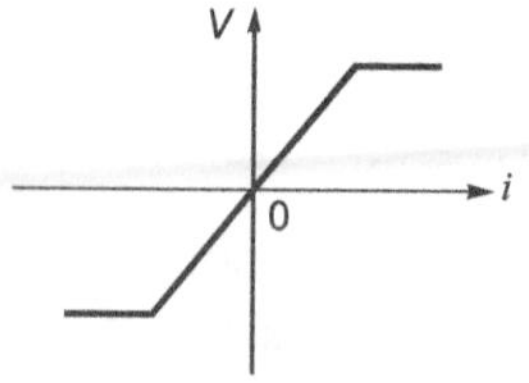

(a) non-linear, active, non-bilateral
(b) linear, active, non-bilateral
(c) non-linear, passive, bilateral
(d) non-linear, passive, non-bilateral

Q.118 If the source impedance is inductive, for maximum transfer of power from the source to the load, the load should be

(a) capacitive
(b) resistive
(c) complex conjugate of complex source impedance
(d) none of the above

Q.119 The equivalent resistance between terminal *A* and *B* for the given circuit is

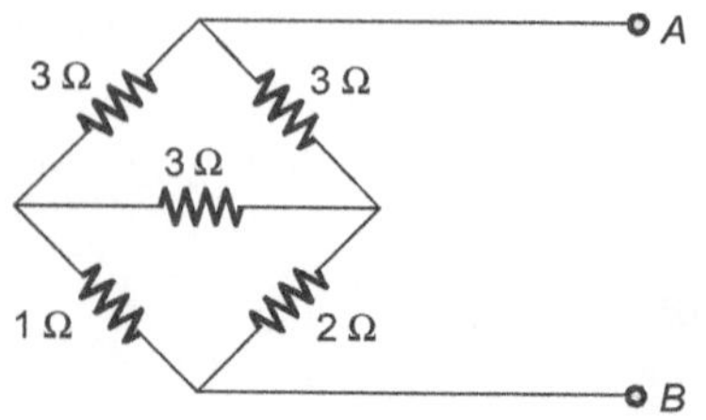

(a) $9\ \Omega$ (b) $\frac{11}{5}\ \Omega$

(c) $3\ \Omega$ (d) $\frac{5}{11}\ \Omega$

Q.120 Assertion (A): Norton's theorem is called dual of the Thevenin's theorem.

Reason (R): If the Thevenin's equivalent voltage source is converted to an equivalent current source, the Norton's equivalent is obtained.

(a) Both A and R are true and R is the correct explanation of A.
(b) Both A and R are true but R is not the correct explanation of A.
(c) A is true but R is false.
(d) A is false but R is true.

Q.121 Assertion (A): In any active network, maximum power transfer to the load takes place when the load impedance is equal to an equivalent impedance of the network as viewed from the terminals of the load.

Reason (R): For finding the maximum power transfer to the load, any network has to be first converted into Thevenin's equivalent across the terminals of the load.

(a) Both A and R are true and R is the correct explanation of A.
(b) Both A and R are true but R is not the correct explanation of A.
(c) A is true but R is false.
(d) A is false but R is true.

Q.122 Assertion (A): The substitution theorem is very useful in the networks consisting the mutual inductances.

Reason (R): By the substitution theorem, the effect of mutual inductances may be replaced by equivalent dependent sources of the same value.

(a) Both A and R are true and R is the correct explanation of A.
(b) Both A and R are true but R is not the correct explanation of A.
(c) A is true but R is false.
(d) A is false but R is true.

Q.123 Consider the following statements related to Thevenin's theorem:

For the application of Thevenin's theorem

1. the circuit should not contain non-linear elements.
2. the circuit may contain unilateral or bilateral elements or both.
3. there should not be magnetic coupling between the load and circuit, to be replaced by Thevenin's theorem.
4. there should not be controlled sources on the load size, controlled from some other part of the circuit.

Which of the above statements is/are true?

(a) 1, 2 and 3 only (b) 2, 3 and 4 only
(c) 1, 3 and 4 only (d) 1 and 4 only

Q.124 Which of the following theorem is applicable to both linear and non-linear networks?

(a) Thevenin's theorem
(b) Superposition theorem
(c) Norton's theorem
(d) Substitution theorem

5. Capacitors and Inductors

Q.125 The equivalent circuit of the capacitor shown below is

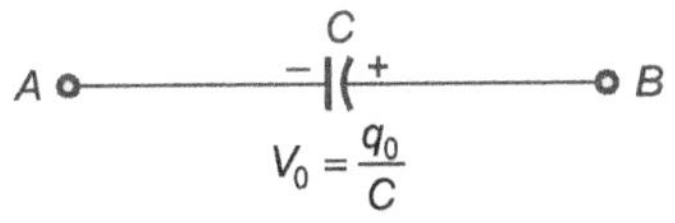

(a) 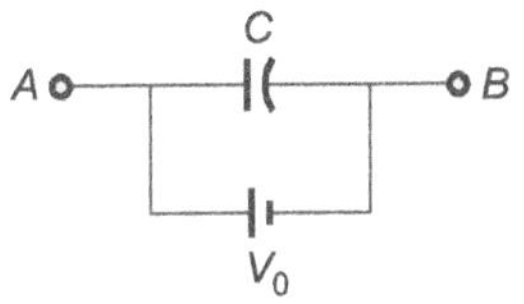

(b) 

(c) 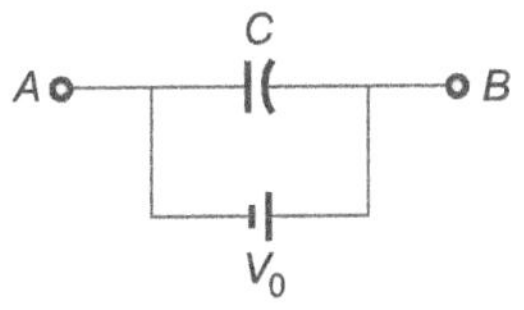

(d) 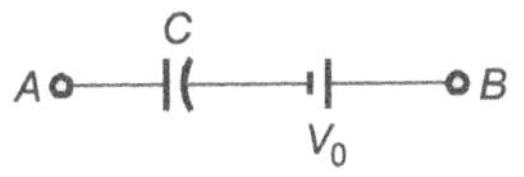

Q.126 The strength of current in 1 Henry inductor changes at a rate of 1 A/sec. The magnitude of energy stored in the inductor after 3 sec is

(a) 2 Joules (b) 3.8 Joules
(c) 4.5 Joules (d) 2.2 Joules

Q.127 The current and voltage profile of an element vs time has been shown in given figure. The element and its value are respectively

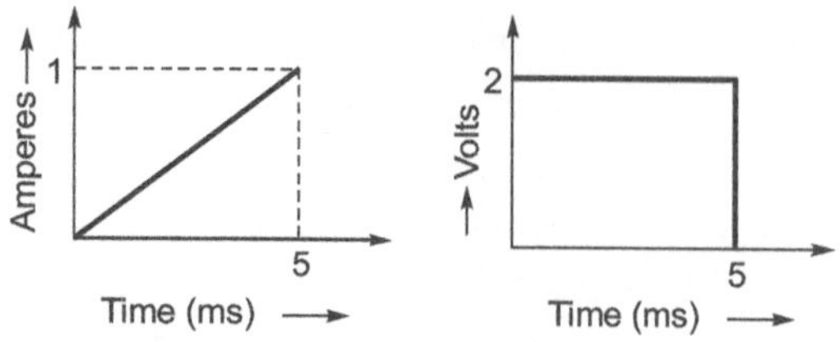

(a) a resistor with $R = 2\ m\Omega$
(b) a capacitor with $C = 2\ \mu F$
(c) an inductor with $L = 0.5$ H
(d) an inductor with $L = 10$ mH

Q.128 The equivalent circuit of the inductor shown below is

(a) 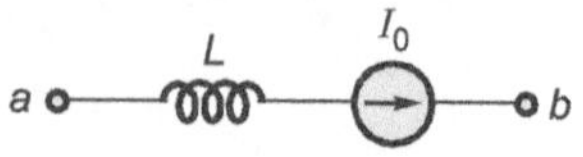

(b) 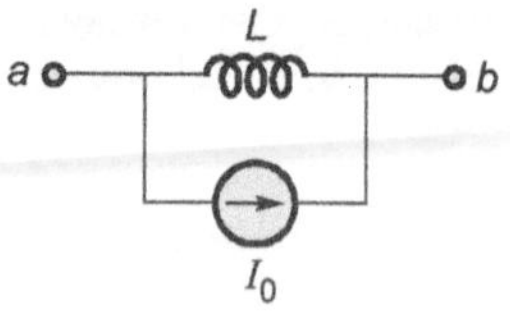

(c) 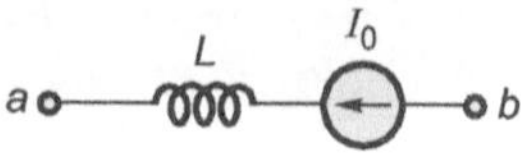

(d) 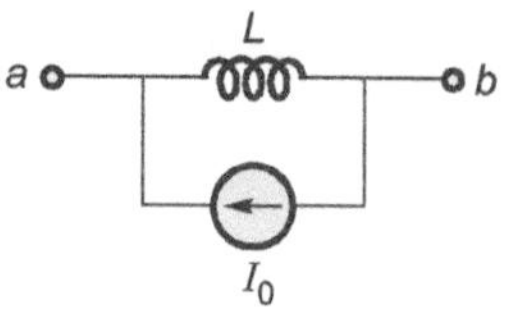

Q.129 Figure shown below exhibits the voltage-time profile of a source to charge a capacitor of 50 μF. The value of charging current in amperes is

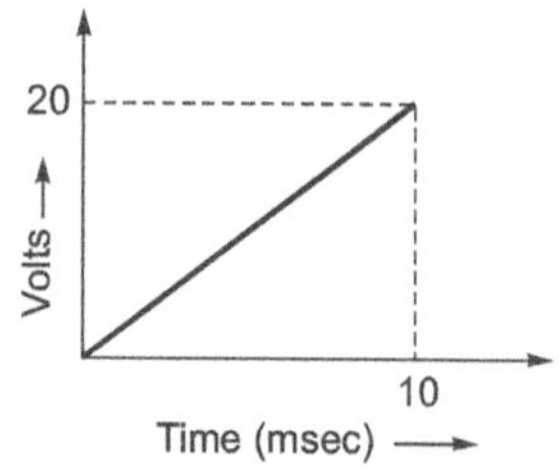

(a) 0.8 (b) 0.5
(c) 0.04 (d) 0.1

Q.130 The equivalent capacitance across the given terminals *A-B* is

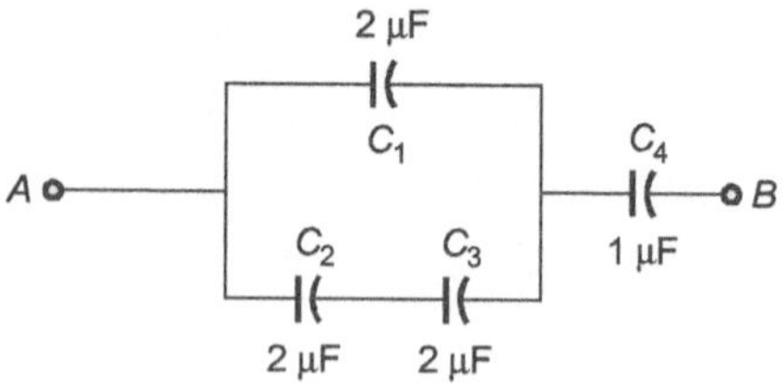

(a) $\frac{1}{2}\mu F$ (b) $\frac{3}{4}\mu F$
(c) $\frac{1}{8}\mu F$ (d) $\frac{4}{3}\mu F$

Q.131 The charging time required to charge the equivalent capacitance between the given terminals *a-b* by a steady direct current of constant magnitude of 10 A is given by

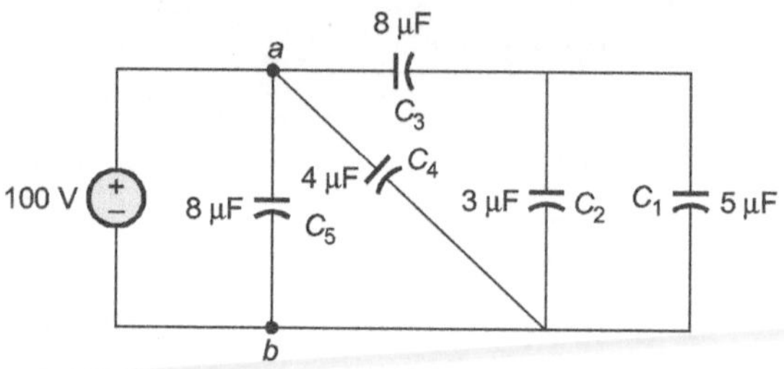

(a) 160 μsec (b) 80 μsec
(c) 21 μsec (d) 45 μsec

Q.132 An ac voltage of 220 V is applied to a pure inductance of 50 H. If the current is 5 A, the instantaneous value of voltage and current will be respectively given by

(a) $v = 622 \sin (314t)$ Volts
$i = 7.07 \cos (314t - 90°)$ Amps
(b) $v = 311 \sin (314t)$ Volts
$i = 14.14 \sin (314t - 90°)$ Amps
(c) $v = 311 \sin (314t)$ Volts
$i = 7.07 \sin (314t - 90°)$ Amps
(d) $v = 622 \sin (314t)$ Volts
$i = 14.14 \cos (314\ t - 90°)$ Amps

Q.133 The voltage and current through a circuit element is $v = 100 \sin (314\ t + 45°)$ volts and $i = 10 \sin (314\ t - 45°)$ amps.
The type of circuit element and its value will be respectively

(a) an inductor with $L = 31.8$ mH
(b) a capacitor with $C = 10$ F
(c) an inductor with $L = 10$ H
(d) a capacitor with $C = 31.8\ \mu F$

Q.134 The equivalent inductance for the inductive circuit shown below at terminal "1–2" is

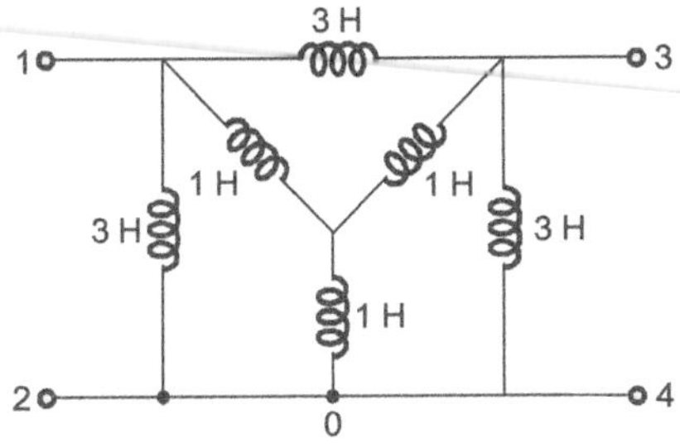

(a) $\frac{1}{4}$ H (b) 1 H
(c) $\frac{3}{4}$ H (d) $\frac{1}{2}$ H

Q.135 There inductances, 1 H, 2 H and L H are in parallel. The maximum value of equivalent inductance will be

(a) $\frac{3}{2}$ H (b) 0 H
(c) 3 H (d) none of these

Q.136 Figure given below shows a parallel combination of 10 mH and 20 mH inductances in parallel. A time varying voltage is applied across the inductances at any time instant and the total current supplied by the time varying source is 25 amps.
The values of instantaneous currents through each of the inductances are given by

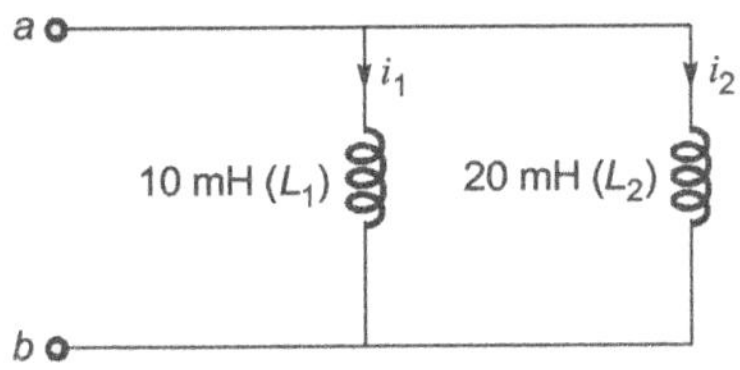

(a) i_1 = 8.33 A and i_2 = 16.67 A
(b) i_1 = 15.56 A and i_2 = 9.44 A
(c) i_1 = 16.67 A and i_2 = 8.33 A
(d) i_1 = 9.44 A and i_2 = 15.56 A

Q.137 The equivalent capacitance between the terminals *a-b* for the capacitive circuit shown below is

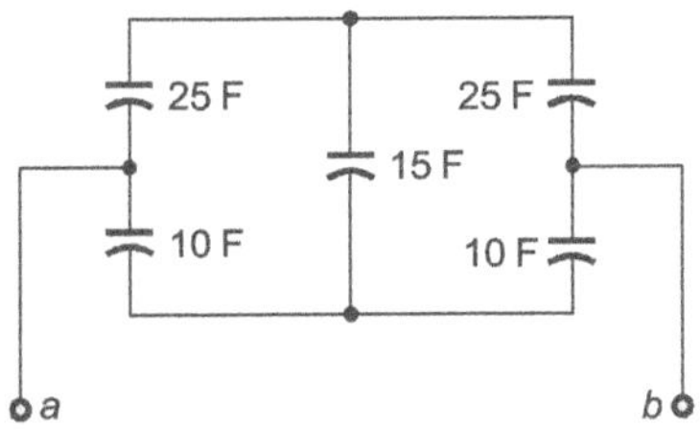

(a) 25.25 F (b) 16.80 F
(c) 12.25 F (d) 17.50 F

Q.138 A voltage $v(t) = e^{-t}\sin t$ is connected to a capacitance C = 2 F. The energy stored in the capacitor for $t = 0$ to $t = 10$ sec is

(a) $2\sin 10 \cdot e^{-20}$ Joules
(b) $\frac{1}{2}\sin^2 10 \cdot e^{-20}$ Joules
(c) $\sin^2 10 \cdot e^{-20}$ Joules
(d) $\frac{1}{2}\sin 100 \cdot e^{-10}$ Joules

Q.139 A sawtooth voltage as shown in figure below is applied between the terminals *A-B* to a capacitor of $C = 30\ \mu F$. The current *i* will be

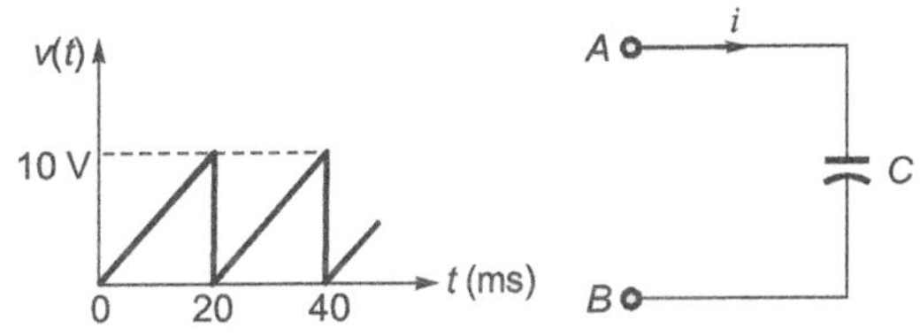

(a) 10.5 mA (b) 50 mA
(c) 0 A (d) 15 mA

Q.140 **Assertion (A):** In steady state condition, the energy consumed is infinite for both inductor and capacitor.
Reason (R): The current *i* in the capacitor and voltage *v* in the inductor are zero in steady state.

(a) Both A and R are true and R is the correct explanation of A.
(b) Both A and R are true but R is not the correct explanation of A.
(c) A is true but R is false.
(d) A is false but R is true.

Q.141 **Assertion (A):** If the voltage across the capacitor being constant, current through it is zero.
Reason (R): On application of dc voltage and with no initial charge, the capacitor first acts as short circuit but as soon as the full charge it retains, the capacitor behaves as open circuit.

(a) Both A and R are true and R is the correct explanation of A.
(b) Both A and R are true but R is not the correct explanation of A.
(c) A is true but R is false.
(d) A is false but R is true.

Q.142 A 10 μF capacitor is charged full by 100 V dc source. It is then connected across an uncharged capacitor of 50 μF capacity. The energy stored in the two capacitors in the final state will be

(a) 1.4 mJ in the former and 8.4 mJ in later
(b) 1.4 mJ in the former and 7 mJ in later
(c) 8.4 mJ in the former and 1.4 mJ in later
(d) 50 mJ in the former and 1.4 mJ in later

Q.143 For the bridge circuit shown below, the charge accumulated in the capacitor under steady state condition is

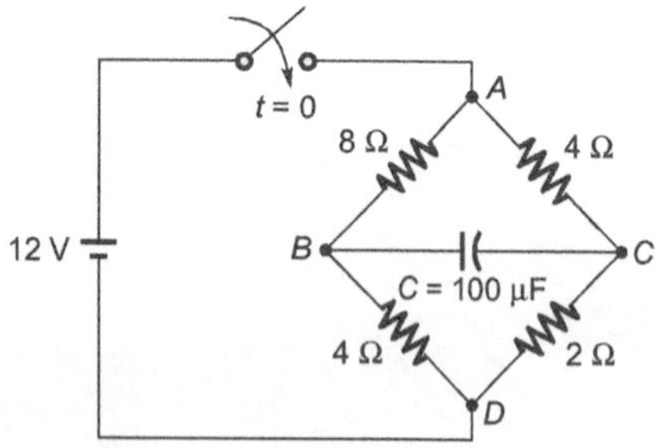

(a) zero (b) 12 μC
(c) infinite (d) 4 μC

Q.144 A current of $i = 300t$ is suddenly applied to an inductance of 1 mH. The induced voltage will be

(a) 300 volts (b) zero
(c) 0.3 volts (d) –0.03 volts

Q.145 A 100 turn coil has an inductance of 6 mH. If the number of turns is increased by 200%, all other quantities remaining the same, the inductance will be

(a) 24 mH (b) 12 mH
(c) 3 mH (d) None of the above

Q.146 A current input, $5\,\delta(t)$ is flow through a capacitor C. The voltage $V_c(t)$, across capacitor is given by

(a) $\frac{5}{C}t$ (b) $5\,C\,u(t)$
(c) $5t$ (d) $\frac{5}{C}u(t)$

Q.147 A voltage waveform $v(t) = 12\,t^2$ is applied across 1 H inductor for $t \geq 0$ with initial current through it being zero. The current through the inductor for $t \geq 0$ is given by

(a) $12\,t$ (b) $4\,t^3$
(c) $24\,t$ (d) $12\,t^3$

Q.148 Which of the following statements associated with capacitor is wrong?

(a) A capacitor resists an abrupt change in the voltage across it in a manner analogous to the way a spring resists abrupt change in its displacement.
(b) A capacitor resists an abrupt change in the current flowing through it.
(c) It is impossible to change the voltage across a capacitor even if the current through the capacitor changes by a finite amount in zero time, for this requires infinite current through the capacitor.
(d) A finite amount of energy can be stored in a capacitor even if the current through the capacitor is zero, such as when the voltage across it is constant.

Q.149 Potential difference across a capacitor of capacitance of 20 μF is increased uniformly from 0 to 240 V in 1 second. The charging current will be

(a) 9.6 mA (b) 1.2 mA
(c) 4.8 mA (d) 12 mA

Q.150 Assertion (A): Lower the self inductance of a coil more the delay in establishing steady current through it.

Reason (R): An inductor opposes a sudden change in current.

(a) Both A and R are true and R is the correct explanation of A.
(b) Both A and R are true but R is not the correct explanation of A.
(c) A is true but R is false.
(d) A is false but R is true.

Q.151 An open coil has

(a) zero resistance and zero inductance
(b) infinite resistance and infinite inductance
(c) infinite resistance and zero inductance
(d) zero resistance and infinite inductance

Q.152 Consider the following statements regarding an inductor:

1. An inductor is a sort of short circuit to dc.
2. An inductor is a sort of open circuit to dc.
3. An inductor never dissipates energy but only stores it.

4. A finite amount of energy can be stored in an inductor even if voltage across it is zero, such as when the current through it is constant.

Which of the above statements is/are not correct?

(a) 2 only (b) 3 and 4 only
(c) 2 and 4 only (d) 2, 3 and 4 only

Q.153 Assertion (A): A capacitor can store charge.
Reason (R): The plates of the condenser are conductive.
(a) Both A and R are true and R is the correct explanation of A.
(b) Both A and R are true but R is not the correct explanation of A.
(c) A is true but R is false.
(d) A is false but R is true.

Q.154 Assertion (A): It is impossible to change the current through an inductor by finite amount in zero time.
Reason (R): The current through an inductor can be changed by finite amount in zero time only when an infinite voltage is applied across it.
(a) Both A and R are true and R is the correct explanation of A.
(b) Both A and R are true but R is not the correct explanation of A.
(c) A is true but R is false.
(d) A is false but R is true.

6. First order RL and RC Circuits

Q.155 The poles of the impedance $Z(s)$ for the network shown in figure below will be real and coincident if

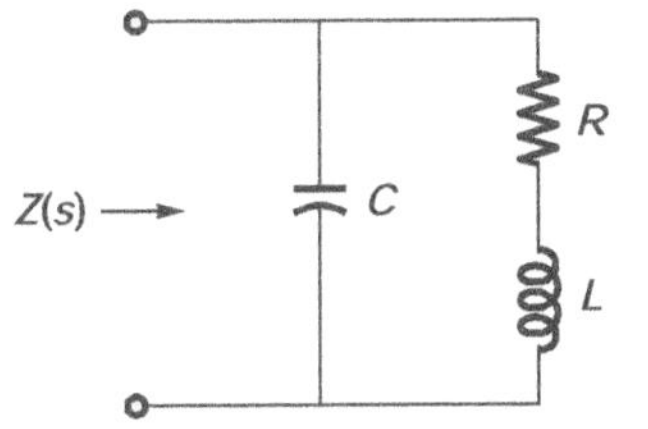

(a) $R = 2\sqrt{\dfrac{L}{C}}$ (b) $R = \dfrac{1}{2}\sqrt{\dfrac{L}{C}}$
(c) $R = 2\sqrt{\dfrac{C}{L}}$ (d) $R = 4\sqrt{\dfrac{L}{C}}$

Q.156 In the complex frequency $S = \sigma + j\omega$, ω has the unit of radian/sec and σ has the unit of
(a) rad (b) neper/sec
(c) radian/sec (d) neper

Q.157 A system is described by the transfer function $H(s) = \dfrac{1}{(s+2)}$. The value of its step response at a very large time will be close to
(a) ∞ (b) zero
(c) $\dfrac{1}{2}$ (d) -1

Q.158 At a certain current, the energy stored in iron cored coil is 1000 J and its copper loss is 2000 W. The time constant (in second) of the coil is
(a) 0.25 (b) 0.5
(c) 1.0 (d) 2.0

Q.159 A ramp voltage $v_i(t) = 100t$ V is applied to a differentiator circuit with $R = 5$ kΩ and $C = 4$ μF. The maximum output voltage is
(a) 10 V (b) 2 V
(c) 50 V (d) 0.2 V

Q.160 The Laplace transform of the waveform below is

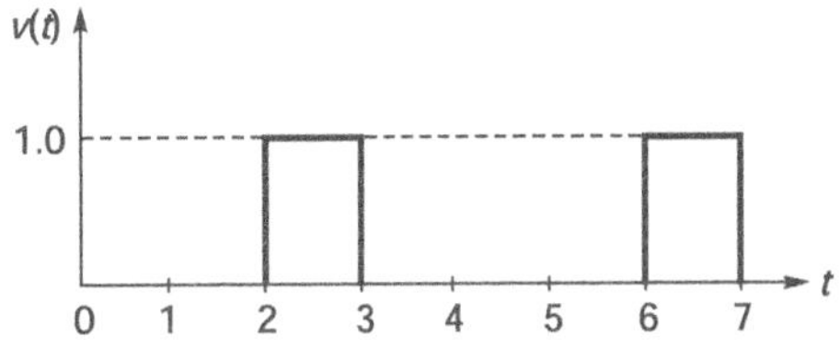

(a) $\dfrac{1}{s}(1 - e^{-2s} + e^{-3s} - e^{-7s})$

(b) $\dfrac{1}{s}(e^{-2s} + e^{-3s} - e^{-6s} + e^{-7s})$

(c) $\dfrac{1}{s}(1 + e^{-2s} - e^{-3s} + e^{-7s})$

(d) none of the above

Q.161 The initial and final values of $\phi(t) = [1 + e^{-t}(\sin t + \cos t)$ are given by
(a) 1, 1 (b) 0, 1
(c) 1, 2 (d) 2, 1

Q.162 The Laplace transform of the ramp function shown below is

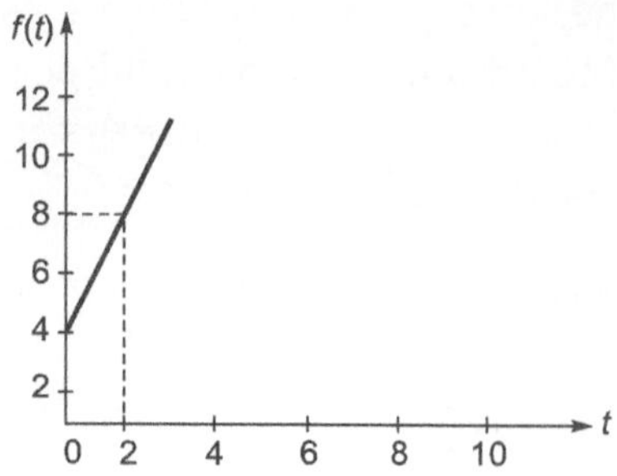

(a) $\frac{2}{s}\left(2+\frac{1}{s}\right)$ (b) $\left(\frac{1}{s}+\frac{2}{s^2}\right)$

(c) $\left(\frac{2}{s}+\frac{4}{s^2}\right)$ (d) $\frac{2}{s}\left(1+\frac{1}{s}\right)$

Q.163 The poles and zeros of the transfer function $\frac{I_2(s)}{I_1(s)}$ for the circuit shown below are located at

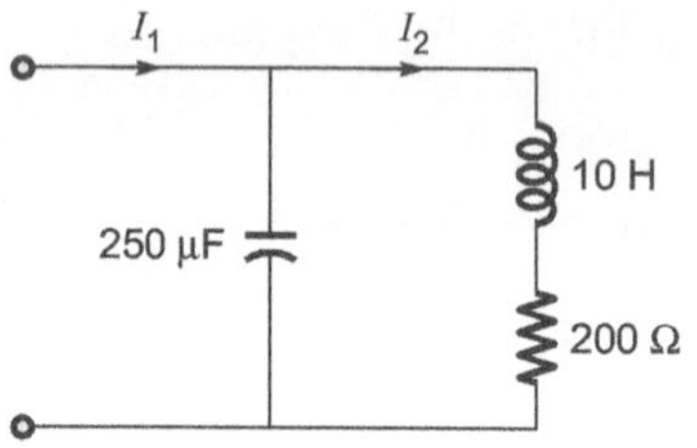

(a) Poles: $s=-2\pm j\sqrt{3}$;
Zeros : $s=10\pm j10\sqrt{3}$

(b) No Poles
Zeros: $s=-2\pm j2\sqrt{3}$

(c) Poles: $s=-10\pm j10\sqrt{3}$;
No zeros

(d) Poles: $s=-10\pm j10\sqrt{3}$;
Zeros: $s=-2\pm j2\sqrt{3}$

Q.164 A capacitor of 0.5 F is initially charged to 1 volt and is subjected to discharge at $t=0$ across a *LR* series circuit where L = 1 H; R = 2 Ω. The current $i(t)$ for $t>0$ is

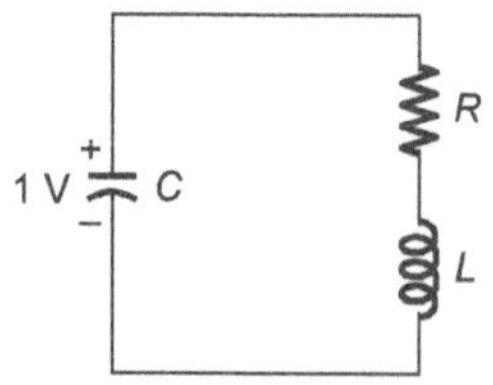

(a) $e^{-t}\cos t\, u(t)$ (b) $2\, e^{-t}\sin t\, u(t)$
(c) $2\, e^{-t}\cos t$ (d) $e^{-t}\sin t\, u(t)$

Q.165 Laplace transform analysis gives
(a) time domain response only
(b) frequency domain response only
(c) both (a) and (b)
(d) none of the above

Q.166 The Laplace transform of the derivative of the signal $f(t)=e^{-t}\,u(t)$ is

(a) $\frac{s}{s+2}$ (b) $\frac{-1}{s+1}$

(c) $\frac{s}{s+1}$ (d) $s+1$

Q.167 For a series *R-C* circuit excited by a dc voltage of 20 V, and with time-constant τ seconds, the voltage across the capacitor at time $t=\tau$ is given by
(a) $20\,(1-e^{1})$ volts (b) $(20-e^{-1})$ volts
(c) $20\,(1-e^{-1})$ volts (d) $20\,e^{-1}$ volts

Q.168 The Laplace-transform equivalent of a given network will have $\frac{7}{5}$ F capacitor replaced by

(a) $\frac{7}{5}s$ (b) $\frac{5}{7s}$

(c) $\frac{5}{7}s$ (d) $\frac{7}{5s}$

Q.169 An RC circuit has a capacitor $C=2$ μF in series with a resistance R = 1 MΩ. The time of 6 secs will be equal to
(a) three times of time constant
(b) two times of time constant
(c) one times of time constant
(d) none of the above

Q.170 The Laplace transform of the gate function shown below is

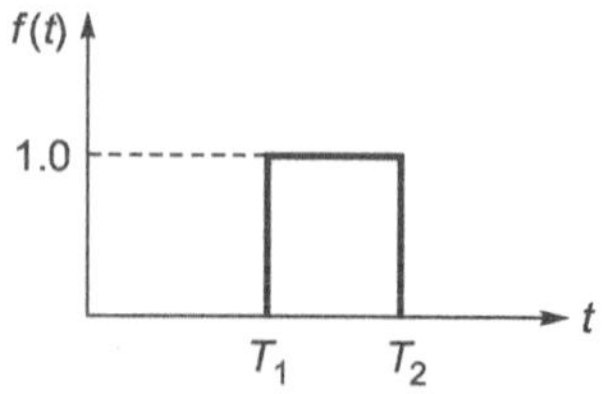

(a) $\left[e^{-T_1 s}-\frac{e^{-T_2 s}}{s}\right]$ (b) $\left[\frac{e^{T_1 s}}{s}-\frac{e^{T_2 s}}{s}\right]$

(c) $\left[\frac{e^{-T_2 s}}{s}-\frac{e^{-T_1 s}}{s}\right]$ (d) $\left[\frac{e^{-T_1 s}}{s}-\frac{e^{-T_2 s}}{s}\right]$

Q.171 Assuming initial condition to be zero, the current $i(t)$ in the circuit shown below is

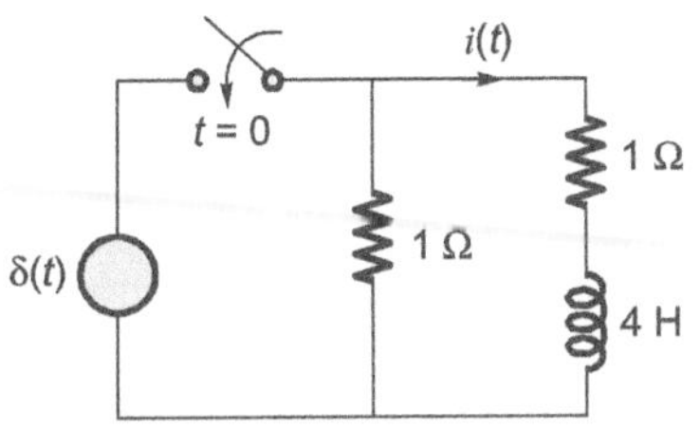

(a) e^{-t} (b) $0.25\ e^{-0.5t}$
(c) $2\ e^{-t}$ (d) $0.5\ e^{-0.5t}$

Q.172 A 10 volts step voltage is applied across a RC series circuit at t = 0 having R = 100 Ω and C = 100 μF.

The values of $i(t)$ and $\frac{di(t)}{dt}$ at t = 0 are respectively given by

(a) –1 A and 10 A/s
(b) 0 A and –5 A/s
(c) 0.1 A and –10 A/s
(d) 0.1 A and 20 A/s

Q.173 If the input voltage $V_1(t) = 10\ e^{-2t}$ V then, the output voltage $V_2(t)$ for the circuit shown below is

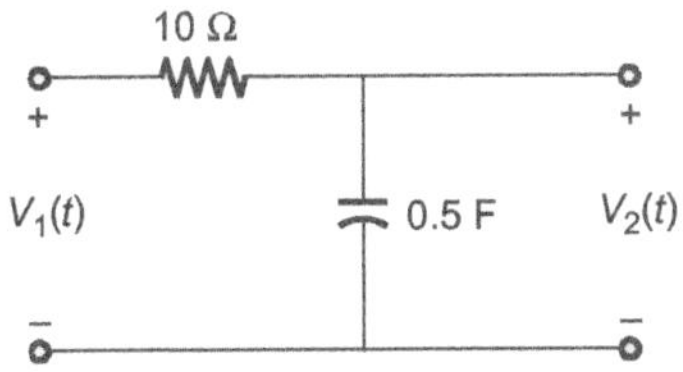

(a) $1.11\ (e^{-0.2t} - e^{-2t})$ (b) $(e^{-0.2t} + e^{-2t})$
(c) $0.5\ (e^{-0.2t} + e^{-2t})$ (d) $(e^{-2t} - e^{-0.2t})$

Q.174 In the circuit shown below, the switch 'S' is opened at t = 0. Prior to that, switch was closed.

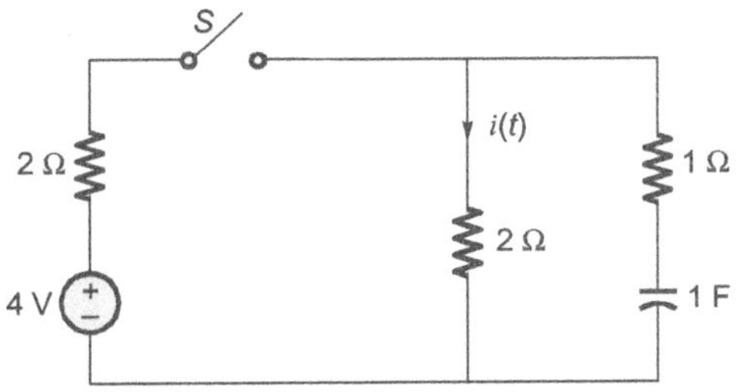

The current $i(t)$ is at $t = 0^+$ is

(a) $\frac{1}{3}$ A (b) 1 A
(c) $\frac{3}{2}$ A (d) $\frac{2}{3}$ A

Q.175 A rectangular pulse of duration T and magnitude A has the Laplace transform

(a) $\frac{A}{s}$ (b) $\frac{A}{s}e^{-Ts}$
(c) $\frac{A}{s}e^{Ts}$ (d) $\frac{A}{s}[1 - \exp(-Ts)]$

Q.176 The time constant of the circuit shown below is

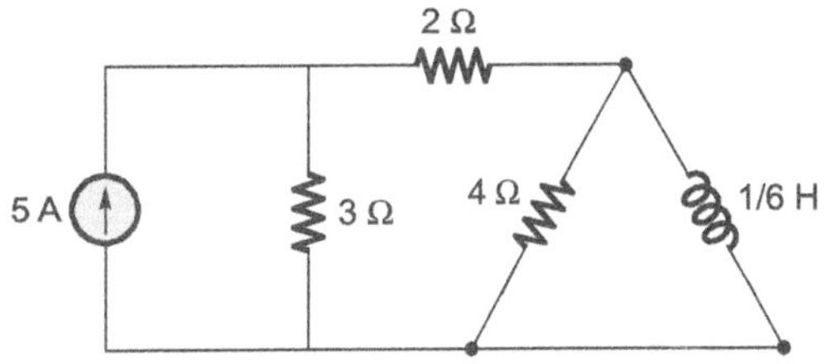

(a) 0.25 sec (b) 0.50 sec
(c) 0.075 sec (d) $\frac{1}{8}$ sec

Q.177 **Assertion (A):** Laplace transform is preferred for solving networks involving higher order differential equations.
Reason (R): The classical method for solving differential equations of higher order is quite cumbersome.

(a) Both A and R are true and R is the correct explanation of A.
(b) Both A and R are true but R is not the correct explanation of A.
(c) A is true but R is false.
(d) A is false but R is true.

Q.178 **Assertion (A):** In a purely capacitive circuit, the current wave is more distorted than the voltage wave.
Reason (R): The harmonics in the current wave are increased in proportion to their frequency numbers.

(a) Both A and R are true and R is the correct explanation of A.
(b) Both A and R are true but R is not the correct explanation of A.
(c) A is true but R is false.
(d) A is false but R is true.

Q.179 The time constant of the circuit shown below is

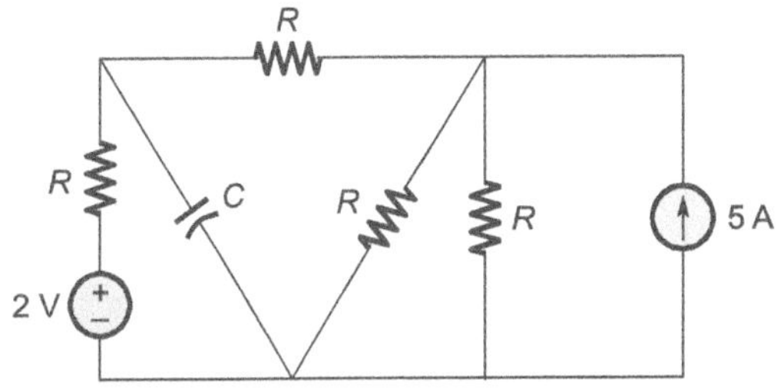

(a) $\frac{RC}{2}$ (b) $\frac{3}{2}RC$

(c) $\frac{5}{3}RC$ (d) $\frac{3}{5}RC$

7. Second Order RLC Circuits

Q.180 A current is given by

$$I(s) = \frac{(s+2)(s+4)}{(s+1)(s+3)s}$$

The initial and final values of currents are respectively

(a) 1 A and $\frac{8}{3}$ A (b) 0 A and $\frac{2}{3}$ A

(c) 1 A and $\frac{2}{3}$ A (d) $\frac{8}{3}$ A and 1 A

Q.181 The initial value theorem does not hold good for which of the following functions?

(a) Ramp function (b) Step function

(c) Delta function (d) Hyperbolic function

Q.182 For the parallel RLC circuit to be underdamped

(a) $R < 2\sqrt{\frac{L}{C}}$ (b) $R > 2\sqrt{\frac{L}{C}}$

(c) $R > \frac{1}{2}\sqrt{\frac{L}{C}}$ (d) $R < \frac{1}{2}\sqrt{\frac{L}{C}}$

Q.183 For the circuit shown below, the switch K is closed at $t = 0$.

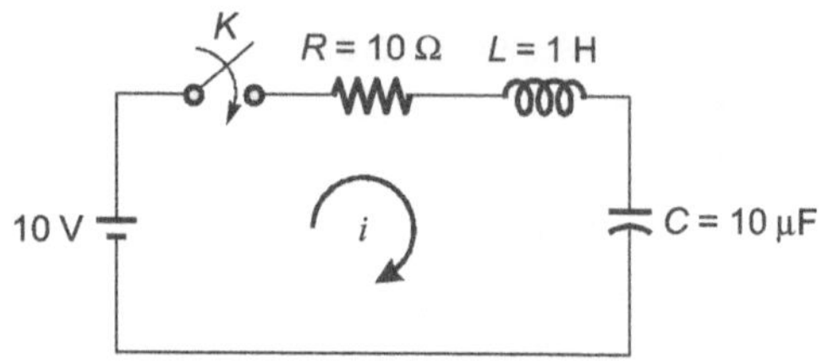

The value of the current and their derivatives for the above circuit at $t = 0$ were foundout and matched as given below:

Current & their derivatives	Values
1. $i(0^+)$	Zero A
2. $\frac{di(0^+)}{dt}$	10 A/sec
3. $\frac{d^2i(0^+)}{dt^2}$	–100 A/sec^2

Which of the above are correctly matched?

(a) 1 and 2 only (b) 2 and 3 only

(c) 1 and 3 only (d) 1, 2 and 3

Q.184 The energy in a network in Laplace domain is given by

$$E(s) = \frac{(s+8)}{(s+1)(s+3)(s+5)}$$

The initial and final values of power are respectively

(a) 1 unit and 0 unit (b) 0 unit and 1 unit

(c) both 1 unit (d) both 0 unit

Q.185 A voltage is given by

$$V(s) = \frac{8s-17}{(s+1)(s-4)}$$

The value of $V(t \to \infty)$ is

(a) $\frac{17}{4}$ volt (b) infinite

(c) zero volt (d) None of these

Q.186 The voltage across a capacitor is given by

$$V_c(s) = \left(\frac{s+1}{s^3+s^2+s+1}\right)$$

If the capacitor has the value of 2 F, initial value of current through it (at $t = 0^+$) will be

(a) 1 A (b) 0 A

(c) 2 A (d) None of these

Q.187 The impedance $Z(s)$ in the circuit shown below is

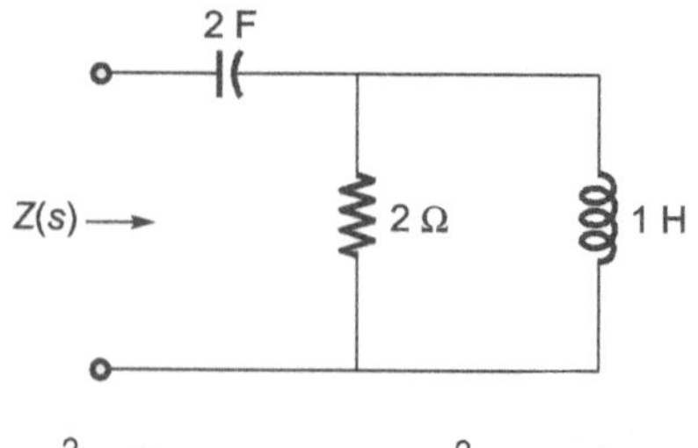

(a) $\frac{s^2+2}{s^2+2s+1}$ (b) $\frac{s^2+s+1}{s^2+s}$

(c) $\frac{2s^2+1}{s^2+s}$ (d) $\frac{4s^2+s+2}{2s^2+4s}$

Q.188 The voltage V_{c1}, V_{c2} and V_{c3} across the capacitors in the circuit shown below under steady-state, are respectively

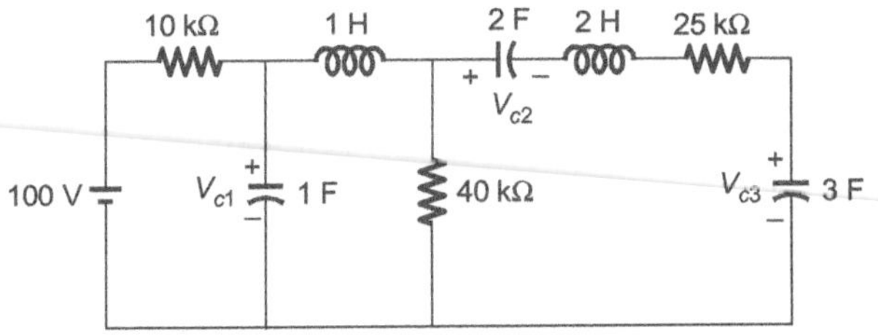

(a) 80 V, 48 V, 32 V (b) 20 V, 12 V, 8 V
(c) 80 V, 32 V, 48 V (d) 20 V, 8 V, 12 V

Q.189 The value of $Z(s)$ in the circuit shown below is given by

$$Z(s) = \frac{s^2 + 1.5s + 1}{s(s+1)}$$

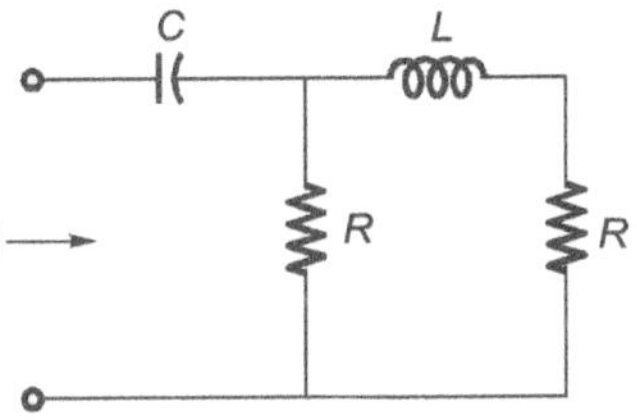

The values of R, L and C are respectively given by

(a) $1\Omega, \frac{1}{2}\text{H}, 2\text{F}$ (b) $\frac{1}{2}\Omega, 2\text{H}, \frac{1}{2}\text{F}$

(c) $1\,\Omega, 2\,\text{H}, 1\,\text{F}$ (d) $\frac{1}{2}\Omega, \frac{1}{2}\text{H}, 1\text{F}$

Q.190 If the value of current $i(t)$ for the circuit shown below is $i(t) = -20\,e^{-2t}$, then the voltage source $v(t)$ will be given by

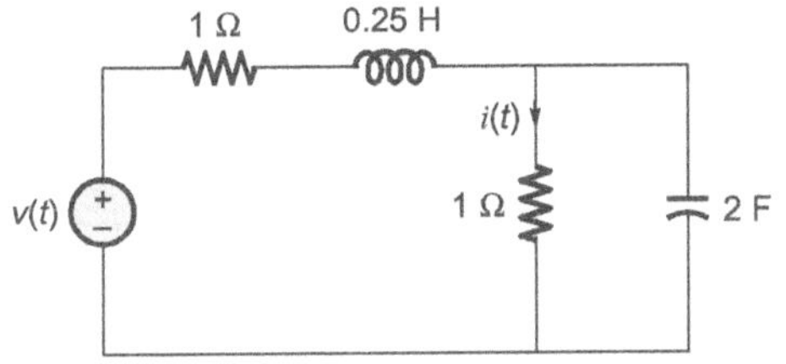

(a) $10\,e^{-2t}$ (b) $20\,e^{-2t}$
(c) $20\,e^{2t}$ (d) $-15\,e^{-2t}$

Q.191 The circuit shown in figure is critically damped.

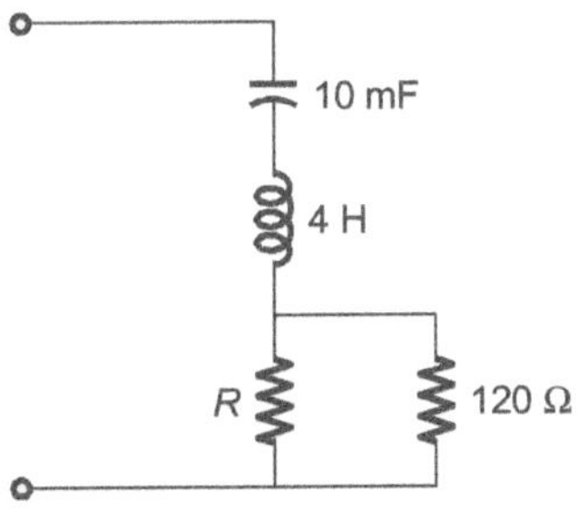

The value of R is
(a) 40 Ω (b) 180 Ω
(c) 120 Ω (d) 60 Ω

Q.192 The voltage across a passive element in an electric circuit is given by $\frac{d^2v}{dt^2} + v = t$. It is given that $V(0^+) = 1$; $\frac{dv}{dt}(0^+) = -2$.

The value of $v(t)$ is
(a) $(1 - \cos t - 3\sin t)$
(b) $(t + \cos t - 3\sin t)$
(c) $(\cos t + 3\sin t)$
(d) $(t - \cos t + 3\sin t)$

Q.193 A voltage is given by $V(s) = \frac{1}{s(s+a)}$.

If $V(\infty) = 2$ volts, then $V(1)$ is
(a) $(1 - e^{-1})$ volts (b) e^{-1} volts
(c) $2\left(1 - \frac{1}{e^{0.5}}\right)$ volts (d) $(1 - e^{-0.5})$ volts

Q.194 Pole of a network is frequency at which
(a) network function becomes zero
(b) network function becomes infinite
(c) network function becomes unity
(d) none of the above

Q.195 Transient current in an R-L-C circuit oscillatory when

(a) $R = 0$ (b) $R > 2\sqrt{\frac{L}{C}}$

(c) $R < 2\sqrt{\frac{L}{C}}$ (d) $R = 2\sqrt{\frac{L}{C}}$

Q.196 Which of the following conditions are necessary for the validity of initial value theorem $\underset{s\to\infty}{Lt}\, sF(s) = \underset{t\to 0}{Lt}\, f(t)$?
(a) $f(t)$ and its derivative $f'(t)$ must have Laplace transform.
(b) If the Laplace transform of $f(t)$ is $F(s)$, the $\underset{s\to 0}{Lt}\, sF(s)$ must exist.
(c) Only $f(t)$ must have Laplace transform.
(d) Both (a) and (b).

Q.197 The poles and zeros of the transfer function $\frac{V_c(s)}{V_i(s)}$ for the circuit shown below are located at

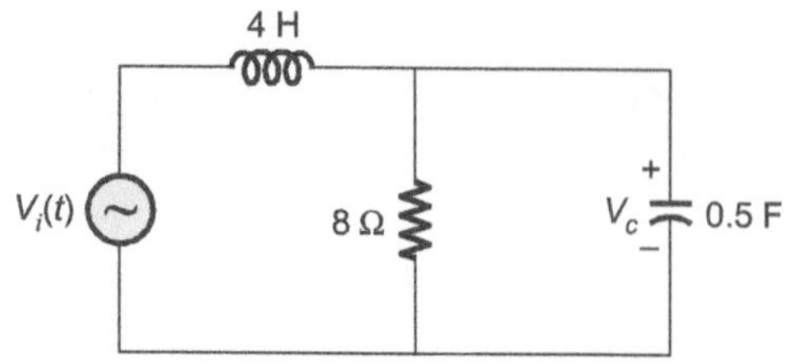

(a) Zeros at $s = -1,\ -j\sqrt{3}$

Poles at $s = \frac{-1}{8} \pm j\sqrt{\frac{31}{8}}$

(b) No zero

Poles at $s = \frac{-1}{8} \pm j\sqrt{\frac{31}{8}}$

(c) Zeros at $s = -1,\ \frac{-1}{2} \pm -j\sqrt{\frac{3}{4}}$

Poles at $s = -1 \pm j\sqrt{31}$

(d) Zeros at $s = \frac{-1}{8} \pm j\sqrt{\frac{31}{8}}$

No poles

Q.198 Assertion (A): Laplace transformation finds its application in solving the transient behaviour of the electric circuits.

Reason (R): The transient response of an electrical circuit can best be described by a linear differential equation.

(a) Both A and R are true and R is the correct explanation of A.
(b) Both A and R are true but R is not the correct explanation of A.
(c) A is true but R is false.
(d) A is false but R is true.

Q.199 The final value of $f(t) = e^{-t}(\sin 2t + \cos 5t)$ is

(a) ∞ (b) 1
(c) –1 (d) 0

Q.200 At $t = 0^+$ with zero initial condition, the voltage across 20 Ω resistor is

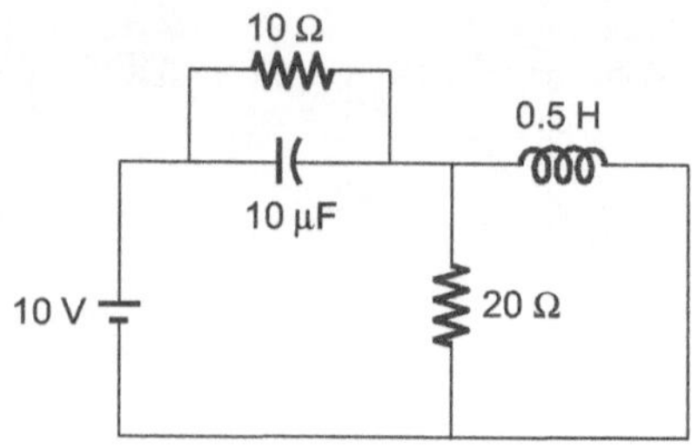

(a) 0 volt (b) 10 volt
(c) 2 volt (d) None of these

Q.201 Assuming zero-initial condition $v_c(t)$ in the given circuit will be given by

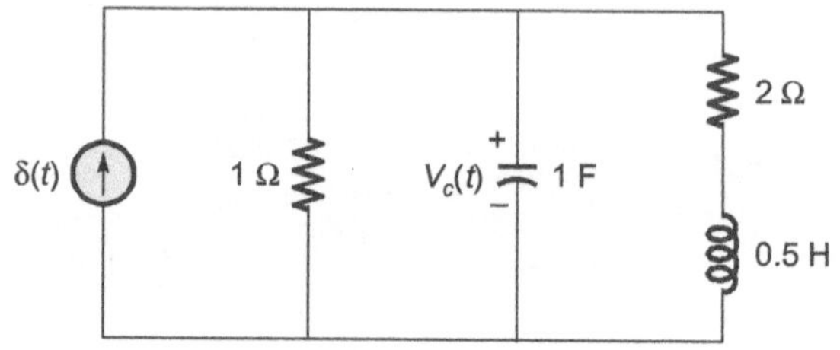

(a) $(e^{-3t} + 2e^{-2t})$ V (b) $(e^{-3t} - e^{-2t})$ V
(c) $(-e^{-3t} + 2e^{-2t})$ V (d) $(e^{-2t} - 2e^{-3t})$ V

Q.202 The initial and final value of current through the resistor R in a series RLC circuit with unit step input and zero initial condition are respectively

(a) 0 A, 0 A (b) 0 A, $\frac{1}{R}$ A
(c) $\frac{1}{R}$ A, 0 A (d) infinite in both cases

Q.203 The value of current through R at $t = \infty$ is

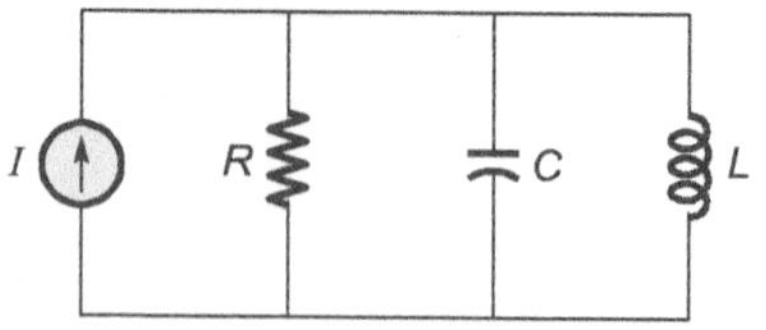

(a) I (b) ∞
(c) zero (d) None of these

Q.204 Consider the following statements for a series RLC circuit excited with a voltage $v(t)$.

1. For $\xi = 0$, there is no-loss in the circuit.
2. The damping ratio of the circuit is independent of resistance R of the circuit.
3. The response of the circuit is oscillator if ξ value is more than unity.

Which of the above statements is/are true?

(a) 2 and 3 (b) 1, 2 and 3
(c) 2 only (d) 1 only

8. Sinusoidal Steady-State Analysis

Q.205 The phasor diagram shown in figure below is for a two-element series circuit having

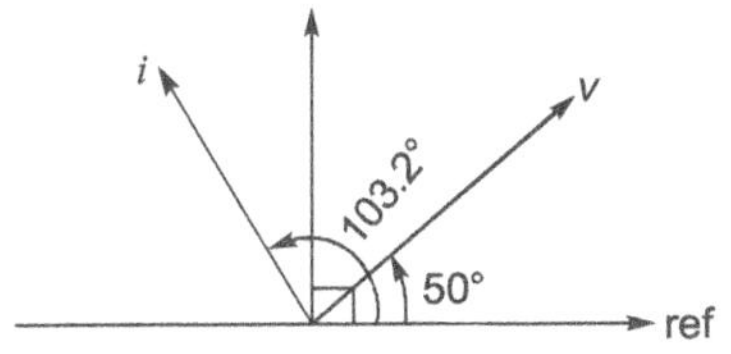

(a) R and C, with $\tan\theta = 4.2635$
(b) R and L, with $\tan\theta = 0.2345$
(c) R and C, with $\tan\theta = 1.3367$
(d) R and L, with $\tan\theta = 1.1918$

Q.206 For the circuit shown below, the voltage across the 3 Ω resistor V_3 is

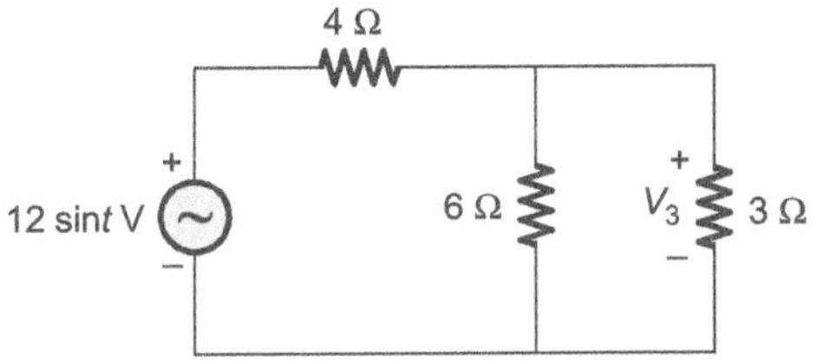

(a) 3 sin$t$ V (b) 4 sin$t$ V
(c) 6 sin$t$ V (d) 12 sin$t$ V

Q.207 A part of a circuit shown below consists of a resistor, a capacitor and an inductor. At steady state, $i_R(t) = 10 \sin t$ and $v_L(t) = 5 \cos t$. The rms value of the current through the capacitor is

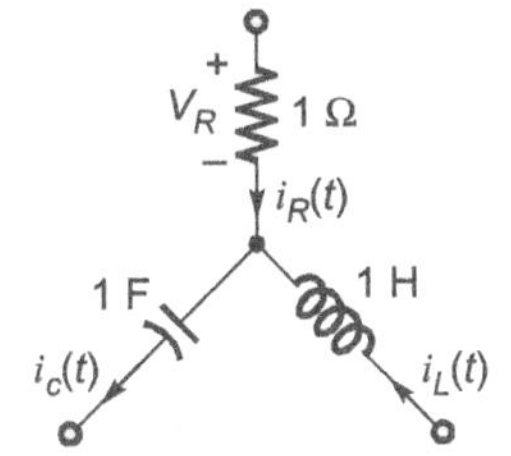

(a) $\sqrt{125}$ A (b) $15\sqrt{2}$ A
(c) $\frac{5}{\sqrt{2}}$ A (d) $\frac{15}{\sqrt{2}}$ A

Q.208 In the circuit shown below, currents I and I_1 are respectively

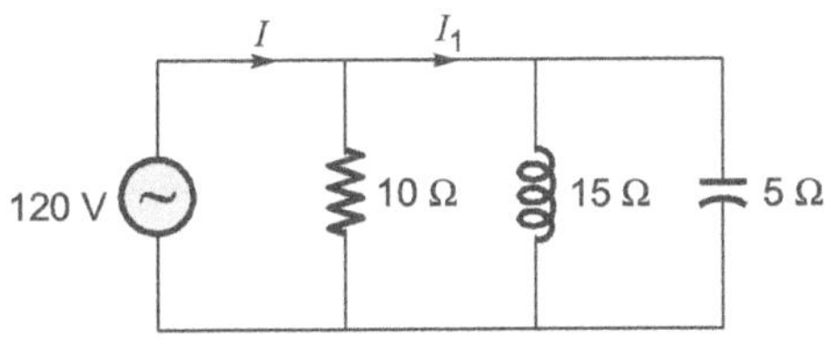

(a) 20 A, 16 A (b) 44 A, 32 A
(c) 34.2 A, 32 A (d) 28 A, 16 A

Q.209 What should be the value of C for the circuit shown below such that the input power factor is unity for any frequency f of the source?

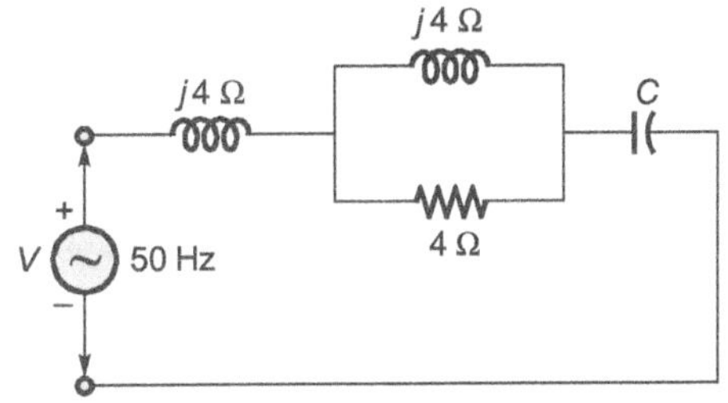

(a) 127 μF (b) 530 μF
(c) 33 μF (d) 268 μF

Q.210 In a series R-L circuit, the current and voltages are given by
$i = \cos(314t - 20°)$, $v = 10\cos(314t + 10°)$.
The values of R and L are respectively
(a) 8.66 Ω, 22.5 mH (b) 22.5 Ω, 8.66 mH
(c) 8.66 Ω, 15.9 mH (d) 22.5 Ω, 15.9 mH

Q.211 A series R-L circuit has resistance and reactance of 15 Ω and 10 Ω respectively. What should be the value of capacitor which when connected across the series combination in parallel, the system attains unity p.f.? (use f = 50 Hz)
(a) 98 μF (b) 65 μF
(c) 76 μF (d) None of these

Q.212 The input p.f. of the circuit shown below is

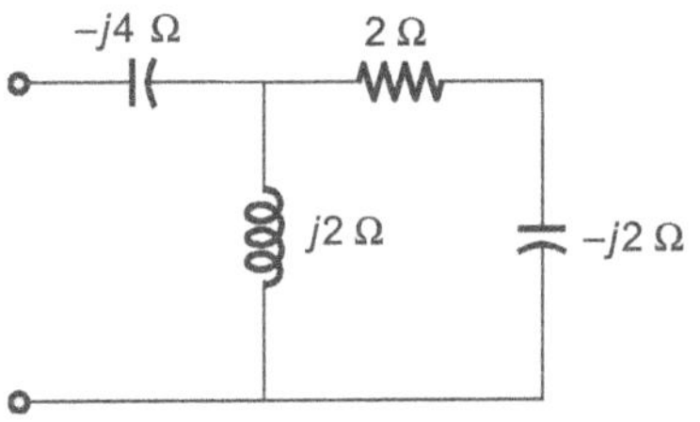

(a) 0.6 (lag) (b) 0.707 (lag)
(c) 0.6 (lead) (d) 0.707 (lead)

Q.213 The input admittance of the circuit shown below is

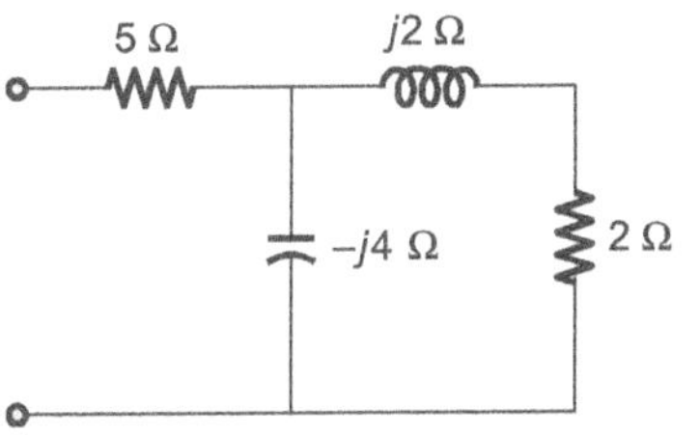

(a) 9 Ω (b) $\frac{1}{3}$ Ω

(c) $\frac{1}{9}$ Ω (d) 3 Ω

Q.214 The driving point impedance

$Z(s) = \frac{s^4 - s^3 + 2s^2}{s+5}$ does not represents a passive one port network because

(a) in the numerator, there is a missing term.

(b) the degree of numerator is 4 while that of denominator is 1.

(c) lowest degree of numerator is 2 while that in denominator is zero.

(d) all of the above.

Q.215 The driving point admittance of the network shown below is

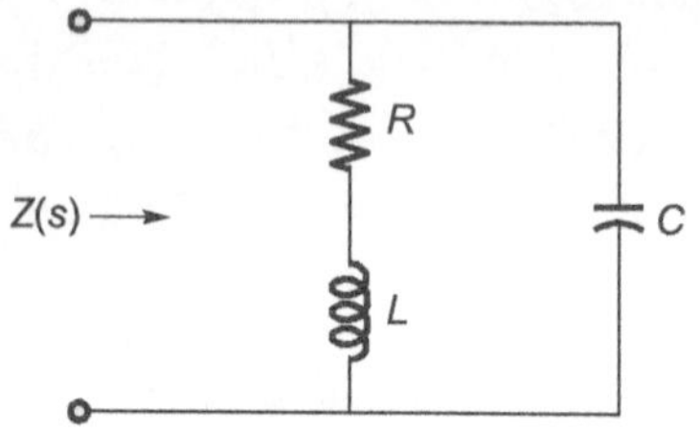

(a) $\frac{1+RCs+LCs^2}{R+Ls}$ (b) $\frac{R+Ls}{1+RCs+LCs^2}$

(c) $\frac{s/L}{s^2+\frac{R}{L}s+\frac{1}{LC}}$ (d) $\frac{s/C+R}{\frac{1}{LC}+\frac{1}{RC}s+s^2}$

Q.216 Figure shown below represents a d.c. 10 V voltage source as well as a 10 V a.c. 50 Hz voltage source while Z_1 and Z_2 are two unknown impedances. When Z_1 is connected across 10 V d.c. source, it draws a current of 10 A while it draws a current of 8 A if connected across 10 V a.c. source.

Now, when Z_2 is connected across 10 V d.c. source, it does not draw any current while it draws a current of 5 A if connected across 10 V a.c. source.

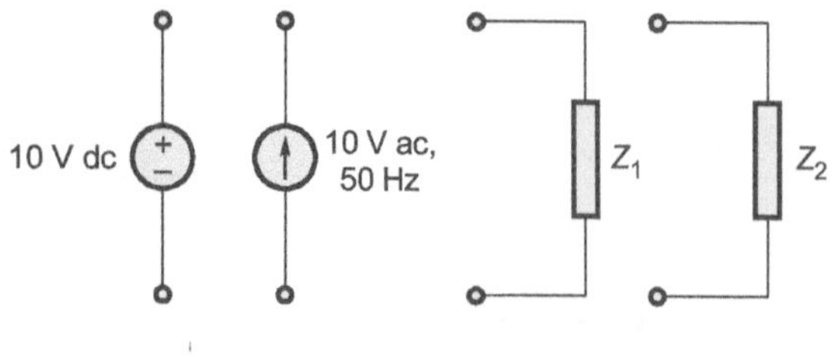

What are the elements in Z_1 and Z_2?

(a) $Z_1 = (1 - j0.25)$ Ω, $Z_2 = j2$ Ω

(b) $Z_1 = (1 + j0.75)$ Ω, $Z_2 = -j2$ Ω

(c) $Z_1 = (1 - j0.25)$ Ω, $Z_2 = -j2$ Ω

(d) $Z_1 = (1 + j0.75)$ Ω, $Z_2 = j2$ Ω

Q.217 The current through the capacitor in the given circuit is

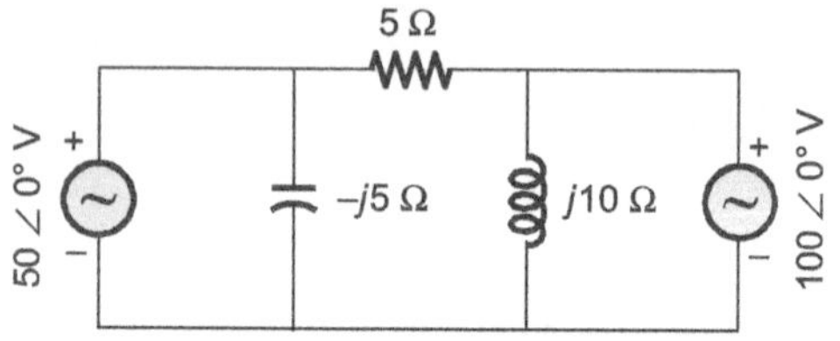

(a) $j30$ A (b) $j\angle{-90°}$ A

(c) $j20\angle{-90°}$ A (d) $j10$ A

Q.218 A parallel RLC circuit is said to be underdamped when

(a) $R < 2\sqrt{\frac{L}{C}}$ (b) $R > \frac{1}{2}\sqrt{\frac{L}{C}}$

(c) $R > 2\sqrt{\frac{L}{C}}$ (d) $R < \frac{1}{2}\sqrt{\frac{L}{C}}$

Q.219 **Assertion (A):** A coil, when connected across 230 V dc supply, will draw more current in comparison to that when connected across 230 V ac supply.

Reason (R): The inductance of the coil opposes the flow of alternating current (not that of direct current).

(a) Both A and R are true and R is the correct explanation of A.

(b) Both A and R are true but R is not the correct explanation of A.

(c) A is true but R is false.

(d) A is false but R is true.

Q.220 A two terminal black box contains one of the R-L-C elements. When the black box is connected to a 220 V ac supply, the current through the source is I. When a capacitance of 0.1 F is inserted in series between the source and the box, the current through the source is $2I$. The element is

(a) a resistance

(b) an inductance

(c) a capacitance

(d) it is not possible to determine the element

Q.221 A system function has a pole at $s = 0$ and zero at $s = -1$. The constant multiplier is unity. For an excitation cos$t$, the steady state response is given by

(a) $\sqrt{2}\cos\left(t+\frac{\pi}{4}\right)$ (b) $\sqrt{2}\cos\left(t-\frac{\pi}{4}\right)$

(c) cos$t$ (d) $\frac{1}{\sqrt{2}}\cos\left(t+\frac{\pi}{4}\right)$

Q.222 For the a.c. circuit given below, what is the value of I?

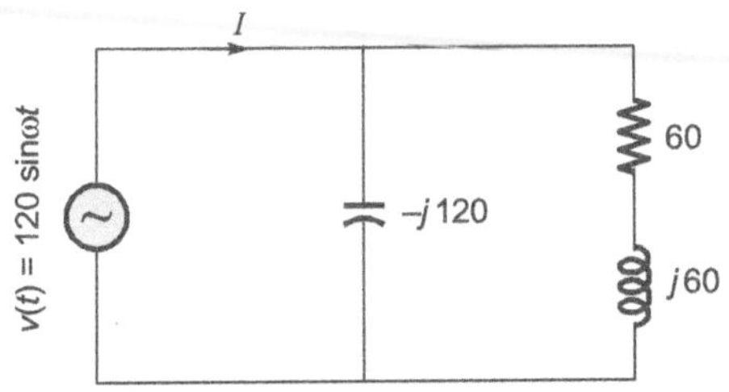

(a) $1 + j1$ (b) $1 + j0$
(c) $0 + j0$ (d) $0 - j1$

Q.223 Consider the following pole-zero diagram of a system, $Z(s) = \frac{V(s)}{I(s)}$:

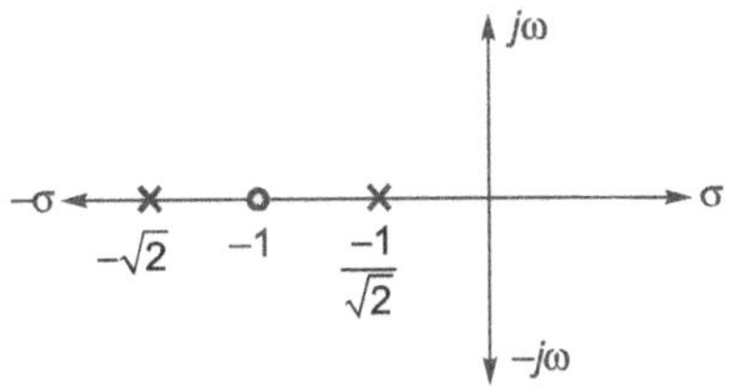

What will be the magnitude of the voltage phasor for $i(t)$ = sin$t$?

(a) $\frac{3}{2}$ (b) 1

(c) $\frac{1}{3}$ (d) $\frac{2}{3}$

Q.224 For the circuit shown below, $I_S = 5\cos t = 5\angle 0°$ A then, the phasor current through the inductor I_L is given by

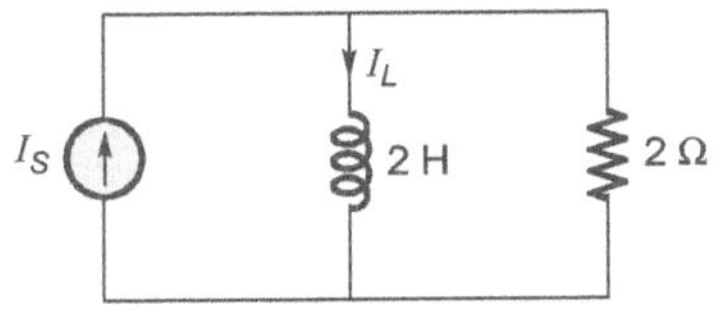

(a) $5\angle 45°$ A (b) $\frac{5}{\sqrt{2}}\angle -45°$ A

(c) $5\angle -45°$ A (d) $10\sqrt{2}\angle 45°$ A

Q.225 A coil takes a current of $1\angle 60°$ A (lag) from a 100 V, 50 Hz supply. The resistance of the coil is

(a) 100 Ω (b) 230 Ω
(c) 50 Ω (d) None of these

9. AC Power Analysis

Q.226 The value of $\left(\frac{I_{max}}{I_{rms}}\right)$ for the waveform shown below is

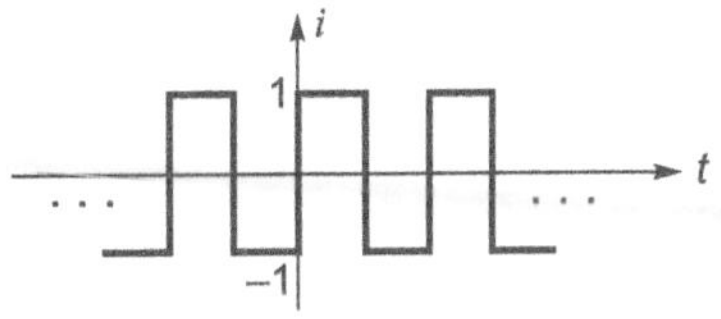

(a) $\sqrt{2}$ (b) 1.11
(c) 1 (d) 2.22

Q.227 For the idealized full wave rectifier system the average and rms value of the voltage is

(a) $\frac{V_m}{\pi}, \frac{V_m}{\sqrt{2}}$ respectively

(b) $\frac{V_m}{\pi}, \frac{V_m}{2}$ respectively

(c) $\frac{2}{\pi}V_m, \frac{V_m}{\sqrt{2}}$ respectively

(d) $\frac{2}{\pi}V_m, \frac{V_m}{2}$ respectively

Q.228 The instantaneous power delivered to the 5 Ω resistor at $t = 0$ in the given circuit is

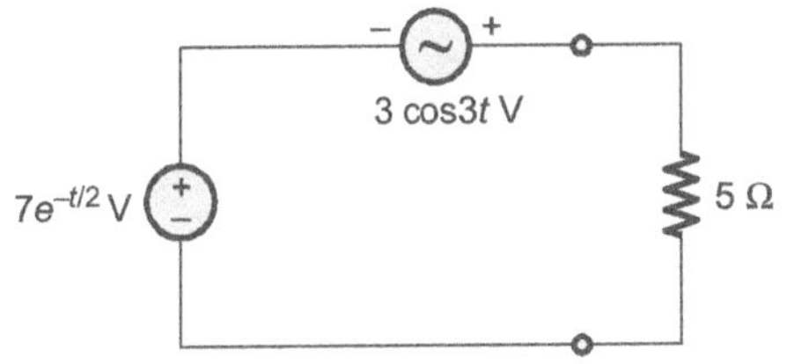

(a) 105 watts (b) 20 watts
(c) 15 watts (d) 35 watts

Q.229 For a sinusoidal waveform, the ratio of average to rms value is

(a) $\frac{\sqrt{2}\pi}{2}$ (b) $\frac{\pi}{2\sqrt{2}}$

(c) $\frac{2\sqrt{2}}{\pi}$ (d) $\frac{2}{\sqrt{2}\pi}$

Q.230 An impedance of $(3 + j4)\ \Omega$ is connected in parallel with a resistance of $10\ \Omega$. The ratio of power loss in these parallel circuits is
(a) 1 : 5 (b) 6 : 5
(c) 3 : 2 (d) 5 : 6

Q.231 A 0.2 HP induction motor runs at an efficiency of 85%. If the operating power factor is 0.8 lag, the reactive power taken by the motor is
(a) 109.28 VAR (b) 125 VAR
(c) 88 VAR (d) 131.65 VAR

Q.232 In an a.c. circuit, $v = 100 \sin(\omega t + 30°)$ V, $i = 5 \sin(\omega t - 30°)$ A. The apparent and reactive power in the circuit are respectively
(a) 250 VA, 217 VAR (b) 188 VA, 206 VAR
(c) 125 VA, 105 VAR (d) None of these

Q.233 The inductive reactance in series with Z in the circuit shown below has a value of $25\ \Omega$. If the voltage drop across Z is 179 volts, the power dissipated in the circuit is 320 W.

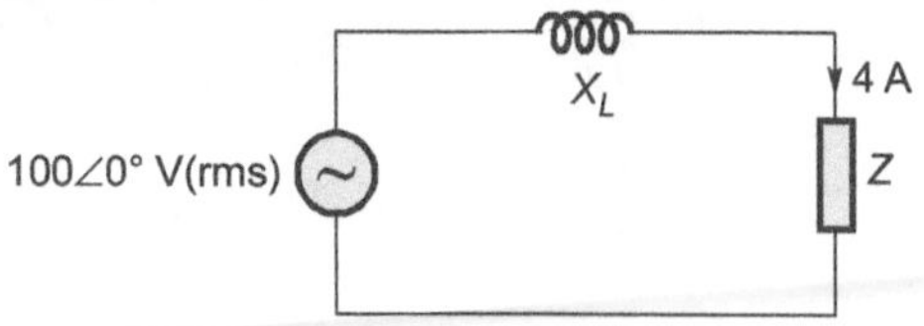

The reactive part of Z is
(a) $10\ \Omega$ (b) $25\ \Omega$
(c) $40\ \Omega$ (d) $56\ \Omega$

Q.234 The values of current I_1 and I_2 in the circuit shown below are given by:
$I_1 = (0.8 + j0.2)$ A and $I_2 = -(1 - j0.5)$ A

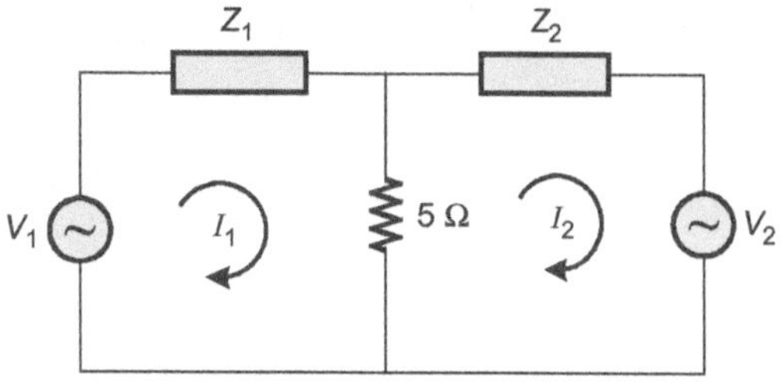

The total power lost in the $5\ \Omega$ resistor is
(a) 6.25 W (b) 1.25 W
(c) 2.25 W (d) 15.75 W

Q.235 The power factor of the circuit shown below is

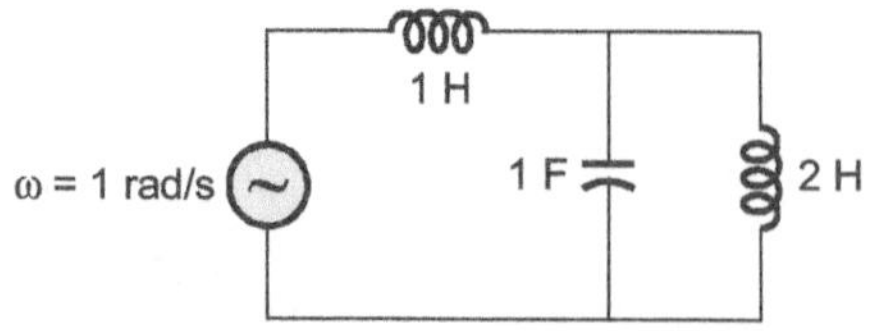

(a) unity (b) 0.6 lag
(c) zero (d) none of these

Q.236 A non-alternating periodic waveform is shown below. It's from factor is

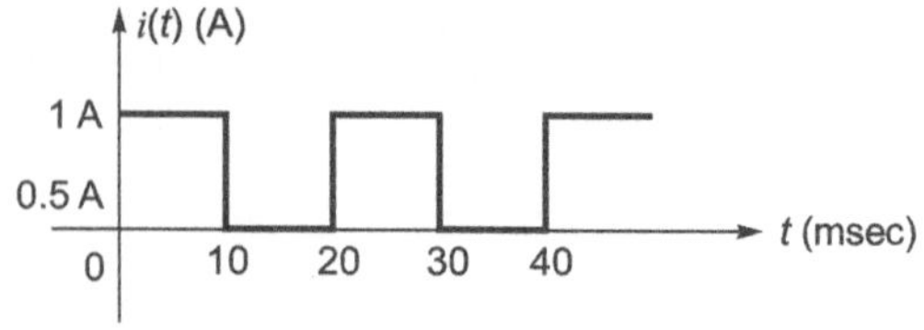

(a) $2\sqrt{2}$ (b) $\frac{1}{\sqrt{2}}$
(c) $\sqrt{2}$ (d) 2

Q.237 The voltage and current in an a.c. circuit are given by

$$v = 100 \sin 314t;\ i = 10 \sin 314t$$

The average power in the circuit is
(a) 225 W (b) 500 W
(c) 250 W (d) 600 W

Q.238 The expression for instantaneous power dissipated in a resistor R connected across a sinusoidal alternating voltage source represented as $v = V_m \sin \omega t$ is given by
(a) $\frac{V_m^2}{2R}$ (b) $\frac{V_m^2}{2R} \cos \omega t$
(c) $\frac{V_m^2}{2R}(1 - \sin 2\omega t)$ (d) $\frac{V_m^2}{2R}(1 - \cos 2\omega t)$

Q.239 The power delivered by the two voltage sources of 6 V and 24 V as shown below will be

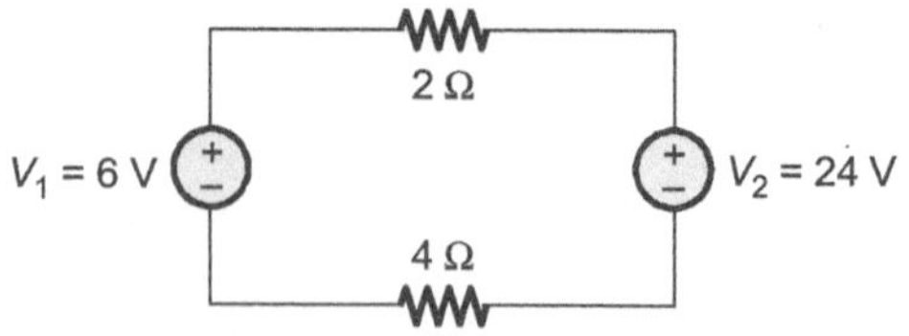

(a) 72 W and 0 W (b) 18 W and 72 W
(c) 0 W and 72 W (d) None of these

Q.240 The power loss in watts in the resistor R shown below is

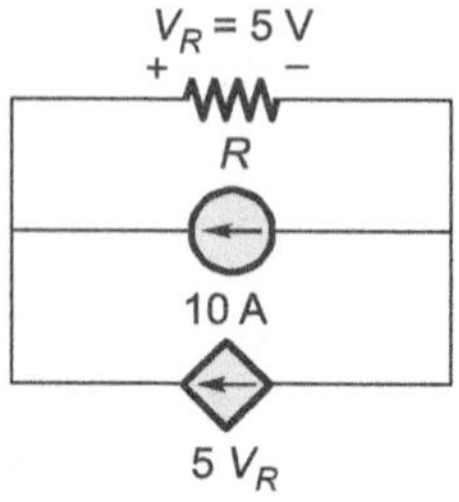

(a) 175 watts (b) 25 watts
(c) 225 watts (d) 62.5 watts

Q.241 The root mean square value of the current wave given by $i = (50 + 30 \sin t)$ A will be
(a) 56.50 A (b) 58.66 A
(c) 54.30 A (d) 52.20 A

Q.242 **Assertion (A):** In an alternating circuit, if the current is $(4 - j3)$ amperes and the impressed voltage is $(100 - j50)$ volts, the reactive power in the circuit is 100 VA.
Reason (R): True power is the real part of (VI).
(a) Both A and R are true and R is the correct explanation of A.
(b) Both A and R are true but R is not the correct explanation of A.
(c) A is true but R is false.
(d) A is false but R is true.

Q.243 In a purely resistive circuit, the average power P_{av} is _______ the peak power P_{max}.
(a) double (b) non-half of
(c) one-fourth of (d) equal to

Q.244 The power consumed in an inductive circuit will be
(a) $VI \cos \phi$ (b) $VI \sin \phi$
(c) VI (d) none of these

Q.245 A series circuit containing passive elements has the following current and applied voltage:

$$v = A\sin\left(\omega t + \frac{\pi}{4}\right), i = B\sin\left(\omega t - \frac{\pi}{6}\right)$$

The circuit elements
(a) may be resistance and inductance
(b) may be inductance, capacitance and resistance
(c) either (a) or (b)
(d) may be resistance and capacitance

10. Magnetically Coupled Circuits

Q.246 For an ideal transformer
(a) both z and y-parameter exist
(b) neither z nor y-parameter exist
(c) z-parameter exist, but not the y-parameters
(d) y-parameters exist, but not the z-parameters

Q.247 Two inductors of values L_1 and L_2 are coupled by a mutual inductance M. By interconnection of the two elements, one can have a minimum inductance of
(a) $L_1 + L_2 - M$ (b) $L_1 + L_2$
(c) $L_1 + L_2 - 2M$ (d) $L_1 + L_2 + M$

Q.248 When two coupled coils of equal self inductance are connected in series in one way, the net inductance is 12 mH and when they are connected in the other way, the net inductance is 4 mH. The maximum value of net inductance when they are connected in parallel in a suitable way is
(a) 6 mH (b) 4 mH
(c) 3 mH (d) 2 mH

Q.249 For the ideal transformer shown below, which of the following options is true?

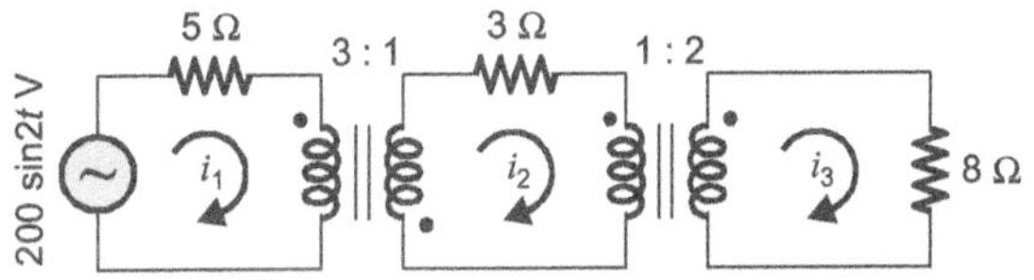

(a) $i_1 = 4 \sin 2t$ A, $i_2 = -12 \sin 2t$ A, $i_3 = 4 \sin 2t$ A
(b) $i_1 = -6 \sin 2t$ A, $i_2 = -4 \sin 2t$ A, $i_3 = 4 \sin 2t$ A
(c) $i_1 = 4 \sin 2t$ A, $i_2 = -12 \sin 2t$ A, $i_3 = -6 \sin 2t$ A
(d) None of the above

Q.250 What the greatest mutual inductance that an air-core transformer can have with the self-inductances of 0.3 and 0.7 H respectively?
(a) 0.500 H (b) 0.210 H
(c) 0.350 H (d) 0.458 H

Q.251 An air-core transformer has primary and secondary currents of $i_1 = 0.2$ A and $i_2 = 0.4$ A that produces a flux of $\phi_{m1} = 100\ \mu$Wb, $\phi_{11} = 250\ \mu$Wb and $\phi_{12} = 300\ \mu$Wb. The turns ratio is $N_1 : N_2 = 25 : 40$. What are the values of L_1 and L_2 in mH ?
(a) 20 mH, 62 mH (b) 43.75 mH, 20 mH
(c) 20 mH, 320 mH (d) 43.75 mH, 62 mH

Q.252 In an ideal transformer with an open-circuited secondary has inductances of $L_1 = 20$ mH, $L_2 = 32$ mH and $M = 13$ mH. The values of primary and secondary voltages when the primary current is increased at the rate of 0.4 kA/s will be respectively

(a) 8 V, –5.2 V (b) ±4 V, 5.2 V
(c) 8 V, ±5.2 V (d) ±4 V, ±5.2 V

Q.253 The input inductance of the circuit for $\omega = 1$ rad/s is

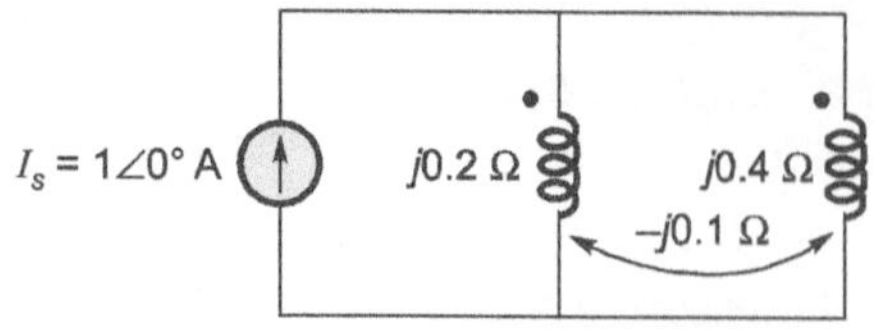

(a) 1.25 H (b) 0.625 H
(c) 0.175 H (d) 0.55 H

Q.254 In a certain magnetic circuit, a current of 1 A produces a flux of 1 Wb. If the linear dimensions of the magnetic circuit are doubled, then for producing the same flux, the current would be

(a) $\frac{1}{4}$ A (b) 1 A
(c) $\frac{1}{2}$ A (d) 2 A

Q.255 Mutual inductance between two magnetically coupled coils depends on the
(a) permeability of the core material
(b) number of turns of the coils
(c) cross-sectional area of their common core
(d) all of the above

Q.256 The overall inductance of two coils connected in series, with mutual inductance opposing self-inductance is L_1; with mutual inductance aiding self-inductance the overall inductance is L_2. The mutual inductance M is given by

(a) $(L_1 - L_2)$ (b) $\frac{1}{2}(L_1 - L_2)$
(c) $\frac{1}{4}(L_2 - L_1)$ (d) $\frac{1}{4}(L_1 - L_2)$

Q.257 In case all the flux from the current in coil-1 links with coil-2, the coefficient of coupling will be
(a) 2.0 (b) 1.0
(c) 0.5 (d) zero

Q.258 In series circuit shown below, for series resonant frequency of 1 rad/s, the value of coupling coefficient K will be

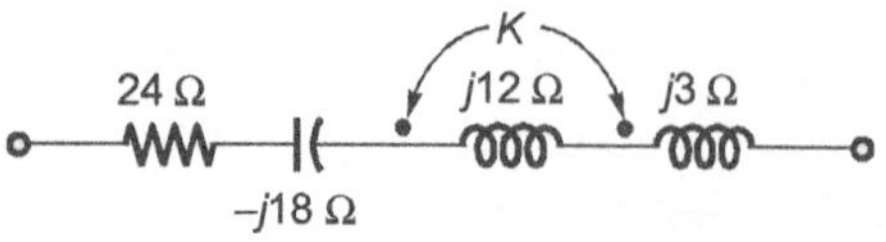

(a) 1.0 (b) 0.5
(c) 0.25 (d) zero

Q.259 The equivalent inductance for the following inductive circuit is

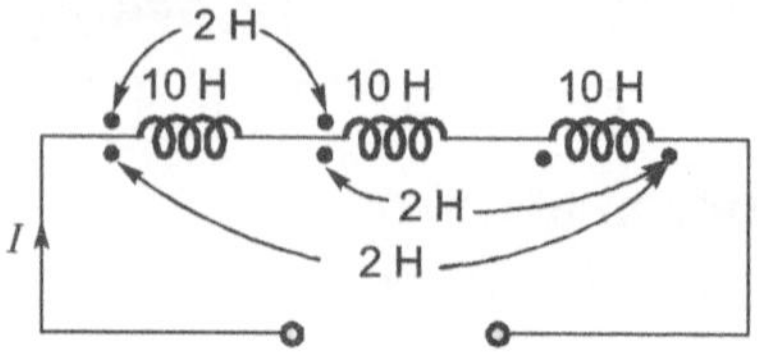

(a) 28 H (b) 30 H
(c) 32 H (d) 34 H

Q.260 The resonant frequency of the given series circuit is

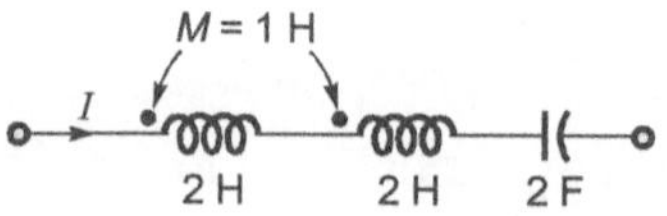

(a) $\frac{1}{2\pi\sqrt{3}}$ Hz (b) $\frac{1}{4\pi\sqrt{3}}$ Hz
(c) $\frac{1}{4\pi\sqrt{2}}$ Hz (d) $\frac{1}{2\pi\sqrt{2}}$ Hz

Q.261 A 3 H inductor has 1000 turns. What should be the number of turns to increase the inductance to 5 H?
(a) 2000 turns (b) 1625 turns
(c) 1290 turns (d) 2560 turns

Q.262 What is the equivalent inductance between the terminals A and B for the circuit shown below? (Given, M = 1 H)

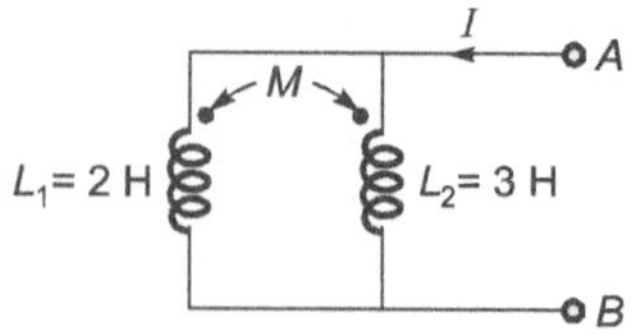

(a) $\frac{5}{3}$ H (b) $\frac{-5}{7}$ H
(c) 0 H (d) $\frac{4}{5}$ H

Q.263 Assertion (A): To determine the relative polarity of the induced voltage in the coupled coil, the coils are marked with dots.
Reason (R): One each coil, a dot is placed at the terminals which are instantaneously of the same polarity on the basis of mutual inductance alone.

(a) Both A and R are true and R is the correct explanation of A.
(b) Both A and R are true but R is not the correct explanation of A.
(c) A is true but R is false.
(d) A is false but R is true.

Q.264 Assertion (A): When a current changes in a circuit, the magnetic flux linking the same circuit changes and an emf is induced in the circuit.
Reason (R): Mutual inductances is the bilateral property of the linked circuit.
(a) Both A and R are true and R is the correct explanation of A.
(b) Both A and R are true but R is not the correct explanation of A.
(c) A is true but R is false.
(d) A is false but R is true.

Q.265 Match **List-I (Connection)** with **List-II (Equivalent inductance)** and select the correct answer using the codes given below the lists:

List-I | **List-II**

A. 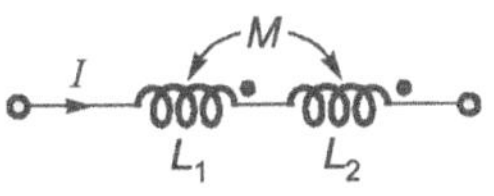

1. $\dfrac{L_1L_2 - M^2}{L_1 + L_2 + 2M}$

B. 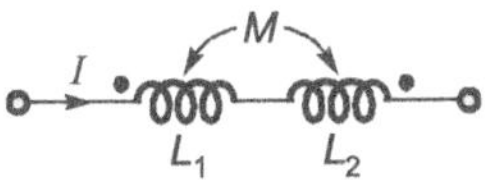

2. $\dfrac{L_1L_2 - M^2}{L_1 + L_2 - 2M}$

C. 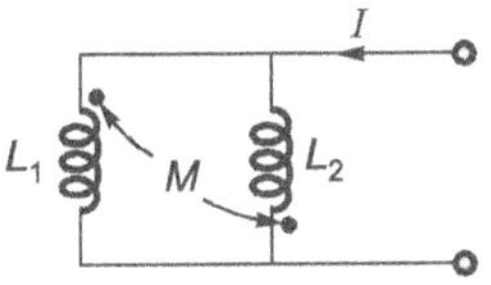

3. $L_1 + L_2 + 2M$

D. 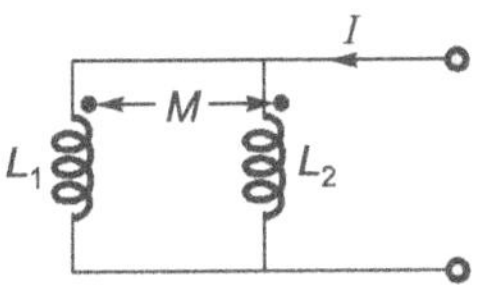

4. $L_1 + L_2 - 2M$

Codes:

	A	B	C	D
(a)	4	3	1	2
(b)	3	4	2	1
(c)	4	3	2	1
(d)	3	4	1	2

Q.266 In an ideal transformer, the mutual inductance = 10 H; number of primary and secondary turns are 50 and 200. The value of primary current required to produce a flux of 0.5 Wb to link with the secondary coil is
(a) 5 A (b) 10 A
(c) 20 A (d) 30 A

Q.267 Two impedances Z_1 and Z_2 are connected in series with the primary and secondary winding of an ideal transformer where the primary coil has $j2\ \Omega$ and the secondary coil has $j3\ \Omega$ reactances as shown in figure below. The mutual inductance of the transformer for $\omega = 100$ rad/sec will be

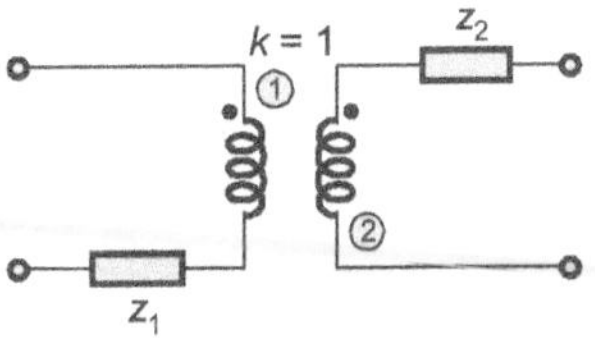

(a) 2.45 H (b) 245 mH
(c) 24.5 mH (d) 0.245 H

11. Frequency Response and Resonance

Q.268 In a double tuned circuit, consisting of two magnetically coupled, identical high Q tuned circuits, at the resonance frequency of either circuit, the amplitude response has
(a) a dip always
(b) a peak always
(c) neither a peak nor a dip
(d) either a peak or dip

Q.269 A network function contains only poles whose real-parts are zero or negative. The network is
(a) always stable.
(b) stable, if the $j\omega$-axis poles are simple.
(c) stable, if the $j\omega$-axis poles are at most of multiplicity 2.
(d) always unstable.

Q.270 The minimum amount of hardware required to make a high pass filter is
(a) a resistance, a capacitance and an inductance.
(b) a resistance, a capacitance and an op-amp.
(c) a resistance and a capacitance.
(d) a resistance, an inductance and an op-amp.

Q.271 The Q-factor of a series RLC circuit having $R = 10\ \Omega$, $L = 0.01$ H and $C = 100\ \mu$F is
(a) 10^3 (b) 1
(c) 10^6 (d) none of these

Q.272 The input frequency of the given circuit for unity power factor operation is

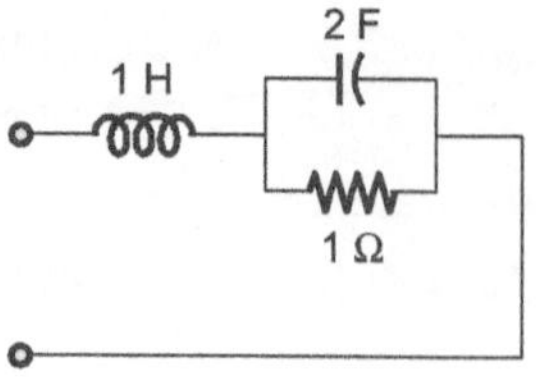

(a) 1 rad/sec (b) 0.5 rad/sec
(c) 1.5 rad/sec (d) 0.25 ard/sec

Q.273 When X_L is equal to X_C, then
(a) $Z = R$ (b) $Z = X_C$
(c) $Z = X_L$ (d) None of these

Q.274 If the quality factor of a series resonant circuit is increased then, its selectivity will
(a) decrease
(b) increase
(c) remains same
(d) may increase or decrease

Q.275 The steady state transfer function $H(j\omega)$ of the network shown below is

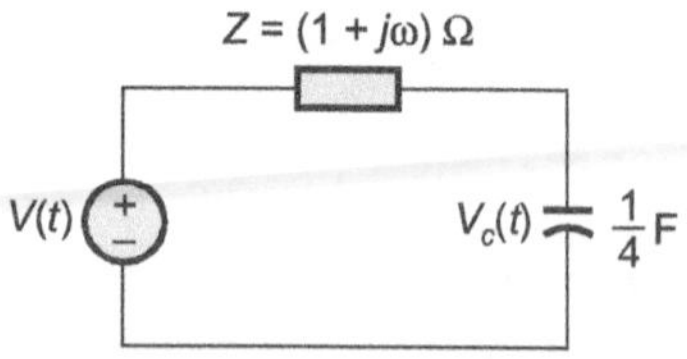

(a) $\dfrac{4}{(4+\omega^2-j\omega)}$ (b) $\dfrac{1}{(4+\omega-j\omega)}$

(c) $\dfrac{4}{(4-\omega^2+j\omega)}$ (d) $\dfrac{-1}{(4+\omega^2-j\omega)}$

Q.276 The transfer function of the circuit shown below is

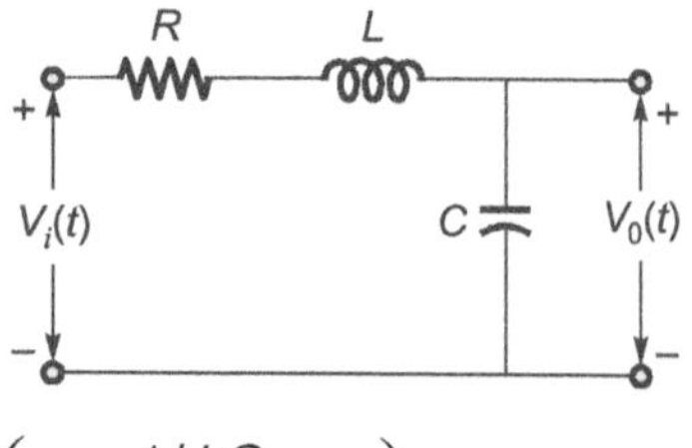

(a) $\left(\dfrac{1/LC}{s^2+\dfrac{1}{RC}s+\dfrac{1}{LC}}\right)$

(b) $\left(\dfrac{1/L}{s^2+\dfrac{1}{RC}s+\dfrac{1}{LC}}\right)$

(c) $\left(\dfrac{1/C}{s^2+\dfrac{R}{L}s+\dfrac{1}{LC}}\right)$

(d) $\left(\dfrac{1/LC}{s^2+\dfrac{R}{L}s+\dfrac{1}{LC}}\right)$

Q.277 The figure of merit for a capacitor in parallel with a resistance R is given by

(a) $\dfrac{1}{\omega RC}$ (b) $\dfrac{R}{\omega C}$

(c) ωRC (d) $\dfrac{\omega C}{R}$

Q.278 A series RLC circuit consist of resistance of 10 ohms, an inductance of 0.1 H and a capacitance of 0.001 μF. The frequency at resonance is
(a) 10^5 Hz (b) 15920 Hz
(c) 10^7 Hz (d) 920 Hz

Q.279 Which of the following is not true for a series RLC resonant circuit?
(a) The maximum voltage across capacitor occurs at f_c below $f_0 = \dfrac{1}{2\pi\sqrt{LC}}$.
(b) A circuit with a flat frequency response curve will be more selective.
(c) A circuit with a tall and narrow frequency response curve will be less selective.
(d) Both (b) and (c).

Q.280 The half-power frequencies of a series resonant circuit where the resonant frequency is 150×10^3 Hz and the bandwidth is 75 kHz will be respectively given by
(a) 117 kHz and 192 kHz
(b) 106 kHz and 181 kHz
(c) 125 kHz and 200 kHz
(d) 88 kHz and 163 kHz

Q.281 For the tank circuit shown below, the circulating current at resonance is given by

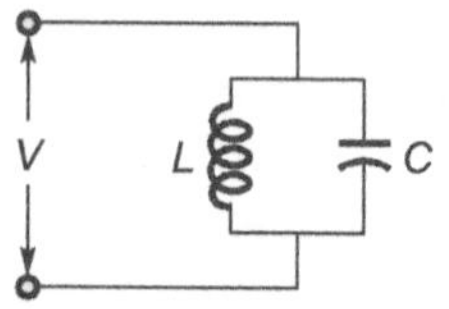

(a) 0 A (b) 1 A

(c) $V \cdot \sqrt{\frac{C}{L}}$ A (d) None of these

Q.282 The value of R_C in the circuit shown below to yield resonance will be

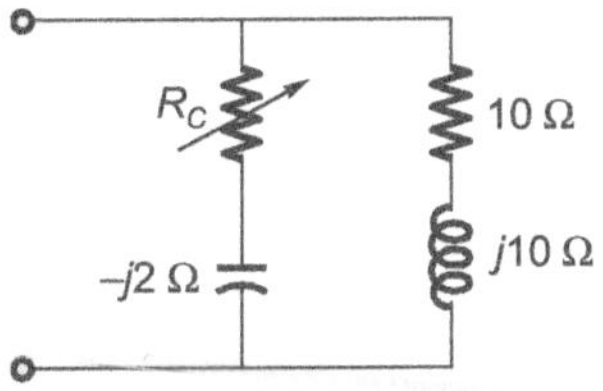

(a) 1 Ω (b) 3 Ω

(c) $\frac{6}{5}$ Ω (d) 6 Ω

Q.283 A resonating circuit has 10 Ω resistance. If the supply is 10 V, the power at half power frequency will be

(a) 2.5 W (b) 5 W
(c) 10 W (d) 20 W

Q.284 For the circuit shown below, what are the values of R_1 and R_2 so that the circuit will resonate at all frequencies?

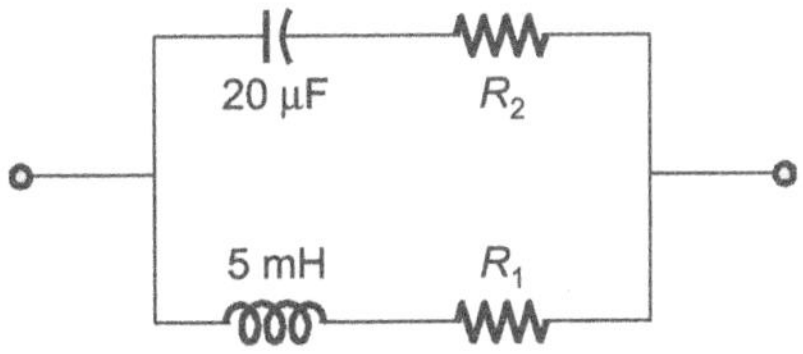

(a) $R_1 = R_2 = 22.5\ \Omega$
(b) $R_1 = 15.8\ \Omega$, $R_2 = 22.5\ \Omega$
(c) $R_1 = R_2 = 15.8\ \Omega$
(d) $R_1 = 22.5\ \Omega$, $R_2 = 15.8\ \Omega$

Q.285 The value of current I_1 in the circuit shown below is

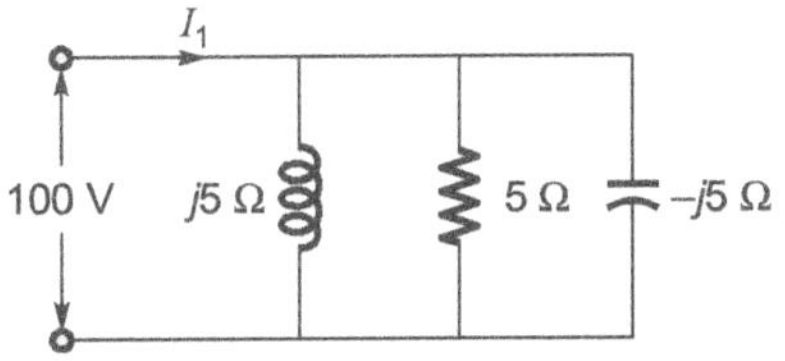

(a) 0 A (b) 60 A
(c) 20 A (d) 40 A

Q.286 The transfer function $\frac{V_0(s)}{V_i(s)}$ of the network shown below is

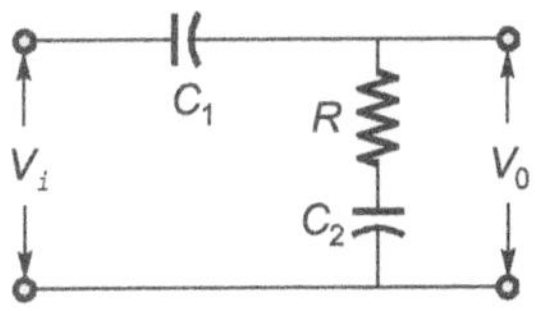

(a) $\frac{RC_1C_2s}{RC_2s + C_1C_2}$ (b) $\frac{C_1(1+RC_2s)}{C_1+C_2+RC_1C_2s}$

(c) $\frac{1+RC_2s}{R+C_1C_2s}$ (d) $\frac{C_2s}{C_1+C_2+RC_1C_2s}$

Q.287 The poles and zeros of the transfer function $\frac{I_2(s)}{I_1(s)}$ for the circuit shown below are located at

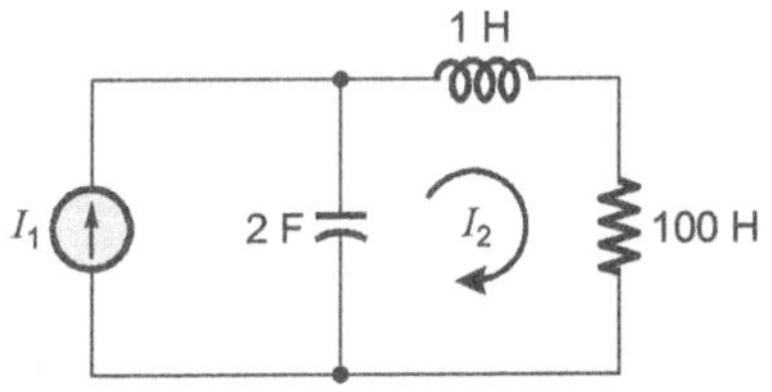

(a) No zeros
Poles: 0, –200
(b) Zeros: –1, –50
Poles: 0, –100
(c) No zeros
Poles: –1, –50
(d) Zeros: 0, –200
Poles: –1, –50

Q.288 The output of a linear system for a unit step input is given by $t^2 e^{-t}$. The transfer function is given by

(a) $\frac{2}{(s+1)^3}$ (b) $\frac{s}{(s+1)^3}$

(c) $\frac{2}{s(s+1)^2}$ (d) $\frac{2s}{(s+1)^3}$

Q.289 A constant voltage but variable frequency ac source feeds L and C in parallel as shown below:

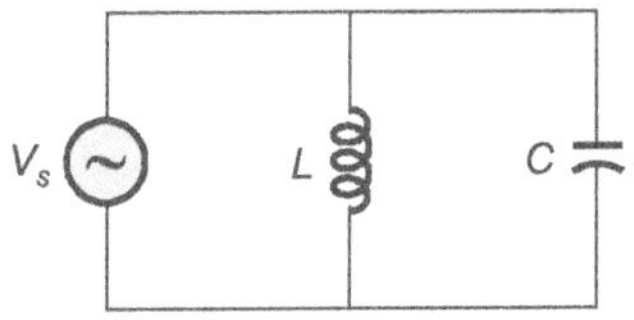

The impedance seen by source is Z

1. Z is zero when $f = 0$.
2. Z is zero when f = infinity.
3. Z is infinite when $f = 0$.
4. Z is infinite when f = infinity.
5. resonant frequency, $\omega_r = \frac{1}{\sqrt{LC}}$ rad/s.

6. $\omega_r = \sqrt{\frac{L}{C}}$ rad/s.

7. $\omega_r = \sqrt{\frac{C}{L}}$ rad/s.

From above, the correct answer is

(a) 3, 4, 7 (b) 1, 2, 6
(c) 3, 4, 5 (d) 1, 2, 5

Q.290 Match **List-I (Types of filters)** with **List-II (Attenuation band)** and select the correct answer using the codes given below the lists:

List-I	List-II
A. Low pass	1. $0 \to f_2, f_1 \to \infty$
B. High pass	2. $f_c \to \infty$
C. Band pass	3. $f_1 \to f_2$
D. Band stop	4. $0 \to f_c$

Codes:

	A	B	C	D
(a)	2	4	3	1
(b)	1	2	3	4
(c)	2	4	1	3
(d)	4	2	1	3

Q.291 The transfer function $\frac{V_0(s)}{V_i(s)} = \frac{5s}{s^2+5s+25}$ is for an active

(a) low pass filter (b) band pass filter
(c) high pass filter (d) all pass filter

Q.292 Match **List-I (Transfer function)** with **List-II (Type of filter)** and select the correct answer using the codes given below the lists:

List-I	List-II
A. $\frac{1}{RCs+1}$	1. APF
B. $\frac{1-RCs}{1+RCs}$	2. HPF
C. $\frac{RCs}{1+RCs}$	3. LPF
D. $\frac{s/C}{s^2+\frac{1}{RC}s+\frac{1}{LC}}$	4. BPF

Codes:

	A	B	C	D
(a)	1	2	3	4
(b)	3	4	2	1
(c)	3	1	2	4
(d)	4	3	1	2

Q.293 Assertion (A): It is necessary that the tuning coil should be of high resistance and low inductance.

Reason (R): High value of Q-factor means not only higher magnification but also a higher selectivity of the tuning coil.

(a) Both A and R are true and R is the correct explanation of A.
(b) Both A and R are true but R is not the correct explanation of A.
(c) A is true but R is false.
(d) A is false but R is true.

Q.294 Voltage transfer function of a simple RC differentiator has

(a) a finite zero and a pole at infinity
(b) a finite zero and a pole at the origin
(c) a zero at the origin and a finite pole
(d) a zero at infinite and a finite pole

12. Two Port Network

Q.295 The two port matrix of an 1 : *n* ideal transfer is $\begin{bmatrix} 1/n & 0 \\ 0 & n \end{bmatrix}$. It describes the transfer in terms of its

(a) *z*-parameters
(b) *h*-parameters
(c) transmission-parameters
(d) *y*-parameters

Q.296 Assertion (A): Simple resistors, inductors and capacitors are linear elements.

Reason (R): The resistances, inductances and capacitances do not change with a change in applied voltage or the circuit current.

(a) Both A and R are true and R is the correct explanation of A.
(b) Both A and R are true but R is not the correct explanation of A.
(c) A is true but R is false.
(d) A is false but R is true.

Q.297 A two port network is not reciprocal if

(a) $BC - AD = -1$
(b) $g_{12} + g_{21} = 0$
(c) $Z_{12} = Z_{21}$
(d) $h_{12} + h_{21} = 1$

Q.298 Assertion (A): Unilateral elements offer varying impedances with variations in flow of current while bilateral elements offer same impedance irrespective of flow of current.

Reason (R): If the magnitude of the current passing through an element is affected due to change in the polarity of the applied voltage, the element is called unilateral element while if the current magnitude remains the same even if the applied emf's polarity is changed, it is called a bilateral element.

(a) Both A and R are true and R is the correct explanation of A.
(b) Both A and R are true but R is not the correct explanation of A.
(c) A is true but R is false.
(d) A is false but R is true.

Q.299 The parameters Z_{11} and Z_{22} of the given *T*-network shown below are

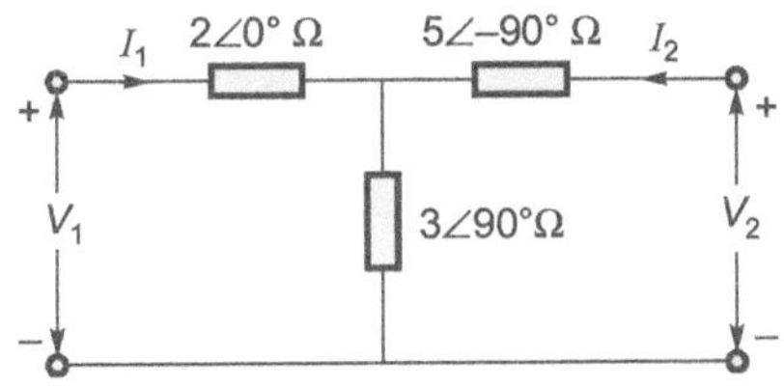

(a) $j3\ \Omega$ and $-j2\ \Omega$
(b) $(2 + j3)\ \Omega$ and $-j2\ \Omega$
(c) $j3\ \Omega$ and $(2 + j3)\ \Omega$
(d) $(2 + j3)\ \Omega$ and $j3\ \Omega$

Q.300 Match **List-I** with **List-II** and select the correct answer using the codes given below the lists:

List-I	List-II
A. h_{11}	1. $1/Z_{22}$
B. h_{12}	2. $-Z_{21}/Z_{22}$
C. h_{22}	3. $\Delta Z/Z_{22}$
D. h_{21}	4. Z_{12}/Z_{22}

Codes:

	A	B	C	D
(a)	3	1	4	2
(b)	1	2	3	4
(c)	3	4	1	2
(d)	1	4	2	3

Q.301 In a two port network, Y_{11} is given by

(a) $Z_{11}/\Delta Y$ (b) $Z_{22}/\Delta Z$
(c) $Z_{22}/\Delta Y$ (d) $Z_{22}/\Delta Z$

Q.302 The value of $Z_{21}(s)$ in the circuit shown below is

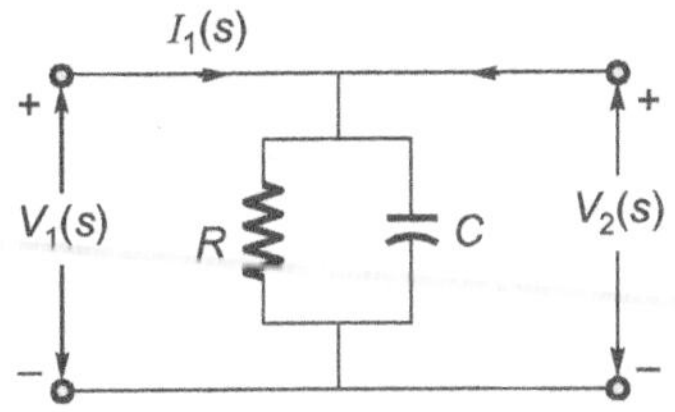

(a) $\dfrac{s}{\left[s+\dfrac{1}{RC}\right]}$ (b) $\dfrac{1}{C\left[s+\dfrac{1}{RC}\right]}$

(c) $\dfrac{1/c}{[s+RC]}$ (d) $\dfrac{RCs}{[1+1/RC]}$

Q.303 A two port network has zero initial conditions. At $t = 0$; an impulse voltage $\delta(t)$ is applied at the input producing a voltage $(e^{-2t} + e^{-3t})\ u(t)$ volts at the output. If the output voltage required is $te^{-2t}u(t)$ then, the input voltage will be

(a) $(e^{-2t} - 0.5e^{-2.5t})\ u(t)$ volts
(b) $(e^{-2.5t} - 0.5e^{-2t})\ u(t)$ volts
(c) $(e^{-2t} + e^{-2.5t})\ u(t)$ volts
(d) $(e^{-2.5t} - e^{-2t})\ u(t)$ volts

Q.304 The *Z*-parameter of the two port network shown below is given by

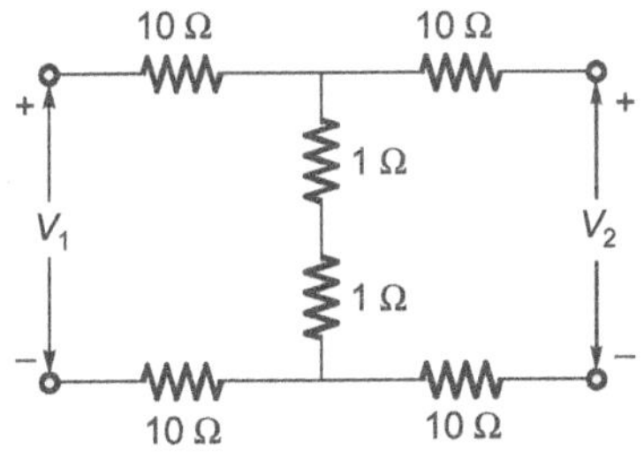

(a) $\begin{bmatrix} 12 & 10 \\ 10 & 12 \end{bmatrix} \Omega$ (b) $\begin{bmatrix} 12 & 2 \\ 2 & 12 \end{bmatrix} \Omega$

(c) $\begin{bmatrix} 22 & 2 \\ 2 & 22 \end{bmatrix} \Omega$ (d) $\begin{bmatrix} 22 & 1 \\ 1 & 22 \end{bmatrix} \Omega$

Q.305 When two networks are connected in series at the output and in parallel at the input, then

(a) *T*-parameters of the respective networks are added.
(b) *h*-parameters of the respective networks are added.
(c) *g*-parameters of the respective networks are added.
(d) *z*-parameters of the respective networks are added.

Q.306 The z-parameters of a circuit are given by $\begin{bmatrix} 4 & 1 \\ 3 & 3 \end{bmatrix}$. Which of the following is true regarding the T-parameters of the given circuit?

(a) $A = \frac{4}{3}, B = 3, C = \frac{1}{3}, D = 1$

(b) $A = \frac{-4}{3}, B = 1, C = \frac{1}{3}, D = 3$

(c) $A = \frac{4}{3}, B = 3, C = 1, D = \frac{1}{3}$

(d) $A = \frac{-4}{3}, B = -3, C = \frac{-1}{3}, D = -1$

Q.307 For the two port network shown below, match **List-I (Z-parameters)** with **List-II (Values in Ω)** and select the correct answer using the codes given below the lists:

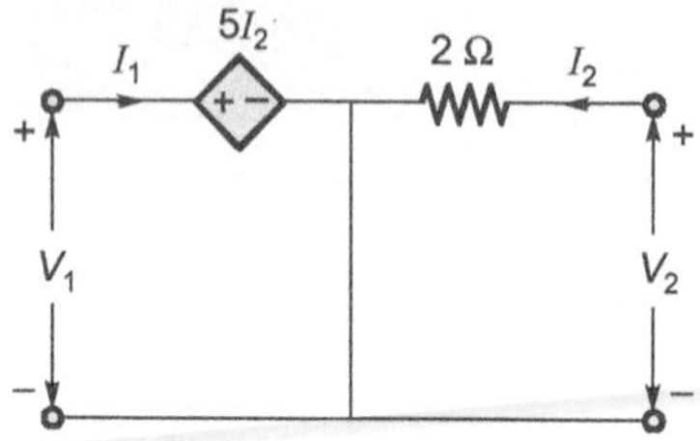

List-I	List-II
A. Z_{11}	1. 2
B. Z_{12}	2. 0
C. Z_{22}	3. 5
D. Z_{21}	4. 1

Codes:

	A	B	C	D
(a)	2	3	2	4
(b)	3	1	3	4
(c)	1	2	3	4
(d)	2	3	1	2

Q.308 For a π-network with admittances Y_A and Y_B as shunt arms and Y_C as series arm, the values of Y_{11}, Y_{21} and Y_{22} will be respectively

(a) $(Y_A + Y_C), Y_B, Y_C$

(b) $Y_A, -Y_C, Y_B$

(c) $(Y_A + Y_C), -Y_C, (Y_B + Y_C)$

(d) $Y_A, (Y_B + Y_C), -Y_B$

Q.309 Match the **List-I (Network parameters)** with **List-II (Dimensions)** and select the correct answer using the codes given below the lists:

List-I	List-II
A. g_{11}	1. Impedance
B. g_{12}	2. Admittance
C. g_{22}	3. Current ratio
D. g_{21}	4. Voltage ratio

Codes:

	A	B	C	D
(a)	2	4	1	3
(b)	2	3	1	4
(c)	1	3	2	4
(d)	1	4	3	2

Q.310 The Y-parameter of the network shown below in matrix form is

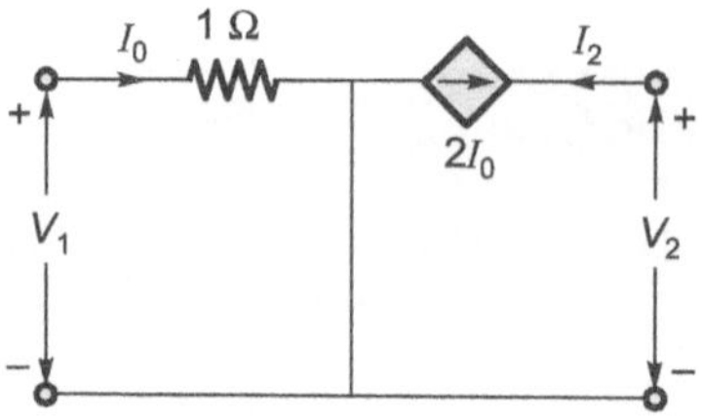

(a) $\begin{bmatrix} 1 & 0 \\ -1 & 0 \end{bmatrix}$ (b) $\begin{bmatrix} 0 & -1 \\ 2 & 0 \end{bmatrix}$

(c) $\begin{bmatrix} 1 & 0 \\ -2 & 0 \end{bmatrix}$ (d) $\begin{bmatrix} -1 & -2 \\ 0 & 0 \end{bmatrix}$

Q.311 The h-parameters (h_{11}, h_{12}) of the two port network shown in given figure are respectively

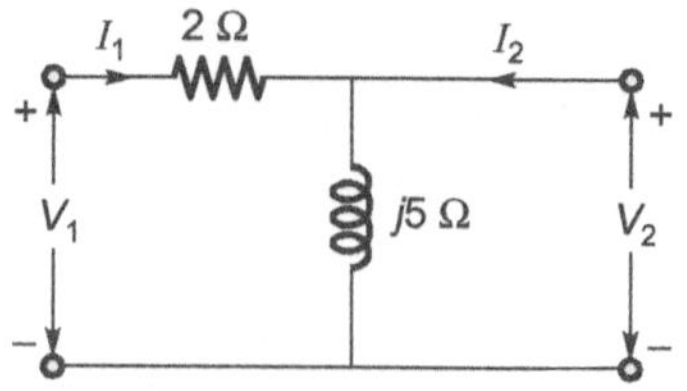

(a) 2, 1 (b) 0, 2
(c) 2, 0 (d) 2, 5

Q.312 A two port network is said to be symmetrical, if

(a) $A = C$ (b) $Z_{11} = Z_{21}$
(c) $h_{12} = h_{21}$ (d) $g_{11}g_{22} - 1 = g_{12}g_{21}$

Q.313 What is the voltage gain of the ideal transformer shown below?

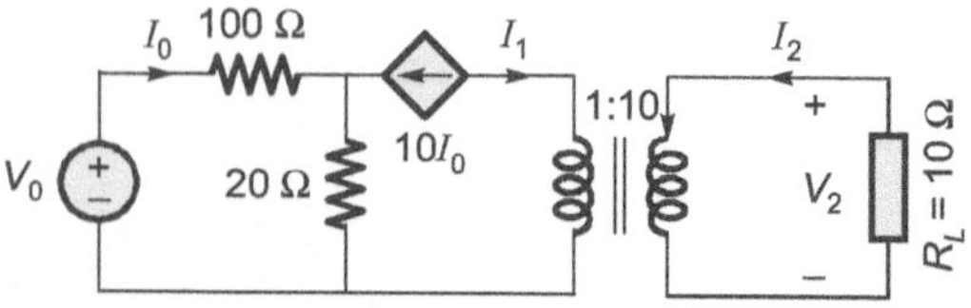

(a) –1.25 (b) –3.125
(c) 1.25 (d) 3.125

13. Network Synthesis

Q.314 Which of the following is a positive real function?

(a) $\frac{(s+1)(s+2)}{(s^2+1)^2}$ (b) $\frac{s^4+s^2+1}{(s+1)(s+2)(s+3)}$

(c) $\frac{(s-1)(s+2)}{(s^2+1)}$ (d) $\frac{(s-1)}{(s^2-1)}$

Q.315 A network function can be completely specified by:
(a) Real parts of poles
(b) Poles and zeros
(c) Real parts of zeros
(d) Poles, zeros and a scalar factor

Q.316 Which of the following property does not relates to LC impedance or admittance functions?
(a) There must be either a pole or a zero at infinity or origin.
(b) This is the ratio of even to odd or odd to even polynomial.
(c) Poles and zeros lie on negative real axis and they alternate.
(d) The highest or lowest powers of numerator or denominator polynomials differ by unity.

Q.317 If the impulse response is realisable by delaying it appropriately and is bounded for bounded excitation, then the system is said to be
(a) causal but unstable
(b) non-causal but stable
(c) causal and stable
(d) non-causal and unstable

Q.318 A network-A has zeros only in the left-half of the *s*-plane while the network-B has zeros only in the right-half of the *s*-plane. The network-A and Network-B will be respectively called
(a) a stable function and an unstable function.
(b) a non-minimum phase function and an unstable function.
(c) an all pass function and a non-minimum phase function.
(d) a minimum phase function and a non-minimum phase function.

Q.319 An L-C admittance or impedance function
(a) has no pole or zero at the origin or infinity.
(b) is an odd rational function.
(c) has simple poles and zeros in the left half of the *s*-plane.
(d) has all zeros on the negative real axis of the *s*-plane.

Q.320 If $F_1(s)$ and $F_2(s)$ are positive real functions, then which of the following are positive real function?

(a) $\frac{1}{F_1(s)}$ and $\frac{1}{F_2(s)}$ (b) $\frac{F_1(s)\cdot F_2(s)}{F_1(s)+F_2(s)}$

(c) Both (a) and (b) (d) Neither (a) nor (b)

Q.321 For the pole-zero plot shown below, the network function is

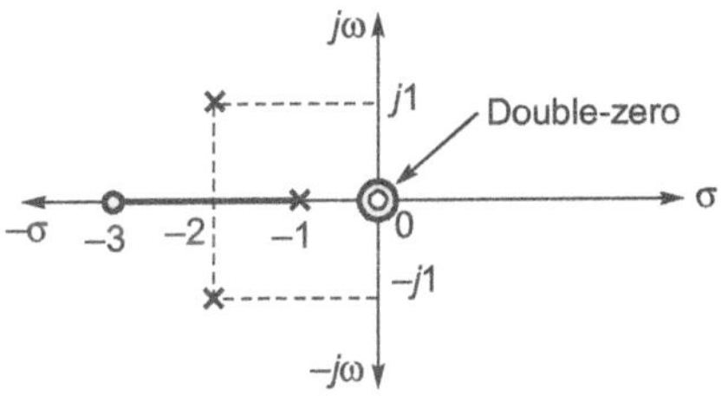

(a) $\frac{(s+1)(s^2+4s+5)}{s^2(s+3)}$

(b) $\frac{s^2(s+3)}{(s+1)(s^2+4s+5)}$

(c) $\frac{s^2(s+2+j)(s+2-j)}{(s+1)(s+3)}$

(d) $\frac{s^2(s+1)}{(s+3)(s+2+j)(s+2-j)}$

Q.322 A stable system must have
(a) zero or negative real part for poles and zeros.
(b) atleast one pole or zero lying in the right-half *s*-plane.
(c) positive real part for any pole or zero.
(d) negative real part for all poles and zero.

Q.323 Which of the following is not a positive real function?
(a) $F(s) = Ls$ (L = Inductance)
(b) $F(s) = K/s$ (K = Constant)
(c) $F(s) = Rs$ (R = Resistance)
(d) $F(s) = \left(\frac{s+1}{s^2+2}\right)$

Q.324 Which of the following is not a Hurwitz polynomial?

(a) $(s + 1)(s^2 + 2s + 3)$
(b) $(s + 3)(s^2 + s - 2)$
(c) $(s^3 + 3s)(1 + 2/s)$
(d) $(s + 1)(s + 2)(s + 3)$

Q.325 A Hurwitz polynomial has

(a) only zeros in the left half of *s*-plane
(b) only poles in the left half of *s*-plane
(c) zero anywhere in *s*-plane
(d) poles anywhere in *s*-plane

Q.326 The function $Z(s) = \dfrac{s(s+3)}{(s+2)(s+4)}$ represent a

(a) R-L function (b) L-C function
(c) R-C function (d) R-L-C network

Q.327 The circuit shown in the figure is a

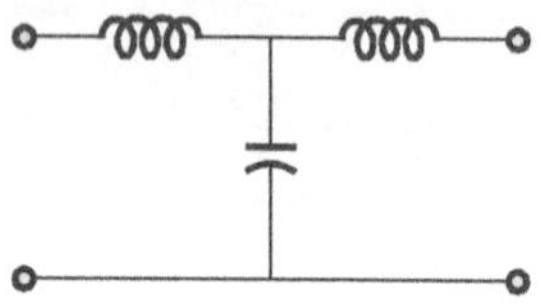

(a) low-pass filter (b) high-pass filter
(c) band-pass filter (d) integrator

Q.328 The driving point admittance of the network shown below is

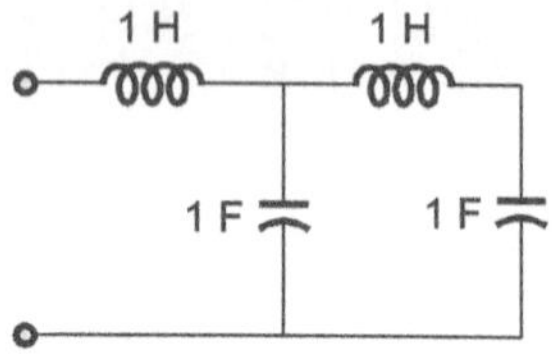

(a) $\left[\dfrac{s^4+3s^2+1}{s(s^2+2)}\right]$ (b) $\left[s+\dfrac{s^2+1}{s(s^2+2)}\right]$

(c) $\left[\dfrac{s^3+2s^2}{s^4+3s^2+1}\right]$ (d) $\left[\dfrac{s^3+2s^2+3}{s^2+2}\right]$

Q.329 Which of the following is a PRF (positive real function)?

(a) $\dfrac{8-s}{s+1}$ (b) $\dfrac{s^2-1}{-s+8}$

(c) $\dfrac{4s+1}{s+2}$ (d) Both (a) and (b)

Q.330 A function $F(s)$ is defined as $F(s)\ \dfrac{s+a}{s^2+bs+c}$.

The necessary condition required for $F(s)$ to be a positive real function is

(a) $a < b$; $a, b, c > 0$
(b) $|a| \geq |b|$; $a, b, c \leq 0$
(c) $|a| > |b|$; $a, b, c < 0$
(d) $a \leq b$; $a, b, c \geq 0$

Q.331 What is the range of values of *m* in $P(s)$, so that $P(s)$ is a Hurwitz polynomial?

$$P(s) = (2s^4 + s^3 + ms^2 + s + 2)$$

(a) $m > 4$ (b) $m > 2$
(c) $m > 1.5$ (d) $m > 0$

Q.332 Consider the impedance functions:

1. $F_1(s) = \dfrac{10(s^2+4)(s^2+6)}{(s^2+1)(s^2+9)}$

2. $F_2(s) = \dfrac{5s(s^2+4)}{(s^2+1)(s^2+3)}$

Out of the above impedance function which are driving point LC immittances of LC network?

(a) 1, 2 (b) 2 only
(c) 1 only (d) None of these

Q.333 Driving point impedance $Z(s) = \dfrac{s(s^2+1)}{s^2+4}$ is not realizable because the

(a) number of zeros is more than the number of poles.
(b) poles and zeros lie on the imaginary axis.
(c) poles and zeros do not alternate on imaginary axis.
(d) poles and zeros are not located on the real axis.

Q.334 Cauer and Foster forms of realizations are used only for

(a) driving point reactance functions
(b) transfer reactance functions
(c) driving point impedance functions
(d) transfer impedance functions

Q.335 The driving point impedance function

$$Z(s) = \frac{s^5+5s^3+3s}{s^4+3s^2+1}$$

can be realized as

(a) R-L-C network (b) R-L network
(c) L-C network (d) R-C network

Q.336 Match **List-I (Network functions)** with **List-II (Type of functions)** and select the correct answer using the codes given below the lists:

List-I

A. $\dfrac{(s+1)(s+4)(s+8)}{s(s+2)(s+6)}$

B. $\dfrac{5(s^2+1)(s^2+9)}{s(s^2+4)}$

C. $\dfrac{10(s+1)(s+3)}{(s+2)(s+6)}$

List-II

1. LC immittance function
2. RC admittance function
3. RL admittance function
4. RLC network function

Codes:

	A	B	C
(a)	3	1	2
(b)	1	2	4
(c)	3	4	2
(d)	3	1	2

Q.337 Assertion (A): Foster's type-I and Foster's type-II networks are equivalent networks.

Reason (R): Foster's type-I and Foster's type-II networks are dual of each other.

(a) Both A and R are true and R is the correct explanation of A.
(b) Both A and R are true but R is not the correct explanation of A.
(c) A is true but R is false.
(d) A is false but R is true.

Q.338 Assertion (A): The realization of one port LC networks can be done in two configurations (commonly known as Cauer-I and Cauer-II forms) using the continued fraction expansion of the driving point impedance.

Reason (R): The basic form of the Cauer realization being a ladder type network.

(a) Both A and R are true and R is the correct explanation of A.
(b) Both A and R are true but R is not the correct explanation of A.
(c) A is true but R is false.
(d) A is false but R is true.

14. Network Topology

Q.339 A possible tree of the topology equivalent of the network shown below is

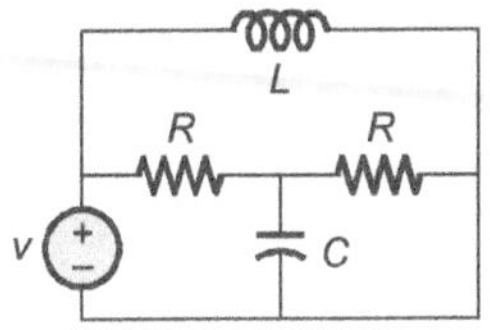

(a) 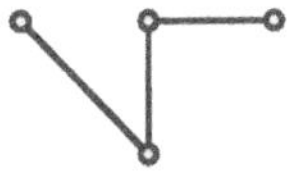 (b) 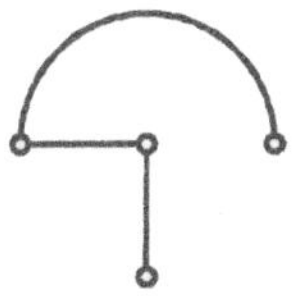

(c) neither (a) nor (b) (d) either (a) or (b)

Q.340 In the incidence matrix of a graph

(a) algebraic sum of the column entries is zero
(b) the determinant value is zero provided the loop is a closed one
(c) both (a) and (b)
(d) none of the above

Q.341 The number of twigs and links in the graph obtained for the given network will be

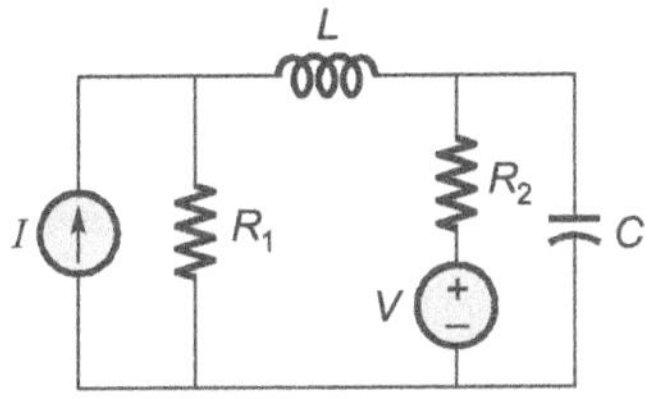

(a) 2, 2 (b) 3, 2
(c) 2, 3 (d) 1, 2

Q.342 Consider the following statements concerned with the incidence matrix of a graph:

1. Determinant of the incident matrix of a closed loop is non-zero.
2. Algebraic sum of the row entires of an incidence matrix is zero.
3. If branch j in the graph is not incident at node i, $a_{ij} = 0$.

Which of the above statements is/are true?

(a) 1, 2, 3 (b) 1 only
(c) 2 and 3 only (d) 3 only

Q.343 The number of edges in a complete graph of n-vertices is

(a) $\dfrac{n(n-1)}{2}$ (b) n
(c) $n(n-1)$ (d) $(n-1)$

Q.344 The number of branches and nodes in the graph are respectively

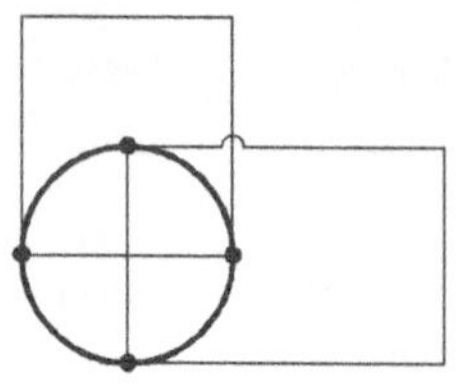

(a) 6, 10 (b) 10, 5
(c) 10, 10 (d) 5, 10

Q.345 Assertion (A): When all the elements (i.e. resistors, inductors, capacitors etc.) in a network are replaced by lines with circles or dots at both ends, the configuration is then called the graph of the network.

Reason (R): The graph of any network is drawn by keeping all the points of intersection of two or more branches and representing the network elements of lines, voltages and current sources by their internal impedances.

(a) Both A and R are true and R is the correct explanation of A.
(b) Both A and R are true but R is not the correct explanation of A.
(c) A is true but R is false.
(d) A is false but R is true.

Q.346 The graph of a network has *N* number of nodes and *B* is the number of branches in it.
Match the **List-I** with **List-II** and select the correct answer using the codes given below the lists:

List-I	List-II
A. Number of chords	1. $N + 1$
B. Rank of the graph	2. $B + 1 - N$
C. No. of *f*-tie sets	3. $N - 1$
D. Number of twigs	4. $N - B + 1$

Codes:

	A	B	C	D
(a)	4	3	3	1
(b)	1	2	2	3
(c)	2	3	2	3
(d)	2	1	2	4

Q.347 Two graphs are said to be isomorphic graph if their incidence matrices
(a) have a determinant value of zero.
(b) are identical.
(c) have a determinant value of unity.
(d) are transpose of each other.

Q.348 Assertion (A): The analysis of network is done using Kirchhoff's laws when the values of each element of the network remains constant for all the values of time.

Reason (R): The analysis of a network is done using graph theory when the values of each element of the network are changed with time.

(a) Both A and R are true and R is the correct explanation of A.
(b) Both A and R are true but R is not the correct explanation of A.
(c) A is true but R is false.
(d) A is false but R is true.

Q.349 Which of the following doesn't represent a possible tree of the coupled circuit shown below?

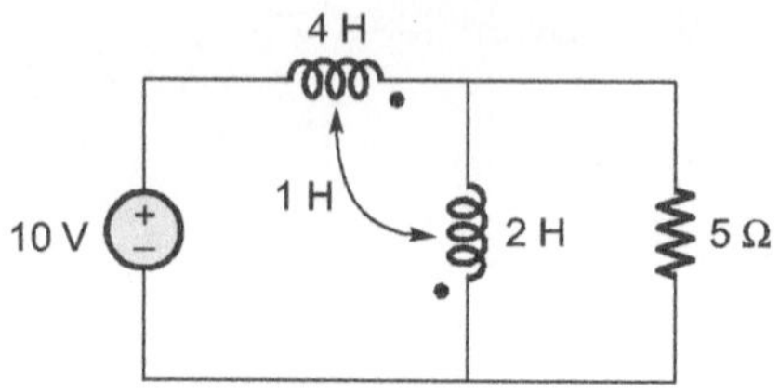

(a)  (b) 

(c) 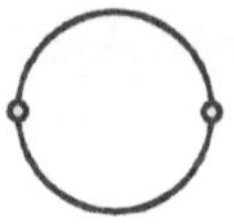 (d) 

Q.350 For the network shown below, the number of links and twigs of it's graph will be respectively

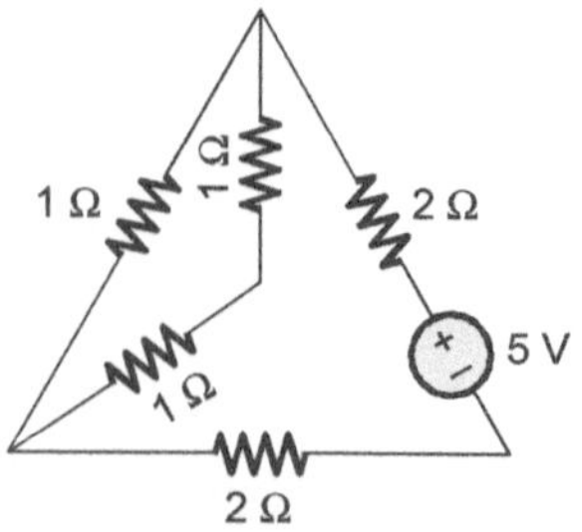

(a) 2, 2 (b) 3, 2
(c) 3, 3 (d) 2, 3

Q.351 A fully connected network has *N* number of nodes. The possible number of trees for this network will be
(a) $N^{(N-2)}$ (b) ${}^{N}C_2$
(c) $N(N-1)/2$ (d) None of these

Q.352 The cut-set matrix corresponding to the links or chords of a tree of a given graph is represented by $[Q_c]$ and it's corresponding tie-set matrix of tree branches is givon by $[B_t]$. The relation between $[Q_c]$ and $[B_t]$ which holds true is:

(a) $[Q_c]^T = -[B_t]$ (b) $[Q_c] = [B_t]$
(c) $[B_t] = -[Q_c]$ (d) $[B_t]^T = -[Q_c]$

Q.353 The incidence matrix of a graph is given by

$$[A_i] = \begin{bmatrix} -1 & 0 & 0 & +1 & -1 & 0 \\ +1 & -1 & 0 & 0 & 0 & -1 \\ 0 & +1 & -1 & 0 & +1 & 0 \\ 0 & 0 & +1 & -1 & 0 & +1 \end{bmatrix}$$

The possible number of trees are

(a) 16 (b) 12
(c) 8 (d) 20

Q.354 The dual of a series RL circuit is a

(a) parallel RL circuit
(b) series RC circuit
(c) parallel RC circuit
(d) series RC circuit

Q.355 Consider the following statements in relation to a given fixed tree of a network:

1. Link currents form an independent set.
2. Branch currents form an independent set.
3. Link voltage form an independent set.
4. Branch voltages form an independent set.

Which of the statements given above is/are correct?

(a) 1, 2 and 4 (b) 1, 2 and 3
(c) 3 and 4 (d) 2, 3 and 4

15. Poly Phase AC Circuits

Q.356 Which of the following is not true for a three-phase system compared to 1-ϕ system?

(a) Domestic power and commercial power can be provided from the same three-phase supply.
(b) Voltage regulation of three-phase system is poorer then single phase system.
(c) Three-phase induction motors are self starting while single-phase induction motors are not.
(d) For a given size of the frame, three phase generator provides more output.

Q.357 In a balanced three-phase star connected system, the phase voltages

(a) leads the respective line voltage by 90°.
(b) are in phase with the respective line voltage.
(c) lags the respective line voltage by 30°.
(d) leads the respective line voltages by 30°

Q.358 Three identical resistances are connected in delta against a balanced three-phase voltage supply. If one of the resistances be removed, the reduction in the power will be

(a) 50% (b) 66.67%
(c) 0.00% (d) 33.33%

Q.359 In a delta connected system, the line current are

(a) 30° ahead of respective phase currents.
(b) 90° behind the respective phase currents.
(c) 30° behind the respective phase currents.
(d) in phase with the respective phase currents.

Q.360 A balanced *Y*-connected load is supplied from a balanced 3-phase 400 V system. If the current in each phase is 15 A and lags the phase voltage by 30°, the total power will be

(a) 15 kW (b) 9 kW
(c) 12 kW (d) 6 kW

Q.361 In a balanced three-phase circuit $V_{AN} = 1500 \angle 20°$ V and $V_{CN} = 1500 \angle -100°$ V. The value of V_{BN} will be

(a) $1500 \angle -120°$ V (b) $\frac{1500}{\sqrt{3}} \angle 220°$ V
(c) $1500 \angle 140°$ V (d) none of these

Q.362 The value of the rms line voltage V_L for the below circuit if the rms line voltage is 100 V will be

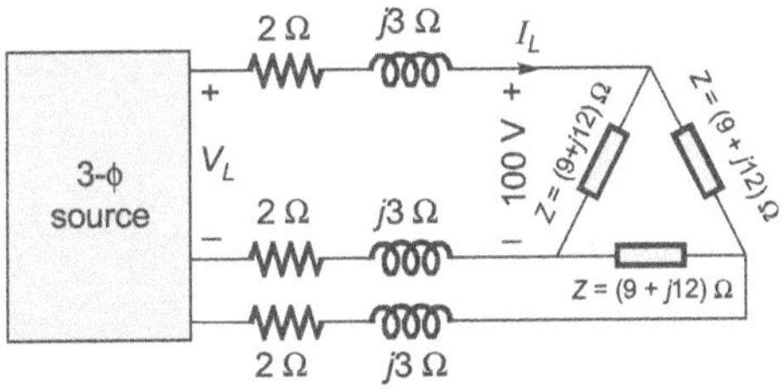

(a) 172 V (b) 125 V
(c) 225 V (d) 150 V

Q.363 Two wattmeters, both have reading of 3 kW when connected for the two-wattmeter method with current coils in lines *A* and *B* of a 600 V *ABC* circuit having a balanced Δ load. The Δ phase impedance will be

(a) resistive with $Z_P = 125\ \Omega$
(b) inductive with $Z_P = 180\ \Omega$
(c) resistive with $Z_P = 180\ \Omega$
(d) capacitive with $Z_P = 200\ \Omega$

Q.364 Three capacitances, each of $C/3$ Farads are connected in delta. Their equivalent star value for each capacitance is
(a) $3C$ (b) C
(c) $\frac{2}{3}C$ (d) $2C$

Q.365 The instantaneous value of currents in both phases B and C of a 3-phase balanced system are –20 A. For a phase sequence of ABC, the instantaneous value of current in phase A is
(a) –40 A (b) 20 A
(c) 40 A (d) –20 A

Q.366 Three equal resistances connected in star take a line current of 10 A when fed from 400 V, 50 Hz source. If the load resistances are reconnected in delta, the line current would be
(a) $10\sqrt{3}$ A (b) 10 A
(c) $\frac{10}{\sqrt{3}}$ A (d) 30 A

Q.367 The minimum number of wattmeters required to measure the real power in an n-phase system with unbalanced load is
(a) $n - 1$ (b) n
(c) $n + 1$ (d) $n - 2$

Q.368 The input to the 3-phase, 50 Hz circuit shown in figure below is 100 V. For a phase sequence of ABC, the wattmeters would read

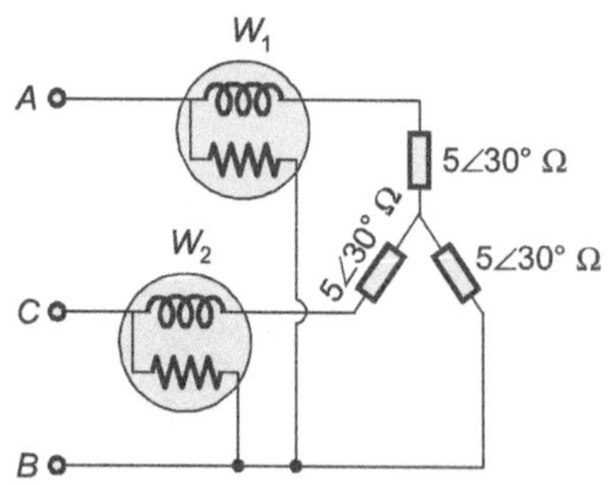

(a) $W_1 = 866$ W and $W_2 = 866$ W
(b) $W_1 = 577.35$ W and $W_2 = 1154.70$ W
(c) $W_1 = 1154.72$ W and $W_2 = 577.36$ W
(d) $W_1 = 1732$ W and $W_2 = 0$ W

Q.369 Three-phase power can be measured by two-wattmeter method in case of
(a) unbalanced load and unbalanced source only
(b) balanced load and balanced source only
(c) balanced load only
(d) both (a) and (b)

Q.370 In the case of power measured by two-wattmeter method in a balanced 3-phase system with a pure capacitive load
(a) one wattmeter will indicate zero and other some non-zero value.
(b) both the wattmeters will indicate the same value but of opposite sign.
(c) both the wattmeters will indicate zero.
(d) both the wattmeters will indicate the same value with same sign.

Q.371 Measurement of power factor of a 3-phase system by two-wattmeter method can be obtained in case of
(a) balanced source with 3-phase, 3-wire unbalanced load.
(b) balanced load and balanced source.
(c) balanced source with 3-phase, 3-wire unbalanced load.
(d) balanced source with 3-phase, 4-wire unbalanced load.

Q.372 Two wattmeters used to measure power of a 3-phase balanced load reads W_1 and W_2. The reactive power drawn by the load is
(a) $W_1 + W_2$ (b) $W_1 - W_2$
(c) $\sqrt{3}(W_1 + W_2)$ (d) $\sqrt{3}(W_1 - W_2)$

Q.373 A three-phase 220 V supply is applied to a balanced Δ-connected three phase load. The phase current being $I_{ab} = 10\angle{-30°}$A as shown in the given figure. The total power received by the Δ-load is

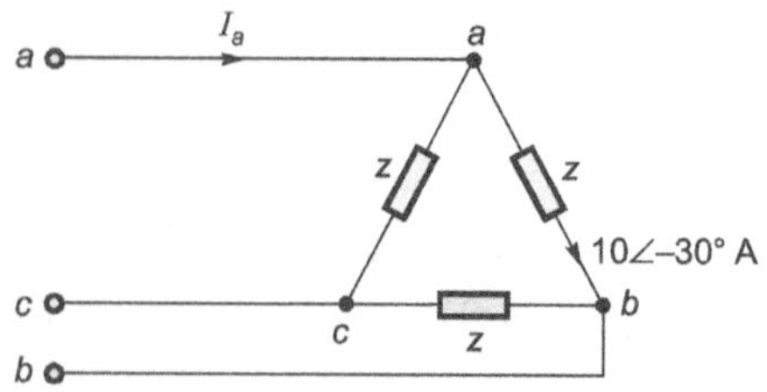

(a) 2250 W (b) 6232 W
(c) 5715 W (d) –2200 W

Q.374 The magnitude of line current in the circuit shown below is

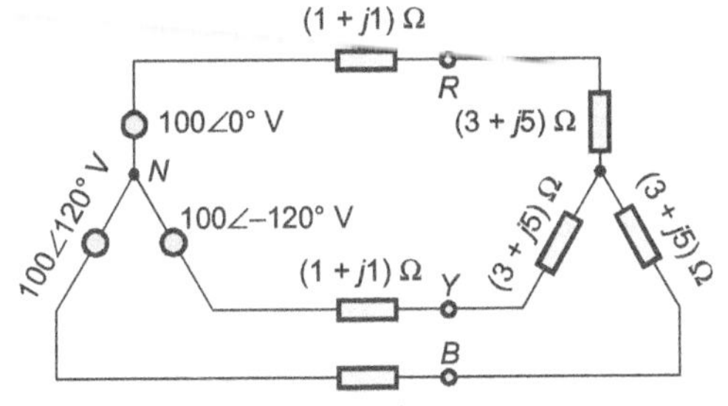

(a) 15 A (b) $50\sqrt{2}$ A

(c) 0 A (d) $\frac{50}{\sqrt{13}}$ A

Q.375 A three-phase, 400 V ac system is connected across a delta connected load having load resistance per phase of 200 Ω. If one of the phases of the delta connected load experiences an open circuit, then power consumed by the load will be

(a) 2400 W (b) 1600 W

(c) 800 W (d) 400 W

Q.376 Assertion (A): Generated voltage or current wave have no odd harmonics.

Reason (R): The field system and armature coils of the generators are all symmetrical and so generates mostly symmetrical voltage wave.

(a) Both A and R are true and R is the correct explanation of A.
(b) Both A and R are true but R is not the correct explanation of A.
(c) A is true but R is false.
(d) A is false but R is true.

Q.377 Assertion (A): RMS value of line voltage is more than $\sqrt{3}$ times of the rms value of phase voltage.

Reason (R): In a star-connected system, the line voltage contains no triple frequency harmonic content.

(a) Both A and R are true and R is the correct explanation of A.
(b) Both A and R are true but R is not the corroct explanation of A.
(c) A is true but R is false.
(d) A is false but R is true.

Q.378 Assertion (A): In a balanced 3-phase star connected system, line voltages are 30° ahead of the respective phase voltages.

Reason (R): The line voltages are $\sqrt{3}$ times of the respective phase voltages.

(a) Both A and R are true and R is the correct explanation of A.
(b) Both A and R are true but R is not the correct explanation of A.
(c) A is true but R is false.
(d) A is false but R is true.

Q.379 Match the **List-I (Power factor)** with **List-II (Wattmeter readings)** for the measurement of power in a three-phase system using two-wattmeter method and select the correct answer using the codes given below the lists:

List-I	List-II
A. 0.5 lag	1. $W_1 = W_2$
B. Unity	2. $W_1 = 2W_2$
C. Zero	3. $W_1 > 0, W_2 = 0$
D. 0.866 lag	4. $W_1 > 0, W_2 < 0$

Codes:

	A	B	C	D
(a)	1	2	4	3
(b)	3	4	1	2
(c)	3	1	4	2
(d)	1	3	2	4

■■■■

Answers Electrical Circuits

1. (c)	2. (a)	3. (c)	4. (a)	5. (d)	6. (a)	7. (c)	8. (c)
9. (d)	10. (c)	11. (c)	12. (c)	13. (d)	14. (c)	15. (d)	16. (d)
17. (b)	18. (c)	19. (a)	20. (d)	21. (c)	22. (d)	23. (b)	24. (d)
25. (c)	26. (d)	27. (a)	28. (a)	29. (a)	30. (b)	31. (a)	32. (a)
33. (d)	34. (d)	35. (a)	36. (d)	37. (c)	38. (d)	39. (b)	40. (a)
41. (a)	42. (c)	43. (b)	44. (d)	45. (c)	46. (a)	47. (c)	48. (b)
49. (d)	50. (c)	51. (c)	52. (b)	53. (d)	54. (b)	55. (c)	56. (c)
57. (d)	58. (c)	59. (a)	60. (d)	61. (b)	62. (d)	63. (c)	64. (a)
65. (b)	66. (d)	67. (c)	68. (b)	69. (a)	70. (d)	71. (c)	72. (a)
73. (b)	74. (a)	75. (d)	76. (d)	77. (b)	78. (a)	79. (b)	80. (a)
81. (a)	82. (c)	83. (b)	84. (c)	85. (b)	86. (c)	87. (c)	88. (d)
89. (d)	90. (b)	91. (c)	92. (b)	93. (d)	94. (a)	95. (d)	96. (d)
97. (c)	98. (d)	99. (b)	100. (c)	101. (b)	102. (c)	103. (d)	104. (c)
105. (d)	106. (c)	107. (a)	108. (b)	109. (d)	110. (c)	111. (a)	112. (b)
113. (c)	114. (d)	115. (c)	116. (b)	117. (c)	118. (a)	119. (b)	120. (a)
121. (d)	122. (a)	123. (c)	124. (d)	125. (d)	126. (c)	127. (d)	128. (b)
129. (d)	130. (b)	131. (a)	132. (c)	133. (a)	134. (b)	135. (d)	136. (c)
137. (d)	138. (c)	139. (d)	140. (d)	141. (a)	142. (b)	143. (a)	144. (c)
145. (a)	146. (d)	147. (b)	148. (b)	149. (c)	150. (d)	151. (c)	152. (a)
153. (b)	154. (a)	155. (a)	156. (b)	157. (c)	158. (c)	159. (b)	160. (d)
161. (d)	162. (a)	163. (c)	164. (d)	165. (a)	166. (c)	167. (c)	168. (b)
169. (a)	170. (d)	171. (b)	172. (c)	173. (a)	174. (b)	175. (d)	176. (c)
177. (a)	178. (a)	179. (d)	180. (a)	181. (b)	182. (c)	183. (d)	184. (a)
185. (b)	186. (c)	187. (d)	188. (a)	189. (c)	190. (a)	191. (d)	192. (b)
193. (c)	194. (b)	195. (c)	196. (d)	197. (b)	198. (a)	199. (d)	200. (b)
201. (c)	202. (a)	203. (c)	204. (d)	205. (c)	206. (b)	207. (d)	208. (a)
209. (b)	210. (c)	211. (a)	212. (d)	213. (c)	214. (d)	215. (a)	216. (b)
217. (d)	218. (b)	219. (a)	220. (b)	221. (b)	222. (b)	223. (d)	224. (b)
225. (c)	226. (c)	227. (c)	228. (b)	229. (c)	230. (b)	231. (d)	232. (a)
233. (c)	234. (d)	235. (c)	236. (c)	237. (b)	238. (d)	239. (c)	240. (a)
241. (c)	242. (d)	243. (b)	244. (a)	245. (c)	246. (a)	247. (c)	248. (c)

249. (c)	250. (d)	251. (d)	252. (c)	253. (c)	254. (c)	255. (d)	256. (c)
257. (b)	258. (c)	259. (b)	260. (b)	261. (c)	262. (a)	263. (a)	264. (b)
265. (d)	266. (b)	267. (c)	268. (b)	269. (b)	270. (c)	271. (b)	272. (b)
273. (a)	274. (b)	275. (c)	276. (d)	277. (c)	278. (b)	279. (d)	280. (a)
281. (c)	282. (d)	283. (b)	284. (c)	285. (c)	286. (b)	287. (a)	288. (d)
289. (d)	290. (c)	291. (b)	292. (c)	293. (d)	294. (c)	295. (c)	296. (a)
297. (d)	298. (a)	299. (b)	300. (c)	301. (d)	302. (b)	303. (a)	304. (c)
305. (c)	306. (a)	307. (d)	308. (c)	309. (b)	310. (c)	311. (a)	312. (d)
313. (b)	314. (b)	315. (d)	316. (c)	317. (b)	318. (d)	319. (c)	320. (c)
321. (b)	322. (a)	323. (d)	324. (b)	325. (b)	326. (a)	327. (a)	328. (c)
329. (c)	330. (d)	331. (a)	332. (d)	333. (c)	334. (a)	335. (c)	336. (d)
337. (d)	338. (a)	339. (c)	340. (c)	341. (a)	342. (d)	343. (a)	344. (b)
345. (a)	346. (c)	347. (b)	348. (b)	349. (c)	350. (d)	351. (a)	352. (d)
353. (a)	354. (c)	355. (a)	356. (b)	357. (c)	358. (d)	359. (c)	360. (b)
361. (c)	362. (a)	363. (c)	364. (b)	365. (c)	366. (d)	367. (a)	368. (b)
369. (d)	370. (b)	371. (b)	372. (d)	373. (c)	374. (d)	375. (b)	376. (d)
377. (d)	378. (b)	379. (c)					

Explanations

1. (c)

We know that,

$$I = \frac{P}{V}$$

∴ Current through the circuit,

$$I = \frac{60}{20} = 3\text{ A}$$

Applying KVL in the loop, we have

$$20 = 4I + V + 12$$

or,
$$V = 20 - 12 - 4I$$
$$= 20 - 12 - 4 \times 3$$
$$= -4 \text{ volts}$$

∴ Power absorbed by unknown element

$$= VI = -4 \times 3 = -12\text{ W}$$

Hence, power supplied by it = 12 W

2. (a)

Energy dissipated in $R = \int_0^{\infty} p\,dt$ where, $p = vi$

$$\therefore \quad \frac{d}{dt}(\text{Energy}) = p = vi$$

Since power is constant, therefore

Power, $P = \frac{\text{Energy}}{\text{Time}} = \frac{100\text{ J}}{1 \times 60\text{ sec}} = \frac{5}{3}\text{ W}$

Hence, $v = \frac{P}{I_2} = \frac{5/3}{0.1} = \frac{50}{3}$ Volts

Applying KCL at node v, we have

$$-I_1 + I_2 + \frac{50/3}{50} - i = 0$$

or, $-0.05 + 0.1 + 0.333 - i = 0$

or, $i = 0.383\text{ A} \approx 0.38\text{ A}$

3. (c)

$$W = QV = 120 \times 3 = 360\text{ J}$$

4. (a)

Maximum heat produced is

$$P = \frac{V^2}{R_{eq}}$$

For P to be maximum, R_{eq} should be minimum which is minimum if all four coils are connected in parallel.

5. (d)

$$\text{Energy} = \text{Power} \times \text{time}$$

$$\therefore \quad W = \Sigma P.t = \sum \frac{V^2}{R} \times t$$

$$= \left[\frac{20^2}{5}\times 3+\frac{30^2}{5}\times(6-3)+\frac{10^2}{5}\times(9-6)\right.$$

$$\left.+\frac{20^2}{5}\times(12-9)+\frac{10^2}{5}\times(15-12)\right]$$

$$= \frac{3}{5}[400+900+100+400+100]$$

$$= \frac{5700}{5} = 1140 \text{ J}$$

Hence, the heat consumed by the resistor = 1140 J.

6. (a)

The current through the capacitor is

$$i = \frac{dq}{dt} = \frac{dq}{dv}\cdot\frac{dv}{dt}$$

Now, $\frac{dq}{dv} = (1+v^2)$ and $\frac{dv}{dt} = \cos t$

$$\therefore \quad i = (1+v^2)\cdot \cos t$$

$$= (1+\sin^2 t)\cdot \cos t$$

9. (d)

Given signal is a ramp signal ($f(t) = At$), which is neither an energy signal nor a power signal.

11. (c)

Power dissipated by the resistor is

$$P = I^2 R = 3^2 \times 20 = 180 \text{ W}$$

Now, energy dissipated by the resistor in one hour (or 60 min)

$$= [180 \times 60] \text{ watt-min}$$

$\therefore$ Energy dissipated by the resistor in one minutes

$$= \left[\frac{180\times 60}{60}\right]\frac{\text{W-min}}{\text{min}} = 180 \text{ watt}$$

12. (c)

We know that,

$$R \propto \frac{l}{A}$$

and $$L = \frac{N^2\mu A}{l} \text{ or } L \propto \frac{N^2}{l}$$

when coil is cut into two equal halves,

$$l' = \frac{l}{2} \text{ and } N' = \frac{N}{2}$$

(Area of cross-section = constant)

So, $$\frac{R'}{R} = \frac{l'}{l} = \frac{l/2}{l} \text{ or } \left[R' = \frac{R}{2}\right]$$

Also, $$\frac{L'}{L} = \left(\frac{N'}{N}\right)^2 \times \left(\frac{l}{l'}\right)$$

$$= \left(\frac{N/2}{N}\right)^2 \times \left(\frac{l}{l/2}\right)$$

$$= \frac{1}{4}\times 2 = \frac{1}{2}$$

So, $$\left[L' = \frac{L}{2}\right]$$

Hence, new values of resistance and inductance which are reconnected in parallel is

$$R_{eq} = R' || R' = \frac{R'}{2} = \frac{R}{4}\,\Omega$$

and $$L_{eq} = L' || L' = \frac{L'}{2} = \frac{L}{4}\text{H}$$

13. (d)

We have; $P_1 = I_1^2 R$

or, $I_1 = \sqrt{\frac{P_1}{R}}$; $P_1 = 1$ W

$P_2 = I_2^2 R$

or, $I_2 = \sqrt{\frac{P_2}{R}}$; $P_2 = 4$ W

When both the sources are present, net current through R will be

$$I = (I_2 - I_1)$$

[as polarity of $|V_1|$ is reverse]

So, power loss in R is

$$P = I^2 R = (I_2 - I_1)^2 R$$

$$= \left(\sqrt{\frac{P_2}{R}} - \sqrt{\frac{P_1}{R}}\right)^2 \times R = \left(\sqrt{P_2} - \sqrt{P_1}\right)^2$$

$$= \left(\sqrt{4} - \sqrt{1}\right)^2 = (2-1)^2 = 1 \text{ Watt}$$

15. (d)

$$VA = \sqrt{(VAr)^2 + (W)^2}$$

16. (d)

Susceptance is the imaginary part of admittance,

$$Y = \frac{1}{Z} = \frac{1}{R + jX_L} = \frac{R - jX_L}{R^2 + X_L^2}$$

$$= \frac{R - jX_L}{Z^2}$$

or, $$Y = \left[\frac{R}{Z^2} - j\frac{X_L}{Z^2}\right] = [G - jS]$$

Here, G = Conductance

$$= \frac{R}{Z^2} \text{ mho}$$

and S = Susceptance

$$= \frac{X_L}{Z^2} \text{ Simen}$$

17. (b)

$$P = VI^* = (a + jb)\cdot(c + jd)^*$$
$$= (a + jb)\,(c - jd)$$
$$= (ac + bd) + j\,(bc - ad)$$
$$= P + jQ$$

Here, P = Active power

and Q = Reactive power

18. (c)

Power, $$P = \frac{V^2}{R} \text{ or } V^2 = PR$$

Since voltage ratings of both bulbs are same, therefore

$$PR = \text{constant}$$

or, $P_A R_A = P_B R_B$

or, $100 \times R_A = 200 \times R_B$

or, $R_A = 2 R_B$

Thus, $R_A > R_B$

19. (a)

$$S = VI^* = (10\angle 15°)(2\angle -45°)^*$$
$$= 20\angle 15 + 45° = 20\angle 60°$$
$$= 20\,(\cos 60° + j \sin 60°)$$
$$= 20\left(\frac{1}{2} + j\frac{\sqrt{3}}{2}\right)$$
$$= (10 + j10\sqrt{3}) = (10 + j17.32)$$
$$= P + jQ$$

20. (d)

Given, power consumed is

$$P = I^2 R_{eq}$$

or, $$I = \sqrt{\frac{P}{R_{eq}}} = \sqrt{\frac{10}{5}} = \sqrt{2}\text{ A}$$

Also, $$I = \frac{V}{|Z|} \text{ or } \sqrt{2} = \frac{(50/\sqrt{2})}{|Z|}$$

or, $|Z| = 25\ \Omega$

or, $\sqrt{X_L^2 + 15^2} = 25$

or, $X_L = \sqrt{25^2 - 15^2} = 20\ \Omega$

Hence, p.f. of given circuit is

$$\cos\phi = \frac{R_{eq}}{|Z|} = \frac{15}{25} = \frac{3}{5} = 0.6 \text{ (lag)}$$

22. (d)

Carbon resistor and semiconductors have non-linear relationship between V and I. Hence, Ohm's law is not applicable. Also, these are not bilateral.

23. (b)

$$I \propto \frac{1}{R} \propto \frac{A}{\rho l} \propto \frac{A}{l}$$

When l is reduced. I will be increased and vice-versa.

24. (d)

$$R_1 + R_2 = 4.5 \quad \text{...(i)}$$

and $$\frac{R_1 R_2}{R_1 + R_2} = 1 \text{ or } R_1 R_2 = 4.5$$

$\therefore$ $$(R_1 - R_2)^2 = (R_1 + R_2)^2 - 4 R_1 R_2$$
$$= (4.5)^2 - 4 \times 4.5 = \frac{9}{4}$$

or, $$R_1 - R_2 = \frac{3}{2} = 1.5 \quad \text{...(ii)}$$

On solving equations (i) and (ii), we get

$R_1 = 3\ \Omega$ and $R_2 = 1.5\ \Omega$

or, $R_1 = 1.5\ \Omega$ and $R_2 = 3\ \Omega$

26. (d)

For series connection,

$$R_{eq} = R_1 + R_2$$

or, $$\frac{R_{eq}}{V_r^2} = \frac{R_1}{V_r^2} + \frac{R_2}{V_r^2}$$

or, $$\frac{1}{P_{eq}} = \frac{1}{P_1} + \frac{1}{P_2} \text{ or } P_{eq} = \frac{P_1 P_2}{P_1 + P_2}$$

Given, $P_1 = P_2 = 1000$ W

$\therefore$ $P_{eq} = 500$ Watt

27. (a)

Energy consumption per week is

$$W = 100 \times 10 \times 7$$
$$= 7000 \text{ kW-hr}$$
$$= 7 \text{ units}$$

30. (b)

For a constant C.S., current will remain constant for all values of loads.

31. (a)

Since the two dielectrics fill space equally, therefore area becomes $A/2$.

Now, $$C = \frac{\varepsilon A}{d} = \frac{\varepsilon_0 \varepsilon_r A}{d}$$

So, $$C_1 = \frac{\varepsilon_0 \varepsilon_1 A}{2d}$$

and $$C_2 = \frac{\varepsilon_0 \varepsilon_2 A}{2d}$$

Hence, $$C_{eq} = C_1 + C_2 = \frac{\varepsilon_0 A(\varepsilon_1 + \varepsilon_2)}{2d}$$

33. (d)

Assertion is false because a network in which the circuit elements like resistance, inductance etc. cannot be physically separated for analysis purposes, is called a distributed network. The best example of a distributed network is a transmission line. Reason is true because most of the electric networks are lumped in nature.

34. (d)

On combining the resitances in parallel, equivalent circuit will reduce as shown in below.

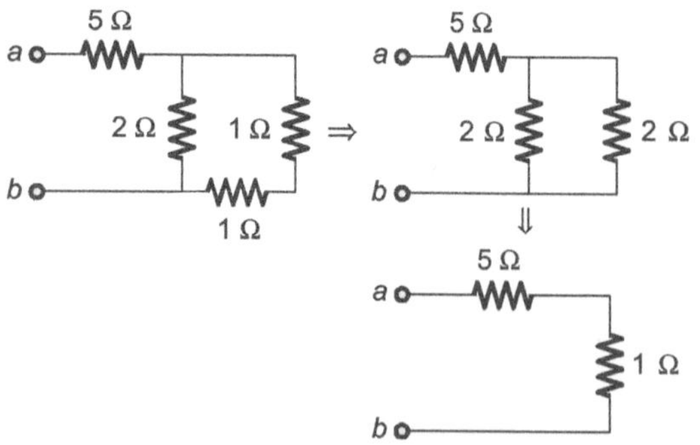

Hence, $R_{a\text{-}b} = 6\ \Omega$

35. (a)

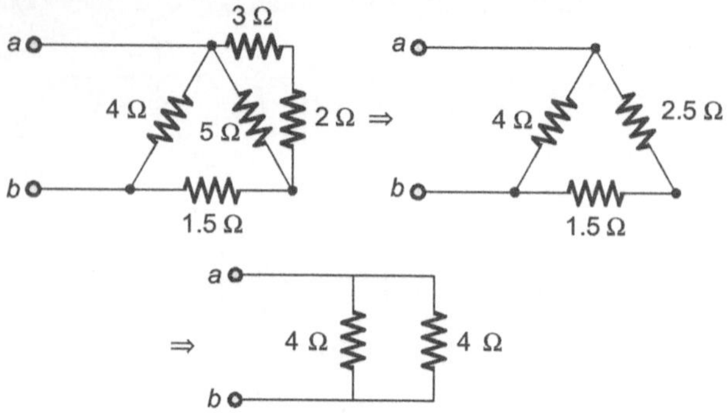

Hence, equivalent resistance is

$$R_{a\text{-}b} = R_{eq} = 4 \| 4 = 2\ \Omega$$

36. (d)

The equivalent resistance of the given circuit is

$$R = R_{eq} = \left(\frac{mr}{2} + \frac{r}{m}\right) = \left(\frac{m^2 r + 2r}{2m}\right)$$

Now, $$I = \frac{V}{R}$$

Hence, for current I to be maximum, R should be minimum.

i.e. $$\frac{dR}{dm} = 0$$

or, $$\frac{2m\,[2mr + 0] - (m^2 r + 2r) \times 2}{(2m)^2} = 0$$

or, $$2\,m\,(2\,mr) - 2\,(m^2 r + 2r) = 0$$

or, $$4\,m^2 - 2\,m^2 - 4 = 0$$

or, $$2\,m^2 - 4 = 0$$

or $$m = \sqrt{2}$$

37. (c)

Converting the current sources into equivalent voltage sources, the circuit is reduced as shown below.

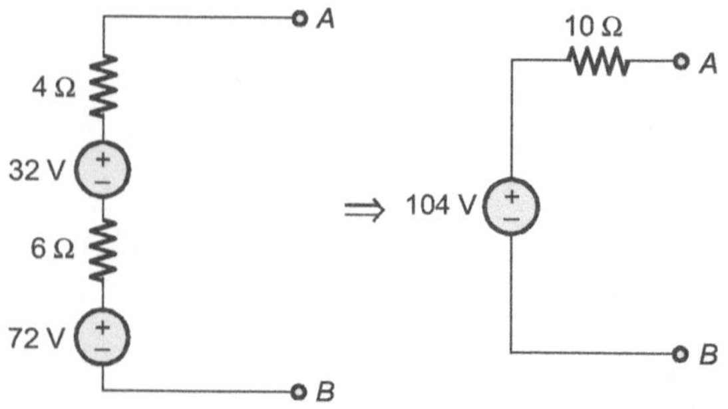

38. (d)

Equivalent voltage in the given circuit is

$$V = 6 + 20 - 10 = 16 \text{ volt}$$

And equivalent resistance is

$$R = 1 + 5 + 2 = 8\ \Omega$$

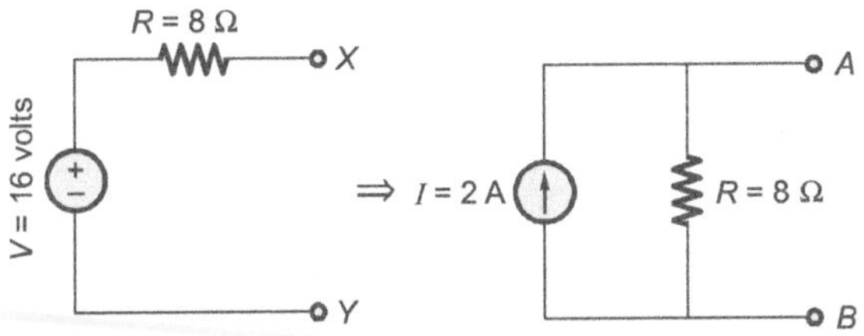

39. (b)

Converting all the voltage sources into equivalent current sources using source transformation, the equivalent circuit will be reudced as shown below.

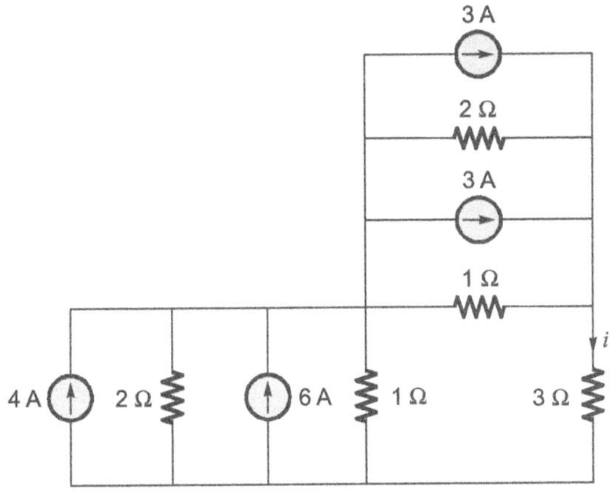

On combining current source and resistances,

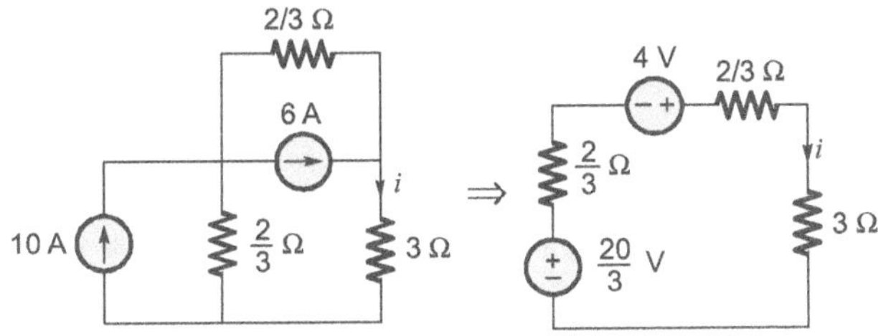

Hence, current,

$$i = \left(\frac{\frac{20}{3}+4}{\frac{2}{3}+\frac{2}{3}+3}\right) = \left(\frac{32/3}{13/3}\right) = 2.46 \text{ A}$$

40. (a)

Total power drawn from the circuit

$$= 2 \times 100 = 200 \text{ watts}$$

Hence, supply current is

$$I = \frac{200}{220} = 0.91 \text{ A}$$

Let R be the series resistance to be inserted in the circuit such that the voltage across the bulbs is 220 V.

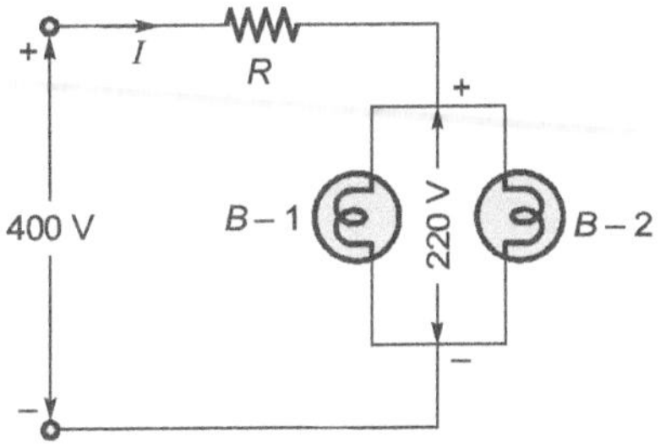

Applying KVL, we get

$$400 = IR + 220$$

or, $$IR = 400 - 220 = 180$$

or, $$R = \frac{180}{I} = \frac{180}{0.91} = 197.8\ \Omega \approx 198\ \Omega$$

41. (a)

Converting the star-connected resistors into Δ equivalent, the given circuit is reduced as shown below.

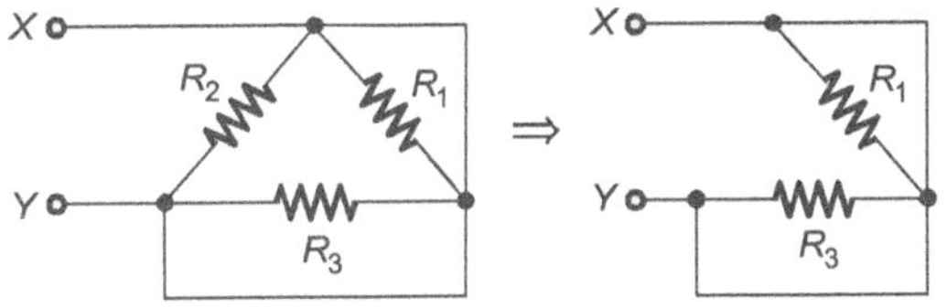

Here, $$R_1 = 1 + 0 + \frac{1 \times 0}{1} = 1\ \Omega$$

$$R_2 = 1 + 1 + \frac{1 \times 1}{0} = \infty\ \Omega$$

(i.e. open circuit)

$$R_3 = 1 + 0 + \frac{1 \times 0}{1} = 1\ \Omega$$

Hence, equivalent resistance between terminals X and Y is

$$R_{eq} = R_{XY} = \frac{0 \times R_1}{0 + R_1} + \frac{0 \times R_3}{0 + R_3} = 0 + 0 = 0\ \Omega$$

42. (c)

KCL is based on law of conservation of charge while KVL is based on law of conservation of energy.

43. (b)

The current through the resistors will purely depend upon the voltage source of 2 V connected across them. Let the currents i_1, i_2 and i_3 respectively flowing through the resistors of 1 Ω, 2 Ω and 5 Ω.

Then, $i_1 = \frac{2}{1} = 2\text{ A}$

$i_2 = \frac{2}{2} = 1\text{ A}$ and $i_3 = \frac{2}{5} = 0.4\text{ A}$

Hence, the total power absorbed by the circuit is

$$W = i_1^2 \times 1 + i_2^2 \times 2 + i_3^2 \times 5$$
$$= (2)^2 \times 1 + (1)^2 \times 2 + (0.4)^2 \times 5$$
$$= 6.8 \text{ watts}$$

44. (d)

Let the current flowing through 5 Ω resistor be I'. Applying KCL, we get

$$5 + I = I'$$

Applying KVL, we get

$$8 = 5I' + 5 \text{ or } I' = \frac{3}{5}\text{ A}$$

So, $I = I' - 5 = \frac{3}{5} - 5 = 0.6 - 5$
$= -4.4$ A

45. (c)

Nodes = 10 (*a, b, c, d, e, f, g, h, k, p*)
Junction points = 3(*b, e, h*)

46. (a)

Applying KCL at the node connecting 0.5 V current source and 1 A current source, we get

$$1 + I = 0.5\text{ V}$$

For $v = 2$ volt, $1 + I = 1$
or, $I = 0$ A

47. (c)

Using KVL in the loop, we have

$V_1 = 10 + 15 = 25$ volts
Also, $V_2 = -25$ volts
$\therefore$ $V_1 + V_2 = 25 - 25 = 0$ volts

48. (b)

Let the required voltage be V.
Then, voltage across 10 Ω resistor

$$= \left(\frac{10}{30+10}\right)\text{V} = 45$$

or, $V = \frac{45 \times 40}{10} = 180$ volts

Hence, required voltage is
$V = 180$ volts

49. (d)

Using Ohm's law,

$$R_T = R_1 + R_2 = \frac{10}{2.5} = 4\,\Omega$$

Given, $V_1 = 2V_2$
As I is same therefore,
$R_1 = 2R_2$
$\therefore$ $R_T = R_1 + R_2 = 2R_2 + R_2 = 3R_2$
So, $R_T = 3R_2 = 4\,\Omega$
or, $R_2 = \frac{4}{3} = 1.33\,\Omega$
and $R_1 = 2R_2 = \frac{8}{3}\,\Omega = 2.66\,\Omega$

50. (c)

We know that,
$R_\Delta = 3\,R_Y$
or, $R_Y = \frac{R_\Delta}{3} = R = \frac{10}{3}\,\Omega$

51. (c)

On combining the current sources, the circuit is reduced as shown below.

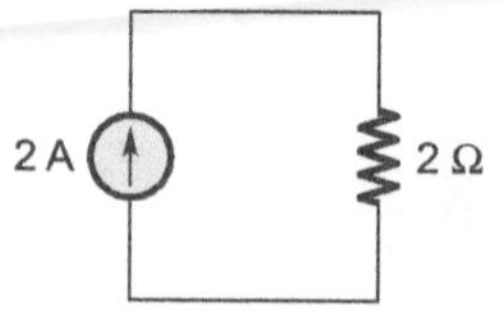

Hence, voltage drop across the 2 Ω resistor
= 2 × 2 = 4 volts

52. (b)

Using KCL,
$I_2 = 6 - 2 = 4$ A
Also, $2 + I_3 = \frac{1}{2}$
or, $I_3 = \frac{1}{2} - 2 = -1.5$ A
and $I_4 = I_2 - I_3 = 4 - (-1.5) = 5.5$ A
Also, $I_1 = 6$ A
$\therefore I_1 + I_2 + I_3 + I_4 = 6 + 4 - 1.5 + 5.5$
$= 15.5 - 1.5 = 14$ A

53. (d)

$$\vec{Z}_{BC} = \left[10 + j10 + \frac{10 \times (j10)}{5}\right]$$
$$= (10 + j10 + j20)$$
$$= (10 + j30)\,\Omega$$

54. (b)

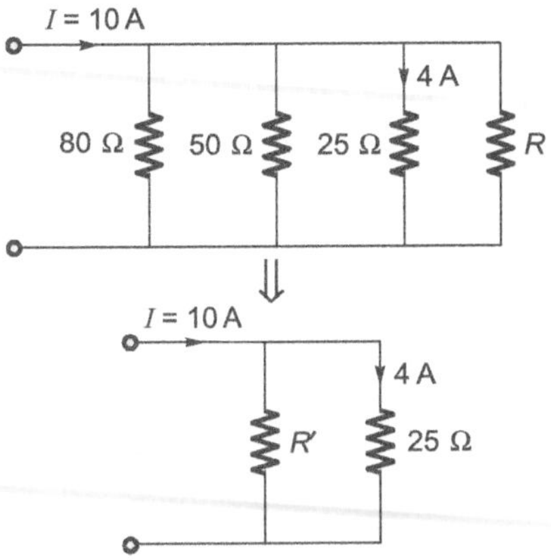

Given, $I_{25\,\Omega} = 4\text{ A}$

Here, $R' = 80 || 50 || R$

$$= \left(\frac{80\times 50}{130}\right) || R = \frac{400}{13} || R$$

$$= \frac{\frac{400}{13}R}{\frac{400}{13}+R} = \frac{400R}{13R+400}$$

Hence, $I_{25\,\Omega} = 4\text{ A} = \left[\frac{R'}{R'+25}\right]I$

or, $$4 = \left[\frac{\frac{400R}{13R+400}}{\frac{400R}{13R+400}+25}\right]\times 10$$

or, $$4 = \frac{400R\times 10}{400R+13\times 25R+400\times 25}$$

$$= \frac{16R\times 10}{16R+13R+400} = \frac{16R\times 10}{29R+400}$$

or, $29R + 400 = 40R$ or $11R = 400$

or, $R = 36.36\ \Omega$

55. (c)

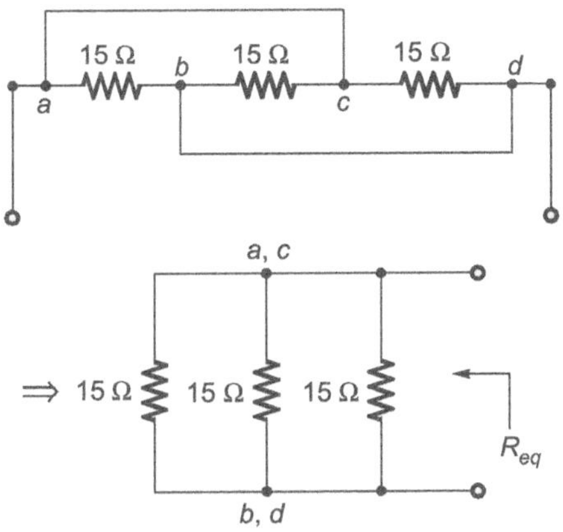

Here, a and c are at equipotential.

Also, b and d are at equipotential.

$$\therefore \quad R_{eq} = \frac{15}{3} = 5\ \Omega$$

56. (c)

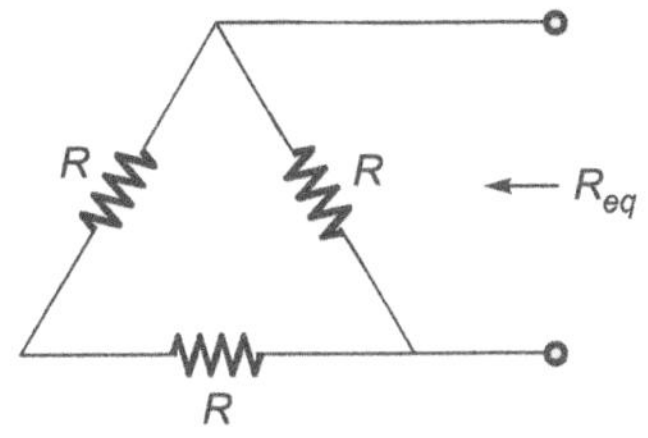

$$R_{eq} = 2R || R = \frac{2}{3}R$$

57. (d)

Number of different node-pair voltage = Number of KCL equations = $n - 1 = 10 - 1 = 9$

59. (a)

$$n = 4$$

Number of loops = $b - (n-1) = 3$

or, $b = 3 + (n-1)$

$= 3 + 4 - 1$

$= 3 + 3 = 6$

Hence, number of branches = 6

Statement-2 is for KCL.

62. (d)

Number of junction point for the given circuit is 6.

63. (c)

Using source transformation, the current source is converted to a voltage source as shown below.

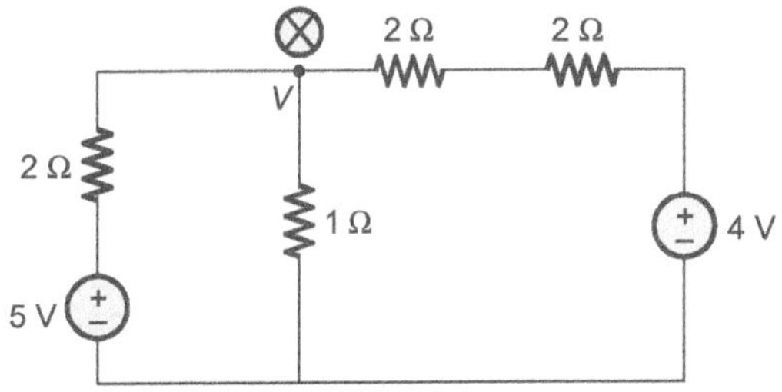

Applying nodal analysis at node (×), we get

$$\frac{V-5}{2} + \frac{V}{1} + \frac{V-4}{4} = 0$$

or, $V = 2$ volts

$\therefore$ Current through $1\ \Omega$ resistor $= \frac{V}{1} = \frac{2}{1} = 2\text{ A}$

64. (a)

Number of loop equation

$$= b-(n-1)$$
$$= 8-(4-1)$$
$$= 8-3 = 5$$

65. (b)

Using source transformation, first we convert the current source into an equivalent voltage source

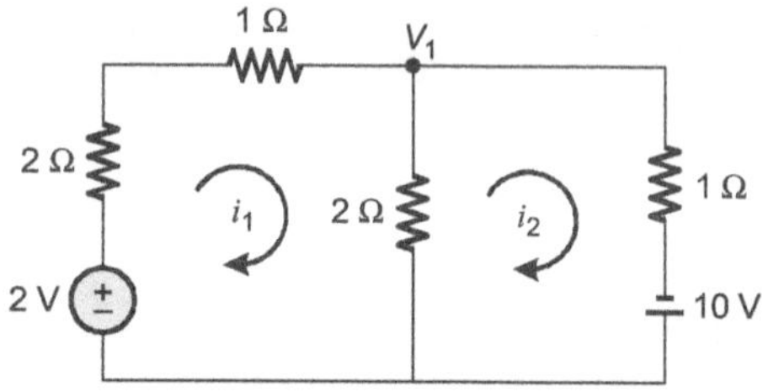

Applying KVL in loop-1, we get

$-2 + 3\,i_1 + (i_1 - i_2)\cdot 2 = 0$

or, $\quad 5\,i_1 - 2\,i_2 = 2 \quad$ ...(*i*)

Applying KVL in loop-2, we get

$2\,(i_2 - i_1) + i_2 - 10 = 0$

or, $\quad 2\,i_1 - 3\,i_2 = 10 \quad$ ...(*ii*)

On solving equations (*i*) and (*ii*), we get

$$i_1 = 2.36 \text{ A} \text{ and } i_2 = 4.91 \text{ A}$$

Hence, the current through the battery of 10 V is

$$i_2 = 4.91 \text{ A}$$

66. (d)

Applying KVL in the loops, we get

$$5\,i_1 + (i_1 - i_2)\,10 = V_1$$

or, $\quad 15\,i_1 - 10\,i_2 = V_1 \quad$ ...(*i*)

Also, $\quad 2\,i_2 + i_2 \times 1 + (i_2 - i_1) \times 10 = 0$

or, $\quad -10i_1 + 13i_2 = 0 \quad$ ...(*ii*)

On solving equations (*i*) and (*ii*), we get

$$i_2 = \frac{10}{95} V_1$$

Also, $\quad V_2 = i_2 \times 1 = i_2 = \dfrac{10}{95} V_1$

or, $\quad \dfrac{V_2}{V_1} = \dfrac{10}{95} = \dfrac{2}{19}$

67. (c)

The equivalent resistance across the 5 A current source is

$$R_{eq} = 5||5 + 2 = \frac{5}{2} + 2$$
$$= 2.5 + 2 = 4.5\ \Omega$$

$\therefore \quad V = R_{eq} \times 5 = 4.5 \times 5 = 22.5$ volts

68. (b)

Number of nodes, $n = 5$

Number of branchs, $b = 6$

Hence, number of KCL equation

$$= n-1 = 5-1 = 4$$

and, number of KVL equations $= b-(n-1)$

$$= 6-(5-1) = 6-4 = 2$$

70. (d)

Applying KVL in the loop,

$$5 = 2i - 2i + i \text{ or } i = 5 \text{ A}$$

$\therefore$ Value of dependent source

$$= 2i = 2 \times 5 = 10 \text{ volt}$$

71. (c)

The given circuit can be reduced by combining resistors and current sources as shown below.

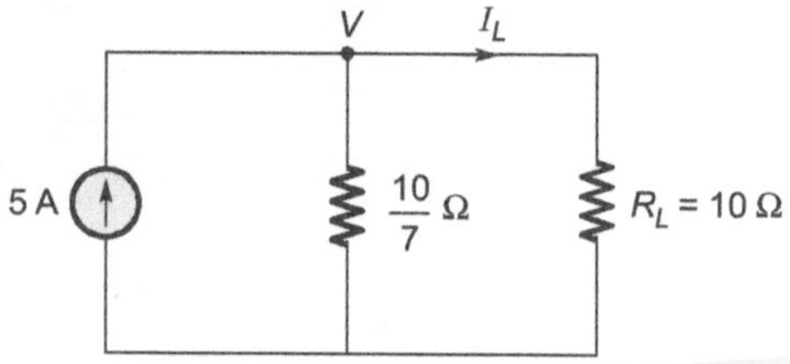

Let the node voltage be V volts.

Applying nodal analysis, we have

$$5 + \frac{V}{(10/7)} + \frac{V}{10} = 0$$

or, $\quad 5 + \left(\dfrac{7}{10} + \dfrac{1}{10}\right) V = 0$

or, $\quad 5 + \dfrac{8}{10} V = 0$

or, $\quad 5 + \dfrac{4}{5} V = 0$

or, $\quad \dfrac{4}{5} V = -5$

or, $\quad V = \dfrac{-25}{4}$ Volt

So, $\quad I_L = \dfrac{V}{10} = \dfrac{-25/4}{10} = \dfrac{-5}{8}$ A

75. (d)

Usind nodal analysis,

$$-2 + \frac{V}{2} + i = 0$$

or, $\quad i = \left(-2 + \dfrac{4}{2}\right) = -2 + 2 = 0$ A

$\therefore \quad i = 0$ A

76. (d)

Applying nodal analysis at the given node, we have

$$-I_x - 3 + \frac{V - 2I_x}{1} = 0$$

or, $3I_x = V - 3$ ...(i)

Also, $10 = 2I_x + V$

or, $V = 10 - 2I_x$ ...(ii)

Putting V from equation (ii) into equation (i),

$$3I_x = 10 - 2I_x - 3$$

or, $5I_x = 7$

or, $I_x = \frac{7}{5} = 1.4$ A

77. (b)

Let the node voltage be V' volts.

Applying nodal analysis at the given node, we have

$$-3 + \frac{V'}{1} + \frac{V'+3}{1} = 0$$

or, $2V' = 0$

or, $V' = 0$ volts

Now, $V' = -3 + V_1$

or, $0 = -3 + V_1$

or, $V_1 = 3$ volts

78. (a)

The given circuit can be redrawn as shown below (By combining current sources).

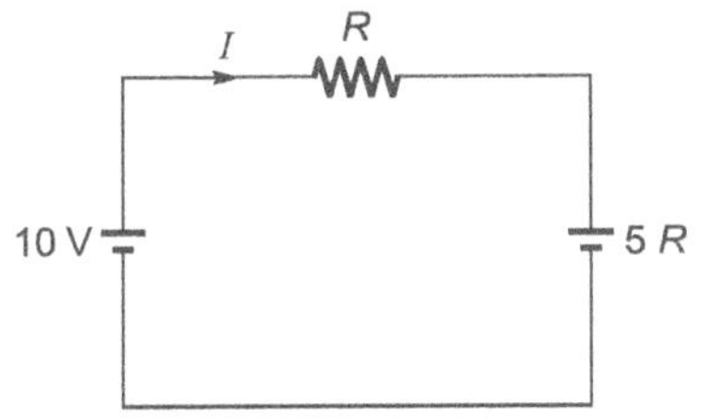

Now, $I = 0$ (Given)

i.e. $\frac{10 - 5R}{R} = 0$

or, $R = 2\ \Omega$

79. (b)

Node potential is always higher than the other voltages appearing in the equations otherwise a negative value for current would result.

82. (c)

We have,

$$\frac{V_1 - e_1}{R_1} + \frac{V_1}{R_5} + \frac{V_1}{R_2} - \frac{V_2 - e_2}{R_2} = 0$$

or, $V_1\left[\frac{1}{R_1} + \frac{1}{R_2} + \frac{1}{R_5}\right] - \frac{V_2}{R_2} = \frac{e_1}{R_1} - \frac{e_2}{R_2}$ ...(i)

Also, $\frac{V_2 - e_2}{R_2} + \frac{V_2 - e_3}{R_3} + \frac{V_2 + e_4}{R_4} - \frac{V_1}{R_2} = 0$

or, $V_2\left[\frac{1}{R_2} + \frac{1}{R_3} + \frac{1}{R_4}\right] - \frac{V_1}{R_2} = \frac{e_2}{R_2} + \frac{e_3}{R_3} - \frac{e_4}{R_4} = 0$ ...(ii)

83. (b)

Let the node voltage be V volts.

Using nodal analysis,

$$-2 + \frac{V}{2} + \frac{V}{3} = 0$$

or, $\frac{5V}{6} = 2$ or $V = \frac{12}{5}$ volts

So, $I = \frac{V}{2} = \frac{12}{10} = \frac{6}{5}$ A

85. (b)

$\sum_{i=1}^{b} V_i(t)\, i_i(t) = 0$ means the total sum of instantenous power consumed in various elements in the branches of a network is zero, which is the statement of Tellegen's theorem.

86. (c)

For finding the Thevenin's equivalent resistance, 8 A current source will be open circuited while V volt voltage source will be short circuited.

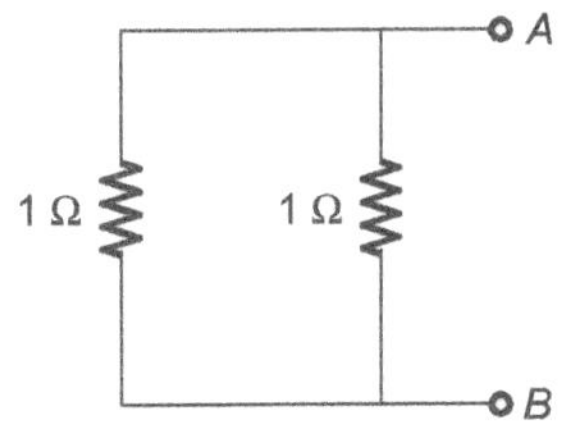

Hence, $R_{eq} = R_{Th} = 1 \,||\, 1 = \frac{1}{2}\ \Omega$

88. (d)

Applying KVL, we get

$$V_2 = 4(I_1 + I_2) - 10\,V_1 \qquad ...(i)$$

and $$V_1 = 2I_1 + 4(I_1 + I_2) - 10\,V_1$$

or, $$11\,V_1 = 6I_1 + 4I_2$$

or, $$V_1 = \frac{6I_1 + 4I_2}{11} \qquad ...(ii)$$

Putting the value of V_1 from equation (*ii*) into equation (*i*), we get

$$V_2 = 4(I_1 + I_2) - 10\left(\frac{6I_1 + 4I_2}{11}\right)$$

$$= I_1\left(4 - \frac{60}{11}\right) + I_2\left(4 - \frac{40}{11}\right)$$

or, $$V_2 = \frac{-16}{11}I_1 + \frac{4}{11}I_2$$

or, $$V_2 = Z_{21}I_1 + Z_{22}I_2$$

Hence, $$Z_{22} = \frac{4}{11}\ \Omega$$

90. (b)

For the given circuit, R_{eq} across the given terminals = R.

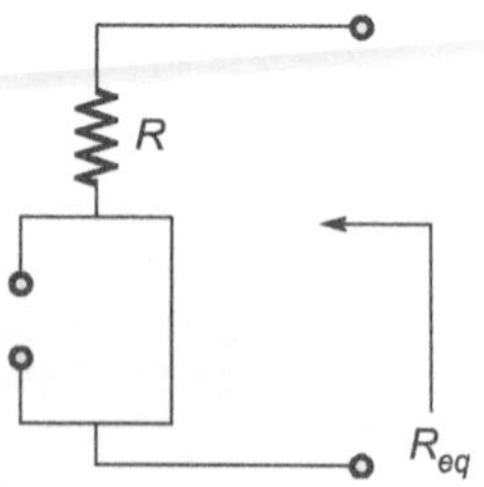

Also, $$V_{oc} = V_{Th} = V$$

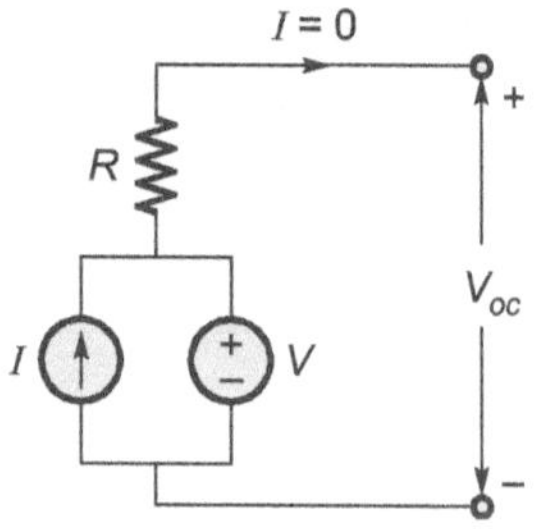

91. (c)

When $R = \infty$, $V_{oc} = V = 5$ volts

When $R = 0$, $I_{sc} = 2.5$ A

$\therefore$ $$R_{int} = \frac{V_{oc}}{I_{sc}} = \frac{5}{2.5} = 2\ \Omega$$

Hence, the Thevenin's equivalent circuit will be drawn as shown below.

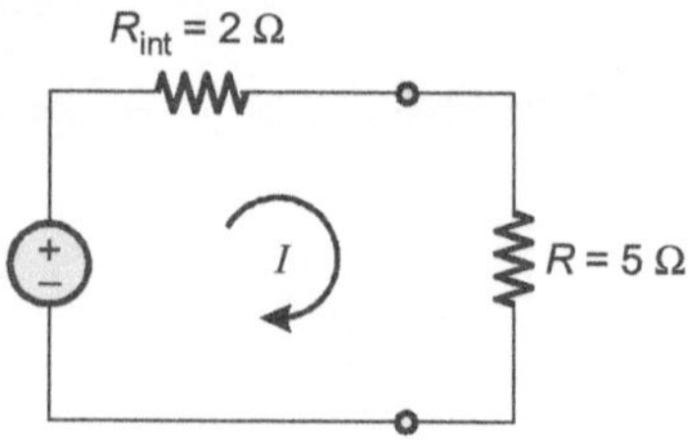

For $R = 5\ \Omega$, $$I = \frac{V_{oc}}{R_{int} + R} = \frac{5}{2+5} = \frac{5}{7}\text{ A}$$

$$= 0.71\text{ A}$$

93. (d)

An ideal voltage source is never short circuited. Hence, we should not use superposition theorem. The voltage across the 5 Ω resistor will be 10 volts. Hence, current through the 5 Ω resistor

$$= \frac{10}{5} = 2\text{ A}$$

94. (a)

The given circuit can be replaced by an equivalent circuit as shown below:

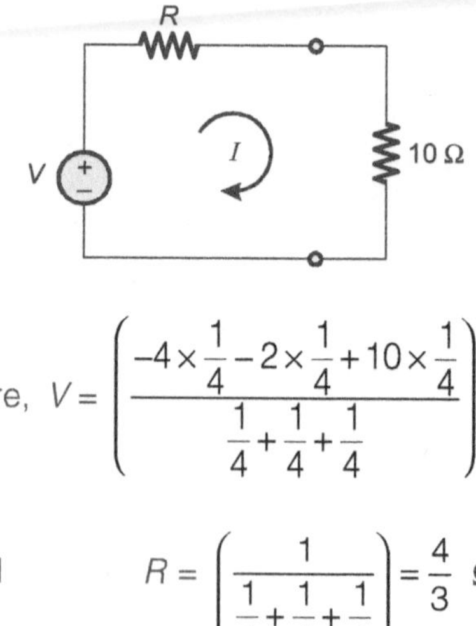

Here, $$V = \left(\frac{-4\times\frac{1}{4} - 2\times\frac{1}{4} + 10\times\frac{1}{4}}{\frac{1}{4}+\frac{1}{4}+\frac{1}{4}}\right) = \frac{4}{3}\text{ volt}$$

and $$R = \left(\frac{1}{\frac{1}{4}+\frac{1}{4}+\frac{1}{4}}\right) = \frac{4}{3}\ \Omega$$

Hence, $$I = \left(\frac{V}{R+10}\right) = \left(\frac{4/3}{4/3+10}\right)$$

$$= \frac{4}{3}\times\frac{3}{34} = \frac{2}{17}\text{ A}$$

95. (d)

For finding V_{oc}, the terminal X-Y is open circuited. Thus, current through voltage source (V_s) becomes zero. i.e. $I_s = 0$ due to which $0.5I_s$ voltage source gets short circuited.

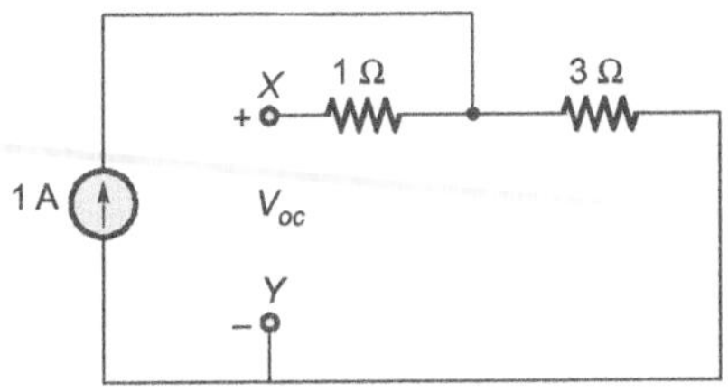

Thus, $V_{oc} = 3 \times 1 = 3$ volt

96. (d)

For finding R_N across diode, the independent voltage sources will be short circuited.

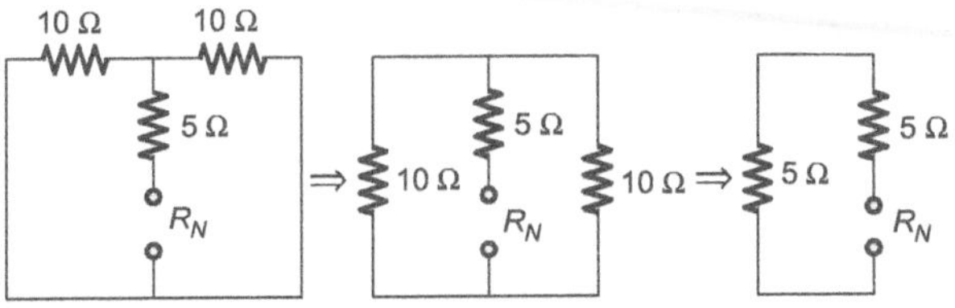

Hence, $R_N = 5 + 5 = 10\ \Omega$

97. (c)

Since the op-amp is ideal, the inverting terminal will be virtually grounded.

Hence, $V_i = I_1 R_1$

or, $\dfrac{V_i}{I_1} = R_1$

or, $\dfrac{I_1}{V_i} = \dfrac{1}{R_1}$ = Input admittance

98. (d)

Here, we apply superposition theorem.

Taking the 20 A current source only:

$$I_1 \text{ (due to 20 A C.S)} = \left(\frac{4}{2+6+4}\right) \times 20 = 6.67 \text{ A}$$

Taking the 10 A current source only:

$$I_2 \text{ (due to 10 A C.S)} = \left(\frac{6}{6+2+4}\right) \times 10 = 5 \text{ A}$$

Hence, the net current through the 2 Ω resistor

$$= I_1 - I_2 = (6.67 - 5) \text{ A} = 1.67 \text{ A}$$

100. (c)

Given circuit can be reduced as shown below.

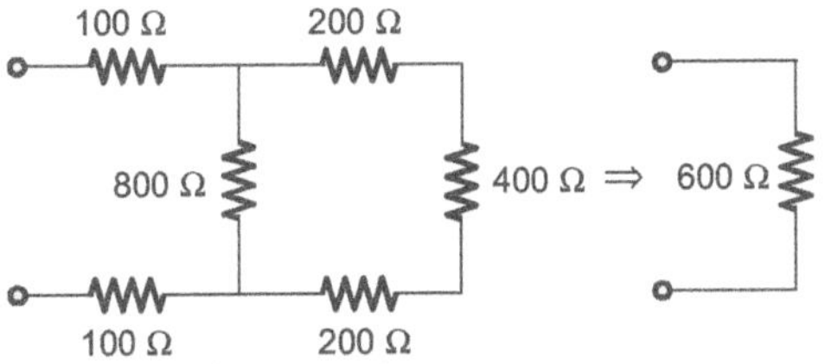

$\therefore$ $R_{eq} = 600\ \Omega$

101. (b)

$$P_{max} = \frac{V^2}{4R_L} = \frac{(12)^2}{4 \times 2} = \frac{144}{8} = 18 \text{ W}$$

102. (c)

Here, $V_{X\text{-}Y} = V_{oc} = 1 \times I_1 = I_1$

Applying KVL in the loop, we have

$I_1 \times 1 - 2 + 2I_1 - 0.5 I_1 = 0$

or, $I_1 = 0.8$ A

Hence, $V_{oc} = 0.8$ V

When the termical X-Y is short circuited then,

$I_1 = 0$

$\therefore$ $I_{sc} = \dfrac{2}{2} = 1\text{ A} = I_N$

So, $R_N = \dfrac{V_{oc}}{I_{sc}} = \dfrac{0.8}{1} = 0.8\ \Omega$

103. (d)

For maixmum power transfer,

$Z_L = Z_S^*$

or, $Z_L = 6 - j4$

$= \sqrt{6^2 + 4^2} \angle -\tan^{-1}\dfrac{4}{6}$

or, $Z_L = 7.211 \angle -33°.69\ \Omega$

105. (d)

Given, $V_{oc} = 5$ V, $I_{sc} = 10$ mA

$\therefore$ $R_{Th} = \dfrac{V_{oc}}{I_{sc}} = \dfrac{5}{10 \times 10^{-3}} = 0.5\text{ k}\Omega$

$= 500\ \Omega$

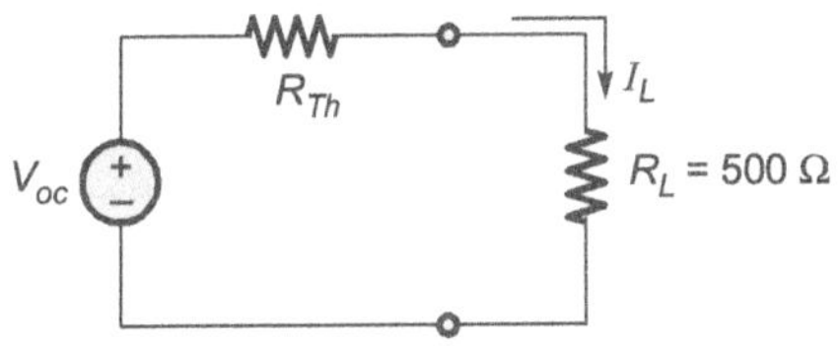

Hence, the current through the load is

$$I_L = \frac{V_{oc}}{R_{Th} + R_L} = \frac{5}{500 + 500}$$

$$= \frac{5}{1000} \text{ A} = 5 \text{ mA}$$

107. (a)

$$Z_{Th} = \frac{(6 - j8) \times (j8)}{6 - j8 + j8} = (10.67 + j8)\ \Omega$$

Also, V_{oc} across Z_L is

$$V_{oc} = \left(\frac{10\angle 0°}{6 - j8 + j8}\right) \times (j8) = 13.33 \angle 90° \text{ V}$$

According to maximum power transfer theorem,

$$Z_L = Z_{Th}^* = (10.67 - j8)\ \Omega$$
$$= (R_L - jX_L)\ \Omega$$

Hence, $P_{max} = \dfrac{V_{oc}^2}{4R_L} = \dfrac{13.33^2}{4 \times 10.67} = 4.16$ watt

108. (b)

Let us apply superposition theorem.

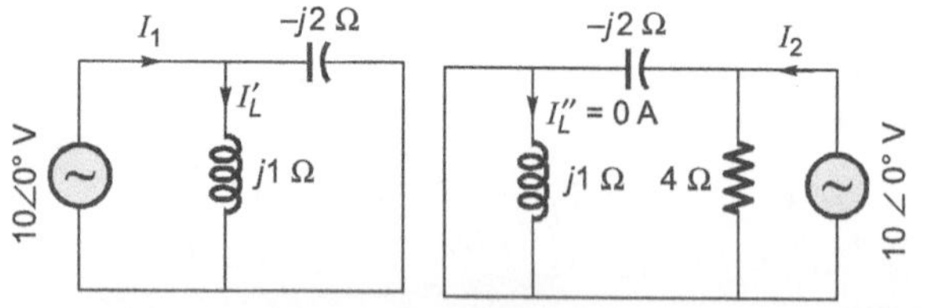

Here, $I_L' = \dfrac{10\angle 0°}{j1} = 10\angle -90°$ A

and $I_L'' = 0$ A

$\therefore$ $I_L = I_L' + I_L''$
$= 10\angle -90° + 0 = -j10$ A

111. (a)

$$V_{oc} = V_{Th} = \frac{10}{20} \times 10 = 5 \text{ volt}$$

(Using voltage divider rule)

R_{Th} across,

$$AB = \left(\frac{10 \times 10}{20} + 5\right) = 10\ \Omega$$

119. (b)

Converting 3 Ω delta connected resistors into star-equivalent, the circuit is reduced as shown below.

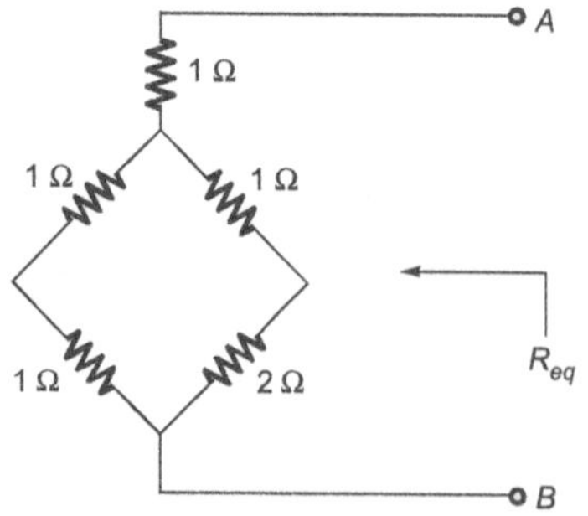

$\therefore$ $R_{eq} = (2 || 3) + 1 = \dfrac{6}{5} + 1 = \dfrac{11}{5}\ \Omega$

121. (d)

$Z_L = Z_s^*$ for maximum power transfer.

Hence, assertion is a false statement.

125. (d)

Due to initial condition, at $t = 0$ capacitor will act as a constant voltage source (at $t = 0$, capacitor acts as short-circuit). Hence, option (d) is correct.

126. (c)

Given, $L = 1$ H; $\dfrac{di}{dt} = 1$ A/sec

or, $di = dt$

or, $\int di = \int dt$

or, $i = t$

$\therefore$ Current in the inductor after 3 sec. is

$I = 3$ A

Hence, energy stored after 3 sec.

$$= \frac{1}{2}LI^2 = \frac{1}{2} \times 1 \times 3^2$$
$$= \frac{9}{2} = 4.5 \text{ Joules}$$

127. (d)

⇒ Since V is not proportional to R therefore, the element can't be a resistor.

⇒ At $t = 5$ ms, even if $i \neq 0$, the element behaves as a short circuit therefore, the element can't be a capacitor (since at $t = 0$ only capacitor behaves as short circuit).

⇒ The current at $t = 0$ is zero and at $t = 5$ ms voltage across the element is zero therefore, the element must be an inductor (at $t = 0$, an inductor acts as open circuit and at $t = \infty$ it behaves as short circuit).

From the given voltage and current profile, we have

$$\frac{di}{dt} = \frac{1}{5 \times 10^{-3}} = 0.2 \times 10^3 \text{ A/sec}$$

and $v(\text{initial}) = 2$ volt

$\therefore$ $v = L\dfrac{di}{dt}$

or, $L = \dfrac{v}{\left(\dfrac{di}{dt}\right)} = \dfrac{2}{0.2 \times 10^3} = 10$ mH

129. (d)

From given figure,

$$\frac{dv}{dt} = \frac{20}{10} = 2\frac{V}{ms} = 2 \times 10^3 \text{ V/sec}$$

$\therefore$ Charging current,

$$i_c = C\frac{dv}{dt} = 50 \times 10^{-6} \times 2 \times 10^3 = 0.1 \text{ A}$$

130. (b)

The equivalent combination of C_2 and C_3

$$= \frac{C_2 \times C_3}{C_2 + C_3} = 1 \, \mu\text{F}$$

The equivalent combination of this 1 μF and $C_1 = 2 \, \mu\text{F}$ is

$$C_1 + 1 \, \mu\text{F} = 3 \, \mu\text{F}$$

Hence, the equivalent capacitance between terminals A and B is

$$C_{AB} = \frac{3 \times C_4}{3 + C_4} = \frac{3 \times 1}{3 + 1} = \frac{3}{4} \, \mu\text{F}$$

131. (a)

Equivalent capacitance between terminals a-b is

$$C_{eq} = C_{a\text{-}b}$$

$$= \left[\frac{(C_1 + C_2)C_3}{C_1 + C_2 + C_3} \,||\, C_4\right] || \, C_5 = 16 \, \mu\text{F}$$

$$\therefore \quad Q_{net} = C_{a\text{-}b} \times V$$

$$= 16 \times 10^{-6} \times 100$$

$$= 1600 \, \mu\text{C}$$

Hence, the charging time required is

$$t = \frac{Q_{net}}{I} = \frac{1600 \times 10^{-6}}{10}$$

$$= 160 \, \mu\text{sec}$$

132. (c)

Max. value of current,

$$I_m = \sqrt{2} I = \sqrt{2} \times 5 = 7.07 \text{ A}$$

$$\omega = 2\pi f = 2\pi \times 50$$

$$= 314 \text{ rad/sec}$$

$$\therefore \quad i = I_m \sin \omega t = 7.07 \sin 314t \text{ Amps}$$

Also, $V_m = \sqrt{2} V = \sqrt{2} \times 220$

$$= 311 \text{ volts}$$

Assuming voltage as the reference phasor,

$$v = 311 \sin 314t \text{ Volt}$$

and $i = 7.07 \sin (314t - 90)° \text{ Amps}$

133. (a)

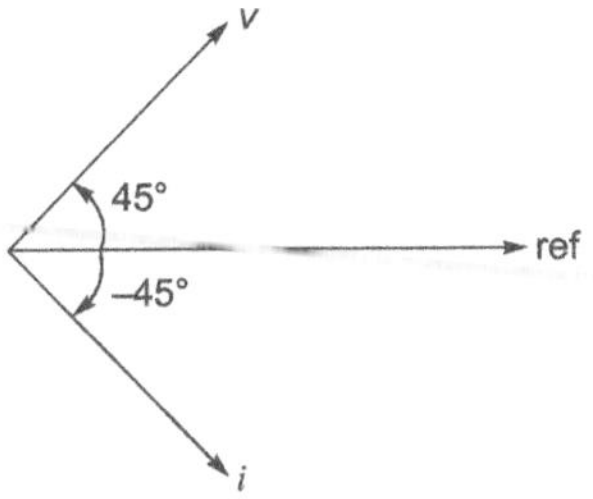

The phase difference between v and i is

$$\phi = (45° + 45°) = 90°$$

Since v leads i therefore, the circuit element is an inductor.

$$\therefore \quad X_L = \frac{V_m}{I_m} = \frac{100}{10} = 10 \, \Omega = 2\pi f L$$

or, $$L = \frac{X_L}{2\pi f} = \frac{10}{2\pi \times 50} \text{ H} = 31.8 \text{ mH}$$

134. (b)

Converting the internal star connected inductance to an equivalent delta, the circuit reduces as shown below.

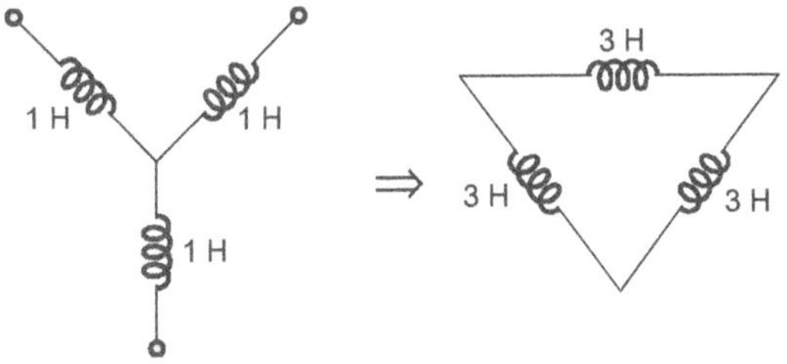

Hence, equivalent circuit becomes as shown below.

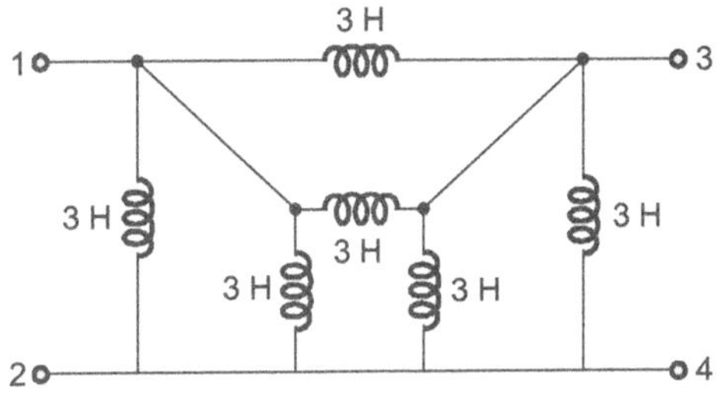

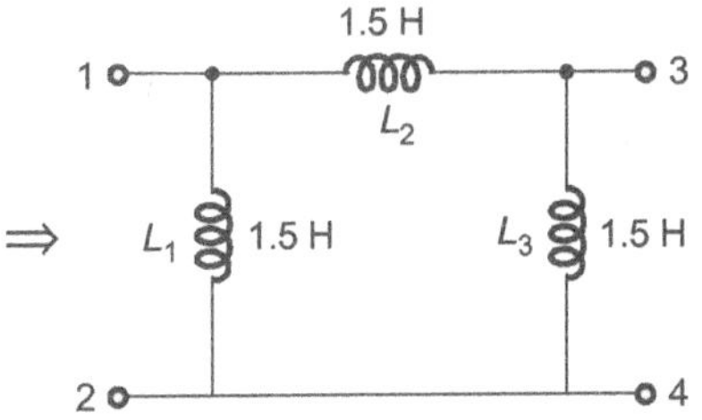

Hence, $L_{1\text{-}2} = L_1 \,||\, (L_2 + L_3)$

$$= \frac{1.5 \times 3}{1.5 + 3} = 1 \text{ H}$$

135. (d)

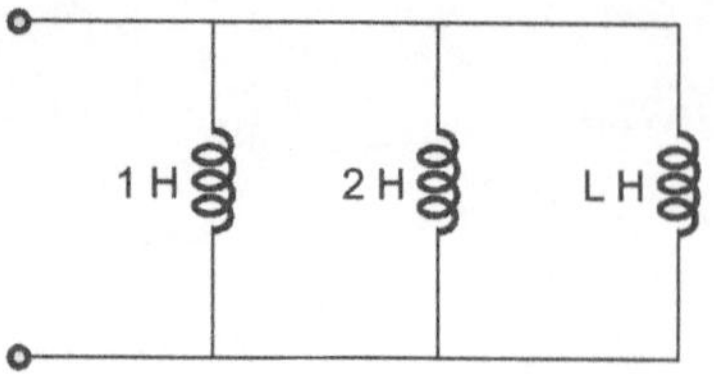

Equivalent inductance,

$$L_{eq} = (1 \| 2 \| L) = \left(\frac{2}{3} \| L\right)$$

$$= \left(\frac{\frac{2}{3}L}{L + \frac{2}{3}}\right) = \left(\frac{2/3}{1 + \frac{2}{3L}}\right)$$

L_{eq} will be maximum of $L \to \infty$.

Hence, $(L_{eq})_{max} = \frac{2}{3}$ H

136. (c)

Let the applied voltage be v.

$\therefore \quad v = L_1 \frac{di_1}{dt}$ and $v = L_2 \frac{di_2}{dt}$

or, $\quad i_1 = \frac{1}{L_1} \int v dt$ and $i_2 = \frac{1}{L_2} \int v dt$

$\therefore \quad \frac{i_1}{i_2} = \frac{L_2}{L_1} = \frac{20}{10} = 2$

or, $\quad i_1 = 2 i_2 \qquad ...(i)$

Also, $\quad i_1 + i_2 = 25 \qquad ...(ii)$

Solving equations (*i*) and (*ii*), we get

$i_1 = 16.67$ A and $i_2 = 8.33$ A

137. (d)

The given circuit can be redrawn as shown below.

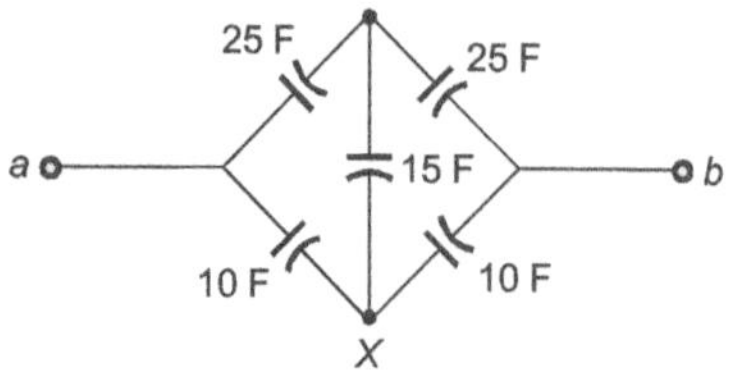

The star at point X can be replaced by equivalent delta having delta brances as C_1, C_2 and C_3 as shown below.

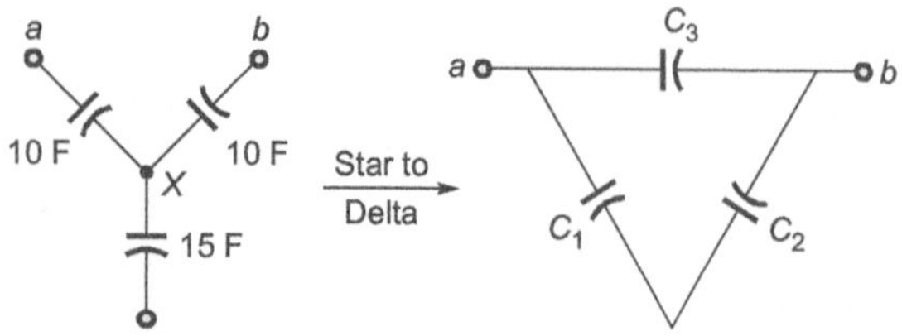

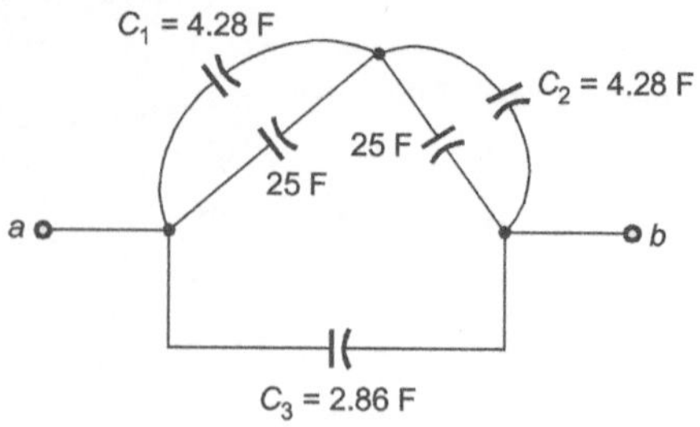

Hence, $C_1 = \frac{10 \times 15}{10 + 15 + 10} = 4.28$ F

$C_2 = \frac{15 \times 10}{15 + 10 + 10} = 4.28$ F

$C_3 = \frac{10 \times 10}{10 + 10 + 15} = 2.86$ F

Combining the parallel capacitances, the equivalent circuit reduces as shown below.

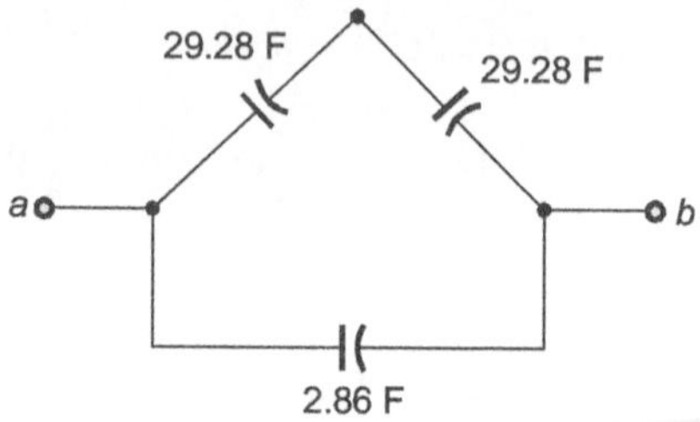

Hence, equivalent capacitance between terminals *a* and *b* is

$$C_{a\text{-}b} = \left(\frac{29.28 \times 29.28}{29.28 + 29.28} + 2.86\right)$$

$$= 17.5 \text{ F}$$

138. (c)

Capacitive energy $= \frac{1}{2} C \left[v(t)^2 - v(t_0)^2\right]$

Given, $\quad v(t) = \sin t \cdot e^{-t}$

$\therefore \quad v(10) = (\sin 10) \cdot e^{-10}$ V

and $\quad v(0) = 0$ V

$\therefore \quad \text{Energy} = \frac{1}{2} \times 2[\sin 10 \cdot e^{-10} - 0]^2$

$= \sin^2 10 \cdot e^{-20}$ Joules

139. (d)

Given waveform is a periodic function having a period of 20 ms.

Here, $\quad \frac{dv}{dt} = \frac{10}{20 \times 10^{-3}} = 0.5 \times 10^3$ V/sec

$\therefore \quad i = C\frac{dv}{dt} = 30 \times 10^{-6} \times 0.5 \times 10^3$

$= 15$ mA

140. (d)

Assertion is false because energy consumed is zero in steady state since capacitor acts as open circuit ($i = 0$) and inductor acts as short-circuit ($v = 0$) in steady state.

142. (b)

Given, $C_1 = 10\,\mu F$

and $C_2 = 50\,\mu F$; $V = 100$ volt

Initial energy stored in the capacitor of $10\,\mu F$ is

$$W_{in} = \frac{1}{2}C_1V^2$$
$$= \frac{1}{2}\times 10\times 10^{-6}\times (100)^2$$
$$= 0.05 \text{ Joules}$$

The initial charge,

$$Q_{in} = CV = 10 \times 10^{-6} \times 100$$
$$= 0.001 \text{ Coulomb}$$

When this $10\,\mu F$ capacitor is connected across the unchanged capacitor, let the final voltage be V' volt.

Then, $W_1 = \frac{1}{2}C_1V'^2$

and $W_2 = \frac{1}{2}C_2V'^2$

Again $V' = \frac{Q}{C}$

where, Q is the final charge and C is the final capacitance.

Now, $Q = Q_{in} = 0.001$ Columb

and $C_{eq} = C = C_1 + C_2$

$$= 10 + 50 = 60\,\mu F$$

Hence, final voltage

$$V' = \frac{0.001}{60\times 10^{-6}} = 16.67 \text{ V}$$

Thus, $W_1 = \frac{1}{2}C_1V'^2$

$$= \frac{1}{2}\times 10\times (16.67)^2\times 10^{-6}$$
$$= 0.0014 \text{ J} = 1.4 \text{ mJ}$$

and $W_2 = \frac{1}{2}C_2V'^2$

$$= \frac{1}{2}\times 50\times 10^{-6}\times (16.67)^2$$
$$= 0.007 \text{ J} = 7 \text{ mJ}$$

143. (a)

Under steady state condition, the capacitor will acts as open circuit.

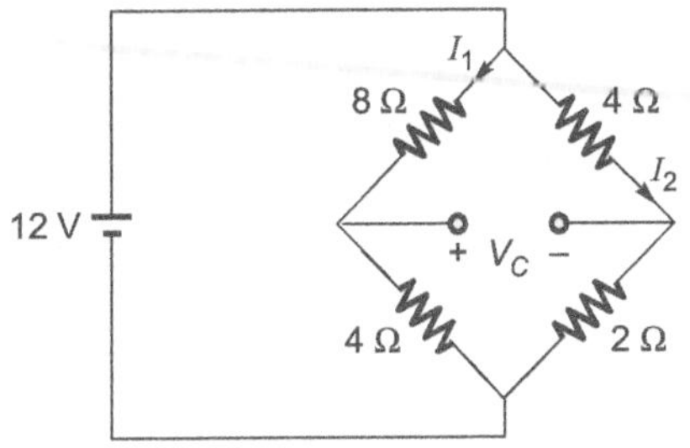

$$I_1 = \frac{12}{8+4} = 1\text{ A}\,;\; I_2 = \frac{12}{4+2} = 2\text{ A}$$

Let the voltage across capacitor be V_C in steady state.

Then, $V_C = -8I_1 + 4I_2$

$$= -8\times 1 + 4\times 2$$
$$= -8 + 8 = 0$$

Hence, charge accumulated in the capacitor is

$$Q = CV_C = (100\,\mu F)\times 0 = 0$$

144. (c)

Induced voltage,

$$V = L\frac{di}{dt} = 1\times 10^{-3}\times \frac{d}{dt}(300t)$$
$$= 10^{-3}\times 300 = 0.3 \text{ volt}$$

145. (a)

$$L = \frac{N^2\mu_0\mu_r A}{l}$$

$\therefore$ $L \propto N^2$

Since N is increased by 200% i.e. N is doubled therefore, new value of L is

$$L' = 4\times L$$
$$= 4\times 6 = 24 \text{ mH}$$

146. (d)

$$i_c(t) = C\frac{dV_c(t)}{dt}$$

or, $V_c(t) = \frac{1}{C}\int i(t)dt = \frac{5}{C}\int \delta(t)dt$

Now, $\int \delta(t)\,dt = u(t)$

So, $V_c(t) = \frac{5}{C}\cdot u(t)$

147. (b)

We know that,

$$v(t) = L\frac{di(t)}{dt}$$

or, $$i(t) = \frac{1}{L}\int v(t)dt$$

or, $$i(t) = \frac{1}{1}\int 12t^2 dt = 12\left[\frac{t^3}{3}\right] = 4\,t^3$$

148. (b)

Statement (b) holds true for an inductor not for a capacitor.

149. (c)

$$I_C = C\frac{dv}{dt} = 20\times10^{-6}\times\left(\frac{240-0}{1}\right) = 4.8\text{ mA}$$

150. (d)

Lower the self inductance of a coil less the delay in establishing steady current through it because lower will be the inductive reactance which will offer less opposition to the flow of current. Hence, assertion is false.

151. (c)

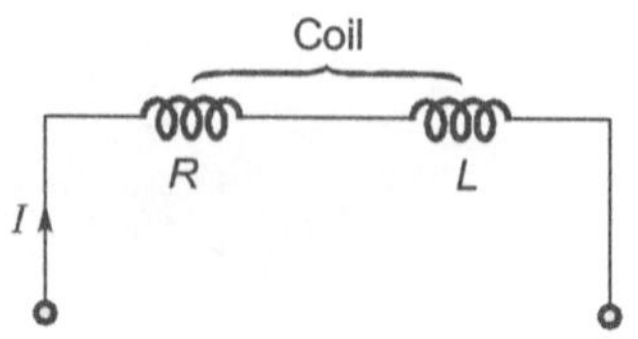

A coil acts as open circuit i.e $I = 0$ if $R = \infty$ or $L = 0$ (so that $X_L = 0\ \Omega$).

152. (a)

An inductor acts as open circuit to dc (as $f = 0$ for dc so that $X_L = 0\ \Omega$).

155. (a)

The impedance $Z(s)$ of the given network is

$$Z(s) = \left[\frac{(R+LS)\times\frac{1}{CS}}{R+LS+\frac{1}{CS}}\right] = \left[\frac{R+LS}{RCS+S^2LC+1}\right]$$

$$= \left[\frac{S+\frac{R}{L}}{S^2+\frac{R}{L}S+\frac{1}{LC}}\right]$$

The poles are at:

$$S = \frac{-R}{2L} \pm j\sqrt{\frac{1}{LC}-\left(\frac{R}{2L}\right)^2}$$

$$= \frac{-R}{2L} \pm j\sqrt{\frac{1}{LC}-\frac{R^2}{4L^2}}$$

Thus, the poles will be coincidence if

$$\left[\frac{1}{LC}-\frac{R^2}{4L^2}\right] = 0 \text{ or } R = 2\sqrt{\frac{L}{C}}$$

157. (c)

$$H(s) = \frac{C(s)}{R(s)} = \frac{1}{(s+2)}$$

Since, $R(s) = \frac{1}{s}$ therefore,

$$C(s) = \frac{1}{s(s+2)} = \frac{A}{s}+\frac{B}{(s+2)}$$

$$A = +\frac{1}{2},\quad B = -\frac{1}{2}$$

$\therefore$ $$C(s) = \frac{1}{2}\left[\frac{1}{s}-\frac{1}{s+2}\right]$$

or, $$C(t) = \frac{1}{2}[1-e^{-2t}]\,u(t)$$

For a very large time (i.e. $t\to\infty$),

$$C(t\to\infty) = \frac{1}{2} = 0.5$$

158. (c)

Time constant, $$\tau = \frac{L}{R} = \frac{I^2L}{I^2R}$$

Given, $$\frac{1}{2}LI^2 = 1000\text{ J}$$

or, $$LI^2 = 2000$$

and $$I^2R = 2000\text{ W}$$

So, $$\tau = \frac{I^2L}{I^2R} = \frac{2000}{2000} = 1\text{ sec}$$

159. (b)

Maximum output voltage of RC differentiator circuit $= RC\frac{dV_i}{dt}$

$$= 5\times10^3\times4\times10^{-6}\times100$$
$$= 20\times10^{-1} = 2\text{ volt}$$

160. (d)

The given waveform can be expressed as

$$v(t) = u(t-2) - u(t-3) + u(t-6) - u(t-7)$$

Taking Laplace transform,

$$V(s) = \frac{1}{s}e^{-2s} - \frac{1}{s}e^{-3s} + \frac{1}{s}e^{-6s} - \frac{1}{s}e^{-7s}$$

or, $$V(s) = \frac{1}{s}(e^{-2s} - e^{-3s} + e^{-6s} - e^{-7s})$$

161. (d)

$$\phi(t) = 1 + e^{-t}(\sin t + \cos t)$$
$$= 1 + e^{-t}\sin t + e^{-t}\cos t$$

or, $$\phi(s) = \frac{1}{s} + \frac{1}{(s+1)^2+1} + \frac{s+1}{(s+1)^2+1}$$

$$= \left[\frac{1}{s} + \frac{s+2}{s^2+2s+2}\right]$$

∴ $$\phi(0) = \underset{s\to\infty}{Lt}\ s\,\phi(s)$$

$$= \underset{s\to\infty}{Lt}\left[1 + \frac{s(s+2)}{s^2+2s+2}\right]$$

$$= \underset{s\to\infty}{Lt}\left[1 + \frac{1+2/s}{1+2/s+2/s^2}\right]$$

$$= 1 + 1 = 2$$

and $$\phi(\infty) = \underset{s\to 0}{Lt}\ s\,\phi(s)$$

$$= \underset{s\to 0}{Lt}\left[1 + \frac{s(s+2)}{s^2+2s+2}\right] = 1$$

162. (a)

The equation of the given ramp function is

$$f(t) = (2t+4)\,u(t) \qquad [\text{using } y = mx + c]$$
$$= 2\,(t+2)$$

∴ $$F(s) = 2\left[\frac{1}{s^2} + \frac{2}{s}\right] = \frac{2}{s^2} + \frac{4}{s} = \frac{2}{s}\left(2 + \frac{1}{s}\right)$$

163. (c)

The given circuit can be redrawn as shown below.

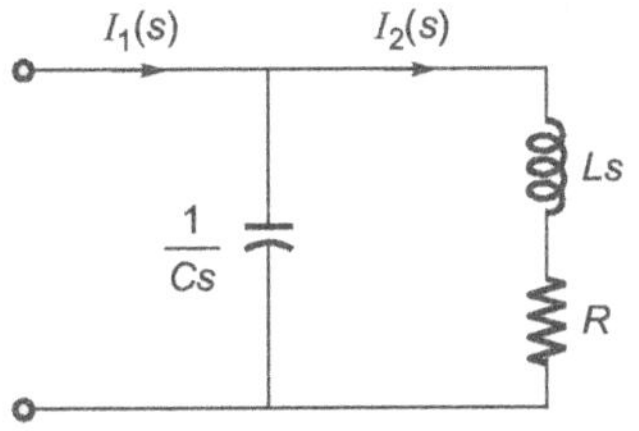

Here, $$I_2(s) = \left(\frac{1/Cs}{1/Cs + Ls + R}\right)I_1(s)$$

or, $$\frac{I_2(s)}{I_1(s)} = \frac{\left(\frac{1}{250\times10^{-6}s}\right)}{\left[\frac{1}{250\times10^{-6}s} + 10s + 200\right]}$$

or, $$\frac{I_2(s)}{I_1(s)} = \left(\frac{400}{s^2+20s+400}\right)$$

Hence, poles are at:

$$s^2 + 20s + 400 = 0$$

or, $$s = -10 \pm j10\sqrt{3}$$

There is no zero.

164. (d)

In Laplace domain, the given circuit can be drawn as shown below.

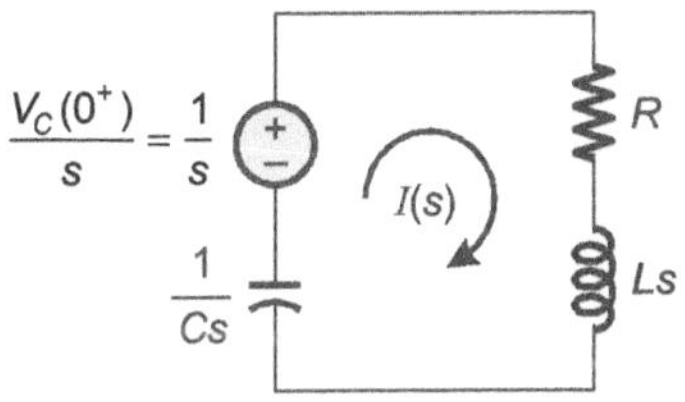

∴ $$\frac{1}{s} = I(s)\left[R + Ls + \frac{1}{Cs}\right]$$

or, $$I(s) = \frac{C}{LCs^2 + RCs + 1}$$

or, $$I(s) = \frac{0.5}{0.5s^2 + s + 1} = \frac{1}{s^2+2s+2}$$

$$= \frac{1}{(s+1)^2+1}$$

or, $$i(t) = 1\,e^{-t}\sin t\,u(t)$$

166. (c)

The derivative of function $f(t)$ is

$$f'(t) = e^{-t}\delta(t) + u(t)(-e^{-t})$$
$$= -e^{-t}u(t) + e^{-t}\delta(t)$$

∴ $$\{f'(t)\} = F'(s) = -\frac{1}{s+1} + 1 = \frac{s}{s+1}$$

167. (c)

$$V_c(t) = V(1 - e^{-t/\tau}) \text{ volts}$$
$$V = 20 \text{ volt} \quad \text{(given)}$$

At $t = \tau$, $$V_c(t)\big|_{t=\tau} = 20\,(1 - e^{-1}) \text{ volts}$$

168. (b)

$$X_c(s) = \frac{1}{Cs} = \frac{1}{\frac{7}{5}s} = \frac{5}{7s}$$

169. (a)

Time constant of series RC circuit is

$$\tau = RC = 1 \times 10^6 \times 2 \times 10^{-6}$$
$$= 2 \text{ secs}$$

Hence, time of 6 seconds will be equal to 3τ = 3 times of time constant.

170. (d)

From the given waveform, we have

$$f(t) = u(t - T_1) - u(t - T_2)$$

$$\therefore \quad F(s) = \left[\frac{1}{s}e^{-T_1 s} - \frac{1}{s}e^{-T_2 s}\right]$$

171. (b)

The given circuit in Laplace domain is shown below.

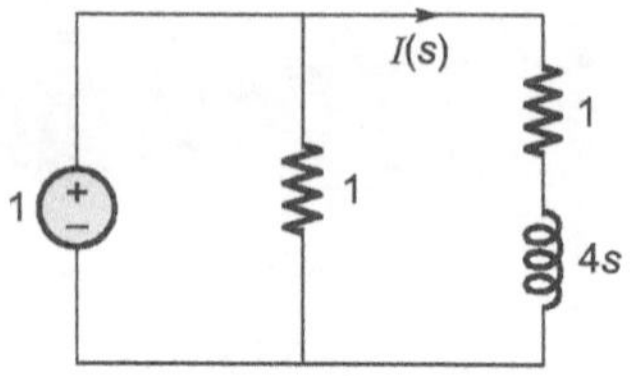

Here, $I(s) = \left(\frac{1}{4s+2}\right) \cdot V(s)$

(Using current divider theorem)

$$= \frac{1/4}{s+1/2} \times 1 = \frac{0.25}{s+0.5}$$

or, $i(t) = 0.25\, e^{-0.5t}\, u(t)$

172. (c)

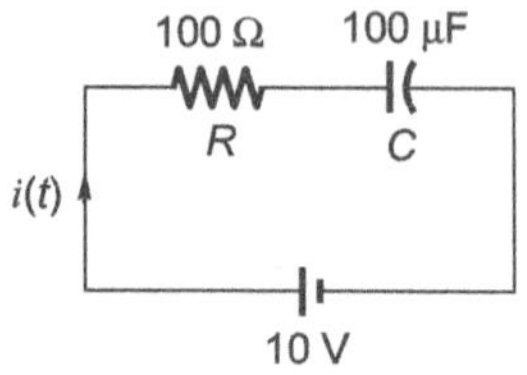

At $t = 0$, capacitor acts as short-circuit.

So, $i(0^+) = \frac{10}{100} = 0.1\text{ A}$

Time constant of given circuit is

$$\tau = \text{RC} = 100 \times 100 \times 10^{-6}$$
$$= 10^{-2} \text{ sec} = 0.01 \text{ sec}$$

With zero initial condition, current in the series RC circuit is

$$i(t) = \frac{V}{R} e^{-t/\tau}$$

$$= \frac{10}{100} e^{-t/0.01} = 0.1 e^{-100t}$$

Hence, $\frac{di}{dt} = -10\, e^{-100t}$

$$i(0^+) = 0.1 \text{ A}$$

$$\therefore \quad \frac{di(0^+)}{dt} = -10 \text{ A/sec}$$

173. (a)

Given circuit is a low-pass filter having transfer function

$$\frac{V_2(s)}{V_1(s)} = \frac{1}{1+RCs} = \frac{1}{1+10\times 0.5s}$$

$$= \frac{1}{5s+1} = \frac{0.2}{s+0.2}$$

When, $V_1(t) = 10\, e^{-2t}$

then, $V_1(s) = \frac{10}{(s+2)}$

So, $V_2(s) = \frac{2}{(s+0.2)(s+2)}$

$$= \frac{A}{s+0.2} + \frac{B}{s+2}$$

Here, $A = \left.\frac{2}{s+2}\right|_{s=-0.2} = \frac{2}{1.98} = 1.11$

and $B = \left.\frac{2}{s+0.2}\right|_{s=-2} = -1.11$

So, $V_2(s) = \frac{1.11}{s+0.2} - \frac{1.11}{s+2}$

or, $V_2(s) = 1.11\,(e^{-0.2t} - e^{-2t})$

174. (b)

For $t = 0^-$, switch S is closed. Hence, in steady state capacitor will act as open circuit.

Thus, $V_c(0^-) = V_c(0^+) = 2$ volts

The circuit at $t = 0^+$ is shown below where, capacitor acts as a constant voltage source.

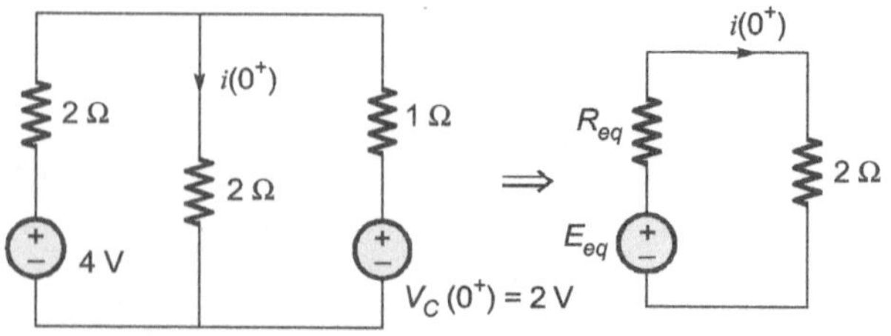

Using source transformation,

$$E_{eq} = \frac{4 \times \frac{1}{2} + 2 \times 1}{\frac{1}{2} + 1} = \frac{2+2}{3/2} = \frac{8}{3}\text{ V}$$

and $R_{eq} = \frac{1}{1+\frac{1}{2}} = \frac{2}{3}\,\Omega$

Hence, $i(0^+) = \left(\frac{E_{eq}}{R_{eq}+2}\right)$

$$= \frac{8/3}{2/3+2} = \frac{8}{8} = 1\text{ A}$$

175. (d)

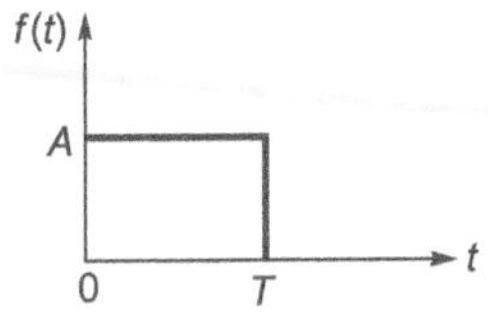

$$f(t) = A\,u(t) - A\,u(t-T)$$

$\therefore$ $f(s) = \frac{A}{s} - \frac{A}{s}e^{-Ts} = \frac{A}{s}(1-e^{-Ts})$

176. (c)

R_{eq} across inductor $= \frac{5\times4}{9} = \frac{20}{9}\,\Omega$

(5 A C.S open circuited)

$\therefore$ Time constant of RL circuit is

$$\tau = \frac{L}{R_{eq}} = \frac{1}{6}\times\frac{9}{20} = \frac{3}{40}\text{ sec}$$

$$= 0.075\text{ sec}$$

179. (d)

For finding time constant of the circuit 5 A current source is open circuited and 2 V voltage source is short circuited.

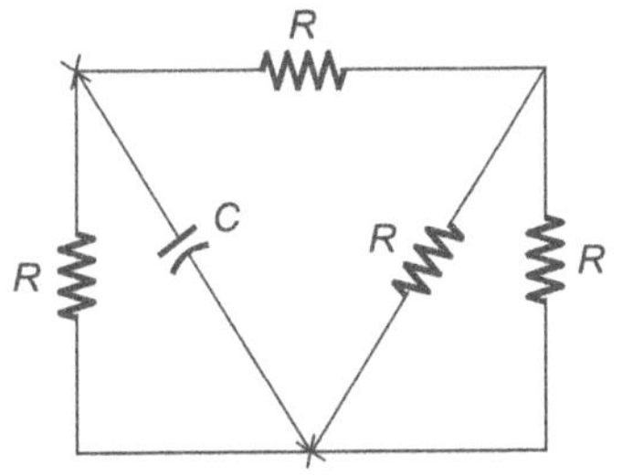

$\tau = R_{eq}C_{eq}$ = Time constant

$$R_{eq} = \left(\frac{3R}{2}\right)||R = \left(\frac{\frac{3R}{2}\times R}{\frac{3R}{2}+R}\right) = \frac{3}{5}R$$

$$C_{eq} = C$$

$\therefore$ $\tau = \frac{3}{5}RC = 0.6\,RC$

180. (a)

$$i(0^+) = \underset{s\to\infty}{Lt}\,[sI(s)]$$

$$= \underset{s\to\infty}{Lt}\, s\cdot\frac{(s+2)(s+4)}{(s+1)(s+3)\cdot s} = 1\text{ A}$$

Also, $i(\infty) = \underset{s\to0}{Lt}\,[sI(s)]$

$$= \underset{s\to0}{Lt}\, s\cdot\frac{(s+2)(s+4)}{(s+1)(s+3)\cdot s} = \frac{8}{3}\text{ A}$$

182. (c)

Characteristic equation of a parallel RLC circuit is

$$s^2 + \frac{1}{RC}s + \frac{1}{LC} = 0$$

Here, $\omega_n = \frac{1}{\sqrt{LC}}$ and $2\xi\omega_n = \frac{1}{RC}$

or, $\xi = \frac{1}{2RC\,\omega_n} = \frac{\sqrt{LC}}{2RC} = \frac{1}{2R}\sqrt{\frac{L}{C}}$

For underdamped circuit,

$$\xi < 1 \text{ or } \frac{1}{2R}\sqrt{\frac{L}{C}} < 1$$

or, $R > \frac{1}{2}\sqrt{\frac{L}{C}}$

183. (d)

At $t = 0^+$, inductor will act as open circuit therefore, $i(0^+) = 0$ A.

Applying KVL in the loop, we have

$$10 = iR + L\frac{di}{dt} + \frac{1}{C}\int i\,dt$$

At $t = 0^+$, $\frac{1}{C}\int i\,dt = v_c(0^+) = 0$ volt

So, $10 = i(0^+)R + L\frac{di(0^+)}{dt}$

or, $\frac{di(0^+)}{dt} = \frac{10 - i(0^+)\times10}{1} = 10\text{ A/sec}$

Now, $10 = i(t)R + L\frac{di}{dt}$

Differentiating the above equation w.r.t time, t we have

$$0 = R\frac{di(t)}{dt} + L\frac{d^2i(t)}{dt^2}$$

$$\text{or,} \quad \frac{d^2 i(t)}{dt^2} = \frac{-R\frac{di(t)}{dt}}{L}$$

$$\frac{d^2 i(0^+)}{dt^2} = \frac{-10\times 10}{1} = -100 \text{ A/s}^2$$

184. (a)

We know that,

$$P(t) = \frac{d(\text{Energy})}{dt}$$

$$\text{or,} \quad P(s) = s\,E(s)$$

$$\text{Hence,} \quad P(0) = \underset{s\to\infty}{Lt}[sP(s)]$$

$$= \underset{s\to\infty}{Lt}\left[\frac{s^2(s+8)}{(s+1)(s+3)(s+5)}\right]$$

$$= 1 \text{ unit}$$

$$\text{and} \quad P(\infty) = \underset{s\to 0}{Lt}[sP(s)]$$

$$= \underset{s\to 0}{Lt}\left[\frac{s^2(s+8)}{(s+1)(s+3)(s+5)}\right]$$

$$= 0 \text{ unit}$$

185. (b)

$$V(s) = \frac{8s-17}{(s+1)(s-4)} = \frac{A}{s+1} + \frac{B}{s-4}$$

$$\text{Here,} \quad A = \left.\frac{8s-17}{s-4}\right|_{s=-1} = 5$$

$$\text{and} \quad B = \left.\frac{8s-17}{s+1}\right|_{s=4} = 3$$

$$\text{So,} \quad V(s) = \frac{5}{s+1} + \frac{3}{s-4}$$

$$\text{or,} \quad V(t) = 5e^{-t} + 3e^{4t}$$

$$\therefore \quad V(\infty) = 0 + \infty = \infty$$

186. (c)

At $t = 0^+$, $s \to \infty$

Hence, in N^r of $V_c(s)$, $s >> 1$

Also, in D^r of $V_c(s)$, at $s \to \infty$, $s^3 >> s^2 >> s >> 1$

$$\text{i.e.} \quad \underset{s\to\infty}{Lt}[V_c(s)] = \frac{s}{s^3} = \frac{1}{s^2}$$

$$\text{Now,} \quad I_c(s) = \frac{V_c(s)}{(1/Cs)} = Cs\,V_c(s)]$$

$$\therefore \quad I_c(0) = \underset{s\to\infty}{Lt}[sI_C(s)]$$

$$= \underset{s\to\infty}{Lt}[s\cdot CsV_c(s)]$$

$$= \underset{s\to\infty}{Lt}[s^2 C\cdot V_c(s)]$$

$$= \underset{s\to\infty}{Lt}\left[s^2 C\times\frac{1}{s^2}\right] = C = 2\text{ A}$$

Hence, initial value of current through the capacitor = 2 A.

187. (d)

The given circuit in Laplace domain is shown below.

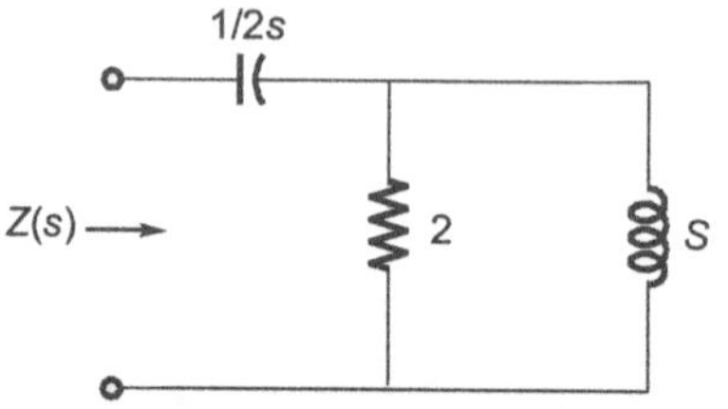

$$Z(s) = \left(\frac{2\times s}{2+s} + \frac{1}{2s}\right) = \left(\frac{2s}{s+2} + \frac{1}{2s}\right)$$

$$= \frac{4s^2+s+2}{2s(s+2)} = \left(\frac{4s^2+s+2}{2s^2+4s}\right)$$

188. (a)

Under steady state condition inductors will act as short circuit while capacitors will act as open circuit.

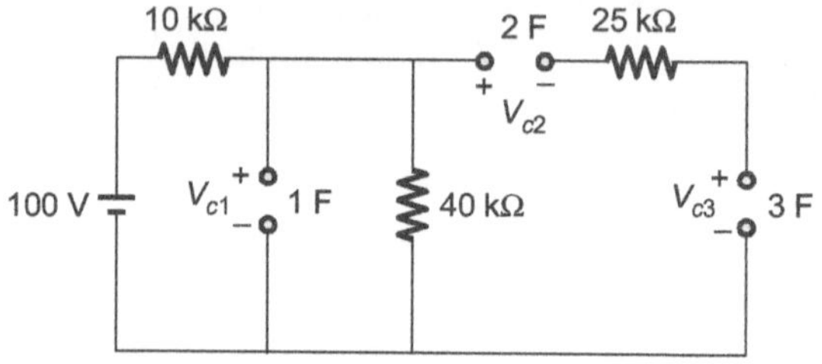

$$\text{Hence,} \quad V_{c1} = \left(\frac{40}{40+10}\right)\times 100 = \frac{40}{50}\times 100$$

$$= 80 \text{ volts}$$

$$\text{Now,} \quad V_{c1} = V_{c2} + V_{c3}$$

$$\text{or,} \quad V_{c2} + V_{c3} = 80 \qquad \ldots(i)$$

$$\text{Also,} \quad V_c \propto \frac{1}{C}$$

$$\therefore \quad \frac{V_{c2}}{V_{c3}} = \frac{3}{2}$$

$$\text{or,} \quad V_{c2} = \frac{3}{5}V_{c1} = \frac{3}{5}\times 80 = 48 \text{ volts}$$

$$\text{and} \quad V_{c3} = 80 - V_{c2}$$

$$= 80 - 48 = 32 \text{ volts}$$

$$\text{Hence,} \quad V_{c1} = 80\text{ V},\ V_{c2} = 48\text{ V},\ V_{c3} = 32\text{ V}$$

189. (c)

$$Z(s) = \frac{(R+Ls)\times R}{2R+Ls}+\frac{1}{Cs}$$

$$= \frac{RCs(R+Ls)+2R+Ls}{Cs(Ls+2R)}$$

$$= \frac{RLCs^2+(R^2C+L)s+2R}{LCs\left(s+\frac{2R}{L}\right)}$$

Given, $Z(s) = \frac{[s^2+1.5s+1]}{s(s+1)}$

On comparing the two values of $Z(s)$, we get

$$R = 1\ \Omega;\quad \frac{2R}{L} = 1$$

or, $L = 2\text{ H and } \frac{2R}{LC} = 1$

or, $C = \frac{2}{2} = 1\text{ F}$

190. (a)

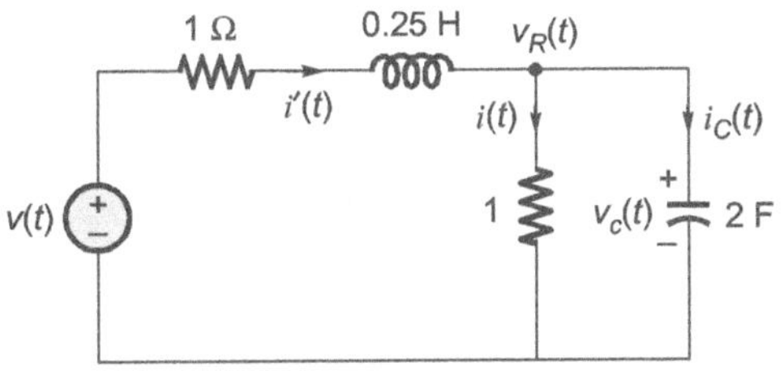

$$v_R(t) = i(t)\times 1 = -20\,e^{-2t} = v_c(t)$$

$\therefore$ $i_c(t) = C\frac{dv_c(t)}{dt} = 2\times[40e^{-2t}]$

$$= 80\,e^{-2t}$$

Now, $i'(t) = i(t) + i_c(t)$

$$= -20e^{-2t} + 80\,e^{-2t} = 60\,e^{-2t}$$

So, $v(t) = i'(t)\times 1+0.25\frac{di'(t)}{dt}+v_R(t)$

$$= 60\,e^{-2t} + 0.25\times(-120e^{-2t}) + (-20e^{-2t})$$

or, $v(t) = (60-30-20)\,e^{-2t} = 10e^{-2t}$

191. (d)

$$R_{eq} = R' = \frac{R\times 120}{R+120} = \left(\frac{120R}{120+R}\right)$$

For an RLC series circuit to be critically damped,

$$R_{eq} = R' = \sqrt{\frac{4L}{C}}$$

or, $\frac{120R}{R+120} = \sqrt{\frac{4\times 4}{10\times 10^{-3}}} = 40$

or, $120\,R = 40\,(R+120)$

or, $3\,R = R + 120 \text{ or } R = 60\ \Omega$

192. (b)

Given,

$$\frac{d^2v}{dt^2}+v = t$$

Taking Laplace transform on both sides,

$$[s^2\,V(s) - sV(0^+) - V'(0^+)] + V(s) = \frac{1}{s^2}$$

or, $V(s)\,(1+s^2) - sV(0^+) - V'(0^+) = \frac{1}{s^2}$

or, $V(s)\,(1+s^2) - s + 2 = \frac{1}{s^2}$

or, $V(s) = \frac{\left(\frac{1}{s^2}+s-2\right)}{(1+s^2)} = \frac{s^3-2s^2+1}{s^2(1+s^2)}$

or, $V(s) = \frac{A}{s^2}+\frac{Bs+C}{s^2+1}$

Using partial fraction expansion,

$A = 1,\quad B = 1,$

$C = -3$

$\therefore$ $V(s) = \frac{1}{s^2}+\frac{s}{s^2+1}-\frac{3}{s^2+1}$

So, $V(t) = (t+\cos t - 3\sin t)$

193. (c)

Given, $V(\infty) = \underset{s\to 0}{Lt}\,[sV(s)] = \underset{s\to 0}{Lt}\left[\frac{1}{s+a}\right]$

$$= \frac{1}{a} = 2$$

or, $a = 0.5$

Thus, $V(s) = \frac{1}{s(s+a)} = \frac{1}{s(s+0.5)}$

$$= 2\left(\frac{1}{s}+\frac{1}{s+0.5}\right)$$

or, $v(t) = 2(1-e^{-0.5t})$

Hence, $v(1) = 2(1-e^{-0.5})$

195. (c)

Characteristic equation of series RLC circuit is

$$s^2 + \frac{R}{L}s + \frac{1}{LC} = 0$$

Here, $\omega_n = \frac{1}{\sqrt{LC}}$

and $2\xi\omega_n = \frac{R}{L}$

or, $\xi = \frac{R}{2L} \times \sqrt{LC} = \frac{R}{2}\sqrt{\frac{C}{L}}$

For oscillatory response,

$$\xi < 1 \quad \text{or} \quad \frac{R}{2}\sqrt{\frac{C}{L}} < 1$$

or, $R < 2\sqrt{\frac{L}{C}}$

197. (b)

Applying KCL at the given node in *s*-domain, we have

$$\frac{V_c(s) - V_i(s)}{4s} + \frac{V_c(s)}{8} + \frac{V_c(s)}{2/s} = 0$$

or, $V_c(s)\left[\frac{1}{4s} + \frac{1}{8} + \frac{s}{2}\right] = \frac{V_i(s)}{4s}$

or, $\frac{V_c(s)}{V_i(s)} = \frac{2}{4s^2 + s + 2}$

Hence, there is no-zero and poles are at

$$4s^2 + s + 2 = 0$$

or, $s = \frac{-1 \pm \sqrt{1-32}}{2 \times 4} = \frac{-1}{8} \pm \frac{j\sqrt{31}}{8}$

198. (a)

Because the transient response of an electrical circuit can best be described by a differential equation hence, Laplace transformation finds its application in solving the transient behaviour of the electric circuits.

199. (d)

$$f(t) = e^{-t}\sin 2t + e^{-2t}\cos 5t$$

$\therefore$ $F(s) = \frac{2}{(s+1)^2 + 4} + \frac{s+2}{(s+2)^2 + 25}$

$$= \left[\frac{2}{s^2 + 2s + 5} + \frac{s+2}{s^2 + 4s + 29}\right]$$

$\therefore$ Find value of $f(t)$ is

$$f(\infty) = \underset{s \to 0}{Lt}\,[sF(s)]$$

$$= \underset{s \to 0}{Lt}\left[\frac{2s}{s^2 + 2s + 5} + \frac{s^2 + 2s}{s^2 + 4s + 29}\right] = 0$$

200. (b)

At $t = 0^+$, capacitor will act as short circuit while inductor an open circuit

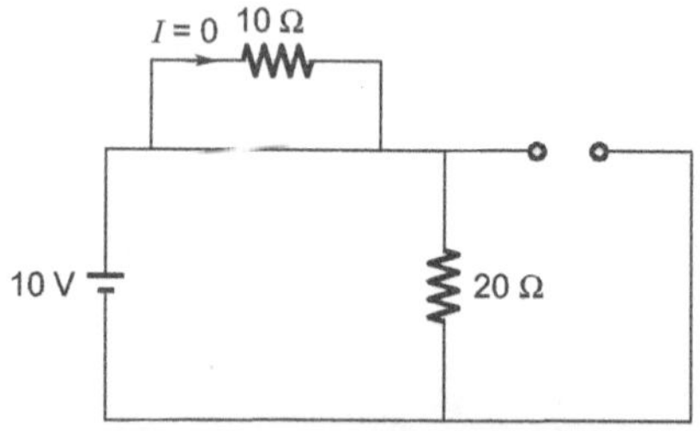

$\therefore$ $V_{20\,\Omega} = 10$ volt

201. (c)

Applying node analysis,

$$-1 + \frac{V_c(s)}{1} + \frac{V_c(s)}{(1/s)} + \frac{V_c(s)}{2 + 0.5s} = 0$$

or, $V_c(s)\left[1 + s + \frac{2}{s+4}\right] = 1$

or, $V_c = \frac{s+4}{(s+3)(s+2)} = \frac{A}{s+3} + \frac{B}{s+2}$

Here, $A = \frac{(s+4)}{(s+3)(s+2)} \cdot (s+3)\Big|_{s=-3} = -1$

$B = \frac{(s+4)}{(s+3)(s+2)} \cdot (s+2)\Big|_{s=-2} = 2$

$\therefore$ $V_c(s) = \frac{-1}{s+3} + \frac{2}{s+2}$

or, $V_c(t) = (-e^{-3t} + 2e^{-2t})$ V

202. (a)

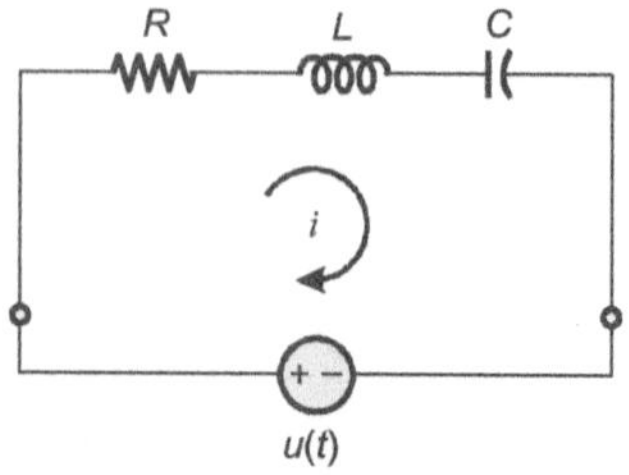

At $t = 0$, inductor act as open circuit

$\therefore$ $i(0^+) = 0$ A

At $t = \infty$, capacitor act as open circuit

$\therefore$ $i(\infty) = 0$ A

203. (c)

At $t = \infty$, inductor will act as short circuit and capacitor open circuit

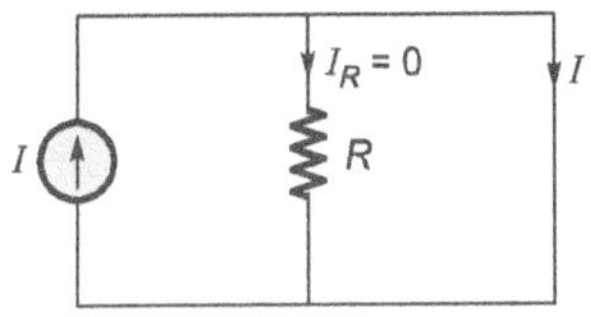

$\therefore \quad I_R = 0$ A

204. (d)

$$s^2 + \frac{R}{L}s + \frac{1}{LC} = 0 \quad \text{(characteristic equation)}$$

Therefore, $\quad \xi = \frac{R}{2}\sqrt{\frac{C}{L}}$

If $R = 0$, $\xi = 0$ (vice-versa) and for $\xi < 1$, response will be oscillatory.

205. (c)

It is clear that i leads v by 53.2° (< 90°). Hence, the two series elements must be R and C.

$\therefore \quad \tan\theta = \tan 53.3°$

$\tan 45° < \tan 53.3° < \tan 60°$

or, $\quad 1 < \tan 53.3° < 1.732$

Hence, $\tan 53.3° = 1.3367$

206. (b)

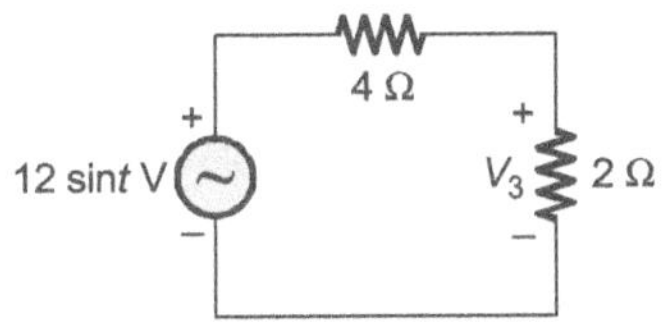

Using voltage division rule, we have

$$V_3 = \frac{2}{6} \times 12 \sin t = 4 \sin t \text{ V}$$

207. (d)

Using KCL at the given node,

$$i_c(t) = i_R(t) + i_L(t)$$

$$= 10 \sin t + \frac{1}{L}\int v_L(t)\, dt$$

$$= 10 \sin t + \frac{1}{1}\int 5 \cos t\, dt$$

$$= 10 \sin t + 5 \sin t$$

or, $\quad i_c(t) = 15 \sin t$ A

Hence, rms value of

$$i_c(t) = \frac{15}{\sqrt{2}} \text{ A}$$

208. (a)

From given figure,

$$I_R = \frac{120}{10} = 12 \text{ A}$$

$$I_L = \frac{120}{15} = 8 \text{ A}$$

and $\quad I_C = \frac{120}{5} = 24$ A

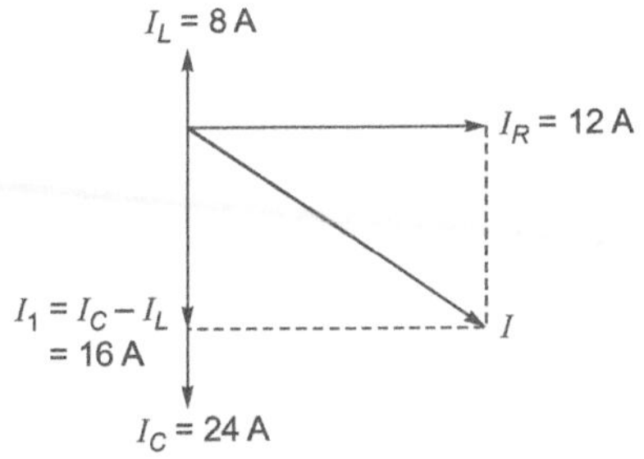

Using phasor diagram shown below, the currents I and I_1 are respectively

$$I_1 = I_C - I_L = 24 - 8 = 16 \text{ A}$$

and $\quad I = \sqrt{I_R^2 + (I_C - I_L)^2}$

$$= \sqrt{12^2 + 16^2} = 20 \text{ A}$$

209. (b)

The input impedance Z of given circuit is

$$Z = \left(\frac{-j}{\omega C}\right) + j4 + \left(\frac{j4 \times 4}{4 + j4}\right)$$

$$= \frac{-j}{\omega C} + j4 + \frac{j4}{1 + j}$$

$$= \frac{-j}{\omega C} + j4 + j2 + 2$$

$$= \frac{-j}{\omega C} + j6 + 2$$

For p.f. to be unity,

$$Img(z) = 0$$

i.e. $\quad 6 - \frac{1}{\omega C} = 0 \quad \text{or} \quad \omega C = \frac{1}{6}$

or, $\quad C = \frac{1}{6\omega} = \frac{1}{6 \times 314} = 530\ \mu\text{F}$

(Here, $\omega = 2\pi f = 2\pi \times 50 = 100\,\pi = 314$ rad/s)

210. (c)

Here, i lags v by 30°

Also, $\quad Z = \frac{V_m}{I_m} = \frac{10}{1} = \sqrt{R^2 + (314L)^2}$

or, $R^2 + (314L)^2 = 100$ ...(*i*)

Also, $\tan 30° = 0.577 = \dfrac{314L}{R}$ ...(*ii*)

Solving, equation (*i*) and (*ii*),

$L = 15.9 \text{ mH};$

$R = 8.66\ \Omega$

211. (a)

The situation is shown in figure below with it's phasor diagram.

Given, $R = 15\ \Omega$ and $X_L = 10\ \Omega$

So, $|Z| = \sqrt{R^2 + X_L^2} = 18.03\ \Omega$

$\therefore \quad Z = 18.03 \angle \theta\ \Omega \quad \left(\tan\theta = \dfrac{X_L}{R}\right)$

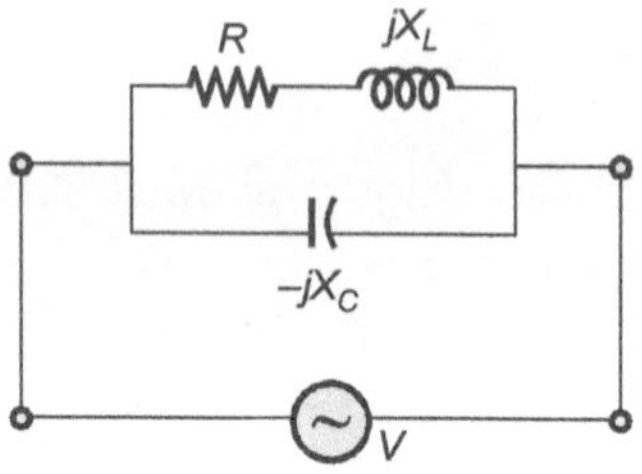

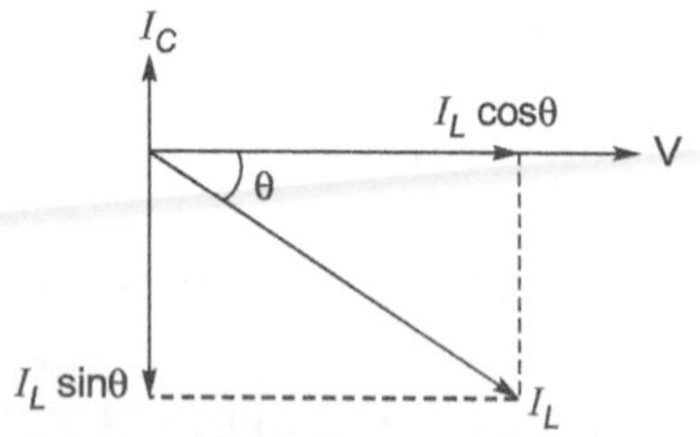

For unity p.f. operation,

$I_C = I_L \sin\theta$

$$\frac{V}{X_C} = \frac{V}{Z} \cdot \frac{X_L}{Z}$$

or, $X_C = \dfrac{Z^2}{X_L} = \dfrac{R^2 + X_L^2}{X_L} = \dfrac{325}{10} = 32.5\ \Omega$

or, $X_C = \dfrac{1}{2\pi f C}$

or, $C = \dfrac{1}{2\pi f X_C} = \dfrac{1}{2\pi \times 50 \times 32.5} = 98\ \mu\text{F}$

212. (d)

$$Z_{in} = -j4 + \frac{j2(2 - j2)}{j2 + 2 - j2}$$

$$= -j4 + \frac{j4 + 4}{2} = -j4 + 2 + j2 = 2 - j2$$

or, $Z_{in} = 2.828\angle{-45°}\ \Omega$

Hence, input p.f. = cos45° = 0.707 (lead)

213. (c)

$$Z_{in} = 5 + \frac{(-j4) \times (2 + j2)}{(2 + j2 - j4)}$$

$$= 5 + \left(\frac{8 - j8}{2 - j2}\right) = 5 + \frac{4(1 - j)}{(1 - j)}$$

$$= 9\ \Omega$$

Hence, $Y_{in} = \dfrac{1}{Z_{in}} = \dfrac{1}{9}\ \Omega$

215. (a)

$$Z(s) = \frac{(R + Ls) \times \frac{1}{Cs}}{R + Ls + \frac{1}{Cs}}$$

$$= \frac{R + Ls}{1 + RCs + LCs^2}$$

$$\therefore \quad Y(s) = \frac{1}{Z(s)} = \frac{1 + RCs + LCs^2}{R + Ls}$$

216. (b)

When Z_1 is connected across d.c. source, it draws 10 A current.

$\therefore \quad R_1 = \dfrac{10 \text{ V}}{10 \text{ A}} = 1\ \Omega$

Again, when Z_1 is connected across a.c. source, it draws 8 A.

$\therefore \quad Z_1 = \dfrac{10}{8} = 1.25\ \Omega$

Thus, $X_1 = \sqrt{Z_1^2 - R_1^2} = \sqrt{1.25^2 - 1}$

$$= \sqrt{(1.25 + 1) \times (1.25 - 1)}$$

$$= \sqrt{2.25 \times 0.25} = 0.75\ \Omega$$

Thus, $Z_1 = (1 + j0.75)\ \Omega$

Now, when Z_2 is connected across 10 V d.c. source, it doesn't draw any current. Hence, Z_2 must be a capacitor which will act as open circuit for d.c. source.

Also, Z_2 draws 5 A when connected to 10 V a.c. source.

$\therefore \quad I = \dfrac{V}{X_C}$

or, $X_C = \dfrac{V}{I} = \dfrac{10}{5} = 2\ \Omega$

Thus, $Z_2 = -j2\ \Omega$

217. (d)

Since two voltage sources are present, we apply. Superposition theorem to find current through the capacitor i.e. current through $-j5\ \Omega$ reactance.

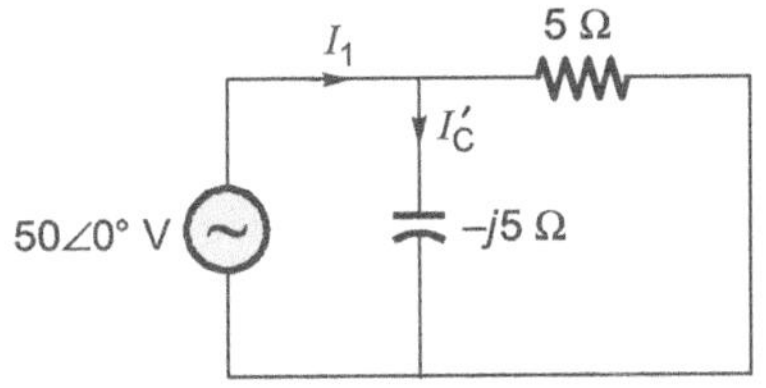

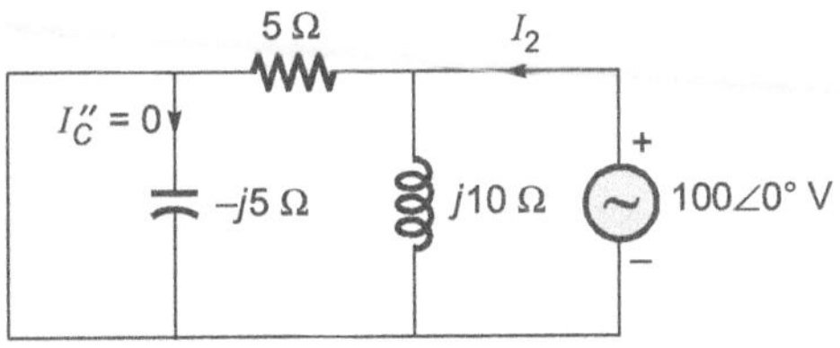

Here, $I_C' = \dfrac{50\angle 0^\circ}{-j5} = 10\angle 90^\circ$ A

and $I_C'' = 0$ A

So, $I_C = I_C' + I_C''$

$= 10\angle 90^\circ \text{ A} = j10 \text{ A}$

218. (b)

The characteristic equation of a parallel RLC circuit is

$$s^2 + \frac{1}{RC}s + \frac{1}{LC} = 0$$

Also, $s^2 + 2\xi\omega_n s + \omega_n^2 = 0$

Thus, $\omega_n = \dfrac{1}{\sqrt{LC}}$

and $2\xi\omega_n = \dfrac{1}{RC}$

$\therefore \quad \xi = \dfrac{1}{2R}\sqrt{\dfrac{L}{C}}$

for underdamped circuit, $\xi < 1$

$$\therefore \quad \frac{1}{2R}\sqrt{\frac{L}{C}} < 1 \text{ or } R > \frac{1}{2}\sqrt{\frac{L}{C}}$$

220. (b)

As current is increased therefore, new impedance should reduce. If the unknown element will be inductive, the new impedance will be $(X_L - X_C)\ \Omega$ so that the new value of current will increase.

221. (b)

T.F. $H(s) = \dfrac{s+1}{s}$

$\therefore \quad H(\omega) = \dfrac{1+j\omega}{j\omega} = \dfrac{1+j}{j}$

or, $H(\omega) = \dfrac{\sqrt{2}\angle 45^\circ}{1\angle 90^\circ} = \sqrt{2}\angle -45$

$= \dfrac{V_0(\omega)}{V_i(\omega)}$

Thus, $V_0 = \sqrt{2}\cos(t - 45^\circ)$

$= \sqrt{2}\cos\left(t - \dfrac{\pi}{4}\right)$

222. (b)

$$Z_T = \frac{(-j120)(60+j60)}{(-j120+60+j60)}$$

$$= \frac{(-j120)(1+j)60}{(60-j60)} = \frac{(-j120)(1+j)}{(1-j)}$$

$$Z_T = \frac{(-j120)(1-1+j2)}{2}$$

$$= (-j60)(j2) = 120\ \Omega$$

So, $I = \dfrac{V}{|Z_T|} = \dfrac{120}{120} = 1 \text{ A} = (1 + j0) \text{ A}$

223. (d)

From the pole-zero plot,

$$Z(s) = \frac{s+1}{\left(s+\frac{1}{\sqrt{2}}\right)(s+\sqrt{2})}$$

$$\therefore \quad Z(j\omega) = \frac{1+j}{\left(\frac{1}{\sqrt{2}}+j\right)(\sqrt{2}+j)}$$

(Since sin$t$ $= i(t)$; $\omega = 1$ rad/s)

$$\therefore \quad |Z(j\omega)| = \frac{\sqrt{2}}{\sqrt{\frac{3}{2}}\cdot\sqrt{3}} = \frac{2}{3}$$

so, $|V| = |I|\cdot|Z| = 1\cdot\dfrac{2}{3} = \dfrac{2}{3}$ volt

224. (b)

Given, $I_S = 5\cos t \quad \therefore \ \omega = 1$ rad/s

So, $X_L = j\omega L = j2\ \Omega$

$$\therefore \quad I_L = \left(\frac{2}{2+j2}\right)I_S = \left(\frac{2}{2+j2}\right)\times(5\angle 0°)$$

$$= \frac{5}{1+j} = \frac{5}{\sqrt{2}}\angle -45° \text{ A}$$

225. (c)

$$Z = \frac{|V|}{|I|} = \frac{100}{1} = 100\ \Omega$$

Also, $\phi = -60° \quad \therefore \cos\phi = \frac{R}{Z}$

or, $\cos 60° = \frac{R}{100}$

or, $R = 100\cos 60°$

$$= 100\times\frac{1}{2} = 50\ \Omega$$

226. (c)

For the square wave shown in given figure,

$$I_{rms} = \left(\frac{1}{T}\int_0^T 1^2 dt\right)^{1/2} = 1 \text{ A}$$

and $I_{max} = 1$ A

Hence, $\frac{I_{max}}{I_{rms}} = 1$

228. (b)

At $t = 0$, the given circuit is as shown below.

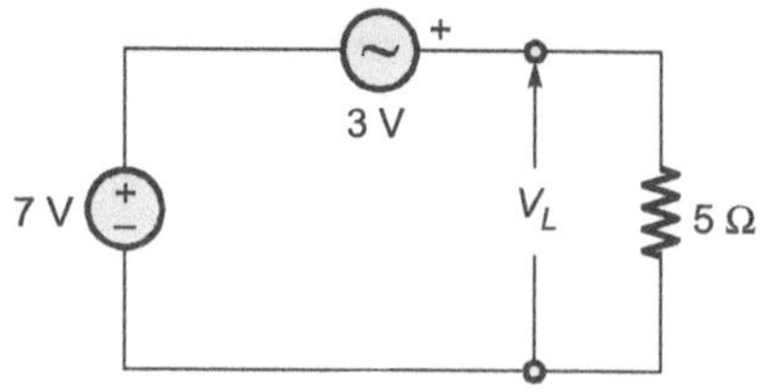

$$V_L = 7 + 3 = 10 \text{ volt}$$

$$\therefore \quad P_L = P_{5\Omega} = \frac{V_L^2}{5} = \frac{100}{5} = 20 \text{ watts}$$

229. (c)

For a sinusoidal waveform,

$$V_{avg} = \frac{2V_m}{\pi} \quad \text{and} \quad V_{rms} = \frac{V_m}{\sqrt{2}}$$

$$\therefore \quad \frac{V_{avg}}{V_{rms}} = \frac{2V_m}{\pi}\times\frac{\sqrt{2}}{V_m} = \frac{2\sqrt{2}}{\pi}$$

230. (b)

$$|Z_1| = \sqrt{3^2+4^2} = 5\ \Omega$$

and $|Z_2| = 10\ \Omega$

Now, $|I_1| = \frac{V}{Z_1} = \frac{V}{5}$

and $|I_2| = \frac{V}{Z_2} = \frac{V}{10}$

Thus, $\frac{|I_1|}{|I_2|} = 2$

$$\therefore \quad \frac{P_1}{P_2} = \frac{I_1^2 R_1}{I_2^2 R_2} = \left(\frac{I_1}{I_2}\right)^2 \times \frac{R_1}{R_2}$$

$$= 4\times\frac{3}{10} = \frac{6}{5}$$

231. (d)

$$P_{out} = 0.2 \text{ HP} = 0.2 \times 746 \text{ W}$$
$$= 149.2 \text{ W}$$

Now, $\eta = \frac{P_{out}}{P_{in}}$

or, $P_{in} = \frac{P_{out}}{\eta} = \frac{149.2}{0.85} = 175.53$ W

Also, p.f. $= \cos\phi = 0.8$

$\therefore \quad \sin\phi = 0.6$

$\therefore$ Reactive power taken by motor is

$$\phi = VI\sin\phi = \frac{P_{in}}{\cos\phi}\times\sin\phi$$

$$= P_{in}\tan\phi$$

$$= \frac{149.2}{0.85}\times\frac{0.6}{0.8} = 131.65 \text{ VAR}$$

232. (a)

$$V_{rms} = \frac{100}{\sqrt{2}} \text{ V}, \quad I_{rms} = \frac{5}{\sqrt{2}} \text{ A}$$

Also, $\phi = 30 - (-30°) = 60°$

$\therefore \quad \cos\phi = 0.5$

$\therefore$ Apparant power,

$$S = V_{rms} I_{rms} = \frac{100}{\sqrt{2}}\times\frac{5}{\sqrt{2}} = 250 \text{ VA}$$

Reactive power,

$$Q = VI\sin\phi = \frac{100}{\sqrt{2}}\times\frac{5}{\sqrt{2}}\times\sin 60°$$

$$= 217 \text{ VAR}$$

233. (c)

$$\cos\phi = \frac{P}{S} = \frac{\text{Active power}}{\text{Apparant power}}$$

$$= \frac{320}{4\times 100} = 0.8$$

Now, power dissipated,

$$P = I^2R$$

or, $$R = \frac{P}{I^2} = \frac{320}{4^2} = 20\ \Omega$$

∴ Resistive part of Z is

$$R = 20\ \Omega$$

Now, $$V_Z = IZ = 179 \text{ volt}$$

or, $$Z = \frac{V_Z}{I} = \frac{179}{4} = 44.75\ \Omega$$

or, $$\sqrt{R^2+X^2} = 44.75$$

or, $$X = \sqrt{(44.75)^2-(20)^2} = 40\ \Omega$$

234. (d)

Power lost in the 5 Ω resistor is

$$P_{5\Omega} = \text{Re}\,\{(I_1-I_2)^2\times 5\}$$

(As current through 5 Ω resistor = $I_1 - I_2$)

$$= \text{Re}\,\{(0.8+j0.2+1-j0.5)\times 5\}$$
$$= \text{Re}\,\{(1.8-j0.3)^2\times 5\}$$
$$= \text{Re}\,\{15.75-j5.4\}$$
$$= 15.75 \text{ watt}$$

235. (c)

Given, ω = 1 rad/sec. So, the circuit can be redrawn as shown below.

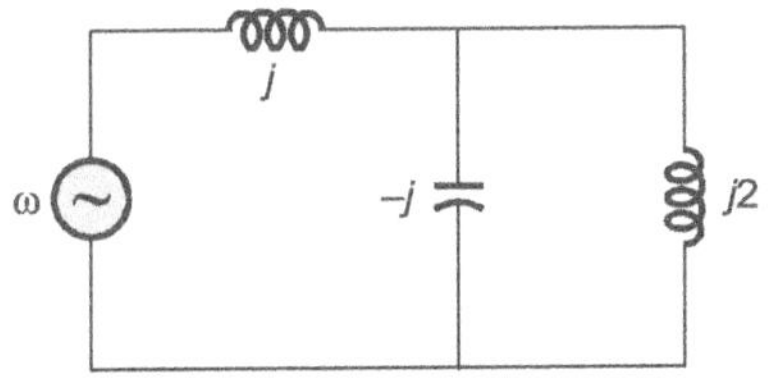

Total impedance of the circuit is

$$Z_T = \frac{(j2)(-j)}{j2-j} + j = -j2+j$$

$$= -j = 1\angle -90°$$

∴ p.f. of the given circuit = cos90° = 0

236. (c)

Here, time period is, T = 20 ms; between 10 ms < t < 20 ms, $i(t)$ = 0 A.

∴ $$i_{rms} = \sqrt{\frac{1}{T}\int i^2 dt}$$

$$= \sqrt{\frac{1}{20\times 10^{-3}}\int_0^{10^{-2}} 1^2 dt}$$

$$= 5\sqrt{2}\times 0.1 = 0.5\sqrt{2} \text{ A}$$

Also, $$i_{avg} = \frac{1}{T}\int_0^{10^{-2}} 1dt = \frac{1}{20\times 10^{-3}}[t]_0^{10^{-2}}$$

$$= 50 \times 0.01 = 0.5 \text{ A}$$

∴ Form factor $$= \frac{i_{rms}}{i_{avg}} = \frac{0.5\sqrt{2}}{0.5} = \sqrt{2}$$

237. (b)

$$V = \frac{V_m}{\sqrt{2}} = \frac{100}{\sqrt{2}} \text{ volt}$$

$$I = \frac{I_m}{\sqrt{2}} = \frac{10}{\sqrt{2}} \text{ A}$$

∴ $$P_{avg} = I^2 R$$

Now, $$R = \frac{V}{I} = Z = \frac{100}{10} = 10\ \Omega$$

(since v and i in phase)

∴ $$P_{avg} = \left(\frac{10}{\sqrt{2}}\right)^2 \times 10 = 500 \text{ watt}$$

238. (d)

$$i = \frac{v}{R} = \frac{V_m \sin\omega t}{R}$$

∴ Instantaneous power,

$$p = v\cdot i = V_m \sin\omega t \cdot \frac{V_m \sin\omega t}{R}$$

$$= \frac{V_m^2}{R}\sin^2\omega t$$

$$= \frac{V_m^2}{R}\left[\frac{1-\cos 2\omega t}{2}\right]$$

$$= \frac{V_m^2}{2R}(1-\cos 2\omega t)$$

239. (c)

Since $V_1 < V_2$, therefore the current will flow from V_2 to V_1 as shown below.

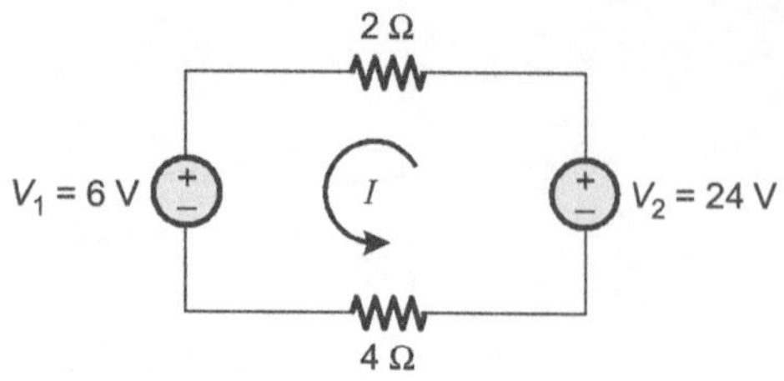

$$I = \frac{V_2 - V_1}{6} = \frac{24-6}{6} = \frac{18}{6} = 3 \text{ A}$$

Hence, the whole power in the circuit will be supplied by V_2 only. Total power delivered by the voltage source V_2 is

$$P_T = I^2 \times (2+4) + 6I$$
$$= 3^2 \times 6 + 6 \times 3 = 54 + 18$$
$$= 72 \text{ watts}$$

240. (a)

$$V_R = (10 + 5V_R) \times R$$

Given, $V_R = 5$ V

So, $5 = (10 + 5 \times 5)$ R

or, $$R = \frac{5}{35} = \frac{1}{7}\,\Omega$$

∴ Power loss in R is

$$P_{loss} = (10 + 5 \times 5)^2 \times \frac{1}{7}$$
$$= 35^2 \times \frac{1}{7} = 175 \text{ watts}$$

241. (c)

$$I_{rms} = \sqrt{(50)^2 + \left(\frac{30}{\sqrt{2}}\right)^2} \text{ A}$$
$$= \sqrt{2500 + \frac{900}{2}} = \sqrt{2950} \text{ A}$$
$$= 54.30 \text{ A}$$

242. (d)

$$V = (100 - j50) \text{ volt,}$$
$$I = (4 - j3) \text{ A}$$

∴ $$S = VI^* = (100 - j50)(4 - j3)^*$$
$$= (100 - j50)(4 + j3)$$
$$= 550 + j100$$

∴ $P = 100$ W and $Q = 100$ VAR

Q is in VAR not in VA. Hence, assertion is not correct.

Also P is the real part of apparent power ($S = VI$).

243. (b)

Instantaneous power in a purely resistive circuit is

$$p = \frac{V_m^2}{2R}(1 - \cos 2\omega t)$$

∴ $$P_{av} = \frac{1}{2\pi}\int_0^{2\pi} p\,dt = \frac{V_m^2}{2R} = \frac{P_{max}}{2}$$

244. (a)

In a purely inductive circuit, power consumed = $Q = VI \sin\phi$. In inductive circuit (having both R and L) power is consumed by resistor R which is $P = VI\cos\phi$.

245. (c)

Phase difference between v and i is

$$\phi = \frac{\pi}{4} - \left(-\frac{\pi}{6}\right) = 45° + 35° = 75°$$

∴ power factor = $\cos\phi = \cos 75°$ lag

Since, i lags v therefore, the circuit elements may be resistance and inductance ($Z = R + jX_L$) or it may be resistance, inductance and capacitance $Z = R + j(X_L - X_C)$, so that p.f. becomes lagging.

247. (c)

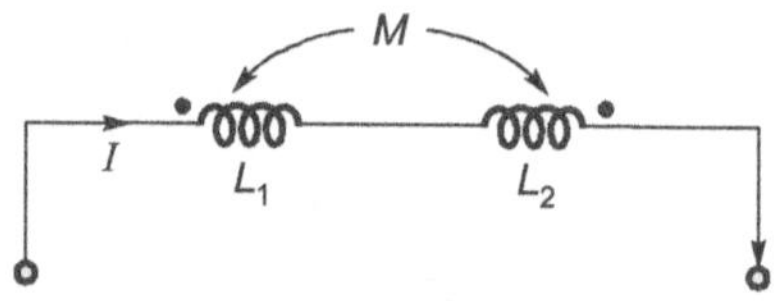

When the effect of mutual inductance is subtractive then, we get minimum value of equivalent inductance, $L_{eq} = (L_1 + L_2 - 2M)$.

248. (c)

Given,

$$L_1 + L_2 + 2M = 12$$

and $L_1 + L_2 - 2M = 4$

Since, $L_1 = L_2 = L$, therefore

$2L + 2M = 12$ or $L + M = 6$ ...(*i*)

and $2L - 2M = 4$ or $L - M = 2$ ...(*ii*)

From above two equations,

$$2L = 8 \text{ or } L = 4 \text{ mH}$$

and $M = L - 2 = 4 - 2 = 2$ mH

To get maximum value in parallel connections,

$$L_{eq} = \frac{L_1L_2 - M^2}{L_1 + L_2 - 2M} = \frac{L^2 - M^2}{2(L-M)}$$

$$= \frac{4^2 - 2^2}{2(4-2)} = \frac{16-4}{2\times 2} = \frac{12}{4} = 3\text{ mH}$$

249. (c)

8 Ω resistance referred to middle circuit

$$= \left(\frac{1}{2}\right)^2 \times 8 = 2\,\Omega.$$

Hence, total resistance in the middle circuit

$= 3 + 2 = 5\,\Omega$

This 5 Ω resistance referred to the first circuit

$= 3^2 \times 5 = 45\,\Omega$

So, $$i_1 = \frac{200\sin 2t}{(45+5)} = 4\sin 2t\text{ A}$$

Now, $$\frac{-i_2}{i_1} = \frac{3}{1} \text{ or } i_2 = -12\sin 2t\text{ A}$$

Also, $$\frac{i_3}{i_2} = \frac{1}{2} \text{ or } i_3 = -6\sin 2t\text{ A}$$

250. (d)

$$M = K\sqrt{L_1L_2}$$

For M to be maximum,

$K = 1$

$\therefore$ $$(M)_{max} = \sqrt{0.3\times 0.7}$$

$$= \sqrt{0.21} = 0.458\text{ H}$$

251. (d)

$$L_1 = \frac{N_1(\phi_{m1} + \phi_{11})}{i_1}$$

$$= \frac{25(100\times 10^{-6} + 250\times 10^{-6})}{0.2}$$

$= 43.75$ mH

Also, $$M = \frac{N_1\phi_{m2}}{i_2} = \frac{N_2\phi_{m1}}{i_1}$$

or, $$\phi_{m2} = \frac{i_2N_2\phi_{m1}}{N_1i_1} = \frac{0.4\times 40\times 100}{25\times 0.2}$$

$= 320\,\mu$Wb

So, $$L_2 = \frac{N_2(\phi_{m2} + \phi_{12})}{i_2}$$

$$= \frac{40(320\times 10^{-6} + 300\times 10^{-6})}{0.4}$$

$= 62$ mH

252. (c)

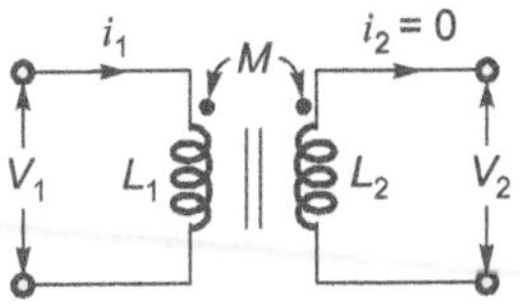

We have:

$$V_1 = \frac{L_1di_1}{dt} \pm \frac{Mdi_2}{dt}$$

(Here, polarity of M is not given in question Hence, it can be +ve or –ve)

and $$V_2 = L_2\frac{di_2}{dt} \pm \frac{Mdi_1}{dt}$$

Since secondary is open-circuited, therefore

$$\frac{di_2}{dt} = 0$$

So, $V_1 = (20\times 10^{-3})\times 0.4\times 10^3$

$= 8$ volt

and $V_2 = \pm(13\times 10^{-3}\times 0.4\times 10^3)$

$= \pm 5.2$ volt

253. (c)

Let the voltage drop across the windings be V volt as shown below.

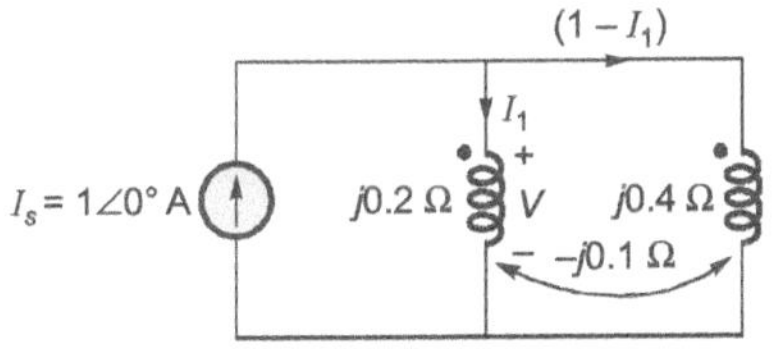

Here, $V = j0.2\,I_1 + j0.1\,(1 - I_1)$

or, $V = j0.1\,I_1 + j0.1$ ...(*i*)

Also, $V = j0.1\,I_1 + j0.4\,(1 - I_1)$

or, $V = j0.4 - j0.3\,I_1$ ...(*ii*)

(In both the equations above, the effect of mutual inductance is positive since current enters both the dots simultaneously).

On solving equation (*i*) and (*ii*), we get

$V = j0.175$ volt

Let the input inductance be L then,

$$j\omega L = \frac{V}{I_S} = \frac{j0.175}{1\angle 0^\circ} = j0.175\,\Omega$$

$\therefore$ $$L = \frac{j0.175}{j\omega} = \frac{j0.175}{j1} = 0.175\text{ H}$$

254. (c)

We know that,

$$\text{Flux} = \frac{MMF}{Rel} = \frac{NI\cdot\mu a}{l}$$

or, $\phi \propto \frac{NIa}{l} \propto \frac{NIr^2}{l} \propto \frac{Ir^2}{l}$

Since $\phi_1 = \phi_2$, therefore

$$\frac{I_2 r_2^2}{l_2} = \frac{I_1 r_1^2}{l_1}$$

or, $$\frac{I_2 \times (2r_1)^2}{2l_1} = \frac{I_1 r_1^2}{l_1} \text{ or } I_2 = \frac{I_1}{2}$$

Hence, the current will become half i.e. $I_2 = \frac{1}{2}$ A.

256. (c)

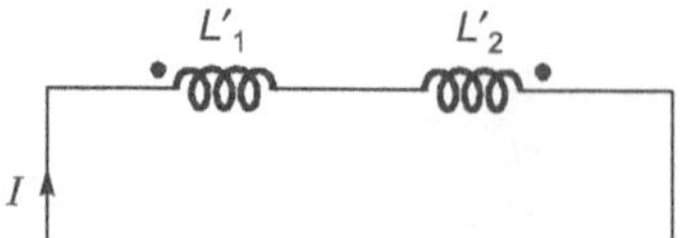

Opposing Circuit

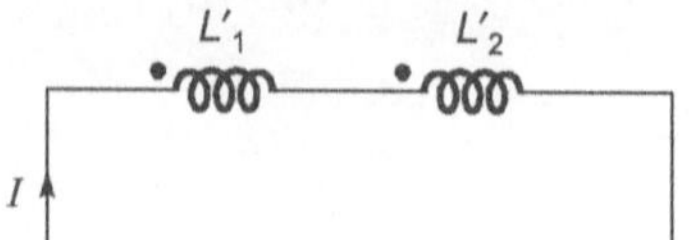

Adding Circuit

Let the self inductances of two coils be L_1' and L_2' respectively.

Then, $L_1 = L_1' + L_2' - 2M$ (For opposing)

and $L_2 = L_1' + L_2' + 2M$ (For adding)

So, $L_2 - L_1 = 4M$

or, $$M = \frac{1}{4}(L_2 - L_1)$$

258. (c)

Using coupling coefficient formula, we have

$$M = K\sqrt{L_1 L_2}$$

Now, for resonance,

$$|(X_L)_{eq}| = |X_C|$$

or, $$|\omega(L_1 + L_2 + 2K\sqrt{L_1 L_2})| = |X_C|$$

or, $$|X_{L_1} + X_{L_2} + 2K\omega\sqrt{L_1 L_2}| = |X_C|$$

or, $$|j12 + j3 + j12K| = |-j18|$$

or, $$15 + 12K = 18$$

or, $$K = \frac{3}{12} = 0.25$$

259. (b)

$$L_{eq} = (10 + 2 - 2) + (10 + 2 - 2) + (10 + 2 - 2)$$
$$= 30 \text{ H}$$

260. (b)

$$L_{eq} = L + L + 2M$$
$$= 2 + 2 + 2 \times 1 = 6 \text{ H}$$

$$\therefore \quad f_0 = \frac{1}{2\pi\sqrt{L_{eq}C}} = \frac{1}{2\pi\sqrt{6 \times 2}} = \frac{1}{4\pi\sqrt{3}} \text{ Hz}$$

261. (c)

We know that,

$$L = \frac{N^2 \mu A}{l} \text{ or } L \propto N^2$$

$$\therefore \quad \frac{L_2}{L_1} = \left(\frac{N_2}{N_1}\right)^2$$

or, $$N_2 = N_1 \cdot \sqrt{\frac{L_2}{L_1}} = 1000 \cdot \sqrt{\frac{5}{3}} \approx 1290 \text{ turns}$$

262. (a)

Due to same polarity, the effect of mutual inductance will be positive.

$$\therefore \quad L_{eq} = \left(\frac{L_1 L_2 - M^2}{L_1 + L_2 - 2M}\right) = \frac{2 \times 3 - 1^2}{2 + 3 - 2 \times 1}$$
$$= \frac{5}{5 - 2} = \frac{5}{3} \text{ H}$$

264. (b)

The reason for assertion is that, emf is induced due to Faraday's law of electromagnetic induction. Reason (R) and assertion (A) are individually true.

266. (b)

$N_1 = 50$, $N_2 = 200$, $M = 10$ H (Given)

Now, $$M = \frac{N_2 \phi_{12}}{I_1} = 200 \times \frac{0.5}{I_1}$$

or, $$10 = \frac{100}{I_1} \text{ or } I_1 = 10 \text{ A}$$

267. (c)

For an ideal transformer, $K = 1$

$$\therefore \quad M = \sqrt{L_1 L_2}$$

or, $$X_M = \sqrt{X_1 X_2}$$

or, $$X_M = \sqrt{j2 \times j3} = j2.45\ \Omega$$

But, $$X_M = 2\pi f M$$

$$\therefore \quad M = \frac{X_M}{\omega} = \frac{2.45}{100}$$
$$= 2.45 \times 10^{-2} \text{ H} = 24.5 \text{ mH}$$

268. (b)

Since quality factor $Q = \frac{\omega_0}{\Delta\omega}$ and Q is high therefore, the resonant frequency ω_0 will be high. Hence, the amplitude response will have a peak always.

269. (b)

Poles are at $S = \sigma \pm j\omega$ (σ can be zero or negative). The system will be stable if the poles at $j\omega$-axis are simple.

270. (c)

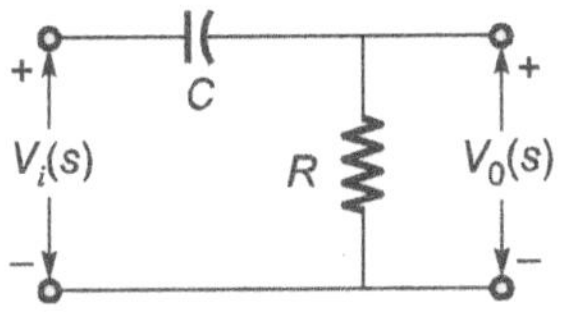

Figure above shows a high-pass filter having a transfer function,

$$\left[\frac{V_0(s)}{V_i(s)} = \frac{RCs}{1+RCs}\right]$$

271. (b)

$$\omega_0 = \frac{1}{\sqrt{LC}} = \frac{1}{\sqrt{0.01 \times 100 \times 10^{-6}}}$$
$$= 10^3 \text{ rad/sec}$$

Also, bandwidth

$$\Delta\omega = \frac{R}{L} = \frac{10}{0.01} = 10^3$$

∴ Quantity factor,

$$Q = \frac{\omega_0}{\Delta\omega} = \frac{10^3}{10^3} = 1$$

272. (b)

Input impedance,

$$Z_{in} = \left[\frac{\frac{R}{j\omega C}}{\frac{1}{j\omega C}+R} + j\omega L\right]$$

$$= \left[\frac{R}{1+j\omega RC} + j\omega L\right]$$

or, $$Z_{in} = \frac{1}{1+j2\omega} + j\omega = \frac{1-j2\omega}{1+4\omega^2} + j\omega$$

$$= \frac{1-j2\omega + j\omega(1+4\omega^2)}{1+4\omega^2}$$

For unity p.f. operation, $Img(Z_{in}) = 0$

∴ $-2\omega + \omega(1+4\omega^2) = 0$

or, $4\omega^3 - \omega = 0$

or, $4\omega^2 - 1 = 0$ (since $\omega \neq 0$)

∴ $\omega = \frac{1}{2}$ rad/s

or, $\omega = \omega_n = 0.5$ rad/s

274. (b)

$Q \propto$ selectivity

275. (c)

Transfer function,

$$H(j\omega) = \frac{V_C(j\omega)}{V(j\omega)}$$

Here, $$V_C(j\omega) = \frac{V(j\omega)\cdot Z_C(j\omega)}{Z(j\omega)+Z_C(j\omega)}$$

$$= \frac{\left(\frac{1}{j\omega \times 0.25}\right)V(j\omega)}{\left(1 + j\omega + \frac{1}{j\omega \times 0.25}\right)}$$

$$= \left(\frac{\frac{4}{j\omega}}{1 + j\omega + \frac{4}{j\omega}}\right)\cdot V(j\omega)$$

or, $$H(j\omega) = \frac{4}{(j\omega)^2 + (j\omega + 4)} = \frac{4}{4-\omega^2+j\omega}$$

∴ $$H(j\omega) = \frac{4}{(4-\omega^2+j\omega)}$$

276. (d)

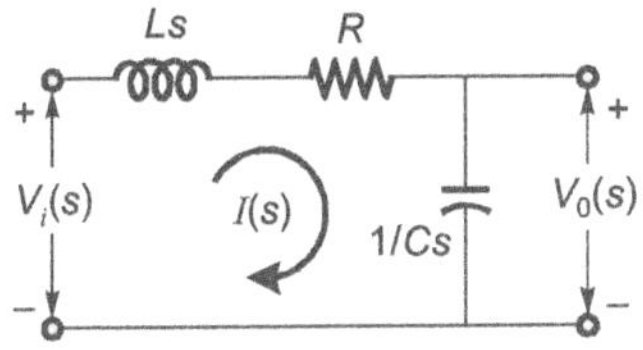

Here, $$V_i(s) = \left[R + Ls + \frac{1}{Cs}\right]$$

and $$V_0(s) = I(s) \times \frac{1}{Cs}$$

∴ $$\frac{V_0(s)}{V_i(s)} = \frac{1}{LCs^2 + RCs + 1}$$

$$= \left(\frac{1/LC}{s^2 + \frac{R}{L}s + \frac{1}{LC}}\right)$$

277. (c)

Figure of merit = Q-factor

$$= \frac{|I_C|}{|I_R|} = \omega RC$$

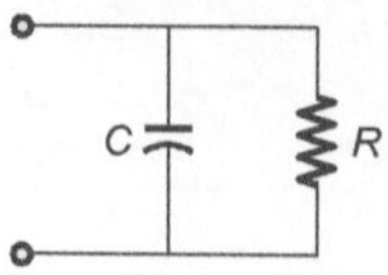

278. (b)

$$f_0 = \frac{1}{2\pi\sqrt{LC}} = \frac{1}{2\pi\sqrt{10^{-1}\times 10^{-9}}}$$

$$= \frac{10^5}{2\pi} = 15915.49 \text{ Hz} \approx 15920 \text{ Hz}$$

279. (d)

Selectivity $\propto Q$ and $Q \propto \frac{1}{\Delta\omega}$.

When $\Delta\omega$ increases, selectivity decreases and vice-versa.

Also, $V_c|_{max}$ occurs at,

$$f_c = \frac{1}{2\pi}\sqrt{\frac{1}{LC} - \frac{R^2}{2L^2}} < f_0$$

280. (a)

Given, $\Delta f = 75 \text{ kHz} = f_2 - f_1$ ...(i)

and $f_0 = \sqrt{f_1 f_2} = 150 \text{ kHz}$ ...(ii)

On solving equations (i) and (ii), we get

$f_1 = 117$ kHz and $f_2 = 192$ kHz

281. (c)

At resonance, $X_L = X_C$

$\therefore$ $I_L = I_C$ (for parallel resonant circuit)

Hence, $I_L = I_C = \frac{V}{X_L}$ or $\frac{V}{X_C}$

or, $$I_L = \frac{V}{2\pi f_0 L} = \frac{V}{2\pi \times \frac{1}{2\pi\sqrt{LC}} \times L}$$

or, $$I_L = V\sqrt{\frac{C}{L}} \text{ A} = I_C$$

($I_L = I_C$ = circulating current)

282. (d)

Here, $$Y = \left[\frac{1}{10+j10} + \frac{1}{R_C - j2}\right]$$

$$= \left[\frac{10-j10}{200} + \frac{R_C + j2}{R_C^2+4}\right]$$

At resonance,

$$Img[Y] = 0$$

or, $$\left[\frac{-10}{200} + \frac{2}{R_C^2+4}\right] = 0 \text{ or } R_C = 6\ \Omega$$

283. (b)

At resonance,

$$I_0 = \frac{V}{R} = \frac{10}{10} = 1 \text{ A}$$

Now, power at half power frequency

$$= \frac{1}{2}P_{max} = \frac{1}{2} \times I_0^2 R$$

$$= \frac{1}{2} \times 1^2 \times 10 = 5 \text{ W}$$

284. (c)

For the given circuit to resonate at all frequency,

$$R_1 = R_2 = \sqrt{\frac{L}{C}}$$

so that, $$f_0 = \frac{1}{2\pi\sqrt{LC}}$$

Note: Resonant frequency of given circuit is

$$f_0 = \frac{1}{2\pi\sqrt{LC}}\sqrt{\frac{R_1^2 - L/C}{R_2^2 - L/C}}$$

Thus, $$R_1 = R_2 = \sqrt{\frac{5\times 10^{-3}}{20\times 10^{-6}}} = 15.8\ \Omega$$

285. (c)

Since $X_L = X_L$, therefore given circuit will be at resonance

$$\therefore \quad I_1 = \frac{100}{5} = 20 \text{ A}$$

286. (b)

The given circuit in s-domain is shown below.

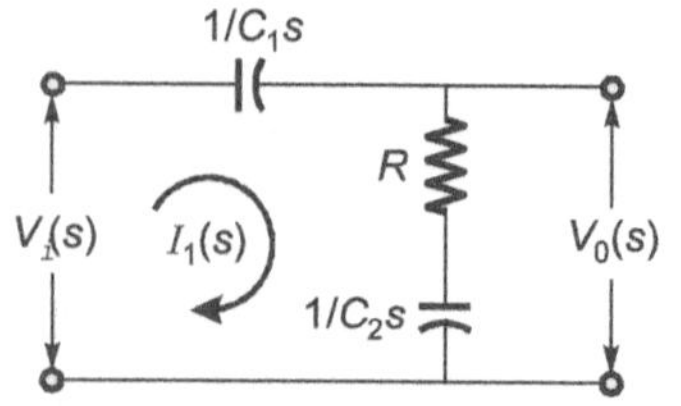

Now, $$V_i(s) = I_1(s)\left[R + \frac{1}{C_1 s} + \frac{1}{C_2 s}\right]$$

and $$V_0(s) = I_1(s)\left[R + \frac{1}{C_2 s}\right]$$

So, $$\frac{V_0(s)}{V_i(s)} = \left(\frac{R+\frac{1}{C_2 s}}{R+\frac{1}{C_1 s}+\frac{1}{C_2 s}}\right)$$

$$= \frac{(RC_2 s+1)C_1 s}{(RC_1 C_2 s^2 + C_1 s + C_2 s)}$$

$$= \frac{C_1(1+RC_2 s)}{C_1+C_2+RC_1 C_2 s}$$

or, $$\frac{V_0(s)}{V_i(s)} = \left(\frac{C_1(1+RC_2 s)}{C_1+C_2+RC_1 C_2 s}\right)$$

287. (a)

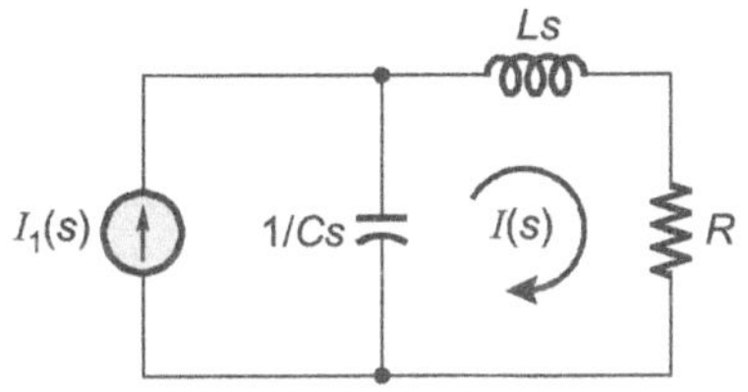

$$I_2(s) = I_1(s)\left(\frac{1/Cs}{R+Ls+\frac{1}{Cs}}\right)$$

or, $$\frac{I_2(s)}{I_1(s)} = \left(\frac{1/Cs}{R+Ls+\frac{1}{Cs}}\right)$$

or, $$\frac{I_2(s)}{I_1(s)} = \frac{1}{s^2 LC + RCs + 1}$$

$$= \frac{1}{2s^2 + 200s + 1}$$

Thus, there is no-zero for T.F. $\frac{I_2(s)}{I_1(s)}$

Poles are at:

$2s^2 + 200s + 1 = 0$

or, $$s = \frac{-200 \pm \sqrt{200^2 - 8}}{2}$$

or, $$s = -100 \pm \sqrt{10^4 - 2}$$

$$= -100 \pm 99.99$$

$$\approx 0 \text{ or } -200$$

Hence, poles are at: $s = 0, -200$

288. (d)

Given, $C(t) = t^2 e^{-t}$

$\therefore$ $$C(s) = \frac{2}{(s+1)^3}$$

Also, $$R(s) = \frac{1}{s}$$

$\therefore$ $$\frac{C(s)}{R(s)} = \text{T.F.} = \frac{2s}{(s+1)^3}$$

289. (d)

Resonant frequency,

$$\omega_r = \frac{1}{\sqrt{LC}} \text{ rad/sec}$$

Also, $$Z = \frac{j\omega L \times \frac{1}{j\omega C}}{j\omega L + \frac{1}{j\omega C}} = \frac{j\omega L}{1-\omega^2 LC}$$

$$= \frac{jL}{\left(\frac{1}{\omega} - \omega LC\right)}$$

When $f = 0$, $\omega = 0$ and $z = 0$.

When $f = \infty$, $\omega = \infty$ and $z = 0$.

291. (b)

When $s \to 0$, $H(0) \to 0$

When $s \to \infty$, $H(\infty) \to 0$

Hence, given T.F. is for a band pass filter.

292. (c)

Putting $s \to 0$ and $s \to \infty$, we can find the type of the filter.

293. (d)

$$Q = \frac{\omega L}{R} \text{ (for a coil)}$$

For Q to be high, R should be low and L should be high.

Hence, assertion is false.

294. (c)

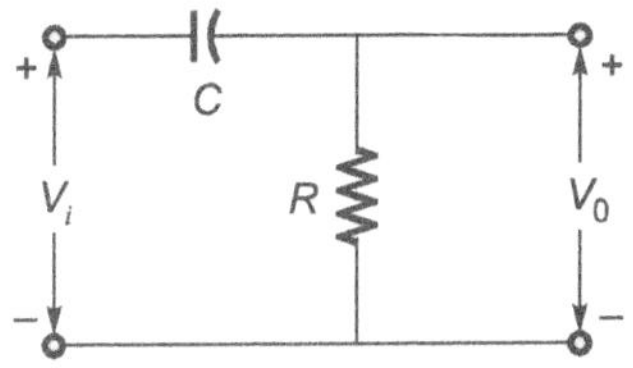

(RC differentiator)

$$\text{T.F} = \frac{V_0(s)}{V_i(s)} = \left(\frac{RCs}{1+RCs}\right)$$

Hence, it has a zero at $s = 0$ and a pole at

$$s = -\frac{1}{RC}.$$

295. (c)

An ideal transfer having turns ratio of 1 : n is shown below.

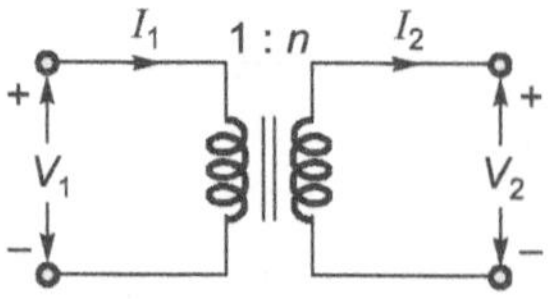

or

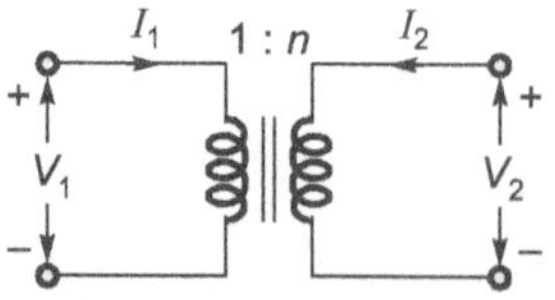

Here, $\frac{V_1}{V_2} = \frac{1}{n}$ or $V_1 = \frac{1}{n}V_2$ ...(1)

Also, $\frac{(-I_2)}{I_1} = \frac{1}{n}$ or $I_1 = -n(I_2)$

or, $I_1 = n(-I_2)$ ...(2)

Comparing equations (1) and (2) with standard transmission-parameter equations,

$V_1 = AV_2 - BI_2$ and $I_1 = CV_2 - DI_2$, we get:

$$\begin{bmatrix} A & B \\ C & D \end{bmatrix} = \begin{bmatrix} 1/n & 0 \\ 0 & n \end{bmatrix}$$

299. (b)

Given, $z_1 = 2\angle 0°\ \Omega$, $z_2 = \angle -90°\ \Omega$

and $z_3 = 3\angle 90°\ \Omega$

Applying KVL,

$$V_1 = I_1(2\angle 0° + 3\angle 90°) + (3\angle 90°)I_2$$

or, $V_1 = (2+j3)I_1 + (j3)I_2$ ...(1)

and $V_2 = I_2(5\angle -90° + 3\angle 90°) + (3\angle 90°)I_1$

$= (3j)I_1 + (-j5 + j3)I_2$

or, $V_2 = (j3)I_1 + (-j2)I_2$ ...(2)

Comparing equations (1) and (2) with the standard z-parameter equations,

$$z_{11} = (2+j3)\ \Omega$$

and $z_{22} = -j2\ \Omega$

300. (c)

Z-parameter equations are:

$$V_1 = Z_{11}I_1 + Z_{12}I_2$$
$$V_2 = Z_{21}I_1 + Z_{22}I_2$$

and h-parameter equations are:

$$V_1 = h_{11}I_1 + h_{12}V_2$$
$$I_2 = h_{21}I_1 + h_{22}V_2$$

Converting the Z-parameter equations into h-parameter equations, we have:

$$V_1 = \frac{(Z_{11}Z_{22} - Z_{12}Z_{21})}{Z_{22}}I_1 + \frac{Z_{12}}{Z_{22}}V_2$$

$$= \frac{\Delta Z}{Z_{22}}I_1 + \frac{Z_{12}}{Z_{22}}V_2$$

and $V_2 = \frac{-Z_{21}}{Z_{22}}I_1 + \frac{1}{Z_{22}}V_2$

301. (d)

$$[Z] = \begin{bmatrix} Z_{11} & Z_{12} \\ Z_{21} & Z_{22} \end{bmatrix} \quad \therefore\ [y] = [Z]^{-1}$$

$$= \frac{1}{\Delta Z}\begin{bmatrix} Z_{22} & -Z_{12} \\ -Z_{21} & Z_{11} \end{bmatrix} = \begin{bmatrix} y_{11} & y_{12} \\ y_{21} & y_{22} \end{bmatrix}$$

Here, $y_{11} = \frac{Z_{22}}{\Delta Z}$

302. (b)

Input admittance, $Y(s) = \left[\frac{1}{R} + Cs\right]$

Also, $I_1(s) = V_1(s) \cdot Y_{in}(s)$

or, $I_1(s) = V_1(s)\left[\frac{1}{R} + Cs\right]$

As $V_1(s) = V_2(s)$ therefore,

$$\frac{V_2(s)}{I_1(s)} = \frac{1}{\left[\frac{1}{R} + Cs\right]}$$

$$= \frac{1}{C\left[s + \frac{1}{RC}\right]} = Z_{12}(s)$$

303. (a)

Given:

Input voltge,

$$v_1(t) = \delta(t)$$

$\therefore$ $V_1(s) = 1$

and output,

$$v_2(t) = (e^{-2t} + e^{-3t})\,u(t)$$

$\therefore$ $V_2(s) = \frac{1}{s+2} + \frac{1}{s+3} = \frac{2s+5}{(s+2)(s+3)}$

Now, impulse response = T.F. = $\frac{V_2(s)}{V_1(s)}$

$$= \frac{2s+5}{(s+2)(s+3)} = H(s)$$

Given, output $v_2'(t) = te^{-2t}$

or, $$\frac{1}{(s+2)^2} = V_2'(s)$$

Let, input required = $v_1'(t)$

Now, $$\frac{V_2'(s)}{V_1'(s)} = H(s)$$

or, $$V_1'(s) = \frac{V_2'(s)}{H(s)}$$

$$= \frac{1}{(s+2)^2} \times \frac{(s+2)(s+3)}{(2s+5)}$$

or, $$V_1'(s) = \frac{s+3}{(s+2)(2s+5)}$$

$$= \frac{A}{s+2} + \frac{B}{s+5/2}$$

Using partial fraction expansion, $A = 1$ and $B = -0.5$.

So, $$V_1'(s) = \frac{1}{s+2} - \frac{0.5}{s+2.5}$$

or, $$v_1'(t) = (e^{-2t} - 0.5e^{-2.5t})u(t) \text{ volts}$$

= Required input voltage

304. (c)

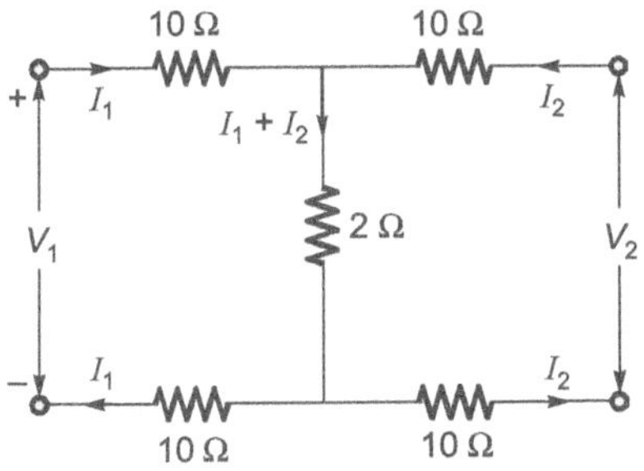

Applying KVL in the two loops,

$$V_1 = 10I_1 + 2(I_1+I_2) + 10I_1$$

or, $$V_1 = 22I_1 + 2I_2$$

$$(\therefore Z_{11} = 22\ \Omega, Z_{12} = 2\ \Omega)$$

Also, $$V_2 = 10I_2 + 2(I_1+I_2) + 10I_2$$

or, $$V_2 = 2I_1 + 22I_2$$

$$(\therefore Z_{21} = 2\ \Omega, Z_{22} = 22\ \Omega)$$

$\therefore$ $$[Z] = \begin{bmatrix} 22 & 2 \\ 2 & 22 \end{bmatrix} \Omega$$

306. (a)

Given, $$V_1 = 4I_1 + I_2 \quad ...(1)$$

$$V_2 = 3I_1 + 3I_2$$

or, $$I_1 = \frac{V_2}{3} + (-I_2) \quad ...(2)$$

Putting the value of $I_1 = \frac{V_2}{3} + (-I_2)$ in equation (1), we have:

$$V_1 = 4\left[\frac{V_2}{3} + (-I_2)\right] + I_2$$

or, $$V_1 = \frac{4}{3}V_2 + 3(-I_2) \quad ...(3)$$

Using equations (2) and (3), we get:

$$\begin{bmatrix} A & B \\ C & D \end{bmatrix} = \begin{bmatrix} 4/3 & 3 \\ 1/3 & 1 \end{bmatrix}$$

307. (d)

Applying KVL in the two loops, we have:

$$V_1 = 0I_1 + 5I_2$$

$\therefore$ $$Z_{11} = 0\ \Omega, Z_{12} = 5\ \Omega$$

and $$V_2 = 2I_2 + 0I_1$$

$\therefore$ $$Z_{21} = 0\ \Omega, Z_{22} = 2\ \Omega$$

$\therefore$ $$\begin{bmatrix} Z_{11} & Z_{12} \\ Z_{21} & Z_{22} \end{bmatrix} = \begin{bmatrix} 0 & 5 \\ 0 & 2 \end{bmatrix}$$

308. (c)

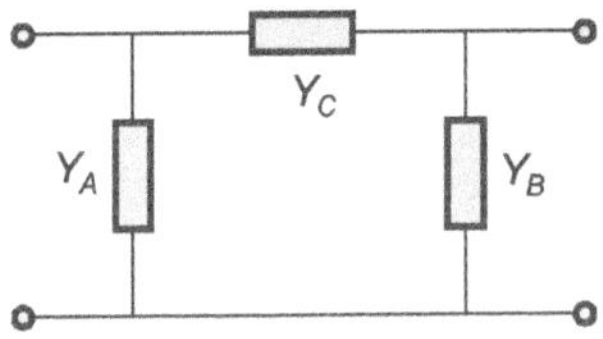

$$[Y] = \begin{bmatrix} Y_A + Y_C & -Y_C \\ -Y_C & Y_B + Y_C \end{bmatrix}$$

310. (c)

Here, $I_1 = I_0$ and $I_2 = -2I_0$

Applying KVL, we get:

$V_1 = I_0$ or $V_1 = I_0 = I_1$ or $I_1 = V_1 + 0\ V_2$

Also, $I_1 = y_{11}V_1 + y_{12}V_2$

So, $y_{11} = 1$ ℧ and $y_{12} = 0$ ℧

Now, $I_2 = -2I_0 = -2V_1 + 0V_2$

Also, $I_2 = y_{21}V_1 + y_{22}V_2$

So, $y_{21} = -2$ ℧ and $y_{22} = 0$ ℧

Hence, $$[y] = \begin{bmatrix} 1 & 0 \\ -2 & 0 \end{bmatrix}$$

311. (a)

$$h_{11} = \left.\frac{V_1}{I_1}\right|_{V_2=0} = 2\,\Omega \text{ and } h_{12} = \left.\frac{V_1}{V_2}\right|_{I_1=0} = 1$$

313. (b)

Here, $I_1 = -10\,I_0$

Applying KVL at input,

$$V_0 = 100I_0 + 20\,(I_0 - I_1)$$
$$= 100I_0 + 20\,(I_0 + 10I_0)$$

or, $V_0 = 100I_0 + 220I_0$

or, $I_0 = -\frac{V_0}{320}$ A

Also, $I_1 = -10I_0 = -\frac{V_0}{32}$ A

For an ideal transformer,

$$I_2 = -\frac{I_1}{n}$$

or, $\frac{I_2}{I_1} = -\frac{1}{n} = -10$

or, $I_2 = -10I_1 = \frac{10V_0}{32}$

Thus, $V_2 = -R_L I_2 = -10 \times \frac{10V_0}{32} = -\frac{100}{32} V_0$

$\therefore$ $\frac{V_2}{V_0} = -\frac{100}{32} = -3.125$

314. (b)

$\frac{(s+1)(s+2)}{(s^2+1)^2}$ has double poles at $s = \pm j$.

Hence, not a PRF.

$\frac{(s-1)(s+2)}{(s^2+1)}$ has a zero (s = 1) in RH s-plane.

Hence, not a PRF.

$\frac{(s-1)}{(s^2-1)}$ has a pole and a zero in RH s-plane.

Hence, not a PRF.

Option (b) is a PRF as all poles and zeros lie in LH s-plane.

317. (b)

Since system is bounded for bounded excitation therefore, it will be stable.

Since the system is delayed therefore, output may exist before the input is applied.

Hence, the system will be non-causal.

321. (b)

Here, zeros are at: s = 0, 0, –3

Poles are at $s = -1, -2 \pm j1$

or, poles are at $(s + 2)^2 + 1 = 0$ and at $s = -1$

or, $s^2 + 4s + 5 = 0$ and at $s = -1$

323. (d)

$$F(s) = \frac{s+1}{s^2+2} = \frac{s+a}{s^2+bs+c}$$

Here, $a = 1$, $b = 0$, $c = 2$

Thus, $a, b, c \geq 0$ but $b < a$

For PRF, $b \geq a$ therefore, option (d) is not a PRF.

324. (b)

For option (a):

$$F_1(s) = (s + 1)(s^2 + 2s + 3)$$
$$= s^3 + 3s^2 + 5s + 3$$

Since all coefficients of $F_1(s)$ are positive and no term is missing, therefore it is a Hurwitz polynomial.

Now, $F_2(s) = (s + 3)\,(s^2 + s - 2)$
$= s^3 + 4s^2 + s - 6$

Since $F_2(s)$ has a negative coefficient, therefore it is not a Hurwitz polynomial.

Similarly, we can check for options (c) and (d).

325. (b)

A Hurwitz polynomial must have no poles and zeros in the RH s-plane.

326. (a)

The pole-zero plot of $Z(s)$ is shown below.

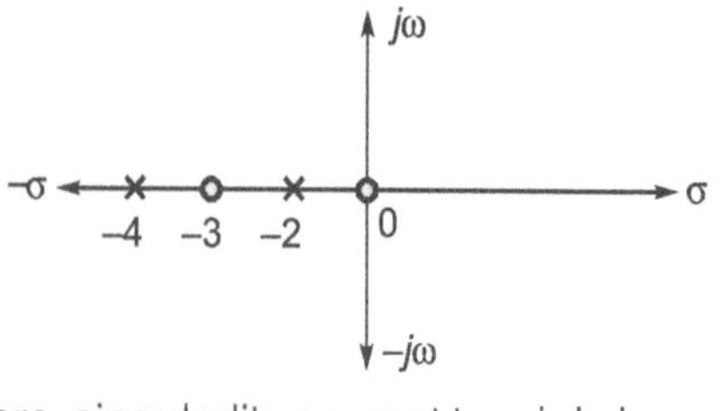

Here, singularity nearest to origin is a zero while near infinity is a pole. Also, pole-zero alternates on negative real axis.

Therefore, $Z(s)$ will represent a R-L impedance function.

327. (a)

For $s \to 0$, L acts as short circuit and C acts as open circuit.

Thus, $V_0 = V_i$

For $s \to \infty$, L acts as open circuit and C acts as short circuit.

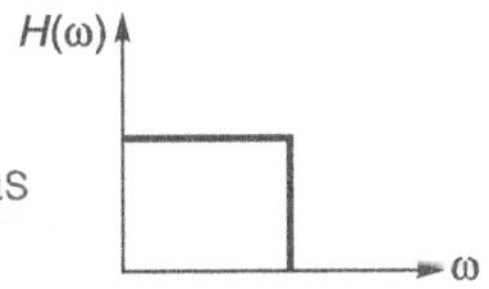

$\therefore \quad V_0 = 0$

Hence, given circuit will act as a LPF.

328. (c)

Converting the given network in Laplace domain, the driving point impedance is

$$Z(s) = \left[\frac{\left(s+\frac{1}{s}\right)\times\frac{1}{s}}{s+\frac{2}{s}} + s\right]$$

$$= \left[\frac{s^2+1}{s^2}\times\frac{s}{s^2+2}\right] + s = \left[\frac{s^2+1}{s(s^2+2)} + s\right]$$

$$= \frac{s^2+1+s^2(s^2+2)}{s(s^2+2)} = \frac{s^4+3s^2+1}{s(s^2+2)}$$

$\therefore$ Driving point admittance $Y(s) = \dfrac{s(s^2+2)}{s^4+3s^2+1}$

329. (c)

$\dfrac{8-s}{s+1}$ is not a PRF since all the coeficients of numerator polynomial are not positive and there is a zero in RH s-plane.

$\dfrac{s^2-1}{-s+8}$ is not a PRF due to the same reason (zeros are at $s = \pm 1$).

$\dfrac{4s+1}{s+2}$ is a PRF since pole is at $s = -\dfrac{1}{4}$ and zero at $s = -2$.

Also, all coefficients of N^r and D^r are positive.

And $\text{Re}\, F(j\omega) = \dfrac{4\omega^2+2}{\omega^2+4}$ and $\text{Re}\, F(j\omega) \geq \forall\, \omega$

331. (a)

Given $P(s) = 2s^4 + s^3 + ms^2 + s + 2$

$= (2s^4 + ms^2 + 2) + (s^3 + s) = M(s) + N(s)$

Continued fraction expansion of $M(s)/N(s)$ is obtained as follows:

$s^3 + s)\, 2s^4 + ms^2 + 2\, (2s$

$\quad 2s^4 + 2s^2$

$\quad (m-2)s^2 + 2)\; s^3 + s \;\; (s/(m-2)$

$\quad\quad s^3 + \dfrac{2s}{m-2}$

$\quad\quad \left(1 - \dfrac{2}{m-2}\right)s$

The quotients of the continued fraction expansion would be positive only when $m > 2$ and

$1 - \dfrac{2}{m-2} > 0$ or $m - 4 > 0$ or, $m > 4$.

Hence for $m > 4$, $P(s)$ will be Hurwitz.

332. (d)

The pole-zero plots of $F_1(s)$ and $F_2(s)$ are shown below.

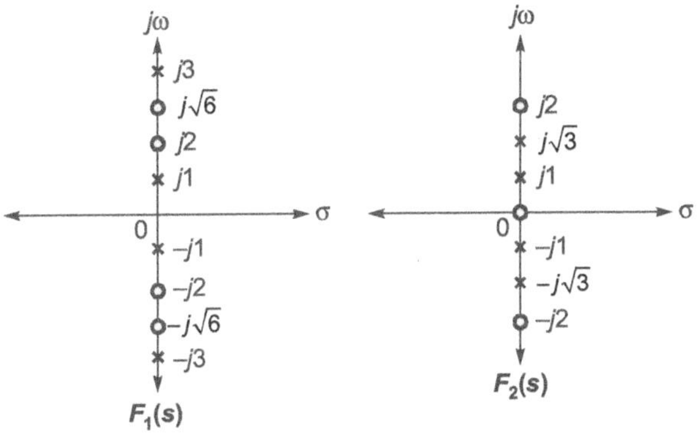

In $Z_1(s)$: Poles and zeros do not alternate. Also, there is no zero-or a pole at origin.

In $Z_2(s)$: Poles and zeros do not alternate.

Hence, neither of $Z_1(s)$ or $Z_2(s)$ represents a LC immittance function.

335. (c)

The continued fraction expansion is shown below

$s^4 + 3s^2 + 1)\; s^5 + 3s^3 + 3s\; (s \to Z_1 = L_1 s \Rightarrow L_1 = 1\text{ H}$

$\quad s^5 + 3s^3 + s$

$\quad 2s^3 + 2s)\; s^4 + 3s^2 + 1\; (s/2 \to Y_2 = C_2 s \Rightarrow C_2 = 1/2\text{ F}$

$\quad\quad s^4 + s^2$

$\quad\quad 2s^2 + 1)\; 2s^3 + 2s\; (s \to Z_3 = L_3 s \Rightarrow L_3 = 1\text{ H}$

$\quad\quad\quad 2s^3 + s$

$\quad\quad\quad s)\; 2s^2 + 1\; (2s \to Y_4 = C_4 s \Rightarrow C_4 = 2\text{ F}$

$\quad\quad\quad\quad 2s^2$

$\quad\quad\quad\quad 1)\; s\; (s$

$\quad\quad\quad\quad\quad s$

$\quad\quad\quad\quad\quad \times$

Hence, $Z(s)$ can be realized as a L-C network.

337. (d)

Foster's type-I and type-II networks are not equivalent networks.
Hence, assertion is false.

339. (c)

Given network has 4 number of nodes and 5 number of branches.
Hence, it's tree will be as shown below.

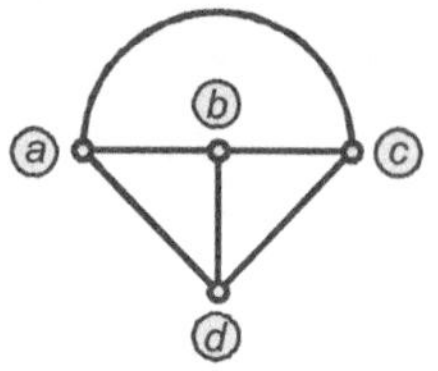

341. (a)

For the given network, number of nodes, $n = 3$ and number of branches = 4 = b.
Hence, twigs = n – 1 = 3 – 1 = 2
and links = $b - (n-1) = 4 - (3-1) = 2$

343. (a)

Number of edges = $^nC_2 = \dfrac{n!}{2!(n-2)!} = \dfrac{n(n-1)}{2}$.

344. (b)

$n = 5$ and $b = 10$

348. (b)

Assertion is true because if the values of each element is changed w.r.t. time, then every time a fresh KVL and KCL equations are to be written for analysis of the network.
Reason is also true but, the correct explanation of assertion is as explained above.
Hence, Reason (R) is not the correct explanation of (A).

349. (c)

The graph of the given network is shown below.

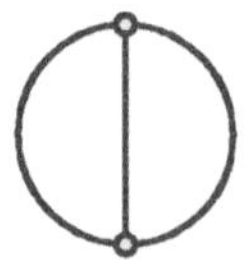

A tree is circuit less, therefore option (c) is wrong.

350. (d)

Number of nodes, $n = 3$ and number of branches, $b = 4$.
So, number of twigs = 3 – 1 = 2
Number of links = $b - (n-1) = 4 - (3-1) = 2$

352. (d)

The cut-set matrix is always equal to negative of transpose of the tie-set matrix.
Hence, $[Q_c] = -[B_t]^T$ or $[B_t]^T = -[Q_c]$

353. (a)

Since it is not said whether the graph is a fully connected or not, therefore we use the general result for finding the number of trees which is equal to det $[AA^T]$ where, A = reduced incidence matrix. By deleting one of the rows, say the bottom most row, the reduced incidence matrix is

$$[A] = \begin{bmatrix} -1 & 0 & 0 & +1 & -1 & 0 \\ +1 & -1 & 0 & 0 & 0 & -1 \\ 0 & +1 & -1 & 0 & +1 & 0 \end{bmatrix}$$

and
$$[A^T] = \begin{bmatrix} -1 & +1 & 0 \\ 0 & -1 & +1 \\ 0 & 0 & -1 \\ +1 & 0 & 0 \\ -1 & 0 & +1 \\ 0 & -1 & 0 \end{bmatrix}$$

$$\text{Hence, det } [A][A^T] = \begin{vmatrix} 3 & -1 & -1 \\ -1 & 3 & -1 \\ -1 & -1 & 3 \end{vmatrix} = 16$$

∴ Number of trees = 16

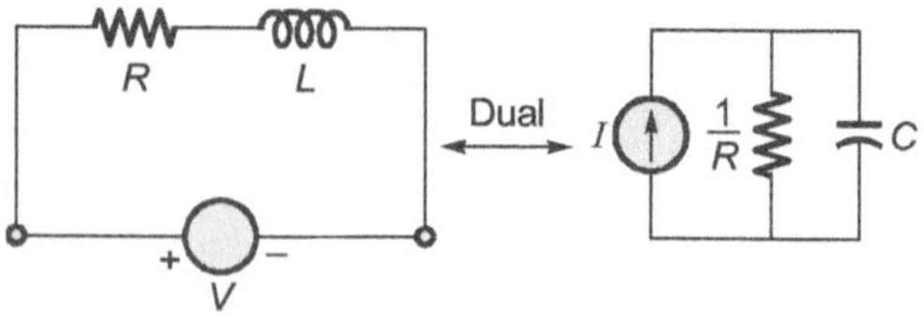

357. (c)

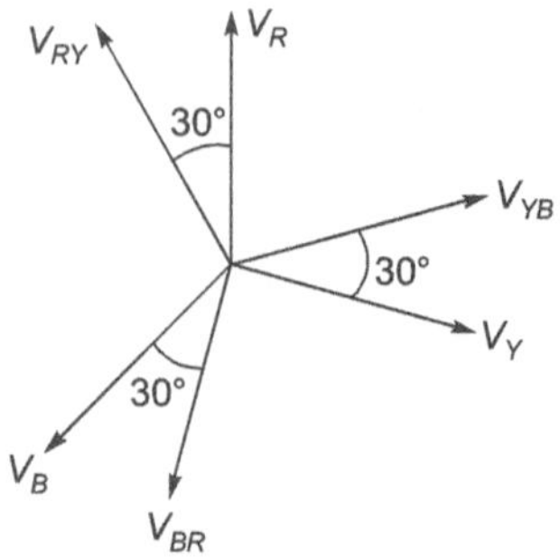

Here, line-to-line voltage leads the respective phase voltage by 30°.

358. (d)

When all the resistance are present,

$$P_\Delta = \frac{3V_{ph}^2}{R} = \frac{3V_L^2}{R}$$

(Since $V_{ph} = V_L$ for Δ connection)

When one of the phase resistance is removed,

$$P_{0.\Delta} = \frac{2V_L^2}{R}$$

$$\therefore \text{ \% reduction in power} = \left(\frac{\frac{3V_L^2}{R} - \frac{2V_L^2}{R}}{\frac{3V_L^2}{R}}\right) \times 100$$

$$= 33.33\%$$

359. (c)

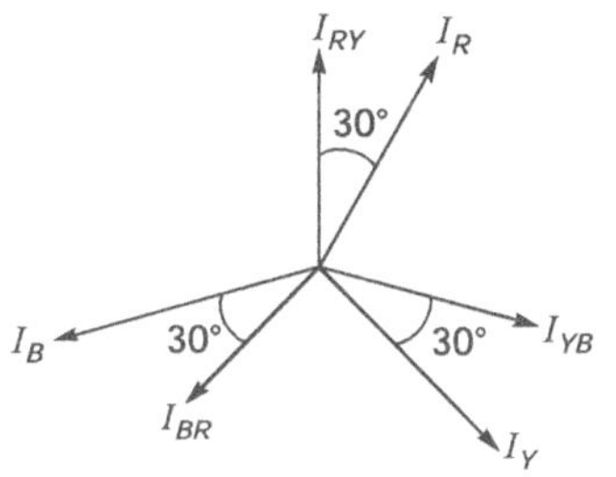

360. (b)

$$\phi = 30° \qquad (I \text{ lags } V \text{ by } 30°)$$

$$\therefore \quad \cos\phi = \cos 30° \text{ (lag)}$$

$\therefore$ Total power,

$$P = \sqrt{3}\, V_L I_L \cos\phi$$

$$= \sqrt{3} \times 400 \times 15 = 9 \text{ kW}$$

361. (c)

Since V_{CN} lags V_{AN} by 120°, therefore phase sequence will be ACB.

Here, V_{BN} will lead V_{AN} by 120° but, will have the same magnitude.

So, $\quad V_{BN} = 1500\angle 20° + 120°$ V

$= 1500\angle 140°$ V

362. (a)

Converting the Δ-connected load in star equivalent, we have

$$Z_Y = \frac{9 + j12}{3} = (3 + j4)\ \Omega$$

So, line to neutral voltage for equivalent Y-load

$= 100/\sqrt{3} = 57.74$ V

$\therefore$ Line current,

$$I_L = \frac{100/\sqrt{3}}{|Z_Y|} = \frac{100/\sqrt{3}}{5}$$

$$= 20/\sqrt{3} = 11.55 \text{ A}$$

So, $\quad V_L = I_L\, |Z_{\text{line}} + Z_Y|$

$= 11.55\,|2 + j3 + 3 + j4|$

$= 11.55\,|5 + j7| = 99.3$ V

$\therefore$ Line voltage at source $= \sqrt{3} \times 99.3 = 172$ V

363. (c)

We have: $\theta = \tan^{-1}\left[\frac{\sqrt{3}(W_1 - W_2)}{W_1 + W_2}\right] = 0°$

(Since $W_1 = W_2 = 3$ kW)

or, $\quad \cos\theta = 1$

Hence, the load will be purely resistive.

Now, $\quad P = 3V_P I_P \cos\theta$

or, $\quad I_P = \frac{P}{3V_P \cos\theta} = \frac{3000 + 3000}{3 \times 600 \times 1}$

$= 3.33$ A

So, $\quad R_\Delta = Z_P = \frac{V_P}{I_P}$

$= \frac{600}{3.33} = 180\ \Omega$

364. (b)

We know that,

$$Z_Y = \frac{Z_\Delta}{3}$$

or, $\quad \frac{1}{\omega C_Y} = \frac{1}{3\omega C_\Delta}$

or, $\quad C_Y = 3C_\Delta$

Given, $\quad C_\Delta = \frac{C}{3}$

$\therefore \quad C_Y = 3 \times \frac{C}{3} = C$ Farad

365. (c)

Since the system is balanced, therefore

$I_A + I_B + I_C = 0$

or, $\quad I_A = (I_B + I_C)$

$= -(-20 - 20) = 40$ A

366. (d)

$$\frac{I_L(\text{star})}{I_L(\text{delta})} = \frac{1}{3}$$

or, $I_L(\text{delta}) = 3 \times I_L(\text{star})$
$= 30$ A

368. (b)

Wattmeter readings are:

$$W_1 = V_L I_L \cos(30 + \phi)$$

and $W_2 = V_L I_L \cos(30 - \phi)$

Here, $V_L = 100$ V, $V_{ph} = \frac{100}{\sqrt{3}}$ V

and $\phi = 30°$

$\therefore$ $I_{ph} = I_L = \frac{V_{ph}}{|Z|} = \frac{100}{\sqrt{3} \times 5} = \frac{20}{\sqrt{3}}$ A

Now, $W_1 = 100 \times \frac{20}{\sqrt{3}} \cos(30 + 30°)$

$= \frac{1000}{\sqrt{3}}$ W $= 577.35$ Watt

and $W_2 = 100 \times \frac{20}{\sqrt{3}} \cos(30 - 30°)$

$= \frac{2000}{\sqrt{3}}$ W $= 1154.70$ Watt

370. (b)

$$W_1 = V_L I_L \cos(30° + \phi)$$

and $W_2 = V_L I_L \cos(30° - \phi)$

When the load is purely capacitance, then $\phi = -90°$.

So, $W_1 = V_L I_L \cos(30° - 90°)$

$= V_L I_L \sin 30° = \frac{V_L I_L}{2}$

and $W_2 = V_L I_L \cos(30° + 90°)$

$= -V_L I_L \sin 30° = \frac{-V_L I_L}{2}$

Hence, $|W_1| = |W_2|$ but $W_1 = -W_2$

372. (d)

We know that:

$$(W_1 - W_2) = V_L I_L \sin\phi = Q_{1-\phi}$$

So, $Q_{3-\phi} = \sqrt{3}(W_1 - W_2)$

373. (c)

$I_{ab} = 10\angle{-30°}$ A

So, $I_{bc} = 10\angle{-150°}$ A

and $I_{ca} = 10\angle{-270°}$ A

Phase current,

$I_a = I_{ab} - I_{ca} = (10\angle{-30°} - 10\angle{-270°})$ A

$= 10[\cos 30° - j\sin 30° - \cos 270° + j\sin 270°]$

$= 10[0.866 - j0.5 - 0 - j1]$

$= 8.66 - j15 = 17.32\angle{-60°}$ A

$\therefore$ $I_a = 17.32\angle{-60°}$ A

$\therefore$ Total power received

$= \sqrt{3} V_L I_L \cos\phi_L$

$= 3 V_P I_P \cos\phi_{ph}$

$= 3 \times 220 \times 10 \times \cos 30°$

$= 5715$ W

374. (d)

Impedance per phase

$= 1 + j1 + 3 + j5$

$= (4 + j6)\ \Omega$

$\therefore$ $|I_{line}| = \frac{100}{\sqrt{4^2 + 6^2}} = \frac{100}{\sqrt{52}} = \frac{50}{\sqrt{13}}$ A

375. (b)

Initially, in closed-Δ, power consumed by the load is

$$P_1 = 3 V_{ph} I_{ph} = 3 \times 400 \times \frac{400}{200}$$

$= 2400$ W

Let, P_2 be the power consumed in open delta.

Then, $P_2 = \frac{400^2}{200} + \frac{400^2}{200} = 1600$ W

379. (c)

$$W_1 = V_L I_L \cos(30 + \phi)$$

and $W_2 = V_L I_L \cos(30 - \phi)$

By finding ϕ from $\cos\phi$, we can find the values of W_1 and W_2.

■■■■

UNIT

4

Electrical & Electronic Measurements

1. Introduction

Q.1 In present day measurement systems
(a) direct methods are commonly used
(b) use of direct methods is limited but indirect methods are commonly used
(c) both direct and indirect methods are commonly used
(d) none of the above

Q.2 The use of electronic instruments is becoming more extensive because they have
(a) the capability to respond to signals from remote places
(b) a fast response and compatibility with digital computers
(c) both (a) and (b)
(d) a high sensitivity but less reliability

Q.3 A thermometer is calibrated 150°C to 200°C. The accuracy is specified within ±0.25 percent of instrument span. The maximum static error is
(a) ± 0.225°C (b) ± 0.250°C
(c) ± 0.125°C (d) ± 0.175°C

Q.4 A set of independent current measurements were recorded as 10.03, 10.10, 10.08 and 10.11 A. The average range of error is
(a) ±0.04 (b) ±0.03
(c) ±0.05 (d) ±0.02

Q.5 Two resistors R_1 and R_2 are connected in series with R_1 = 28.7 Ω and R_2 = 3.624 Ω. The total resistance to the appropriate number of significant figures can be written as
(a) 32.324 Ω (b) 32.4 Ω
(c) 32.32 Ω (d) 32.3 Ω

Q.6 A wheatstone bridge requires a change of 8 Ω in the unknown arm of the bridge to produce a change in deflection of 3 mm of the galvanometer. The deflection factor (scale factor) is
(a) 0.375 mm/Ω (b) 2.67 Ω/mm
(c) 24 Ω-mm (d) can't be determined

Q.7 A moving coil voltmeter has a uniform scale with 100 divisions, the full scale reading is 200 V and 1/10 of a scale division can be estimated with a fair degree of certainty.
The resolution of the instrument in volt is
(a) 2 (b) 0.01
(c) 0.2 (d) 1

Q.8 A multimeter having a sensitivity of 2 kΩ/V is used for the measurement of voltage across a circuit having an output resistance of 10 kΩ. The open circuit voltage of the circuit is 6 volts. The percentage error in multimeter reading when it is set to 10 volt is given by
(a) 33% low (b) 22% high
(c) 33% high (d) 22% low

Q.9 A 50 V range voltmeter is connected across the terminals *X* and *Y* of the circuit shown in figure below. The voltage across the terminals are measured both under open circuit and loaded conditions.

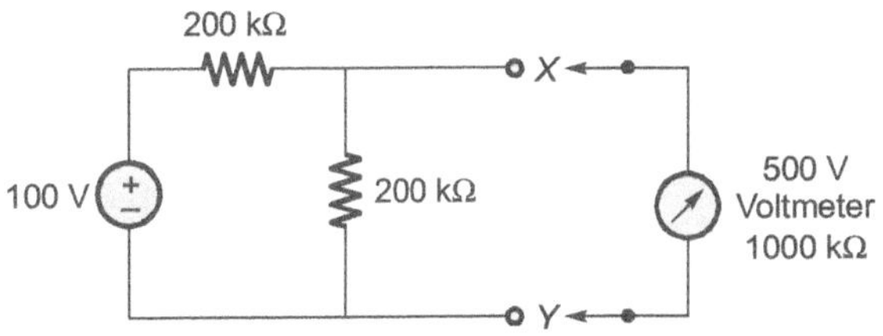

The accuracy in the measurement of voltage across the terminals *X* and *Y* in percent is
(a) 9.1% (b) 90.9%
(c) 45.45% (d) 4.54%

Q.10 The number of significant figures in the two resistors having resistances 4 × 10^6 kΩ and 0.345 kΩ are respectively
(a) 6 and 3 (b) 2 and 4
(c) 1 and 4 (d) 1 and 3

Q.11 An ammeter reads 6.7 A and the true value of current is 6.5 A. The correction factor is
(a) 0.97 (b) 0.20
(c) 0.03 (d) none of these

Q.12 A pressure gauge is calibrated from 0 - 50 kN/m^2. It has a uniform scale with 100 scale divisions. One fifth of a scale division can be read with certainty. The gauge has a
(a) threshold of 0.1 kN/m^2
(b) dead zone of 0.2 kN/m^2
(c) resolution of 0.1 kN/m^2
(d) both (a) and (b)

Q.13 The resistance of a circuit is found by measuring current flowing and the power fed into the circuit. If the limiting errors in the measurement of power and current are respectively ±1.5% and ±1.0%, then the limiting error in the measurement of resistance will be
(a) ±4% (b) ±3.5%
(c) ±2.5% (d) ±0.5%

Q.14 A resistance is determined by voltmeter-ammeter method. The voltmeter reads 100 V with an uncertainty of ±12 V and the ammeter reads 10 A with an uncertainty of ±2 A. The uncertainty in the measurement of resistance is
(a) 1.56 Ω (b) 0.26 Ω
(c) 3.25 Ω (d) 2.33 Ω

Q.15 The resistance of an unknown resistor is determined by wheatstone bridge. The solution for the unknown resistance is stated as:

$$R_4 = \frac{R_1 R_2}{R_3}$$

where limiting values of various resistances are:

R_1 = 500 Ω ± 1%,
R_2 = 615 Ω ± 1%,
R_3 = 100 Ω ± 0.5%

The limiting error of the unknown resistor in ohm is
(a) ± 76.88 Ω (b) ± 66.66 Ω
(c) ± 3.07 Ω (d) ± 98.76 Ω

Q.16 A Wattmeter has a full scale range of 2500 W. It has an error ±1% of true value. What would be the range of reading if true power is 1250 W?
(a) 1225 W – 1275 W
(b) 1245 W – 1255 W
(c) 1200 W – 1300 W
(d) 1237.5 W – 1262.5 W

Q.17 In a permanent magnet moving coil ammeter, the deflection of the pointer is proportional to product of flux density of magnetic field produced by the permanent magnet and the current in the moving coil. If the strength of the permanent magnet becomes 95% of the original, the meter gives errorneous reading resulting into error. This error can be classified as:
(a) gross error (b) systematic error
(c) random error (d) none of these

Q.18 A 0 - 10 A ammeter has a guaranteed accuracy of 1 percent of full-scale deflection. The limiting error while reading 2.5 A is
(a) 1% (b) 2%
(c) 4% (d) none of these

Q.19 Dynamic response consists of
(a) two parts, one steady state and the other transient state response
(b) only transient state response
(c) only steady state response
(d) steady state and transient frequency response

Q.20 Match **List-I (Quantity)** with **List-II (Dimensions)** and select the correct answer using the codes given below the lists:

List-I	List-II
A. Emf	1. $[MT^{-2}I^{-1}]$
B. Magnetic flux density	2. $[M^{-1}L^{-2}T^{2}I^{2}]$
C. Magnetizing force	3. $[ML^{2}T^{-3}I^{-1}]$
D. Reluctance	4. $[L^{-1}I]$

Codes:

	A	B	C	D
(a)	2	1	3	4
(b)	3	1	2	4
(c)	3	2	4	1
(d)	3	1	4	2

Q.21 The symbol nano (n) and Giga (G) are respectively multiplied by a factor of
(a) 10^{-9} and 10^{12} (b) 10^{-9} and 10^{9}
(c) 10^{9} and 10^{-9} (d) 10^{12} and 10^{-9}

Q.22 The most stable primary atomic standards for frequency is
(a) Caesium beam standard
(b) Hydrogen maser standard
(c) quartz standard
(d) Rubidium vapour standard

Q.23 If an analog PMMC 0 - 10 A ammeter is provided with no controlling mechanism and the moving parts are free to rotate, what will be the reading of the instrument if a current of 1 A (d.c.) is passed through the moving coil (The torque produced is sufficient to overcome the frictional losses)?
(a) 1 A
(b) 10 A
(c) the pointer will continuously rotate
(d) the pointer will remain stationary

Q.24 Match **List-I (Damping Mechanism)** with **List-II (Applications)** and select the correct answer using the codes given below the lists:

List-I
A. Fluid friction damping
B. Electromagnetic damping
C. Eddy current damping
D. Air friction damping

List-II
1. Electrostatic instruments
2. Galvanometers
3. PMMC instruments
4. Moving iron instruments

Codes:

	A	B	C	D
(a)	1	2	3	4
(b)	2	1	3	4
(c)	4	3	2	1
(d)	4	2	3	1

Q.25 **Assertion (A):** The frictional torque exerts a considerable influence on the performance of an indicating instrument.
Reason (R): If the weight of moving parts is large, the frictional torque will be small.
(a) Both A and R are true and R is a correct explanation of A.
(b) Both A and R are true but R is not a correct explanation of A.
(c) A is true but R is false.
(d) A is false but R is true.

Q.26 **Assertion (A):** Torque/Weight ratio of an instrument is an index of its performance, the higher the ratio, the better will be its performance.
Reason (R): The ratio of deflecting torque to frictional torque is a measure of reliability of the instrument indications and of the inherent quality of the design.
(a) Both A and R are true and R is a correct explanation of A.
(b) Both A and R are true but R is not a correct explanation of A.
(c) A is true but R is false.
(d) A is false but R is true.

Q.27 Improper setting of range of multimeter leads to an error called
(a) random error (b) limiting error
(c) instrumental error (d) observational error

Q.28 Match **List-I (Dynamic characteristics of an Instrument)** with **List-II (Definition)** and select the correct answer using the codes given below the lists:

List-I
A. Speed of response
B. Fidelity
C. Lag
D. Dynamic Error

List-II
1. the delay in the response of an instrument to changes in the measured variable.
2. the difference between the true value of a quantity changing with time and the value indicated by the instrument.
3. the degree to which an instrument indicates the changes in the measured variable.
4. the rapidity with which an instrument responds to changes in the measured quantity.

Codes:

	A	B	C	D
(a)	4	1	3	2
(b)	4	3	1	2
(c)	4	2	1	3
(d)	1	4	2	3

Q.29 The difference between the expected value of the variable and the measured variable is termed as

(a) absolute error (b) random error
(c) instrumental error (d) gross error

Q.30 Errors that occur even when all the gross and systematic errors are taken care of are called
(a) environmental errors
(b) instrumental errors
(c) limiting errors
(d) random errors

Q.31 Limiting errors are
(a) manufacturer's specifications of accuracy
(b) manufacturer's specifications of instrumental error
(c) environmental errors
(d) random errors

Q.32 Consider the following statements about the null type instruments used in measurement and instrumentation:
1. Null type instruments are more accurate than deflection type instruments.
2. Null type instruments can be highly sensitive as compared with deflection types instruments.
3. Null type instruments are more suited for measurements under dynamic conditions.
4. A DC potentiometer and a Wattmeter are null type instruments.

Which of the above statements are correct?
(a) 2, 3 and 4 (b) 1, 2, 3 and 4
(c) 1 and 2 (d) 1, 2 and 3

Q.33 **Assertion (A):** To eliminate the parallax error, pointer and the scale of the instrument are kept in the same plane.
Reason (R): Parallax errors arise on account of pointer and the scale not being in the same plane.
(a) Both A and R are true and R is a correct explanation of A.
(b) Both A and R are true but R is not a correct explanation of A.
(c) A is true but R is false.
(d) A is false but R is true.

Q.34 Match **List-I** with **List-II** and select the correct answer using the codes given below the lists:

List-I	List-II
A. Instrumental errors	**1.** Parallax error
B. Observational error	**2.** Human mistakes
C. Residual error	**3.** Loading effect
D. Gross errors	**4.** Random error

Codes:

	A	B	C	D
(a)	3	4	2	1
(b)	3	1	4	2
(c)	4	3	1	2
(d)	2	3	1	4

Q.35 Match **List-I** with **List-II** and select the correct answer using the codes given below the lists:

List-I
A. Four terminal
B. Atomic clock
C. Quartz crystal
D. Ceramic core totoid shape device

List-II
1. Oscillator frequency
2. Time device
3. Inductance
4. Resistance

Codes:

	A	B	C	D
(a)	3	4	1	2
(b)	4	2	1	3
(c)	4	1	3	2
(d)	1	2	3	4

Q.36 A random noise generator produces a signal
(a) whose amplitude varies randomly
(b) which has no periodic frequency
(c) has an unpredictable power spectrum
(d) none of the above

Q.37 Which of the following has the highest accuracy?
(a) standard resistance
(b) standard inductance
(c) standard capacitance
(d) standard mutual inductance

Q.38 In a.c. circuits, the connection of measuring instruments causes loading errors which may effect
(a) only the magnitude of the quantity being measured.
(b) only the phase of the quantity being measured.
(c) both the magnitude and phase of the quantity being measured.
(d) magnitude, phase and also the waveform of the quantity being measured.

Q.39 A zero to 300 V voltmeter has an error of ±2% of the full-scale deflection. If the true voltage is 30 V, then the range of readings on this voltmeter would be
(a) 20 V to 40 V (b) 24 V to 36 V
(c) 29.4 V to 30.6 V (d) 29.94 V to 30.06 V

Q.40 A resistor of 10 kΩ with 5% tolerance is connected in series with a 5 kΩ resistor of 10% tolerance. What is the tolerance limit of the series network?
(a) 5% (b) 6.67%
(c) 10% (d) 8.33%

Q.41 Match **List-I** with **List-II** and select the correct answer using the codes given below the lists:

List-I
A. Precision
B. Accuracy
C. Resolution
D. Static

List-II
1. The smallest change in the input quantity which can be detected with its certainty.
2. Closeness of reading with its true value.
3. Measure of reproducibility of the measurements.
4. Ratio of infinitesimal change sensitivity in output to infinitesimal change in input.

Codes:

	A	B	C	D
(a)	3	2	1	4
(b)	2	3	1	4
(c)	2	3	4	1
(d)	3	2	4	1

Q.42 The total current, $I = I_1 + I_2$ in a circuit is measured as $I_1 = 150 \pm 1$ A, $I_2 = 250 \pm 2$ A, where the limits of error are given as standard deviation. Current I is measured as
(a) (400 ± 3) A (b) (400 ± 2.24) A
(c) (400 ± 1/5) A (d) (400 ± 1) A

Q.43 A set of independent current measurements taken by four observers was recorded as: 117.02 mA, 117.11 mA, 117.08 mA and 117.03 mA
What is the range of error?
(a) ± 0.045 (b) ± 0.054
(c) ± 0.065 (d) ± 0.056

2. Measurement of Resistance

Q.44 In order to reduce skin effect in resistance standards when they are used on high frequency a.c.
(a) adjacent turns of the resistance coil should carry current in the opposite direction
(b) they should be provided with four terminals
(c) the diameter of the resistance wire should be small
(d) none of the above

Q.45 The condition for a resistor to have the same value of resistance at medium frequencies as with d.c. is
(a) $CR^2 = \omega L$ (b) $CR^2 = 2L$
(c) $CR^2 = L$ (d) $CR^2 = 2\omega L$

Q.46 A resistor R has an effective inductance of L and a distributed capacitance of C. Its time constant at medium frequencies is
(a) $\frac{L}{R} - CR$ (b) $\frac{L}{R}$
(c) CR (d) $CR - \frac{L}{R}$

Q.47 A 100 Ω resistor has a time constant of 30 ns. What is the value of capacitor to be connected in parallel with the resistor to make its phase angle equal to zero?
(a) 30 pF (b) 300 nF
(c) 30 nF (d) 300 pF

Q.48 **Assertion (A):** Manganin is an alloy of copper, manganese and nickel and is universally used as resistance material for precision resistors and for resistance measuring apparatus.
Reason (R): The foremost property of manganin is that it has almost zero temperature coefficient of resistance near about room temperature. Moreover, it shows no ageing effect.
(a) Both A and R are true and R is a correct explanation of A.
(b) Both A and R are true but R is not a correct explanation of A.
(c) A is true but R is false.
(d) A is false but R is true.

Q.49 A resistance R is measured using the connections shown in figure below.

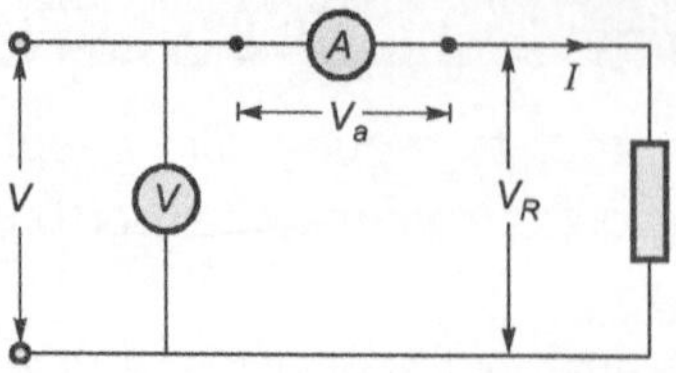

The current measured is 10 A on a range 10 A and the voltage measured is 125 V on a range of 150 V. The scale of the ammeter and voltmeter are uniform, the total scale divisions of ammeter are 100 and that of voltmeter are 150. The scales of these instruments are such that 1/10 of a scale can be distinguished. The constructional error of the ammeter is ±0.3% and that of voltmeter is ±0.4%. The resistance of ammeter is 0.25 Ω. The unknown resistance R will be specified as:

(a) R = 12.50 + 0.4%
(b) R = 12.25 + 0.867%
(c) R = 12.50 + 0.867%
(d) R = 12.25 + 0.467%

Q.50 Each of the ratio arms of a laboratory type wheatstone bridge has guaranteed accuracy of ±0.05%, while the standard arm has a guaranteed accuracy of ±0.1%. The ratio arms are both set at 1 kΩ and and bridge is balanced with standard arm adjusted to 3154 Ω. The upper and the lower limits of the unknown resistance, based upon the guaranteed accuracies of the known bridge arms are respectively

(a) 3217 Ω and 3091 Ω
(b) 3160.3 Ω and 3091.7 Ω
(c) 3160.3 Ω and 3091 Ω
(d) 3160.3 Ω and 3147.7 Ω

Q.51 A highly sensitive galvanometer can detect a current as low as 0.1 nA. This galvanometer is used in a wheatstone bridge as a detector as shown in figure below. The resistance of galvanometer is negligible.

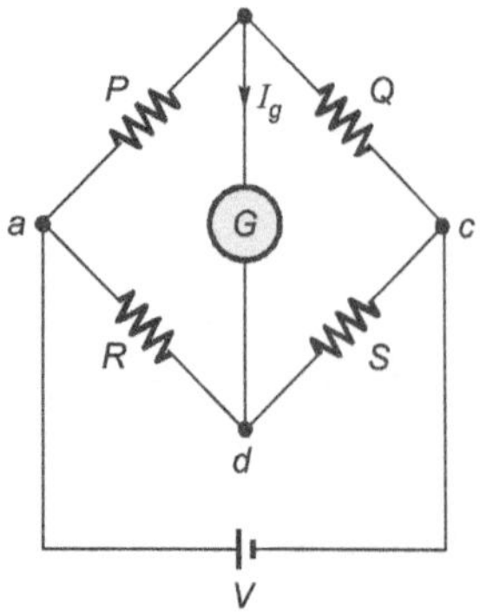

Each arm of the bridge has a resistance of 1 kΩ and the input voltage applied to bridge is 20 V. Neglecting the resistance of the galvanometer as compared with the internal resistance of bridge, the smallest change in the resistance which can be detected is:

(a) 20 μΩ (b) 15 μΩ
(c) 2 μΩ (d) 1 μΩ

Q.52 A cable is tested by loss of charge method using a ballistic galvanometer with following results:

Discharged immediately after electrification, deflection 200 divisions.

Discharged after 20*s* and after electrification:

(*i*) deflection, 100 divisions,
(*ii*) when in parallel with a resistance of 10 MW, deflection 50 divisions.

What is the insulation resistance of the cable?

(a) 20 MΩ (b) 30 MΩ
(c) 40 MΩ (d) 10 MΩ

Q.53 Consider the following statements about the bridge measurement:

1. The measurement accuracy is directly related to the accuracy of the bridge component as well as that of the null indicator used.
2. The basic dc bridge used for accurate measurement of resistance is called wheatstone's bridge.
3. When the bridge is in balanced condition, current flows through the galvanometer, causing a deflection of its pointer.
4. A more sensitive galvanometer deflects by a greater amount for the same current.
5. The sensitivity of the bridge measurement can only be expressed in mm/μA.

Which of these statements is/are correct?

(a) 2 only (b) 1, 3, 4 and 5
(c) 2 and 4 (d) 1, 2 and 5

Q.54 The value of unknown resistance R_x for the wheatstone bridge shown below is

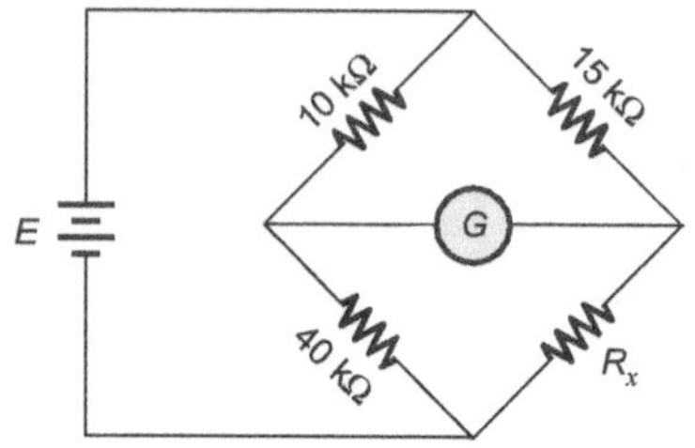

(a) 40 kΩ (b) 26.67 kΩ
(c) 3.75 kΩ (d) 60 kΩ

Q.55 For the Kelvin's bridge shown in figure below, the ratio of R_a to R_b is 1000 Ω, R_1 is 5 Ω and $R_1 = 0.5\ R_2$.

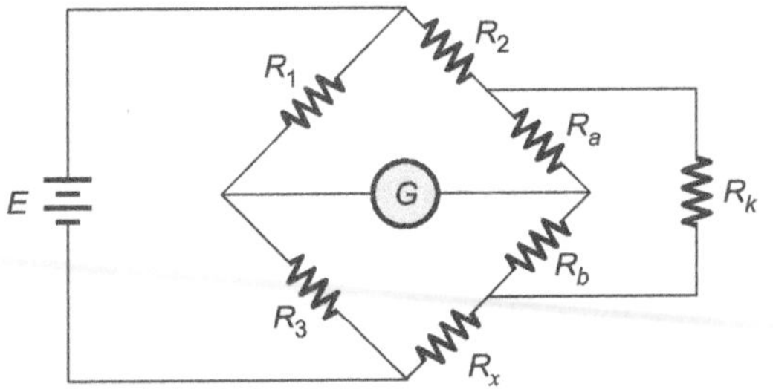

The value of R_x is
(a) 10 mΩ (b) 5 mΩ
(c) 100 mΩ (d) 50 mΩ

Q.56 **Assertion (A):** Low resistances are provided with four terminals.
Reason (R): By providing four terminals, the effect of thermo-electric emfs are eliminated.
(a) Both A and R are true and R is a correct explanation of A.
(b) Both A and R are true but R is not a correct explanation of A.
(c) A is true but R is false.
(d) A is false but R is true.

Q.57 What is the insulation resistance of a cable in which the voltage falls from 100 to 10 V in 20 seconds and cable capacitance is 300 pF?
(a) 22 MΩ (b) 29 GΩ
(c) 25 MΩ (d) 18 GΩ

Q.58 Match **List-I (Instruments)** with **List-II (Quantity being measured)** and select the correct answer using the codes given below the lists:

List-I	List-II
A. Kelvin's double bridge	1. High R
B. Wheatstone bridge	2. Frequency (ω)
C. Megger	3. Low R
D. Wein's bridge	4. Medium R

Codes:

	A	B	C	D
(a)	4	1	3	2
(b)	1	3	2	4
(c)	3	4	2	1
(d)	3	4	1	2

Q.59 The value of resistance of an earthing electrode depends upon:
(a) shape and material of electrode
(b) depth to which electrode is driven into earth
(c) specific resistance of soil
(d) all of the above

Q.60 Which of the following method is not used for the measurement of resistance less than one ohm?
(a) Ohm-meter method
(b) Voltmeter-ammeter method
(c) Potentiometer method
(d) both (a) and (b)

Q.61 Match **List-I (Methods of resistance measurement)** with **List-II (Applications)** and select the correct answer using the codes given below the lists:

List-I
A. Carey-Foster slide-wire Bridge
B. Loss of charge method
C. Direct deflection method
D. Wheatstone bridge method

List-II
1. Localizing the cable faults.
2. Volume and surface resistivity measurement.
3. Leakage resistance of capacitor.
4. Difference between the standard and the unknown resistances.

Codes:

	A	B	C	D
(a)	1	4	3	2
(b)	4	2	3	1
(c)	4	3	2	1
(d)	1	2	3	4

Q.62 The measurement of an unknown resistance is done by using voltmeter ammeter methods using two arrangements as shown below.

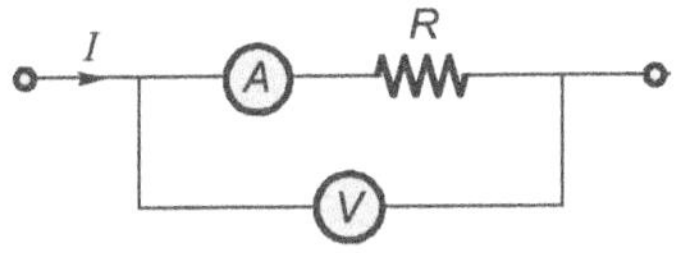

Arrangement-1

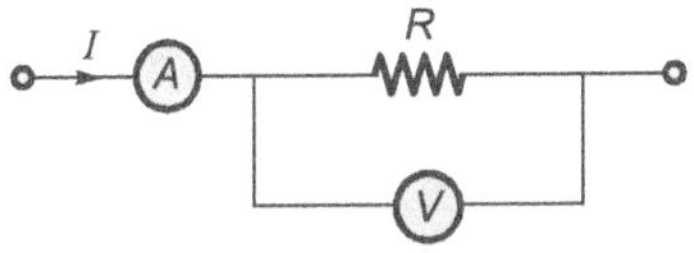

Arrangement-2

If the resistances of voltmeter and milliammeter are 500 Ω and 1 Ω respectively and voltmeter reads 20 V and milliammeter 100 mA, then the value of unknown resistances in arrangements-1 and 2 will be respectively
(a) 200 Ω and 289 Ω (b) 199 Ω and 333 Ω
(c) 200 Ω and 333 Ω (d) 199 Ω and 289 Ω

Q.63 **Assertion (A):** The use of wheatstone bridge is limited to the measurement of resistances ranging from a few ohm to several megaohm.
Reason (R): The upper limit is set by the reduction in sensitivity to unbalance caused by high resistance values while the lower limit is set by the resistance of the connecting leads and by contact resistance at the binding posts.
(a) Both A and R are true and R is a correct explanation of A.
(b) Both A and R are true but R is not a correct explanation of A.
(c) A is true but R is false.
(d) A is false but R is true.

Q.64 **Assertion (A):** The "Kelvin double bridge" can be used for the accurate measurement of low value of resistance.
Reason (R): The effect of thermo-electric emfs can be eliminated by making another measurement with the battery connections reversed.
(a) Both A and R are true and R is a correct explanation of A.
(b) Both A and R are true but R is not a correct explanation of A.
(c) A is true but R is false.
(d) A is false but R is true.

Q.65 An unknown resistance is measured by "substitution method". First a standard known resistance of 100 Ω is connected in series with a circuit having a rheostat and a galvanometer. The battery voltage is 10 V and the setting of the rheostat is 500 Ω and the galvanometer shows a deflection of 60°. After this test, the battery voltage goes down to 9 V and when the unknown resistance is substituted for the known resistance, the galvanometer again shows a deflection of 60° with the same setting of the rheostate. The value of unknown resistance is:
(a) 90 Ω (b) 100 Ω
(c) 54 Ω (d) 40 Ω

Q.66 A wheatstone bridge is used for measuring the value of change of resistance of a strain gauge which forms one of the arms of the bridge. All the arms of the bridge including the strain gauge have a resistance of 100 Ω each. The maximum allowable power dissipation from the strain gauge is 250 mW.
If a source of 20 V is available, then what is the value of series resistance to be connected between the source and the bridge to limit the input voltage of the bridge to permissible limit?
(a) 100 Ω (b) 50 Ω
(c) 25 Ω (d) 150 Ω

Q.67 **Assertion (A):** The determination of voltage by a potentiometer is quite dependent of the source resistance.
Reason (R): A potentiometer is an instrument designed to measure an unknown voltage by comparing it with a known voltage.
(a) Both A and R are true and R is a correct explanation of A.
(b) Both A and R are true but R is not a correct explanation of A.
(c) A is true but R is false.
(d) A is false but R is true.

Q.68 Which of the following is not true regarding a potentiometer used in measurement of an unknown voltage source?
(a) Since no current flows, power consumed in potentiometer circuit is zero.
(b) It is the standard for the calibration of voltmeters and ammeters.
(c) It's accuracy is very low.
(d) It can be used to determine current simply by measuring the voltage produced by the unknown current passing through a known standard resistance.

Q.69 The instrument normally used to check the insulation resistance is
(a) ohm-meter (b) multimeter
(c) megger (d) tong-tester

Q.70 The absolute measurement of resistance is done by
(a) Bridge method (b) Lorenz method
(c) Rayleigh method (d) Ohm's law method

Q.71 A meggar is usually a
(a) moving iron type instrument
(b) hot-wire type instrument
(c) electrostatic type instrument
(d) moving coil type instrument

Q.72 A shunt type ohmmeter is not suitable for high resistance measurement because
(a) scale is highly cramped for high resistance values.
(b) battery cannot supply the necessary current for proper meter detection.
(c) full scale value of meter may be exceeded.
(d) a very low resistance of the meter would short the high unkonwn resistance.

Q.73 Consider the following statements:
1. Megger is, essentially, a mega ohmmeter.
2. For measurement of a resistance by a potentiometer standardisation of the potentiometer is not necessary at all.
3. An ohmmeter is basically an ammeter.
4. A 100 kΩ resistance can be measured by ammeter-voltmeter method.

Which of the above statements is/are not correct?
(a) Only 4 (b) 1, 2 and 3
(c) 2 and 4 (d) 3 and 4

Q.74 In the circuit shown in figure, if the ammeter indicates 1 A and the voltmeter having an internal resistance of 1 kΩ indicates 100 V, then the value of R would be

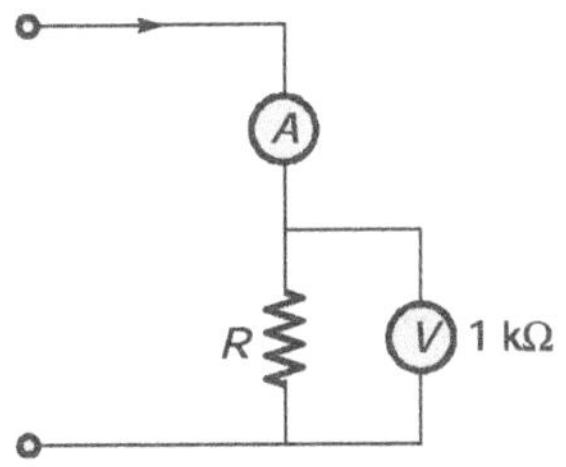

(a) 111.11 Ω (b) 105.2 Ω
(c) 100 Ω (d) 90.9 Ω

Q.75 The current flowing through the galvanometer (G) having an internal resistance of 1.25 kΩ as shown in figure below is

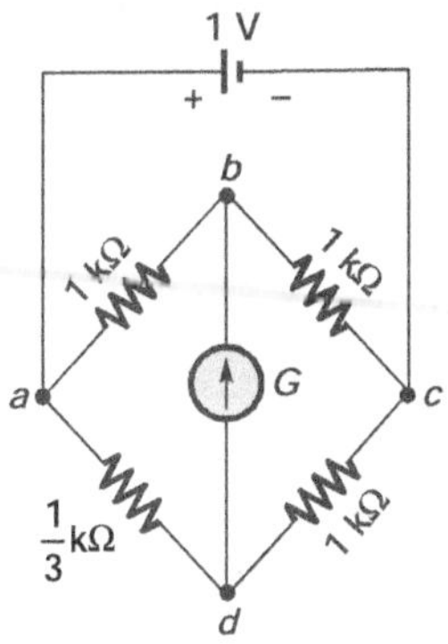

(a) $\frac{1}{3}$ mA (b) $\frac{1}{5}$ mA
(c) $\frac{1}{8}$ mA (d) 0 mA

Q.76 The accuracy of Kelvin's double bridge for the measurement of low resistance is high because the bridge
(a) use two pairs of resistance arms
(b) has medium value resistance in the ratio arms
(c) uses a low resistance link between standard and test resistances
(d) uses a null indicating galvanometer

3. AC Bridges

Q.77 A bridge circuit works at a frequency of 3 kHz. Which detector can be used for detecting the null conditions in the bridge?
(a) Vibration galvanometer and tunable amplifiers.
(b) Vibration galvanometer and head phones.
(c) Headphones and tunable amplifiers.
(d) Vibration galvanometer, headphones and tunable amplifiers.

Q.78 Inductance comparison bridge uses
(a) One L and one C
(b) One L
(c) One C
(d) none of the above

Q.79 The equations under conditions of balance for a bridge measuring resistance R_1 and inductance L_1 of a choke coil are

$$R_1 = \frac{R_2 R_3}{R_4} \text{ and } L_1 = R_2 R_3 C_4$$

In order to achieve convergence of balance

(a) R_2 and C_4 should be chosen as variable
(b) R_2 and R_3 should be chosen as variable
(c) R_3 and C_4 should be chosen as variable
(d) R_4 and C_4 should be chosen as variable

Q.80 In measurement of resistance by "Carey Foster bridge" no error is introduced due to

(a) connecting leads
(b) thermoelectric emfs
(c) contact resistance
(d) all of the above

Q.81 A capacitance comparison bridge as shown in figure below is used to measure a capacitive impedance at a frequency of 2 kHz. The bridge constants at balance are:

C_3 = 100 μF, R_1 = 10 kΩ, R_2 = 50 kΩ, R_3 = 100 kΩ.

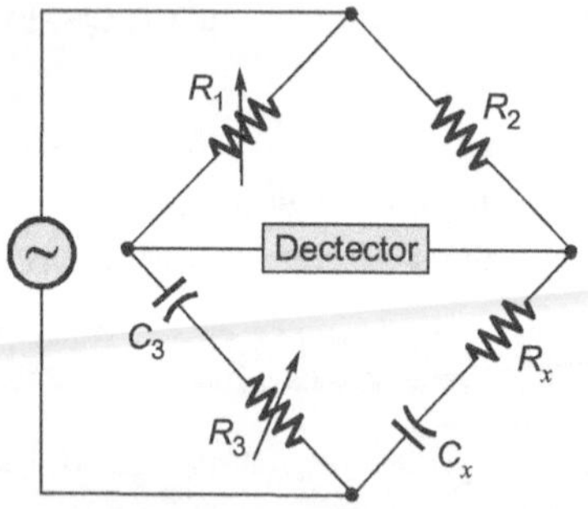

The equivalent series circuit of the unknown impedance will be represented as:

(a) 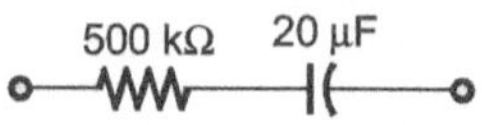

(b) 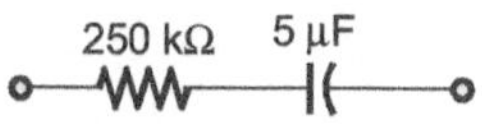

(c) 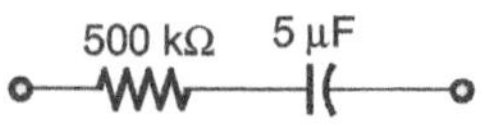

(d) 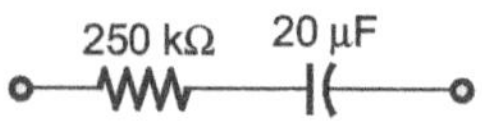

Q.82 A Wien bridge circuit consists of the following:

R_1 = 4.7 kΩ, C_1 = 5 nF
R_2 = 20 kΩ, C_2 = 10 nF
R_3 = 10 kΩ
R_4 = 100 kΩ

The frequency of the circuit is

(a) 2.25 kHz (b) 1.50 kHz
(c) 3.75 kHz (d) 3.28 kHz

Q.83 An ac bridge uses a detector in the form of

(a) ammeter (b) voltmeter
(c) headphones (d) wattmeter

Q.84 Maxwell's bridge is used to measure Q-factor in the range

(a) 1 - 10 (b) 30 - 50
(c) 50 - 75 (d) 75 - 100

Q.85 Anderson bridge is used to measure unknown

(a) resistance (b) frequency
(c) inductance (d) capacitance

Q.86 Hay's bridge is used to measure an inductance of

(a) low *Q* (b) medium *Q*
(c) high *Q* (d) very high *Q*

Q.87 To measure precise inductance from a few μH to several henries, which of the following bridge is used?

(a) Maxwell's bridge
(b) Hay's bridge
(c) Maxwell-Wien bridge
(d) Anderson bridge

Q.88 Wagner's ground is used to

(a) eliminate earth from the bridge network capacitance
(b) measure capacitance
(c) measure resistance
(d) measure inductance

Q.89 An ac bridge with terminals *ABCD* has in
Arm *AB*-a resistance of 800 Ω in parallel with a capacitor of 0.5 μF,
Arm *BC*-a resistance of 400 Ω in series with a capacitor of 1 μF.
Arm *CD*-a resistance of 1000 Ω,
Arm *DA*-a pure resistance *R*.
What is the value of R required to produce balance?

(a) 0.25 kΩ (b) 1 kΩ
(c) 1.5 kΩ (d) 0.5 kΩ

Q.90 Consider the following statements regarding the use of alternating current bridge methods for measurement of electrical quantities:

1. Alternating current bridge circuits find applications in communication systems and complex electronic circuits.

2. Head phones are extremely useful for power and low audio frequency ranges.
3. Tuneable amplifier detectors are the most versatile of the detectors.
4. An a.c. bridge is balanced if either the phase angle or the magnitude condition of the bridge arm impedances is satisfied.

Which of the above statements is/are correct?
(a) 1, 2, 3 and 4 (b) 1 and 2
(c) 1, 3 and 4 (d) 1 and 3

Q.91 **Assertion (A):** The Wien's bridge may be difficult to balance unless the waveform of the applied voltage is sinusoidal.
Reason (R): Wien bridge is frequency sensitive.
(a) Both A and R are true and R is a correct explanation of A.
(b) Both A and R are true but R is not a correct explanation of A.
(c) A is true but R is false.
(d) A is false but R is true.

Q.92 Match **List-I (Bridge circuits)** with **List-II (Applications)** and select the correct answer using the codes given below the lists:

List-I
A. Maxwell's inductance - capacitance bridge
B. Anderson's bridge
C. Hay's bridge
D. Wien's bridge

List-II
1. Accurate value of capacitance in terms of inductance.
2. As a Notch filter.
3. Wide range of inductance at power and audio frequencies.
4. Inductance of a coil with large time constant.

Codes:

	A	B	C	D
(a)	3	4	2	1
(b)	4	1	3	2
(c)	3	1	4	2
(d)	3	2	4	1

Q.93 A four arm a.c. bridge *a*, *b*, *c*, *d* has the following impedances:
Arm ***ab*** : $z_1 = 200\angle 60°\Omega$ (inductive impedance)
Arm ***ad***: $z_2 = 400\angle -60°\Omega$ (capacitive impedance)
Arm ***bc*** : $z_3 = 300\angle 0°\ \Omega$ (Purely resistive)

The value of unknown impedance z_4 in arm *cd* will be
(a) $200\angle -45°\ \Omega$ (b) $600\angle 30°\ \Omega$
(c) $275\angle 45°\ \Omega$ (d) none of these

Q.94 Consider the wheatstone bridge circuit shown below.

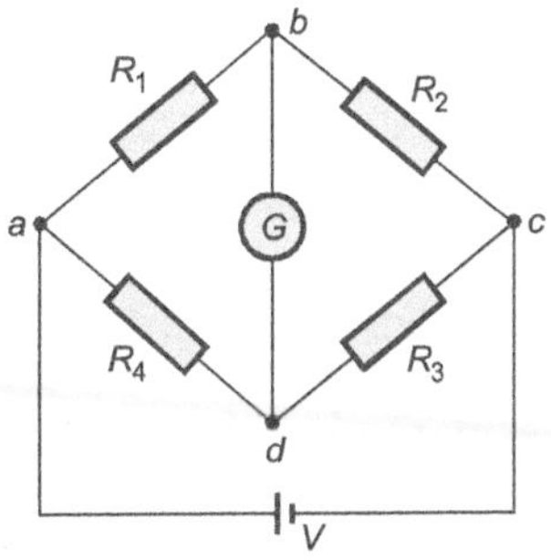

What resistance range must resistor R_3 have in order to measure unknown resistor in the range 1 - 100 kΩ? (Given, R_1 = 1 k and R_2 = 10 k).
(a) 10 kΩ to 1 MΩ (b) 0.5 kΩ to 50 kΩ
(c) 10 kΩ to 0.5 MΩ (d) 1 kΩ to 100 kΩ

Q.95 The ac bridge shown below is used to measure an unknown impedance. The bridge is in balanced condition. The impedance *Z* is

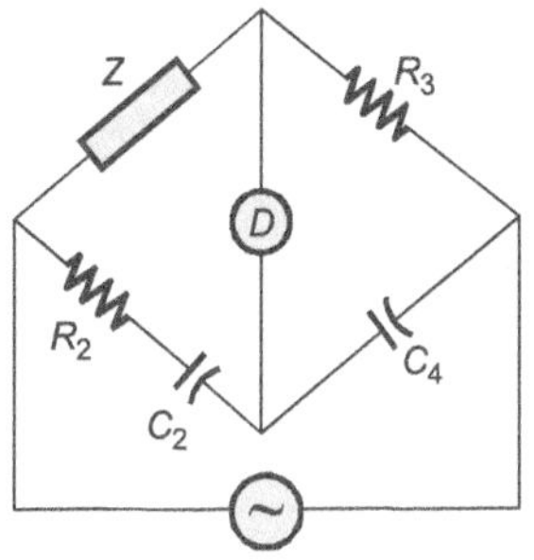

(a) a parallel combination of a resistor and a capacitor.
(b) a series combination of a resistor and a capacitor.
(c) a parallel combination of a resistor and an inductor.
(d) a series combination of a resistor and an inductor.

Q.96 Hay bridge is a modification of
(a) Anderson's bridge
(b) De-Sauty's bridge
(c) Maxwell's bridge
(d) None of the above

Q.97 In Wien bridge, the output frequency is determined by
(a) R-L combination
(b) R-C combination
(c) R-L-C combination
(d) R-L-L combination

Q.98 De-Sauty bridge is used for the measurement of
(a) capacitance and power factor
(b) power factor and dissipation factor
(c) power factor and quality factor
(d) dissipation factor and capacitance

Q.99 The shielding of bridge elements is done to reduce the effects of
(a) distributed capacitance
(b) stray capacitance
(c) leakage current
(d) external magnetic field

Q.100 **Assertion (A):** If each component in a bridge has a defining screen connected to one end, a very high accuracy in measurement is made possible by the addition of a Wagner earthing device.
Reason (R): Wagner earthing device avoids all the error caused due to inter-capacitance between high and low arms of the bridge.
(a) Both A and R are true and R is a correct explanation of A.
(b) Both A and R are true but R is not a correct explanation of A.
(c) A is true but R is false.
(d) A is false but R is true.

Q.101 **Assertion (A):** The balance conditions obtained in finding the unknown values of inductance and resistance in a Hay's bridge is independent of frequency.
Reason (R): Hay's bridge is not suitable for measurement of coils having Q less than 10 and for these applications a Maxwell's bridge is more suited.
(a) Both A and R are true and R is a correct explanation of A.
(b) Both A and R are true but R is not a correct explanation of A.
(c) A is true but R is false.
(d) A is false but R is true.

Q.102 Match **List-I (Detectors)** with **List-II (Frequency range)** and select the correct answer using the codes given below the lists:

List-I
A. Vibrant galvanometers
B. Headphones
C. Tuneable amplifier detectors
D. Galvanometer

List-II
1. 250 Hz - 4 kHz
2. DC
3. 5 - 1000 Hz
4. 10 Hz - 100 kHz

Codes:

	A	B	C	D
(a)	4	3	1	2
(b)	3	2	4	1
(c)	4	2	3	1
(d)	3	1	4	2

Q.103 Consider the bridge circuit shown in figure below:

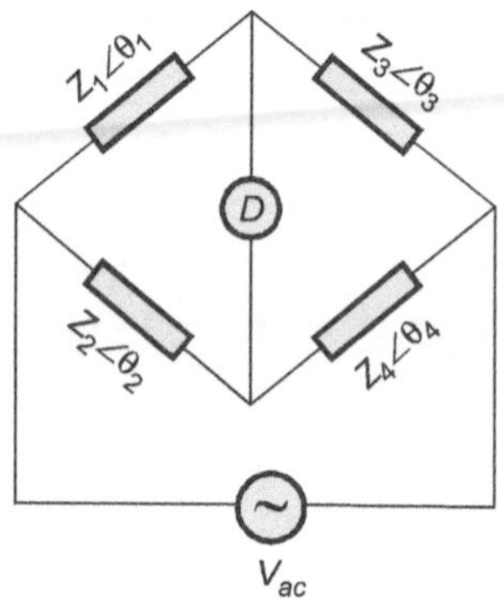

Match **List-I (Values of impedance angle θ_1)** with **List-II (Elements present in Z_1)** and select the correct answer using the codes given below the lists:

List-I	List-II
A. $-90° < \theta_1 < 0°$	1. Purely resistive
B. $-90°$	2. R and L in series
C. $0° < \theta_1 < 90°$	3. Purely capacitive
D. $0°$	4. R and C in series
	5. Purely inductive

Codes:

	A	B	C	D
(a)	2	3	4	1
(b)	4	3	2	1
(c)	4	5	2	1
(d)	2	5	4	3

Q.104 A Schering bridge has the following constants:
Arm *AB*: Capacitor of 1 μF in parallel with 1.2 kΩ resistance.
Arm *AD* : Resistance of 4.7 kΩ
Arm *BC* : Capacitor of 1 μF
Arm *CD* : Unknown capacitor C_x and R_x
If the frequency of supply is 0.5 kHz, the dissipation factor will be
(a) 3.76 (b) 2.25
(c) 4.25 (d) 4.50

Q.105 **Assertion (A):** D.C. and A.C. bridges are examples of direct measuring instruments.
Reason (R): Direct measuring instruments are most commonly used in engineering practice because they are the most simple and inexpensive.
(a) Both A and R are true and R is a correct explanation of A.
(b) Both A and R are true but R is not a correct explanation of A.
(c) A is true but R is false.
(d) A is false but R is true.

Q.106 A vibration galvanometer is tuned
(a) by changing its damping constant
(b) by attaching weight to the vibrating coil
(c) by changing the length and tension of vibrating coil
(d) all of the above

Q.107 In the Maxwell bridge shown in figure below, the values of resistance R_x and inductance L_x of a coil are to be calculated after balancing the bridge. The component values are shown in the figure at balance. The values of R_x and L_x will be respectively

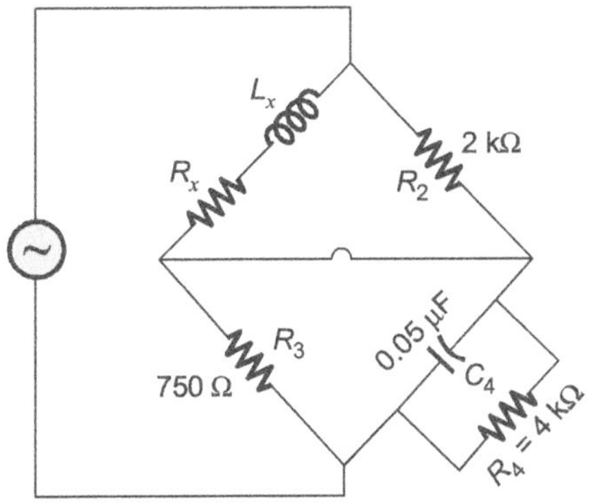

(a) 375 Ω, 75 mH (b) 75 Ω, 150 mH
(c) 37.5 Ω, 75 mH (d) 75 Ω, 75 mH

Q.108 The bridge circuit shown in figure

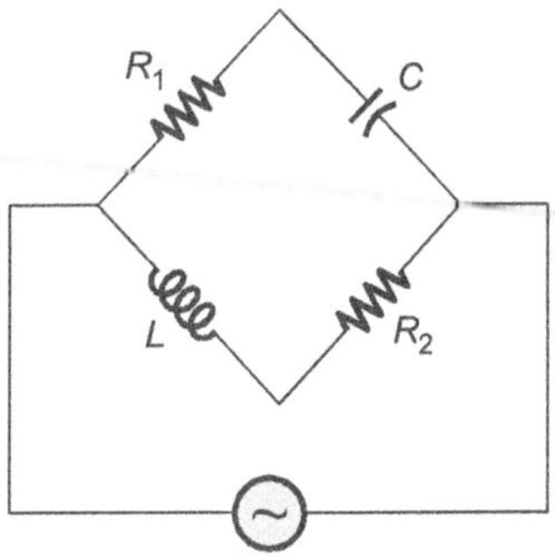

(a) cannot be balanced.
(b) can be balanced but, the frequency of excitation must be known.
(c) can be balanced for only one frequency.
(d) can be balanced at any frequency.

Q.109 A Wien-bridge is used to measure the frequency of the input signal. However, the input signal has 10% third harmonic distortion. Specifically, the signal is
$2\sin 400\pi t + 0.2\sin 1200\,\pi t$ (with t in secs)
With this input the balance will
(a) lead to a null indication and setting will correspond to a frequency of 200 Hz.
(b) Lead to a null indication and setting will correspond to a frequency of 260 Hz.
(c) lead to a null indication and setting will correspond to a frequency of 400 Hz.
(d) not lead to null indication.

Q.110 Wagner earth devices in AC bridge circuits are used for
(a) shielding all the bridge elements from external magnetic field
(b) eliminating the effect of stray capacitance
(c) minimizing the effect of inter-component capacitance
(d) eliminating all the node to earth capacitances

4. Electromechanical Indicating Instruments

Q.111 A moving coil instrument has the following data:
Number of turns = 100
Width of the coil = 20 mm
Depth of the coil = 30 mm
Flux density in the gap = 0.1 Wb/m^2
If the control spring constant is 2×10^{-6} N-m/degree, then the deflection of the pointer when the coil carries a current of 10 mA is

(a) $\frac{\pi}{3}$ radians (b) $\frac{\pi}{2}$ radians
(c) $\frac{\pi}{6}$ radians (d) $\frac{\pi}{4}$ radians

Q.112 A PMMC uses a
(a) taut bond
(b) moving coil
(c) electrodynamometer
(d) moving iron type

Q.113 A moving iron movement uses
(a) radial vane (b) electrodynamometer
(c) ribbon (d) moving coil

Q.114 A 1 mA meter movement with an internal resistance of 100 Ω is to be converted into a 0 - 100 mA. The value of shunt resistant required is
(a) 0.11 Ω (b) 1.10 Ω
(c) 1.09 Ω (d) 1.01 Ω

Q.115 Which of the following is not a desired characteristic of shunts used in PMMC instruments?
(a) The resistance of the shunt should not vary with time.
(b) It must be manufactured using manganin or copper.
(c) It should have a low thermal emf.
(d) It should carry the current without excessive temperature rise.

Q.116 To select the range, a multirange ammeter uses a
(a) double pole double throw switch
(b) single pole double throw switch
(c) simple switch
(d) make before break type switch

Q.117 A D'Arsonval movement is
(a) taut band
(b) PMMC
(c) electrodynamometer
(d) moving iron type

Q.118 What is the value of multiplier resistance on the 50 V range of a dc voltmeter that uses a 500 μA meter movement with an internal resistance of 1 kΩ?
(a) 109 kΩ (b) 199 kΩ
(c) 99 kΩ (d) 119 kΩ

Q.119 Loading effect in a voltmeter can be avoided by
(a) using high voltage range
(b) using a high sensitive voltmeter
(c) using an accurate and precise instrument
(d) using a low sensitive voltmeter

Q.120 Assertion (A): Gravity control is not suited for indicating instruments in general and portable instruments in particular.
Reason (R): Gravity control is used in only vertically mounted instruments.
(a) Both A and R are true and R is a correct explanation of A.
(b) Both A and R are true but R is not a correct explanation of A.
(c) A is true but R is false.
(d) A is false but R is true.

Q.121 Assertion (A): The ratio of frictional torque to deflecting torque is a measure of reliability of the instrument indications and of the inherent quality of the design.
Reason (R): Torque/weight ratio of an instrument is an index of its performance, the higher the ratio, the better will be its performance.
(a) Both A and R are true and R is a correct explanation of A.
(b) Both A and R are true but R is not a correct explanation of A.
(c) A is true but R is false.
(d) A is false but R is true.

Q.122 Figure given below shows a circuit for testing of a galvanometer.
When R_3 = 350 Ω, the galvanometer deflection is 140 mm.
When R_3 = 1080 Ω, the galvanometer deflection is 50 mm.

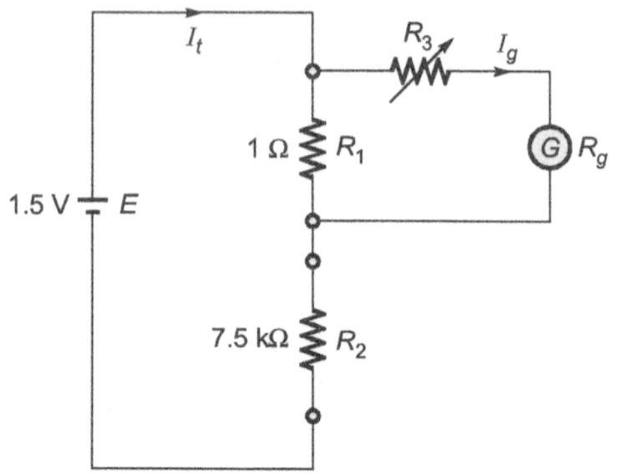

The resistance of galvanometer is

(a) 44 Ω (b) 34 Ω
(c) 64 Ω (d) 54 Ω

Q.123 Match **List-I (Parts of analog instruments)** with **List-II (Materials used in construction)** and select the correct answer using the codes given below the lists:

List-I	List-II
A. Control springs	1. Jewel
B. Permanent magnets	2. Aluminium
C. Pointers	3. Alnico
D. Bearings	4. Phosphor bronze

Codes:

	A	B	C	D
(a)	4	1	3	2
(b)	4	1	2	3
(c)	2	3	4	1
(d)	4	3	2	1

Q.124 In a D'Arsonval galvanometer, an iron core is usually used between the permanent magnet pole faces.
This is used so that

(a) moment of inertia of moving parts becomes smaller.
(b) flux density in the air gap becomes high thereby a large deflecting torque is produced.
(c) the effect of stray magnetic fields is reduced.
(d) none of the above

Q.125 What is the value of the multiplier resistor for a 10 V rms ac range on the voltmeter for the rectifier circuit shown below?

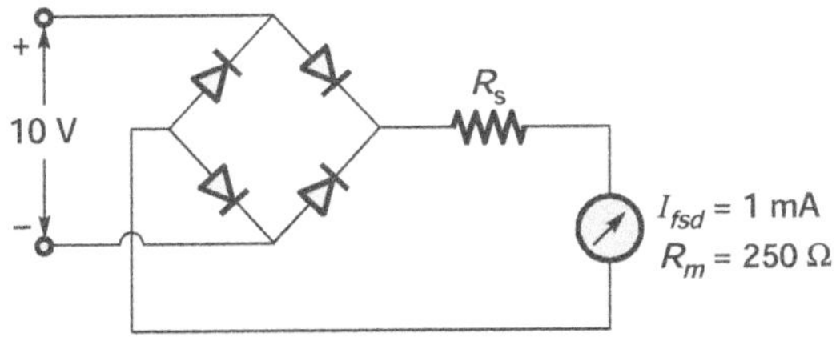

(a) 6.25 kΩ (b) 2.25 kΩ
(c) 8.75 kΩ (d) 4.75 kΩ

Q.126 The sensitivity of a voltmeter is defined as

(a) Ω/V (b) V/Ω
(c) I/Ω (d) Ω/I

Q.127 A diode used as rectifier in ac voltmeter should have

(a) high forward current and low reverse current.
(b) high forward current and high reverse current.
(c) low forward current and high reverse current.
(d) low forward current and low reverse current.

Q.128 The ac voltmeter using PMMC measures

(a) true RMS voltage (b) peak voltage
(c) average voltage (d) instantaneous voltage

Q.129 The ohm per volt rating on ac ranges as compared to the same rating on dc ranges is

(a) less (b) more
(c) equal (d) none of these

Q.130 Consider the following statements about moving coil instruments:

1. These instruments are used for both ac and dc.
2. The scale is uniformly divided.
3. The weight-to-torque ratio is high which gives a high accuracy.
4. The error due to stray magnetic fields are more compared to moving iron and electrodynamometer type instruments.
5. The cost of these instruments is higher compared to moving iron instruments.

Which of these statements are correct?

(a) 1, 2 and 4 only (b) 2, 3 and 4 only
(c) 4 and 5 only (d) 2, 3 and 5 only

Q.131 **Assertion (A):** In case of moving iron instruments, accurate reading of current is not possible at the lower end.
Reason (R): The scale of moving iron instruments is cramped at the lower end.

(a) Both A and R are true and R is a correct explanation of A.
(b) Both A and R are true but R is not a correct explanation of A.
(c) A is true but R is false.
(d) A is false but R is true.

Q.132 The inductance of a moving iron instrument is given by:

$$L = (10 + 5\theta - \theta^2)\ \mu\text{H},$$

where θ is the deflection in radian from zero position.
The spring constant is 12×10^{-6} N-m/rad.

The deflection of the pointer for a current of 5 A is
(a) 2.56 radian (b) 96.8 degree
(c) 6.25 radian (d) 85 degree

Q.133 The coil of a 150 V moving iron voltmeter has a resistance of 400 Ω and an inductance of 0.75 H. The current consumed by the instrument when placed on a 150 V d.c. supply is 0.05 A. What is the capacitance of the capacitor necessary to eliminate the frequency error?
(a) 38.6 μF (b) 66 μF
(c) 45.5 nF (d) 22.4 μF

Q.134 A rectifier type instrument uses a bridge rectifier and has its scale calibrated in terms of rms values of a sine wave. It indicates a voltage of 2.22 V when measuring a voltage of triangular waveshape. What is the percentage error in the measurement of voltage?
(a) –3.9% (b) +2.22%
(c) –2.2% (d) +3.9%

Q.135 Match **List-I (Meter type)** with **List-II (Major uses)** and select the correct answer using the codes given below the lists:

List-I
A. Thermo couple
B. Electrostatic
C. Electrodynamometer
D. Moving Iron

List-II
1. Measurement of Radio frequency a.c. signals.
2. Measurement of high voltages.
3. Precise measurement of a.c. current and voltage at power frequencies.
4. Widely used in indicator type applications such as panels.

Codes:

	A	B	C	D
(a)	3	2	1	4
(b)	4	3	2	1
(c)	1	4	3	2
(d)	1	2	3	4

Q.136 Which of the following instruments can be used for both a.c. and d.c.?
1. Thermocouple instruments
2. Rectifier type instruments
3. PMMC (d' Arsonval)
4. Electrodynamometer
5. Moving Iron

(a) 2, 4 and 5 only (b) 1, 2, 3 and 5 only
(c) 1, 2, 4 and 5 only (d) 1, 2 and 4 only

Q.137 **Assertion (A)** : The power requirement of electrostatic instruments are quite high.
Reason (R) : The current drawn on a.c. by the electrostatic instruments are fairly small.
(a) Both A and R are true and R is a correct explanation of A.
(b) Both A and R are true but R is not a correct explanation of A.
(c) A is true but R is false.
(d) A is false but R is true.

Q.138 Swamping resistance is connected
(a) in series with the ammeters to reduce errors on account of friction.
(b) in series with the shunt to reduce temperature error in shunted ammeters.
(c) in series with the meter and have a negligible resistance co-efficient in order to reduce temperature errors in shunted ammeters.
(d) in series with meter and have high resistance temperature co-efficient in order to reduce errors in ammeter.

Q.139 Horizontally mounted moving iron instruments use
(a) air friction damping
(b) electromagnetic damping
(c) fluid friction damping
(d) eddy current damping

Q.140 An astatic movement in an electrodynamometer type of instrument is used for
(a) providing eddy current damping.
(b) eliminating errors on account of external magnetic fields.
(c) increasing the operating torque of the instruments.
(d) eliminating error on the account of low torque to weight ratio.

Q.141 The sensitivity of a PMMC instrument is 10 kΩ/V. If this instrument is used in a rectifier type voltmeter with half waves rectification, the sensitivity would be
(a) 4.5 kΩ/V (b) 9 kΩ/V
(c) 22.2 kΩ/V (d) 10 kΩ/V

Q.142 Assertion (A): In an electrodynamometer type instrument, the X/R ratio of both fixed and moving coils should be equal.

Reason (R): With equal X/R ratio of both the moving and fixed coils, errors on account of change in frequency are prevented.

(a) Both A and R are true and R is a correct explanation of A.
(b) Both A and R are true but R is not a correct explanation of A.
(c) A is true but R is false.
(d) A is false but R is true.

Q.143 A 10 A electrodynamic ammeter is controlled by a spring having a constant of 0.1×10^{-6} N-m/degree. The full scale deflection is 110°. The mutual inductance at 0° deflection is 2 μH and the change in mutual inductance is linear with deflection.

What is the inductance of the instrument when measuring a current of 10 A?

(a) 1.15 μH (b) 3.65 μH
(c) 2.21 μH (d) 2.36 μH

Q.144 Assertion (A): Hot wire instrument is an electrothermic type of instrument.

Reason (R): Hot wire instruments are used for measurement of current at high frequencies without any significant errors.

(a) Both A and R are true and R is a correct explanation of A.
(b) Both A and R are true but R is not a correct explanation of A.
(c) A is true but R is false.
(d) A is false but R is true.

Q.145 The thermocouple type of instruments can be used to measure

(a) d.c. current and voltages only
(b) a.c. current only
(c) both a.c. and d.c. values of current only
(d) both a.c. and d.c. values of voltage and current

Q.146 Which of the following is not an electrothermic type of instrument?

(a) Bolometer
(b) PMMC instruments
(c) Thermocouple instruments
(d) Hot wire instruments

Q.147 For a certain dynamometer ammeter the mutual inductance M varies with deflection θ (expressed in degrees) as:

$$M = 5 \sin(\theta - 30°) \text{ mH}$$

What is the value of deflecting torque produced by a direct current of 50 mA corresponding to a deflection of 60°?

(a) 9.9 μN-m (b) 8.6 μN-m
(c) 10.8 μN-m (d) 12.5 μN-m

Q.148 Assertion (A): The rectifier type of instruments are calibrated for r.m.s. values of sinusoidal voltages or currents.

Reason (R): In order that a rectifier type instrument should read true rms value of the sinusoidal input, the scale markings are made with using a multiplying factor of 2.22 in case of instruments using full wave rectifier and by 1.11 in case of instruments using half wave rectifiers.

(a) Both A and R are true and R is a correct explanation of A.
(b) Both A and R are true but R is not a correct explanation of A.
(c) A is true but R is false.
(d) A is false but R is true.

Q.149 Consider the following statements related to analog instruments:

1. A transfer instrument is one which has different accuracy for d.c. and a.c.
2. Electro-dynamometer instruments fall in the category of transfer instruments.
3. Electro-dynamometer type of instruments are used as Watt-meters, varmeters and with some modification as power factor meters and frequency meters.
4. PMMC instruments are less accurate than moving iron and electrodynamometer type instruments.
5. Moving iron instruments are more costly than PMMC instruments.

Which of the above statements is/are not true?

(a) 1, 4 and 5 (b) 3 and 4
(c) 2 and 3 (d) 2, 4 and 5

Q.150 A current of $i(t) = 0.5 + 0.3 \sin \omega t - 0.2 \sin 3\,\omega t$ is passed through the ammeters A_1, A_2 and A_3 connected in series as shown below.

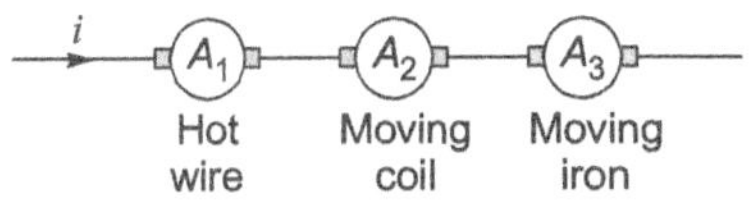

The reading of ammeters A_1, A_2 and A_3 will be respectively
(a) 489 mA, 500 mA and 489 mA
(b) 560 mA, 560 mA and 500 mA
(c) 489 mA, 560 mA and 560 mA
(d) 560 mA, 500 mA and 560 mA

Q.151 Error due to hysteresis is predominant in
(a) moving coil instruments
(b) electrodynamometer type instruments
(c) moving iron instruments
(d) hot -wire instruments

Q.152 The developed torque in an electrostatic voltmeter is given by

(a) $I^2 \frac{dM}{d\theta}$ (b) $VI \cdot \frac{dC}{d\theta}$

(c) $\frac{1}{2} I^2 \frac{dL}{d\theta}$ (d) $\frac{1}{2} V^2 \frac{dC}{d\theta}$

Q.153 The instrument whose deflection is given by the expression $\theta \propto I^2 \frac{dM}{d\theta}$ is known as
(a) electrodynamic type
(b) repulsive type
(c) electrostatic type
(d) attraction type

Q.154 A current $i = (10 + 10 \sin t)$ amperes is passed through an ideal moving iron type ammeter. Its reading will be
(a) 0 A (b) 10 A
(c) $\sqrt{150}$ A (d) $10\sqrt{2}$ A

Q.155 Match **List-I (Instruments)** with **List-II (Application)** and select the correct answer using the codes given below the lists:

List-I
A. Electro-dynamometer instrument
B. Thermocouple instrument
C. Ramp generator
D. Weston standard cell

List-II
1. True rms value meter
2. Transfer instrument between a.c. and d.c.
3. Time base of CRO
4. Standard of electromotive force (Emf)

Codes:

	A	B	C	D
(a)	4	1	3	2
(b)	4	3	1	2
(c)	2	1	3	4
(d)	2	3	1	4

Q.156 In the statement "the wattmeter commonly used for power measurement at commercial frequencies is of the *X*-type. This meter consists of two coil systems, the fixed system being the *Y*-coil and the moving system being the *Z*-coil". *X*, *Y* and *Z* stands respectively for
(a) dynamometer, voltage and current
(b) dynamometer, current and voltage
(c) induction, voltage and current
(d) induction, current and voltage

Q.157 A high frequency ac signal is applied to a PMMC instrument. If the rms value of the ac signal is 2 V, then the reading of the instrument will be
(a) 0 (b) 2 V
(c) $\sqrt{2}$ V (d) $4\sqrt{2}$ V

Q.158 Assertion (A): Rectifier instruments are indispensable for measurements in communication circuits.
Reason (R): Rectifier instruments can be used at frequencies well above the range of other ac instruments of the ordinary type (from about 20 Hz to 20 kHz).
(a) Both A and R are true and R is a correct explanation of A.
(b) Both A and R are true but R is not a correct explanation of A.
(c) A is true but R is false.
(d) A is false but R is true.

Q.159 Assertion (A): Thermocouple instruments are used for measurements of currents from power frequencies to 100 MHz.
Reason (R): In thermocouple instruments, the upper limit of frequency is determined by skin effect and stray capacitance errors and depends on whether the instrument is used as an ammeter or a voltmeter and its current rating.
(a) Both A and R are true and R is a correct explanation of A.

(b) Both A and R are true but R is not a correct explanation of A.
(c) A is true but R is false.
(d) A is false but R is true.

Q.160 Assertion (A): In portable instruments pointer and scale need not to have the boldness and contrast as required in case of switchboard and panel instruments.

Reason (R): In portable instruments readings are taken from short distance.

(a) Both A and R are true and R is a correct explanation of A.
(b) Both A and R are true but R is not a correct explanation of A.
(c) A is true but R is false.
(d) A is false but R is true.

5. Measurement of Power & Energy

Q.161 Which one of the following meters is an induction type instrument?
(a) Ammeter (b) Voltmeter
(c) Energymeter (d) Multimeter

Q.162 The error caused by pressure coil inductance in a wattmeter is compensated by connecting a
(a) capacitor across a part of series resistance in the pressure coil circuit.
(b) capacitor in series with the series resistance in the pressure coil circuit.
(c) capacitor across a part of a series inductor in the pressure coil circuit.
(d) capacitor in series with the inductor in the pressure coil circuit.

Q.163 Power at RF is measured using
(a) directly (b) $P = E^2/R$ or $P = I^2 R$
(c) wattmeter (d) voltmeter

Q.164 Creeping in a single phase induction type energy meter may be due to
(a) overvoltage
(b) vibration
(c) neither overvoltage nor vibration
(d) overvoltage or vibration

Q.165 In a single phase induction type energy meter, the lag adjustment is done
(a) to bring the pressure coil flux in phase with the applied voltage.
(b) to make the current coil flux to lag 90° behind the applied voltage.
(c) to make the pressure coil flux to lag 90° behind the applied voltage.
(d) none of the above

Q.166 A wattmeter has a current coil of 0.1 Ω resistance and a pressure coil of 6500 Ω respectively. The wattmeter shows a reading of 12 A at 250 V with unity power factor.
The percentage errors, due to resistance only with each of the two methods of connections shown below will be respectively

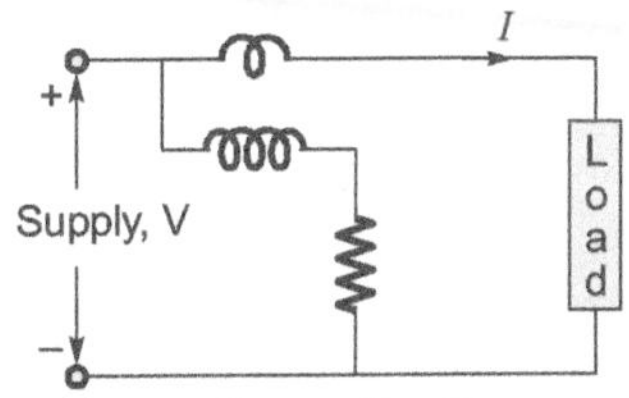

Connection (*i*)

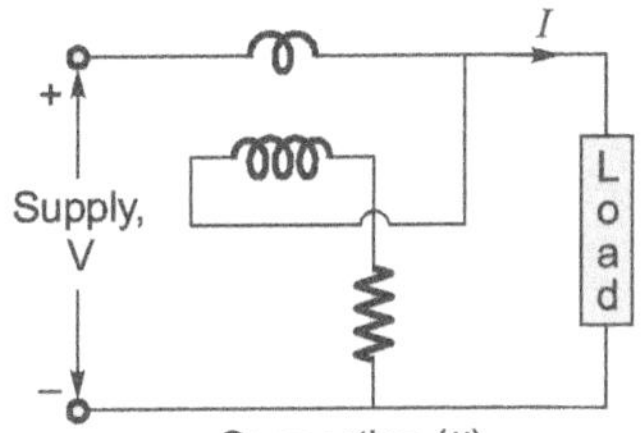

Connection (*ii*)

(a) 0.26% in connection (*i*) and 0.16% in connection (*ii*)
(b) 0.32% in connection (*i*) and 0.48% in connection (*ii*)
(c) 0.16% in connection (*i*) and 0.26% in connection (*ii*)
(d) 0.48% in connection (*i*) and 0.32% in connection (*ii*)

Q.167 The power factor in a 3-phase circuit using two wattmeter method is given by

(a) $\cos\left[\cot^{-1}\dfrac{(P_1+P_2)}{\sqrt{3}(P_1-P_2)}\right]$

(b) $\cos\left[\tan^{-1}\dfrac{\sqrt{3}(P_1+P_2)}{(P_1-P_2)}\right]$

(c) $\sin\left[\tan^{-1}\dfrac{(P_1+P_2)}{\sqrt{3}(P_1-P_2)}\right]$

(d) $\cos\left[\cot^{-1}\dfrac{\sqrt{3}(P_1-P_2)}{\sqrt{3}(P_1+P_2)}\right]$

Q.168 A wattmeter reads 5.54 kW when its current coil is connected in the red phase and its voltage coil is connected between the neutral and red phase of a symmetrical 3-phase system supplying a balanced load of 30 A at 400 V. If the phase sequence is RYB, then what will be the reading of the instrument if the current coil remains unchanged and voltage coil get connected between blue and yellow phases?

(a) 6.5 kW (b) 7.2 kVar
(c) 11.08 kW (d) 5.8 kVar

Q.169 Match **List-I (Power factor)** with **List-II (Wattmeter observations in two-wattmeter method)** and select the correct answer using the codes given below the lists:

	List-I		List-II
A.	Zero leading	1.	$W_1 - 2W_2 = 0$
B.	0.866 lagging	2.	W_1 = negative value, W_2 = positive value
C.	Unity	3.	$W_1 + W_2 = 0$
D.	0.3 lagging	4.	$W_1 - W_2 = 0$

Codes:

	A	B	C	D
(a)	4	1	3	2
(b)	3	2	4	1
(c)	3	1	4	2
(d)	4	1	3	2

Q.170 Phantom leading for testing of energy meter is used

(a) for meters having low current ratings.
(b) to reduce power loss during testing.
(c) to isolate the current and potential circuits.
(d) to test meters having a large current ratings for which loads may not be available in the laboratory. This also reduce power losses during testing.

Q.171 The braking torque provided by a permanent magnet in a single phase meter is proportional to the

(a) square of the flux of the permanent magnet
(b) speed of the meter
(c) both (a) and (b)
(d) neither (a) nor (b)

Q.172 The meter constant of a 230 V, 10 A watthour meter is 1800 revolutions per kWh. The meter is tested at half load and rated voltage and unity power factor. The meter is found to make 80 rev. in 138 seconds. The meter error at half load is

(a) 0.817% fast (b) 0.562% slow
(c) 0.727% fast (d) 0.625% slow

Q.173 Assertion (A): Induction type energy meters are universally used for measurement of energy in domestic and industrial a.c. circuits.

Reason (R): Induction type of meters possess lower friction and higher torque/weight ratio.

(a) Both A and R are true and R is a correct explanation of A.
(b) Both A and R are true but R is not a correct explanation of A.
(c) A is true but R is false.
(d) A is false but R is true.

Q.174 Match **List-I** with **List-II** and select the correct answer using the codes given below the lists:

List-I
A. Braking system
B. Driving system
C. Moving system
D. Registering system

List-II
1. Aluminium disc
2. Train of reduction gears
3. Shading bands
4. Electro-magnets
5. Permanent magnet

Codes:

	A	B	C	D
(a)	5	3	4	2
(b)	5	4	1	2
(c)	1	4	5	3
(d)	2	3	4	1

Q.175 A 230 V, single phase, watt hour meter has a constant load of 4 A passing through it for 6 hours at unity at unity power factor. The meter disc makes 2208 revolutions during this period. If the number of revolutions made by the meter are 1472 when operating at 230 V and 5 A for 4 hours, then power factor of the load would be

(a) 0.6 lag (b) 0.707 lag
(c) 0.866 lag (d) 0.8 lag

Q.176 The declared constant of a 5 A, 220 V d.c. watthour meter is 3275 revolutions per kWh. The speed of the disc at full load is

(a) 8 rps (b) 3 rps
(c) 1 rps (d) 2 rps

Q.177 In an induction type energy meter, an increase in temperature results in
(a) a decrease in torque produced by all shading bands.
(b) a decrease in the resistance of the eddy current paths.
(c) an increase in the angle of lag 'α' of the eddy currents.
(d) an increase in angle of lag between V and potential coil flux ϕ_P.

Q.178 In electrodynamometer type wattmeters, current coils designed for carrying heavy currents use stranded wire or laminated conductors
(a) to reduce eddy current losses in conductors
(b) to reduce iron losses
(c) to reduce hysteresis losses
(d) all of the above

Q.179 In an electrodynamometer wattmeter, the inductance of pressure coil circuit produces error
(a) which is constant irrespective of the power factor of the load.
(b) which is higher at low power factors.
(c) which is lower at low power factors.
(d) all of the above

Q.180 A dynamometer wattmeter measures power in a 50 Hz, single phase circuit without error, at all power factors. The resistance of the voltage coil and its series resistance are 400 Ω and 10,000 Ω respectively. The series resistance has a distributed self-capacitance equivalent to a shunt capacity of 20 pF.
What is the self inductance of the pressure coil?
(a) 5 mH (b) 4 mH
(c) 3 mH (d) 2 mH

Q.181 Assertion (A): For measuring the power factor of the load in case of a two-wattmeter method, load may be balanced or unbalanced.
Reason (R): For measuring the power in a three-phase circuit, load may be balanced or unbalanced, source may be balanced or unbalanced but there must be three-phase load connected to three-phase source via three-phase wires.
(a) Both A and R are true and R is a correct explanation of A.
(b) Both A and R are true but R is not a correct explanation of A.
(c) A Is true but R is false.
(d) A is false but R is true.

Q.182 The voltage and current in a circuit are given by,
$v(t) = 5 + 2\sin(\omega t + 30°) + 10\sin(3\omega t - 60°)$
and $i(t) = 10 + 5\sin\omega t + 8\sin(2\omega t - 30°)$
The total power consumed in the circuit is
(a) 49.64 watt (b) 54.33 watt
(c) 28.64 watt (d) none of these

Q.183 In an electrodynamometer type wattmeter
1. moving coil is iron cored.
2. fixed coil is air cored.
3. air friction damping is used.
4. the fixed coil is connected across the load.
5. the moving coil is connected in series with the load.
Which of the above statements is/are true?
(a) 2 and 3 only (b) 1, 3, 4 and 5 only
(c) 2, 4 and 5 only (d) 3 and 5 only

Q.184 Assertion (A): A wattmeter must be used for the measurement of power in a.c. circuit instead of merely an ammeter and a voltmeter.
Reason (R): Wattmeter measures average active and reactive powers.
(a) Both A and R are true and R is a correct explanation of A.
(b) Both A and R are true but R is not a correct explanation of A.
(c) A is true but R is false.
(d) A is false but R is true.

Q.185 Assertion (A): An electrodynamometer type wattmeter reads low when the load power factor is leading.
Reason (R): The effect of pressure coil inductance is to increase the phase angle between load current and pressure coil current when the load power factor is leading.
(a) Both A and R are true and R is a correct explanation of A.
(b) Both A and R are true but R is not a correct explanation of A.
(c) A is true but R is false.
(d) A is false but R is true.

Q.186 The ratio of measured power to true power in case of an electrodynamometer type wattmeter is given by (β = internal angle of pressure coil circuit)

(a) $1 + \tan\phi \cdot \tan\beta$ (b) $\dfrac{\cos\phi}{\cos\phi \cdot \cos(\phi - \beta)}$

(c) $\dfrac{1}{1+\tan\phi \cdot \tan\beta}$ (d) $\cos\phi \cdot \cos(\phi - \beta)$

Q.187 Match **List-I** with **List-II** and select the correct answer using the codes given below the lists:

List-I

A. Creeping
B. Lag adjustment
C. Friction compensation
D. Temperature compensation

List-II

1. Saturable magnetic shunt
2. Shading loop
3. Shading bands
4. Diametrically opposite holes in the disc
5. Mutemp

Codes:

	A	B	C	D
(a)	3	2	5	1
(b)	1	3	2	5
(c)	4	2	5	3
(d)	4	3	2	5

Q.188 **Assertion (A):** Phantom loading of energy meter consists of supplying the fixed coil circuit from a circuit of required normal voltage, and the moving coil circuit from a separate low voltage supply.

Reason (R): When the current rating of a meter under test is high, a test with actual loading arrangement would involve a considerable waste of power due to which phantom or fictious loading is done.

(a) Both A and R are true and R is a correct explanation of A.
(b) Both A and R are true but R is not a correct explanation of A.
(c) A is true but R is false.
(d) A is false but R is true.

Q.189 A dynamometer wattmeter with its voltage coil connected across the load side of the instrument reads 250 W. If the load voltage be 200 V, what power is being taken by the load?

(The voltage coil has a resistance of 2000 Ω)

(a) 250 W (b) 270 W
(c) 230 W (d) can't be determined

Q.190 The power flowing in a 3-ϕ, 3-wire, balanced load system is measured by the two-wattmeter method. The reading on wattmeter –*A* is 3 kW and on wattmeter –*B* is –1 kW. The factor of the load is

(a) 0.382 (b) 0.277
(c) 0.356 (d) 0.289

Q.191 Holes are filled on the opposite side of the disc of an induction type energy meter to

(a) avoid creep on no-load
(b) balance the disc
(c) dissipate heat energy due to eddy currents
(d) increase the deflecting torque

Q.192 Two-wattmeter method is employed to measure power in a 3-phase balanced system with the current coils connected in the A and C lines. The phase sequence is ABC. If the wattmeter with its current coil in A-phase line reads zero, then the power factor of the 3-phase load will be

(a) zero lagging (b) zero leading
(c) 0.5 lagging (d) 0.5 leading

Q.193 The meter constant of a single-phase 240 V induction watt-hour meter is 400 rev/kWh. The speed of the meter disc for a current of 10 A at 0.8 p.f. lagging will be

(a) 12.8 rpm (b) 16.02 rpm
(c) 18.2 rpm (d) 21.1 rpm

6. Cathode Ray Oscilloscope (CRO)

Q.194 Post deflection acceleration is used to

(a) enhance the intensity of the beam
(b) repel the electron beam
(c) increase the velocity of the electron beam
(d) focus the beam

Q.195 The function of the synchronization section in CRO is

(a) to adjust the intensity control.
(b) to start the horizontal sweep at the same relative point on the vertical signal.
(c) to control the gain of the amplifier.
(d) to match the horizontal sweep rate with the frequency of the vertical signal.

Q.196 The distance between two peaks measured on the X-axis is 2 cm, at 1 ms/div. The frequency of the signal is
(a) 5 kHz (b) 50 Hz
(c) 1 kHz (d) 500 Hz

Q.197 A dual trace CRO uses
(a) one electron gun (b) two electron gun
(c) two pairs of HDPs (d) none of these

Q.198 A sampling CRO is used for
(a) high frequencies
(b) very low frequencies
(c) very high frequencies
(d) low frequencies

Q.199 A time base selector basically consists of
(a) LC oscillator (b) RC oscillator
(c) crystal oscillator (d) wein bridge oscillator

Q.200 Trigger pulses in the CRO are used
(a) to generate low voltage required for the CRT
(b) to synchronise the input with the time base generator
(c) to generator high voltage required for the CRT
(d) to synchronise the input and the vertical amplifier

Q.201 The waveform shown in figure below is observed on the screen of an oscilloscope.

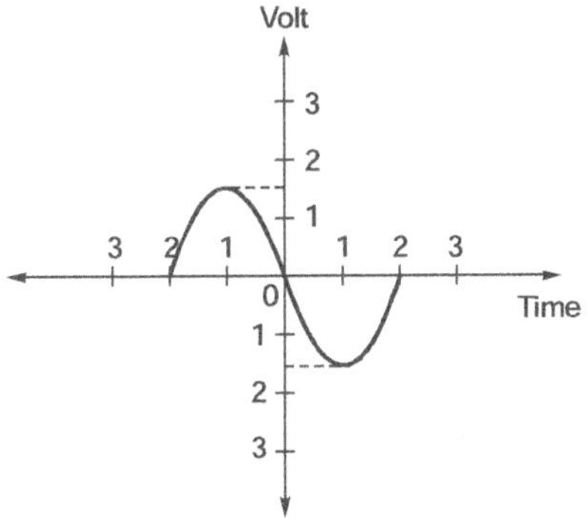

If the vertical attenuation is set to 0.5 V/div, what is the peak-to-peak amplitude of the given signal?
(a) 0.5 V (b) 1.5 V
(c) 2.0 V (d) 3.0 V

Q.202 If the time/div control is set to 2 μs/div for the waveform of previous problem (Q.No. 201), then what is the frequency of the signal?
(a) 125 kHz (b) 75 kHz
(c) 100 kHz (d) 155 kHz

Q.203 The most common approaches used by oscilloscope manufacturers for delayed sweep operations are
1. Intensified sweep, where the delayed sweep acts as a positional magnifier.
2. Normal triggering sweep after the desired time delay, which is set from the panel controls.
3. A delay plus trigger mode, where a visual indication, such as light, indicates that the delay time has elapsed and the sweep is ready to be triggered.

Which of the above statements is/are correct?
(a) 1, 2 and 3 (b) 2 and 3
(c) 2 only (d) 3 only

Q.204 A Lissajous pattern on an oscilloscope is stationary and has 5 horizontal tangencies and 2 vertical tangencies. The frequency of horizontal input is 1000 Hz. The frequency of vertical input is
(a) 0.4 kHz (b) 1.5 kHz
(c) 3.5 kHz (d) 2.5 kHz

Q.205 A sinusoidal input is applied to the vertical input of an oscilloscope starting at $t = 0$. Match **List-I (Lissajous patterns)** with **List-II (Phase shifts between vertical and horizontal inputs)** and select the correct answer using the codes given below the lists:

List-I | **List-II**

A. 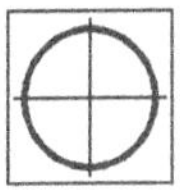 1. 0°

B.  2. 120°

C. 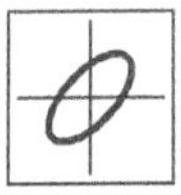 3. 90°

4. 45°

Codes:

	A	B	C
(a)	2	1	3
(b)	3	2	4
(c)	3	1	4
(d)	1	4	3

Q.206 If the distance of screen from a CRT to centre of deflection plates is 15 cm. The length of deflection plates is 2 cm, the distance between plates is 1 m and the accelerating voltage is 500 V, the deflection sensitivity is
(a) 33.2 V/cm (b) 0.03 cm/V
(c) 66.4 V/cm (d) 0.015 cm/V

Q.207 During the retrace time, the electrons forming the horizontal beam
(a) move from right to the left on the screen
(b) move from left to right on the screen
(c) move from top to bottom of screen
(d) move from bottom to top of screen

Q.208 Post acceleration is needed in a CRO if the frequency of the signal is
(a) more than 1 MHz (b) more than 10 Hz
(c) less than 1 MHz (d) more than 10 MHz

Q.209 For the Lissajous pattern shown below, if the horizontal frequency is 51 Hz, then the vertical frequency will be

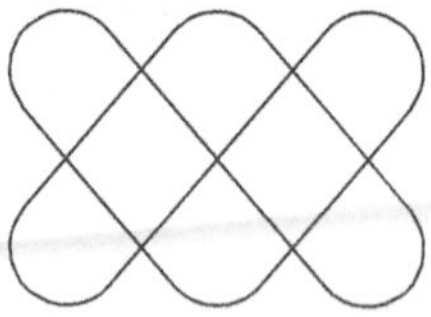

(a) 34 Hz (b) 153 Hz
(c) 76.5 Hz (d) none of these

Q.210 A CRT has an anode voltage of 2000 V and parallel deflecting plates 2 cm long and 5 mm apart. The screen is 30 cm from the centre of the plates. If the input voltage is applied to the deflecting plates through amplifier having an overall gain of 100, the input voltage required to deflect the beam through 3 cm would be
(a) 0.5 V (b) 2 V
(c) 2.5 V (d) 1 V

Q.211 What is the maximum velocity of the beam of electrons in a CRT having a cathode-anode voltage of 800 V?
Assume the electrons to leave the cathode with zero velocity.
(a) 16.8 km/s (b) 22.2 km/s
(c) 25 m/s (d) 86 km/s

Q.212 The function of a delayed-sweep feature in many laboratory-quality oscilloscope is
(a) to magnify a selected portion of the undelayed sweep
(b) to measure waveform jitter or rise time
(c) both (a) and (b)
(d) none of the above

Q.213 **Assertion (A):** Lissajous pattern is used for accurate measurement of frequency of a signal.
Reason (R): The signal, whose frequency is to be measured, is applied to *X*-plates.
(a) Both A and R are true and R is a correct explanation of A.
(b) Both A and R are true but R is not a correct explanation of A.
(c) A is true but R is false.
(d) A is false but R is true.

Q.214 Match **List-I (Various controls in a CRO)** with **List-II (Methods)** and select the correct answer using the codes given below the lists:
List-I
A. Intensity control
B. Positioning control
C. Focus control
List-II
1. By changing the potential of the focusing anode
2. By applying a small dc voltage to the deflecting plates
3. By varying the cathode grid potential
Codes:

	A	B	C
(a)	2	3	1
(b)	3	2	1
(c)	1	2	3
(d)	1	3	2

Q.215 Consider the following statements related to a cathode ray oscilloscope:
1. The signal to be examined are usually applied to the horizontal deflection plates.
2. The vertical deflection plates are fed by a sweep voltage that provides a time base.
3. The horizontal amplifier receives an input from the saw tooth sweep generator which is triggered by the synchronizing amplifier.
4. Electrostatic focusing technique is superior to magnetostatic focusing technique.

Which of the above statements is/are not true?
(a) only 3 (b) 1 and 4
(c) 2, 3 and 4 (d) 1, 2 and 4

Q.216 If the bandwidth of an oscilloscope is given as direct current to 100 MHz, what is the fastest rise time a sine wave can have to be accurately reproduced by the instrument?
(a) 3.5 ns (b) 35 ns
(c) 7 ns (d) 0.35 ns

Q.217 Assertion (A): Electrostatic focusing technique works on the principle of "double concave lens."
Reason (R): There is difference in potential between focusing anode and the two accelerating anodes due to which a non-uniform field exists on each of the two ends of the focusing anode.
(a) Both A and R are true and R is a correct explanation of A.
(b) Both A and R are true but R is not a correct explanation of A.
(c) A is true but R is false.
(d) A is false but R is true.

Q.218 Assertion (A): The dual trace oscilloscope has two separate vertical input channels.
Reason (R): The dual trace oscilloscope cannot capture two fast transient events, as it cannot switch quickly enough between traces.
(a) Both A and R are true and R is a correct explanation of A.
(b) Both A and R are true but R is not a correct explanation of A.
(c) A is true but R is false.
(d) A is false but R is true.

Q.219 An electronic switch is used in a
(a) dual trace CRO (b) dual beam CRO
(c) single beam CRO (d) sampling CRO

Q.220 A signal of 10 mV at 75 MHz is to be measured. Which of the following instrument can be used?
(a) VTVM
(b) Cathode ray oscilloscope
(c) Moving iron voltmeter
(d) Digital multimeter

Q.221 Spectrum analyser is a combination of
(a) Narrow band superheterodyne receiver and CRO
(b) Signal generator and CRO
(c) Oscillator and wave analyser
(d) VTVM and CRO

Q.222 Which of the following statements correctly represents the post acceleration in a cathode-ray tube?
(a) It provides deflection of the beam.
(b) It increase the brightness of the trace if the signal frequency is higher than 10 MHz.
(c) It accelerates the beam before deflection.
(d) It increase the brightness of the trace of low frequency signal.

Q.223 To the y-input of a CRO a signal defined by 10 sin $100t$ is applied. To the x-input, the signal 10 cos $100t$ is fed. The gain for both x-channel and y-channel is same. The screen shows
(a) a sinusoidal signal
(b) a straight line
(c) a ellipse
(d) a circle

Q.224 A Lissajous pattern, as shown in figure below, is observed on the screen of a CRO when voltages of frequencies f_x and f_y are applied to the x and y-plates respectively. The frequency ratio $f_x : f_y$ is equal to

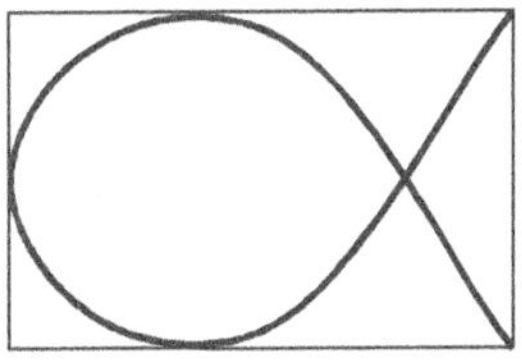

(a) 3 : 2 (b) 1 : 2
(c) 2 : 3 (d) 2 : 1

Q.225 A CRO screen has ten divisions on the horizontal scale. If a voltage signal 5 sin($314t$ + 45°) is examined with a time base setting of 5 m sec/div, then the number of cycles of signal displayed on the screen will be
(a) 0.5 (b) 2.5
(c) 5 (d) 10

Q.226 Calculate the maximum velocity of the beam of electrons in a CRT having a cathode-anode voltage of 1000 V. Assume the electrons to leave the cathode with zero velocity.
(Charge of electron = 1.60×10^{-19} C and mass of electron = 9.1×10^{-31} kg)
(a) 0.1875×10^6 m/s (b) 0.1875×10^7 m/s
(c) 0.1875×10^8 m/s (d) 0.1875×10^9 m/s

7. Transducers

Q.227 A thermocouple is to be used to measure in the range of 700°C to 800°C. Select the pair that would be most suitable for the applications.
(a) iron-constantan
(b) platinum – platinum + rhodium
(c) copper-constantan
(d) chromel-alumel

Q.228 The transducer suited for high temperature and fast response time is a
(a) RTD (b) Thermistor
(c) Thermocouple (d) Bimetallic strip

Q.229 Rotameter is a flowmeter based on
(a) variable area (b) variable length
(c) variable pressure (d) none of these

Q.230 In a thermocouple element, heat energy transferred to the hot junction is converted into electrical energy by
(a) Hall effect (b) Faraday's effect
(c) Johnson's effect (d) Seebeck effect

Q.231 A resistance strain gauge with a factor of 2 is cemented to a steel member, which is subjected to a strain of 1×10^{-6}.
If the original resistance value of the gauge is 130 Ω, what is the change is resistance?
(a) 153 mΩ (b) 260 μΩ
(c) 65 μΩ (d) 140 mΩ

Q.232 A variable reluctance type inductive transducer has a coil of inductance 2500 μH when the target mode of ferromagnetic material is 1 mm away from the core. What is the value of inductance when a displacement of 0.04 mm is applied to the target in a direction moving it towards the core?
(a) 260 μH (b) 250 μH
(c) 104 μH (d) 604 μH

Q.233 While of the following is not a pair of active transducers?
(a) Accelerometer and thermocouple
(b) Piezoelectric transducer and tachnogenerator
(c) Photo-voltaic cell and thermopile
(d) LVDT and potentiometer

Q.234 An LVDT has an input voltage of 6.3 V and an output voltage of 5.2 V. It's range is ±0.5 m. What is the change in output voltage produced for a core movement going from +0.45 m to –0.30 m?
(a) –7.80 V (b) –4.68 V
(c) –2.20 V (d) –3.00 V

Q.235 A platinum resistance thermometer has a resistance of 180 Ω at 20° C. What is it's resistance at 60° C? (α_{20} = 0.00392)
(a) 205 Ω (b) 208 Ω
(c) 225 Ω (d) 229 Ω

Q.236 RTD is a/an
(a) active transducer
(b) passive transducer
(c) inductive transducer
(d) capacitive transducer

Q.237 A thermistor is inherently suitable for temperature measurement involving
(a) high span but low sensitivity
(b) low span but high sensitivity
(c) high span and high sensitivity
(d) low span and low sensitivity

Q.238 While selecting a transducer for a particular application
(a) only the output characteristics should be considered.
(b) only the transfer characteristics should be considered.
(c) only the input characteristics should be considered.
(d) input, output and transfer characteristic should be considered.

Q.239 Capacitive transducers are generally used for
(a) transient measurement
(b) dynamic measurement
(c) static measurement
(d) both static and dynamic measurement

Q.240 A resistance wire strain gauge uses a soft iron wire of small diameter. The gauge factor is +4.2. Neglecting the piezoelectric effects, the Poisson's ratio is
(a) 2.6 (b) 0.625
(c) 1.6 (d) 2.1

Q.241 Ceramic materials used for piezo-electric transducers

(a) are basically made of barium titanate.
(b) do not have piezo-electric properties in their original state but these properties are produced by special polarizing treatment.
(c) are polycrystalline in nature.
(d) all of the above

Q.242 The rotor of a synchro-transmitter
(a) uses a salient pole type of construction
(b) uses a cylindrical pole type of construction
(c) may use either salient or cylindrical type of construction
(d) none of the above

Q.243 Match **List-I (Type)** with **List-II (Operation)** and select the correct answer using the codes given below the lists:

List-I
A. Bimaterial
B. Orifice
C. Bourdon tube
D. Hydrometer

List-II
1. Specific gravity to displacement
2. Temperature to displacement
3. Pressure to displacement
4. Velocity to pressure

Codes:

	A	B	C	D
(a)	4	3	2	1
(b)	2	4	3	1
(c)	3	1	2	4
(d)	2	3	4	1

Q.244 **Assertion (A):** A piezoelectric crystal acts as an inverse transducer.
Reason (R): An inverse transducer is defined as a device which converts an electrical quantity into a non-electrical quantity.
(a) Both A and R are true and R is a correct explanation of A.
(b) Both A and R are true but R is not a correct explanation of A.
(c) A is true but R is false.
(d) A is false but R is true.

Q.245 Consider the following statements related to resistance potentiometer for the measurement of displacement:
1. They are very useful for the measurement of large amplitudes of displacement.
2. The frequency response of wire wound potentiometers is limited.
3. They require a large force to move their sliding contacts.

Which of the above statements is/are correct?
(a) 1 and 3 only (b) 2 only
(c) 2 and 3 only (d) 1, 2 and 3

Q.246 A resistance, wire strain gauge with a gauge factor of 2 is bonded to a steel structural member subjected to a stress of 100 MN/m^2. The modulus of elasticity of steel is 200 GN/m^2. What is the percentage change in the value of the gauge resistance due to the applied stress?
(a) 1% (b) 0.01%
(c) 0.1% (d) 0.001%

Q.247 Which of the following is not true regarding Thermocouple used for measuring the temperature?
(a) They are cheaper than the resistance thermometers.
(b) Thermocouples follow the temperature changes with a small time lag and as such are suitable for recording comparatively rapid changes in temperature.
(c) Thermocouples are very convenient for measuring the temperature at one particular point in a piece of apparatus.
(d) Thermocouples are very accurate and used for precision work.

Q.248 With the differential outputs
1. accuracy in increased.
2. output is more affected by external magnetic fields.
3. sensitivity is reduced.
4. the effects of changes in supply voltage and frequency are reduced.
5. the effective variations due to temperature changes are increased.

Of these the correct statements is/are
(a) 3, 4 and 5 (b) 1 and 4
(c) 2 and 3 (d) 2, 3 and 4

Q.249 The output voltage of a LVDT is 1.5 V at maximum displacement. At a load of 0.5 MΩ, the deviation from linearity is maximum and it is ±0.003 V from a straight line through origin. What is the percentage linearity at the given load?

(a) ± 0.2% (b) ± 0.01%
(c) ± 0.15% (d) can't be determined

Q.250 A piezo-electric crystal having dimensions of 5 mm × 5 mm × 1.5 mm and a voltage sensitivity of 0.055 V-m/N is used for force measurement. If the voltage developed is 100 V, then force acting on it is

(a) 20 Newton (b) 60 Newton
(c) 30 Newton (d) 40 Newton

Q.251 A Hall effect transducer is used for the measurement of a magnetic field of 0.5 Wb/m^2. The 2 mm thick slab is made of Bismuth for which the Hall's co-efficient is -1×10^{-6} V-m/(A-Wbm^{-2}) and the current is 3 A. The output voltage developed is

(a) –0.25 mV (b) –0.20 μV
(c) –0.45 mV (d) –0.75 mV

Q.252 The lower limit of useful working range of a transducer is determined by

(a) cross sensitivity
(b) dynamic response
(c) by transducer error and noise
(d) minimum useful input level

Q.253 Measurement of power at high frequencies and thermal conductivity can be done with the help of

(a) thermistor (b) LVDT
(c) thermocouple (d) none of these

Q.254 **Assertion (A):** Thermistor acts as an ideal temperature transducer.

Reason (R): Thermistor exhibits a very high negative temperature coefficient of resistance.

(a) Both A and R are true and R is a correct explanation of A.
(b) Both A and R are true but R is not a correct explanation of A.
(c) A is true but R is false.
(d) A is false but R is true.

Q.255 Match **List-I (Class of transducers)** with **List-II (Applications)** and select the correct answer using the codes given below the lists:

Lst-I	List-II
A. Hall effect pickup	1. Gas flow
B. Pirani gauge	2. Magnetic flux
C. Differential transformer	3. Position
D. Resistance strain gauge	4. Torque

Codes:

	A	B	C	D
(a)	2	1	4	3
(b)	2	1	3	4
(c)	3	2	4	1
(d)	1	2	3	4

Q.256 **Assertion (A):** The strain gauge should have a high value of gauge factor.

Reason (R): A high value of gauge factor indicates a small change in resistance for a particular strain resulting in high sensitivity.

(a) Both A and R are true and R is a correct explanation of A.
(b) Both A and R are true but R is not a correct explanation of A.
(c) A is true but R is false.
(d) A is false but R is true.

Q.257 **Assertion (A):** LVDT is a frictionless device.

Reason (R): The transducer possesses a high sensitivity which is typically about 40 V/mm.

(a) Both A and R are true and R is a correct explanation of A.
(b) Both A and R are true but R is not a correct explanation of A.
(c) A is true but R is false.
(d) A is false but R is true.

Q.258 The voltage sensitivities of Barium titanate and quartz are respectively 12×10^{-3} V-m/N and 50×10^{-3} V-m/N. Their respective permittivities are 12.5×10^{-9} F/m and 40.6×10^{-12} F/m. What are their charge sensitivities?

(a) 1.04×10^{6} C/N, 1.23×10^{9} C/N
(b) 150 pC/N, 2 pC/N
(c) 20 pC/N, 50 pC/N
(d) none of the above

Q.259 Dummy strain gauge is used in conjunction with the main strain gauge to

(a) Calibrate the system
(b) Compensate temperature effects
(c) improve sensitivity
(d) reduce strain on the main gauge

Q.260 Which one of the following transducers can be used for measurement of pressure as high as 10, 000 atmosphere?
(a) Mcleod gauge (b) Pirani gauge
(c) Bridgman gauge (d) Knudsen gauge

Q.261 Rochelle salt is a crystalline material used in producing
(a) velocity transducer
(b) photoelectric transducer
(c) piezoelectric transducer
(d) differential transformer transducer

Q.262 Match **List-I (Instruments)** with **List-II (Measurand)** and select the correct answer using the codes given below the lists:

List-I	List-II
A. Mcload gauge	1. Temperature
B. Turbine meter	2. Pressure
C. Pyrometer	3. Flow
D. Synchros	4. Displacement

Codes:

	A	B	C	D
(a)	1	4	2	3
(b)	2	3	1	4
(c)	1	3	2	4
(d)	2	4	1	3

Q.263 Measurement of flow, thermal conductivity and liquid level using thermistors make use of
(a) resistance decrease with temperature
(b) resistance increase with temperature
(c) self-heating phenomenon
(d) change of resistivity

Q.264 Which of the following can be used/modified for measurement of angular speed?
1. LVDT 2. Magnetic pick-up
3. Tacho-generator 4. Strain gauge
Select the correct answer using the code given below.
(a) Only 1 and 2 (b) Only 2 and 3
(c) Only 3 (d) Only 2, 3 and 4

8. Instrument Transformers

Q.265 Turn compensation is used in current transformer primarily for reduction of
(a) phase angle error
(b) both ratio and phase angle errors
(c) ratio error, reduction in phase angle error is incidental
(d) none of the above

Q.266 The ratio marked on the instrument transformers is
(a) nominal ratio
(b) turns ratio
(c) transformation ratio
(d) none of the above

Q.267 A 1000/5 A, 50 Hz current transformer has a secondary burden comprising a non-inductive impedance of 1.6 Ω. The primary winding has one turn. Neglecting the leakage reactance and assuming the iron loss in the core to be 1.5 W at full-load, the ratio error at full load is (the magnetizing mmf is 100 A)
(a) –1.52% (b) +1.61%
(c) –3.61% (d) +2.25%

Q.268 The ratio and phase angle error in potential transformers may be reduced by
(a) increasing the exciting current
(b) increasing the resistance and leakage reactance in the transformer
(c) by not employing turns compensation
(d) None of the above

Q.269 The size of potential transformers
(a) is smaller than that of power transformers of the same VA rating.
(b) is much greater than that of power transformers of the same VA rating because they are designed for low ratio and phase angle errors which require large size cores and winding conductors.
(c) is the same as that of power transformers of the same VA rating.
(d) none of the above

Q.270 In current transforms there are errors caused because
1. there is always a magnetic leakage and consequently the primary flux linkages are not equal to the secondary flux linkages.
2. the flux density in the core is not a linear function of the magnetizing force i.e. the transformer core becomes saturated.

3. the transformer input must have a component which supplies the core losses and I^2R losses of transformer windings due to flow of current I_0.
4. there is some exciting mmf required by the primary winding to produce flux and therefore, the transformers draws a magnetizing current I_m.

Which of the above statements is/are correct?

(a) 1, 3 and 4 (b) 2, 3 and 4
(c) 1, 2 and 4 (d) 1, 2, 3 and 4

Q.271 Wilson compensation is used to

(a) reduce the phase angle error of a current transformer.
(b) reduce the ratio error of a current transformer.
(c) reduce the phase angle error of a potential transformer.
(d) reduce the ratio error of a potential transformer.

Q.272 Assertion (A): The secondary winding circuit of a current transformer is never open circuited while its primary winding is energised.

Reason (R): The eddy current and hysteresis losses would be very high under these conditions and due to this the transformer may be overheated and completely damaged. Moreover, the core may become permanently magnetized and this gives appreciable ratio and phase angle errors.

(a) Both A and R are true and R is a correct explanation of A.
(b) Both A and R are true but R is not a correct explanation of A.
(c) A is true but R is false.
(d) A is false but R is true.

Q.273 A 250 : 5, current transformer is used along with an ammeter. If ammeter reading is 2.7A, the line current is

(a) 13.5 A (b) 135 A
(c) 0.054 A (d) 167 A

Q.274 Assertion (A): Instrument transformers can be used in both a.c. and d.c. circuits.

Reason (R): The rating of low range meter can be fixed irrespective of the value of high voltage or current to be measured.

(a) Both A and R are true and R is a correct explanation of A.
(b) Both A and R are true but R is not a correct explanation of A.
(c) A is true but R is false.
(d) A is false but R is true.

Q.275 A current transformer has turns ratio 1 : 399 and is rated at 200/5 A. The core loss component is 3 A and magnetizing component is 8 A, under full load conditions. The percentage ratio and phase angle error (degree) under full load condition if secondary circuit power factor is 0.8 leading will be respectively

(a) 0.3713% and 0.1562°
(b) 0.2752% and 0.2568°
(c) 0.3713% and 0.2355°
(d) none of the above

Q.276 The ratio and phase angle error of instrument transformers depends upon

(a) core loss component of no-load current I_0.
(b) magnetizing component of no-load current I_0.
(c) both core loss and magnetizing component of no-load current I_0.
(d) neither the core loss nor the magnetizing component of no-load current.

Q.277 If the secondary burden of a current transformer is 30 VA and secondary current is 5 A, then the impedance of the connected load will be

(a) 2.5 Ω (b) 0.6 Ω
(c) 2.4 Ω (d) 1.2 Ω

Q.278 A PT is a device which is

(a) electrically coupled
(b) electro-magnetically coupled
(c) electro-statically coupled
(d) conductively coupled

Q.279 The problem of electro-static coupling in a transformer is accurate at

(a) high frequency
(b) power frequency
(c) low frequency
(d) high load on the transformer

Q.280 Usually a CT has
(a) no overload capacity
(b) lower power overload capacity than PT
(c) the same overload capacity as a PT
(d) a higher overload capacity than a PT

Q.281 Capacitive potential transformers are used
(a) for keeping the value of transformation ratio constant irrespective of the burden by making certain adjustment.
(b) for primary winding phase voltages above 100 kV.
(c) because they are cheaper than the electro-magnetic transformers above a certain voltage range.
(d) all of the above

Q.282 **Assertion (A):** The secondary winding of a P.T. can be open-circuited without any damage being caused either to the operator or to the transformer.
Reason (R): The potential transformer may be considered as parallel transformer with its secondary winding operating under open circuit conditions whereas the current transformer may be thought as a series transformer under virtual short circuit conditions.
(a) Both A and R are true and R is a correct explanation of A.
(b) Both A and R are true but R is not a correct explanation of A.
(c) A is true but R is false.
(d) A is false but R is true.

Q.283 A comparison between a potential transformer and a power transformer has been done as shown below.

S.No.	Characteristics	Potential transformer	Power transformer
1.	Cost	More	Less
2.	Error in measurement	More	Less
3.	Overloading	2-3 times	Around 30 times
4.	Efficiency	Less	More

Which of the above are the correct comparison among the two said transformers?
(a) 2 and 3 only (b) 1, 2, 3 and 4
(c) 1, 3 and 4 only (d) 1, 2 and 3 only

9. Miscellaneous

Q.284 In d.c. potentiometer measurements, a second reading is often taken after reversing the polarities of the d.c. supply and the unknown voltage, and the average of the two readings is taken. This is with a view to eliminate the effect of
(a) ripples in the d.c. supply
(b) stray magnetic fields
(c) stray thermal emf's
(d) erroneous standardisation

Q.285 A slide wire potentiometer is used to measure voltage across a d.c. circuit. The voltage is 1.2 V as read by the potentiometer. Across the same points 20,000 Ω/V voltmeter reads 0.6 V on its 5 V range. Input resistance between the two points is
(a) 20,000 Ω (b) 100,000 Ω
(c) 50,000 Ω (d) 120,000 Ω

Q.286 If Q_e is the effective Q of the coil, C is the resonant capacitance and C_d is the distributed capacitance, then the true Q in a Q-meter will be
(a) $Q_e[(C+C_d)/C]$ (b) $Q_e[C/(C+C_d)]$
(c) $Q_e[C_d/(C+C_d)]$ (d) $Q_e[(C+C_d)/C_d]$

Q.287 In an analog data acquisition unit, what is the correct sequence of the blocks starting from the input?
(a) Transducer-Recorder-Signal conditioner
(b) Transducer-Signal conditioner-Recorder
(c) Signal conditioner-Transducer-Recorder
(d) Signal conditioner-Recorder-Transducer

Q.288 A single slide wire is used for the measurement of current in a circuit. The voltage drop across a standard resistance of 1.0 Ω is balanced at 70 cm. What is the magnitude of the current, if the standard cell having an emf of 1.45 volts is balanced at 50 cm?
(a) 3.09 A (b) 2.64 A
(c) 2.03 A (d) 1.45 A

Q.289 **Assertion (A):** No power is consumed in the circuit containing the unknown emf when the potentiometer is balanced.
Reason (R): The determination of voltage by a potentiometer is quite independent of the source resistance.

(a) Both A and R are true and R is a correct explanation of A.
(b) Both A and R are true but R is not a correct explanation of A.
(c) A is true but R is false.
(d) A is false but R is true.

Q.290 Consider the following statements related to a DC potentiometer:
1. It can be used for the calibration of all voltmeters, ammeters and wattmeters.
2. The potentiometer can be used for the measurement of current, power and resistance.
3. Since the potentiometer is a d.c. device, the instruments to be calibrated must be d.c. moving iron or electro-dynamometer types.
4. DC potentiometers are based on comparison methods and have less degree of accuracy compared to deflection type voltage measuring instruments.

Which of the above statements is/are not correct?
(a) 4 only (b) 3 only
(c) 3 and 4 only (d) 2, 3 and 4 only

Q.291 Assertion (A): The value of resistance (R) is indirectly determined by measuring the value of *Q*.
Reason (R): The value of effective resistance with a.c. differs from its d.c. value on account of "Skin effect" and "Eddy current losses".
(a) Both A and R are true and R is a correct explanation of A.
(b) Both A and R are true but R is not a correct explanation of A.
(c) A is true but R is false.
(d) A is false but R is true.

Q.292 Assertion (A): In a practical storage factor meter, the shunt resistance to which the oscillator delivers current should be in mega-ohm (MΩ).
Reason (R): The storage factor meter is an instrument which is designed to measure the value of *Q* directly and as such is very useful in measuring the characteristics of coils and capacitors.
(a) Both A and R are true and R is a correct explanation of A.
(b) Both A and R are true but R is not a correct explanation of A.
(c) A is true but R is false.
(d) A is false but R is true.

Q.293 A series RLC circuit is used for measurement of inductance of a coil using resonant method. Resonance conditions are produced by varying the capacitance. A thermoelectric type ammeter is used in series with the circuit. Also an electronic voltmeter is connected across the tuning capacitor.
(a) The condition of resonance are more reliably indicated when the voltmeter connected across the capacitor reads maximum.
(b) The conditions of resonance are more reliably indicated when the ammeter reads maximum.
(c) The conditions of resonance are indicated when both ammeter and voltmeter read minimum.
(d) The conditions of resonance are the same when both ammeter and voltmeter read maximum.

Q.294 The resonant frequency of an RLC series circuit is 1.5 MHz with the tuning capacitor set at 150 pF. The bandwidth is 10 kHz. The effective resistance of the circuit is
(a) 14.75 Ω (b) 29.5 Ω
(c) 4.7 Ω (d) 9.4 Ω

Q.295 Which of the following is not true regarding a flux meter?
(a) The industrial form of flux meter is portable
(b) There is no control springs
(c) The air friction damping is negligible
(d) The flux meter is more sensitive and accurate than the ballistic galvanometer

Q.296 A distortion analyser measures the total
(a) average power (b) rms power
(c) peak power (d) dc power

Q.297 A spectrum analyser is used to display
(a) frequency band spectrum
(b) amplitude
(c) time
(d) phase

Q.298 The self capacitance of a coil is measured by using a quality factor meter.

- The first measurement is at f_1 = 1 MHz and C_1 = 500 pF.
- The second measurement is at f_2 = 2 MHz and C_2 = 110 pF.

The distributed capacitance of the coil is given by

(a) 5 pF (b) 10 pF
(c) 20 pF (d) 35 pF

Q.299 Calculate the value of the self capacitance when the following measurements are performed:

f_2 = 2 MHz and C_1= 500 pF;
f_2 = 6 MHz and C_2= 50 pF

(a) 2.25 pF (b) 4.25 pF
(c) 5.50 pF (d) 6.25 pF

Q.300 *Q*-factor is defined as

(a) $\frac{\text{Reactance}}{\text{Resistance}}$ (b) $\frac{\text{Resistance}}{\text{Reactance}}$
(c) $\frac{\text{Resistance}}{\text{Impedance}}$ (d) $\frac{\text{Impedance}}{\text{Resistance}}$

Q.301 The shunt resistance R_{sh} used in *Q*-meter has a value of

(a) 0.02 Ω to 0.04 Ω (b) 2 Ω to 4 Ω
(c) 0.2 Ω to 0.4 Ω (d) 2 kΩ to 4 kΩ

Q.302 A *Q*-meter is used to measure

(a) voltage (b) inductance
(c) power (d) resistance

Q.303 A frequency meter is used to measure

(a) frequency (b) ratio
(c) time interval (d) phase

Q.304 Schmitt trigger used in digital measurement time converts input to

(a) Sawtooth wave (b) Sine wave
(c) Square wave (d) Pulses

Q.305 A certain fluxmeter has the following constants:

Air-gap flux density = 0.05 Wb/m^2
Turns on moving coil = 40
Area of moving coil = 750 mm^2

If the flux linking a 10 turn search coil of 200 mm^2 area connected to the fluxmeter is reversed in a uniform field of 0.5 Wb/m^2, the deflection of the fluxmeter would be

(a) 76.5° (b) 66.2°
(c) 53.5° (d) 44.7°

Q.306 In an unshunted fluxmeter, the sensitivity is dependent upon

(a) resistance of moving coil
(b) resistance of the search coil
(c) resistance of both the moving coil and the search coil
(d) resistance of neither the moving coil nor the search coil

Q.307 **Assertion (A):** If higher voltage ranges have to be measured a precision potential divider, called a "Volt-ratio box", is used.

Reason (R): A volt-ratio box provides multiple ranges.

(a) Both A and R are true and R is a correct explanation of A.
(b) Both A and R are true but R is not a correct explanation of A.
(c) A is true but R is false.
(d) A is false but R is true.

Q.308 Consider the following statements related to data Acquisition systems:

1. Analog data acquisition systems are used when the physical quantity being monitored has a narrow bandwidth.
2. Digital data acquisition systems are used when wide frequency width is required or when lower accuracies can be tolerated.
3. Digital data acquisition systems are in general, more complex than analog systems, both in terms of instrumentation involved and the volume and complexity of the data they can handle.
4. Time division multiplexing is especially useful when telemetering slow-changing, low bandwidth data.
5. Examples of data usually sent by FDM systems are vibration, acceleration and dynamic strain.

Of these, the correct statements is/are

(a) 2, 3 and 4 (b) 1, 2 and 3
(c) 3, 4 and 5 (d) 1, 2 and 5

Q.309 **Assertion (A):** Conversion to digitized systems is widely used today.

Reason (R): In digitized system, the circuits are low cost, accurate and simple to implement.

(a) Both A and R are true and R is a correct explanation of A.

(b) Both A and R are true but R is not a correct explanation of A.
(c) A is true but R is false.
(d) A is false but R is true.

Q.310 In a reed type frequency meter all the reeds
(a) have different natural frequency
(b) have same natural frequency
(c) have different natural frequencies but the difference in natural frequencies of adjacent reeds is ± 0.5 Hz
(d) none of the above

Q.311 In signal generators
(a) energy is created.
(b) energy is generated.
(c) energy is converted from a simple d.c. source into a.c. energy at some specific frequency.
(d) all of the above

Q.312 Match **List-I (Wave analysers)** with **List-II (Applications)** and select the correct answer using the codes given below the lists:

List-I
A. Hetrodynane wave analyser
B. Harmonic distortion analyser
C. Spectrum analyser

List-II
1. Study of non-linear behaviour of circuit elements.
2. Frequency measurement in the MHz range.
3. Study of information about bandwidth, effects of different types of modulation and spurious signal generation.

Codes:

	A	B	C
(a)	3	2	1
(b)	2	1	3
(c)	2	3	1
(d)	1	2	3

Q.313 A potentiometer that is accurate to ± 0.0001 V (standard deviation) is used to measure current through a standard resistance of 0.1 ± 0.1 Ω% (standard deviation). The voltage across the resistance is measured to be 0.2514 V. What is the percentage accuracy of measurement of current?
(a) ±0.203%
(b) ±0.155%
(c) ±0.265%
(d) ±0.108%

Q.314 Fluxmeter is a special type of ballistic galvanometer having
(a) very high electromagnetic damping and control torque.
(b) very high electromagnetic damping and very small control torque.
(c) very small electromagnetic damping and very high control torque.
(d) very small electromagnetic damping and control torque.

Q.315 **Assertion (A):** The Q-meter measures the Q-factor of a coil when the circuit is in resonance.
Reason (R): The Q-factor of a coil depends only on its inductance and not on its resistance.
(a) Both A and R are true and R is a correct explanation of A.
(b) Both A and R are true but R is not a correct explanation of A.
(c) A is true but R is false.
(d) A is false but R is true.

Q.316 **Assertion (A):** Rayleigh current balance is considered to be the most satisfactory method for absolute measurement of current.
Reason (R): Rayleigh current balance method of absolute measurement of current does not need accurate measurement of dimensions, measurement of earth magnetic field horizontal component and determination of the tension constant of a suspension wire.
(a) Both A and R are true and R is a correct explanation of A.
(b) Both A and R are true but R is not a correct explanation of A.
(c) A is true but R is false.
(d) A is false but R is true.

Q.317 Harmonic distortion percentage is accurately measured by
(a) VTVM
(b) Harmonic distortion analyser
(c) CRO
(d) VOM

Q.318 A comparison between a ballistic galvanometer and fluxmeter is given below. Which of the following characteristics are correctly matched?

S.No.	Characteristics	Ballistic Galvanometer	Flux meter
1.	Sensitivity	Less	More
2.	Accuracy	Less	More
3.	Controlling torque	High	Very small
4.	Electromagnetic damping	Less	Heavy

(a) 1 and 2 only (b) 1, 2, 3 and 4
(c) 2, 3 and 4 only (d) 3 and 4 only

10. Digital Meters

Q.319 An average reading VTVM uses one diode connected in sereis with an external resitance. A high value of series resistance is used so that the instrument should have
(a) a high input impedance
(b) a linear *v-i* characteristics
(c) low power consumption
(d) all of the above

Q.320 Electronic voltmeters which use rectifiers employ negative feedback. This is done
(a) to increase the overall gain
(b) to improve stability
(c) to overcome non-linearity of diodes
(d) none of the above

Q.321 A voltmeter with a broad bandwidth has
(a) low noise level and high sensitivity
(b) high noise level and high sensitivity
(c) low noise level and low sensitivity
(d) high noise level and low sensitivity

Q.322 A $4\frac{1}{2}$ digit voltmeter is used for voltage measurements. What is its resolution?
(a) 0.01% (b) 0.001%
(c) 6.25% (d) 3.125%

Q.323 An integrator contains a 100 kΩ and 1 μF capacitor. If the voltage applied to the integrator input is 1 V, what voltage magnitude will be present at the output of the integrator after 1 second?
(a) 100 V (b) 10 V
(c) 1 V (d) 0.1 V

Q.324 Measurement in a ramp type DVM is performed during the
(a) negative slope (b) position slope
(c) both of the slope (d) none of these

Q.325 The principle of voltage to time conversion is used in
(a) dual slope type DVM
(b) successive approximation type DVM
(c) integrating type DVM
(d) none of the above

Q.326 SAR type DVM uses the principle of
(a) voltage to current conversion
(b) voltage to binary conversion
(c) voltage to frequency conversion
(d) voltage to time conversion

Q.327 Measurement by dual slope DVM is performed during
(a) rising slope
(b) falling slope
(c) rising and falling slope
(d) none of the slopes

Q.328 **Assertion (A):** The dual slope technique has excellent noise rejection.
Reason (R): Noise and superimposed ac are averaged out in the process of integration.
(a) Both A and R are true and R is a correct explanation of A.
(b) Both A and R are true but R is not a correct explanation of A.
(c) A is true but R is false.
(d) A is false but R is true.

Q.329 A $3\frac{1}{2}$ digit DVM is used for measuring voltage. If the range of the meter is 10 - 100 V, then the resolution of the meter would be
(a) 10% (b) 0.1%
(c) 1% (d) 0.01%

Q.330 Match **List-I (Type of DVM's)** with **List-II (Characteristics)** and select the correct answer using the codes given below the lists:

List-I
A. Potentiometric-type DVM
B. Intergrating-type DVM
C. Dual slop-type DVM
D. Ramp-type DVM

List-II
1. Require excellent characteristics in the linearity of the ramp

2. Input filters are usually required
3. Self-balancing potentiometer
4. Excellent noise rejection

Codes:

	A	B	C	D
(a)	1	3	2	4
(b)	3	2	4	1
(c)	1	4	2	3
(d)	3	1	4	2

Q.331 Consider the following statements related to ramp type DVM:

1. It uses the principle of voltage to time conversion.
2. The ramp technique circuit is easy to design and its cost is less.
3. The output pulse can be transmitted over long feeder lines.
4. The error due to noise are very less in ramp type DVM's.

Which of the above statements is/are correct?

(a) 1 and 3 only (b) 2 and 3 only
(c) 3 and 4 only (d) 3 only

Q.332 A DVM measures
(a) peak value
(b) rms value
(c) peak-to-peak value
(d) average value

Q.333 Which of the following is a correct statement to a DVM?
(a) Common mode rejection is the same for ac and dc voltages.
(b) It does not have series mode error.
(c) Its accuracy depends on source impedance.
(d) Use of shunts for dc current measurements increases its accuracy.

Q.334 Accuracy of DVM is specified as
(a) percentage of the actual reading
(b) percentage of the full-scale reading
(c) number of least significant digits
(d) none of the above

Answers Electrical & Electronic Measurements

1. (b)	2. (c)	3. (c)	4. (a)	5. (d)	6. (b)	7. (c)	8. (a)
9. (b)	10. (d)	11. (a)	12. (c)	13. (b)	14. (d)	15. (a)	16. (d)
17. (b)	18. (c)	19. (a)	20. (d)	21. (b)	22. (b)	23. (c)	24. (a)
25. (c)	26. (a)	27. (d)	28. (b)	29. (a)	30. (d)	31. (a)	32. (c)
33. (a)	34. (b)	35. (b)	36. (a)	37. (a)	38. (d)	39. (b)	40. (b)
41. (a)	42. (b)	43. (a)	44. (c)	45. (b)	46. (a)	47. (d)	48. (a)
49. (b)	50. (d)	51. (a)	52. (d)	53. (c)	54. (d)	55. (a)	56. (c)
57. (b)	58. (d)	59. (d)	60. (a)	61. (c)	62. (b)	63. (a)	64. (b)
65. (d)	66. (a)	67. (d)	68. (c)	69. (c)	70. (c)	71. (d)	72. (a)
73. (a)	74. (a)	75. (c)	76. (a)	77. (c)	78. (c)	79. (d)	80. (d)
81. (a)	82. (d)	83. (c)	84. (a)	85. (c)	86. (c)	87. (d)	88. (a)
89. (b)	90. (d)	91. (a)	92. (c)	93. (d)	94. (a)	95. (d)	96. (c)
97. (b)	98. (d)	99. (b)	100. (c)	101. (d)	102. (d)	103. (b)	104. (a)
105. (d)	106. (c)	107. (a)	108. (d)	109. (d)	110. (d)	111. (c)	112. (b)

113. (a)	114. (d)	115. (b)	116. (d)	117. (b)	118. (c)	119. (c)	120. (b)
121. (a)	122. (d)	123. (d)	124. (b)	125. (c)	126. (a)	127. (a)	128. (c)
129. (a)	130. (d)	131. (a)	132. (b)	133. (c)	134. (a)	135. (d)	136. (c)
137. (d)	138. (c)	139. (a)	140. (b)	141. (a)	142. (a)	143. (c)	144. (b)
145. (d)	146. (b)	147. (c)	148. (c)	149. (a)	150. (d)	151. (c)	152. (d)
153. (a)	154. (c)	155. (c)	156. (b)	157 (a)	158. (b)	159. (a)	160. (a)
161. (c)	162. (a)	163. (b)	164. (d)	165. (c)	166. (d)	167. (a)	168. (b)
169. (c)	170. (d)	171. (c)	172. (a)	173. (a)	174. (b)	175. (d)	176. (c)
177. (a)	178. (a)	179. (b)	180. (d)	181. (d)	182. (b)	183. (a)	184. (c)
185. (a)	186. (a)	187. (d)	188. (d)	189. (c)	190. (b)	191. (a)	192. (c)
193. (a)	194. (c)	195. (b)	196. (d)	197. (a)	198. (c)	199. (c)	200. (b)
201. (b)	202. (a)	203. (a)	204. (d)	205. (c)	206. (b)	207. (a)	208. (d)
209. (c)	210. (d)	211. (a)	212. (c)	213. (c)	214. (b)	215. (d)	216. (a)
217. (a)	218. (b)	219. (a)	220. (b)	221. (a)	222. (b)	223. (d)	224. (c)
225. (b)	226. (c)	227. (b)	228. (c)	229. (a)	230. (d)	231. (b)	232. (c)
233. (d)	234. (a)	235. (b)	236. (b)	237. (b)	238. (d)	239. (b)	240. (c)
241. (d)	242. (a)	243. (b)	244. (a)	245. (d)	246. (c)	247. (d)	248. (b)
249. (a)	250. (c)	251. (d)	252. (c)	253. (a)	254. (a)	255. (b)	256. (c)
257. (b)	258. (b)	259. (b)	260. (c)	261. (c)	262. (b)	263. (a)	264. (b)
265. (c)	266. (a)	267. (c)	268. (d)	269. (b)	270. (d)	271. (a)	272. (a)
273. (b)	274. (d)	275. (c)	276. (c)	277. (d)	278. (b)	279. (a)	280. (d)
281. (d)	282. (a)	283. (c)	284. (c)	285. (b)	286. (b)	287. (b)	288. (c)
289. (b)	290. (a)	291. (a)	292. (d)	293. (a)	294. (c)	295. (d)	296. (b)
297. (a)	298. (c)	299. (d)	300. (a)	301. (a)	302. (b)	303. (c)	304. (d)
305. (a)	306. (d)	307. (a)	308. (c)	309. (a)	310. (c)	311. (c)	312. (b)
313. (d)	314. (b)	315. (c)	316. (a)	317. (b)	318. (d)	319. (d)	320. (c)
321. (d)	322. (a)	323. (b)	324. (a)	325. (a)	326. (b)	327. (b)	328. (a)
329. (c)	330. (d)	331. (b)	332. (d)	333. (a)	334. (b)		

Explanations

1. (b)

Measurement by direct methods are not always possible, feasible and practicable. In most of the cases direct method of measurements are inaccurate because they involve human factors. Moreover, they are less sensitive. Due to these reasons use of direct method of measurement is limited while indirect methods are commonly used.

3. (c)

Span of the thermometer

$= 200 - 150 = 50°C$

$\therefore$ Maximum static error

$= \pm 0.25\%$ of $50°C$

$$= \pm\frac{0.25}{100}\times 50 = \pm 0.125°C$$

4. (a)

Average value of current is

$$I_{avg} = \frac{10.03+10.10+10.11+10.8}{4}$$

$= 10.08$ A

Maximum value of error is

$\varepsilon_{max} = I_{max} - I_{avg}$

$= 10.11 - 10.08 = 0.03$ A

Minimum value of error is

$\varepsilon_{min} = I_{avg} - I_{min}$

$= 10.08 - 10.03 = 0.05$ A

$\therefore$ Average range of error is

$$\varepsilon_{avg} = \pm\left[\frac{\varepsilon_{max}+\varepsilon_{min}}{2}\right]$$

$$= \pm\left[\frac{0.03+0.05}{2}\right] = \pm 0.04 \text{ A}$$

5. (d)

Given, $R_1 = 28.7\ \Omega$ (three significant figures)

$R_2 = 3.624\ \Omega$ (four significant figures)

For series connection,

$R_{total} = R_1 + R_2$

$= 28.7 + 3.624$

$= 32.324\ \Omega$

Since one of the resistance (i.e. R_1) is accurate only to three significant figures, therefore the result of R_{total} has to be reduced to three significant figures. Hence, total resistance,

$R_{total} = 32.3\ \Omega$

6. (b)

$$\text{Sensitivity} = \frac{\text{Output}}{\text{Input}} = \frac{3}{8}\text{ mm}/\Omega$$

Deflection factor,

$$D = \frac{1}{\text{Sensitivity}} = \frac{8}{3}\ \Omega/\text{mm}$$

$= 2.67\ \Omega$/mm

7. (c)

$$1 \text{ scale division} = \frac{200}{100} = 2\text{ volt}$$

$$\text{Now, resolution} = \frac{1}{10}\text{th scale division}$$

$$= \frac{1}{10}\times 2 = 0.2 \text{ volt}$$

8. (a)

The circuit diagram for the measurement of voltage using multimeter is shown below:

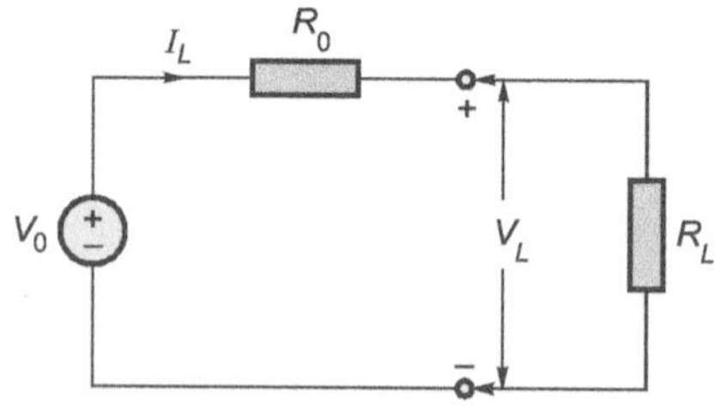

Given, multimeter (voltmeter) reading = 10 V.

$\therefore$ Input resistance of voltmeter,

R_L = Sensitivity × Voltmeter reading

$= 2 \times 10^3 \times 10 = 20\ k\Omega$

Also, $V_0 = 6$ volts

= Open circuit voltage

and $R_0 = 10\ k\Omega$

= Output resistance of circuit

Let the current in the circuit be I_L.

Now, $V_L = I_L R_L$ and $V_0 = I_L(R_0 + R_L)$

So, $$\frac{V_L}{V_0} = \frac{I_L R_L}{I_L(R_0+R_L)}$$

or, $$V_L = \frac{V_0}{\left(1+\frac{R_0}{R_L}\right)}$$

$$= \frac{6}{\left(1+\frac{10}{20}\right)} = \frac{6}{1.5} = 4 \text{ volts}$$

So, $(V_L)_{measured}$ = 4 volts and $(V_L)_{True}$ = 6 volts

∴ % error in voltage reading

$$= \frac{(V_L)_{measured} - (V_L)_{True}}{(V_L)_{True}} \times 100$$

$$= \left(\frac{4-6}{6}\right) \times 100 = -33.33\%$$

$$\approx -33\% \approx 33\% \text{ low}$$

9. (b)

The given circuit can be reduced to its Thevenin's equivalent circuit as shown below.

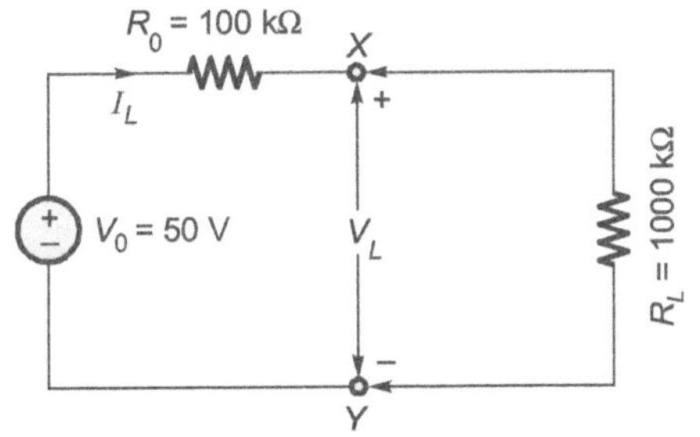

True value of voltage across terminals X-Y is

$$(V_L)_{True} = V_0 = 50 \text{ Volt}$$

Measured value of voltage across terminals X-Y is

$$(V_L)_{measured} = \left(\frac{R_L}{R_0+R_L}\right)V_0$$

$$= \left(\frac{1000}{1000+100}\right) \times 50$$

$$= \frac{1000}{1100} \times 50 = 45.45 \text{ volts}$$

$$\therefore \text{Loading error} = \left[\frac{(V_L)_{measured} - (V_L)_{True}}{(V_L)_{True}}\right] \times 100$$

$$= \left(\frac{45.45-50}{50}\right) \times 100 = -9.1\%$$

Accuracy in measurement of voltage

= 100 – loading error (in %)

= 100 – 9.1 = 90.9%

11. (a)

$$\text{Error in ammeter reading} = \frac{I_m - I_T}{I_T} = \varepsilon_r$$

or, $$\frac{I_m}{I_T} = 1 + \varepsilon_r$$

or, $$\frac{I_T}{I_m} = \frac{1}{1+\varepsilon_r}$$

$$= \text{Correction factor} = \frac{6.5}{6.7} = 0.97$$

12. (c)

$$1 \text{ scale division} = \frac{50}{100} \text{kN/m}^2 = 0.5 \text{ kN/m}^2$$

$$\therefore \quad \text{resolution} = \frac{1}{5}\text{th of scale division}$$

$$= \frac{1}{5} \times 0.5 = 0.1 \text{ kN/m}^2$$

13. (b)

$$R = \frac{P}{I^2}$$

∴ Given, $\varepsilon_P = \pm 1.5\%$, $\varepsilon_I = \pm 1.0\%$

∴ Limiting error in the measurement of resistance is $\varepsilon_R = (\varepsilon_P + 2\varepsilon_I) = \pm(1.5 + 2 \times 1)\%$

$= \pm 3.5\%$

14. (d)

Resistance, $R = \frac{V}{I}$ (Using V-I method)

$$\frac{\delta R}{\delta V} = \frac{1}{I} \text{ and } \frac{\delta R}{\delta I} = \frac{-V}{I^2}$$

Given, $W_V = \pm 12$ V

$W_I = \pm 2$ A

V = 100 Volt

and I = 10 A

∴ Uncertainty in the measurement of resistance is

$$W_R = \sqrt{\left(\frac{\delta R}{\delta V}\right)^2 \times W_V^2 + \left(\frac{\delta R}{\delta I}\right)^2 \times W_I^2}$$

$$= \sqrt{\left(\frac{1}{I}\right)^2 \times W_V^2 + \left(\frac{V}{I^2}\right)^2 \times W_I^2}$$

$$= \sqrt{\left(\frac{1}{10}\right)^2 \times 12^2 + \left(\frac{100}{100}\right)^2 \times 4}$$

$$= \sqrt{\frac{144}{100} + 4} = \sqrt{\frac{544}{100}} = 2.33\ \Omega$$

15. (a)

The limiting error in percentage for the unknown resistor is

$$\varepsilon_r = \pm(\varepsilon_1 + \varepsilon_2 + \varepsilon_3)$$
$$= \pm(1 + 1 + 0.5)\% = \pm 2.5\%$$

Also, true value of resistance

$$R_4 = \frac{500 \times 615}{100} = 3075\ \Omega$$

So, limiting error of the unknown resistor in ohm

$$= \pm\frac{2.5}{100} \times 3075\ \Omega$$
$$= \pm 76.875\ \Omega \approx \pm 76.88\ \Omega$$

16. (d)

$$\text{Error} = \pm\frac{1}{100} \times 1250 = \pm 12.5\ \text{W}$$

∴ Range of reading

$$= (1250 \pm 12.5)\ \text{W}$$
$$= 1237.5\ \text{W to } 1262.5\ \text{W}$$

17. (b)

The error occurs due to shortcoming of the instrument because the strength of the permanent magnet becomes 95%. Thus, this error falls in the category of systematic error.

18. (c)

Error at desired scale

$$= \left(\frac{\text{Full scale error} \times \text{Full scale value}}{\text{Desired value}}\right)$$

$$= \frac{1\% \times 10}{2.5} = 4\%$$

20. (d)

Emf, $$e = \frac{\text{Work done}}{\text{Charge}} = \frac{F \cdot x}{q}$$

∴ $$[e] = \frac{[MLT^{-2}][L]}{[IT]} = [ML^2T^{-3}I^{-1}]$$

Magnetic flux density, $B = \dfrac{\phi}{A}$

Now, $$e = N\frac{d\phi}{dt}$$

∴ $$\phi = \frac{e.t}{N}$$

or, $$[\phi] = [ML^2T^{-3}I^{-1}][T] = [ML^2T^{-2}I^{-1}]$$

So, $$[B] = \frac{[\phi]}{[A]} = \frac{[ML^2T^{-2}I^{-1}]}{[L^2]}$$

$$= [MT^{-2}I^{-1}]$$

Also, magnetizing force,

$$H = \frac{\text{mmf}}{\text{length}} = \frac{NI}{l}$$

∴ $$[H] = \frac{[I]}{[L]} = [IL^{-1}]$$

And, reluctance $= \dfrac{\text{mmf}}{\text{Flux}} = \dfrac{NI}{\phi} = Rl$

∴ $$[Rl] = \frac{[I]}{[ML^2T^{-2}I^{-1}]} = [M^{-1}L^{-2}T^2I^2]$$

23. (c)

The PMMC ammeter should indicate the value of current when the deflecting torque (T_d) = controlling torque (T_c).

Since, $T_c = 0$, therefore the pointer will continuously rotate due to T_d and will not stop at a certain value of current on scale due to absence of balancing torque/force (T_c or F_c).

25. (c)

If the weight of moving parts is large, the frictional torque will be large. Hence, reason is false.

27. (d)

The errors introduced by the observer are called observational errors. Improper setting of range of multimeter will lead to observational error.

32. (c)

Since the response of null type instrument is slower, therefore they are not suited for dynamic conditions. Hence, statement-3 is false.

A Wattmeter is an indicating type instrument while a DC potentiometer is a null type instrument.

34. (b)

- Loading effect is one of the cause of Instrumental error.
- Parallax error fall in the category of observational error.
- Residual error is also called Random error.
- Gross error is mainly due to human mistakes in recording, reading and calculating measurement results.

39. (b)

Given, $\frac{\delta V}{V} = 2\%$

$\therefore \quad \delta V = \frac{2}{100} \times 300 = 6 \text{ V}$

Hence, the range of reading

$= 30 \pm 6 = 24 \text{ V to } 36 \text{ V}$

40. (b)

Tolerance for 10 kΩ resistor

$$= \frac{5}{100} \times 10 \text{ k}\Omega = 0.5 \text{ k}\Omega$$

Tolerance for 5 kΩ resistor

$$= \frac{10}{100} \times 5 \text{ k}\Omega = 0.5 \text{ k}\Omega$$

$\therefore$ Total tolerance = 0.5 + 0.5 = 1 kΩ

Total resistance of series network

= 10 + 5 = 15 kΩ

Thus, tolerance limit $= \frac{1}{15} \times 100 = 6.67\%$

42. (b)

Since the limits of error are given as standard deviation, therefore error in current measurement is given by

$$\sigma_I = \sqrt{\left(\frac{\delta I}{\delta I_1}\right)^2 \cdot \sigma_{I_1}^2 + \left(\frac{\delta I}{\delta I_2}\right) \cdot \sigma_{I_2}^2}$$

Now, $\quad I = I_1 + I_2$

$\therefore \quad \frac{\delta I}{\delta I_1} = \frac{\delta I}{\delta I_2} = 1 \text{ and } \sigma_{I_1} = 1, \sigma_{I_2} = 2$

$\therefore \quad \sigma_I = \sqrt{1^2 . 1^2 + 1^2 \cdot 2^2}$

$= \sqrt{5} = 2.24 \text{ A}$

Hence, $\quad I = (400 \pm 2.24) \text{ A}$

43. (a)

Average current (I_{avg})

$$= \frac{117.02 + 117.11 + 117.08 + 117.03}{4}$$

$= 117.06 \text{ mA}$

As $\quad I_{max} = 117.11 \text{ mA}$

and $\quad I_{min} = 117.02 \text{ mA}$

$\therefore$ Range of error

$$= \pm \frac{(I_{max} - I_{avg}) + (I_{avg} - I_{min})}{2}$$

$$= \pm \frac{0.05 + 0.04}{2} = \pm 0.045 \text{ mA}$$

44. (c)

- To reduce the error in standard resistance due to stray inductance, the adjacent wires should carry current in opposite direction so as to reduce the resultant magnetic field.
- To reduce the error due to contact (lead) resistance, a four terminal construction is used.
- To reduce the skin effect when used at high frequency a.c., small diameter wires are used.

45. (b)

At high and medium frequencies, a resistor can be represented as shown in figure below.

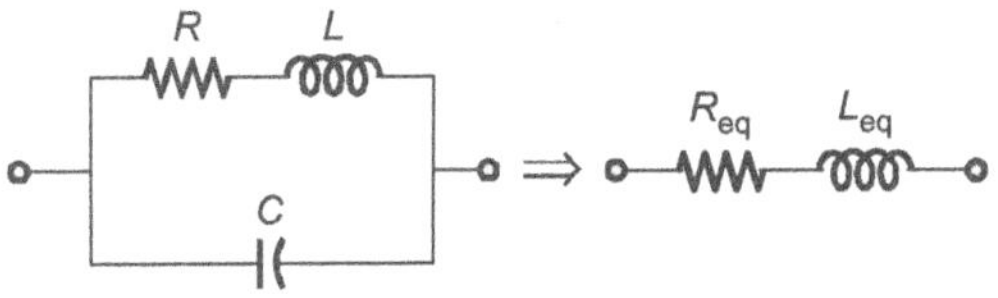

The equivalent impedance of the above shown resistor at an angular frequency ω rad/s is

$$Z_{eq} = \frac{\left(\frac{1}{j\omega C}\right)(R + j\omega L)}{R + j\omega L + \left(\frac{1}{j\omega C}\right)}$$

$$= \frac{R + j\omega[(1 - \omega^2 LC) - CR^2]}{1 + \omega^2 C(CR^2 - 2L)}$$

$$= R_{eq} + jX_{eq}$$

Here, $\quad R_{eq} = \frac{R}{1 + \omega^2 C(CR^2 - 2L)}$

and $\quad X_{eq} = \frac{\omega [L(1 - \omega^2 LC) - CR^2]}{1 + \omega^2 C(CR^2 - 2L)}$

If the resistance is to remain independent of frequency i.e. to have the same value of resistance at medium frequencies as with d.c, we have:

$$CR^2 = 2L$$

so that, $R_{eq} = R$

46. (a)

At medium frequencies, we have:

$$L_{eq} = \frac{L - CR^2}{1+\omega^2 C(CR^2 - 2L)}$$

$$R_{eq} = \frac{R}{1+\omega^2 C(CR^2 - 2L)}$$

∴ Time constant,

$$\tau = \frac{L_{eq}}{R_{eq}} = \frac{L - CR^2}{R} = \frac{L}{R} - CR$$

47. (d)

Let the the required capacitance be C_1.

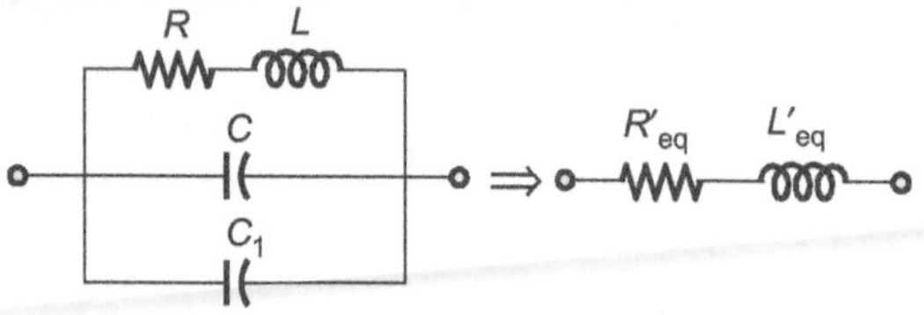

Without the external capacitance C_1,

$$\tan\phi = \frac{X_{eq}}{R_{eq}} = \frac{\omega(L - CR^2)}{R}$$

$$= \omega\left(\frac{L}{R} - CR\right)$$

(Using the values obtained in Q.No. 45)

Now, when an external capacitance C_1 is connected in parallel, then new phase angle

$$= \tan\phi' = \omega\left[\frac{L}{R} - (C_1 + C)R\right]$$

For the phase angle to be zero, $\tan\phi' = 0$

$$\therefore \quad \frac{L}{R} - (C_1 + C)\,R = 0$$

or, $$\left(\frac{L}{R} - CR\right) - (C_1 R) = 0$$

Now, $\frac{L}{R} - CR$ = time constant = τ = 30 ns (Given)

So, $30 \times 10^{-9} - C_1 \times 100 = 0$

or, $$C_1 = \frac{30 \times 10^{-9}}{100} = 0.3 \text{ nF} = 300 \text{ pF}$$

49. (b)

Reading error of voltmeter

$$= \pm\frac{\text{Scale divisions that can be distinguised}}{\text{Total scale divisions}} \times 100$$

$$= \pm\frac{\frac{1}{10}}{150} \times 100 = \pm\frac{1}{15}\% = \pm 0.067\%$$

Similarly, reading error of ammeter

$$= \pm\frac{\frac{1}{10}}{100} \times 100 = \pm 0.1\%$$

Thus, total error of voltmeter

$$\varepsilon_V = \pm 0.4 \pm 0.067 = \pm 0.467\%$$

and total error of ammeter

$$\varepsilon_I = \pm 0.3 \pm 0.1 = \pm 0.4\%$$

Now, unknown resistance, $R = \frac{V}{I}$

Hence, total systematic error in measurement

$$= \varepsilon_V + \varepsilon_I = \pm 0.467 \pm 0.4$$
$$= \pm 0.867\%$$

Measured value of resistance is

$$R_m = \frac{125}{10} = 12.5\ \Omega$$

For the arrangement shown in given figure, ammeter is near to the unknown test resistance (R), therefore

$$\text{error} = \frac{R_a}{R_T} = \frac{R_m - R_T}{R_T}$$

or, $$R_T = R_m - R_a$$

∴ True value of resistance,

$$R_T = R_m - R_a = 12.5 - 0.25$$
$$= 12.25\ \Omega$$

Hence, the value of R can be specified as:

$$R = 12.25 \pm 0.867\%$$

50. (d)

Value of unknown resistance,

$$R = \frac{P}{Q}S$$

$$= \frac{1000}{1000} \times 3154 = 3154\ \Omega$$

Now, percentage error in determination of R is

$$\varepsilon_r = (\pm 0.05 \pm 0.05 \pm 0.1) = \pm 0.2\%$$

∴ Limiting values of R

$$= (3154 \pm 0.2\%)$$
$$= (3154 \pm 6.308)\ \Omega$$
$$= 3147.7\ \Omega \text{ to } 3160.3\ \Omega$$

51. (a)

Let the resistance R be changed to $R + \Delta R$. The Thevenin's equivalent circuit across terminals bd is shown below.

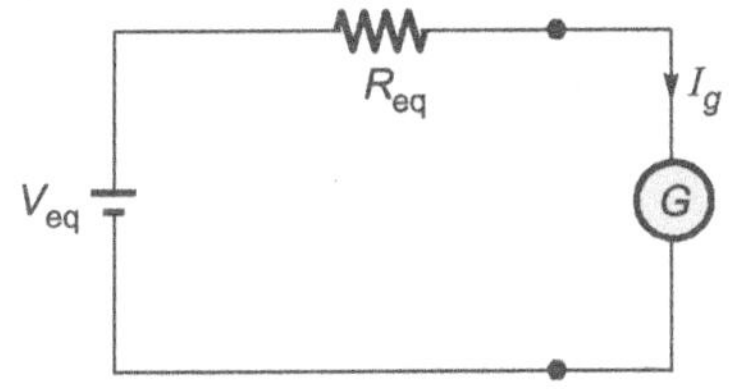

Here, $$V_{eq} = V\left[\frac{R+\Delta R}{R+\Delta R+S} - \frac{P}{P+Q}\right]$$

Since, $P = Q = R = S$, therefore

$$V_{eq} = V\left[\frac{R+\Delta R}{R+\Delta R+R} - \frac{R}{R+R}\right]$$

$$= V\left[\frac{R+\Delta R}{2R+\Delta R} - \frac{1}{2}\right]$$

Also, $$R_{eq} = \frac{(R+\Delta R)S}{R+\Delta R+S} + \frac{PQ}{P+Q}$$

$$= \frac{(R+\Delta R)R}{2R+\Delta R} + \frac{R}{2}$$

Considering $\Delta R << R$, we have

$$V_{eq} = V\left[\frac{2R+2\Delta R-2R-\Delta R}{2(2R+\Delta R)}\right]$$

$$= V\left[\frac{\Delta R}{2(2R+\Delta R)}\right] \approx V\frac{\Delta R}{4R}$$

and, $$R_{eq} \approx \frac{R}{2} + \frac{R}{2} = R$$

So, $$I_g = \frac{V_{eq}}{R_{eq}} \quad \text{(Since, } R_g = 0\text{)}$$

or, $$I_g = \frac{V\Delta R}{4R^2} \text{ or } \Delta R = \frac{4R^2 I_g}{V}$$

Given, $I_g = 0.1$ nA, $R = 1$ kΩ, $V = 20$ volt

$$\therefore \quad \Delta R = \frac{4\times(10^3)^2\times(0.1\times10^{-9})}{20}$$

$$= 20 \times 10^{-6}\ \Omega = 20\ \mu\text{A}$$

52. (d)

Let R be the insulation resistance of the cable and R' be the equivalent resistance of the parallel combination of insulation resistance R and 10 MΩ resistance.

Now, $$R = \frac{0.4343t}{C\log_{10}\left(\frac{V}{V_C}\right)}$$

$$= \frac{0.4343t}{C\log_{10}\left(\frac{\theta}{\theta_C}\right)} \quad \text{(Since } V \propto \theta\text{)}$$

Given, $t = 20s$, $\theta = 200$ divisions, $\theta_c = 100$ divisions.

$$\therefore \quad R = \frac{0.4343t}{C\log_{10}\left(\frac{200}{100}\right)} = \frac{0.4343\times30}{C\log_{10}2} \quad ...(1)$$

Again, for $\theta = 200$ and $\theta_c = 50$, we have:

$$R' = \frac{0.4343\times30}{C\log\left(\frac{200}{50}\right)}$$

$$= \frac{0.4343\times30}{2C\log_{10}2} \quad ...(2)$$

Dividing equation (1) by (2), we get:

$$\frac{R}{R'} = 2 \text{ or } R = 2R' \quad ...(3)$$

Also, $$R' = \frac{10R}{R+10} \quad ...(4)$$

($R' = R \,||\, 10$ MΩ)

Equating equations (3) and (4), we get:

$$\frac{R}{2} = \frac{10R}{R+10}$$

or, $R^2 + 10R = 20R$

or, $R^2 - 10R = 0$ or $R = 10$ MΩ

53. (c)

- The measurement accuracy is directly related to the accuracy of the bridge component and not to that of the null indicator used.
- When the bridge is in balanced condition, no current flows through the galvanometer.
- The sensitivity of the bridge measurement can be expressed in mm/μA or degree/mA or radians/mA.

54. (d)

At balance condition, we have:

$$\frac{40K}{R_x} = \frac{10K}{15K}$$

or, $$R_x = \frac{15}{10}\times40K = 60\text{ k}\Omega$$

55. (a)

For the given Kelvin's bridge, we have:

$$\frac{R_x}{R_2} = \frac{R_b}{R_a}$$

Given, $\frac{R_a}{R_b} = 1000,\ R_1 = 0.5\ R_2 = 5\ \Omega$

$$\therefore \quad \frac{R_x}{10} = \frac{1}{1000}$$

or, $$R_x = \frac{10}{1000} = \frac{1}{100}\ \Omega = 0.01\ \Omega = 10\ \text{m}\Omega$$

56. (c)

Low resistances are provided with four terminals in order that the resistance value becomes definite irrespective of the nature of the contacts at the current terminals.

57. (b)

Using loss of charge method measurement of insulation resistance of cable, we have:

$$R_i = \frac{0.4343t}{C\log_{10}\left(\frac{V}{V_C}\right)}$$

Given, $t = 20\ s,\ V = 100$ volt, $V_c = 10$ volt and $C = 300$ pF

$$\therefore \quad R_i = \frac{0.4343\times 20}{300\times 10^{-12}\log_{10}\left(\frac{100}{10}\right)}$$

$$= 0.02895\times 10^{12}\ \Omega = 28.95\ \text{G}\Omega \approx 29\ \text{G}\Omega$$

60. (a)

- Ohm-meter method is used for measurement of medium resistance ($1 < R < 100\ \text{k}\Omega$).
- Volt-ammeter method is used for measurement of low resistance and medium resistance both.
- Potentiometer method is used for measurement low resistance ($R \le 1\Omega$).

62. (b)

Arrangement-1:

Measured value of resistance,

$$R_m = \frac{V}{I} = \frac{20}{100\times 10^{-3}} = 200\ \Omega$$

Also, $R_m = R_a + R = 200$

or, $R = 200 - R_a = 200 - 1 = 199\ \Omega$

Arrangement-2:

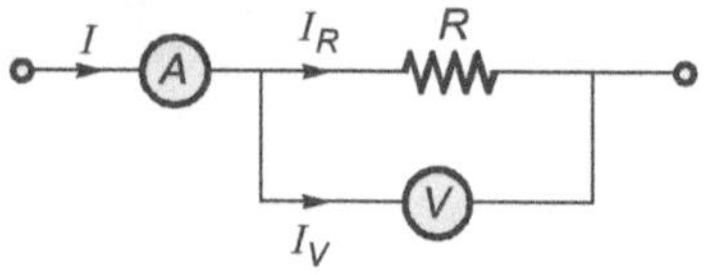

Here, $$I_V = \frac{V}{R_V} = \frac{20}{500} = 40\ \text{mA}$$

$$\therefore \quad I_R = I - I_V = 100 - 40 = 60\ \text{mA}$$

So, $$R = \frac{V}{I_R} = \frac{20}{60}\times 10^3 = 333.33\ \Omega \approx 333\ \Omega$$

64. (b)

The "Kelvin double bridge" can be used for the accurate measurement of low value of resistance because it can eliminate the effect of connecting leads and contact resistance due to which the error caused in the measurement of low resistance can be eliminated.

The reason is also a true statement but, not a correct explanation of assertion.

65. (d)

The arrangement shown in figure below.

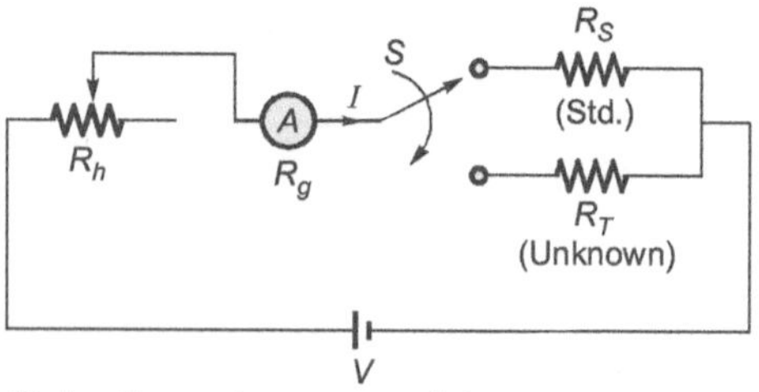

Let R_T be the unknown resistance.

Here, $R_g = 0\ \Omega$ (as not given)

Given:

For $V_1 = 10$ volt, $\theta_1 = 60°$, $R_s = 100\ \Omega$, $R_h = 500\ \Omega$

For $V_2 = 9$ volt, $\theta_2 = 60°$, $R_h = 500\ \Omega$, $R_T = ?$

The deflection in the galvanometer, $\theta \propto I$

Since, $\theta_1 = \theta_2$, therefore $I_1 = I_2$

Now, $$\frac{V_1}{R_h + R_s} = \frac{V_2}{R_h + R_T}$$

or, $$R_T = \left(\frac{V_2}{V_1}(R_h + R_s) - R_h\right)$$

or, $R_T = \left(\frac{9}{10}(500+100)-500\right)$

$= 0.9 \times 600 - 500$

$= 540 - 500 = 40\ \Omega$

Thus, unknown value of resistance,

$R_T = 40\ \Omega$

66. (a)

Given, the resistance of strain gauge,

$R = 100\ \Omega$

Let I be the current through each arm under balanced conditions.

Then, $I^2R = P$

or, $I = \sqrt{\frac{P}{R}} = \sqrt{\frac{250\times10^{-3}}{100}}$

$= \sqrt{25\times10^{-4}} = 0.05\text{ A}$

$= 50\text{ mA}$

Voltage applied across the bridge =

$2IR = 2\times50\times10^{-3}\times100$

$= 10\text{ volt}$

∴ Voltage across the series resistor

$= 20 - 10 = 10\text{ V} = V_s$

and current through the series resistor

$= 2I = 2\times0.05 = 0.1\text{ A} = I_s$

So, resistance of series resistor,

$R_s = \frac{V_s}{I_s} = \frac{10}{0.1} = 100\ \Omega$

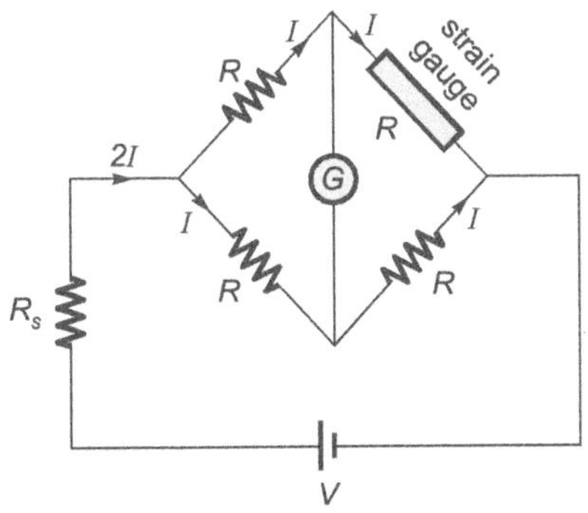

67. (d)

Since no power is consumed in the circuit containing the unknown emf when the potentiometer is balanced (null condition) therefore the determination of voltage by a potentiometer is quite independent of the source resistance. Hence, assertion is a false statement. Reason is a true statement.

68. (c)

Since the unknown voltage is measured using the comparison method, potentiometer has high degree of accuracy in the measurement of unknown voltage.

72. (a)

In a shunt ohm-meter, meter current is given by

$$I_m = \left(\frac{R}{R+R_P}\right)I_{f_s}$$

where, R = unknown resistance

Also, $R_P = \frac{R_1R_m}{R_1+R_m}$,

which is a very low value.

When unknown resistance (R) is in very high range, there is hardly any change in I_m due to which scale is highly cramped for reading high resistance value.

73. (a)

- Standardization of the potentiometer is done for converting the meter scale in terms of voltage. Hence, it is not necessary for the measurement of resistance. Hence, statement-2 is true.
- A 100 kΩ resistance lies in the high scale range, therefore it can't be measured by ammeter-voltmeter method. Thus, statement-4 is not true.
- Meggar is, essentially, a mega ohmmeter because it is used for the measurement of high value of resistance (insulation resistance).
- An ohmmeter is basically an ammeter because it reads the value of current and, then giving the value of unknown resistance by measuring the ratio V/I.

74. (a)

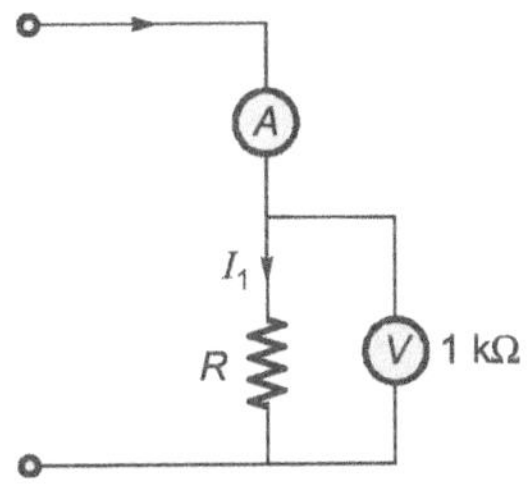

Here, $V = I_1R$

But, $I_1 = \left(\dfrac{10^3}{10^3 + R}\right) I$

Given, $I = 1$ A and $V = 100$ volt

$\therefore \quad 100 = \left(\dfrac{10^3}{10^3 + R}\right) R$ or $R = 111.11\ \Omega$

75. (c)

The Thevenin's equivalent resistance across the galvanometer (G) is

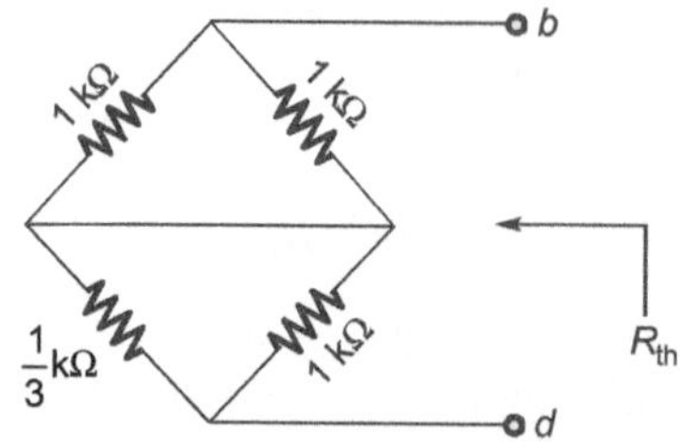

$$R_{th} = (1\,k \,||\, 1\,k) + \left(1\,k \,||\, \frac{1}{3}\,k\right)$$

$$= 0.5\,k + 0.25\,k = 0.75\ k\Omega$$

The Thevenin's equivalent voltage across the galvanometer is

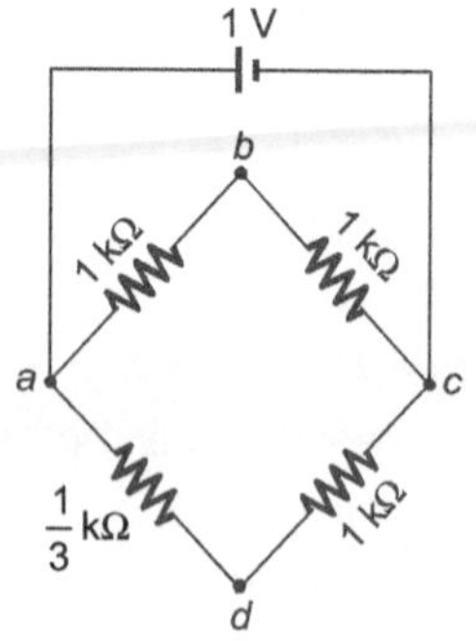

$$V_{th} = V_{bd} = (V_b - V_d)$$

$$= \left[\left(\frac{1}{1+1}\right) \times 1 - \left(\frac{1}{1+\frac{1}{3}}\right) \times 1\right] V$$

$$= \left(0.5 - \frac{3}{4}\right) V = -0.25 \text{ volt}$$

Hence, the Thevenin's equivalent circuit across the galvanometer will be represented as shown below.

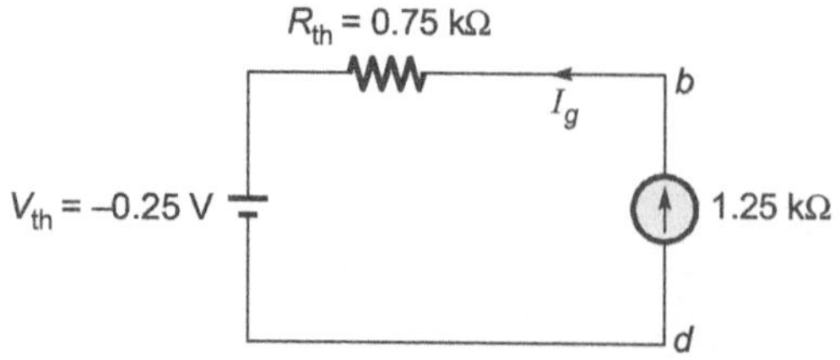

Thus, $-0.25 = (-I_g)\,(0.75\,k + 1.25\,k)$

or, $I_g = \dfrac{0.25}{2 \times 10^3}\ \text{A} = \dfrac{1}{8}\ \text{mA}$

Hence, current through galvanometer,

$$I_g = \frac{1}{8}\ \text{mA}$$

76. (a)

Kelvin bridge overcomes the difficulties that arise in a wheatstone bridge on account of the resistance of the leads and the contact resistances while measuring low value resistances. The Kelvin double bridge incorporates the idea of a second set of ratio arms hence, the name double bridge and the use of four terminal resistor for the low resistance arms. The second set of ratio arms eliminate the effect of connecting lead resistance.

77. (c)

For 3 kHz frequency headphone (Range – 250 Hz to 4 kHz) and tuned amplifier detector (Range – 10 Hz to 100 kHz) will be used.

79. (d)

The bridge circuit for the given balance condition indicates that of a Maxwell's inductance capacitance bridge as shown below.

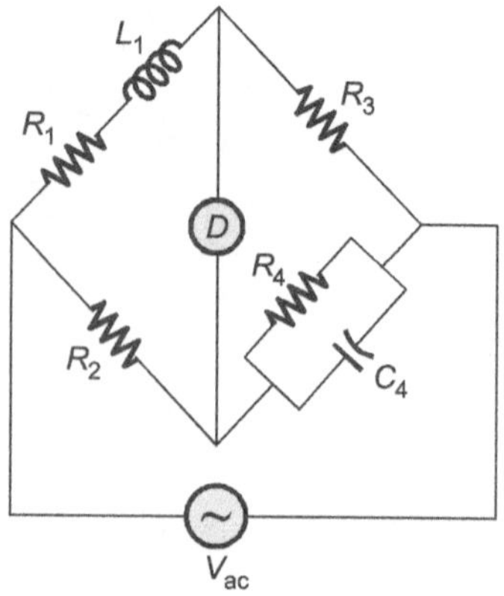

For sliding balance, R_4 and C_4 will be varied since they are not common in the equations of R_1 and L_1.

In order to achieve convergence to balance point, R_4 and C_4 will be varied since they are in the same arm of the bridge.

80. (d)

"Carey Foster slide wire bridge" is a special type of bridge circuit which is used for the measurement of medium resistance by comparing with a standard known resistance. This bridge circuit will give no error due to connecting lead resistances, thermoelectric emfs and contact resistances.

81. (a)

At balance condition,

$$Z_1 Z_x = Z_2 Z_3$$

or, $$R_1\left(R_x - \frac{j}{\omega C_x}\right) = R_2\left(R_3 - \frac{j}{\omega C_3}\right)$$

or, $$\left(R_1 R_x - j\frac{R_1}{\omega C_x}\right) = \left(R_2 R_3 - j\frac{R_2}{\omega C_3}\right)$$

Separating real and imaginary parts, we have:

$$R_x = \frac{R_2 R_3}{R_1} \text{ and } C_x = \frac{R_1}{R_2}\cdot C_3$$

Given, R_1 = 10 kΩ, R_2 = 50 kΩ, R_3 = 100 kΩ, C_3 = 100 μF

Thus, $$R_1 = \frac{50\times 100}{10}\text{ k}\Omega = 500\text{ k}\Omega$$

and $$C_x = \frac{10}{50}\times 100\ \mu\text{F} = 20\mu\text{F}$$

82. (d)

The frequency measured by the Wien bridge is:

$$f = \frac{1}{2\pi\sqrt{R_1 R_2 C_1 C_2}}$$

$$\therefore\quad f = \frac{1}{2\pi\sqrt{4.7\times 10^3\times 20\times 10^3\times 5\times 10^{-9}\times 10\times 10^{-9}}}$$

$$= \frac{1}{2\pi\sqrt{5\times 10^{-10}}\times 4.7}$$

$$= \frac{10^5}{2\pi\sqrt{5\times 4.7}} = \frac{10^5}{2\pi\sqrt{23.5}}$$

$$= 0.0328\times 10^5\text{ Hz} = 3.28\text{ kHz}$$

84. (a)

Maxwell's bridge is used to measure Q-factor of medium Q-coils i.e. $1 < Q < 10$

85. (c)

Anderson bridge is used for the measurement of low Q-coils i.e. measurement of low inductance.

86. (c)

Hay's bridge is used for the measurement of inductance of high Q-coils ($Q > 10$).

87. (d)

Since, $Q = \frac{\omega L}{R}$, therefore for an inductance measurement from a few μH to several henries, Q will be very less (< 1). Thus, Anderson bridge will be used to measure this precise inductance.

88. (a)

Wagner's ground connection device is used to remove all the earth capacitances from the bridge network and inter arm capacitances in the bridge circuit.

89. (b)

The bridge circuit will be drawn as shown below:

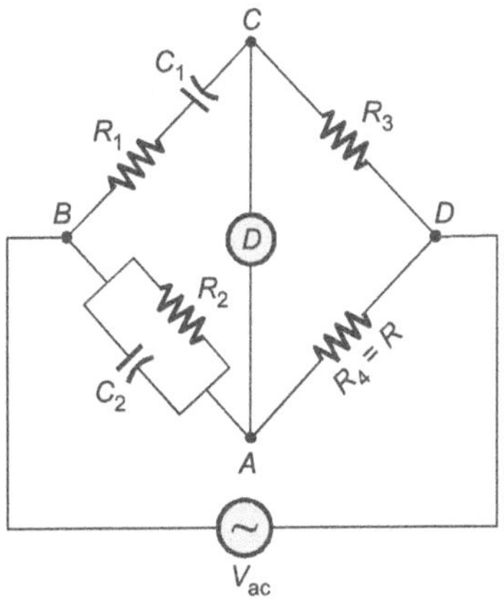

The given bridge configuration is of Wien bridge.

Here, R_1 = 400 Ω, C_1 = 1 μF
R_2 = 800 Ω, C_2 = 0.5 μF
R_3 = 1000 Ω, and $R_4 = R$ = ?

At balance condition, $Z_1 Z_4 = Z_2 Z_3$

or, $$\left(R_1 + \frac{1}{j\omega C_1}\right)R = \frac{R_2 R_3}{(1 + j\omega C_2 R_2)}$$

or, $R(1 + j\omega C_1 R_1 + j\omega R_2 C_2 - \omega^2 C_1 R_1 C_2 R_2)$
$= j\omega C_1 R_2 R_3$

Separating imaginary parts, we have:

$$R(R_1 C_1 + R_2 C_2) = C_1 R_2 R_3$$

or, $$\frac{R_3}{R} = \frac{R_1}{R_2} + \frac{C_2}{C_1}$$

Putting the values, we get:

$$\frac{1000}{R} = \frac{400}{800} + \frac{0.5}{1} = 1$$

or, $R = 1000\ \Omega = 1\ \text{k}\Omega$

Hence, the value of R required to produce balance is

$$R = 1000\ \Omega = 1\ \text{k}\Omega$$

90. (d)

- Vibration galvanometers are extremely useful for power and low audio frequency ranges while headphones are used as detectors at frequencies of 250 Hz and over upto 3 or 4 kHz. Hence, statement-2 is false.
- Tuneable amplifier detectors are the most versatile of the detectors as the transistor amplifier can be tuned electrically and thus can be made to respond to a narrow bandwidth at the bridge frequency.
- An a.c. bridge is balanced only if both the phase angle and magnitude conditions are satisfied. Hence, statement-4 is not true.

92. (c)

- Hay's bridge is used to measure inductance of high Q coils $\left(Q = \frac{\omega L}{R}\right)$. i.e. inductance of coil with large time constant $\left(\frac{L}{R}\right)$.
- Wien's bridge is used as a Notch filter.
- Maxwell's inductance capacitance bridge measures wide range of inductance at power (50 Hz) and audio frequencies (20 - 20 kHz).
- Though, Anderson's bridge is used to measure inductance of low Q-coils, it can also be used to measure accurate value of capacitance in terms of inductance.

93. (d)

The bridge circuit can be drawn as shown below.

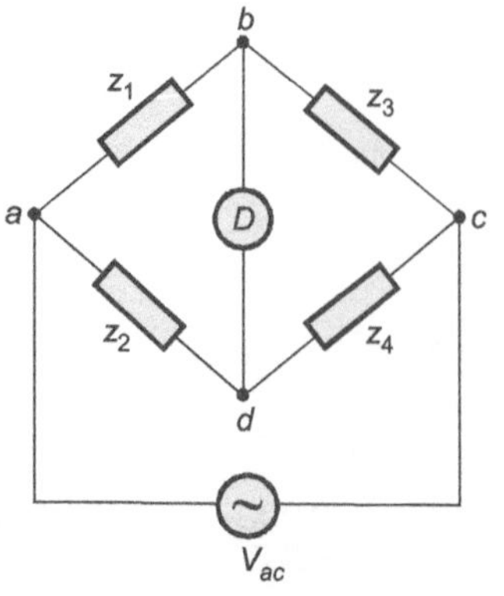

At balance,

$$|Z_1||Z_4| = |Z_2||Z_3| \text{ or } |Z_4| = \frac{|Z_2||Z_3|}{|Z_1|}$$

or, $$|Z_4| = \frac{400 \times 300}{200} = 600\ \Omega$$

Also, $\theta_1 + \theta_4 = \theta_2 + \theta_3$

or, $$\theta_4 = (\theta_2 + \theta_3) - \theta_1$$
$$= (-60° + 0) - 60° = -120°$$

Thus, unknown impedance,

$$Z_4 = 600\angle -120°\ \Omega$$

94. (a)

At balance condition,

$$\frac{R_1}{R_4} = \frac{R_2}{R_3} \text{ or } R_3 = \frac{R_2}{R_1}R_4$$

Given, $R_1 = 1\ \text{k}\Omega$, $R_2 = 10\ \text{k}\Omega$

and $R_4 = 1$ to $100\ \text{k}\Omega$

For $R_4 = 1\ \text{k}\Omega$, $R_3 = \frac{10}{1} \times 1 = 10\ \text{k}\Omega$

For $R_4 = 100\ \text{k}\Omega$, $R_3 = \frac{10}{1} \times 100$

$$= 1000\ \text{k}\Omega = 1\ \text{M}\Omega$$

∴ Required range of $R_3 = 10\ \text{k}\Omega$ to $1\ \text{M}\Omega$

95. (d)

At balance condition,

$$Z \times \frac{1}{j\omega C_4} = R_3 \times \left(R_2 + \frac{1}{j\omega C_2}\right)$$

or, $$Z = jR_3\omega C_4\left(R_2 + \frac{1}{j\omega C_2}\right)$$

$$= jR_2R_3\omega C_4 + R_3\frac{C_4}{C_2}$$

or, $$Z = R_3 \cdot \frac{C_4}{C_2} + jR_2R_3\omega C_4$$

$$= R_{eq} + jX_{eq} = R_{eq} + j\omega L_{eq}$$

Here, $R_{eq} = R_3 \cdot \frac{C_4}{C_2}$ and $L_{eq} = R_2R_3C_4$

Thus, Z is a series combination of a resistor and an inductor.

99. (b)

Due to presence of stray capacitances there is error in the measurement of unknown quantity. Therefore, shielding of bridge is necessary.

100. (c)

Wagner earthing device removes all the earth capacitances from the bridge network. Hence, reason is not correct.

101. (d)

In Hay's bridge, at balance condition,

$$L_1 = \left[\frac{R_2R_3C_4}{1+(1/Q)^2}\right]$$

and $$R_1 = \left[\frac{\omega^2R_2R_3R_4C_4^2}{1+(1/Q)^2}\right]$$

Since $Q = \dfrac{1}{\omega R_4C_4}$ and depends on frequency of source, therefore values of R_1 and L_1 depends on frequency of source.

104. (a)

The bridge circuit is shown below:

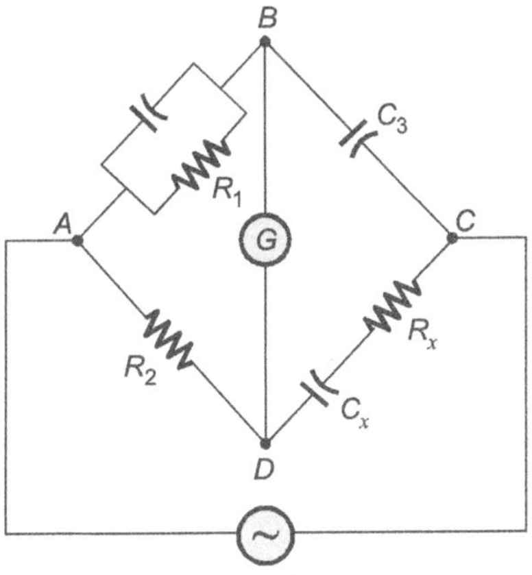

Given, R_1 = 1.2 kΩ, C_1 = 1 μF, R_2 = 4.7 kΩ, C_3 = 1 μF, f = 0.5 kHz = 500 Hz

At balance, $Z_1Z_x = Z_2Z_3$

or, $$\left(\frac{R_1 \times \dfrac{1}{j\omega C_1}}{R_1 + \dfrac{1}{j\omega C_1}}\right) \times \left(R_x + \frac{1}{j\omega C_x}\right) = \frac{R_2}{j\omega C_3}$$

or, $$\left(R_x + \frac{1}{j\omega C_x}\right) = \left(\frac{R_2}{j\omega C_3}\right) \times \left(\frac{1+j\omega R_1C_1}{R_1}\right)$$

$$= \frac{R_2}{j\omega R_1C_3} + \frac{R_2C_1}{C_3}$$

Separating real and imaginary parts,

$$R_x = \frac{R_2C_1}{C_3} \text{ and } \frac{1}{C_x} = \frac{R_2}{R_1C_3}$$

or, $$C_x = \frac{R_1C_3}{R_2}$$

Thus, $$R_x = \frac{4.7\times10^3 \times 1\times10^{-6}}{1\times10^{-6}} = 4.7 \text{ k}\Omega$$

and $$C_x = \frac{1.2}{4.7}\times10^{-6} = 0.255\ \mu\text{F}$$

∴ The dissipation factor,

$$D = \omega R_xC_x$$
$$= 2\pi \times 500 \times 4.7 \times 10^3 \times 0.255 \times 10^{-6}$$
$$= 3.765$$

105. (d)

D.C. and A.C. bridges are examples of comparison type instruments because these instruments measure the unknown quantity by comparison with a standard. Hence, assertion is a false statement. Reason is a true statement.

107. (a)

At balance, we have:

$$R_x = \frac{R_2R_3}{R_4} = \frac{2000\times750}{4000} = 375\ \Omega$$

and $$L_x = R_2R_3C_4$$
$$= 2000 \times 750 \times 0.05 \times 10^{-6}$$
$$= 75 \text{ mH}$$

109. (d)

Since the signal contains 3rd harmonic frequency term, therefore Wien-bridge will not lead to null indication since it is used to measure only fundamental frequency.

111. (c)

The deflecting torque,

$$T_d = NBAI = GI$$

Given, $N = 100$, $A = 20 \times 30 \times 10^{-6}$ m^2

$A = 600\times 10^{-6}$ m^2

$B = 0.1$ Wb/m^2,

$K = 2\times 10^{-6}$ N-m/degree,

$I = 10$ mA

∴ $$T_d = 100 \times 0.1 \times 600 \times 10^{-6} \times 10 \times 10^{-3}$$
$$= 6 \times 10^{-5} \text{ N-m}$$

At equilibrium,

$$T_c = T_d$$

or, $$K\theta = T_d \text{ or } \theta = \frac{T_d}{K}$$

$$\therefore \quad \theta = \frac{6\times10^{-5}}{2\times10^{-6}} = 30° = \frac{\pi}{6} \text{ radian}$$

Thus, the deflection of the pointer = $\frac{\pi}{6}$ radian.

112. (b)

A PMMC uses a moving coil and the magnet is permanent.

114. (d)

Given, $R_m = 100\ \Omega$, $I_m = 1$ mA, $I = 100$ mA

Shunt multiplying factor,

$$m = \frac{I}{I_m} = \frac{100}{1} = 100$$

∴ Required value of shunt resistance is

$$R_{sh} = \frac{R_m}{m-1} = \frac{100}{100-1} = \frac{100}{99}$$

$$= 1.01\ \Omega$$

115. (b)

Shunts used in PMMC instruments are made up of manganin or constantan since they have very low temperature coefficient of resistance.

118. (c)

The sensitivity of 500 μA meter movement is given by,

$$S = \frac{1}{I_m} = \frac{1}{500\ \mu\text{A}} = 2\ \text{k}\Omega/\text{V}$$

The value of the multiplier resistance can be obtained as

$$R_s = (S \times \text{range} - R_m)$$

$$= (2\ \text{k}\Omega/\text{V} \times 50\ \text{V} - 1\ \text{k}\Omega)$$

$$= 100\ \text{k}\Omega - 1\ \text{k}\Omega = 99\ \text{k}\Omega$$

119. (c)

Higher the sensitivity of the voltmeter, lower the loading effect and vice-versa.

120. (b)

Gravity control is not suited for indicating instruments in general and portable instruments in particular because for using this type of control, the instruments must be mounted in level position otherwise there will be a very serious zero error. Gravity control is used in only vertically mounted instruments. Hence, reason is also a true statement but, not a correct explanation of assertion.

121. (a)

The ratio of frictional torque to deflecting torque is a measure of reliability of the instrument because lower is this ratio, higher the deflecting torque and hence higher is the torque to weight ratio. Higher is the torque to weight ratio, higher is the sensitivity of the instrument.

Hence, both assertion and reason are true and reason is a correct explanation of assertion.

122. (d)

Since the value of R_1 is very small and the variable resistor R_3 is in a circuit connected in parallel with R_1, therefore the variation of R_3 will have no effect on the total current I_t. However, variation in R_3 will change the galvanometer current (I_g).

When $R_3 = 350\ \Omega$, the galvanometer current,

$$I_{g_1} = \left(\frac{1}{1+350+R_g}\right)I_t \quad \ldots(i)$$

When $R_3 = 1080\ \Omega$, the galvanometer current,

$$I_{g_2} = \left(\frac{1}{1+1080+R_g}\right)I_t \quad \ldots(ii)$$

From equations (*i*) and (*ii*), we have:

$$\frac{I_{g_1}}{I_{g_2}} = \left(\frac{1+1080+R_g}{1+350+R_g}\right)$$

$$= \left(\frac{1081+R_g}{351+R_g}\right)$$

Now, $\theta \propto I_g$, therefore,

$$\frac{I_{g_1}}{I_{g_2}} = \frac{14}{5}$$

$$\therefore \quad \frac{14}{5} = \left(\frac{1081+R_g}{351+R_g}\right)$$

or, $$4914 + 14R_g = 5405 + 5R_g$$

or, $$R_g = \frac{491}{9} = 54.55\ \Omega \approx 54\ \Omega$$

Thus, the required galvanometer resistance = 54 Ω.

125. (c)

The dc sensitivity,

$$S_{dc} = \frac{1}{I_{fsd}} = \frac{1}{1\times10^{-3}} = 1\,\text{k}\Omega/\text{V}$$

For the full-wave rectifier meter,

$$S_{ac} = 0.9\,S_{dc}$$
$$= 0.9 \times 1\,\text{k}\Omega/\text{V} = 0.9\,\text{k}\Omega/\text{V}$$

∴ The multiplier resistor required is

$$R_s = (S_{ac} \times \text{Range} - R_m)$$
$$= (0.9\,\text{k}\Omega/\text{V} \times 10\,\text{V} - 250)$$
$$= (900 \times 10 - 250) = 8750\,\Omega = 8.75\,\text{k}\Omega$$

127. (a)

A diode used as rectifier in ac voltmeter should have high forward current and low reverse current due to which silicon diodes are generally preferred.

129. (a)

Ohm per volt (Ω/V) = sensitivity

Since $S_{ac} = 0.9\,S_{dc}$ or $S_{dc} = 1.11\,S_{ac}$,

hence, $S_{dc} > S_{ac}$

Therefore, ohm per volt rating on ac ranges as compared to the same rating on dc ranges is less.

130. (d)

- PMMC instruments are used only for dc.
- Since $\theta \propto I$, therefore scale of PMMC instruments is linear.
- The torque to weight ratio is high i.e. weight to torque ratio is less due to which high accuracy is obtained in the measured value of current or voltage.
- Since, PMMC instruments uses a very high value of magnetic field (0.5 Wb/m^2), the errors due to stray magnetic fields are small compared to MI and electrodynamometer instruments.
- Cost of PMMC instruments is higher than MI instruments.

Thus, only statements 2 and 5 are correct.

131. (a)

Moving iron instruments can't give accurate reading at lower end of scale because the moving iron instruments have cramped scale at lower ends of the scale. Hence, both assertion and reason are true and reason is a correct explanation of assertion.

132. (b)

Given, $L = (10 + 5\theta - \theta^2)\,\mu\text{H}$, $I = 5\text{A}$,

$K = 12 \times 10^{-6}$ N-m/rad

$$\therefore \quad \frac{dL}{d\theta} = (5 - 2\theta)\,\mu\text{H/rad}$$

In moving iron instruments, the deflection of pointer is given by

$$\theta = \frac{I^2}{2K}\frac{dL}{d\theta}$$
$$= \frac{25}{2\times12\times10^{-6}}(5-2\theta)\times10^{-6}$$

or, $24\,\theta = 25(5 - 2\theta)$

or, $240 = 125 - 50\theta$ or $\theta = \dfrac{125}{74}$

$= 1.69$ radian

or, $\theta = 1.69 \times \dfrac{180}{\pi}$ degrees $= 96.8°$

133. (c)

In order that there is no frequency error, value of capacitance to be connected across the series multiplier resistance R_s is

$$C = 0.41\frac{L}{R_s^2}$$

Now, total resistance of instrument circuit

$$= \frac{150}{0.05} = 3000\,\Omega$$

Given, resistance of coil,

$$R_m = 400\,\Omega$$

∴ Series multiplier resistance,

$$R_s = 3000 - 400 = 2600\,\Omega$$

Thus, $C = 0.41 \times \dfrac{0.75}{(2600)^2}$

$= 0.0455 \times 10^{-6}\,\text{F} = 45.5\,\text{nF}$

134. (a)

The meter uses a full wave rectifier circuit and it indicates a value of 2.22 V. The form factor for full wave rectifier sinusoidal waveform is 1.11.

∴ Average value of voltage is:

$$V_{avg} = \frac{\text{rms value of voltage reading}}{\text{form factor}}$$
$$= \frac{2.22}{1.11} = 2\text{ volt}$$

For a triangular waveform, peak value of voltage is

$$V_m = 2\,V_{avg} = 4\text{ V}$$

∴ rms value of voltage is

$$V = \frac{V_m}{\sqrt{3}} = \frac{4}{\sqrt{3}} = 2.31\text{ volt}$$

$$\therefore\ \%\text{ error} = \left(\frac{2.22-2.31}{2.31}\right)\times 100 = -3.9\%$$

137. (d)

Assertion is a false statement because the power requirement of electrostatic instruments are quite low.

139. (a)

Moving iron instruments uses air-friction damping due to low operating magnetic field.

141. (a)

For half-wave rectifier meter,

$$S_{ac} = 0.45\,S_{dc}$$
$$= 0.45 \times 10\text{ k}\Omega/\text{V}$$
$$= 4.5\text{ k}\Omega/\text{V}$$

143. (c)

For an electrodynamometer ammeter, deflection, θ is given by:

$$\theta = \frac{I^2}{K}\frac{dM}{d\theta} \quad\text{or}\quad \frac{dM}{d\theta} = \frac{K\theta}{I^2}$$

or,
$$\frac{dM}{d\theta} = \frac{0.1\times10^{-6}\times110}{(10)^2}$$
$$= 0.11\times10^{-6}\text{ H/rad}$$
$$= 0.11\times10^{-6}\times\frac{\pi}{180}\text{ H/degree}$$
$$= 0.00192\ \mu\text{H/degree}$$

∴ Final inductance = Initial inductance + change in inductance

$$= 2 + 0.00192 \times 110 = 2.21\ \mu\text{H}$$

144. (b)

Hot wire instrument is an electrothermic type of instrument because the electrothermic instruments indicate the value of current when some element of its circuit is heated by flow of current. Hot wire instrument is one of them where current to be measured is passed through a fine platinum iridium wire and when the wire is heated it expands and the deflection of the pointer indicates the measured value of current.

Reason is also a true statement but, not a correct explanation of assertion.

147. (c)

For an electrodynamometer ammeter,

$$T_d = I^2\frac{dM}{d\theta}$$

Given, $I = 50$ mA, $M = 5\sin(\theta - 30°)$ mH, $\theta = 60°$

$$\therefore\quad \frac{dM}{d\theta} = 5\cos(\theta-30°)\text{ mH/deg}$$

Now,
$$\left.\frac{dM}{d\theta}\right|_{\theta=60°} = 5\cos(60°-30°)\text{ mH/deg}$$
$$= 5\cos30°\text{ mH/deg}$$

$$\therefore\quad T_d = (50\times10^{-3})^2\times5\cos30°\times10^{-3}\text{ N-m}$$
$$= 1.0825\times10^{-5}\text{ N-m} \approx 10.8\ \mu\text{N-m}$$

149. (a)

- A transfer instrument is one that may be calibrated with a d.c. source and then used without modification to measure a.c. This requires the transfer type instrument to have same accuracy for both d.c. and a.c.
- The electrodynamometer type of instruments is a kind of transfer instrument.
- Statement-3 is correct.
- PMMC instruments are more accurate than *MI* and *EDM* instruments because of higher torque to weight ratio in them.
- PMMC instruments are more costly than moving iron instruments.

Hence, statements 1, 4 and 5 are not correct.

150. (d)

Hot-wire ammeter (A_1) and moving iron ammeter (A_3) will read rms value of current, while moving coil ammeter (A_2) will read average value of current.

Given, $i(t) = 0.5 + 0.3\sin\omega t - 0.2\sin3\omega t$

Now,
$$i_{rms} = \sqrt{(0.5)^2+\left(\frac{0.3}{\sqrt{2}}\right)^2+\left(\frac{0.2}{\sqrt{2}}\right)^2}$$
$$= 0.5612\text{ A} \approx 0.560\text{ mA}$$

Also, $i_{avg} = 0.5\text{ A} = 500\text{ mA}$

Thus, A_1 and A_3 both will read 560 mA while A_2 will read 500 mA.

151. (c)

In moving-iron instrument due to presence of moving-iron disc hysteresis loss occurs in it which results in error in the meter reading.

152. (d)

In an electrostatic voltmeter, the deflecting torque or the developed torque is given by

$$T_d = \frac{1}{2}V^2\frac{dC}{d\theta}$$

where, dC = change in capacitance
and $d\theta$ = change in deflection

154. (c)

Moving iron type ammeter will read rms value of current.

Given, $i = (10 + 10 \sin t)$ A

$$\therefore \quad i_{rms} = \sqrt{10^2 + \left(\frac{10}{\sqrt{2}}\right)^2}$$

$$= \sqrt{100+50} = \sqrt{150}\text{ A}$$

156. (b)

In a dynamometer type wattmeter, the fixed coil is current coil and moving coil is a voltage coil or pressure coil.

157. (a)

Since PMMC reads average value of voltage and current, therefore for high frequency ac signal reading will be zero.

161. (c)

An energymeter is an induction and an integrating type instrument.

162. (a)

To reduce the error in the wattmeter reading due to pressure coil inductance, the wattmeters are compensated by connecting a capacitor in parallel with a portion of series resistance (multiplier) as shown below.

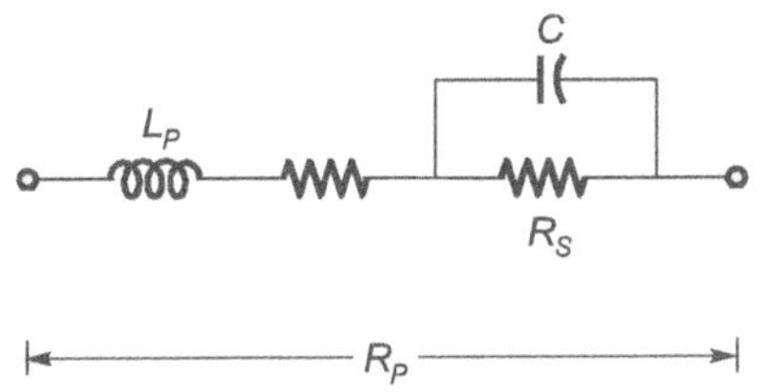

163. (b)

At AF (audio frequency) and RF (radio frequency) it is easier to measure voltage, current and impedance, than to measure power. Direct measurement of power is not done in this range. Here, power is calculated from the equation $P = E^2/R$ or $P = I^2 R$.

165. (c)

For the energy meter to record true energy the pressure coil should be highly inductive (i.e. $\Delta = 90°$). Thus, for the angle between the shunt magnet flux (ϕ_p) and the supply voltage to be 90°, a lagging coil is placed between the potential coil and the rotating aluminium disc. Lag adjustment sometimes is called power factor or quadrature or inductive load adjustment.

166. (d)

$$\text{True power} = VI\cos\phi$$
$$= 250 \times 12 \times 1 = 3000\text{ W}$$

For connection (i), the error is caused due to power loss in the current coil.

Power loss in current coil

$$= I^2 r_a = (12)^2 \times 0.1$$
$$= 14.4 \text{ watt}$$

$$\therefore \quad \%\text{ error} = \frac{14.4}{3000}\times 100 = 0.48\%$$

For connection (ii), error is caused due to power loss in pressure coil.

Power loss in pressure coil

$$= \frac{V^2}{R_P} = \frac{(250)^2}{6500} = 9.6 \text{ watt}$$

$$\therefore \quad \%\text{ error} = \frac{9.6}{3000}\times 100 = 0.32\%$$

167. (a)

From two-wattmeter method of measurement of power, we have:

$$W_1 + W_2 = \sqrt{3}\,V_L I_L \cos\phi$$

and $$W_1 - W_2 = V_L I_L \sin\phi$$

$$\therefore \quad \frac{W_1 - W_2}{W_1 + W_2} = \frac{\tan\phi}{\sqrt{3}}$$

or, $$\tan\phi = \sqrt{3}\,\frac{(W_1 - W_2)}{(W_1 + W_2)}$$

or, $\cot\phi = \dfrac{W_1+W_2}{\sqrt{3}(W_1-W_2)}$

or, $\phi = \cot^{-1}\left[\dfrac{W_1+W_2}{\sqrt{3}(W_1-W_2)}\right]$

$\therefore$ p.f. $= \cos\phi$

$$= \cos\left[\cot^{-1}\frac{(W_1+W_2)}{\sqrt{3}(W_1-W_2)}\right]$$

(where, $W_1 = P_1$ and $W_2 = P_2$)

168. (b)

Let V and I be the phase voltage and phase current respectively.

Phase voltage, $V = \dfrac{400}{\sqrt{3}}$ Volt

Phase current, $I = 30$ A

When the current coil is connected in the red phase and it's voltage coil is connected between the neutral and red phase as shown in figure below, the wattmeter will read the power in one phase.

$\therefore$ $W_1 = VI\cos\phi = 5.54\times10^3$ Watt

or, $\cos\phi = \dfrac{5.54\times10^3}{\left(\dfrac{400}{\sqrt{3}}\times30\right)} = 0.8$

Thus, $\sin\phi = 0.6$

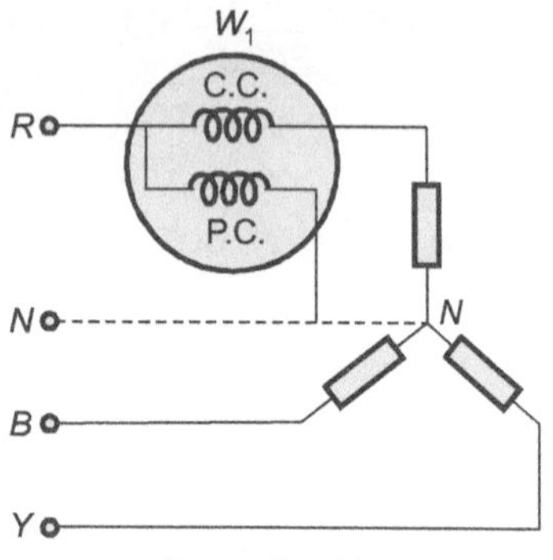

Connection (*i*)

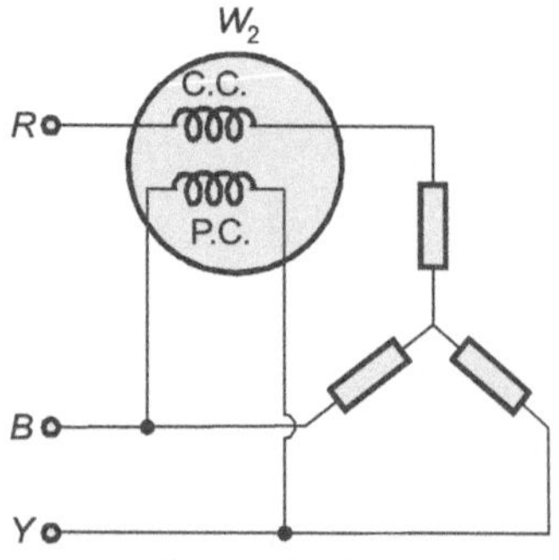

Connection (*ii*)

Now, when the current coil is connected in the red phase and the pressure coil circuit is connected across the yellow and the blue phases as shown in above figure (connection (*ii*)), the reading of the wattmeter is

$$W_2 = V_{YB}I_R\cos(90°-\phi)$$

(Angle between V_{BY} and $I_R = (90° - \phi)$ using phasor diagram)

$$= \sqrt{3}\,VI\sin\phi$$

Here, $VI\sin\phi$ = reactive power per phase

$\therefore$ $W_2 = Q = \sqrt{3}\times\dfrac{400}{\sqrt{3}}\times30\times0.6$

$= 7200$ Var $= 7.2$ kVar

169. (c)

From two-wattmeter measurement of power, we have:

$W_1 = V_LI_L\cos(30+\phi)$

and $W_2 = V_LI_L\cos(30-\phi)$

- When $\cos\phi = 0$ leading, $\phi = -90°$

$\therefore$ $W_1 = V_LI_L\cos(30-90°) = \dfrac{V_LI_L}{2}$

and $W_2 = V_LI_L\cos(30+90°) = \dfrac{-V_LI_L}{2}$

Thus, $W_1 = -W_2$ or $W_1 + W_2 = 0$

- When $\cos\phi = 0.866$ lagging, $\phi = 30°$

$\therefore$ $W_1 = V_LI_L\cos(30°+30°) = \dfrac{V_LI_L}{2}$

and $W_2 = V_LI_L\cos(30°-30°) = V_LI_L$

Thus, $W_2 = 2W_1$ or, $W_1 - 2W_2 = 0$

- When $\cos\phi = 1$, $\phi = 0°$

$\therefore$ $W_1 = V_LI_L\cos 30°$

and $W_2 = V_LI_L\cos 30°$

Thus, $W_1 = W_2$

- When $\cos\phi = 0.3$ lagging, $\phi = 72.54°$

$\therefore$ $W_1 = V_LI_L\cos(30°+72.54°)$

$= -$ve value

and $W_2 = V_LI_L\cos(30°-72.54°)$

$= +$ve value

170. (d)

When the energy meter is tested under high current rating, there is lot of waste of power. To reduce this unwanted power loss, Phantom or Fictitious loading is done.

171. (c)

Braking torque provided by the permanent magnet is given by

$$T_B \propto N\phi_m^2 d$$

Where, N = speed of the disc in rpm

ϕ_m = maximum value of flux of the permanent magnet

d = distance of the permanent magnet from the centre of the revolving disc

172. (a)

Given, energy meter constant,

K = 1800 revolutions per kWh

Actual energy or true energy consumed by the energy meter at half load during 138 seconds is:

$$W_T = VI\cos\phi \times t \times 10^{-3}\text{ kWh}$$

$$= 230 \times 5 \times 1 \times \frac{138}{3600} \times 10^{-3}\text{ kWh}$$

$$= 44.08 \times 10^{-3}\text{ kWh}$$

Also, measured energy or energy recorded is

$$W_m = \frac{\text{Number of revolutions made by disc}}{\text{Energy meter constant}}$$

$$= \frac{80}{1800} = 44.4 \times 10^{-3}\text{ kWh}$$

Hence, % error $= \left(\frac{W_m - W_T}{W_T}\right) \times 100$

$$= \left(\frac{44.44 - 44.08}{44.08}\right) \times 100$$

$$= 0.8167$$

Since, error is +ve, therefore meter runs fast.

174. (b)

There are four main parts of the operating mechanism of an induction type energy meter:

- **Braking system:** A permanent magnet positioned near the edge of the aluminium disc forms the braking system.
- **Driving system:** The driving system of the meter consists of two series and shunt magnet (i.e. electro-magnets).
- **Moving system:** The moving system consists of an aluminium disc mounted on a light alloy shaft.
- **Registering system:** The function of a registering or counting mechanism is to record continuously a number which is proportional to the revolutions made by the moving system which is achieved with the help of train of reduction gears.

175. (d)

Energy supplied $= VI\cos\phi \times t \times 10^{-3}$ kWh

$= 230 \times 4 \times 1 \times 6 \times 10^{-3}$

$= 5.52$ kWh

Energy meter constant,

$$K = \frac{\text{Number of revolutions}}{\text{Energy supplied}}$$

$$= \frac{2208}{5.52} = 400\text{ rev/kWh}$$

Energy consumed when the meter makes 1472 revolutions

$$= \frac{1472}{400} = 3.68\text{ kWh}$$

Now, energy consumed

$= VI\cos\phi \times t \times 10^{-3}$ kWh

$= 230 \times 5 \times \cos\phi \times 4 \times 10^{-3}$ kWh

$= 3.68$ kWh (As obtained above)

or, $$\cos\phi = \frac{3.68}{230 \times 5 \times 4 \times 10^{-3}} = 0.8$$

Hence, power factor of the load = 0.8

176. (c)

Here, $\cos\phi = 1$ (for d.c. watthour meter)

Energy supplied to the energy meter in kWh

$= VI\cos\phi \times t \times 10^{-3}$

$= 220 \times 5 \times 1 \times 10^{-3} = 1.1$ kWh

∴ Energy supplied to the energy meter per minute $= \frac{1.1}{60}$ kWh

Given, energy meter constant,

K = 3275 revolutions/kWh

∴ Disc speed in rpm = Energy supplied per minute × Energy meter constant

$$= \frac{1.1}{60} \times 3275\text{ rpm} = 60.04\text{ rpm}$$

$$= \frac{60.04}{60}\text{ rps} = 1.00069\text{ rps} \approx 1\text{ rps}$$

177. (a)

An increase in temperature is accompanied by a rise in resistance of all copper and aluminium parts.

180. (d)

Given, $R_P = 400\ \Omega$,
$R_S = 10,000\ \Omega$
$C_C = 20$ pF

Self inductance of the pressure coil is given by

$$L_P = C_C \times R_S^2$$
$$= 20 \times 10^{-12} \times (10^4)^2$$
$$= 20 \times 10^{-4} = 2 \text{ mH}$$

181. (d)

Measurement of power factor with the help of two-wattmeter method can be done only if load is balanced. Hence, assertion is not true. Reason is a correct statement.

182. (b)

Total power consumed in the circuit
= Average active power

$$= 5\times10+\frac{1}{2}\times2\times5\cos(30°-0°)+0$$

$$= 50 + 5\cos 30° = 50+5\times\frac{\sqrt{3}}{2} = 54.33 \text{ watt}$$

183. (a)

In an electrodynamometer type wattmeter:

- Both fixed and moving coils are air cored.
- Air friction damping is used due to low value of operating magnetic field.
- The moving coil is connected across the load.
- The fixed coil is connected is series with the load.

184. (c)

A wattmeter must be used for the measurement of power in a.c. circuit instead of merely an ammeter and a voltmeter because the power factor can't be measured with the help of voltmeter and ammeter.

Wattmeter measures average active power not average reactive power. Hence, reason is not a correct statement.

185. (a)

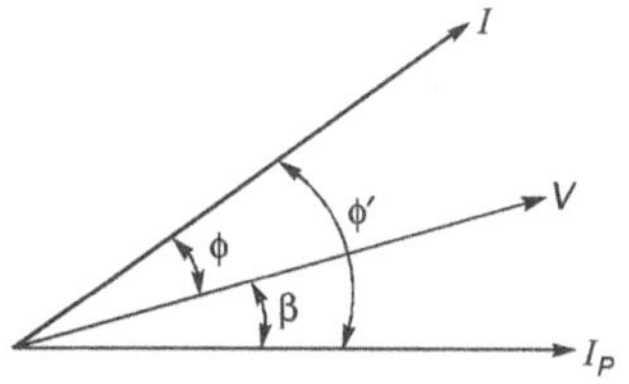

In the above phasor diagram

I = load current

and I_P = Pressure coil current

$\phi' = (\phi + \beta)$

We know that, measured power is

$$W_m \propto \left[\frac{VI\cos\beta\cdot\cos(\phi+\beta)}{R_S}\right]$$

At leading power factor of load, angle $(\phi + \beta)$ is increased, therefore $\cos(\phi + \beta)$ is reduced. Hence, wattmeter reads low at leading p.f. load.

186. (a)

We have: $$\frac{W_T}{W_m} = \left[\frac{\cos\phi}{\cos\phi\cdot\cos(\phi-\beta)}\right]$$

$$= \left[\frac{1}{1+\tan\phi\cdot\tan\beta}\right]$$

or, $$\frac{W_m}{W_T} = (1+\tan\phi\cdot\tan\beta)$$

188. (d)

Phantom loading consists of supplying the pressure coil circuit (or moving coil circuit) from a circuit of required normal voltage, and the current circuit (or fixed coil circuit) from a separate low voltage supply. Hence, assertion is not true.

Reason is a correct statement.

189. (c)

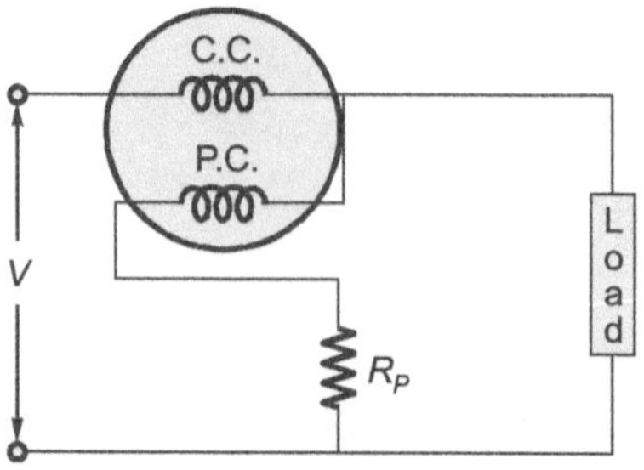

Power loss in the voltage coil $= \dfrac{200^2}{2000} = 20$ watt

Instrument reading = 250 watt

$\therefore$ Actual load = 250 − 20

= 230 watt

190. (b)

$$\tan\phi = \sqrt{3}\left(\frac{W_A - W_B}{W_A + W_B}\right) = \sqrt{3}\left(\frac{3+1}{3-1}\right)$$

$$= \sqrt{3}\times\frac{4}{2} = 2\sqrt{3}$$

$\therefore$ $1 + \tan^2\phi = 1 + 12 = 13$

or, $\sec^2\phi = 13$

or, $\cos\phi = \frac{1}{\sqrt{13}} = 0.277$

191. (a)

Cripping is due to over compensation provided to the running friction. To avoid it holes are made on opposite side of the disc of the energy meter.

192. (c)

When power factor = 0.5 (ϕ = 60°), then one wattmeter will read zero. The load will be lagging in nature.

$W_1 = V_L I_L \cos(30 + \phi)$

and $W_2 = V_L I_L \cos(30 - \phi)$

For $\phi = 60°$, $W_1 = 0$ watt

193. (a)

Meter constant,

K = 400 revolutions per kWh

$$= 400 \times \frac{240\times 10\times 0.8}{1000} \text{ rev/hr}$$

= 768 revolution per hour

$$= \frac{768}{60} \text{ rpm} = 12.8 \text{ rpm}$$

Hence, speed of meter disc = 12.8 rpm

194. (c)

Post deflection acceleration is used to increase the velocity of the electron beam.

196. (d)

Let the signal be as shown below.

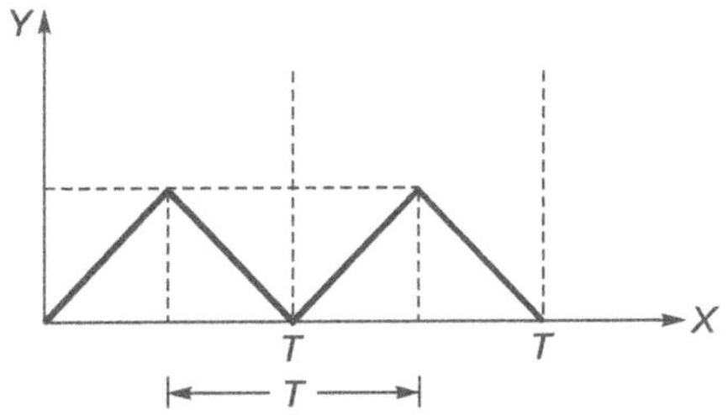

Given, T corresponds to 2 cm

= 2 division on X-axis

Since, 1 division = 1 ms on X-axis

therefore, T = 2 ms

$$\therefore \quad f = \frac{1}{T} = \frac{1}{2\times 10^{-3}} = 500 \text{ Hz}$$

197. (a)

A dual trace CRO uses two pairs of vertical deflecting plates and one electron gun.

201. (b)

Here, no. of divisions occupied for $V_{P\text{-}P} = 3$.

$$V_{P\text{-}P} = \left(\frac{\text{Volts}}{\text{div}}\right)\times(\text{No. of divisions})$$

$= 0.5 \times 3$

$= 1.5$ volt

202. (a)

For the waveform of problem-7, the time period of the signal is calculated using the equation

$$T = \left(\frac{\text{time}}{\text{div}}\right)\times\left(\frac{\text{No. of division}}{\text{Cycle}}\right)$$

or, $T = 2\ \mu\text{s} \times 4 = 8\ \mu\text{s}$

$\therefore$ Frequency,

$$f = \frac{1}{T} = \frac{1}{8\times 10^{-6}} = 125 \text{ kHz}$$

204. (d)

We have:

$$\frac{f_y}{f_x} = \frac{\text{No. of horizontal tangencies}}{\text{No. of vertical tangencies}}$$

or, $\frac{f_y}{f_x} = \frac{5}{2}$

or, $f_y = \frac{5}{2}f_x = \frac{5}{2}\times 1000$

= 2500 Hz = 2.5 kHz

206. (b)

Given, L = 15 cm, l_d = 2 cm, d = 1 cm

and E_a = 500 V

We know that deflection sensitivity is

$$S = \frac{Ll_d}{2dE_a} = \frac{15\times 2}{2\times 1\times 500}$$

= 0.03 cm/V

208. (d)

Post acceleration is needed to increase the brightness of the beam if the frequency of the signal is more than 10 MHz.

209. (c)

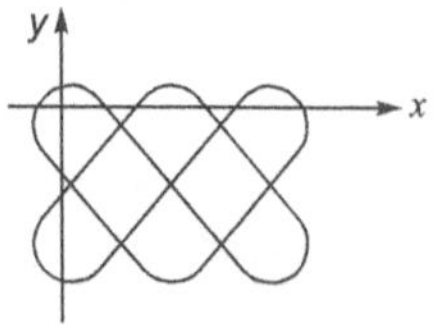

No. of horizontal tangencies = 6

No. of vertical tangencies = 4

Now, $\frac{f_y}{f_x} = \frac{\text{No. of horizontal tangencies}}{\text{No. of vertical tangencies}}$

$= \frac{6}{4} = \frac{3}{2}$

Given, $f_x = 51$ Hz

$\therefore$ $f_y = \frac{3}{2}f_x = \frac{3}{2} \times 51 = 76.5\text{ Hz}$

210. (d)

Deflection, $D = \frac{L l_d E_d}{2dE_a}$

$\therefore$ Applied voltage to deflecting plates,

$$E_d = \frac{2dE_aD}{Ll_d}$$

Given, $d = 5$ mm, $E_a = 2000$ V,
$D = 3$ cm, $L = 30$ cm,
$l_d = 2$ cm

$\therefore$ $E_d = \frac{2 \times 5 \times 10^{-3} \times 2000 \times 3 \times 10^{-2}}{30 \times 10^{-2} \times 2 \times 10^{-2}}$

$= 100$ V

Thus, input voltage required for a deflection of

3 cm $= \frac{E_d}{\text{gain}} = \frac{100}{100} = 1\text{ volt}$

211. (a)

The velocity of electron is given by

$$v_{ax} = \sqrt{\frac{2eE_a}{m}}$$

Here, m = mass of electron = 9.1×10^{-31} kg and charge on electron = e = 1.6×10^{-19} C

$\therefore$ $v_{ax} = \sqrt{\frac{2 \times 1.6 \times 10^{-19} \times 800}{9.1 \times 10^{-31}}}$

$= 16.8 \times 10^6$ m/s

$= 16.8$ km/s

213. (c)

The signal, whose frequency is to be measured, is applied to vertical deflecting plates i.e. Y-plates. Hence, reason is a false statement.

215. (d)

- The signal to be examined are usually applied to the vertical deflection plates. Hence, statement-1 is not correct.
- The horizontal deflection plates are fed by a sweep voltage that provides a time base. Hence, statement-2 is false.
- Statement-3 is correct.
- Since, magnetostatic focusing technique has higher sensitivity than the electrostatic focusing technique therefore, magnetostatic focusing technique is superior to electrostatic focusing technique.

Thus, statements 1, 2 and 4 are not correct.

216. (a)

We know that,

Bandwidth × Rise time = 0.35

or, Rise time $= \frac{0.35}{\text{Bandwidth}}$

Given, BW = 100 MHz

$\therefore$ Rise time $= \frac{0.35}{100 \times 10^6}$

$= 0.35 \times 10^{-8}$ sec

$= 3.5$ ns

217. (a)

On account of difference of potential between focusing anode and the two accelerating anodes, a non-uniform field exists on each of the two ends of the focusing anode. The equipotential surfaces, thus, form a "Double concave lens."

218. (b)

Both are true statement but reason is not correct. Since a dual trace oscilloscope has one electron guns or beams and two separate vertical input channels.

220. (b)

At high frequencies, CRO is used to measure the voltage.

223. (d)

Given, $y(t) = 10 \sin 100t$

$= 10 \cos (100t - 90°)$

and $x(t) = 10 \cos 100t$

Since, the phase difference between the given two signals is 90°, therefore the screen will display a circle.

224. (c)

$$\frac{f_x}{f_y} = \frac{\text{No. of vertical tangents}}{\text{No. of horizontal tangents}} = \frac{2}{3}$$

225. (b)

Given, time base setting = 5 m sec/div

For 10 division on horizontal scale,

time = 50 m sec

Given, $\omega = 314$ rad/sec

or, $2\pi f = 314$ or $f = 50$ Hz

∴ Number of cycles of signal displayed on the screen

$= 50 \times 10^{-3} \times 50$

$= 2.5$ cycles

226. (c)

Work done by the electric field = eV

Now, kinetic energy of electron $= \frac{1}{2} mv^2 = eV$

or, $$v = \sqrt{\frac{2eV}{m}} = \sqrt{\frac{2 \times 1.6 \times 10^{-19} \times 1000}{9.1 \times 10^{-31}}}$$

$= 0.1875 \times 10^8$ m/s

227. (b)

Material used	Temperature range
Iron – Constantan	–200 to 900°C
Platinum – Platinum + Rhodium	0 to 1500°C
Copper – Constantan	–200 to 400°C
Chromel – Alumel	–200 to 900°C

228. (c)

Thermocouple can measure temperature in range of –270°C to 2700°C with high speed of response and with quite good accuracy.

231. (b)

Given, $G_f = 2$, Strain $= 1 \times 10^{-6}$, $R = 130\ \Omega$

We know that, $$G_f = \frac{\Delta R / R}{\text{Strain}}$$

or, $\Delta R = G_f \times \text{strain} \times R$

$= 2 \times 1 \times 10^{-6} \times 130$

$= 260\ \mu\Omega$

232. (c)

Given, inductance with air gap length of 1 mm is

$L = 2500\ \mu$H

Length of air-gap when a displacement is applied to the target

$= 1.00 - 0.04 = 0.96$ mm

Now, inductance is inversely proportional to the length of air-gap.

Therefore, inductance (L) with a gap length of 0.96 mm

$= L + \Delta L$

$$= 2500\ \mu\text{H} \times \frac{1}{0.96\ \text{mm}} = 2604\ \mu\text{H}$$

Thus, change in inductance is

$\Delta L = 2604\ \mu\text{H} - 2500\ \mu\text{H} = 104\ \mu\text{H}$

233. (d)

LVDT and potentiometer are passive transducers because they require an external supply for giving output.

234. (a)

Since 0.5 m core displacement produces 5.2 V, therefore 0.45 m core movement will produce

$$\frac{0.45 \times 5.2}{0.5} = 4.68 \text{ volt}$$

Similarly, a –0.30 m core movement produces

$$\frac{-0.30 \times -5.2}{-0.5} = -3.12 \text{ V}$$

∴ Change in output voltage

$= -3.12\text{ V} - 4.68\text{ V} = -7.8\text{ V}$

235. (b)

We know that, resistance at any temperature (T) is

$R = R_0 [1 + \alpha (T - T_0)]$

$= R_0 [1 + \alpha \Delta T]$

∴ $R = 180 [1 + 0.00392 (60 - 20)]$

$= 180 [1 + 0.1568]$

$= 208.22\ \Omega \approx 208\ \Omega$

236. (b)

RTD is a passive transducer because it requires an external supply for giving output.

240. (c)

We know that, gauge factor

$$G_f = 1 + 2\gamma \quad \text{(where, } \gamma = \text{Poisson's ratio)}$$

$$\therefore \quad \gamma = \frac{G_f - 1}{2} = \frac{4.2-1}{2} = \frac{3.2}{2} = 1.6$$

244. (a)

Both Assertion and Reason are true and reason (R) is a correct explanation of assertion (A) because when a voltage is applied to a peiezoelectric crystal (across its surfaces), it changes its dimension causing a mechanical displacement.

246. (c)

We know that, modulus of elasticity,

$$Y = \frac{\text{Stress}}{\text{Strain}}$$

$$\therefore \quad \text{strain} = \frac{\text{Stress}}{Y} = \frac{100 \times 10^6}{200 \times 10^9}$$

$$= 0.5 \times 10^{-3} = 500 \times 10^{-6}$$

$$= 500 \text{ micro strain}$$

Now, $G_f = \dfrac{\Delta R / R}{\text{strain}}$

or, $\dfrac{\Delta R}{R} = G_f \times \text{strain}$

$$= 2 \times 500 \times 10^{-6} = 10^{-3}$$

$$\therefore \text{ \% change in resistance} = \left(\frac{\Delta R}{R} \times 100\right)\%$$

$$= (10^{-3} \times 100)\% = 0.1\%$$

247. (d)

Thermocouples have a low accuracy and hence they cannot be used for precision work.

248. (b)

Differential output will increase sensitivity.
Due to differential output, output will be less affected by external magnetic fields and variation due to temperature changes will be reduced.
Hence, only statements 1 and 4 are correct.

249. (a)

$$\text{\%age linearity} = \pm\frac{0.003}{15} \times 100 = \pm 0.2\%$$

250. (c)

The applied pressure is, $P = \dfrac{V_0}{gt}$

Here, V_0 = voltage developed = 100 V
t = thickness = 1.5 mm
g = 0.055 V-m/N

$$\therefore \quad P = \frac{100}{0.055 \times 1.5 \times 10^{-3}} \text{ N/m}^2$$

$$= 1.2 \text{ MN/m}^2$$

Thus, force acting on the crystal

$$= P.A = 1.2 \times 10^6 \times 5 \times 5 \times 10^{-6}$$

$$= 30 \text{ N}$$

251. (d)

Output voltage developed is

$$V_H = \frac{R_H B I}{t}$$

Here, R_H = Hall's coefficient
$= -1 \times 10^{-6}$ V-m/(A-Wbm^{-2})
$B = 0.5$ Wb/m^2
$I = 3$ A
t = width of specimen = 2 mm

$$\therefore \quad V_H = \frac{-1 \times 10^{-6} \times 3 \times 0.5}{2 \times 10^{-3}}$$

$$= -0.75 \times 10^{-3} \text{ V} = -0.75 \text{ mV}$$

254. (a)

Due to very high negative temperature co-efficient of resistance, thermistor acts as an ideal temperature transducer.

256. (c)

Gauge factor,

$$G_f = \frac{\Delta R / R}{\Delta L / L}$$

or, $G_f \propto \Delta R \propto$ change in resistance

The gauge factor of the strain gauge should be high so that for a large change in resistance for a particular strain results in high sensitivity. Hence reason is a false statement.
High value of gauge factor is required for excellent and reproducible result in the measurement of strain gauge output which is the required correct reason for assertion.

257. (b)

Both assertion and reason are true. However, the reason for LVDT to be a frictionless device is that there is no physical contact between the movable core and coil structure.

258. (b)

We know that charge sensitivity,

$$d = \varepsilon g$$

Given, for Barium titanate,

$$g_b = 12 \times 10^{-3} \text{ V-m/N}$$

and $\varepsilon_b = 12.5 \times 10^{-9}$ F/m

Also, for quartz,

$$g_q = 50 \times 10^{-3} \text{ V-m/N}$$

and $\varepsilon_q = 40.6 \times 10^{-12}$ F/m

So, charge sensitivity of Barium titanate is

$$d_b = \varepsilon_b g_b$$
$$= 12 \times 10^{-3} \times 12.5 \times 10^{-9} \text{ C/N}$$
$$= 150 \text{ pC/N}$$

and charge sensitivity of quartz is

$$d_q = \varepsilon_q g_q$$
$$= 40.6 \times 10^{-12} \times 50 \times 10^{-3} \text{ C/N}$$
$$= 2030 \times 10^{-15} \text{ C/N} \approx 2 \text{ pC/N}$$

263. (a)

Thermistors have negative temperature coefficient of resistance i.e. resistance decreases with temperature due to which it is used for the measurement of flow, thermal conductivity and liquid level.

264. (b)

LVDT and strain-gauge measures linear displacement.

266. (a)

The ratio marked on the instrument transformer is the nominal ratio (K_n) which is defined as

$$K_n = \frac{\text{Rated primary winding current}}{\text{Rated secondary winding current}}$$

(for a C.T.)

$$\text{or,} \quad K_n = \frac{\text{Rated primary winding voltage}}{\text{Rated secondary winding voltage}}$$

(for a P.T.)

267. (c)

Nominal ratio, $K_n = \dfrac{1000}{5} = 200$

In absence of any other data, the turns ratio is assumed to be equal to the nominal ratio.

$\therefore$ Turns ratio, $n = K_n = 200$.

Primary winding turns,

$$N_S = nN_P = 200$$

Secondary impedance = 1.6 Ω

Secondary induced voltage,

$$E_S = 5 \times 1.6 = 8 \text{ V}$$

$\therefore$ Primary induced voltage,

$$E_P = \frac{E_S}{n} = \frac{8}{200} \text{ V}$$

Loss component of exciting current,

$$I_e = \frac{\text{Iron loss}}{E_P} = \frac{1.5}{8/200}$$
$$= 37.5 \text{ A}$$

The magnetizing component current,

$$I_m = \frac{\text{Magnetizing mmf}}{N_P}$$
$$= \frac{100}{1} = 100 \text{ A}$$

As the burden is purely resistive and therefore, the secondary load angle is zero or $\delta = 0°$.

Now, actual ratio,

$$R = n + \frac{I_m \sin\delta + I_e \cos\delta}{I_s}$$
$$= 200 + \frac{37.5}{5} = 207.5$$

$$\therefore \text{ Ratio error } = \left(\frac{K_n - R}{R}\right) \times 100$$
$$= \left(\frac{200 - 207.5}{207.5}\right) \times 100 = -3.61\%$$

268. (d)

The ratio and phase angle error in a P.T. may be reduced by reducing I_m and I_e, by reducing the resistance and leakage reactance and using turns compensation.

Hence, none of the option is correct.

271. (a)

Wilson compensation is used for reducing the phase angle error of a current transformer.

273. (b)

We have, $\dfrac{I_1}{I_2} = \dfrac{250}{5}$

Since ammeter is in secondary of C.T., therefore,

$$I_2 = 2.7 \text{ A}$$

$$\therefore \quad \frac{I_1}{2.7} = \frac{250}{5} \quad \text{or} \quad I_1 = 135 \text{ A}$$

So, line current is 135 A.

274. (d)

Instrument transformers can be used only for a.c. circuits and not for d.c. circuits. This is the only disadvantage of an instrument transformer.
Hence, assertion is a false statement.

275. (c)

Given, $I_c = 3$ A, $I_m = 8$ A,

$\cos\phi = 0.8$ leading

$\delta = -\cos^{-1}0.8 = -36.86°$

(Negative due to leading p.f.)

$$n = \frac{N_s}{N_p} = \frac{399}{1}$$

$$k_n = \frac{I_p}{I_s} = \frac{2000}{5} = 400$$

= Nominal ratio

Now, actual ratio is

$$R = n + \frac{I_m \sin\delta + I_c \cos\delta}{I_s}$$

where, $I_s = 5$ A (full load)

$$= 399 + \frac{8\sin(-36.86°) + 3 \times 0.8}{5}$$

$$= 398.52$$

$$\therefore \text{ \% ratio error} = \left(\frac{k_n - R}{R}\right) \times 100$$

$$= \left(\frac{400 - 398.52}{398.52}\right) \times 100$$

$$= 0.3713\%$$

Also, phase angle error is

$$\theta = \frac{180}{\pi}\left[\frac{I_m \cos\delta - I_c \sin\delta}{nI_s}\right] \text{degree}$$

$$= \frac{180}{\pi}\left[\frac{8 \times 0.8 - 3 \times \sin(-36.86°)}{399 \times 5}\right]$$

$$= 0.2355°$$

277. (d)

Secondary burden = 30 VA and
Secondary current = 5 A

$$\therefore \text{ Secondary voltage} = \frac{30}{5} = 6 \text{ volt}$$

Hence, impedance of the connected load

$$= \frac{6}{5} = 1.2\ \Omega$$

285. (b)

Input resistance between two points

$= 20000 \times 5 = 100,000\ \Omega$

288. (c)

Voltage drop per unit length

$$= \frac{1.45}{50} = 0.029 \text{ V/cm}$$

Voltage drop across 70 cm length

$= 0.029 \times 70 = 2.03$ V

$$\therefore \text{ Current through the resistor} = \frac{2.03}{1\Omega} = 2.03 \text{ A}$$

289. (b)

Both assertion (A) and reason (R) are individually correct statements.
The current reason for assertion is that, no power is consumed in the circuit containing the unknown emf when the potentiometer is balanced because the basic principle of operation of a potentiometer is balance or null condition.

290. (a)

All the three statements 1, 2 and 3 are true.
Statements-4 is not true because the measurements using comparison methods are capable of a high degree of accuracy since the result obtained does not depend upon the actual deflection of a pointer, as is the case in deflection methods, but depends only upon the accuracy with which the voltage of the reference source is known.
Hence, only statement-4 is not correct.

291. (a)

The value of resistance (R) varies in a complex manner with frequency due to skin effect and eddy current losses, therefore it is indirectly determined by measuring the value of *Q*.
Hence, both assertion (A) and reason (R) are true and reason is a correct explanation of assertion.

292. (d)

- A "*Q*-meter" is also called a "storage factor meter".
- The shunt resistance to which the oscillator delivers current in a practical "storage factor

meter" (Q-meter) should be in milli-ohms. By keeping R_{sh} in milli-ohms (very low value), it introduces almost no resistance into the oscillator circuit and therefore represents a voltage source of magnitude E, with a very small internal resistance. If this input voltage E is kept constant, the voltage appearing across the capacitor is Q-times E and voltage connected across the capacitor can be calibrated to read the value of Q directly.

Hence, assertion is not a correct statement.
Reason is a correct statement.

294. (c)

Given, $f_0 = 1.5$ MHz, $C = 150$ pF,
BW $= \Delta f = 10$ kHz

We know that, quality factor of series RLC resonant circuit is

$$Q = \frac{f_0}{\Delta f} = \frac{1}{\omega_0 RC}$$

or, $$Q = \frac{f_0}{\Delta f} = \frac{1}{2\pi f_0 RC} \quad \text{or } R = \frac{1}{2\pi f_0^2 C}$$

or, $$R = \frac{10^4}{2\pi \times (1.5\times 10^6)^2 \times 150 \times 10^{-12}}$$
$$= 4.7\ \Omega$$

295. (d)

The flux meter is less sensitive and accurate than the ballistic galvanometer.
Hence, option (d) is not correct.

296. (b)

A distortion analyser measures the total harmonic power (rms power) present in the test wave rather the distortion caused by each component. The simplest method is to suppress the fundament frequency by means of a high pass filter whose cut-off frequency is a little above the fundament frequency. This high pass allows only the harmonics to pass and the total harmonic distortion can then be measured.

297. (a)

A spectrum analyser provides a calibrated graphical display on its CRT, with frequency on the horizontal axis and amplitude (voltage) on the vertical axis.

298. (c)

The distributed capacitance of the coil is given by

$$C_s = \left(\frac{C_1 - 4C_2}{3}\right) = \left(\frac{500 - 4\times 110}{3}\right)$$
$$= \left(\frac{500 - 410}{3}\right) = \frac{60}{3} = 20$$

$\therefore \quad C_s = 20$ pF

299. (d)

Given, $f_2 = 3f_1$

$$\therefore \quad \frac{1}{2\pi\sqrt{L(C_2 + C_s)}} = \frac{3}{2\pi\sqrt{L(C_1 + C_s)}}$$

or, $(C_1 + C_s) = 9(C_2 + C_s)$
or, $(C_1 + C_s) = (9C_2 + 9C_s)$

or, $$C_s = \left(\frac{C_1 - 9C_2}{8}\right) = \left(\frac{500 - 9\times 50}{8}\right)$$
$$= \left(\frac{500 - 450}{8}\right) = \frac{50}{8} = 6.25 \text{ pF}$$

Hence, required self capacitance = 6.25 pF.

300. (a)

For a series RLC resonant circuit, we have:

$$Q\text{-factor} = \frac{\omega_0 L}{R} = \frac{X_L}{R}$$
$$= \frac{1}{\omega_0 RC} = \frac{X_C}{R}$$

301. (a)

The shunt resistance R_{sh} used in Q-meter has a value in milli-ohms i.e. from 20 mΩ to 40 mΩ.

305. (a)

Constant of fluxmeter

$$G = NBA = 40 \times 0.05 \times 750 \times 10^{-6}$$
$$= 1500 \times 10^{-6}$$

Flux linking with the search coil

$$= 0.5 \times 200 \times 10^{-6}$$
$$= 100 \times 10^{-6} \text{ Wb}$$

Change in flux linking the coil

$$= 2 \times 100 \times 10^{-6} = 200 \times 10^{-6}$$

Now, the change in flux $\phi = \left(\frac{G}{N}\right)\theta$

$\therefore$ Deflection, $\theta = \frac{N\phi}{G} = \frac{10 \times 200 \times 10^{-6}}{1500 \times 10^{-6}}$ rad

$= \frac{4}{3}$ rad $= \frac{4}{3} \times \frac{180}{\pi}$ degree

$= 76.39$ degree

306. (d)

The value of resistance of either the fluxmeter (moving coil) or the search coil does not affect the meter sensitivity.
However, the value of total circuit resitance should not exceed about 20 ohm otherwise the damping will not be adequate.

309. (a)

Conversion to digitized systems is widely used today because the complex circuits used are low cost, accurate and simple to implement.
Hence, both assertion and reason are true and reason is a correct explanation of assertion.

310. (c)

The natural frequency of vibration of the reeds in a reed type frequency meter depends upon their weights and dimensions. Since the reeds have different weights and sizes, their natural frequencies of vibration are different. The reeds are arranged in ascending order of natural frequency, the difference is usually 0.5 Hz.
Thus, the natural frequency of first reed may be 47 Hz, of the second 47.5 Hz, of the next 48 Hz and so on.

313. (d)

Current, $I = \frac{V}{S} = \frac{0.2514}{0.1} = 2.514$ A

The fractional standard deviation of current is

$$\frac{\delta I}{I} = \sqrt{\left(\frac{\delta V}{V}\right)^2 + \left(\frac{\delta S}{S}\right)^2}$$

$$= \sqrt{\left(\frac{0.0001}{0.2514}\right)^2 + (0.001)^2}$$

$= 0.00108$

$\therefore$ Percentage accuracy of measurement of current $= \pm 0.00108 \times 100 = \pm 0.108$ %.

315. (c)

Reason (R) is a false statement because in a series RLC resonant circuit (Q-meter operates at series resonance), Q of the coil is given by:

$$Q = \frac{\omega_0 L}{R}$$

Here, ω_0 = Resonant frequency of the circuit

318. (d)

A ballistic galvanometer is more accurate and more sensitive than a fluxmeter.
Hence, characteristics 1 and 2 are not correctly matched.

320. (c)

Electronic voltmeters which use rectifier employ negative feedback and uses a high open loop gain to overcome the non-linearity of the rectifier diodes.

322. (a)

Number of full digits is $n = 4$

$\therefore$ Resolution $= \frac{1}{10^n} = \frac{1}{10^4} = 0.0001$

$= 0.01\%$

323. (b)

An Op-amp integrator is shown below.

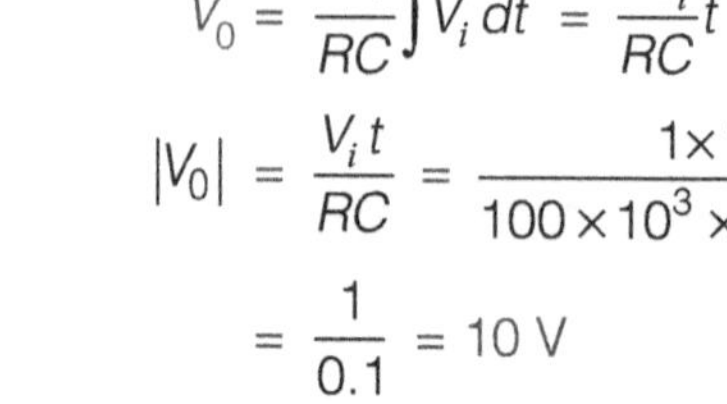

The output voltage is given by

$$V_0 = \frac{-1}{RC}\int V_i\,dt = \frac{-V_i}{RC}t$$

or, $|V_0| = \frac{V_i t}{RC} = \frac{1 \times 1}{100 \times 10^3 \times 1 \times 10^{-6}}$

$= \frac{1}{0.1} = 10$ V

328. (a)

The dual slope technique uses an integrator instead of an ADC circuit due to which it has excellent noise rejection because noise is averaged out by the positive and negative ramps using the process of integration.

Hence, both assertion (A) and reason (R) are true and reason (R) is a correct explanation of assertion (A).

329. (c)

We know that,

Sensitivity (s) = Resolution $\times (f_s)_{min}$

Since range of meter is 10 - 100 V, therefore lowest full scale of the meter, $(f_s)_{min} = 10$ V.

Also, resolution $= \dfrac{1}{10^n}$

Here, n = Number of full digits = n

$\therefore$ Resolution $= \dfrac{1}{10^3}$

Hence, sensitivity $= \dfrac{1}{10^3} \times 10 = \dfrac{1}{100} = 0.01$

$= 0.01 \times 100\ \% = 1\ \%$

331. (b)

- The principle of voltage to time conversion is used in dual slope type DVM not in ramp type DVM.
  Hence, statement-1 is false.
- Statement-2 and 3 are correct.
- The error due to noise can be large in case of ramp type DVM if noise is superimposed on the input signal due to which input filters are required in this type of converter.
  Hence, statement-4 is false.
  Thus, only statement-2 and 3 are correct.

■■■■

UNIT 5

Control Systems

1. Introduction

Q.1 The Laplace transform of a transportation lag of 10 seconds is

(a) $\frac{1}{s+10}$ (b) exp (10s)

(c) $\exp\left(\frac{-s}{10}\right)$ (d) exp (–10s)

Q.2 As compared to an open loop system, a closed loop system is
(a) more accurate but less stable
(b) less accurate but more stable
(c) more accurate as well as more stable
(d) less accurate as well as less stable

Q.3 Linear system obeys
(a) reciprocity theorem
(b) principle of maximum power transfer
(c) both (a) and (b)
(d) none of the above

Q.4 The most commonly used input signal(s) in control system is/are
(a) ramp or velocity function
(b) step function
(c) accelerating function
(d) all of the above

Q.5 Consider the following statements regarding control systems:
1. In open loop control system, the control action depends upon the desired output.
2. All control systems operated by present timing mechanism are open loop.
3. In a closed loop control system feedback signal is usually negative.
4. In open-loop control system stability cannot be ensured.

Which of the above statements are correct?
(a) 1, 2, 3 and 4 (b) 1 and 4 only
(c) 1, 2 and 3 only (d) 2 and 3 only

Q.6 **Assertion (A):** Closed loop control systems are known as feedback control systems.
Reason (R): In closed loop control systems, the control action is dependent on the desired output.
(a) Both A and R are true and R is a correct explanation of A.
(b) Both A and R are true but R is not a correct explanation of A.
(c) A is true but R is false.
(d) A is false but R is true.

Q.7 As compared to a closed loop system, an open loop system is
(a) slow and more reliable
(b) fast and more reliable
(c) slow and less reliable
(d) slow and more reliable

Q.8 Match **List-I** with **List-II** and select the correct answer using the codes given below the lists:

List-I
A. Immersion rod
B. Air conditioner
C. Automatic washing machine
D. Automatic control of traffic

List-II
1. Open loop control system
2. Closed loop control system
3. Neither a closed loop nor an open loop system

Codes:

	A	B	C	D
(a)	3	1	2	2
(b)	1	1	1	3
(c)	3	1	1	2
(d)	1	2	1	1

Q.9 **Assertion (A):** The control system which operates on the time basis is an open loop system.

Reason (R): A field control d.c. motor is an example of open loop system.
(a) Both A and R are true and R is the correct explanation of A.
(b) Both A and R are true but R is not the correct explanation of A.
(c) A is true but R is false.
(d) A is false but R is true.

Q.10 A ramp function occurs at $t = a$. It's Laplace transform is
(a) e^{-as} (b) $\frac{e^{-as}}{s}$
(c) $\frac{e^{as}}{s^2}$ (d) None of these

Q.11 Match **List-I (Laplace transforms)** with **List-II (Time functions)** and select the correct answer using the codes given below the lists:

List-I	List-II
A. $\frac{a^2}{s(s^2+a^2)}$	1. $(1 - at)\,e^{-at}$
B. $\frac{1}{(s-a)^2}$	2. $\cos at$
C. $\frac{s}{s^2+a^2}$	3. $1 - \cos at$
D. $\frac{s}{(s+a)^2}$	4. $t\,e^{at}$

Codes:

	A	B	C	D
(a)	1	4	2	3
(b)	3	2	4	1
(c)	3	4	2	1
(d)	1	2	4	3

Q.12 A standard signal used for comparison in a closed-loop control system is called the
(a) step input (b) reference input
(c) error input (d) ramp input

Q.13 The difference between the output response and the reference signal is known as the _____ signal.
(a) actuating (b) bias
(c) velocity (d) none of these

Q.14 The Laplace transform of unity function is
(a) 1 (b) zero
(c) s (d) $1/s$

Q.15 The phenomena of 'limit cycles' and 'jump resonance' are observed in
(a) linear systems
(b) distributed systems
(c) non-linear systems
(d) discrete time systems

Q.16 The Laplace transform of $e^{-2t} \sin 3\omega t$ is
(a) $\frac{3s}{(s+2)^2+3\omega^2}$ (b) $\frac{3\omega}{(s+2)^2+9\omega^2}$
(c) $\frac{3\omega}{(s-2)^2+9\omega^2}$ (d) $\frac{3s}{(s-2)^2+3\omega^2}$

Q.17 A control system is said to be robust when
(a) it has low sensitivities
(b) it is stable over a wide range of parameter variation
(c) both (a) and (b)
(d) neither (a) nor (b)

Q.18 Which of the following is not correctly matched? (CLCS = Closed Loop Control System and OLCS = Open Loop Control System)

System	Type
(a) A missile launching system	CLCS
(b) A field controlled DC motor	OLCS
(c) Radar tracking system	OLCS
(d) Armature controlled DC motor	CLCS

Q.19 Which of the following is not true regarding an open loop control system?
(a) Highly affected by non-linearities.
(b) Open loop systems are inaccurate.
(c) These are not reliable.
(d) The effect of parameter variation and internal noise is less.

Q.20 As compared to an open loop control system, a closed loop control system
(a) is faster and has more bandwidth.
(b) is faster and has less bandwidth.
(c) is slower and has more bandwidth.
(d) is slower and has less bandwidth.

Q.21 **Assertion (A):** The cost for making a closed loop system is more compared to an open loop system.
Reason (R): A closed loop system is relatively more complex than an open loop system.

(a) Both A and R are true and R is a correct explanation of A.
(b) Both A and R are true but R is not a correct explanation of A.
(c) A is true but R is false.
(d) A is false but R is true.

2. Transfer Function

Q.22 The type and order of the system described by the open loop transfer function $\frac{s(s+4)}{(s+1)(s+8)}$ are respectively
(a) 1 and 2 (b) 0 and 3
(c) 1 and 3 (d) 0 and 2

Q.23 Non-minimum phase transfer function is defined as the transfer function
(a) which has zeros only in the left-half *s*-plane
(b) which has poles in the right-half *s*-plane
(c) which has zeros in the right-half *s*-plane
(d) which has poles in the left-half *s*-plane

Q.24 **Assertion (A):** If the number of zeros are less than the number of poles (i.e. $Z < P$), we say that there are zeros at infinity and the order of such zeros is P-Z.
Reason (R): The value of the transfer function becomes zero for *s* tends to zero.
(a) Both A and R are true and R is a correct explanation of A.
(b) Both A and R are true but R is not a correct explanation of A.
(c) A is true but R is false.
(d) A is false but R is true.

Q.25 The transfer function of the network shown below is

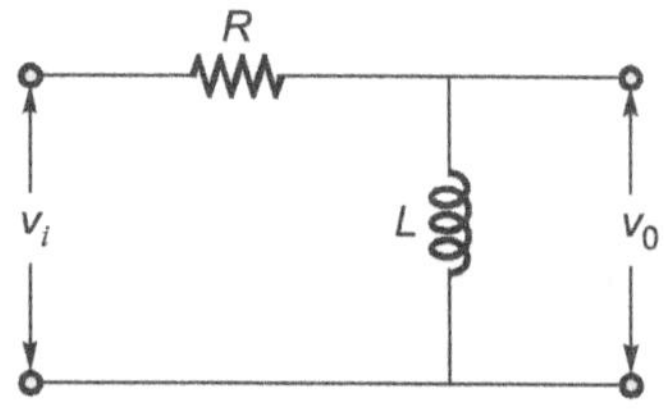

(a) $\frac{s}{R+sL}$ (b) $\frac{sL}{R+sL}$
(c) $\frac{Rs}{R+sL}$ (d) $\frac{1}{R+sL}$

Q.26 Consider the following statements regarding the properties of transfer function:
1. The transfer function is independent of the input of the system.
2. The transfer function is defined for linear time-variant system.
3. The transfer function of a continuous data system is expressed only as a function of the complex variable s.
4. All the initial conditions are not necessarily set to zero for finding the transfer function of a given system.

Which of the above statements is/are not true?
(a) 1, 3 and 4 only (b) 2, 3 and 4 only
(c) 1 and 3 only (d) 1 and 4 only

Q.27 The differential equation of a control system having input $x(t)$ and output $y(t)$ is given as

$$\frac{d^2y(t)}{dt^2}+9\frac{dy(t)}{dt}+20y(t)=20x(t)$$

The output response of the system for unit step input is given by
(a) $-1+5e^{-4t}-4e^{-5t}$ (b) $1-e^{-4t}-e^{-5t}$
(c) $1-5e^{-4t}+4e^{-5t}$ (d) none of these

Q.28 The step response of a system is given by
$c(t) = 1 + 0.25\,e^{-50t} - 1.25e^{-10t}$
The steady state gain of the transfer function in time constant form will be
(a) 1 (b) 7
(c) –5 (d) –7

Q.29 The poles and zeros of the transfer function $\frac{V_2(s)}{V_1(s)}$ for the network shown below are located at

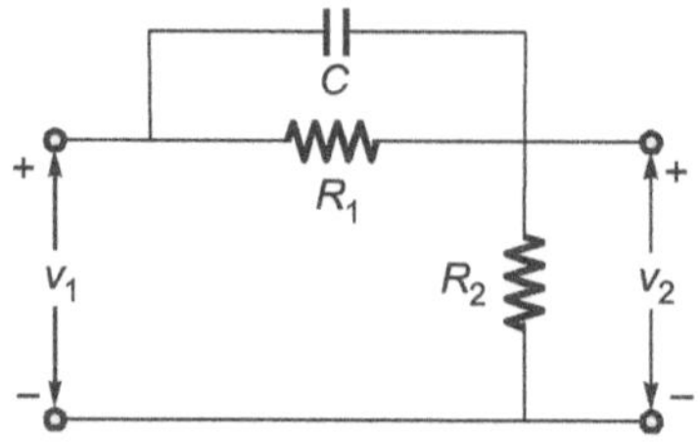

(a) $\frac{-1}{R_2C}$ and $-\left(\frac{R_1+R_2}{R_1R_2C}\right)$ respectively
(b) $-\left(\frac{R_1+R_2}{R_1R_2C}\right)$ and $\frac{-1}{R_1C}$ respectively

(c) $\frac{-1}{R_1C}$ and $-\frac{1}{R_2C}$ respectively

(d) $\left(\frac{R_1+R_2}{R_1R_2C}\right)$ and $-\frac{1}{R_2C}$ respectively

Q.30 The pole-zero configuration of a transfer function is shown below:

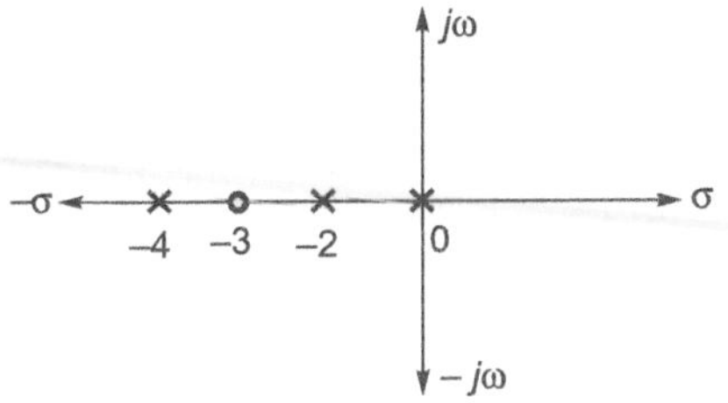

If the value of the transfer function at $s = 1$ is 3.2, the gain factor K is

(a) 7 (b) −16

(c) 6 (d) 12

Q.31 The type of a system denotes the number of

(a) zeros of infinity (b) poles of origin

(c) poles of infinity (d) zeros of origin

Q.32 Match **List-I (Roots in s-plane)** with **List-II (Corresponding impulse response)** and select the correct answer using the codes given below the lists:

List-I

A.

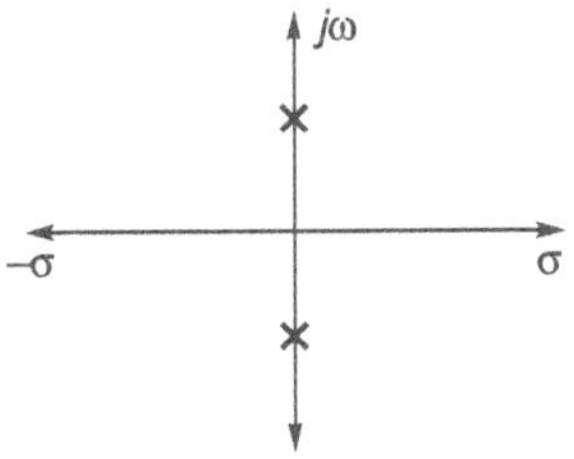

B.

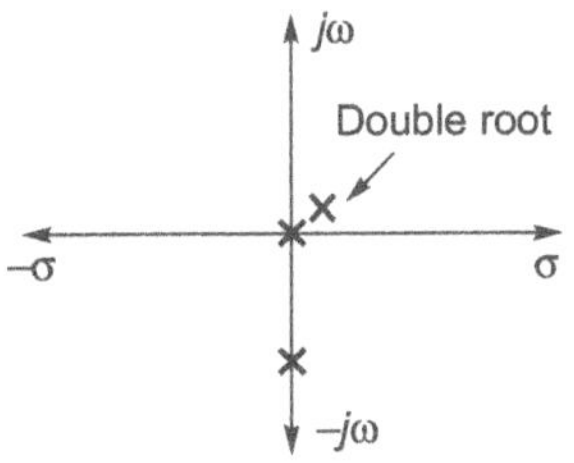

C.

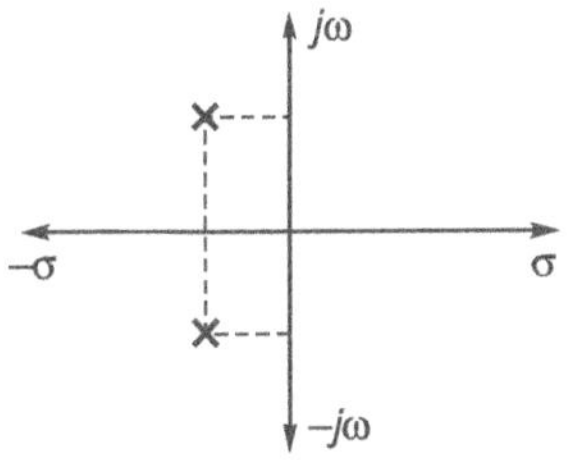

D.

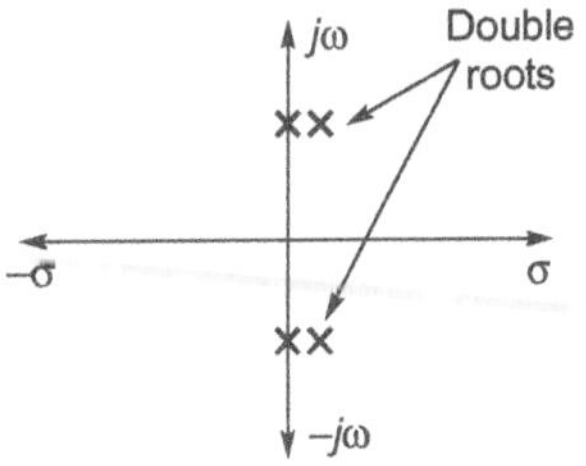

List-II

1.

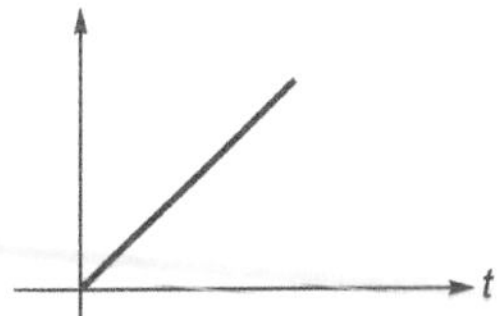

2.

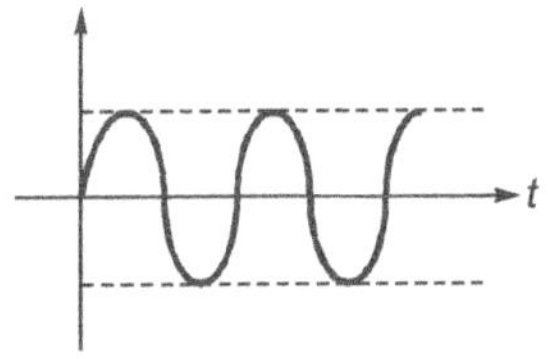

3.

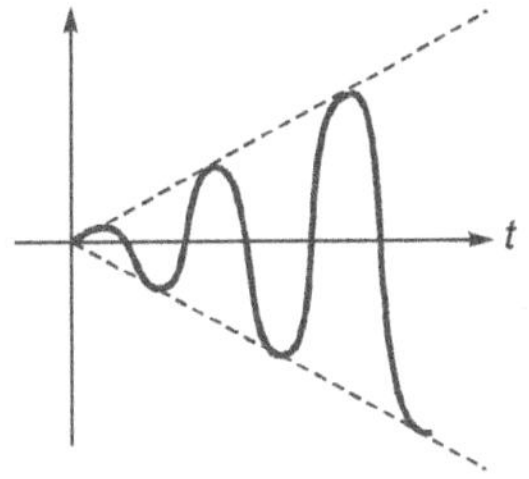

4.

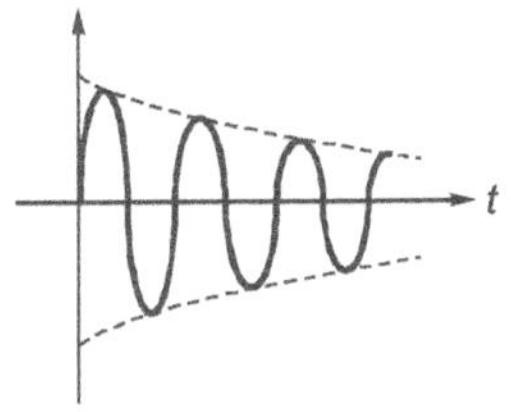

Codes:

	A	B	C	D
(a)	2	1	4	3
(b)	1	2	4	3
(c)	2	4	3	1
(d)	1	2	3	4

Q.33 The transfer function of a system is the ratio of

(a) Fourier transform of the input variable to that of output variable.

(b) Laplace transform of the output variable to that of input variable.

(c) Fourier transform of the output variable to that of input variable.

(d) Laplace transform of the input variable to that of output variable.

Q.34 The transfer function is applicable to
(a) linear and time-variant systems
(b) linear systems only
(c) linear and time-invariant systems
(d) non-linear systems only

Q.35 The DC gain of the system represented by the following transfer function is

$$H(s) = \frac{(s+2)\,(s+5)}{(s+3)\,(s+8)}$$

(a) 0.416 (b) 8
(c) 5 (d) 0.832

Q.36 The unit impulse response of a control system is given by $c(t) = -te^{-t} + e^{-t}$. Its transfer function is

(a) $\frac{-1}{(s+1)^2}$ (b) $\frac{s}{(s+1)}$
(c) $\frac{1}{(s+1)^2}$ (d) $\frac{s}{(s+1)^2}$

Q.37 The impulse response of an initially relaxed linear system is $e^{-3t}\,u(t)$. To produce a response of $t\,e^{-3t}\,u(t)$, the input must be equal to

(a) $3\,e^{-t}\,u(t)$ (b) $\frac{1}{3}e^{-3t}u(t)$
(c) $e^{-3t}\,u(t)$ (d) $e^{-t}\,u(t)$

Q.38 A linear time invariant system, initially at rest when subjected to a unit step input gave response $c(t) = t\,e^{-t}\,(t \geq 0)$. The transfer function of the system is

(a) $\frac{s}{(s+1)^2}$ (b) $\frac{1}{s(s+1)^2}$
(c) $\frac{1}{(s+1)^2}$ (d) $\frac{1}{s(s+1)}$

Q.39 A linear time invariant system having input $r(t)$ and output $y(t)$ is represented by the differential equation

$$2\frac{d^2y(t)}{dt^2} + \frac{dy(t)}{dt} + 5y(t) = r(t) + 2r(t-1)$$

The transfer function of the given system is represented as

(a) $\frac{2e^{s}+1}{2s^2+s+5}$ (b) $\frac{1+e^{-s}}{s^2(2s^2+s+5)}$
(c) $\frac{1+2e^{-s}}{s^2(2s^2+s+5)}$ (d) $\frac{2e^{-s}+1}{2s^2+s+5}$

Q.40 The open loop transfer function of a system is

$$G(s)H(s) = \frac{K(s+10)}{s(s+8)(s+16)(s+72)}$$

A closed loop pole will be located at $s = -12$, when the value of K is
(a) 4355 (b) 5760
(c) 9600 (d) 9862

Q.41 The singularities of a function are the points in the s-plane at which the function or its derivates
(a) have a non-zero value
(b) have a zero value
(c) do not exist
(d) have a negative value

Q.42 **Assertion (A):** The final value theorem cannot be applied to a function given by $F(s) = \frac{a}{s^2+a^2}$.

Reason (R): The function $s\,F(s)$ has two poles on the imaginary axis of s-plane.
(a) Both A and R are true and R is a correct explanation of A.
(b) Both A and R are true but R is not a correct explanation of A.
(c) A is true but R is false.
(d) A is false but R is true.

Q.43 When two time constant elements are cascaded non-interactively then, the overall transfer function of such an arrangement
(a) is never the product of two individual transfer functions.
(b) is always the product of two individual transfer functions.
(c) is always the sum of two individual transfer functions.
(d) may or may not be the product of two individual transfer functions.

Q.44 The unit impulse response of a unity feedback control system whose open loop transfer function

$G(s) = \frac{2s+1}{s^2}$ is given by

(a) $te^{-t}\,u(t)$ (b) $(1-t)\,e^{-t}\,u(t)$
(c) $(2-t)\,e^{-t}\,u(t)$ (d) none of these

Q.45 The impulse response of a linear time invariant system is a unit step function. The transfer function of this system would be

(a) 1 (b) $\frac{1}{s^2}$

(c) s (d) $\frac{1}{s}$

Q.46 Consider the function $F(s) = \frac{5}{s(s^2+s+2)}$,

where $F(s)$ is the Laplace transform of $f(t)$.

$\lim_{t \to \infty} f(t)$ is equal to

(a) 5 (b) 2.5

(c) infinity (d) zero

Q.47 If $\mathcal{L}\{f(t)\} = F(s)$, then $\mathcal{L}\{f(t)\, e^{at}\}$ is equal to

(a) $\frac{F(s)}{s-a}$ (b) $e^{-as} F(s)$

(c) $F(s-a)$ (d) $e^{as} F(s)$

Q.48 The transfer function of high-pass RC network is

(a) $\frac{RC}{1+RCs}$ (b) $\frac{1}{1+RCs}$

(c) $RCs\,(1 + RCs)$ (d) $\frac{RCs}{1+RCs}$

Q.49 For the given OP-AMP circuit shown below, the transfer function $\left|\frac{V_0(s)}{V_i(s)}\right|$ is

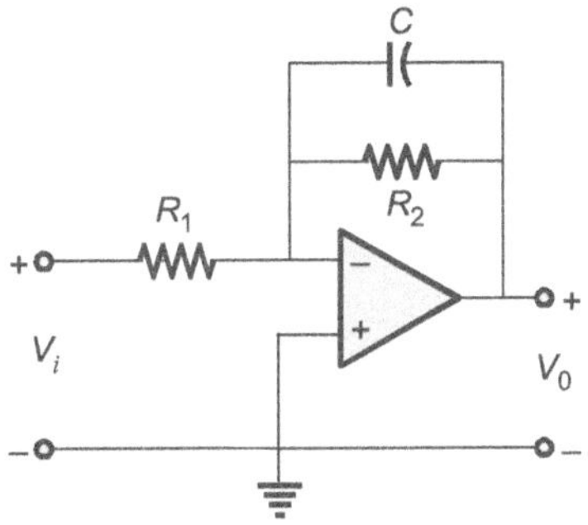

(a) $\frac{R_2}{R_1(1+R_2Cs)}$ (b) $\frac{1}{1+R_1Cs}$

(c) $\frac{R_1}{R_2(1+R_1Cs)}$ (d) $\frac{R_1}{(1+R_2Cs)}$

Q.50 The pole- zero configuration of a transfer function is shown in figure below.

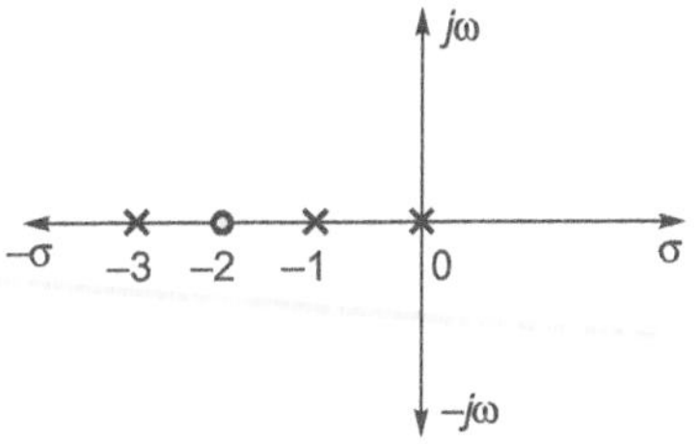

The value of the transfer function at $s = 2$ is found to be 4.

The transfer function of the given system is:

(a) $\frac{(s+2)}{(s+1)(s+3)}$ (b) $\frac{30(s+2)}{s(s+1)(s+3)}$

(c) $\frac{10(s+1)(s+3)}{s(s+2)}$ (d) $\frac{15(s+2)}{s(s+1)(s+3)}$

3. Block Diagrams

Q.51 The transfer function of the control system represented by the block diagram shown below is

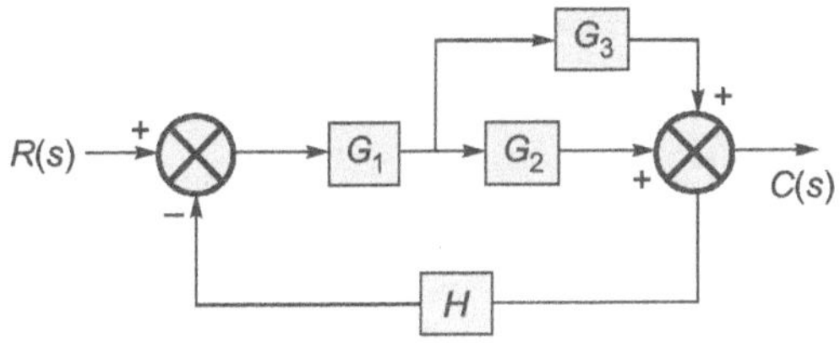

(a) $\frac{G_1G_2 - G_1G_3}{1+G_1H+G_2H}$

(b) $\frac{G_1G_2 + G_2G_3}{1+G_1H+G_3H}$

(c) $\frac{G_1G_2 + G_1G_3}{1+G_1G_2H+G_1G_3H}$

(d) $\frac{G_1G_2 + G_2G_3}{1+G_1G_2H+G_1G_3H}$

Q.52 Which one of the following block diagrams in options given is equivalent to the below shown block diagram?

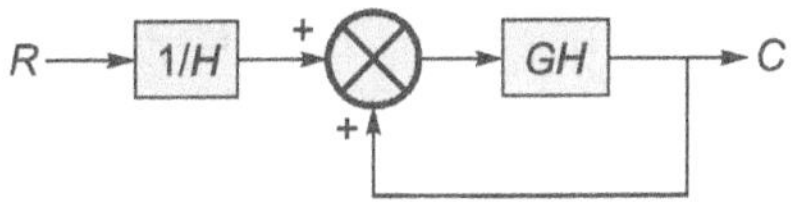

(a) 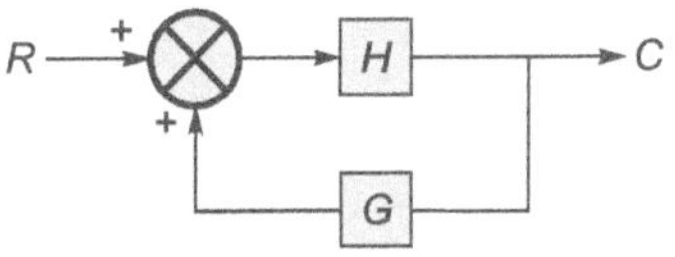

(b) 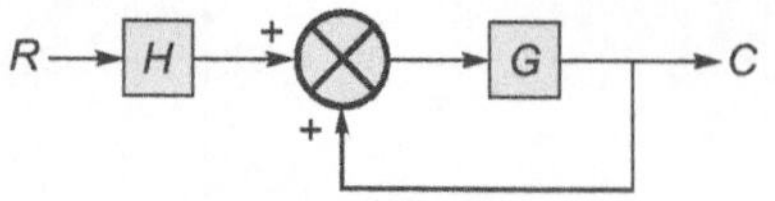

(c) 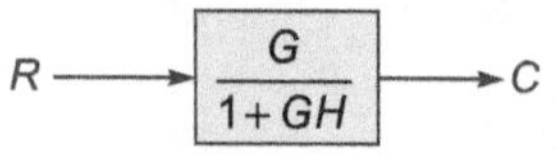

(d) None of the above

Q.53 Consider the following statements:

1. Transfer function is an expression which relates output to input in s-domain.
2. Transfer function gives information about the internal structure of the system.
3. A system can be represented by the block diagram if the transfer function of the system is known.
4. Block diagram is the flow of system variables from one block to another block represented by a single line.

Which of the above statements is/are true?

(a) 3 and 4 only (b) 1, 3 and 4 only
(c) 1, 2 and 4 only (d) 1, 2, 3 and 4

Q.54 The ratio $C(s)/R(s)$ for the system shown in figure below is

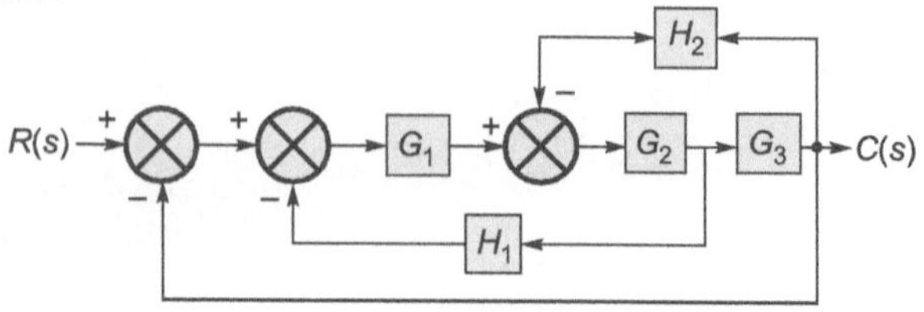

(a) $\dfrac{G_1G_2G_3}{1+G_1G_2G_3+G_1G_2H_1+G_2G_3H_2}$

(b) $\dfrac{1+G_1G_2G_3}{1+G_1G_2H_2+G_2G_3H_1+G_1G_2G_3}$

(c) $\dfrac{G_1G_2G_3}{1-G_1G_2G_3+G_1G_2H_2+G_2G_3H_1}$

(d) $\dfrac{1+G_1G_2G_3}{1+G_1G_2G_3+G_1G_2H_1+G_1G_3H_2}$

Q.55 The transfer function $\dfrac{C(s)}{R(s)}$ for the system described by the block diagram shown below is given by:

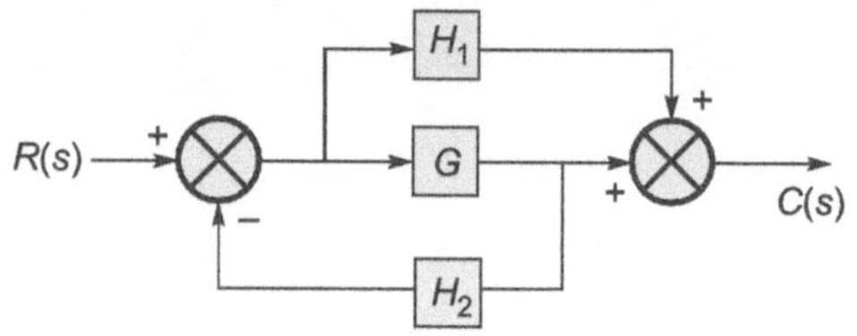

(a) $\dfrac{1+H_1}{1-GH_2}$ (b) $\dfrac{1+G}{1+GH_1}$

(c) $\dfrac{G+H_1}{1+GH_2}$ (d) None of these

Q.56 The transfer matrix for the multi input-multi output (MIMO) system represented by the block diagram shown below is given by

$$G(s)=\begin{bmatrix} G_{11} & G_{12} \\ G_{21} & G_{22} \end{bmatrix}$$

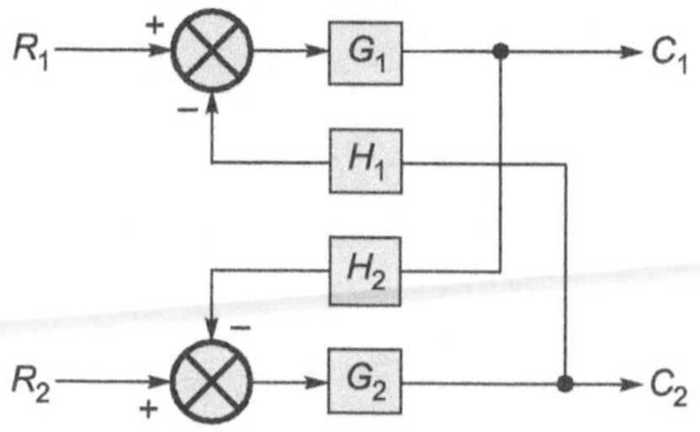

Match **List-I** with **List-II** and select the correct answer using the codes given below the lists:

List-I	List-II
A. G_{11}	1. $\dfrac{-G_1G_2H_1}{1-G_1G_2H_1H_2}$
B. G_{12}	2. $\dfrac{G_2}{1-G_1G_2H_1H_2}$
C. G_{21}	3. $\dfrac{G_1}{1-G_1G_2H_1H_2}$
D. G_{22}	4. $\dfrac{-G_1G_2H_2}{1-G_1G_2H_1H_2}$

Codes:

	A	B	C	D
(a)	2	4	1	3
(b)	3	1	4	2
(c)	2	1	4	3
(d)	3	4	1	2

Q.57 Consider the three block diagrams A, B and C shown below.

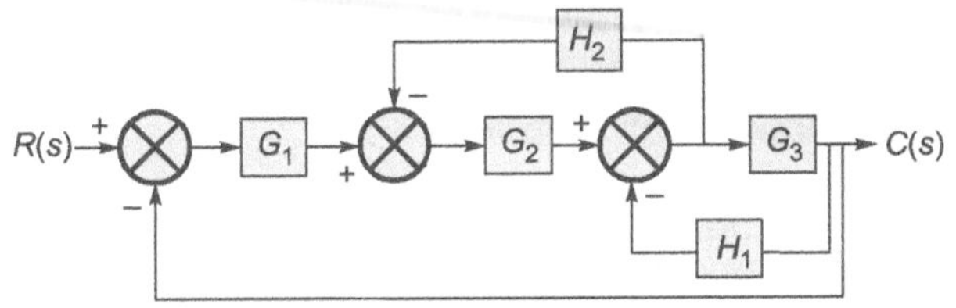

Block Diagram A

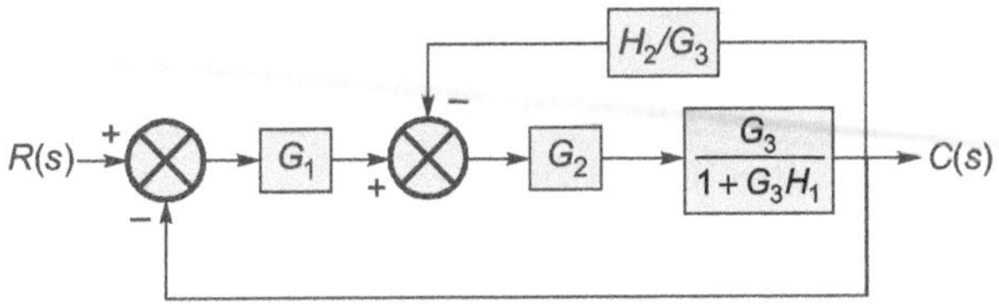

Block Diagram B

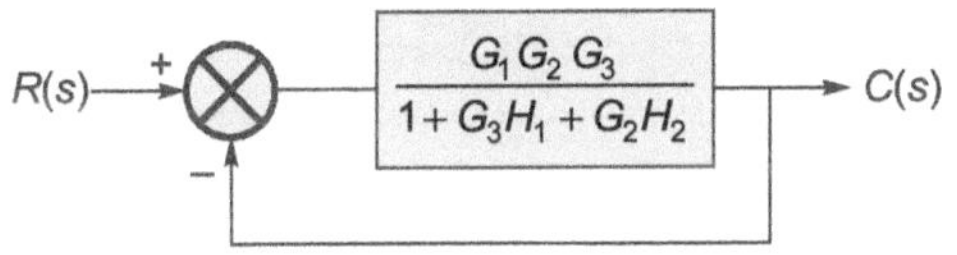

Block Diagram C

Which one of the following statements is correct in respect of the above block diagrams?

(a) A, B and C are equivalent
(b) Only A and B are equivalent
(c) Only B and C are equivalent
(d) Only A and C are equivalent

Q.58 The poles of the transfer function $C(s)/R(s)$ of the system represented by the block diagram shown below are located at

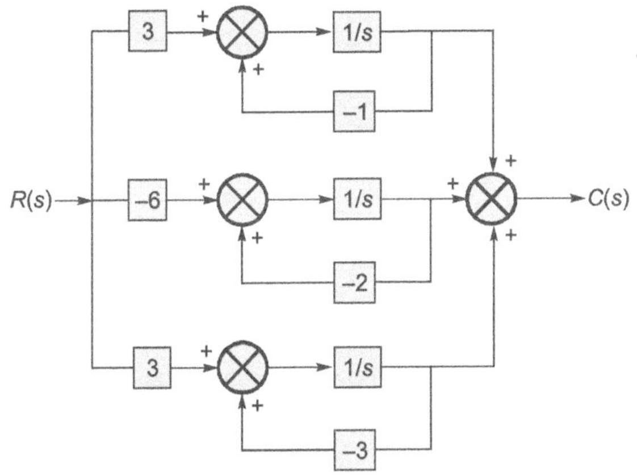

(a) 0, –2 and –4 (b) –1, –2, and –4
(c) –2, –3, and –4 (d) –1, –2 and –3

Q.59 The transfer function $\dfrac{C(s)}{R(s)}$ for the block diagram shown below is

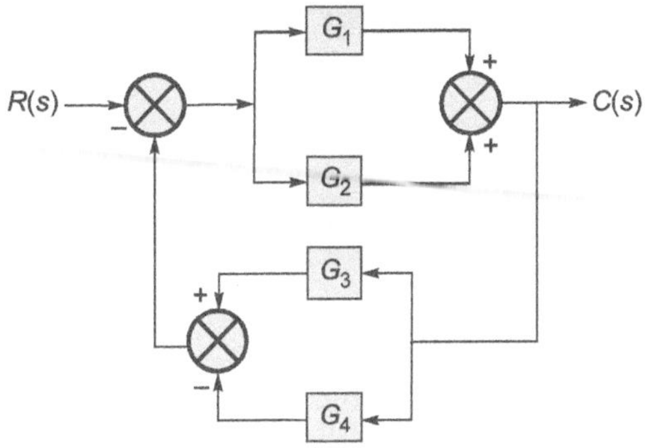

(a) $\dfrac{(G_1+G_2)G_3G_4}{1-G_1G_4-G_2G_4+(G_1+G_2)G_3}$

(b) $\dfrac{G_3G_4}{1-G_1G_4-G_2G_4+(G_1+G_2)G_3}$

(c) $\dfrac{G_1+G_2}{1+G_1G_3-G_1G_4+G_2G_3-G_2G_4}$

(d) $\dfrac{1+G_1G_2}{1+G_1G_3-G_1G_4+G_2G_3-G_2G_4}$

Q.60 The value of C_1/R_1 for the system described by the block diagram shown in figure below is

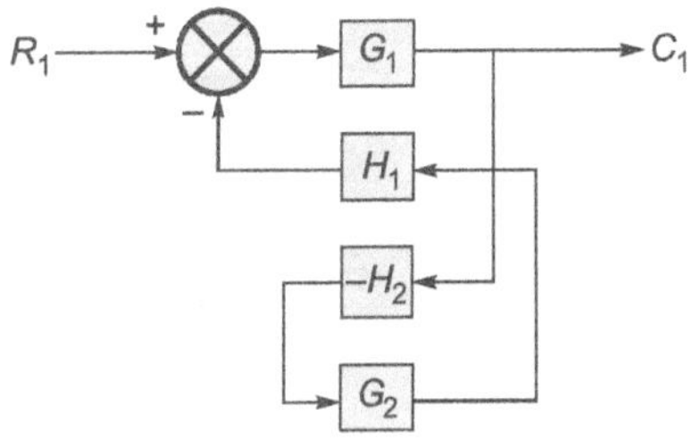

(a) $\dfrac{-G_2H_2}{1-G_1G_2H_1H_2}$ (b) $\dfrac{G_1}{1-G_1G_2H_1H_2}$

(c) $\dfrac{G_1H_2G_2}{1+G_1G_2H_1H_2}$ (d) None of these

Q.61 The closed loop transfer function of the system shown below is

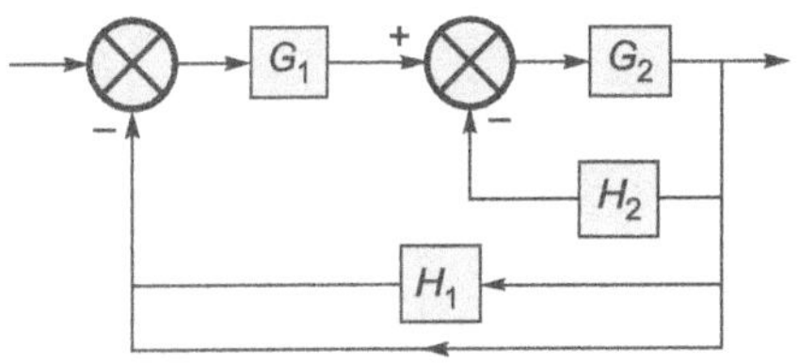

(a) $\dfrac{G_1G_2}{1+(G_1+G_2+H_1)H_2}$

(b) $\dfrac{1+G_1G_2}{1+(G_1+H_2-G_1H_1)G_2}$

(c) $\dfrac{G_1G_2}{1+(G_1+G_2+G_1H_2)G_2}$

(d) $\dfrac{G_1G_2}{1+(G_1+H_2+G_1H_1)G_2}$

Q.62 **Assertion (A):** Block diagram reduction technique is time consuming.
Reason (R): At every stage modified block diagram has to be drawn.
(a) Both A and R are true and R is a correct explanation of A.
(b) Both A and R are true but R is not a correct explanation of A.
(c) A is true but R is false.
(d) A is false but R is true.

Q.63 The closed loop transfer function of the system whose block diagram is given in figure below when $R_1 = 0$ is given by

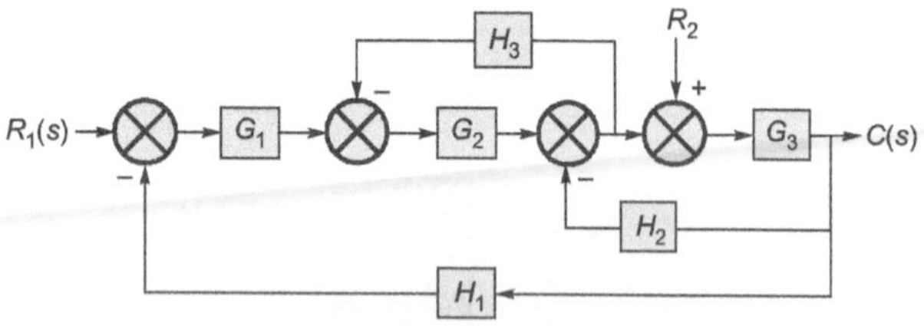

(a) $\dfrac{G_3(1+G_2H_3)}{1+G_1G_2G_3H_1-G_3H_2+G_2H_3}$

(b) $\dfrac{G_2(1+G_3H_3)}{1+G_1G_2G_3H_2+G_2H_1+G_3H_3}$

(c) $\dfrac{G_3(1+G_2H_3)}{1+G_1G_2G_3H_1+G_3H_2+G_2H_3}$

(d) $\dfrac{G_2(1+G_3H_3)}{1+G_1G_2G_3H_2-G_2H_1+G_3H_3}$

Q.64 The equivalent transfer function of three parallel blocks given by

$G_1(s)=\dfrac{1}{s}$, $G_2(s)=\dfrac{1}{s+3}$, $G_3(s)=\dfrac{s+1}{s+5}$ is

(a) $\dfrac{s^2+2s+1}{(s+3)(s+5)}$ (b) $\dfrac{3s^2+16s+15}{s(s+3)(s+5)}$

(c) $\dfrac{4s^2+5s+6}{s(s+3)(s+5)}$ (d) $\dfrac{s^2+11s+10}{s(s+3)(s+5)}$

Q.65 Three blocks having transfer functions,

$$G_1(s)=\frac{1}{s+1},\ G_2=\frac{s-1}{s+4},\ G_3(s)=\frac{-s}{s^2+2}$$

are cascaded.
The equivalent transfer function is

(a) $\dfrac{s^4-2s+6}{(s+1)(s+4)(s^2+2)}$

(b) $\dfrac{(s-1)}{(s+1)(s+4)(s^2+2)}$

(c) $\dfrac{2s^4+4s^3-2s+1}{(s+1)(s+4)(s^2+1)}$

(d) $\dfrac{s(1-s)}{(s+1)(s+4)(s^2+2)}$

Q.66 Which of the following options is not true regarding the block diagram reduction techniques?
(a) When two or more summing points are directly connected they can be interchanged.
(b) When blocks are connected in series, transfer function of such blocks get multiplied.
(c) When blocks are in parallel, transfer function of blocks get added algebraically.
(d) Shifting a take-off point before a summing point does not require any modification in the block diagram reduction technique.

Q.67 The equivalent closed loop transfer function of the block diagram shown below is:

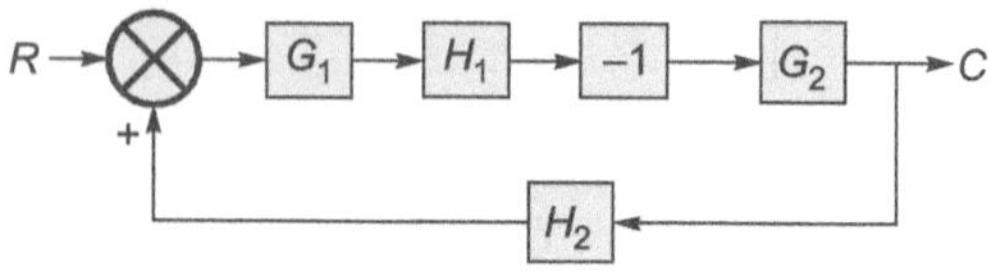

(a) $\dfrac{-G_1G_2H_2}{1-G_1G_2H_1H_2}$ (b) $\dfrac{G_1G_2H_1H_2}{1+G_1G_2H_1H_2}$

(c) $\dfrac{-G_1G_2H_1}{1+G_1G_2H_1H_2}$ (d) $\dfrac{G_1G_2H_2}{1+G_1G_2H_1H_2}$

Q.68 Consider the systems shown in figure below:

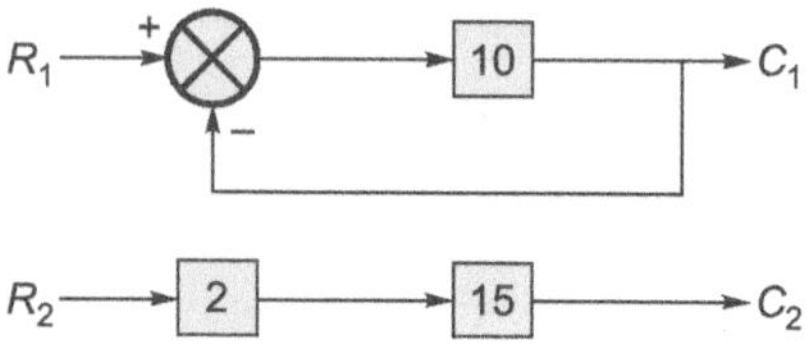

If the forward path gain is reduced by 5% in each system, then the variation in C_1 and C_2 will be approximately respectively

(a) 5% and 1% (b) 10% and zero%
(c) 1% and 5% (d) 10% and 5%

4. Signal Flow Graphs

Q.69 Mason's rule is applied to
(a) signal flow graph
(b) rotational system
(c) block diagram reduction technique
(d) both (a) and (b)

Q.70 **Assertion (A):** Signal flow graph is applicable to the linear systems.
Reason (R): Signal flow graph method of finding the transfer function of a system is very simple and does not require any reduction technique.
(a) Both A and R are true and R is a correct explanation of A.
(b) Both A and R are true but R is not a correct explanation of A.
(c) A is true but R is false.
(d) A is false but R is true.

Q.71 Consider the following three cases of signal flow graph and their corresponding transfer functions:

Signal flow graph — Transfer function

1. 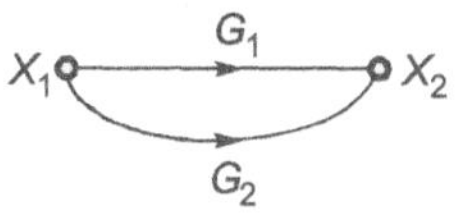

$\frac{X_2}{X_1} = G_1 + G_2$

2. 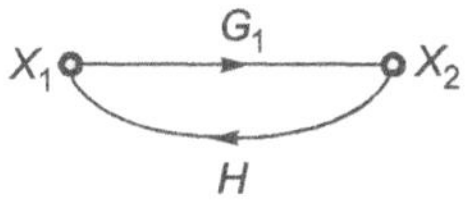

$\frac{X_2}{X_1} = \frac{G}{1+GH}$

3. 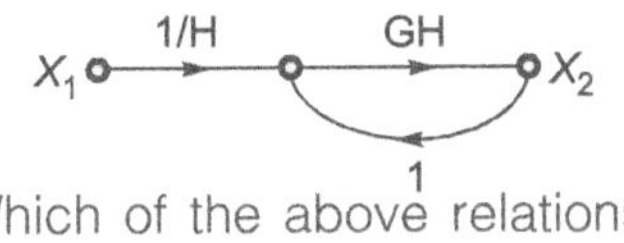

$\frac{X_2}{X_1} = \frac{G}{1+GH}$

Which of the above relations is/are correctly matched?
(a) 1 and 2 (b) 2 and 3
(c) 1 only (d) 1 and 4

Q.72 The signal flow graph shown below has M number of forward path and N number of individual loops.

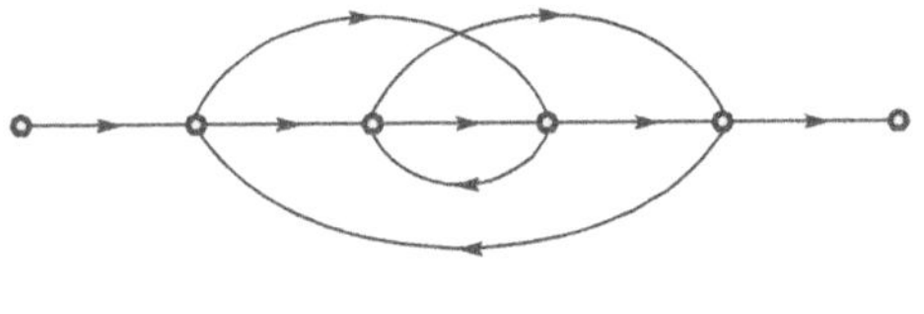

What are the values of M and N?
(a) $M = 3$ and $N = 4$
(b) $M = 4$ and $N = 6$
(c) $M = 3$ and $N = 6$
(d) $M = 4$ and $N = 3$

Q.73 The signal flow graph for a certain feedback control system is given below.

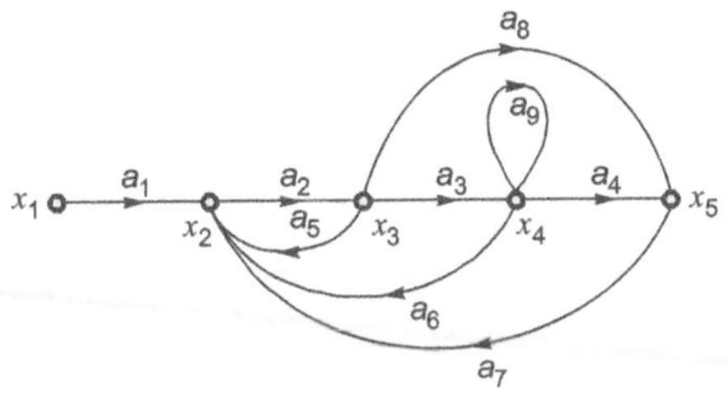

Now, consider the following set of equations for the nodes:
1. $x_2 = a_1x_1 + a_5x_3 + a_6x_4 + a_7x_5$
2. $x_3 = a_2x_2$
3. $x_4 = a_3x_3$
4. $x_5 = a_4x_4 + a_8x_3$

Which of the above equations are correct?
(a) 1, 2 and 3 (b) 1, 2 and 4
(c) 1, 3 and 4 (d) 2, 3 and 4

Q.74 The gain $\frac{C(s)}{R(s)}$ of the signal flow graph shown below is

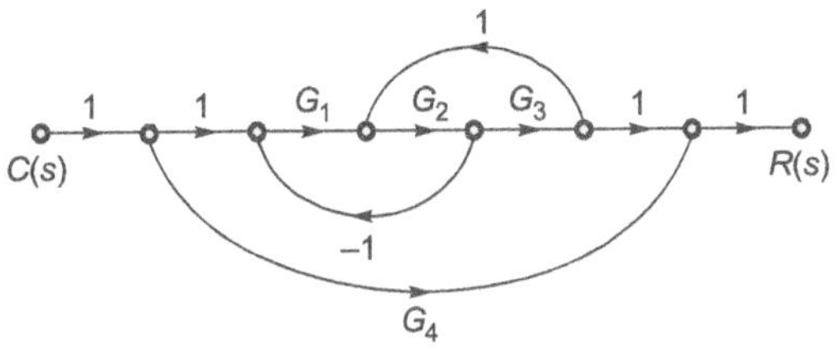

(a) $\dfrac{G_1G_2G_3 + G_4(1 - G_2G_3)}{1 + G_1G_2 + G_2G_3}$

(b) $\dfrac{G_1G_2G_3 + G_1G_2G_4 - G_2G_3G_4}{1 + G_1G_2 - G_2G_3}$

(c) $\dfrac{G_2G_3(G_1 - G_4)}{1 + G_1G_2 + G_2G_3}$

(d) $\dfrac{G_1G_2G_3 + G_4(1 + G_1G_2 - G_2G_3)}{1 + G_1G_2 - G_2G_3}$

Q.75 The number of forward paths and the number of non-touching loop pairs for the signal flow graph shown in the figure below are respectively

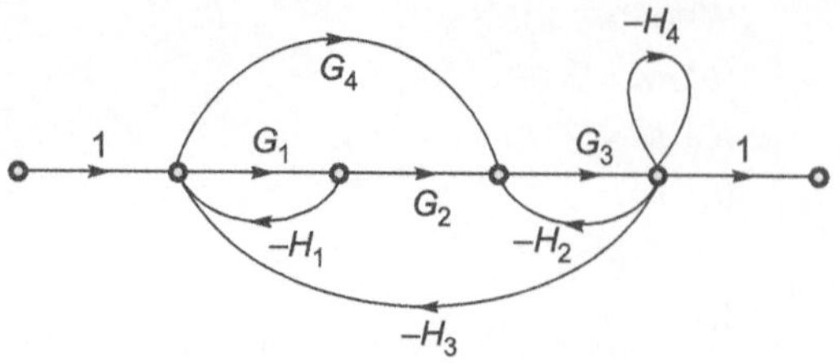

(a) 3, 2 (b) 2, 3
(c) 2, 2 (d) 3, 3

Q.76 The signal flow graph of the RLC circuit of figure (i) is shown in figure (ii).

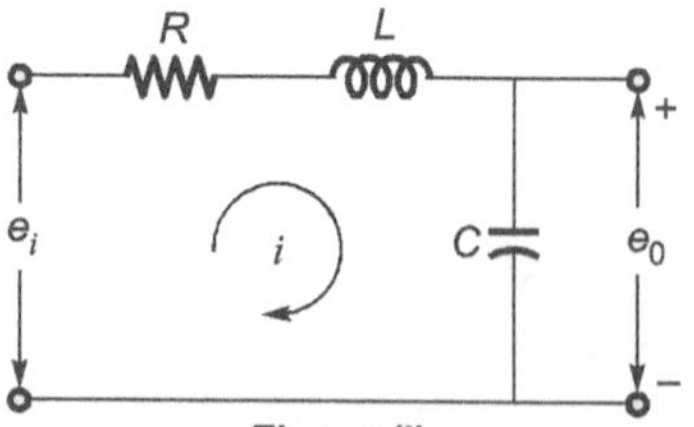

Figure (i)

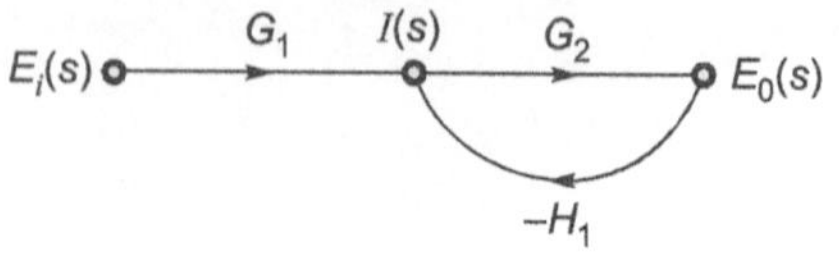

Figure (ii)

The values of G_1 and H_1 are respectively

(a) $\frac{1}{R+sL}, \frac{1}{R+sL}$ (b) $\frac{1}{Cs}, \frac{1}{R+sL}$

(c) $-\frac{1}{R+sL}, \frac{1}{Cs}$ (d) $\frac{1}{R+sL}, -\frac{1}{R+sL}$

Q.77 The closed loop transfer function C/R for the signal graph shown below is

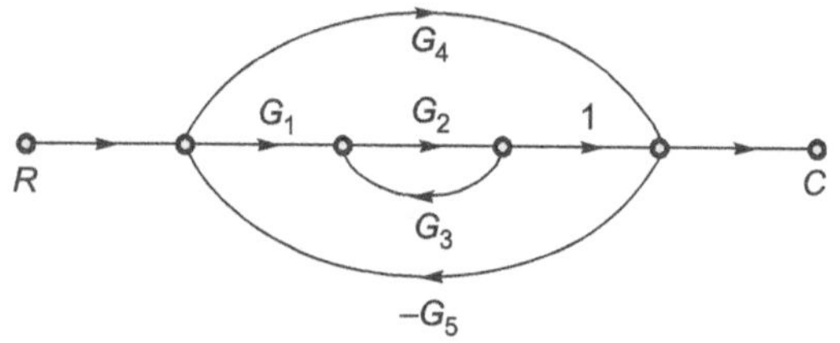

(a) $\dfrac{G_1G_2+G_4}{1+G_2G_3+G_1G_5+G_4G_5}$

(b) $\dfrac{G_1G_2}{1-G_2G_3+G_1G_2G_5+G_4G_5}$

(c) $\dfrac{G_1G_2+G_4(1-G_2G_3)}{1-G_2G_3+G_1G_2G_5+G_4G_5}$

(d) none of the above

Q.78 Consider the following statements with regards to the signal flow graph shown below.

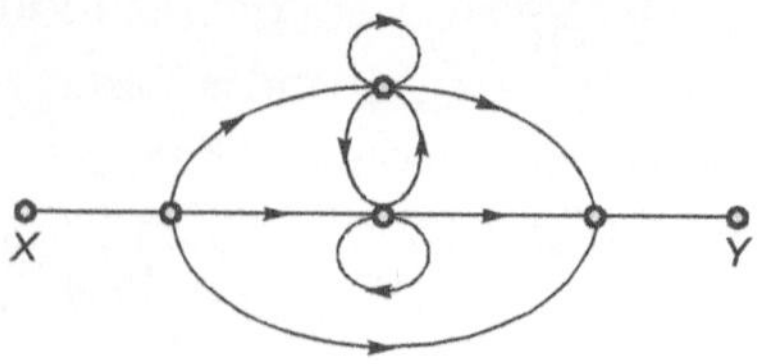

1. The number of forward paths are 2.
2. The number of forward paths are 3.
3. The number of loops are 3.
4. The number of loops are 5.
5. The number of non-touching loop pairs is 1.
6. The number of non-touching loop pairs is nil.

Which of these statements are correct?

(a) 1, 3 and 6 (b) 2, 4 and 6
(c) 2, 3 and 5 (d) 1, 4 and 5

Q.79 Consider the following statements related to signal flow graph (SFG) representation of a system:

1. In signal flow graph, the systems are represented by separate nodes.
2. During construction of SFG all variables, summing points and take off points are represented by separate nodes.
3. If a summing point is placed before a take-off point, the summing and take off points are represented by separate nodes.
4. If a summing point is placed after take off point, the summing and take off points are represented by a single node.
5. SFG method is applicable to linear systems only.

Which of these statements are not true?

(a) 2, 4 and 5 (b) 3 and 4
(c) 2, 3 and 4 (d) 3, 4 and 5

Q.80 The signal flow graph of the differential equation given by

$$\frac{d^2y}{dt^2}+10\frac{dy}{dt}-5y=3x$$

will be represented as

(a)

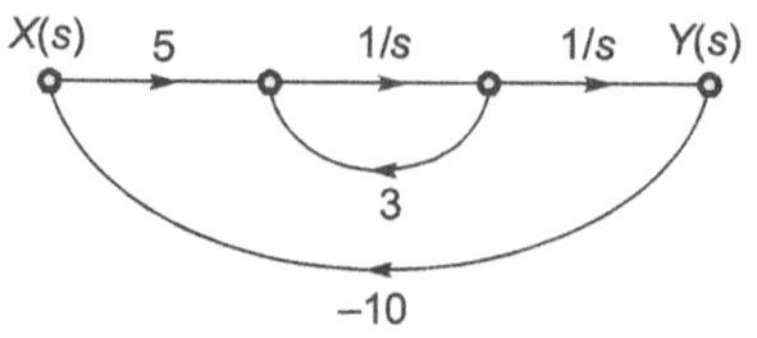

(b)

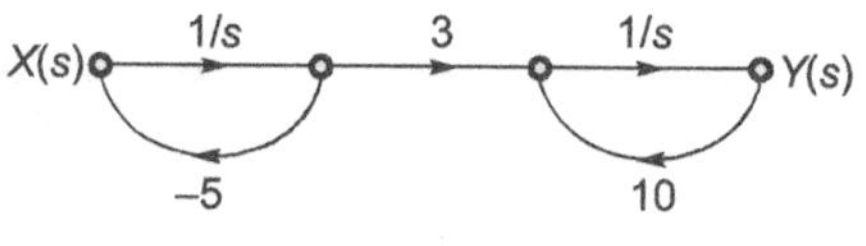

(c) 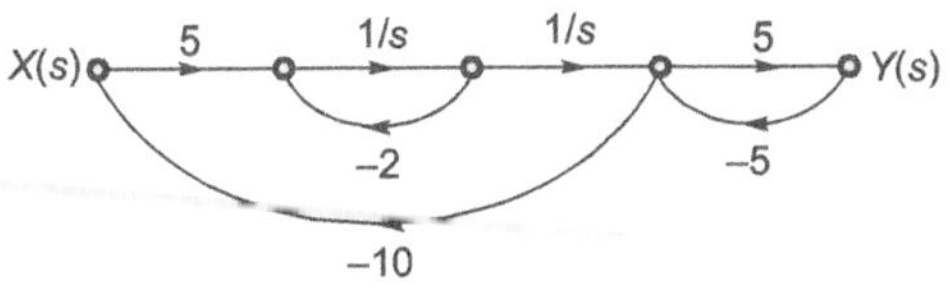

(d) 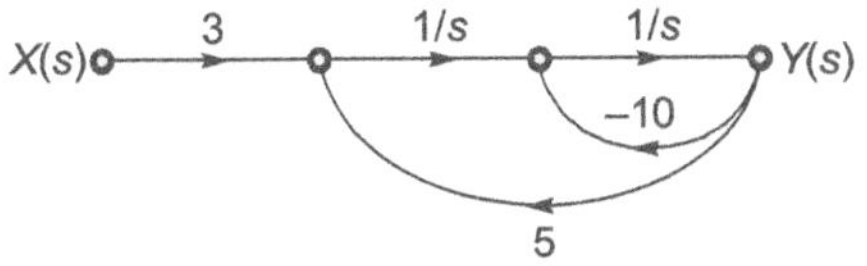

Q.81 The transfer function of the signal flow graph shown below is

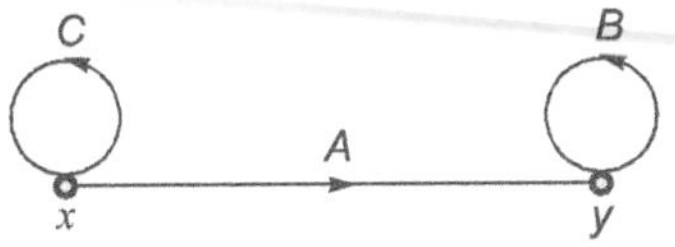

(a) $\dfrac{A}{(1-B)(1-C)}$ (b) $\dfrac{1}{(1-B-C)}$

(c) $\dfrac{A}{(1-B)}$ (d) $\dfrac{A}{(1-B)(C-1)}$

Q.82 The transfer function of the signal flow graph shown in figure below is

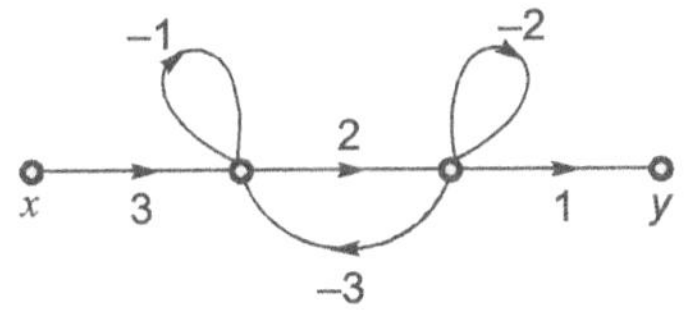

(a) $\dfrac{-3}{2}$ (b) $\dfrac{5}{6}$

(c) $\dfrac{1}{2}$ (d) $\dfrac{-11}{7}$

Q.83 Consider the block diagram shown in figure below.

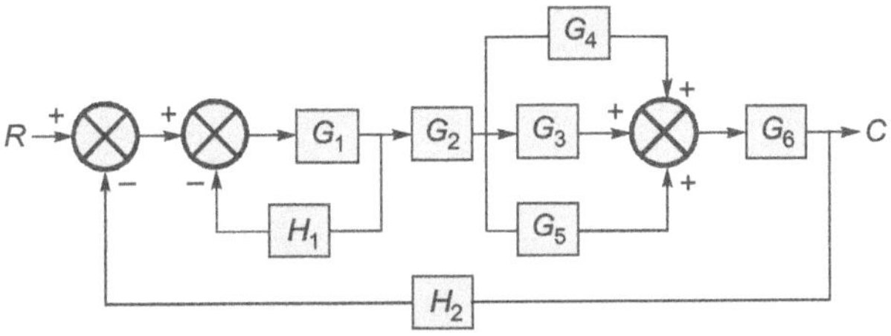

The signal flow graph of the above block diagram will be represented as:

(a) 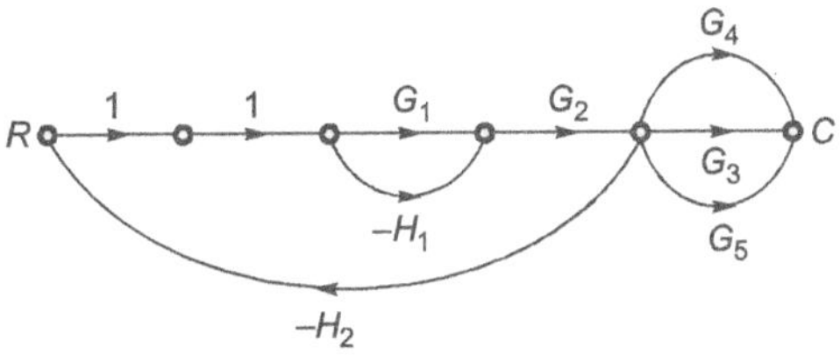

(b) 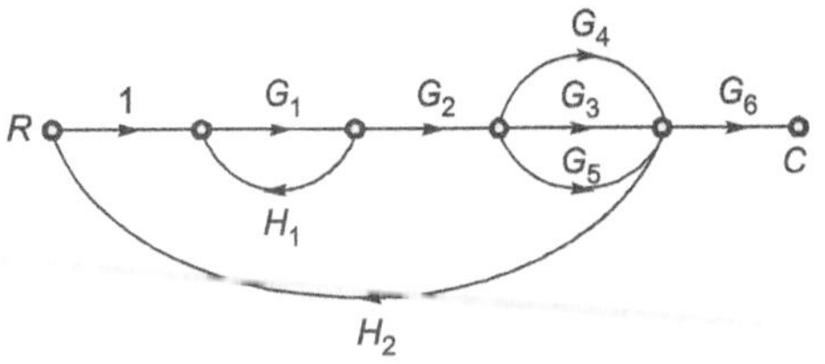

(c) 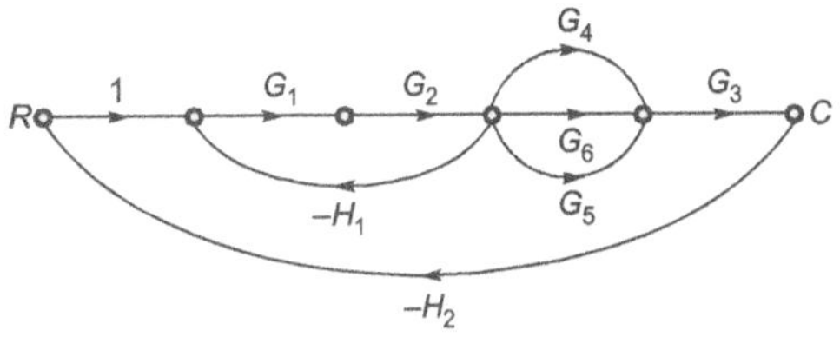

(d) 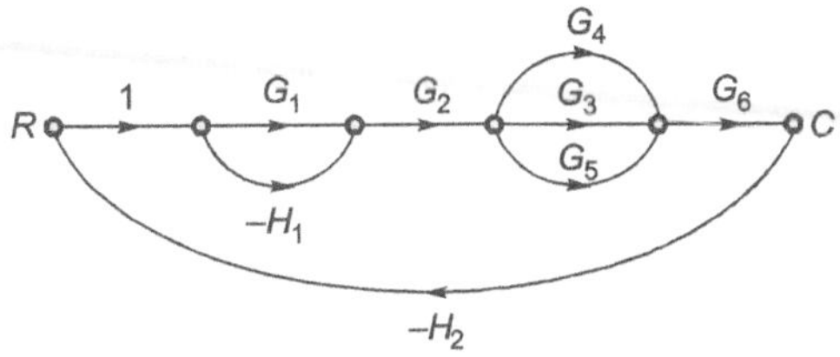

Q.84 In the previous question, the transfer function C/R will be given by:

(a) $\dfrac{G_1G_2G_3G_6 + G_1G_2G_4G_6 + G_1G_2G_5G_6}{1+G_1H_1 + G_1G_2G_3H_2 + G_1G_2G_4H_2 + G_1G_2G_5H_2}$

(b) $\dfrac{G_1G_2G_3 + G_1G_2G_4 + G_1G_2G_5G_6}{1+G_1H_1 + G_1G_2H_2 + G_3G_4H_2G_4}$

(c) $\dfrac{G_1G_2G_3 + G_1G_2G_3G_4 + G_1G_2G_3G_6}{1+G_1G_2H_1 + G_1G_2G_3H_2 + G_1G_2G_3G_4H_2 + G_1G_2G_3G_5H_2}$

(d) $\dfrac{G_1G_2G_3G_6 + G_1G_2G_4G_6 + G_1G_2G_5G_6}{1-G_1H_1 - G_1G_2G_3H_2 + G_1G_2G_4H_2 + G_1G_2G_5H_2}$

5. Feedback Characteristics

Q.85 In the system shown in figure below, the sensitivity of the closed loop transfer function w.r.t. parameter a is

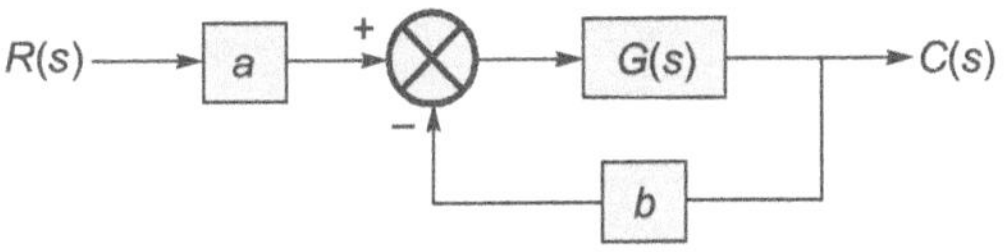

(a) $\dfrac{1}{1+aG(s)}$ (b) 1

(c) $\dfrac{1}{1+ab}$ (d) $\dfrac{1}{ab}$

Q.86 For the system shown in figure below, $S_{G_1}^{T}$ is equal to (where, T = closed loop transfer function)

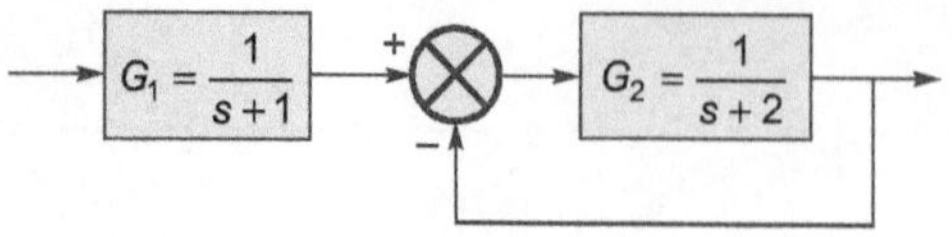

(a) −2.5 (b) 0
(c) −1.5 (d) 1

Q.87 In control system, the feedback
(a) reduces the error and increases the sensitivity of the system to parameter variation.
(b) increases both the error and sensitivity of the system to the parameter variation.
(c) reduces the error and the sensitivity of the system to parameter variation.
(d) increases both the error and sensitivity of the system to parameter variation.

Q.88 For the closed loop transfer function shown below, $\frac{E(s)}{R(s)}$ is given by

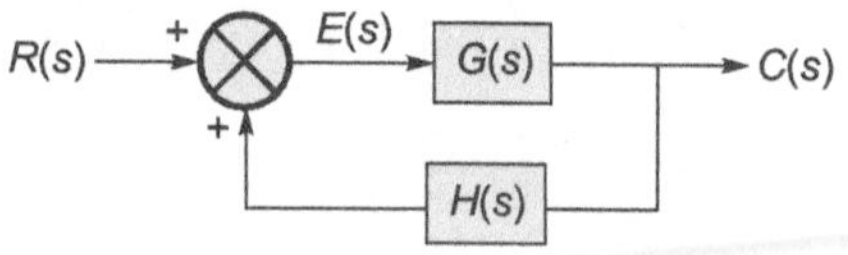

(a) $\frac{G(s)}{1+G(s)H(s)}$ (b) $\frac{1}{1+G(s)H(s)}$
(c) $\frac{1}{1-G(s)H(s)}$ (d) $\frac{G(s)}{1-G(s)H(s)}$

Q.89 Due to positive feedback
(a) stability of the system is increased but, gain is reduced.
(b) both stability and gain of the system are increased.
(c) both stability and gain of the system are reduced.
(d) stability of the system is reduced but, gain is increased.

Q.90 Consider the following statements:
1. Closed loop system is less sensitive to variations in feedback path parameters than variations in forward path parameters.
2. Feedback can improve the stability or may be harmful to stability if it is not properly designed and applied.
3. Feedback controls the time response of the system by adjusting the location of poles.

Which of the these statements are correct?
(a) 2 and 3 only (b) 1 and 2 only
(c) 2 only (d) 1, 2 and 3

Q.91 An unity feedback control system has a transfer function of $\frac{Ks+b}{s^2+as+b}$. The open loop transfer function of this system is:

(a) $\frac{1}{s^2+as-Ks}$ (b) $\frac{Ks+b}{s^2-s(K-a)}$
(c) $\frac{Ks-b}{s^2-s(K-a)}$ (d) $\frac{s+b}{s^2+as-K}$

Q.92 **Assertion (A):** Feedback in control system improves the time response of the system.
Reason (R): Time constant of an open loop system is less than the time constant of a closed loop system.
(a) Both A and R are true and R is a correct explanation of A.
(b) Both A and R are true but R is not a correct explanation of A.
(c) A is true but R is false.
(d) A is false but R is true.

Q.93 For the unity feedback system shown below, if the input to the system is a unit-step signal, then the integral square error $I_e = \int_0^{\infty} e^2(t)dt$ for infinite value of gain k will be

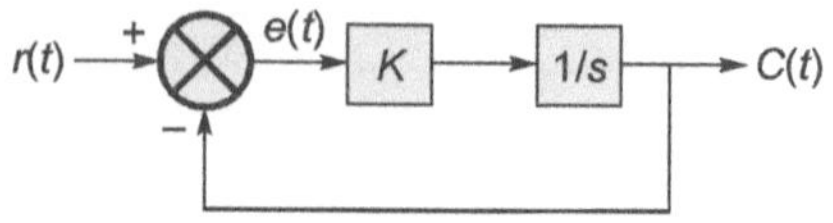

(a) 1.0 (b) −0.25
(c) 0.5 (d) 0

Q.94 Introduction of feedback to a system
(a) increases the stability
(b) decreases the stability
(c) has no effect on stability
(d) none of the above

Q.95 In a speed control system, output rate feedback is used to
(a) limit the speed of motor
(b) limit the acceleration of motor
(c) reduce the damping of the system
(d) increase the gain margin

Q.96 The effect of tacho-meter feedback in a control system is to reduce
(a) only time constant
(b) only gain
(c) damping
(d) both gain and time constant

6. Modelling of Control Systems

Q.97 The electrical time-constant of an armature-controlled dc servomotor is
(a) larger than mechanical time-constant
(b) not related to mechanical time-constant
(c) equal to mechanical time-constant
(d) smaller than mechanical time-constant

Q.98 A mass spring system under equilibrium condition is shown in figure below.

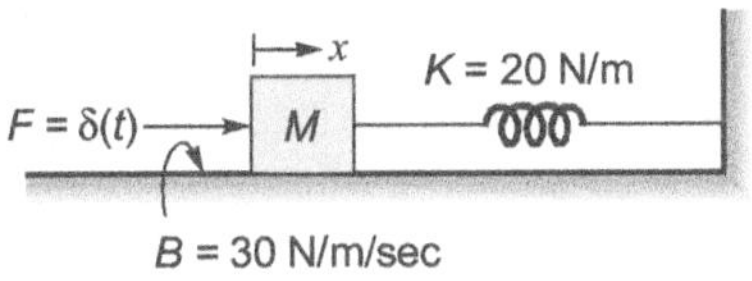

The output displacement $x(t)$ for M = 10 kg is given by
(a) $(e^{-t} - e^{-2t})$ (b) $10\,(e^{-t} - e^{-2t})$
(c) $0.1\,(e^{-t} - e^{-2t})$ (d) $5\,(e^{-t} + e^{-2t})$

Q.99 Two wheels are mechanically coupled as shown in figure below.

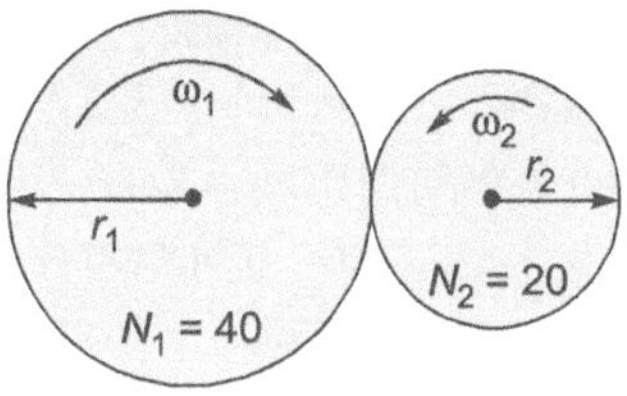

Here, ω_1 and ω_2 = angular velocities (in rad/s)
T_1 and T_2 = Torque (in N-m)
r_1 and r_2 = radius of the wheels (in cm)
θ_1 and θ_2 = angular displacements (in degrees)
N_1 and N_2 = No. of teeths on wheels
It is given that ω_1 = 10 rad/s, N_1 = 40, N_2 = 20, T_1 = 5 N-m, r_1 = 20 cm and θ_1 = 20°.
Match **List-I** with **List-II** and select the correct answer using the codes given below the lists:

List-I	List-II
A. ω_2	1. 40
B. T_2	2. 20
C. θ_2	3. 10
D. r_2	4. 2.5

Codes:

	A	B	C	D
(a)	2	1	4	3
(b)	2	4	1	3
(c)	2	3	4	1
(d)	1	2	3	4

Q.100 The poles and zeros of the transfer function $\frac{X_2(s)}{X_1(s)}$ for the mechanical system shown below are respectively

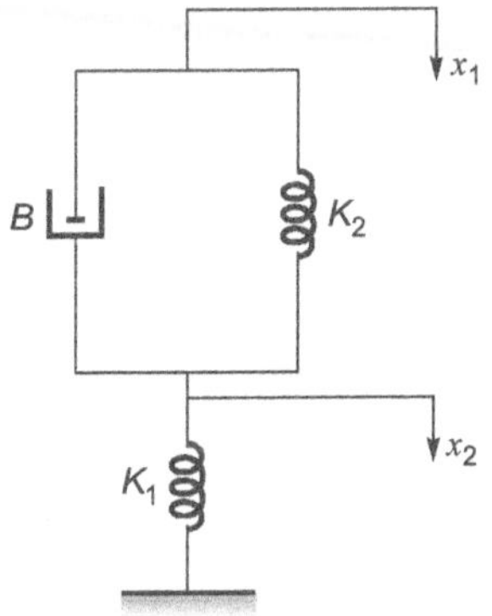

(a) $-\left[\frac{K_1+K_2}{B}\right]$ and $\frac{-K_2}{B}$

(b) $\frac{-K_1}{B}$ and $-\left[\frac{K_1+K_2}{B}\right]$

(c) $-\left[\frac{K_1+K_2}{B}\right]$ and $\frac{-K_1}{B}$

(d) $\frac{-K_2}{B}$ and $-\left[\frac{K_1+K_2}{B}\right]$

Q.101 Consider the mechanical system shown in figure below.

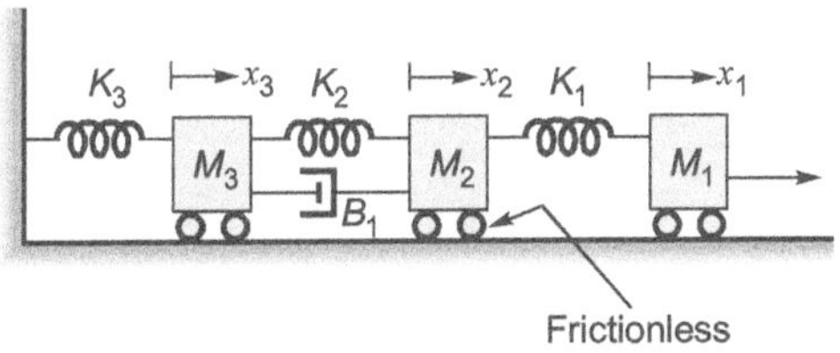

Which one of the following circuits shows the force-voltage analogous electrical circuit for the mechanical diagram shown above?

(a) 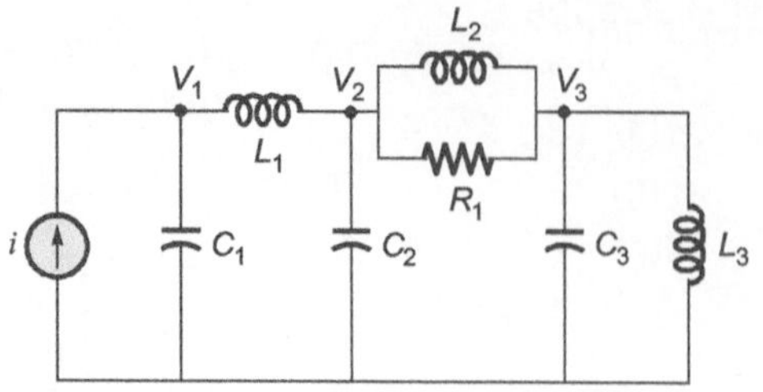

(b) 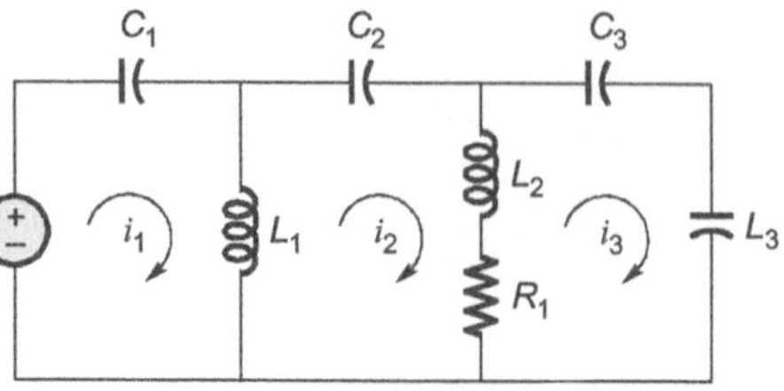

(c) 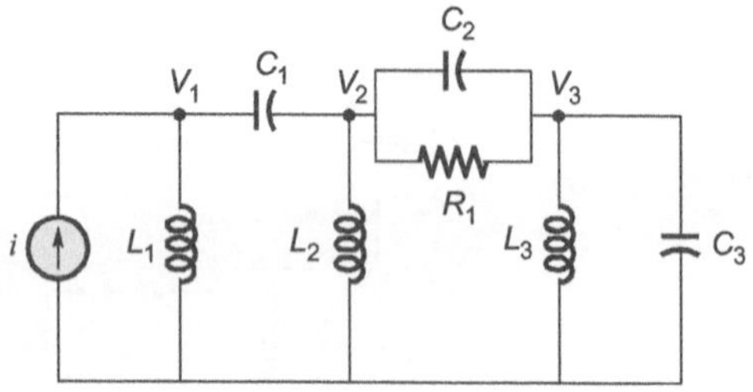

(d) 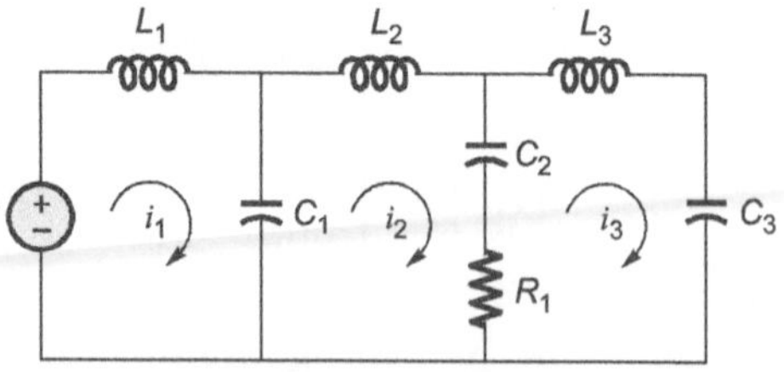

Q.102 Match **List-I** with **List-II** and select the correct answer using the codes given below the lists:

List-I	**List-II**
A. Torque	**1.** Capacitance
B. Moment of inertia	**2.** Inductance
C. Angular displacement	**3.** Current
D. Elastance	**4.** Flux linkage

Codes:

	A	B	C	D
(a)	4	2	3	1
(b)	4	3	2	1
(c)	3	1	4	2
(d)	3	2	4	1

Q.103 Consider the mechanical system shown below:

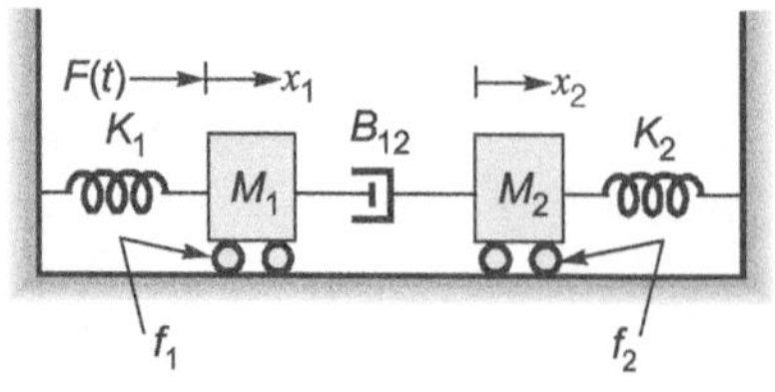

The system is described by the differential equations:

(a) $F(t) = M_1 \dfrac{d^2x_1}{dt^2} + f_1 \dfrac{dx_1}{dt} + B_{12} \dfrac{d(x_2 - x_1)}{dt} + K_2 x_2$

$B_{12} \dfrac{d}{dt}(x_2 - x_1) = M_2 \dfrac{d^2x_2}{dt^2} + f_2 \dfrac{dx_2}{dt} + K_1 x_1$

(b) $F(t) = M_1 \dfrac{d^2x_1}{dt^2} + f_1 \dfrac{dx_1}{dt} + B_{12} \dfrac{d}{dt}(x_2 - x_1) + K_1 x_1$

$B_{12} \dfrac{d}{dt}(x_1 - x_2) = M_2 \dfrac{d^2x_2}{dt^2} + f_2 \dfrac{dx_2}{dt} + K_1 x_1$

(c) $F(t) = M_1 \dfrac{d^2x_1}{dt^2} + f_1 \dfrac{dx_1}{dt} + B_{12} \dfrac{d}{dt}(x_1 - x_2) + K_1 x_1$

$B_{12} \dfrac{d}{dt}(x_1 - x_2) = M_2 \dfrac{d^2x_2}{dt^2} + f_2 \dfrac{dx_2}{dt} + K_2 x_2$

(d) none of the above

Q.104 Which of the following is not a desirable property of a servomotor in feedback control systems?
(a) Response should be fast.
(b) Torque-speed characteristic should be linear.
(c) It should be reversible.
(d) Relation between electrical control system and rotor speed should be non-linear.

Q.105 Consider the following statements related to A.C. servomotors compared to D.C. servomotor used in control system applications.
1. It has less efficiency.
2. It is maintenance free.
3. It gives smooth operation.
4. It gives high power output.

Which of these statements is/are true regarding an A.C. servomotor?
(a) 1, 3, and 4 (b) 4 only
(c) 1, 2 and 3 (d) 2, 3 and 4

Q.106 A field controlled DC servomotor is a
(a) open loop system with a large time constant.
(b) closed loop system with a small time constant.
(c) open loop system with a small time constant.
(d) closed loop system with a large time constant.

Q.107 Match **List-I (Control system components)** with **List-II (Principle of operation)** and select the correct answer using the codes given below the lists:

List-I

A. Synchronous
B. Potentiometer
C. Tachometer

List-II

1. Mechanical displacement into electrical energy.
2. Angular position of a shaft into an electrical signal.
3. Electrical output proportional to speed.

Codes:

	A	B	C
(a)	3	1	2
(b)	2	1	3
(c)	1	2	3
(d)	3	2	1

Q.108 An a.c. servomotor
(a) is noise free and has smaller R to X ratio.
(b) is noisy and has higher R to X ratio.
(c) is noise free and has larger R to X ratio.
(d) is noisy and has smaller R to X ratio.

Q.109 Match **List-I (Control System Components)** with **List-II (Applications)** and select the correct answer using the codes given below the lists:

List-I

A. Stepper Motors
B. Synchros
C. Servomotors
D. Potentiometer

List-II

1. Radars, electromechanical actuators.
2. Displacement transducer
3. Printers, Watches
4. Detectors and encoders

Codes:

	A	B	C	D
(a)	1	2	3	4
(b)	3	2	4	1
(c)	1	4	3	2
(d)	3	4	1	2

Q.110 Match **List-I (Types of friction)** with **List-II (Corresponding frictional force vs Velocity curve)** and select the correct answer using the codes given below the lists:

List-I — List-II

A. Static friction — 1.
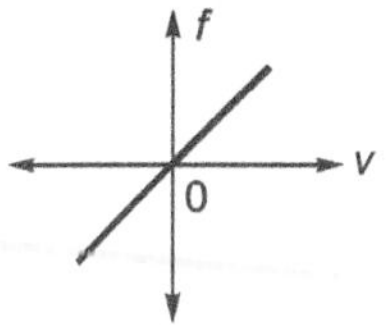

B. Coulomb friction — 2.
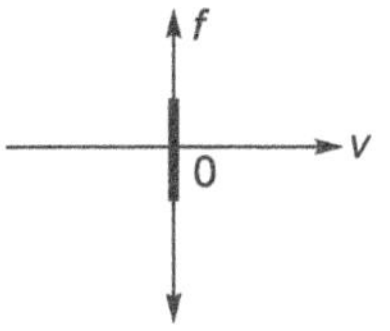

C. Viscous friction — 3.
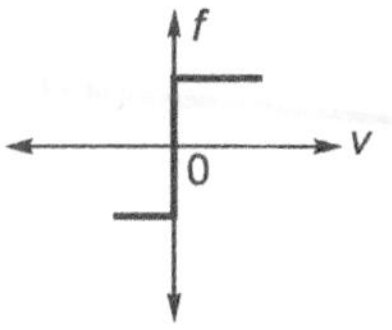

Codes:

	A	B	C
(a)	1	2	3
(b)	2	3	1
(c)	2	1	3
(d)	3	2	1

Q.111 **Assertion (A):** D'Alembert's principle states that "for any body, the algebraic sum of externally applied forces and the forces resisting motion in any given direction is zero."
Reason (R): D'Alembert principle is used in writing the equation of motion of mechanical systems.
(a) Both A and R are true and R is a correct explanation of A.
(b) Both A and R are true but R is not a correct explanation of A.
(c) A is true but R is false.
(d) A is false but R is true.

Q.112 **Assertion (A):** When two time constant elements are cascaded interactively, the overall transfer function of such an arrangement is the product of two individual transfer functions.
Reason (R): There is loading effect present in such an arrangement.
(a) Both A and R are true and R is a correct explanation of A.
(b) Both A and R are true but R is not a correct explanation of A.
(c) A is true but R is false.
(d) A is false but R is true.

Q.113 Assertion (A): An a.c. servomotor used in control system applications must have high starting torque and low inertia.
Reason (R): A large value of rotor diameter and small axial length gives a higher value of starting torque and a low value of inertia.
(a) Both A and R are true and R is a correct explanation of A.
(b) Both A and R are true but R is not a correct explanation of A.
(c) A is true but R is false.
(d) A is false but R is true.

Q.114 Assertion (A): When the rotor of the synchro is inclined along any one of the stator axis completely, then this position is called electrical zero position.
Reason (R): When the rotor of the synchro is inclined along any one of the stator axis completely, then maximum voltage will be induced in that winding and almost zero voltage will be induced in the other two windings.
(a) Both A and R are true and R is a correct explanation of A.
(b) Both A and R are true but R is not a correct explanation of A.
(c) A is true but R is false.
(d) A is false but R is true.

Q.115 Assertion (A): Stepper motors can be used for speed control.
Reason (R): The motor speed is proportional to the rate of command pulses.
(a) Both A and R are true and R is a correct explanation of A.
(b) Both A and R are true but R is not a correct explanation of A.
(c) A is true but R is false.
(d) A is false but R is true.

Q.116 The stator of the synchros is made of
(a) laminated silicon steel
(b) cast iron
(c) stainless steel
(d) none of the above

Q.117 A synchro is used to
(a) accelerate a rotating shaft
(b) convert an angular position of shaft into an electrical signal
(c) convert linear motion into angular position
(d) amplify low frequency signals

Q.118 A synchro transmitter-receiver unit is a
(a) two phase ac device
(b) 3-phase ac device
(c) dc device
(d) single phase ac device

Q.119 The transfer function of a tachometer is of the form
(a) Ks (b) $\frac{K}{s}$
(c) $\frac{K}{1+s}$ (d) $\frac{K}{s(1+s)}$

Q.120 In force-current analogy stiffness constant K is analogous to
(a) inductance
(b) reciprocal of inductance
(c) capacitance
(d) reciprocal of capacitance

Q.121 In liquid level systems, the units of capacitance of a tank is
(a) m^2 (b) m^3
(c) $\frac{m^2}{s}$ (d) $m^{-2}\,s$

Q.122 In force-voltage analogy, viscous friction constant B is analogous to
(a) resistance
(b) reciprocal of resistance
(c) inductance
(d) capacitance

7. Time Domain Analysis of Control Systems

Q.123 The solution of the differential equation
$$\frac{d^2y}{dt^2}+2\frac{dy}{dt}+2y=3 \text{ is}$$
(a) overdamped (b) critically damped
(c) oscillatory (d) underdamped

Q.124 A second order system is said to be critically damped if the damping factor (ξ) is
(a) $\xi = 0.707$ (b) $\xi < 1$
(c) $\xi = 1$ (d) $\xi > 1$

Q.125 The ratio of damped frequency to natural frequency of the given system having damping ratio ξ is

(a) ξ (b) $\xi\omega_n$

(c) $\sqrt{1-\xi^2}$ (d) $\frac{1}{\xi\omega_n}$

Q.126 In time domain specification, the time delay is the time required for the response to reach
(a) 25% of final value
(b) 50% of the final value
(c) 10% of the final value
(d) 63.3% of the final value

Q.127 For the second order prototype system, when the undamped natural frequency increases, the maximum overshoot of the output
(a) stays the same (b) decreases
(c) doubles (d) increases

Q.128 Damping in a control system is a function of

(a) gain (b) $\frac{1}{\sqrt{\text{gain}}}$

(c) $\frac{1}{\text{gain}}$ (d) $\sqrt{\text{gain}}$

Q.129 Damping ratio is defined as the ratio of
(a) undamped natural frequency to critical damping.
(b) critical damping to actual damping.
(c) damped natural frequency to actual damping.
(d) actual damping to critical damping.

Q.130 A linear second-order system with the transfer function $G(s) = \frac{49}{s^2 + 16s + 49}$ is initially at rest and is subjected to a step input signal.

The response of the system will exhibit a peak overshoot of
(a) 25% (b) 0%
(c) 2% (d) none of these

Q.131 The closed-loop transfer function of a system is

$$\frac{C(s)}{R(s)} = \frac{10(s+1)}{(s+2)(s+5)}$$

Which of the following statements is not true?
(a) The system is non-oscillatory.
(b) The closed-loop frequency response of this system exhibits a resonance peak.
(c) The damping ratio of the closed-loop poles is less than unity.
(d) The unit step response exhibits overshoot due to the presence of a zero.

Q.132 The steady state error of a control system can be reduced by
(a) increasing time constant of the system only.
(b) increasing gain constant of the system only.
(c) increasing both the time constant and gain of the system.
(d) none of the above.

Q.133 The system shown in figure below has unit step input.

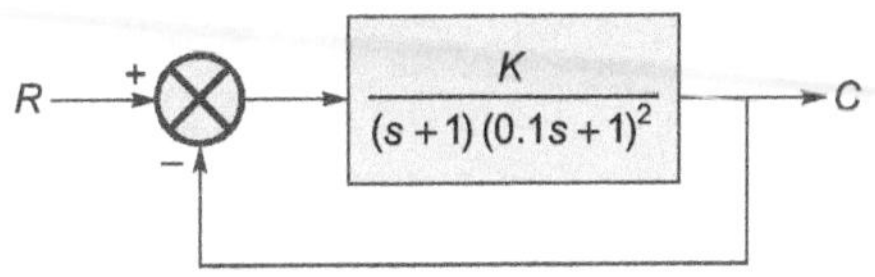

In order that the steady state error is 0.1, the value of K required should be
(a) 9.0 (b) 10.0
(c) 0.1 (d) 20.0

Q.134 The steady-state error coefficients for a system are $K_p = \infty$, K_v = finite constant and $K_a = 0$. The type of the system is
(a) 4 (b) 2
(c) 1 (d) 0

Q.135 The steady-state error of a feedback control system with an acceleration input becomes finite in a
(a) type-3 system (b) type-2 system
(c) type-1 system (d) type-0 system

Q.136 Consider the unity feedback system shown below:

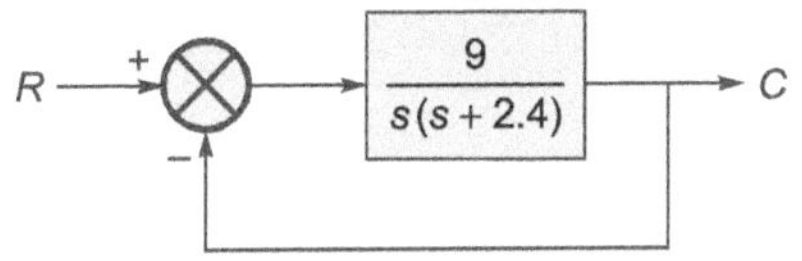

The settling time of the resulting second order system for 2% tolerance band will be
(a) 2.25 seconds (b) 3.5 seconds
(c) 2.65 seconds (d) 3.33 seconds

Q.137 For a control system, the Laplace transform of error signal $e(t)$ is given by $\frac{8(s+3)}{s(s+10)}$. The steady state value of the error will be
(a) 0.8 (b) 0.3
(c) 2.4 (d) 3.2

Q.138 For a type one system, the steady-state error due to step input is equal to
(a) zero (b) 0.75
(c) infinite (d) 0.50

Q.139 For an unity feedback control system with $G(s) = \frac{K}{s(s+8)}$, the value of K for damping ratio of 0.5 is
(a) 16 (b) 64
(c) 8 (d) 32

Q.140 The damping ratio of a system having the characteristic equation $s^2 + 2s + 8 = 0$ is
(a) 0.300 (underdamped)
(b) 0.353 (underdamped)
(c) 1.24 (overdamped)
(d) 1.02 (overdamped)

Q.141 The closed-loop transfer function of a unity-feedback system is given by $\frac{C(s)}{R(s)} = \frac{1}{(1+Ts)}$.
The steady state error to a unit ramp input is:
(a) Infinite (b) $\frac{1}{T}$
(c) 1 (d) T

Q.142 The peak overshoot of step-input response of an underdamped second-order system is an indication of
(a) rise time of the system
(b) settling time of the system
(c) damping ratio of the system
(d) natural frequency of the system

Q.143 Consider the position control system shown below:

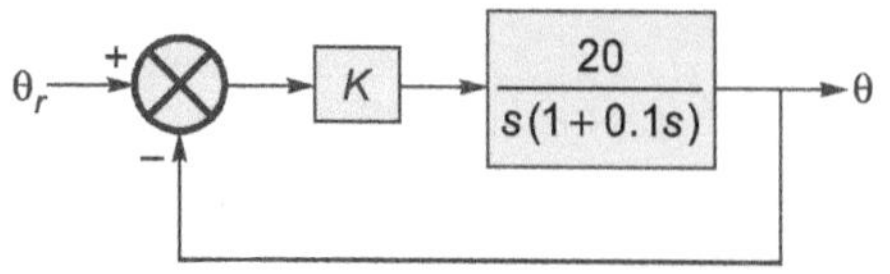

The value of K such that the steady state error is 20° for input $\theta_r = 300t$ rad/sec, is
(a) 42.97 (b) 58.62
(c) 0.75 (d) none of these

Q.144 Assertion (A): With the increase in bandwidth of the system the response of the system becomes fast.
Reason (R): Damping ratio of the system decreases with the increase in bandwidth.
(a) Both A and R are true and R is a correct explanation of A.
(b) Both A and R are true but R is not a correct explanation of A.
(c) A is true but R is false.
(d) A is false but R is true.

Q.145 For the stable system described by the block diagram shown below, Match **List-I (Static error coefficients)** with **List-II (Values)** and select the correct answer using the codes given below the lists:

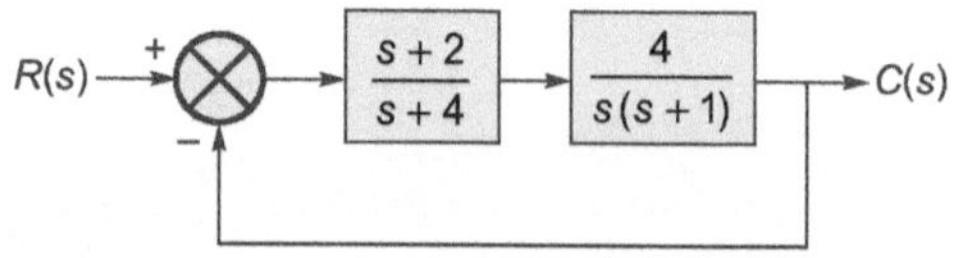

List-I	List-II
A. K_a	1. ∞
B. K_v	2. 0
C. K_p	3. 2
	4. −1

Codes:

	A	B	C
(a)	2	1	4
(b)	1	4	3
(c)	2	1	3
(d)	2	3	1

Q.146 The unit step response of a system is given by $c(t) = 1 + 0.25\,e^{-50t} - 1.25\,e^{-10t}$
The given system is
(a) underdamped (b) critically damped
(c) overdamped (d) undamped

Q.147 Match **List-I (Transfer function of systems)** with **List-II (Nature of damping)** and select the correct answer using the codes given below the lists:

List-I	List-II
A. $\frac{12}{s^2+8s+12}$	1. Underdamped
B. $\frac{16}{s^2+8s+16}$	2. Overdamped
C. $\frac{20}{s^2+8s+20}$	3. Undamped
D. $\frac{4}{s^2+4}$	4. Critically damped

Codes:

	A	B	C	D
(a)	2	4	1	3
(b)	1	2	4	3
(c)	2	1	4	3
(d)	1	4	3	2

Q.148 The block diagram of an electronic pacemaker is given in figure below.

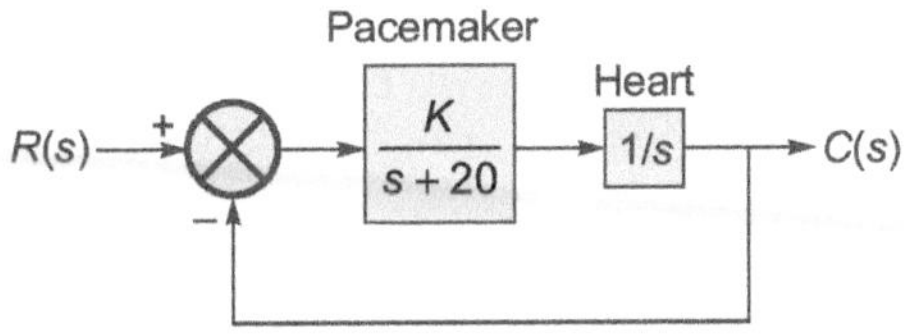

What is the value of K for which the steady-state error to a unit ramp input is 0.02?

(a) 10 (b) 100
(c) 10^3 (d) 1

Q.149 Which one of the following equations gives the steady-state error for a unity feedback system excited by $r(t) = 2 + 5t + 2t^2$?

(a) $\frac{2}{1+K_p}+\frac{1}{K_v}+\frac{4}{K_a}$ (b) $\frac{2}{1+K_p}+\frac{5}{K_v}+\frac{4}{K_a}$

(c) $\frac{1}{K_p}+\frac{1}{K_v}+\frac{1}{K_a}$ (d) $\frac{1}{1+K_p}+\frac{5}{K_v}+\frac{2}{K_a}$

Q.150 An unity feedback control system with closed loop transfer function is given by

$$\frac{C(s)}{R(s)}=\frac{Ks+b}{s^2+as+b}$$

The steady state error due a unit ramp input response is

(a) $\frac{b}{a-k}$ (b) $\frac{a}{b}+\frac{K}{b}$

(c) $\frac{a}{b}-K$ (d) $\frac{a}{b}-\frac{K}{b}$

Q.151 The second order approximation using dominant pole concept for the transfer function

$$\frac{C(s)}{R(s)}=\frac{5}{(s+10)(s^2+s+1)}$$ is

(a) $\frac{5}{s^2+s+1}$ (b) $\frac{5}{(s+10)}$

(c) $\frac{0.5}{s^2+s+1}$ (d) $\frac{1}{s^2+s+1}$

Q.152 The unit step response of the system represented by the block diagram shown below is

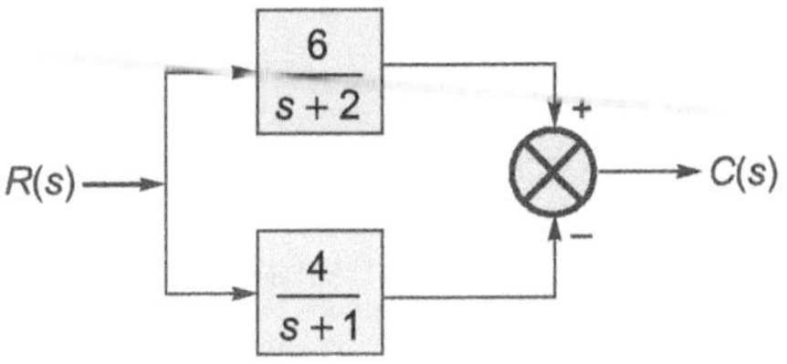

(a) $-1-3e^{-2t}+4e^{-t}$ (b) $1+3e^{-2t}+4e^{-t}$
(c) $-1+3e^{-2t}-4e^{-t}$ (d) none of the above

Q.153 For the given series RLC circuit, damping ratio (ξ) and undamped natural frequency (ω_n) are respectively given by

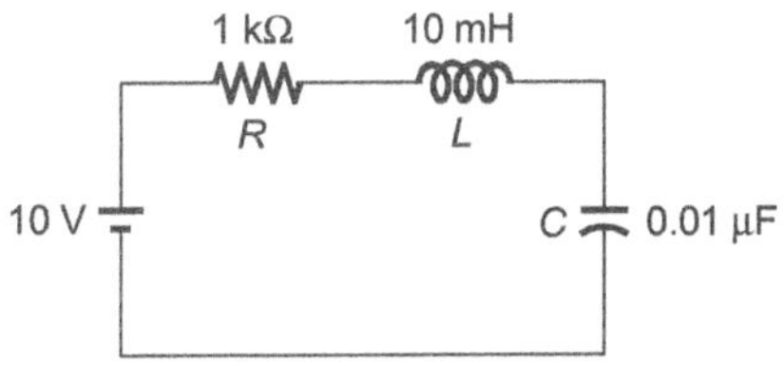

(a) 0.75 and 10 k-rad/sec
(b) 0.5 and 100 k-rad/sec
(c) 0.75 and 100 k-rad/sec
(d) 0.5 and 10 k-rad/sec

Q.154 For the unity feedback control system shown below, by what factor the amplifier gain K should be multiplied so that the damping ratio is increased from 0.3 to 0.9?

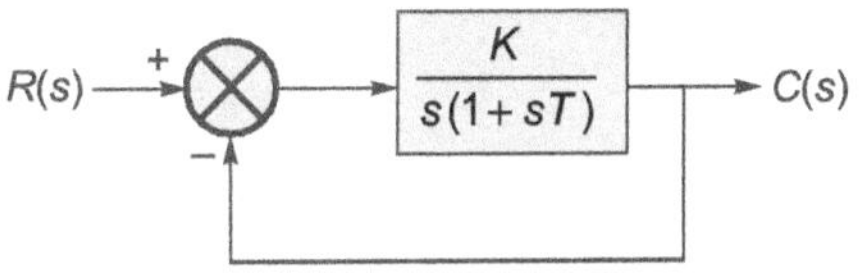

(a) $\frac{1}{9}$ (b) 3

(c) $\frac{1}{3}$ (d) 9

Q.155 A system is shown in figure below.

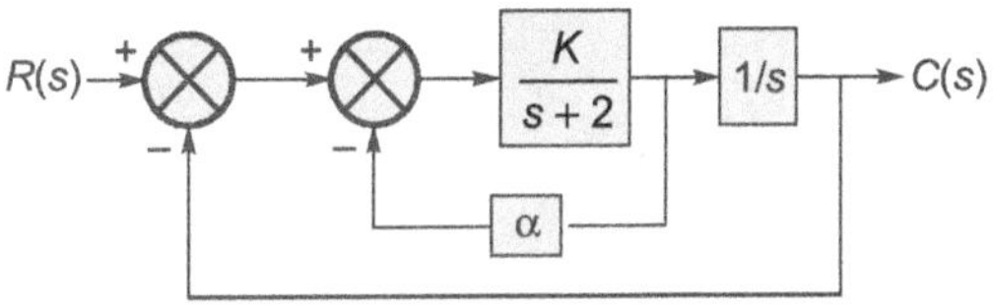

What are the values of K and α respectively so that the system has a damping ratio of 0.7 and an undamped natural frequency ω_n of 4 rad/sec?

(a) 16, 0.50 (b) 8,0.225
(c) 16,0.225 (d) 8,0.50

Q.156 For type-2 system, the steady state error due to parabolic input is
(a) finite (b) zero
(c) infinite (d) indeterminate

Q.157 The open loop transfer function of a unity-feedback control system with unit step input is given by:

$$G(s) = \frac{1}{s(s+1)}$$

The settling time of the system for 5% tolerance band is,
(a) 8 sec (b) 6 sec
(c) 4 sec (d) 4.5 sec

Q.158 The integral of a ramp function becomes
(a) a unit-impulse function
(b) a gate function
(c) a unit-step function
(d) a parabolic function

Q.159 Consider a unity feedback control system as shown in figure below:

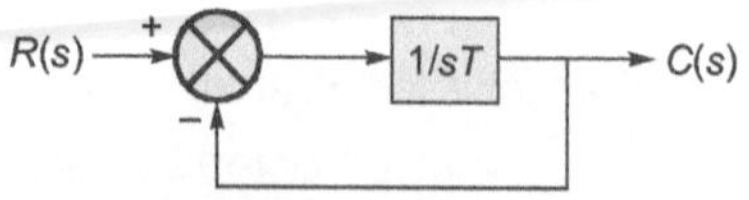

The error signal $e(t)$ for a unit ramp input for $T \geq 0$ is
(a) T (b) $T - e^{-t/T}$
(c) $T(1 - e^{-t/T})$ (d) $t + T \cdot e^{-t/T}$

Q.160 A second order system has a closed loop transfer function $\frac{C(s)}{R(s)} = \frac{25}{s^2 + 8s + 25}$.

If the system is initially at rest and subjected to a unit step input at t = 0, the third peak in the response will occur at

(a) $\frac{2\pi}{3}$ sec. (b) π sec.

(c) $\frac{5\pi}{3}$ sec. (d) π sec.

Q.161 A closed servo system is represented by the differential equation

$$\frac{d^2c}{dt^2} + 8\frac{dc}{dt} = 64e$$

where c is the displacement of the output shaft, r is the displacement of the input shaft and $e = r - c$.
The damping ratio and natural frequency for this system are respectively
(a) 0.5 and 8 rad/s (b) 0.75 and 4 rad/s
(c) 0.5 and 4 rad/s (d) 0.75 and 8 rad/s

Q.162 The poles of a closed loop control system are located at $-1 \pm j$.
For the given system match **List-I (Time domain specifications)** with **List-II (Values)** and select the correct answer using the codes given below the lists:

List-I
A. Damping ratio
B. Undamped natural frequency (rad/s)
C. Damping factor
D. Settling time (in secs)

List-II
1. 4.0
2. 1.0
3. 0.707
4. 1.414

Codes:

	A	B	C	D
(a)	3	1	2	4
(b)	3	4	2	1
(c)	4	2	1	3
(d)	4	1	2	3

Q.163 Consider the block diagram of a control system shown below.

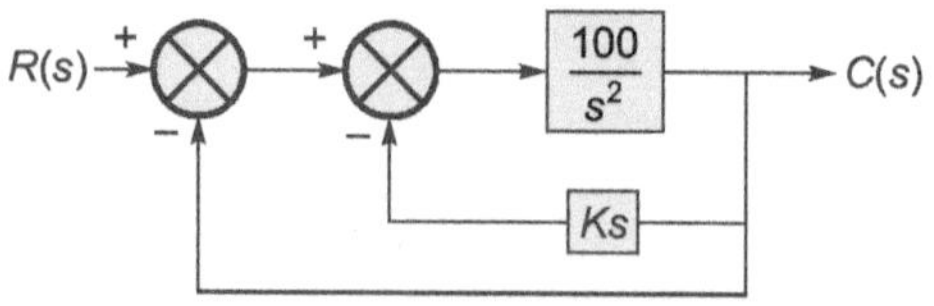

What is the value of gain k such that the system is non oscillatory yet has the lowest possible settling time to step input?
(a) 0 (b) 0.5
(c) 0.2 (d) none of these

Q.164 **Assertion (A) :** The "dominant pole" concept is used to control the dynamic performance of the system, whereas the "insignificant poles" are used for the purpose of ensuring that the controller transfer function can be realized by physical components.

Reason (R) : If the magnitude of the real part of a pole is atleast 5 to 10 times that of a dominant pole or a pair of complex dominant poles, the pole may be regarded as Insignificant in so far as the transient response is concerned.
(a) Both A and R are true and R is a correct explanation of A.
(b) Both A and R are true but R is not a correct explanation of A.
(c) A is true but R is false.
(d) A is false but R is true.

Q.165 **Assertion (A):** The step-function input represents an instantaneous change in the reference input.
Reason (R): If the input is an angular position of a mechanical shaft, a step input represents the sudden rotation of the shaft.
(a) Both A and R are true and R is a correct explanation of A.
(b) Both A and R are true but R is not a correct explanation of A.
(c) A is true but R is false.
(d) A is false but R is true.

Q.166 Consider the following statements related to second order control system with unit step input:
1. The right-half s-plane corresponds to negative damping.
2. Negative damping gives a response that doesn't grows in magnitude without bound with time.
3. The imaginary axis corresponds to zero damping.
4. The maximum overshoot is often used to measure the relative stability of a control system.
5. Higher the damping in the system larger is the peak overshoot in the transient response of the system.

Which of the above statements is/are correct?
(a) 1 and 3 (b) 1, 4 and 5
(c) 2, 3 and 4 (d) 1, 3 and 4

Q.167 Which of the following is not true regarding addition of a pole to the forward-path transfer function?
(a) Maximum overshoot of the closed-loop system is increased.
(b) Rise time of the step response is increased.
(c) Stability of the system is increased.
(d) Bandwidth of the system is reduced.

Q.168 For a second-order prototype system, when the undamped natural frequency is increased, the maximum overshoot of the output will
(a) increase
(b) remains the same
(c) decrease
(d) increase or decrease depending on the value of damping factor of the system

Q.169 Consider a unity feedback system shown in figure below where, $K > 0$ and $a > 0$.

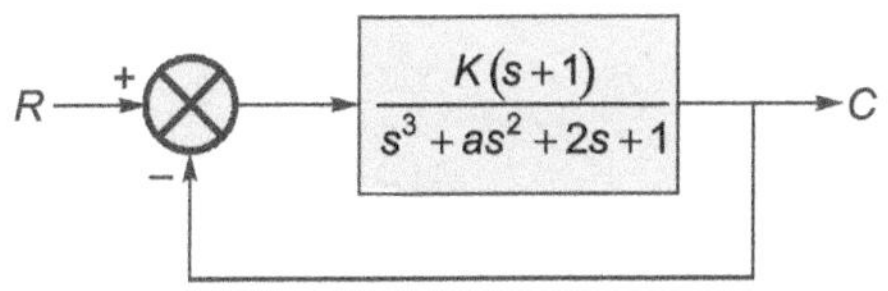

What are the values of k and a respectively so that the system oscillates at a frequency of 2 rad/sec?
(a) 3, 0.50 (b) 2, 0.50
(c) 0.2, 0.50 (d) 2, 0.75

Q.170 The transient response of a system is mainly due to
(a) internal forces (b) stored energy
(c) friction (d) external forces

Q.171 A second order system has a damping ratio ξ and undamped natural frequency of oscillation ω_n respectively. The settling time at 5% of tolerance band of the system is
(a) $\frac{2}{\xi\omega_n}$ (b) $\frac{4}{\xi\omega_n}$
(c) $\frac{3}{\xi\omega_n}$ (d) $\frac{1}{\xi\omega_n}$

Q.172 The unit impulse response of a linear time invariant second order system is
$$g(t) = 10e^{-8t}\sin 6t \qquad (t \geq 0)$$
The natural frequency and the damping ratio of the system are respectively
(a) 10 rad/sec and 0.6
(b) 10 rad/sec and 0.8
(c) 5 rad/sec and 0.6
(d) 6 rad/sec and 0.8

Q.173 A quantity whose magnitude has a definite repeating time cycle is called a
(a) transient
(b) steady state periodic
(c) steady state aperiodic
(d) transient state periodic

Q.174 In a second order underdamped system, the value of peak overshoot in per unit terms is

(a) $M_p = \exp\left(\dfrac{\pi\xi}{\sqrt{1-\xi^2}}\right)$

(b) $M_p = -\exp\left(\dfrac{\pi\xi}{\sqrt{1-\xi^2}}\right)$

(c) $M_p = \exp\left(\dfrac{-\pi\xi}{\sqrt{1-\xi^2}}\right)$

(d) $M_p = \left(\dfrac{\pi\xi}{\sqrt{1-\xi^2}}\right)$

Q.175 A first order system has a time constant of 20*s*. It is subjected to a step input. The settling time of the output is assumed to be the time it reaches 95% of its final steady state value. The settling time of the system is
(a) 100 sec (b) 60 sec
(c) 20 sec (d) infinite

Q.176 In second order system, the frequency of damped oscillations is 18 rad/s. The value of damping factor is 0.866. The natural frequency of oscillations is:
(a) 15.6 rad/s (b) 19.3 rad/s
(c) 36 rad/s (d) 9 rad/s

8. Stability Analysis of Linear Control Systems

Q.177 The presence of non-linearities in a control system tends to introduce
(a) instability (b) steady state error
(c) transient error (d) all of these

Q.178 Which one of the following is not a property of the Liapunov function?
(a) It is a unique function for a given system.
(b) It is positive definite, atleast in the neighbourhood of the origin.
(c) It is a scalar function.
(d) It's time derivative is non-positive.

Q.179 The number of sign changes in the Routh's array indicates the number of roots lying in the
(a) origin of s-plane
(b) left half of s-plane
(c) right half of s-plane
(d) centre of the s-plane

Q.180 **Assertion (A):** A system is said to be stable if the impulse response approaches zero for sufficiently large time.
Reason (R): If the impulse response approaches infinity for sufficiently large time, the system is said to be unstable
(a) Both A and R are true and R is a correct explanation of A.
(b) Both A and R are true but R is not a correct explanation of A.
(c) A is true but R is false.
(d) A is false but R is true.

Q.181 The open loop transfer function of a unity feedback system is $\dfrac{K}{s(1+0.4s)(1+0.25s)}$.
What is the range of value of K so that the closed loop system is absolutely stable?
(a) $K > 6.5$ (b) $0 < K < 4.25$
(c) $0 < K < 9.8$ (d) $0 < K < 6.5$

Q.182 **Assertion (A):** If the system is stable we can determine the relative stability by the settling time of the system.
Reason (R): If the settling time is less than that of the other system then the system is said to be relatively more stable.
(a) Both A and R are true and R is a correct explanation of A.
(b) Both A and R are true but R is not a correct explanation of A.
(c) A is true but R is false.
(d) A is false but R is true.

Q.183 If the unit step response of a system is given by $c(t) = -1.5 + e^t + 0.5e^{-2t}$, then the system will be
(a) stable (b) unstable
(c) relatively stable (d) marginally stable

Q.184 The characteristic equation of a servo system is given by

$$a_0s^4 + a_1s^3 + a_2s^2 + a_3s + a_4 = 0$$

Now, consider the following conditions required to be satisfied by the coefficient of the above characteristic equation for the system to be stable:

1. $a_1 > 0$ and $a_4 > 0$
2. $a_1 < 0$ and $a_4 > 0$
3. $a_1a_2 < a_0a_3$
4. $a_1a_2 > a_0a_3$
5. $a_0a_3^2 + a_1^2a_4 < a_1a_2a_3$
6. $a_0a_3^2 + a_1^2a_4 > a_1a_2a_3$

Which of the above conditions holds true for the given system to be stable?

(a) 2, 3 and 6 (b) 1, 3 and 5
(c) 1, 4 and 5 (d) 2 and 4

Q.185 The open loop transfer function of a unity feedback control system is given by

$$G(s) = \frac{K}{(s+2)(s+4)(s^2+6s+25)}$$

The oscillating frequency of the given system is

(a) 2.25 rad/sec (b) 6.60 rad/sec
(c) 3.82 rad/sec (d) 4.06 rad/sec

Q.186 The characteristic equation of a system is given by $s^3 + s + 2 = 0$.
What are the number of roots in the right half s-plane and on the $j\omega$-axis respectively?

(a) 3, 0 (b) 2, 1
(c) 1, 2 (d) 2, 0

Q.187 Which of the following is the correct statement?
A minimum phase network is one whose transfer function has

(a) zeros in the right hand s-plane and poles in the left hand s-plane.
(b) zeros and poles in the left hand s-plane.
(c) zeros in the left hand s-plane and poles in the right hand s-plane.
(d) arbitrary distribution of zeros and poles in the s-plane.

Q.188 For a discrete-time system to be unstable, all the poles of the z-transfer function should lie

(a) on the left-half of z-plane
(b) on the right-half of z-plane
(c) outside the circle of unit radius
(d) within a circle of unit radius

Q.189 Consider the following characteristic equation of a system:

$$s^3 + 2Ks^2 + (K+2)s + 4 = 0$$

Which one of the following is correct?

(a) The system Is stable for all positive values of K.
(b) The system is unstable for all values of K.
(c) The system is stable for values of $K > 0.73$.
(d) The system is stable for value of $K < 0.73$.

Q.190 For which of the following values of K, the feedback system shown in the below figure is stable?

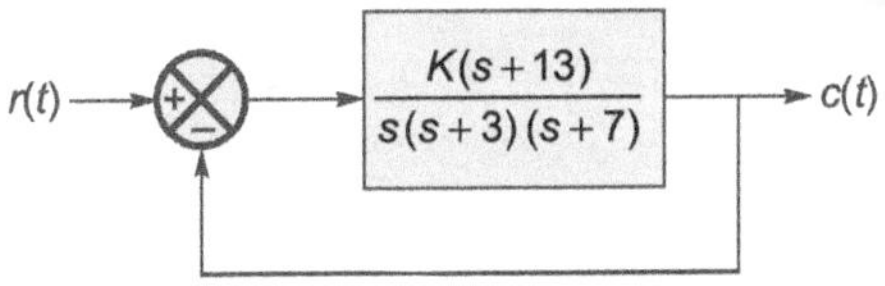

(a) $K < 0$ (b) $K > 0$
(c) $0 < K < 54$ (d) $0 < K < 70$

Q.191 What is the range of K for which the open loop transfer function

$$G(s) = \frac{Ke^{-s}}{s(s^2+5s+9)}$$

represents an unstable closed loop system?

(a) $0 < K < 7.5$ (b) $K < 5$
(c) $7.5 < K < 0$ (d) None of the above

Q.192 What is the range of values of K ($K > 0$) such that the characteristic equation

$$s^3 + 3(K+1)s^2 + (7K+5)s + (4K+7) = 0$$

has roots more negative than $s = -1$?

(a) $\infty > K > 0.53$ (b) $-2.5 < K < 0.53$
(c) $\infty > K > 0.25$ (d) $1 > K > 0.25$

Q.193 A closed loop control system is shown below:

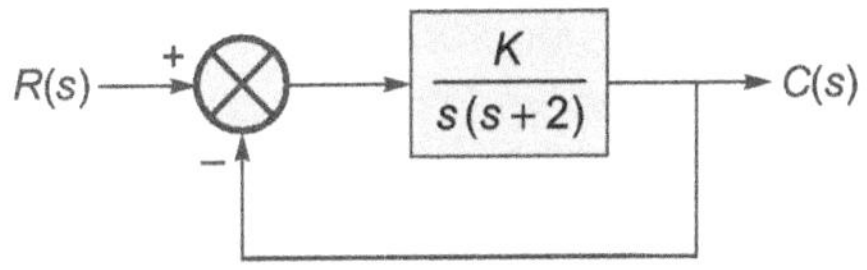

Match **List-I (Different values of K)** with **List-II (Roots of Characteristic Equation)** and select the correct answer using the codes given below the lists:

List-I	List-II
A. $0 < K < 1$	1. Complex conjugate
B. $K = 0$	2. Real and equal

C. $K = 1$
D. $K > 1$

3. Open loop poles
4. Real and distinct

Codes:

	A	B	C	D
(a)	4	2	3	1
(b)	4	3	2	1
(c)	1	2	3	4
(d)	1	3	2	4

Q.194 Consider the following statements regarding the characteristic equation of a system given by:

$$s^4 + 5s^3 + 2s + 10 = 0$$

1. The system is unstable.
2. The system is stable.
3. Number of roots with zero real part = 0.
4. Number of roots with positive real part = 4
5. Number of roots with negative real part = 2.

Which of the above statements are correct?

(a) 2 and 3
(b) 1 and 3
(c) 2, 3 and 4
(d) 1, 3 and 5

Q.195 The transfer function of a system is $\frac{(1+2s)}{(1-s)}$.

The system is then which one of the following?

(a) Non-minimum phase function
(b) Low-pass system
(c) Second-order system
(d) None of the above

Q.196 Assertion (A): For a system to be stable, the roots of the characteristic equation should not lie in the right half of the s-plane or on the imaginary axis.

Reason (R): A system is said to be stable if the zeros of the closed loop transfer function are all in the left half of *s*-plane.

(a) Both A and R are true and R is a correct explanation of A.
(b) Both A and R are true but R is not a correct explanation of A.
(c) A is true but R is false.
(d) A is false but R is true.

Q.197 Assertion (A): Location of the poles of a transfer function in the s-plane affects greatly the transient response of the system and hence stability.

Reason (R): The poles that are close to the imaginary axis in the left-half s-plane give rise to transient responses that will decay relatively slowly, whereas the poles that are far away from the axis (relative to the dominant poles) correspond to fast-decaying time responses.

(a) Both A and R are true and R is a correct explanation of A.
(b) Both A and R are true but R is not a correct explanation of A.
(c) A is true but R is false.
(d) A is false but R is true.

Q.198 Assertion (A): Routh-Hurwitz criterion doesn't provides the information about the number of roots that lie on the $j\omega$-axis while it tests whether any of the roots of the characteristic equation lie in the right-half s-plane or not.

Reason (R): The Routh-Hurwitz criterion represents a method of determining the location of zeros of a polynomial with constant real coefficients with respect to the left half and right half of the s-plane, without actually solving for the zeros.

(a) Both A and R are true and R is a correct explanation of A.
(b) Both A and R are true but R is not a correct explanation of A.
(c) A is true but R is false.
(d) A is false but R is true.

Q.199 Consider the following statements related to Routh-Hurwitz criterion of determining the stability of a system:

1. It is applicable if the characteristic equation has real coefficients, complex terms or exponential functions of s.
2. The criterion can be applied to any stability boundaries in complex plane, such as the unit circle in the z-plane.
3. The number of changes of signs in the elements of the first column equals the number of roots with negative real parts.

Which of these statements is/are not correct?

(a) 2 and 3 only
(b) 1, 2 and 3
(c) 1 only
(d) 1 and 2 only

Q.200 The characteristic equation of a linear discrete-data control system is given as:

$$F(z) = z^3 + z^2 + z + K = 0,$$

Where K is a real constant.

What is the range of values of K so that the system is stable?

(a) $0 < K < \infty$ (b) $0 < K < 1$

(c) $-3 < K < \infty$ (d) $-\frac{1}{3} < K < 1$

Q.201 The characteristic equation of a linear time-invariant discrete-data system is given by:

$F(z) = a_n z^n + a_{n-1} z^{n-1} + ... + a_1 z + a_0 = 0$

where all the coefficients are real.

According to "Jury's stability criterion", there are certain necessary conditions which must be satisfied for the above discrete-data system to have no roots on or outside the unit circle.

Which of the following necessary condition does not holds true for the discrete-data system to have all roots inside the unit circle according to "Jury's stability criterion"?

(a) $|a_0| > |a_n|$

(b) $F(1) > 0$

(c) $F(-1) > 0$ (if n = even integer)

(d) $F(-1) < 0$ (if n = odd integer)

Q.202 The first two rows of Routh's tabulation of a third-order system are

s^3	2	2
s^2	4	4

Select the correct answer from the following choices.

(a) The given system has one root in the right-half s-plane.

(b) The given system has two roots on the $j\omega$-axis at $s = j$ and $-j$ while the third root is in the left-half s-plane.

(c) The given system has two roots on the $j\omega$-axis at $s = j2$ and $-j2$ while the third root is in the left-half s-plane.

(d) The given system has two roots on the $j\omega$-axis at $s = j2$ and $s = -j2$ while the third root is in the right-half s-plane.

Q.203 The characteristic equation of a second order discrete-data system is given by:

$$F(z) = z^2 + z + 0.25 = 0$$

The above system is

(a) stable (b) marginally stable

(c) unstable (d) asymptotically stable

Q.204 The z-transform of a signal is given by:

$$C(z) = \frac{z^{-1}(1-z^{-4})}{(1-z^{-2})^2}$$

Its final value is

(a) $\frac{1}{4}$ (b) zero

(c) -1 (d) infinity

Q.205 The dynamic response of a system is given by the following differential equation:

$$a_3\frac{d^3c(t)}{dt^3} + a_2\frac{d^2c(t)}{dt^2} + a_1\frac{dc(t)}{dt} + a_0c(t)$$

$$= b_2\frac{d^2r(t)}{dt^2} + b_1\frac{dr(t)}{dt} + b_0r(t)$$

where $c(t)$ and $r(t)$ are respectively the output and the input and a_3, a_2, a_1, a_0, b_2, b_1 and b_0 are constants. It is a

(a) non-linear system

(b) linear system

(c) linear time variant system

(d) linear time invariant system

Q.206 The transfer function of a system is:

$$G(s) = \frac{100e^{-ST}}{s(s+10)}$$

The system

(a) is a linear system.

(b) is a non-linear system.

(c) has a transportation lag.

(d) none of the above.

9. The Root Locus Technique

Q.207 Which of the following option is correct? The root locus is the path of the roots of the characteristic equation traced out in the s-plane

(a) as the output of the system is changed

(b) as the sensitivity is changed.

(c) as the system parameter is changed

(d) as the input of the system is changed

Q.208 Number of intersection of the asymptotes of the complete root loci is

(a) three (b) unknown

(c) two (d) one

Q.209 The open-loop transfer function of a unity feedback control system is given by:

$$G(s) = \frac{K(s+2)}{s(s^2+2s+2)}$$

The centroid and the angle of root locus asymptotes are respectively

(a) $-\frac{2}{3}$ and $+60°, -60°$

(b) zero and + 90°, –90°

(c) zero and +120°, –120°

(d) $-\frac{2}{3}$ and $+90°, -90°$

Q.210 Considering the root locus diagram for a system with

$$G(s)H(s) = \frac{K(s+5)}{s(s+2)(s+4)(s^2+2s+2)},$$

the meeting point of the asymptotes on the real axis occurs at

(a) –0.75 (b) –1.25

(c) –1.2 (d) –0.65

Q.211 Assertion (A): Root locus is a graphical method in which roots of the characteristic equation are plotted in s-plane for the different value of parameter.

Reason (R): The locus of the roots of the characteristic equation when gain is varied from zero to infinity is called root locus.

(a) Both A and R are true and R is a correct explanation of A.

(b) Both A and R are true but R is not a correct explanation of A.

(c) A is true but R is false.

(d) A is false but R is true.

Q.212 Assertion (A) : An addition of real pole at $s = -p_0$ in the transfer function $G(s)H(s)$ of a control system results in the increase of stability margin.

Reason (R) : An addition of real pole at $s = -p_0$ in the transfer function $G(s)H(s)$ will make the resultant root loci bend towards the right.

(a) Both A and R are true and R is a correct explanation of A.

(b) Both A and R are true but R is not a correct explanation of A.

(c) A is true but R is false.

(d) A is false but R is true.

Q.213 The open-loop transfer function of a closed loop control system is given as:

$$G(s)H(s) = \frac{K}{s(s^2+6s+10)}$$

Which of the following statements is correct about the root locus of the above system?

(a) Two number of branches of root locus will terminate at infinity.

(b) For $K = 40$, the system is marginally stable.

(c) The centroid of the given system is located at (–1, 0).

(d) The root locus intersects the $j\omega$-axis at $s = \pm j3.16$.

Q.214 The root locus of $s(s-1) + K(s+1) = 0$ is a circle. The co-ordinates of the centre and the radius (in units) of this circle are respectively

(a) (–2, 0) and $\sqrt{3}$ units

(b) (–1, 0) and $\sqrt{2}$ units

(c) (–2, 0) and $\sqrt{2}$ units

(d) (–1, 0) and $\sqrt{3}$ units

Q.215 What is the open-loop transfer function of a unity feedback control system having root locus shown in the following figure?

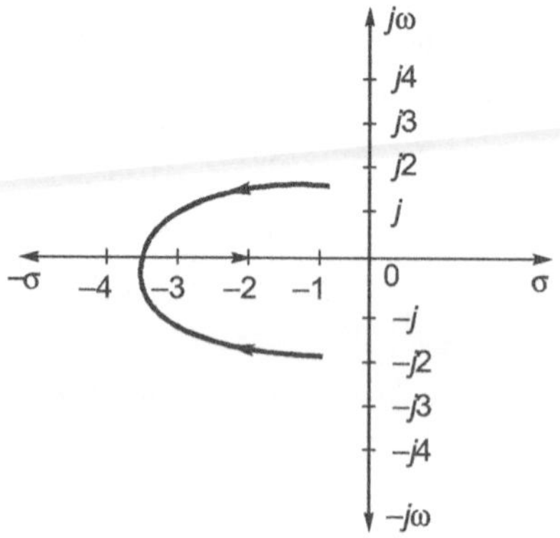

(a) $\frac{K(s+2)}{(s+1)(s+3)}$ (b) $\frac{K}{(s+2)(s^2+2s+2)}$

(c) $\frac{K(s+2)}{s^2+2s+2}$ (d) $\frac{K(s+4)}{(s+2)(s+1)}$

Q.216 A control system has $G(s)H(s) = \frac{K(s+1)}{s^2(s+3.6)}$ $(0 < K < \infty)$. The number of breakaway point in the root locus diagram is/are

(a) one (b) two

(c) three (d) none of these

Q.217 The open loop transfer function of a unity feedback control system is given by

$$G(s)H(s) = \frac{K(s+1)}{s(s+2)(s+4)(s+5)}$$

The breakaway point in its root locus will lie between
(a) 0 and –1 (b) beyond –5
(c) –4 and –5 (d) –2 and –4

Q.218 The characteristic equation of a closed loop control system is given by:

$$s^2 + 2s + 10 + K(s^2 + 6s + 10) = 0$$

The angle of asymptotes for the root loci for $K \geq 0$ are given by
(a) 180°, 360° (b) 90°, 270°
(c) 90°, 180° (d) none of these

Q.219 The breakaway point in the root loci plot for the loop transfer function $G(s) = \dfrac{K}{s(s^2 + 6s + 12)}$ is
(a) –1.5 (b) –2
(c) –1 (d) –2.2

Q.220 Consider the following statements pertaining to root locus techniques:
1. Routh-Hurwitz criterion may be used to find the intersections of the root loci on the imaginary axis.
2. Breakaway points on the root loci of an equation corresponds to multiple-order roots of the equation.
3. The breakaway points on the root loci of $1 + KG(s)H(s) = 0$ must satisfy

$$\frac{d}{ds}\{G(s)H(s)\} = 0$$

4. n root loci arrive or depart a breakaway point at $\dfrac{180}{n}$ degrees.

Which of these statements are correct?
(a) 1, 2 and 3 (b) 2, 3 and 4
(c) 1, 2, 3 and 4 (d) 1, 2 and 4

Q.221 **Assertion (A):** From the root sensitivity standpoint, a system should not be operated at the breakaway points.
Reason (R): The root sensitivity at the breakaway points is zero.
(a) Both A and R are true and R is a correct explanation of A.
(b) Both A and R are true but R is not a correct explanation of A.
(c) A is true but R is false.
(d) A is false but R is true.

Q.222 Consider the following statements:
1. All the rules and properties for the construction of root loci in the s-plane can be applied to the construction of root loci of discrete data system in the z-plane without any modification.
2. The breakaway points of the root loci must always be on the real axis.
3. There is only one intersection of the asymptotes of the complex root loci.
4. Adding a pole to $Q(s)/P(s)$ has the general effect of pushing the root loci to the left, whereas adding a zero pushes the loci to the right.

Which of the above statements is/are not correct related to root locus technique in control system?
(a) 2 and 4 (b) 1, 2, 3 and 4
(c) 3 and 4 (d) 2, 3 and 4

Q.223 For the loop transfer function:

$$G(s)H(s) = \frac{Ks}{(s+1)(s^2+1)}$$

the angle of departure at $s = j$ for $K > 0$ is
(a) 90° (b) 75°
(c) 155° (d) 135°

Q.224 Which of the following is not true regarding the existence of breakaway point in root locus technique?
(a) A breakaway point exists between two zeros if the region between them is a part of root locus.
(b) If we see the root locus between two poles on real axis, then there is a breakaway point between these two poles.
(c) A breakaway point exists to the left of a pole if the region to the left of it on real axis is a part of root locus.
(d) A break away point exists to the left of a zero on real axis if the region to the left of it is a part of root locus.

Q.225 The open loop transfer function of a system is given by

$$G(s)H(s) = \frac{K(s+1)^2(s+5)}{s(s+10)}$$

The number of loci terminating at infinity is

(a) 3 (c) 2
(c) 1 (d) 0

Q.226 A unity feedback system has open-loop transfer function,

$$G(s) = \frac{K(s+\alpha)}{s^2(s+\beta)}; \beta > \alpha > 0$$

The valid root-loci for this system is

(a) 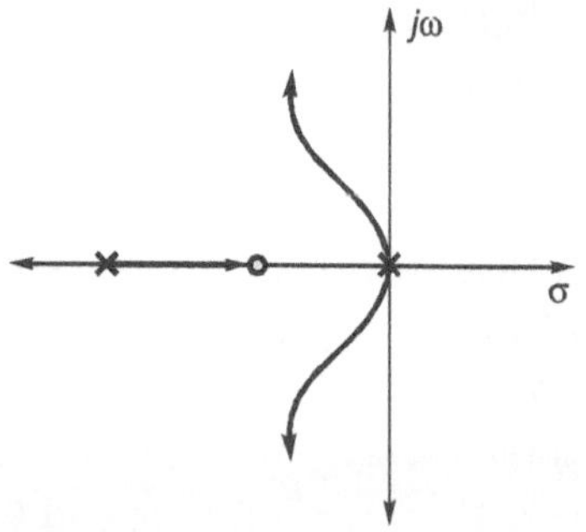

(b) 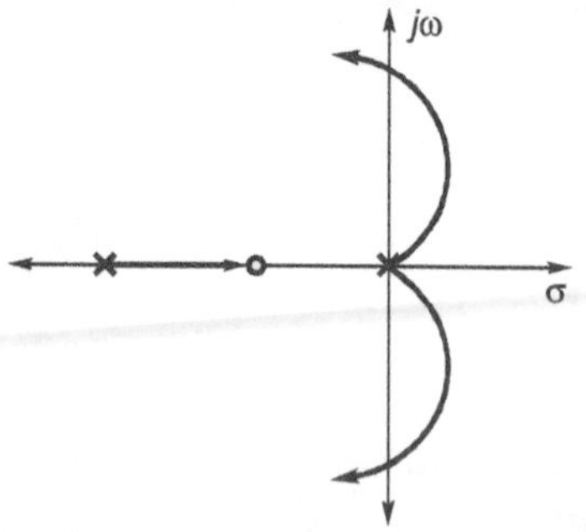

(c) 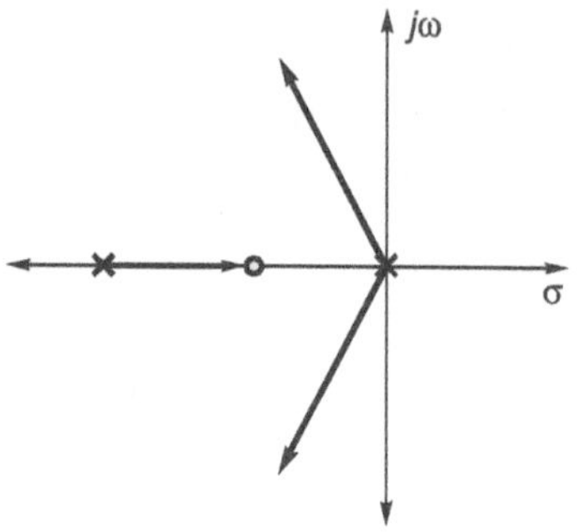

(d) 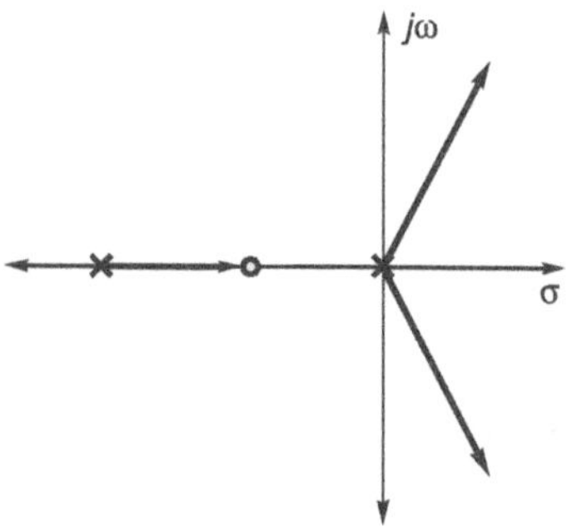

10. Frequency Domain Analysis of Control Systems

Q.227 The Bode diagram approach is the most commonly used method for the analysis and synthesis of
(a) non-linear feedback control system only
(b) open-loop system only
(c) linear feedback control system only
(d) all of the above

Q.228 Nichol's chart is a plot of
(a) magnitude in decibels vs log ω
(b) magnitude vs phase angle in the polar coordinates
(c) magnitude in decibels vs phase shift on rectangular coordinates
(d) none of the above

Q.229 The initial slope of Bode plot for a transfer function having no poles at origin is
(a) –20 dB/decade (b) 0 dB/decade
(c) +10 dB/decade (d) –10 dB/decade

Q.230 The gain margin of a control system having the loop transfer function $G(s)H(s) = \frac{\sqrt{2}}{s(s+1)}$ is
(a) 0 dB (b) 6 dB
(c) 3 dB (d) ∞

Q.231 Gain crossover frequency is the frequency at which the gain of $G(j\omega)$ is
(a) 100 dB (b) 0 dB
(c) 5 dB (d) 10 dB

Q.232 For a control system having gain margin of –10 dB, the magnitude of $GH(s)$ for 180° phase shift is
(a) 10 dB (b) $-\frac{1}{10}$ dB
(c) $\frac{1}{10}$ dB (d) –10 dB

Q.233 If the Nyquist plot of the loop transfer function G(s)H(s) of a closed-loop system does not encloses the (–1, j0) point in the $G(s)H(s)$ plane, the gain margin of the system will be
(a) infinity (b) less than zero
(c) greater than zero (d) zero

Q.234 A system with phase margin close to zero or gain margin close to unity is

(a) unstable (b) highly stable
(c) oscillatory (d) relatively stable

Q.235 If the transfer function of a first-order system in $G(s) = \frac{10}{s+0.2}$, then the time constant of this system will be

(a) 10 secs (b) $\frac{1}{50}$ secs
(c) 5 secs (d) 50 secs

Q.236 The polar plot of $G(j\omega) = \frac{1}{j\omega(1+j\omega T)}$

(a) crosses the positive imaginary axis
(b) crosses the negative real axis
(c) crosses the negative imaginary axis
(d) none of the above

Q.237 If the bandwidth of the system is increased, then

(a) both rise time and damping ratio will decrease
(b) rise time will increase but damping ratio will decrease
(c) rise time will decrease but damping ratio will increase
(d) both rise time and damping ratio will increase.

Q.238 Polar plot of $G(s) = \frac{1}{1+Ts}$ is a

(a) straight line (b) semicircle
(c) parabola (d) circle

Q.239 Consider the following statements related to frequency domain analysis:

1. The cut-off rate is the slope of the log magnitude curve near the cut-off frequency.
2. The bandwidth is defined as the band of frequencies lying between 3 dB points.
3. Higher the value of resonant frequency of the system, slower is the time response.
4. The magnitude of resonant peak gives the information about the relative stability of the system.

Which of these statements is/are not correct?

(a) 1, 3 and 4 (b) 1, 2 and 3
(c) 2 and 3 (d) 3 and 4

Q.240 The inverse polar plot of the open loop transfer function, $G(s) = \frac{1+sT}{sT}$ will be represented by

(a)
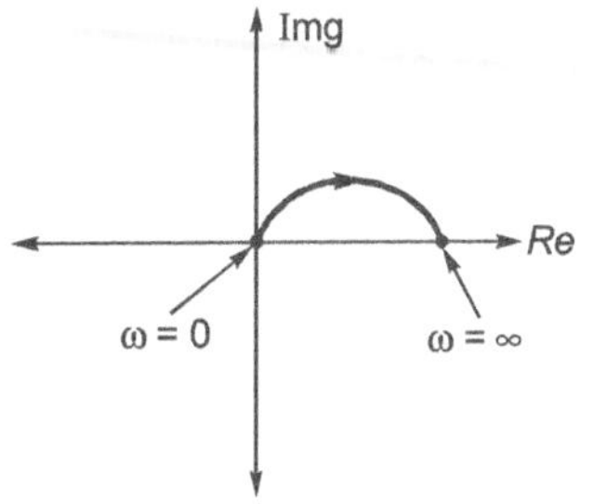

(b)
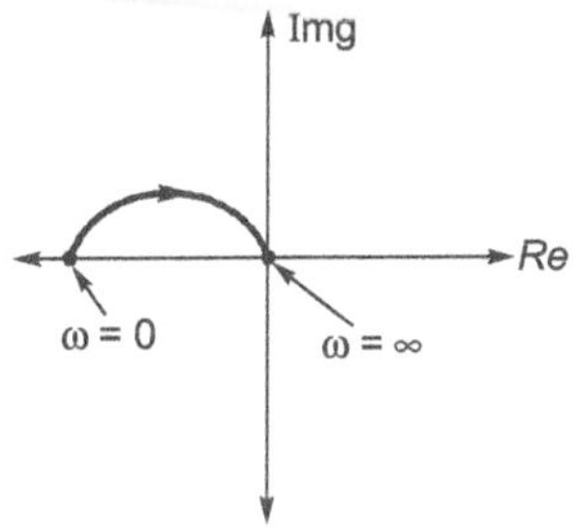

(c)
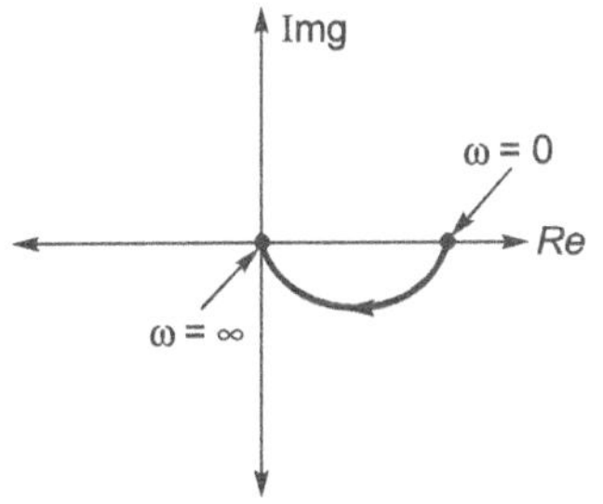

(d)
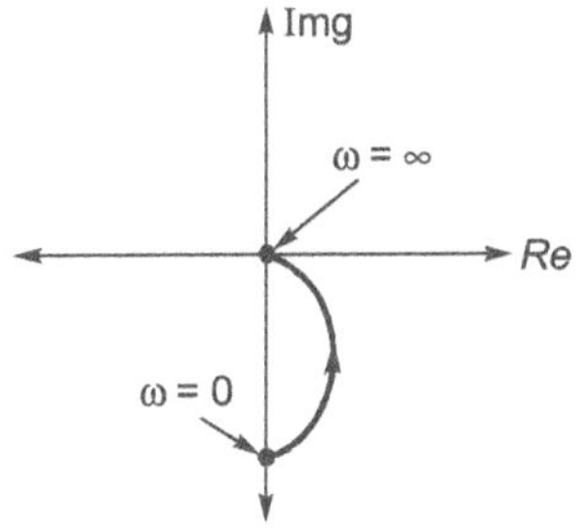

Q.241 The constant *M*-circles corresponding to the magnitude (*M*) of the closed loop transfer function of a linear system for value of *M* less than one lie in the *G*-plane and to the

(a) lower side of the $M = -j1$ line
(b) upper side of the $M = \pm j1$ line
(c) left of the $M = 1$ line
(d) right of the $M = 1$ line

Q.242 The slope of the line due to $\frac{1}{(j\omega)^2}$ factor in magnitude part of Bode plot is

(a) –40 db per octave
(b) –20 db per octave
(c) –12 db per octave
(d) –4 db per octave

Q.243 The magnitude plot of the open loop transfer function $G(s)$ of a certain system is shown in figure below.

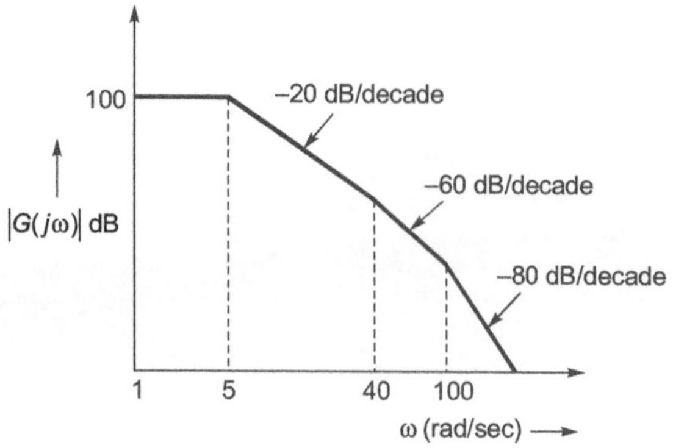

If the system is of minimum phase type, then the open-loop transfer function $G(s)$ will be given by:

(a) $\frac{10^3}{(1+0.01s)\,(1+0.025s)^2\,(1+0.2s)}$

(b) $\frac{10^5}{(1+0.01s)\,(1+0.025s)^2\,(1+0.2s)}$

(c) $\frac{100}{(1+0.1s)\,(1+0.02s)\,(1+0.25s)^2}$

(d) $\frac{10^3(1+0.01s)}{(1+0.025s)^2(1+0.2s)}$

Q.244 A unity feedback system has the open loop transfer function,

$$G(s)H(s) = \frac{K}{s(s+5)\,(s+10)}$$

For the given system to be marginally stable, the value of gain cross-over frequency would be

(a) 7.07 rad/sec (b) 0.707 rad/sec
(c) 0.0707 rad/sec (d) 1.414 rad/sec

Q.245 The transfer function of a lag-lead network is given by

$$G_c(j\omega) = K\frac{\left(s+\frac{1}{T_1}\right)\left(s+\frac{1}{T_2}\right)}{\left(s+\frac{\beta}{T_1}\right)\left(s+\frac{1}{\beta T_2}\right)}$$

The phase of $G_c(j\omega)$ will become zero at

(a) $\omega = \frac{1}{T_1\sqrt{\beta}}$ rad/s (b) $\omega = \frac{1}{T_2}$ rad/s

(c) $\omega = \frac{1}{T_2\sqrt{\beta}}$ rad/s (d) $\omega = \frac{1}{\sqrt{T_1 T_2}}$ rad/s

Q.246 For the system shown below, what is the range of value of K for stability?

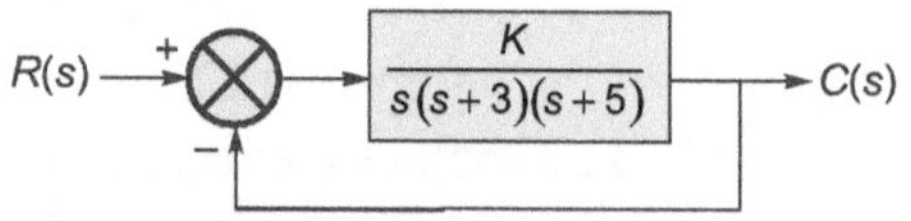

(a) $0 < K < \frac{1}{120}$ (b) $K < \frac{1}{120}$

(c) $0 < K < 120$ (d) $K > 120$

Q.247 Match **List -I (Polar plot of systems)** with **List-II (Transfer functions)** and select the correct answer using the codes given below the lists:

List-I

A.
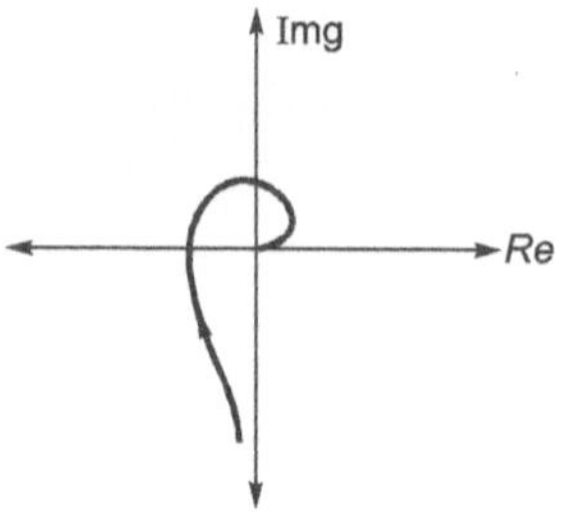

B.
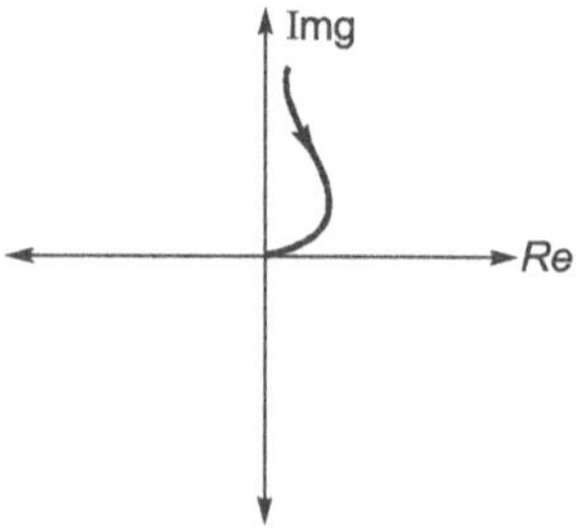

C.
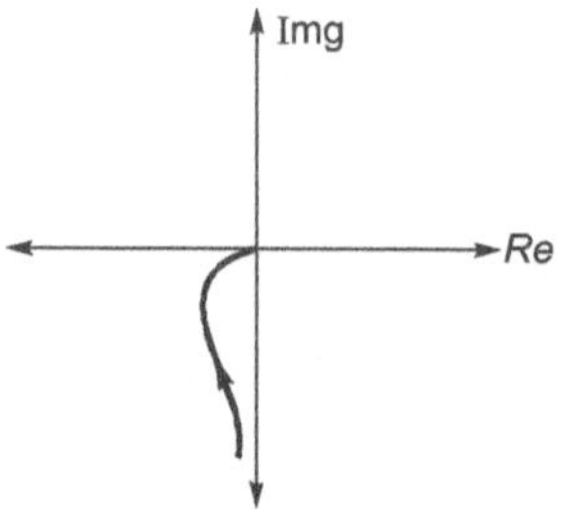

D. 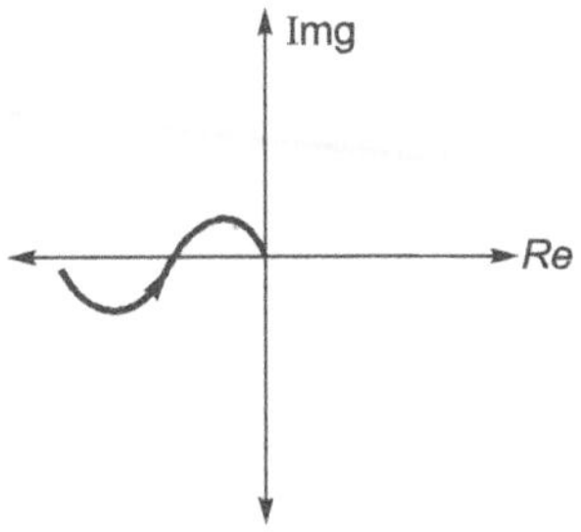

List-II

1. $\dfrac{1}{s(1+T_1 s)}$

2. $\dfrac{1}{s(1+T_1 s)(1+T_2 s)(1+T_3 s)}$

3. $\dfrac{(1+T_0 s)}{s^2(1+T_1 s)(1+T_2 s)}$

4. $\dfrac{1}{s^3(1+Ts)}$

Codes:

	A	B	C	D
(a)	3	1	2	4
(b)	2	4	1	3
(c)	2	1	4	3
(d)	1	2	3	4

Q.248 Assertion (A): The Nyquist criterion represents a method of determining the location of the characteristic equation roots with respect to the left half and the right half of the *s*-plane.

Reason (R): The Nyquist criterion does not give the exact location of the characteristic equation roots.

(a) Both A and R are true and R is a correct explanation of A.
(b) Both A and R are true but R is not a correct explanation of A.
(c) A is true but R is false.
(d) A is false but R is true.

Q.249 Which of the following is not true?

(a) Increasing ξ decreases BW and increases t_r
(b) Increasing ω_n increases BW and decreases t_r
(c) The larger the bandwidth, the faster the system will respond
(d) BW is inversely proportional to ω_n

Q.250 If the Nyquist plot cuts the negative real axis at a distance of 0.25, then the gain margin and phase margin of the system will be respectively

(a) 0.25 and 0° (b) 4.0 and –180°
(c) –0.25 and 180° (d) 4.0 and 0°

Q.251 Match **List-I (Different values of damping ratio)** with **List-II (Values of resonant peaks)** and select the correct answer using the codes given below the lists:

List-I	List-I
A. $\xi = 0$	1. More than unity
B. $\xi = 0.707$	2. Not defined
C. $\xi = 0.85$	3. Unity
D. $\xi = 0.5$	4. Infinite

Codes:

	A	B	C	D
(a)	1	3	2	4
(b)	4	1	2	3
(c)	4	3	2	1
(d)	1	2	3	4

Q.252 Assertion (A) : Gain margin alone is adequate to indicate relative stability when system parameters other than the loop gain are subject to variation.

Reason (R) : Gain margin is the amount of gain in dB that can be added to the loop before the closed-loop system becomes unstable.

(a) Both A and R are true and R is a correct explanation of A.
(b) Both A and R are true but R is not a correct explanation of A.
(c) A is true but R is false.
(d) A is false but R is true.

Q.253 Assertion (A): Phase margin is measured at the gain crossover frequency.

Reason (R): Gain crossover frequency can be easily determined from Bode plot than from the Nyquist plot.

(a) Both A and R are true and R is a correct explanation of A.
(b) Both A and R are true but R is not a correct explanation of A.
(c) A is true but R is false.
(d) A is false but R is true.

Q.254 If the magnitude of closed loop transfer function of a system is given by *M*, then the centre and radius of *M*-circle will be respectively given by:

(a) $\left(\frac{-M^2}{M^2-1}, 0\right)$ and $\left(\frac{M^2}{M^2-1}\right)$

(b) $\left(\frac{M^2}{M^2-1}, 0\right)$ and $\left(\frac{M^2}{M^2-1}\right)$

(c) $\left(\frac{-M^2}{M^2-1}, 0\right)$ and $\left(\frac{1}{M^2-1}\right)$

(d) $\left(\frac{M^2}{M^2-1}, 0\right)$ and $\left(\frac{M}{M^2-1}\right)$

Q.255 The open loop transfer function of a unity feedback system is given by

$$G(s) = \frac{25}{s(s+1)(s+10)}$$

The value of ω_{pc} is

(a) $\sqrt{10}$ rad/sec (b) $\sqrt{5}$ rad/sec

(c) $2\sqrt{5}$ rad/sec (d) $\frac{1}{\sqrt{2}}$ rad/sec

Q.256 If the gain of the open-loop system is increased to four times, the gain margin of the system

(a) remains same
(b) becomes half
(c) becomes one-fourth
(d) none of the above

Q.257 A minimum phase system with no zeros has a phase angle of −190° at gain cross-over frequency. The system is

(a) stable
(b) unstable
(c) conditionally stable
(d) marginally stable

Q.258 Phase crossover frequency is defined at

(a) $\angle G(j\omega)\, H(j\omega) = -180°$
(b) $\angle G(j\omega)\, H(j\omega) = 0°$
(c) $\angle G(j\omega)\, H(j\omega) = 90°$
(d) $\angle G(j\omega)\, H(j\omega) = 180°$

Q.259 The Bode plot for a transfer function is shown below:

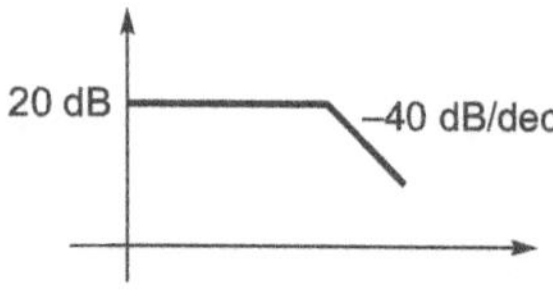

The steady state error corresponding to a parabolic input is

(a) zero (b) 0.01
(c) infinity (d) none of these

Q.260 The frequency at which the Nyquist plot crosses the negative real axis is known as

(a) natural frequency
(b) damped natural frequency
(c) gain crossover frequency
(d) phase crossover frequency

Q.261 A unity feedback system has a forward path gain of $G(s) = \frac{K}{s(s+5)(s+10)}$

The phase angle of the Nyquist plot for $\omega \to \infty$ is

(a) 0° (b) −90°
(c) −180° (d) 90°

Q.262 The open loop transfer function of a system is given as $G(s)H(s) = \frac{K}{s(1+as)}$, $\left(a < \frac{1}{K}\right)$.

The Bode plot of this system is represented as:

(a)

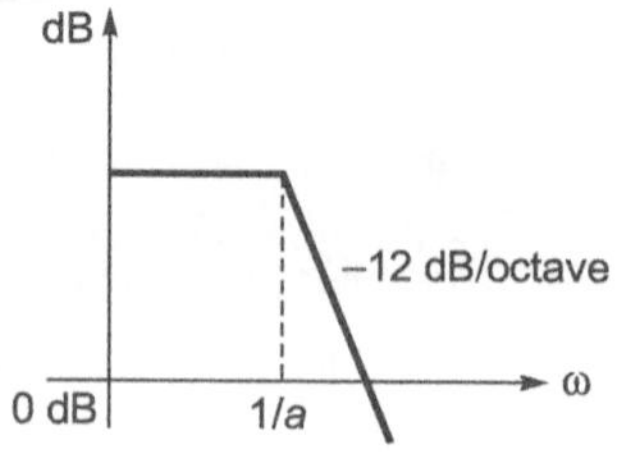

(b)

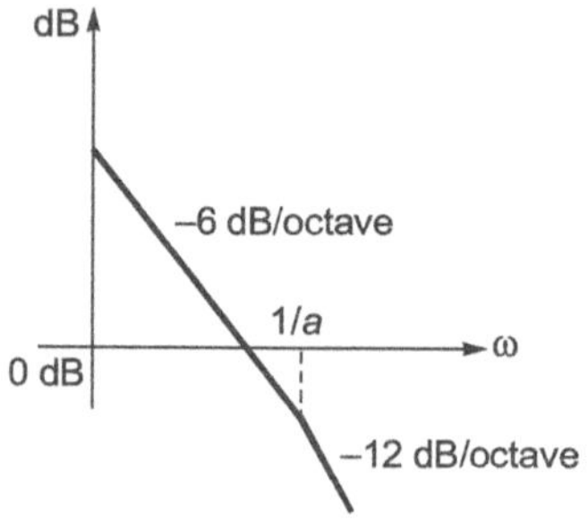

(c)

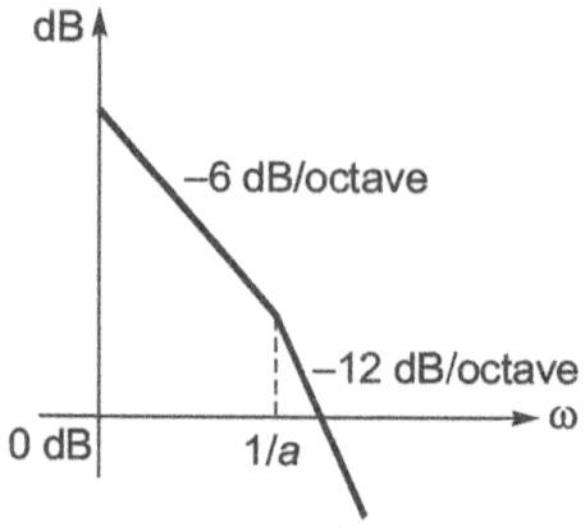

(d) none of the above

Q.263 A second order system has

$$M(j\omega) = \frac{100}{100 - \omega^2 + 10\sqrt{2}j\omega}$$

It's M_p (peak magnitude) will be approximately

(a) 0.5 (b) 1

(c) $\sqrt{2}$ (d) 2

Q.264 The closed loop transfer function of a control system is given by $\frac{C(s)}{R(s)} = \frac{1}{1+s}$.

For the input $r(t) = \sin t$, the steady state value of $c(t)$ is equal to

(a) $\frac{1}{\sqrt{2}}\cos t$ (b) 1

(c) $\frac{1}{\sqrt{2}}\sin t$ (d) $\frac{1}{\sqrt{2}}\sin(t - \pi/4)$

Q.265 A system has transfer function equal to $\frac{1-s}{1+s}$.

Its gain at ω = 1 rad/s is

(a) 1 (b) 0

(c) –1 (d) none of these

Q.266 The gain margin of a unity negative feedback system having forward path transfer function $\frac{K}{s(1+Ts)}$ is

(a) infinity (b) KT

(c) 1 (d) zero

Q.267 The characteristic equation of a closed loop control system is given by $s^2 + 4s + 16 = 0$. The resonant frequency (in radian/sec) of the system is

(a) 2 (b) $2\sqrt{3}$

(c) 4 (d) $2\sqrt{2}$

Q.268 The gain margin (in dB) of a system having the open-loop transfer function $G(s)\,H(s) = \frac{\sqrt{2}}{s(1+s)}$ is

(a) 0 (b) 3

(c) 6 (d) infinity

Q.269 Nichol's chart is useful for detailed study and analysis of

(a) closed loop frequency response

(b) open loop frequency response

(c) closed loop and open loop frequency responses

(d) none of the above

Q.270 The forward path transfer function of an unity feedback system is given by

$$G(s) = \frac{1}{(1+s)^2}$$

What is the phase margin for this sytem?

(a) –180° (b) 0°

(c) 90° (d) 180°

Q.271 The closed loop system having the open loop transfer function, $G(s)H(s) = \frac{1}{s(1+2s)(1+s)}$ is

(a) unstable (b) stable

(c) marginally stable (d) conditionally stable

Q.272 The nyquist plot for a unity feedback control system having open loop transfer function $G(s) = \frac{K(1-s)}{s+1}$ is shown in figure below:

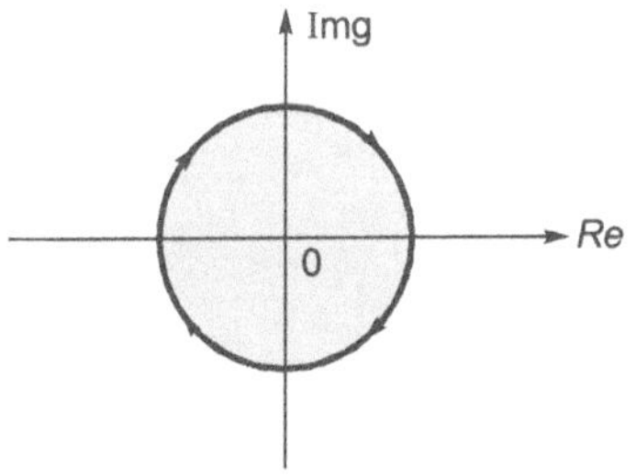

The system is

(a) stable for all values of K

(b) stable for $K > 1$

(c) unstable for $K > 1$ and stable for $K < 1$

(d) unstable for $K < 1$

Q.273 A unity feedback control system has a forward path transfer function of $\frac{e^{-Ts}}{s(1+s)}$. Its phase value will be zero at a frequency of ω_1. Which one of the following equation should be satisfied for it?

(a) $\omega_1 = \cot(T\omega_1)$ (b) $\omega_1 = \tan(T\omega_1)$

(c) $T\omega_1 = \cos(\omega_1)$ (d) $T\omega_1 = \tan(\omega_1)$

Q.274 What is the approximate value of gain margin in the Nyquist diagram shown below?

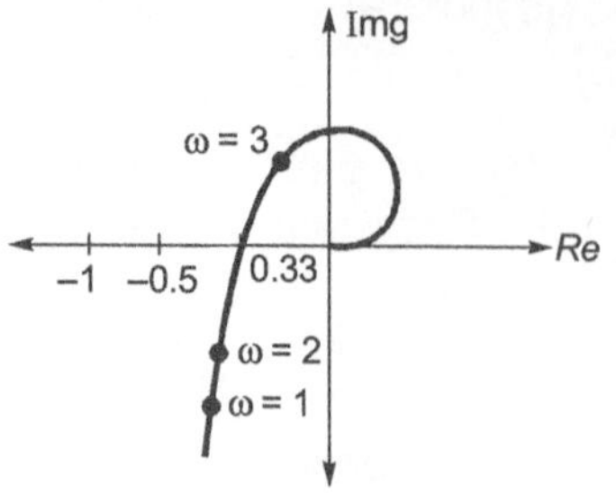

(a) 0.67 (b) 3.0
(c) 1.0 (d) 0.33

Q.275 The magnitude-frequency response of a control system is given in figure below:

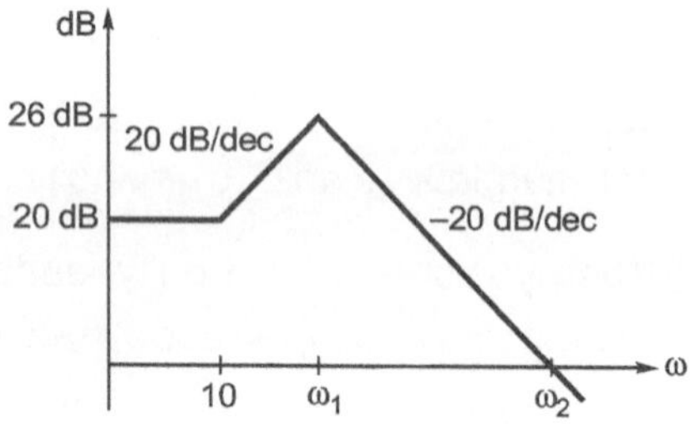

The values of ω_1 and ω_2 are respectively
(a) 100 and 400 (b) 20 and 400
(c) 20 and 200 (d) 10 and 200

Q.276 Match **List-I** with **List-II** and select the correct answer using the codes given below the lists:

List-I
A. Resonant peak
B. Damped natural frequency (rad/s)
C. Resonant frequency (rad/s)
D. Peak overshoot

List-II
1. $\sqrt{\omega_n^2 - 2\xi^2\omega_n^2}$
2. $e^{\frac{-\xi\pi}{\sqrt{1-\xi^2}}}$
3. $\dfrac{1}{\xi\sqrt{4-4\xi^2}}$
4. $\sqrt{\omega_n^2 - \xi^2\omega_n^2}$

Codes:

	A	B	C	D
(a)	3	4	1	2
(b)	3	1	4	2
(c)	2	4	1	3
(d)	2	1	4	3

Q.277 A second order underdamped system has a damping ratio of 0.8. It is subjected to a sinusoidal input of unit amplitude. It has resonant peak of
(a) 108% (b) 92%
(c) 20% (d) none of these

Q.278 Bandwidth, a frequency domain concept is indication of
(a) rise time in time domain
(b) settling time in time domain
(c) steady state error in time domain
(d) all of the above

11. Industrial Controllers & Compensators

Q.279 A first order dynamic linear system with a proportional controller exhibits an offset to a unit step input. The offset can be reduced by
(a) increasing the proportional gain
(b) adding derivative mode
(c) adding integral mode
(d) decreasing the porportional gain

Q.280 Addition of zeros in a transfer function causes
(a) lag compensator
(b) lead compensator
(c) lead-lag compensator
(d) none of the above

Q.281 A phase lead compensation network
(a) speeds up the dynamic response
(b) decreases the system bandwidth
(c) reduces the steady state error
(d) is applied when error constants are specified

Q.282 Match **List-I (Type of controllers)** with **List-II (Operation)** and select the correct answer using the codes given below the lists:

List-I
A. Hydraulic controller
B. Pneumatic controller
C. Electronic controller

List-II
1. Fire and explosion of proof operation
2. Flexible operation
3. High torque, high speed operation

Codes:

	A	B	C
(a)	3	1	2
(b)	3	2	1
(c)	1	2	3
(d)	1	3	2

Q.283 Consider the following statements related to compensators used in control system applications:

1. When transient response is satisfactory and steady state characteristics are not satisfactory, normally lag compensation is employed.
2. Phase lag network is used to increase system stability and maintain velocity gain constant.
3. The capacitance is not used to fabricate a lag network.

Which of these statements is/are correct?

(a) 1 and 3 only (b) 1, 2 and 3
(c) 1 only (d) 1 and 2 only

Q.284 The transfer function of a phase-lead controller is given by

(a) $\frac{1+bTs}{1+Ts}$; $b<1,\ T>0$

(b) $\frac{1-bTs}{1+Ts}$; $b<1,\ T>0$

(c) $\frac{1-bTs}{1+Ts}$; $b>1,\ T>0$

(d) $\frac{1+bTs}{1+Ts}$; $b>1,\ T>0$

Q.285 The input to a controller is

(a) servo-signal (b) desired variable signal
(c) sensed-signal (d) error signal

Q.286 Addition of a pole to the closed-loop transfer function results in

(a) reduction in rise time
(b) reduction in maximum overshoot
(c) increase in bandwidth
(d) increase in stability of the system

Q.287 Which of the following properties is/are not exhibited by a phase lead network?

1. The velocity constant is usually increased.
2. The slope of the magnitude curve is reduced at the gain cross over frequency.
3. The bandwidth is reduced.
4. The response becomes faster.

(a) 1, 2, 3 and 4 only (b) 2 and 3 only
(c) 3 only (d) 1 only

Q.288 **Assertion (A):** If the phase shift decreases rapidly near the gain cross over frequency, phase lead compensation becomes ineffective.
Reason (R): The additional phase lead at gain crossover frequency is added to a much smaller phase angle than that at the old gain crossover frequency.

(a) Both A and R are true and R is a correct explanation of A.
(b) Both A and R are true but R is not a correct explanation of A.
(c) A is true but R is false.
(d) A is false but R is true.

Q.289 **Assertion (A) :** The phase lag network provides an attenuation of 20 log a (a < 1) at high frequencies having transfer function equal to

$$\left(\frac{1+aTs}{1+Ts}\right).$$

Reason (R): The lag network allows to pass high frequencies and low frequencies are attenuated.

(a) Both A and R are true and R is a correct explanation of A.
(b) Both A and R are true but R is not a correct explanation of A.
(c) A is true but R is false.
(d) A is false but R is true.

Q.290 **Assertion (A):** Introduction of phase lag network in forward path increases the phase shift.
Reason (R): A phase lag network has pole nearer to the imaginary axis as compared to zero.

(a) Both A and R are true and R is a correct explanation of A.
(b) Both A and R are true but R is not a correct explanation of A.
(c) A is true but R is false.
(d) A is false but R is true.

Q.291 Match **List-I (Type of controllers)** with **List-II (Alternate names)** and select the correct answer using the codes given below the lists:

List-I

A. Two-position controller
B. Proportional controller
C. Integral controller
D. Derivative controller

List-II

1. Rate controller
2. Reset controller
3. An amplifier with adjustable gain
4. ON-OFF controller

Codes:

	A	B	C	D
(a)	4	2	3	1
(b)	4	3	2	1
(c)	3	4	2	1
(d)	3	4	1	2

Q.292 The open loop transfer function of a three action controller (PID controller) is given by:

(a) $K_p\left[1+\frac{1}{sT_i}+sT_d\right]$ (b) $\left[K_p+\frac{1}{sT_i}+sT_d\right]$

(c) $K_p[1+sT_d]+\frac{1}{sT_i}$ (d) $K_p\left[1+\frac{1}{sT_i}\right]+sT_d$

Q.293 The action of a PD controller is to

(a) increase the rise time and steady state error of the system.
(b) reduce the rise time and increase the steady error of the system.
(c) maintain a constant value of rise time and reduce the steady state error of the system.
(d) reduce the rise time and maintain a constant value of steady state error of the system.

Q.294 For the derivative feedback control shown in figure below, the damping ratio is equal to 0.7.

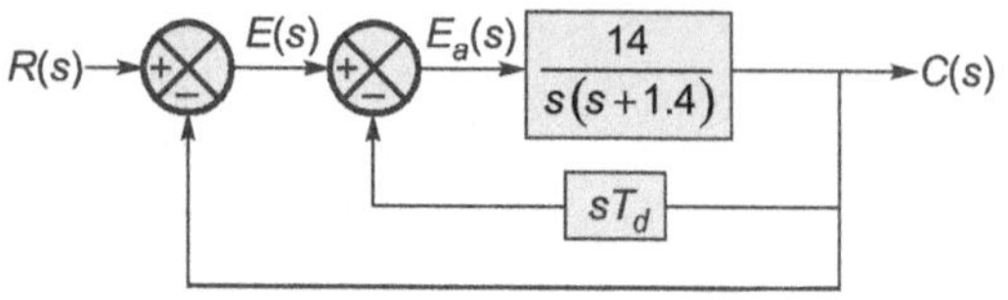

The value of derivative gain constant is

(a) 0.125 (b) 0.382
(c) 0.274 (d) 0.225

Q.295 The block diagram of an integral control is shown in figure below.

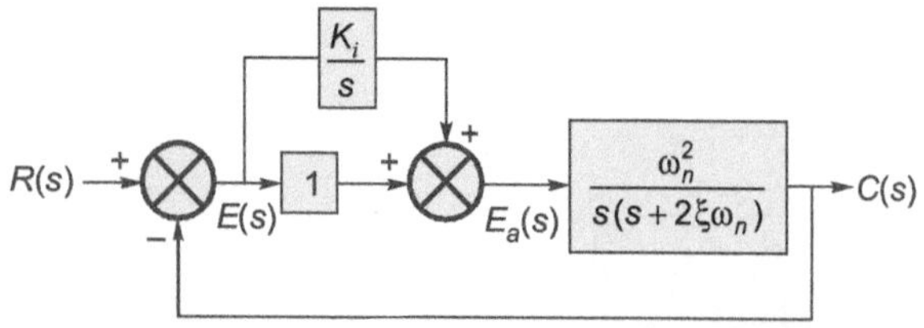

The steady state error in the above system for a parabolic input is

(a) $\left(\frac{2\xi}{\omega_n}+K_i\right)$ (b) $\frac{2\xi}{\omega_n}$

(c) zero (d) $\frac{2\xi}{K_i\omega_n}$

Q.296 A system employing proportional plus error rate control is shown in figure below.

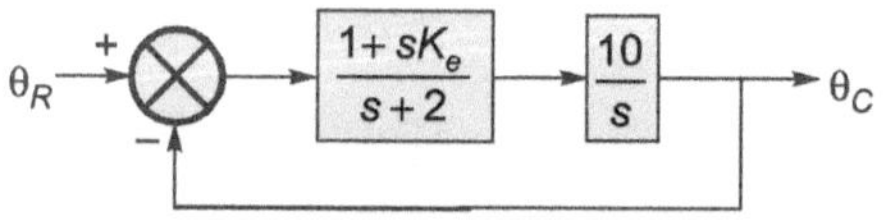

The value of error rate control (K_e) and 2% settling time for a damping ratio of 0.5 are respectively

(a) 0.116 and 2.53 sec
(b) 0.265 and 0.116 sec
(c) 0.116 and 0.265 sec
(d) 0.265 and 2.53 sec

Q.297 A feedback control system employing output-rate damping is shown in figure below.

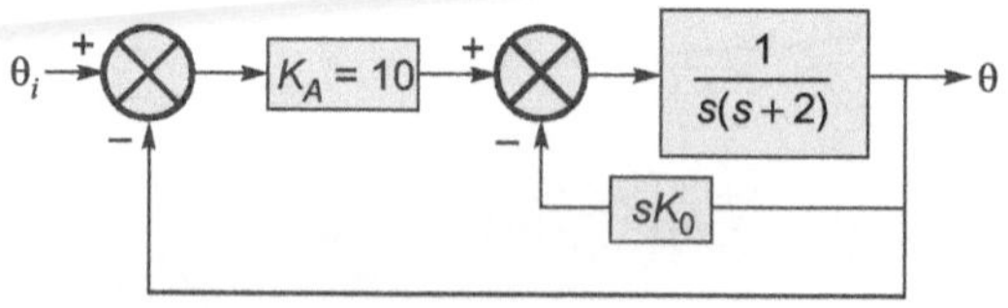

What is the steady state error (in radian) for the above system resulting from an unit ramp input if the damping ratio is 0.6?

(a) 0.16 rad (b) 0.38 rad
(c) 0.22 rad (d) 0.45 rad

Q.298 In the control system shown below, the controller which can give infinite value of steady-stae error to a ramp input is

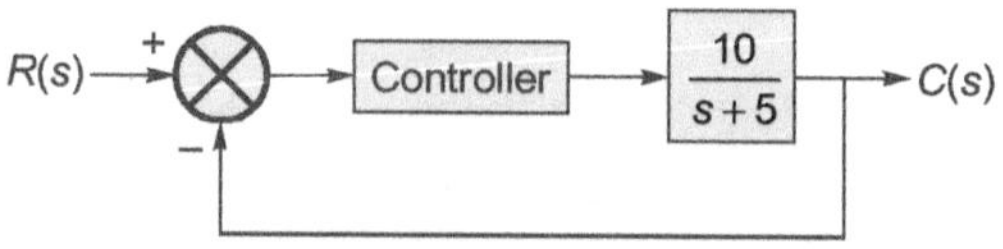

(a) derivative plus proportional type
(b) derivative type
(c) proportional type
(d) both (a) and (b)

Q.299 The transfer function of the OP-AMP circuit shown below is given by:

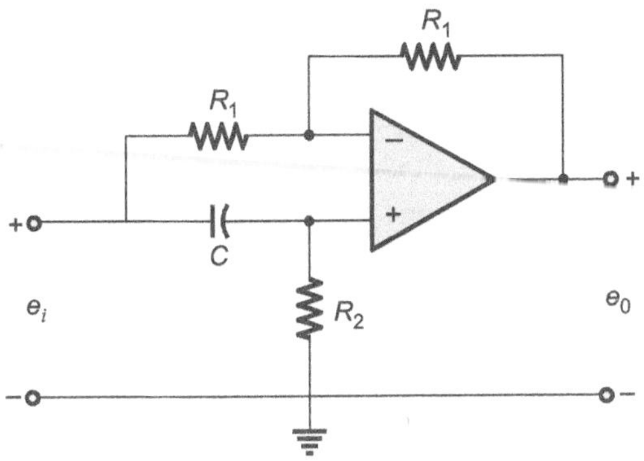

(a) $\frac{R_2 Cs}{1+R_2 Cs}$ (b) $\frac{1-R_1 Cs}{1+R_1 Cs}$

(c) $\frac{R_2 Cs-1}{R_2 Cs+1}$ (d) $\frac{R_2 Cs}{1+R_1 Cs}$

Q.300 The transfer function of a phase lead compensator is given by $\left[\frac{1+Ts}{1+\alpha Ts}\right]$, where $\alpha < 1$ and $T > 0$.
Which of the following is not a correct value of maximum phase shift provided by such a compensator?

(a) $\sin^{-1}\left[\frac{\alpha+1}{\alpha-1}\right]$ (b) $\tan^{-1}\left[\frac{1-\alpha}{2\sqrt{\alpha}}\right]$

(c) $\cos^{-1}\left[\frac{2\sqrt{\alpha}}{\alpha+1}\right]$ (d) $\sin^{-1}\left[\frac{1-\alpha}{1+\alpha}\right]$

Q.301 Assertion (A): Stepping motors are compatible with modern digital sytems.
Reason (R): The nature of command is in the forms of pulses.
(a) Both A and R are true and R is a correct explanation of A.
(b) Both A and R are true but R is not a correct explanation of A.
(c) A is true but R is false.
(d) A is false but R is true.

Q.302 Consider the control sytem shown in figure below employing a proportional compensator.

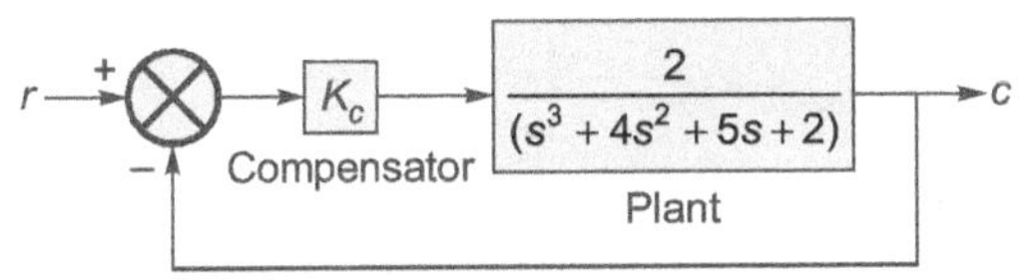

If the steady-state error has to be less than two percent for a unit step input, then the value of K_c would be

(a) 0 (b) 9
(c) 6 (d) can't be determined

Q.303 Match **List-I (Controllers)** with **List-II (Transfer functions)** and select the correct answer using the codes given below the lists:

List-I	List-II
A. PID controller	1. $\frac{K_1 s}{K_2 s}$
B. PD controller	2. $\frac{K_1 s+K_2}{K_3 s}$
C. PI controller	3. $\frac{K_1 s^2+K_2 s+K_3}{K_4 s}$
D. P controller	4. $\frac{K_1 s+K_2}{K_3}$

Codes:

	A	B	C	D
(a)	2	3	4	1
(b)	3	4	1	2
(c)	3	4	2	1
(d)	3	1	4	2

Q.304 The transfer function of a system is given as:

$$\frac{C(s)}{R(s)} = \frac{1+s}{1+0.2s}$$

It represents a
(a) lag network (b) lead network
(c) lag-lead network (d) proportional network

Q.305 Integral feedback control
(a) reduces feedback time
(b) improves transient response
(c) improves steady state response
(d) none of the above

Q.306 The industrial controller having the best steady state accuracy is
(a) a derivative controller
(b) an integral controller
(c) a rate feedback controller
(d) a proportional controller

Q.307 The transfer function of a phase lead controller is $\frac{1+3Ts}{1+Ts}$. The maximum value of phase shift provided by this controller is
(a) 90° (b) 60°
(c) 45° (d) 30°

Q.308 With regard to the filtering property, the lead compensator and the lag compensators are respectively
(a) low pass and high pass filters
(b) high pass and low pass filters
(c) both high pass filters
(d) both low pass filters

Q.309 In industrial control system, which one of the following methods is most commonly used in designing a system for meeting performance specifications?
(a) The transfer function is first determined and then either a lead compensation or a lag compensation is implemented.
(b) The transfer function is first determined and PID controllers are implemented by mathematically determining PID constants.
(c) PID controllers are implemented without the knowledge of the system parameters using "Ziegler Nichol's method."
(d) PID controllers are implemented using "Ziegler Nichol's method" after determining the system transfer function.

Q.310 Match **List-I (Control system components)** with **List-II (Applications)** and select the correct answer using the codes given below the lists:

List-I	List-II
A. Lag	1. Increases bandwidth
B. Lead	2. Attenuation
C. Lag-lead	3. Increases damping factor
D. Rate	4. Second order

Codes:

	A	B	C	D
(a)	3	1	4	2
(b)	2	4	1	3
(c)	3	4	1	2
(d)	2	1	4	3

12. State Variable Analysis

Q.311 Which one of the following information is necessary to formulate the problem of control systems optimization?
(a) System state equation, output equation, control vector and the performance index only.
(b) System state equation, output equation, control vector, constraints of the problem, the performance index and system parameters.
(c) System state equation, output equation, control vector and constraints of the problem only.
(d) System state equation, output equation and the control vector only.

Q.312 For arbitrary pole placement, the following combination is necessary.
(a) State feedback and stability
(b) Output feedback and observability
(c) Output feedback and controllability
(d) State feedback and controllability

Q.313 For a linear time invariant system, an optimal controller can be designed if
(a) the system is unstable but observable
(b) the system is controllable and observable
(c) the system is stable and unobservable
(d) the system is uncontrollable but stable

Q.314 The eigen values of linear system are the location of
(a) zeros of the system
(b) poles of the system
(c) both (a) and (b)
(d) finite poles and zeros

Q.315 If the system matrix of a linear time invariant continuous system is given by $A = \begin{bmatrix} 0 & 1 \\ -2 & -5 \end{bmatrix}$, its characteristic equation will be given by
(a) $s^2 + 5s + 2 = 0$ (b) $s^2 - 2s - 5 = 0$
(c) $s^2 + s + 3 = 0$ (d) $s^2 + 2s + 5 = 0$

Q.316 The state-variable description of a linear autonomous system is $\overline{X} = A\overline{X}$, where $\overline{X}$ is a two-dimensional state vector and A is a matrix given by $A = \begin{bmatrix} 0 & 5 \\ 5 & 0 \end{bmatrix}$.
The poles of the system are located at
(a) –5 and +5 (b) $-5j$ and $+5j$
(c) –5 and –5 (d) +5 and +5

Q.317 Assertion (A): The eigen values of a linear continuous-data time invariant system controls the stability of the system.

Reason (R): The roots of the characteristic equation are the same as the eigen values of system matrix A of the state equations.

(a) Both A and R are true and R is a correct explanation of A.
(b) Both A and R are true but R is not a correct explanation of A.
(c) A is true but R is false.
(d) A is false but R is true.

Q.318 The vector matrix differential equation of a system is given by

$$\dot{x} = \begin{bmatrix} 0 & 1 \\ -2 & -3 \end{bmatrix} x$$

The state transition matrix of the system is

(a) $\begin{bmatrix} -e^{-t} + 2e^{-2t} & 2e^{-t} + 2e^{-2t} \\ 2e^{-t} - e^{-2t} & e^{-t} - e^{-2t} \end{bmatrix}$

(b) $\begin{bmatrix} -e^{-t} + 2e^{-2t} & e^{-t} - e^{-2t} \\ 2e^{-t} + 2e^{-2t} & 2e^{-t} - e^{-2t} \end{bmatrix}$

(c) $\begin{bmatrix} 2e^{-t} - e^{-2t} & e^{-t} - e^{-2t} \\ -2e^{-t} + 2e^{-2t} & -e^{-t} + 2e^{-2t} \end{bmatrix}$

(d) $\begin{bmatrix} -2e^{-t} + 2e^{-2t} & e^{-t} - e^{-2t} \\ -e^{-t} + 2e^{-2t} & 2e^{-t} - e^{-2t} \end{bmatrix}$

Q.319 The system matrix of a continous time system is given by:

$$A = \begin{bmatrix} -5 & 1 & 0 \\ 0 & -5 & 1 \\ 0 & 0 & -5 \end{bmatrix}$$

The characteristic equation is

(a) $s^3 + 12s^2 - 70s + 120 = 0$
(b) $s^3 - 17s^2 + 10s + 50 = 0$
(c) $s^3 + s^2 + 2s + 20 = 0$
(d) $s^3 + 15s^2 + 75s + 125 = 0$

Q.320 A system is described by the following equations:

$$\dot{x}(t) = \begin{bmatrix} -1 & 1 \\ 0 & -2 \end{bmatrix} \dot{x}(t) + \begin{bmatrix} 1 & 0 & 1 \\ 0 & 1 & 1 \end{bmatrix} u(t)$$

$$y(t) = \begin{bmatrix} 1 & 2 \\ 1 & 0 \\ 1 & 1 \end{bmatrix} x(t)$$

The transfer function of the system is

(a) $\dfrac{s^3 - 3s^2 - 9s + 2}{s^2 + 3s + 2}$ (b) $\dfrac{s^2 + 3s + 2}{s^3 - 3s - 9s + 2}$

(c) $\dfrac{1}{s^2 + 3s + 2}$ (d) $\dfrac{3s^2 + 9s^2 + 2s + 1}{s^3 + s^2 + 3s + 2}$

Q.321 The state variable description of an autonomous system is, $\dot{X} = AX$ where X is a two-dimensional vector and A is a matrix given by $A = \begin{bmatrix} \sigma & -\omega \\ \omega & \sigma \end{bmatrix}$.

The eigen values of A are

(a) $j\omega$ and $\sigma + j\omega$
(b) $\sigma + j\omega$ and $\sigma - j\omega$
(c) $\sigma - j\omega$ and $\sigma + j\omega$
(d) $\sigma - j\omega$ and $j\omega$

Q.322 The system equations are given by

$$\dot{x}(t) = \begin{bmatrix} 0 & 1 \\ -2 & -3 \end{bmatrix} x(t) + \begin{bmatrix} 0 \\ 1 \end{bmatrix} u(t)$$

$$y(t) = [1 \quad 0]\, x(t)$$

The transfer function of the above system is

(a) $\dfrac{1}{s^2 + 3s + 2}$ (b) $\dfrac{2}{s^2 + 3s + 2}$

(c) $\dfrac{1}{s^2 + s + 1}$ (d) $\dfrac{2}{s^2 + s + 2}$

Q.323 Consider the following statements related to state space analysis of control systems:

1. The zeros of the system can be obtained from eigen value of the system matrix.
2. A system is said to be observable if every state x_0 can be exactly determined from the measurement of the output 'y' over a finite interval of time $0 \le t \le t_f$.
3. The process by which transfer function changes to state diagram or state equations is called decomposition of the transfer function.
4. The state space techniques can be applied to linear and time invariant systems only.

Which of the above statements are correct?

(a) 1, 3 and 4 (b) 1, 2 and 3
(c) 2 and 3 (d) 2, 3 and 4

Q.324 Consider the system shown in figure below:

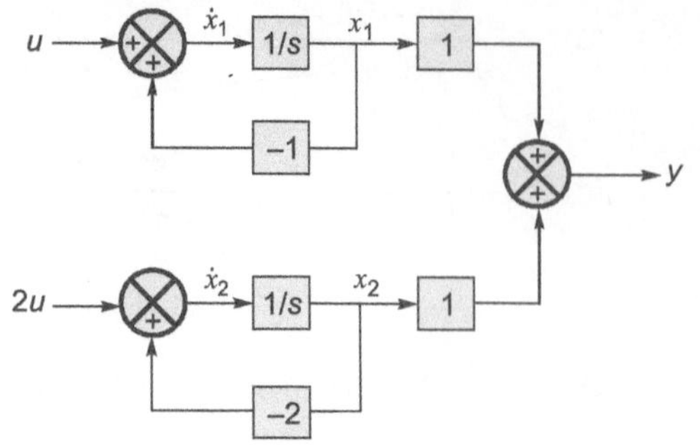

The system is

(a) controllable but unobservable

(b) uncontrollable but observable

(c) uncontrallable and unobservable

(d) controllable and observable

Q.325 Consider the following matrix:

$$\dot{x} = \begin{bmatrix} 0 & 1 \\ -6 & -5 \end{bmatrix},\ x(0) = \begin{bmatrix} 1 \\ 0 \end{bmatrix}$$

$x(t)$ is given by

(a) $\begin{bmatrix} 2e^{-2t} - e^{-3t} \\ 5(e^{-2t} - e^{-3t}) \end{bmatrix}$ (b) $\begin{bmatrix} 3e^{-2t} - 2e^{-3t} \\ -6(e^{-2t} - e^{-3t}) \end{bmatrix}$

(c) $\begin{bmatrix} -e^{-2t} + 2e^{-3t} \\ e^{-2t} - e^{-3t} \end{bmatrix}$ (d) $\begin{bmatrix} 2e^{-2t} - e^{-3t} \\ 2(e^{-3t} - e^{-2t}) \end{bmatrix}$

Q.326 The transfer function of the system shown below is

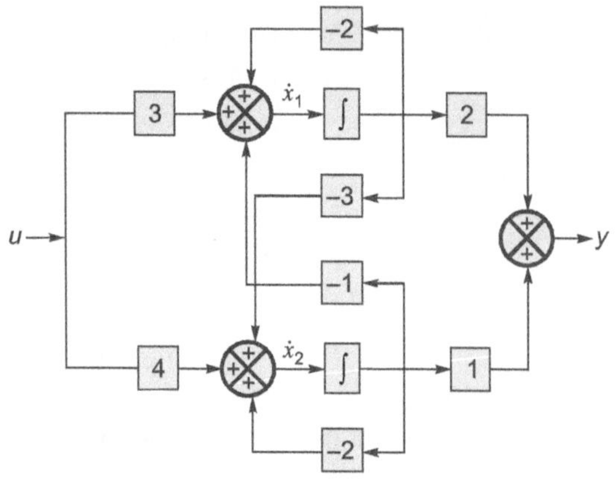

(a) $\dfrac{10s+3}{s^2+4s+1}$ (b) $\dfrac{1}{s^2+4s+1}$

(c) $\dfrac{s+1}{s^2+4s+1}$ (d) none of these

Q.327 The state equation for the circuit shown below is

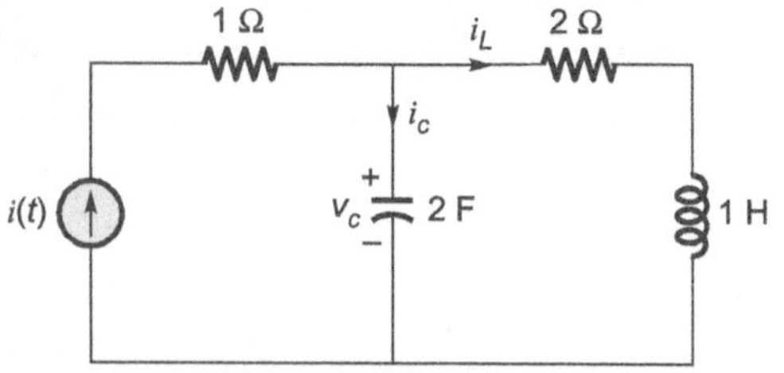

(a) $\begin{bmatrix} \dot{v}_C \\ \dot{i}_L \end{bmatrix} = \begin{bmatrix} 0.5 & 0 \\ 0 & 2 \end{bmatrix} \begin{bmatrix} v_C \\ i_L \end{bmatrix} + \begin{bmatrix} -1 \\ 0.5 \end{bmatrix} i(t)$

(b) $\begin{bmatrix} \dot{v}_C \\ \dot{i}_L \end{bmatrix} = \begin{bmatrix} 0.5 & -0.5 \\ 1 & 2 \end{bmatrix} \begin{bmatrix} v_C \\ i_L \end{bmatrix} + \begin{bmatrix} 0.5 \\ -1 \end{bmatrix} i(t)$

(c) $\begin{bmatrix} \dot{v}_C \\ \dot{i}_L \end{bmatrix} = \begin{bmatrix} 0 & -0.5 \\ 1 & 2 \end{bmatrix} \begin{bmatrix} v_C \\ i_L \end{bmatrix} + \begin{bmatrix} 0.5 \\ 0 \end{bmatrix} i(t)$

(d) $\begin{bmatrix} \dot{v}_C \\ \dot{i}_L \end{bmatrix} = \begin{bmatrix} 0.5 & -0.5 \\ 1 & 0 \end{bmatrix} \begin{bmatrix} v_C \\ i_L \end{bmatrix} + \begin{bmatrix} -1 \\ 0.5 \end{bmatrix} i(t)$

Q.328 The state space representation of the system represented by the SFG shown below is

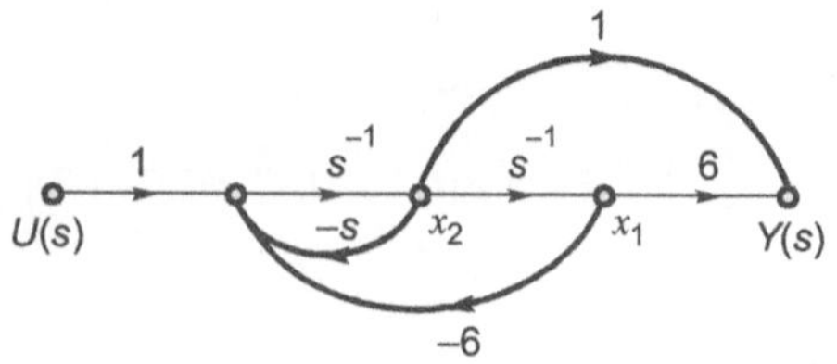

(a) $\dot{x} = \begin{bmatrix} 0 & 1 \\ -6 & -5 \end{bmatrix} x + \begin{bmatrix} 0 \\ 1 \end{bmatrix} u(t)\ ;\ y(t) = \begin{bmatrix} 6 & 1 \end{bmatrix} x$

(b) $\dot{x} = \begin{bmatrix} 1 & 0 \\ -5 & -6 \end{bmatrix} x + \begin{bmatrix} 0 \\ 2 \end{bmatrix} u(t)\ ;\ y(t) = \begin{bmatrix} 6 & 0 \end{bmatrix} x$

(c) $\dot{x} = \begin{bmatrix} 1 & -5 \\ 0 & -6 \end{bmatrix} x + \begin{bmatrix} 1 \\ 1 \end{bmatrix} u(t)\ ;\ y(t) = \begin{bmatrix} 1 & 6 \end{bmatrix} x$

(d) $\dot{x} = \begin{bmatrix} 1 & -6 \\ -5 & 0 \end{bmatrix} x + \begin{bmatrix} 1 \\ 0 \end{bmatrix} u(t)\ ;\ y(t) = \begin{bmatrix} 0 & 6 \end{bmatrix} x$

Q.329 The state variable representation of a system is given by:

$$\dot{X} = \begin{bmatrix} 0 & 1 & 0 \\ 0 & 0 & 1 \\ -6 & -11 & -6 \end{bmatrix} X + \begin{bmatrix} 0 \\ 0 \\ 2 \end{bmatrix} U$$

$$Y = [1\ 0\ 0]\, X$$

The system is
(a) neither controllable nor observable
(b) controllable but not observable
(c) uncontrollable but observable
(d) both controllable and observable

Q.330 The zeros of following system are located at

$$X = \begin{bmatrix} 0 & 1 \\ -20 & -9 \end{bmatrix} X + \begin{bmatrix} 0 \\ 1 \end{bmatrix} u$$

$$Y = [-17 \quad -5]\, X + [1]\, u$$

(a) –4 and – 5
(b) –1 and –4
(c) –1 and –3
(d) –3 and –5

Answers Control Systems

1.	(d)	2.	(a)	3.	(c)	4.	(b)	5.	(d)	6.	(a)	7.	(c)	8.	(d)
9.	(b)	10.	(d)	11.	(c)	12.	(b)	13.	(a)	14.	(d)	15.	(c)	16.	(b)
17.	(c)	18.	(c)	19.	(d)	20.	(a)	21.	(a)	22.	(d)	23.	(c)	24.	(c)
25.	(b)	26.	(d)	27.	(c)	28.	(a)	29.	(b)	30.	(d)	31.	(b)	32.	(a)
33.	(b)	34.	(c)	35.	(a)	36.	(d)	37.	(c)	38.	(a)	39.	(d)	40.	(b)
41.	(c)	42.	(a)	43.	(b)	44.	(c)	45.	(d)	46.	(b)	47.	(c)	48.	(d)
49.	(a)	50.	(b)	51.	(c)	52.	(d)	53.	(b)	54.	(a)	55.	(c)	56.	(b)
57.	(a)	58.	(d)	59.	(c)	60.	(b)	61.	(d)	62.	(a)	63.	(c)	64.	(b)
65.	(d)	66.	(d)	67.	(c)	68.	(c)	69.	(a)	70.	(b)	71.	(c)	72.	(a)
73.	(b)	74.	(d)	75.	(c)	76.	(a)	77.	(c)	78.	(c)	79.	(b)	80.	(d)
81.	(a)	82.	(c)	83.	(d)	84.	(a)	85.	(b)	86.	(d)	87.	(c)	88.	(c)
89.	(d)	90.	(a)	91.	(b)	92.	(c)	93.	(d)	94.	(d)	95.	(c)	96.	(c)
97.	(d)	98.	(c)	99.	(b)	100.	(a)	101.	(d)	102.	(c)	103.	(c)	104.	(d)
105.	(c)	106.	(a)	107.	(b)	108.	(c)	109.	(d)	110.	(b)	111.	(b)	112.	(d)
113.	(c)	114.	(a)	115.	(a)	116.	(c)	117.	(b)	118.	(d)	119.	(a)	120.	(b)
121.	(a)	122.	(a)	123.	(d)	124.	(c)	125.	(c)	126.	(b)	127.	(a)	128.	(b)
129.	(d)	130.	(b)	131.	(c)	132.	(c)	133.	(a)	134.	(c)	135.	(b)	136.	(d)
137.	(c)	138.	(a)	139.	(b)	140.	(b)	141.	(d)	142.	(c)	143.	(a)	144.	(b)
145.	(d)	146.	(c)	147.	(a)	148.	(c)	149.	(b)	150.	(d)	151.	(c)	152.	(a)
153.	(b)	154.	(a)	155.	(c)	156.	(a)	157.	(b)	158.	(d)	159.	(c)	160.	(c)
161.	(a)	162.	(b)	163.	(c)	164.	(b)	165.	(a)	166.	(d)	167.	(c)	168.	(b)
169.	(d)	170.	(b)	171.	(c)	172.	(b)	173.	(b)	174.	(c)	175.	(b)	176.	(c)
177.	(d)	178.	(a)	179.	(c)	180.	(b)	181.	(d)	182.	(a)	183.	(b)	184.	(c)
185.	(d)	186.	(d)	187.	(b)	188.	(c)	189.	(c)	190.	(d)	191.	(c)	192.	(a)

193. (b)	194. (d)	195. (d)	196. (c)	197. (a)	198. (d)	199. (b)	200. (b)
201. (a)	202. (b)	203. (a)	204. (c)	205. (d)	206. (c)	207. (c)	208. (d)
209. (b)	210. (a)	211. (a)	212. (d)	213. (d)	214. (b)	215. (c)	216. (a)
217. (d)	218. (d)	219. (b)	220. (c)	221. (c)	222. (a)	223. (d)	224. (c)
225. (c)	226. (a)	227. (d)	228. (c)	229. (b)	230. (d)	231. (b)	232. (a)
233. (c)	234. (d)	235. (c)	236. (d)	237. (a)	238. (b)	239. (c)	240. (a)
241. (d)	242. (c)	243. (b)	244. (a)	245. (d)	246. (c)	247. (b)	248. (b)
249. (d)	250. (d)	251. (c)	252. (d)	253. (b)	254. (a)	255. (a)	256. (c)
257. (b)	258. (a)	259. (c)	260. (d)	261. (d)	262. (b)	263. (b)	264. (d)
265. (a)	266. (a)	267. (d)	268. (d)	269. (a)	270. (d)	271. (b)	272. (c)
273. (a)	274. (b)	275. (b)	276. (a)	277. (d)	278. (b)	279. (c)	280. (b)
281. (d)	282. (a)	283. (c)	284. (d)	285. (d)	286. (b)	287. (c)	288. (a)
289. (c)	290. (d)	291. (b)	292. (a)	293. (d)	294. (c)	295. (d)	296. (a)
297. (b)	298. (d)	299. (c)	300. (a)	301. (a)	302. (d)	303. (c)	304. (b)
305. (c)	306. (b)	307. (d)	308. (b)	309. (d)	310. (d)	311. (c)	312. (d)
313. (b)	314. (b)	315. (a)	316. (b)	317. (a)	318. (c)	319. (d)	320. (a)
321. (b)	322. (a)	323. (c)	324. (d)	325. (b)	326. (a)	327. (c)	328. (a)
329. (d)	330. (c)						

Explanations

1. (d)

$$f(t) = \delta(t-10)$$

$$\therefore \quad F(s) = 1 \cdot e^{-10s}$$

$$= e^{-10s} = \exp(-10s)$$

(since $\mathcal{L}\{f(t-a)\} = F(s) \cdot e^{-as}$)

5. (d)

- There is no-control action in open loop systems. Hence, statement 1 is false.
- Generally, all control systems operated by present timing mechanism are open loop because they are more stable than closed loop control system. Hence, statement-2 is true.
- Due to various advantages of negative feedback systems (except reduction in gain), feedback signal is usually negative in a closed loop control system. Hence, statement-3 is true.
- In open-loop control systems stability can be ensured because they are more stable than closed-loop control system.

Thus, only statements 2 and 3 are true.

9. (b)

The control system which operates on the time basis is an open loop system. For example, in automatic control of traffic, the time for each lamp (red, yellow and green) is fixed. The operation of each lamp does not depends upon the density of the traffic but depends upon the fixed time. Hence, assertion is true.

Reason is also true but, it is not the correct explanation of assertion.

10. (d)

Ramp function occurring at $t = 0$ is denoted by $r(t) = t\,u(t)$.

Ramp function occurring at $t = a$ will be denoted by $r(t-a) = (t-a)\,u(t-a)$.

$$\therefore \quad \mathcal{L}\{r(t-a)\} = \mathcal{L}(t-a)\,u(t-a)\}$$
$$= \frac{1}{s^2}\cdot e^{-as}$$

14. (d)

The laplace transform of unity function i.e.

$$\mathcal{L}[u(t)] = \frac{1}{s}$$

16. (b)

$$\mathcal{L}\{e^{-at}\sin bt\} = \frac{b}{(s+a)+b^2}$$

Here, $a = 2$ and $b = 3\omega$

$$\therefore \quad \mathcal{L}\{e^{-2t}\sin 3\omega t\} = \mathcal{L}\{e^{-2t}\sin 3\omega t\}$$
$$= \left[\frac{3\omega}{(s+2)^2+9\omega^2}\right]$$

19. (d)

The effect of parameter variation and internal noise is more in an open loop system.

21. (a)

Due to presence of feedback in a closed loop system, it is more complex and hence, costly compared to an open loop system.

22. (d)

$$G(s) = \frac{s(s+4)}{(s+1)(s+8)}$$

There is no open loop pole at origin. Hence, type of system = 0.

Highest power of s of the characteristic equation $1+G(s) = 0$ will be 2.

Hence, order of given system = 2.

24. (c)

For $Z < P$,

$$\text{Let,} \quad \text{T.F.} = \frac{(s+a)}{(s+b)(s+c)}$$
$$= \frac{\left(1+\frac{a}{s}\right)}{s\left(1+\frac{b}{s}\right)\left(1+\frac{c}{s}\right)}$$

Here, number of zeros = 1 and no. of poles = 2

$$\therefore \quad P - Z = 1$$

When $s \to \infty$, transfer function becomes zero. Thus, there is one zero $(P - Z = 1)$ at infinity. Thus, assertion is true.

Since value of transfer function becomes zero as $s \to \infty$, therefore, reason is false.

25. (b)

Let i be the current in the given circuit.

$$\text{Then,} \quad v_i = iR + L\frac{di}{dt}$$
$$\text{or,} \quad V_i(s) = I(s)\,[R + Ls]$$
$$\text{or,} \quad I(s) = \frac{v_i(s)}{R+Ls}$$
$$\text{Also,} \quad V_0(s) = (Ls)\,I(s) = \left(\frac{Ls}{R+Ls}\right)V_i(s)$$
$$\text{or,} \quad \frac{V_0(s)}{V_i(s)} = \left(\frac{Ls}{R+Ls}\right) = \text{Transfer function}$$

26. (d)

The transfer function is defined only for linear time invariant system. All the initial conditions are set to zero for finding the transfer function of a system. Hence, only statements 1 and 4 are not true.

27. (c)

Given differential equation is

$$\frac{d^2y(t)}{dt^2} + \frac{9dy(t)}{dt} + 20\,y(t) = 20\,x(t)$$

or, $s^2Y(s) + 9s\,Y(s) + 20\,Y(s) = 20\,X(s)$

$$\text{or,} \quad \frac{Y(s)}{X(s)} = \frac{20}{s^2+9s+20}$$
$$\text{Given,} \quad X(s) = \frac{1}{s}$$
$$\therefore \quad Y(s) = \frac{20}{s(s^2+9s+20)}$$
$$= \frac{4}{s} + \frac{B}{s+4} + \frac{C}{s+5}$$

Residues are: $A = 1$, $B = -5$ and $C = 4$

$$\therefore \quad Y(s) = \frac{1}{s} - \frac{5}{s+4} + \frac{4}{s+5}$$
$$\text{or,} \quad y(t) = (1 - 5e^{-4t} + 4\,e^{-5t})$$

28. (a)

$$\text{Given,} \quad c(t) = 1 + 0.25\,e^{-50t} - 1.25e^{-10t}$$

or, $$C(s) = \frac{1}{s} + \frac{0.25}{(s+50)} - \frac{1.25}{(s+10)}$$

$$= \frac{500}{s(s+50)(s+10)}$$

Now, $$R(s) = \frac{1}{s}$$

$\therefore$ $$\text{T.F} = \frac{C(s)}{R(s)} = \frac{500}{(s+50)(s+10)}$$

$$= \frac{500}{500\left(1+\frac{1}{50}s\right)\left(1+\frac{s}{10}\right)}$$

$$= \frac{1}{\left(1+\frac{s}{500}\right)\left(1+\frac{s}{10}\right)}$$

Thus, $\frac{C(s)}{R(s)}$ is in time constant form having steady state gain = 1.

29. (b)

For given network, total input impedance is

$$Z(s) = \left(R_2 + \frac{R_1 \times \frac{1}{Cs}}{R_1 + \frac{1}{Cs}}\right) = \left(R_2 + \frac{R_1}{1+R_1 Cs}\right)$$

$$= \left(\frac{R_1 + R_2 + R_1 R_2 Cs}{1+R_1 Cs}\right)$$

$\therefore$ Total current in circuit is,

$$I(s) = \frac{V_1(s)}{Z(s)}$$

So, $$V_2(s) = I(s) \times R_2 = \frac{V_1(s)}{Z(s)} R_2$$

$$= \frac{V_1(s)\left[1+R_1 Cs\right]}{\left[R_1 + R_2 + R_1 R_2 Cs\right]} \times R_2$$

or, $$\frac{V_2(s)}{V_1(s)} = \frac{R_2(1+R_1 Cs)}{(R_1 + R_2 + R_1 R_2 Cs)}$$

= Transfer function

Hence, pole is at

$$s = -\frac{(R_1 + R_2)}{R_1 R_2 C} = -\left(\frac{1}{R_1 C} + \frac{1}{R_2 C}\right)$$

and zero is at $s = -\frac{1}{R_1 C}$.

30. (d)

From given pole-zero plot, transfer function is

$$G(s) = \frac{K(s+3)}{s(s+2)(s+4)}$$

Given, $$G(1) = 3.2 = \frac{K \times 4}{1 \times 3 \times 5}$$

or, $$K = \frac{3.2 \times 15}{4} = 12$$

So, gain factor $K = 12$

31. (b)

The type of a system denotes the no. of poles of origin.

32. (a)

For A: Two conjugate poles on imaginary axis i.e. it's impulse response = $\sin at$ (marginally stable).

For B: Two poles at origin i.e. impulse response = $t\,u(t)$ (unstable system).

For C: Two complex conjugate poles with negative real parts. So, its impulse response = $e^{-at}\sin bt$ (stable system).

For D: Two imaginary complex double roots. So, impulse response = $t \sin bt$. (unstable system).

34. (c)

Transfer function is applicable to only linear and time-variant systems.

35. (a)

Given transfer function is

$$H(s) = \frac{(s+2)(s+5)}{(s+3)(s+8)}$$

Converting above transfer function in time-constant form, we get:

$$H(s) = \frac{2 \times 5\left(1+\frac{s}{2}\right)\left(1+\frac{s}{5}\right)}{3 \times 8\left(1+\frac{s}{3}\right)\left(1+\frac{s}{8}\right)}$$

Hence, dc gain is

$$K = \frac{2 \times 5}{3 \times 8} = \frac{10}{24} = 0.416$$

36. (d)

Given, $c(t) = -te^{-t} + e^{-t}$ (impulse response)

We know that,

$\mathcal{L}$\{Impulse response\} = Transfer function ($H(s)$)

$$\therefore \quad H(s) = \frac{-1}{(s+1)^2} + \frac{1}{(s+1)}$$

$$= \frac{1}{s+1}\left[\frac{-1}{s+1} + 1\right]$$

$$= \frac{1}{(s+1)} \times \frac{s}{(s+1)} = \frac{s}{(s+1)^2}$$

37. (c)

We know that, impulse response $= \frac{C(s)}{R(s)} =$ Transfer fuction.

For given system,

$$\frac{C(s)}{R(s)} = \frac{1}{(s+3)}$$

Now, $c(t) = t\,e^{-3t}$

$$\therefore \quad C(s) = \frac{1}{(s+3)^2}$$

So, $R(s) = C(s) \times (s+3)$

$$= \frac{1}{(s+3)^2} \times (s+3) = \frac{1}{(s+3)}$$

or, $r(t)$ = Required input $= e^{-3t}\,u(t)$

38. (a)

Given, $c(t) = t\,e^{-t}$

$$\therefore \quad C(s) = \frac{1}{(s+1)^2}$$

Now, $r(t) = u(t)$ (unit step input)

$$\therefore \quad R(s) = \frac{1}{s}$$

So, $$\text{T.F} = \frac{C(s)}{R(s)} = \frac{s}{(s+1)^2}$$

39. (d)

Taking Laplace transform of the given differential equation on both sides, we get

$2s^2\,Y(s) + sY(s) + 5\,Y(s) = R(s) + 2e^{-s}\,R(s)$

or, $$\frac{Y(s)}{R(s)} = \left(\frac{1+2e^{-s}}{2s^2+s+5}\right)$$

40. (b)

$$K = \frac{\sum(\text{distance from poles})}{\sum(\text{distance from zeros})}$$

$$= \frac{12 \times 4 \times 4 \times 60}{2} = 5760$$

42. (a)

The final value theorem is valid only if $sF(s)$ does not has any poles on the $j\omega$ axis and in the right half of the s-plane. Hence, both A and R are true and R is a correct explanation of A.

44. (c)

Given, O.L.T.F. $= \frac{2s+1}{s^2}$

$$\therefore \quad \text{C.L.T.F.} = \frac{C(s)}{R(s)} = \frac{G(s)}{1+G(s)H(s)}$$

$$= \left[\frac{\frac{2s+1}{s^2}}{1+\frac{2s+1}{s^2}}\right] \quad (H(s)=1)$$

or, $$\frac{C(s)}{R(s)} = \frac{2s+1}{(s+1)^2} = \frac{2s}{(s+1)^2} + \frac{1}{(s+1)^2}$$

For unit impulse response, $R(s) = 1$

$$\therefore \quad C(s) = 2\left[\frac{(s+1)}{(s+1)^2} - \frac{1}{(s+1)^2}\right] + \frac{1}{(s+1)^2}$$

or, $$C(s) = 2\left[\frac{1}{(s+1)} - \frac{1}{(s+1)^2}\right] + \frac{1}{(s+1)^2}$$

or, $c(t) = 2\,(e^{-t} - te^{-t}) + te^{-t}$

$= 2e^{-t} - te^{-t} = (2-t)\,e^{-t}$

Thus, impulse response, $c(t) = (2-t)\,e^{-t}\,u(t)$

45. (d)

Impulse response,

$c(t) = u(t)$

We now that, $\mathcal{L}$(impulse response) = Transfer function $= \frac{C(s)}{R(s)}$

$$\therefore \quad \frac{C(s)}{R(s)} = \text{T.F.} = \frac{1}{s}$$

46. (b)

Using final value theorem

$$\lim_{t\to\infty} f(t) = \lim_{s\to 0}[sF(s)]$$

$$= \lim_{s\to 0}\left[\frac{5}{s^2+s+2}\right] = \frac{5}{2} = 2.5$$

48. (d)

A high pass RC network can be represented as shown below.

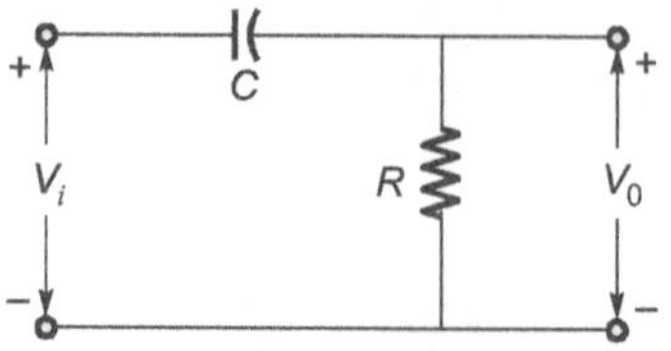

Its transfer function is $\left[\frac{V_0(s)}{V_i(s)} = \frac{RCs}{1+RCs}\right]$.

49. (a)

Transforming the given circuit in Laplace domain, the circuit becomes as shown below.

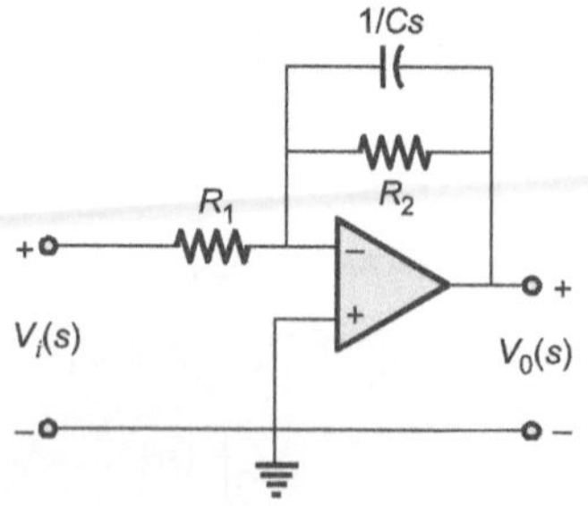

Net impedance of feedback path is

$$Z_2(s) = \frac{R_2 \times \frac{1}{Cs}}{R_2 + \frac{1}{Cs}} = \frac{R_2}{1+R_2Cs}$$

and $Z_1(s) = R_1$

∴ For inverting op-amp,

$$\frac{V_0(s)}{V_i(s)} = \frac{-Z_2(s)}{Z_1(s)} = -\frac{\frac{R_2}{1+R_2Cs}}{R_1} = \frac{-R_2}{R_1(1+R_2Cs)}$$

$$\therefore \quad \left|\frac{V_0(s)}{V_i(s)}\right| = \frac{R_2}{R_1(1+R_2Cs)}$$

50. (b)

Let system gain be K.

$$\therefore \quad \text{T.F.} = \frac{C(s)}{R(s)} = H(s) = \frac{K(s+2)}{s(s+1)(s+3)}$$

(Using given pole- zero plot)

Given, $H(2) = 4$

or, $\frac{K\times 4}{2\times 3\times 5} = 4$ or $K = 30$

$$\therefore \quad H(s) = \frac{30(s+2)}{s(s+1)(s+3)}$$

51. (c)

On combining the blocks G_2 and G_3 which are in parallel, the given block diagram will be reduced as shown below.

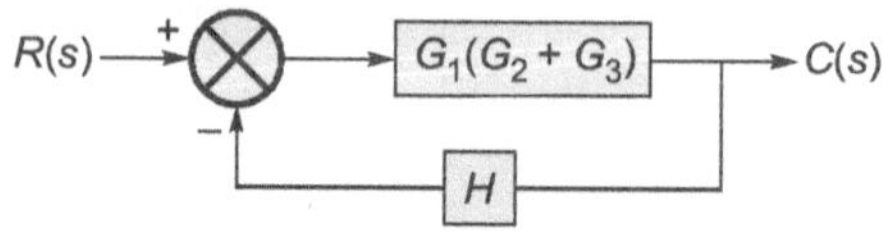

∴ Transfer function,

$$\frac{C(s)}{R(s)} = \left[\frac{G_1(G_2+G_3)}{1+G_1H(G_2+G_3)}\right]$$

$$= \left[\frac{G_1G_2+G_2G_3}{1+G_1G_2H+G_1G_3H}\right]$$

52. (d)

The T.F. of the given block diagram can be reduced to: $\frac{C}{R} = \left(\frac{G}{1-GH}\right)$

Its block diagram is shown below, where a closed loop feedback system is transformed into an unity feedback system and vice-versa.

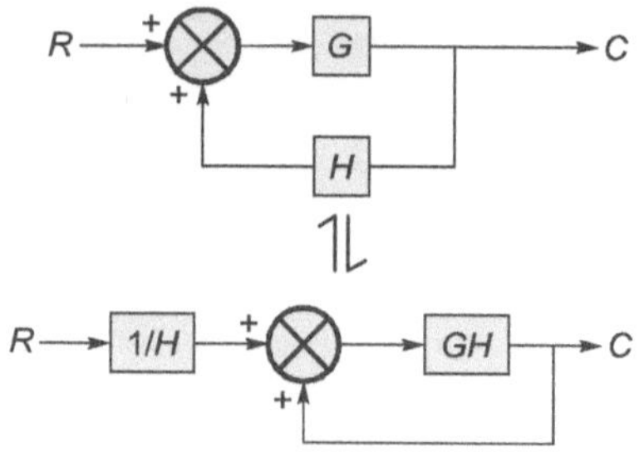

53. (b)

Transfer function does not gives information about the internal structure of the system. Hence, staement-2 is false.

54. (a)

On shifting the take-off point beyond the block G_3 and combining blocks G_2 and G_3, the given block diagram is reduced as shown below.

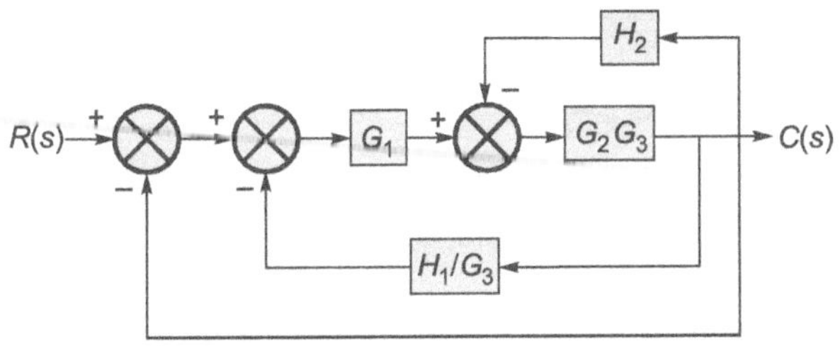

The block diagram is further reduced as shown below.

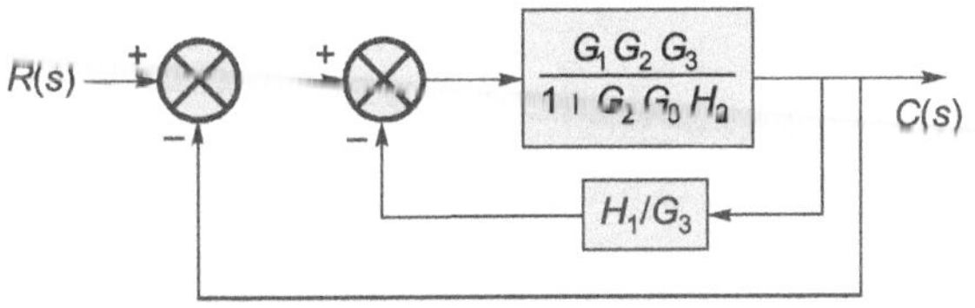

Reducing the above block diagram,

$$\frac{C(s)}{R(s)} = \left(\frac{G_1 G_2 G_3}{1 + G_2 G_3 H_2 + G_1 G_2 H_1 + G_1 G_2 G_3}\right)$$

55. (c)

On shifting the take-off point beyond block G, we have the reduced block diagram as shown below:

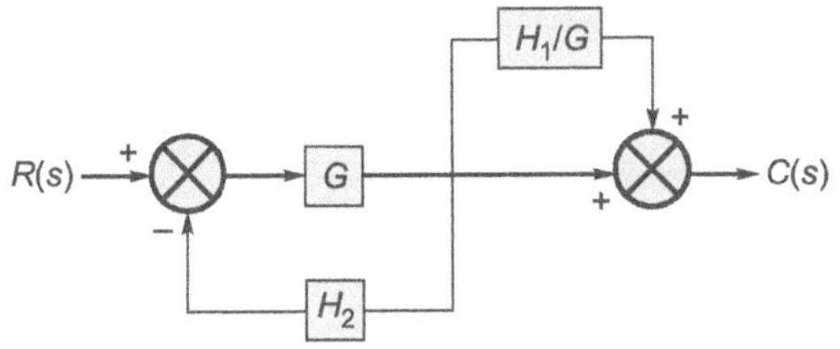

On further reducing the above block diagram, we get the block diagram as shown below.

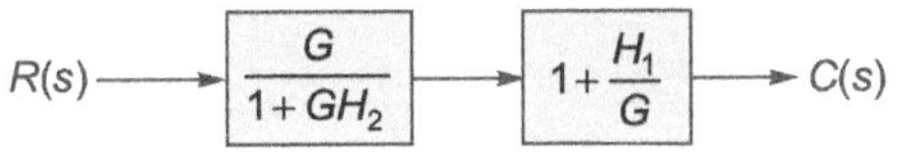

Thus, $\frac{C(s)}{R(s)} = \left(\frac{G + H_1}{1 + GH_2}\right)$

56. (b)

Here, $G_{11} = \left.\frac{C_1}{R_1}\right|_{R_2 = C_2 = 0}$

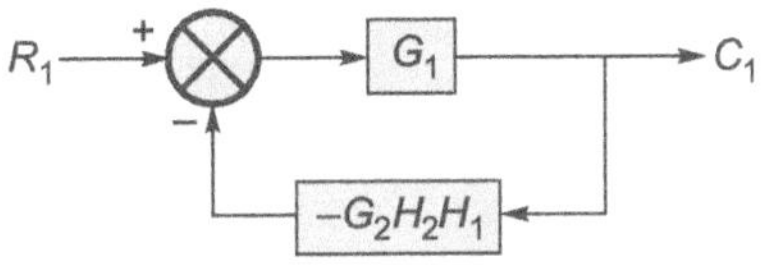

$\therefore$ $G_{11} = \left[\frac{G_1}{1 - G_1 G_2 H_1 H_2}\right]$

Also, $G_{22} = \left.\frac{C_2}{R_2}\right|_{R_1 = C_1 = 0}$

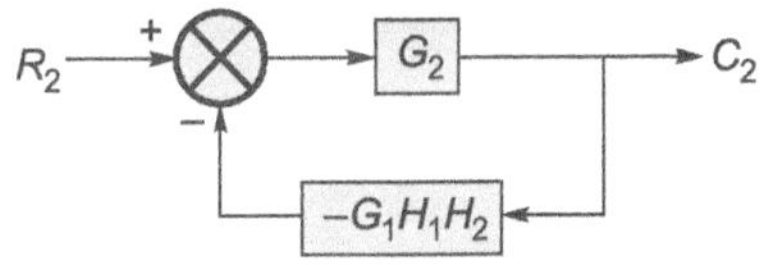

$\therefore$ $G_{22} = \left[\frac{G_2}{1 - G_1 G_2 H_1 H_2}\right]$

Also, $G_{12} = \left.\frac{C_1}{R_2}\right|_{R_1 = C_2 = 0}$

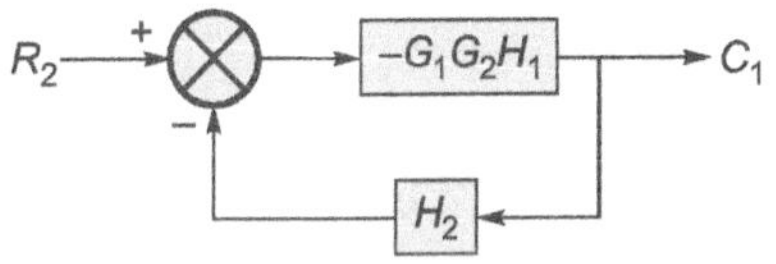

$\therefore$ $G_{12} = \left[\frac{-G_1 G_2 H_1}{1 - G_1 G_2 H_1 H_2}\right]$

Also, $G_{21} = \left.\frac{C_2}{R_1}\right|_{R_2 = C_1 = 0}$

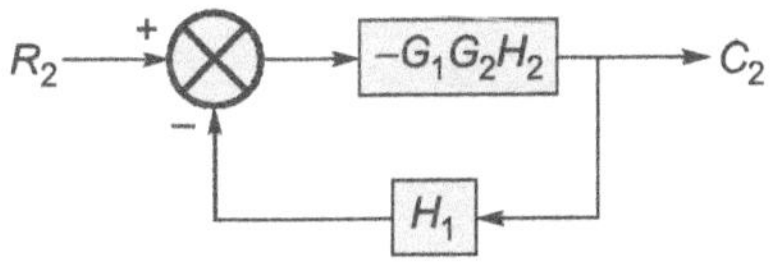

$\therefore$ $G_{21} = \left[\frac{-G_1 G_2 H_2}{1 - G_1 G_2 H_1 H_2}\right]$

57. (a)

Shifting the take-off point after block G_3 in block diagram-A, the reduced block diagram will be as shown below.

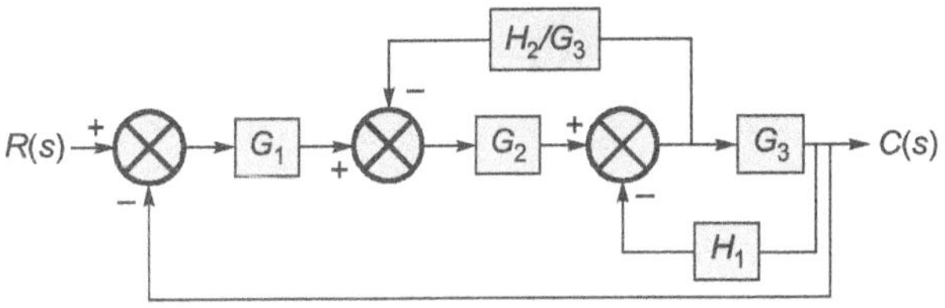

Reducing further, the block diagram is reduced as shown below.

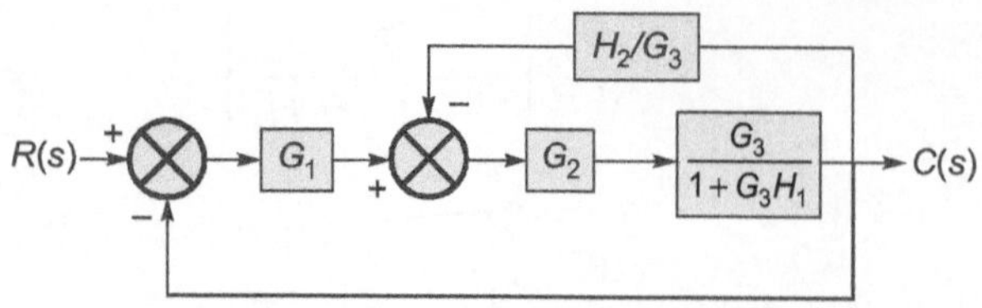

Thus, the block diagrams A and B are equivalent to each other.

On further reducing the block diagram-B, we get the block diagram-C. Hence, all the three block diagrams A, B and C are equivalent to each other.

58. (d)

The given block diagram can be reduced as shown below.

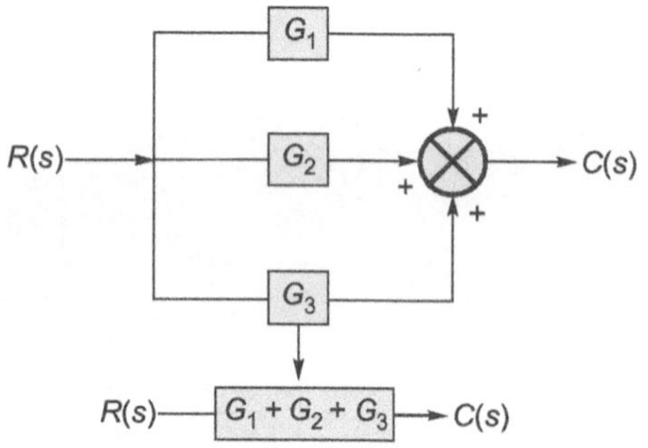

Here, $G_1 = \frac{3}{s+1}$, $G_2 = \frac{-6}{s+2}$

and $G_3 = \frac{3}{s+3}$

$$\therefore \quad \frac{C(s)}{R(s)} = G_1 + G_2 + G_3$$

$$= \frac{3}{s+1} - \frac{6}{s+2} + \frac{3}{s+3}$$

$$= \frac{3(s+2)(s+3) - 6(s+1)(s+3) + 3(s+1)(s+2)}{(s+1)(s+2)(s+3)}$$

Hence, the poles of given system are:

$s = -1, -2, -3$

59. (c)

Combining the blocks in parallel, the block diagram is reduced as shown below.

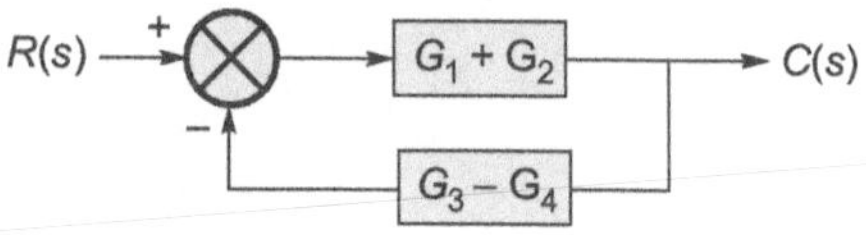

$$\text{Thus,} \quad \frac{C(s)}{R(s)} = \frac{G_1 + G_2}{1 + (G_1 + G_2)(G_3 - G_4)}$$

$$= \frac{G_1 + G_2}{1 + G_1G_3 - G_1G_4 + G_2G_3 - G_2G_4}$$

60. (b)

The given block diagram can be reduced as shown below.

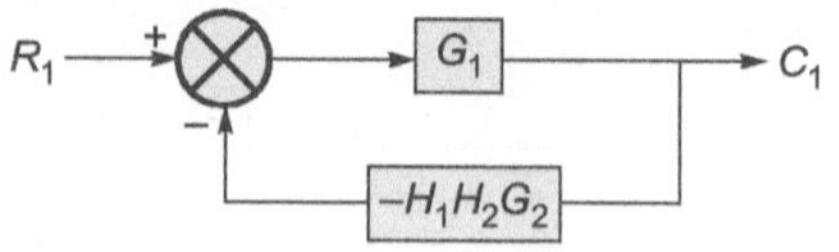

$$\text{Thus,} \quad \frac{C_1}{R_1} = \left(\frac{G_1}{1 - G_1G_2H_1H_2}\right)$$

61. (d)

Reducing the given block diagram, we get the block diagram as shown below.

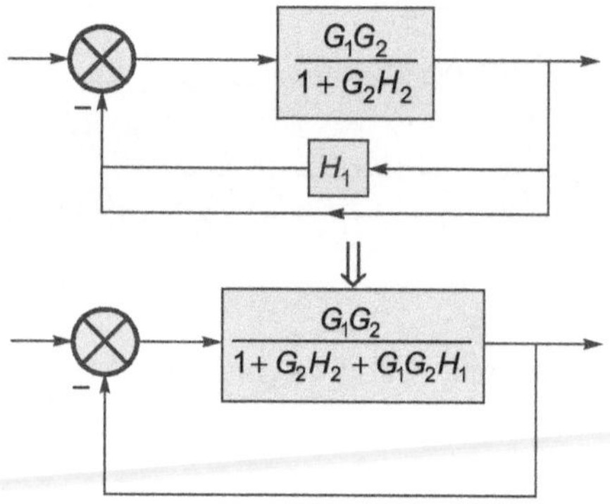

Thus, closed loop transfer function is

$$T(s) = \frac{G_1 G_2}{1 + G_1G_2 + G_2H_2 + G_1G_2H_1}$$

$$= \frac{G_1 G_2}{1 + (G_1 + H_2 + G_1H_1)G_2}$$

63. (c)

When $R_1 = 0$, shifting the summing point between G_1 and G_2 after G_3 the block diagram is reduced as shown below.

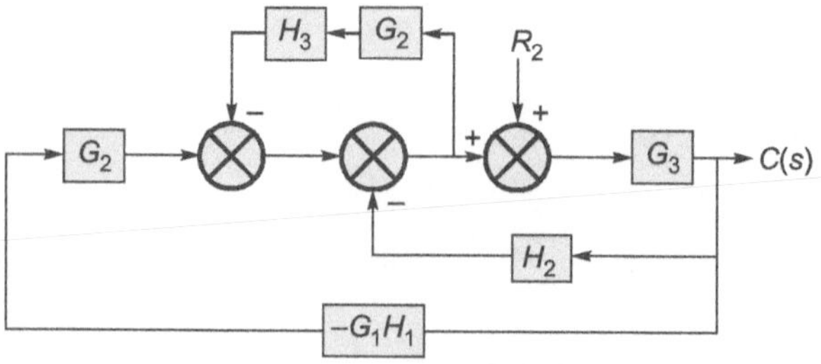

Interchanging the two summing points (after G_2), and on simplifying, the above block diagram is reduced as shown below.

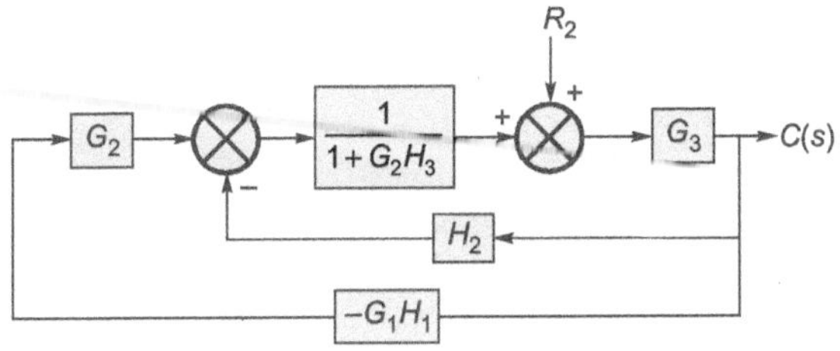

Shifting the summing point after the block, the above block diagram is reduced as shown below.

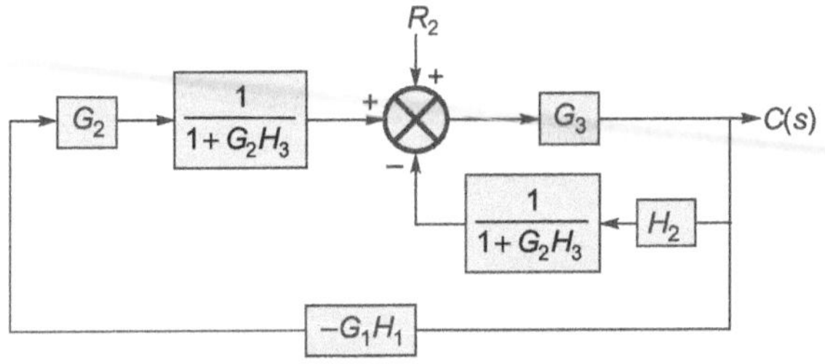

On simplifying above block diagram, we have the reduced block diagram as shown below.

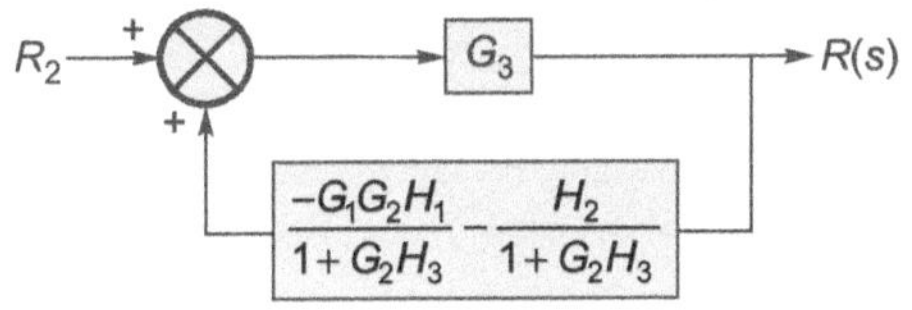

$$\text{Thus, } \frac{C(s)}{R_2(s)} = \frac{G_3}{1-\left[\dfrac{-G_1G_2H_1 - H_2}{1+G_2H_3}\right]G_3}$$

$$= \frac{G_3(1+G_2H_3)}{1+G_1G_2H_3H_1 + G_3H_2 + G_2H_3}$$

64. **(b)**

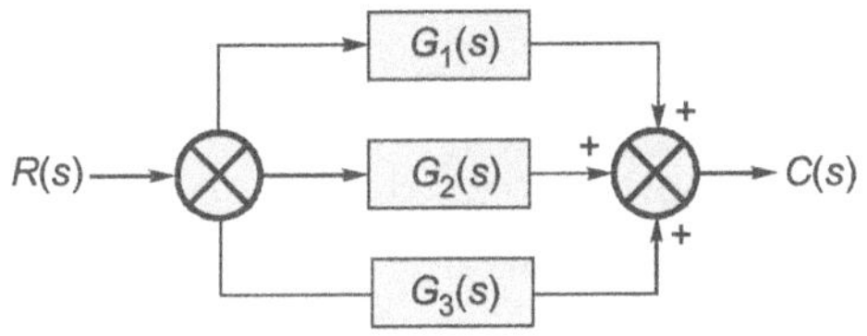

The equivalent T.F. is

$$\frac{C(s)}{R(s)} = G_1(s) + G_2(s) + G_3(s)$$

$$= \frac{1}{s} + \frac{1}{s+3} + \frac{s+1}{s+5}$$

$$\text{or, } \frac{C(s)}{R(s)} = \frac{(s+3)(s+5)+5(s+5)+s(s+3)}{s(s+3)(s+5)}$$

$$= \frac{s^2+8s+15+s^2+5s+s^2+3s}{s(s+3)(s+5)}$$

$$= \frac{3s^2+16s+15}{s(s+3)(s+5)}$$

65. **(d)**

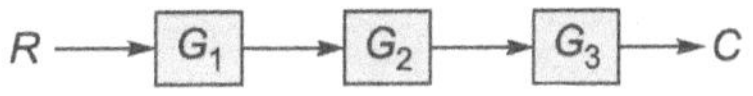

$$\text{T.F.} = \frac{C(s)}{R(s)} = G_1G_2G_3$$

$$= \frac{1}{(s+1)} \times \left(\frac{s-1}{s+4}\right) \times \left(\frac{-s}{s^2+2}\right)$$

$$= \frac{-s(s-1)}{(s+1)(s+4)(s^2+2)}$$

$$= \frac{s(1-s)}{(s+1)(s+4)(s^2+2)}$$

66. **(d)**

Associate law:

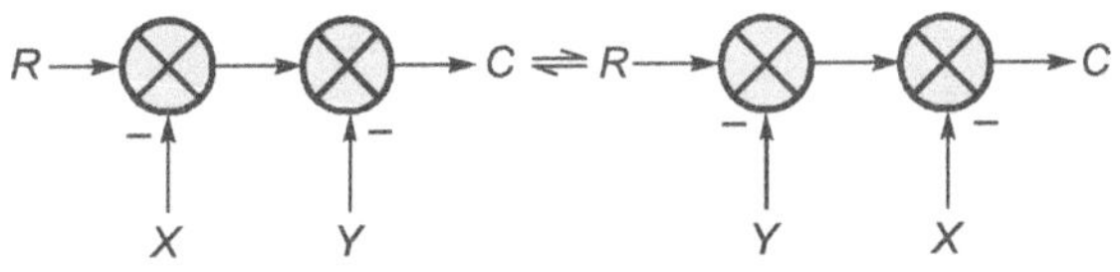

Blocks in series:

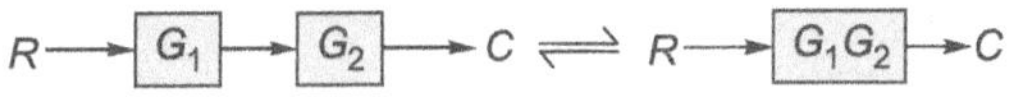

Blocks in parallel:

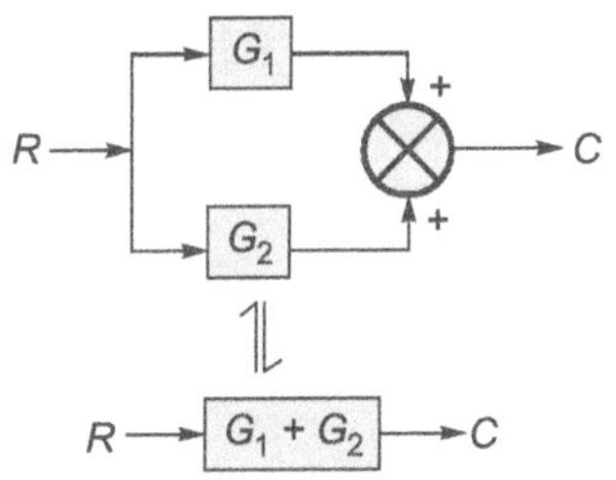

Shifting of take-off point before summing point:

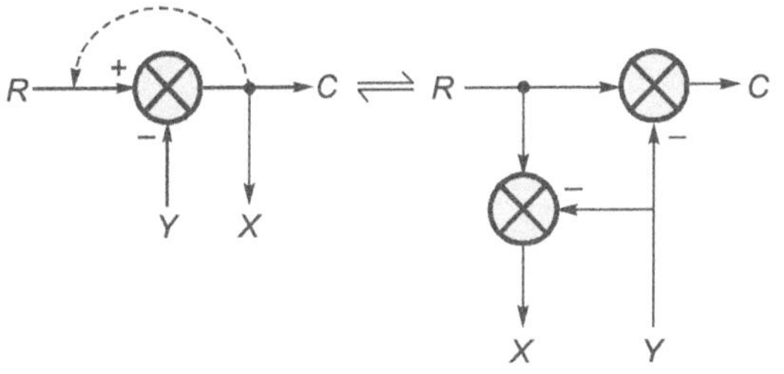

Hence, option (d) is false.

67. **(c)**

Open loop gain $= -G_1G_2H_1$

$$\therefore \quad \text{CLTF} = \frac{C(s)}{R(s)} = \frac{-G_1G_2H_1}{1-(-G_1G_2H_1)(H_2)}$$

$$= \left(\frac{-G_1G_2H_1}{1+G_1G_2H_2}\right)$$

68. (c)

For system-1,

$$\frac{C_1}{R_1} = \frac{10}{1+10} = \frac{10}{11}$$

For system-2,

$$\frac{C_2}{R_2} = 30 \text{ or } C_2 = 30\,R_2$$

When forward path is reduced by 5%, then in system-2 C_2 will reduce by 5%.

However, when forward path is reduced by 5% in system-1, then

$$\left(\frac{C_1}{R_1}\right)_{new} = \frac{9.5}{1+9.5} = \frac{9.5}{10.5} = 0.904 \approx 1$$

Thus, C_1 reduces by approximately 1%.

69. (a)

Mason's rule or Mason's gain formula is applied to signal flow graph for finding the transfer function.

70. (b)

Signal flow graph is applicable to the linear systems because output is proportional to gain of the system.

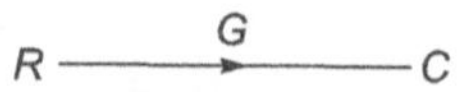

Here, $\frac{C}{R} = G$ or $C \propto G$

Reason (R) is also true but is not a correct explanation of assertion (A).

71. (c)

For SFG-2, $\frac{X_2}{X_1} = \frac{G}{1-GH}$ (Loop = G_1H)

For SFG-3, $\frac{X_2}{X_1} = \frac{G}{1-GH}$ (Loop = GH)

For SFG-1, $\frac{X_2}{X_1} = G_1 + G_2$ (No loop)

Hence, only 1 is correctly matched.

74. (d)

Gain of forward paths are:

$$P_1 = G_1G_2G_3 \text{ and } P_2 = G_4$$

Here, $\Delta_1 = 1$

and $\Delta_2 = (1 + G_1G_2 - G_2G_3)$

Individual loops are:

$$L_1 = -G_1G_2$$

and $L_2 = G_2G_3$

Non-touching loop = Nil

$$\therefore \quad \frac{C}{R} = \frac{P_1\Delta_1 + P_2\Delta_2}{\Delta}$$

$$= \frac{G_1G_2G_3 + G_4(1 + G_1G_2 - G_2G_3)}{1 + G_1G_2 - G_2G_3}$$

75. (c)

Forward paths are:

$$P_1 = G_1G_2G_3 \text{ and } P_2 = G_3G_4$$

Gain product of non-touching loop pairs are:

$$L_1L_2 = G_1H_1H_4$$

and $L_1L_3 = G_1G_3H_1H_2$

Thus, there are two forward paths and two non-touching loop pairs.

76. (a)

Converting the given circuit in s-domain and writing the equations, we get:

$$I(s) = \frac{E_i(s) - E_0(s)}{R + sL} = \frac{E_i(s)}{R+sL} - \frac{E_0(s)}{R+sL} \quad ...(1)$$

Also, from given SFG, we have:

$$I(s) = G_1E_i(s) - H_1E_0(s) \quad ...(2)$$

On comparing equations (1) and (2), we have

$$G_1 = \frac{1}{R+sL} \text{ and } H_1 = \frac{1}{R+sL}$$

77. (c)

Number of forward paths are two.

i.e., $P_1 = G_1G_2$, $\Delta_1 = 1$

and $P_2 = G_4$, $\Delta_2 = (1 - G_2G_3)$

Individuals loops are:

$$L_1 = G_2G_3,\ L_2 = -G_1G_2G_5$$

$$L_3 = -G_4G_5$$

Non-touching loop pairs = Nil

$$\therefore \quad \frac{C}{R} = \frac{\sum P_i\Delta_i}{\Delta} = \frac{P_1\Delta_1 + P_2\Delta_2}{\Delta}$$

$$= \frac{G_1G_2 + G_4(1 - G_2G_3)}{1 - (G_2G_3 - G_1G_2G_5 - G_4G_5)}$$

$$= \frac{G_1G_2 + G_4(1 - G_2G_3)}{1 - G_2G_3 + G_4G_5 + G_1G_2G_5}$$

79. (b)

If a summing point is placed before a take-off point, the summing and take-off points are represented by a single node. Hence, statement-3 is not correct.

Also, if summing point is placed after take off point, the summing and take-off points are represented by separate nodes. Thus, statement-4 is also false.

80. (d)

Given, $\frac{d^2y}{dt^2}+10\frac{dy}{dt}-5y=3x$

Taking Laplace transform on both sides, we get:

$s^2Y(s)+10s\,Y(s)-5Y(s)=3X(s)$

or, $$\frac{Y(s)}{X(s)}=\frac{3}{s^2+10s-5}=\frac{\frac{3}{s^2}}{1+\frac{10}{5}-\frac{5}{s^2}}$$

$$=\frac{\frac{3}{s^2}}{1-\left(\frac{5}{s^2}-\frac{10}{s}\right)}=\frac{P_1\Delta_1}{\Delta}$$

Here, forward path = P_1 (Grain product = $\frac{3}{s^2}$)

Since, highest degree of s in denominator of P_1 is 2, therefore we require two integrators in forward path.

Also, $\Delta=1-\left(\frac{5}{s^2}-\frac{10}{5}\right)$, therefore there will be two individual loops in the SFG.

These two loops are: $L_1=\frac{5}{s^2}$ and $L_2=\frac{-10}{s}$.

Thus, we can draw the SFG of given differential equation. Finally, we will get SFG as in option (d).

81. (a)

Forward path is: $P_1=A$; $\Delta_1=1$

Loops are: $L_1=B$, $L_2=C$

Product of non-touching loop = $L_1L_2=BC$

∴ Transfer function,

$$\frac{Y(s)}{U(s)}=\frac{P_1\Delta_1}{\Delta}=\frac{A}{1-(B+C)+BC}$$

$$=\frac{A}{B(C-1)-(C-1)}$$

$$=\frac{A}{(B-1)(C-1)}=\frac{A}{(1-B)(1-C)}$$

82. (c)

Here, number of forward path

= 1 ($P_1=3\times2\times1=6$) and $\Delta_1=1$

Number of loops = 3

Loop gains are: $L_1=-1$, $L_2=-2$, $L_3=6$

Product of non-touching loop gain

$=L_1L_2=-1\times-2=2$

∴ $$\frac{y}{x}=\frac{P_1\Delta_1}{\Delta}$$

So, $$\frac{y}{x}=\frac{6\times1}{1-(-1-2-6)+2}$$

$$=\frac{6}{12}=\frac{6}{12}=\frac{1}{2}$$

84. (a)

Here, $P_1=G_1G_2G_3G_6,\ \Delta_1=1$

$P_2=G_1G_2G_4G_6,\ \Delta_2=1$

$P_3=G_1G_2G_5G_6,\ \Delta_3=1$

Individual loops are:

$L_1=G_1H_1$

$L_2=-G_1G_2G_2H_2$

$L_3=-G_1G_2G_4H_2$

$L_4=-G_1G_2G_5H_2$

∴ $$\frac{C}{R}=\frac{P_1\Delta_1+P_2\Delta_2+P_3\Delta_3}{\Delta}$$

$$=\frac{G_1G_2G_3G_6+G_1G_2G_4G_6+G_1G_2G_5G_6}{1+G_1H_1+G_1G_2G_3H_2+G_1G_2G_4H_2+G_1G_2G_5H_2}$$

85. (b)

The closed loop transfer function is given by

$$\frac{C(s)}{R(s)}=M(s)=\frac{aG(s)}{1+bG(s)}$$

∴ $$S_a^{M(s)}=\frac{a}{M(s)}\cdot\frac{\partial}{\partial a}[M(s)]$$

Now, $$\frac{a}{M(s)}=\left[\frac{1+bG(s)}{G(s)}\right]$$

and $$\frac{\partial}{\partial a}[M(s)]=\frac{\partial}{\partial a}\left[\frac{aG(s)}{1+bG(s)}\right]=\left[\frac{G(s)}{1+bG(s)}\right]$$

∴ $$S_a^{M(s)}=\left[\frac{1+bG(s)}{G(s)}\right]\times\left[\frac{G(s)}{1+bG(s)}\right]=1$$

86. (d)

The CLTF is

$$T(s)=\frac{G_1G_2}{1+G_2}$$

Now, $S_{G_1}^T = \frac{G_1}{T} \cdot \frac{\partial T(s)}{\partial G_1}$

Here, $\frac{\partial T(s)}{\partial G_1} = \frac{(1+G_2) \cdot G_2 - G_1 G_2 . 0}{(1+G_2)^2} = \frac{G_2}{1+G_2}$

$\therefore \quad S_{G_1}^T = \frac{G_1}{\left(\frac{G_1 G_2}{1+G_2}\right)} \times \frac{G_2}{1+G_2} = 1$

88. (c)

We have: $E(s) = R(s) + C(s)\,H(s)$ ...(*i*)

Also, $C(s) = E(s)\,G(s)$ ...(*ii*)

From (*i*) and (*ii*), we have:

$E(s) = R(s) + E(s)\,G(s)\,H(s)$

or, $E(s)\,[1 - G(s)H(s)] = R(s)$

or, $\frac{E(s)}{R(s)} = \left[\frac{1}{1 - G(s)H(s)}\right]$

90. (a)

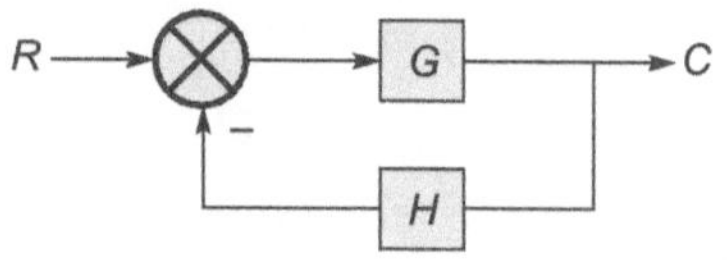

For the closed loop control system shown above,

let, $H = \frac{C}{R}$ = transfer function.

Then, $S_H^T = \frac{-GH}{1+GH}$

and $S_G^T = \frac{1}{1+GH}$

Here, $|S_H^T| > |S_G^T|$

Thus, closed loop system is more sensitive to variation in feedback path parameters than variation in forward path parameters. Hence statement-1 is false. Thus, statements 2 and 3 are only correct.

91. (b)

Given, $\frac{G(s)}{1+G(s)} = \frac{Ks + b}{s^2 + as + b}$

$\therefore \quad G(s) = \frac{Ks + b}{s^2 + s(a - K)}$

$= \frac{Ks + b}{s^2 - s(K - a)}$

92. (c)

Time constant of open loop system is more than that of closed loop system. Lesser is the time constant of the system, faster is its time response.

93. (d)

Given, $R(s) = \frac{1}{s}$

$\therefore \quad \frac{C(s)}{R(s)} = \frac{\frac{K}{s}}{1 + \frac{K}{s}} = \frac{K}{s+K}$

Also, $\frac{E(s)}{R(s)} = \frac{1}{1 + G(s)H(s)} = \frac{s}{s+K}$

or, $E(s) = \left(\frac{s}{s+K}\right) \times \frac{1}{s} = \frac{1}{s+K}$

$\therefore \quad e(t) = e^{-Kt}$

$\therefore \quad I_e = \int_0^\infty e^2(t)\,dt = \int_0^\infty e^{-2Kt}dt$

$= \left[\frac{e^{-2Kt}}{-2K}\right]_0^\infty = -\frac{1}{2K}[0 - 1] = \frac{1}{2K}$

Thus, when $K = \infty$, $I_e = 0$.

94. (d)

Introduction of feedback to a system will increase the stability if negative feedback is introduced while it will reduce the stability if positive feedback is introduced.

98. (c)

The differential equation for the given system is

$$F = \delta(t) = \frac{Md^2x}{dt^2} + \frac{Bdx}{dt} + Kx$$

or, $F = 10\frac{d^2x}{dt^2} + 30\frac{dx}{dt} + 20x = \delta(t)$

Taking Laplace transform on both sides, we get:

$(10s^2 + 30s + 20)\,X(s) = 1$

or, $X(s) = \frac{1}{10s^2 + 30s + 20}$

or, $X(s) = \frac{0.1}{s^2 + 3s + 2} = \frac{0.1}{(s+1)(s+2)}$

or, $X(s) = \frac{K_1}{s+1} + \frac{K_2}{s+2}$

Here, $K_1 = \frac{0.1}{1} = 0.1,\ K_2 = -0.1$

$\therefore$ $X(s) = 0.1\left(\frac{1}{s+1} - \frac{1}{s+2}\right)$

or, $x(t) = 0.1\,(e^{-t} - e^{-2t})$

99. (b)

We know that,

$$\frac{r_1}{r_2} = \frac{N_1}{N_2} = \frac{\omega_2}{\omega_1} = \frac{T_1}{T_2} = \frac{\theta_2}{\theta_1}$$

$$\therefore \quad r_2 = r_1 \times \frac{N_2}{N_1} = 20 \times \frac{20}{40} = 10 \text{ cm}$$

$$\omega_2 = \omega_1 \times \frac{N_1}{N_2} = 10 \times \frac{40}{20} = 20 \text{ rad/s}$$

$$T_2 = T_1 \times \frac{N_2}{N_1} = 5 \times \frac{20}{40} = 2.5 \text{ N-m}$$

$$\theta_2 = \theta_1 \times \frac{N_1}{N_2} = 20 \times \frac{40}{20} = 40°$$

100. (a)

The differential equation for the given mechanical system is

$$K_1 X_2 = B\frac{d}{dt}(x_1 - x_2) + K_2(x_1 - x_2)$$

Taking Laplace transform on both sides, we get:
$K_1X_2(s) = BsX_1(s) - BsX_2(s) + K_2X_1(s) - K_2X_2(s)$
or, $X_2(s)\,[Bs + K_2 + K_1] = X_1(s)\,[Bs + K_2]$

or, $$\frac{X_2(s)}{X_1(s)} = \frac{Bs + K_2}{Bs + K_1 + K_2}$$

Hence, pole is at

$$s = -\left(\frac{K_1 + K_2}{B}\right)$$

and zero is at

$$s = \frac{-K_2}{B}$$

101. (d)

We have, $$F = M\frac{d^2x}{dt^2} + B\frac{dx}{dt} + Kx$$

and $$V = L\frac{d^2q}{dt^2} + R\frac{dq}{dt} + \frac{q}{c}$$

or, $$V = L\frac{di}{dt} + iR + \frac{1}{C}\int i\,dt$$

Hence, for *F-V* analogy, we have:

$M \leftrightarrow L$; $B \leftrightarrow R$; $K \leftrightarrow \frac{1}{C}$; $x \leftrightarrow i$

Thus, option (d) will be the correct circuit.

102. (c)

For mechanical rotational system, we have:

$$T = J\frac{d^2\theta}{dt^2} + B\frac{d\theta}{dt} + K\theta$$

$$= J\frac{d\omega}{dt} + B\omega + K\int \omega\, dt \qquad ...(i)$$

For electrical system, we have:

$$I = C\frac{d^2\phi}{dt^2} + \frac{1}{R}\frac{d\phi}{dt} + \frac{\phi}{L}$$

$$= C\frac{dv}{dt} + \frac{1}{R}V + \frac{1}{L}\int v\,dt \qquad ...(ii)$$

Comparing equations (*i*) and (*ii*), we have:

$T \leftrightarrow I$; $J \leftrightarrow C$; $B \leftrightarrow \frac{1}{R}$; $K \leftrightarrow \frac{1}{L}$ or $\frac{1}{K} \leftrightarrow L$; $\theta \leftrightarrow \phi$

Here, $\frac{1}{K}$ = Elastance

103. (c)

The free-body diagram for the two masses will be shown as below.

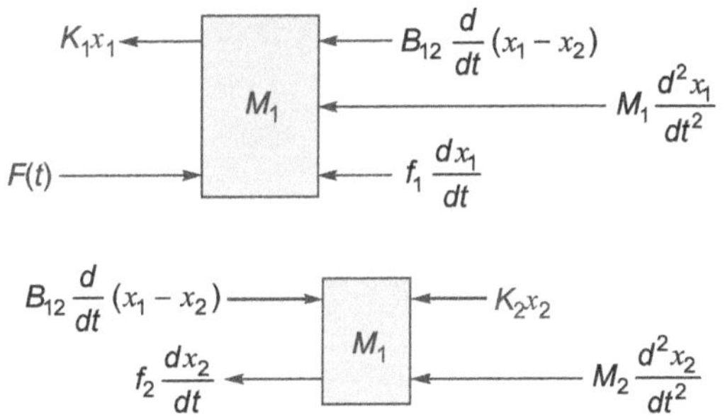

Hence, the differential equations for the two masses can be written as follows:

$$F(t) = M_1\frac{d^2x_1}{dt^2} + f_1\frac{dx_1}{dt} + B_{12}\frac{d}{dt}(x_1 - x_2) + K_1x_1$$

and $$B_{12}\frac{d}{dt}(x_1 - x_2) = M_2\frac{d^2x_2}{dt^2} + f_2\frac{dx_2}{dt} + K_2x_2$$

104. (d)

Relation between electrical control system and rotor speed should be linear.

105. (c)

An A.C. servomotor doesn't require brushes compared to a D.C. servomotor. So, it gives maintenance free operation. It's efficiency is less due to low power output but, gives smooth operation.

111. (b)

Both assertion and reason are true. However, the correct reason is that all the forces in the direction of reference direction are considered as positive and the forces opposite to the reference direction are taken as negative due to which the algebraic sum of forces is zero.

112. (d)

When two time constant elements are cascaded non-interactively, then the overall transfer function is the product of two individual TFs.
Hence, assertion is false.

113. (c)

Reason is a false statement because small diameter and large axial length will give a lower value of inertia of motor.

115. (a)

Stepper motors can be used for speed control because the motor speed is proportional to the rate of command pulses.

120. (b)

$$F = M\frac{d^2x}{dt^2} + B\frac{dx}{dt} + Kx$$

and $$I = C\frac{d^2\phi}{dt^2} + \frac{1}{R}\frac{d\phi}{dt} + \frac{\phi}{L}$$

So, $K \leftrightarrow \frac{1}{L}$.

121. (a)

If q = flow; A = Area of cross-section and h = height of water in tank.

Then, $$\theta = A\frac{dh}{dt} \quad ...(i)$$

Also, $$i = C\frac{dv}{dt} \quad ...(ii)$$

On comparing equations (*i*) and (*ii*), we have

$$C\text{(Farad)} = A \equiv m^2$$

122. (a)

$$F = M\frac{d^2x}{dt^2} + B\frac{dx}{dt} + Kx \quad ...(i)$$

$$V = L\frac{d^2q}{dt^2} + R\frac{dq}{dt} + \frac{q}{C} \quad ...(ii)$$

On comparing equations (*i*) and (*ii*), we have

$$B \leftrightarrow R$$

123. (d)

Taking Laplace transform on both sides of the given differential equation, we get

$$(s^2 + 2s + 2)\, Y(s) = 3$$

or, $$Y(s) = \left[\frac{3}{s^2 + 2s + 2}\right]$$

Characteristic equation is

$$s^2 + 2s + 2 = 0$$

Also, $$s^2 + 2\xi\omega_n s + \omega_n^2 = 0$$

Here, $$\omega_n = \sqrt{2} \text{ rad/sec}$$

and $$2\xi\omega_n = 2$$

or, $$\xi = \frac{2}{2\sqrt{2}} = \frac{1}{\sqrt{2}} = 0.707$$

Since, $\xi < 1$, therefore given system is underdamped.

124. (c)

If $\xi < 1$, system is underdamped.
If $\xi > 1$, system is overdamped.
If $\xi = 1$, system is critically damped.

125. (c)

$$\frac{\text{Damped natural frequency}}{\text{Natural frequency}} = \frac{\omega_d}{\omega_n}$$

$$= \frac{\omega_n\sqrt{1-\xi^2}}{\omega_n} = \sqrt{1-\xi^2}$$

126. (b)

Delay time (t_d) is the time required for the response curve to reach 50% of the final value.

127. (a)

Undamped natural frequency = ω_n
and maximum overshoot,

$$M_p = e^{\frac{-\xi\pi}{\sqrt{1-\xi^2}}}$$

Since M_p is independent of ω_n, therefore with increase in ω_n, M_p remains constant.

128. (b)

The general relation between ξ and gain K is

$$\xi \propto \frac{1}{\sqrt{K}}$$

or, $$\text{Damping ratio} \propto \frac{1}{\sqrt{\text{Gain of the system}}}$$

129. (d)

$$\text{Damping ratio} = \frac{\text{Actual damping}}{\text{Critical damping}} = \frac{\xi\omega_n}{\omega_n} = \xi$$

130. (b)

Here, $$\omega_n = 7 \text{ rad/s and } \xi = \frac{16}{2\times 7}$$

$$= \frac{8}{7} = 1.14$$

Since $\xi > 1$, therefore the system is overdamped. Hence, there will be no peak overshoot in the output response of the system.

131. (c)

Given, $$\frac{C(s)}{R(s)} = \frac{10(s+1)}{(s+2)(s+5)}$$

It's characteristic equation is $(s + 2)(s + 5) = 0$

or, $$s^2 + 7s + 10 = 0$$

or, $$s^2 + 2\xi\omega_n s + \omega_n^2 = 0$$

Here, $\omega_n = \sqrt{10}$ r/s, $2\xi\omega_n = 7$ or $\xi = \frac{7}{2\sqrt{10}} = 1.1$

Thus, damping ratio $\xi > 1$.

Hence, option (c) is false.

132. (c)

Time constant of the system,

$$\tau = \frac{1}{\xi\omega_n}$$

When τ is increased means damping is reduced, thus steady state error reduces.

Also, when gain of the system increases, then damping reduces and therefore, steady state error reduces.

$$\text{Steady sate error} \propto \xi \propto \frac{1}{\sqrt{K}}$$

133. (a)

$$G(s)H(s) = \frac{K}{(s+1)(0.1s+1)^2}$$

Since input is unit step, therefore steady state error is

$$e_{ss} = \lim_{s\to 0}\left(\frac{1}{1+K_p}\right) = 0.1 \quad \text{(Given)}$$

Here, $$K_p = \lim_{s\to 0}[G(s)H(s)]$$

$$= \lim_{s\to 0}\left[\frac{K}{(s+1)(0.1s+1)^2}\right] = K$$

$\therefore$ $$0.1 = \frac{1}{1+K} \text{ or, } 1 + K = 10 \text{ or } K = 9$$

134. (c)

We have, $$K_p = \lim_{s\to 0}[G(s)H(s)] = \infty$$

$$K_v = \lim_{s\to 0}[sG(s)H(s)]$$

$$= \text{finite constant}$$

and $$K_a = \lim_{s\to 0}[s^2G(s)H(s)] = 0$$

From above values of K_p, K_v and K_a we conclude that the type of the system should be one.

135. (b)

Steady state error with an acceleration input having an amplitude of A is given by

$$e_{ss} = \frac{A}{K_a}$$

where, $$K_a = \lim_{s\to 0}[s^2G(s)H(s)]$$

Hence, if the type of the system = 2, then K_a = some non-zero value or finite value due to which we will get some finite value of K_a.

136. (d)

The characteristic equation of the given closed loop system is

$$1+\frac{9}{s(s+2.4)} = 0 \text{ or } s^2 + 2.4s + 9 = 0$$

Comparing above equation with

$s^2 + 2\xi\omega_n s + \omega_n^2 = 0$, we have

$\omega_n = 3$ rad/s and $2\xi\omega_n = 2.4$ or $\xi\omega_n = 1.2$

Thus, setting time for 2% tolerance band is

$$t_s = 4T = \frac{4}{\xi\omega_n}$$

or, $$t_s = \frac{4}{1.2} = 3.33 \text{ sec.}$$

137. (c)

Given, $$E(s) = \frac{8(s+3)}{s(s+10)}$$

$\therefore$ $$e(\infty) = \lim_{s\to 0}[sE(s)]$$

(Using final value theorem)

or, $$e(\infty) = \lim_{s\to 0}\left[\frac{8s(s+3)}{s(s+10)}\right]$$

$$= \lim_{s\to 0}\left[\frac{8(s+3)}{(s+10)}\right] = \frac{8\times 3}{10} = 2.4$$

138. (a)

$$e_{ss} = \frac{A}{1+K_p}$$

where, $$K_p = \lim_{s\to 0}[G(s)H(s)]$$

(Here, A = magnitude of step input)
Since system is type-1, therefore $K_p = \infty$

$\therefore$ $$e_{ss} = \frac{A}{1+K_p} = \frac{A}{1+\infty} = 0$$

139. (b)

Characteristic equation is

$$1+\frac{K}{s(s+8)} = 0 \text{ or } s^2 + 8s + K = 0$$

Here, $\omega_n = \sqrt{K}$ rad/s

and $2\xi\omega_n = 8$ or $\omega_n = \dfrac{8}{2\xi}$

Given, $\xi = 0.5$

So, $$\omega_n = \sqrt{K} = \frac{8}{2\times 0.5} = 8$$

or, $K = 64$

140. (b)

Given, $s^2 + 2s + 8 = 0$

Here, $\omega_n = \sqrt{8}$ rad/s $= 2\sqrt{2}$ rad/sec
and $2\xi\omega_n = 2$

or, $$\xi = \frac{2}{2\omega_n} = \frac{1}{2\sqrt{2}} = \frac{0.707}{2} = 0.353$$

Since, $\xi < 1$, therefore
system is underdamped.

141. (d)

Given, $$\frac{C(s)}{R(s)} = \frac{1}{1+Ts} = \frac{G(s)}{1+G(s)}$$

(Since $H(s) = 1$)

$\therefore$ $$G(s) = \frac{1}{Ts}$$

For a unit ramp input,

Here, $$e_{ss} = \frac{1}{K_v}$$

$$K_v = \lim_{s\to 0}[s\,G(s)]$$

$$= \lim_{s\to 0}\left[s\times\frac{1}{Ts}\right] = \frac{1}{T}$$

$\therefore$ $$e_{ss} = \frac{1}{(1/T)} = T \text{ units}$$

142. (c)

If damping of system increases, peak overshoot M_p decreases and vice-versa.

$$\left[M_p \propto \frac{1}{\text{Damping of the system}}\right]$$

143. (a)

Input = $300t$ rad/sec = ramp input of magnitude $A = 300$.
For ramp input, steady state error is

$$e_{ss} = \frac{A}{K_a} = \frac{300}{K_a}$$

where, $$K_V = \lim_{s\to 0}[s\,G(s)\,H(s)]$$

Now, from given block diagram, we have:

$$G(s)H(s) = \frac{20K}{s(1+0.1s)}$$

$\therefore$ $$K_V = \lim_{s\to 0}\left(\frac{20sK}{s(1+0.1s)}\right) = 20K$$

So, $$\text{error} = e_{ss} = \frac{300}{20K} = 20° \text{ (Given)}$$

or, $$\frac{15}{K} = 20\times\frac{\pi}{180}$$

or, $K = \dfrac{15\times180}{20\pi} = \dfrac{135}{\pi} = 42.97$

So, value of gain is

$$K = 42.97$$

144. (b)

When BW is increased, the system response becomes fast due to fall in rise time (t_r).

With the increase in bandwidth of the system, damping ratio (ξ) decreases.

Thus, both assertion and reason are true but, reason is not the correct explanation of assertion.

145. (d)

We have,

$$G(s)H(s) = \frac{4(s+2)}{s(s+1)(s+4)}$$

$$\therefore \quad K_a = \lim_{s\to0}[s^2 G(s)H(s)]$$

$$= \lim_{s\to0}\left[\frac{4s(s+2)}{(s+1)(s+4)}\right] = 0$$

Also, $K_v = \lim_{s\to0}[sG(s)H(s)]$

$$= \lim_{s\to0}\left[\frac{4(s+2)}{(s+1)(s+4)}\right] = \frac{4\times2}{1\times4} = 2$$

and $K_p = \lim_{s\to0}[G(s)H(s)]$

$$= \lim_{s\to0}\left[\frac{4(s+2)}{s(s+1)(s+4)}\right] = \infty$$

146. (c)

Given, $c(t) = 1 + 0.25\,e^{-50t} - 1.25\,e^{-10t}$

$$\therefore \quad C(s) = \frac{1}{s} + \frac{0.25}{(s+50)} - \frac{1.25}{(s+10)}$$

or, $C(s) = \dfrac{500}{s(s+10)(s+50)}$

$$\therefore \quad \frac{C(s)}{R(s)} = \frac{500}{(s+10)(s+50)}$$

$$\left(\text{Since, } R(s) = \frac{1}{s}\right)$$

or, $\dfrac{C(s)}{R(s)} = \dfrac{500}{s^2+60s+500}$

$$= \frac{\omega_n^2}{s^2+2\xi\omega_n s+\omega_n^2}$$

Thus, $\omega_n = \sqrt{500}$ rad/sec

and $2\xi\omega_n = 60$ or $\xi\omega_n = 30$

Thus, $\xi = \dfrac{30}{\sqrt{500}} = \dfrac{3}{\sqrt5} = 1.34$

Since, $\xi > 1$, therefore given system is overdamped.

147. (a)

Characteristic equations are:

- $s^2 + 8s + 12 = 0$

  Here, $\omega_n = \sqrt{12}$ and $2\xi\omega_n = 8$

  or, $\xi = \dfrac{4}{\sqrt{12}} = 1.5$ ($\xi > 1 \therefore$ overdamped)

- $s^2 + 8s + 16 = 0$

  Here, $\omega_n = 4$ and $2\xi\omega_n = 8$

  or, $\xi = 1$ (critically damped)

- $s^2 + 8s + 20 = 0$

  Here, $\omega_n = \sqrt{20}$

  and $2\xi\omega_n = 8$

  or, $\xi = 0.894$ ($\xi < 1 \therefore$ underdamped)

- $s^2 + 4 = 0$

  or, $\omega_n = 2$

  and $\xi = 0$ ($\therefore$ undamped)

148. (c)

Here, $G(s)H(s) = \dfrac{K}{s(s+20)}$

As the input is a unit ramp function, therefore

$$e_{ss} = \frac{1}{K_v} = 0.02$$

Now, $K_v = \lim_{s\to0}[s\,G(s)H(s)]$

$$= \lim_{s\to0}\left[\frac{k}{s+20}\right] = \frac{k}{20}$$

$$\therefore \quad 0.02 = \frac{20}{K}$$

or, $K = \dfrac{20}{0.02} = 1000$

149. (b)

Given, $r(t) = 2 + 5t + 2t^2$

$$\therefore \quad R(s) = \left(\frac{2}{s} + \frac{5}{s^2} + 2\cdot\frac{2}{s^3}\right) = \left(\frac{2}{s} + \frac{5}{s^2} + \frac{4}{s^3}\right)$$

$$\therefore \quad e_{ss} = \lim_{s\to 0}\left[\frac{s\,R(s)}{1+G(s)H(s)}\right]$$

$$= \lim_{s\to 0}\left[\frac{\left(2+\frac{5}{s}+\frac{4}{s^2}\right)}{1+G(s)\,H(s)}\right]$$

$$= \lim_{s\to 0}\left[\frac{2}{1+G(s)H(s)}+\frac{5}{s+sG(s)H(s)}+\frac{4}{s^2+s^2G(s)H(s)}\right]$$

or, $$e_{ss} = \left[\frac{2}{1+K_p}+\frac{5}{K_v}+\frac{4}{K_a}\right]$$

150. (d)

Given, $$\frac{C(s)}{R(s)} = \frac{G(s)}{1+G(s)} = \frac{Ks+b}{s^2+as+b}$$

$$\therefore \quad G(s) = \text{Open loop T.F} = \frac{Ks+b}{s^2+s(a-K)}$$

$$\therefore \quad K_v = \lim_{s\to 0}[sG(s)]$$

$$= \lim_{s\to 0}\left[\frac{s(Ks+b)}{s^2+s(a-K)}\right]$$

$$= \lim_{s\to 0}\left[\frac{Ks+b}{s+(a-K)}\right] = \frac{b}{a-K}$$

So, $$e_{ss} = \frac{1}{K_v} = \left(\frac{a-K}{b}\right) = \left(\frac{a}{b}-\frac{K}{b}\right)$$

151. (c)

Given, $$\frac{C(s)}{R(s)} = \frac{5}{(s+10)(s^2+s+1)}$$

In time constant form,

$$\frac{C(s)}{R(s)} = \frac{5}{10(1+0.1s)(1+s+s^2)}$$

$$= \frac{0.5}{(1+0.1s)(1+s+s^2)}$$

Using dominant pole concept, the given transfer function can be approximated to

$$\frac{C(s)}{R(s)} = \left(\frac{0.5}{s^2+s+1}\right)$$

152. (a)

From given block diagram, we have

$$\frac{C(s)}{R(s)} = \frac{6}{s+2}-\frac{4}{s+1} = \frac{2(s-1)}{(s+1)(s+2)}$$

Given, $r(t) = 1$

$$\therefore \quad R(s) = \frac{1}{s}$$

So, $$C(s) = \left[\frac{6}{s+2}-\frac{4}{s+1}\right]\cdot\frac{1}{s}$$

$$= \frac{6}{s(s+2)}-\frac{4}{s(s+1)}$$

or, $$C(s) = 6\left[\frac{1}{2s}-\frac{1}{2(s+2)}\right]-4\left[\frac{1}{s}-\frac{1}{s+1}\right]$$

or, $$C(s) = \frac{-1}{s}-\frac{3}{s+2}+\frac{4}{s+1}$$

$$\therefore \quad c(t) = (4e^{-t}-3e^{-2t}-1)$$

= Required step response

153. (b)

The characteristic equation of series RLC circuit is given by:

$$s^2+\frac{R}{L}s+\frac{1}{LC} = 0$$

Comparing with $s^2+2\xi\omega_n s+\omega_n^2 = 0$, we have:

$$\omega_n = \frac{1}{\sqrt{LC}} \text{ and } \xi = \frac{R}{2}\sqrt{\frac{C}{L}}$$

$$\therefore \quad \omega_n = \frac{1}{\sqrt{10^{-2}\times 10^{-8}}} = 10^5 \text{ rad/sec}$$

= 100 k-rad/sec

and $$\xi = \frac{10^3}{2}\sqrt{\frac{10^{-8}}{10^{-2}}} = \frac{10^3}{2}\sqrt{10^{-6}}$$

$$= \frac{1}{2} = 0.5$$

154. (a)

Characteristic equation of given system is

$$1+G(s)\,H(s) = 0$$

or, $$1+\frac{K}{s(1+sT)} = 0$$

or, $$s^2T+s+K = 0$$

or, $$s^2+\frac{s}{T}+\frac{K}{T} = 0$$

Here, $$\omega_n = \sqrt{\frac{K}{T}} \text{ and } \xi = \frac{1}{2\sqrt{KT}}$$

For $\xi_1 = 0.3$, gain $= K_1$

For $\xi_2 = 0.9$, gain $= K_2$

$$\therefore \quad \frac{\xi_1}{\xi_2} = \frac{1}{2\sqrt{K_1 T}} \times 2\sqrt{K_2 T} = \sqrt{\frac{K_2}{K_1}}$$

or, $$\frac{0.3}{0.9} = \sqrt{\frac{K_2}{K_1}} \text{ or } \frac{K_2}{K_1} = \frac{1}{9}$$

So, $$K_2 = \frac{1}{9} K_1$$

155. (c)

From given block diagram, the closed loop transfer function is

$$\frac{C(s)}{R(s)} = \frac{K}{s^2 + (2 + K\alpha)s + K}$$

Characteristic equation is,

$$s^2 + (2 + K\alpha)\, s + K = 0$$

Here, $\omega_n = \sqrt{K}$

and $2\xi\omega_n = (2 + K\alpha)$

Given, $\omega_n = 4$ rad/sec $\quad \therefore K = 16$

Also, $\xi = 0.7$

$\therefore \quad 2 \times 0.7 \times 4 = (2 + 16\alpha)$

or, $(2 + 16\alpha) = 5.6$

or, $$\alpha = \frac{3.6}{16} = 0.225$$

156. (a)

For parabolic input, $e_{ss} = \dfrac{A}{K_a}$ (A = constant)

$$K_a = \lim_{s \to 0} [s^2 G(s) H(s)]$$

= Finite for type-2 system

So, e_{ss} = Some finite value

157. (b)

Characteristic equation is: $1 + G(s)\, H(s) = 0$

or, $s^2 + s + 1 = 0$

Here, $2\xi\omega_n = 1$ or $\xi\omega_n = \dfrac{1}{2} = 0.5$

For 5% tolerance band, settling time is

$$t_s = 3T = \frac{3}{\xi\omega_n} = \frac{3}{0.5} = 6 \text{ sec}$$

159. (c)

CLTF is, $$\frac{C(s)}{R(s)} = \frac{1}{1+Ts}$$

Given, $R(s) = \dfrac{1}{s^2}$ (unit ramp input)

$$\therefore \quad C(s) = \frac{1}{s^2(1+Ts)} = \frac{A}{s} + \frac{B}{s^2} + \frac{1}{1+Ts}$$

Here, $A = -T, B = 1, C = T^2$

So, $$C(s) = \left(\frac{-T}{s} + \frac{1}{s^2} + \frac{T^2}{1+Ts}\right)$$

$$= \left(-\frac{T}{s} + \frac{1}{s^2} + T \cdot \frac{1}{s + \frac{1}{T}}\right)$$

Taking inverse Laplace transform on both sides, we get:

$$c(t) = (-T + t + Te^{-t/T})$$

$\therefore$ Error signal,

$$e(t) = c(t) - r(t)$$
$$= c(t) - t = -T + T e^{-t/T}$$
$$= T(1 - e^{-t/T})$$

160. (c)

Given, $$\frac{C(s)}{R(s)} = \frac{25}{s^2 + 8s + 25}$$

$$= \frac{\omega_n^2}{s^2 + 2\xi\omega s + \omega_n^2}$$

Here, $\omega_n = 5$ rad/s, $2\xi\omega_n = 8$

or, $$\xi = \frac{8}{2 \times 5} = 0.8$$

Peak time, $$t_p = \frac{n\pi}{\omega_d} = \frac{n\pi}{\omega_n\sqrt{1-\xi^2}}$$

For 3rd peak, $n = 5$

(3rd overshoot is the third peak)

So, $$t_p = \frac{5\pi}{5\sqrt{1-0.8^2}} = \frac{5\pi}{5 \times 0.6} = \frac{\pi}{0.6} = \frac{5\pi}{3}$$

161. (a)

Given, $$\frac{d^2c}{dt^2} + 8\frac{dc}{dt} = 64\,(r - c)$$

or, $s^2 C(s) + 8s\, C(s) = 64\, [R(s) - C(s)]$

or, $$\frac{C(s)}{R(s)} = \frac{64}{s^2 + 8s + 64}$$

$$= \frac{\omega_n^2}{s^2 + 2\xi\omega_n s + \omega_n^2}$$

Hence, $\omega_n = 8$ rad/sec

and $2\xi\omega_n = 8$ or $\xi = 0.5$

162. (b)

Given, poles are at $s = -1 \pm j$

∴ Characteristic equation is

$$(s + 1)^2 + 1 = 0$$

or, $$s^2 + 2s + 2 = 0$$

Comparing above equation with

$$s^2 + 2\xi\omega_n s + \omega_n^2 = 0$$

we have, $\omega_n = \sqrt{2}$ rad/sec

and $2\xi\omega_n = 2$

or, $$\xi = \frac{2}{2\omega_n} = \frac{1}{\sqrt{2}} = 0.707$$

∴ Damping ratio, $\xi = \frac{1}{\sqrt{2}} = 0.707$

Undamped natural frequency,

$$\omega_n = \sqrt{2} \text{ rad/sec} = 1.414 \text{ rad/sec.}$$

Damping factor $= \xi\omega_n = 1$

Settling time (For 2% tolerance band) is

$$t_s = 4T = \frac{4}{\xi\omega_n} = 4 \text{ seconds}$$

Damped natural frequency is

$$\omega_d = \omega_n\sqrt{1-\xi^2} = \sqrt{2}\sqrt{1-\frac{1}{2}}$$

$$= 1 \text{ rad/sec}$$

163. (c)

Forward gain,

$$G(s) = \frac{\frac{100}{s^2}}{1+\frac{100Ks}{s^2}} = \frac{100}{s^2 + 100Ks}$$

$$\therefore \quad \frac{C(s)}{R(s)} = \frac{100}{s^2 + 100Ks + 100}$$

$$= \frac{\omega_n^2}{s^2 + 2\xi\omega_n s + \omega_n^2}$$

Here, $\omega_n = 10$ rad/s and $2\xi\omega_n = 100K$

or, $$K = \frac{2\xi \times 10}{100} = \frac{\xi}{5}$$

For no-oscillation and to have lowest possible settling time $\xi = 1$.

$$\therefore \quad K = \frac{1}{5} = 0.2$$

164. (b)

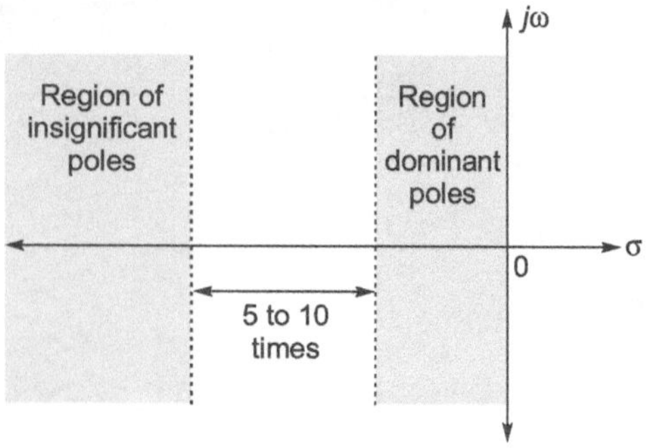

Both assertion and reason are true. However, the correct explanation of assertion is that the poles having real part magnitude 5 to 10 times of the magnitude of real part of dominant poles doesn't affect the transient response (dynamic performance) of the system to a great extent due to which they are termed as insignificant poles and hence are neglected for transient stability study purpose.

166. (d)

Negative damping gives a response that grows in magnitude without bound in time. Hence, statement-2 is false.

Since $M_p \propto \frac{1}{\xi}$, therefore higher is the damping in the system, less is the peak overshoot of the system. Hence, statement-5 is not correct.

167. (c)

With addition of a pole to the forward-path transfer function, bandwidth will reduce.

Since $BW \propto \frac{1}{\text{Rise time}}$, therefore rise time will increase.

Since rise time $\propto \frac{1}{\text{Stability}}$, therefore stability will decrease.

168. (b)

We know that, maximum overshot is

$$M_p = e^{\frac{-\xi\pi}{\sqrt{1-\xi^2}}}$$

M_p is independent of ω_n. So $M_p \propto \xi$ only.

169. (d)

Given, $G(s)H(s) = \dfrac{K(s+1)}{s^3 + as^2 + 2s + 1}$

Now, $1 + G(s)H(s) = 0$

or, $s^3 + as^2 + (2 + K)s + (1 + K) = 0$

Routh's array is

s^3	1	$(2+K)$
s^2	a	$1+K$
s^1	$\dfrac{a(2+K)-(1+K)}{a}$	0
s^0	$1+K$	

For oscillation, we have:

$$\frac{a(2+K)-(1+K)}{a} = 0 \text{ or } a(2+K) = (1+K)$$

Also, oscillating frequency will be given by roots of $A_2(s) = 0$.

i.e. $as^2 + 1 + K = 0$ or $s = \pm j\sqrt{\dfrac{1+K}{a}} = \pm j\omega$

Given, $\omega = \sqrt{\dfrac{1+K}{a}} = 2 \text{ rad/sec}$

We see that only option (d) is matching.

i.e. for $K = 2$ and $a = 0.75$,

$$\omega = \sqrt{\frac{1+2}{0.75}} = \sqrt{\frac{3}{0.75}} = \sqrt{4} = 2 \text{ rad/sec.}$$

171. (c)

For 2% tolerance band,

$$t_s = 4T = \frac{4}{\xi\omega_n}$$

For 5% tolerance band,

$$t_s = 3T = \frac{3}{\xi\omega_n}$$

172. (b)

Given, $c(t) = g(t) = 10e^{-8t}\sin 6t$

Here, $\omega_d = \omega_n\sqrt{1-\xi^2} = 6$

and $\xi\omega_n = 8$

or, $\sqrt{\omega_n^2 - (\xi\omega_n)^2} = 6$ or $\sqrt{\omega_n^2 - 64} = 6$

or, $\omega_n = \sqrt{36+64} = \sqrt{100}$

$= 10 \text{ rad/sec}$

So, $\xi\omega_n = 8$ or $\xi = \dfrac{8}{10} = 0.8$

175. (b)

Since the settling time = time when output reaches 95% of its final value, therefore tolerance band = 5%.

For 5% tolerance band $t_s = 3T = 3 \times 20 = 60$ sec

176. (c)

We know that, damped natural frequency,

$$\omega_d = \omega_n\sqrt{1-\xi^2}$$

Given, $\omega_d = 18$ rad/s, $\xi = 0.866 = \dfrac{\sqrt{3}}{2}$

$\therefore$ $\omega_n = \dfrac{\omega_d}{\sqrt{1-\xi^2}} = \dfrac{18}{\sqrt{1-\frac{3}{4}}}$

$= 18 \times 2 = 36$ rad/s

179. (c)

Number of sign changes in the first column of Routh's array indicates the number of closed loop poles or roots of characteristic equation lying in the right half of s-plane.

180. (b)

Both assertion and reason are individually true. The correct reason for assertion is BIBO stability criteria.

181. (d)

The characteristic equation is,

$$1 + G(s)H(s) = 0$$

or, $1 + \dfrac{K}{s(1+0.4s)(1+0.25s)} = 0$

or, $s^3 + 6.5s^2 + 10s + 10K = 0$

The Routh's array is:

s^3	1	10
s^2	6.5	10
s^1	$\dfrac{65-10K}{6.5}$	
s^0	$10K$	

For absolute stability, we have:

$$10K > 0 \text{ and } \frac{65-10K}{6.5} > 0$$

or, $K > 0$ and $K < 6.5$

So, $0 < K < 6.5$

(Range of K for stability)

182. (a)

The settling time is inversely proportional to the real part of the dominant roots. If the settling time is less compared to other system, then the system will be relatively more stable.

183. (b)

Given, step response:

$$c(t) = -1.5 + e^{t} + 0.5e^{-2t}$$

or, $$C(s) = \frac{-1.5}{s} + \frac{1}{s-1} + \frac{0.5}{(s+2)}$$

$$= \frac{3}{s(s-1)(s+2)}$$

∴ Transfer function is

$$\frac{C(s)}{R(s)} = \frac{3}{(s-1)(s+2)} \quad \left(\because R(s) = \frac{1}{s}\right)$$

Since one of the closed loop pole ($s = 1$) lies in RH s-plane, therefore given system is unstable.

184. (c)

The Rouths array is formed as follows:

s^4	a_0	a_2	a_4
s^3	a_1	a_3	
s^2	$\frac{a_1a_2 - a_0a_3}{a_1}$	$\frac{a_1a_4}{a_1}$	
s^1	$\frac{a_3\left[\frac{a_1a_2 - a_0a_3}{a_1}\right] - \frac{a_1^2a_4}{a_1}}{\left(\frac{a_1a_2 - a_0a_3}{a_1}\right)}$		
s^0	a_4		

For the system to be stable, there should not be any sign change in the first column of Routh's array.

Therefore, $a_1 > 0$ and $a_4 > 0$ ...(*i*)

Also, $a_1a_2 - a_0a_3 > 0$ or, $a_1a_2 > a_0a_3$ ...(*ii*)

and $a_1a_2a_3 - a_0a_3^2 - a_1^2a_4 > 0$

or, $a_0a_3^2 + a_1^2a_4 < a_1a_2a_3$ ...(*iii*)

Hence, above conditions must hold for the given system to be stable.

185. (d)

The characteristic equation of given system is:

$$1 + G(s) = 0$$

or, $(s+2)(s+4)(s^2+6s+25) + K = 0$

or, $s^4 + 12s^3 + 69s^2 + 198s + (200 + K) = 0$

Routh's array is

s^4	1	69	$(200+K)$
s^3	12	198	
s^2	52.5	$(200+K)$	
s^1	$\left(198 - \frac{12(200+K)}{52.5}\right)$		
s^0	$(200+K)$		

Given system will be stable when,

$200 + K > 0$ or $K > -200$

Also, $198 - \frac{12(200+K)}{52.5} > 0$ or $K < 666.25$

Thus, $-200 < K < 666.25$ (For stability)

System will oscillate when $K = K_{mar} = 666.25$

Auxiliary equation is : $52.5s^2 + (200 + K) = 0$

or, $52.5s^2 = -866.25$ or, $s^2 = -16.5$

or, $s = \pm j4.06 = \pm j\omega$

∴ Oscillating frequency, $\omega = 4.06$ rad/sec

186. (d)

Routh's array is formed as shown below,

s^3	1	1
s^2	0	2
s^1		
s^0		

Replacing ε in place of 0 in 1st column, we have:

s^3	1	1
s^2	ε	2
s^1	$\frac{-(2-\varepsilon)}{\varepsilon}$	
s^0	2	

To check for sign change:

$$\lim_{\varepsilon \to 0} \varepsilon = +0 \text{ and } \lim_{\varepsilon \to 0} -\left(\frac{2-\varepsilon}{\varepsilon}\right) = -\infty$$

Thus, there are two sign changes in 1st column of Routh's array (+0 to −∞ and −∞ to + 2).

Hence, number of roots in RH s-plane = 2.

Number of roots in LH s-plane = 1.

Number of roots on the $j\omega$-axis = 0.

188. (c)

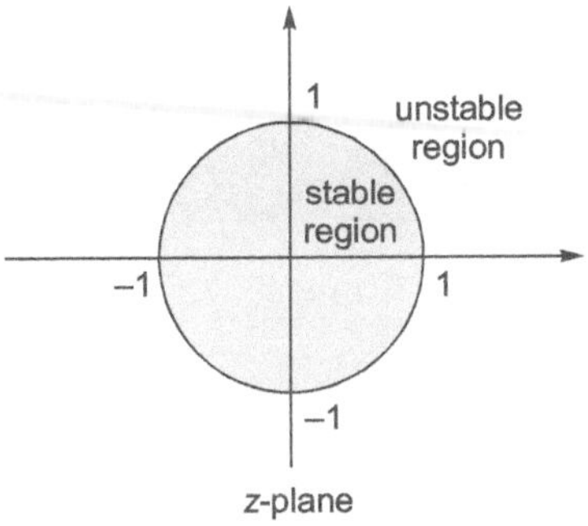

189. (c)

Given characteristic equation is,

$$s^2 + 2Ks^2 + (K + 2)s + 4 = 0$$

Routh's array is:

s^3	1	$(K+2)$
s^2	$2K$	4
s^1	$\left(\dfrac{K^2+2K-2}{K}\right)$	
s^0	4	

For stability, $2K > 0$ or $K > 0$

Also, $\dfrac{K^2+2K-2}{K} > 0$ or $K^2 + 2K - 2 > 0$

Now, $K^2 + 2K - 2 = 0$ or, $K = 0.73, -2.73$

For $K > 0.73$, $K^2 + 2K - 2 > 0$

(since K should be > 0)

Hence, system will be stable if $K > 0.73$.

190. (d)

The characteristic equation is

$$1 + G(s)H(s) = 0$$

or, $$1+\frac{K(s+13)}{s(s+3)(s+7)} = 0$$

or, $s^3 + 10s^2 + (21 + K)s + 13K = 0$

s^3	1	$(21+K)$
s^2	10	$13K$
s^1	$\dfrac{10(21+K)-13K}{10}$	
s^0	$13K$	

For stability, $13K > 0$ or $K > 0$

Also, $\dfrac{10(21+K)-13K}{10} > 0$ or $210 - 3K > 0$

or, $K < 70$

Hence, $0 < K < 70$ (For stability).

191. (c)

For low frequency, $e^{-s} = (1 - s)$

$\therefore$ $$G(s) = \frac{K(1-s)}{s(s^2+5s+9)}$$

The characteristic equation is

$$1 + G(s) = 0$$

or, $s(s^2 + 5s + 9) + K(1 - s) = 0$

or, $s^3 + 5s^2 + (9 - K)s + K = 0$

The Routh's array is

s^3	1	$(9-K)$
s^2	5	K
s^1	$\left(\dfrac{45-6K}{5}\right)$	
s^0	K	

For stability, $K > 0$ and $\dfrac{45-6K}{5} > 0$ or $K < 7.5$

$\therefore$ $0 < K < 7.5$ (for stability)

$0 > K > 7.5$ (for unstability)

192. (a)

Putting $s = (z - 1)$ in the given characteristic equation, we get:

$(z-1)^3 + 3(K+1)(z-1)^2 + (7K+5)(z-1) + (4K+7) = 0$

or, $z^3 + 3Kz^2 + (K+2)z + 4 = 0$

Routh's array is:

z^3	1	$(K+2)$
z^2	$3K$	4
z^1	$\dfrac{3K(K+2)-4}{3K}$	
z^0	4	

For stability, $\dfrac{3K(K+2)-4}{3K} > 0$ and $3K > 0$

i.e. $3K^2 + 6K - 4 > 0$ and $K > 0$

Now, roots of $3K^2 + 6K - 4 = 0$ are : $K = -2.5, 0.53$

Since $K > 0$ $\therefore$ $K = 0.53$ ($K = -2.5$ neglected)

For $K > 0.53$, $3K^2 + 6K - 4 > 0$

Hence, the range of value of K is

$$0.53 < K < \infty \text{ or } \infty > K > 0.53$$

193. (b)

The characteristic equation is

$$1 + G(s)H(s) = 0 \text{ or } 1+\frac{K}{s(s+2)} = 0$$

or, $s^2 + 2s + K = 0$

or, $s = -1 \pm \sqrt{1-K}$

When, $0 < K < 1$, roots are real and distinct.

When, $K = 0$, roots are $s = 0, -2$ (Poles of open loop T.F. $G(s)\,H(s)$).

When, $K = 1$, roots are $s = -1, -1$ i.e. real and equal.

When, $K > 1$, roots are complex conjugate.

194. (d)

The Routh's array is

$$\begin{array}{c|ccc} s^4 & 1 & 0 & 10 \\ s^3 & 5 & 2 & \\ s^2 & -2/5 & 10 & \\ s^1 & 127 & & \\ s^0 & 10 & & \end{array}$$

Since there are two sign changes in first column of Routh's array, therefore the system is unstable.

Number of roots with positive real part = 2

Number of roots with negative real part = 2

Number of roots with zero real part = 0

195. (d)

Since one pole ($s = 1$) lies in RH s-plane, therefore the system is unstable.

196. (c)

Poles of the closed loop transfer function is the characteristic equation of a closed loop control system. For the system to be stable all roots of characteristic equation should lie in LH s-plane. Hence reason is false.

198. (d)

Routh-Hurwitz criterion tests whether any roots of the characteristic equation lie in the right-half s-plane as well as the number of roots that lie on the $j\omega$-axis. Therefore, assertion is false.

199. (b)

- Routh-Hurwitz criterion is not applicable if characteristic equation has coefficients which are exponential or complex.
- It is not applied in z-plane because z-plane having unit circle is the stability boundary of discrete-data system.
- Statement-3 is also false because no. of sign change in 1st column of Routh's array indicates the number of roots with positive real part or roots lying in RH s-plane.

200. (b)

Given, $F(z) = z^3 + z^2 + z + k = 0$

Using bilinear transformation, we put $z = \dfrac{1+r}{1-r}$ in the above characteristic equation.

Thus, $\left(\dfrac{1+r}{1-r}\right)^3 + \left(\dfrac{1+r}{1-r}\right)^2 + \left(\dfrac{1+r}{1-r}\right) + K = 0$

or, $(1-K)r^3 + (1+3K)\,r^2 + 3(1-K)r + 3 + K = 0$

Routh's array is:

$$\begin{array}{c|cc} r^3 & (1-K) & 3(1-K) \\ r^2 & (1+3K) & (3+K) \\ r^1 & \dfrac{8K(1-K)}{(1+3K)} & 0 \\ r^0 & (3+K) & \end{array}$$

For the system to be stable, we have:

$1 - K > 0$ or $K < 1$ ...(*i*)

$1 + 3K > 0$ or $K > -\dfrac{1}{3}$ ...(*ii*)

$3 + K > 0$ or $K > -3$ ...(*iii*)

From equation (*i*), (*ii*) and (*iii*) the range of values of K for stability is, $0 < K < 1$.

201. (a)

The necessary conditions for stability are:

$F(1) > 0$

$F(-1) > 0$, if n = even integer

$F(-1) < 0$, if n = odd integer

and $|a_0| < a_n$

202. (b)

Routh's array is

$$\begin{array}{c|cc} s^3 & 2 & 2 \\ s^2 & 4 & 4 \\ s^1 & 0 & 0 \end{array}$$

Since Routh's array ends abruptly, therefore we construct auxiliary equation for finding other elements of Routh's array.

$A_2(s) = 4s^2 + 4$

$\therefore \quad \dfrac{dA_2(s)}{ds} = 8s$

Hence, new Routh's array is

s^3	2	2
s^2	4	4
s^1	8	0
s^0	4	

Since there is no sign change in 1st column of Routh's array, therefore given system will be marginally stable provided there is no repeated poles on $j\omega$-axis.

Thus, $A_2(s) = 0$ or $4s^2 + 4 = 0$

or, $s = \pm j$ (distinct imaginary roots)

Hence, the given system is marginally stable having one root in LH s-plane and two roots on the $j\omega$-axis at $s = \pm j$.

203. (a)

For a second order discrete- data system given by: $F(z) = a_2z^2 + a_1z + a_0 = 0$

to be stable, the necessary and sufficient conditions are:

$$F(1) > 0$$
$$F(-1) > 0 \text{ and } |a_0| < a_2$$

Here, $F(z) = z^2 + z + 0.25$

So, $a_0 = 0.25,\ a_1 = 1,\ a_2 = 1$

Thus, $F(1) = 1^2 + 1 + 0.25 = 2.25 > 0$

$F(-1) = 1 - 1 + 0.25 = 0.25 > 0$

and $|a_0| = 0.25 < a_2 = 1$

Since all the conditions are satisfied, therefore given system is stable.

204. (c)

Given, $C(z) = \dfrac{z^4 - 1}{z(z^2 - 1)^2}$

Applying final value theorem,

$$\lim_{n\to\infty} C(n) = \lim_{z\to\infty} \{ZC(z)\}$$

$$= \lim_{z\to 0} \frac{z(z^4 - 1)}{z(z^2 - 1)^2} = -1$$

207. (c)

The root locus is defined as location of closed loop poles when system gain k is varied from zero to infinity.

208. (d)

Centroid is the intersection point of asymptotes of the complete root loci which lie on the real axis.

209. (b)

$$G(s) = \frac{K(s+2)}{s(s^2+2s+2)}$$

Number of open loop poles = $P = 3$

and number of open loop zeros = $Z = 1$

Poles are at: $s = -1 \pm j$, 0 and zero is at $s = -2$

$$\therefore \quad \text{Centroid} = \frac{\Sigma(\text{real part of OL poles}) - (\text{real part of OL zeros})}{(P - Z)}$$

$$= \frac{(-1-1-0)-(-2)}{3-1} = \frac{-2+2}{2}$$

$$= 0$$

Here, $P - Z = 2$

So, angle of asymptotes are given by

$$\theta = \left(\frac{2q+1}{P-Z}\right) \times 180^\circ \quad (q = 0, 1)$$

Here, $\theta_1 = \dfrac{180^\circ}{2} = 90^\circ$

and $\theta_2 = \dfrac{3}{2} \times 180^\circ = 270^\circ$

$\therefore\ \theta_1 = 90^\circ$ (anticlockwise) and $\theta_2 = +270^\circ$

or, -90° (clockwise)

210. (a)

Centroid ≡ meeting point of the asymptotes on the real axis

$$\equiv \frac{\Sigma(\text{real part of OL poles}) - (\text{real part of OL zeros})}{(P - Z)}$$

Here, $P = 5,\ Z = 1$

$\therefore \quad P - Z = 5 - 1 = 4$

Poles are at: $s = 0, -2, -4, -1 \pm j1$

Zero is at: $s = -5$

$$\therefore \quad \text{Centroid} = \frac{(0-2-4-1-1)-(-5)}{4}$$

$$\equiv \frac{-8+5}{4} \equiv \frac{-3}{4} \equiv -0.75$$

212. (d)

Addition of real pole to an open loop T.F. decreases the stability because root locus shifts towards right of s-plane. Hence, assertion is a false statement.

213. (d)

Given, $G(s)\,H(s) = \dfrac{K}{s(s^2+6s+10)}$

Here, $P = 3$ and $Z = 0$

$\therefore$ $P - Z = 3$

Number of branches terminating at infinity = $3 = P - Z$.

Number of branches terminating at zero = $Z = 0$.

Poles are at $s = 0$ and $s = -3 \pm j1$.

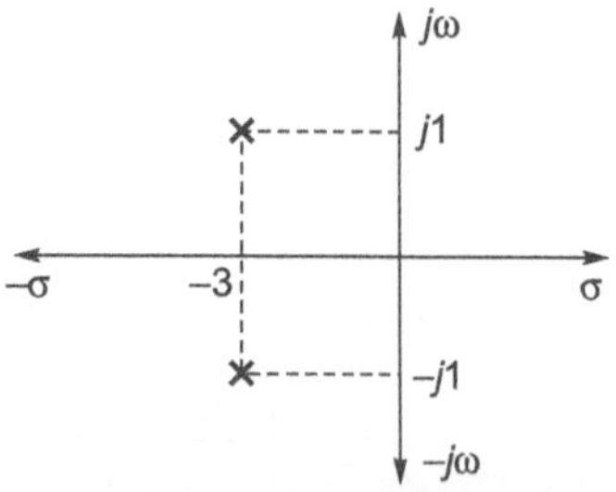

Since, three number of poles have to terminate at infinity, therefore the two complex poles will terminate at infinity due to which there will be intersection of root locus branches with $j\omega$-axis.

Now, characteristic equation is

$$1 + G(s)H(s) = 0$$

or, $s^3 + 6s^2 + 10s + K = 0$

Routh array is

s^3	1	10
s^2	6	K
s^1	$\dfrac{60-K}{6}$	
s^0	K	

For stability, $0 < K < 60$.

Now, $K_{marg} = 60$ (For marginal stability)

and $A_2(s) = 6s^2 + K = 0$

or, $6s^2 + K_{marg} = 0$

or, $s = \pm j\sqrt{10} = \pm j3.16$

Hence, root locus intersects $j\omega$-axis at $s = \pm j3.16$.

Also, centroid $= \dfrac{(-3-3-0)-0}{3}$

$= -2 \equiv (-2, 0)$

214. (b)

Given, $s(s-1) + K(s+1) = 0$

or, $1 + \dfrac{K(s+1)}{s(s-1)} = 0$

or, $1 + G(s)H(s) = 0$

$\therefore$ $G(s)H(s) = \dfrac{K(s+1)}{s(s-1)} = \dfrac{K(s+a)}{s(s+b)}$

It's centre is,

$$C \equiv (-a, 0) \equiv (-1, 0)$$

and radius is,

$$r = \sqrt{a(a-b)} = \sqrt{1[1-(-1)]}$$

$$= \sqrt{2} \text{ units}$$

215. (c)

Branches of root locus either terminate at infinity or zero. Here, one branch is terminating at $s = -2$, therefore there is a zero at $s = -2$. Since two branches are meeting between $s = -3$ and $s = -4$, therefore there is a breakaway point between $s = -3$ and -4. Hence, there must be poles at $s = (-1 \pm j)$.

Therefore, open loop T.F. is

$$G(s)H(s) = \frac{K(s+2)}{(s+1)^2+1} = \frac{K(s+2)}{s^2+2s+2}$$

Here, $P = 2$ and $Z = 1$.

Thus, one branch terminates at $s = -2$ (zero) and one branch terminates at infinity ($P - Z = 1$).

216. (a)

Given, $G(s)H(s) = \dfrac{K(s+1)}{s^2(s+3.6)}$

Here, $P = 3$, $Z = 1$

$\therefore$ Number of branches terminating at infinity $= P - Z = 2$.

Number of branches terminating at zero = $Z = 1$.

$$\text{Centroid} \equiv \frac{(0+0-3.6)-(-1)}{2}$$

$$\equiv \frac{-2.6}{2} = -1.3$$

Angle of asymptotes are: $\theta_1 = 90°$, $\theta_2 = 270°$

For breakaway point, $1 + G(s)H(s) = 0$

or, $1 + \dfrac{K(s+1)}{s^2(s+3.6)} = 0$

or, $K = \dfrac{-(s^3+3.6s^2)}{s+1}$

$\therefore$ $\dfrac{dK}{ds} = 0$

or, $s^3 + 3.3s^2 + 3.6s = 0$

or, $s(s^2 + 3.3s + 3.6) = 0$

Hence, $s = 0$ and $s = \dfrac{-3.3 \pm \sqrt{3.3^2 - 4 \times 1 \times 3.6}}{2}$

$\therefore$ $s = 0$ and $s = (-1.65 \pm j0.936)$

The root locus diagram is shown below.

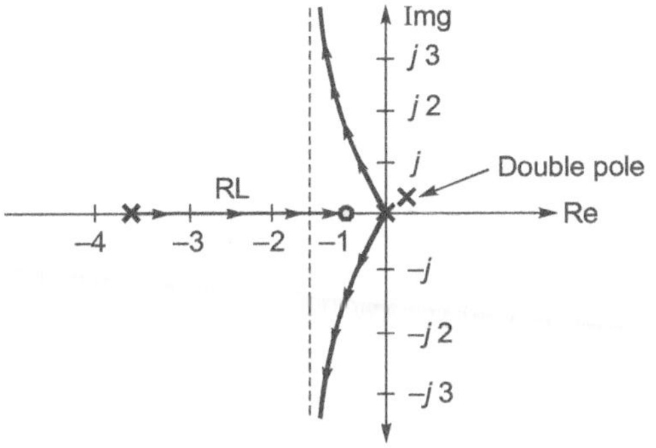

Here, $(-1.65 \pm j0.936)$ is not a part of root locus. Thus, it has only one breakaway point $s = 0$.

217. (d)

Given, $G(s)H(s) = \dfrac{K(s+1)}{s(s+2)(s+4)(s+5)}$

Here, $P = 4$, $Z = 1$, $P - Z = 3$

No. of branches of RL terminating at zero = 1.

No. of branches of RL terminating at infinity = 3.

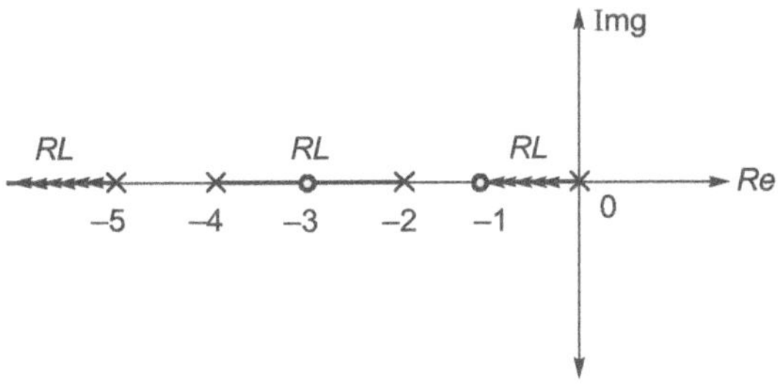

Since out of the given option, real axis between $s = -2$ and $s = -4$ is only a part of root locus, therefore breakaway point will lie between –2 and –4.

218. (d)

Given, $s^2 + 2s + 10 + K(s^2 + 6s + 10) = 0$

or, $$1 + \frac{K(s^2 + 6s + 10)}{s^2 + 2s + 10} = 0$$

or, $1 + G(s)H(s) = 0$

Thus, $$G(s)H(s) = \frac{K(s^2 + 6s + 10)}{s^2 + 2s + 10}$$

Here, number of open loop poles, $P = 2$.

Number of open loop zero, $Z = 2$ $\therefore$ $P - Z = 0$.

Angle of asymptotes are given by:

$$\theta = \frac{(2q+1)\,180°}{(P - Z)}$$

Since $P - Z = 0$, therefore there are no angle of asymptotes for the root locus of the given system.

219. (b)

Given, $G(s) = \dfrac{K}{s(s^2 + 6s + 12)}$

$1 + G(s) = 0$

or, $s(s^2 + 6s + 12) + K = 0$

or, $K = -(s^3 + 6s^2 + 12s)$

For B.A. points, $\dfrac{dK}{ds} = 0$ or $3s^2 + 12s + 12 = 0$

or, $s^2 + 4s + 4 = 0$

or, $(s + 2)^2 = 0$

or, $s = -2, -2$

For given O.L.T.F., $P = 3$, $Z = 0$, $P - Z = 3$

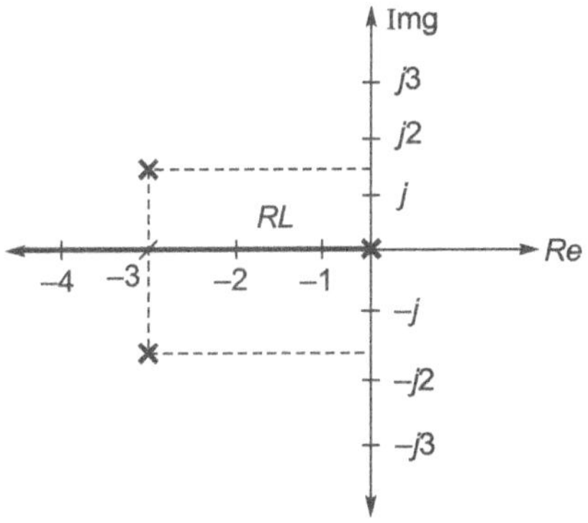

Poles are at $s = 0$ and $s = -3 \pm j\sqrt{3} = -3 \pm j1.732$

Since entire –ve real axis is a part of root locus, therefore $s = -2$ is a valid breakaway point.

221. (c)

From the root sensitivity, standpoint, a system should not be operated at the breakaway points because root sensitivity at breakaway point is infinite.

i.e. $$S_K^s = \frac{K}{s}\frac{ds}{dK} = \infty$$

at breakaway points,

(Since $\dfrac{dK}{ds} = 0$ at B.A. points)

Here, α = closed loop poles(s)

and β = system gain (K)

Thus, $S_{\beta}^{\alpha} = \frac{\beta}{\alpha}\frac{d\alpha}{d\beta}$

or, $S_K^s = \frac{K}{s}\frac{ds}{dK}$

222. (a)

There is possibility of complex B.A. point. Also, when a pole is added to OLTF, root locus pushes to right and reverse happens when *a* zero is added.

223. (d)

Given, $G(s)H(s) = \frac{Ks}{(s+1)(s^2+1)}$

Here, $P = 3$, $Z = 1$, $P - Z = 2$

No. of branches of RL terminating at zero = 1.

No. of branches of RL termination at infinity = 2.

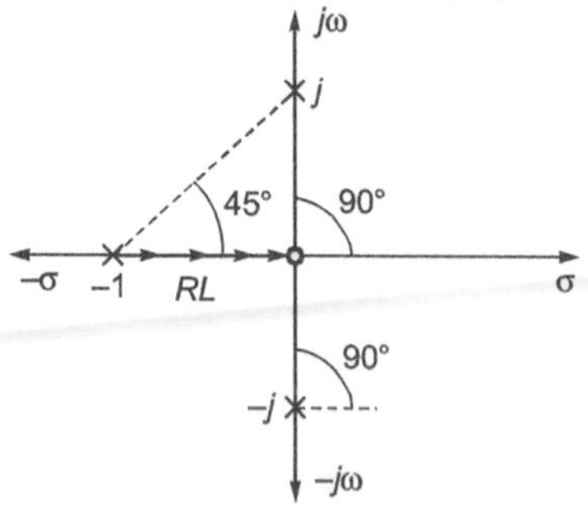

Here, two branches of RL has to terminate at infinity.

Thus, the two imaginary poles will terminate at infinity.

Hence, we need to find angle of departure for which we join all poles and zeros with the imaginary pole at $s = j1$.

Now, $\phi = \Sigma(\phi_z - \phi_P)$

Here, $\Sigma\phi_z = 90°$

and $\Sigma\phi_P = 90° + 45° = 135°$

Thus, $\phi = (90° - 135°) = -45°$

∴ Angle of departure is,

$$\phi_D = 180° + \phi$$
$$= 180° - 45° = 135°$$

225. (c)

Here, $P = 2$, $Z = 3$ and $Z > P$

∴ No. of branches of RL terminating at infinity

$$= Z - P = 3 - 2 = 1$$

226. (a)

To find whether root locus intersects the imaginary axis or not, we construct Routh's array.

Characteristic equation is

$$s^3 + \beta s^2 + Ks + K\alpha = 0$$

Routh's array is:

s^3	1	K
s^2	β	$K\alpha$
s^1	$\frac{(\beta-\alpha)K}{\beta}$	
s^0	$K\alpha$	

For $\beta > \alpha > 0$, root-loci doesn't intersect the $j\omega$-axis and is stable for all positive values of K (no sign change in first column for $\beta > \alpha > 0$ and $K > 0$).

Also, $P = 3$, $Z = 1$, $P - Z = 2$

Thus, two branches of RL will terminate at infinity while one branch will terminate at a zero.

Therefore, possible root locus diagram will be represented by option (a).

229. (b)

With no-pole at origin, initial slope will be 0 dB/decade.

For n-pole at origin, initial slope will be $-20\,n$ dB/decade.

230. (d)

Given, $G(j\omega)H(j\omega) = \frac{\sqrt{2}}{(j\omega)(j\omega+1)}$

Now, $|G(j\omega)H(j\omega)| = \frac{\sqrt{2}}{\omega\sqrt{1+\omega^2}}$

and $\angle G(j\omega)H(j\omega) = \phi = -90° - \tan^{-1}\omega$

At $\omega = \omega_{pc}$, $\phi = -180°$

or, $-90° - \tan^{-1}\omega_{pc} = -180°$

or, $\tan^{-1}\omega_{pc} = 90°$

or, $\omega_{pc} = \infty$ rad/sec

So, $|X| = |G(j\omega)H(j\omega)|_{\omega=\omega_{pc}} = 0$

∴ $\text{G.M.} = \frac{1}{|X|} = \frac{1}{0} = \infty$

231. (b)

At gain crossover frequency (ω_{gc}), $|G(j\omega)| = 1$.

$\therefore$ Gain in dB $= 20 \log |G(j\omega)|$
$= 20 \log 1 = 0$ dB

232. (a)

Given, $\text{G.M.} = -10 \text{ dB} = 20 \log \dfrac{1}{|X|}$

or, $\log \dfrac{1}{|X|} = \dfrac{-1}{2}$

or, $\dfrac{1}{|X|} = 10^{-1/2}$

or, $|X| = \sqrt{10}$
$= |G(j\omega)H(j\omega)|_{\text{at } \omega=\omega_{pc}}$

$\therefore$ $20 \log |X| = 20 \log \sqrt{10}$
$= 10 \log 10 = 10$ dB

233. (c)

Since the critical point $(-1 + j0)$ is not enclosed, therefore magnitude of OLTF, $|X|$ at $\omega = \omega_{pc}$ will be less than unity.

Therefore, $\text{G.M.} = 20 \log\left(\dfrac{1}{|X|}\right)$ will be positive and the system will be stable.

235. (c)

Given, $G(s) = \dfrac{10}{s+0.2} = \dfrac{10}{0.2(1+5s)}$

$= \dfrac{50}{1+5s} = \dfrac{50}{1+Ts}$

Hence, time constant $= T = 5$ seconds

236. (d)

Given, $G(j\omega) = \dfrac{1}{j\omega(1+j\omega T)}$

or, $G(s) = \dfrac{1}{s(1+Ts)}$

Given system is of type-1 and order-2, therefore its polar plot will be as shown below.

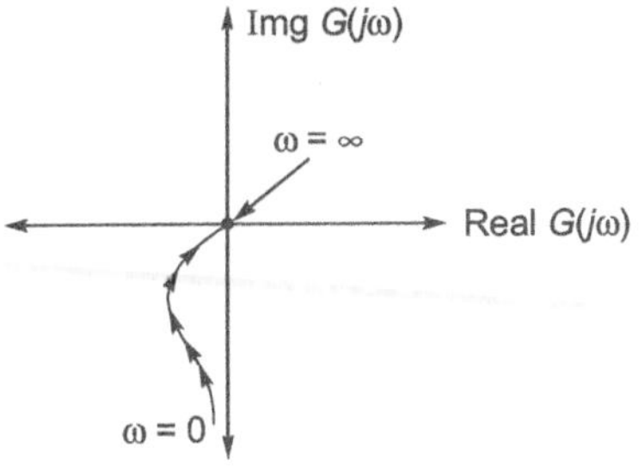

Hence, the polar plot will neither cross the real axis nor the imaginary axis.

237. (a)

We know that, $\text{BW} \propto \dfrac{1}{\text{Rise time}}$

Thus, if BW is increased, rise time will decrease and the system response will become fast. Due to faster response of system, damping ratio will decrease.

238. (b)

Given, $G(s) = \dfrac{1}{1+Ts}$

It is a type-0 and order-1 system, therefore its polar-plot will be as shown below.

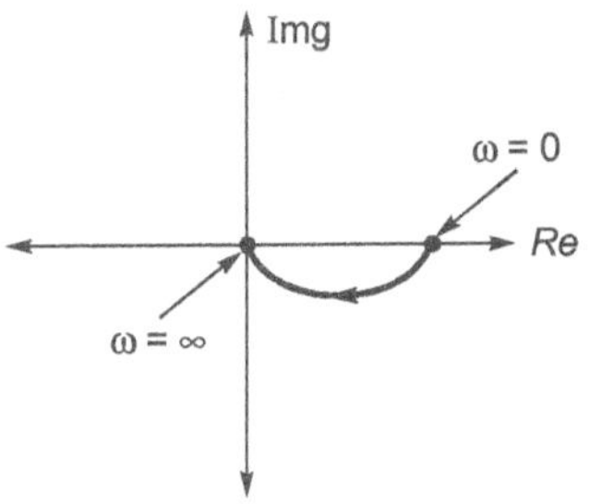

Hence, the polar plot will be semicircular in shape.

239. (c)

The bandwidth is defined as the band of frequencies lying between –3 dB points. Hence, statement-2 is false.
Higher the value of resonant frequency of the system, faster is the time response. Hence, statement-3 is false.

240. (a)

Given, $G(s) = \dfrac{1+sT}{sT}$

The inverse polar plot of $G(j\omega)$ is the polar plot of $\frac{1}{G(j\omega)}$.

$$\therefore \quad \frac{1}{G(j\omega)} = \frac{j\omega T}{1+j\omega T}$$

Thus, $$|G(j\omega)^{-1}| = \frac{\omega T}{\sqrt{1+(\omega T)^2}}$$

and $$\angle G(j\omega)^{-1} = 90° - \tan^{-1}\omega T$$

ω	$\lvert G(j\omega)^{-1}\rvert$	$\angle G(j\omega)^{-1}$
0	0	90°
∞	1	0°

Hence, inverse polar plot will be as shown below.

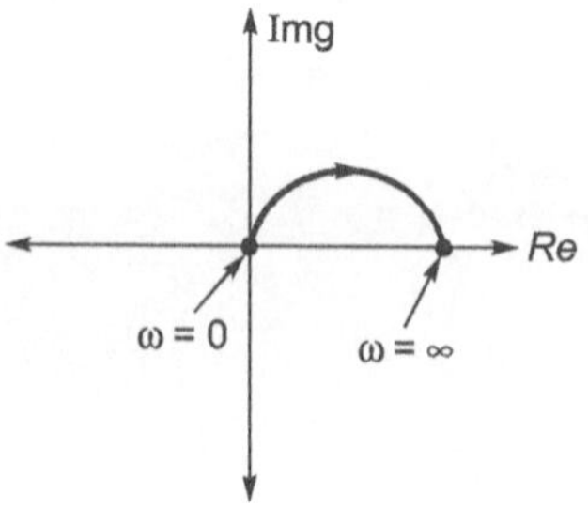

242. (c)

$\frac{1}{(j\omega)^2}$ represents type-2 factor which will have a slope of –40 db/decade.

We know that,

–20 db/decase = –6 db/octave

∴ –40 db/decase = –12 db/octave

243. (b)

Since, the system is of minimum phase type, it has no poles or zeros in the right hand side of s-plane.

⇒ There will be three corner frequencies at:

$$\omega_1 = 5 \text{ rad/s}$$

$$\therefore T_1 = \frac{1}{\omega_1} = 0.2 \text{ sec.}$$

(one pole added at $\omega_1 = 5$ rad/s)

$$\omega_2 = 40 \text{ rad/s}$$

$$\therefore T_2 = \frac{1}{\omega_2} = 0.025 \text{ sec}$$

(two pole added at $\omega_1 = 40$ rad/s)

and $$\omega_3 = 100 \text{ rad/s}$$

$$\therefore T_3 = \frac{1}{\omega_3} = 0.01 \text{ sec.}$$

(two poles added at $\omega_3 = 100$ rad/s)

So, open loop T.F. is

$$G(s) = \frac{K}{(1+0.2s)(1+0.025s)^2(1+0.01s)}$$

Now, 20 log K = 100

$$\therefore \quad K = 10^5$$

Thus,

$$G(s) = \frac{10^5}{(1+0.2s)(1+0.025s)^2(1+0.01s)}$$

244. (a)

For marginally stable system,

$$\text{P.M} = \text{G.M} = 0 \text{ and } \omega_{pc} = \omega_{gc}$$

Now, phase margin = (180° + ϕ)

where, $$\phi = |\angle G(j\omega)H(j\omega)|_{\omega_{pc}}$$

Given, P.M = 0° (for marginally stable)

$$\therefore \quad 0° = 180° + \angle G(j\omega)H(j\omega)$$

or, $$\angle G(j\omega)H(j\omega) = -180°$$

Given, $$G(s)H(s) = \frac{K}{s(s+5)(s+10)}$$

or, $$\angle G(j\omega)H(j\omega) = -90° - \tan^{-1}\frac{\omega}{5} - \tan^{-1}\frac{\omega}{10}$$

$$\therefore \quad -90° - \tan^{-1}\frac{\omega}{5} - \tan^{-1}\frac{\omega}{10} = -180°$$

or, $$\tan^{-1}0.2\,\omega + \tan^{-1}0.1\,\omega = 90°$$

or, $$\tan^{-1}\left(\frac{0.2\omega + 0.1\omega}{1-(0.2\omega)(0.1\omega)}\right) = 90°$$

or, $$1 - 0.02\,\omega^2 = 0$$

or, $$\omega = \frac{1}{\sqrt{0.02}} = \frac{10}{\sqrt{2}} = 7.07 \text{ rad/sec}$$

Thus, $$\omega_{pc} = \omega_{gc} = \omega = 7.07 \text{ rad/sec}$$

246. (c)

Given, $$G(s)H(s) = \frac{K}{s(s+3)(s+5)}$$

$$\therefore \quad G(j\omega)H(j\omega) = \frac{K}{(j\omega)(3+j\omega)(5+j\omega)}$$

$$= \frac{K}{-8\omega^2 + j\omega(-\omega^2+15)} \quad ...(1)$$

The OLTF has no-poles in RH s-plane. Thus, for stability the $-1 + j0$ (critical point) point should not be encircled by Nyquist plot.

For finding the point of intersection of Nyquist plot with the negative real axis,

$$\text{Img}\,[G(j\omega)\,H(j\omega)] = 0$$

or, $$-\omega^2 + 15 = 0$$

or, $$\omega = \sqrt{15} \text{ rad/sec}$$

Putting, $\omega = \sqrt{15}$ rad/s in equation (1), we have

$$G(j\omega)H(j\omega) = \frac{K}{-8 \times (\sqrt{15})^2} = \frac{-K}{120}$$

For system to be stable, $\frac{K}{120} < 1$ or $K < 120$

So, $0 < K < 120$ (for stability)

249. (d)

BW $\propto \omega_n$. Hence, option (d) is not correct.

For a prototype second order system,

$$\text{BW} = \omega_n\left[(1-2\xi^2) + \sqrt{\xi^4 - 4\xi^2 + 2}\right]$$

Thus, $$\text{BW} \propto \omega_n \propto \frac{1}{\xi}$$

250. (d)

Here, $|X| = 0.25$ (Given)

$$\therefore \quad \text{G.M.} = \frac{1}{|X|} = \frac{1}{0.25} = 4$$

Also, $\phi = -180°$

$$\therefore \quad \text{P.M.} = 180 + \phi = 180° - 180° = 0°$$

251. (c)

Resonant peak is given by,

$$M_r = \frac{1}{2\xi\sqrt{1-\xi^2}}$$

- For $\xi = 0$, $M_r = \infty$
- For $\xi = 0.707 = \frac{1}{\sqrt{2}}$, $M_r = 1$
- For $\xi = 0.5$ i.e., $\xi < \frac{1}{\sqrt{2}}$, $M_r > 1$
- For $\xi = 0.85$, $\xi > \frac{1}{\sqrt{2}}$, no M_r or not defined

In frequency domain analysis,

$$\omega_r = \omega_n\sqrt{1-2\xi^2}$$

Thus, for ω_r to be real, $\xi < \frac{1}{\sqrt{2}}$. Hence for $\xi > \frac{1}{\sqrt{2}}$ there is no M_r or M_r is not defined.

252. (d)

Gain margin alone is inadequate to indicate relative stability when system parameters other than the loop gain are subject to variation. Due to variation in some parameters, phase shift of the system can vary which may results instability. Hence, we must also consider the concept of phase margin for determining the relative stability of the system.

255. (a)

Given,

$$G(j\omega)\,H(j\omega) = \frac{25}{(j\omega)(1+j\omega)(10+j\omega)} = \frac{2.5}{(j\omega)(1+j\omega)(1+0.1j\omega)}$$

$$\therefore \quad \phi = \angle G(j\omega)\,H(j\omega) = -90° - \tan^{-1}\omega - \tan^{-1}(0.1\,\omega)$$

At $\omega = \omega_{pc}$, $\phi = -180°$, therefore

$$-180° = -90° - \tan^{-1}\omega - \tan^{-1}(0.1\,\omega)$$

or, $$\omega = \omega_{pc} = \frac{1}{\sqrt{0.1}} = \sqrt{10} \text{ rad/sec}$$

= Phase crossover frequency

256. (c)

$$\text{G.M.} \propto \frac{1}{\text{Open loop gain}}$$

$\therefore$ When open-loop gain becomes 4-times, GM will become one-fourth.

257. (b)

Given, at ω_{pc},

$$\phi = -190°$$

$$\therefore \quad \text{Phase margin} = 180° + \phi = 180° - 190° = -10°$$

Since, phase margin is negative, therefore system is unstable.

259. (c)

Initial slope of the system = 0 dB/decade
Therefore, given system is a type-0 system.
For type-0 system, for parabolic input, steady state error is infinite since,

$$K = \lim_{s\to 0}[s^2G(s)H(s)] = 0$$

260. (d)

At $\omega = \omega_{gc}$, $\phi = -180°$

i.e. Nyquist plot crosses the negative real axis.

261. (d)

$$G(j\omega)\,H(j\omega) = \frac{K}{(j\omega)(5+j\omega)(10+j\omega)}$$

$$\therefore \angle G(j\omega)\,H(j\omega) = -90° - \tan^{-1}\frac{\omega}{5} - \tan^{-1}\frac{\omega}{10} = \phi$$

At $\omega \to \infty$, $\phi = -90° - 90° - 90° = -270° = +90°$

262. (b)

Given, $$G(s)\,H(s) = \frac{K}{s(1+as)}$$

or, $$G(j\omega)\,H(j\omega) = \frac{K}{(j\omega)(1+ja\omega s)}$$

Initial slope = –20 dB/decade = –6 dB/octave
(Since Type = 1)

Corner frequency is, $\omega_c = \frac{1}{a}$ rad/sec

It will cross 0 dB line at $\omega = K$ and $K < \frac{1}{a}$.

So, the Bode plot will be as shown below.

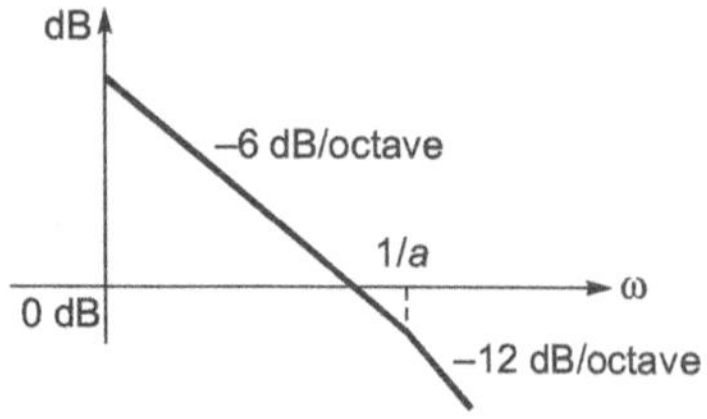

263. (b)

$$\text{Peak magnitude} = 1+\exp\left(\frac{-\xi\pi}{\sqrt{1-\xi^2}}\right)$$

Given, $2\xi\omega_n = 10\sqrt{2}$, $\omega_n = 10$ rad/s

$$\therefore \quad \xi = \frac{1}{\sqrt{2}}$$

So, peak magnitude = $1 + e^{-\pi} \approx 1$

264. (d)

Given, $$\frac{C(s)}{R(s)} = \frac{1}{1+s}$$

Thus, $$H(j\omega) = \frac{C(j\omega)}{R(j\omega)} = \frac{1}{1+j\omega}$$

Here, $\omega = 1$ rad/s

$$\therefore \quad |H(j\omega)| = \frac{1}{\sqrt{1+\omega^2}} = \frac{1}{\sqrt{1+1}} = \frac{1}{\sqrt{2}}$$

and $$\angle H(j\omega) = -\tan^{-1}\omega = -\tan^{-1}(1) = -45° = -\frac{\pi}{4}$$

$$\therefore \quad c(t) = \frac{1}{\sqrt{2}}\sin\left(t-\frac{\pi}{4}\right)$$

265. (a)

Given transfer function represents an all pass filter which has gain of 1 at all frequencies.

266. (a)

$$G(j\omega)\,H(j\omega) = \frac{K}{(j\omega)(1+j\omega T)}$$

$$\therefore \quad |G(j\omega)H(j\omega)| = \frac{K}{\omega\sqrt{1+(\omega T)^2}}$$

$$\angle G(j\omega)\,H(j\omega) = -90° - \tan^{-1}\omega T = \phi$$

At $\omega = \omega_{pc}$,

$$\phi = -180°$$

$\therefore -90° - \tan^{-1}\omega_{pc} = -180°$ or, $\omega_{pc} = \infty$ rad/s

$$\therefore \quad |G(j\omega)\,H(j\omega)|_{\omega=\omega_{pc}} = X = 0$$

$$\therefore \quad \text{G.M.} = \frac{1}{|X|} = \frac{1}{0} = \infty$$

267. (d)

Given, $s^2 + 4s + 16 = 0$ or $s^2 + 2\xi\omega_n s + \omega_n^2 = 0$

$$\therefore \quad \omega_n = 4 \text{ rad/s}$$

and $2\xi\omega_n = 4$ or $\xi = \frac{1}{2}$

$\therefore$ Resonant frequency,

$$\omega_r = \omega_n\sqrt{1-2\xi^2} = 4\sqrt{1-2\times\frac{1}{4}} = 4\sqrt{\frac{1}{2}} = \frac{4}{\sqrt{2}} = 2\sqrt{2} \text{ rad/sec}$$

268. (d)

Given,

$$G(j\omega)\,H(j\omega) = \frac{\sqrt{2}}{(j\omega)\,(1+j\omega)}$$

$$\therefore \quad |G(j\omega)\,H(j\omega)| = \frac{\sqrt{2}}{\omega\sqrt{1+\omega^2}}$$

and $\angle G(j\omega)\,H(j\omega) = -90° - \tan^{-1}\omega = \phi$

At $\omega = \omega_{pc}$, $\phi = -180°$

$\therefore\ -90° - \tan^{-1}\omega_{pc} = -180°$ or $\omega_{pc} = \infty$

$$\therefore \quad \text{G.M.} = \frac{1}{|G(j\omega)\,H(j\omega)|_{\omega=\omega_{pc}}} = \frac{1}{0} = \infty$$

269. (a)

Nichol's chart consists of magnitude and phase angle of a closed loop system represented as a family of circle known as M and N-circle.

270. (d)

$$\text{Given,} \quad G(j\omega) = \frac{1}{(1+j\omega)^2}$$

$$\therefore \quad |G(j\omega)| = \frac{1}{(\sqrt{1+\omega^2})^2} = \frac{1}{1+\omega^2}$$

At $\omega = \omega_{gc}$, $|G(j\omega)| = 1$

$$\therefore \quad \frac{1}{1+\omega_{gc}^2} = 1 \text{ or } \omega_{gc} = 0 \text{ rad/sec}$$

Now, $\angle G(j\omega) = -2\tan^{-1}\omega = \phi$

At $\omega = \omega_{gc} = 0$ rad/s

$$\phi = -2 \times 0 = 0°$$

$$\therefore \quad \text{P.M.} = 180° + \phi = 180° + 0° = 180° = \pi$$

271. (b)

Given,

$$G(j\omega)\,H(j\omega) = \frac{1}{(j\omega)\,(1+j2\omega)\,(1+j\omega)}$$

$$= \left[\frac{-3}{(1+4\omega^2)(1+\omega^2)} - j\frac{1-2\omega^2}{\omega(1+4\omega^2)(1+\omega^2)}\right]$$

For finding ω_{gc}, Img $[G(j\omega)H(j\omega)] = 0$

$$\text{or,} \quad 1-2\omega_{pc}^2 = 0 \text{ or } \omega_{pc} = \frac{1}{\sqrt{2}} = 0.707 \text{ rad/s}$$

Now,

$$|G(j\omega)\,H(j\omega)|_{\omega_{pc}} = \left|\frac{-3}{\left(1+\frac{4}{2}\right)\left(1+\frac{1}{2}\right)}\right| = \left|\frac{-3}{3\times\frac{3}{2}}\right|$$

$$= \left|-\frac{2}{3}\right| = 0.66$$

Thus, Nyquist plot will cut real axis at –0.66.

The given system: $G(s)H(s) = \dfrac{1}{s(1+2s)(1+s)}$ is of type-1 and order-3.

Therefore, Nyquist plot will be as shown below:

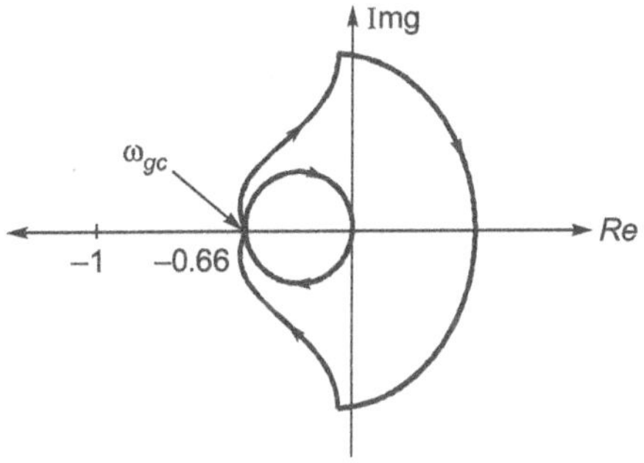

Now, from above Nyquist plot, N = 0 = Number of encirclements.

Also, P = number of open loop poles in RH s-plane = 0

Now, $N = P - Z$

or, $0 = 0 - Z$ or $Z = 0$

Thus, no. of closed loop poles in RH s-plane = 0.

Hence, system is stable.

272. (c)

$$\text{Given,} \quad G(s)H(s) = \frac{K(1-s)}{(s+1)}$$

$$\text{or,} \quad G(j\omega)H(j\omega) = \frac{K(1-j\omega)}{(1+j\omega)}$$

$$\therefore \quad |G(j\omega)\,H(j\omega)| = K\sqrt{\frac{1+\omega^2}{1+\omega^2}} = K$$

and $\angle G(j\omega)\,H(j\omega) = -2\tan^{-1}\omega$

ω	$\|G(j\omega)\,H(j\omega)^{-1}\|$	$\angle G(j\omega)\,H(j\omega)$
0	K	0°
1	K	–90°
∞	K	–180°

Thus, the radius of Nyquis plot will be K.

Now, no.of open loop poles in RH s-plane = 0.
When $K > 1$, no. of encirclement, $N = -1$
When $K < 1$, no. of encirclement, $N = 0$
Thus, for $K > 1$, $N = P - Z$
or, $-1 = 0 - Z$ or $Z = 1$
∴ System is unstable.
For $K < 1$, $N = P - Z$ or $0 = 0 - Z$ or $Z = 0$
∴ System is stable.

273. (a)

Given, $G(s)\,H(s) = \dfrac{e^{-Ts}}{s(1+s)}$

$$\therefore \quad G(j\omega)\,H(j\omega) = \frac{e^{-j\omega T}}{(j\omega)(1+j\omega)} = \frac{\cos\omega T - j\sin\omega T}{(j\omega)(1+j\omega)}$$

or, $\angle G(j\omega)\,H(j\omega) = -\omega T - 90° - \tan^{-1}\omega$

At $\omega = \omega_1$, $\angle G(j\omega)\,H(j\omega) = 0°$

$\therefore -\omega_1 T - 90° - \tan^{-1}\omega_1 = 0$

or, $\tan^{-1}\omega_1 = -(\omega_1 T + 90°)$

or, $\omega_1 = -\tan(90° + \omega_1 T) = \cot\omega_1 T$

$\therefore \quad \omega_1 = \cot(\omega_1 T)$

274. (b)

At $\omega = \omega_{gc}$, $|X| = 0.33$

(Using Nyquist diagram)

$$\therefore \quad \text{G.M.} = \frac{1}{|X|} = \frac{1}{0.33} = \frac{100}{33} \approx 3$$

275. (b)

Number of decade change from first break frequency

10 rad/s ω_1 is $= \dfrac{26-20}{20} = \dfrac{6}{20}$ dec.

Now, $\log\dfrac{\omega_1}{10} = \dfrac{6}{20} \Rightarrow \omega_1 = 20$ rad/sec

Now, no. of decade change from break frequency

ω_1 to $\omega_2 = \dfrac{26}{20} = 1.3$

$\therefore \quad \log\dfrac{\omega_2}{\omega_1} = 1.3$ or $\log\dfrac{\omega_2}{20} = 1.3$

or, $\omega_2 = 400$ rad/s

277. (d)

Resonant frequency is given by

$$\omega_r = \omega_n\sqrt{1-2\xi^2}$$

For resonant peak, $1 - 2\xi^2 > 0$

or, $\xi < \dfrac{1}{\sqrt{2}}$ or $\xi < 0.707$

Since $\xi > 0.707$, therefore there is no M_r (resonant peak).

279. (c)

One of the disadvantage of a proportional controller is that it exhibits a permanent residual error at its operating point known as offset and given by $e_{ss} = \left(\dfrac{A}{1+K_p}\right)$. The offset can be reduced by adding integral mode.

280. (b)

Pole-zero plot for lead and lag compensators are shown below.

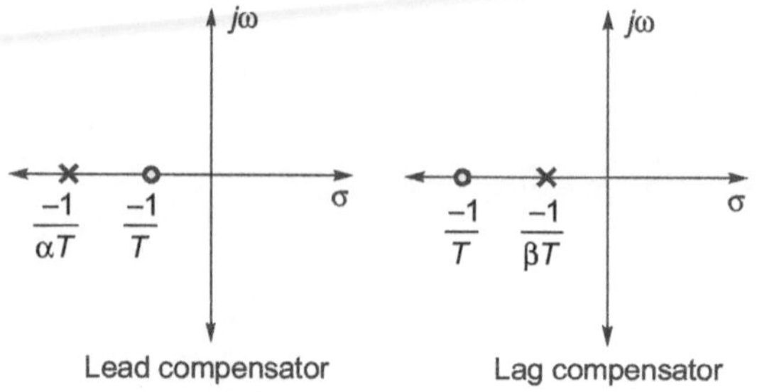

Since zero is close to origin in a lead compensator, therefore addition of zero in a transfer function will cause lead compensator.

283. (c)

Lag compensator is used to reduce steady-state error, therefore statement-1 is true.
As lag compensator reduces steady state error therefore, statement-2 is false.
Statement-3 is also false because capacitance is used to fabricate a lag network.

284. (d)

For lead compensator, zero is near to origin and pole is near to infinity as shown below. So, constant b will be more than unity and time constant T will be greater than zero.

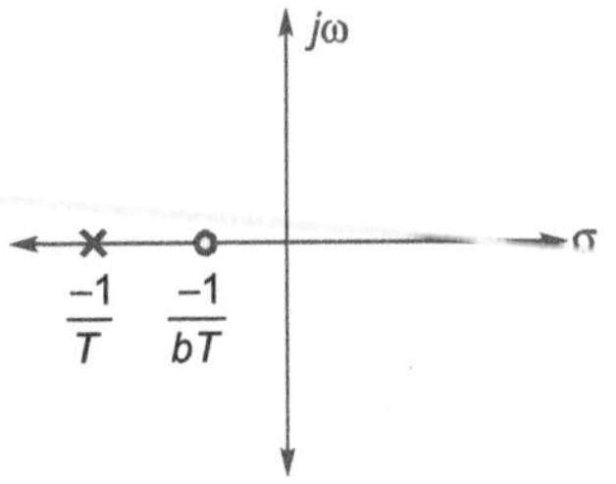

285. **(d)**

For any type of controller, the input is always an error signal.

287. **(c)**

Due to a phase lead network, bandwidth is increased. Hence, option (d) is correct.

289. **(c)**

Here, reason (R) is false because the lag network allows to pass low frequencies and high frequencies are attenuated.

290. **(d)**

The transfer function of a phase-lag network is:

$$H(s) = \frac{1+aTs}{1+Ts} \; (a<1)$$

The pole-zero plot is shown below:

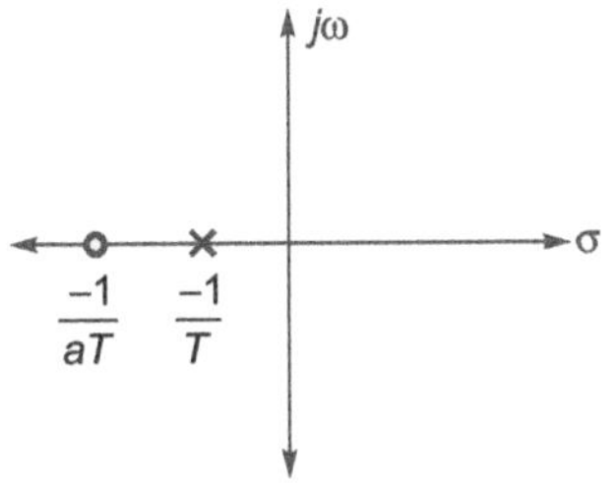

Here, pole is nearer to the imaginary axis and origin compared to zero.

Hence, reason (R) is true.

Introduction of phase lag network in forward path reduces the phase shift. Hence, assertion (A) is a false statement.

292. **(a)**

A PID controller is shown below.

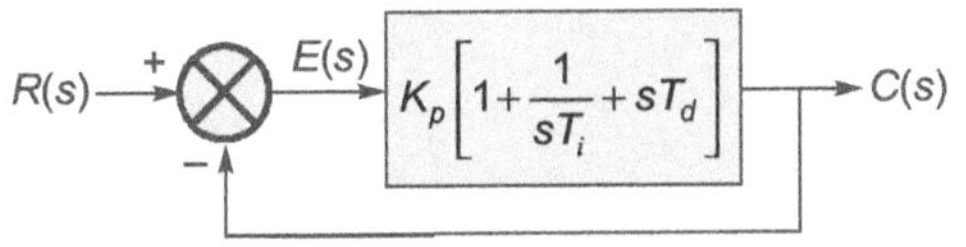

Here, $G(s)$ = O.L.T.F = $K_p\left[1+\frac{1}{sT_i}+sT_d\right]$

293. **(d)**

Due to PD controller, a zero is added to the forward path which results in a lead compensator. Therefore, steady state error remais the same while rise time of the system is reduced (i.e. stability is increased).

294. **(c)**

For the given system, characteristic equation is

$$\left(1+\frac{14}{s(s+1.4)+14sT_d}\right)=0$$

or, $s^2 + 1.4s + 14s\,T_d + 14 = 0$

or, $s^2 + (1.4 + 14T_d)\,s + 14 = 0$

Comparing with $s^2 + 2\xi\omega_n s + \omega_n^2 = 0$, we have:

$\omega_n = \sqrt{14}$ rad/s and $2\xi\omega_n = (1.4 + 14T_d)$

or, $2\times 0.7\times\sqrt{14} = 1.4 + 14T_d$

($\because \xi = 0.7$, given)

or, $$\left(\frac{1.4\sqrt{14}-1.4}{14}\right) = T_d$$

or, $$T_d = \left(\frac{0.1\sqrt{14}-1.4}{14}\right)$$

$$= \left(\frac{\sqrt{14}-1}{10}\right) = \frac{2.74}{10}$$

$$= 0.274$$

295. **(d)**

For the given system, forward path transfer function is

$$\left(\frac{K_i}{s}+1\right)\left(\frac{\omega_n^2}{s(s+2\xi\omega_n)}\right) = \frac{(K_i+s)\omega_n^2}{s^2(s+2\xi\omega_n)}$$

For parabolic input, steady state error is

$$e_{ss} = \frac{1}{K_a}$$

where, $$K_a = \lim_{s\to 0}[s^2 G(s)H(s)]$$

$$= \lim_{s\to 0}\left[\frac{(K_i+s)\omega_n^2}{s+2\xi\omega_n}\right] = \frac{K_i\,\omega_n^2}{2\xi\omega_n}$$

$$= \frac{K_i\omega_n}{2\xi}$$

$\therefore$ $$e_{ss} = \frac{1}{K_a} = \frac{2\xi}{K_i\omega_n}$$

296. **(a)**

The forward path gain of the given system is:

$$G(s) = \frac{10(1+sK_e)}{s(s+2)}$$

$\therefore$ Characteristic equation is

$$1 + G(s)H(s) = 0 \text{ or } 1+\frac{10(1+sK_e)}{s(s+2)} = 0$$

or, $s^2 + (2 + 10\,K_e)s + 10 = 0$

Here, $\omega_n = \sqrt{10}$ rad/s

and $2\xi\omega_n = (2 + 10K_e)$

Given, $\xi = 0.5$

So, $2\times 0.5\times\sqrt{10} = (2 + 10K_e)$

or, $$K_e = \frac{\sqrt{10}-2}{10} = \left(\frac{3.16-2}{10}\right)$$

$$= \frac{1.16}{10} = 0.116$$

Also, 2% settling time

$$= 4T = \frac{4}{\xi\omega_n} = \frac{4}{0.5\times\sqrt{10}}$$

$$= 2.53 \text{ sec}$$

297. **(b)**

The characteristic equation of the given system is

$$1+G(s)\,H(s) = 0$$

Here, $$G(s)\,H(s) = \frac{10}{s(s+2)+K_0 s}$$

$$= \frac{10}{s(s+K_0+2)}$$

$\therefore$ Characteristic equation is

$$1+\frac{10}{s(s+K_0+2)} = 0$$

or, $s^2 + (K_0 + 2)s + 10 = 0$

Comparing with $s^2 + 2\xi\omega_n s + \omega_n^2 = 0$ we have:

$$\omega_n = \sqrt{10} \text{ rad/s}$$

and $2\xi\omega_n = K_0 + 2$

Given, $\xi = 0.6$

$\therefore$ $2\times 0.6\times\sqrt{10} = K_0 + 2$

or, $K_0 = 1.792$

For a unit ramp input, $e_{ss} = \dfrac{1}{K_v}$

where, $K_v = \lim\limits_{s\to 0}[sG(s)H(s)]$

$$= \lim_{s\to 0}\left[\frac{10s}{s(s+K_0+2)}\right] = \frac{10}{K_0+2}$$

$\therefore$ $$e_{ss} = \frac{K_0+2}{10} = \frac{1.792+2}{10} = \frac{3.792}{10}$$

$$= 0.3792 \approx 0.38 \text{ radian}$$

298. **(d)**

Case-1: For integral type

$$G(s)H(s) = \frac{10}{T_i\,s(s+5)}$$

$\therefore$ $$K_v = \lim_{s\to 0}[sG(s)H(s)] = \frac{10}{5T_i}$$

$\therefore$ $$e_{ss} = \frac{T_i}{2}$$

Case-2: For derivative type

$$G(s)\,H(s) = \frac{10sT_d}{(s+5)}$$

$\therefore$ $$K_v = \lim_{s\to 0}[sG(s)H(s)]$$

$$= \lim_{s\to 0}\left[\frac{10s^2T_d}{s+5}\right] = 0$$

$\therefore$ $$e_{ss} = \frac{1}{K_v} = \infty$$

Case-3: For proportional type

$$G(s)\,H(s) = \frac{10K_p}{s+5}$$

$\therefore$ $$K_v = \lim_{s\to 0}[sG(s)H(s)]$$

$$= \lim_{s\to 0}\left[\frac{10K_p}{s+5}\right] = 2K_p$$

$\therefore$ $$e_{ss} = \frac{1}{K_v} = \frac{1}{2K_p}$$

Case-4: For proportional plus derivative type

$$G(s)H(s) = \frac{10K_p(1+T_d s)}{s+5}$$

$\therefore$ $$K_v = \lim_{s\to 0}[sG(s)H(s)]$$

$$= \lim_{s\to 0}\left[\frac{10sK_p(1+T_d s)}{s+5}\right] = 0$$

$\therefore$ $$e_{ss} = \frac{1}{K_v} = \infty$$

299. (c)

Redrawing the given circuit in s-domain, we have:

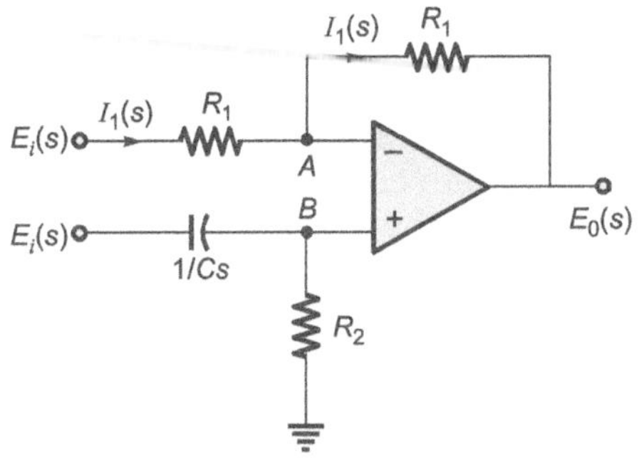

Now, $$V_B(s) = \frac{R_2}{\left(R_2 + \frac{1}{Cs}\right)} E_i(s)$$

From virtual ground concept,

$$V_A(s) = V_B(s) = \frac{R_2}{\left(R_2 + \frac{1}{Cs}\right)} E_i(s)$$

$$= \frac{R_2 Cs}{1 + R_2 Cs} E_i(s)$$

Now, $$I_1(s) = \frac{E_i(s) - V_A(s)}{R_1} = \frac{V_A(s) - E_0(s)}{R_1}$$

or, $$2V_A(s) = E_i(s) + E_0(s)$$

or, $$E_0(s) = 2\,V_A(s) - E_i(s)$$

or, $$E_0(s) = 2\left[\frac{R_2 Cs}{1 + R_2 Cs} E_i(s)\right] - E_i(s)$$

or, $$\frac{E_0(s)}{E_i(s)} = \left[\frac{R_2 Cs - 1}{R_2 Cs + 1}\right]$$

301. (a)

Since the nature of command is in the form of pulses, therefore stepping motors are compatible with modern digital systems.

302. (d)

For given system,

$$G(s)H(s) = \frac{2K_c}{(s^3 + 4s^2 + 5s + 2)}$$

With proportional controller,

$$K_p = \lim_{s \to 0} [G(s)H(s)]$$

$$= \lim_{s \to 0} \left[\frac{2K_c}{s^3 + 4s^2 + 5s + 2}\right] = K_c$$

$$\therefore \quad e_{ss} = \frac{1}{(1 + K_p)} = \frac{1}{(1 + K_c)}$$

(for a unit step input)

Given, e_{ss} = less than 2% of input

i.e. $e_{ss} < 0.02$

Now, characteristic equation is:

$1 + G(s)H(s) = 0$

or, $s^3 + 4s^2 + 5s + 2 + 2K_c = 0$

Routh's array is:

s^3	1	5
s^2	4	$2 + 2K_c$
s^1	$\frac{18 - 2K_c}{4}$	
s^0	$2 + 2K_c$	

For stability of given system,

$2 + 2K_c > 0$ or $K_c > -1$

and $\frac{18 - 2K_c}{4} > 0$ or $K_c < 9$

Hence, $-1 < K_c < 9$ (for stability) ...(1)

Now, $e_{ss} = \frac{1}{1 + K_c} < 0.02$ (given) ...(2)

From equations (1) and (2), we conclude that the minimum value of steady state error = 0.1 when $K_c = 9$.

Also, for all value of K_c between –1 and 9, value of e_{ss} will be more than 0.02 i.e. more than 2% of input.

Therefore, there is no such value of K_c for given system i.e. can not be determined.

304. (b)

Given, $$\frac{C(s)}{R(s)} = \frac{1 + s}{1 + 0.2s}$$

The pole-zero plot is shown below.

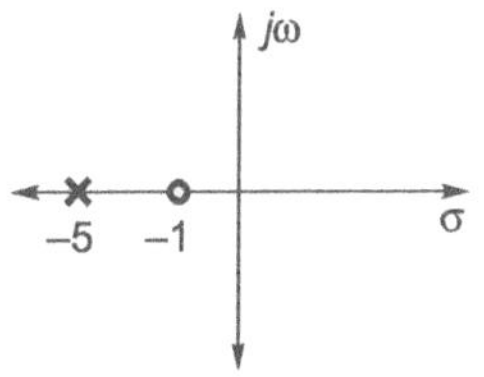

Since zero is near to origin, therefore it represents a lead network.

307. (d)

Given, $$H(s) = \frac{1 + 3Ts}{1 + Ts} = \frac{1 + T_1 s}{1 + \alpha T_1 s}$$

Here, $T_1 = 3T$ and $\alpha T_1 = T$

$$\therefore \quad \alpha = \frac{T}{T_1} = \frac{T}{3T} = \frac{1}{3}$$

$$\therefore \quad \sin\phi_m = \frac{1-\alpha}{1+\alpha} = \frac{1-\frac{1}{3}}{1+\frac{1}{3}} = \frac{2}{4} = \frac{1}{2}$$

or, $\phi_m = 30° =$ maximum phase shift

313. (b)

An optimal controller for an LTI system can be designed provided the system is both controllable and observable.

314. (b)

Eigen value are given by $|sI - A| = 0$, which is the location of poles.

315. (a)

Characteristic equation is,

$$|sI - A| = 0 \text{ or } \left| \begin{bmatrix} s & 0 \\ 0 & s \end{bmatrix} - \begin{bmatrix} 0 & 1 \\ -2 & -5 \end{bmatrix} \right| = 0$$

$$\text{or, } \begin{vmatrix} s & -1 \\ 2 & s+5 \end{vmatrix} = 0 \text{ or } s(s+5) + 2 = 0$$

or, $s^2 + 5s + 2 = 0$

316. (b)

Poles of given system are given by

$$|sI - A| = 0 \text{ or } \begin{vmatrix} s & -5 \\ -5 & s \end{vmatrix} = 0$$

or, $s^2 + 25 = 0$ or $s = \pm j5$

318. (c)

We have:

$$(sI - A) = \begin{bmatrix} s & 0 \\ 0 & s \end{bmatrix} - \begin{bmatrix} 0 & 1 \\ -2 & -3 \end{bmatrix} = \begin{bmatrix} s & -1 \\ 2 & s+3 \end{bmatrix}$$

$$\therefore |sI - A| = \begin{vmatrix} s & -1 \\ 2 & s+3 \end{vmatrix} = s(s+3) + 2$$

$$= s^2 + 3s + 2 = (s+1)(s+2)$$

$$\therefore \quad \phi(s) = (sI - A)^{-1}$$

$$= \begin{bmatrix} \frac{s+3}{(s+1)(s+2)} & \frac{1}{(s+1)(s+2)} \\ \frac{2}{(s+1)(s+2)} & \frac{s}{(s+1)(s+2)} \end{bmatrix}$$

$$\therefore \quad \phi(t) = \mathcal{L}^{-1}\{\phi(s)\} = \mathcal{L}^{-1}(sI - A)^{-1}$$

$$= \begin{bmatrix} 2e^{-t} - e^{-2t} & e^{-t} - e^{-2t} \\ -2e^{-t} + 2e^{-2t} & -e^{-t} + 2e^{-2t} \end{bmatrix}$$

319. (d)

The characteristic equation is $|sI - A| = 0$.

$$\text{Now, } (sI - A) = \begin{bmatrix} s & 0 & 0 \\ 0 & s & 0 \\ 0 & 0 & s \end{bmatrix} - \begin{bmatrix} -5 & 1 & 0 \\ 0 & -5 & 1 \\ 0 & 0 & -5 \end{bmatrix}$$

$$= \begin{bmatrix} s+5 & -1 & 0 \\ 0 & s+5 & -1 \\ 0 & 0 & s+5 \end{bmatrix}$$

Now, $|sI - A| = 0$

or, $(s+5)[(s+5)^2 + 0] + 1(0-0) + 0 = 0$

or, $(s+5)^3 = 0$ or $s^3 + 15s^2 + 75s + 125 = 0$

320. (a)

Transfer function is given by

$$\frac{Y(s)}{U(s)} = C[sI - A]^{-1} B$$

$$\text{Given, } A = \begin{bmatrix} -1 & 1 \\ 0 & -2 \end{bmatrix}, B = \begin{bmatrix} 1 & 0 & 1 \\ 0 & 1 & 1 \end{bmatrix}$$

$$\text{and } C = \begin{bmatrix} 1 & 2 \\ 1 & 0 \\ 1 & 1 \end{bmatrix}$$

$$\text{Now, } (sI - A) = \begin{bmatrix} s+1 & 1 \\ 0 & s+2 \end{bmatrix}$$

$$\therefore \quad (sI - A)^{-1} = \frac{1}{s^2 + 3s + 2} \begin{bmatrix} s+2 & 1 \\ 0 & s+1 \end{bmatrix}$$

$$\therefore \quad C[sI - A]^{-1} = \begin{bmatrix} 1 & 2 \\ 1 & 0 \\ 1 & 1 \end{bmatrix} \begin{bmatrix} s+2 & 1 \\ 0 & s+1 \end{bmatrix} \times \frac{1}{s^2 + 3s + 2}$$

$$= \frac{1}{s^2 + 3s + 2} \begin{bmatrix} s+2 & 2s+3 \\ s+2 & 1 \\ s+2 & s+2 \end{bmatrix}$$

$$\text{So, T.F.} = \frac{Y(s)}{U(s)} = C[sI - A]^{-1} B$$

$$= \frac{1}{s^2 + 3s + 2} \begin{bmatrix} s+2 & 2s+3 \\ s+2 & 1 \\ s+2 & s+2 \end{bmatrix} \begin{bmatrix} 1 & 0 & 1 \\ 0 & 1 & 1 \end{bmatrix}$$

$$= \left[\frac{s^3 - 3s^2 - 9s + 2}{s^2 + 3s + 2} \right]$$

321. (b)

Eigen values of A are given by

$$|sI - A| = 0 \text{ or } \begin{vmatrix} s-\sigma & \omega \\ -\omega & s-\sigma \end{vmatrix} = 0$$

or, $(s-\sigma)^2 + \omega^2 = 0$

or, $s = \sigma \pm j\omega$

or, $s = \sigma + j\omega$ and $s = \sigma - j\omega$

322. (a)

Given, $A = \begin{bmatrix} 0 & 1 \\ -2 & -3 \end{bmatrix}, B = \begin{bmatrix} 0 \\ 1 \end{bmatrix}, C = [1 \quad 0]$

Transfer function of the given system is

$$\frac{Y(s)}{U(s)} = C(sI - A)^{-1} B$$

Now, $(sI - A) = \begin{bmatrix} s & -1 \\ 2 & s+3 \end{bmatrix}$

$\therefore$ $|sI - A| = s(s+3) + 2$
$= s^2 + 3s + 2$

Now, $(sI - A)^{-1} = \dfrac{1}{s^2 + 3s + 2} \begin{bmatrix} s+3 & 1 \\ -2 & s \end{bmatrix}$

Now, $C(sI - A)^{-1} B = [1 \quad 0] \begin{bmatrix} s+3 & 1 \\ -2 & s \end{bmatrix} \begin{bmatrix} 0 \\ 1 \end{bmatrix} \times \dfrac{1}{(s^2 + 3s + 2)}$

$$= [s+3 \quad 1] \begin{bmatrix} 0 \\ 1 \end{bmatrix} \times \frac{1}{s^2 + 3s + 2}$$

$$= \frac{1}{s^2 + 3s + 2}$$

Hence, $\dfrac{Y(s)}{U(s)} = \dfrac{1}{s^2 + 3s + 2}$

323. (c)

The poles of the system can be obtained from eigen values of the system matrix. Hence, statement-1 is false.

State space techniques can be applied to linear or non-linear, time variant or time invariant systems.

Hence, statement-4 is false.

324. (d)

From given block diagram, the state equations can be written as:

$$\dot{x}_1 = -x_1 + u \text{ and } \dot{x}_2 = -2x_2 + 2u$$

In matrix form,

$$\begin{bmatrix} \dot{x}_1 \\ \dot{x}_2 \end{bmatrix} = \begin{bmatrix} -1 & 0 \\ 0 & -2 \end{bmatrix} \begin{bmatrix} x_1 \\ x_2 \end{bmatrix} + \begin{bmatrix} 1 \\ 2 \end{bmatrix} u$$

Also, $y = x_1 + x_2$

In matrix form,

$$y = [1 \quad 1] \begin{bmatrix} x_1 \\ x_2 \end{bmatrix}$$

Thus, $A = \begin{bmatrix} -1 & 0 \\ 0 & -2 \end{bmatrix}, B = \begin{bmatrix} 1 \\ 2 \end{bmatrix}, C = [1 \quad 1]$

Now, $Q_C = [B \quad AB]$

and $Q_0 = [C^T \quad A^T C^T]$

Here, $AB = \begin{bmatrix} -1 & 0 \\ 0 & -2 \end{bmatrix} \begin{bmatrix} 1 \\ 2 \end{bmatrix} = \begin{bmatrix} -1 \\ -4 \end{bmatrix}$

and $A^T C^T = \begin{bmatrix} -1 & 0 \\ 0 & -2 \end{bmatrix} \begin{bmatrix} 1 \\ 1 \end{bmatrix} = \begin{bmatrix} -1 \\ -2 \end{bmatrix}$

So, $Q_C = [B \quad AB] = \begin{bmatrix} 1 & -1 \\ 2 & -4 \end{bmatrix}$

$\therefore$ $|Q_C| = \begin{vmatrix} 1 & -1 \\ 2 & -4 \end{vmatrix} = -4 + 2 = -2 \neq 0$

and $Q_0 = [C^T \quad A^T C^T] = \begin{bmatrix} 1 & -1 \\ 1 & -2 \end{bmatrix}$

$\therefore$ $|Q_0| = \begin{vmatrix} 1 & -1 \\ 1 & -2 \end{vmatrix} = -2 + 1 = -1 \neq 0$

Since, $|Q_C| \neq 0$ and $|Q_0| \neq 0$, therefore given system is both controllable and observable.

325. (b)

Given,

$$A = \begin{bmatrix} 0 & 1 \\ -6 & -5 \end{bmatrix}, x(0) = \begin{bmatrix} 1 \\ 0 \end{bmatrix}$$

Now, $(sI - A)^{-1} = \dfrac{1}{s^2 + 5s + 6} \begin{bmatrix} s+5 & 1 \\ -6 & s \end{bmatrix}$

$\therefore$ $\phi(t) = \mathcal{L}^{-1} (sI - A)^{-1}$

$$= \mathcal{L}^{-1} \begin{bmatrix} \dfrac{s+5}{s^2 + 5s + 6} & \dfrac{1}{s^2 + 5s + 6} \\ \dfrac{-6}{s^2 + 5s + 6} & \dfrac{s}{s^2 + 5s + 6} \end{bmatrix}$$

$$= \mathcal{L}^{-1} \begin{bmatrix} \dfrac{s+5}{(s+2)(s+3)} & \dfrac{1}{(s+2)(s+3)} \\ \dfrac{-6}{(s+2)(s+3)} & \dfrac{s}{(s+2)(s+3)} \end{bmatrix}$$

$$= \mathcal{L}^{-1}\begin{bmatrix} \left(\frac{3}{s+2}-\frac{2}{s+3}\right) & \left(\frac{1}{s+2}-\frac{1}{s+3}\right) \\ -6\left(\frac{1}{s+2}-\frac{1}{s+3}\right) & \left(\frac{3}{s+3}-\frac{1}{s+2}\right) \end{bmatrix}$$

$$\text{or, } \phi(t) = \begin{bmatrix} (3e^{-2t}-2e^{e^{-3t}}) & (e^{-2t}-e^{-3t}) \\ -6(e^{-2t}-e^{-3t}) & (3e^{-3t}-e^{-2t}) \end{bmatrix}$$

So, $x(t) = \phi(t)\cdot x(0)$

= state transition equation

$$= \begin{bmatrix} (3e^{-2t}-2e^{-3t}) & (e^{-2t}-e^{-3t}) \\ -6(e^{-2t}-e^{-3t}) & (3e^{-3t}-e^{-2t}) \end{bmatrix}\begin{bmatrix}1\\0\end{bmatrix}$$

$$= \begin{bmatrix} 3e^{-2t}-2e^{-3t} \\ -6(e^{-2t}-e^{-3t}) \end{bmatrix}$$

326. (a)

From given block diagram, the state equations can be written as:

$$\dot{x}_1 = -2x_1 - x_2 + 3u$$

$$\dot{x}_2 = -3x_1 - 2x_2 + 4u$$

Also, output equation is

$$y = 2x_1 + x_2$$

In matrix form, we have:

$$\begin{bmatrix}\dot{x}_1\\\dot{x}_2\end{bmatrix} = \begin{bmatrix}-2 & -1\\-3 & -2\end{bmatrix}\begin{bmatrix}x_1\\x_2\end{bmatrix} + \begin{bmatrix}3\\4\end{bmatrix}u$$

and $y = [2 \;\; 1]\, x(t)$

$$\text{Now, } (sI - A) = \begin{bmatrix}s+2 & 1\\3 & s+2\end{bmatrix}$$

$$\therefore \quad (sI-A)^{-1} = \frac{1}{s^2+4s+1}\begin{bmatrix}s+2 & -1\\-3 & s+2\end{bmatrix}$$

∴ Transfer function,

$$\frac{Y(s)}{U(s)} = C(sI-A)^{-1}$$

$$= [2 \;\; 1]\begin{bmatrix}s+2 & -1\\-3 & s+2\end{bmatrix}\begin{bmatrix}3\\4\end{bmatrix}\times\frac{1}{s^2+4s+1}$$

$$= \frac{10s+3}{s^2+4s+1}$$

327. (c)

Let us select the state variables as v_C and i_L. Applying KVL in the mesh-2, we have:

$$v_C = 2i_L + L\frac{di_L}{dt}$$

$$\text{or, } \frac{di_L}{dt} = -2i_L + v_C \qquad ...(i)$$

Also, by applying KCL at the given node, we get:

$$-i(t) + 2\frac{dv_C}{dt} + i_L = 0$$

$$\text{or, } \frac{dv_C}{dt} = \frac{1}{2}i(t) - \frac{1}{2}i_L \qquad ...(ii)$$

From equations (*i*) and (*ii*), state equations in matrix form can be written as:

$$\begin{bmatrix}\dot{v}_C\\\dot{i}_L\end{bmatrix} = \begin{bmatrix}0 & -0.5\\1 & -2\end{bmatrix}\begin{bmatrix}v_C\\i_L\end{bmatrix} + \begin{bmatrix}0.5\\0\end{bmatrix}i(t)$$

328. (a)

The state equations from the given signal flow graph can be written as:

$$\dot{x}_1 = x_2 \text{ and } \dot{x}_2 = -6x_1 - 5x_2 + u(t)$$

In matrix form,

$$\begin{bmatrix}\dot{x}_1\\\dot{x}_2\end{bmatrix} = \begin{bmatrix}0 & 1\\-6 & -5\end{bmatrix}\begin{bmatrix}x_1\\x_2\end{bmatrix} + \begin{bmatrix}0\\1\end{bmatrix}u(t)$$

Also, output is

$$y(t) = 6x_1 + x_2$$

In matrix form

$$y(t) = [6 \;\; 1]\begin{bmatrix}x_1\\x_2\end{bmatrix}$$

329. (d)

$$Q_C = [B \;\; AB \;\; A^2B]$$

$$\text{Here, } A = \begin{bmatrix}0 & 1 & 0\\0 & 0 & 1\\-6 & -11 & -6\end{bmatrix}, \; B = \begin{bmatrix}0\\0\\2\end{bmatrix}$$

$$AB = \begin{bmatrix}0 & 1 & 0\\0 & 0 & 1\\-6 & -11 & -6\end{bmatrix}\begin{bmatrix}0\\0\\2\end{bmatrix} = \begin{bmatrix}0\\2\\-12\end{bmatrix}$$

$$A^2B = \begin{bmatrix}0 & 1 & 0\\0 & 0 & 1\\-6 & -11 & -6\end{bmatrix}\begin{bmatrix}0\\2\\-12\end{bmatrix} = \begin{bmatrix}2\\-12\\50\end{bmatrix}$$

$$\therefore \quad Q_C = \begin{bmatrix}0 & 0 & 2\\0 & 2 & -12\\2 & -12 & 50\end{bmatrix}$$

So, $|Q_C| = 0 - 0 + 2(0-4)$

$= -8 \neq 0$

Hence, the system is controllable.

Also, $Q_0 = [C^T \;\; A^TC^T \;\; (A^T)^2\, C^T]$

Here, $C^T = \begin{bmatrix} 1 \\ 0 \\ 0 \end{bmatrix}, A^T = \begin{bmatrix} 0 & 0 & -6 \\ 1 & 0 & -11 \\ 0 & 1 & -6 \end{bmatrix}$

$$A^TC^T = \begin{bmatrix} 0 & 0 & -6 \\ 1 & 0 & -11 \\ 0 & 1 & -6 \end{bmatrix}\begin{bmatrix} 1 \\ 0 \\ 0 \end{bmatrix} = \begin{bmatrix} 0 \\ 1 \\ 0 \end{bmatrix} \text{ and}$$

$$(A^T)^2 C^T = \begin{bmatrix} 0 & 0 & -6 \\ 1 & 0 & -11 \\ 0 & 1 & -6 \end{bmatrix}\begin{bmatrix} 0 \\ 1 \\ 0 \end{bmatrix} = \begin{bmatrix} 0 \\ 0 \\ 1 \end{bmatrix}$$

$$\therefore \quad Q_0 = \begin{bmatrix} 1 & 0 & 0 \\ 0 & 1 & 0 \\ 0 & 0 & 1 \end{bmatrix}$$

So, $|Q_C| = \begin{vmatrix} 1 & 0 & 0 \\ 0 & 1 & 0 \\ 0 & 0 & 1 \end{vmatrix} = 1(1-0) - 0 + 0$

$= 1 \neq 0$

Since $|Q_0| \neq 0$, therefore given system is observable.

330. (c)

T.F., $\dfrac{Y(s)}{U(s)} = C(sI - A)^{-1} B + D$

$$= \frac{C\, Adj(sI - A)\, B + |sI - A| D}{|sI - A|}$$

Hence, zeros are at:

$C.Adj\,(sI-A)B + |sI - A|D = 0$

Now, $(sI - A) = \begin{bmatrix} s & -1 \\ 20 & s+9 \end{bmatrix}$

Co-factor of $(sI - A) = \begin{bmatrix} s+9 & -20 \\ 1 & s \end{bmatrix}$

$\therefore \; Adj\,(sI - A) = \begin{bmatrix} s+9 & 1 \\ -20 & s \end{bmatrix}$

Also, $|sI - A| = s^2 + 9s + 20$

So, zeros are at:

$$[-17 \;\; -5]\begin{bmatrix} s+9 & 1 \\ -20 & s \end{bmatrix}\begin{bmatrix} 0 \\ 1 \end{bmatrix} - (s^2 + 9s + 20)\cdot 1 = 0$$

or, $-17 - 5s + s^2 + 9s + 20 = 0$

or, $s^2 + 4s + 3 = 0$ or, $(s + 1)(s + 3) = 0$

or, $s = -1, -3$

■■■■

UNIT 6

Electrical Machines

1. Magnetic Circuits

Q.1 What is the inductance of a coil in which a current of 0.1A increasing at the rate of 0.5 A/s represents a power flow of 0.5 watt?

(a) 0.1 H (b) $\frac{1}{2}$H
(c) 10^2 H (d) 10 H

Q.2 A coil of 100 turns is wound on a torroidal magnetic core having a reluctance of 10^4 AT/Wb. When the coil current is 5 A and is increasing at the rate of 200 A/s, the voltage across the coil would be (assume coil resistance to be 2 Ω)

(a) 210 volt (b) 230 volt
(c) 115 volt (d) Zero volt

Q.3 The field winding of a dc electro-magnet is wound with 960 turns and has a resistance of 50 Ω. The excitation voltage is 230 V and the magnetic flux linking the coil is 5 mWb. The energy stored in the magnetic field is

(a) 22 Joules (b) 11 Joules
(c) 19 Joules (d) 29 Joules

Q.4 The flux in a magnetic core is alternating sinusoidally at a frequency of 600 Hz. The maximum flux density is 2 Tesla and the eddy current loss is 15 Watts. What would be the eddy current loss in the core if the frequency is raised to 800 Hz and the maximum flux density is reduced to 1.5 Tesla?

(a) 12 Watts (b) 25.25 Watts
(c) 15 Watts (d) 18 Watts

Q.5 Two inductances of 15 mH and 25 mH are connected in series such that their fluxes oppose each other. They are so placed that the coefficient of coupling is 0.8. The value of mutual inductance between them will be

(a) 15.5 mH (b) 18 mH
(c) 12.6 mH (d) None of these

Q.6 The property of a material which opposes the production of magnetic flux in it is known as:

(a) mmf (b) permeance
(c) permittivity (d) reluctance

Q.7 Air exhibits

(a) ferromagnetism (b) paramagnetism
(c) diamagnetism (d) antiferromagnetism

Q.8 The laws of electromagnetic induction are summarized in the following equation:

(a) $e = L\frac{di}{dt}$ (b) $e = iR$

(c) $e = -\frac{d\psi}{dt}$ (d) None of these

Q.9 EMF induced in a coil rotating in a uniform magnetic field will be maximum when the

(a) rate of cutting flux by the coil sides is minimum.
(b) flux linking with the coil is maximum.
(c) rate of change of flux linkage is minimum.
(d) none of the above.

Q.10 The emf induced in a conductor rotating in a bipolar field is

(a) both dc and ac (b) dc only
(c) ac only (d) none of these

Q.11 **Assertion (A) :** In an electric circuit, the current is due to the presence of electromotive force.
Reason (R) : In a magnetic circuit, the magnetic flux is due to the presence of a magnetomotive force.

(a) Both A and R are true and R is a correct explanation of A.
(b) Both A and R are true but R is not a correct explanation of A.
(c) A is true but R is false.
(d) A is false but R is true.

Q.12 **Assertion (A):** Leakage flux has a path dominated through the surrounding air.

Reason (R): Leakage flux may be defined as that flux which does not follow the intended path in a magnetic circuit.

(a) Both A and R are true and R is the correct explanation of A
(b) Both A and R are true but R is not the correct explanation of A
(c) A is true but R is false
(d) A is false but R is true

Q.13 A conductor 20 cm long moves at right angle to its length at a constant speed of 30 m/s in a uniform magnetic field of flux density 1.2 T. The emf induced in case the conductor motion is normal to the field flux is

(a) 0 volt (b) 28.8 volt
(c) 7.2 volt (d) 14.4 volt

Q.14 A magnetic circuit with relative permeability of 100 has a core cross-sectional area of 5 cm^2 and mean core length of 25 cm. The coil has 120 turns with an mmf of 1000 AT. The magnetic core flux is

(a) 0.75 mWb (b) 1 mWb
(c) 0.05 mWb (d) 0.25 mWb

Q.15 Match **List-I** (Magnetic quantities) with **List-II** (Units) and select the correct answer using the codes given below the lists:

List-I	List-II
A. Permeability	1. Wb
B. Magnetic field intensity	2. Wb/m^2
C. Magnetic flux	3. H/m
D. Magnetic flux density	4. Amp.-turns/m

Codes:

	A	B	C	D
(a)	4	3	1	2
(b)	3	4	1	2
(c)	4	2	3	1
(d)	3	1	2	4

Q.16 The mutual inductance between two closely coupled coils is 1 H. Now the turns of one coil is decreased to half and those of the other is doubled. The new value of mutual inductance would be

(a) $\frac{1}{4}$ H (b) 2 H
(c) $\frac{1}{2}$ H (d) 1 H

Q.17 **Assertion (A):** A magnetic circuit is not operated deep into saturation zone.

Reason (R): Large exciting current can put extra overhead on the source supplying power to the coil.

(a) Both A and R are true and R is a correct explanation of A.
(b) Both A and R are true but R is not a correct explanation of A.
(c) A is true but R is false.
(d) A is false but R is true.

Q.18 "In all cases of electromagnetic induction, an induced voltage will cause a current to flow in a closed circuit in such a direction that the magnetic field which is caused by that current will oppose the change that produces the current", is the original statement of

(a) Lenz's Law
(b) Fleming's law of induction
(c) Ampere's circuital law
(d) Faraday's law of magnetic induction

Q.19 Iron is removed from the iron cored coil so that the coil becomes air-cored coil. Inductance of this air-cored coil will

(a) increase
(b) decrease
(c) remain the same
(d) increase or decrease depending upon the coil configuration

Q.20 A coil with 60 turns wound over a ferromagnetic core having relative permeability of 400 has an inductance of 50 mH. If coil turns are doubled and the core is replaced by a new ferromagnetic core having relative permeability 600, the new inductance would be

(a) 0.15 H (b) 0.3 H
(c) 1.5 H (d) 3 H

Q.21 An iron-cored choke, with 2 mm air-gap length, takes 2 A when fed from a constant-voltage source of 230 V. If its air-gap length is increased to 10 mm, then the magnetic flux produced by the choke would

(a) remains constant and the current would increase.
(b) decrease and the current would increase.

(c) decrease and the current would also decrease.
(d) remain constant and the current would decrease.

Q.22 A coil of 1000 turns is wound on a core. A current of 1 A flowing through the coil creates a core flux of 1 mWb. What is the energy stored in the magnetic field?

(a) 1 J
(b) $\frac{1}{4}$ J
(c) 2 J
(d) $\frac{1}{2}$ J

Q.23 **Assertion (A) :** To determine the relative polarity of the induced voltage in the coupled coil, the coils are marked with dots.

Reason (R) : When the currents through each of the mutually coupled coils are going away from the dot or towards the dot, the mutual inductance is negative while for the case when the current through the coil is leaving the dot for one coil and entering the other, the mutual inductance is positive

(a) Both A and R are true and R is a correct explanation of A.
(b) Both A and R are true but R is not a correct explanation of A.
(c) A is true but R is false.
(d) A is false but R is true.

Q.24 A circular metallic disc is placed in a vertical magnetic field of constant induction in the downward direction. If the disc is rotated in a horizontal plane, the emf induced will be

(a) zero
(b) increasing radially in the outward direction
(c) constant and independent of disc size
(d) decreasing radially in the outward direction

Q.25 Match **List-I (Electric Circuit)** with **List-II (Magnetic Circuit)** and select the correct answer using the codes given below the lists:

List-I
A. Current
B. Conductivity
C. Electric field intensity
D. Current density

List-II
1. Magnetic flux density
2. Magnetic field intensity
3. Magnetic flux
4. Permeability

Codes:

	A	B	C	D
(a)	2	4	3	1
(b)	3	4	2	1
(c)	2	3	4	1
(d)	3	1	2	4

Q.26 A coil of 300 turns, wound on a core of non-magnetic material, has an inductance of 10 mH. The average value of emf induced in the coil when a current of 5 A is reversed in 8 milliseconds is

(a) 6.25 V
(b) 12.5 V
(c) 0 V
(d) 2.2 V

Q.27 One 800-turn flat coil with an area of 5×10^{-2} m^2 is rotating in a magnetic field of flux density 60 mWb/m^2 at 1500 rpm. The value of induced emf in the coil if the plane of the coil is parallel to the field is

(a) 375 V
(b) 288 V
(c) 392 V
(d) 377 V

2. Transformer

Q.28 Which of the following is not true relating to an ideal transformer?

(a) The iron loss in an ideal transformer is zero
(b) The winding resistance has a zero value
(c) The leakage reactance has a non-zero value
(d) The magnetizing current is zero

Q.29 The emf per turn of a single phase 10 kVA, 2200/220 V, 50 Hz transformer is 10 V. The net cross-sectional area of core for a maximum flux density of 1.5 T is

(a) 300 cm^2
(b) 200 cm^2
(c) 100 cm^2
(d) 25 cm^2

Q.30 A 1 kVA, 400 Hz transformer is desired to be used at a frequency of 60 Hz. The kVA rating of the transformer at this reduced frequency is

(a) 82.5 VA
(b) 550 VA
(c) 150 VA
(d) 500 VA

Q.31 A 2200 / 250 V transformer takes 0.5 A at a p.f. of 0.3 on open-circuit. The magnetising component of no-load primary current is

(a) 0.150 A (b) 0.477 A
(c) 0.795 A (d) 0.352 A

Q.32 A 50 MVA, 76.2/33 kV, 1-ϕ, 50 Hz, two-winding transformer with tap changer has percentage impedance of (0.5 + j7.0). What tapping must be used to maintain rated voltage at the secondary on full load at 0.8 lagging power factor?
(a) 3.8 % down (b) 4.6 % up
(c) 4.6 % up (d) 4.6 % down

Q.33 The ohmic drop of a transformer is 1% while the reactance drop is 5% of the voltage. The regulation of the transformer at 0.8 power factor leading and 0.8 power factor lagging are respectively
(a) 2.2 % and –3.8% (b) –3.4% and 4.6 %
(c) –2.2 % and 3.8% (d) 3.4 % and –4.6 %

Q.34 A 1 kV, 50 Hz supply to a transformer results in 650 W hysteresis loss and 400 W eddy current loss. If both the applied voltage and frequency are doubled, the new core loss would be
(a) 2900 W (b) 2628 W
(c) 3120 (d) 2600 W

Q.35 A transformer designed for operation of 60 Hz supply is working on 50 Hz supply system without changing its voltage and current ratings. When compared with full-load efficiency at 60 Hz, the transformer efficiency on full load at 50 Hz will
(a) increase by a factor of 1.2
(b) increase marginally
(c) remain unaltered
(d) decrease marginally

Q.36 For a core-type power transformers, both primary and secondary windings have circular coil sections, because this section
(a) has the strongest mechanical shape.
(b) results in less core material and, therefore, less core loss.
(c) is easier to wind.
(d) require minimum conductor material and, therefore, less I^2R loss.

Q.37 A 10 kVA, 400 V/ 200 V, 1-phase transformer with a percentage resistance of 3% and percentage reactance of 6% is supplying a current of 50 A to a resistive load. The value of the load voltage is
(a) 390 V (b) 194 V
(c) 196 V (d) 192 V

Q.38 The full-load voltage rogulation of a power transformer is zero when power factor of the load is near
(a) unity and leading (b) unity and lagging
(c) zero and leading (d) zero and lagging

Q.39 Match **List-I (Type of transformers)** with **List-II (Applications)** and select the correct answer using the codes given below the lists:

List-I
A. Distribution transformer
B. Audio frequency transformer
C. Power transformer
D. Pulse transformer

List-II
1. At generating stations
2. Thyristor firing circuits
3. Impedance matching
4. At different city locations

Codes:

	A	B	C	D
(a)	4	3	1	2
(b)	3	4	2	1
(c)	4	2	1	3
(d)	3	1	2	4

Q.40 A 3-phase delta-star transformer has secondary to primary turns ratio per phase of 5. For a primary line current of 10 A, the secondary line current would be
(a) 3.464 A (b) 50 A
(c) 1.155 A (d) 1.633 A

Q.41 A 40 kVA transformer has a core loss of 400 W and a full-load copper loss of 800 W. The proportion of full-load at maximum efficiency is
(a) 62.3% (b) 50%
(c) 100% (d) 70.7%

Q.42 The necessary conditions for parallel operation of two 1-phase transformers is that these should have the same
1. kVA rating 2. voltage regulation
3. polarity 4. voltage ratio
5. efficiencies 6. X/R ratio

From these, the correct answer is
(a) 1, 3, 5, 6 (b) 3, 4, 5, 6
(c) 1, 2, 3, 4 (d) 2, 3, 4, 6

Q.43 A 400 V/ 100 V, 10 kVA, two-winding transformer is reconnected as an autotransformer across a suitable voltage. The maximum rating of such a transformer could be
(a) 12.5 kVA (b) 50 kVA
(c) 8.75 kVA (d) 15 kVA

Q.44 A 3-phase star-delta transformer has secondary to primary turns ratio per phase of 5. For a primary voltage of 400 V, the secondary voltage would be
(a) 2000 V (b) 3464 V
(c) 1154.7 V (d) 46.2 V

Q.45 A 5 kVA, 250/125 V, 1-phase transformer has leakage impedance of (0.02 + *j*0.08) pu. Its value in ohm referred to H.V. side is
(a) 0.25 + *j*1 (b) 0.02 + *j*0.08
(c) 0.0625 + *j*0.25 (d) 0.2 + *j*0.8

Q.46 The voltage regulation of a transformer at full load and 0.8 pf lagging is 2.5%. The voltage regulation at full load 0.8 pf leading will be
(a) zero (b) –2.5%
(c) 2.5% (d) –0.9%

Q.47 In a transformer, the winding is tapped in the middle
(a) to eliminate the axial forces on the windings.
(b) to provide a mechanical balance to the windings.
(c) to reduce the insulation level of the windings.
(d) to avoid the radial forces on the windings.

Q.48 When compared to power transformer, a distribution transformer has
(a) high %age impedance and low I^2R loss to core-loss ratio.
(b) low %age impedance and high I^2R loss to core-loss ratio.
(c) high %age impedance and high I^2R loss to core-loss ratio.
(d) low %age impedance and low I^2R loss to core-loss ratio.

Q.49 Voltage applied to the primary of a transformer is kept constant but its frequency is decreased. Under this operation
(a) magnetizing current and core-loss current both decreases.
(b) magnetizing current increases but core-loss current decreases.
(c) magnetizing current and core-loss current both increases.
(d) magnetizing current decreases but core-loss current increases.

Q.50 A 3 : 1 transformer has impedance of (1 + *j*5) Ω on the L.V. side and (9 + *j*45) Ω on the H.V. side. The total equivalent impedance at the H.V. terminal is
(a) 2 + *j*10 Ω (b) 8 + *j*40 Ω
(c) 10 + *j*50 Ω (d) 18 + *j*90 Ω

Q.51 For a single-phase transformer, r_e = total equivalent resistance, x_e = total equivalent leakage reactance, P_c = core loss. The load current at which maximum efficiency occurs is
(a) $\frac{P_c}{x_e}$ (b) $\sqrt{\frac{P_c}{x_e}}$
(c) $\frac{P_c}{r_e}$ (d) $\sqrt{\frac{P_c}{r_e}}$

Q.52 Short-circuit test on a single-phase transformer gave the following data:
30 V at 50 Hz, 20 A, p.f. = 0.2 lag.
If s.c. test is performed on 30 V, 25 Hz, then short-circuit current
(a) decreases at a p.f. > 0.2
(b) increases at a p.f. < 0.2
(c) decreases at a p.f. < 0.2
(d) increases at a p.f. > 0.2

Q.53 A 20 kVA, 2300/230 V, two winding transformer is to be used as an autotransformer to give 2300 V/2530 V. Its rating will be
(a) 200 kVA with conducted kVA = 20
(b) 220 kVA with conducted kVA = 200
(c) 200 kVA with conducted kVA = 180
(d) 220 kVA with conducted kVA = 20

Q.54 A 400/200 V, 50 Hz transformer operates at a flux density of 1.2 T when energised from its H.V. side. For this transformer, linear dimensions of core are doubled while the number of turns are halved on both its H.V. and L.V. side, then its flux density would be

(a) 1.2 T (b) 3.6 T
(c) 0.6 T (d) 4.8 T

Q.55 Match **List-I** with **List-II** and select the correct answer using codes given below the lists:

List-I
A. Regulation of ideal transformer
B. Eddy current loss
C. Magnetic flux
D. Efficiency of ideal transformer

List-II
1. 100%
2. current
3. 0%
4. Laminations

Codes:

	A	B	C	D
(a)	3	2	4	1
(b)	1	2	4	3
(c)	3	4	2	1
(d)	1	4	2	3

Q.56 **Assertion (A):** Now a days cold rolled steel is used almost exclusively inspite of its high cost.
Reason (R) : Cold rolled has high permeability and lower losses at a given flux density.
(a) Both A and R are true and R is a correct explanation of A.
(b) Both A and R are true but R is not a correct explanation of A.
(c) A is true but R is false.
(d) A is false but R is true

Q.57 **Assertion (A):** A transformer on short-circuit takes some energy from the mains even if the output power is zero.
Reason (R): At short-circuit, the energy input is required to meet no-load copper losses in primary and iron losses in the core.
(a) Both A and R are true and R is a correct explanation of A.
(b) Both A and R are true but R is not a correct explanation of A.
(c) A is true but R is false.
(d) A is false but R is true.

Q.58 Which of the following 3-phase connections of a transformer causes interference to the nearby communication system?
(a) Star-delta (b) Star-star
(c) Delta-star (d) Delta-delta

Q.59 When two transformers of different kVA ratings are connected in parallel, they will divide the total load in proportion to their respective kVA ratings when their
(a) equivalent impedances are equal
(b) efficiencies are equal
(c) equivalent impedances are inversely proportional to their respective ratings
(d) none of the above

Q.60 If the transformers operated in parallel are not connected with regard to polarity
(a) the transformers will not share the load in proportion to their kVA ratings.
(b) dead short-circuit will take place.
(c) a circulating current will exist on no-load.
(d) the transformer of lower rating will be out of operation.

Q.61 Maximum transient currents flow through a transformer winding when it is switched on with secondary ______ circuited and when the input voltage wave is passing through the ____ value.
(a) short, zero (b) open, maximum
(c) open, zero (d) short, maximum

Q.62 A 400 V/200 V/200 V, 50 Hz three-winding transformer is connected as shown in the figure below.

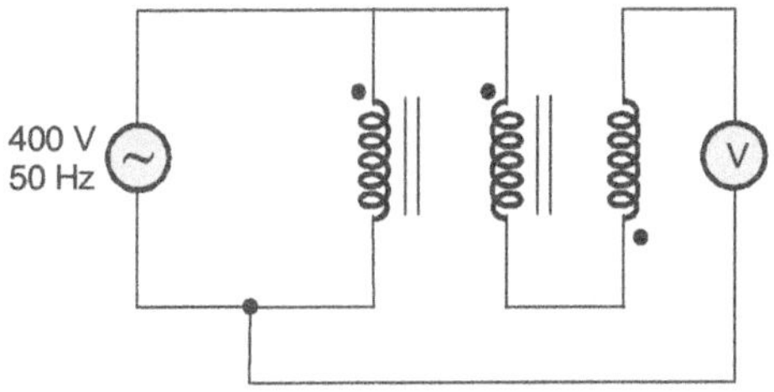

The reading of the voltmeter V will be
(a) –400 V (b) 800 V
(c) –600 V (d) 0 V

Q.63 Scott-connected transformers can convert:
(*i*) 3 to 2 phases (*ii*) 2 to 3 phases
(*iii*) 2 to 4 phases (*iv*) 3 to 4 phases
From the above correct answer is
(a) (*ii*) only (b) (*i*) and (*ii*)
(c) (*iii*) and (*iv*) only (d) (*i*) and (*iv*) only

Q.64 The part of transformer which is subjected to maximum heating is
(a) winding (b) oil
(c) frame (d) core

Q.65 Malfunctioning of the Buchholz relay may be caused due to
(a) heavy external short circuit
(b) excessive overheating
(c) dropping of oil level below the relay level during operation
(d) improper breather action

Q.66 The leakage flux of primary and secondary windings can be reduced to the minimum by
(a) employing low permeability magnetic material core.
(b) winding primary and secondary coils one upon the other coaxially.
(c) increasing the number of turns.
(d) winding primary and secondary coils on separate limbs.

Q.67 Two transformers, each having iron loss of P_i watts and full-load copper loss of P_C, are put to back to back test and full-load current is allowed to flow through the secondaries. The total input will be
(a) $P_i + P_c$ (b) $2P_i$
(c) $2(P_i + P_c)$ (d) $2P_C$

Q.68 Consider the following statements related to a transformer supplying an inductive load:
1. The secondary terminal voltage is less than the secondary induced emf.
2. The power factor on primary side will be lower then that of load.
3. The secondary current results in equivalent primary current in phase opposition.

Out of the above statements:
(a) 1 and 2 are correct
(b) 2 and 3 are correct
(c) only 3 is correct
(d) 1, 2 and 3 are correct

Q.69 A single-phase, 3000/1000 V, 10 kVA transformer has its leakage impedance of 2%. The voltage required to be applied to the h.v. side for performing short-circuit test at rated current is
(a) 120 V (b) 60 V
(c) 140 V (d) 20 V

Q.70 A 2/1 ratio two-winding transformer is connected as an auto-transform to a given voltage ratio 2/3. The kVA rating as a two-winding transformer compared to an auto-transformer will be
(a) 0.67 (b) 1.5
(c) 0.5 (d) 0.33

Q.71 A 1 MVA, 50/40 kV, single-phase auto-transformer is fully loaded. The current in the common section of the winding will be
(a) 5 A (b) 45 A
(c) –15 A (d) –5 A

Q.72 Following statements are made regarding the short-circuit test on a 1-phase transformer:
1. It is performed at rated current.
2. It helps in the calculation of voltage regulation.
3. It gives ohmic loss on the side instruments are connected.
4. It is performed at rated voltage.
5. L.V. side is short circuited.
6. It gives the core loss.
7. H.V. side is short circuited

From the above statements, the correct one are
(a) 2, 4, 5, 6 (b) 2, 3, 4, 7
(c) 1, 2, 5 (d) 1, 2, 6, 7

Q.73 A 50 Hz transformer having equal hysteresis and eddy current losses at rated excitation is operated at 45 Hz at 90% of rated voltage. Compared to the rated operating point, the core loss under this condition
(a) reduces by 10% (b) reduces by 19%
(c) reduces by 14.5% (d) remains unchanged

Q.74 In a transformer, $r_{e2} + jx_{e2}$ = equivalent leakage impedance (z_{e2}). Its maximum voltage regulation is
(a) r_{e2} at $\frac{r_{e2}}{z_{e2}}$ lagging p.f.
(b) x_{e2} at $\frac{x_{e2}}{z_{e2}}$ lagging p.f.
(c) z_{e2} at $\frac{r_{e2}}{z_{e2}}$ leading p.f.
(d) z_{e2} at $\frac{r_{e2}}{z_{e2}}$ lagging p.f.

Q.75 Transformer maximum efficiency, for a constant load current occurs at
(a) unity p.f. (b) zero p.f. leading
(c) zero p.f. lagging (d) any p.f.

Q.76 A bank of three-identical single-phase 250 kVA, 11 kV/230 V transformer is used to provide 400 V low-tension supply from a 11 kV, 3-phase sub-station. The effective kVA rating of the bank will be

(a) 250 (b) $250\sqrt{3}$
(c) 500 (d) 750

Q.77 The full-load copper-loss and iron-loss of a transformer are 6400 W and 5000 W respectively. The copper-loss and iron-loss at half load will be respectively
(a) 3200 W and 2500 W
(b) 3200W and 5200 W
(c) 1600 W and 1250 W
(d) 1600 W and 5000 W

Q.78 The use of higher flux density in the transformer design
(a) reduces the weight per kVA
(b) increases the weight per kVA
(c) has no relation with the weight of transformer
(d) increases the weight per kW

Q.79 Consider the following statements relating to the constructional features of a large power transformer:
1. The conservator is used to protect level of oil in the transformer tank.
2. The bushing is used to protect transformer insulation against lightning overvoltages.
3. The buchholz relay is an over current relay.
4. Silica gel is used to absorb moisture.

Which of these statements are correct?
(a) 1, 2, 3 and 4 (b) 2, and 3
(c) 1 and 4 (d) 1, 2, and 4

Q.80 Consider the following statements related to an ideal transformer:
1. Voltages are transferred in the reverse ratio.
2. Impedances are transferred in the reverse ratio squared.
3. Power and volt-amperes remain unchanged on either sides.
4. Currents are transferred in the direct ratio.

Which of these statements are correct?
(a) 1, 2, 3 and 4 (b) 1 and 4
(c) 3 only (d) 2 and 3

Q.81 **Assertion (A):** In open circuit test of a transformer, the wattmeter reading can be taken as equal to transformer core loss.
Reason (R) : The ohmic loss during open circuit is only about 9 to 16 percent of the full load primary loss.
(a) Both A and R are true and R is a correct explanation of A.
(b) Both A and R are true but R is not a correct explanation of A.
(c) A is true but R is false.
(d) A is false but R is true.

Q.82 **Assertion (A) :** The distribution transformers should be designed to have a low value of leakage impedance.
Reason (R) : For a transformer of large voltage regulation, the voltage at the consumer's terminals will fall appreciably with the increase in load-this has a detrimental effect on the operation of the fluorescent tube, T.V. sets, refrigerator motors etc.
(a) Both A and R are true and R is a correct explanation of A.
(b) Both A and R are true but R is not a correct explanation of A.
(c) A is true but R is false.
(d) A is false but R is true.

Q.83 In a 400V, 50 Hz transformer, the total iron loss is 2.5 kW. When the supply voltage and the frequency is reduced to 200 V and 25 Hz respectively the corresponding loss is 850 watts. The eddy current and hysteresis loss at normal voltage and frequency in kW will be respectively
(a) 0.9, 1.6 (b) 1.2, 1.3
(c) 1.6, 0.9 (d) 1.3, 1.2

Q.84 If P_1 and P_2 be the iron and copper losses of a transformer on full-load, what is the ratio of P_1 and P_2 such that maximum efficiency occurs at 75% of full load?
(a) 0.2385 (b) 0.5252
(c) 0.7575 (d) 0.5625

Q.85 A single phase, 1100/220 V transformer under test gave the following test results :
OC test : 1100 V, 0.5 A, 55 W
SC test : 10 V, 80 A, 400 W on LV side, HV short-circuited.

The efficiency of the transformer when secondary supplies 100 A at 0.8 pf lag is
(a) 96.28% (b) 97.50%
(c) 92.63% (d) 95.44%

Q.86 A two winding transformer is
(a) electrically linked (b) conductively linked
(c) inductively linked (d) not linked at all

Q.87 The primary winding of a 220/6 V, 50 Hz transformer is energised from 110 V, 60 Hz supply. The secondary output voltage will be
(a) 6.0 V (b) 2.5 V
(c) 3.6 V (d) 3.0 V

Q.88 A two-winding single phase transformer has a voltage regulations of 4.5 % at full-load and unity power factor. At full-load and 0.80 power factor lagging load the voltage regulation will be
(a) 4.5 %
(b) more than 4.5 %
(c) less than 4.5 %
(d) 4.5 % or more than 4.5 %

Q.89 A 1 : 5 step-up transformer has 120 V across the primary and 600 ohms resistance across the secondary. Assuming 100% efficiency, the primary current equals
(a) 0.2 Amp. (b) 20 Amp.
(c) 10 Amp. (d) 5 Amp.

3. Direct Current Machine

Q.90 Consider the following statements about the field windings used for various types of dc machines:
1. For dc shunt machines, large number of turns of small cross-section are used.
2. For dc series machine, small number of turns of large cross-section are used.
3. For dc compound machine, both shunt (thick wire) and series (thin wire) field windings are used.

Out of the above statements the correct statements is/are:
(a) 3 only (b) 1, 2 and 3
(c) 1 and 2 only (d) None of these

Q.91 A 6-pole dc machine has 300 conductors and flux per pole is 15 mWb. If the machine is driven at 1800 rpm, then the average emf induced per conductor will be
(a) 3.6 V (b) 2.7 V
(c) 9 V (d) 162 V

Q.92 A commutator with a diameter of 50 cm rotates at 1000 rpm. For a brush width of 1.5 cm, the time of commutation is
(a) 573 μsec (b) 256 msec
(c) 625 μsec (d) 448 msec

Q.93 A 24 kW, 250 V, 1600 rpm separately-excited d.c. generator has armature circuit resistance of 0.1 Ω. The machine is first run at rated speed and the field winding current is adjusted to give an open circuit voltage of 260 V. Now, when the generator is loaded to deliver its rated current, the speed of the driving motor is found to be 1500 r.p.m. Assuming flux to remain constant, the terminal voltage of the generator under these conditions is
(a) 250 volt (b) 226 volt
(c) 238 volt (d) 234 volt

Q.94 **Assertion (A):** DC series motors are suitable for buses, trolleys, hoists and trains.
Reason (R): DC series motors are variable speed motors.
(a) Both A and R are true and R is a correct explanation of A.
(b) Both A and R are true but R is not a correct explanation of A.
(c) A is true but R is false.
(d) A is false but R is true.

Q.95 **Assertion (A):** Armature resistance control method of speed control of dc motors is wasteful and inefficient.
Reason (R): Large amount of power is wasted in the control resistances, specially in case of shunt motors.
(a) Both A and R are true and R is a correct explanation of A.
(b) Both A and R are true but R is not a correct explanation of A.
(c) A is true but R is false.
(d) A is false but R is true.

Q.96 Match **List-I (Tests on DC machines)** with **List-II (Results)** and select the correct answer using the codes given below the lists:

List-I
A. Field test
B. Swinburne's test
C. Hopkinson test
D. Retardation test
List-II
1. Stray losses in case of dc series machines
2. Constant losses
3. Performance of machines regarding commutation and temperature rise
4. Stray losses in case of dc shunt machines

Codes:

	A	B	C	D
(a)	1	3	2	4
(b)	4	1	2	3
(c)	1	2	3	4
(d)	4	3	2	1

Q.97 At a certain speed and flux, the voltage generated by a dc generator is 230 volts. If the speed is increased by 20% and the flux is simultaneously reduced by 10%, the voltage generated will be
(a) decreased by 8%
(b) increased by 10%
(c) decreased by 20%
(d) increased by 8%

Q.98 LAP winding is employed in a dc machine of
(a) high current and low voltage rating
(b) low current and high voltage rating
(c) high current and high voltage rating
(d) low current and low voltage rating

Q.99 Consider the following parts of a dc machine:
1. Pole core
2. Brushes
3. Yoke
4. Armature core
Which of the above parts are not subjected to iron loss?
(a) 2 only
(b) 1 and 3 only
(c) 1, 2 and 3 only
(d) 2 and 3 only

Q.100 Pole shoe of a dc machine is laminated for the purpose of
(a) decreasing hysteresis loss
(b) decreasing eddy current loss
(c) manufacturing ease
(d) decreasing both eddy current and hysteresis loss

Q.101 In a dc motor, the windage loss is proportional to
(a) square of flux density
(b) supply voltage
(c) square of supply voltage
(d) square of armature speed

Q.102 The principle of dynamically induced emf is utilized in
(a) transformer
(b) generator
(c) thermocouple
(d) choke

Q.103 The direction of induced emf in case of a dc machines can be determined using
(a) Fleming's left hand rule
(b) Faraday's law of magnetic induction
(c) Fleming's right hand rule
(d) Biot-Savart's law

Q.104 In a dc machine
(a) the current and emf in armature conductors are unidirectional while those at the terminals are alternating.
(b) the current and emf in armature conductors and at the terminals are unidirectional.
(c) the emf in armature conductors and at the terminals is alternating while current is unidirectional.
(d) the current and emf in armature conductors are alternating while those at the terminals are unidirectional.

Q.105 In a dc machine, the brushes are physically placed in the
(a) interpolar axis and electrically connected to the coils lying in polar axis.
(b) polar axis and electrically connected to the coils lying in the polar axis.
(c) interpolar axis and electrically connected to the coils lying in interpolar axis.
(d) polar axis and electrically connected to the coils lying in the interpolar axis.

Q.106 A lap wound dc machine has 400 conductors and 8 poles. The voltage induced per conductor is 2 V. The machine generates a voltage of
(a) 800 V
(b) 400 V
(c) 200 V
(d) 100 V

Q.107 In dc machines, the air gap flux distribution in space at no load is
(a) Rectangular
(b) Triangular
(c) Flat topped
(d) Pulsating

Q.108 In a dc machine without any brush shift, the shift of the magnetic neutral axis showing to armature reaction is
(a) against the direction of rotation for both the generator and the motor.
(b) in the direction of rotation for the generator and against the direction of rotation for the motor.
(c) in the direction of rotation for both the generator and the motor.
(d) against the direction of rotation for the generator and in the direction of motor.

Q.109 In a dc machine, the number of mechanical and electrical degrees will be same when the number of poles on the machine is
(a) 8 (b) 4
(c) 2 (d) 1

Q.110 Which of the following does not constitute the constant losses occurring in a dc machine?
(a) Mechanical loss
(b) Brush contact power loss
(c) Series field loss
(d) Eddy current loss

Q.111 A series motor is drawing a load current of 1 A from the supply lines. If now the load is reduced such that the current drawn is halved, the speed of the machine (neglecting the saturation and armature resistance) would be
(a) unchanged
(b) reduced by 100%
(c) increased by 100%
(d) reduced by 50%

Q.112 Consider the following statements regarding the magnetization characteristics of a dc generator:
1. It represents the variation of field flux with field current.
2. It is non-linear from zero to rated field current.
3. it always starts from the origin.
4. Variation of open-circuit terminal voltage with field current is independent of speed.
5. It helps to determine the critical field resistance at a given speed.

Which of the above statements are correct?
(a) 1, 2 and 5 only (b) 1 and 5 only
(c) 2, 3 and 4 only (d) 3 and 5 only

Q.113 In swinburne's method of testing dc machine, the shunt machine is run as a
(a) generator at full load at rated speed and rated voltage.
(b) generator at no-load at near rated speed and rated voltage.
(c) motor at no-load at rated speed and rated voltage.
(d) generator at no-load at rated speed and rated terminal voltage.

Q.114 A 200 V dc shunt motor delivers an output of 17 kW with an input of 20 kW. The field winding resistance is 50 Ω and the armature-circuit resistance is 0.04 Ω. Maximum efficiency will be obtained when total armature ohmic losses are equal to
(a) 5.23 kW (b) 3.68 kW
(c) 3.20 kW (d) 2.63 kW

Q.115 Match **List-I (Method of speed control of DC motor)** with **List-II (Applications/ Characteristics)** and select the correct answer using codes given below the lists:

List-I
A. Rheostatic control
B. Use of diverter resistance
C. Field control method
D. Armature voltage control

List-II
1. Speed control of series motor
2. Poor motor efficiency
3. Speed below base speed
4. Speed above base speed

Codes:

	A	B	C	D
(a)	2	1	4	3
(b)	1	2	3	4
(c)	1	2	4	3
(d)	2	4	3	1

Q.116 Take the normal values as unity, the approximate values of current and speed of a dc shunt motor under the conditions of **List-I** are given in **List-II**. Match the two lists and select the correct answer using the codes given below the lists:

List-I
A. Terminal voltage halved, field current and power output constant.

B. Terminal voltage halved, field current and load torque constant.
C. Field flux halved, armature voltage halved and power output constant.
D. Field flux doubled, armature voltage and load torque constant.

List-II

	Armature current	Speed
1.	1.0	0.5
2.	0.5	0.5
3.	2.0	1.0
4.	2.0	0.5

Codes:

	A	B	C	D
(a)	1	2	3	4
(b)	4	1	3	2
(c)	1	3	4	2
(d)	4	3	2	1

Q.117 A dc shunt motor runs at a no-load speed of 1140 rpm. At full-load, armature reaction weakens the main flux by 5% whereas the armature circuit voltage drops by 10%. The motor full load speed in rpm is
(a) 1200 (b) 1000
(c) 1203 (d) 1080

Q.118 A 440 V dc shunt motor has no-load ($I_a = 0$) speed of 2000 rpm. It is running at 1000 rpm at full load torque, reduced armature voltage and fullfield. If load torque is reduced to 50% of rated value with armature voltage and field voltage held constant at previous values, the speed increases to 1050 rpm. The armature voltage drop at full load (neglecting armature reaction effect) is
(a) 14 V (b) 16 V
(c) 19 V (d) 22 V

Q.119 A cumulatively compounded dc motor delivers rated load power at rated speed. If the series field is short circuited, then
(a) I_a increases and ω_m decreases.
(b) I_a remains constant but ω_m increases.
(c) both armature current I_a and speed ω_m increases.
(d) both armature current I_a and speed ω_m decreases.

Q.120 A dc shunt generator has full-load voltage regulation of 10% at rated speed of 1000 rpm. If it is now driven at 1250 rpm, then its voltage regulation at full load would
(a) remain unchanged
(b) be less than 10%
(c) be 12.5%
(d) be more than 10%

Q.121 In a loaded dc generator, if the brushes are given a shift from the interpolar axis in the direction of rotation, then the commutation will
(a) improve with fall of terminal voltage V_t.
(b) deteriorate with rise in V_t.
(c) improve with rise in V_t.
(d) deteriorate with fall of V_t.

Q.122 A 220 V dc machine has an armature resistance of 1 Ω. If the full-load current is 20 A, the difference in the induced voltages when the machine is running as a motor and as a generator is
(a) 50 V (b) 0 V
(c) 40 V (d) 20 V

Q.123 A dc series motor drives a load. The motor is unsaturated and the load torque is proportional to speed squared. For a speed of 400 rpm, motor line current is 20 A. For a speed of 800 rpm, the line current will be
(a) 40 A (b) 25.25 A
(c) 60 A (d) 20 A

Q.124 The windings of a large dc series motor and their functions are given below. Match **List-I (Windings)** with **List-II (Functions)** and select the correct answer using the codes given below the lists:

List-I
A. Compensating winding
B. Commutating pole winding
C. Series field winding
D. Shunt field winding

List-II
1. to generate working flux
2. to avoid runaway speed at no-load
3. to avoid sparking
4. to avoid field distortion under the pole

Codes:

	A	B	C	D
(a)	1	2	3	4
(b)	4	3	1	2
(c)	4	3	2	1
(d)	4	2	1	3

Q.125 A dc generator works on the principle of the production of
(a) statically induced emf
(b) dynamically induced emf
(c) both (a) and (b)
(d) neither (a) nor (b)

Q.126 A 4-pole lap wound armature has 144 slots with two coil sides per slot, each coil having two turns. If the flux per pole is 20 mWb and the armature rotates at 720 rpm, the induced emf will be
(a) 99 V (b) 182 V
(c) 138 V (d) 125 V

Q.127 The armature of a 2-pole, 200 V generator has 400 conductors and runs at 300 rpm. If the number of turns in each field coil is 1200, what would be the average value of the emf induced in each coil on breaking the field, if the flux dies away completely in 0.1 sec?
(a) 400 V (b) 1 kV
(c) 1.5 kV (d) 1.2 kV

Q.128 A dc series generator , having an external characteristic which is a straight line through zero to 50 V at 200 A is connected as a booster between a station bus-bar and a feeder of 0.3 Ω resistance as shown in figure below.

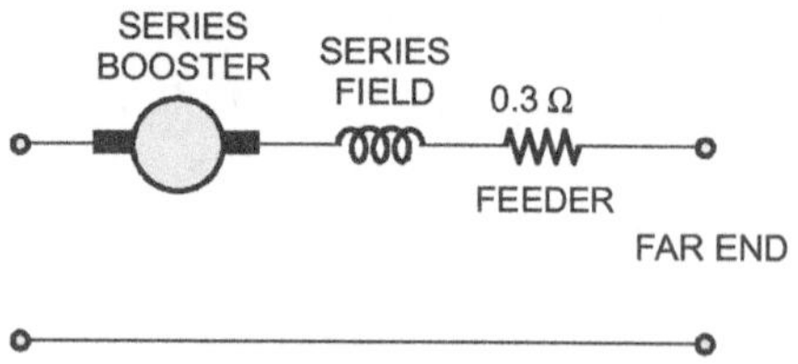

What is the voltage difference between the station bus-bar and the far end of the feeder at a current of 50 A?
(a) 1.5 V (b) –2.5 V
(c) 0.5 V (d) – 1.5 V

Q.129 A 25 kW, 250 V shunt generator delivers rated current at rated voltage. On removal of load, the terminal voltage rises to 275 V. The voltage regulation is
(a) 0.1 per unit (b) – 0.09 per unit
(c) – 0.1 per unit (d) 0.09 per unit

Q.130 A shunt generator is to be converted into a level compound generator by addition of a series field winding. From a test on the machine which shunt excitation only, it is found that the shunt current is 4.1 A to give 440 V on no-load and 5.8 A to give the same voltage when the machine supplies its full load of 200 A. A shunt field winding has 1,200 turns per pole. The number of series turns required per pole for the required purpose will be
(a) 6 (b) 7
(c) 9 (d) 11

Q.131 **Assertion (A) :** The generator with more drooping characteristic carries small load.
Reason (R) : The generator having more drooping characteristic delivers less current than the current delivered by the generator having less drooping characteristic.
(a) Both A and R are true and R is a correct explanation of A.
(b) Both A and R are true but R is not a correct explanation of A.
(c) A is true but R is false.
(d) A is false but R is true.

Q.132 The condition required for the parallel operation of DC generators is
(a) same polarity
(b) equal terminal voltage
(c) both (a) and (b)
(d) parallel operation is possible even without (a) and (b)

Q.133 **Assertion (A) :** In retardation test or running down test of dc machines, machine under test is speeded up slightly above its normal speed and supply to the armature is cut-off. Consequently, the armature slows down and its kinetic energy is utilized to meet the rotational losses.
Reason (R) : Retardation test is used for finding out the stray losses of shunt wound dc machines.
(a) Both A and R are true and R is a correct explanation of A.

(b) Both A and R are true but R is not a correct explanation of A.
(c) A is true but R is false.
(d) A is false but R is true.

Q.134 Consider the following statements related to dc machines:
1. For the development of electromagnetic torque, the armature mmf wave must be stationary in space.
2. The effect of current in the armature coils is to set up an armature mmf that is at an angle of 90° to the brush axis.
3. With the increases in the reluctance of the pole tips, the magnitude of armature cross flux is reduced and the distortion of the resultant flux density wave is minimised.
4. For improving the commutation process, interpoles are used while for neutralizing the effect of armature reaction, compensation windings are used.

Which of the above statements is/are true?
(a) 1, 2 and 4 only (b) 1 and 4 only
(c) 2, 3 and 4 only (d) 1, 3 and 4 only

Q.135 The internal characteristic of a differentially compound generator is the plot of
(a) armature current against terminal voltage.
(b) load current against internal induced emf.
(c) load current against terminal voltage.
(d) armature current against internal induced emf.

Q.136 A 230 V dc series motor develops its rated output at 1500 rpm while taking 20 A. Armature and series field resistances are 0.3 Ω and 0.2 Ω respectively. Neglecting saturation, what should be the external resistance required to be added to obtain rated torque at starting?
(a) 8.08 Ω (b) 11.0 Ω
(c) 6.25 Ω (d) 9.8 Ω

Q.137 The maximum efficiency of a dc machine is obtained when variable armature circuit loss is
(a) equal to core loss
(b) equal to copper loss
(c) equal to fixed loss
(d) equal to iron loss

Q.138 Consider the following statements about the use of starters in dc motors:
1. Starter in a dc motor is used to start the motor.
2. In case of a three-point starter, the field circuit is connected across the hold coil.
3. A three-point starter can be used where wide range of speed control, by shunt field control is required.
4. The four-point starter permits the change of field current by field rheostat, without affecting the holding coil current.

Which of the statements given above are correct ?
(a) 1 and 2 (b) 2 and 3
(c) 4 only (d) 1, 2 and 4

Q.139 Match **List-I (DC Generators)** with **List-II (Voltage Regulation)** and select the correct answer using codes given below the lists:

List-I	List-II
A. Overcompounded	1. Zero
B. Differentially-compounded	2. Positive
C. Shunt	3. Negative
D. Level Compounded	4. Poor

Codes:

	A	B	C	D
(a)	3	4	2	1
(b)	1	2	4	3
(c)	3	2	4	1
(d)	1	4	2	3

Q.140 Armature-circuit resistance control method controls the speed of dc machines
(a) above base speed only
(b) below base speed only
(c) either above or below base speed
(d) both above and below base speed

Q.141 In a shunt dc machine, the armature and field winding resistance are respectively
(a) high and low (b) low and high
(c) of lower values (d) of higher values

Q.142 Back emf of a d.c. motor is a function of which of the following factors?
1. Speed 2. Armature current
3. Field flux 4. Residual magnetism

Select the correct answer using the codes given below:

(a) 2 and 3 (b) 1 and 4
(c) 3 and 4 (d) 1 and 3

Q.143 Four types of d.c. generators of constant speed are considered **(List-I)**. Their external characteristics at constant speed are given in **List-II**. Match **List-I (Type of d.c. generator)** with **List-II (External Characteristics)** and select the correct answer using the codes given below the lists:

List-I
A. Separately excited
B. Series excited
C. Shunt excited
D. Over-compound excited

List-II

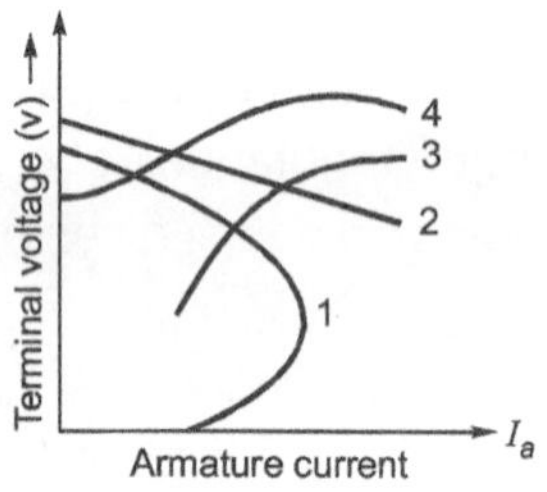

Codes:

	A	B	C	D
(a)	2	3	1	4
(b)	1	4	2	3
(c)	1	3	2	4
(d)	2	4	1	3

Q.144 A belt-driven cumulatively-compounded dc generator is delivering power to the dc mains. If the belt snaps then the machine will run as a
(a) cumulative compound motor in the same direction.
(b) differentially compound motor in the same direction.
(c) cumulative compound motor in the opposite direction.
(d) differentially compound motor in the opposite direction.

Q.145 The brush-axis of a bipolar dc motor is rotated by 90°. The effect of this rotation on the back emf E_b and the torque developed T_d would be such that
(a) both E_b and T_d are unchanged
(b) E_b is zero, but T_d is unchanged
(c) E_b is unchanged, but T_d is zero
(d) both E_b and T_d are zero

Q.146 Match **List-I (DC Motors)** with **List-II (Characteristics)** labelled 1, 2, 3 and 4 and select the correct answer using codes given below lists:

List-I
A. Shunt motor
B. Series motor
C. Cumulative compound motor
D. Differentially compound motor

List-II

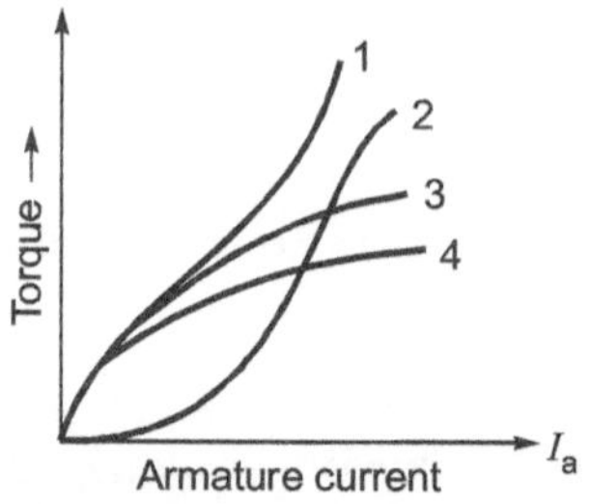

Codes:

	A	B	C	D
(a)	3	2	1	4
(b)	4	1	2	3
(c)	3	1	2	4
(d)	4	2	1	3

Q.147 If load on a d.c. shunt motor is increased, its speed is decreased primarily due to
(a) increase in its flux
(b) decrease in back emf
(c) increase in armature current
(d) increase in brush drop

Q.148 If the field circuit of a loaded shunt motor is suddenly opened
(a) it would race to almost infinite speed.
(b) it would draw abnormally high armature current.
(c) circuit breaker or fuse will open the circuit before too much damage is done.
(d) torque developed by the motor would be reduced to zero.

Q.149 It is possible to increase the field flux and at the same time, increase the speed a d.c. motor provided its _______ is held constant.
(a) applied voltage
(b) torque
(c) armature circuit resistance
(d) armature current

Q.150 Which of the following quantity maintains the same direction whether d.c. machine runs as a generator or as a motor?
(a) Induced emf (b) Armature current
(c) Field current (d) Supply current

Q.151 Sparkless commutation can be achieved by employing
(a) compensating winding
(b) interpoles
(c) high resistance carbon brushes
(d) any one or more of the above methods

Q.152 An over-compounded dc generator is supplying 800 A at full-load on 500 V. The resistance of each conductor is 0.02 Ω. The percentage of compounding is
(a) 12.8 % (b) 6.4 %
(c) 3.2 % (d) 1.6 %

Q.153 The electric braking system employed with dc shunt motors, in which the armature connections are reversed so that the motor tends to run in opposite direction is called the ____ braking.
(a) plugging (b) regenerative
(c) dynamic (d) rheostatic

Q.154 Regenerative braking
(a) can be easily applied to dc series motors.
(b) can be easily applied to dc shunt motors.
(c) can be used for stopping the motor.
(d) can't be used when the load on motor has overhauling characteristics.

Q.155 Performance of large dc machines regarding commutation and temperature rise etc. at full load can be checked by
(a) brake test (b) Swinburne's test
(c) Hopkinson's test (d) running down test

Q.156 A dc shunt generator has a speed of 800 rpm when delivering 20 A to the load at the terminal voltage of 220 V. If the same machine is run as a motor it takes a line current of 20 A from 220 V supply. The speed of the machine as a motor will be
(a) more than 800 rpm
(b) less than 800 rpm
(c) 800 rpm
(d) may be more or less than 800 rpm

Q.157 A 4-pole lap wound dc shunt motor rotates at a speed of 1500 rpm, has a flux of 0.4 mWb and the total number of conductors are 1000. What is the value of emf?
(a) 20 volt (b) 0.1 volt
(c) 100 volt (d) 10 volt

Q.158 The speed-torque characteristic of a DC series motor are approximately similar to those of the ______motor.
(a) universal (b) synchronous
(c) two-phase (d) dc shunt

Q.159 The speed of a d.c. shunt motor at no-load is
(a) 5 to 10% of full-load
(b) 35 to 40 % of full-load
(c) 15 to 20% of full-load
(d) 25 to 30% of full-load

4. Synchronous Machine

Q.160 Match **List-I (Name of test)** with **List-II (Result)** and select the correct code given below the lists:

List-I
A. Open Circuit and short circuit test
B. Open circuit and zero power factor test
C. Slip test
D. Maximum lagging current test

List-II
1. Leakage reactance
2. Direct axis synchronous reactance
3. Quadrature axis synchronous reactance
4. Ratio of direct axis synchronous reactance to quadrature axis synchronous reactance

Codes:

	A	B	C	D
(a)	1	2	4	3
(b)	1	2	3	4
(c)	2	1	4	3
(d)	2	1	3	4

Q.161 The phasor diagram of a synchronous machine connected to an infinite bus is shown below. The machine is acting as a

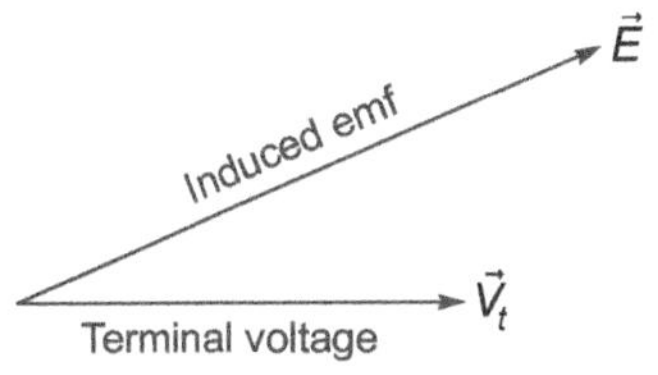

(a) generator and operating at a lagging pf.
(b) generator and operating at a leading pf.
(c) motor and operating at a leading pf.
(d) motor and operating at a lagging pf.

Q.162 A 6-pole alternator with 36 slots carries a 3-phase distributed winding. Each coil is short-pitched by one slot. The winding factor is given by

(a) $\frac{\cot 15^\circ}{3\sqrt{2}}$ (b) $\frac{\cot 15^\circ}{4}$
(c) $\frac{\cot 15^\circ}{2\sqrt{2}}$ (d) $\frac{\cot 15^\circ}{6}$

Q.163 A winding is distributed in the slots along the air-gap periphery
1. to add mechanical strength to the winding.
2. to reduce the amount of conductor material required.
3. to reduce the harmonics in the generated emf.
4. to reduce the size of the machine.
5. for full utilization of iron and conductor materials.

From these, the correct answer is
(a) 1, 3, 4, 5 (b) 1, 2, 4, 5
(c) 1, 2, 3, 5 (d) 1, 3, 5

Q.164 Which of the following graphs represents the speed-torque characteristic of a synchronous motor?

(a) 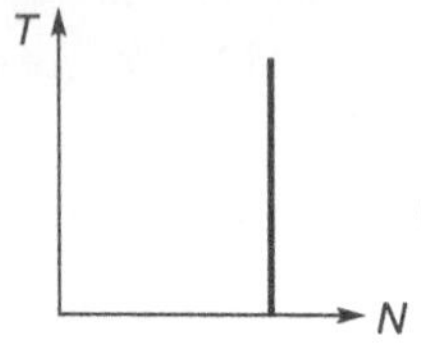

(b) 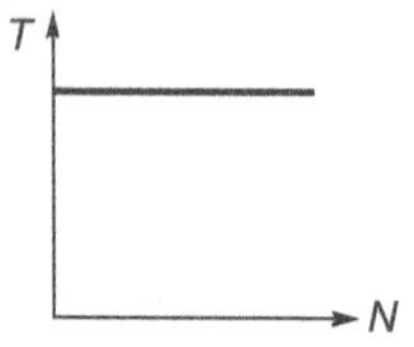

(c) 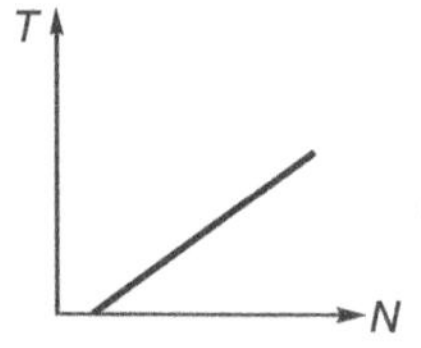

(d) 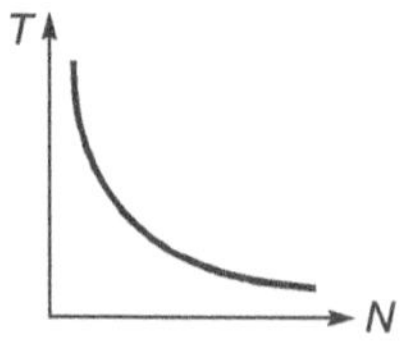

Q.165 In figure the characteristic that corresponds to the variation of synchronous reactances of synchronous machine with the field current is

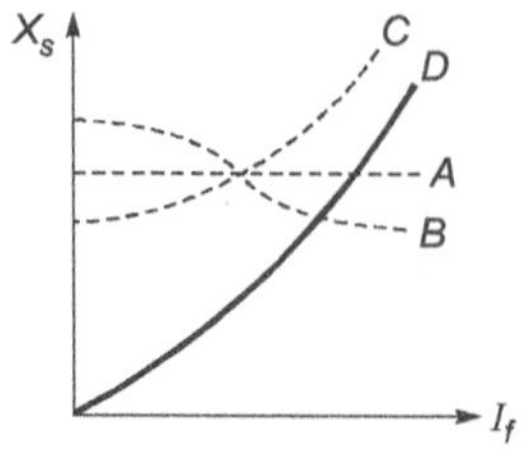

(a) Curve *A* (b) Curve *B*
(c) Curve *C* (d) Curve *D*

Q.166 The four methods of calculating voltage regulation of a 3-phase alternator are:
1. Emf method
2. Saturated synchronous reactance method
3. New ASA method
4. Mmf method

The correct sequence in the ascending order of the values of regulation obtained by these methods is
(a) 4, 3, 1, 2 (b) 4, 3, 2, 1
(c) 3, 4, 2, 1 (d) 3, 4, 1, 2

Q.167 The flux per pole in a synchronous motor with the field circuit ON and the stator disconnected from the supply is found to be 25 mWb. When the stator is connected to the rated supply with the field excitation unchanged, the flux per pole in the machine is found to be 20 mWb while the motor is running on no-load. Assuming no-load losses to be zero, the no-load current drawn by the motor from the supply
(a) is zero
(b) lags the supply voltage
(c) is in phase with the supply voltage
(d) leads the supply voltage

Q.168 Which of the following method would give a lower than actual value of regulation of an alternator?
(a) zpf method (b) mmf method
(c) Emf method (d) ASA method

Q.169 A synchronous motor is operating on no-load at unity power factor. If the field current is increased, the pf will become
(a) leading and the current will increase.
(b) lagging and the current will decrease.
(c) leading and the current will decrease.
(d) lagging and the current will increase.

Q.170 In a round-rotor alternator, reactive power is maximum at a load angle of
(a) 180° (b) 45°
(c) 90° (d) 0°

Q.171 The power-angle characteristic of a salient machine contains a second harmonic term, because
(a) of the slot harmonics.

(b) the frequency has a second harmonic component.
(c) of the difference in reluctance of air gap in the direct and quadrature axis.
(d) the air-gap flux density has a second harmonic component.

Q.172 If the excitation is kept constant in an alternator, then the excitation voltage variation with speed would be
(a) first linear and then horizontal
(b) first curved and then linear
(c) linear
(d) first linear and then curved upwards

Q.173 **List-I** and **List-II** pertain to the operation of salient-pole synchronous machine at rated speed. Match the two lists using the codes given below the lists:

List-I	List-II
A. Generator at lagging pf	1. $V_t \cos\delta < E_f$
B. Motor at lagging pf	2. $V_t \cos\delta > E_f$
C. Generator at leading pf	3. $V_t \cos\delta = E_f$
D. Motor at leading pf	

Codes:

	A	B	C	D
(a)	1	2	3	2
(b)	2	1	1	2
(c)	2	1	1	3
(d)	1	2	2	1

Q.174 When a 3-phase synchronous motor is running above synchronous speed, the damper winding produces
(a) reluctance torque
(b) damping torque
(c) induction motor torque
(d) inductor generator torque

Q.175 A 3-phase induction motor draws 1 MVA at a pf of 0.8 lag. A synchronous condenser is connected in parallel to draw an additional 750 kVA at 0.6 pf lead. The pf of the total load supplied by the mains is
(a) unity (b) 0.707 lead
(c) 0.707 lag (d) zero

Q.176 In a synchronous generator operating at zero-power factor leading, the effect of armature reaction is
(a) magnetizing
(b) cross-magnetizing
(c) demagnetizing
(d) partially cross magnetizing & demagnetizing

Q.177 Potier reactance of an alternator is almost the same as
(a) leakage reactance of field winding
(b) field winding reactance
(c) total armature reactance
(d) armature leakage reactance

Q.178 The impedance value of a generator is 0.2 pu on a base value of 11 kV, 50 MVA. The impedance value for a base of 22 kV, 150 MVA is
(a) 0.15 pu (b) 0.24 pu
(c) 0.20 pu (d) 0.30 pu

Q.179 The relative speed between the magnetic fields of stator and rotor under steady state operation is zero for
(a) a dc machine
(b) an induction machine
(c) a synchronous machine
(d) all of the above machines

Q.180 The result of a 'slip test' for determining direct-axis (X_d) and quadrature axis (X_q) reactances of a star-connected, salient pole alternator are given below:
Line values: V_{max} = 2820 V; V_{min} = 2800 V; I_{max} = 360 A; I_{min} = 275 A
Hence, the two reactances will be
(a) X_d = 4.5 Ω and X_q = 5.92 Ω
(b) X_d = 10.25 Ω and X_q = 7.78 Ω
(c) X_d = 5.92 Ω and X_q = 4.5 Ω
(d) X_d = 7.78 Ω and X_q = 10.25 Ω

Q.181 Match **List-I** with **List-II** and select the correct answer using codes given below the lists:

List-I
A. Synchronous motor input when the excitation is reduced to zero
B. zero power factor method
C. Reactance voltage
D. Deceleration due to speed reversal

List-II
1. Commutating poles
2. Plugging
3. Reluctance Power
3. Voltage regulation

Codes:

	A	B	C	D
(a)	1	4	3	2
(b)	3	4	1	2
(c)	1	2	3	4
(d)	3	4	2	1

Q.182 **Assertion (A):** Distributed winding results in efficient cooling.

Reason (R): With distributed winding high current density in the copper can be adopted due to even distribution of copper over the surface of the armature resulting in even distribution of copper loss.

(a) Both A and R are true and R is a correct explanation of A.
(b) Both A and R are true but R is not a correct explanation of A.
(c) A is true but R is false.
(d) A is false but R is true.

Q.183 **Assertion (A):** The synchronous condenser requires more copper for field winding in comparison with a synchronous motor for equal armature voltage and current ratings.

Reason (R): The synchronous condenser is a synchronous motor operated with heavy excitation i.e. heavy field current.

(a) Both A and R are true and R is a correct explanation of A.
(b) Both A and R are true but R is not a correct explanation of A.
(c) A is true but R is false.
(d) A is false but R is true.

Q.184 Consider the following statements related to synchronous machines:

1. Pull-out torque of a synchronous motor increases with the decrease in excitation.
2. A small short-circuit ratio indicates small armature reaction in an alternator.
3. Under short-circuit conditions, the armature reaction in the alternator is having magnetising effect.
4. Synchronous reactance can be determined from open-circuit and short-circuit tests.

From these, the correct statements are:

(a) 2, 3 and 4 (b) 2 and 3
(c) 1, 2, 3 and 4 (d) 4 only

Q.185 A 4-pole star-connected alternator running at 1500 rpm have 72 slots with 10 conductors per slot. The number of turns per phase and slot angle is

(a) 120 turns and 20 degrees
(b) 100 turns and 10 degrees
(c) 120 turns and 10 degrees
(d) 100 turns and 20 degrees

Q.186 To eliminate r^{th} harmonic from the induced emf in a phase of a synchronous machine the pitch of the coil must be

(a) $\frac{r}{r-1}$*th* fraction of full pitch
(b) $\frac{2r-1}{r}$*th* fraction of full pitch
(c) $\frac{r-1}{r}$*th* fraction of full pitch
(d) $\frac{2r}{r+1}$*th* fraction of full pitch

Q.187 When an alternator designed for operation at 60 Hz is operated at 50 Hz?

(a) Operating voltage will increase in the ratio of 1.2.
(b) Operating voltage will reduce in the ratio of 5/6.
(c) kVA rating will increase in the ratio of 1.2.
(d) Operating voltage will reduce in the ratio of $\left(\frac{5}{6}\right)^2$.

Q.188 In a synchronous generator

(a) the armature mmf lags behind the air-gap flux and air-gap flux leads the field mmf.
(b) the armature mmf leads the air-gap flux and air-gap flux leads the field mmf.
(c) the armature mmf leads the air-gap flux and the air-gap flux lags behind the field mmf.
(d) the armature mmf lags behind the air-gap flux and air-gap flux lags behind the field mmf.

Q.189 The regulation obtained by synchronous impedance method is always higher than actual value because

(a) synchronous reactance is assumed constant while it is not.

(b) effective armature resistance is assumed constant.
(c) field current is increased to give short-circuit current about twice of full-load current.
(d) synchronous impedance is assumed variable while it is not.

Q.190 Modern alternators are designed to have poor regulation as it
(a) increases the efficiency of operation
(b) increases the value of short-circuit current
(c) limits the value of short-circuit current
(d) both (a) and (b)

Q.191 An alternator with higher value of SCR has
(a) poor voltage regulation and higher stability limit.
(b) better voltage regulation and lower stability limit.
(c) poor voltage regulation and lower stability limit.
(d) better voltage regulation and higher stability limit.

Q.192 A synchronous generator has its field winding on the rotor and armature winding on the stator. When running under steady state conditions its air gap field is
(a) stationary w.r.t. stator.
(b) rotating at synchronous speed against the direction of rotor rotation.
(c) rotating at synchronous speed in the direction of rotor rotation.
(d) rotating at synchronous speed w.r.t. rotor.

Q.193 When does a synchronous motor operate with lagging power factor current?
(a) While it is heavily loaded
(b) While it is overexcited
(c) While it is critically loaded
(d) While it is underexcited

Q.194 A 3-phase synchronous motor connected to an infinite bus-bar, with constant excitation, is driving a certain load and operating at leading power factor. If the shaft load is reduced
(a) the load angle will decrease.
(b) power factor will decrease.
(c) both (a) and (b)
(d) neither (a) nor (b)

Q.195 A 3-phase, star-connected, 11 kV, 165 MVA alternator has synchronous reactance of 1 Ω per phase. What is the short circuit current in the alternator at the instant of short circuit?
(a) 17.96 kA (b) 22.22 kA
(c) 6.30 kA (d) None of these

Q.196 A star-connected synchronous motor rated 187 kVA, 3-ϕ, 2300 V, 47 A, 50 Hz, 187.5 rpm has an effective resistance of 1.5 Ω and a synchronous reactance of 20 Ω per phase. What is the internal power developed by the motor when it is operating at rated current and 0.8 power factor leading?
(a) 100 kW (b) 125 kW
(c) 140 kW (d) 155 kW

Q.197 Two identical 2 MVA alternators operate in parallel. The governor of first machine is such that the frequency drops uniformly from 50 Hz on no-load to 47.5 Hz on full-load. The corresponding uniform speed drop of the second machine is 50 Hz to 48 Hz. How will they share a load of 3 MW?
(a) $P_1 = \frac{5}{3}$ MW, $P_2 = \frac{4}{3}$ MW
(b) $P_1 = \frac{4}{3}$ MW, $P_2 = \frac{5}{3}$ MW
(c) $P_1 = \frac{1}{3}$ MW, $P_2 = \frac{8}{3}$ MW
(d) $P_1 = \frac{8}{3}$ MW, $P_2 = \frac{1}{3}$ MW

Q.198 A 100 MVA synchronous generator operates on full load at a frequency of 50 Hz. The load is suddenly reduced to 50 MW. Due to time lag in governor system, the steam valve begins to close after 0.4 second. If the inertia constant of the machine is: $H = 5$ kW-s/kVA, then the amount of excess energy in MJ transferred to the rotating parts will be
(a) 50 MJ (b) 100 MJ
(c) 20 MJ (d) None of these

Q.199 What is the distribution factor for a uniformly distributed 3-phase winding having a phase spread of 120°?
(a) 0.955 (b) 0.880
(c) 0.967 (d) 0.827

Q.200 What is the pitch factor for the following winding? 36 stator slots, 4 poles, coil-span 1 to 8.
(a) cos 20° (b) > cos 20°
(c) < cos 20° (d) None of these

Q.201 An alternator is running at synchronous speed. Its field is now energised from an ac source at rated frequency, then which of the following option will hold true?
(a) Core loss will decrease.
(b) Copper loss will decrease.
(c) Efficiency will decrease.
(d) None of the above.

Q.202 Assertion (A): In a 3-phase supply, floating neutral is undesirable.
Reason (R): Floating neutral in a 3-phase supply system may give rise to high voltage across the load.
(a) Both A and R are true and R is a correct explanation of A.
(b) Both A and R are true but R is not a correct explanation of A.
(c) A is true but R is false.
(d) A is false but R is true.

Q.203 Assertion (A): Line currents of 3-phase 3-wire system cannot contain zero-sequence current whether it is delta connected or star connected.
Reason (R): Fourth wire is necessary for the flow of zero sequence currents in the lines of the 3-phase system.
(a) Both A and R are true and R is a correct explanation of A.
(b) Both A and R are true but R is not a correct explanation of A.
(c) A is true but R is false.
(d) A is false but R is true.

Q.204 A synchronous motor, connected to an infinite bus, is working at a leading power factor. Its excitation emf E_f and terminal voltage V_f are related as under
(a) $E_f < V_t$ and E_f lags V_t
(b) $E_f > V_t$ and E_f leads V_t
(c) $E_f < V_t$ and E_f leads V_t
(d) $E_f > V_t$ and E_f lags V_t

Q.205 In a synchronous motor, hunting can be reduced to minimum possible by
(a) providing damper winding in the rotor pole faces.
(b) using a flywheel.
(c) designing the motor for adequate synchronising power.
(d) any of the above method.

Q.206 Which of the following methods is employed for starting a 3-phase synchronous motor?
(a) Star-delta starter
(b) resistance starter in the stator circuit
(c) Damper winding in conjunction with a star-delta starter or an auto-transformer starter.
(d) Damper winding

Q.207 Match **List-I (Synchronous machine operation)** with **List-II (Phasor diagram)** and select the correct answer using the codes given below the lists:

List-I
A. As an alternator
B. At the moment of synchronisation
C. As a synchronous motor
D. At the moment of losing synchronism

List-II

1. 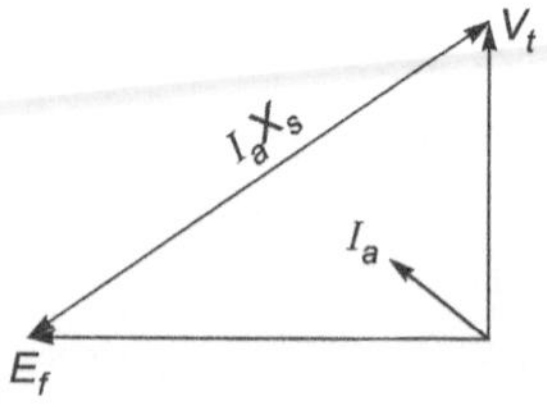

2. 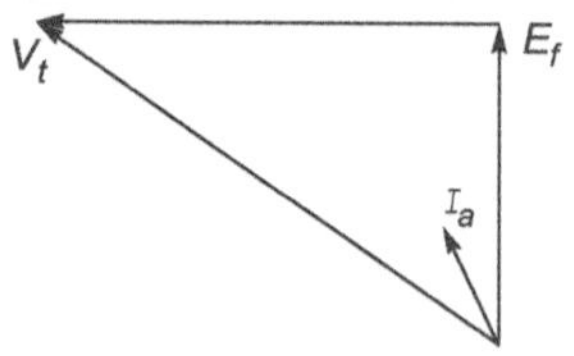

3. 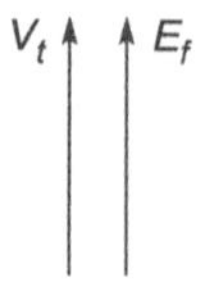

4. 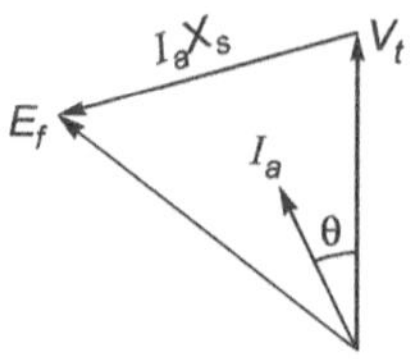

Codes:

	A	B	C	D
(a)	3	1	2	4
(b)	4	3	2	1
(c)	3	2	1	4
(d)	3	4	2	1

Q.208 A synchronous machine is operating at constant load and at unity power factor. If its excitation is increased, then it operates as:

1. a motor at a leading pf
2. a generator at a leading pf
3. a motor at a lagging pf
4. a generator at a lagging pf
5. absorber of reactive power
6. exporter of reactive power

From these, the correct answer is

(a) 1, 4, 6 (b) 2, 3, 5
(c) 1, 3, 6 (d) 3, 4, 5

Q.209 Match **List-I (Methods of full-load regulation of 25 MVA alternator at 0.6 pf lagging)** with **List-II (% regulation)** and select the correct answer using the codes given below the lists:

List-I	List-II
A. zpf	1. 32%
B. mmf	2. 18%
C. emf	3. 13%

Codes:

	A	B	C
(a)	1	2	3
(b)	2	3	1
(c)	3	2	1
(d)	1	3	2

Q.210 **Assertion (A):** A compromise between good starting torque and good damping effect is made for obtaining satisfactory operation of a synchronous motor.

Reason (R): For good starting torque, damper bars should have low resistance but, for large damping effect near synchronous speed, the damper bars should have high resistance.

(a) Both A and R are true and R is a correct explanation of A.
(b) Both A and R are true but R is not a correct explanation of A.
(c) A is true but R is false.
(d) A is false but R is true.

Q.211 **Assertion (A):** The synchronising-power coefficient is a measure of the stiffness of electromagnetic coupling between stator and rotor fields.

Reason (R): Too large stiffness of coupling means that the motor tends to follow closely, the variation of speed caused by the disturbance in electric power supply.

(a) Both A and R are true and R is a correct explanation of A.
(b) Both A and R are true but R is not a correct explanation of A.
(c) A is true but R is false.
(d) A is false but R is true.

Q.212 The current from the stator of an alternator is taken out to the external load circuit through

(a) commutator segments
(b) solid connections
(c) slip rings
(d) carbon brushes

Q.213 An alternator is delivering rated current at rated voltage and 0.8 power-factor lagging case. If it is required to deliver rated current at rated voltage and 0.8 power factor leading, the required excitation will be:

(a) more or less (b) the same
(c) more (d) less

Q.214 Inverted V-curve of a synchronous motor is the plot of

(a) armature current against the field current for a given load
(b) armature current against the power factor for a given load
(c) power factor against the field current for a given load
(d) power factor against armature current for a given load

5. Three-Phase Induction Machine

Q.215 A 3-phase, 50 Hz induction motor has a full-load speed of 1440 rpm. The number of poles of the motor is

(a) 4 (b) 5
(c) 6 (d) None of these

Q.216 A properly shunted centre-zero galvanometer is connected in the rotor circuit of a 6-pole, 50 Hz would-rotor induction motor. If the galvanometer makes 90 complete oscillations in one minute, the speed of rotor would be

(a) 1000 rpm (b) 740 rpm
(c) 880 rpm (d) 970 rpm

Q.217 A 10-pole, 3-phase, 50 Hz, delta connected induction motor takes 120 kW during blocked-rotor test at rated voltage and frequency. If the stator ohmic loss is equal to rotor ohmic loss during blocked rotor test, then the starting torque if DOL starting is used for this motor will be

(a) 822.56 N-m (b) 954.93 N-m
(c) 525.76 N-m (d) 730 N-m

Q.218 A 3-phase squirrel cage induction motor has maximum torque equal to twice the full-load torque. The per phase rotor resistance and per phase standstill reactance referred to stator are 0.2 Ω and 2 Ω respectively (Stator impedance is neglected). What is the ratio of motor starting toque to its full-load torque, if the motor is started by direct-on-line starter?

(a) 0.396 (b) 0.194
(c) 0.198 (d) 0.400

Q.219 A 10 kW, 400 V, 3-phase induction motor draw a current of 5 times its full-load current at rated voltage and at standstill developing a starting toque of 1.5 times its full-load torque. An auto-transformer is installed to reduce the starting current and to give full-load torque at starting. Neglecting exciting current of auto-transformer, the voltage applied to the motor terminals at the time of starting with auto transformer is

(a) 220 volts (b) 241.8 volts
(c) 326.6 volts (d) 215.9 volts

Q.220 A 3-phase, 20 kW, 400 V, 1470 rpm, 50 Hz squirrel cage induction motor develops a torque of 100 N-m at a speed of 1400 rpm. If the motor is connected to a 30 Hz supply; for keeping the same air-gap flux, the supply voltage and the speed for the same load torque would respectively be

(a) 240 V, 750 rpm (b) 240 V, 800 rpm
(c) 240 V, 780 rpm (d) 240 V, 840 rpm

Q.221 In a 3-phase squirrel cage induction motor, applied voltage is reduced to x times its rated value. Then, its starting line current from mains and starting torque by auto-transformer starting would reduce respectively from their corresponding direct switching values by a factor of

(a) x, x^2 (b) x, x
(c) x^2, x^2 (d) x^2, x

Q.222 A 3-phase IM with stator in delta draws a line current of 10 A. When connected in star, it would draw a line current of

(a) 10 A (b) $\frac{10}{\sqrt{3}}$ A
(c) 30 A (d) $\frac{10}{3}$ A

Q.223 The power factor of a delta-connected 3-phase, 50 kW, IM is 0.4 when delivering 35% of its rated load. If the stator is reconnected in star, then its

(a) pf is worsened, stator current increases
(b) pf is improved, stator current increases
(c) pf remains unchanged, stator current decreases
(d) pf is improved, stator current decreases.

Q.224 A 3-phase, star-connected SRIM is fed from 400 V, 50 Hz source. Stator to rotor effective turns ratio is 2. At a rotor speed of 1440 rpm, the rotor induced emf per phase would be

(a) 8.0 V (b) 4.62 V
(c) 9.24 V (d) 46.2 V

Q.225 A decrease in the number of poles of a 3-phase IM results in

(a) increase in maximum pf
(b) no-change in maximum pf
(c) decrease in maximum pf
(d) cannot be determined

Q.226 The diameter of the circle diagram for a 3-ϕ IM is given by (V_1 = stator phase voltage)

(a) $\frac{V_1}{X_2}$ (b) $\frac{V_1}{R_1 + R_2}$
(c) $\frac{V_1}{X_1 + X_2}$ (d) $\frac{V_1}{X_1}$

Q.227 The complete circle diagram of a 3-phase induction motor can be drawn with the help of

(a) Blocked-rotor test and stator-resistance test
(b) stator resistance test and Running-light test
(c) Running-light test and Blocked-rotor test
(d) Running-light test, Blocked-rotor test and stator resistance test

Q.228 A 3-phase induction requires starter

(a) because the motor does not posses an inherent starting torque
(b) to increase the starting torque
(c) because the induced secondary voltage and hence the currents in the windings are abnormally high during starting

(d) because there is no back emf induced in the stator winding during starting.

Q.229 If the auto-transformer tapping K is more than unity, then the starting torque in a 3-phase squirrel cage induction motor when started with this auto-transformer starter would be (Te_{st} = starting torque)

(a) $K^2 \times T_{est}$ (b) $\frac{1}{K} \times T_{est}$

(c) $K \times T_{est}$ (d) $\frac{1}{K^2} \times T_{est}$

Q.230 The following starting method for a 3-phase squirrel cage induction motor is inferior in view of the poor starting torque per ampere of the line-current drawn:

(a) Auto-transformer method of starting
(b) Impedance starting
(c) Direct-on-line starting
(d) Star-delta starting

Q.231 The stator of a 3-phase, 6-pole slip ring induction is connected to 50 Hz source but its rotor is energised from 20 Hz source. The rotor would run at a speed of

(a) 600 rpm, or 1400 rpm
(b) 1600 rpm or, 400 rpm
(c) 600 rpm or, 400 rpm
(d) 1600 rpm or, 600 rpm

Q.232 A slip ring induction motor drives a constant torque load. If the supply voltage reduces to $1/\sqrt{2}$ times its previous stator voltage, then slip and current get modified by factors of

(a) 2 and 2 respectively
(b) $\sqrt{2}$ and 2 respectively
(c) $\sqrt{2}$ and $\frac{1}{\sqrt{2}}$ respectively
(d) 2 and $\sqrt{2}$ respectively

Q.233 Match **List-I (Type of motor)** with **List-II (Application)** and select the correct answer using the codes given below the lists:

List-I
A. Single phase shaded motor
B. 3-phase squirrel cage induction motor
C. Synchronous motor
D. DC series motor

List-II
1. Condenser
2. Hair dryer
3. Cranes
4. Centrifugal pump

Codes:

	A	B	C	D
(a)	4	2	1	3
(b)	2	4	1	3
(c)	4	3	2	1
(d)	2	3	1	4

Q.234 Assertion (A): Induction generators are widely used for generating power for commercial supply
Reason (R): Induction generators are simple and rugged in construction, cheaper in initial cost and easy in maintenance.

(a) Both A and R are true and R is a correct explanation of A.
(b) Both A and R are true but R is not a correct explanation of A.
(c) A is true but R is false.
(d) A is false but R is true.

Q.235 Assertion (A): The torque-sped characteristic of an induction motor in operating region closely resembles with that of a dc shunt motor.
Reason (R): In an induction motor, even at full load, slip doesn't exceed 5% i.e. the speed varies from 95% to very close to synchronous speed for different loads.

(a) Both A and R are true and R is a correct explanation of A.
(b) Both A and R are true but R is not a correct explanation of A.
(c) A is true but R is false.
(d) A is false but R is true.

Q.236 A squirrel cage induction motor has a full-load slip of 0.05. The motor starting current at rated voltage is 6 times its full-load current. What is the percentage tapping on the auto-transformer starter which will give full-load torque at start?

(a) 74.50 (b) 33.33
(c) 68.90 (d) 86.20

Q.237 A 10 kW, 420 V, 3-phase, 4-pole, 50 Hz, delta connected squirrel cage induction motor gave the following data for blocked rotor test:

210 V, 20 A, 5 kW

Stator core loss at rated voltage and frequency is 300 Watts. The d.c. resistance measured between any two stator terminals is 0.6 Ω. The air-gap power at rated voltage and frequency with rotor blocked is:

(a) 21 kW (b) 16 kW
(c) 18 kW (d) 24 kW

Q.238 Increased air-gap length in a three-phase induction motor

1. increases the slip at which maximum torque occurs.
2. reduce the maximum torque while starting torque is increased.
3. results in poor power factor at no-load and also on full load.
4. increases the magnetizing current drawn by the motor.
5. reduced the mutual flux in the air-gap.

Of these, the correct statements are:

(a) 1, 5 and 3 (b) 2, 3 and 4
(c) 1, 5 and 4 (d) 3 and 4

Q.239 A 3-phase induction motor is designed to operate at 50 Hz and *V* volts. Its breakdown torque occurs at a slip of 0.2 and its stator impedance is negligible. If the motor is made to operate from *V* volts, 45 Hz, the ratio of starting current at 45 Hz operation to their corresponding values at 50 Hz operation would be

(a) 1.111 (b) 0.989
(c) 1.106 (d) 1.262

Q.240 A 10 kW, 50 Hz, 3-phase induction motor develops rated torque at 1440 rpm. If the load torque is reduced to half, then assuming linear torque-slip characteristics near the operating range, the new speed of the motor would be

(a) 1420 rpm (b) 1500 rpm
(c) 1480 rpm (d) 1470 rpm

Q.241 The ratio of maximum torque to starting torque in a 3-phase induction motor neglecting stator impedance is given by (s_m = slip at maximum torque)

(a) $\dfrac{s_m}{0.5(S_m^2+1)}$ (b) $0.5\left(s_m+\dfrac{1}{s_m}\right)$

(c) $\left(0.5s_m+\dfrac{1}{0.5s_m}\right)$ (d) $\dfrac{0.5s_m}{(1+s_m^2)}$

Q.242 A 4-pole, 3-phase slip-ring induction motor is used as a frequency changer. Its stator is excited from 3-phase, 50 Hz supply. A load requiring 3-phase, 20 Hz supply is connected to the star-connected rotor through the slip-rings of the motor.
At what speed the prime mover should drive the rotor of this motor?

(a) 2100 rpm against the direction of stator field
(b) 900 rpm against the direction of stator field
(c) 2100 rpm in the direction of stator field
(d) none of the above

Q.243 **Assertion (A):** In large high speed induction motors it is not essential to use semi-closed slots to reduce the reluctance.
Reason (R): Large high speed induction motors have large pole pitch, so the large number of conductors per pole and, therefore, can have desired flux density in the air gap without an excessive magnetizing current.

(a) Both A and R are true and R is a correct explanation of A.
(b) Both A and R are true but R is not a correct explanation of A.
(c) A is true but R is false.
(d) A is false but R is true.

Q.244 In a squirrel cage induction motor high starting torque is achieved by using

(a) high resistance in series with the rotor circuit.
(b) double cage rotor
(c) short circuited rotor winding across slip rings
(d) none of the above

Q.245 In double cage induction motor the inner cage has

(a) high resistance and high leakage reactance
(b) high resistance and low leakage reactance
(c) low resistance and low leakage reactance
(d) low resistance and high leakage reactance

Q.246 The rotor slots in a 3-phase induction motor are kept inclined to avoid.

(a) Crawling (b) Skewing
(c) Cogging (d) none of these

Q.247 Presence of 5th harmonics in induction motor causes

(a) Cogging
(b) Small reverse braking torque
(c) Crawling
(d) Hunting

Q.248 Match **List-I** with **List-II** and select the correct answer using the codes given below the lists:

List-I	List-II
A. Power factor	1. space harmonics
B. Cogging	2. rotor resistance
C. Starting torque	3. air gap
D. Crawling	4. harmonic synchronous torque

Codes:

	A	B	C	D
(a)	3	4	2	1
(b)	4	3	2	1
(c)	4	1	2	3
(d)	3	1	2	4

Q.249 The power scale of circle diagram of an induction motor is determined from ______ test data only.
(a) stator resistance (b) short-circuit
(c) slip (d) open-circuit

Q.250 The equivalent circuit per phase of a three phase transformer is similar to that of a three-phase induction motor but the transformer does not develop any torque. This is due to
(a) absence of air-gap
(b) low supply voltage
(c) non-fulfilment of condition of space condition of winding
(d) none of the above

Q.251 The power factor of a lightly loaded induction motor is quite low because
(a) of the current due to air gap
(b) the current drawn is largely a magnetizing component due to laminated core
(c) the current drawn is largely an energy component due to laminated core
(d) the current drawn is largely a magnetising current due to air gap.

Q.252 In a 3-phase induction motor, the maximum torque
(a) varies as rotor circuit resistance
(b) varies inversely as rotor circuit resistance
(c) is independent of rotor circuit resistance
(d) is constant

Q.253 In a 3-phase induction motor, the starting torque will be maximum when
(a) $R_2 = X_2^2$ (b) $R_2 = \frac{1}{X_2}$
(c) $R_2 = \sqrt{X_2}$ (d) $\frac{1}{R_2} = \frac{1}{X_2}$

Q.254 The maximum possible speed of a 3-phase squirrel cage induction motor running at a slip of 4% is
(a) 3000 rpm (b) 960 rpm
(c) 1440 rpm (d) 2880 rpm

Q.255 Which of the following statements about a squirrel cage induction motor is incorrect?
(a) It has high starting torque.
(b) It is provided with cage rotor which is simple, robust and almost instrucible in construction.
(c) It has no commutator and no slip ring.
(d) Almost 90 percent of induction motors are of this type.

Q.256 The rotor of a 3-phase induction motor rotates in the same direction as that of stator rotating field. This can be explained by
(a) Faraday's law of electromagnetic induction.
(b) Fleming's right hand rule
(c) Lenz's law
(d) Newton's third law of motion

Q.257 Consider the following statements related to a three-phase induction motor:
1. An auto-transformer with x% tap is used to start an induction motor. Supply current is reduced to x/100 of the value under direct-on-line.
2. The double cage induction motor has comparatively higher copper loss.
3. The speed of an induction motor can't be controlled as easily and efficiently as that of a dc shunt motor.
4. Deep bar or double cage construction is used for improving the running performance of the motor.

Out of these, the statements which are not true are:
(a) 1, 2 and 4 (b) 2 and 3
(c) 1, 3 and 4 (d) 3 and 4

Q.258 **Assertion (A):** Start-delta starting is one of the popular method for starting a three-phase induction motor.

Reason (R): The starting current is reduced while the starting torque is increased when a star-delta starter is used to start a three-phase induction motor.

(a) Both A and R are true and R is a correct explanation of A.
(b) Both A and R are true but R is not a correct explanation of A.
(c) A is true but R is false.
(d) A is false but R is true.

Q.259 If rotor input of a SCIM running with a slip of 10% is 100 kW, gross power developed by its rotor is ______ kW.

(a) 10 (b) 90
(c) 99 (d) 80

Q.260 The fractional slip of an induction motor is the ratio of

(a) rotor cu-loss/rotor input
(b) stator cu-loss/stator input
(c) rotor cu-loss/rotor output
(d) rotor cu-loss/stator cu loss

Q.261 The torque developed by a 3-phase induction motor depends on

(a) speed, frequency and number of poles.
(b) voltage, current and stator impedance.
(c) synchronous speed, rotor speed and frequency.
(d) rotor emf, rotor current and rotor p.f.

Q.262 In case of a 3-phase induction motor, shaft power is 2700 W and mechanical losses are 180 W. At a slip of 4%, the rotor ohmic loss will be

(a) 115.2 W (b) 120 W
(c) 108 W (d) 105 W

Q.263 A 3-phase IM is driving a constant torque load of 1 pu at 5% slip. It has a maximum torque of 2 pu at 10% slip. Torque-speed variation in stable zone is assumed to be linear. For the stable operation of the motor, the minimum pu supply voltage is approximately

(a) 0.25 (b) 0.50
(c) 0.707 (d) 0.80

Q.264 In a 3-phase slip-ring induction motor, 3-phase balanced supply is given to the rotor and stator winding is short-circuited. The rotor would

(a) run at half the synchronous speed
(b) run in the direction of rotating field
(c) not run
(d) run against the direction of rotating field

Q.265 A prime-mover drives a 6-pole, 3-phase induction frequency converter. The converter is connected to 60 Hz, 3-phase supply on the primary. If the prime-mover speed is 3000 rpm, the frequencies of the possible outputs from the converter are

(a) 180 Hz, 210 Hz (b) 176 Hz, 86 Hz
(c) 90 Hz, 210 Hz (d) 120 Hz, 60 Hz

Q.266 A variable frequency induction motor is operated at constant *V/f* ratio. It has the following features over wide speed range regarding motor magnetising current (I_m) and maximum torque (T_m).

(a) Constant I_m, Constant T_m
(b) Variable I_m, Constant T_m
(c) Constant I_m, Variable T_m
(d) Variable I_m, Variable T_m

Q.267 Which of the following is not the advantage of an induction generator?

(a) When short circuited it delivers little or no sustained power.
(b) It delivers only leading current
(c) It is simple and rugged in construction, cheaper in cost and needs little maintenance.
(d) It does not hunt or drop out of synchronism.

Q.268 **Assertion (A):** Running light test of a three-phase induction motor is similar to the open-circuit test of a transformer.

Reason (R): Running light test gives the fixed losses occurring in a three-phase induction motor.

(a) Both A and R are true and R is a correct explanation of A.
(b) Both A and R are true but R is not a correct explanation of A.
(c) A is true but R is false.
(d) A is false but R is true.

Q.269 The frequency of the rotor current in a 3-phase, 50 Hz, 4-pole induction motor at full load speed is about

(a) 50 Hz (b) 0 Hz
(c) 20 Hz (d) 2 Hz

Q.270 The power factor of a squirrel cage induction motor is
(a) low at heavy load only
(b) low at light and heavy load both
(c) low at light load only
(d) low at rated load only

Q.271 The synchronous speed for a 3-phase, 6-pole induction motor is 1200 rpm. If the number of poles is now reduced to 4 with the frequency remaining constant, the rotor speed with a slip of 5% will be
(a) 1500 rpm (b) 1750 rpm
(c) 1690 rpm (d) 1710 rpm

6. Fractional KW Motors

Q.272 A 125 W, 4-pole, 110 V, 50 Hz single phase induction motor delivers rated output at a slip of 6%. The total copper loss at full load is 25 Watts. Rotational loss may be assumed to be 25 Watts. Neglecting stator copper loss, the full-load efficiency of the motor would be:
(a) 71.43% (b) 88.62%
(c) 75% (d) 90.08%

Q.273 The stepping angle for a 3-stack, 16-tooth variable reluctance motor is:
(a) 22.5° (b) 15°
(c) 7.5° (d) 3.75°

Q.274 Consider the following statements related to an universal motor:
1. An universal motor runs on d.c. as well as on a.c.
2. No extra winding is required in a universal motor for ac operation.
3. The direction of rotation of universal motor can be reversed by reversing either the field or armature leads, but not both.
4. Universal motors are designed for operation at low speeds (less than 3500 rpm).

Of these, the correct statements are:
(a) 2, 3 and 4 (b) 1 and 2
(c) 1, 2 and 4 (d) 1 and 3

Q.275 An 8-pole, 50 Hz, single phase induction motor is running at 690 rpm. It's slip w.r.t. backward and forward fields respectively are:
(a) 8%, 192% (b) 192%, 8%
(c) 8%, 200% (d) 192%, 6%

Q.276 Single phase ac motors are classified on the basis of
(a) rpm (b) construction
(c) starting method (d) both (b) and (c)

Q.277 In an induction motor
(a) stator current, stator flux and rotor emf are in phase
(b) rotor current and rotor flux are in phase and both lag behind the rotor emf by 90°
(c) both (a) and (b)
(d) none of the above

Q.278 The torque developed by a single phase induction motor drops to zero at
(a) synchronous speed
(b) a speed slightly above synchronous speed
(c) a speed slightly below synchronous speed
(d) none of the above

Q.279 In a single phase induction motor
(a) main winding is placed on stator and auxiliary winding on rotor.
(b) auxiliary winding is placed on stator and main winding on rotor.
(c) both the windings are placed on rotor.
(d) both the windings are placed on stator.

Q.280 In a comparative study of the torque-slip characteristic of a balanced poly-phase induction motor and that of a single phase induction motor, it is found that for zero slip
(a) the torque for the former is zero and has a non-zero negative value for the later.
(b) the torque has non-zero positve value on both of the characteristics.
(c) the torque is zero on both of the characteristics.
(d) the torque has a non-zero positive value on the former and has a non-zero negative value on the later.

Q.281 Double revolving field theory is based on the idea that pulsating field produced in single phase motors can be resolved into two components of ____ its amplitude and rotating in____ direction with synchronous sped.
(a) $\frac{1}{\sqrt{2}}$ times of, same
(b) half, opposite

(c) $\frac{1}{\sqrt{2}}$ times of, opposite
(d) half, same

Q.282 A single phase induction motor is
(a) self starting
(b) not self starting
(c) self starting with the help of an auxiliary winding
(d) none of the above

Q.283 The capacitor in a capacitor-start induction run ac motor is connected in series with
(a) compensating winding
(b) starting winding
(c) squirrel cage winding
(d) running winding

Q.284 Which of the following statements associated with shaded pole induction motor is incorrect?
(a) It has high operating efficiency.
(b) It has squirrel cage type rotor.
(c) It is one of the simplest and cheapest single phase induction motor.
(d) It has laminated salient poles each provided with its own exciting coil.

Q.285 Assertion (A) : In a single phase induction motor, while the motor is connected to single phase ac supply and is rotating, the two fields (stator and rotor field) combine to form a resultant rotating field revolving at a synchronous speed.
Reason (R) : The rotor field created by rotor currents acts at right angle to the stator field and also lags the stator field by 90° electrical degrees in time phase.
(a) Both A and R are true and R is a correct explanation of A.
(b) Both A and R are true but R is not a correct explanation of A.
(c) A is true but R is false.
(d) A is false but R is true.

Q.286 Assertion (A) : Linear induction motor wide use in industry and electric traction.
Reason (R) : Linear induction motors are cheaper in initial as well as in maintenance costs.
(a) Both A and R are true and R is a correct explanation of A.
(b) Both A and R are true but R is not a correct explanation of A.
(c) A is true but R is false.
(d) A is false but R is true.

Q.287 In comparison to a resistance start motor, the capacitor start motor has
1. high starting torque
2. high running torque
3. larger phase difference between two winding currents
4. low starting current

Out of these, the correct answer is
(a) 1, 2, 3 and 4 (b) 3 and 4
(c) 1, 3 and 4 (d) 1 and 4

Q.288 The direction of rotation of a shaded pole induction motor
(a) can be reversed by interchanging main winding terminals.
(b) can be reversed by interchanging the supply terminals.
(c) can be reversed by open-circuiting the shading rings.
(d) can't be reversed unless there is a provision for shifting of shading coil from one half to the other half of the pole.

Q.289 In various types of split-phase, 1-phase induction motors, the starting torques produced are in the following descending order.
(a) Capacitor-split, resistance-split, shaded pole
(b) Resistor-split, capacitor-split, shaded pole
(c) Shaded pole, resistor-split, capacitor split
(d) Capacitor-split, resistor-split, shaded pole

Q.290 Which of the following statements associated with an ac series motor is incorrect?
(a) Its torque-speed characteristic is similar to that of a dc series motor.
(b) Its torque-speed characteristic is similar to that of a dc shunt motor.
(c) Its power factor decreases with the increase in load torque.
(d) Its speed falls with the increase in load torque.

Q.291 Speed of universal motor is
(a) proportional to frequency of supply.

(b) proportional to square of frequency of supply.
(c) independent of frequency of supply.
(d) inversely proportional to frequency of supply.

Q.292 A single phase reluctance motor
(a) has salient pole rotor structure and runs at super-synchronous speed.
(b) has salient pole rotor structure and runs at sub-synchronous speed.
(c) has non-salient pole rotor structure and runs at synchronous speed.
(d) has salient pole rotor structure and runs at synchronous speed.

Q.293 The torque of a reluctance motor can be effectively decreased by
(a) decreasing reluctance of the magnetic circuit along the quadrature axis
(b) increasing reluctance of the magnetic circuit along the direct axis
(c) increasing the ratio of quadrature axis reluctance to direct axis reluctance
(d) decreasing the ratio of quadrature axis reluctance to direct axis reluctance

Q.294 In a single phase hysteresis motor
(a) starting torque is caused by only eddy current while running torque is caused by only hysteresis.
(b) starting as well as running torques are caused by both eddy current and hysteresis.
(c) starting torque is caused by both eddy current and hysteresis while the running torque is caused by hysteresis.
(d) starting as well as running torques both are caused by only hysteresis.

Q.295 **Match List-I (Type of motors)** with **List-II (Application)** and select the correct answer using the codes given below the lists:

List-I
A. Shaded pole motor
B. Reluctance motor
C. Universal motor
D. Single phase IM

List-II
1. Ceiling fan
2. Food mixture
3. Timing and control purposes
4. Toys

Codes:

	A	B	C	D
(a)	4	3	2	1
(b)	2	3	1	4
(c)	3	4	2	1
(d)	2	1	3	4

Q.296 Which of the following single phase motor has no teeth or winding?
(a) Universal motor (b) Hysteresis motor
(c) Split-phase motor (d) Reluctance motor

Q.297 **Match List-I (Type of motors)** with **List-II (Characteristics)** and select the correct answer using the codes given below the lists:

List-I
A. Shaded Pole motor
B. Hysteresis motor
C. PMDC motor
D. Capacitor start motor

List-II
1. Least noise
2. Highest starting torque
3. Lowest speed
4. Used in cassette tape recorder

Codes:

	A	B	C	D
(a)	3	4	2	1
(b)	1	4	2	3
(c)	3	1	4	2
(d)	1	2	3	4

Q.298 **Assertion (A):** When a ceiling fan using a capacitor run motor is switched on, it hums but does not run. When it is driven by some external means it runs in the direction in which it was made to run.
Reason (R): The capacitor connected in series with the auxiliary winding is short circuited.
(a) Both A and R are true and R is a correct explanation of A.
(b) Both A and R are true but R is not a correct explanation of A.
(c) A is true but R is false.
(d) A is false but R is true.

Q.299 The rotor of a stepper motor has
(a) a commutator and no winding
(b) a commutator and brushes only
(c) a commutator and no brushes
(d) neither a commutator, brushes nor winding

Q.300 A stepping motor has four concentric coils on its 4-pole stator. It has an unwound variable reluctance motor with 5 teeth. The step angle for this motor is

(a) 36° (b) 18°
(c) 9° (d) 72°

Q.301 A single phase induction motor is running at N rpm. Its synchronous speed is N_s. If its slip w.r.t. forward field is s. What is the slip w.r.t. the backward field?

(a) s (b) $-s$
(c) $(1 - s)$ (d) $(2 - s)$

Q.302 Stepper motors are mostly used for

(a) high power requirements
(b) very high speed of operation
(c) very low speed of operation
(d) control system applications

Q.303 Consider the following types of 1-ϕ motors:

1. Capacitor start induction motor.
2. Capacitor start and run induction motor.
3. Permanent split capacitor motor.
4. Shaded pole motor.

The correct sequence of these in the ascending order of magnitude of starting torque is

(a) 1, 2, 3, 4 (b) 2, 3, 4, 1
(c) 3, 4, 2, 1 (d) 4, 3, 2, 1

Q.304 In a stepper motor the angular displacement

(a) cannot be readily interfaced with micro computer based controller.
(b) cannot be precisely controlled.
(c) can be precisely controlled.
(d) cannot be used for positioning of work tables and tools in NC machines.

Q.305 A hysteresis motor

(a) is a constant speed motor.
(b) is not a self-starting motor.
(c) cannot be run in reverse speed.
(d) needs dc excitation.

Q.306 The most suitable servomotor for low power application is

(a) a dc series motor.
(b) a dc shunt motor.
(c) an ac series motor.
(d) an ac two-phase induction motor.

Q.307 A stepper motor is

(a) a single-phase ac motor.
(b) a dc motor.
(c) a multi-phase motor.
(d) a two-phase motor.

7. Basic Concept of Rotating Electrical Machines

Q.308 A linear magnetic circuit has flux linkages of 1.2 Wb turns when a current of 10 A flows through its coil. The energy stored in the magnetic field of this coil, in Joules is

(a) 2.4 (b) 1.2
(c) 12 (d) 6

Q.309 The electromagnetic force and/or torque, developed in any physical system, acts in such a direction as to tend to

(a) increase the magnetic stored energy at constant flux.
(b) increase the magnetic stored energy at constant current.
(c) decrease the magnetic stored energy at constant flux.
(d) decrease the magnetic stored energy at constant mmf.

Q.310 Singly-and doubly-excited magnetic systems are respectively

(a) reluctance motor and synchronous motors
(b) dc shunt machines and solenoids
(c) synchronous motors and moving-iron instruments
(d) loud-speakers and tachometers

Q.311 A rotating electromechanical energy conversion device has uniform air-gap. If δ is the space angle between the axis of stator field and rotor field, then the average torque developed is proportional to (a and b are constants)

(a) $a \sin \delta + B \sin 2\delta$ (b) $a \sin \delta$
(c) $a \sin 2\delta$ (d) δ

Q.312 Armature winding is one in which working

(a) flux is produced by the working emf.
(b) emf is produced by the working flux.
(c) flux is produced by field current.
(d) emf is produced by the leakage flux.

Q.313 In ac rotating machines, the generated or speed emf

(a) lags ϕ by 90°
(b) leads ϕ by 90°
(c) is in phase with the working flux ϕ
(d) lags ϕ by 180°

Q.314 A 3-phase, 4-pole alternator has 48 stator slots carrying a 3-phase distributed winding. Each coil of the winding is short chorded by one slot pitch. The belt-factor of the winding is given by

(a) $\frac{\cos 7.5}{16}$ (b) $\frac{\sin 7.5}{8}$
(c) $\frac{1}{8 \sin 7.5}$ (d) $\frac{\cos 7.5}{8}$

Q.315 Consider the following statements about the chorded coils:
1. Chorded coil leads to reduction in harmonics.
2. In chorded coils, coil span may be less or more than 180° electrical.
3. Chorded coils require more conductors material.
4. Chorded coils reduce the magnitude of output voltage.
5. Chorded coils give better generated emf waveform.
6. In chorded coils, coil span is less than 180° electrical.

From these, the correct statements are:
(a) 1, 2, 4 and 6 (b) 4, 5 and 6
(c) 1, 4, 5 and 6 (d) 2, 3 and 4

Q.316 For eliminating 3rd harmonic from the phase emf generated in an alternator, the coil span in terms of full-pitch would be

(a) $\frac{5}{4}$ (b) $\frac{4}{3}$
(c) $\frac{5}{3}$ (d) $\frac{2}{3}$

Q.317 If the dimensions of all the parts of a synchronous generator, and the number of field and armature turns are doubled, then the generated voltage will change by a factor of
(a) 2 (b) 8
(c) 4 (d) 1

Q.318 A 6-pole machine is rotating at a speed of 1200 rpm. This speed in electrical radians per sec and mechanical radian per sec are respectively given by

(a) $60\pi, 20\pi$ (b) $\frac{40\pi}{3}, 40\pi$
(c) $40\pi, 120\pi$ (d) $120\pi, 40\pi$

Q.319 In an alternator, frequency per revolution is equal to
(a) speed in rps
(b) twice the number of poles
(c) number of pole-pairs
(d) number of poles

Q.320 A three-phase alternator is wound with a 60 degree phase spread armature winding and develops 300 kVA. If the armature is reconnected utilizing all the coils for single-phase operation with a phase spread of 180 degree, the new rating of the machine will be
(a) 300 kVA (b) 200 kVA
(c) 100 kVA (d) 250 kVA

Q.321 A synchronous machine has full-pitch coils having coil-span of 12 slots. For eliminating third harmonic, the coil-span should be
(a) 10 slots (b) 9 slots
(c) 8 slots (d) 6 slots

Q.322 The 5th harmonic mmf wave, produced by 3-phase currents flowing in 3-phase balanced winding, rotates w.r.t fundamental field (N_s = speed of fundamental mmf wave) at a speed of

(a) $\frac{5}{4}N_s$ (b) $\frac{5}{6}N_s$
(c) $\frac{4}{5}N_s$ (d) $\frac{6}{5}N_s$

Q.323 Match **List-I (Power losses)** with **List-II (Dependent upon)** and select the correct answer using the codes given below the lists:

List-I	List-II
A. Stray load loss	1. Load
B. Brush constant loss	2. Value of flux
C. Hysteresis loss	3. Rotor rotation
D. No-load rotational loss	4. Square of load

Codes:

	A	B	C	D
(a)	3	1	4	2
(b)	4	1	2	3
(c)	3	4	2	1
(d)	4	1	3	2

Q.324 Output rating of
1. dc generators is in kVA.
2. ac motor is in kW.
3. ac generator is in kVA.
4. dc motors is in kW.
5. induction motors is in h.p.
6. synchronous motors is in kVA.

From these, the correct statements are:
(a) 2, 3 and 4 (b) 1, 3, 4 and 5
(c) 1, 4 and 5 (d) 2, 3, 5 and 6

Q.325 If P_i be the iron loss and P_0 be the ohmic loss on full load, then which of the following conditions has to be satisfied to obtain maximum efficiency at 2/3 full load?
(a) $P_0 = \frac{2}{3}P_i$ (b) $P_0 = \frac{3}{2}P_i$
(c) $P_0 = \frac{4}{9}P_i$ (d) $P_0 = \frac{9}{4}P_i$

Q.326 If ϕ_m is the maximum value of flux due to any one phase, the resultant flux in 3-phase and 2-phase ac machines would be given by
(a) $2\,\phi_m$ and $1.5\,\phi_m$; both rotating
(b) $1.5\,\phi_m$, rotating and ϕ_m, standstill
(c) $1.5\,\phi_m$, and ϕ_m; both standstill
(d) $1.5\,\phi_m$, and ϕ_m; both rotating

Q.327 Resistance of one conductor for a dc machine is 10 mΩ. What would be the armature circuit resistance if the dc machine with 640 conductors and 8 poles is wave-connected ?
(a) 3.2 Ω (b) 2.4 Ω
(c) 1.6 Ω (d) 1.2 Ω

Q.328 A 6-pole, 50 Hz, 3-phase induction motor has a stator bore diameter of 1.2 meters. Its stator is cut and laid out flat, so that rotating mmf wave now becomes a travelling mmf wave. What is the linear velocity of the travelling mmf wave in meters per second ?
(a) 56.6 (b) 62.8
(c) 36.4 (d) None of these

Q.329 **Assertion (A) :** In a rotating electrical machine, having 2 poles on the stator and 4 poles on the rotor, the net electromagnetic torque developed is zero.
Reason (R) : In all rotating electrical machines, the number of rotor poles must be equal to the number of stator poles for development of electromagnetic torque.
(a) Both A and R are true and R is a correct explanation of A.
(b) Both A and R are true but R is not a correct explanation of A.
(c) A is true but R is false.
(d) A is false but R is true.

Q.330 **Assertion (A) :** Exciting winding is one which produces working flux.
Reason (R) : Armature winding in one in which the working emf is induced by the working flux.
(a) Both A and R are true and R is a correct explanation of A.
(b) Both A and R are true but R is not a correct explanation of A.
(c) A is true but R is false.
(d) A is false but R is true.

Q.331 **Assertion (A) :** At the time of winding design of three-phase alternators, the attention is mainly directed for the attenuation of triplen harmonics, by adopting a chording angle of 60° electrical.
Reason (R): Three-phase alternators are invariably star-connected to eliminate triplen harmonics from the line e.m.f.
(a) Both A and R are true and R is a correct explanation of A.
(b) Both A and R are true but R is not a correct explanation of A.
(c) A is true but R is false.
(d) A is false but R is true.

Q.332 The induced emf in the armature conductors of a d.c motor is
(a) sinusoidal (b) trapezoidal
(c) rectangular (d) alternating

Q.333 The principle of operation of a 3-phase induction motor is most similar to that of a
(a) synchronous motor
(b) repulsion-star induction motor
(c) transformer with a shorted secondary
(d) capacitor-star induction motor

Q.334 The magnetizing current drawn by transformers and induction motors is the cause of their ______ power factor.
(a) zero (b) lagging
(c) unity (d) leading

Q.335 If in an alternator, chording angle for fundamental flux wave is α, its value of 5th harmonic will be
(a) 5α (b) $\frac{\alpha}{5}$
(c) 25α (d) $\frac{\alpha}{25}$

Answers Electrical Machines

1. (d)	2. (a)	3. (b)	4. (c)	5. (a)	6. (d)	7. (b)	8. (c)
9. (d)	10. (c)	11. (b)	12. (a)	13. (c)	14. (d)	15. (b)	16. (d)
17. (a)	18. (a)	19. (b)	20. (b)	21. (a)	22. (d)	23. (c)	24. (a)
25. (b)	26. (b)	27. (d)	28. (c)	29. (a)	30. (c)	31. (b)	32. (d)
33. (c)	34. (a)	35. (d)	36. (d)	37. (b)	38. (c)	39. (a)	40. (c)
41. (d)	42. (d)	43. (b)	44. (c)	45. (a)	46. (d)	47. (a)	48. (b)
49. (c)	50. (d)	51. (d)	52. (d)	53. (b)	54. (a)	55. (c)	56. (a)
57. (c)	58. (b)	59. (c)	60. (b)	61. (c)	62. (d)	63. (b)	64. (a)
65. (c)	66. (b)	67. (c)	68. (d)	69. (b)	70. (d)	71. (a)	72. (c)
73. (c)	74. (d)	75. (a)	76. (d)	77. (d)	78. (a)	79. (c)	80. (c)
81. (c)	82. (a)	83. (c)	84. (d)	85. (a)	86. (c)	87. (d)	88. (b)
89. (d)	90. (c)	91. (b)	92. (a)	93. (d)	94. (b)	95. (a)	96. (c)
97. (d)	98. (a)	99. (c)	100. (b)	101. (d)	102. (b)	103. (c)	104. (d)
105. (d)	106. (d)	107. (c)	108. (b)	109. (c)	110. (c)	111. (c)	112. (b)
113. (c)	114. (d)	115. (a)	116. (b)	117. (d)	118. (d)	119. (b)	120. (d)
121. (a)	122. (c)	123. (a)	124. (b)	125. (b)	126. (c)	127. (d)	128. (b)
129. (a)	130. (d)	131. (a)	132. (c)	133. (a)	134. (d)	135. (d)	136. (b)
137. (c)	138. (c)	139. (a)	140. (b)	141. (b)	142. (d)	143. (a)	144. (b)
145. (d)	146. (a)	147. (b)	148. (c)	149. (d)	150. (c)	151. (d)	152. (c)
153. (a)	154. (b)	155. (c)	156. (b)	157. (d)	158. (a)	159. (a)	160. (c)
161. (a)	162. (b)	163. (d)	164. (a)	165. (b)	166. (b)	167. (d)	168. (b)
169. (a)	170. (d)	171. (c)	172. (c)	173. (d)	174. (d)	175. (a)	176. (a)
177. (d)	178. (a)	179. (d)	180. (c)	181. (b)	182. (a)	183. (a)	184. (d)
185. (c)	186. (c)	187. (b)	188. (d)	189. (a)	190. (c)	191. (d)	192. (a)
193. (d)	194. (c)	195. (a)	196. (c)	197. (b)	198. (c)	199. (d)	200. (a)
201. (c)	202. (c)	203. (a)	204. (b)	205. (d)	206. (c)	207. (d)	208. (a)
209. (b)	210. (c)	211. (a)	212. (b)	213. (d)	214. (c)	215. (a)	216. (d)

217. (b)	218. (a)	219. (c)	220. (b)	221. (c)	222. (d)	223. (d)	224. (b)
225. (a)	226. (c)	227. (d)	228. (c)	229. (d)	230. (b)	231. (a)	232. (d)
233. (b)	234. (d)	235. (a)	236. (a)	237. (c)	238. (d)	239. (c)	240. (d)
241. (b)	242. (a)	243. (a)	244. (b)	245. (d)	246. (c)	247. (b)	248. (a)
249. (b)	250. (c)	251. (d)	252. (c)	253. (d)	254. (d)	255. (a)	256. (c)
257. (a)	258. (c)	259. (b)	260. (a)	261. (d)	262. (b)	263. (c)	264. (d)
265. (c)	266. (a)	267. (b)	268. (b)	269. (d)	270. (c)	271. (d)	272. (a)
273. (c)	274. (c)	275. (b)	276. (d)	277. (c)	278. (c)	279. (d)	280. (a)
281. (b)	282. (c)	283. (b)	284. (a)	285. (a)	286. (b)	287. (c)	288. (d)
289. (a)	290. (b)	291. (c)	292. (d)	293. (d)	294. (c)	295. (a)	296. (b)
297. (c)	298. (c)	299. (d)	300. (b)	301. (d)	302. (d)	303. (d)	304. (c)
305. (a)	306. (b)	307. (c)	308. (d)	309. (c)	310. (a)	311. (b)	312. (b)
313. (a)	314. (c)	315. (c)	316. (d)	317. (b)	318. (d)	319. (c)	320. (b)
321. (c)	322. (d)	323. (b)	324. (a)	325. (d)	326. (d)	327. (c)	328. (b)
329. (a)	330. (a)	331. (d)	332. (a)	333. (c)	334. (b)	335. (b)	

Explanations

1. (d)

Power flow through the coil is: $p = ei$

$$= \left(L\frac{di}{dt}\right)\cdot i = Li\cdot\frac{di}{dt}$$

or,
$$L = \frac{p}{\left(i\frac{di}{dt}\right)} = \frac{0.5}{0.1\times 0.5} = 10 \text{ Henery}$$

2. (a)

Voltage across the coil $= L\frac{di}{dt} + IR$

Given, number of turns of coil $= N = 100$

$$\frac{di}{dt} = 200 \text{ A/s}$$

Reluctance, $Rl = 10^4$ AT/Wb

∴ Inductance of coil,

$$L = \frac{N^2\mu_0\mu_r A}{l} = \frac{N^2}{Rl}$$

$$= \frac{100^2}{10^4} = 1 \text{ H}$$

So, voltage across the coil

$$= 1 \times (200) + 5 \times 2$$
$$= 210 \text{ volt}$$

3. (b)

Current through the field winding,

$$I = \frac{\text{Excitation voltage}}{\text{Field winding resistance}}$$

$$= \frac{230}{50} = 4.6 \text{ A}$$

Given, $\phi = 5$ mWb

∴ Inductance of coil,

$$L = \frac{N\phi}{I} = \frac{960\times 0.005}{4.6}$$
$$= 1.043 \text{ H}$$

So, energy stored in magnetic field

$$= \frac{1}{2}LI^2 = \frac{1}{2}\times 1.043\times(4.6)^2$$
$$= 11.04 \text{ J} \approx 11 \text{ Joules}$$

4. (c)

We know that eddy current loss,

$$P_e \propto f^2 B_m^2$$

$$\therefore \quad \frac{P_{e2}}{P_{e1}} = \left(\frac{f_2}{f_1}\right)^2 \times \left(\frac{B_{m2}}{B_1}\right)^2$$

or, $$P_{e2} = P_{e1} \times \left(\frac{f_2}{f_1}\right)^2 \times \left(\frac{B_{m2}}{B_{m1}}\right)^2$$

$$= 15 \times \left(\frac{1.5}{2}\right)^2 \times \left(\frac{800}{600}\right)^2 = 15 \text{ W}$$

5. (a)

$$M = K\sqrt{L_1 L_2}$$

$$= 0.8\sqrt{15 \times 10^{-3} \times 25 \times 10^{-3}}$$

$$= 0.8 \times 10^{-3}\sqrt{15 \times 25}$$

$$= 15.5 \text{ mH}$$

6. (d)

Resistance in electrical circuit corresponds to reluctance in magnetic circuit. Reluctance opposes the production of magnetic flux in a magnetic circuit and resistance opposes the flow of current in electrical circuit.

8. (c)

From Faraday's law of electromagnetic induction,

$$e = -\frac{d\psi}{dt}$$

(minus sign is due to Lenz's law).

9. (d)

$$e = L\frac{di}{dt}$$

Hence, emf induced is maximum when di/dt is maximum.

10. (c)

As the field is bipolar, therefore the induced emf will be alternating in nature.

11. (b)

$$I = \frac{\text{EMF}}{R} \text{ (for an electric circuit)}$$

$$\phi = \frac{\text{MMF}}{\text{Redactance}}$$

(for a magnetic circuit)

13. (c)

Given, $\theta = 90°$

$\therefore$ emf induced $= B\, l\, v \sin\theta$

$= 1.2 \times 0.2 \times 30 \times \sin 90°$

$= 7.2$ volt

14. (d)

Given, $\mu_r = 100$, $a = 5$ cm^2, $l = 25$ cm, $N = 120$ turns, NI = MMF = 1000 AT

We know that,

$$\text{Flux} = \frac{\text{MMF}}{Rl}$$

$$= \frac{NI}{\left(\frac{l}{\mu_0 \mu_r A}\right)} = \frac{NI \cdot \mu_0 \mu_r A}{l}$$

or, $$\text{Flux} = \frac{1000 \times 4\pi \times 10^{-7} \times 100 \times 5 \times 10^{-4}}{25 \times 10^{-2}}$$

$$= \frac{4\pi}{5} \times 10^{-4} = 0.25 \text{ mWb}$$

15. (b)

$$H = \frac{NI}{l}\left(\frac{\text{Ampere-turn}}{\text{metre}}\right)$$

$$\phi = B.A.$$

or, $$B = \frac{\phi}{A} \text{ (Wb/m}^2\text{)}$$

ϕ has a unit of Wb; permeability, μ has unit of Henery/m.

16. (d)

Since, $$L = \frac{N^2 \mu A}{l}$$

or, $L \propto N^2$

Since, turns of one coil is decreased to half, therefore its new inductance,

$$L_1' = \frac{L_1}{4}$$

Since, turns of other coil is doubled, therefore its new value of inductance,

$$L_2' = 4L_2$$

Now, mutual inductance,

$$M = k\sqrt{L_1 L_2}$$

So, new value of mutual inductance is,

$$M' = K\sqrt{L_1' L_2'}$$

$$\therefore \quad M' = M.\sqrt{\frac{L_1' \cdot L_2'}{L_1 L_2}} = 1 \cdot \sqrt{\frac{\frac{1}{4}L_1 \times 4L_2}{L_1 L_2}}$$

$$= 1 \text{ H}$$

19. **(b)**

We know that inductance,

$L \propto \mu$ (Permeability) $\propto \mu_0 \mu_r \propto \mu_r$. Since relative permeability of iron is more than that of air ($\mu_r = 1$) therefore, with iron-cored coil L will be more while with air-cored coil L will be less.

20. **(b)**

$$L = \frac{N^2 \mu_0 \mu_r A}{l} \text{ or } L \propto N^2 \mu_r$$

$$\therefore \quad \frac{L_2}{L_1} = \left(\frac{N_2}{N_1}\right)^2 \times \left(\frac{\mu_{r_2}}{\mu_{r_1}}\right)$$

or, $$\frac{L_2}{L_1} = (2)^2 \times \frac{600}{400} = 6$$

or, $L_2 = 6L_1$

or, $L_2 = 6 \times 50 \times 10^{-3} = 0.3$ H

21. **(a)**

Since length of air-gap is increased, therefore, reluctance offered to the magnetic circuit will increase.

Also, $$\text{Flux} = \frac{MMF}{\text{Reluctance}} = \frac{NI}{\text{Reluctance}}$$

Hence, to maintain the constant flux, the choke will draw more current so that the net mmf is increased.

22. **(d)**

$$L = \frac{N\phi}{I} = \frac{1000 \times 10^{-3}}{1} = 1 \text{ H}$$

$$\therefore \text{ Energy stored} = \frac{1}{2}LI^2 = \frac{1}{2} \times 1 \times 1^2$$

$$= \frac{1}{2} \text{ Joule}$$

24. **(a)**

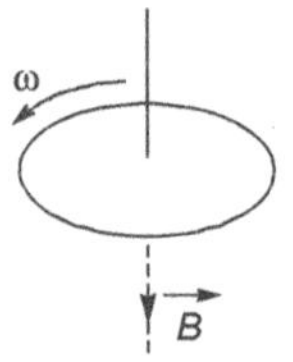

Since v and B will be parallel, therefore emf induced = 0.

26. **(b)**

Rate of change of current is

$$\frac{di}{dt} = \frac{5-(-5)}{8 \times 10^{-3}} = 1250 \text{ A/s}$$

$\therefore$ Emf induced in the coil is

$$e = L\frac{di}{dt}$$

$$= 0.01 \times 1250 = 12.5 \text{ V}$$

27. **(d)**

Emf induced, $e = N\phi\omega$

Here, $\phi = B \times A$

$= 60 \times 10^{-3} \times 5 \times 10^{-2}$

$= 3 \times 10^{-3}$ Wb

And, $$\omega = \frac{2\pi}{60}N = \frac{2\pi \times 1500}{60} = 50\,\pi \text{ rad/s}$$

Given, N = Number of turns = 800

So, emf induced,

$e = 800 \times 3 \times 10^{-3} \times 50\,\pi$

$= 377$ volt

28. **(c)**

In an ideal transformer there is no-loss (either in winding or core). The leakage reactance is also zero (as no-voltage drop).

Magnetizing current is zero due to infinite permeability of core.

29. **(a)**

Maximum flux,

$$\phi_m = \frac{\text{Emf per turn}}{4.44f}$$

$$= \frac{10}{4.44 \times 50} = 0.045 \text{ Wb}$$

$\therefore$ New cross-sectional area of core,

$$A = \frac{\phi_{max}}{B_{max}} = \frac{0.045}{1.5} = 0.03 \text{ m}^2 = 300 \text{ cm}^2$$

30. **(c)**

kVA rating of transformer,

$S \propto VI$

or, $S \propto V$ (I = constant)

Since $V \propto f$, therefore, $S \propto f$

$$\therefore \quad \frac{S_{60}}{S_{400}} = \frac{60}{400} \times 1000 = 150 \text{ VA}$$

31. (b)

Given, $I_0 = 0.5$ A

$\cos\phi_0 = 0.3$ (lag)

$\therefore$ $I_c = I_0 \cos\phi_0 = 0.5 \times 0.3 = 0.15$ A

and $I_m = \sqrt{I_0^2 - I_C^2} = \sqrt{0.5^2 - 0.15^2}$

$= 0.477$ A

I_m can also be calculated as follows

or, $I_m = I_0 \sin\phi_0 = 0.5 \times \sqrt{1 - 0.5^2}$

$= 0.477$ A

32. (d)

Given, $Z = (0.5 + j7.0)\%$

$= (R + jX)\%$

$\therefore$ Regulation at 0.8 pf lag is

$\epsilon_R = (\%R\cos\phi + \%X\sin\phi)$

$= 0.5 \times 0.8 + 7 \times 0.6 = 4.6\ \%$

Hence, voltage to be raised, for maintaining rated voltage = 4.6%.

Thus, tap setting required on the HV side = –4.6% or 4.6 % down.

33. (c)

$$\epsilon_R = (\%V_R)\cos\phi \pm (\%V_x)\sin\phi$$

At 0.8 pf lag,

$\epsilon_R = 1 \times 0.8 + 5 \times 0.6$

$= 0.8 + 3 = 3.8\ \%$

At 0.8 pf lead,

$\epsilon_R = 1 \times 0.8 - 5 \times 0.6$

$= 0.8 - 3 = -2.2\ \%$

34. (a)

Since, $\frac{V}{f}$ = constant

$\therefore$ ϕ_m = constant

or, B_m = constant

$\therefore$ $P_e \propto V^2$

$\therefore$ $\frac{Pe_2}{Pe_1} = (2)^2$

or, $P_{e2} = 4P_{e1} = 2 \times 800 = 1600$ W

Also, $P_h \propto f$

$\therefore$ $\frac{P_{h2}}{P_{h1}} = \frac{f_2}{f_1} = 2$

or, $P_{h2} = 2P_{h1} = 2 \times 650 = 1300$ W

$\therefore$New core loss $= P_{e2} + P_{h2}$

$= 1600 + 1300$

$= 2900$ W

35. (d)

We know that, eddy current loss $P_e \propto V^2$ and hysteresis loss,

$$P_h \propto \frac{V^{1.6}}{f^{0.6}}$$

When V = constant, P_e = constant

and $P_h \propto \frac{1}{f^{0.6}}$

When frequency is increased from 50 Hz to 60 Hz, P_h will increase. Thus, core loss will increase marginally, therefore efficiency will decrease marginally.

36. (d)

Primary and secondary windings of a core type transformer have circular coil section to reduce the mean length of each turn so that it require minimum conductor material and thus have less copper loss which will increase the efficiency.

37. (b)

$\cos\phi = 1$ and $\sin\phi = 0$(since the load is resistive) voltage regulation,

$\epsilon_R = R_{pu}\cos\phi + X_{pu}\sin\phi$

$= 0.03 \times 1 + 0.06 \times 0$

$= 0.03$ pu

or, $\frac{E_2 - V_2}{V_2} = 0.03$

or, $\frac{E_2}{V_2} = 1.03$

or, $V_2 = \frac{E_2}{1.03} = \frac{200}{1.03} = 194$ volt

Hence, load voltage = 194 volt

38. (c)

Voltage regulation of a power transformer is zero when $\cos\phi = 0$ (leading).

40. (c)

Given, $\frac{N_{Y(P)}}{N_{\Delta(P)}} = 5 = \frac{I_{\Delta(P)}}{I_{Y(P)}}$

or, $I_{\Delta(P)} = 5 \cdot I_{Y(P)}$ ...(*i*)

Also, $I_{\Delta(P)} = \frac{I_{\Delta(L)}}{\sqrt{3}} = \frac{10}{\sqrt{3}}$ A ...(*ii*)

From equation (*i*) and (*ii*),

$$I_{Y(P)} = \frac{10}{5\sqrt{3}} = \frac{2}{\sqrt{3}} = 1.155 \text{ A}$$

Now, in star connection,

$$I_{Y(P)} = I_{Y(L)} = 1.155 \text{ A}$$

41. (d)

Proportion of full load at maximum efficiency

$$= \sqrt{\frac{P_i}{P_{cu(fl)}}} = \sqrt{\frac{400}{800}}$$

$$= \frac{1}{\sqrt{2}} = 0.707 = 70.7\%$$

43. (b)

Given, $a_{2W} = 4$ and $S_{2W} = 10$ kVA

Maximum rating of the autotransformer is

$$(S_a)_{max} = (S_{2W} + a_{2W} \cdot S_{2W})$$
$$= 10 + 4 \times 10 = 50 \text{ kVA}$$

44. (c)

Given, $\dfrac{N_{\Delta(P)}}{N_{Y(P)}} = 5 = \dfrac{V_{\Delta(P)}}{V_{Y(P)}}$

Also, $V_{Y(P)} = \dfrac{400}{\sqrt{3}}$ V

$\therefore$ $V_{\Delta(P)} = 5 \times \dfrac{400}{\sqrt{3}} = \dfrac{2000}{\sqrt{3}}$

$= 1154.7$ volt

For delta connection,

$$V_{\Delta(L)} = V_{\Delta(P)} = 1154.7 \text{ volt}$$

45. (a)

Given, $Z_{pu} = (0.02 + j0.08)$ pu

$$Z_{base(HV)} = \frac{(250)^2}{5 \times 10^3} = 12.5\ \Omega$$

$\therefore$ $(Z_{actual})_{HV} = Z_{pu} \times Z_{base(HV)}$

$= (0.02 + j0.08) \times 12.5$

$= (0.25 + j1)\ \Omega$

46. (d)

At 0.8 p.f. lag,

$$\in_R = (R_{pu}\cos\phi + X_{pu}\sin\phi)$$

or, $0.025 = 0.8\,R_{pu} + 0.6\,X_{pu}$

Also, at 0.8 p.f. lead,

$$\in_R = 0.8\,R_{pu} - 0.6\,X_{pu}$$

Hence, possible value of $\in_R = -0.9\%$

(since at leading p.f., $\in_R$ can't be ± 2.5% or zero.)

48. (b)

% Z of distribution transformer is less (around 10%).

Flux density, B_m is less for a distribution transformer compared to a power transformer. Thus, core loss is less.

Hence, $\dfrac{I^2R \text{ loss}}{\text{core loss}}$ is more.

49. (c)

$$\phi_m \propto \frac{V}{f} \propto I_m$$

Since, V = constant and f is decreased, therefore there will be overfluxing in the transformer. Hence, it will draw more magnetizing current to provide this increased flux.

Also, $P_e \propto V^2$

$\therefore$ P_e = Constant (As V = constant)

and $P_h \propto \dfrac{V^{1.6}}{f^{0.6}} \propto \dfrac{1}{f^{0.6}}$

As $f\downarrow$, $P_h\uparrow$

Hence, core loss will increase. Therefore, I_c will also increase (As core loss = VI_c).

50. (d)

Given, $a = 3$

$\therefore$ Total equivalent impedance referred to the H.V. side

$= (9 + j45) + (3)^2(1 + j5)$

$= 9 + j45 + 9 + j45$

$= (18 + j90)\ \Omega$

51. (d)

At η_{max},

Copper loss = Iron loss

or, $I_m^2 r_e = P_c$

or, $I_m = \sqrt{\dfrac{P_C}{r_e}}$

52. (d)

$$I_{SC} = \frac{V}{|Z_{eq}|} = \frac{V}{\sqrt{R_{eq}^2 + X_{eq}^2}}$$

As frequency decrease from 50 Hz to 25 Hz, then X_{eq} decreases.

So, I_{SC} (short circuit current) increases.

Also, at short circuit,

$$\text{p.f.} = \cos\phi_{sc} = \frac{R_{eq}}{\sqrt{R_{eq}^2 + X_{eq}^2}}$$

Thus, cosϕ is also increased due to decrease In X_{eq}.

53. (b)

Given, $$a_{auto} = \frac{253}{230} (> 1)$$

Now, $$\frac{S_{auto}}{S_{2w}} = \left(\frac{a_{auto}}{a_{auto}-1}\right) = \frac{253}{23} = 11$$

$$\therefore \quad S_{auto} = 11 \times 20 = 220 \text{ kVA}$$

Also,

$$\text{Conducted kVA} = \frac{1}{a_{auto}} \times S_{auto}$$

$$= \frac{230}{253} \times 220 = 200 \text{ kVA}$$

54. (a)

$$V = \sqrt{2}\pi f \phi_m N$$

$$\therefore \quad \phi_m \propto \frac{V}{fN} \text{ or } B_m \propto \frac{V}{fNA}$$

$$\therefore \quad \frac{B_{m_2}}{B_{m_1}} = \frac{V_2}{V_1} \times \frac{f_1 N_1 A_1}{f_2 N_2 A_2}$$

$$= \left(\frac{400}{200}\right) \times \left(\frac{50}{50}\right) \times \left(\frac{N}{N/2}\right) \times \left(\frac{A}{4A}\right)$$

$$= 2 \times 2 \times \frac{1}{4} = 1$$

$$\therefore \quad B_{m2} = B_{m1} = 1.2\text{T}$$

57. (c)

Reason is false because at open-circuit, the energy input is required to meet no-load copper losses in primary and iron losses in the core.

59. (c)

For proportional load sharing, the pu impedances of the two transformers on their own respective bases should be equal.

i.e. $$Z_{1(pu)} = Z_{2(pu)}$$

or, $$\frac{Z_1}{Z_{1(Base)}} = \frac{Z_2}{Z_{2(Base)}} \text{ or } \frac{Z_1}{Z_2} = \frac{Z_{1(Base)}}{Z_{2(Base)}}$$

or $$\frac{Z_1}{Z_2} = \frac{S_{2(Rated)}}{S_{1(Rated)}} \text{ or } Z \propto \frac{1}{S_{(Rated)}}$$

61. (c)

When transformer is switched at an instant when $V = 0$, flux in the core rises from $-\phi_m$ to $+\phi_m$ in half cycle (doubling effect) and calls for huge magnetizing current in the primary winding of the transformer.

62. (d)

Due to the polarities (dot marking) in the three windings,

$$V = -200 - 200 + 400 = 0 \text{ volt}$$

63. (b)

Scott-connection converts 2-phase to 3-phase connections and vice-versa.

68. (d)

Due to the voltage drop in secondary winding, the secondary induced emf will be more than secondary terminal voltage ($E_2 > V_2$).

Using phasor diagram of transformer we can conclude that *pf* on primary side will be lower than that of load.

Also, I_1 and I_2 will be out of phase.

69. (b)

Given, $Z_{eq} = 0.02$ pu,

Rated current $= 1 \text{ pu} = I_{sc}$

$\therefore$ Applied voltage $= V_{sc} = I_{sc} \cdot Z_{eq}$

$= 1 \times 0.02 = 0.02$ pu

$= 0.02 \times 3000 = 60$ V

70. (d)

Given, $$a_{auto} = \frac{3}{2} (>1)$$

$$\therefore \quad \frac{S_{auto}}{S_{2w}} = \frac{a_{auto}}{(a_{auto}-1)} = \frac{\frac{3}{2}}{\frac{3}{2}-1} = 3$$

$$\therefore \quad \frac{S_{2w}}{S_{auto}} = \frac{1}{3} \text{ or } S_{2W} = 0.33\, S_{auto}$$

71. (a)

Current drawn by primary is

$$I_1 = \frac{1 \times 10^6}{50 \times 10^3} = 20 \text{ A}$$

and load current,

$$I_2 = \frac{1\times 10^6}{40} = 25\,\text{A}$$

Hence, current in common section of the winding

$$= I_2 - I_1 = 25 - 20 = 5\text{ A}$$

73. (c)

$$f_2 = 45\text{ Hz} = 90\%\text{ of }50\text{ Hz}$$

Here, $\frac{V}{f} = \frac{90\%\text{ of }V}{90\%\text{ of }50\text{ Hz}}$

$= \text{constant (for both cases)}$

∴ $B_m = \text{constant}$

Now, $P_e \propto f^2 \quad (B_m = \text{constatnt})$

∴ $\frac{P_{e_2}}{P_{e_1}} = (0.9)^2$

or, $P_{e_2} = (0.9)^2 P_{e_1}$ and $P_h \propto f$

∴ $\frac{P_{h2}}{P_{h1}} = 0.9$ or $P_{h2} = 0.9\,P_{h1}$

Since $P_{e1} = P_{h1} = P$

Therefore, $P_{e2} = 0.81\,P_1$ and $P_{h2} = 0.9\,P$

∴ Total core loss,

$$P_2 = (0.9 + 0.81)P = 1.71\,P$$

Also, initial core loss $= P_1 = 2P$

∴% reduction in core loss

$$= \left(\frac{2P - 1.71P}{2P}\right)\times 100 = 14.5\%$$

74. (d)

$$\epsilon_R = z_{pu}\cos(\theta_{eq} - \phi)$$

Maximum voltage regulation occurs at lagging p.f. when $\theta_{eq} = \phi$.

Hence, $(\epsilon_R)_{max} = Z_{pu} = Z_{e2}$,

At $\cos\phi = \frac{r_{e2}}{z_{e2}}$ (lagging)

75. (a)

$$\eta = \left(\frac{VI\cos\phi}{VI\cos\phi + I^2R_{eq} + P_i}\right)$$

When, $I = \text{constant}$,

η is maximum at $\cos\phi = 1$.

76. (d)

kVA rating of bank of transformer = thrice of single-phase rating = 3 × 250 = 750 kVA

77. (d)

Copper loss $\propto S^2$

and iron loss = constant (at all loads)

∴ for half load, copper loss

$$= \frac{1}{4}\times 6400 = 1600\text{ W}$$

and iron loss = 5000 W

78. (a)

Copper weight $\propto \frac{1}{\sqrt{B_m}}$ and core weight $\propto \frac{1}{B_m}$.

If $B_m\uparrow$, weight of transformer per kVA reduces.

81. (c)

During open circuit test, exciting current is about 2 to 6 percent of the full load current i.e. the ohmic loss in the primary varies from 0.04 to 0.36 percent of the full-load primary ohmic loss due to which copper loss can be ignored. Hence, reason is false.

83. (c)

Since v/f ratio = constant in both case, therefore B_m = constant.

∴ $P_h \propto f$ or $P_h = Af$ (say)

and $P_e \propto f^2$ or $P_e = Bf^2$ (say)

∴ Total loss, $P_i = P_h + P_e = Af + Bf^2$

Now, at 50 Hz and 400 V,

$$50\,A + (50)^2 B = 2500 \quad ...(i)$$

At 25 Hz and 200 V,

$$25\,A + (25)^2 B = 850 \quad ...(ii)$$

Solving equation (*i*) and (*ii*), we have

$$B = 0.64 \text{ and } A = 18$$

Hence, at 400 V, 50 Hz (normal V and f),

$P_e = Bf^2 = 0.64 \times 50^2$

$= 1600\text{ W} = 1.6\text{ kW}$

and $P_h = Af = 18 \times 50 = 900\text{ W}$

$= 0.9\text{ kW}$

84. (d)

We know that, copper loss $\propto S^2$

∴ At 75% of full-load, copper loss

$$= x^2(P_{cu})_{fl}$$

$$= \left(\frac{3}{4}\right)^2 \times P_2 = \frac{9}{16}P_2$$

At maximum efficiency (given at 75% of *fl*),

copper loss = iron loss

or, $\frac{9}{16}P_2 = P_1$ or $\frac{P_1}{P_2} = \frac{9}{16} = 0.5625$

85. (a)

$$\text{Output} = V_2 I_2 \cos\phi = 220 \times 100 \times 0.8 = 17600\text{ W} = 17.6\text{ kW}$$

Iron loss = 55 W = 0.055 kW

Copper loss at 100 A of load current

$$= 400 \times \left(\frac{100}{80}\right)^2 = 625\text{ W} = 0.625\text{ kW}$$

∴ Transformer efficiency,

$$\eta = \left(\frac{\text{output}}{\text{output + losses}}\right) \times 100$$

$$= \left(\frac{17.6}{17.6 + 0.055 + 0.625}\right) \times 100 = 96.28\%$$

87. (d)

Turns ratio will remain same.

$$\therefore \quad \frac{V_2}{110} = \frac{6}{220}$$

or, V_2 = secondary voltage

$$= 110 \times \frac{6}{220} = 3\text{ volts}$$

88. (b)

$$\%R = V_r \cos\phi + V_x \sin\phi$$

At upf ($\cos\phi = 1$); $\%R = 0.045 = V_r$

or, $V_r = 0.045$ pu

At 0.8 pf lag,

$$\%R = 0.8 \times 0.04 + V_x \times 0.6 = 0.036 + 0.6\, V_x = (3.6\%) + 0.6\,(\%V_x)$$

Hence, %*R* will be more than 4.5% at 0.8 p.f. lag condition.

89. (d)

η = 100% i.e. losses are zero

$$\therefore \quad V_2 = 5\,V_1 = 600\text{ V}$$

$$\text{or,} \quad I_2 = \frac{V_2}{R_2} = \frac{600}{600} = 1\text{ A}$$

$$\frac{I_1}{I_2} = 5 \Rightarrow I_1 = 5I_2 = 5\text{ A}$$

90. (c)

For dc shunt machine, field current is very less, therefore for required field AT/Pole, number of turns of field winding is increased. Thus, *R* should be high for low I_f.

For dc series machine field winding current is equal to armature current which is of larger value. Therefore, number of turns of field winding is reduced to get the required field AT/Pole. As it has to handle large current, therefore resistance of field winding should be less by keeping the winding thick. Statement-3 is wrong.

91. (b)

Given, speed,

$$N = 1800\text{ rpm}$$

$$n = \frac{1800}{60}\text{ rps} = 30\text{ rps}$$

Average emf induced per conductor

$$= nP\phi = 30 \times 6 \times 15 \times 10^{-3} = 2.7\text{ V}$$

92. (a)

Peripheral velocity of commutator is

$$v = \pi DN = \pi \times 50 \times \frac{1000}{60}\text{ cm/s}$$

Now, brush width = $v \times$ time of commutation

or, time of commutation

$$= \frac{1.5 \times 60}{\pi \times 50 \times 1000} = 0.573\text{ ms} = 573\text{ μs}$$

93. (d)

We know that,

$$E_a \propto N\phi$$

or, $E_a \propto N$ (Since ϕ = constant)

$$\therefore \quad \frac{E_{a2}}{E_{a1}} = \frac{N_2}{N_1} = \frac{1500}{1600}$$

$$\text{or,} \quad E_{a2} = \frac{15}{16}E_{a1} = \frac{15}{16} \times 260 = 243.75\text{ V}$$

Now, rated armature current is

$$I_a = \frac{24 \times 10^3}{250} = 96\text{ A}$$

So, the terminal voltage is

$$V = E_{a2} - I_a R_a$$
$$= (243.75 - 96 \times 0.1)$$
$$= 234.15 \text{ volt} \approx 234 \text{ volt}$$

94. (b)

The reason of assertion is that dc series motor develop high torque at start due to which these motors are used for hard to start loads. Reason is also correct because these are variable speed motors.

97. (d)

We know that,

$$E_g \propto N\phi$$

$$\therefore \quad \frac{E_{g2}}{E_{g1}} = \frac{N_2\phi_2}{N_1\phi_1}$$

$$\text{or,} \quad \frac{E_{g2}}{E_{g1}} = \left(\frac{1.2N_1}{N_1}\right) \times \left(\frac{0.9\phi_1}{\phi_1}\right) = 1.08$$

$$\therefore \quad E_{g2} = 108\text{ \% of } E_{g1}$$

Thus, voltage generated is increased by 8%.

98. (a)

For LAP winding, number of parallel path = number of poles.
Thus, resistance is less compared to wave winding. Hence, current rating is more and voltage rating is less compared to wave winding.

101. (d)

Windage loss is a rotational loss (mechanical loss) and is due to air friction. It is directly proportional to square of speed (N).

106. (d)

Emf generated,

$$E_a = (\text{voltage induced per conductor}) \times \frac{Z}{A}$$

Here, $P = A = 8$
= Number of parallel paths
(due to LAP winding)

$$\therefore \quad E_a = 2 \times \frac{400}{8} = 100 \text{ volts}$$

107. (c)

Armature mmf wave is triangular while air-gap mmf is flat-topped.

108. (b)

Due to armature reaction, the MNA rotates by angle β (brush shift angle) in the direction of rotation of generator and by angle β is opposite to the direction of rotation of motor for sparkless commutation.

109. (c)

$$\theta_{ele} = \frac{P}{2}\theta_{mech}$$

Hence, for $P = 2$

$$\theta_{ele} = \theta_{mech}$$

110. (c)

Series field copper loss is a variable loss.

111. (c)

If the armature reaction is ignored then,

$$E_a \approx \text{Constant} \approx V$$

Now, $E_a \propto \phi N$

Since saturation is neglected,

therefore $\phi \propto I_a$

and $E_a = \text{Constant}$

$$\therefore \quad N_2 I_{a2} = N_1 I_{a1}$$

$$\text{or,} \quad N_2 = N_1 \times \frac{I_{a1}}{I_{a2}}$$

$$= N_1 \times \frac{I}{1/2I} = 2N_1$$

or, $N_2 = 2N_1$

Hence, speed becomes two times, i.e. increases by 100%.

112. (b)

- Magnetization characteristic is also called saturation characteristic or no-load characteristic or open-circuit characteristic. It represents the variation of open-circuit terminal voltage (V_{oc}) with I_f. Since, $V_{oc} \propto \phi\omega_m$, therefore it also represents variation of field flux (ϕ) with field current (I_f).
  Hence, statement-1 is correct.
- It is linear from zero to rated speed (air-gap line). Hence, statement-2 is false.
- If residual magnetism is present (in shunt, series and compound generators), then it will not start from origin, Hence, statement-3 is false.

- V_{oc} vs I_f depends on speed as $V_{oc} \propto \phi\,\omega_m$. Hence, statement-4 is false.
- Slope of air-gap line of the open-circuit characteristic gives the value of critical field resistance at a given speed since $R_f \propto \omega_m$. Hence, statement-5 is correct.

Thus, only statements-1 and 5 are correct.

113. (c)

Swinburne's test is carried out on dc machine to measure the no-load machine losses with the help of which efficiency of dc machine can be calculated.

In this test, the machine, whether it is a generator or a motor is run as a no-load shunt motor at rated speed and with rated terminal voltage.

114. (d)

$$\text{Field circuit loss} = \frac{V^2}{R_f} = \frac{200^2}{50} = 800 \text{ Watt}$$

Also,
$$I_f = \frac{V_f}{R_f} = \frac{200}{50} = 4 \text{ A}$$

$$\text{Output} = 17 \text{ kW} = E_a I_a$$
$$\text{Input} = VI_L = 20 \text{ kW}$$

or,
$$I_L = \frac{20\times10^3}{200} = 100 \text{ A}$$

$\therefore$
$$I_a = I_L - I_f = 100 - 4 = 96 \text{ A}$$

$\therefore$ Armature copper loss $= I_a^2 R_a$

$$= 96^2 \times 0.04 = 368.44 \text{ Watt}$$

$$\text{Total losses} = 20 \text{ kW} - 17 \text{ kW} = 3 \text{ kW} = 3000 \text{ Watt}$$

For maximum efficiency,

Armature ohmic loss = constant loss(P_k)

$\therefore$
$$P_k = \text{Total loss} - I_a^2 R_a = 3000 - 368.44 = 2631.36 \text{ watt} \approx 2.63 \text{ kW}$$

Hence, at maximum efficiency, armature ohmic loss = 2.63 kW.

116. (b)

For a dc shunt motor, speed $\omega_m \propto \dfrac{V}{\phi}$

(V = terminal voltage and ϕ = field flux)

Also, Torque $T_e \propto \phi I_a$ and $\phi \propto I_f$

$\therefore$
$$T_e \propto I_f I_a$$

Also,
$$P_{out} = E_a I_a$$

or,
$$P_{out} \propto \phi\,\omega_m I_a \propto I_f \omega_m I_a$$

Case-I:

$$V' = 0.5\,V,\ I_f' = I_f$$

and
$$P_{out}' = P_{out}$$

Thus,
$$\omega_m' \propto \frac{0.5\,V}{I_f}$$

$\therefore$
$$\omega_m' = 0.5\,\omega_m$$

Also,
$$P_{out}' \propto \phi'\omega_m' I_a' = \text{constant}$$

So,
$$\frac{P_{out}'}{P_{out}} = \frac{I_f \times 0.5\,\omega_m \times I_a'}{I_f \times \omega_m \times I_a}$$

or,
$$I_a' = \frac{I_a}{0.5} = 2I_a$$

Thus, new value of speed and current are:

$$\omega_m' = 0.5 \text{ pu and } I_a' = 2 \text{ pu}$$

Case-II:

$$V' = 0.5\,V,\ I_f' = I_f = \text{constant},$$

$$T_e' = T_e = \text{constant}$$

New speed,
$$\omega_m' \propto \frac{0.5\,V}{I_f}$$

or,
$$\omega_m' = 0.5\,\omega_m = 0.5 \text{ pu}$$

Also,
$$T_e' = T_e$$

or,
$$I_f' I_a' = I_f I_a$$

or,
$$I_a' = I_a = 1 \text{ pu}$$

$\therefore$ New speed,

$$\omega_m' = 0.5 \text{ pu and new current}$$

$$I_a' = 1 \text{ pu}$$

Case-III:

$$I_f' = 0.5\,I_f,\ V' = 0.5\,V$$

and
$$P_{out}' = \text{constant} = P_{out}$$

$\therefore$ New Speed,

$$\omega_m' \propto \frac{0.5\,V}{0.5\,I_f}$$

or,
$$\omega_m' = \omega_m = 1 \text{ pu}$$

$$P_{out}' = \text{constant} = P_{out}$$

or,
$$I_f' \omega_m' \times I_a' = I_f\,\omega_m I_a$$

or, $I'_a = \frac{\omega_m}{\omega'_m} \times \frac{I_f}{I'_f} = \frac{1}{1} \times \frac{1}{0.5} = 2$ pu

∴ New speed,

$\omega'_m = 1$ pu and

New current,

$I'_a = 2$ pu

Case-IV:

$I'_f = 2I_f,\ V' = V = \text{constant}$

and $T'_e = T_e = \text{constant}$

∴ New speed,

$$\omega'_m \propto \frac{V}{2I_f}$$

or, $\omega'_m = 0.5\,\omega_m = 0.5$ pu

Also, $T'_e = T_e = \text{constant}$

or, $I'_f I'_a = I_f I_a$

or, $I'_a = \frac{I_f}{I'_f} \times I_a$

$= \frac{I_f}{2I_f} \times I_a = 0.5\,I_a = 0.5$ pu

Hence, new current = 0.5 pu

and new speed = 0.5 pu

117. (d)

Speed of dc shunt motor is given by

$$\omega_m = \frac{V - I_a R_a}{k\phi}$$

Given, $\phi_2 = 0.95\,\phi_1$ (due to armature reaction)

Also, $I_a R_a$ drop = 0.1 pu = 0.1 V_1 pu

∴ New speed motor is

$$\omega_{m_2} \propto \frac{V - 0.1V_1}{0.95\,\phi_1} \,\alpha\, \frac{0.9\,V_1}{0.95\,\phi_1}$$

Also, $\omega_{m_1} \propto \frac{V_1}{\phi_1}$

So, $\frac{\omega_{m_2}}{\omega_{m_1}} = \frac{0.90}{0.95}$

or, $\omega_{m_2} = 1140 \times \frac{0.90}{0.95} = 1080$ rpm

118. (d)

Given, $\omega_0 = 2000$ rpm, $V_0 = E_{a0} = 440$ V,

$\phi_0 = \phi_{fl}$

$\omega_1 = 1000$ rpm, $T_1 = T_{fl}$;

$V_1 = V_{rated}$, $\phi_1 = \phi_{fl}$, $I_{a1} R_a = ?$

$\omega_2 = 1050$ rpm, $T_2 = 0.5\,T_{fl}$,

$V_2 = V_{rated}$, $\phi_2 = \phi_{fl}$

We know that,

$$E_a \propto \phi\omega$$

∴ $\frac{E_{a1}}{E_{a0}} = \frac{\omega_1}{\omega_0}$ $(\phi_1 = \phi_2 = \phi_{fl})$

or, $E_{a1} = 440 \times \frac{1000}{2000} = 220$ volt

∴ $V_{rated} = E_{a1} + I_{a1} R_a$

or, $V_{rated} = 220 + I_{a1} R_a$ ...(*i*)

Again, $E_a \;\alpha\; \phi\omega_m$

∴ $\frac{E_{a2}}{E_{a0}} = \frac{\omega_2}{\omega_0}$

or, $E_{a2} = 440 \times \frac{1050}{2000} = 231$ volt

∴ $V_{rated} = E_{a2} + I_{a2} R_a$

or, $V_{rated} = 231 + I_{a2} R_a$ ...(*ii*)

Also, $T \propto \phi I_a$

or, $T \propto T_a$ (ϕ = constant)

∴ $\frac{T_2}{T_1} = \frac{I_{a2}}{I_{a1}}$

or, $I_{a2} = 0.5\,I_{a1}$ ...(*iii*)

Equating equations (*i*) and (*ii*),

$220 + I_{a1} R_a = 231 + I_{a2} R_a$

or, $I_{a1} R_a - I_{a2} R_a = 11$

Using equation (*iii*), we have;

$I_{a1} R_a - 0.5\,I_{a1} R_a = 11$

or, $I_{a1} R_a = 22$ volts

Hence, armature voltage drop at full load

= 22 volts

119. (b)

In a cumulatively compounded dc motor, net flux is given by $\phi \propto (N_f I_f + N_{se} I_{se})$.

When the series field is short circuited then, $N_{se} I_{se} = 0$ and , therefore flux will decrease due to which speed (ω_m) will increase $\left(\omega_m \propto \frac{V}{\phi}\right)$.

Since load power = constant

i.e. $E_a I_a = \text{constant}$

or, $\phi\omega_m I_a = \text{constant}$

or, $I_a = \text{constant}$

(since ϕ decreases and ω_m increases so, $\phi\omega_m$ = constant).

Hence, I_a = constant and ω_m increases

120. (d)

Voltage regulation of dc shunt generator is

$$\epsilon_R = \left(\frac{V_{NL}-V_{FL}}{V_{FL}}\right) = \left(\frac{E_{a0}-V}{V}\right)$$

For generator,

$$V = E_a - I_a R_a$$

As speed is increased from 1000 rpm to 1250 rpm, E_{a0} will increase since $E_a \propto \phi N$. Hence, voltage regulation will increase and becomes more than 10%.

121. (a)

Since brushes are given a shift in the direction of rotation for dc generator, therefore commutation will improve. However, due to brush shift, armature reaction will have the demagnetizing effect which will reduce the induced emf E_a. Due to which terminal voltage will fall. ($V = E_a - I_a R_a$).

122. (c)

For motor:

$$E_{am} = V - I_a R_a$$

For generator:

$$E_{ag} = V + I_a R_a$$

$$\therefore \quad E_{ag} - E_{am} = 2 I_a R_a$$

$$= 2 \times 20 \times 1 = 40 \text{ Volt}$$

123. (a)

Given, $T \propto N^2$

Also, $T \propto \phi I_a$

or, $T \propto I_{a2}$

(Since, $\phi \propto I_a$ in unsaturated region)

Hence, $$\left(\frac{N_2}{N_1}\right)^2 = \left(\frac{I_{a2}}{I_{a1}}\right)^2 \text{ or } \frac{I_{a2}}{I_{a1}} = \frac{N_2}{N_1}$$

or, $$I_{a2} = \frac{800}{400} \times 20 = 40 \text{ A}$$

125. (b)

A dc generator works on the principle of dynamically induced emf i.e. whenever flux is cut by the conductor, dynamically induced emf is produced in it according to the laws of electromagnetic induction, which will cause a flow of current in the conductor if the circuit is closed.

126. (c)

Given, $\phi = 20 \times 10^{-3}$ Wb ;

$A = P = 4$ (for LAP winding) ;

$N = 720$ rpm

Z = Number of slots × number of coil sides per slot × number of turns in each coil

$= 144 \times 2 \times 2 = 576$

∴ Emf induced,

$$E_a = \frac{NP\phi z}{60\,A}$$

$$= \frac{720 \times 4 \times 20 \times 10^{-3} \times 576}{60 \times 4} = 138.24 \text{ V}$$

127. (d)

Number of parallel path = 2

(whether LAP or wave winding)

Flux per pole, $$\phi = \frac{E_a \times 60}{NZ} \times \frac{A}{P}$$

$$= \frac{200 \times 60}{300 \times 400} \times \frac{2}{2} = 0.1 \text{ Wb}$$

∴ Emf induced in each field coil,

$$e = N_f \times \frac{d\phi}{dt} = 1200 \times \left(\frac{0.1-0}{0.1}\right)$$

$$= 1200 \text{ V} = 1.2 \text{ kV}$$

128. (b)

If the series generator is operated on initial straight portion of the characteristic, it gives voltage approximately proportional to the current.

Therefore, $V \propto I$

$$\therefore \quad \frac{V_2}{V_1} = \frac{I_2}{I_1}$$

or, $$V_2 = 50 \times \frac{50}{200} = 12.5 \text{ volt}$$

Also, voltage drop in feeder = 50 × 0.3 = 15 V

Hence, voltage difference between the station bus-bar and far end of the feeder

$= 12.5 - 15 = -2.5$ V

129. (a)

Given, $V_{FL} = 250$ V, $V_{NL} = 275$ V

∴ % voltage regulation

$$= \left(\frac{V_{NL}-V_{FL}}{V_{FL}}\right) \times 100$$

$$= \left(\frac{275-250}{250}\right) \times 100 = 10\% = 0.1 \text{ pu}$$

130. (d)

At no-load, AT required to generate a voltage of 440 V is

$$AT_0 = 4.1 \times 1200 = 4920$$

At full-load, AT required to generator a voltage of 440 V is

$$AT_{FL} = 5.8 \times 1200 = 6960$$

Hence, additional AT required at full-load

$$= AT_{FL} - AT_0$$
$$= 6960 - 4920 = 2040$$

Given, full-load current,

$$I_{FL} = 200 \text{ A}$$

∴ Number of series field turns required

$$= \frac{2040}{200} = 10.2 \text{ turns}$$
$$\approx 11 \text{ turns}$$

134. (d)

Armature mmf wave is directed along the brush axis. Therefore, statement-2 is false. For development of steady state torque in rotating electrical machines both stator and rotor mmf wave should be stationary to each other. Since stator mmf wave (field mmf wave) is stationary, therefore rotor mmf wave (armature mmf wave) must also be stationary.

135. (d)

Internal characteristic: E_a vs I_a or I_a vs E_a
External characteristic: V vs I_L or I_L vs V

136. (b)

Motor torque, $T_e = k\phi I_a$
Since magnetic saturation is neglected,

$$\phi \propto I_a \qquad \therefore T_e \propto I_a^2$$

Now, I_a = Rated armature current
= 20 A = Constant
(As T_e = constant required)

At starting, $E_a = 0$ (as $N = 0$)

Let the required external resistance be R_{ext}.

$$\therefore \quad V = I_a(R_a + R_{se} + R_{ext}) \quad (\text{Since } E_a = 0)$$

$$\text{or,} \quad R_{ext} = \frac{V}{I_a} - (R_a + R_{se}) \quad (\text{Since } E_a = 0)$$

$$= \frac{230}{20} - 0.5 = 11\ \Omega$$

137. (c)

Fixed loss = constant loss

At η_{max}, armature variable circuit loss
= constant loss (or fixed loss)

138. (c)

Starter in a dc motor is used to limit the high starting current. Hence, statement-1 is false.

In a 3-point starter, field circuit is connected in series with the hold coil.

A three-point starter can't be used where wide rage of speed control is required. It is used below base speed.

141. (b)

Armature resistance, R_a is around 1 Ω.
Shunt field winding has resistance lying between 100 – 250 Ω.

142. (d)

$$E_a \propto \phi\omega_m \propto \phi N$$

Hence, E_a is a function of flux (ϕ) and speed (N).

144. (b)

When the belt snaps, the machine now starts drawing power from the supply and the direction of current in the series winding reverses due to which it starts behaving as a differentially compound motor and runs in the same direction to maintain the constant speed of the machine.

145. (d)

By rotating the brush axis by 90°, net flux will be zero i.e. $\phi = \phi_f - \phi_{ar} = 0$ (or, $F = F_f\ F_{ar}$).
Hence, both E_b and T_d becomes zero.

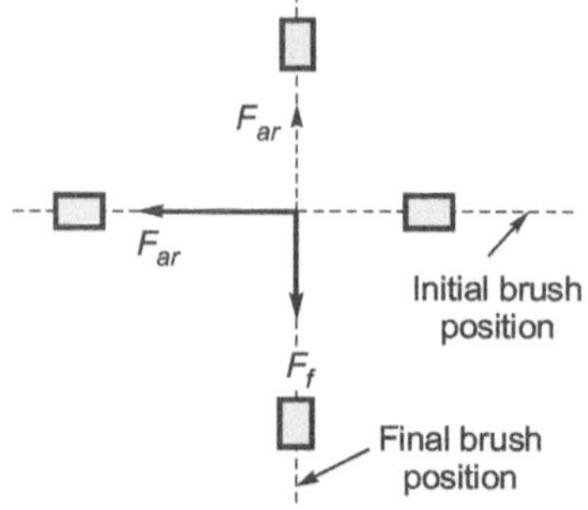

147. (b)

When load increases, I_a increases so that $E_b = (V - I_aR_a)$ falls and therefore, speed is decreased ($\omega_m \propto E_b$).

149. (d)

$$\omega_m = \frac{V - I_a R_a}{k\phi}$$

If ϕ is increased and I_a = constant, then ω_m can be increased by increasing V.

152. (c)

For an over compounded dc generator, percentage of compounding = voltage regulation

$$\%\epsilon_R = \left(\frac{E_a - V}{V}\right) \times 100$$

$$= \left(\frac{I_a R_a}{V}\right) \times 100$$

$$= \left(\frac{800 \times 0.02}{500} \times 100\right) = \frac{16}{5} = 3.2\%$$

153. (a)

When the armature connections are reversed than torque reverses this is known as plugging.

156. (b)

$$E_g = V + I_a R_a$$

and $E_m = V - I_a R_a$

Now, $E_a \propto N$

$$\therefore \quad \frac{E_m}{E_g} = \frac{V - I_a R_a}{V + I_a R_a} < 1$$

or, $\frac{N_m}{N_g} < 1$ or $N_m < N_g$ or $N_m < 800$ rpm

157. (d)

$$E_a = \frac{NP\phi Z}{60A} \quad (A = P \text{ for LAP winding})$$

$$\therefore \quad E_a = \frac{1500 \times 0.4 \times 10^{-3} \times 1000}{60} = 10 \text{ volt}$$

159. (a)

$$N \propto \frac{E_b}{\phi} \propto \frac{(V - I_a R_a)}{\phi}$$

As E_b is less at no-load, therefore N is less.

161. (a)

As $|\vec{E}_f| > |\vec{V}|$ and $\vec{E}_f$ leads $\vec{V}$, therefore machine is acting as a generator and operating at lagging pf.

162. (b)

Number of slots/pole/phase,

$$m = \frac{36}{6 \times 3} = 2$$

Slots/Pole, $n = \frac{36}{6} = 6$

Slot angle, $\beta = \frac{180°}{6} = 30°$

$$\therefore \quad K_d = \frac{\sin\left(2 \times \frac{30°}{2}\right)}{2\sin\left(\frac{30°}{2}\right)} = \frac{1}{4 \sin 15°}$$

Also, $K_C = \cos\frac{\alpha}{2} = \cos\frac{30°}{2} = \cos 15°$

So, $K_W = K_C \cdot K_d = \frac{\cot 15°}{4}$

164. (a)

For a synchronous motor, $N = N_s$ for all value of torque.

167. (d)

Since after connecting the stator to supply, flux per pole is reduced to 20 mWb from initial value of 25 mWb, therefore effect of armature reaction is demagnetizing i.e. $F_f > F_r$. Due to which motor is over-excited and runs at leading power factor.

168. (b)

Mmf method is called optimistic method because it gives voltage regulation lower than the actual value.

169. (a)

Motor becomes over-excited and starts operating at leading pf with the increase in field current. Here, I_a will increase.

170. (d)

$$Q_{out} = \left(\frac{VE_f}{X_S}\cos\delta - \frac{V^2}{X_S}\right) \quad (\text{Neglecting } r_a)$$

Thus, Q_{out} is maximum when $\delta = 0°$.

171. (c)

$$P = \underbrace{\frac{VE_f}{X_d}\sin\delta}_{\text{Excitation Power}} + \underbrace{\frac{V^2}{2}\left(\frac{1}{X_q} - \frac{1}{X_d}\right)\sin 2\delta}_{\text{Reluctance Power}}$$

172. (c)

$$X_s \propto f \text{ and } X_s \propto E_f$$

$\therefore$ $E_f \propto f$

Now, $f \propto N_S$

$\therefore$ $E_f \propto N_S$

or, $E_f = kN_S$ (k = constant)

174. (d)

Slip becomes negative if motor runs above N_S. Hence, the damper winding produces induction generator torque.

175. (a)

$$S_{IM} = [(10^6 \times 0.8) + j(10^6 \times 0.6)]$$
$$= (800 \times 10^3 + j600 \times 10^3)$$

Also, $S_{SC} = (750 \times 0.6 - j750 \times 0.8) \times 10^3$
$$= (450 \times 10^3 - j600 \times 10^3)$$

$\therefore$ Q_{Total} = 600 kVAR − 600 kVAR = 0

Thus, $\cos\phi = 1$

176. (a)

for zero-pf lead operation, synchronous generator will be under excited. Hence, $F_f < F_r$ due to which effect of armature reaction will be magnetizing.

178. (a)

$$Z_{pu}(\text{new}) = Z_{pu}(\text{old}) \times \frac{MVA_{b(\text{new})}}{MVA_{b(\text{old})}} \times \left(\frac{KV_{b(\text{old})}}{KV_{b(\text{new})}}\right)^2$$

$$= 0.2 \times \frac{150}{50} \times \left(\frac{11}{22}\right)^2$$

$$= 0.6 \times \frac{1}{4} = \frac{0.3}{2} = 0.15 \text{ pu}$$

180. (c)

$$X_d = \frac{V_{max}}{I_{min}} \text{ and } X_q = \frac{V_{min}}{I_{max}}$$

or, $$X_d = \frac{2820/\sqrt{3}}{275} \text{ and } X_q = \frac{2800/\sqrt{3}}{360}$$

or, $X_d = 5.92\ \Omega$ and $X_q = 4.5\ \Omega$

183. (a)

A synchronous condenser is an over excited synchronous motor operating on no-load and improves the pf of the system.

Since it is over-excited ($F_f > F_r$), therefore it requires more copper for field winding.

184. (d)

Pull-out torque α E_f. Hence, statement-1 is false. Short-circuit ratio is not related to armature reaction.

Statement-4 is correct as,

$$X_S = \frac{V_{OC}}{I_{SC}}$$

Under short-circuit condition, effect of armature reaction will be demagnetising.

185. (c)

$$\text{Slot angle} = \frac{180°}{(\text{Number of slots / pole})}$$

$$= \frac{180°}{\left(\frac{72}{4}\right)} = 10°$$

Total number of conductors,

$$Z = 72 \times 10 = 720$$

$\therefore$ Number of turns per phase $= \dfrac{720}{2 \times 3} = 120$

186. (c)

To eliminate n^{th} harmonic,

$$\frac{n\alpha}{2} = 90° \text{ or } \alpha = \frac{180°}{n}$$

$\therefore$ Coil span with short-pitched winding

$$= 180° - \frac{180°}{n} = \left(\frac{n-1}{n}\right) \times 180$$

$$= \left(\frac{n-1}{n}\right) \times \text{full-pitch}$$

$$= \left(\frac{r-1}{r}\right) \times \text{full-pitch} \quad (\text{Here } r = n)$$

187. (b)

$$\frac{V}{f} = \text{constant}$$

$\therefore$ $$\frac{V_2}{f_2} = \frac{V_1}{f_1} \text{ or } V_2 = \frac{f_2}{f_1} V_1 = \frac{50}{60} V_1$$

or, $$V_2 = \frac{5}{6} V_1$$

190. (c)

With higher value of X_S or Z_S, voltage regulation becomes poor but, short circuit current

$$I_{SC} = \frac{1}{Z_{pu}} \text{ or } \frac{1}{X_S(\text{pu})} \text{ is reduced.}$$

191. (d)

$$\text{SCR} = \frac{1}{X_{S(sat)}} \propto \frac{\text{Air gap}}{N_{ph}}$$

SCR ↑, $X_{S(sat)}$ ↓, voltage regulation improves, air gap ↑, stability ↑ due to increase in moment of inertia of machine.

192. (a)

Here, stator mmf is stationary in space.
Rotor mmf rotates at N_S opposite to the direction of rotor rotation (N_S).
So, air gap field remains stationary in space i.e. w.r.t. stator.

194. (c)

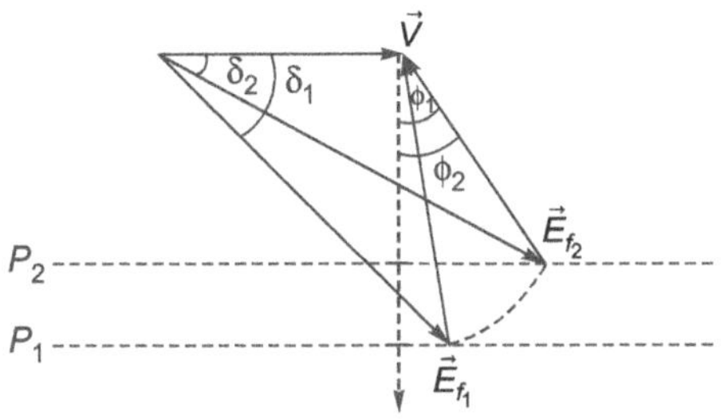

Here, $|\vec{E}_{f_1}| > |\vec{E}_{f_2}|$, $P_1 > P_2$, $\delta_1 > \delta_2$ and $\phi_1 < \phi_2$ or, $\cos\phi_1 > \cos\phi_2$.
Thus, with reduction in shaft load, cosϕ and δ both are reduced.

195. (a)

Here, sub-transient reactance = X_S = 1 Ω
∴ Short-circuit current at the instant of fault

$$= 2\sqrt{2} \times \frac{\text{Phase voltage}}{X_S}$$

$$= 2\sqrt{2} \times \frac{11000/\sqrt{3}}{1}$$

$$= 17962.9 \text{ A} \approx 17.96 \text{ kA}$$

196. (c)

Given, V_L = 2300 volt,
I_L = 47 A, cos ϕ = 0.8 (leading)

Power input to motor = $P_{in} = \sqrt{3}\, V_L I_L \cos\phi$

$$= \sqrt{3} \times 2300 \times 47 \times 0.8$$
$$= 149.78 \text{ kW}$$

Total copper loss

$$= 3I^2R = 3 \times (47)^2 \times 1.5$$
$$= 9.94 \text{ kW}$$

∴ Internal power developed

$$= P_{in} - 3I^2R = 149.78 - 9.94$$
$$= 139.84 \text{ kW} \approx 140 \text{ kW}$$

197. (b)

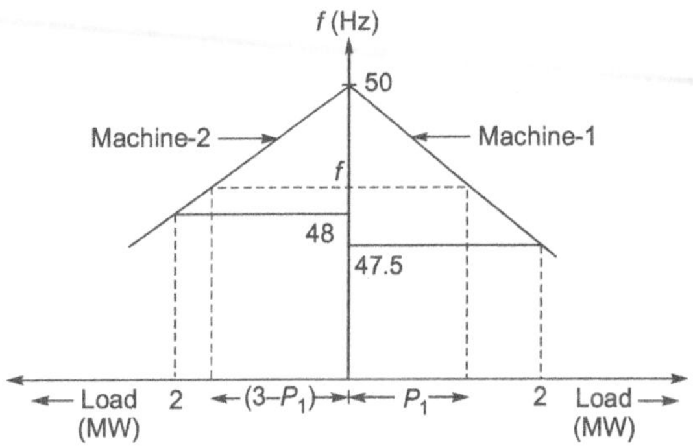

Let the common frequency at 3 MW be f.
Equating slopes, we get:

$$\frac{50-f}{3-P_1} = \frac{50-48}{2} = 1$$

or, $$f = 47 + P_1 \quad ...(i)$$

and $$\frac{50-47.5}{2} = \frac{50-f}{P_1}$$

or, $$f = 50 - 1.25\, P_1 \quad ...(ii)$$

From equations (*i*) and (*ii*), we have

$$47 + P_1 = 50 - 1.25\, P_1$$

or, $$P_1 = \frac{3}{2.25} = \frac{4}{3} \text{ MW}$$

and $$P_2 = 3 - \frac{4}{3} = \frac{5}{3} \text{ MW}$$

Hence, $$P_1 = \frac{4}{3} \text{ MW and } P_2 = \frac{5}{3} \text{ MW}$$

198. (c)

Excess power input to the generator shaft before the steam valve begins to close = 50 MW.
∴ Excess energy transferred to the rotating parts in 0.4 second = 50× 10^6 × 0.4 J = 20 MJ.

199. (d)

$$K_d = \frac{\sin\left(\frac{\text{Phase spread}}{2}\right)}{\left(\frac{\text{Phase spread}}{2}\right)}$$

$$= \frac{\sin\left(\frac{120°}{2}\right)}{\left(60 \times \frac{\pi}{180}\right)} = \frac{3}{\pi}\sin 60°$$

$$= \frac{3}{\pi} \times \frac{\sqrt{3}}{2} = 0.827$$

200. (a)

Pole pitch = 9 slots and coil-span = 7 slots

β = slot pitch angle

$$= \frac{180°}{(\text{Number of slots / Pole})}$$

$$= \frac{180°}{\left(\frac{36}{4}\right)} = 20°$$

∴ Chording angle,

α = 2 slot-pitches = 2 × 20° = 40°

Hence, pitch factor

$$= K_c = \cos\frac{\alpha}{2} = \cos\frac{40°}{2} = \cos 20°$$

201. (c)

When the field winding on the rotor is excited with alternating current, a pulsating field is created which can be resolved into two components rotating in opposite direction at synchronous speed.

The relative speed between backward rotating field and stationary armature will be zero. However, the relative speed between forward field component and armature will be $2N_s$. Thus, frequency of emf generated in the armature = $2f$. Hence, core loss in the armature at double frequency $2f$ would be more due to which efficiency will fall.

204. (b)

In a synchronous motor, E_f always lags V_t by δ.

Since motor is overexcited, therefore $|\vec{E}_f| > |\vec{V}_t|$.

207. (d)

In generator action, E_f leads V_t and in motor action, V_t leads E_f.

208. (a)

For generator action, if excitation is increased (over-excited), it starts operating at lagging pf and exports lagging VAR. So, for generator statements 4 and 6 holds true.

For motor action, if excitation is increased (over excited), it starts operating at leading pf and takes leading VAR (exports lagging VAR) from the system. Hence, for motor action, statements-1 and 6 will hold true.

209. (b)

EMF method is also called "pessimistic method" because voltage regulation found is more than actual.

MMf method is also called "optimistic method" because voltage regulation found is less than actual.

210. (c)

For good starting torque, the damper bars should have high resistance and for large damping effect near synchronous speed, the damper bars should have low resistance. Hence, reason is false.

211. (a)

$$S_p = \frac{dP}{d\delta} = P_{max}\cos\delta$$

(δ = angle between stator and rotor field axis.)

Higher the value of S_p, higher is the stability of the machine.

213. (d)

At leading p.f., alternator becomes under-excited, therefore

$$F_f < F_r$$

Hence, excitation will decrease.

214. (c)

V-curve: I_a vs I_f for a given load

Inverted V-curve: $\cos\phi$ vs I_f for a given load.

215. (a)

$$N = \frac{120f}{P} \text{ or } P = \frac{120f}{N}$$

$$= \frac{120 \times 50}{1440} = \frac{25}{6} = 4\frac{1}{6} \text{ Poles}$$

Since the number of poles must be even and a whole number, therefore the number of poles must be 4.

216. (d)

Rotor frequency,

$$f_2 = sf_1 = \frac{90}{60} = 1.5 \text{ Hz}$$

$$\therefore \text{ Slip, } \quad s = \frac{f_2}{f_1} = \frac{1.5}{50} = 0.03$$

So, rotor speed,

$$N_r = (1 - s)N_s$$

$$= (1 - 0.03) \times \frac{120 \times 50}{6} = 970 \text{ rpm}$$

217. (b)

During blocked rotor test, power input to a 3-phase induction motor appears almost entirely as I^2R loss in both stator and rotor windings.

Since, stator ohmic loss = rotor ohmic loss

$$\therefore \text{ Rotor } I^2R \text{ loss} = \frac{120}{2} = 60 \text{ kW} = 3I_2^2 r_2$$

Now,

$$\omega_{Sm} = \frac{2\pi}{60} \times N_S = \frac{2\pi}{60} \times \frac{120 \times 50}{10}$$

$$= 20\,\pi \text{ rad/sec}$$

So, starting torque,

$$T_{st} = \frac{3}{\omega_{Sm}} \times I_2^2 r_2 = \frac{60 \times 10^3}{20\,\pi}$$

$$= 954.93 \text{ N-m}$$

218. (a)

$$S_m = \frac{r_2}{x_2} = \frac{0.2}{2} = 0.1$$

Now,

$$\frac{T_{st}}{T_{em}} = \left(\frac{2}{S_m + \frac{1}{S_m}} \right)$$

$$= \frac{2}{(0.1 + 10)} = \frac{2}{10.1} \quad \ldots(i)$$

Given, $\quad T_{em} = 2T_{fl} \quad \ldots(ii)$

From equations (*i*) and (*ii*), we have

$$\frac{T_{st}}{T_{fl}} = \frac{4}{10.1} = 0.396$$

219. (c)

Since torque, $T_e \propto V^2$

$$\therefore \quad T_{st_1} = 1.5\, T_{fl} \propto (400)^2$$

Also, $\quad T_{st_2} = T_{fl} \propto (V_1)^2$

So, $$\frac{T_{st_2}}{T_{st_1}} = \frac{1}{1.5} = \left(\frac{V_1}{400}\right)^2$$

or, $$V_1 = \frac{400}{\sqrt{1.5}} = 326.6 \text{ Volts}$$

220. (b)

For same air-gap flux,

$$\frac{V}{f} = \text{constant}$$

or, $$\frac{V_2}{30} = \frac{400}{50} \text{ or } V_2 = 240 \text{ volt}$$

For same load torque,

$$sf = \text{constant}$$

or, $\quad sN_s = \text{constant}$

or, $\quad N_s - N_r = \text{constant}$

$$\therefore \quad N_{S_2} - N_{r_2} = N_{S_1} - N_{r_1}$$

Here, $$N_{S_1} = \frac{120 \times 50}{4} = 1500 \text{ rpm},$$

$$N_{r_1} = 1400 \text{ rpm}$$

$$N_{S_2} = \frac{120 \times 30}{4} = 900 \text{ rpm}$$

$$\therefore \quad 900 - N_{r_2} = 1500 - 1400$$

or, $\quad N_{r_2} = 800$ rpm

222. (d)

$$\frac{I_{s(\text{line})_Y}}{I_{s(\text{line})_\Delta}} = \frac{1}{3}$$

or, $$I_{s(\text{line})_Y} = \frac{10}{3} \text{ A}$$

223. (d)

We know that,

$$\frac{I_{s(\text{line})_Y}}{I_{s(\text{line})_\Delta}} = \frac{1}{3}$$

$\therefore \quad I_{s(line)Y} = \dfrac{I_{s(line)\Delta}}{3}$

Thus, stator current will decrease.

Also, $\quad Z_Y = \dfrac{Z_\Delta}{3}$

$\therefore \quad Z_Y < Z_\Delta$ (Impedances)

So, $\quad (\cos\phi)_Y > (\cos\phi)_\Delta$

224. (b)

$$\frac{E_1}{E_2} = 2 \text{ or } E_2 = \frac{E_1}{2} = 200 \text{ V}$$

Rotor induced emf

$$= sE_2 = \frac{1}{25} \times 200$$

$$= 8 \text{ volt} \quad \text{(line value)}$$

Here, slip $\quad s = \dfrac{1500 - 1440}{1500}$

$$= \frac{60}{1500} = \frac{1}{25}$$

$\therefore$ Rotor induced emf per phase

$$= \frac{8}{\sqrt{3}} = 4.62 \text{ volt}$$

225. (a)

$$(\cos\phi)_{max} = \frac{R_2'}{R_2'^2 + (S_m X_2')^2}$$

Now, $\quad f = \dfrac{NP}{120}$

$\therefore$ When $P\downarrow$, $f\downarrow$ and hence $X_2'\downarrow$

Thus, when $f\downarrow$, $X_2'\downarrow$ and $(\cos\phi)_{max}$ increases

228. (c)

A 3-phase induction motor requires starter to limit the starting current because the secondary induced emf at starting = E_2 (S = 1) which is high as a result of which high starting current flows which may damage the winding.

231. (a)

Speed of rotor field w.r.t. stator = $N_r \pm$ speed of rotor field w.r.t. rotor

or, $\quad \left(\dfrac{120 \times 50}{6}\right) = N_r \pm \left(\dfrac{120 \times 20}{6}\right)$

or, $\quad 1000 = N_r \pm 400$

or, $\quad N_r = 1000 \pm 400$

$\quad = 1400 \text{ rpm or } 600 \text{ rpm}$

232. (d)

$$T = \frac{3}{\omega_{s_m}} \times \frac{V_1^2}{R_2'} s$$

or, $\quad T \propto V_1^2 s$

Now, $\quad T = \text{constant} \quad \therefore T_2 = T_1$

or, $\quad V_2^2 s_2 = V_1^2 s_1$

or, $\quad s_2 = \left(\dfrac{V_1}{V_2}\right)^2 s_1 \text{ or } s_2 = 2s_1$

$\therefore$ Slip increases 2-times

Also, $\quad T = \dfrac{I_2'^2}{\omega_{s_m}} \times \dfrac{R_2'}{s} = \text{constant}$

or, $\quad I_2'^2 \propto s$

$$\frac{I_2'}{I_1'} = \sqrt{\frac{s_2}{s_1}} = \sqrt{2}$$

$\therefore$ Current increases $\sqrt{2}$ times

234. (d)

Induction generators are not used for generating power for commercial supply. A 3-ϕ IM becomes an induction generator when N_s becomes less than N_r (i.e. slip becomes negative). It draws reactive power from supply and at the same time delivers active power to the supply mains. Hence, assertion is false.

236. (a)

Given, $\quad I_{SC} = 6I_{fl}$

$\therefore \quad \dfrac{T_s}{T_{fl}} = x^2 \left(\dfrac{I_{SC}}{I_{fl}}\right)^2 \times S_{fl}$

or, $\quad 1 = x^2 (6)^2 \times 0.05$

or, $\quad x = \sqrt{\dfrac{1}{1.8}} = 0.745 = 74.5\%$

237. (c)

With the stator in delta,

Resistance per phase $= \dfrac{R \times 2R}{3R} = 0.6$

or, $\quad R = 0.9\ \Omega$

$\therefore$ Effective stator resistance per phase

$$= 1.2 \times 0.9 = 1.08\ \Omega$$

Power input at rated voltage during blocked rotor test

$$= 5 \times \left(\frac{420}{210}\right)^2 = 20\text{ kW}$$

Stator current at rated voltage during blocked rotor test

$$= 20 \times \frac{420}{210} = 40\text{ A}$$

$\therefore$ Air-gap power, P_g

$$= 20 \times 10^3 - \left[3 \times \left(\frac{40}{\sqrt{3}}\right)^2 \times 1.8 + 300\right]$$

$= 17972$ W $= 17.97$ kW $\simeq 18$ kW

238. (d)

$$S_m = \frac{R_2'}{X_2'}$$

$\Rightarrow$ As air gap is increased, X_2' and X_1' (leakage reactances) are increased so that S_m is reduced, Te_m and T_{st} are also reduced.

$\Rightarrow$ Mutual flux remains constant as now, motor draw more magnetizing current to keep this flux constant.

$\Rightarrow$ $\cos\phi = \dfrac{R_2'}{\sqrt{R_2'^2 + (sX_2')^2}}$ is decreased.

239. (c)

Given, $S_m = \dfrac{R_2}{X_2} = 0.2$

At 50 Hz, leakage reactance $= X_2$

At 45 Hz, leakage reactance $= 0.9\,X_2$

(Since, $X \propto f$)

Let I_{s_1} and I_{s_2} be the starting currents at 50 Hz and 45 Hz respectively.

Then, $$I_{s_1} = \frac{V}{\sqrt{R_2^2 + (X_2)^2}}$$

and $$I_{s_2} = \frac{V}{\sqrt{R_2^2 + (0.9X_2)^2}}$$

$$\therefore \quad \frac{I_{s_2}}{I_{s_1}} = \sqrt{\frac{R_2^2 + X_2^2}{R_2^2 + (0.9X_2)^2}}$$

$$= \sqrt{\frac{S_m^2 + 1}{S_m^2 + 0.9^2}} = \sqrt{\frac{0.2^2 + 1}{0.2^2 + 0.81}}$$

$$= \sqrt{\frac{1.04}{0.85}} = \sqrt{1.223} = 1.106$$

240. (d)

Slip, $s_1 = \dfrac{1500 - 1440}{1500} = 0.04$

For T-N characteristic to be linear,

$$T_e = \frac{3V^2}{\omega_{sm}} \cdot \frac{s}{R_2} \quad \text{or} \quad T_e \propto s$$

$$\therefore \quad \frac{T_{e_2}}{T_{e_1}} = \frac{s_2}{s_1}$$

or, $$\frac{1}{2} = \frac{s_2}{0.04} \quad \text{or} \quad s_2 = 0.02$$

Hence, new motor speed

$$= 1500 \times (1 - 0.02) = 1470\text{ rpm}$$

241. (b)

$$\frac{T_{st}}{T_{em}} = \frac{2}{s_m + \frac{1}{s_m}} = \frac{2s_m}{s_m^2 + 1}$$

or, $$\frac{T_{em}}{T_{st}} = 0.5\left(s_m + \frac{1}{s_m}\right)$$

242. (a)

Speed of rotor field w.r.t. stator

$$= \frac{120f_1}{P} = \frac{120 \times 50}{4} = 1500\text{ rpm}$$

Speed of rotor field w.r.t. rotor structure

$$= \frac{120f_2}{P} = \frac{120 \times 20}{4} = 600\text{ rpm}$$

Let rotor speed $= N_r$

Then, $N_r \pm 600 = 1500$ rpm

For +ve sign, $N_r = 900$ rpm

(in the direction of stator field)

For –ve sign, $N_r = 2100$ rpm

(against the direction of stator field)

244. (b)

For SCIM, we use double cage rotor for getting improved performance during starting (i.e. low I_{st} and high T_{st}).

245. (d)

Inner cage has low R and high X.
Outer cage has high R and low X.

246. (c)

Rotor slots are skewed to avoid the magnetic interlocking phenomenon of stator and rotor slots called Cogging.

247. (b)

⇒ Cogging is due to magnetic interlocking of stator and rotor slots.

⇒ Crawling is due to space harmonics (mainly 7th harmonic dominant).

⇒ Due to 5th harmonic mmf wave, which is a negative sequence quantity and, therefore it rotates at one fifth speed of synchronous speed in opposite direction of fundamental mmf wave, a small reverse braking torque will act on the motor.

251. (d)

Due to presence of air gap in a three-phase induction motor, it draws large value of magnetizing current to establish the required value of flux in the air-gap. Due to this reason the power factor of 3-phase IM is quite low.

252. (c)

$$T_{max} = \frac{3}{2\omega_{sm}} \times \frac{V_1^2}{X_2'}$$

∴ T_{max} is independent of rotor resistance.

253. (d)

$$S_{T\max} = \frac{R_2}{X_2}$$

During starting,

$$S = S_{T\max} = 1$$

$$\therefore \quad \frac{R_2}{X_2} = 1 \text{ or } \frac{1}{R_2} = \frac{1}{X_2} \text{ or } R_2 = X_2$$

254. (d)

$$N_r = (1-s)\,N_s$$

$$= (1-0.04) \times \frac{120 \times 50}{P}$$

$$= 0.96 \times \frac{6000}{P} = \frac{5760}{P}$$

For speed to be maximum, maximum number of poles = 2.

$$\therefore \quad N_r = \frac{5760}{2} = 2880 \text{ rpm}$$

255. (a)

The starting torque of a SCIM is low because there is no provision of inserting an external resistance in the rotor circuit during starting because the rotor windings are short circuited. Hence, a specially designed rotor (double cage rotor) is used in this type of motor to get improved starting performance.

257. (a)

⇒ For auto-transformer starting,

$$I_{S(line)} = x^2 I_{SC} = x^2 \times (I_s)_{DOL}$$

∴ Supply current will reduce by $(x/100)^2$

⇒ Deep bar or double cage construction is used for improving the running performance of motor.

⇒ Due to use of two rotor bars, a double cage IM has comparatively higher copper loss.

258. (c)

$$(I_{s(\text{line})})_Y = \frac{1}{3}(I_{S(\text{line})})_\Delta$$

$$(T_{st})_{\text{star-delta}} = \frac{1}{3} \times (T_{st})_{\text{DOL}\,\Delta}$$

Hence, starting current is reduced but, there is reduction in starting torque. Therefore, reason is false.

259. (b)

$$\text{Gross power} = P_d = (1-s)\,P_g$$
$$= (1-0.1) \times 100$$
$$= 90 \text{ kW}$$

262. (b)

$$P_{sh} = 2700 \text{ W}, P_{rot} = 180 \text{ W}$$

$$\therefore \quad P_d = 2880 \text{ W } (= P_{sh} + P_{rot})$$

Now, $$\frac{P_{cu}}{P_d} = \frac{1-s}{s}$$

or, $$P_{cu} = 2880 \times \left(\frac{0.04}{0.96}\right) = 120 \text{ Watt}$$

263. (c)

$T \alpha \frac{V^2 s}{R^2}$ or, $V^2 s$ = constant (for constant toque)

$$\therefore \quad \left(\frac{V_2}{V_1}\right)^2 = \frac{s_1}{s_2}$$

$$\text{or,} \quad V_2 = V_1\sqrt{\frac{s_1}{s_2}} = 1\sqrt{\frac{1}{2}} = 0.707 \text{ pu}$$

264. (d)

Rotor will run against the direction of rotating field i.e. at slip speed to reduce the relative speed (since stator can't run therefore it will rotate the rotor).

265. (c)

The possible synchronous speeds are:
$N_s = 3000 \pm 1200 = 4200$ rpm or 1800 rpm

For 4200 rpm, $f = \frac{6 \times 4200}{120} = 210$ Hz

For 1800 rpm, $f = \frac{6 \times 1800}{120} = 90$ Hz

266. (a)

Since, V/f = constant, flux ϕ = constant, therefore I_m = constant. Maximum torque will remain constant in constant V/f ratio speed control method of 3-ϕ IM.

268. (b)

Running light test (no-load test) of a 3-ϕ IM is similar to open circuit test of a transformer because at no-load the rotor circuit acts as open-circuit when rated voltage and frequency is supplied to stator of the IM.

269. (d)

Since slip is very less at full load (2-5%), therefore possible rotor frequency should be 2 Hz.

270. (c)

As load on a 3-ϕ squirrel cage induction motor is increased, power factor is improved (due to decrease in slip).

$$\cos\phi = \frac{R_2}{\sqrt{R_2^2 + (sX_2)^2}}$$

As slip decreases, $\cos\phi$ increases.

271. (d)

Given,

$$N_{S_1} = 1200 \text{ rpm}, P_1 = 6$$

$$s_2 = 0.05, P_2 = 4$$

Now, $$f_1 = \frac{N_{S_1}P_1}{120} = \frac{1200 \times 6}{120} = 60 \text{ Hz}$$

Rotor frequency,

$$f' = sf_1 = 0.05 \times 60 = 3 \text{ Hz}$$

Now, $$N_{S_2} = \frac{120 \times 60}{4} = 1800 \text{ rpm}$$

$\therefore$ N = speed $= (1 - s_2)\, N_{S_2}$

$$= (1 - 0.05) \times 1800$$

$$= 1710 \text{ rpm}$$

272. (a)

$$P_{out} = 125 \text{ Watts}, P_{rot} = 25 \text{ Watts}$$

Total copper loss; P_{cu} = 25 Watt

$\therefore$ Input power to motor,

$$P_{in} = P_{out} + \text{Losses}$$

$$= P_{out} + (P_{cu} + P_{rot})$$

$$= 125 + (25 + 25) = 175 \text{ Watt}$$

$\therefore$ Efficiency of motor,

$$\eta = \frac{P_{out}}{P_{in}} \times 100$$

$$= \frac{125}{175} \times 100 = 71.43\%$$

273. (c)

Number of stacks or phases, $n = 3$
Number of teeth or poles, $P = 16$
$\therefore$ Stepping angle,

$$\alpha = \frac{360}{nP} = \frac{360}{3 \times 16} = 7.5^\circ$$

274. (c)

⇒ The compensating winding is required in an universal motor for ac operation to neutralize reactance voltage.

⇒ Universal motors are designed for operation at high speeds (> 3500 rpm).

275. (b)

$$N_s = \frac{120 \times 50}{8} = 750 \text{ rpm}$$

Forward slip,

$$s = \frac{N_s - N}{N_s} = \frac{750 - 690}{750}$$

$$= 0.08 = 8\%$$

Backward slip $= 2 - s = 2 - 0.08$

$$= 1.92 = 192\%$$

278. (c)

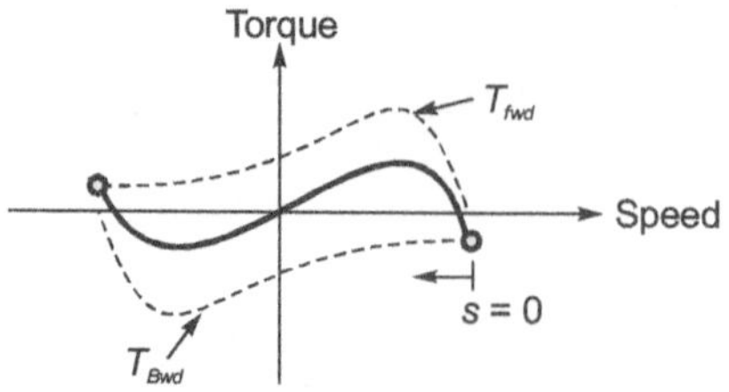

From above T-ω characteristic, $T = 0$ at speed = 0 and slightly below N_s.

280. (a)

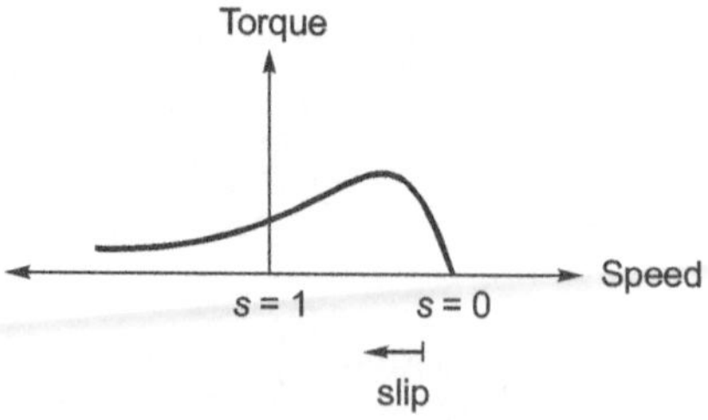

(T-N char of 3-ϕ IM)

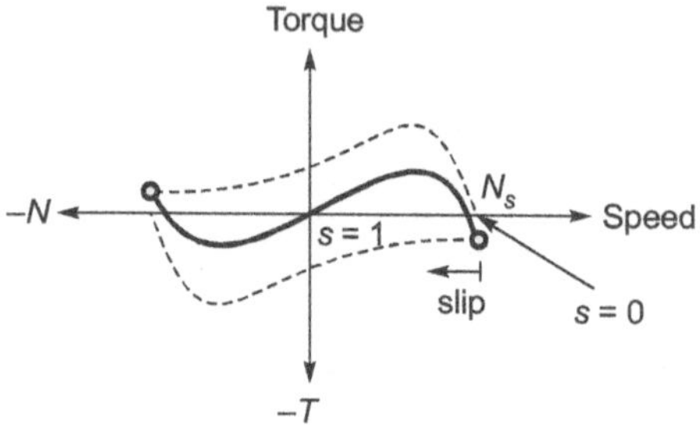

(T-N char of 1-ϕ IM)

From T-N char. of 3-ϕ IM, we conclude that:

For, $s = 0$, $T = 0$

From T-N char. of 1-ϕ IM, we conclude that:

For, $s = 0$, T = small –ve value

282. (c)

A single-phase IM is not self starting, therefore an auxilary winding is used to start this motor.

283. (b)

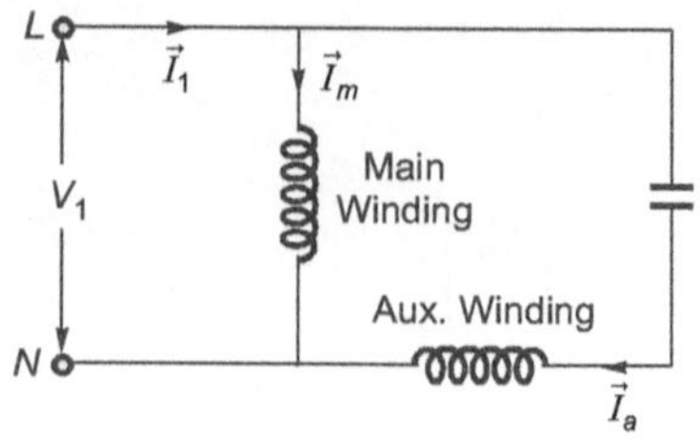

Here, C is connected in series with auxiliary winding (or starting winding).

286. (b)

Assertion is true because linear motion induction motors have linear motion due to which they are wide used in industry and electric traction.

289. (a)

Shaded pole motor has smallest starting torque and speed (1 - 2 rpm). T_{st} of capacitor-split phase IM > T_{st} of resistance-split phase IM.

293. (d)

Torque developed in a reluctance motor is given by

$$T_e \propto (Rl_q - Rl_d)$$

If $\frac{Rl_q}{Rl_d}$ is less, it means $Rl_q < Rl_d$ and thus, T_e is less.

294. (c)

In a 1-ϕ hysteresis motor,

eddy current torque, $T_{eddy} \propto s$

and hysteresis torque = constant

During starting, slip is high, therefore torque developed is due to both eddy current and hysteresis. However, during running condition slip is very low due to which eddy current torque will be very less. Therefore, torque is only due to hysteresis during running condition.

297. (c)

Capacitor start motor has highest starting toque. Hysteresis motor gives noise less operation.

298. (c)

The capacitor might be open circuited due to which no starting torque is developed and motor fails to start.

300. (b)

Step size/angle,

$$\Delta\theta = \frac{360°}{m N_r}$$

Given, N_r = Number of rotor teeths = 5

m = Number of phase/stacks = 4

$$\therefore \quad \Delta\theta = \frac{360°}{5\times 4} = 18°$$

308. (d)

$$\text{Energy stored} = \frac{1}{2}LI^2 = W$$

Now, $$L = \frac{N\phi}{I}$$

$$\therefore \quad W = \frac{1}{2}\left(\frac{N\phi}{I}\right)I^2 = \frac{1}{2}\cdot(N\phi)\cdot I$$

$$= \frac{1}{2}\times 1.2\times 10 = 6 \text{ Joules}$$

310. (a)

In synchronous-motor, dc excitation is given to rotor for flux generation while it's stator is connected is 3-ϕ supply. Therefore, it has doubly-excited system. Reluctance motor is an excitation-less motor ie. it has singly-excited magnetic system.

311. (b)

Average torque developed is

T_d = (stator mmf)·(rotor mmf)·sinδ

or, $T_d \propto a \sin\delta$

(where, a = constant and δ = torque angle)

313. (a)

In ac rotating machines like synchronous generators/motor/induction motors, working flux (ϕ) leads generated emf by 90°. However, in static device like a transformer working flux (ϕ) lags the applied voltage by 90°.

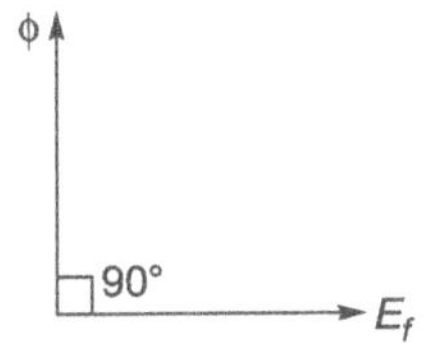

314. (c)

$$m = \text{number of slots/pole /phase} = \frac{48}{4\times 3} = 4$$

$$\text{Number of slots/pole} = \frac{48}{4} = 12$$

$\therefore$ slot-pitch angle,

$$\beta = \frac{180°}{12} = 15° \text{ ele}$$

$$\therefore \quad k_d = \left(\frac{\sin\frac{m\beta}{2}}{m\sin\frac{\beta}{2}}\right) = \frac{\sin\left(\frac{4\times 15}{2}\right)}{4\sin\frac{15}{2}}$$

$$= \frac{\sin 30°}{4\sin 7.5} = \frac{1}{8\sin 7.5}$$

316. (d)

For eliminating n^{th} harmonic,

$$\frac{n\alpha}{2} = 90° \text{ or } \alpha = \frac{180°}{n}$$

$\therefore$ Coil span of short-pitched winding

$$= 180° - \frac{180°}{n}$$

$$= \left(\frac{n-1}{n}\right)\times \text{Full-pitch}$$

For eliminating 3rd harmonic, coil-span

$$= \frac{2}{3}\times \text{Full-pitch}$$

317. (b)

Emf induced per phase $\alpha\ f\phi N_{ph}$

Here, ϕ = flux per pole

$$= \frac{2}{P}B_P LD$$

(L = Axial length of armature and D = diameter)

$$\therefore \quad E_{ph}\ \alpha\ N_{ph}LD\ (fB_m = \text{constant})$$

When, $N'_{ph} = 2N_{ph}$, $L' = 2L$ and $D' = 2D$

then, $(E_{ph})_{new}\ \alpha\ (2\times 2\times 2)(N_{ph}LD)\ \alpha\ 8N_{ph}$

Thus, generated voltage will increase by a factor of 8.

318. (d)

Given, N = 1200 rpm

$$\therefore \quad \omega = \frac{2\pi}{60}N = \frac{2\pi}{60}\times 1200$$

$$= 40\pi \text{ mech.rad/sec}$$

We know that,

$$\theta_{ele} = \frac{P}{2}\cdot\theta_{mech} = \frac{6}{2}\times 40\pi$$
$$= 120\,\pi \text{ elec-rad/sec}$$

319. (c)

$$f = \frac{NP}{120} \text{ cycles/sec}$$

or, $$\frac{\text{frequency}}{\text{rev}} = \text{Number of pole pairs}$$

320. (b)

kVA rating, $\alpha\, E_f \propto k_d$

$$\therefore \quad \frac{(kVA)_{new}}{(kVA)_{old}} = \frac{(k_d)_{new}}{(k_d)_{old}} = \frac{\left(\frac{\sin 90°}{90°}\right)}{\left(\frac{\sin 30°}{30°}\right)} = \frac{2}{3}$$

$$\left(\text{Since, } k_d = \frac{\sin\left(\frac{\text{phase spread}}{2}\right)}{\left(\frac{\text{phase spread}}{2}\right)}\right)$$

$$\therefore\ (kVA)_{new} = \frac{2}{3}\times 300 = 200 \text{ kVA}$$

321. (c)

To eliminate 3rd harmonic,

$$\frac{3\alpha}{2} = 90°$$

or, $\alpha = 60°$

$\therefore$ coil span $= 180° - 60° = 120°$

Now, 180° elec occupy 12 slots.

$$\therefore\ 120° \text{ elec occupy } \frac{12}{180}\times 120 = 8 \text{ slots}$$

Hence, coil span for eliminating 3rd harmonic = 8 slots.

322. (d)

5th harmonic is a negative sequence quantity which rotates at a speed of $\frac{1}{5}N_s$ w.r.t observer in space while the fundamental field rotates at a speed of N_s in space.

The direction of rotation of 5th harmonic mmf is opposite to that of fundamental mmf wave. Hence, speed of fifth harmonic w.r.t fundamental mmf wave

$$= N_s - \left(-\frac{N_s}{5}\right) = \frac{6}{5}N_s$$

323. (b)

Hysteresis loss, $P_h \propto B_m^x f$

Brush constant loss $= V_D I_L \propto I_L$

Stray load loss $\propto I_L^2$

No-load rotational loss (friction and windage loss) is due to rotor rotation.

325. (d)

At maximum efficiency, copper loss = iron loss

or, $x^2 P_0 = P_i$ (x = fraction of full load)

At $\frac{2}{3}$ full-load,

$$\left(\frac{2}{3}\right)^2 P_0 = P_i \text{ or } P_0 = \frac{9}{4}P_i$$

326. (d)

If n be the number of phases then,

magnitude of resultant flux $= \frac{n}{2}\phi_m$

(n = 3 for 3-phase and n = 2 for 2-phase)

$\therefore$ For 3 phase,

$$\phi_{res} = \frac{3}{2}\phi_m = 1.5\,\phi_m$$

and for 2 phase,

$$\phi_{res} = \frac{2}{2}\phi_m = \phi_m$$

Both the resultant flux will be rotating.

327. (c)

For wave-winding, number of conductors per parallel path

$$= \frac{Z}{2} = \frac{640}{2} = 320$$

$\therefore$ Armature circuit resistance,

$$r_a = \frac{320\times 10\times 10^{-3}}{2} = 1.6\,\Omega$$

328. (b)

Speed of rotating mmf wave = $\left(\frac{2}{P}f\right)$ rps.

As in one revolution, a peripheral distance of πD meter is traversed, therefore speed of travelling

$$\text{mmf wave} = \left(\frac{2f}{P}(\pi D)\right) \text{m/s}$$

$$= 2f\left(\frac{\pi D}{P}\right) \text{m/s}$$

$$= [2f\ (\text{Pole-pitch})]\ \text{m/s}$$

Now, $D = 1.2$ m, $P = 6$

$\therefore$ Speed of travelling mmf wave is

$$v = 2 \times 50 \times \left(\frac{\pi \times 1.2}{6}\right)$$

$$= 62.83 \text{ m/s}$$

331. (d)

At the time of winding design of 3-phase alternators, the attention is mainly directed for the attenuation of non-triplen harmonics (i.e. 5th and 7th harmonics) because the triplen harmonics (3rd, 9th etc) can be automatically eliminated from line emf if alternator is connected in star.

335. (b)

For n^{th} harmonic,

$$\text{Chording angle} = \frac{360°}{n}$$

Given, for fundamental flux wave,

Chording angle $= \alpha = 360°$

$\therefore$ For 5th harmonic,

$$\text{Chording angle} = \alpha_5 = \frac{360°}{5} = \frac{\alpha}{5}$$

■■■■

UNIT 7

Power Systems

1. Performance of Transmission Lines, Line Parameters & Corona

Q.1 A sample of copper wire has a resistance of 50 Ω at 10 °C. The maximum operating temperature of the wire if its resistance is to increase by at most 10 percent will be:
(Given, the temperature coefficient at 10°C is $\alpha = 0.00409°C^{-1}$).
(a) 19.55 °C (b) 22.22 °C
(c) 34.45 °C (d) 42.67 °C

Q.2 A single-phase, two wire transmission line, 15 km long, is made up of round conductors, each 0.8 cm radius, separated from each other by 40 cm. The equivalent diameter of a fictitious hollow, thin-walled conductor having the same inductance as the original line is given by
(a) 1.557 cm (b) 1.246 cm
(c) 1.6 cm (d) 0.623 cm

Q.3 If the conductor radius increases, inductance of the line,
(a) increases
(b) decreases
(c) remains same
(d) may increase or decrease

Q.4 For an underground cable, the surge impedance is generally taken as
(a) 40 Ω (b) 80 Ω
(c) 100 Ω (d) 400 Ω

Q.5 Consider the following phenomenon occuring in a power system:
1. Corona loss 2. Charging current
3. Ferranti effect 4. Dielectric loss
5. Skin effect
Which of the above are associated with transmission lines?
(a) 1, 3 and 5 (b) 2, 3, 4 and 5
(c) 3, 2 and 4 (d) 1, 2, 3 and 5

Q.6 The corona loss or the critical disruptive voltage in a transmission line at a temperature (t) and pressure (b) can be foundout by using "air-density correction factor (δ)" given by
(a) $\delta = \left[\dfrac{273+t}{3.92b}\right]$ (b) $\delta = 3.92\,(t+273)\,b$
(c) $\delta = \left[\dfrac{3.92b}{273+t}\right]$ (d) $\delta = \left[\dfrac{273+t}{b}\right]$

Q.7 A three-phase, 50 Hz, 400 kV, transmission line is 300 km long. The line inductance is 0.97 mH/km per phase and capacitance is 0.0115 μF/km per phase. Assuming a lossless line, the surge impedance of the line will be
(a) 196.44 Ω (b) 286.62 Ω
(c) 250 Ω (d) 290.43 Ω

Q.8 The resistance and reactance of a short line are equal. At zero regulation the load will be
(a) upf (b) zpf
(c) 0.707 lag (d) 0.707 lead

Q.9 Shunt reactors are needed
(a) to boost receiving end voltage under light loads.
(b) to boost receiving end voltage under heavy loads.
(c) to bring down receiving end voltage at light loads.
(d) to bring down receiving end voltage under heavy loads.

Q.10 Operating transmission lines close to their maximum power transfer capability will impair
(a) stability
(b) steady state power flow
(c) voltage profile
(d) transmission efficiency

Q.11 The transmission efficiency increases with
(a) decrease in pf and increase in voltage level
(b) decrease in pf and decrease in voltage level
(c) increase in pf and Increase in voltage level
(d) increase in pf and decrease in voltage level

Q.12 A short line with reactance of 20 Ω and negligible resistance operates with a sending end voltage of 132 kV and a receiving end voltage of 126 kV. The maximum power that can be transmitted with this voltage profile is
(a) 400 MW (b) 832 MW
(c) 1000 MW (d) 132 MW

Q.13 The surge impedance of a 100 km line is 390 Ω. If the line is extended to 200 km, the value of surge impedance is
(a) 195 Ω
(b) 780 Ω
(c) depends on operating frequency
(d) 390 Ω

Q.14 The surge impedance loading of a 220 kV overhead line is approximately
(a) 121 MW (b) 1.21 MW
(c) 12.1 MW (d) 121,000 MW

Q.15 A lossless line terminated with its surge impedance has a
(a) sag of voltage at midpoint
(b) swell of voltage at midpoint
(c) flat voltage profile
(d) uniform voltage dips along length of line

Q.16 A 220 kV, 20 km long, 3-phase line has following parameters:
$A = D = 0.96\angle 3°$; $B = 55\angle 65°$ Ω/phase, $C = 0.5 \times 10^{-4}\angle 80°$ ℧/phase.
It's charging current is
(a) $\frac{11}{\sqrt{3}}$ A (b) 11 A
(c) 220 A (d) $\frac{220}{\sqrt{3}}$ A

Q.17 The economic size of conductor is determined by
(a) Kelvin's law (b) Kirchoff's law
(c) Faraday's law (d) All of these

Q.18 Skin effect depends upon
(a) type of conductor material
(b) shape of the conductor
(c) both (a) and (b)
(d) none of the above

Q.19 Proximity effect depends upon
(a) size of the conductor
(b) resistivity of material
(c) permeability of the material
(d) all of the above

Q.20 The overall diameter '*D*' of a standard conductor is given in terms of diameter of each strand '*d*' and number of layers '*n*' as
(a) $D = 2(n-1)d$ (b) $(2n-1)/d$
(c) $D = 2(2n-1)d$ (d) $D = 2(n-D)d$

Q.21 In a transmission system, the weight of copper used is proportional to
(a) E^2 (b) E
(c) $\frac{1}{E^2}$ (d) $\frac{1}{E}$

Q.22 As the spacing between the phase conductors are increased, match **List-I** with **List-II** and select the correct code given below the lists:

List-I	List-II
A. Line resistance	1. Decrease
B. Line inductance	2. Increase
C. Line capacitance	3. Remains same

Codes:

	A	B	C
(a)	1	2	2
(b)	1	3	2
(c)	3	1	2
(d)	3	2	1

Q.23 Which of the following conductor spirality effect is ignored?
(a) Magnetic (b) Non-magnetic
(c) Both (a) and (b) (d) None of these

Q.24 Skin effect is ignored in which supply system?
(a) AC (b) DC
(c) Both (d) None of these

Q.25 For a good voltage profile under no load conditions a long line needs
(a) shunt capacitors at receiving end
(b) shunt reactors at receiving end
(c) shunt resistance at receiving end
(d) shunt capacitor at sending end

Q.26 Corona has can be reduced by use of hollow conductor because
(a) the current density is reduced.
(b) the eddy current in conductor is eliminated.
(c) for a given cross-section, the radius is increased.
(d) of better ventilation in the conductor.

Q.27 The charging current in a transmission line
(a) lags the voltage by 90°
(b) leads the voltage by 45°
(c) leads the voltage by 90°
(d) leads the voltage by 180°

Q.28 Which of the following relationships is not valid for a short transmission line?
(a) $I_S = I_R$ (b) $B = Z = C$
(c) $A = D = 1$ (d) None of these

Q.29 Sag of an overhead line is proportional to the square of
(a) tension (b) span
(c) temperature (d) weight of conductor

Q.30 The per unit "regulation" of a short transmission line having a per unit resistance voltage drop v_r and a per unit reactance voltage drop of v_x at rated current and, with power factor cosθ, is given by
(a) $(v_r + v_x)\cos\theta$ (b) $(v_r + v_x)\sin\theta$
(c) $v_r\cos\theta + v_x\sin\theta$ (d) $v_r\sin\theta + v_x\cos\theta$

Q.31 The charging reactance of 100 km line is 500 Ω, the charging reactance of 200 km length line will be
(a) 1000 Ω (b) 250 Ω
(c) 500 Ω (d) 1255 Ω

Q.32 When bundle conductors are used in place of single conductors, the effective inductance and charging current will respectively
(a) decrease and increase
(b) increase and decrease
(c) decrease and remain unaffected
(d) remain unaffected and increase

Q.33 A hollow conductor is at a potential *V*. The potential at any point inside the hollow conductor is
(a) zero (b) V/2
(c) V (d) None of these

Q.34 **Assertion (A):** Feeders are designed mainly from the point of view of the voltage drop in them.
Reason (R): Current loading of the feeder remains the same along its length.
(a) Both A and R are true and R is a correct explanation of A.
(b) Both A and R are true but R is not a correct explanation of A.
(c) A is true but R is false.
(d) A is false but R is true.

Q.35 **Assertion (A):** Heavier the conductor smaller will be the sag.
Reason (R): Sag is directly proportional to the square of the span.
(a) Both A and R are true and R is a correct explanation of A.
(b) Both A and R are true but R is not a correct explanation of A.
(c) A is true but R is false.
(d) A is false but R is true.

Q.36 The propagation constant of a transmission line is given as:
(a) $j\sqrt{\dfrac{L}{C}}$ (b) $j\sqrt{LC}$
(c) $j\sqrt{C/L}$ (d) $j\omega\sqrt{LC}$

Q.37 The SIL of a line is proportional to
(a) V^2 (b) V
(c) $\sqrt{V}$ (d) $V^{1/3}$

Q.38 Load compensation can be achieved by
(a) installing the compensating equipment near the source
(b) installing the compensating equipment near the load
(c) either (a) or (b)
(d) both (a) and (b)

Q.39 The characteristic impedance of the line depends upon:
(a) the characteristic of the line per unit length
(b) length of the line
(c) radius and spacing between conductors
(d) all of the above

Q.40 Consider the two configurations of conductors as shown below.

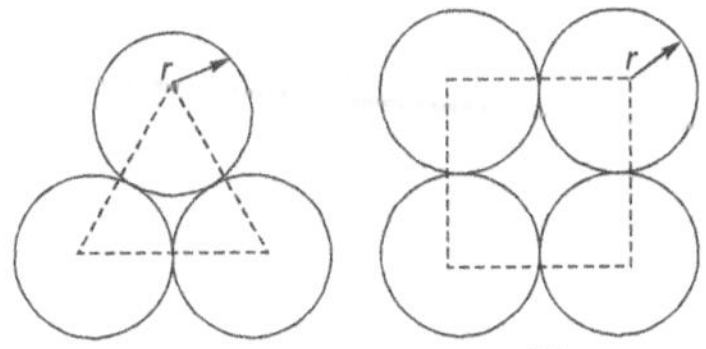

(a) (b)

The geometric mean radius of the above two configurations in terms of the radius r are respectively given by

(a) $\sqrt[3]{r' \times 2r \times 2r}$ and $\sqrt[4]{r' \times 2r \times 2r \times \sqrt{2}r}$

(b) $\sqrt[3]{r' \times r \times r}$ and $\sqrt[4]{r' \times r \times 2r \times 2r}$

(c) $\sqrt[3]{r' \times 2r \times 2r}$ and $\sqrt[4]{r' \times 2r \times 2r \times 2\sqrt{2}r}$

(d) $\sqrt[3]{r' \times r \times 2r}$ and $\sqrt[4]{r' \times r \times r \times \sqrt{2}r}$

Q.41 A single-phase transmission line of impedance $j0.8$ ohm supplies a resistive load of 500 A at 300 V. The sending end power factor is

(a) unity (b) 0.8 lagging
(c) 0.8 leading (d) 0.6 lagging

Q.42 A 50 Hz 3-phase transmission line of length 100 km has a capacitance of $0.03/\pi$ μF/km. It is represented as a π-model. The shunt admittance at each end of the transmission line will be

(a) $150 \times 10^{-6} \angle 90$ mho
(b) $100 \times 10^{-6} \angle 90°$ mho
(c) $50 \times 10^{-6} \angle 90°$ mho
(d) $\frac{10^6}{150} \angle 90°$ mho

Q.43 A rectangular voltage wave is impressed on a loss-free overhead line, with the far end of the line being open-circuited. On reaching the end of this line

(a) both the current and the voltage waves are reflected with positive sign.
(b) both the current and the voltage waves are reflected with negative sign.
(c) the current wave is reflected back with positive sign, but the voltage wave with negative sign.
(d) the current wave is reflected back with negative sign, but the voltage wave with positive sign.

Q.44 A three-phase, 50 Hz transmission line has a capacitance of line to neutral $C_n = 0.01$ μF/km. The voltage of the line is 100 kV. The charging current per kilometer of line is

(a) $\frac{314}{\sqrt{3}}$ A/km (b) $\frac{314}{\sqrt{3} \times 10^3}$ A/km

(c) $\frac{\sqrt{3} \times 10^3}{314}$ A/km (d) $\sqrt{3} \times 10^3$ A/km

Q.45 The breakdown strength of air at STP is 21 kV/cm. Its breakdown strength at 30°C and 72 cm of the *Hg* will be

(a) 21.25 kV/cm (b) 20.2 kV/cm
(c) 23 kV (d) 19.5 kV/cm

Q.46 Consider the plot of an unknown parameter (X) of the transmission line vs temperature in °C as shown below.

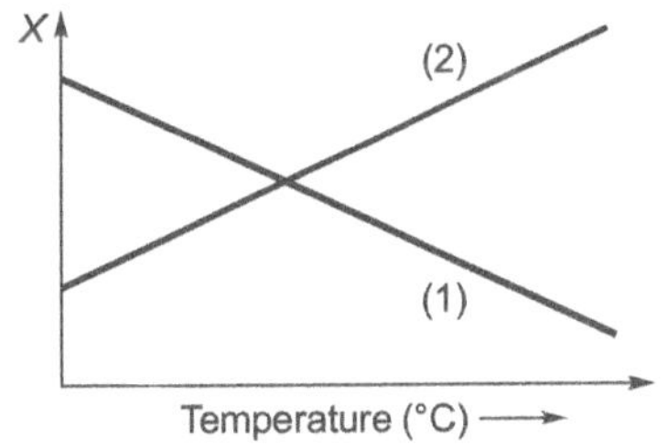

If the unknown parameter is taken as Sag and tension of the transmission line respectively then, the resulting variation of "Sag vs temperature" and "tension vs temperature" will be respectively represented by the straight line (s)

(a) 1 and 2 (b) 1 only
(c) 2 and 1 (d) 2 only

Q.47 The parameter A of a nominal-T circuit of a transmission line for which $R = 10\ \Omega$, $X = 20\ \Omega$ and $Y = 400$ μS for each phase will be given by

(a) $(996 + j2) \times 10^{-3}$
(b) $(0.996 + j200) \times 10^3$
(c) $(996 + j200) \times 10^{-3}$
(d) $(0.996 + j2) \times 10^3$

Q.48 A transmission line has an electrical line length of 6°. If the velocity of propagation is 3×10^8 m/sec then, for a 50 Hz system the length of the line in km will be

(a) 80 km (b) 100 km
(c) 180 km (d) 250 km

Q.49 In a short line, maximum power transfer occurs when δ is (here θ is angle of impedance of the line and δ = power angle):

(a) $> \theta$ (b) $= \theta$
(c) $< \theta$ (d) $= 90°$

Q.50 A line has an electrical line length of 8°. At 60 Hz, the ratio of line length to wave length is

(a) 0.01 (b) 1.02
(c) 0.022 (d) 3.46

Q.51 The corona loss on a particular system at 50 Hz is 1.2 kW/km/phase. The corona loss at 60 Hz will be

(a) 1.2 kW/km/phase
(b) 0.996 kW/km/phase
(c) 1.44 kW/km/phase
(d) 1.36 kW/km/phase

Q.52 The Sag for a span of 350 m, if the ultimate tensile strength of conductor is 5000 kg, factor of safety 2 and weight of conductor is 0.550 kg/m is

(a) 3.37 m (b) 6.12 m
(c) 1.2 m (d) 10.8 m

Q.53 For any fixed degree of inductive shunt compensation, additional series capacitive compensation

(a) increases the effective length of line
(b) increases virtual surge impedance of line
(c) decreases virtual surge impedance loading of the line
(d) none of the above

Q.54 In a matrix form, the equation of a 4-terminal network representing a transmission line is given by

$$\begin{bmatrix} V_S \\ I_S \end{bmatrix} = \begin{bmatrix} A & B \\ C & D \end{bmatrix} \begin{bmatrix} V_R \\ I_R \end{bmatrix}$$

The two networks considered are as shown below:

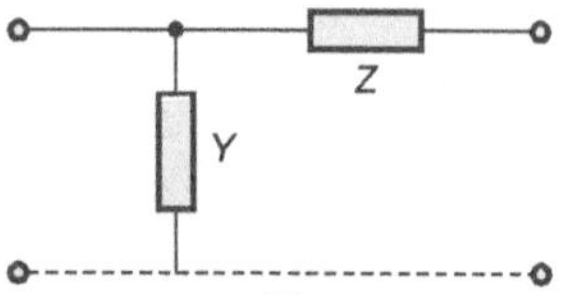

(A)

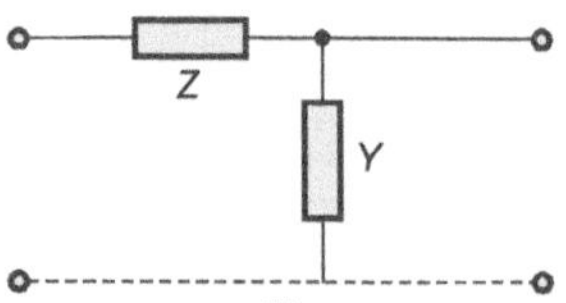

(B)

The possible transfer matrix for the networks (A) and (B) could be

(*i*) $\begin{bmatrix} 1 & YZ \\ 0 & Z \end{bmatrix}$ (*ii*) $\begin{bmatrix} Y & 0 \\ YZ & 1 \end{bmatrix}$

(*iii*) $\begin{bmatrix} 1 & Z \\ Y & 1+YZ \end{bmatrix}$ (*iv*) $\begin{bmatrix} 1+YZ & Z \\ Y & 1 \end{bmatrix}$

The correct combination for the two networks (A) and (B) would be

(a) (*i*) and (*ii*) (b) (*i*) and (*iii*)
(c) (*ii*) and (*iv*) (d) (*iii*) and (*iv*)

Q.55 The self-inductance of a long-cylindrical conductor due to its internal flux linkages is K H/m. If the diameter of the conductor is doubled, then the self-inductance of the conductor due to its internal flux linkages would be

(a) 0.5 K H/m (b) K H/m
(c) 1.414 K H/m (d) 4 K H/m

Q.56 Consider the following statements:

1. Equivalent-T circuit of a long line is preferred over equivalent π-circuit.
2. The nature of reactive power compensation is different for peak load and off-peak load conditions.
3. Ferranti effect is significant only on medium and long lines.

Which of the statements given above are correct?

(a) 1 and 2 (b) 1 and 3
(c) 2 and 3 (d) 1, 2 and 3

Q.57 Two arrangements of conductors are proposed for a 3-phase transmission line: one with equilateral spacing of 4 m and the other a flat with 4 m distance between the conductors as shown in the figure below.

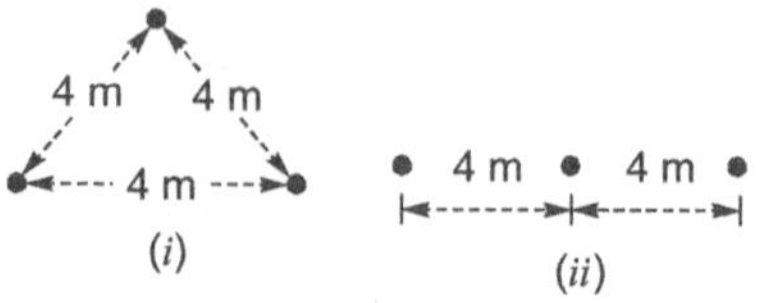

The conductor diameter in each case is 2 cm. Assuming that the line is transposed in both cases, which one of the following statements would be true?

(C_n = Capacitance in F/m line to neutral and L = Inductance in H/m per phase)

(a) $C_{n1} = C_{n2}$ and $L_1 > L_2$
(b) $C_{n1} > C_{n2}$ and $L_1 < L_2$
(c) $C_{n1} < C_{n2}$ and $L_1 > L_2$
(d) $C_{n1} > C_{n2}$ and $L_1 = L_2$

Q.58 A medium line with parameters *A*, *B*, *C* and *D* is extended by connecting a short line of impedance *Z* in series. The overall *ABCD* parameters of the series combination will be

(a) $\begin{bmatrix} A & AZ \\ C+DZ & D \end{bmatrix}$ (b) $\begin{bmatrix} A+BZ & B \\ C+DZ & D \end{bmatrix}$

(c) $\begin{bmatrix} A & AZ+B \\ C & CZ+D \end{bmatrix}$ (d) $\begin{bmatrix} AZ & B \\ C/Z & D \end{bmatrix}$

Q.59 For equilateral spacing of conductors of an untransposed 3-phase line, we have
(a) balanced receiving end voltage and no communication interference.
(b) unbalanced receiving end voltage and no communication interference.
(c) balanced receiving end voltage and communication interference.
(d) unbalanced receiving end voltage and communication interference.

Q.60 The reflection coefficient for the transmission line shown in figure at point *P* is

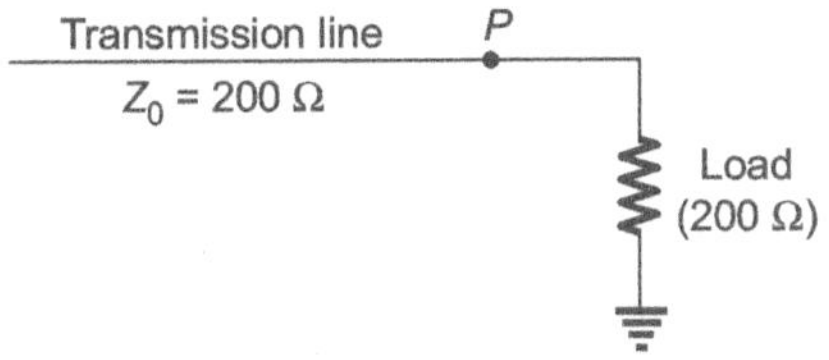

(a) +1 (b) −1
(c) zero (d) 0.5

Q.61 For a single-phase overhead line having solid copper conductors of diameter 1 cm, spaced 60 cm between centers, the inductance in mH/km is
(a) $0.05 + 0.2\, ln\, 60$
(b) $0.2\, ln\, 60$
(c) $0.05 + 0.2\, ln\left(\frac{60}{0.5}\right)$
(d) $0.2\, ln\left(\frac{60}{0.5}\right)$

Q.62 If the length of a wire of resistance *R* is uniformly stretched to *n* times its original value, its new resistance will be

(a) nR (b) $\frac{R}{n}$
(c) n^2R (d) $\frac{R}{n^2}$

Q.63 A single-phase 2-wire system supplies a load of 1000 kW. The voltage between the conductors is 11000 V. Return wire is earthed. Now, a third wire is added in the system and supply made is 3-phase. For same power factor and same loss in the line the transmitted power will be
(a) 2450 kW (b) 3100 kW
(c) 3000 kW (d) 1500 kW

Q.64 A lossless radial transmission line with surge impedance loading
(a) takes negative VAR at sending end and zero VAR at receiving end.
(b) takes positive VAR at sending end and zero VAR at receiving end.
(c) has flat voltage profile and unity power factor at all points along it.
(d) has sending end voltage higher than receiving end voltage and unity power factor at sending end.

Q.65 The L/C ratio for 132 kV and 400 kV line are typically 160×10^3 and 62.5×10^1 respectively. The natural 3-phase loading for the two lines will be respectively
(a) 108.9 and 2560 MW
(b) 44 and 2560 MW
(c) 44 and 6400 MW
(d) 6400 and 44 MW

Q.66 The time taken for a square wave to travel a 600 km long overhead transmission line is
(a) 6 sec (b) 1 sec
(c) 0.02 sec (d) 0.002 sec

Q.67 For a synchronous generator connected to an infinite bus through a transmission line, how are the change of voltage (ΔV) and the change of frequency (Δf) related to the active power (P) and the reactive power (Q)?
(a) ΔV is proportional to P and Δf to Q
(b) ΔV is proportional to Q and Δf to P
(c) Both ΔV and Δf are proportional to P
(d) Both ΔV and Δf are proportional to Q

Q.68 A transmission line of 100 km length has certain values of parameters *A*, *B*, *C*, *D*. If the length is made 210 km the parameter
(a) *C* increases and *D* decreases
(b) *C* and *D* both decreases
(c) *C* and *D* both increases
(d) *C* decreases and *D* increases

Direction for Questions (69 to 73):

Each of the following question consists of two statements, one labelled the **'Assertion (A)'** and the other labelled the **'Reason (R)'**. Examine the two statements carefully and decide if the Assertion (A) and Reason (R) are individually true and if so whether the Reason (R) is correct explanation of the Assertion (A). Select your answers to these questions using the codes given below:

Codes:
(a) Both A and R are true and R is the correct explanation of A.
(b) Both A and R are true but R is not a correct explanation of A.
(c) A is true but R is false.
(d) A is false but R is true.

Q.69 **Assertion (A):** Proximity effect affects the current distribution.
Reason (R): Proximity effect results in an increase in the resistance of the conductor and decrease in self reactance.

Q.70 **Assertion (A):** Corona reduces the effects of transients produced by lightning and other causes.
Reason (R): Charges induced on the line by lightning or other causes are partially dissipated as a corona loss.

Q.71 **Assertion (A):** Transposition of conductors in a transmission line is necessary.
Reason (R): Corona losses are reduced by transposition of conductors.

Q.72 **Assertion (A):** The insulation of modern EHV lines is designed based on the lightning voltage.
Reason (R): With the developments in the design of relay breaker systems, it is possible to control and minimize switching over voltages.

Q.73 **Assertion (A):** EHV transmission lines make use of bundled conductors.
Reason (R): Bundled conductors reduce the line inductance per phase and increase the capacitance.

Q.74 The voltage stability analysis is carried out by which power flow based method?
(a) $P-V$ curves
(b) $Q-V$ curves
(c) Both (a) and (b)
(d) None of these

Q.75 Match the items given in **List-I** and those in **List-II** and select your answer to these questions using the codes given below the lists:

List-I
A. Skin effect
B. Proximity effect
C. Ferranti effect
D. Corona
E. Surge impedance

List-II
1. Increase in resistance but decrease in self inductance.
2. Increase in ac resistance.
3. Causes power loss in the line.
4. Owing to voltage drop across line inductance due to flow of a charging current.
5. Square root of ratio of line impedance and shunt admittance.

Codes:

	A	B	C	D	E
(a)	1	2	3	4	5
(b)	5	4	1	2	3
(c)	3	2	5	4	1
(d)	2	1	4	3	5

Q.76 Match the items given in **List-I (Type of conductors)** and those in **List-II (Application)** and select your answer to these questions using the codes given below the lists:

List-I
A. All Aluminium (AA)
B. ACSR
C. Copper
D. Galvanised steel

List-II
1. Restricted use
2. LV lines with short spans
3. Stray wires, earth wires, guard wires
4. LV and HV lines with large spans

Codes:

	A	B	C	D
(a)	1	2	3	4
(b)	2	4	1	3
(c)	3	1	4	2
(d)	4	3	2	1

Q.77 Match the items given in **List-I (Methods)** and those in **List-II (Line constants)** and select your answer to these questions using the codes given below the lists:

List-I

A. Simple series impedance method
B. End condenser method
C. Nominal-T method
D. Nominal-π method

List-II

1. $A = D = \left(1 + \frac{YZ}{2}\right)$

   $B = Z$ and $C = Y\left(1 + \frac{YZ}{4}\right)$

2. $A = D = \left(1 + \frac{YZ}{2}\right)$

   $B = Z\left(1 + \frac{YZ}{4}\right)$ and $C = Y$

3. $A = D = 1$; $B = Z$ and $C = 0$
4. $A = (1 + YZ)$; $B = Z$; $C = Y$; $D = 1$

Codes:

	A	B	C	D
(a)	1	2	3	4
(b)	3	4	2	1
(c)	2	3	4	1
(d)	4	1	3	2

2. Compensation Techniques, Voltage Profile Control and Load Frequency Control

Q.78 Series capacitive compensation on EHV transmission lines is used to
(a) reduce the line loading
(b) improve the protection of the line
(c) reduce the voltage profile
(d) improve the stability of the system

Q.79 Phase modifier is normally installed in the case of
(a) short transmission line
(b) medium length lines
(c) long length lines
(d) for all length lines

Q.80 For the stable operation of interconnected system, the passive element that can be used as an interconnecting element is
(a) resistor (b) capacitor
(c) reactor (d) any of these

Q.81 For a synchronous phase modifier the load angle is
(a) 0° (b) 25°
(c) 30° (d) None of these

Q.82 A synchronous phase modifier supplies
(a) both active and reactive power
(b) both lagging and leading reactive power
(c) inductive reactive power only
(d) none of the above

Q.83 Series capacitors on transmission lines are of little use when
(a) the load VAR requirement is small
(b) the load VAR requirement is large
(c) the load VAR requirement is fluctuating
(d) none of the above

Q.84 Full load compensation in a line requires
(a) shunt reactors (b) transformers
(c) shunt capacitors (d) series capacitors

Q.85 Use of additional shunt capacitor can be made for increasing the capability of line as it
(a) increases phase shift β
(b) increases α
(c) reduces surge impedance Z_0
(d) all of the above

Q.86 Shunt capacitance in an EHV line is restored to
(a) improve the stability
(b) reduce fault level
(c) improve the voltage
(d) none of the above

Q.87 The reactive power transfer over a line mainly depends on
(a) power angle δ (b) $|V_S| - |V_R|$
(c) V_S (d) V_R

Q.88 For synchronous generator connected to a power station, increase in excitation will
(a) decrease reactive power generation
(b) increase reactive power generation
(c) decrease active power generation
(d) increase active power generation

Q.89 A power system needs injection of VARs at
(a) off-peak load
(b) peak load
(c) both peak and off-peak load
(d) full load

Q.90 Shunt capacitors in a substation
(a) consume lagging VARs
(b) deliver lagging VARs
(c) consume active power
(d) deliver active power

Q.91 During load shedding
(a) system voltage is reduced
(b) system frequency is reduced
(c) system loads are switched off
(d) system power factor is changed

Q.92 Flow fo active power through a line in a power system may be controlled by
(a) tap changing transformer
(b) phase shifting transformer
(c) synchronous phase modifier
(d) any of the above

Q.93 **Assertion (A):** The best location for the pf correction equipment to be installed is the generating end.
Reason (R): Generators are relieved of carrying excessive current due to poor power factor.
(a) Both A and R are true and R is the correct explanation of A.
(b) Both A and R are true but R is not a correct explanation of A.
(c) A is true but R is false.
(d) A is false but R is true.

Q.94 Which of the following is not a sub-component of load-frequency controller?
(a) speed changer
(b) linkage mechanism
(c) exciter
(d) hydraulic amplifier

Q.95 Unit of *'R'* and *'B'* in load frequency control have a unit of
(a) MW-s and MVAr/Hz respectively
(b) Hz/MW and MW/Hz respectively
(c) MVA/Hz and MVA/Hz respectively
(d) MVAr/Hz and MW-s respectively

Q.96 Shunt capacitors are used to raise the power factor of the load of 100 kW from 0.8 lag to unity, the supply being 3-phase at 10 kV. In star bank, the capacitance per phase required will be
(a) 0.8 μF (b) 2.4 μF
(c) 1.6 μF (d) 3.2 μF

Q.97 A shunt reactor of 100 MVAR is operated at 98% of its rated voltage and at 96% of its rated frequency. The reactive power absorbed by the reactor is
(a) 98 MVAR (b) 100.04 MVAR
(c) 96.04 MVAR (d) 100.06 MVAR

Q.98 Consider the following statements:
1. In case of transmission system, for improving the power factor, a synchronous condenser can be installed either at load end or source end.
2. A phase advancer can be used to improve the power factor of an induction motor.
3. Schrage motor and Compensated induction motors are used for improving the power factor of the system.

Which of the above statements are true?
(a) 1 and 2 (b) 2 only
(c) 2 and 3 (d) 1, 2 and 3

Q.99 Consider the following statements about the various methods of load frequency control in an interconnected power system:
1. In "Flat Frequency Control", the frequency of the system is maintained constant by regulating only one station-A and without any regulation of station-B.
2. In "Flat Tie-Line Control", the increase in load of an area is met by increasing the generation in that area and thus power flow in the tie line is kept constant irrespective of load demands.
3. "Tie-line Bias Control" is the modification of "flat frequency control" and "flat Tie-line control".

Which of the above statement(s) is/are true?
(a) 2 and 3 only (b) 2 only
(c) 1, 2 and 3 (d) 1 and 2 only

Q.100 **Assertion (A):** The higher the load with smaller power factor the greater is the voltage variation.
Reason (R): The voltage variation at a load is an indication of the unbalance between the reactive power generated and absorbed by that load.
(a) Both A and R are true and R is the correct explanation of A.
(b) Both A and R are true but R is not a correct explanation of A.
(c) A is true but R is false.
(d) A is false but R is true.

Q.101 The speed of a 200 MW alternator decreases by 4% from no-load to full-load. It's rated capacity is 200 MW. The speed regulation of alternator in Hz/MW will be
(a) 0.02 (b) 0.03
(c) 0.04 (d) 0.01

Q.102 Two alternators rated 100 MW and 200 MW have a governor droop characteristics of 4% from no load to full load. They are connected in parallel to share a load of 250 MW. The load shared by the machines will be respectively
(a) 98.26 MW, 151.74 MW
(b) 66.72 MW, 183.56 MW
(c) 83.33 MW, 166.67 MW
(d) 57.44 MW, 192.56 MW

Q.103 A turbine generator set has a regulation constant 5% on the generator rating of 100 MVA, 50 Hz. The generator frequency decreases by 0.02 Hz. The increase in turbine output power for steady-state operation is
(a) 0.8 MW (b) 8 MW
(c) 3.6 MW (d) 0.2 MW

Q.104 Whenever there is change in frequency of the system due to change in the load demand, the component of the load frequency controller which senses this change in load is
(a) speed changer
(b) hydraulic amplifier
(c) linkage mechanism
(d) speed governor

Q.105 When the speed governor fails to sense any variation/change in the load demand the resulting load frequency control is termed as
(a) proportional controller operation
(b) integral controller operation
(c) free governor operation
(d) none of the above

Q.106 A synchronous generator is supplying load of 300 kW at 0.6 pf lagging. If the pf is increased to unity, the additional kW required for the same kVA loading is
(a) 700 kW (b) 300 kW
(c) 200 kW (d) 500 kW

Q.107 Consider the following statements:
1. Steady-state analysis of a power system deals with small step changes in the load demand.
2. Dynamic analysis of a power system deals with large step changes in the load demand.
3. Steady-state analysis gives the result in time domain while dynamic analysis gives the result in frequency domain.

Which of the above statement(s) is/are true?
(a) 1, 2 and 3 (b) 2 and 3 only
(c) 1 and 2 only (d) 3 only

Q.108 What is the correct expression for Area Control Error (ACE) for an interconnected 2-area power system, if ΔP_t is incremental tie line power deviation, Δf is frequency deviation and b is frequency bias coefficient?
(a) $ACE = b\,\Delta P_t + \Delta f$
(b) $ACE = \Delta P_t + b\,\Delta f$
(c) $ACE = \frac{1}{b}\Delta P_t + \Delta f$
(d) $ACE = \Delta P_t + \frac{1}{b}\Delta f$

Q.109 Two generators rated at 200 MW and 400 MW are operating in parallel. Both the governors have a droop of 4%, when the total load is 300 MW. They share the load as (suffix '1' is used for generator of 200 MW and suffix '2' is used for generator of 400 MW):
(a) P_1 = 100 MW and P_2 = 200 MW
(b) P_1 = 150 MW and P_2 = 150 MW
(c) P_1 = 200 MW and P_2 = 100 MW
(d) P_1 = 200 MW and P_2 = 400 MW

Q.110 A 500 MW generator has a speed regulation of 4%. If the frequency drops by 0.12 Hz with an unchanged reference, the increase in turbine power will be

(a) 10 MW (b) 30 MW
(c) 45 MW (d) 20 MW

3. Distribution Systems, Cables & Insulators

Q.111 If r is the the radius of the conductor and R the the radius of the sheath of the cable, the cable operates stably from the view point of dielectric strength if:

(a) $\frac{r}{R} > 1.0$ (b) $\frac{r}{R} < 0.368$
(c) $\frac{r}{R} < 1.0$ (d) $\frac{r}{R} > 2.718$

Q.112 The insulation resistance of a cable of 10 km length is 1 MΩ. The resistance for a length of 40 km is

(a) 1 MΩ (b) 4 MΩ
(c) 0.25 MΩ (d) Can't be determined

Q.113 Sheath losses can be eliminated by

(a) transposing cable along with cross bonding
(b) transposing cables only
(c) cross bonding
(d) none are effective

Q.114 Armouring protect cables against

(a) earth fault
(b) capacitance leakage
(c) mechanical damage
(d) stray magnetic fields

Q.115 Match **List-I (Type of Cables)** with **List-II (Operatng Voltages)** and select the correct answer using the code given below the lists:

List-I	List-II
A. LT Cable	1. Upto 11,000 V
B. EHT Cable	2. Upto 33,000 V
C. ST Cable	3. Upto 1000 V
D. HT Cable	4. Upto 66,000 V

Code:

	A	B	C	D
(a)	3	4	2	1
(b)	1	2	4	3
(c)	1	3	2	4
(d)	3	4	1	2

Q.116 Consider the following statements:

1. AAC (All Aluminium conductors) are universaly employed for feeders as well as distributors.
2. The conductor size for a feeder is mainly governed by the permissible voltage drop in the line.

Which of the above statement(s) is/are true?

(a) 1 only (b) 1 and 2
(c) none (d) 2 only

Q.117 Which of the following distribution systems is the most economical?

(a) Single phase AC system
(b) DC system
(c) Three phase, 4-wire AC system
(d) Three phase, 3-wire ac system

Q.118 For high ovltage transmission lines, conductors are suspended from towers so as to

(a) increase clearance from ground
(b) reduce wind and snow effect
(c) take care of increase in length
(d) reduce clearance from ground

Q.119 The function of steel wire in an ACSR conductor is to

(a) compensate for skin effect
(b) take care of surges
(c) provide additional mechanical strength
(d) reduce inductance

Q.120 The string efficiency of insulators can be increased by

(a) reducing the number of strings
(b) increasing the number of strings
(c) correct grading of insulators of various capacitances
(d) changing the orientation of strings

Q.121 The insulating materials most commonly used in a cable is

(a) PVC (b) impregnated paper
(c) VIR (d) All of the above

Q.122 If the frequency of a transmission system is changed from 50 Hz to 100 Hz, the string efficiency will

(a) increase
(b) decrease
(c) remains unchanged
(d) may increase or decrease

Q.123 The insulation thickness on the conductor in cables depends upon
(a) voltage (b) power factor
(c) reactive power (d) current

Q.124 Which of the following distribution system is the simplest and have lowest installation cost?
(a) inter connected system
(b) radial system
(c) ring main system
(d) none of the above

Q.125 Three insulating materials with same maximum working stress and permittivities 2.5, 3.0, 4.0 are used in a single core cable. The location of the materials w.r.t. the core of the cable will be:
(a) 2.5, 3.0, 4.0 (b) 3.0, 2.5, 4.0
(c) 4.0, 3.0, 2.5 (d) 4.0, 2.5, 3.0

Q.126 A 13.2 kV single-conductor cable has an outside diameter of 10 cm. The conductor radius that must be withstood by the insulating material in the most economical (optimal-ratio) configuration will be given by:
(a) 1.718 cm (b) 0.367 cm
(c) 2.718 cm (d) 1.840 cm

Q.127 In a test of a three-conductor cable, the three conductors are first bunched together, and the capacitance between the bunched conductors and the sheath is found to be C_A. Then, two of the conductors are bunched with the sheath, and the capacitance between these and the third conductor is found to be C_B. The value of C_S (Capacitance between conductor and sheath) and C_C (Capacitance between phase conductors) will be respectively given by:

(a) $3C_A$ and $\frac{1}{2}\left(C_A - \frac{1}{3}C_B\right)$

(b) $\frac{C_A}{3}$ and $\frac{1}{2}\left(C_B - \frac{1}{3}C_A\right)$

(c) $3\,C_A$ and $\frac{1}{2}\left(C_B - \frac{1}{3}C_A\right)$

(d) $\frac{C_A}{3}$ and $\frac{1}{2}\left(C_A - \frac{1}{3}C_B\right)$

Q.128 Three insulating materials *A*, *B* and *C* have breakdown strengths of 250 kV/cm, 200 kV/cm and 150 kV/cm with respective permittivities of 2.5, 3.0 and 3.5. They are used in a single core cable. The location of the materials with respect to the outermost surface of the cable will be:
(a) *A*, *B* and *C* (b) *A*, *C* and *B*
(c) *C*, *B* and A (d) *B*, *C* and *A*

Q.129 The capacitance per kilometer of a three wire cable are 0.90 μF between the three bunched conductors and the sheath, and 0.40 μF between one conductor and the other two connected to the sheath. The line-to-ground capacitance of a 20 km length of this calbe will be
(a) 11 μF (b) 0.45 μF
(c) 4.5 μF (d) 9 μF

Q.130 **Assertion (A) :** The capacitance of a cable is more than the capacitance of an overhead transmission line.
Reason (R) : The electrical permittivity of the cable insulation is 3 to 5 times greate than air insulation around the conductors of an overhead line.
(a) Both A and R are true and R is the correct explanation of A.
(b) Both A and R are true but R is not the correct explanation of A.
(c) A is true but R is false.
(d) A is false but R is true.

Q.131 The diameter (for economic operation) of a single core cable metal sheathed for a working voltage of 80 kV if the dielectric strength of the insulating material is 60 kv/cm will be given by
(a) 7.23 cm (b) 3.61 cm
(c) 2.66 cm (d) 6.32 cm

Q.132 The dielectric loss in a cable is not a function of
(a) capacitance of the cable
(b) frequency of supply
(c) square of the voltage
(d) radius of the conductor

Q.133 Which of the following is not a characteristic of a ring main distributor?
(a) Better flexibility
(b) Greater voltage regulation
(c) More power loss
(d) Greater reliability

Q.134 Match **List-I** with **List-II** and select the correct answer using the codes given below the lists:

List-I

A. Poles

B. Cross-arms

C. Guys and stays

D. Continuous earth wire on top of the tower

List-II

1. to support line conductors at a safe distance from ground.
2. to keep conductors at a safe distance from each other and from the pole.
3. to protect the line against the lightning.
4. to stay overhead lines at angle and terminal positions.

Codes:

	A	B	C	D
(a)	1	2	3	4
(b)	2	1	4	3
(c)	1	2	4	3
(d)	3	4	1	2

Q.135 For an existing AC transmission line, the string efficiency is 80%, if DC voltage is supplied for the same set up, the string efficiency will be

(a) < 80% (b) 80%
(c) 90% (d) 100%

Q.136 In a three unit insulator string, voltage across the lowest unit is 17.5 kV and string efficiency is 84.28%. The total voltage across the string will be equal to

(a) 8.825 kV (b) 44.25 kV
(c) 88.25 kV (d) 442.5 kV

4. Generating Power Stations

Q.137 A hydrograph indicates

(a) the discharge at any time during the period under consideration.
(b) the maximum and minimum run-off during the period.
(c) the average run-off during the period
(d) all of the above

Q.138 Which is the conventional source of energy?

(a) Solar
(b) Radio-active substances
(c) Geothermal
(d) Wind

Q.139 The power output from a hydro-electric power plant doesn't depend on

(a) discharge (b) head
(c) type of turbine (d) system efficiency

Q.140 A station operates with a load factor and plant capacity factor of 0.75, with a maximum demand of 100 MW. The reserve capacity is

(a) 100 MW (b) 75 MW
(c) 0 MW (d) 133.33 MW

Q.141 The area under daily load curve gives

(a) energy generated during peak hours
(b) energy generated during off-peak hours
(c) energy generated per day
(d) energy generated to supply peak load

Q.142 Within the boiler, the steam has highest temperature in

(a) superheater (b) water tubes
(c) water walls (d) water drum

Q.143 In which part of the thermal power plant, the steam pressure is less than that of atmosphere?

(a) Boiler (b) Turbine
(c) Superheater (d) Condenser

Q.144 An industrial consumer has a load of 1500 kW at a 0.8 p.f. lag for 12 hours and 1000 KW upf for 12 hours during the day. The daily load factor is

(a) 0.833 (b) 0.666
(c) 1.5 (d) 0.8

Q.145 In a nuclear reactor, chain reaction is controlled by

(a) Iron rods (b) Cadmium rods
(c) Graphite rods (d) Brass rods

Q.146 In a nuclear power station using boiling water reactor, water is used as

(a) a moderator but not as a coolant
(b) a coolant but not as a moderator
(c) both moderator and coolant
(d) neither maoderator nor coolant

Q.147 The yearly load duration curve of a power plant is a straight line. The maximum load is 750 MW and the minimum load is 600 MW. The utilization factor and the plant capacity factor are respectively

(a) 0.80, 0.56 (b) 0.83, 0.75
(c) 0.90, 0.78 (d) 0.75, 0.83

Q.148 Out of the following plant categoriess:

1. Nuclear 2. Run or river
3. Pump storage 4. Gas turbines

The peak load power plants aro

(a) 2 and 3 (b) 1 and 3
(c) 3 only (d) 3 and 4

Q.149 Electrostatic precipitator is installed between

(a) induced fan and chimney
(b) air preheater and induced fan
(c) economiser and air preheater
(d) boiler furnace and economiser

Q.150 The efficiency of a nuclear power plant is less than that of a conventional fuel fired thermal plant because of

(a) less injection of heat in the condenser
(b) higher temperature conditions
(c) higher pressure conditions
(d) low pressure and temperature conditions

Q.151 A generating station which has a high investment cost and low operating cost is usually operated as a

(a) peak load station
(b) base load station
(c) medium load station
(d) none of the above

Q.152 Match **List-I (Types of power plant)** with **List-II (Basic feature)** and select the correct answer using the codes given below the lists:

List-I
A. Nuclear
B. Thermal
C. Hydro
D. Pumped storage

List-II
1. Maximum capital cost
2. Maximum operating cost
3. Minimum operating cost
4. Least starting time

Codes:

	A	B	C	D
(a)	1	2	3	4
(b)	3	1	4	2
(c)	2	4	1	3
(d)	4	3	2	1

Q.153 The surge tank is a component of a

(a) Thermal PP
(b) Nuclear PP
(c) Pumped storage PP
(d) Hydro PP

Q.154 Match **List-I (Type of turbine)** with **List-II (Water discharge)** and select the correct answer using the codes given below the lists:

List-I	List-II
A. Pelton	1. High
B. Francis	2. Medium
C. Kaplan	3. Low

Codes:

	A	B	C
(a)	1	2	3
(b)	3	2	1
(c)	1	3	2
(d)	3	1	2

Q.155 Maximum efficiency of a modern coal-fired steam-raising thermal power plant is restricted to about 0.35 (a low value), mainly because of

(a) low alternator efficiency
(b) high energy loss in boilers
(c) low steam turbine mechanical efficiency
(d) high energy loss from turbine exhaust to condenser

Q.156 The maximum demand on a generating station is 20 MW, a load factor of 75%, a plant capacity factor of 50%, and a plant use factor of 80%. What is the maximum energy that could be produced daily if the plant were in use all the time?

(a) 720 MWh (b) 225 MWh
(c) 360 MWh (d) 180 MWh

Q.157 The choice of number and size of units in a station are governed by best compromise between

(a) a plant load factor and capacity factor
(b) plant capacity factor and plant-use factor
(c) plant load factor and use factor
(d) none of the above

Q.158 A peak load station

(a) should have a low operating cost
(b) should have a low capital cost
(c) can have a operating cost high
(d) both (b) and (c)

Q.159 The load pattern for 24 hours is shown below.

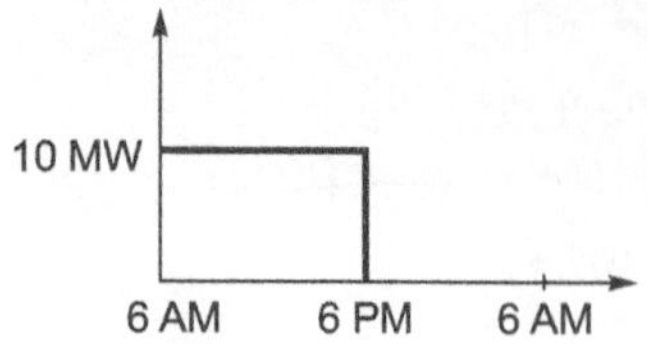

The factor is

(a) 1.0 (b) 0.6
(c) 0.5 (d) 2.0

Q.160 A hydel plant supplied by a river with a discharge of 3000 m^3/sec at a head of 30 m, with a plant efficiency of 79% develops a power of

(a) 423.12 MW (b) 345.2 MW
(c) 823.7 MW (d) 697.5 MW

Q.161 A coal fired steam power station working at a plant load factor of 80% has 500 MW generating unit, the heat content of the coal is 2 kW-h/kg and the overall efficiency is 40%. If the coal loading capacity of a train is 2000 metric tons, then the number of trains required daily for plant will be

(a) 1 (b) 102
(c) 6 (d) 24

Q.162 The modern trend in electric power generation is

(a) to have a large number of small size thermal plants located at different places.
(b) to have a large size thermal plants near load centres.
(c) to have a large size thermal plants located near coal fields.
(d) none of the above

Q.163 A super critical boiler is one which operates above the pressure and temperature of the following values

(a) 100 kg/cm^2 and 540°C
(b) 218 kg/cm^2 and 540° C
(c) 100 kg/cm^2 and 373° C
(d) 218 kg/cm^2 and 373° C

Q.164 In a steam power plant heat from the flue gases is recovered in

(a) a condenser
(b) a chimney
(c) economiser and air preheater
(d) a desuper heater

Q.165 A generating station supplies peak loads of 100, 125 and 50 MW to three groups of consumers. The station load factor is 0.5 and diversity factor is 1.5. The maximum demand on the station and the average demand are respectively

(a) 165.45 MW and 89.66 MW
(b) 183.33 MW and 89.66 MW
(c) 165.45 MW and 91.66 MW
(d) 183.33 MW and 91.66 MW

Q.166 The daily load pattern on a plant is as follows: 50 MW for 3 hrs; 75 MW for 16 hrs; 100 MW for 2 hrs and no load for 3 hrs.
The load factor of the plant is

(a) 38.62% (b) 49.37%
(c) 64.58% (d) 72.86%

Q.167 A generating station has a connected load of 600 MW. The maximum demand is 380 MW. The number of units generated per annum is 2100×10^6. The demand factor and load factor are respectively given by:

(a) 63% and 63% (b) 63% and 72%
(c) 72% and 63% (d) 72% and 68%

Q.168 A hydroelectric station has to operate with a mean head of 300 m and is supplied from a reservoir which drains a catchment area of 500 sq. km with an annual rainfall of 150 cm. 72% of rainfall can be utilized. Assuming there is a loss of 2% in penstock, turbine efficiency of 92% and generator efficiency of 95%, What is the power that can be developed by the station?

(a) 62.55 MW (b) 38.24 MW
(c) 52 MW (d) 43.14 MW

Q.169 Match **List-I** with **List-II** and select the correct answer using the codes given below the lists:

List-I	List-II
A. Moderator	1. Lithium
B. Reflector	2. Cadmium
C. Regulating rod	3. High grade graphite
D. Coolant	4. Heavy water

Codes:

	A	B	C	D
(a)	3	1	4	2
(b)	4	3	2	1
(c)	2	4	3	1
(d)	1	2	3	4

Q.170 An interconnected system has the following power plants:

1. Nuclear	2. Steam
3. Diesel	4. Gas turbine
5. Hydro with storage	6. Run-off rive
7. Pumped storage	

Which of the above three power plants are used extensively for supplying peak loads?

(a) 1, 2, 3 (b) 3, 4, 7
(c) 4, 5, 6 (d) 2, 3, 5

Q.171 A generating station has a maximum demand of 20 MW and a connected load of 40 MW. The generated units are 4.36×10^7 per annum. The demand factor is

(a) 2 (b) 2/3
(c) 1/2 (d) 1/3

Q.172 Match **List-I (Classification of head)** with **List-II (Type of turbine)** and select the correct answer using the codes given below the lists:

List-I

A. Low head, 2 - 15 m
B. Medium head, 15 - 70 m
C. High head, 70 - 500 m
D. Very high head > 500 m

List-II

1. Propeller of Kaplan
2. Kaplan or Francis
3. Pelton
4. Francis or Pelton

Codes:

	A	B	C	D
(a)	1	3	4	2
(b)	4	2	1	3
(c)	1	2	4	3
(d)	4	3	1	2

Q.173 **Assertion (A) :** Load factor and diversity factor play an important part in the cost of the supply of electrical energy.

Reason (R) : Lower the values of diversity factor and load factor, lower will be the overall cost per unit generated.

(a) Both A and R are true and R is the correct explanation of A.
(b) Both A and R are true but R is not the correct explanation of A.
(c) A is true but R is false.
(d) A is false but R is true.

Q.174 **Assertion (A) :** Availability of storage may convert an uneconomical hydro-project into an economical one.

Reason (R) : Storage increases the firm capacity of the power plant and increase in firm capacity of hydro-plant means dispensing with the alternative steam capacity.

(a) Both A and R are true and R is the correct explanation of A.
(b) Both A and R are true but R is not the correct explanation of A.
(c) A is true but R is false.
(d) A is false but R is true.

Q.175 Match **List-I (Type of Consumer)** with **List-II (Demand Factor)** and select the correct answer using the codes given below the lists:

List-I

A. Domestic and Commercial lightning upto 1 kW
B. Domestic power
C. Industrial power
D. Tube well

List-II

1. 0.5
2. 05. – 0.8
3. 0.5 – 0.7
4. 0.9 – 0.95

Codes:

	A	B	C	D
(a)	1	2	3	4
(b)	2	3	4	1
(c)	3	1	2	4
(d)	4	3	1	2

Q.176 A 120 MW power station delivers 110 MW for 4 hours, 60 MW for 10 hours and is shut down for the rest of the day. For general maintenance, it is shunt down for 60 days per annum. The annual load factor of the station is

(a) 28.88% (b) 36.11%
(c) 52.67% (d) 44.21%

Q.177 Match **List-I** with **List-II** and select the correct answer using the codes given below the lists:

List-I

A. Load-duration curve
B. Load curve
C. Integrated load-duration curve
D. Hydrograph

List-II
1. Load demand arranged in a descending order of magnitude versus time in hours
2. Discharge (m^3/sec) vs time
3. Load demand versus time
4. Plot of the cumulative integration of area under the load curve starting at zero loads to the particular load.

Codes:

	A	B	C	D
(a)	1	2	3	4
(b)	3	1	2	4
(c)	4	3	2	1
(d)	1	3	4	2

Q.178 A plant having maximum demand more than the installed capacity will have utilization factor
(a) less than 100% (b) equal to 100%
(c) more than 100% (d) none of these

Q.179 If capacity factor = load factor, then
(a) utilization factor is zero
(b) utilization factor is non-zero
(c) utilization factor is equal to one
(d) none of the above

Q.180 When compared to a thermal plant, the operating cost and capital cost of a hydro-electric plant are respectively
(a) high and low (b) low and high
(c) both high (d) both low

Q.181 It is necessary to heat water before feeding it back to the boiler as because of this
(a) the dissolved oxygen and carbon dioxide which would otherwise corrode boiler are removed in the feed water heater.
(b) thermal stresses due to cold water entering the boiler drum are avoided.
(c) some other impurities carried by steam and condensate due to corrosion in boiler and condenser are precipitated outside the boiler.
(d) all of the above

5. Fault Analysis

Q.182 When all the three phases are short circuited it gives rise to
(a) asymmetrical fault current
(b) symmetrical fault current
(c) zero fault current
(d) both (a) and (b)

Q.183 The zero sequence reactance will be more than the positive sequence reactance, if the equipment is:
(a) Generator (b) Transformer
(c) Transmision line (d) Both (b) and (c)

Q.184 The sequence networks for a three-phase faults are connected in
(a) series (b) parallel
(c) series opposition (d) none of the above

Q.185 Match **List-I (Symbol for 3-phase transformer connections)** with **List-II (Zero-sequence equivalent circuits)** and select the correct answer using the codes given below the lists:

List-I

A. 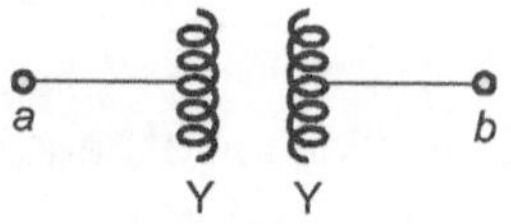

B. 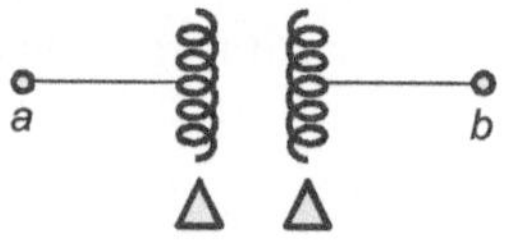

C. 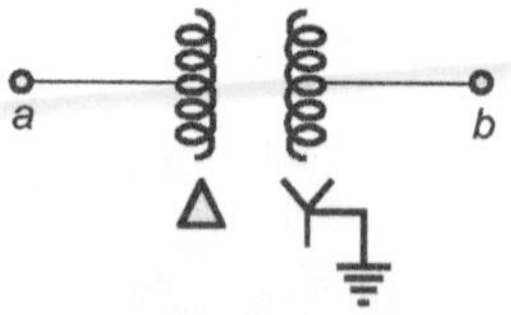

D. 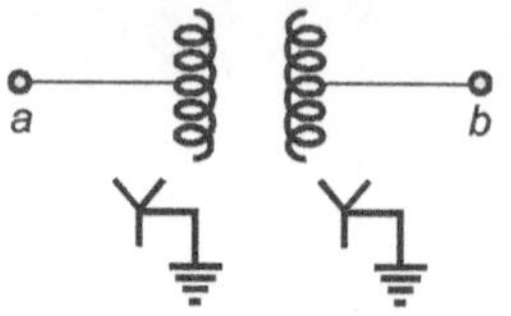

List-II

1. 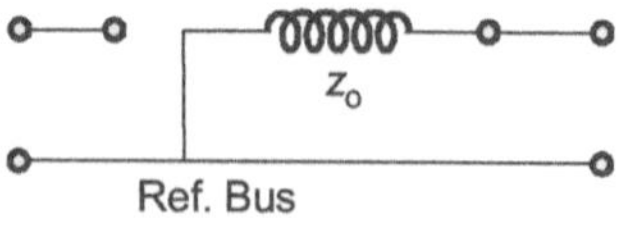

2. 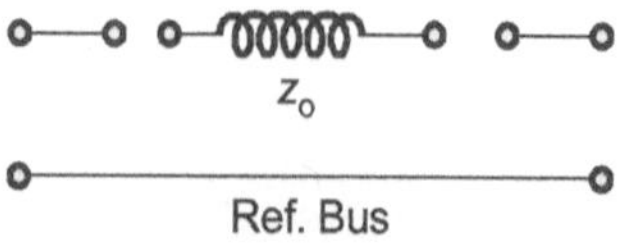

3. 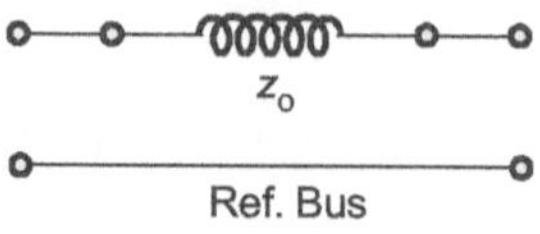

4. 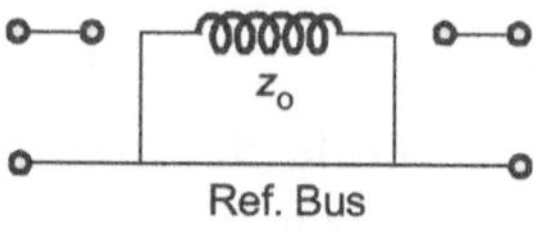

Codes:

	A	B	C	D
(a)	2	4	1	3
(b)	3	4	2	1
(c)	1	2	3	4
(d)	2	1	3	4

Q.186 The base impedance and base voltage for a given power system are 10 Ω and 400 V respectively. The base kVA and the base current will be respectively
(a) 16000 kVA and 40 A
(b) 4 kVA and 16 A
(c) 16 kVA and 40 A
(d) 4 kVA and 400 A

Q.187 The per-unit admittance Y_{pu} of a power system in terms of the base voltage V_{base} and the base voltamperes $(VA)_{base}$ is given by
(a) $\frac{YV_{base}^2}{(VA)_{base}}$ pu (b) $\frac{Y(VA)_{base}}{V_{base}^2}$ pu
(c) $\frac{V_{base}^2}{Y(VA)_{base}}$ pu (d) $\frac{(VA)_{base}}{YV_{base}^2}$ pu

Q.188 In a line to ground fault on a system, the fault current is 1500 A. The zero sequence current is
(a) 1500 A (b) 0 A
(c) 750 A (d) 500 A

Q.189 The fault current, in a double line to ground fault on a system depends on
(a) z_1, z_2 and z_0
(b) z_1 and z_2 only
(c) z_2 and z_0 only
(d) does not depend on any of them

Q.190 There may be a zero sequence components in
1. line voltage of star circuit
2. phase voltage of star circuit
3. line current of delta circuit
4. phase current of delta circuit

Select the correct answer from the given options:
(a) 1, 3 and 4 (b) 2 and 4 only
(c) 2 only (d) 1, 2 and 4

Q.191 A 10 kVA, 400 V/200V single phase transformer with 10% impedance draws a steady short circuit line current of
(a) 50 A (b) 150 A
(c) 250 A (d) 350 A

Q.192 A single line to ground fault occurs at the terminals of an unloaded generator operating at rated voltage where,
$X_1 = X_2 = 0.2$ pu ; $X_o = 0.1$ pu ; $X_n = 0.05$ pu
The fault current in pu is
(a) 6.125 pu (b) 1.625 pu
(c) –0.225 pu (d) 4.615 pu

Q.193 The following sequence currents were measured during a fault:
$I_{a1} = -j3.05$ pu; $I_{a2} = 0.68$ pu; $I_{ao} = j2.37$ pu
The type of fault is
(a) 3-ϕ fault (b) L-G fault
(c) L-L-G fault (d) L-L fault

Q.194 In a practical power system the most severe fault is the
(a) LLG fault (b) LG fault
(c) Both (a) and (b) (d) 3-ϕ fault

Q.195 In the network as shown below, the marked parameters are pu impedances. The bus admittance matrix of the network is

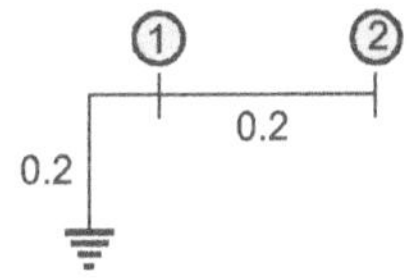

(a) $\begin{bmatrix} 10 & -5 \\ -5 & 5 \end{bmatrix}$ (b) $\begin{bmatrix} 5 & -5 \\ -5 & 10 \end{bmatrix}$
(c) $\begin{bmatrix} -10 & 5 \\ 5 & -5 \end{bmatrix}$ (d) $\begin{bmatrix} -5 & 5 \\ 5 & -10 \end{bmatrix}$

Q.196 A 30 MVA, 11 kV, 3-phase alternator has $X_d'' = 0.25$ pu, $z_2 = 0.35$ pu and $z_0 = 0.1$ pu. Its neutral is solidly grounded. A single line to ground fault occurs at the terminals of the alternator. The fault current will be
(a) – j1575 A (b) – j3150 A
(c) – j6755 A (d) – j4725 A

Q.197 A single phase transformer is rated as 2.5 kVA, 11/0.4 kV. If its leakage reactance is 0.96 Ω when referred to LV side, then it's leakage reactance in pu will be
(a) 0.015 (b) 0.125
(c) 0.625 (d) 1.125

Q.198 **Assertion (A):** Unsymmetrical faults can be analyzed only using symmetrical components.

Reason (R): Unsymmetrical fault currents are highly unbalanced.

(a) Both A and R are ture and R is the correct explanation of A.
(b) Both A and R ar true but R is not the correct explanation of A.
(c) A is true but R is false.
(d) A is false but R is true.

Q.199 A three-phase transformer having zero-sequence impedance of z_0 and neutral grounded with impedance z_n has the zero-sequence network as shown in the below figure.

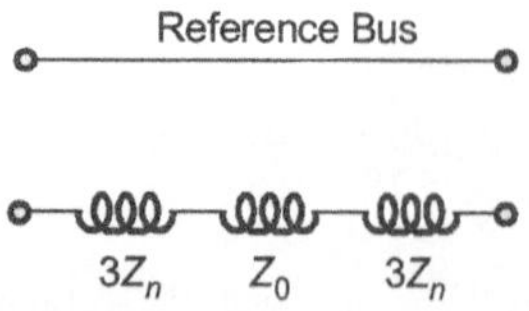

The connections of its windings are
(a) star-delta
(b) star-star
(c) delta-star with neutral grounded
(d) star-star with neutrals grounded

Q.200 A single-phase, 10 kVA, 200 V generator has an internal impedance Z_g of 2 Ω. Using the ratings of the generator as base values, the generated per-unit voltage required to produce full-load current under short-circuit condition will be
(a) 1 pu (b) 0.25 pu
(c) 0.5 pu (d) 1.25 pu

Q.201 A 345 kV transmission line has a series impedance of (4 + j 60) Ω and a shunt admittance of $j2 \times 10^{-3}$ S. Using 100 MVA and the line voltage as base values, the per-unit impedance and per-unit admittance of the line will be respectively
(a) $(3.36 + j50.4) \times 10^{-3}$ pu and $j2.66$ pu
(b) $(3.36 + j50.4) \times 10^{-3}$ pu and $j2.38$ pu
(c) $(50.4 + j3.36) \times 10^{-3}$ pu and $j2.38$ pu
(d) $(50.4 + j3.36)$ pu and $j2.66$ pu

Q.202 In the given figure, a three phase short circuit occurs at point F.

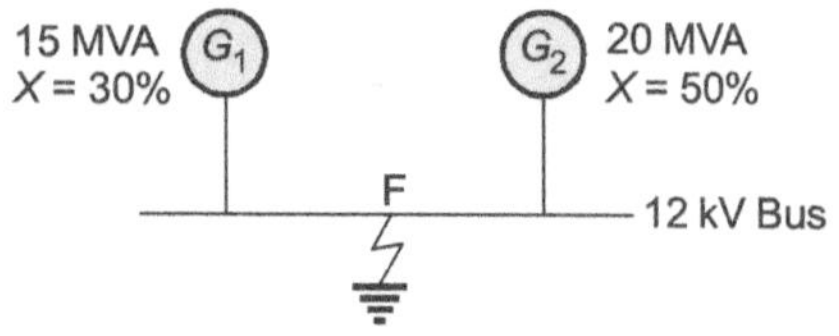

The fault current will be
(a) 4330 A (b) 9000 A
(c) 6435 A (d) 10 kA

Q.203 A three-phase 10 MVA, 11 kV alternator has an internal reactance of 5%. The series reactor needed to restrict the SC current of alternator to 8 times of full load current is
(a) 10 Ω (b) 9.075 Ω
(c) 0.9075 Ω (d) 90.75 Ω

Q.204 The line currents of a three phase supply are: $I_R = 3 + j5$ A ; $I_Y = 2 + j2$ A ; $I_B = -2 - j1$ A
The zero sequence current is:
(a) $1 + j2$ A (b) $5 + j7$ A
(c) $1 + j4$ A (d) $-2 - j1$ A

Q.205 The one line diagram of a system is shown in figure below:

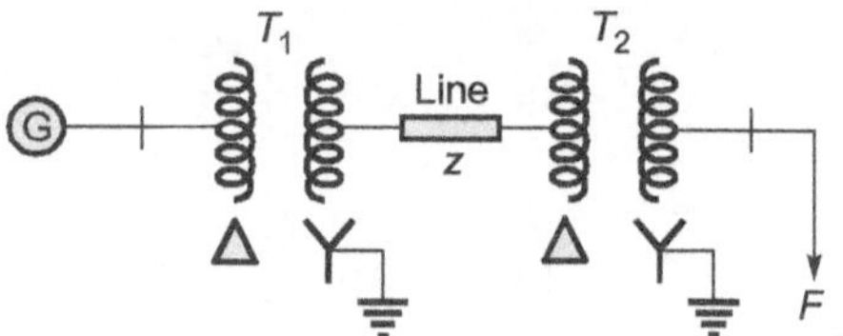

The zero sequence network is

(a)
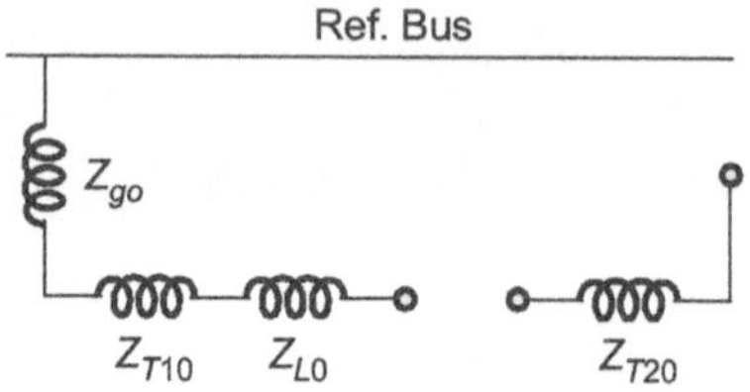

(b)
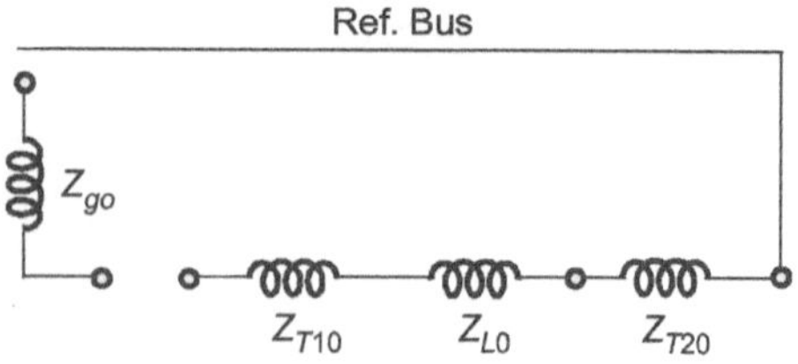

(c)
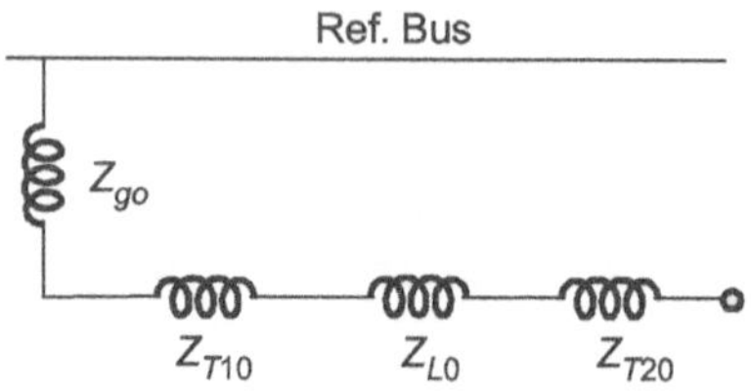

(d)
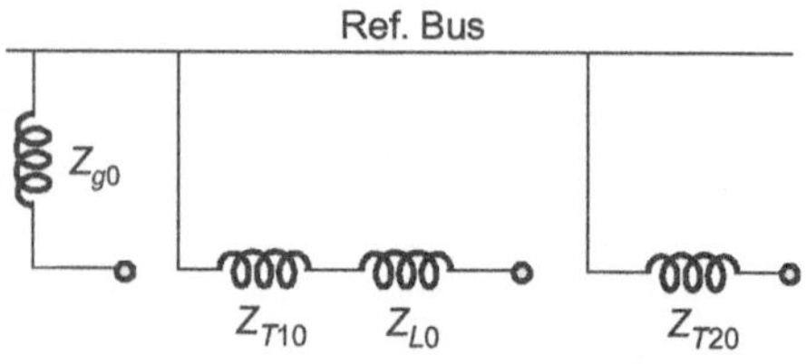

Q.206 In an unloaded generator operating at rated voltage, it is observed that the magnitude of fault current for a L-G fault is same as a three phase fault. The reactances of the generator are $x_1 = x_2 = 0.15$ pu; $x_0 = 0.04$ pu. For the given condition, the value of the neutral grounding reactance is

(a) 0.367 pu (b) 0.0367 pu
(c) 6.667 pu (d) 0.1 pu

Q.207 A small station has a bus bar divided into three sections. Each section is connected to a tie bar with a reactance rated 5 MVA, reactance 0.1 pu as shown in figure below. A generator of rating 5 MVA and reactance 0.2 pu is connected to each section of the bus bar. The short circuit capacity of the breaker if a 3-ϕ fault takes place on one of the section will be

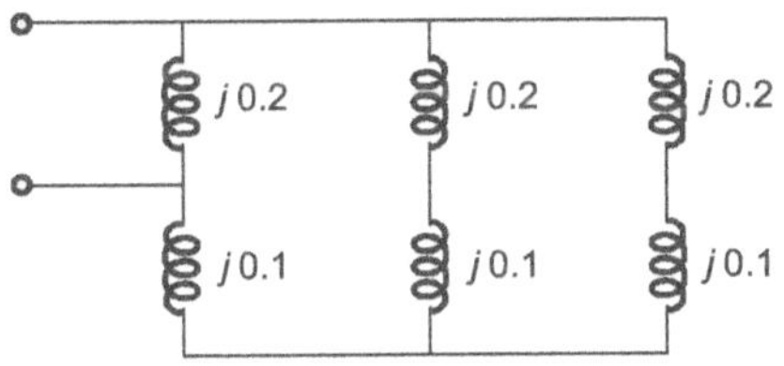

(a) 45 MVA (b) 90 MVA
(c) 10 MVA (d) 5 MVA

Q.208 Consider the following statements:

1. The positive and negative sequence components of voltage in a balanced 3-ϕ system are zero and only zero sequence component is present.
2. The zero sequence current flows only where neutral return path is available.
3. For a line-to-line fault, the sequence networks are connected in series.

Which of the above statement(s) is/are true?

(a) 1 and 3 only (b) 2 and 3 only
(c) 1 and 2 only (d) 2 only

Q.209 The bus admittance matrix of the network shown in the given figure for which the marked parameters are per unit impedance is

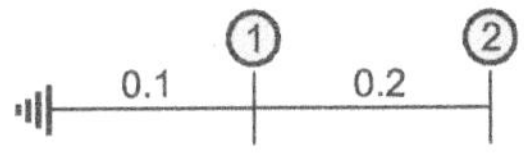

(a) $\begin{bmatrix} 0.3 & -0.2 \\ -0.2 & 0.2 \end{bmatrix}^{-1}$ (b) $\begin{bmatrix} 0.3 & 0.2 \\ 0.2 & 0.2 \end{bmatrix}$

(c) $\begin{bmatrix} 0.3 & -0.2 \\ -0.2 & 0.2 \end{bmatrix}$ (d) $\begin{bmatrix} 15 & -5 \\ -5 & 5 \end{bmatrix}$

Q.210 In an unbalanced 3-phase system, the currents are measured as: I_a = zero, $I_b = 6\angle 60°$ A and $I_c = 6\angle -120°$ A. The corresponding sequence components will be:

	I_{a_0}	I_{a_1}	I_{a_2}
(a)	zero	$(3 - j\sqrt{3})$ A	$(-3 + j\sqrt{3})$ A
(b)	zero	$(-3 + j\sqrt{3})$ A	$(3 - j\sqrt{3})$ A
(c)	zero	$(-9 + j3\sqrt{3})$ A	$(9 - j3\sqrt{3})$ A
(d)	zero	$(9 - j3\sqrt{3})$ A	$(-9 + j3\sqrt{3})$ A

Q.211 When a 50 MVA, 11 kV, 3-phase generator is subjected to a 3-phase fault, the fault current is $-j5$ pu. When it is subjected to a line-to-line fault, the positive sequence current is $-j4$ pu. The positive and negative sequence reactances are respectively

(a) $j0.2$ and $j0.05$ pu
(b) $j0.2$ and $j0.25$ pu
(c) $j0.25$ and $j0.25$ pu
(d) $j0.05$ and $j0.05$ pu

Q.212 A 50 MVA, 11 kV, 3-ϕ synchronous generator is subjected to a 3-phase short circuit fault and the fault current is 2000 A. The per unit value of positive sequence reactance is

(a) 1.5 pu (b) 1.312 pu
(c) 1.274 pu (d) 1.516 pu

Q.213 Assertion (A) : In case of machines of different ratings connected to a system, the percentage reactance of each machine is calculated at an arbitrary kVA, know as base kVA.

Reason (R) : The value of the base kVA is quite unimportant and may be equal to the kVA of the largest plant, equal to the total plant capacity in kVA or any arbitrary value.

(a) Both A and R are ture and R is the correct explanation of A.
(b) Both A and R ar true but R is not the correct explanation of A.
(c) A is true but R is false.
(d) A is false but R is true.

Q.214 Match **List-I (Different types of faults)** with **List-II (Causes)** and select the correct answer using the codes given below the lists:

List-I
A. Over current
B. Under voltage
C. Unbalance
D. Surges

List-II
1. Short circuits or failure of alternator's field.
2. Lightning.
3. Leakage due to corona or overload on the supply system.
4. Breaking of one of the phase conductors.

Codes:

	A	B	C	D
(a)	2	3	1	4
(b)	4	1	3	2
(c)	2	3	4	1
(d)	3	1	4	2

Q.215 The zero sequence reactance (in pu) are indicated in the network shown in figure below. The zero sequence driving-point reactance of node-3 will be

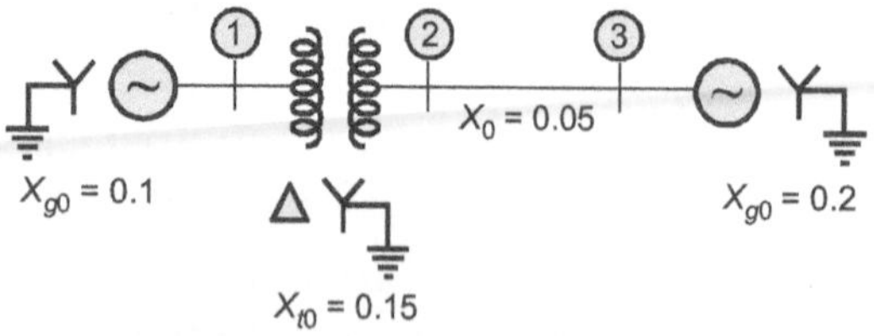

(a) 0.05 pu (b) 0.20 pu
(c) 0.10 pu (d) 0.30 pu

Q.216 Symmetrical faults can be analysed using
(a) Thevenin's theorem on per phase basis
(b) bus admittance matrix
(c) symmetrical components
(d) both (a) and (b)

Q.217 In short circuit studies of a large power system network
(a) the series resistances of lines are neglected while load currents are not neglected.
(b) the series resistances of lines are considered while load currents are neglected.
(c) both series resistances of the lines and load currents are taken into consideration.
(d) both series resistances of lines and load currents are neglected.

6. Load-Flow Study

Q.218 Assumption of decoupling is valid only if
(a) $X_{Line} < R_{Line}$ (b) $X_{Line} = R_{Line}$
(c) Any conditon (d) $R_{Line} << X_{Line}$

Q.219 In a load flow solution $V_i = 1.083\angle 15°$ pu and $V_k = 0.986\angle -2°$ pu. What is the direction of active power (P) and reactive power (Q) flow in the line $i - k$?
(a) P and Q flow from i to k
(b) P flows from i to k and Q flows from k to i
(c) Both flow from k to i
(d) Data is insufficient to determine direction

Q.220 The slack bus has to be a
(a) PV bus (b) PQ bus
(c) QV bus (d) No constraint

Q.221 Providing synchronous condensers at a load bus will control
(a) P at bus (b) Q
(c) p.f. (d) (b) and (c)

Q.222 Generally in load flow solutions the load is modelled as a
(a) constant impedance load
(b) constant power load
(c) constant current load
(d) dynamically varying load

Q.223 At a slack bus
(a) δ is unknown (b) P is known
(c) Q is unknown (d) V is unknown

Q.224 The complex power injections at a bus are as shown below. Two lines are connected drawing powers as shown in figure below. S_x is

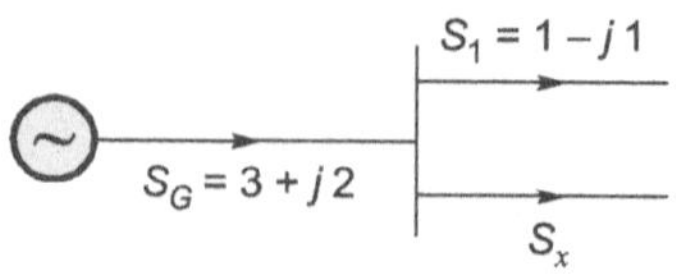

(a) $4 + j1$ (b) $2 + j3$
(c) $2 + j1$ (d) $-2 - j3$

Q.225 When the bus voltage magnitudes at the two ends of a transmission line are equal, if the line is lossless, then S_{ij} is related to S_{ji} by which of the following relationships? (S_{ij} is complex power from bus i to j)

(a) $S_{ij} = S_{ji}$ (b) $S_{ij} \geq S_{ji}$

(c) $S_{ij} = -S_{ji}^*$ (d) $S_{ij} = S_{ji}^*$

Q.226 If $S = VI^*$, P is positive and Q is negative, the network is

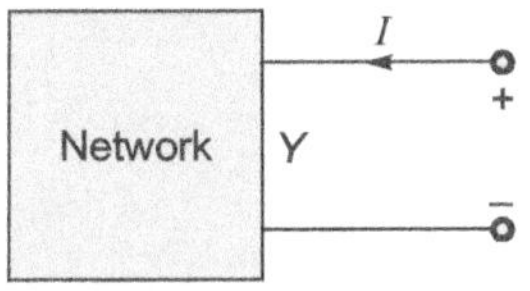

(a) absorbing P and Q
(b) absorbing P and delivering Q
(c) delivering P and absorbing Q
(d) delivering P and Q

Q.227 In a power system with 100 buses, 1 bus is taken as slack bus, 15 buses are *PV* buses and rest are *PQ* buses. The number of variables to be solved by *NR* method in polar coordinates is

(a) 100 (b) 198
(c) 200 (d) 183

Q.228 In the above problem, if equations were formulated in rectangular coordinates, the number of variables to be solved by *NR* method will be

(a) 100 (b) 198
(c) 200 (d) 183

Q.229 In FDLF method the Jacobian matrix is
(a) to be updated once in a few iterations
(b) to be updated at every iteration
(c) is a constant and need to be computed only once
(d) a null matrix

Q.230 The voltage of generator and an infinite bus are given as $0.92\angle 10°$ pu and $1.0\angle 0°$ pu respectively. The generator acts as a
(a) shunt capacitor
(b) shunt coil
(c) data is insufficient
(d) none of the above

Q.231 The time taken for each iteration in *NR* method compared to *GS* method is
(a) more
(b) equal
(c) less
(d) depend upon the size of the power system

Q.232 Y_{BUS} matrix is a
(a) Null matrix (b) Sparse matrix
(c) Full matrix (d) Unity matrix

Q.233 For the power system network shown below, the value of S_{G1}, S_{21}, S_{23} and S_{G3} are respectively given by (all units are in MVA)

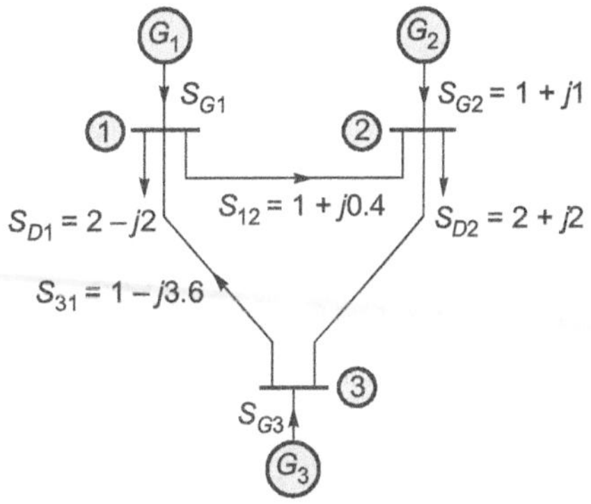

(a) $S_{G1} = (1 + j2.2)$ MVA, $S_{G3} = (2 + j2)$ MVA
$S_{21} = -j1.4$ MVA, $S_{23} = (-1 + j0.4)$ MVA
(b) $S_{G1} = -j1.4$ MVA, $S_{G3} = (1 + j2.2)$ MVA
$S_{21} = (2 + j2)$ MVA, $S_{23} = (-1 + j0.4)$ MVA
(c) $S_{G1} = (2 + j2)$ MVA, $S_{G3} = (1 + j2.2)$ MVA
$S_{21} = (-1 + j0.4)$ MVA, $S_{23} = -j1.4$ MVA
(d) $S_{G1} = (2 + j2)$ MVA, $S_{G3} = (-1 + j0.4)$MVA
$S_{21} = (1 + j2.2)$ MVA, $S_{23} = -j1.4$ MVA

Q.234 For the system shown in figure below, it is desired that $|V_1| = |V_2| = 1$ pu. The line impedance is $j0.05$ pu. If the real power input into the line is 10 pu, the value of Q_{line} in pu will be

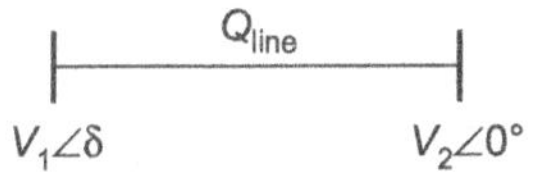

(a) 2.68 pu (b) 8.04 pu
(c) 1.34 pu (d) 5.36 pu

Q.235 A power system consists of 300 buses out of which 20 buses are generator buses, 25 buses are ones with reactive support, 15 buses are with fixed capacitors. Rest are load buses. What is the size of Jacobian in N-R method, solved in polar form?
(a) 510 × 510 (b) 554 × 554
(c) 600 × 600 (d) 560 × 560

Q.236 The model used for transmission lines in load flow studies is

(a) T-model (b) π model
(c) long-line model (d) short-line model

Q.237 A load $S_L = (5.0 + j0.5)$ pu is connected at a bus whose voltage is $1.02\angle{-20°}$ pu. The load is to be converted into an equivalent admittance and absorbed into Y_{bus} matrix. The diagonal element to be added at the bus is:

(a) $\left[\dfrac{5.0+j0.5}{(1.02)^2}\right]$ (b) $\left[\dfrac{5.0+j0.5}{1.02}\right]$

(c) $\left[\dfrac{5.0-j0.5}{(1.02)^2}\right]$ (d) $\left[\dfrac{5.0-j0.5}{1.02}\right]$

Q.238 The connection of a shunt capacitor at bus 'i' of an existing power system, will affect which of the following bus admittance matrix elements?
(a) All the elements
(b) None of the elements
(c) All elements incident on bus i
(d) Only diagonal element of bus i

Q.239 **Assertion (A):** Bus admittance matrix is a sparse matrix.
Reason (R): In load flow studies, one of the buses is taken as a slack bus.
(a) Both A and R are ture and R is the correct explanation of A.
(b) Both A and R ar true but R is not the correct explanation of A.
(c) A is true but R is false.
(d) A is false but R is true.

Q.240 A single line diagram of a power system is shown in the given figure. The diagonal element of the Y_{Bus} matrix are

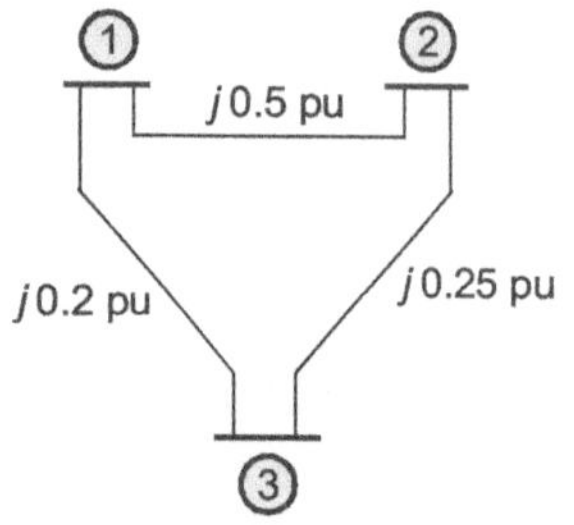

	Y_{11}	Y_{22}	Y_{33}
(a)	$j0.7$ pu	$j0.75$ pu	$j0.45$ pu
(b)	$-j0.7$ pu	-0.75 pu	$-j0.45$ pu
(c)	$-j7$ pu	$-j6$ pu	$-j9$ pu
(d)	$j7$ pu	$-j6$ pu	$-j9$ pu

Q.241 The per-unit impedance of a short transmision line is $j0.06$. The per unit load on the line is $(1 + j0.6)$ pu at a receiving end voltage of $1\angle 0°$ pu. The average reactive power flow over the line is
(a) 56 pu (b) –32 pu
(c) –56 pu (d) 32 pu

Q.242 Two networks are connected by a tie-line. The slack bus is present in the first network. During the Y_{bus} formulation, the tie line data is not considered. The load flow
(a) will converge, but solution is incorrect.
(b) will not converge.
(c) can be solved by deleting row and column in Y_{bus} pertaining to slack bus.
(d) will not be affected.

Q.243 In solution of load flow equations, if slack bus is changed
(a) complex voltages and losses will change
(b) only complex voltages will change, losses will be same
(c) complex voltages do not change, however line losses are affected
(d) choice of slack bus has no effect on solution of load flow equations.

Q.244 A shunt capacitor is connected at bus-2 with reactance of 0.2 pu in the power system network shown below. What is $[Y_{22}]$?
(a) $-j15$ pu (b) $+j25$ pu
(c) $+j15$ pu (d) $-j25$ pu

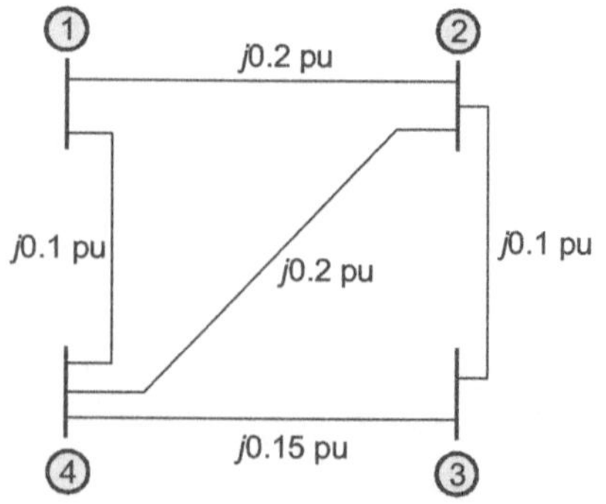

Q.245 Consider the following statements with respect to load flow studies in power system:
1. Bus admittance matrix is a sparse matrix.
2. Gauss-Seidel method is preferred over Newton-Raphson method for load flow studies.

3. One of the buses is taken as slack bus in load flow studies.

Which of the above statements are correct?

(a) 1, 2 and 3 (b) 1 and 2
(c) 1 and 3 (d) 2 and 3

Q.246 The voltage phasor of a circuit is $10\angle 15°$ V and the current phasor is $2\angle -45°$ A. The active and the reactive powers in the circuit are respectively

(a) 10 W and 17.32 VAR
(b) 5 W and 8.66 VAR
(c) 20 W and 60 VAR
(d) $20\sqrt{2}$ W and $10\sqrt{2}$ VAR

Q.247 In N-R and G-S method of load flow study the variables are respectively expressed in

(a) rectangular co-ordiantes and polar co-ordinates.
(b) polar co-ordinates and rectangular co-ordinates.
(c) polar co-ordinates only.
(d) rectangular co-ordinates only.

7. Switch Gear and Protection

Q.248 For an arcing fault where the arc resistance is significant an impedance or admittance relay will

(a) over reach
(b) under reach
(c) the reach of relay is unaffected
(d) the reach will depend on the arcing resistance and may under reach or over reach.

Q.249 Kick fuses are used across relay coils to prevent relay operation during

(a) heavy external faults
(b) inrush current of transformers when they are energized
(c) line to ground faults
(d) bolted faults

Q.250 There is a problem of high transient over voltages while interrupting

(a) capacitive currents
(b) inductive currents
(c) heavy short circuit current
(d) (a) or (c)

Q.251 The most suitable circuit breaker for voltages above 220 kV is

(a) Vaccum CB (b) SF_6 CB
(c) Air blast CB (d) Oil CB

Q.252 Match **List-I** with **List-II** and select the correct answer using the codes given below the lists:

List-I

A. Mho relay
B. Reactance relay
C. Simple impedance relay

List-II

1. Voltage restrained OC relay
2. Direction restrained OC relay
3. Voltage restrained directional relay

Codes:

	A	B	C
(a)	3	1	2
(b)	2	3	1
(c)	3	2	1
(d)	1	2	3

Q.253 Pilot relaying schemes are used for protection of

(a) bus bars
(b) transformers
(c) transmission lines
(d) instrument transfomers

Q.254 Relays A and B have to be coordinated on two adjacent line sections as shown in figure below. Zone-1 and Zone-2 settings are indicated on the diagram.

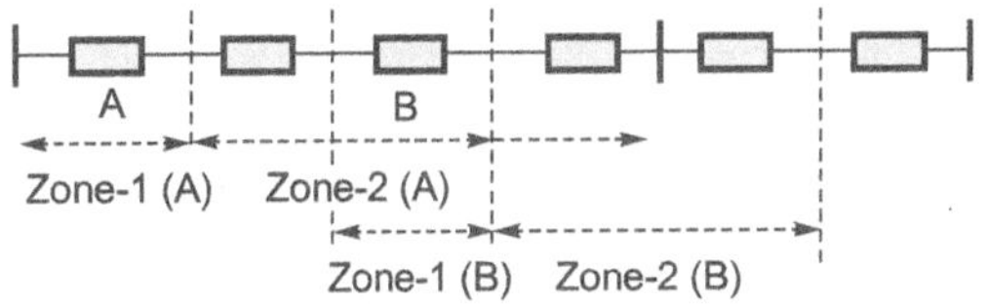

Which of the following indicates correct time setting for zone-2 of relays A and B?

(a) $TZ_{2A1} = 0.6$ s; $TZ_{2B2} = 0.3$ s
(b) $TZ_{2A1} = 0.3$ s; $TZ_{2B2} = 0.6$ s
(c) $TZ_{2A1} = 0.3$ s; $TZ_{2B2} = 0.3$ s
(d) $TZ_{2A1} = 0.1$ s; $TZ_{2B2} = 0.3$ s

Q.255 Identify the relays to be provided with directional and non-directional elements in the system shown in figure below.

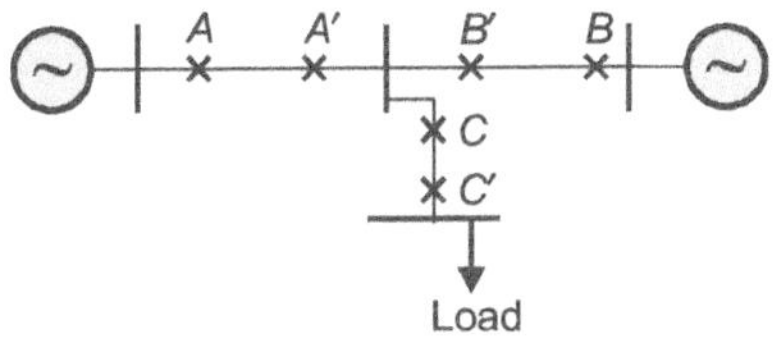

(a) Directional relays : A, B and C
Non-directional relays : A', B' and C'
(b) Directional relays : C and C'
Non-directional relays : A, A', B and B'
(c) Directional relays : A', B' and C
Non-directional relays : A, B and C'
(d) Directional relays : A' and B'
Non-directional relays : A, B, C and C'

Q.256 If the fault current is 2000 A, the relay setting 50% and the CT ratio is 400/5, then the plug setting multiplier will be
(a) 15 (b) 25
(c) 50 (d) None of these

Q.257 The percent bias for a generator protection lies between
(a) 15 to 20% (b) 10 to 15%
(c) 5 to 10% (d) None of these

Q.258 High synchronous reactance is preferred in the present day alternators because one can have
(a) reduced sub-transient currents
(b) reduced hormonic currents
(c) reduced transient currents
(d) high voltage regulation with load

Q.259 A 11/132 kV transformer is connected delta/star. The current transformers on the low voltage side have a ratio 600/5 for circulating current merz price differential relay. The ratio of CT on high voltage side will be
(a) $\frac{5}{50\sqrt{3}}$ (b) $\frac{50\sqrt{3}}{5}$
(c) $\frac{50}{5\sqrt{3}}$ (d) $\frac{50}{5}$

Q.260 Consider the following statements:
1. Coupling capacitor offers low impedance for power signal and high impedance for carrier signal.
2. Line trap offers low impedance for carrier signals and high impedance for power signal.
3. A simple differential relay maloperates when an internal fault occurs due to different CT ratios of the current transformers.

Which of the above statement(s) is/are true?
(a) 1 and 2 only (b) 3 only
(c) 2 and 3 only (d) None of these

Q.261 Assertion (A): A short line fault is less severe as compared to a long line fault.
Reason (R): Restricking voltage is a function of power factor of the system.
(a) Both A and R are ture and R is the correct explanation of A.
(b) Both A and R ar true but R is not the correct explanation of A.
(c) A is true but R is false.
(d) A is false but R is true.

Q.262 Match **List-I (Type of circuit breaker)** with **List-II (Characteristics)** and select the correct answer using the codes given below the lists:

List-I
A. Oil circuit breaker
B. vaccum circuit breaker
C. Air blast circuit breaker
D. SF_6 circuit breaker

List-II
1. Very sensitive to restricking voltage
2. Noiseless operation
3. Fire hazards
4. Least maintenance

Codes:

	A	B	C	D
(a)	1	4	2	3
(b)	2	1	3	4
(c)	3	4	1	2
(d)	1	2	3	4

Q.263 A 3-phase oil circuit breaker has operating voltage = 11 kV, rupturing capacity = 500 MVA and normal current rating = 500 A. Its symmetrical breaking current will be
(a) 26.24 kA (rms) (b) 27.68 kA (rms)
(c) 28.22 kA (rms) (d) 29.45 kA (rms)

Q.264 In the above problem, the rated making current will be
(a) 88.88 kA (b) 66.91 kA
(c) 89.00 kA (d) 55.55 kA

Q.265 Assertion (A): Static relays are superior to electro-magnetic relays.
Reason (R): Electro-magnetic relays are costly than static relays.
(a) Both A and R are ture and R is the correct explanation of A.

(b) Both A and R ar true but R is not the correct explanation of A.
(c) A is true but R is false.
(d) A is false but R is truo.

Q.266 Match **List-I (Power system component)** with **List-II (Applications)** and select the correct answer using the codes given below the lists:

List-I
A. Current transformer
B. Surge absorber
C. Earthing
D. Circuit breaker

List-II
1. To carry full load current continuously
2. To feed the current coils of indicating and metering instruments
3. To minimize the danger due to over-voltage
4. For providing safety to the users

Codes:

	A	B	C	D
(a)	4	3	2	1
(b)	1	3	2	4
(c)	2	3	1	4
(d)	2	3	4	1

Q.267 For the figure shown below, the relays which must be made directional to have selectivity will be

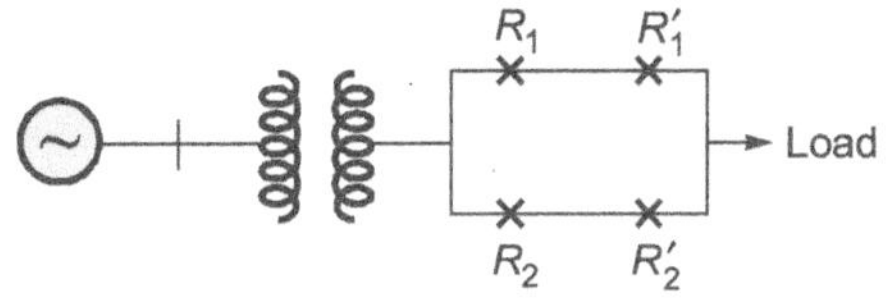

(a) R_1 and R_2 (b) R_1 and R_2'
(c) R_1' and R_2 (d) R_1' and R_2'

Q.268 Current chopping is due to
(a) interruption of heavy capacitive currents
(b) interruption of bolted faults
(c) interruption of low inductive currents
(d) interruption of unity power factor loads

Q.269 The relay most likely to maloperate during power swing is
(a) reactance relay (b) mho relay
(c) impedance relay (d) both (a) and (c)

Q.270 For protection of short lines against ground faults where arc resistance is comparable to the impedance of the line, the preferred relay is
(a) impedance relay (b) reactance relay
(c) admittance relay (d) over current relay

Q.271 A three phase, delta-wye - connected, 30 MVA, 33/11 kv transformer is protected by a differential relay. The CT current ratio on the primary side is 500 : 5 and that on the secondary side is 2000 : 5. The relay current setting for faults drawing upto 200 percent of the rated current will be
(a) 3.31 A (b) 5.30 A
(c) 524.88 A (d) 6.81 A

Q.272 Match **List-I (Type of over current relay)** with **List-II (Operating Characteristic)** and select the correct answer using the codes given below the lists:

List-I
A. Extremely inverse relay
B. Inverse relay
C. IDMT relay
D. Definite time relay

List-II

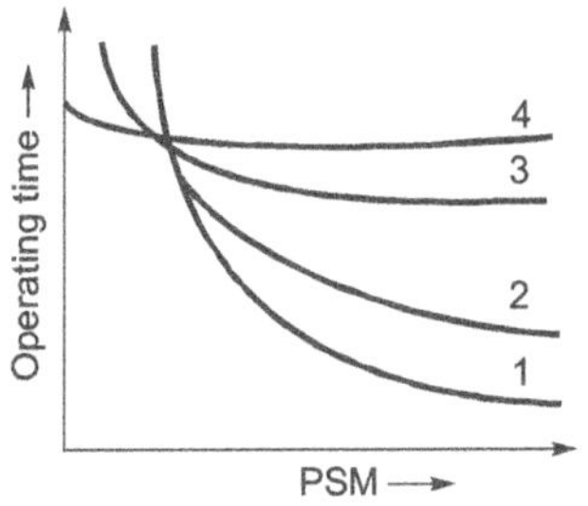

Codes:

	A	B	C	D
(a)	2	1	4	3
(b)	1	2	3	4
(c)	2	3	1	4
(d)	1	2	4	3

Q.273 Three sections of a feeder are provided with circuit breakers CB1, CB2, CB3, CB4, CB5 and CB6. For a fault F as indicated in figure,

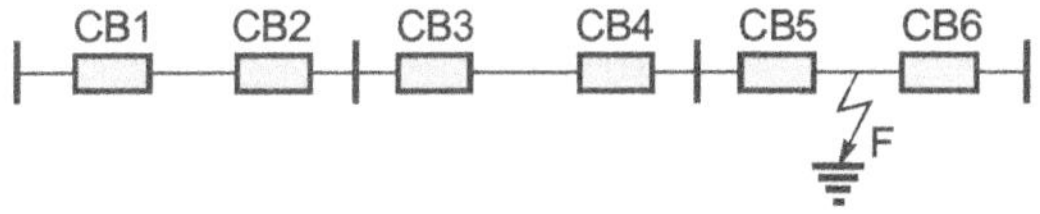

(a) CB5 must be set to trip after CB1 trips.
(b) CB5 must be set to trip after CB3 and CB4 trips.
(c) CB5 must be set to trip after CB2 trips.
(d) CB5 must be set to trip before CB1, CB2, CB3 and CB4 trips.

Q.274 An over current relay of current rating 5 A and setting 150% is connected to the secondary of CT of ratio 300/5. Then, the current in the lines for which the relay pick up is
(a) 450 A (b) 150 A
(c) 300 A (d) 200 A

Q.275 A 3-phase, 33/66 kV transformer is connected Y/Δ and is provided with differential protection scheme. The CT on the low voltage side has a ratio of 300/5. What will be the ratio of the CT on the HV side?
(a) $\frac{50}{5\sqrt{3}}$ (b) $\frac{5}{50\sqrt{3}}$
(c) $\frac{50}{5}$ (d) $\frac{50\sqrt{3}}{5}$

Q.276 Match **List-I (Power system components)** with **List-II (Relaying schemes)** and select the correct answer using the codes given below the lists:

List-I	List-II
A. Feeder	1. Offset Mho relay
B. Transformer	2. Translay relay
C. Generator	3. Ratio relay
D. Transmission line	4. Buchholz relay

Codes:

	A	B	C	D
(a)	2	4	3	1
(b)	4	3	2	1
(c)	1	2	3	4
(d)	2	4	1	3

Q.277 Current chopping is minimum in case of
(a) Vaccum CB (b) Oil CB
(c) SF_6 CB (d) ABCB

Q.278 The neutral of a 10 MVA, 11 kV alternator is earthed through a resistance of 5 Ω. The earth fault relay is set to operate at 0.75 A. The CT's have a ratio of 10^3 : 5. The percentage of the alternator winding unprotected is
(a) 11.8% (b) 15%
(c) 88.2% (d) 85%

Q.279 **Assertion (A):** A fault nearer the source is more severe from the view-point of arc interruption as compared to a fault far from the source.
Reason (R): Lower the value of the inductance and capacitance, the higher will be the natural frequency of oscilaltions and more severe will be the effect of restriking voltage.
(a) Both A and R are ture and R is the correct explanation of A.
(b) Both A and R ar true but R is not the correct explanation of A.
(c) A is true but R is false.
(d) A is false but R is true.

Q.280 In a 66 kV, 3-ϕ, 50 Hz power system the line to ground capacitance is 0.04 μF and the inductance is 4 H. The voltage appearing across the breaker pole when a magnetising current of 4 A is interrupted will be:
(a) 10 kV (b) 40 kV
(c) 15 kV (d) 25 kV

Q.281 In the above problem what is the resistance required to be connected across the contacts to eliminate the restriking voltage?
(a) 50 kΩ (b) 28 kΩ
(c) 5 kΩ (d) 8 kΩ

Q.282 Reactor switching is done with
(a) air blast circuit breaker
(b) vaccum circuit breaker
(c) SF_6 circuit breaker
(d) minimum oil circuit breaker

Q.283 **Assertion (A):** The basic principle of a distance relay involves the comparison of the fault current seen by the relay with the voltage at the relaying point.
Reason (R): Since the impedance is proportional to the length of the transmission line, it is appropriate to use a relay capable of measuring the impedance of a line upto a given point.
(a) Both A and R are ture and R is the correct explanation of A.
(b) Both A and R ar true but R is not the correct explanation of A.
(c) A is true but R is false.
(d) A is false but R is true.

8. Power System Stability

Q.284 A 500-MVA synchronous machine has H_1 = 4.6 MJ/MVA, and a 1500 MVA machine has H_2 = 3.0 MJ/MVA. The two machines operate in parallel in a power station. The equivalent H constant for the two, relative to a 100 MVA base will be
(a) 22 MJ/MVA (b) 45 MJ/MVA
(c) 52 MJ/MVA (d) 68 MJ/MVA

Q.285 The load carrying capability of a long ac line is
(a) limited by conductor size.
(b) limited by stability considerations.
(c) limited by thermal consderations.
(d) decreased by using bundled conductors.

Q.286 Fast acting excitation systems
(a) improve transient stability.
(b) improve steady state stability.
(c) both (a) and (b).
(d) have no effect on stability of the system.

Q.287 A shunt reactor in general
(a) reduces steady state power transfer limit.
(b) improves steady state power transfer limit.
(c) does not affect power transfer.
(d) reduces the reactive power consumption.

Q.288 The angle swings during transients
(a) should not exceed 90°.
(b) can exceed 90° but can't exceed 180°.
(c) can go beyond 180°.
(d) will not be greater than swings under minor disturbances.

Q.289 If the inertia constant H = 9.0 MJ/MVA for a 50 MVA generator, the stored energy is
(a) 9.0 MJ (b) 125 MJ
(c) 450 MJ (d) 5.55 MJ

Q.290 Bundled conductors in a transmission line will
(a) reduce line loading.
(b) improve steady state stability.
(c) improve protection of line.
(d) reduce voltage of buses at the two ends.

Q.291 A 1500 MVA synchronous generator has an inertia constant of 5.0 MJ/MVA. The inertia constant of the machine relative to a 50 MVA base is
(a) 25 MJ/MVA (b) 50 MJ/MVA
(c) 125 MJ/MVA (d) 150 MJ/MVA

Q.292 If the initial acceleration power of a 100 MVA, 50 Hz synchronous generator having H = 5 MJ/MVA is X pu, the initial acceleration in elec-deg/s^2 and the inertia constant in MJ-s/ele-deg respectively will be
(a) 31.4 X, 18 (b) 1800 X, 0.056
(c) X/1800, 0.056 (d) X/31.4, 18

Q.293 The critical clearing angle of a power system is related to
(a) reactive power limit
(b) short circuit limit
(c) steady state stability limit
(d) transient stability limit

Q.294 The electrical stiffness of a synchronous generator connected to a very large system can be increased by
(a) increasing the excitation or the power angle of the machine.
(b) reducing the excitation of the synchronous reactance of the machine.
(c) increasing the synchronous reactance of the machine.
(d) operating the generator at a much lower MW level compared to the steady state limit.

Q.295 The kinetic enrgy stored in the rotor of a 50-MVA, six-pole, 60-Hz synchronous machine is 200 MJ. The input to the machine is 25 MW at a developed power of 22.5 MW. The acceleration of the rotor will be
(a) 1.06 elec-rad/s^2 (b) 2.356 elec-rad/s^2
(c) 2.25 elec-rad/s^2 (d) 1.25 elec-rad/s^2

Q.296 A transmission line is capable of delivering a maximum power of P pu from sending end to receiving end. If the line is compensated by 30%, the maximum power that can transferred is
(a) P pu (b) 1.3 P pu
(c) 1.8 P pu (d) 1.42 P pu

Q.297 Fast acting circuit breakers
(a) improve transient stability
(b) improve steady state stability
(c) improve both transient and steady state stability
(d) do not affect stability

Q.298 The damping power is approximately given by

(a) $P_D = D\frac{d\delta}{dt}$ (b) $P_D = D\frac{d^2\delta}{dt^2}$

(c) $P_D = D\left(\frac{d\delta}{dt}+\frac{d\omega}{dt}\right)$ (d) $P_D = D\frac{d\omega}{dt}$

Q.299 A generator is supplying power to an infinite bus through a double circuit line, when a three phase short circuit occurs at the end of one of the lines. The fault is cleared after some time. The critical clearing time is

(a) more if the fault is cleared by openig a line.
(b) more if the fault is cleared without opening a line.
(c) same in both cases of (a) and (b).
(d) depends on auto recolsure facility provided in the breakers which opens the line.

Q.300 A generating station has a generator *A* rated 250 MVA, *H* = 1.6 pu and generator *B* rated 500 MVA, *H* = 1.0 pu. The pu inertia constant *H* on a base of 100 MVA of the equivalent machine when *A* and *B* are operating in parallel is

(a) 2.6 pu (b) 0.6 pu
(c) 1.625 pu (d) 9.0 pu

Q.301 For the system shown in figure below, the maximum power transferred from the sending end to the receiving end before a fault occurs at point *Q* will be

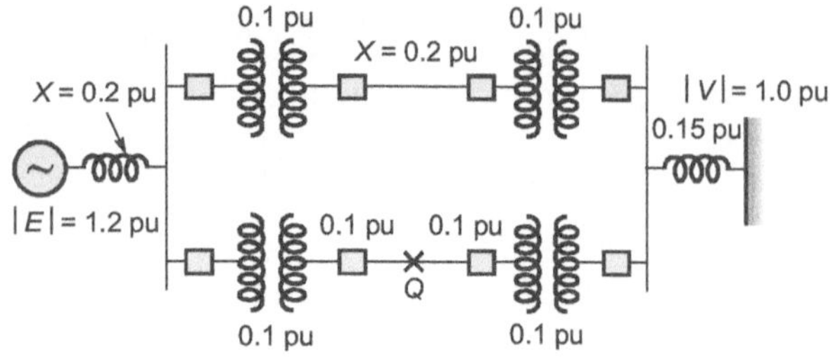

(a) 2.18 pu (b) 1.55 pu
(c) 0.67 pu (d) 0.55 pu

Q.302 A 100 MVA, two pole, 60 Hz generator has a moment of inertia of 50×10^3 kg-m^2. The angular momentum is given by:

(a) 0.125 MJ-s/ele-deg
(b) 0.329 MJ-s/ele-deg
(c) 1.22 MJ-s/ele-deg
(d) 0.625 MJ-s/ele-deg

Q.303 The sending end and receiving end voltages of a transmission line at a 100 MW load are equal to 115 kV. The per phase line impedance is (4 + *j* 7) Ω. The maximum steady state power that can be transmitted over the line will be

(a) 225.2 MW (b) 425 MW
(c) 826.5 MW (d) 650 MW

Q.304 The inertia constant *H* for a 60 Hz, 100 MVA hydroelectric generator is 4.0 MJ/MVA. If the input to the generator is suddenly increased by 20 MVA, What is the acceleration imparted to the rotor?

(a) 128°/s^2 (b) 540°/s^2
(c) 660°/s^2 (d) 426°/s^2

Q.305 The acceleration of the rotor of a 100 MVA, two-pole, 60 Hz generator is 76°/s^2 and remains constant for twelve cycles then, the speed of the rotor at the end of this twelve cycle is

(a) 3600 rpm (b) 3605.25 rpm
(c) 3800.52 rpm (d) 3602.53 rpm

9. Optimal Power System Operation

Q.306 If $\frac{\delta P_L}{\delta P_i}$ for a plant *i* with output P_i and loss in the system being P_L increases, the penalty factor of the system

(a) increases
(b) decreases
(c) does not change
(d) depends on the incremental fuel cost

Q.307 The incremental fuel costs of two plants are given by

$$\frac{dF_1}{dP_1} = 0.01\, P_1 + 2.0 \text{ Rs/MWhr}$$

$$\frac{dF_2}{dP_2} = 0.012\, P_2 + 1.6 \text{ Rs/MWhr}$$

For economic schedule of a load of 180 MW

(a) P_1 = 100 MW; P_2 = 80 MW
(b) P_1 = 90 MW; P_2 = 90 MW
(c) P_1 = 80 MW; P_2 = 100 MW
(d) P_1 = 120 MW; P_2 = 60 MW

Q.308 A generator is supplying a load. An incremental change in load of 4 MW requires generation to be increased by 6 MW. The cost at plant bus is ₹ 30/MWhr. The incremental cost at the receiving end is

(a) ₹ 30/MWhr (b) ₹ 45/MWhr
(c) ₹ 33.4/MWhr (d) ₹ 6.0/MWhr

Q.309 The current distribution factors for the power system network shown below are given by

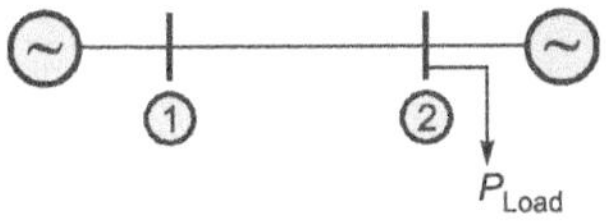

(a) $Na_1 = 0$; $Na_2 = 0$
(b) $Na_1 = 0.5$; $Na_2 = 0.5$
(c) $Na_1 = 1.0$; $Na_2 = 0$
(d) $Na_1 = 0$; $Na_2 = 1.0$

Q.310 The incremental transmission loss of a plant can't be a
(a) negative quantity (b) positive quantity
(c) a zero value (d) non-zero value

Q.311 When generating units are loaded to equal incremental costs, it results in
(a) minimum fuel costs
(b) fuel costs are at a maximum
(c) fuel costs are not affected
(d) maximum loading of generating units

Q.312 Unit of heat rate curve is
(a) Rs-hr (b) Rs/MWh
(c) Rs/hr (d) Million kcal/hr

Q.313 Unit of λ is
(a) Rs/hr (b) Rs/MW
(c) Rs/MWh (d) MW/Rs

Q.314 Cost curves are expressed as:
(a) Rs/million cal
(b) Million cal/hr × Rs/million cal
(c) Rs/hr
(d) Both (b) and (c)

Q.315 The optimization problem is
(a) to allocate load demand among various units such that the cost of generation is maintained constant.
(b) to allocate total load demand among various units such that the cost of generation is minimized.
(c) to allocate total load demand among various units such that the cost of generatin is enormously increased.
(d) to allocate total load demand among various units such that there is no effect with cost of generation.

Q.316 The curve obtained by considering the change in cost of generation to change in real-power generation at various point is
(a) fuel cost curve
(b) input-output curve
(c) incremental cost curve
(d) all of the above

Q.317 In the economic operation of a power system, the effect of increased penalty factor between a generating plant and system load centre is to
(a) decrease the load on the generating plant
(b) increase the load on the plant
(c) hold the plant load constant
(d) decrease the load first and then increase

Q.318 In a two-plant system, the entire load is located at plant-2, which is connected to plant-1 by a transmission line. Plant-1 supplies 100 MW of power with a corresponding transmission loss of 5 MW. The penalty factors for the two plants will be given by:
(a) $L_1 = 0$, $L_2 = 1$ (b) $L_1 = -1.11$, $L_2 = 0$
(c) $L_1 = 1.11$, $L_2 = 1$ (d) $L_1 = 1$, $L_2 = 1.11$

Q.319 A two bus system is shown in figure below. If 100 MW is transmitted from plant-1 to the load, a loss fo 10 MW is incurred. System incremental cost is λ = Rs 30/MWhr. The value of P_1 and P_2 will be

$$\frac{dF_1}{dP_1} = 0.02\,P_1 + 16.0 \text{ Rs/MWhr and}$$

$$\frac{dF_2}{dP_2} = 0.04\,P_2 + 20.0 \text{ Rs/MWhr}$$

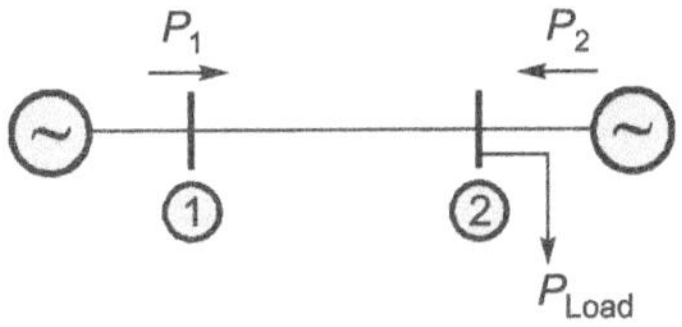

(a) $P_1 = 100$ and $P_2 = 175$ MW
(b) $P_1 = 175$ MW and $P_2 = 100$ MW
(c) $P_1 = 100$ MW and $P_2 = 250$MW
(d) $P_1 = 175$ MW and $P_2 = 250$ MW

Q.320 In the above problem the value of P_{Load} is
(a) 40 MW (b) 394.38 MW
(c) 415.55 MW (d) 10 MW

Q.321 The fuel cost of a unit is given by

$$F = 1.5 + 20P + 0.1P^2 \text{ Rs/hr}$$

where P is the unit output in MW. The incremental fuel cost at an output of 200 MW is

(a) 60 Rs/MWhr (b) 40 Rs/MWhr
(c) 80 Rs/MWhr (d) 20 Rs/MWhr

Q.322 A turbine generator set has a regulation constant of 5% on the generator rating of 100 MVA, 50 Hz. The generator frequency decreases by 0.02 Hz. The increase in trubine output power for steady state operation is

(a) 0.8 MW (b) 8 MW
(c) 3.6 MW (d) 0.2 MW

Q.323 Match **List-I (Loss coefficients)** with **List-II (Values)** for the power system network shown below and select the correct answer using the codes given below the lists:

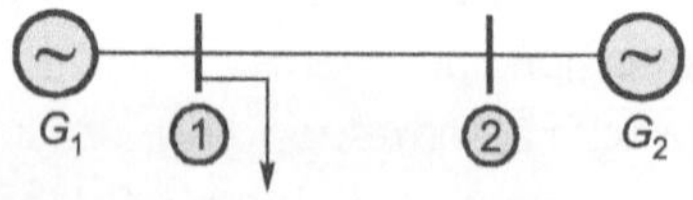

List-I	List-II
A. B_{12}	1. 7×10^{-4} MW^{-1}
B. B_{22}	2. 0.00 MW^{-1}
C. B_{11}	

Codes:

	A	B	C
(a)	1	2	1
(b)	2	1	1
(c)	2	1	2
(d)	1	1	2

Q.324 On a system with two plants, the incremental costs are given by

$$\frac{dF_1}{dP_1} = 0.01\,P_1 + 20 \text{ Rs/MW-hr}$$

$$\frac{dF_2}{dP_2} = 0.015\,P_2 + 22.5 \text{ Rs/MW-hr}$$

The system is running under optimal schedule with $P_1 = P_2 = 100$ MW.

If $\frac{\partial P_L}{\partial P_2} = 0.2$, the penalty factors will be given by

(a) $L_1 = 1.428$; $L_2 = 1.25$
(b) $L_1 = 1$; $L_2 = 1.25$
(c) $L_1 = 1.428$; $L_2 = 1$
(d) $L_1 = 1.25$; $L_2 = 1.428$

Q.325 In the above problem, $\frac{\partial P_L}{\partial P_1}$ has a value of

(a) 0.1 (b) 0.3
(c) –0.1 (d) 0.25

Q.326 The incremental fuel costs (in some appropriate unit) for a power plant consisting of three generating units is given by:

$$IC_1 = 20 + 0.3\,P_1$$
$$IC_2 = 30 + 0.4\,P_2$$
$$IC_3 = 30$$

where P_i is the power in MW generated by unit for i = 1, 2 and 3. Assume that all the three units are operating all the time. Minimum and maximum loads on each unit are 50 MW and 300 MW respectively. If the plant is operating on economic load dispatch to supply the total power demand of 700 MW, the power generated by each unit is

(a) P_1 = 242.86 MW ; P_2 = 157.14 MW and P_3 = 300 MW
(b) P_1 = 157.14 MW ; P_2 = 242.86 MW and P_3 = 300 MW
(c) P_1 = 300.0 MW ; P_2 = 300.0 MW and P_3 = 100 MW
(d) P_1 = 233.3 MW ; P_2 = 233.3 MW and P_3 = 233.4 MW

Q.327 To get an optimal solution to an optimization problem, we will define an objective function as

(a) $C^* = \sum_{i=1}^{n} C_i(P_{Gi})$

(b) $C^* = \sum_{i=1}^{n} C_i\left(P_{G_i} - \lambda\right)$

(c) $C^* = C - \lambda\left\{\sum_{i=1}^{n} P_{G_i} - P_D\right\}$

(d) $C^* = \lambda - C\left\{\sum_{i=1}^{n} P_{G_i} < P_D\right\}$

Q.328 Equality and inequality constraints are respectively

(a) $\sum_{i=1}^{n} P_{Gi} - P_D = 0$ and

$$P_{G_i}(\text{min}) < P_{G_i} < P_{G_i}(\text{max})$$

(b) $\sum_{i=1}^{n} P_{G_i} + P_D = 0$ and

$$P_{G_i}(\text{min}) < P_{G_i} < P_{G_i}(\text{max})$$

(c) $\sum_{i=1}^{n} P_{G_i} - P_D = 0$ and

$$P_{G_i}(\text{min}) > P_{G_i} > P_{G_i}(\text{max})$$

(d) none of the above

Q.329 The method of getting the solution to an optimization problem with neglected transmission losses:
1. does sense the changes in the loads.
2. does not sense the location of the change in load.
3. does sense the changes in the load and the location of changes in the loads.
4. does not sense both the location and the changes in the load.

Which of the above statements is/are true?
(a) 1 and 2 only (b) either 3 or 4
(c) 4 only (d) 3 only

Q.330 Transmission loss is
(a) a function of real-power generation.
(b) independent of real-power generation.
(c) a function of reactive power generation.
(d) a function of bus voltage-magnitude and its angle.

10. HVDC

Q.331 In case of HVDC system, there is
(a) corona loss but no skin effect
(b) skin effect but no corona loss
(c) both skin effect and corona loss
(d) neither corona loss nor skin effect

Q.332 A dc line carries
(a) same power as an ac line
(b) less power than an equivalent ac line
(c) more power than the ac line
(d) both (a) and (c)

Q.333 HVDC transmission is preferred to EHV AC because
(a) HVDC terminal equipment are inexpensive
(b) VAR compensation is not required in HVDC system
(c) system reliability can be improved
(d) harmonic problem is avoided

Q.334 Back-to-back HVDC is used to
(a) increase the transmission capability
(b) decrease line losses
(c) reduce voltage drop
(d) provide stable interconnection

Q.335 HVDC transmission needs
(a) dc filters
(b) ac filters and transformers
(c) ac filters
(d) pulse converters

Q.336 In HVDC system a converter acts as a rectifier when the firing angle α has a value
(a) $0° < \alpha < 120°$ (b) $0° < \alpha < 90°$
(c) $90° < \alpha < 180°$ (d) none of these

Q.337 The capital cost on a HVDC converter is
(a) les than ac substation
(b) same as ac substation
(c) either less or more than ac substation
(d) more than that of ac substation

Q.338 In a dc transmision
(a) it is necessary for the sending and receiving end to be operated in synchronism.
(b) the effects of inductive and capacitive reactances are greater than in ac transmission line of the same rating.
(c) there are no effects due to inductive and capacitive reactances.
(d) power transfer capability is limited by stability considerations.

Q.339 As compared to ac transmission dc transmission is free from
(a) inductance
(b) capacitance
(c) phase dispalcement
(d) all of the above

Q.340 Which of the following is/are correctly matched?

Characteristics	EHV-AC system	HVDC system
1. Corona and radio interference	More	Less
2. Short circuit current	Small	Large
3. Voltage control	Difficult	Easier

(a) 1, 2 and 3 (b) 1 and 3
(c) 2 and 3 (d) 2 only

Q.341 Assertion (A): In HVDC systems, the dc voltage should be as high as possible and the rectifier control angle α should be as low as possible.
Reason (R): Control of dc voltage is expressed by the rectifier control angles α and γ respectively.
(a) Both A and R are ture and R is the correct explanation of A.
(b) Both A and R ar true but R is not the correct explanation of A.
(c) A is true but R is false.
(d) A is false but R is true.

Q.342 If p is the pulse number and n is an integer, what is the order of harmonics on a.c. side and d.c. side of an HVDC converter?
(a) $(np + 1)$ and $(np - 1)$ respectively
(b) $(np - 1)$ and np respectively
(c) $(np + 1)$ and np respectively
(d) $(np \pm 1)$ and np respectively

Q.343 Consider the following statements regarding HVDC power transmission:
1. The modern HVDC systems use 12-pulse converters.
2. DC systems never use ground or sea return
3. Most of present day schemes are two-terminal links.

Which of statements given above is/are correct?
(a) 1, 2 and 3 (b) 1 only
(c) 2 and 3 only (d) 1 and 3 only

Q.344 In the HVDC system, the ac harmonics which gets effectively eliminated with 12 pulse bridge converters are
(a) triplen harmonics
(b) triplen and 5th harmonics
(c) triplen, 5th and 7th harmonics
(d) 5th and 7th harmonics

■■■■

Answers Power Systems

1. (c)	2. (b)	3. (b)	4. (a)	5. (d)	6. (c)	7. (d)	8. (d)
9. (c)	10. (a)	11. (c)	12. (b)	13. (d)	14. (a)	15. (c)	16. (a)
17. (a)	18. (c)	19. (d)	20. (c)	21. (c)	22. (d)	23. (b)	24. (b)
25. (b)	26. (c)	27. (c)	28. (b)	29. (b)	30. (c)	31. (b)	32. (a)
33. (c)	34. (d)	35. (d)	36. (d)	37. (a)	38. (b)	39. (d)	40. (c)
41. (d)	42. (a)	43. (d)	44. (b)	45. (d)	46. (c)	47. (a)	48. (b)
49. (b)	50. (c)	51. (d)	52. (a)	53. (d)	54. (d)	55. (b)	56. (c)
57. (b)	58. (c)	59. (c)	60. (c)	61. (c)	62. (c)	63. (a)	64. (a)
65. (c)	66. (d)	67. (b)	68. (a)	69. (a)	70. (a)	71. (c)	72. (d)
73. (b)	74. (b)	75. (d)	76. (b)	77. (b)	78. (d)	79. (c)	80. (c)
81. (a)	82. (b)	83. (b)	84. (c)	85. (c)	86. (c)	87. (b)	88. (b)
89. (c)	90. (b)	91. (c)	92. (a)	93. (d)	94. (c)	95. (b)	96. (b)
97. (b)	98. (c)	99. (c)	100. (a)	101. (d)	102. (c)	103. (a)	104. (d)
105. (c)	106. (c)	107. (c)	108. (b)	109. (a)	110. (b)	111. (b)	112. (c)
113. (a)	114. (c)	115. (d)	116. (c)	117. (b)	118. (a)	119. (c)	120. (c)
121. (d)	122. (c)	123. (a)	124. (b)	125. (c)	126. (d)	127. (b)	128. (c)
129. (d)	130. (a)	131. (a)	132. (d)	133. (c)	134. (c)	135. (d)	136. (b)
137. (d)	138. (b)	139. (c)	140. (c)	141. (c)	142. (a)	143. (d)	144. (a)
145. (b)	146. (c)	147. (b)	148. (d)	149. (b)	150. (d)	151. (b)	152. (a)
153. (d)	154. (b)	155. (d)	156. (a)	157. (b)	158. (d)	159. (c)	160. (d)

161. (c)	162. (c)	163. (b)	164. (c)	165. (d)	166. (c)	167. (a)	168. (d)
169. (b)	170. (b)	171. (c)	172. (c)	173. (c)	174. (a)	175. (c)	176. (b)
177. (d)	178. (c)	179. (c)	180. (b)	181. (d)	182. (b)	183. (c)	184. (d)
185. (a)	186. (c)	187. (a)	188. (d)	189. (a)	190. (b)	191. (c)	192. (d)
193. (c)	194. (d)	195. (a)	196. (c)	197. (a)	198. (a)	199. (d)	200. (c)
201. (b)	202. (a)	203. (c)	204. (a)	205. (d)	206. (b)	207. (a)	208. (d)
209. (d)	210. (b)	211. (a)	212. (b)	213. (a)	214. (d)	215. (c)	216. (d)
217. (d)	218. (d)	219. (a)	220. (a)	221. (d)	222. (b)	223. (c)	224. (b)
225. (c)	226. (b)	227. (d)	228. (b)	229. (c)	230. (b)	231. (a)	232. (b)
233. (c)	234. (d)	235. (b)	236. (b)	237. (c)	238. (d)	239. (b)	240. (c)
241. (c)	242. (b)	243. (a)	244. (a)	245. (c)	246. (a)	247. (b)	248. (b)
249. (b)	250. (a)	251. (b)	252. (c)	253. (c)	254. (b)	255. (d)	256. (d)
257. (a)	258. (c)	259. (b)	260. (d)	261. (d)	262. (c)	263. (a)	264. (b)
265. (c)	266. (d)	267. (d)	268. (c)	269. (d)	270. (b)	271. (a)	272. (b)
273. (d)	274. (a)	275. (d)	276. (d)	277. (c)	278. (a)	279. (a)	280. (b)
281. (c)	282. (b)	283. (a)	284. (d)	285. (b)	286. (a)	287. (c)	288. (b)
289. (c)	290. (b)	291. (d)	292. (b)	293. (d)	294. (d)	295. (b)	296. (d)
297. (a)	298. (a)	299. (b)	300. (d)	301. (a)	302. (b)	303. (c)	304. (b)
305. (d)	306. (a)	307. (a)	308. (b)	309. (c)	310. (c)	311. (a)	312. (d)
313. (c)	314. (c)	315. (b)	316. (c)	317. (c)	318. (c)	319. (d)	320. (b)
321. (a)	322. (a)	323. (c)	324. (d)	325. (b)	326. (a)	327. (c)	328. (a)
329. (a)	330. (a)	331. (c)	332. (c)	333. (c)	334. (d)	335. (d)	336. (b)
337. (d)	338. (c)	339. (d)	340. (b)	341. (c)	342. (d)	343. (d)	344. (d)

Explanations

1. (c)

Given: $R_1 = 50\ \Omega$ and $R_2 = 110\%$ of $50\ \Omega$.

$$\therefore \quad R_2 = \frac{110}{100} \times 50 = 55\ \Omega$$

$$T_1 = 10\ °C,\ T_2 = ?$$

We know that,

$$R_2 = R_1[1 + \alpha \Delta T]$$

$$\therefore \quad 55 = 50[1 + 0.00409\,(T_2 - 10)]$$

$$\text{or,} \quad T_2 = 34.45\ °C$$

2. (b)

The fictitious conductor is one whose radius is r' and whose diameter is therefore,

$$2r' = 2r\,e^{-1/4} = 2 \times 0.8 \times 0.7788$$
$$= 1.24608 \text{ cm}$$
$$\approx 1.246 \text{ cm}$$

3. (b)

Inductance,

$$L = 2 \times 10^{-7}\, ln\left[\frac{\text{GMD}}{\text{GMR}}\right] \text{H/m/phase}$$

Here, GMR ∝ radius of conductor.

Hence, if radius increases, GMR also increases.

As a result of which inductance will decrease.

4. (a)

For underground cable, $Z_s = 40\ \Omega$

For overhead line, $Z_s = 400\ \Omega$

5. (d)

Dielectric loss occurs in underground cables not in a transmission line.

7. (d)

Surge impedance of the line is given by

$$Z_S = \sqrt{\frac{L}{C}}$$

or, $$Z_S = \sqrt{\frac{0.97 \times 10^{-3}}{0.0115 \times 10^{-6}}} = 290.43\ \Omega$$

8. (d)

Zero regulation of a transmission line occurs at a leading power factor when $\phi = \theta$.

Since, $R = X$

Therefore, $\tan\theta = \frac{X}{R} = 1$ or $\theta = 45°$

$\therefore$ At zero regulation, $\phi = 45°$

Hence, $\cos\phi = \cos 45° = 0.707$ leading

9. (c)

Shunt reactors are used to bring down receiving end voltage during light load or no-load condition to overcome Ferranti effect.

11. (c)

The losses occuring in a transmission line is

$$\text{loss} \propto \frac{1}{(V\cos\phi)^2}$$

As pf is improved and voltage is increased, losses reduces and hence, efficiency of transmission increases.

12. (b)

Maximum power transmitted is

$$P_{max} = \frac{V_S \cdot V_R}{X}$$

$$= \frac{132 \times 126}{20} = 831.6 \text{ MW} \approx 832 \text{ MW}$$

13. (d)

Surge impedance is independent of length of the line.

14. (a)

Surge impedance,

$$Z_s = 400\ \Omega$$

$$\therefore \quad \text{SIL} = \frac{220 \times 220}{400} \text{MW} = 121 \text{MW}$$

16. (a)

Charging current/phase is

$$|I_C| = V_{ph} \cdot Y$$

Here, $Y = C = 0.5 \times 10^{-4}$ ℧/phase

$$\therefore \quad |I_C| = \left(\frac{220}{\sqrt{3}} \times 10^3\right) \times 0.5 \times 10^{-4}$$

$$= \frac{11}{\sqrt{3}} A$$

18. (c)

Skin effect depends upon type of conductor material, shape of conductor, frequency of current and diameter of conductor.

19. (d)

Like skin effect, proximity effect depends on the conductor size, frequency of the supply, resistivity and relative permeability of the material.

22. (d)

Line resistance doesn't depends upon the spacing between conductors. As spacing is increased, GMD will increase which will increase inductance and decrease capacitance

$$L \propto ln\left(\frac{\text{GMD}}{\text{GMR}}\right) \text{ and } C \propto \frac{1}{ln\left(\frac{\text{GMD}}{r}\right)}$$

24. (b)

In case of DC, current is uniformally distributed throughout the conductor surface.

27. (c)

Charging current is the current flowing through capacitance of the transmission line which leads the supply voltage by 90°.

28. (b)

For a short transmission line, $B = Z$ and $C = Y = 0$. i.e. capacitance is neglected due to which $Y = 0$.

29. (b)

Sag, $$S = \frac{Wl^2}{8T}$$

$$\therefore \quad S \propto l^2 \propto (\text{span})^2$$

31. (b)

Charging reactance of a line $\propto \dfrac{1}{\text{(length of the line)}}$

As length of line doubles, charging reactance will reduce to half.

32. (a)

Bundle conductors increases self GMD therefore, inductance will decrease while capacitance will increase. Due to increase in capacitance, charging current will increase.

34. (d)

Feeder are designed mainly from the point of view of their current carrying capacity. Hence, A is false.

35. (d)

Sag, $$S = \frac{Wl^2}{8T}$$

$\therefore$ Sag $\propto$ weight of conductor.

Hence, heavier the conductor, more will be the Sag.

36. (d)

The propagation constant of a transmission line is given by

$$\gamma = \sqrt{YZ}$$
$$= \sqrt{(j\omega C)(j\omega L)} = j\omega\sqrt{LC}$$

39. (d)

$Z_C = \sqrt{\dfrac{L}{C}}$ = characteristic impedance of the line.

Here, L and C will depend upon radius, length of the line as well as spacing (d) between the conductors.

40. (c)

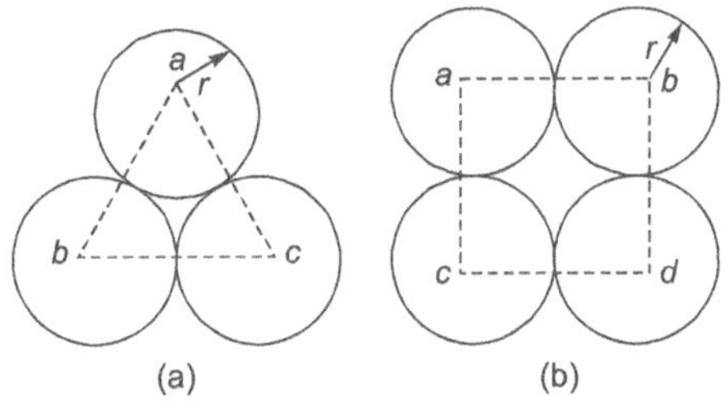

(a) (b)

For conf. (a),

$$D_s = \sqrt[3]{D_{aa}D_{ab}D_{ac}}$$
$$= \sqrt[3]{r' \times 2r \times 2r}$$

For conf. (b),

$$D_s = \sqrt[4]{D_{aa}D_{ab}D_{ac}D_{ad}}$$
$$= \sqrt[4]{r' \times 2r \times 2r \times 2\sqrt{2}\,r}$$

41. (d)

Given, $I_R = 500\angle 0°$ A, $V_R = 300\angle 0°$ V

We know that,

$$V_s = AV_R + BI_R$$
$$= 300\angle 0° + (j0.8)(500\angle 0°)$$
$$= (300 + j400) \text{ volt}$$
$$= 500\angle \tan^{-1}\frac{4}{3} \text{ volt} = 500\angle\phi \text{ volt}$$

Here, $\tan\phi = \dfrac{4}{3}$

$\therefore$ $\cos\phi = \dfrac{3}{5} = 0.6$

Also, $I_s = 500\angle 0°$ A and $V_s = 500\angle\phi$

Hence, V_s leads I_s

Therefore, sending end pf = 0.6 lagging

42. (a)

In π-model, the shunt admittance at each end of the line is $Y/2$.

$$\therefore \quad \frac{Y}{2} = j\omega\frac{C}{2} = j(2\pi f)\cdot\frac{C}{2}$$
$$= j(2\pi \times 50)\times\frac{0.03}{\pi}\times\frac{1}{2}\times 10^{-6}\times 100$$
$$= 150\times 10^{-6}\angle 90° \text{ mho}$$

43. (d)

As the far-end (receiving end) is open-circuited therefore, reflection coefficient of voltage, $\rho_V = +1$ and reflection coefficient of current, $\rho_I = -1$. Hence, reflected current wave will have negative sign whereas the reflected voltage wave will have positive sign.

44. (b)

Charging current per kilometer per phase will be given by:

$$I_C = \frac{V_{ph}}{X_C} = \frac{V_{line}}{\sqrt{3}\times X_C} = \frac{V_{line}}{\sqrt{3}}\times\omega C$$
$$= \frac{100\times 10^3 \times 2\pi\times 50\times 0.01\times 10^{-6}}{\sqrt{3}}$$
$$= \frac{\pi\times 10^{-1}}{\sqrt{3}} \text{ A/km} = \frac{314}{\sqrt{3}\times 10^3} \text{ A/km}$$

45. (d)

The breakdown strength of air at any temperature (t) and pressure (b) is given by:

$$\gamma' = 21\delta = 21\times\left(\frac{3.92b}{273+t}\right)$$

$$= 21\times\left(\frac{3.92\times72}{273+30}\right)\text{kV/cm}$$

$$= 21\times0.93 = 19.5 \text{ kV/cm}$$

46. (c)

When temperature is increased, Sag of transmission line increases while it's tension decreases. Hence, "tension vs temperature" will be represented by straight line (1) and "Sag vs temperature" by straight line (2) respectively.

47. (a)

The parameter A of a nominal-T circuit of a transmission line is given by:

$$A = D = \left(1+\frac{YZ}{2}\right)$$

$$= 1+\left(j\frac{4\times10^{-4}}{2}\right)(10+j20)$$

$$= 1 + j2\times10^{-4}(10+j20)$$

$$= 1 + j20\times10^{-4} - 40\times10^{-4}$$

$$= (1-40\times10^{-4}) + j2\times10^{-3}$$

$$= (0.996 + j2\times10^{-3})$$

$$= (996 + j2)\times10^{-3}$$

48. (b)

Given, $\quad v = 3\times10^8$ m/sec

Electrical line length = $\beta l = \frac{6\times\pi}{180}$ radian

Now, velocity $v = \lambda f$ and $\lambda = \frac{2\pi}{\beta}$

So, velocity $v = \frac{2\pi f}{\beta}$ or $\beta = \frac{2\pi f}{v}$

Hence, $\quad l = \frac{6\pi}{180\beta} = \frac{6\pi}{180\times\left(\frac{2\pi f}{v}\right)}$

$$= \frac{6\pi\times v}{180\times2\pi f} = \frac{3v}{180\times f}$$

$$= \frac{3\times3\times10^8}{180\times50} = 10^5 \text{ m} = 100 \text{ km}$$

49. (b)

For a short transmission line, receiving end power is given by:

$$P_R = \frac{|V_S||V_R|}{|Z|}\cos(\theta_s-\delta) - \frac{|V_R|^2}{|Z|}\cos\theta_s$$

Hence, P_R is maximum when $\theta_s = \delta$.

50. (c)

Given, electrical line length = $\beta l = \frac{8\pi}{180}$ radian

or, $\quad l = \frac{8\pi}{180\beta}$

Now, $\quad \beta = \frac{2\pi}{\lambda}$

$\therefore \quad l = \frac{8\pi}{180\left(\frac{2\pi}{\lambda}\right)}$

or, $\quad \frac{l}{\lambda} = \frac{8\pi}{180\times2\pi} = \frac{4}{180} = 0.022$

$\therefore \quad \frac{l}{\lambda} = 0.022$

51. (d)

We know that,

Corona loss $\propto (f+25)$

$\therefore \quad \frac{P_{C_2}}{P_{C_1}} = \frac{60+25}{50+25} = \frac{85}{75} = \frac{17}{15}$

or, $\quad P_{C_2} = \frac{17}{15}P_{C_1} = \frac{17}{15}\times1.2$

$= 1.36$ kW/km/phase

52. (a)

Tension, $\quad T = \frac{\text{Ultimate strength}}{\text{Factor of safety}}$

$= \frac{5000}{2} = 2500$ kg

Given, $\quad W = 0.55$ kg/m, L = span = 350 m

$\therefore$ Sag, $\quad S = \frac{WL^2}{8T} = \frac{0.55\times(350)^2}{8\times2500}$

$= 3.3687 \approx 3.37$ m

53. (d)

Due to the additional series capacitive compensation, the virtual surge impedance of the line will decrease as $X_{eff} = (X_L - X_{se})$, which in turn decreases the effective length of the line and increases virtual surge impedance loading.

54. (d)

For network-(A),

$$\begin{bmatrix} A & B \\ C & D \end{bmatrix} = \begin{bmatrix} 1 & 0 \\ Y & 1 \end{bmatrix}\begin{bmatrix} 1 & Z \\ 0 & 1 \end{bmatrix} = \begin{bmatrix} 1 & Z \\ Y & 1+YZ \end{bmatrix}$$

For network-(B),

$$\begin{bmatrix} A & B \\ C & D \end{bmatrix} = \begin{bmatrix} 1 & Z \\ 0 & 1 \end{bmatrix}\begin{bmatrix} 1 & 0 \\ Y & 1 \end{bmatrix} = \begin{bmatrix} 1+YZ & Z \\ Y & 1 \end{bmatrix}$$

55. (b)

The internal flux linkage of a long-cylindrical conductor is

$$\phi_{int} = \frac{\mu_0 I}{8\pi} \quad (I = \text{current})$$

Hence, it is independent of the diameter of the conductor.

56. (c)

The result obtained using nominal-T and nominal-π method are same but from practical point of view equivalent π-model is preferred as the need of creating a new bus is avoided. Hence (1) is a wrong statement.

57. (b)

Fig. (*i*) represent symmetrical arrangement whereas Fig. (*ii*) represents unsymmetrical arrangement of conductors.

Now, $L \propto ln\left(\frac{\text{GMD}}{\text{GMR}}\right)$ and $C_n \propto \frac{1}{ln\left(\frac{\text{GMD}}{r}\right)}$

Due to unsymmetrical spacing, GMD will increase which will increase L and will decrease C_n.
Therefore, $C_{n1} > C_{n2}$ and $L_1 < L_2$.

58. (c)

The situation is shown below.

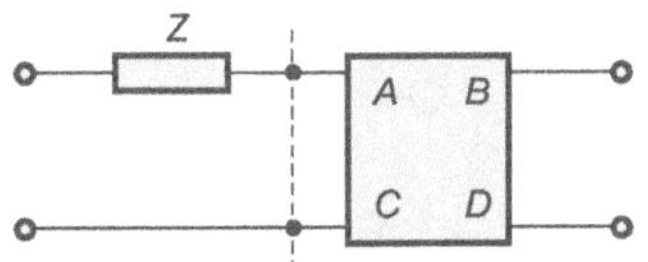

The overall *ABCD* parameter

$$= \begin{bmatrix} 1 & Z \\ 0 & 1 \end{bmatrix}\begin{bmatrix} A & B \\ C & D \end{bmatrix} = \begin{bmatrix} A & AZ+B \\ C & CZ+D \end{bmatrix}$$

59. (c)

As the line is not transposed therefore, there will be communication interference.

60. (c)

Given: $R = 200\ \Omega$ and $Z_0 = 200\ \Omega$
Reflection coefficient of voltage,

$$\rho_V = \left(\frac{R - Z_0}{R + Z_0}\right) = \left(\frac{200-200}{200+200}\right) = 0$$

Reflection coefficient of current,

$$\rho_I = \left(\frac{Z_0 - R}{R + Z_0}\right) = \left(\frac{200-200}{200+200}\right) = 0$$

61. (c)

Given, $r = 0.5$ cm, $D = 60$ cm (mutual distance between center to center).
Inductance per phase is,

$$L = 0.05 + 0.2\, ln\left(\frac{D}{r}\right) \text{ mH/km}$$

$$= 0.05 + 0.2\, ln\left(\frac{60}{0.5}\right) \text{ mH/km}$$

62. (c)

Let the original length = l and original cross-sectional area = A

New length = $l' = nl$

Also, new cross-sectional area = $A' = \frac{A}{n}$

Volume of wire = constant

Now, $R \propto \frac{l}{A}$

$\therefore \quad \frac{R'}{R} = \frac{l'}{l} \times \frac{A}{A'} = \frac{nl}{l} \times \frac{A}{A/n} = n^2$

or, $R' = n^2 R$ = new resistance

63. (a)

In case of single-phase 2-wire system, we have

$P_1 = 1000$ kW
$= 11 \times 10^3 \times I_1 \cos\phi \; (= 3\ VI_1 \cos\phi)$

or, $I_1 \cos\phi = \frac{1000}{11}$

In case of 3-phase system, we have

$P_2 = 3 \times 11 \times I_2 \cos\phi \; (= 3\ VI_2 \cos\phi)$

As losses remains same, therefore

$$2I_1^2R = 3I_2^2R \text{ or } I_2 = \sqrt{\frac{2}{3}}\, I_1$$

So, $$P_2 = 3\times 11\times\sqrt{\frac{2}{3}}\, I_1\cos\phi$$

$$= 3\times 11\times\sqrt{\frac{2}{3}}\times\frac{1000}{11}$$

$$= \sqrt{6}\times 1000 \text{ kW} = 2450 \text{ kW}$$

Hence, $P_2 = 2450$ kW

65. (c)

The natural 3-phase loadings for the two lines are:

$$P_1 = \frac{(132)^2}{\sqrt{160\times 10^3}}\text{ MW}$$

$$= 43.56 \text{ MW} \approx 44 \text{ MW}$$

and $$P_2 = \frac{(400)^2}{\sqrt{62.5\times 10^1}}\text{ MW} = 6400 \text{ MW}$$

66. (d)

Required time, $t = \dfrac{600\times 10^3}{3\times 10^8} = 0.002$ sec

68. (a)

Here, 210 km is a long length line.

C = Admittance of line, which will increase with increase in line length.

With increase in line length, A and D decreases ($A = D = \cosh\gamma l$)

69. (a)

Both (A) and (R) are true and (R) is the correct explanation of (A).

70. (a)

Corona acts as a safety valve and protects the transmission line from direct lightning strokes due to the induced charges on the line. Hence, option (a) is correct.

71. (c)

Transposition of conductor in a transmission line is necessary to reduce the telephonic interference of the transmission line with the communication lines. Hence, (R) is false.

72. (d)

The insulation of modern EHV lines is designed based on the switching over voltages. Hence, (A) is false.

73. (b)

EHV transmission line make use of bundled conductor for increasing self GMD due to which inductive reactance of the line is reduced as a result of which steady-state stability is increased. It also reduces corona loss. Hence, (A) is true. Due to increase in self GMD, C is increased and L is decreased.

74. (b)

Voltage of a system depends upon reactive power not on active power.

78. (d)

With series capacitive compensation, the net transfer reactance of the line will be $X = (X_L - X_C)$ due to which the power transmitted from sending to the receiving end is increased. Thus, it will improve the steady-state stability of the system.

81. (a)

Synchronous phase modifier is an overexcited synchronous motor operating on no-load under wide variation of excitation. For no-load operation, $\delta = 0$.

84. (c)

Full load compensation in a line requires shunt capacitors for improving power factor of the system.

86. (c)

Shunt capacitor is usually connected at load end for improving power factor and voltage.

87. (b)

Reactive power,

$$Q \propto \Delta V$$

or, $$Q \propto [|V_s| - |V_R|]$$

89. (c)

A power system requires a considerable amount of reactive power whether it is at peak load or at off-peak load.

90. **(b)**

Shunt capacitors in a substation fulfills the requirement of lagging reactive power (VARs) of loads. Hence, it delivers lagging VARs. Shunt capacitors are also called static capacitors.

91. **(c)**

Load shedding means switching off some loads.

95. **(b)**

$$(\Delta f)_{\text{steady-state}} = \left(\frac{1}{B_{pu} + \frac{1}{R}}\right)(\Delta P_C - \Delta P_D)$$

Here, $B = \frac{dP_D}{df}$ has unit of MW/Hz and R has unit of Hz/MW.

96. **(b)**

Given, $\cos\phi_1 = 0.8$ lag, $\cos\phi_2 = 1$

or, $\tan\phi_1 = \frac{3}{4}$

Reactive power demand is,

$$Q_1 = \tan\phi_1$$

$$= 100 \times 10^3 \times \frac{3}{4} = 75000 \text{ VAR}$$

For raising the pf from 0.8 lag to unity, the capacitor bank will supply the required VAR to the load. Hence, capacitance per phase is,

$$C = \frac{Q_1}{\omega V^2} = \frac{Q_1}{2\pi f \times V^2}$$

$$= \frac{75000}{2\pi \times 50 \times (10^4)^2} = 2.4\ \mu\text{F}$$

97. **(b)**

Let the rated voltage be V kV.

Then, reactance of the reactor is

$$X = \frac{V^2}{100} \quad \left(\text{since, } X = \frac{V^2}{Q}\right)$$

Now, at 98% of rated voltage and at 96% of rated frequency, the reactive power absorbed will be

$$Q' = \frac{V'^2}{X'} = \frac{(0.98\,V)^2}{(0.96\,X)}$$

$$= \left[\frac{0.98^2}{0.96} \times \frac{V^2}{X}\right] \quad (\text{since, } X \propto f)$$

$$= \left(\frac{0.98^2}{0.96} \times 100\right) \text{MVAR}$$

$$= 100.04 \text{ MVAR}$$

98. **(c)**

A synchronous condenser is always installed at the load (receiving) end in a transmission line. Hence, statement 1 is false.

A schrage motor and a compensated inductor motor operates at a high power factor (approximately unity) and thus improves the power factor of the system.

99. **(c)**

Tie-line bias control is modification of flat frequency control and flat Tie-line control. Here, the system is allowed to follow its normal regulating characteristic directed towards holding normal frequency.

101. **(d)**

Given, $P_{\text{rated}} = 200$ MW

We know that,

$$\Delta f = \left(\frac{1}{B_{pu} + \frac{1}{R}}\right)(\Delta P_C - \Delta P_D)$$

Here, $1/R$ is in pu-MW/Hz or, R is in Hz/pu-MW.

Now, % regulation = 4% of f_{schedule} = 4% of 50 Hz

$$\therefore \quad R = \frac{4}{100} \times 50 = 2 \text{ Hz/pu-MW}$$

$$\therefore \quad R \text{ in Hz/MW} = \frac{2}{P_{\text{rated}}} = \frac{2}{200} = 0.01 \text{ Hz/MW}$$

102. **(c)**

Let the power supplied by 100 MW alternator be x MW and that by 200 MW be $(250-x)$ MW.

$$\therefore \quad \text{Droop in speed} = \frac{4x}{100}\% \quad (\text{for alternator-1})$$

and Droop in speed $= \frac{4}{200}(250-x)\%$

(for alternator-2)

Since the two machines are runnings in parallel therefore, their droop in speed will be equal.

$$\therefore \quad \frac{4x}{100} = \frac{4}{200}(250 - x)$$

or, $x = 83.33$ MW

Hence, the power supplied by the alternators will be

$$P_1 = x = 88.33 \text{ MW}$$

and $$P_2 = 250 - x = 250 - 83.33$$
$$= 166.67 \text{ MW}$$

103. (a)

Here, full-load frequency = 50 Hz and full load output P_1 = 100 MW

No-load frequency = 52.5 Hz

When frequency decreases by 0.02 Hz, new frequency = 49.98 Hz

Let new load = P_2

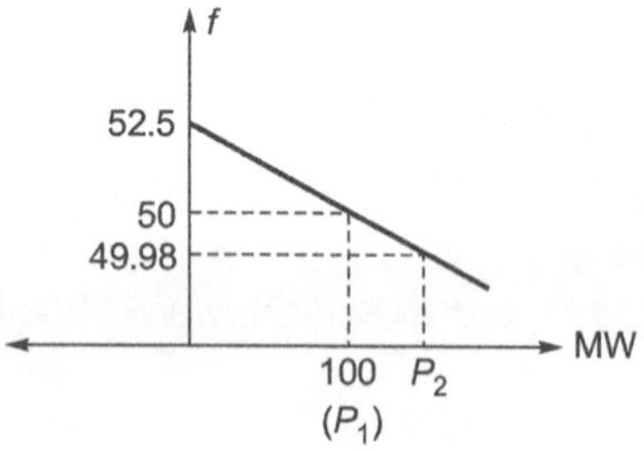

Since, slope is equal therefore,

$$\frac{2.5}{100} = \frac{52.5 - 49.98}{P_2}$$

or, $$P_2 = \frac{2.52 \times 100}{2.5} = 100.8 \text{ MW}$$

∴ Increase in turbine output power is,

$$P_2 - P_1 = 0.8 \text{ MW}$$

104. (d)

First of all, the speed governor senses any change in the load demand and sends a signal to the speed changer.

106. (c)

Given, $\cos\phi_1 = 0.6$ and $\cos\phi_2 = 1$

$$\text{kVA loading} = \frac{\text{kW}}{\cos\phi_1} = \frac{300}{0.6} = 500 \text{ kVA}$$

$$\text{New pf} = 1 = \cos\phi_2$$

Since, kVA remains constant therefore, new kW required = kVA × $\cos\phi_2$ = 500 × 1 = 500 kW

Hence, additional kW loading required

$$= 500 \text{ kW} - 300 \text{ kW}$$
$$= 200 \text{ kW}$$

108. (b)

Unit of frequency bias coefficient is MW/Hz. Hence, option (b) is correct.

109. (a)

Given, Machine-1 → 200 MW, 4% droop
Machine-2 → 400 MW, 4% droop

Now, $P_1 + P_2 = 300$ MW ...(*i*)

As droops are same, therefore,

$$\frac{P_1}{200} = \frac{P_2}{400}$$

$$\Rightarrow \quad P_2 = 2P_1 \text{ or } \frac{P_1}{P_2} = \frac{1}{2}$$

$$\therefore \quad P_1 = \frac{1}{3} \times 300 = 100 \text{ MW}$$

and $$P_2 = 300 - 100 = 200 \text{ MW}$$

110. (b)

Speed regulation,

$$R = 4\% \text{ of } 50 \text{ Hz}$$
$$= \frac{4}{100} \times 50 = 2 \text{ Hz}$$

$$\therefore \quad R = \frac{2 \text{ Hz}}{500 \text{ MW}} = 0.004 \text{ Hz/MW}$$

Given, drop in frequency,

$$\Delta f = -0.12 \text{ Hz}$$

∴ Increase in turbine power is,

$$\Delta P = -\frac{1}{R}\Delta f = \frac{-1}{0.004} \times (-0.12)$$
$$= 30 \text{ MW}$$

∴ Turbine power incremental is,

$$\Delta P = 30 \text{ MW}$$

111. (b)

The cable operates stably from the view point of dielectric strength if, $\frac{R}{r} > e$ or, $\frac{r}{R} < \frac{1}{e}$ or, $\frac{r}{R} < 0.368$.

112. (c)

The insulation resistance of a cable,

$$R_i \propto \frac{1}{(\text{length of cable})}$$

$$\therefore \quad \frac{R_{i_2}}{R_{i_1}} = \frac{10}{40} = \frac{1}{4}$$

or, $$R_{i_2} = \frac{1}{4} \times 1 = 10.25 \text{ M}\Omega$$

116. (c)

ACSR conductors are universally employed for distribution systems (feeders as well as distributors). AAC can be used for distribution purpose provided the spans are small.
Hence, statement-1 is false.
The conductor size for a feeder is mainly governed by current carrying capacity and overall economy. Hence, 2 is also a false statement.

122. (c)

String efficiecy is indopendent of frequency of transmission system.

123. (a)

Higher the operating voltage of the cable, more is the insulation thickness required.

125. (c)

In capacitance grading of cable for same maximum working stress,

$$\varepsilon_1 r = \varepsilon_2 r_1 = \varepsilon_3 r_2$$

Since, $\varepsilon_3 > \varepsilon_2 > \varepsilon_1$ therefore, $r_2 < r_1 < r$
Hence, location of the materials w.r.t. the cable will be 4.0, 3.0, 2.5.

126. (d)

The optimal ratio of the radius of the cable to the radius of the conductor is given by:

$$\frac{R_1}{R_2} = e = 2.718$$

or,
$$\frac{\frac{10}{2}}{R_2} = 2.718 \text{ or } R_2 = 1.84 \text{ cm}$$

127. (b)

For the first test, the equivalent capacitance will be $C_A = 3\,C_S$ or $C_S = \frac{C_A}{3}$

For the second test, the equivalent capacitance will be

$$C_B = C_S + 2C_C \text{ or } C_C = \frac{1}{2}\left(C_B - \frac{1}{3}C_A\right)$$

128. (c)

This is an example of capacitance grading of cables. The dielectric stress can be made uniform such that,

$$\varepsilon_A r_A = \varepsilon_B r_B = \varepsilon_C r_C$$

Here, $\varepsilon_A = 250$ kV/cm
$\varepsilon_B = 200$ kV/cm
and $\varepsilon_C = 150$ kV/cm
Since $\varepsilon_A > \varepsilon_B > \varepsilon_C$ therefore, $r_C > r_B > r_A$
Hence, the location of materials w.r.t. outermost sufrace of cable will be *C*, *B*, *A*.
And the location w.r.t. core of the cable will be *A*, *B* and *C*.

129. (d)

Given, capacitance betwen three bunched conductors and the sheath

$$= 0.90\ \mu\text{F/Km} = 3C_S$$

or, $C_S = 0.3\ \mu$F/Km
Also, capacitance between one conductor and the other two connected to the sheath

$$= 0.40\ \mu\text{F/km} = C_S + 2C_C$$

$$\text{or, } C_C = \frac{0.40 - C_S}{2}$$

$$= \frac{0.4 - 0.3}{2} = 0.05\ \mu\text{F/km}$$

Hence, the line-to ground capacitance is:

$$C_n = C_S + 3\,C_C$$
$$= 0.3 + 3 \times 0.05 = 0.45\ \mu\text{F/km}$$

∴ For 20 km length, required value of

$$C_n = 0.45 \times 20 = 9\ \mu\text{F}$$

130. (a)

Due to small distance between phase conductors, small distance between conductor and sheath and high electrical permittivity of insulation in cables, the capacitance value is more than the capacitance of an overhead line.

131. (a)

For economic operation,

$$\frac{R_1}{R_2} = e$$

Given, $V = 80$ kV, $E_{max} = 60$ kv/cm

Now, $$V = E_{max} R_2\, ln\left(\frac{R_1}{R_2}\right)$$

or, $$80 = 60\, R_2\, ln\, e = 60\, R_2$$

or, $$R_2 = \frac{80}{60} = 1.33 \text{ cm}$$

∴ Diameter of conductor is

$$D_2 = 1.33 \times 2 = 2.66 \text{ cm}$$

∴ Diameter of sheath is

$$D_1 = 2 \times (eR_2)$$
$$= 2 \times (2.718 \times 1.33) = 7.23 \text{ cm}$$

132. (d)

Dielectric loss is

$$P_d = 3V_{ph}^2 \omega C_o \delta \quad \text{(For 3-}\phi\text{)}$$

$$P_d = V_{ph}^2 \omega C_o \delta \quad \text{(For 1-}\phi\text{)}$$

136. (b)

String efficiency,

$$\eta = \left(\frac{\text{voltage across the string}}{n \times \text{voltage across the disc near power conductor}}\right)$$

∴ Total voltage across the string

$$= 0.8428 \times 3 \times 17.5 \text{ kV} = 44.25 \text{ kV}$$

138. (b)

Conventional source of energy are: Water, Coal, Gas, radioactive substances. Non-conventional source of energy are: Wind, Solar energy, Geothermal.

139. (c)

Power output of a hydroelectric plant

$$= \frac{0.736}{75} QWh\eta$$

Here, Q = discharge (m^3/sec)
W = density of water
h = Water head
η = system efficiency

140. (c)

$$\frac{\text{Load factor}}{\text{Plant capacity factor}} = \left(\frac{\text{Average load/Max. demand}}{\text{Avg. load/Plant Capacity}}\right)$$

or, $$\frac{0.75}{0.75} = \frac{\text{Plant Capacity}}{\text{Maximum demand}}$$

or, Plant capacity = Maximum demand
= 100 MW

Hence,

Reserve capacity = Plant capacity − Max. demand
= 100 − 100 = 0 MW

142. (a)

Superheated steam has the highest temperature which comes out from super heater.

143. (d)

Condenser in a steam power plant has lowest operating pressure.

144. (a)

$$\text{Average load} = \frac{1500 \times 12 + 1000 \times 12}{24}$$
$$= \frac{2500}{2} = 1250 \text{ kW}$$

Max. demand = 1500 kW

$$\therefore \quad \text{Load factor} = \frac{\text{Average load}}{\text{Maximum demand}}$$
$$= \frac{1250}{1500} = 0.833$$

145. (b)

Control rods made up of boron, Cadmium are used to control the nuclear chain reaction in a nuclear power plant.

147. (b)

$$\text{Average annual load} = \frac{750 + 600}{2} = 675 \text{ MW}$$

$$\text{Plant capacity factor} = \frac{\text{Average annual load}}{\text{Plant capacity}}$$
$$= \frac{675}{900} = 0.75$$

$$\text{Utilization factor} = \frac{\text{Maximum demand}}{\text{Plant capacity}}$$
$$= \frac{750}{900} = 0.83$$

148. (d)

Pump storage PP and Gas turbine PP are used as peak load power plants because they are quick to start.

150. (d)

Efficiency of a thermal power plant is more than that of a nuclear power plant because the water in the boiler is used at high temperature and high pressure.

154. **(b)**

Higher the head, lower is the water discharge.

155. **(d)**

There is high percentage of loss occuring in condenser due to which a modern coal-fired steam-raising thermal PP has low efficiency.

156. **(a)**

Daily energy generated

$= \text{Average load} \times 24$

$\text{Average load} = \text{MD} \times \text{LF} = 20 \times 0.75$

$= 15 \text{ MWh}$

∴ Daily energy generated

$= 15 \times 24 = 360 \text{ MWh}$

The maximum energy that can be produced daily if the plant is running all the time

$$= \frac{\text{Actual energy generated}}{\text{Plant capacity factor}}$$

$$= \frac{360}{0.5} = 720 \text{ MWh}$$

159. **(c)**

From given load pattern, average load

$$= \frac{100 \times 12}{24} = 50 \text{ MW}$$

Max. demand = 100 MW

$$\therefore \quad \text{Load factor} = \frac{50}{100} = \frac{\text{Average load}}{\text{Max. demand}} = 0.5$$

160. **(d)**

$$\text{Power output} = \left(\frac{0.736}{75} QWh\eta\right) = 9.81\,(Qh\eta)\text{ kW}$$

(Here, W = density of water = 1000 kg/m^3)

$$\therefore \quad P_{out} = 9.81 \times 3000 \times 30 \times 0.79 \text{ kW}$$

$$= 9.81 \times 90 \times 0.79 \text{ MW}$$

$$= 697.5 \text{ MW}$$

161. **(c)**

Power generated by the power station

$= 500 \times 0.8 = 400 \text{ MW}$

Total energy generated in a day

$= 400 \times 10^3 \times 24 \text{ kWh}$

Energy required from coal

$$= \frac{400 \times 10^3 \times 24}{0.4} \text{ kWh}$$

$$= 24000 \times 10^3 \text{ kWh}$$

∴ Number of trains required

$$= \frac{24000 \times 10^3}{2 \times 2000 \times 10^3} = 6$$

162. **(c)**

To reduce the transportation cost to a great extent, modern thermal power plants are located near coal field.

164. **(c)**

Economiser takes feed water and extracts heat from flue gases and therefore, feeds hot water to boiler. Air preheater takes atmospheric air from forced draught fan and extracts heat from flue gases coming out from economiser therefore, it feeds hot air to boiler.

165. **(d)**

$$\text{Diversity factor} = \frac{\text{Sum of individual max. demand}}{\text{Maximum demand on station}}$$

∴ Maximum demand on station

$$= \frac{100 + 125 + 50}{1.5} = 183.33 \text{ MW}$$

$$\text{Also,} \quad \text{Load factor} = \frac{\text{Average load}}{\text{Maximum demand}}$$

or, $\text{Average load} = 0.5 \times 183.33 = 91.66 \text{ MW}$

166. **(c)**

Average demand (load)

$$= \frac{(50 \times 3) + (75 \times 16) + (100 \times 2) + (0 \times 3)}{24}$$

$= 64.58 \text{ MW}$

Max. demand = 100 MW

$$\therefore \quad \text{Load factor} = \frac{64.58}{100}$$

$$= 0.6458 = 64.58\%$$

167. **(a)**

$$\text{Demand factor} = \frac{380}{600} = 0.63 = 63\%$$

$$\text{Avg. demand} = \frac{2100 \times 10^6}{8760} \text{ kW}$$

$$= 239.7 \text{ MW}$$

$$\therefore \quad \text{Load factor} = \frac{239.7}{380} = 0.63 = 63\%$$

168. (d)

Water available in reservoir (annually)

$= 0.72 \times 500 \times 10^6 \times 1.5$

$= 540 \times 10^6 \text{ m}^3$

Average discharge per second

$$= \frac{540 \times 10^6}{8760 \times 3600} = 17.12 \text{ m}^3/\text{sec}$$

h = water head = 300 m

$\eta = 0.98 \times 0.92 \times 0.95 = 0.856$

∴ Power generated is

$$P = \frac{0.736}{75} \times 17.12 \times 1000 \times 300 \times 0.856 \text{ kW}$$

$= 43143 \text{ kW} = 43.14 \text{ MW}$

170. (b)

Diesel, Gas turbine and pumped storage power plants are quick to start and therefore, can be used to supply peak load and can improve load factor of the system.

171. (c)

$$\text{Demand factor} = \frac{\text{Maximum demand}}{\text{Connectedload}} = \frac{20}{40} = \frac{1}{2}$$

173. (c)

(R) is false because higher the value of diversity and load factor, lower will be the overall cost per unit generated.

175. (c)

$$\text{Demand factor} = \left(\frac{\text{Maximum demand}}{\text{Connected load}}\right)$$

176. (b)

Plant capacity = 120 MW

Energy supplied in one day

$= (110 \times 4) + (60 \times 10) = 1040 \text{ MWh}$

Number of working days in a year

$= 365 - 60 = 305$ days

Energy supplied per year

$= 1040 \times 305 = 3,17,200 \text{ MWh}$

∴ Annual load factor,

$$= \frac{\text{MWh supplied / annum}}{\text{max. demand} \times \text{working hrs}}$$

$$= \frac{3,17,200}{120 \times 305 \times 24} \times 100 = 36.11\%$$

178. (c)

$$\text{Plant utilization factor} = \frac{\text{Maximum demand}}{\text{Plant capacity}}$$

Since MD > Installed capacity therefore, PUF > 1 i.e. more than 100%.

179. (c)

Since plant capacity factor = Load factor

$$\therefore \frac{\text{Average load}}{\text{Plant capacity}} = \frac{\text{Average load}}{\text{Maximum demand}}$$

or, Plant capacity = Maximum demand

Now, $$\text{PUF} = \frac{\text{Maximum demand}}{\text{Plant capacity}} = 1$$

180. (b)

Hydro-power plants are used as a base load power plant due to high capital cost and low operating cost.

182. (b)

A three-phase fault is a symmetrical fault.

183. (c)

For transmission line

$z_0 >> z_1$ and $z_1 = z_2$

184. (d)

A three-phase fault is a symmetrical fault where only positive sequence network (current exists).

185. (a)

Using switch diagram, we find the zero sequence network for a transformer.

186. (c)

$$\text{Base current} = \frac{400}{10} = 40 \text{ A}$$

and $$\text{Base kVA} = \frac{40 \times 400}{1000} = 16 \text{ kVA}$$

187. (a)

$$Z_{base} = \frac{V_{base}}{I_{base}} = \frac{V_{base}}{(VA)_{base} / V_{base}}$$

$$= \frac{V_{base}^2}{(VA)_{base}}$$

Now, $Y_{base} = \frac{1}{Z_{base}} = \frac{(VA)_{base}}{V_{base}^2}$

$\therefore$ $Y\text{(in pu)} = \frac{Y}{Y_{base}} = \frac{YV_{base}^2}{(VA)_{base}}$ pu

188. (d)

$$I_{fLG} = 3I_0$$

or, $I_0 = \frac{I_{fLG}}{3} = \frac{1500}{3} = 500$ A

189. (a)

$$I_{f,LLG} = 3I_0$$

Here, I_0 will depend upon z_0, z_1 and z_2 therefore, $I_{f,LLG}$ will also depend upon z_0, z_1 and z_2.

191. (c)

$$I_{SC}\text{(in pu)} = \frac{V\text{(in pu)}}{z_{pu}} = \frac{1}{0.1} = 10 \text{ pu}$$

Now, $I_{base} = \frac{10^4}{400} = \frac{100}{4} = 25$ A

Hence, $I_{SC}\text{(in A)} = 10 \times 25 = 250$ A

192. (d)

The fault current is given by

$$I_{f,LG} = \frac{3E_{ph}}{(z_1 + z_2 + z_0 + 3z_n)}$$

$$= \frac{3 \times 1}{(0.2 + 0.2 + 0.1 + 3 \times 0.05)}$$

$= 4.615$ pu

193. (c)

Since, $I_{a1} + I_{a2} + I_{a0} = -j3.05 + j0.68 + j2.37 = 0$ pu

therefore, the given fault is a double line to ground fault.

195. (a)

$$Y_{11} = \frac{1}{0.2} + \frac{1}{0.2} = 10 \text{ pu}$$

$$Y_{12} = Y_{21} = -\frac{1}{0.2} = -5 \text{ pu}$$

$$Y_{22} = \frac{1}{0.2} = 5 \text{ pu}$$

Hence, bus admittance matrix is

$$[Y] = \begin{bmatrix} 10 & -5 \\ -5 & 5 \end{bmatrix}$$

196. (c)

$$I_{a1} = \frac{E_a}{z_1 + z_2 + z_0} = -j1.43 \text{ pu}$$

Also, $I_{fLG} = 3I_{a1} = -j4.29$ pu

Base current, $I_b = \frac{30 \times 10^3}{\sqrt{3} \times 11} = 1574.6$ A

Hence, $I_{fLG} = -j4.29 \times 1574.6 = -j6755$ A

197. (a)

Let, $V_{base} = 0.4$ kV, $(kVA)_{base} = 2.5$

$$(Z_{base})_{LV} = \frac{(0.4)^2 \times 10^3}{2.5} = 64\ \Omega$$

$\therefore$ Leakage reatance in pu

$$= \frac{0.96}{64} = 0.015 \text{ pu}$$

198. (a)

Unsymmetrical faults (LG, LL and LLG faults) are highly unbalanced due to which we analyze it using symmetrical components i.e. positive, negative and zero-sequence components.

199. (d)

On both primary and secondary sides, the neutral will be ground through z_n due to which the zero-sequence network will result as given in question.

200. (c)

Let, Base voltage = 200 V = 1 pu

Base kVA = 10 kVA = 1 pu

$\therefore$ Base current $= \frac{10,000}{200} = 50$ A $= 1$ pu

The generated voltage required to produce the rated current under short circuit is

$= IZ_g = 50 \times 2 = 100$ V

$= \frac{100}{200}$ pu $= 0.5$ pu

201. (b)

$$Z_{pu} = \frac{Z}{Z_{base}}$$

$\therefore$ $Z_{pu} = (4 + j60) \times \frac{100 \times 10^6}{(345 \times 10^3)^2}$

$= (3.36 + j50.4) \times 10^{-3}$ pu

$$Y_{pu} = (j2\times10^{-3})\times\frac{(345\times10^3)^2}{100\times10^6}$$
$$= j2.38 \text{ pu}$$
$$\left(\text{Since, } Y_{pu} = \frac{Y V_{base}^2}{(VA)_{base}} \text{ pu}\right)$$

202. (a)

Let Base voltage,

$$V_{base} = 12 \text{ kV}$$

Base MVA, $S_{base} = 15$ MVA

Given, $X_1 = 0.3$ pu

Now, X_2 has to be found on common base of 15 MVA.

$$\therefore \quad X_2(\text{new}) = 0.5\times\frac{15}{20}\times\left(\frac{12}{12}\right)^2 = \frac{150}{4}\%$$
$$= \frac{1.5}{4} \text{ pu}$$

$\therefore$ Thevenin equivalent reactance is

$$X_{th} = \frac{0.3\times\frac{1.5}{4}}{0.3+\frac{1.5}{4}} = \frac{1}{6} \text{ pu} = \frac{100}{6}\%$$

$\therefore$ Short circuit MVA

$$= 15\times\frac{100}{(100/6)} = 90 \text{ MVA}$$

So, short circuit current

$$= \frac{90}{\sqrt{3}\times12} \text{ kA} = 4.330 \text{ kA}$$
$$= 4330 \text{ A}$$

203. (c)

Let the required series reactance be $x\ \Omega$

Now, $\quad \text{SC MVA} = \text{Rated MVA}\times\frac{100}{\%Z}$

or, $\quad 80 = 10\times\frac{100}{5+x\%}$

or $\quad \%x = 7.5$

or, $\quad x = 0.075$ pu

Now, $\quad X_{base} = \frac{11^2}{10} = \frac{121}{10}\ \Omega$

Hence, required series reatance

$$= x(\text{pu})\times X_{base}$$
$$= 0.075\times\frac{121}{10} = 0.9075 \text{ pu}$$

204. (a)

We know that, the zero sequence current is

$$I_0 = \frac{1}{3}(I_R + I_Y + I_B)$$
$$= \frac{1}{3}(3+j5+2+j2-2-j1)$$
$$= (1+j2) \text{ A}$$

205. (d)

Using switch diagram, we can easily find the zero-sequence network for the given one line diagram.

Since transfomer-T_1 has $\Delta\text{-}Y_g$ connection therefore, shunt switch on primary side will be closed while series switch on secondary side will be closed.

Since transformer-T_2 has $\Delta\text{-}Y_g$ connection, same thing happens for it as that for transformer-T_1.

206. (b)

Since, $\quad I_{fLG} = I_{f3\phi}$

therefore, the neutral grounding reactance is

$$x_n = \frac{1}{3}(x_1 - x_0) = \frac{1}{3}(0.15-0.04)$$
$$= 0.0367 \text{ pu}$$

207. (a)

The Thevenin equivalent reactance across the given two terminals is

$$X_{th} = \frac{(j0.1+j0.15)\times(j0.2)}{(j0.1+j0.15+j0.2)} = j\frac{1}{9}\text{ pu} = \frac{100}{9}\%$$

Hence, short circuit capacity of the breaker is

$$\text{SC MVA} = \text{Rated MVA}\times\frac{100}{\%X_{th}}$$
$$= 5\times\frac{100}{(100/9)} = 45 \text{ MVA}$$

208. (d)

In a balanced three-phase system, only positive sequence component exists. Hence, 1 is a false statement.

For a line-to-line fault, the sequence networks are connected in series opposition. Hence, 3 is also a false statement.

209. (d)

$$Y_{11} = \frac{1}{0.1} + \frac{1}{0.2} = 10 + 5 = 15 \text{ pu}$$

$$Y_{22} = \frac{1}{0.2} = 5 \text{ pu}$$

and $$Y_{12} = Y_{21} = -\frac{1}{0.2} = -5 \text{ pu}$$

210. (b)

We have,

$$\begin{bmatrix} I_{a_0} \\ I_{a_1} \\ I_{a_2} \end{bmatrix} = \frac{1}{3}\begin{bmatrix} 1 & 1 & 1 \\ 1 & \alpha & \alpha^2 \\ 1 & \alpha^2 & \alpha \end{bmatrix}\begin{bmatrix} I_a \\ I_b \\ I_c \end{bmatrix}$$

$\therefore$ $$I_{a_0} = \frac{1}{3}(I_a + I_b + I_c)$$

$$= \frac{1}{3}(0 + 6\angle 60° + 6\angle -120°) = 0 \text{ A}$$

Also, $$I_{a_1} = \frac{1}{3}(I_a + \alpha I_b + \alpha^2 I_c)$$

or, $$I_{a_1} = \frac{1}{3}[0 + 1\angle 120° \times 6\angle 60° + 1\angle 240° \times 6\angle -120°]$$

$$= \frac{1}{3}(-9 + j3\sqrt{3}) \text{ A}$$

$$= (-3 + j\sqrt{3}) \text{ A}$$

and $$I_{a_2} = \frac{1}{3}(I_a + \alpha^2 I_b + \alpha I_c)$$

$$= \frac{1}{3}[0 + 1\angle 240° \times 6\angle 60° + 1\angle 120° \times 6\angle -120°]$$

$$= \frac{1}{3}(9 - j3\sqrt{3}) = (3 - j\sqrt{3}) \text{ A}$$

211. (a)

For a 3-ϕ fault,

$$I_{f,\,3\text{-}\phi} = \frac{E_{ph}}{x_1} \quad \text{or} \quad x_1 = \frac{1}{-j5} = j0.2 \text{ pu}$$

For a L-L fault,

$$I_{f,LL} = \frac{\sqrt{3}E_{ph}}{x_1 + x_2} \quad \text{or} \quad -j4 = \frac{1}{x_1 + x_2}$$

or, $$x_1 + x_2 = j0.25 \text{ pu}$$

or $$x_2 = j0.25 - j0.2 = j0.05 \text{ pu}$$

Hence, $x_1 = j0.2$ pu and $x_2 = j0.05$ pu

212. (b)

$$I_{f,3\text{-}\phi} = \frac{E_{ph}}{x_1} \quad \text{or} \quad x_1 = \frac{11 \times 10^3/\sqrt{3}}{2000}$$

$$= 3.175\ \Omega$$

Now, $$Z_{base} = \frac{(11)^2}{50} = 2.42\ \Omega$$

$\therefore$ $$x_1 \text{ (in pu)} = \frac{3.175}{2.42} = 1.312 \text{ pu}$$

215. (c)

The zero sequence network of the given power system model is shown below.

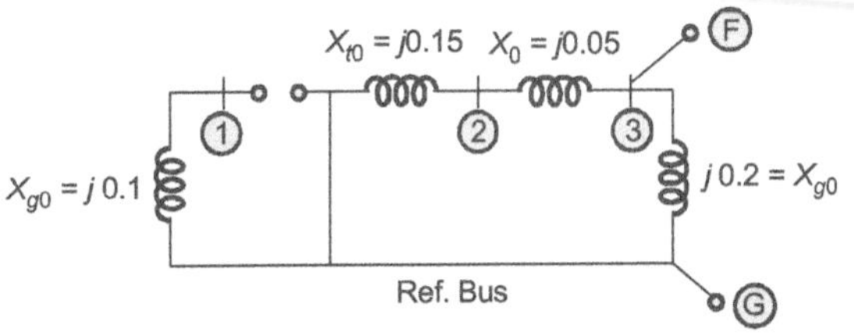

Combining the reactances, the network is reduced as shown below.

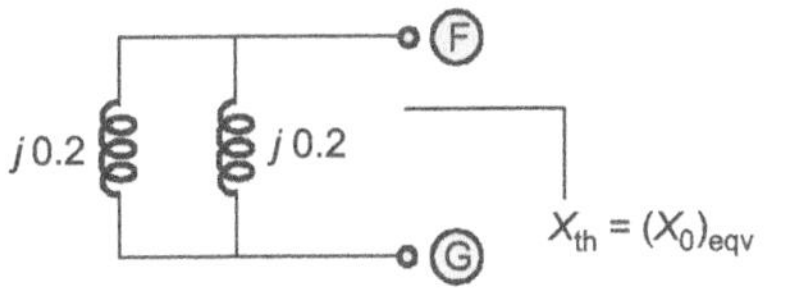

When fault occurs at bus-3 then, the Thevenin's equivalent of the zero sequence network at bus-3 will be

$$X_{th} = (X_0)_{eqv} = (j0.2)\,||\,(j0.2)$$

$$= j0.1 = 10\% = 0.1 \text{ pu}$$

218. (d)

Assumption of decoupling is valid in load flow study only if resistance of the line is negligible in comparison to the reactance of the line.

219. (a)

Since active power flows from leading bus to lagging bus therefore, P will flow from i to k. Since, reactive power flows from high potential bus to low potential bus therefore, Q will flow from i to k.

220. (a)

Out of the 15% of Generator buses in a power system, one bus is taken as a slack bus which take care of losses occuring in the system. At a generator bus P and V are known. Hence, a slack bus can be a PV bus.

221. (d)

A synchronous condeser regulates Q and thus controls power factor of the system.

223. (c)

At a slack bus, $|V|$ and δ are known while P and Q are unknown.

224. (b)

$$S_G = S_1 + S_x$$

or
$$S_x = S_G - S_1$$
$$= 3 + j2 - (1 - j1)$$
$$= (2 + j3)$$

226. (b)

Given, $S = (P - j\,Q)$ .....for given network

Since P is positive, therefore the given network absorbs P.

Since Q is negative, therefore the given network absorbs $-Q$ or delviers $+Q$.

227. (d)

Given, $n = 100$

Slack bus = 1, PV Bus = 15,

PQ bus = 100 – 16 = 84

Size of Jacobian matrix

$$= (2n - 2 - x) \times (2n - 2 - x)$$

Here, x = Number of PV buses

$\therefore$ Size of $[J] = (2 \times 100 - 2 - 15) \times (2 \times 100 - 2 - 15)$

$= 183 \times 183$

Hence, number of variables to be solved by NR method in polar coordinates = 183.

228. (b)

In rectangular co-ordinates, PV buses will be absent for finding size of the Jacobian matrix.

Hence, size of $[J]$ will be

$$(2n - 2) \times (2n - 2) = (200 - 2) \times (200 - 2)$$
$$= 198 \times 198$$

Hence, 198 number of variables are required to be solved.

229. (c)

In Fast decoupled load flow method, the Jacobian matrix is a constant and need to be computed only once. Due to this reason FDLF gives faster result compared to GS and NR method.

230. (b)

Active power will flow from generator to infinite bus. Reactive power (Q) will flow from infinite bus to generator. Hence, generator supplies P and takes $+Q$ therefore, generator will act as a shunt coil.

233. (c)

At bus –1,

$$S_{G1} + S_{31} = S_{D1} + S_{12}$$

or,
$$S_{G1} = S_{D1} + S_{12} - S_{31}$$
$$= (2 + j2) \text{ MVA}$$

Now,
$$S_{21} = -S_{12}^*$$
$$= (-1 + j0.4) \text{ MVA}$$

At bus-2, we have

$$S_{G2} = S_{D2} + S_{21} + S_{23}$$

or,
$$S_{23} = S_{G2} - S_{D2} - S_{21}$$
$$= -j1.4 \text{ MVA}$$

Also,
$$S_{32} = -S_{23}^* = -j1.4 \text{ MVA}$$

At bus-3,
$$S_{G3} = S_{31} + S_{32}$$
$$= (1 + j2.2) \text{ MVA}$$

234. (d)

Let, $V_2 = 1\angle 0°$ pu and $V_1 = 1\angle\delta$ pu

Then,
$$P_1 = P_2 = \frac{|V_1||V_2|}{X}\sin\delta$$

or,
$$10 = \frac{1 \times 1}{0.05}\sin\delta \quad \text{or} \quad \delta = 30°$$

The reactive power is given by

$$Q_1 = \frac{|V_1|^2}{X} - \frac{|V_1||V_2|}{X}\cos\delta$$

$$= \frac{1}{0.05} - \frac{1}{0.05}\cos 30°$$

or, $Q_1 = 2.68 \text{ pu} = -Q_2$

Hence, $Q_{line} = Q_1 - Q_2 = 2Q_1 = 5.36$ pu

235. (b)

If buses with fixed capacitor i.e. takes as PQ bus otherwise PV bus,

$$x = 20 + 25 - 1 = 44$$

Size of Jacobian matrix

$$= (2n - 2 - x) \times (2n - 2 - x)$$
$$= (600 - 2 - 44) \times (600 - 2 - 44)$$
$$= 554 \times 554$$

237. (c)

$$Z_{load} = \frac{V_L^2}{S_L^*}$$

or, $$Y_{load} = \frac{S_L^*}{V_L^2} = \left[\frac{5 - j0.5}{(1.02)^2}\right]$$

239. (b)

Bus admittance matrix is a sparse matrix because most of the elements of $[Y]_{Bus}$ are zero.

240. (c)

The diagonal elements of Y_{Bus} are given by

$$Y_{11} = y_{12} + y_{13}$$

$$= \frac{1}{j0.2} + \frac{1}{j0.5} = -j7 \text{ pu}$$

$$Y_{22} = y_{23} + y_{21}$$

$$= \frac{1}{j0.25} + \frac{1}{j0.5} = -j6 \text{ pu}$$

$$Y_{33} = y_{32} + y_{31}$$

$$= \frac{1}{j0.25} + \frac{1}{j0.2} = -j9 \text{ pu}$$

241. (c)

The sending end voltage is

$$V_S = V_R + IZ$$

$$= 1\angle 0° + (1 + j0.6)(j0.06)$$

or, $$V_S = (0.964 + j0.06) \text{ pu}$$

$$= 0.96581\angle 3.5615° \text{ pu}$$

(Here, $|V_s| = \sqrt{0.964^2 + 0.06^2} = 0.9658$)

The average reactive power flow over the line is given by

$$Q_{avg} = \frac{1}{2X}(|V_S|^2 - |V_R|^2) \text{ p.u.}$$

$$= \frac{1}{2 \times 0.06}(0.9658^2 - 1^2)$$

$$= -0.56 \text{ p.u.}$$

242. (b)

The load flow will not converge due to absence of slack bus in the second network.

243. (a)

Slack bus is also called reference bus where $|V|$ and δ are known while P and Q are unknown. In power system, slack bus takes care of losses occuring in the system. Hence, when slack bus is changed, $|V|$ and δ will change. Also, losses in the system will change.

244. (a)

$$Y_{22} = \left(\frac{1}{j0.2} + \frac{1}{j0.2} + \frac{1}{j0.1}\right) + \frac{1}{-j0.2}$$

$$= -j5 - j5 - j10 + j5$$

$$= -j15 \text{ pu}$$

245. (c)

Newton-Raphson method is preferred over GS method. Hence, statement-2 is false.

246. (a)

The phase angle between voltage and current is

$$\phi = 15° - (-45°) = 60°$$

Hence, power factor = $\cos\phi = \cos 60° = 0.5$ lag (As V leads I)

Also, $$\sin\phi = \sin 60° = \frac{\sqrt{3}}{2}$$

Thus, active power,

$$P = VI\cos\phi$$

$$= 10 \times 2 \times 0.5$$

$$= 10 \text{ W}$$

and reactive power,

$$Q = VI\sin\phi = 10 \times 2\frac{\sqrt{3}}{2}$$

$$= 10\sqrt{3} \text{ VAR} = 17.32 \text{ VAR}$$

248. (b)

The reach of a reactance relay is unaffected due to the arc resistance while an impedance and admittance relay under reaches.

250. (a)

High transient over voltages occurs in power system due to formation of capacitance across the breaker poles. These are also called switching over voltages.

253. (c)

Pilot relaying schemes provide high speed protection for transmission lines.

256. (d)

Relay operating current,

I_{op} = Current setting × CT secondary rated current × CT ratio

$$= \frac{50}{100} \times 5 \times \frac{400}{5} = 200 \text{ A}$$

$$\therefore \text{PSM} = \frac{\text{Fault current}}{\text{Relay operating current}} = \frac{2000}{200} = 10$$

258. (c)

Higher value of synchronous reactance of alternators reduces transient currents.

259. (b)

Since primary VA output = Primary VA input

$$\therefore \sqrt{3} \times 132 \times 10^3 \times I = \sqrt{3} \times 11 \times 10^3 \times 600$$

or, $I = 50$ A

$$\therefore \text{CT ratio on HV side} = \frac{50}{\left(\frac{50}{\sqrt{3}}\right)} = \frac{50\sqrt{3}}{5}$$

260. (d)

Coupling Capacitor offers high impedance for power signal and low impedance for carrier signal (frequency range is 50 - 500 kHz).

Hence, staement-1 is false.

Line trap offers low impedance for power signal and high impedance for carrier signal. Hence, statement-2 is also false.

A simple differential relay maloperates due to an external fault due to which we use a percentage differential relay. Hence, statement-3 is also false.

261. (d)

Restriking voltage, $V_r = V_m\left(1 - \cos\frac{t}{\sqrt{LC}}\right)$

which depends upon V which in turn is a function of p.f. of the system. A short line fault is more severe compared to a long line fault due to high value of V_r for it.

263. (a)

Symmetrical breaking current

$$= \frac{\text{Rupturing capacity}}{\sqrt{3} \times \text{operating voltage}}$$

$$= \frac{500 \times 10^6}{\sqrt{3} \times 11 \times 10^3} = 26.24 \text{ kA (rms)}$$

264. (b)

Rated making current

= 2.55 × symmetrical breaking current

= 2.55 × 26.24 kA

= 66.91 kA

265. (c)

Static relays are superior to electro-magnetic relays because the moving parts and the contacts are eliminated. CTs and PTs employed in static relays are of lesser VA rating. Also, static relays are more accurate and have high speed of operation.

267. (d)

For protection of parallel lines we require non-directional relay at the source end and directional relay at the load end. Hence, relays R_1' and R_2' will be directional relays.

269. (d)

All three relays i.e. reactance, Impedance and mho relay maloperates during power swing but, mho relay has least changes of maloperation.

270. (b)

Since reach of the reactance relay is unaffected due to the arcing fault resistance, therefore reactance relay is preferred.

271. (a)

Primary line current is

$$I_P = \frac{30 \times 10^6}{\sqrt{3} \times 33 \times 10^3} = 524.88 \text{ A}$$

and secondary line current is

$$I_S = 3I_P = 3 \times 524.88 = 1574.64 \text{ A}$$

The CT current on primary side is

$$I_1 = 524.88 \times \frac{5}{500} = 5.249 \text{ A}$$

The CT current on secondary side is

$$I_2 = 1574.64 \times \frac{5}{200} \times \sqrt{3} = 6.818 \text{ A}$$

Hence, the relay current at 200% of the rated is

$= 2(I_2 - I_1) = 2 \times (6.818 - 5.249)$

$= 3.3138 \text{ A} \approx 3.31 \text{ A}$

273. (d)

The selectivity of relays on radial feeder is achieved based on the time of operation of the relay. The time of operation of the relays at various location is so adjusted that the relay farthest from the source will have minimum time of operation and as it approaches towards the source the time of operation is increased. Hence, CB5 must trip before CB1, CB2, CB3 and CB4 trips.

274. (a)

Current rating of relay = 5 A
Plug setting current for 150% of current setting

$$= \frac{150}{100} \times 5 = 7.5 \text{ A}$$

Now, secondary line current = CT ratio × Plug setting current

$$= \frac{300}{5} \times 7.5 = 450 \text{ A}$$

275. (d)

CT connection:

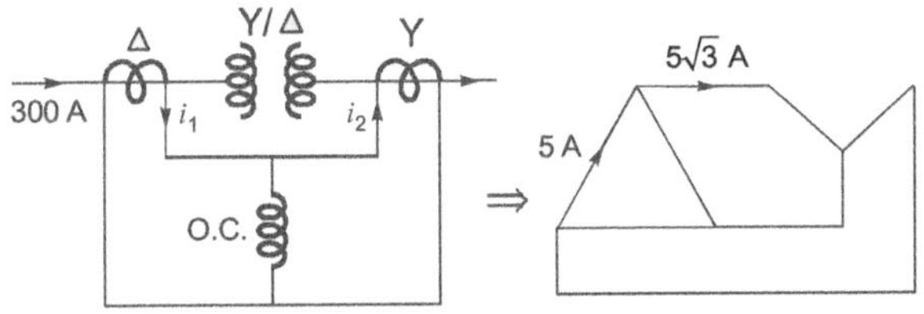

$$\frac{V_2}{V_1} = \frac{I_1}{I_2}$$

$$\frac{66}{33} = \frac{300}{I_2}$$

$$I_2 = 150 \text{ A}$$

CT ratio on H.V. side

$$= \frac{I_2}{i_2} = \frac{150}{5\sqrt{3}} = \frac{150\sqrt{3}}{5 \times 3} = \frac{50\sqrt{3}}{5}$$

276. (d)

Ratio relay is also called distance relay or impedance relay.

278. (a)

$$\text{Pickup value of relay} = \frac{1000 \times 0.75}{5} = 150 \text{ A}$$

Let the % of winding unprotected be x%.
Voltage induced in this section

$$= \frac{11 \times 10^3}{\sqrt{3}} \times \frac{x}{100} \text{ volt}$$

Given, earthing resistance = $r = 5\ \Omega$

$$\text{Now,} \quad 150 = \frac{\left(\frac{11 \times 10^3}{\sqrt{3}} \times \frac{x}{100}\right)}{5}$$

= pickup value of relay

$$\text{or,} \quad x = 11.8\%$$

279. (a)

For a short line fault, L and C values will be lower so that the restriking voltage will be more. Hence, a fault nearer to the source is more severe from the view point of arc interruption compared to a fault far from the source.
Restriking voltage,

$$V_r = V_m\left(1 - \cos\frac{t}{\sqrt{LC}}\right)$$

Also, natural frequency of oscillation $\omega = \frac{1}{\sqrt{LC}}$ will be higher if L and C value are low (i.e. if fault occurs near the source).

280. (b)

Voltage apearing across the breaker pole
= prospective voltage

$$= i_a \cdot \sqrt{\frac{L}{C}} = 4\sqrt{\frac{4}{0.04 \times 10^{-6}}} = 40 \text{ kV}$$

281. (c)

Required resistance is,

$$R = 0.5\sqrt{\frac{L}{C}} = 0.5\sqrt{\frac{4}{0.04 \times 10^{-6}}} = 5 \text{ k}\Omega$$

284. (d)

We know that,

$$G_e H_e = G_1 H_1 + G_2 H_2$$

= Total K.E. stored

$$\text{or,} \quad H_e = \frac{G_1 H_1 + G_2 H_2}{G_e}$$

Here, $G_e = 100$ MVA

$$\text{Hence,} \quad H_e = H = \frac{4.6 \times 500 + 3 \times 1500}{100}$$

= 68 MJ/MVA

286. (a)

Fast field excitation system is one of the method to improve transient stability of the system.

287. (c)

Shunt reactors are used to reduce the effect of Ferranti effect when a medium or a long line is lightly loaded or is on no-load. It doesn't affect the steady state power transfer limit.

289. (c)

$$H = \frac{\text{K.E. stored}}{\text{Rating of machine } (G)}$$

or, K.E. stored $= GH = 50 \times 9 = 450$ MJ

290. (b)

Bundled conductors increases self GMD due to which the series inductive reactance is reduced as a result of which power transferred from sending end to the receiving end is increased. Hence, it improves the steady state stability of the system.

291. (d)

Since inertia constant,

$$H \propto \frac{1}{\text{Rating of machine}}$$

$$\therefore \quad \frac{H_2}{H_1} = \frac{G_2}{G_1}$$

or, $$H_2 = 5 \times \frac{1500}{50} = 150 \text{ MJ/MVA}$$

292. (b)

From swing equation,

$$\delta'' = \frac{P_a}{M} = \frac{X_{pu} \times S}{\left(\frac{GH}{180f}\right)}$$

$$= \frac{180 \times 50 \times X \times 100}{5 \times 100} = 1800\,X \text{ ele-deg/s}^2$$

Also, inertia constant,

$$M = \frac{GH}{180f} \text{ MJ-s/ele-deg}$$

$$= \frac{100 \times 5}{180 \times 50} = \frac{1}{18}$$

$= 0.056$ MJ-s/ele-deg

293. (d)

t_{cr} and δ_{cr} are related to transient stability limit. The fault in a power system must be cleared before δ reaches δ_{cr} and before t_{cr}.

295. (b)

Accelerating power,

$$P_a = 25 - 22.5 = 2.5 \text{ MW}$$

Inertia constant,

$$H = \frac{\text{K.E. stored}}{\text{Rating of machine}}$$

$$= \frac{200}{50} = 4 \text{ MJ/MVA}$$

Also, $$M = \frac{GH}{\pi f} = \frac{50 \times 4}{\pi \times 60} = 1.06 \text{ MJ-s/rad}$$

From swing equation,

$$P_a = M \cdot \frac{d^2\delta}{dt^2}$$

or, $$\frac{d^2\delta}{dt^2} = \frac{P_a}{M} = \frac{2.5}{1.06} = 2.356 \text{ rad/s}^2$$

Hence, acceleration of rotor = 2.356 ele-rad/s^2

296. (d)

$$P_m = \frac{|V_S||V_R|}{X} = P \text{ pu}$$

When the line is compensated by 30% then, new transfer reactance will be

$$X' = 0.7\,X$$

Hence, the new value of maximum power transferred will be

$$P' = \frac{|V_S||V_R|}{0.7X} = 1.42\,P \text{ pu}$$

297. (a)

Fast acting circuit breakers clear the fault in the system before critical celaring time (t_{cr}) and hence improves transient stability of the system.

299. (b)

If the breaker doesn't opens the faulted line then, the fault will be cleared after critical clearing time (t_{cr}).

300. (d)

The total K.E. stored by the machines

$= 250 \times 1.6 + 500 \times 1$

$= 900$ MVA pu

Hence, the pu inertia constant H on a base of 100 MVA is

$$H = \frac{\text{Total K.E. stored}}{\text{Base MVA}} = \frac{900}{100} = 9 \text{ pu}$$

301. (a)

The transfer reactance is

$X_{eq} = 0.2 + (0.4 || 0.4) + 0.15 = 0.55$ pu

Hence, the maximum power transferred before a fault occurs at Q is

$$P_{max} = \frac{1.2 \times 1}{0.55} = 2.18 \text{ pu}$$

302. (b)

The stored kinetic energy is

$$GH = \frac{1}{2} J\omega_m^2$$

$$= \frac{1}{2} \times (50 \times 10^3)\left(\frac{2\pi \times 3600}{60}\right)^2$$

$$= 3553 \text{ MJ}$$

∴ Angular momentum,

$$M = \frac{GH}{\pi f} = \frac{\text{K.E. stored}}{\pi f}$$

$$= \frac{3553}{180 \times 60} \text{ MJ-s/ele-degree}$$

$$= 0.329 \text{ MJ-s/ele-degree}$$

303. (c)

Given, $|V_R| = 115 \times 10^3$ volt

Hence, $$(P_R)_{max} = \frac{|V_R||V_S|}{|Z|} - \frac{|V_R|^2}{|Z|}\cos\theta_s$$

$$= \frac{|V_S||V_S|}{|Z|} - \frac{|V_R|^2 . R}{|Z_s|^2}$$

$$= \left[\frac{115^2}{\sqrt{4^2 + 7^2}} - \frac{115^2 \times 4}{(4^2 + 7^2)}\right] \text{MW}$$

$$= 115^2 [0.124 - 0.0615]$$

$$= 826.56 \text{ MW}$$

304. (b)

Acelerating power,

$$P_a = 20 \text{ MVA}$$

$$M = \frac{GH}{180 f} = \frac{100 \times 4}{180 \times 60}$$

$$= \frac{1}{27} \text{ MJ-s/ele-degree}$$

From swing equation, we have:

$$P_a = M\frac{d^2\delta}{dt^2} \quad \text{or} \quad \frac{d^2\delta}{dt^2} = \frac{P_a}{M}$$

$$= \frac{20}{\left(\frac{1}{27}\right)} = 540 \text{ ele-deg/s}^2$$

$$= 540°/\text{s}^2$$

305. (d)

Time required for twelve cycles

$$= \frac{12}{160} = 0.2 \text{ sec}$$

Now, $$\delta'' = 76°/\text{s}^2 = 76 \times \frac{1}{360} \text{ rev/s}^2$$

$$= \frac{76}{360} \text{ rps/sec}$$

$$= 76 \times \frac{60}{360} \text{ rpm/sec}$$

$$= 12.67 \text{ rpm/sec}$$

Hence, rotor speed at the end of 12 cycles

$$= \left(\frac{120 \times 60}{2}\right) + 12.67 \times 0.2$$

$$= 3602.534 \text{ rpm}$$

306. (a)

Penalty factor of the system is

$$L = \frac{1}{\left(1 - \frac{\partial P_L}{\partial P_i}\right)}$$

when $\frac{\partial P_L}{\delta P_i}$ increases, L will increase.

307. (a)

For economic scheduling of plants, we have:

$$\frac{dF_1}{dP_1} = \frac{dF_2}{dP_2}$$

or, $0.01 P_1 + 2 = 0.012 P_2 + 1.6$

or, $6P_2 - 5P_1 = 200$ ...(1)

Also, $P_1 + P_2 = 180$ ...(2)

On solving equations (1) and (2), we get

$P_1 = 100$ MW

and $P_2 = 80$ MW

308. (b)

We have, $L\frac{dC}{dP} = \lambda$

Here, penalty factor,

$$L = \frac{1}{\left(1-\frac{\partial P_L}{\partial P}\right)} = \frac{1}{\left(1-\frac{2}{6}\right)} = \frac{3}{2}$$

Also, $P_{Loss} = 6 - 4 = 2$ MW

So, $\lambda = L\frac{dC}{dP} = \frac{3}{2}\times 30$

= ₹ 45 per MWhr

309. (c)

Since the load is at plant-2 therefore,

$Na_2 = 0$ and $Na_1 = 1$

310. (c)

The incremental transmission loss of a plant can be positive, a non-zero value but, can't be zero.

311. (a)

To minimize the total fuel cost, the generating units must operate at equal incremental cost of production.

312. (d)

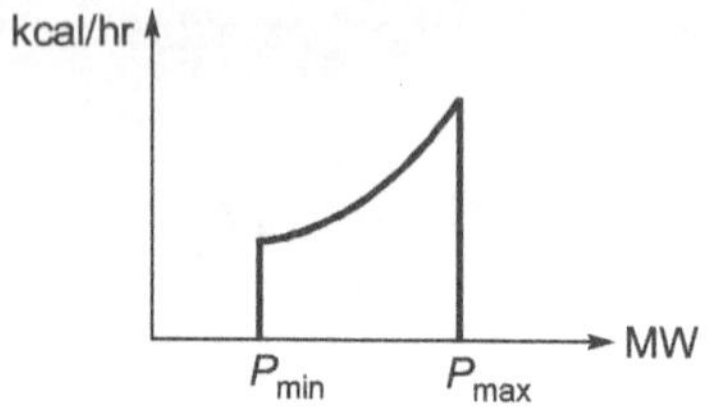

Heat-rate curve

313. (c)

$$\frac{dC_i}{dP_i} = \lambda = \text{Lagangrian multiplier}$$

(C_i= cost of production of ith unit in Rs/hr)

Hence, unit of λ is Rs/MWh.

317. (c)

Penalty factor is, $L_i = \frac{1}{\left(1-\frac{\partial P_{Loss}}{\partial P_i}\right)}$

Hence, it has no effect on plant load.

318. (c)

Since all the load is located at plant-2 therefore, varying P_2 does not affect the transmision loss (P_{loss}).

Hence, $B_{12} = B_{21} = B_{22} = 0$

Thus, $P_{Loss} = 5 = P_1^2 B_{11}$

or, $P_{Loss} = 10^4 B_{11} = 5$

or, $B_{11} = 5\times 10^{-4}$ MW^{-1}

Also, $\frac{\partial P_{Loss}}{\partial P_1} = 2P_1B_{11} = 2\times 100\times 5\times 10^{-4}$

$= 0.1$

Hence, $L_1 = \frac{1}{\left(1-\frac{\partial P_{Loss}}{\partial P_1}\right)} = 1.11$

and $L_2 = \frac{1}{\left(1-\frac{\partial P_{Loss}}{\partial P_2}\right)} = \frac{1}{1-10} = 1$

319. (d)

Since load is at bus-2 therefore,

$B_{12} = B_{21} = B_{22} = 0$

$P_{loss} = B_{11}P_1^2$

or, $10 = B_{11}\times(100)^2$

or, $B_{11} = 0.001$ MW^{-1}

$\therefore$ $\frac{\partial P_{Loss}}{\partial P_1} = 2B_{11}P_1$ and $\frac{\partial P_{Loss}}{\partial P_2} = 0$

$\therefore$ $L_1 = \frac{1}{1-(2\times 0.001\times 100)} = 1.25$

and $L_2 = 1$

Now, for plant-1, we have

$$\frac{dF_1}{dP_1} + \lambda\frac{\partial P_{Loss}}{\partial P_1} = \lambda$$

or, $(0.02P_1 + 16) + 30\,(2\times 0.001\times P_1) = 30$

or, $P_1 = 175$ MW

For plant-2, we have

$$\frac{dF_2}{dP_2} = \lambda$$

or, $0.04\,P_2 + 20 = 30$ or $P_2 = 250$ MW

320. (b)

$P_{Load} = P_1 + P_2 - P_{Loss}$

Now, $P_{Loss} = B_{11}P_1^2 = 0.001\times 175^2$

$= 30.625$ MW

$\therefore$ $P_{Load} = 175 + 250 - 30.625$

$= 394.375$ MW ≈ 394.38 MW

321. (a)

The incremental fuel cost is

$$\frac{dF}{dP} = (20 + 0.2P) \text{ Rs/MWhr}$$
$$= 20 + 0.2 \times 200 = 60 \text{ Rs/MWhr}$$

322. (a)

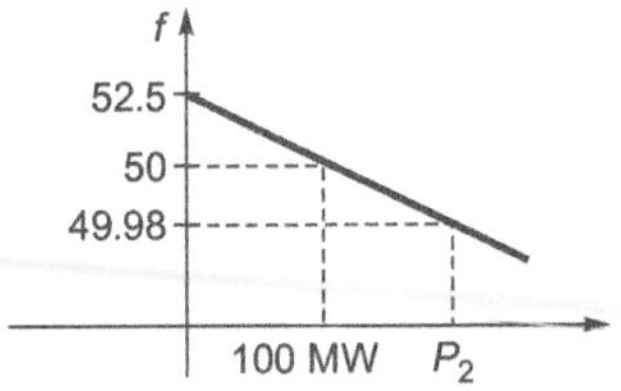

Since slopes are equal therefore,

$$\frac{2.5}{100} = \frac{52.5 - 49.98}{P_2} = \frac{2.52}{P_2}$$

or, $$P_2 = \frac{2.52 \times 100}{2.5} = 100.8 \text{ MW}$$

$\therefore$ $P_2 - P_1$ = increase in turbine output = 0.8 MW

In the given system, a power transmission of 100 MW from generator-2 leads a loss of 7 MW.

323. (c)

Here, $B_{11} = B_{12} = B_{21} = 0$

(since load is connected at plant-1)

and $P_{Loss} = B_{22} P_2^2$

or, $$B_{22} = \frac{P_{Loss}}{P_2^2} = \frac{7}{10^4} = 7 \times 10^{-4} \text{ MW}^{-1}$$

324. (d)

For economic scheduling, we have:

$$L_i \frac{dF_i}{dP_i} = \lambda, \quad \text{where, } L_i = \frac{1}{\left(1 - \frac{\partial P_L}{\partial P_i}\right)}$$

Here, $P_2 = 100$ MW; $\frac{\partial P_L}{\partial P_2} = 0.2$;

$$L_2 = \frac{1}{\left(1 - \frac{\partial P_L}{\partial P_2}\right)}$$

Now, $$L_2 \frac{dF_2}{dP_2} = \lambda$$

or, $$(0.015 \times 100 + 22.5) \times \frac{1}{(1 - 0.2)} = \lambda$$

or, $\lambda = 30$ Rs/MW-hr

$\therefore$ Penalty factor,

$$L_2 = \frac{1}{1 - \frac{\partial P_L}{\partial P_2}} = \frac{1}{1 - 0.2} = 1.25$$

Also, $$L_1 \frac{dF_1}{dP_1} = \lambda$$

or, $L_1 (0.01 \times 100 + 20) = 30$ or $L_1 = 1.428$

325. (b)

We have, $$L_1 = \frac{1}{\left(1 - \frac{\partial P_L}{\partial P_1}\right)} = 1.428$$

or, $$\frac{\partial P_L}{\partial P_1} = 0.3$$

326. (a)

For optimal generation,

$P_3 = 300$ MW (max. load)

Also, $IC_1 = IC_2$

or, $20 + 0.3P_1 = 32 + 0.4\,P_2$ ...(1)

and $P_1 + P_2 + P_3 = 700$

or, $P_1 + P_2 = 400$ MW ...(2)

On solving equations (1) and (2), we get

$P_1 = 242.86$ MW and $P_2 = 157.14$ MW

330. (a)

$$P_{Loss} = B_{11} P_1^2 + 2B_{12} P_1 P_2 + B_{22} P_2^2$$

Hence, it is a function of real-power.

332. (c)

Due to requirement of less number of conductors, a dc line carries more power than a ac line. (losses of the line reduces in HVDC).

333. (c)

HVDC transmission is more reliable than EHV AC because it uses ground or sea return and therefore, in case of fault on a line it can still be used to supply power with the help of healthy lines.

338. (c)

As operating frequency, $f = 0$ therefore, there is no effect of X_L and X_C.

343. (d)

DC systems uses ground or sea return. Hence, statement-2 is false.

UNIT

8

Digital Electronics

1. Basics of Digital Electronics

Q.1 The decimal equivalent of $(101111 \cdot 1101)_2$ is
(a) 52.6789 (b) 38.6254
(c) 47.8125 (d) 51.6328

Q.2 Consider the following statements associated with data representation by 1's and 2's complements:
1. The 2's complement system requires only one arithmetic operation.
2. The 1's complement system requires two arithmetic operations.
3. The 1's complement is often used in logical manipulations for inversion operation.
4. The 2's complement is used only for arithmetic applications.

Which of the statements given above are correct?
(a) 1, 2, 3 and 4 (b) 1, 3 and 4
(c) 2, 3 and 4 (d) 1, 2 and 3

Q.3 Match **List-I** with **List-II** and select the correct answer using the codes given below the lists:

List-I	List-II
A. Error detecting code	1. Excess-3 code
B. Unit distance code	2. Hamming code
C. Reflective code	3. BCD code
D. Weighted code	4. Gray code

Codes:

	A	B	C	D
(a)	3	2	1	4
(b)	4	1	3	2
(c)	1	4	2	3
(d)	2	4	1	3

Q.4 The following two numbers are converted into desired bases x and y respectively.

$$(7 \cdot FD6)_{16} = (7.7726)_x$$
$$(7864)_{10} = (1EB8)_y$$

The values of x and y are respectively
(a) 10 and 16 (b) 8 and 16
(c) 4 and 8 (d) 8 each

Q.5 If $(100110) \times (x)_{10} = (101010110)_2$, then the value of x will be equal to
(a) 16 (b) 5
(c) 14 (d) 9

Q.6 The range of numbers that can be represented using 6-bits in signed binary number representation is
(a) –31 to + 31 (b) –32 to + 31
(c) –32 to + 32 (d) –31 to + 32

Q.7 A decimal number $(22)_{10}$ may be represented by the following ways:
1. $(1\,1\,0\,0\,1)_{GRAY}$ 2. $(1\,0\,1\,1\,0)_2$
3. $(26)_8$ 4. $(16)_{16}$

Which of the above is/are correct representation of $(22)_{10}$?
(a) 1, 2, 3 and 4 (b) 2 and 4
(c) 2, 3 and 4 (d) 3 and 4

Q.8 The 7-bit even parity Hamming code of the binary bits 0101 is
(a) 0 0 1 0 1 0 0 (b) 0 1 0 0 1 0 1
(c) 1 0 1 0 0 1 0 (d) 1 1 0 1 1 0 1

Q.9 If $(2AC9)_{16} = (Z)_7$, then the value of Z is
(a) 43635 (b) 58477
(c) 42587 (d) 38999

Q.10 **Assertion (A):** Digital circuits are often called switching circuits.
Reason (R): Each type of digital circuit obeys a certain set of logic rules.
(a) Both A and R are true and R is the correct explanation of A.
(b) Both A and R are true but R is not the correct explanation of A.
(c) A is true but R is false.
(d) A is false but R is true.

Q.11 Which of the following statements about digital systems are correct?
1. These are easier to design compared to analog systems.

2. These are highly affected by unwanted electrical signals.
3. Accuracy and precision of digital systems are greater when compared with analog systems.
4. The fabrication of digital ICs is complicated and costly than that of analog ICs.

Choose the correct code from the given options:
(a) 3 only (b) 2, 3 and 4
(c) 1 and 3 (d) 1, 2, 3 and 4

Q.12 $(1010.01)_2 - (111.111)_2$ results into
(a) $(0010.011)_2$ (b) $(0101.110)_2$
(c) $(0001.101)_2$ (d) None of these

Q.13 **Assertion (A):** The advantage of performing subtraction by the complement method is reduction in the hardware.
Reason (R): Instead of having separate digital circuits for addition and subtraction, only adding circuits are needed for subtraction by complement method.
(a) Both A and R are true and R is the correct explanation of A.
(b) Both A and R are true but R is not the correct explanation of A.
(c) A is true but R is false.
(d) A is false but R is true.

Q.14 The decimal equivalent of the hexadecimal number $(A0F9.0EB)_{16}$ is
(a) 56423.4321 (b) 58657.0625
(c) 32641.1249 (d) 41209.0572

Q.15 If $100_b = 16_{10}$, then the value of b would be equal to
(a) 6 (b) 4
(c) 3 (d) 8

Q.16 Consider the following series with a missing number:
10000, 121, 100, ?, 24, 22, 20
If in the above series, the same integer is expressed in different number systems, the missing number of the series would be equal to
(a) 49 (b) 13
(c) 31 (d) 25

Q.17 Each of the following arithmetic operations is correct in at least one number system:
(*i*) 1234 + 5432 = 6666 is valid in any number system with base ≥ 6.
(*ii*) $\sqrt{41} = 5$ is valid in a number system with base 6.
(*iii*) $\frac{302}{20} = 12.1$ is valid in a number system with base 5.

Out of the above arithmetic operations:
(a) (*ii*) and (*iii*) are correct
(b) only (*iii*) is correct
(c) only (*ii*) is correct
(d) none is correct

Q.18 Consider the following statements:
1. In a self-complementing code, the code word for the 9's complement of N is the 1's complement of the code word for N.
2. Each one of the 4-bit BCD codes has 6 illegal states.
3. The Gray code cannot be used for mathematical operations.
4. Excess-3 code can be used for mathematical operations.

Which of the statements given above is/are correct?
(a) 1 and 2 (b) 1, 2, 3 and 4
(c) 1, 3 and 4 (d) 2 and 4

Q.19 The ASCII code is a
(a) cyclic code (b) reflective code
(c) weighted code (d) alphanumeric code

Q.20 **Assertion (A):** The 2's complement system has two zeros, a positive 0 (all 0s) and a negative 0 (all 1's).
Reason (R): The complement of the complement gives the original number
(a) Both A and R are true and R is the correct explanation of A.
(b) Both A and R are true but R is not the correct explanation of A.
(c) A is true but R is false.
(d) A is false but R is true.

Q.21 **Assertion (A):** The computer method of division requires successive subtraction.
Reason (R): The computer method of multiplication requires repeated addition.
(a) Both A and R are true and R is the correct explanation of A.

(b) Both A and R are true but R is not the correct explanation of A.
(c) A is true but R is false.
(d) A is false but R is true.

Q.22 **Assertion (A):** Gray code is sometimes called reflective code.
Reason (R): The n least significant bits for 2^n through $(2^{n+1} - 1)$ are the mirror images for 0 through (2^n-1).
(a) Both A and R are true and R is the correct explanation of A.
(b) Both A and R are true but R is not the correct explanation of A.
(c) A is true but R is false.
(d) A is false but R is true.

Q.23 Which of the following 4-bit binary pairs does not forms a Gray code pair?
(a) 0101, 0100 (b) 1110, 1010
(c) 1010, 1001 (d) 1100, 1101

Q.24 Match **List-I (Data representation)** with **List-II (Range with n-bits)** and select the correct answer using the codes given below the lists:

List-I
A. Signed numbers in 1's complement form
B. Signed magnitude
C. Unsigned magnitude
D. Signed numbers in 2's complement form

List-II
1. 0 to $(2^n - 1)$
2. $-(2^{n-1})$ to $+ (2^{n-1} - 1)$
3. $-(2^{n-1})$ to $+ (2^{n-1})$
4. $-(2^{n-1}-1)$ to $+ (2^{n-1} - 1)$

Codes:

	A	B	C	D
(a)	3	3	1	2
(b)	1	4	3	4
(c)	2	4	1	3
(d)	4	4	1	2

Q.25 A number is represented in 2's complement as $(1101)_2$. It's equivalent decimal number is
(a) –3 (b) +5
(c) –2 (d) –5

Q.26 Match **List-I** with **List-II** and select the correct answer using the codes given below the lists:

List-I
A. $(345)_8$
B. $(11011.011)_2$
C. $(47)_{16}$

List-II
1. $(1B{\cdot}6)_{16}$
2. $(229)_{10}$
3. $(107)_8$
4. $(326)_{10}$
5. $(D8.7)_{16}$

Codse:

	A	B	C
(a)	2	5	3
(b)	4	3	2
(c)	2	1	3
(d)	4	1	2

Q.27 Which of the following statements is/are correct in regard to excess-3 code?
(a) It is an unweighted code
(b) It is a self-complementing code
(c) It is a BCD code
(d) All of the above

Q.28 Multiplication of two binary numbers A and B is 11011. If A = 011, then B is
(a) 1001 (b) 1100
(c) 1011 (d) 0101

Q.29 Which of the following statement is not correct in regard to the "GRAY Code"?
(a) It is a cyclic code.
(b) It is a BCD code.
(c) It is a reflective code.
(d) It is a unit distance code.

Q.30 **Assertion (A):** The "XS-3 code," also called "Excess-3 code" is a non-weighted BCD code.
Reason (R): In XS-3 code, each binary code word is the corresponding 8421 code word plus 0011 (3).
(a) Both A and R are true and R is the correct explanation of A.
(b) Both A and R are true but R is not the correct explanation of A.
(c) A is true but R is false.
(d) A is false but R is true.

Q.31 2's complement of the number 1010101 is
(a) 1110011 (b) 0101010
(c) 0101011 (d) 1101010

Q.32 The most negative number that can be represented by 8-bits in 2's complement system is
(a) -128_{10} (b) -64_{10}
(c) -127_{10} (d) -256_{10}

Q.33 The ASCII is an input code which is a
(a) four bit code (b) two bit code
(c) eight bit code (d) seven bit code

Q.34 The number of bits required to represent an eight digit decimal number BCD is
(a) 32 (b) 16
(c) 8 (d) 24

Q.35 An example of self complementing code is
(a) 7421 code (b) Gray code
(c) XS-3 code (d) 8421 code

Q.36 When signed numbers are used in binary arithmetic, then which one of the following notations would have unique representation for zero?
(a) sign magnitude (b) 1's complement
(c) 2's complement (d) 9's complement

Q.37 In the following codes a binary number $Y_1 Y_2 Y_3$ is converted into which code by the circuit given below?

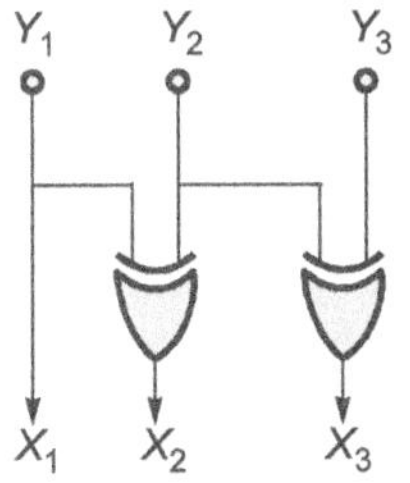

(a) Excess-3 code (b) Gray code
(c) Decimal code (d) BCD code

Q.38 Addition of all gray code to convert decimal (0–9) into gray code is
(a) 129 (b) 108
(c) 69 (d) 53

Q.39 The binary number 110011 is to be converted to gray code. The number of gates and type required are:
(a) 6, AND (b) 6, XNOR
(c) 6, XOR (d) 5, XOR

2. Boolean Algebra & Minimization Techniques

Q.40 Three Boolean operators are
(a) NOR, NAND, NOT
(b) NOT, NAND, OR
(c) NOR, OR, NOT
(d) NOT, OR, AND

Q.41 The only function of a NOT gate is to
(a) stop a signal
(b) act as a universal gate
(c) invert an input signal
(d) none of these

Q.42 Negative logic in a logic circuit is one in which
(a) logic 0 and 1 are represented by the negative and zero voltages respectively.
(b) logic 0 and 1 are represented by zero and positive voltages respectively.
(c) logic 0 voltage level is lower than logic 1 voltage level.
(d) logic 0 voltage level is higher than logic 1 voltage level.

Q.43 Which of the following Boolean expressions is **not** true?
(a) $A \cdot A = A$ (b) $A + \bar{A} = 1$
(c) $A + 1 = A$ (d) $A \cdot \bar{A} = 0$

Q.44 The simplified form of the Boolean expression $\bar{A} + A\bar{B}$ is
(a) $A + \bar{B}$ (b) $\bar{A} + \bar{B}$
(c) $A + B$ (d) $\bar{B} + A$

Q.45 The Boolean expression:

$$Y = AB + (A + B)\,(\bar{A} + B)$$

may be simplified as
(a) $Y = B$ (b) $Y = \bar{A}$
(c) $Y = \bar{B}$ (d) $Y = A$

Q.46 The complement form of $Y = A(B + C)\,(\bar{C} + \bar{D})$ is
(a) $A + BC$ (b) $\bar{A} + \bar{B}\bar{C} + CD$
(c) $A + \bar{B}\bar{C} + C\bar{D}$ (d) $A + BC + CD$

Q.47 The POS form of expression is suitable for circuit using:
(a) NOR (b) AND
(c) XOR (d) NAND

Q.48 Which of the following is/are the correct reason for using Gray code in K-maps?
(a) No other code is available
(b) Gray code is efficient than binary code
(c) Gray code provides cell values that differ in only one bit in adjacent cells
(d) All of the above

Q.49 When eight adjacent 1's are grouped horizontally or vertically the group so found is called
(a) hex (b) quad
(c) pair (d) octet

Q.50 Consider the following statements associated with Boolean algebra:
1. There is nothing like subtraction or division in Boolean algebra.
2. There is possibility of having negative or fractional numbers.
3. A variable has no numerical significance.
4. Logical multiplication is same as the AND operation, and logical addition is the same as the OR operation.
5. There are only three constants within the Boolean system.

Which of the statements given above are correct?
(a) 2, 4 and 5 (b) 1, 3 and 4
(c) 2, 3 and 5 (d) 1, 4 and 5

Q.51 **Assertion (A):** Boolean algebra and Binary number system are different from each other.
Reason (R): There are some basic operations like AND, OR and NOT which are performed in Boolean algebra.
(a) Both A and R are true and R is the correct explanation of A.
(b) Both A and R are true but R is not the correct explanation of A.
(c) A is true but R is false.
(d) A is false but R is true.

Q.52 According to De Morgan's theorem:
1. the NOR gate is equivalent to a bubbled NAND gate.
2. individual variables can be removed form under a NOT sign, and a sum of product from can be transformed to a product-of-sums form.
3. the complement of the product of variables is equal to the sum of their individual complements.
4. the NAND gate is equivalent to a bubbled NOR gate.
5. individual variables can be removed from under a NOT sign, and a product-of-sum form can be transformed to a sum-of-products form.

Which of the above statements about De Morgan's theorem is/are not correct?
(a) 2, 3 and 4 (b) 1 and 5
(c) 1, 4 and 5 (d) 1 and 4

Q.53 The complement of the expression $(\overline{AB}+\bar{A}+AB)$ is equal to
(a) 0 (b) $\bar{A}$
(c) B (d) 1

Q.54 Match **List-I (Boolean expressions)** with **List-II (Minimized expression)** and select the correct answer using the codes given below the lists:

List-I	List-II
A. $AB+A\bar{B}C+B\bar{C}$	1. 0
B. $(\overline{A+\overline{BC}})(A\bar{B}+ABC)$	2. $B+C$
C. $A[B+\bar{C}(\overline{AB+A\bar{C}})]$	3. $AC+B\bar{C}$
D. $A\bar{B}C+B+B\bar{D}+AB\bar{D}+\bar{A}C$	4. AB

Codes:

	A	B	C	D
(a)	3	4	1	2
(b)	1	2	3	4
(c)	3	1	4	2
(d)	2	3	4	1

Q.55 The minimized Boolean expression for the logic diagram shown below is

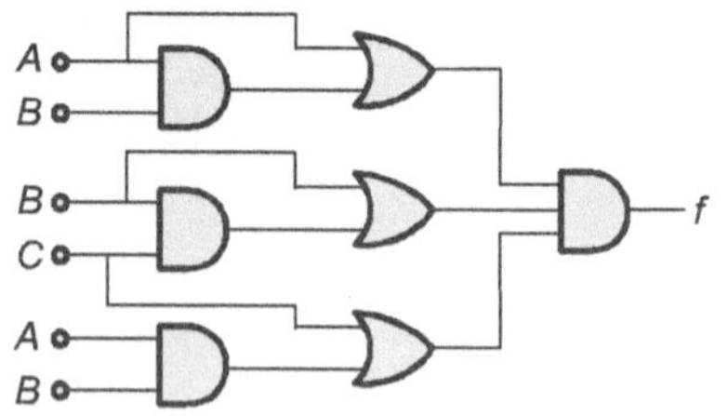

(a) $f = BC$ (b) $f = AB$
(c) $f = A + C$ (d) $f = B + C$

Q.56 **Assertion (A):** The use of K-maps is usually limited to six-variables.
Reason (R): A K-map can not be used for problems involving more than six variables.
(a) Both A and R are true and R is the correct explanation of A.
(b) Both A and R are true but R is not the correct explanation of A.
(c) A is true but R is false.
(d) A is false but R is true.

Q.57 **Assertion (A):** The binary number designations of the rows and columns of the K-map are in Gray code.
Reason (R): The Gray code designation is used in K-map to ensure that two physically adjacent squares are really adjacent i.e. their minterms or maxterms differ by only one variable.
(a) Both A and R are true and R is the correct explanation of A.
(b) Both A and R are true but R is not the correct explanation of A.
(c) A is true but R is false.
(d) A is false but R is true.

Q.58 Consider the following statements:
1. The terms which cannot be combined, further in the tabular method are called "prime implicants."
2. The implicants which will definitely occur in the final expression are called "essential prime implicants."
3. The prime implicant chart is a pictorial representation of the relationships between the prime implicants and the minterms of the expression.

Which of the above statements is/are correct?
(a) 2 and 3 (b) 1 only
(c) 1, 2 and 3 (d) 1 and 2

Q.59 $X_8 X_4 X_2 X_1$ is an 8421 BCD input to a logic circuit whose output is 1 when $X_8 = 0$, $X_4 = 0$ and $X_2 = 1$, or when $X_8 = 0$ and $X_4 = 1$. The simplest output of this logic circuit would be:
(a) $X_1 X_4$ (b) $X_3 \overline{X}_4$
(c) $\overline{X}_2 + X_3$ (d) $X_2 + X_4$

Q.60 The inputs to a computer circuit are 4-bits of the binary number $A_3 A_2 A_1 A_0$. The circuit is required to produce 1 if and only if all of the following conditions hold.
1. The MSB is 1 or any of the other bits are 0.
2. A_2 is 1 or any of the other bits are 0.
3. Any of the 4-bits are 0.

The minimized expression for the output of the computer circuit is
(a) $A_1A_2 + A_0\overline{A}_3 + \overline{A}_0$ (b) $\overline{A}_2\overline{A}_3 + \overline{A}_1 + \overline{A}_0$
(c) $A_0 + \overline{A}_1 + A_2A_3$ (d) $A_0 A_3 + A_1 \overline{A}_2$

Q.61 The minimum number of gate inputs required to implement the minimized output of the expression given by f are:
$$f = \Sigma m(1, 5, 6, 12, 13, 14) + d(2, 4)$$
(a) 9 (b) 7
(c) 10 (d) None of these

Q.62 The minimized expression for the output given by:
$$f = (A+B+C)(\overline{A}+B+\overline{C})(\overline{A}+\overline{B}+\overline{C})(A+\overline{B}+\overline{C})(\overline{A}+\overline{B}+C)$$ is
(a) $AB\overline{C} + \overline{A}BC + \overline{A}B\overline{C}$
(b) $\overline{A}\overline{B}C + A\overline{B}C + \overline{A}B\overline{C}$
(c) $A\overline{B}C + A\overline{B}\overline{C} + \overline{A}B\overline{C}$
(d) $A\overline{B}\overline{C} + \overline{A}\overline{B}C + \overline{A}B\overline{C}$

Q.63 For 4 number of variables, maximum possible self dual expressions are
(a) 65536 (b) 16
(c) 256 (d) None of these

Q.64 For the given Venn diagram, the minimized expression for the shaded area is

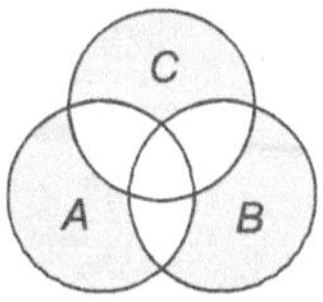

(a) $AB + BC + CA$
(b) $\overline{A}\overline{B}C + AB\overline{C} + A\overline{B}C$
(c) $A\overline{B}\overline{C} + \overline{A}B\overline{C} + A\overline{B}C$
(d) None of these

Q.65 For the digital circuit shown below, the output is true if majority number of inputs are false.

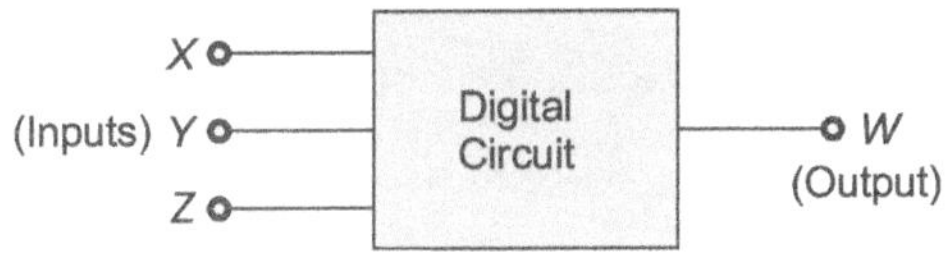

The Boolean expression for the output is given by:
(a) $W = X\overline{Y}Z + \overline{X}YZ + XY\overline{Z} + \overline{X}\overline{Y}Z$
(b) $W = XYZ + X\overline{Y}Z + XY\overline{Z} + \overline{X}YZ$
(c) $W = \overline{X}Y\overline{Z} + X\overline{Y}\overline{Z} + \overline{X}\overline{Y}\overline{Z} + \overline{X}\overline{Y}Z$
(d) $W = \overline{X}\overline{Y}Z + X\overline{Y}\overline{Z} + \overline{X}Y\overline{Z} + \overline{X}YZ$

Q.66 Match **List-I (Logic circuit)** with **List-II (Output)** and select the correct answer using the codes given below the lists:

List-I

A.

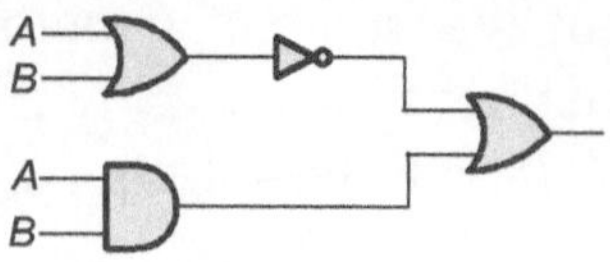

B.

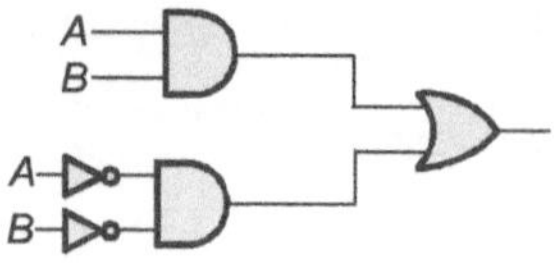

C.

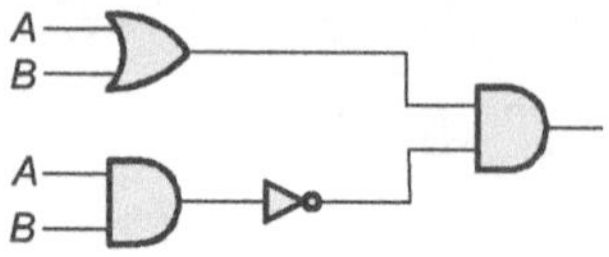

List-II

1. $A\bar{B}+\bar{A}B$
2. $(\bar{A}+B)(A+\bar{B})$
3. AB

Codes:

	A	B	C
(a)	3	3	2
(b)	1	2	1
(c)	2	1	3
(d)	2	2	1

Q.67 Match **List-I (Boolean expression)** with **List-II (Minimized expression)** and select the correct answer using the codes given below the lists:

List-I	List-II
A. $A\cdot(A+B)$	1. $A+B$
B. $AB+BC+\bar{B}C$	2. $AB+C$
C. $\bar{A}B+AB+\bar{A}\bar{B}$	3. A
D. $A+A\bar{B}+\bar{A}B$	4. $\bar{A}+B$

Codes:

	A	B	C	D
(a)	3	2	4	1
(b)	2	4	3	1
(c)	1	2	3	4
(d)	3	4	1	2

Q.68 The minimized output equation of the K-map shown below has

DE \ ABC	000	001	011	010	110	111	101	100
00	0	0	1	1	1	1	1	0
01	0	0	0	0	1	0	1	1
11	1	1	0	0	1	0	0	1
10	0	1	1	1	0	0	0	0

(a) 5 minterms (b) 7 minterms
(c) 6 minterms (d) 9 minterms

Q.69 The simplified expression for the Boolean expression:
$f(A, B, C, D) = \Sigma m(0, 2, 3, 6, 7) + \Sigma d(8, 10, 11, 15)$ is

(a) $\bar{A}C+\bar{B}\bar{D}$ (b) $\bar{A}(C+\bar{D})(\bar{B}+C)$
(c) either (a) or (b) (d) both (a) and (b)

Q.70 The following are some Boolean equations. One of the option which is not true is

(a) $A + 0 = A$ (b) $A + 1 = 1$
(c) $A + A = 1$ (d) $1 + 1 = 0$

Q.71 In the following Boolean expressions given in options which one is not correct?

(a) $A \cdot 0 = 0$
(b) $(A+\bar{A})A = A$
(c) $A + AB + ABC = A$
(d) $A(B + A) = B$

Q.72 The Boolean expression Y given below can be simplified to

$$Y = ABC + A\bar{B} + AB\bar{C} + \bar{A}$$

(a) B (b) C
(c) $\bar{A}$ (d) 1

Q.73 The minimized expression for the given K-map (x : don't care) is

CD \ AB	00	01	11	10
00	0	0	1	1
01	0	x	x	1
11	x	x	1	x
10	1	0	1	1

(a) $A+\bar{B}C$ (b) $B + AC$
(c) $C + AB$ (d) ABC

Q.74 Given the Boolean expression, $\bar{x}+\bar{y}=0$. Which one of the following is TRUE?

(a) $\bar{x}yz = y+\bar{y}$
(b) $\bar{x}\bar{y}\bar{z}+xyz = xy\bar{z}+\bar{x}\bar{y}z$
(c) $\bar{x}y+x\bar{y} = xy+\bar{x}\bar{y}$
(d) $\bar{x}y+x\bar{y}+xz = x\bar{y}+yz$

Q.75 The dual of expression, $f = \overline{\overline{AB} + \bar{A} + AB}$ is

(a) 0 (b) 1
(c) A (d) None of these

Q.76 The max terms of the expression $A(\bar{B} + A)B$ are given by

(a) $\pi\,(0, 1, 3)$ (b) $\pi\,M(1, 2)$
(c) $\pi\,M(0, 3)$ (d) $\pi\,M(0, 1, 2)$

Q.77 The expression, $f = \overline{(A+\bar{B})(C+\bar{D})}$ after demorganization results into

(a) $(A+\bar{B})(C+\bar{D})$ (b) $A\bar{B}+C\bar{D}$
(c) $AB + CD$ (d) None of these

Q.78 The truth table of a circuit is shown below:

Inputs			Output
A	B	C	Z
0	0	0	1
0	0	1	0
0	1	0	1
0	1	1	1
1	0	0	1
1	0	1	1
1	1	0	1
1	1	1	1

The Boolean expression for output Z is

(a) $(\overline{A+B})(\overline{B+\bar{C}})$ (b) $\overline{(\overline{A+B})(\overline{B+\bar{C}})}$
(c) $\overline{(\overline{\bar{A}+B})(\overline{B+C})}$ (d) All of the above

Q.79 A combinational circuit has input A, B and C and its K-map is as shown in figure. The output of the circuit is given by

A \ BC	00	01	11	10
0		1		1
1	1		1	

(a) $(\bar{A}B + A\bar{B})\bar{C}$ (b) $(AB + \bar{A}\bar{B})\bar{C}$
(c) $\bar{A}\bar{B}\bar{C}$ (d) $A \oplus B \oplus C$

3. Logic Gates & Switching Circuits

Q.80 Match **List-I (Logic gates)** with **List-II (Symbols)** and select the correct answer using the codes given below the lists:

List-I **List-II**

A. Positive logic AND gate 1. 

B. Positive logic OR gate 2. 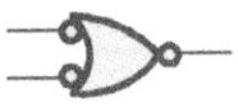

C. Positive logic NAND gate 3. 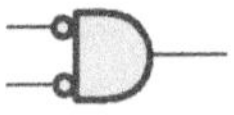

D. Positive logic NOR gate 4. 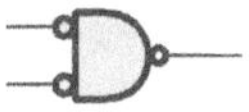

Codes:

	A	B	C	D
(a)	4	3	2	1
(b)	1	4	2	3
(c)	2	4	1	3
(d)	3	2	4	1

Q.81 The minimum number of NAND gate required to implement the logic function given by $f = A + BC\bar{D}$ is

(a) 6 (b) 5
(c) 3 (d) 4

Q.82 **Assertion (A):** When 0s and 1s are interchanged in the truth table, the positive logic AND gate becomes negative logic OR gate, and positive logic NAND gate becomes negative logic NOR gate.

Reason (R): The simple method of converting the logic designation (i.e. from positive to negative logic or vice versa) is that all 0's are replaced with 1s and all 1s with 0s in the truth table.

(a) Both A and R are true and R is the correct explanation of A.
(b) Both A and R are true but R is not the correct explanation of A.
(c) A is true but R is false.
(d) A is false but R is true.

Q.83 Match **List-I (Logic gates)** with **List-II (Applications)** and select the correct answer using the codes given below the lists:

List-I

A. Exclusive OR gate
B. NOT gate
C. Exclusive NOR gate
D. NOR gate

List-II

1. As a controlled inverter
2. As an universal gate
3. Complementation
4. Even-odd parity checker

Codes:

	A	B	C	D
(a)	4	3	1	2
(b)	1	3	4	2
(c)	1	2	4	3
(d)	4	2	1	3

Q.84 **Assertion (A):** The EX-NOR gate can be used as a "one-bit comparator.

Reason (R): The output of the EX-NOR gate is high if the number of high inputs to it is even while the output is low if the number of high inputs to it is odd.

(a) Both A and R are true and R is the correct explanation of A.
(b) Both A and R are true but R is not the correct explanation of A.
(c) A is true but R is false.
(d) A is false but R is true.

Q.85 Consider the following logic circuit:
What is the required input condition (A, B, C) to make the output X = 1, for the above logic circuit?

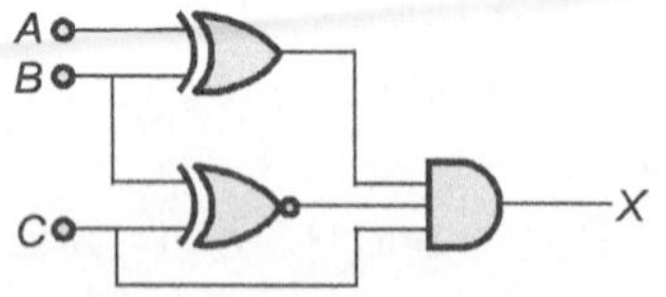

(a) (1, 0, 1) (b) (0, 0, 1)
(c) (1, 1, 1) (d) (0, 1, 1)

Q.86 Match **List-I (Logic expression)** with **List-II (Equivalent expression)** and select the correct answer using the codes given below the lists:

List-I	List-II
A. $\overline{X}$	1. $X \odot 1$
B. $(X+Y)\overline{XY}$	2. $X \odot Y$
C. X	3. $X \oplus 1$
D. $XY+\overline{(X+Y)}$	4. $X \oplus Y$

Codes:

	A	B	C	D
(a)	3	4	1	2
(b)	1	2	3	4
(c)	3	2	1	4
(d)	1	4	3	2

Q.87 If the waveforms A, B and C shown in figure are EX-NORed, the output (Y) waveform will be

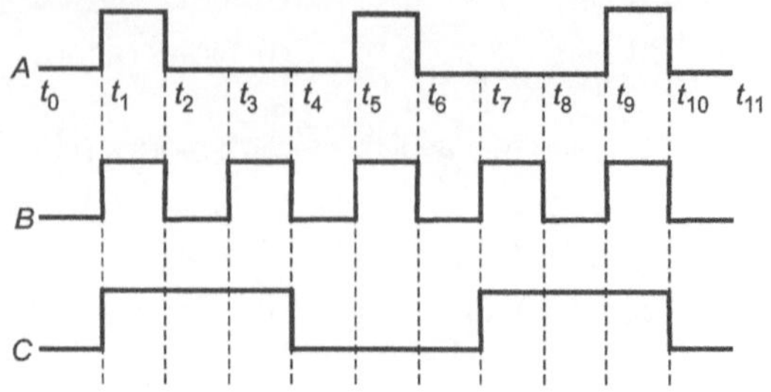

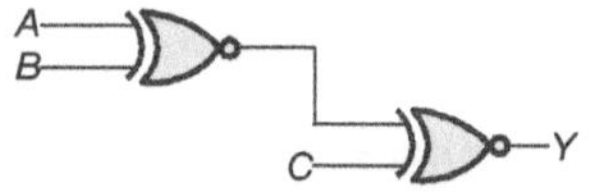

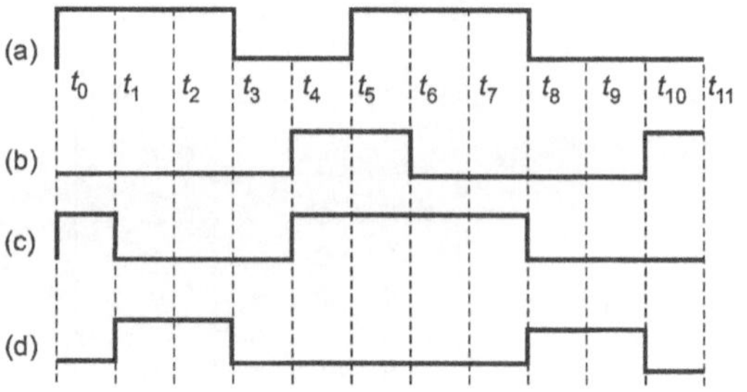

Q.88 Three squares waves X, Y and Z as shown in figure below are to be ANDed. The frequency of ANDed output will be:

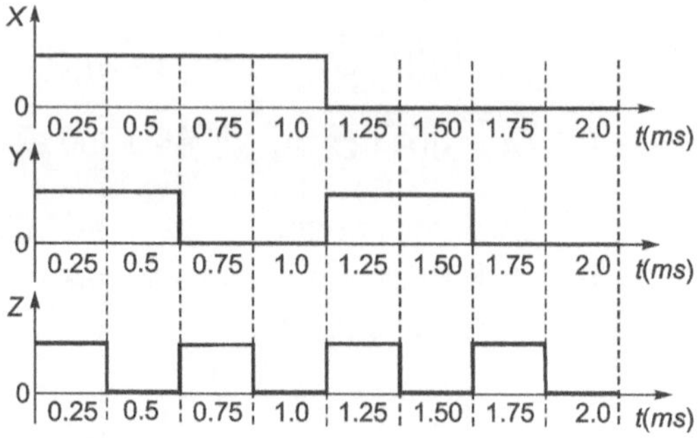

(a) 1.5 KHz (b) 2.0 KHz
(c) 4.0 KHz (d) 0.5 KHz

Q.89 Which of the following option is not correct?

(a) 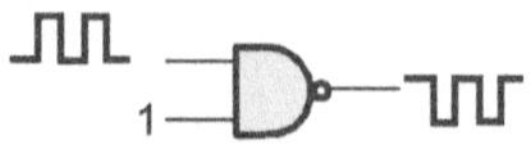

(b) 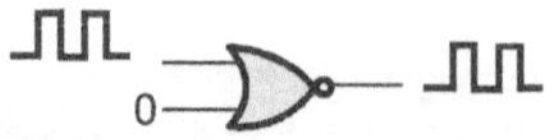

(c) 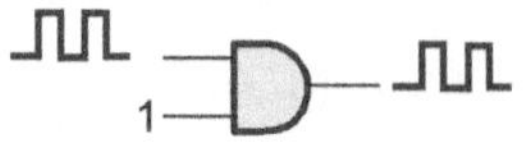

(d) 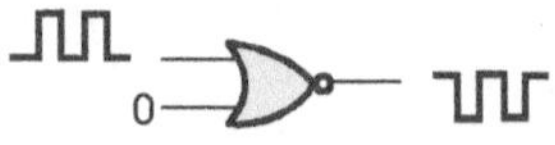

Q.90 Which of the following logic gate follow both commutative and associate law?

(a) AND gate (b) NOR gate
(c) NAND gate (d) None of these

Q.91 For the logic diagram shown below, the output f is

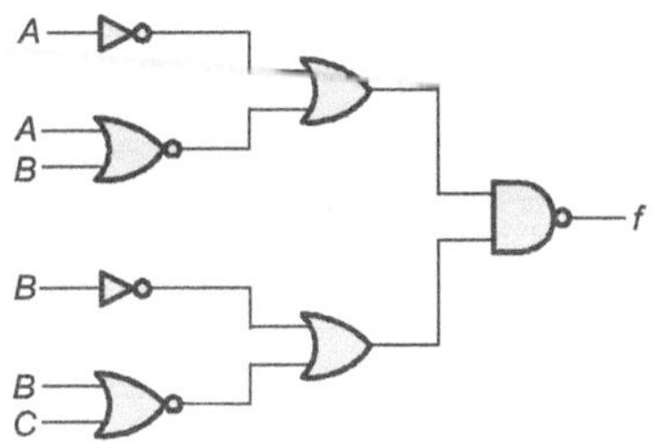

(a) $B + C$ (b) $\bar{B}C$
(c) $A + B$ (d) AB

Q.92 The minimum number of NOR gates requires to implement the expression $f(A, B, C) = \pi M(0, 1, 2, 3, 4, 7)$ is
(a) 5 (b) 4
(c) 7 (d) 6

Q.93 Consider the following statements associated with logic gates:
1. An OR gate in the positive logic system becomes an AND gate in the negative logic system and vice-versa.
2. The dual of an expression is obtained by changing ANDs to ORs, ORs to ANDs,0s to 1s, to 0s and by complementing the variables.
3. NAND and NOR are the only two available universal logic gates.
4. Universal gates do not follow commutative law.
5. On logic diagrams, an inversion bubble at the point where the input is connected to a gate, indicates an active-HIGH input.

Which of the statements given above is/are **not** correct?
(a) 1 and 3 (b) 2 and 5
(c) 1 and 5 (d) 5 only

Q.94 The number of two input NAND gate required to implement an OR gate and an EX-NOR gate are respectively
(a) 2 and 5 (b) 3 and 4
(c) 3 and 5 (d) 2 and 4

Q.95 Three set of simultaneous equations are given by
(i) $\bar{A} + AB = 0$
(ii) $AB = AC$
(iii) $AB + A\bar{C} + CD = \bar{C}D$

The solution of the above three set of simultaneous equations will be
(a) $A = D = 1$ and $B = C = 0$
(b) $B = C = D = 1$ and $A = 0$
(c) $C = D = 1$ and $A = B = 1$
(d) $A = 1$ and $B = C = D = 0$

Q.96 If one of the input to an EX-NOR gate is fixed to logic '1', then output of this gate will be
(a) complement of its input
(b) logic '0'
(c) same as input
(d) logic '1'

Q.97 An AND gate
(a) gives high output only when all inputs are low.
(b) is equivalent to a series switching circuit.
(c) implements logic addition.
(d) is equivalent to a parallel switching circuit.

Q.98 An XOR gate produces output 1 only when two inputs are
(a) equal (b) low
(c) high (d) different

Q.99 Consider the truth table shown below:

Inputs		Output
A	**B**	**f**
1	0	1
0	1	1
1	1	0

The logic gate represented by the above truth table is
(a) OR (b) NOR
(c) EX-OR (d) NAND

Q.100 A logic gate is an electronic circuit which
(a) allows flow of electrons only in one direction
(b) alternates between 0 and 1 values
(c) operates on binary algebra
(d) performs arithmetic and logic functions

Q.101 The output of a logic gate is '0' when both its inputs are different. The gate is
(a) either an EX-OR or NAND gate
(b) either an EX-NOR or NAND gate
(c) either an EX-OR or NOR gate
(d) either an EX-NOR or NOR gate

Q.102 For the diode circuit shown below, both the diodes D_1 and D_2 are ideal.

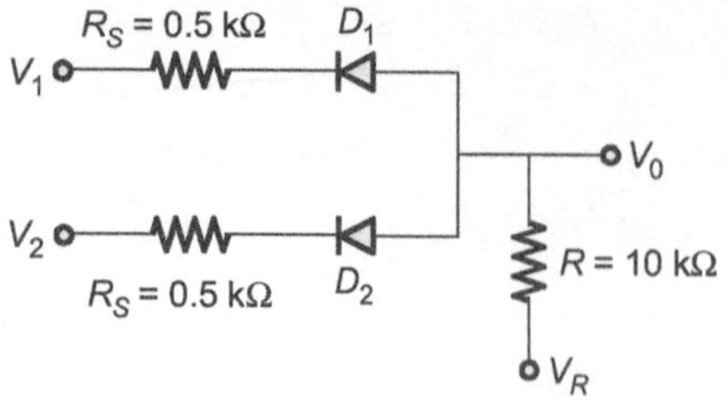

If V_R = +5 V, then match **List-I (Values of V_1 and V_2)** with **List-II (Output voltage V_0)** and select the correct answer using the codes given below the lists:

List-I	List-II
A. V_1 = 0 and V_2 = 5 volt	1. 5 volt
B. V_1 = V_2 = 5 volt	2. 0.12 volt
C. V_1 = V_2 = 0 volt	3. 0.24 volt
	4. 2.5 volt

Codes:

	A	B	C
(a)	3	1	2
(b)	4	1	3
(c)	2	3	4
(d)	3	4	2

Q.103 A circuit has three inputs and one output. The output is 1 if at least two of the three input variables are 1, otherwise it is zero. The minimum number of basic gates required to implement the output (*Y*) are

(a) 3 (b) 4
(c) 7 (d) 8

Q.104 The boolean expression for the output *f* of the digital circuit shown below is

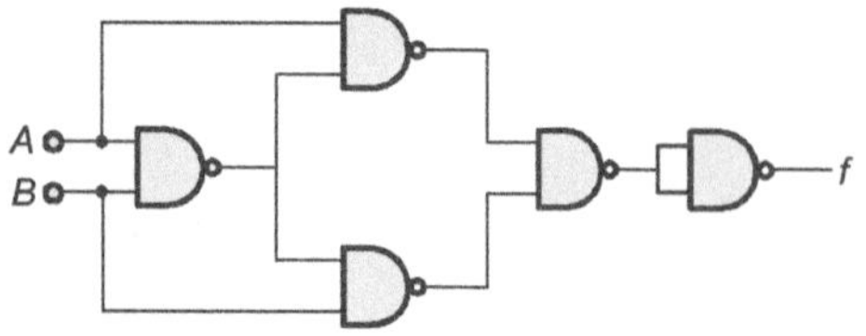

(a) $A\bar{B}+\bar{A}B$ (b) $\overline{AB}$
(c) $\overline{A+B}$ (d) $AB+\bar{A}\bar{B}$

Q.105 In the truth table of an N-input OR gate, in the column for the output of the gate,

(a) the number of zeros is equal to N.
(b) the number of zeros is always 1.
(c) the output is zero if any one of the input is zero.
(d) the number of times a zero occurs is a complex function of N.

Q.106 Consider the following statements associated with logic gates:

1. Logic circuit of any complexity can be realised by using only the three basic gates namely AND, OR and NOT.
2. AND, OR and NOT gates are called universal building blocks.
3. AND/OR/INVERT logic (AOI logic) can be converted to NAND logic or NOR logic.
4. A NAND gate can be used as an inverter by connecting all its input terminals except one, to logic 1 and applying the signal to be inverted to the remaining terminal.

Which of the statements given above is/are correct?

(a) 1, 3 and 4 (b) Only 3
(c) 1 and 3 (d) 1 and 4

Q.107 Assertion (A): When a number of variables are to be X-NORed, a number of two-input X-NOR gates can be used.

Reason (R): Three or more variable X-NOR gates will make the circuit costly.

(a) Both A and R are true and R is the correct explanation of A.
(b) Both A and R are true but R is not the correct explanation of A.
(c) A is true but R is false.
(d) A is false but R is true.

Q.108 If $F = X \odot X \odot X$.........*n* times, then the value of *F* is

(a) *X* when *n* is even (b) Zero when *n* is odd
(c) *X* when n is odd (d) Zero when *n* is even

Q.109 The simplified version of the logic circuit shown below consists of

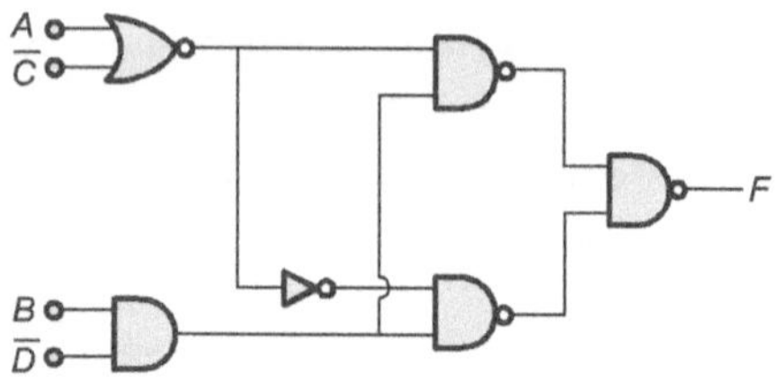

(a) a NAND gate
(b) a AND gate and a NOT gate
(c) an X-OR gate
(d) an OR gate

Q.110 The circuit shown in figure is a ECL OR-AND-INVERTER circuit. The output Z is

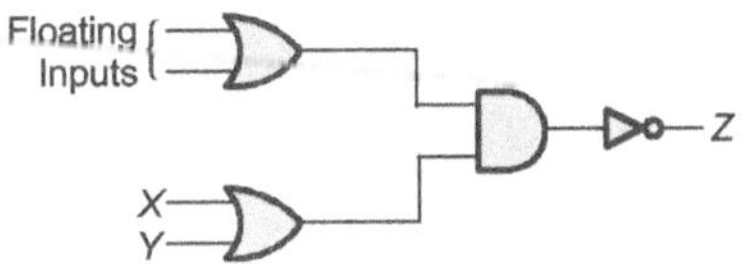

(a) 0 (b) $X+Y$
(c) 1 (d) $\overline{X+Y}$

Q.111 The output of the logic gate shown below is

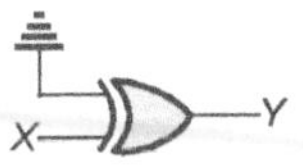

(a) $\overline{X}$ (b) 0
(c) 1 (d) X

Q.112 In the ring oscillator shown below, the propagation delay of each inverter is 100 pico seconds. What is the fundamental frequency of oscillator output?

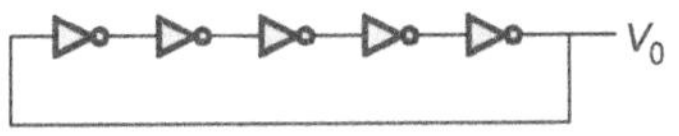

(a) 10 MHz (b) 100 MHz
(c) 1 GHz (d) 2 GHz

Q.113 What is the parity checker output in the below figure if inputs are respectively 1 1 1 1 1 (D_4 D_3 D_2 D_1 D_0) and 1 0 0 0 0 (D_4 D_3 D_2 D_1 D_0)?

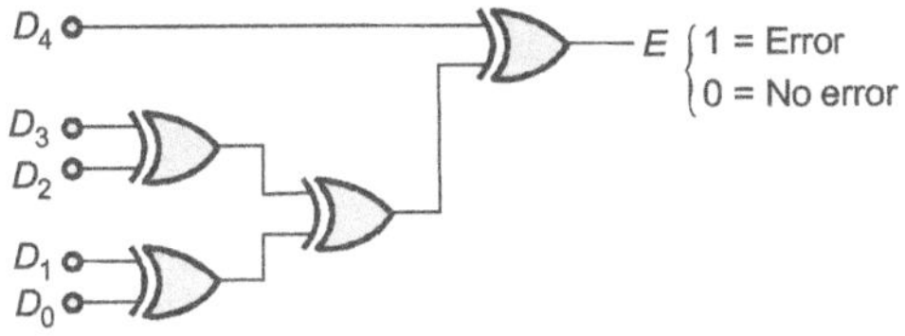

(a) Error, No error (b) No error, Error
(c) No error, Error (d) No error, No error

Q.114 The transistor circuit with inputs A and B and output Y as shown below represents a

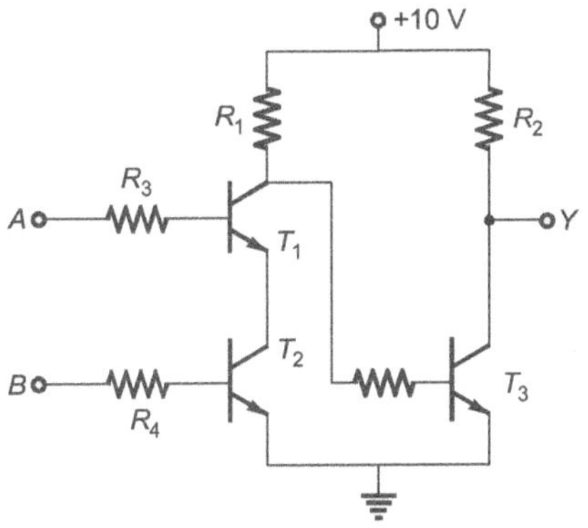

(a) NOR gate (b) NAND gate
(c) EX-OR gate (d) AND gate

4. Combinational Logic Circuits

Q.115 Consider the following statements:

1. When a MUX is used to implement a logic function, the logic variables are applied to the MUX's data inputs.
2. The circuit for a DEMUX is basically the same as that for a decoder.

Of these:
(a) statement-1 is correct while 2 is false.
(b) both the statements are correct.
(c) statement-1 is false while 2 is correct.
(d) both the statements are false.

Q.116 Match **List-I (Combinational Circuits)** with **List-II (Applications)** and select the correct answer using the codes given below the lists:

List-I
A. Decoder
B. Multiplexer
C. Encoder
D. Demultiplexer

List-II
1. Serial to parallel converter
2. Digital to analog converter
3. Parallel to serial converter
4. Analog to digital converter

Codes:

	A	B	C	D
(a)	1	3	2	4
(b)	4	3	2	1
(c)	2	1	4	3
(d)	3	4	1	2

Q.117 The logic function $F(A, B, C) = \Sigma(1, 3, 5, 6)$ can be implemented using:
(a) One 4 × 1 MUX
(b) One 4 × 1 MUX and an OR gate
(c) One 4 × 1 MUX and a AND gate
(d) One 4 × 1 MUX and a NOT gate

Q.118 In a binary adder with two inputs X and Y, the correct set of logical expression for the output S (sum) and C (carry) are respectively
(a) $S = X\overline{Y} + \overline{X}Y$ and $C = XY$
(b) $S = \overline{X}\overline{Y} + XY$ and $C = \overline{X}Y$
(c) $S = X\overline{Y} + \overline{X}Y$ and $C = \overline{X}Y$
(d) $S = \overline{X}\overline{Y} + XY$ and $C = XY$

Q.119 A certain multiplexer can switch one of 32 data inputs to its output. The number of control inputs in this multiplexer is

(a) 32 (b) 16
(c) 5 (d) 8

Q.120 Consider the following statements associated with the use of various adder circuits:

1. A ripple carry adder is a parallel adder in which the carry-out of each full-adder is the carry-in to the next most significant adder.
2. Serial adders are used where speed is more important than circuit minimization.
3. The look-ahead carry adder speeds up the process by eliminating the ripple carry.
4. Serial adders are faster than parallel adders.

Which of the statements given above are correct?

(a) 1, 2 and 4 (b) 2 and 4 only
(c) 1 and 3 only (d) 2, 3 and 4

Q.121 Assertion (A): The multiplexer can be viewed as a function generator.

Reason (R): The multiplexer acts like a digitally controlled multi-position switch.

(a) Both A and R are true and R is the correct explanation of A.
(b) Both A and R are true but R is not the correct explanation of A.
(c) A is true but R is false.
(d) A is false but R is true.

Q.122 Assertion (A): Using a decoder could have advantages over using a multiplexer.

Reason (R): The decoder is more economical in cases where non-trivial, multiple-output expressions of the same input variables are required.

(a) Both A and R are true and R is the correct explanation of A.
(b) Both A and R are true but R is not the correct explanation of A.
(c) A is true but R is false.
(d) A is false but R is true.

Q.123 Assertion (A): The speed of operation of the parallel binary adder is high.

Reason (R): The parallel binary adder is said to generate its output immediately after the inputs are applied.

(a) Both A and R are true and R is the correct explanation of A.
(b) Both A and R are true but R is not the correct explanation of A.
(c) A is true but R is false.
(d) A is false but R is true.

Q.124 A multiplexer has

(a) one data input and number of data outputs
(b) one data output, a number of data inputs, and a number of select inputs
(c) one data output and a number of data inputs
(d) one data output and a number of select inputs

Q.125 Consider the following statements:

1. A decoder with a data input is called a demultiplexer.
2. An octal decoder with active-LOW outputs will output seven LOWs and one HIGH for each combination of inputs.
3. A hexadecimal decoder is sometimes called a 4-line-to-10-line decoder.
4. The Gray code is not a BCD-type code.

Which of the statements given above are not correct?

(a) 1, 3 and 4 (b) 1 and 3 only
(c) 2 and 3 only (d) 2 and 4 only

Q.126 A multiplexer can be used as a

(a) Logic element (b) Counter
(c) Flip-flop (d) 7-segment LED driver

Q.127 The combinational circuit shown below has three data lines D_1, D_2, and D_3 while one select line S.

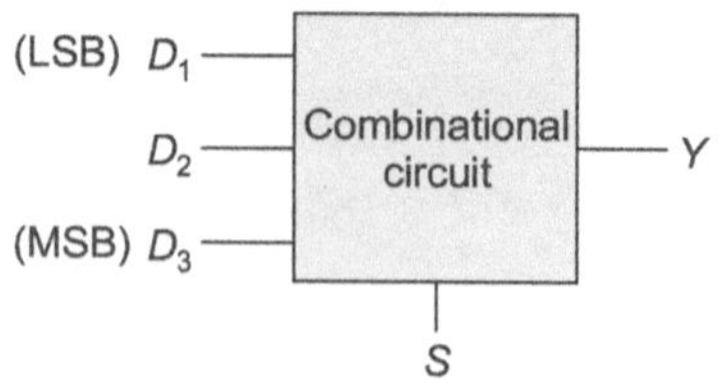

When the control line is high, the circuit is to detect when one of the data lines has '1' on it. No more than one data line will ever have '1' on it. When the control line is low, the circuit will output '0', regardless of what is on the data lines. The output Y of the combinational circuit will be:

(a) $C\bar{D}_1 + CD_2 + D_3$ (b) $\bar{C}(D_1 + D_2) + CD_3$
(c) $CD_1 + CD_2 + \bar{C}D_3$ (d) $C(D_1 + D_2 + D_3)$

Q.128 Figure below shows an automobile alarm circuit used to detect certain undesirable conditions. The three switches are used to indicate the status of the door by the driver's seat, the ignition and the head lights, respectively.

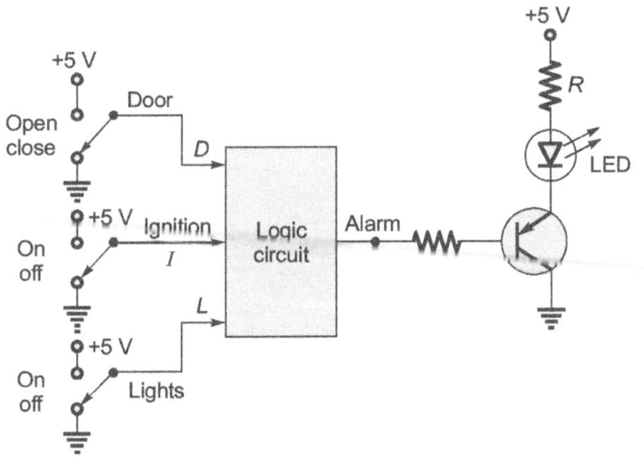

What is the output of the above logic circuit with these three switches (D, I and L) as inputs so that the alarm will be activated whenever either of the following condition exists:
The headlights are ON while the ignition is OFF.
The door is open while the ignition is ON.

(a) $I\bar{L} + D\bar{I}$ (b) $\bar{I}\bar{L} + \bar{D}I$
(c) $\bar{I}\bar{L} + DI$ (d) $IL + DI$

Q.129 A logic circuit takes 4-BCD inputs (A, B, C and D) to give an output F. Output F is '1' if the input is an invalid BCD-code. The number of two input NAND gates required to implement the output Y is

(a) 3 (b) 4
(c) 5 (d) 6

Q.130 A 1-bit full adder takes 15 ns to generate carry out bit while 35 ns for the sum bit. If we design a 4-bit adder using cascade connection of four 1-bit full-adder, then the maximum number of additions that can be performed by this 4-bit adder will be:

(a) 166×10^5/sec (b) 10^7/sec
(c) 125×10^5/sec (d) None of these

Q.131 The number of two-input multiplexers required to implement an EX-NOR and a NAND gate are respectively

(a) 2 each (b) 3 and 2
(c) 2 and 3 (d) 2 and 1

Q.132 The circuit shown below represents a

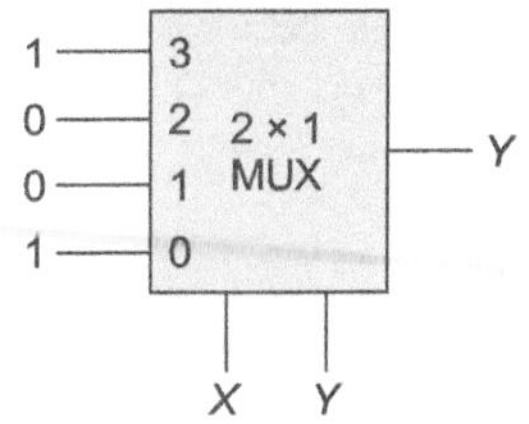

(a) NAND gate (b) exclusive-OR gate
(c) NOR gate (d) exclusive-NOR gate

Q.133 The logic expression $f(A, B, C) = \pi M(0, 3, 5)$ is required to be implemented using a 4×1 MUX as shown below.

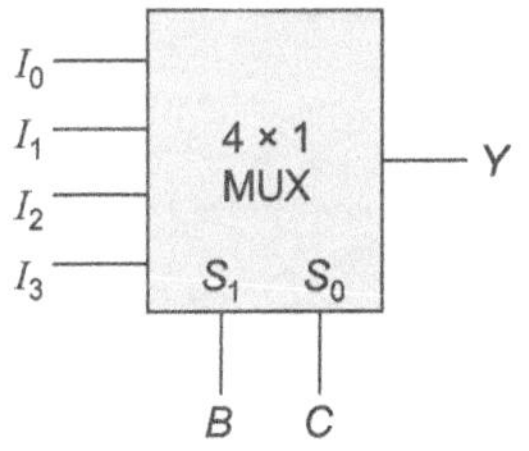

The combinations of the input to this 4×1 MUX circuit would be

(a) $I_0 = A, I_1 = 0, I_2 = 1, I_3 = \bar{A}$
(b) $I_0 = A, I_1 = \bar{A}, I_2 = 1, I_3 = A$
(c) $I_0 = 1, I_1 = A, I_2 = 0, I_3 = \bar{A}$
(d) $I_0 = \bar{A}, I_1 = A, I_2 = 1, I_3 = \bar{A}$

Q.134 A full adder circuit can be implemented using:

1. One 3×8 decoder and two OR gates.
2. One 3×8 decoder, one OR gate and a NOT gate.
3. Two half adders and one OR gate.
4. Two half adders and one NOT gate.
5. Nine NAND/NOR gates.
6. 6 NAND/NOR gates.

Select the correct code from the given options.

(a) 2, 4 and 5 (b) 1, 4 and 6
(c) 1, 3 and 5 (d) 2, 3 and 6

Q.135 To add two 4-bit numbers using parallel adder circuit we require

(a) 4-full adders and 1-OR gate
(b) 8-half adders
(c) either (a) or (b)
(d) neither (a) nor (b)

Q.136 A combinational circuit has three inputs namely A, B and C (A being MSB and C being LSB).

Match **List-I (Combinational Circuit outputs)** with **List-II (Logic expressions)** and select the correct answer using the codes given below the lists:

List-I

A. Full adder carry output
B. Full subtractor borrow output
C. Sum output of full adder or difference output of full subtractor

List-II

1. $ABC+\bar{A}B+\bar{A}\bar{B}C$
2. $AB+A\bar{B}C+\bar{A}BC$
3. $ABC+\bar{A}B\bar{C}+\bar{A}\bar{B}C+A\bar{B}\bar{C}$

Codes:

	A	B	C
(a)	2	1	3
(b)	1	2	3
(c)	1	3	2
(d)	2	3	1

Q.137 The correction to be applied in decimal adder to the generated sum is
(a) 00101 (b) 00110
(c) 01101 (d) 01010

Q.138 The gates required to build a half adder are:
(a) EX-OR gate and NOR gate
(b) Four NAND gate
(c) EX-OR gate and OR gate
(d) EX-OR gate and AND gate

Q.139 The code where all successive numbers differ from their preceding number by single bit is
(a) Binary code (b) Excess-3 code
(c) Gray code (d) BCD code

Q.140 A device which changes from serial data to parallel data is
(a) DEMULTIPLEXER
(b) MULTIPLEXER
(c) COUNTER
(d) FLIP-FLOP

Q.141 A device which converts BCD to Seven Segment is called
(a) Multiplexer (b) Decoder
(c) Demultiplexer (d) Encoder

Q.142 The logic circuit shown below converts a binary code $x_1x_2x_3$ into

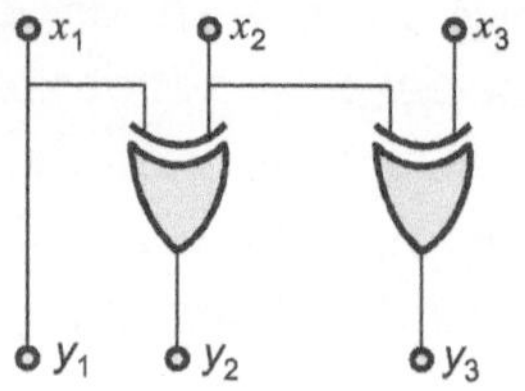

(a) Hamming code (b) BCD code
(c) Excess-3 code (d) Gray code

Q.143 The logic circuit shown below can be minimized to

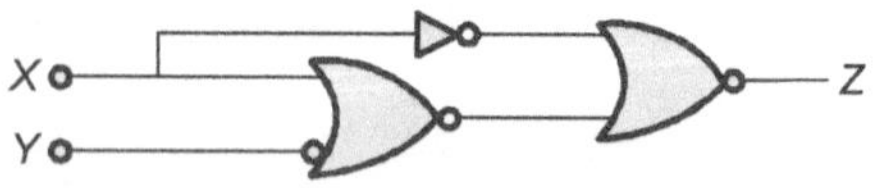

(a) 

(b) 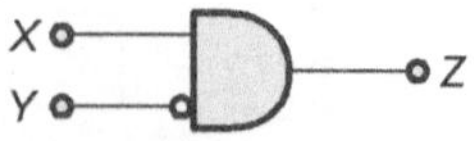

(c) 

(d) 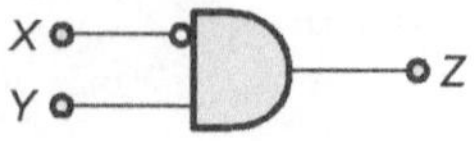

Q.144 When the set of input data to an even parity generator is 0111, the output will be
(a) undefined
(b) 0
(c) 1
(d) depends on the previous input

Q.145 A combinational circuit has 3 inputs *A*, *B*, *C* and output *Y*. *Y* is true for the following input combinations:
A is false, *B* is true
A is false, *C* is true
A, *B*, *C* are False
A, *B*, *C* are true
The simplified expression for output *Y* can be implemented using:
(a) three 2-input NAND gate
(b) one 2-input NAND gate
(c) four 2-input NAND gate
(d) two 2-input NAND gate

Q.146 The logic circuit shown in figure implements.

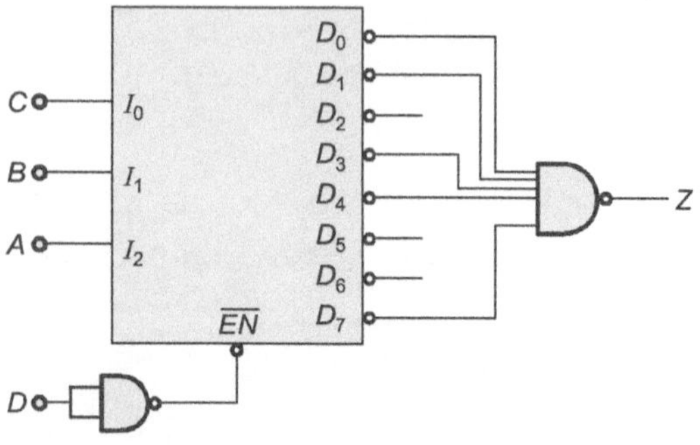

(a) $D(A \odot C+\bar{A}\bar{C})$ (b) $D(B \oplus C+\bar{A}\bar{C})$
(c) $D(B \oplus C+\bar{A}\bar{B})$ (d) $D(B \odot C+\bar{A}\bar{B})$

Q.147 In the TTL circuit In figure, S_2 to S_0 are select lines and X_2 and X_0 are input lines. S_0 and X_0 are LSBs. The output Y is

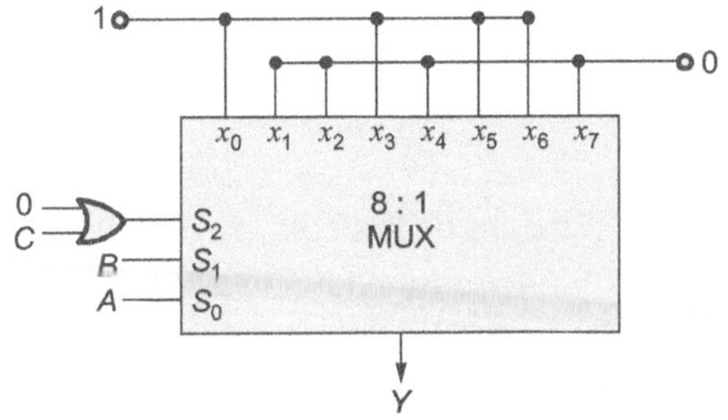

(a) Intermediate (b) $A \oplus B$
(c) $\overline{A \oplus B}$ (d) $C \odot A \oplus B$

Q.148 The network shown in figure implements:

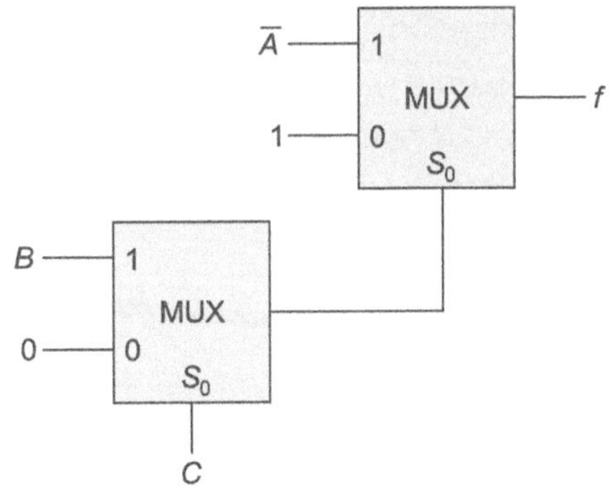

(a) NOR gate (b) NAND gate
(c) XOR gate (d) XNOR gate

Q.149 Match **List-I (Operation)** with **List-II (Associated device)** and select the correct answer using the codes given below the lists:

List-I	List-II
A. Counting	1. ROM
B. Decoding	2. Multiplexer
C. Data selection	3. Demultiplexer
D. Code conversion	4. Register

Codes:

	A	B	C	D
(a)	3	4	2	1
(b)	3	4	1	2
(c)	4	3	1	2
(d)	4	3	2	1

Q.150 Compared to a parallel adder, a serial adder is
(a) faster, and requires more hardware
(b) slower, and requires more hardware
(c) faster, and requires less hardware
(d) slower, and requires less hardware

Q.151 In a 2-bit half adder, the sum and the carry bits are obtained, respectively, by the
(a) NAND and OR gates
(b) AND and NOR gates
(c) NOr and EXCLUSIVE-OR gates
(d) EXCLUSIVE-OR and AND gates

Q.152 Given two binary numbers A and B, an adder circuit can be used to perform the binary subtraction A-B by
(a) first obtaining the complement of B and adding it to A
(b) first obtaining the complement of A and adding it to B
(c) first using a subtraction circuit to subtract B from 0, and then adding the result to A
(d) None of the above

Q.153 A 16-to-1 multiplexer can be used to
(a) convert a 4-bit serial input to a parallel output
(b) convert a 4-bit parallel input to a serial output
(c) convert a 16-bit serial input to a parallel output
(d) convert a 16-bit parallel input to a serial output

5. Sequential Logic Circuits

Q.154 For which of the following flip-flops, the output is clearly defined for all combinations of two inputs?
(a) R-S flip-flop (b) J-K flip-flop
(c) D flip-flop (d) None of these

Q.155 What J-K input condition will always set 'Q^+' upon the occurrence of the active clock transition ?
(a) $J = 1, K = 0$ (b) $J = 1, K = 1$
(c) $J = 0, K = 1$ (d) $J = 0, K = 0$

Q.156 In a J-K flip-flip, toggle means
(a) set $Q = 0$ and $\bar{Q} = 1$
(b) set $Q = 1$ and $\bar{Q} = 0$
(c) change the output to the opposite state
(d) no change in the output

Q.157 The output of S-R flip-flop when $S = 1$, $R = 0$ is
(a) low (b) high
(c) no change (d) high impedance

Q.158 An eight stage ripple counter uses a flip-flop with propagation delay of 75 nano-seconds. The pulse width of the strobe is 50 nano-seconds. The frequency of the input signal which can be used for proper operation of the counter is approximately equal to
(a) 500 MHz (b) 2 MHz
(c) 4 MHz (d) 1.5 MHz

Q.159 The output of a *J-K* flip-flop with asynchronous preset and clear inputs if '1'. The output can be changed to '0' with which one of the following conditions?
(a) By applying $J = 1$, $K = 0$ and using the clock
(b) By applying a synchronous preset input
(c) By applying $J = 1$, $K = 1$ and using the clock
(d) By applying $J = 0$, $K = 0$ and using the clock

Q.160 A *J-K* flip-flop can be implemented using *D* flip-flop connected such that
(a) $D = J\bar{Q} + KQ$ (b) $D = \bar{J}Q + K\bar{Q}$
(c) $D = J\bar{Q} + K\bar{Q}$ (d) $D = J\bar{Q} + \bar{K}Q$

Q.161 For the circuit shown below, both the flip-flops are initially cleared. If the input x is driven by a square wave signal, then the value of Q_1Q_0 in consecutive time intervals indicates that the circuit behaves as a

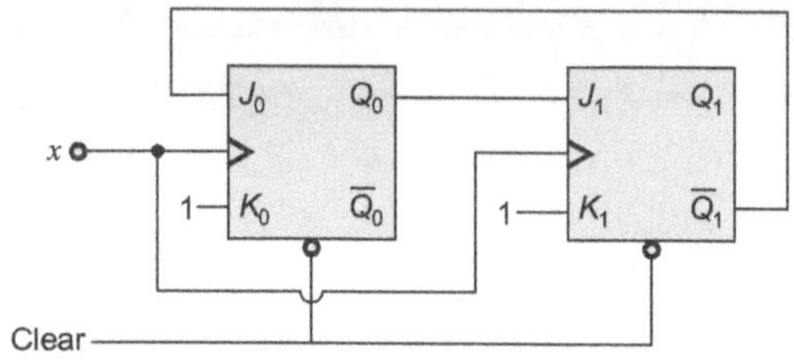

(a) Mod-4 counter (b) Mod-3 counter
(c) Mod-2 counter (d) both (b) and (c)

Q.162 Assertion (A): In general, asynchronous circuits are considerably faster than synchronous circuits.
Reason (R): In an asynchronous circuit, events can occur after one event is completed and there is no need to wait for a clock pulse.
(a) Both A and R are true and R is the correct explanation of A.
(b) Both A and R are true but R is not the correct explanation of A.
(c) A is true but R is false.
(d) A is false but R is true.

Q.163 Consider the following statements:
1. Sequential circuits are always faster than combination circuits.
2. In an asynchronous circuit there is no problem of stability.
3. The logic circuits whose outputs at any instant of time depend only on the input signals present at that time are known as combinational circuits.
4. In a combinational circuit, for a change in the input, the output appears immediately.
5. In a sequential circuits, the output signals are fed back to the input side.

Which of the statement given above are correct?
(a) 2, 3 and 4 (b) 1 and 5 only
(c) 3 and 5 only (d) 1, 2 and 4

Q.164 Assertion (A): A latch is a memory device with the capability of storing one binary digit of information.
Reason (R): A basic latch is made up of cross coupled inverters.
(a) Both A and R are true and R is the correct explanation of A.
(b) Both A and R are true but R is not the correct explanation of A.
(c) A is true but R is false.
(d) A is false but R is true.

Q.165 Assertion (A): The indeterminate condition of the *J-K* flip-flop is permitted in *S-R* flip-flop.
Reason (R): A *J-K* flip-flop has a characteristic similar to that of an *S-R* flip-flop.
(a) Both A and R are true and R is the correct explanation of A.
(b) Both A and R are true but R is not the correct explanation of A.
(c) A is true but R is false.
(d) A is false but R is true.

Q.166 The *J-K* master-slave flip-flops in figure shown below has its J and K inputs tied to $+V_{CC}$, and a series of pulses is applied to its CLK input.

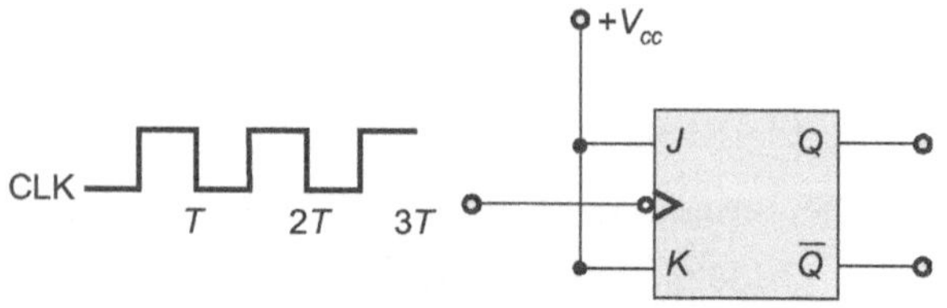

The waveform at Q is given by

(a) 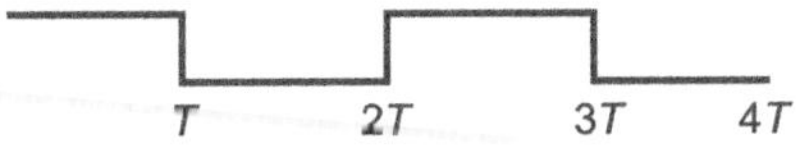

(b) 

(c) 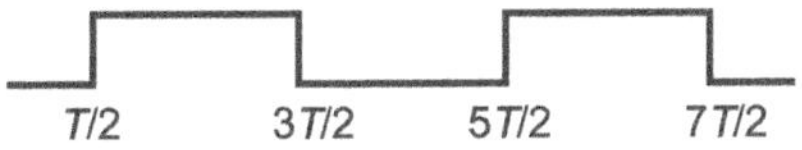

(d) None of the above

Q.167 For the *J-K* flip-flop shown below to function as a divide-by-2 element, the J and K inputs should be equal to

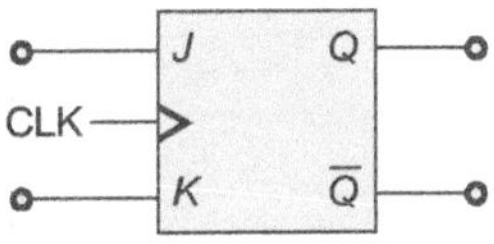

(a) $J = 0, K = 1$ (b) $J = 1, K = 0$
(c) $J = K = 1$ (d) $J = K = 0$

Q.168 Match **List-I** with **List-II** and select the correct answer using the codes given below the lists:

List-I
A. J-K flip-flop
B. *D* flip-flop
C. Master-slave flip-flop
D. *T* flip-flop

List-II
1. Transparent latch
2. Most versatile and widely used of all the flip flops
3. Not widely available as commercial items
4. Pulse triggered flip-flop

Codes:

	A	B	C	D
(a)	2	1	4	3
(b)	2	4	1	3
(c)	3	1	4	2
(d)	4	1	2	3

Q.169 Match **List-I** with **List-II** and select the correct answer using the codes given below the lists:

List-I
A. Propagation delay time
B. Set-up time
C. Hold time

List-II
1. The minimum time for which the control levels need to be maintained constant on the input terminal of flip-flop, in order to enable the flip-flop to respond reliably.
2. The minimum time for which the control signals need to be maintained constant at the input terminal of flip-flop.
3. The time interval between the time of application of the triggering edge and the time at which the output actually makes a transition.

Codes:

	A	B	C
(a)	2	1	3
(b)	1	2	3
(c)	3	1	2
(d)	3	2	1

Q.170 Match **List-I** with **List-II** and select the correct answer using the codes given below the lists:

List-I
A. NOR gate S-R latch
B. NAND gate S-R latch
C. Non-gated latch
D. Clocked and non-edge triggered latch

List-II
1. Asynchronous latch
2. Active-low latch
3. Synchronous latch
4. Active-high latch

Codes:

	A	B	C	D
(a)	2	4	1	3
(b)	4	2	1	3
(c)	4	2	3	1
(d)	1	3	2	4

Q.171 **Assertion (A):** An unclocked flip-flop is called a latch.

Reason (R): A latch is constructed using two cross-coupled NAND gates or NOR gates.

(a) Both A and R are true and R is the correct explanation of A.
(b) Both A and R are true but R is not the correct explanation of A.
(c) A is true but R is false.
(d) A is false but R is true.

Q.172 Consider the following statements associated with flip-flops:
1. Any one type of flip-flop can be converted to any other type by providing a suitable combinational circuit.
2. The inputs to a master-slave flip-flop must change when the clock is HIGH.

3. In a master-slave flip-flop, master is level triggered while slave is edge triggered.
4. Data lock-out flip-flops are nothing but master-slave flip-flops.

Which of the statements given above is/are not correct?

(a) 3 only (b) 1 and 4 only
(c) 1 and 2 only (d) 2 and 3 only

Q.173 For the flip-flop shown below, there is one clock pulse for each bit time.

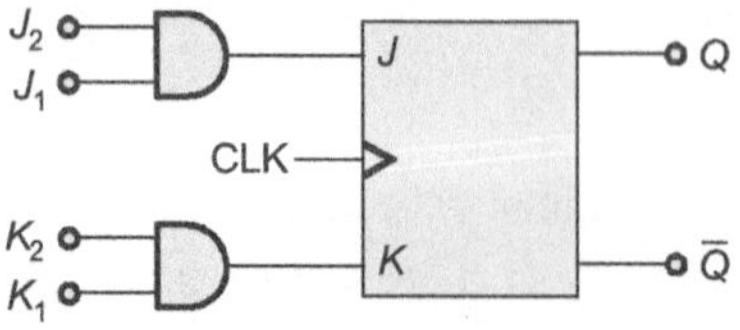

If the following serial data are applied to the flip-flop, then the resulting decimal value of the serial data that appears on the Q output will be (assume that, Q is initially '0')

$J_1 = 01101101$; $J_2 = 10011011$
$K_1 = 01101001$; $K_2 = 11011011$

(a) 14 (b) 24
(c) 18 (d) 20

Q.174 For a flip-flop formed from two NOR gates as shown in figure the unusable state corresponds to

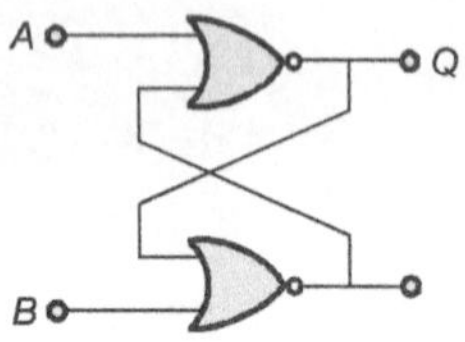

(a) $A = 0, B = 0$ (b) $A = 1, B = 0$
(c) $A = 1, B = 1$ (d) $A = 0, B = 1$

Q.175 The digital circuit as shown below represents to which one of the following?

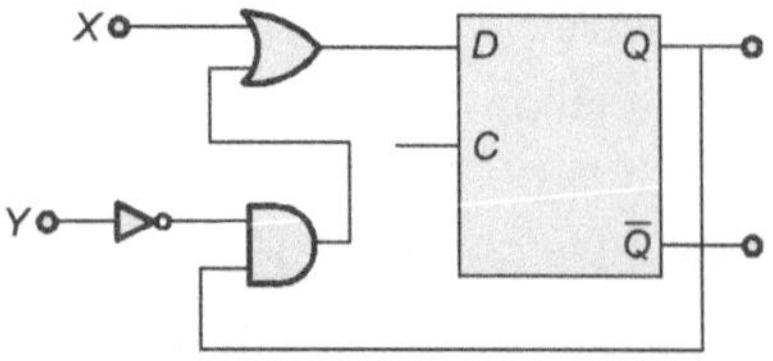

(a) J-K flip-flop (b) S-R flip-flop
(c) T flip-flop (d) Master-slave flip-flop

Q.176 If t_p is the pulse width, Δt is the propagation delay, T is period of pulse train then which of the following condition can avoid the race around condition?

(a) $2\,t_p > \Delta t > T$ (b) $t_p = \Delta t = T$
(c) $2\,t_p < \Delta t < T$ (d) $2t_p < \Delta t > T$

Q.177 Match **List-I (Logic Circuit)** with **List-II (Circuit Realization)** and select the correct answer using the codes given below the lists:

List-I
A. J-K flip-flop
B. T flip-flop
C. D flip-flop

List-II

1. 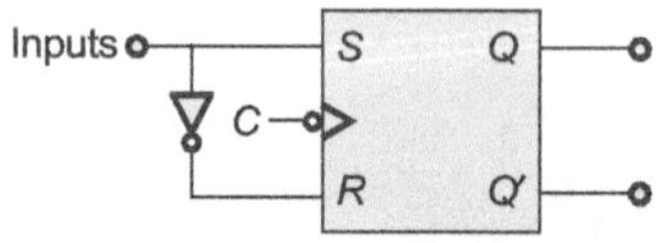

2. 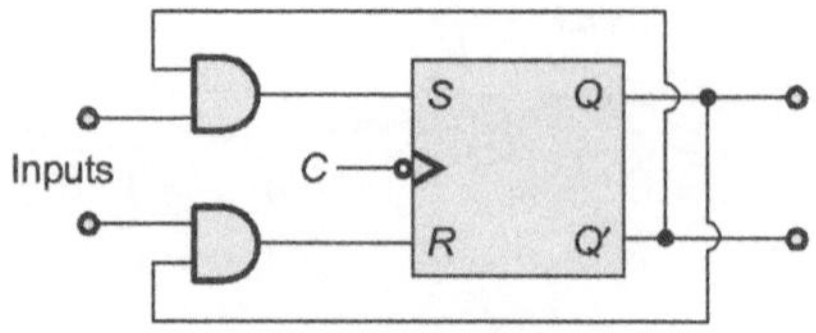

3. 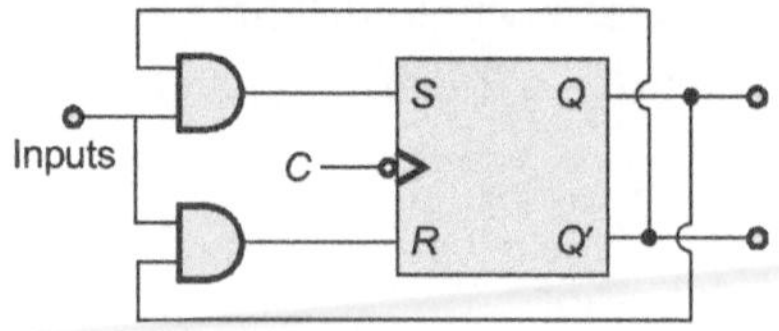

Codes:

	A	B	C
(a)	3	2	1
(b)	1	3	2
(c)	2	1	3
(d)	2	3	1

Q.178 The characteristic equation of a T flip-flop is given by

(a) $Q_{n+1} = T\bar{Q}_n + \bar{T}Q_n$ (b) $Q_{n+1} = T + \bar{Q}_n$
(c) $Q_{n+1} = \overline{T\bar{Q}}_n + TQ_n$ (d) $Q_{n+1} = \overline{TQ_n}$

Q.179 Race around condition always arise in a
(a) combinational circuit
(b) asynchronous circuit
(c) synchronous circuit
(d) digital circuit

Q.180 In a J-K flip-flop we have $J = \bar{Q}$ and $K = 1$ (shown in figure). Assuming the flip-flop was initially cleared and then clocked for 6-pulses, the sequence at the Q output will be

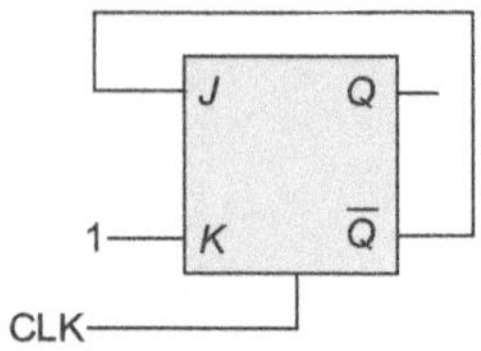

(a) 010000 (b) 110010
(c) 010101 (d) 101010

Q.181 The output of a J-K flip-flop Q_n is '0'. The state of the flip-flop changes when a clock pulse is applied. The possible combination of J_n and K_n inputs could be ('X' denote don't care)
(a) 'X' and '1' respectively
(b) '0' and 'X' respectively
(c) '1' and 'X' respectively
(d) 'X' and '0' respectively

Q.182 When a flip-flop is reset, its output will be
(a) $Q=1; \bar{Q}=0$ (b) $Q=1; \bar{Q}=1$
(c) $Q=0; \bar{Q}=1$ (d) $Q=0; \bar{Q}=0$

Q.183 A *T*-flip-flop can be made from a *J-K* flip-flop by using
(a) an XOR gate (b) a AND gate
(c) an OR gate (d) none of these

Q.184 In a clocked *S-R* flip-flop made with NAND gates, the input combination *S* = 1, *R* = 1, is not permitted because it leads to
(a) a race around condition.
(b) both the outputs being zero simultaneously.
(c) an unpredictable output when the inputs become zero.
(d) the output becoming available before the clock goes low.

Q.185 The *J-K* master-slave flip-flop is different from the *S-R* flip-flop (both made with NAND gates) in that it has
(a) no race around condition
(b) no need of a clock
(c) a valid output when both inputs are 1
(d) a much shorter delay

6. Registers

Q.186 Data can be changed from spacial code to temporal code and vice-versa by using
(a) Counters
(b) Shift registers
(c) A/D converters
(d) Combinational circuits

Q.187 Shifting a register content to left by one bit position is equivalent to
(a) division by two
(b) subtraction by two
(c) multiplication by two
(d) addition by two

Q.188 The content of a 4-bit register is initially 1101. The register is shifted 2 times to the right with the serial input being 1011101.
The content of the register after each shift operations are respectively
(a) 1110 and 0111 (b) 1001 and 1001
(c) 0001 and 1000 (d) 1101 and 1011

Q.189 Match **List-I (Digital Circuits)** with **List-II (Applications)** and select the correct answer using the codes given below the lists:

List-I
A. Dynamic MOS registers
B. Flip-flops
C. Buffer registers
D. Shift registers

List-II
1. Single-bit register
2. Ring counters
3. Pocket calculators
4. Storage of data

Codes:

	A	B	C	D
(a)	4	1	3	2
(b)	3	1	4	2
(c)	3	2	4	1
(d)	4	1	2	3

Q.190 Consider the following statements associated with shift registers:
1. Shift registers are used in digital systems to provide time delays.
2. Dynamic shift registers are made up of MOS inverters.
3. In a bidirectional shift register, data can be fed in serially and data is also shifted out serially.
4. In a shift register, data is said to be in parallel form if the bits are available sequentially.

Which of the statements given above are correct?
(a) 2 and 3 only (b) 1, 3 and 4
(c) 1, 2, 3 and 4 (d) 1 and 2 only

Q.191 Assertion (A): A shift register has a specified sequence of states.

Reason (R): Shift registers are a type of logic circuits closely related to counters.

(a) Both A and R are true and R is the correct explanation of A.
(b) Both A and R are true but R is not the correct explanation of A.
(c) A is true but R is false.
(d) A is false but R is true.

Q.192 Consider the following statements:
1. Shift register is a synchronous device.
2. Level-triggered FFs cannot be used to design a shift register.
3. A shift register is constructed using *D* flip-flops only.
4. In shift registers master-slave FFs are used to avoid timing problems.

Which of the statements given above are correct?
(a) 1, 2, 3 and 4 (b) 2, 3 and 4
(c) 1, 2 and 4 (d) 1, 2 and 3

Q.193 Assertion (A): Parallel-in, parallel-out and serial-in, serial-out shift registers are called " static shift registers".
Reason (R): Each one of the memory elements (i.e. FFs) used to build the parallel-in, parallel-out and serial-in, serial-out shift registers can retain the data bit indefinitely.
(a) Both A and R are true and R is the correct explanation of A.
(b) Both A and R are true but R is not the correct explanation of A.
(c) A is true but R is false.
(d) A is false but R is true.

Q.194 How long will it take to shift an 8-bit number into an 8-bit serial-in parallel-out shift register if the clock is set at 10 MHz?
(a) 100 ns (b) 800 ns
(c) 0 ns (d) 700 ns

Q.195 Assertion (A): Buffer registers do nothing more than storing a binary bit.
Reason (R): Most of the buffer registers use *T* flip-flops.
(a) Both A and R are true and R is the correct explanation of A.
(b) Both A and R are true but R is not the correct explanation of A.
(c) A is true but R is false.
(d) A is false but R is true.

Q.196 For a n-bit shift register match **List-I (Operation)** with **List-II (Number of clock pulse required)** and select the correct answer using the codes given below the lists:

List-I	**List-II**
A. Parallel-in	1. n
B. Serial-out	2. 0
C. Parallel-out	3. $n-1$
D. Serial-in	4. 1

Codes:

	A	B	C	D
(a)	1	2	4	3
(b)	2	1	4	3
(c)	4	3	1	2
(d)	4	3	2	1

Q.197 The flip-flop used in shift registers are generally
(a) *S-R* flip-flop (b) *J-K* flip-flop
(c) *D* flip-flop (d) *T* flip-flop

Q.198 Data can be changed from spacial code to temporal code spacial by using
(a) ADCs and DACs
(b) PISO shift resisters
(c) SIPO shift resisters
(d) Timers

Q.199 The initial contents of the 4-bit serial-in parallel-out right-shift shift register shown in figure is 0110. After three clock pulses applied, the contents of the shift register will be

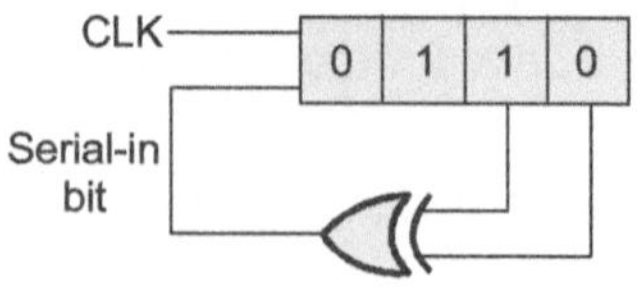

(a) 0000 (b) 0101
(c) 1010 (d) 1111

Q.200 The 8-bit shift-left shift register and *D* flip-flop shown in the figure is synchronized with the same clock. The *D* flip-flop is initially cleared. The circuit acts as

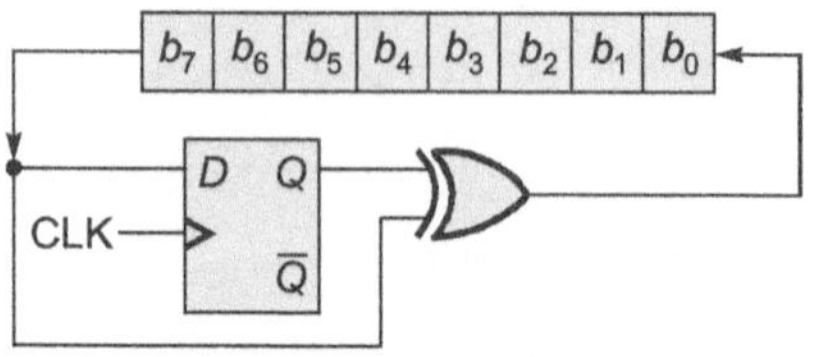

(a) binary to 2's complement converter
(b) binary to gray code converter
(c) binary to 1's complement converter
(d) binary to excess-3 code converter

Q.201 A three bit shift register is shown in the given figure. To have the content '000' again, the number of clock pulses required would be

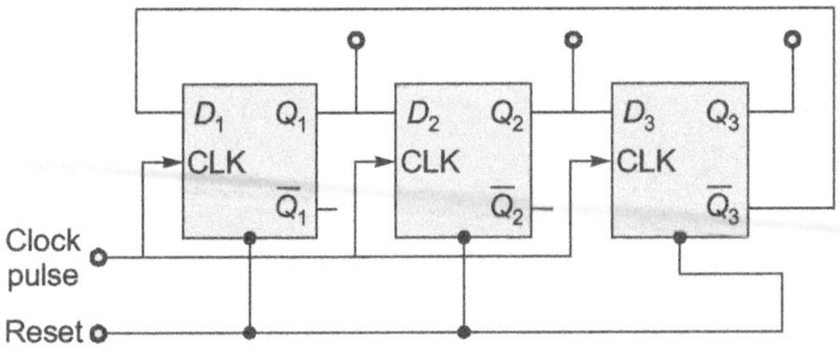

(a) 3 (b) 6
(c) 8 (d) 16

Q.202 The 54/76164 chip is 8-bit serial input-parallel output shift register. The clock is 1 MHz. The time needed to shift a 8-bit binary number into the chip is
(a) 1 μs (b) 2 μs
(c) 8 μs (d) 16 μs

Q.203 Compared to a dynamic MOS shift register, a static shift register
(a) consumes more power and stores data indefinitely.
(b) is faster and consumes more power.
(c) consumes less power and occupies less chip area.
(d) occupies more chip area and saves data for a shorter time.

7. Counters

Q.204 How many flip-flops are required for Mod-16 counter?
(a) 4 (b) 6
(c) 3 (d) 5

Q.205 A ring counter consisting of five flip-flops will have
(a) 32 states (b) infinite states
(c) 5 states (d) 10 states

Q.206 If the input to *T*-flip flop is 100 Hz signal, the final output of the three *T*-flip flops in cascade is
(a) 333 Hz (b) 1000 Hz
(c) 500 Hz (d) 12.5 Hz

Q.207 A 4-bit synchronous counter uses flip-flops with propagation delay times of 20 ns each. The maximum possible time required for change of state will be
(a) 60 ns (b) 20 ns
(c) 30 ns (d) 10 ns

Q.208 How many flip-flops are required to construct a decade counter?
(a) 3 (b) 10
(c) 2 (d) 4

Q.209 If the clock input applied to a cascaded Mod-6 and Mod-4 counter is 48 kHz, the output of the cascaded arrangement would be
(a) 2 kHz (b) 4.8 kHz
(c) 8 kHz (d) 12 kHz

Q.210 Consider the following statements associated with ring counter:
1. Ring counters are constructed by modifying the serial-in, serial-out, shift registers.
2. Ring counter outputs can be used as a sequence of synchronizing pulses.
3. The ring counter is a decimal counter.
4. The keyboard encoder is an example of the application of a shift register used as a ring counter in conjunction with other devices.

Which of the statements given above are correct?
(a) 1, 3 and 4 (b) 1, 2, 3 and 4
(c) 2, 3 and 4 (d) 1, 2 and 4

Q.211 For what minimum value of propagation delay in each FF will a 10-bit ripple counter skip a count when it is clocked at 10 MHz?
(a) 5 ns (b) 100 ns
(c) 10 ns (d) None of these

Q.212 **Assertion (A):** The synchronous counters can operate at much higher frequencies than those that can be used in asynchronous counters.
Reason (R): In a synchronous counter all the FFs change state simultaneously in synchronization with the clock pulse.
(a) Both A and R are true and R is the correct explanation of A.
(b) Both A and R are true but R is not the correct explanation of A.
(c) A is true but R is false.
(d) A is false but R is true.

Q.213 A binary ripple counter is required to count up to $16,383_{10}$. If the clock frequency is 8192 MHz, what is the frequency at the output of the MSB?

(a) 625 kHz (b) 225 kHz
(c) 150 kHz (d) 500 kHz

Q.214 For a counter having n-FFs match **List-I (Counter type)** with **List-II (Number of states)** and select the correct answer using the codes given below the lists:

List-I	List-II
A. Ring counter	1. 2n
B. Johnson counter	2. n
C. Ripple counter	3. 2^n

Codes:

	A	B	C
(a)	3	2	1
(b)	2	1	3
(c)	3	1	2
(d)	2	3	1

Q.215 Assertion (A): A ripple counter suffers from the problem of lock-out.

Reason (R): A counter is said to suffer from the problem of lock-out if it keeps going from one invalid state to another invalid state after successive clock pulses and never returns to a normal state.

(a) Both A and R are true and R is the correct explanation of A.
(b) Both A and R are true but R is not the correct explanation of A.
(c) A is true but R is false.
(d) A is false but R is true.

Q.216 Match **List-I** with **List-II** and select the correct answer using the codes given below the lists:

List-I
A. Ripple counter
B. Ring counter
C. Creeping counter

List-II
1. Twisted ring counter
2. Asynchronous counter
3. Used in SAR type A/D converter

Codes:

	A	B	C
(a)	2	1	3
(b)	1	2	3
(c)	2	3	1
(d)	3	2	1

Q.217 Consider the following statements:

1. A counter is said to be of self-starting type, if it returns to a valid state and counts normally after one or more clock pulses, even if it enters an invalid state.
2. A synchronous counter in which the output of one counter drives the clock input of another counter is called a hybrid counter.
3. 'CLEAR' and 'PRESET' are the two synchronous inputs being used.
4. Decoding errors are present both in synchronous and asynchronous counters.

Which of the above statements is/are not correct?

(a) 1 and 2 (b) 3 only
(c) 2 and 3 (d) 3 and 4

Q.218 A 25 : 1 ripple counter is required. If 4 flip-flops are available on a chip, how many chips are needed?

(a) 25 (b) 2
(c) 1 (d) 3

Q.219 Minimum number of J-K flip-flops needed to construct a BCD counter is

(a) 2 (b) 3
(c) 4 (d) 5

Q.220 The counter shown in the figure is a

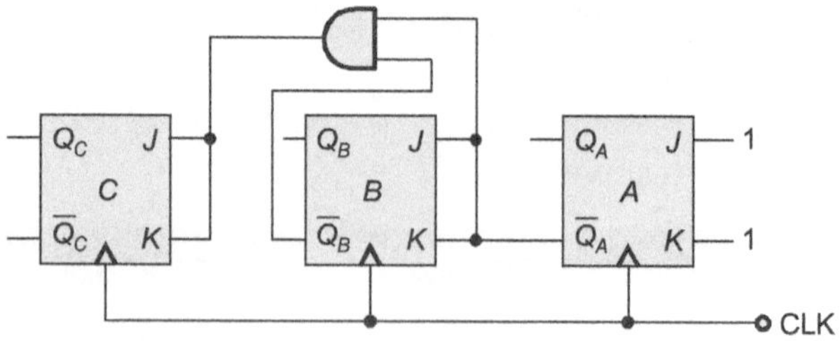

(a) MOD-8 up counter
(b) MOD-8 down counter
(c) MOD-6 up counter
(d) MOD-6 down counter

Q.221 The frequency of the pulse at z in the network shown in figure is

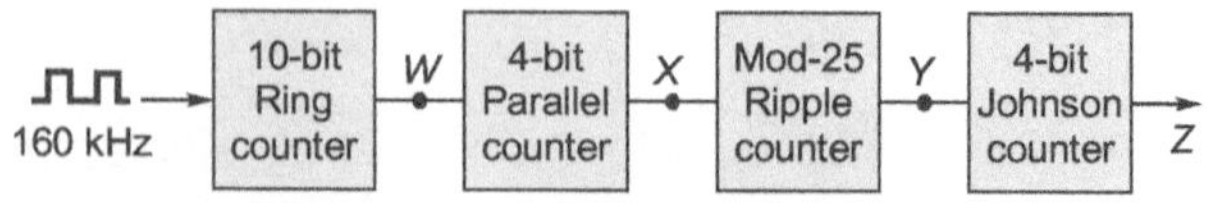

(a) 10 Hz (b) 160 Hz
(c) 40 Hz (d) 5 Hz

Q.222 Match **List-I (Digital circuits)** with **List-II (Type of circuit)** and select the correct answer using the codes given below the lists:

List-I
A. BCD to 7-segment Decoder
B. 4-to-1 multiplexer
C. 4-bit shift register
D. BCD counter

List-II
1. Sequential circuit
2. Combinational circuit
3. Neither sequential nor combinational circuit

Codes:

	A	B	C	D
(a)	2	1	2	1
(b)	3	2	1	3
(c)	2	2	1	1
(d)	2	1	2	3

Q.223 The circuit shown below is a Mod-K counter. The value of K is

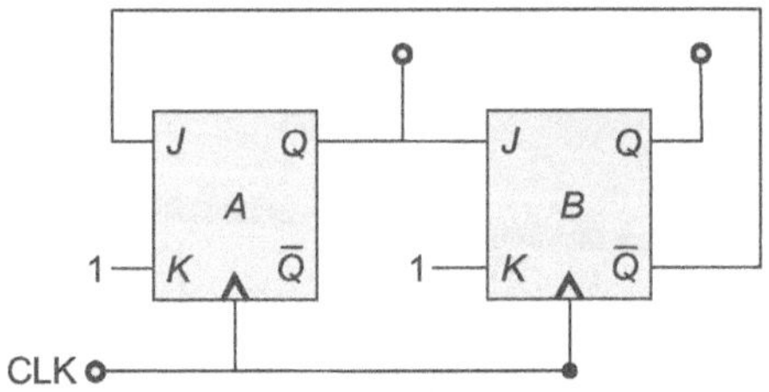

(a) 1 (b) 2
(c) 3 (d) 4

Q.224 In the circuit shown below, the outputs Y_1 and Y_2 for the given initial condition $Y_1 = Y_2 = 1$ and after four-clock pulses will be

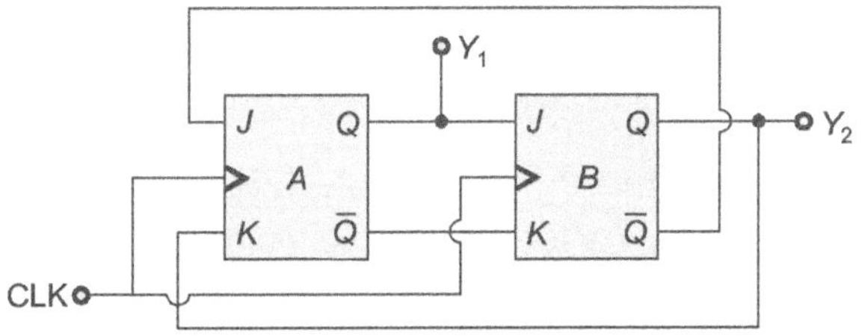

(a) $Y_1 = 1, Y_2 = 0$ (b) $Y_1 = 0, Y_2 = 0$
(c) $Y_1 = 0, Y_2 = 1$ (d) $Y_1 = 1, Y_2 = 1$

Q.225 The decoding error of the counter can be avoided by
(a) increasing propagation delay of flip-flops used in the counter
(b) using very fast logic
(c) using the strobe signal
(d) reducing the propagation delay of flip-flops used in the counter

Q.226 The counter starts off in the '0000' state, and then clock pulses are applied. Some time later the pulses are removed and the counter FFs read '0011'. How many clock pulses have occurred?
(a) 3 (b) 35
(c) 51 (d) Any of these

Q.227 A switch tailoring counter is made by using a single D-flip-flop. The resulting circuit is a
(a) SR flip-flop (b) JK flip-flop
(c) D flip-flop (d) T flip-flop

Q.228 A ripple counter made with J-K flip-flop can be made to count down, rather than up, by
(a) connecting the $\bar{Q}$ output to the input of the next stage, rather than the Q output
(b) making the output transition to occur at the rising, rather than the falling edge of the clock
(c) designing a logic circuit that inverts the output of the counter
(d) finding the two's complement of the output

8. Logic Families

Q.229 In the circuit shown below, let the cut-in voltage of diode be 0.6 V, and be ideal otherwise. If $S = A = 1$ and $B = 0$, the current through diodes D_2 and D_3 are respectively

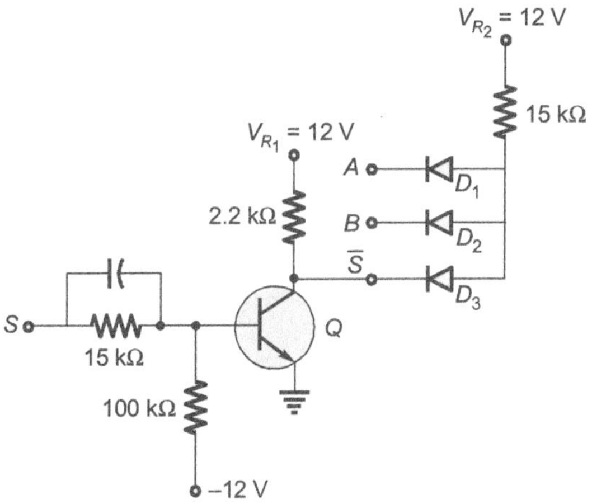

(a) zero and 0.76 mA
(b) zero
(c) 0.76 mA and zero
(d) 0.25 mA and 0.76 mA

Q.230 When an *n-p-n* transistor is used to make an inverter for positive logic, the transistor is forced into saturation when the input is high, because
(a) the large stored charge in the base, in saturation, ensures a fast turn off
(b) the output voltage (V_{CE}) in saturation is very close to zero
(c) the high base current in saturation leads to a smaller turn off time
(d) saturated transistors have a high voltage gain

Q.231 Compared to a passive pull-up (resistor) output stage, the active pull-up(totem-pole) output stage has
(a) longer delays and lower power consumption
(b) shorter delays and lower power consumption
(c) shorter delays and higher power consumption
(d) longer delays and higher power consumption

Q.232 In today's world, the almost exclusive logic family used for making digital circuits is
(a) NMOS (b) RTL
(c) CMOS (d) TTL

Q.233 The logic family which belongs to non-saturated logic is
(a) NMOS (b) TTL
(c) CMOS (d) ECL

Q.234 Emitter-coupled logic (ECL) is the fastest bipolar transistor logic because
(a) it uses current, rather than voltages, as the output variables.
(b) it uses a circuit configuration that prevents the transistors from going into saturation.
(c) it has no *p-n-p* transistors.
(d) it uses differential inputs.

Q.235 The digital logic family which has minimum power dissipation is
(a) RTL (b) TTL
(c) DTL (d) CMOS

Q.236 Which of the following is the fastest logic?
(a) CMOS (b) LSI
(c) ECL (d) TTL

Q.237 The digital logic family which has the lowest propagation delay time among all the logic family members is
(a) CMOS (b) TTL
(c) PMOS (d) None of these

Q.238 In digital ICs, Schottky transistors are preferred over normal transistors because of their
(a) lower power dissipation
(b) lower propagation delay
(c) higher fan out
(d) higher propagation delay

Q.239 Which TTL logic gate is used for wired ANDing?
(a) ECL gates (b) Open collector output
(c) Tristate output (d) Totem pole

Q.240 CMOS circuits are extensively used for ON-chip computers mainly because of their extremely
(a) high noise immunity
(b) low cost
(c) large packing density
(d) low power consumption

Q.241 The circuit shown in the figure represents a

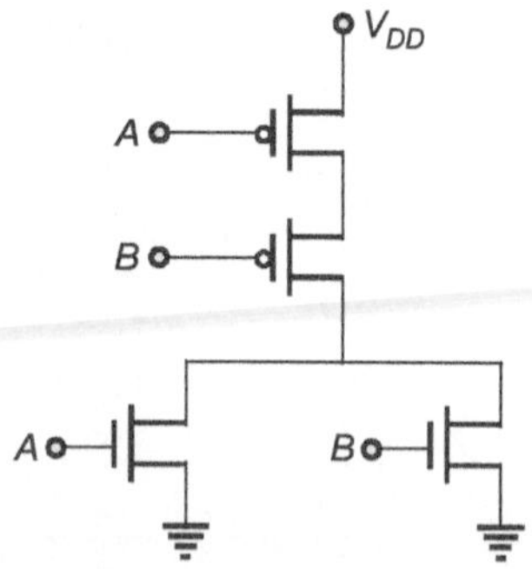

(a) AND gate (b) NOR gate
(c) NAND gate (d) OR gate

Q.242 CMOS logic has the property of
(a) decreased area
(b) high noise margin
(c) low static power dissipation
(d) increased capacitance and delay

Q.243 An emitter coupled logic (ECL)
1. has low noise margin of about 250 mV.
2. need inverters.
3. has higher cost.
4. has higher fanout of about 25.
5. has high output impedance.

Which of the above characteristic is possessed by an emitter-coupled logic?
(a) 1, 3 and 4 (b) 2 and 5
(c) 1, 2, 3 and 5 (d) 2, 3 and 4

Q.244 Assertion (A): By connecting a schottky barrier diode between the base and the collector of each transistor in TTL fan-out is increased.

Reason (R): In schottky TTL the transistors are not allowed to go into saturation.

(a) Both A and R are true and R is the correct explanation of A.
(b) Both A and R are true but R is not the correct explanation of A.
(c) A is true but R is false.
(d) A is false but R is true.

Q.245 Match **List-I (Logic family)** with **List-II (Characteristics)** and select the correct answer using the codes given below the lists:

	List-I		List-II
A.	CMOS	1.	Fastest
B.	ECL	2.	Highest figure of merit
C.	RTL	3.	Highest fan-out
D.	HTL	4.	Lowest noise-margin

Codes:

	A	B	C	D
(a)	3	1	4	2
(b)	3	4	1	2
(c)	4	3	2	1
(d)	4	2	3	1

Q.246 Match **List-I (Logic family)** with **List-II (Basic circuit)** and select the correct answer using the codes given below the lists:

List-I

A. Complementary metal oxide semiconductor logic
B. Transistor-transistor logic
C. Resistor transistor logic
D. Emitter-coupled logic

List-II

1. NOR or OR gate
2. NOR gate
3. NAND gate
4. NOT gate

Codes:

	A	B	C	D
(a)	2	4	3	1
(b)	1	2	4	3
(c)	2	1	3	4
(d)	4	3	2	1

Q.247 Assertion (A): Both NMOS and PMOS are more economical than CMOS.

Reason (R): Both NMOS and PMOS have greater packing density than that of CMOS.

(a) Both A and R are true and R is the correct explanation of A.
(b) Both A and R are true but R is not the correct explanation of A.
(c) A is true but R is false.
(d) A is false but R is true.

Q.248 Consider the following statements:

1. Modern MOSFET circuitry is constructed using PMOS devices.
2. The ECL is mainly used in supercomputers.
3. A tri-state gate utilizes the advantages of high speed of operation of the totem-pole configuration and wire ANDing of the open-collector configuration.
4. HTL are quite useful in the industrial environment where the noise level is usually high due to the presence of motors, high voltage switches, etc.

Which of the statements given above is/are not correct?

(a) 1 and 4 (b) 2 and 4
(c) 1 only (d) 3 only

Q.249 Assertion (A): The figure of merit of an IC family should be high.

Reason (R): The figure of merit has the unit of energy.

(a) Both A and R are true and R is the correct explanation of A.
(b) Both A and R are true but R is not the correct explanation of A.
(c) A is true but R is false.
(d) A is false but R is true.

Q.250 Match **List-I (Logic families)** with **List-II (Power dissipation per gate in mW)** and select the correct answer using the codes given below the lists:

	List-I		List-II
A.	ECL	1.	50
B.	CMOS	2.	0.01
C.	IIL	3.	0.1
D.	TTL	4.	10

Codes:

	A	B	C	D
(a)	1	2	4	3
(b)	1	2	3	4
(c)	2	1	4	3
(d)	2	4	3	1

Q.251 For LOW power schottky TTL,
I_{IL}(max) = –0.36 mA ; I_{IH}(max) = 20 μA
I_{OL}(max) = 8 mA ; I_{OH}(max) = –400 μA
The fan-out of this LOW power schottky TTL is
(a) 22 (b) 20
(c) either (a) or (b) (d) neither (a) nor (b)

Q.252 The voltage levels which should be used to measure the rise time and fall time for a signal going from 0.2 V to 3.9 V will be respectively
(a) 0.59 V and 2.24 V (b) 0.26 V and 2.24 V
(c) 0.59 V and 3.71 V (d) 0.26 V and 3.71 V

Q.253 For the circuit shown in figure, $V_{BE(sat)}$ = 0.8 V, V_γ = 0.5 V, $V_{CE(sat)}$ = 0.2 V, the drop across the conducting diode is 0.7 V and V_γ(diode) = 0.6 V. The inputs of this switch (*A*, *B*, *C*) are obtained from the outputs of similar gates. The given circuit will function as a

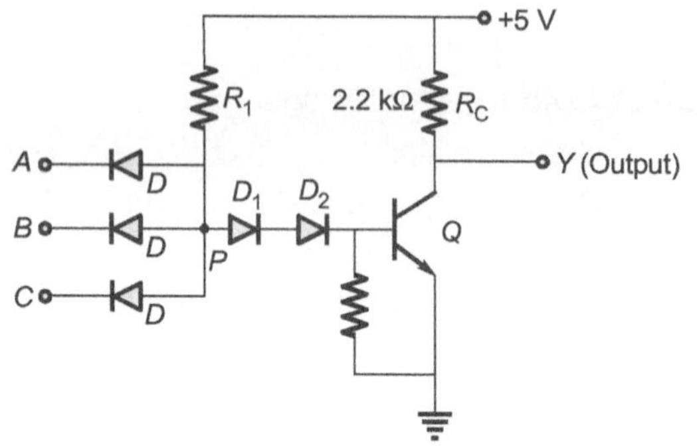

(a) NOR gate (b) AND gate
(c) OR gate (d) NAND gate

Q.254 In positive logic, the transistor gate in the circuit below is a

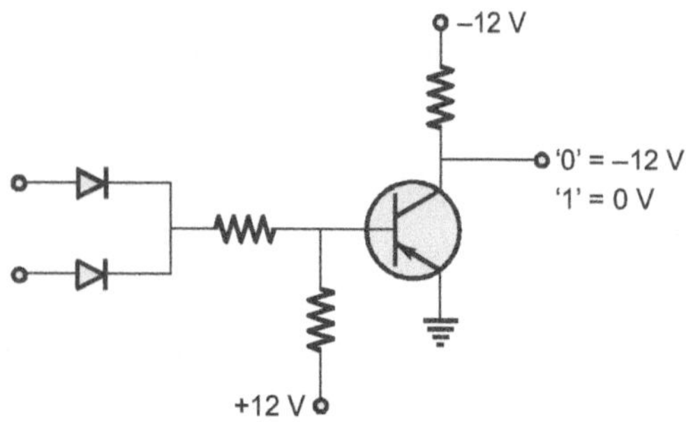

(a) AND gate (b) NOR gate
(c) NAND gate (d) OR gate

Q.255 The problem of "current hogging" is associated with
(a) DCTL gates (b) DTL gates
(c) ECL gates (d) CMOS gates

Q.256 In the given figure, the LED emits light when

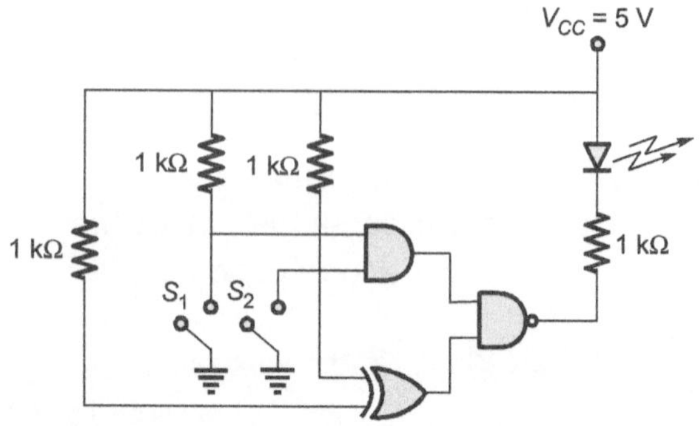

(a) both switches are closed
(b) both switches are open
(c) only one switch is closed
(d) LED does not emit light irrespective of the switch positions

Q.257 In the circuit shown in figure the output *z* is

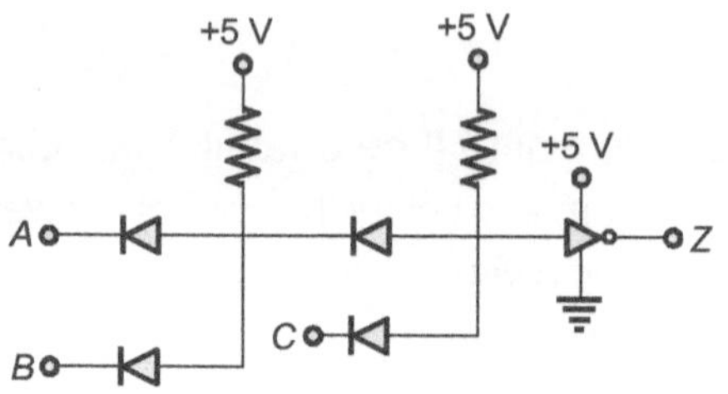

(a) $AB + \bar{C}$ (b) $\overline{AB}C$
(c) $\overline{ABC}$ (d) ABC

Q.258 Consider the CMOS circuit shown in figure. It acts as a

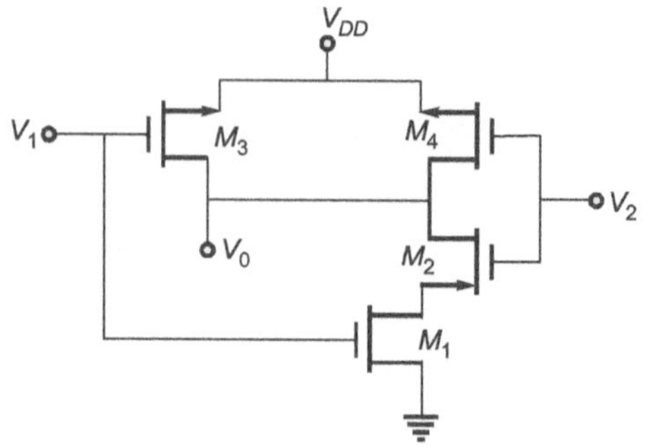

(a) Negative NAND (b) Positive NAND
(c) Negative NOR (d) Positive NOR

Q.259 For a negative logic gate, the given circuit represents a

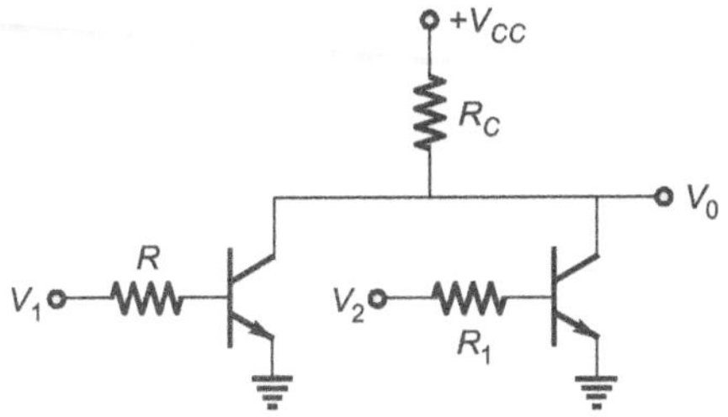

(a) AND gate (b) OR gate
(c) NAND gate (d) NOR gate

9. A/D and D/A Converters

Q.260 The speed of conversion is maximum in
(a) Dual-slope A/D converter
(b) Successive-approximation A/D converter
(c) Counter ramp A/D converter
(d) Parallel-comparator A/D converter

Q.261 The A/D converter whose conversion time is independent of the number of bits is
(a) counter type
(b) Flash type
(c) Dual slope type
(d) Successive approximation type

Q.262 In successive-approximation A/D converter, offset voltage equal to 1/2 LSB is added to the D/A converter's output. This is done to
(a) reduce the maximum quantization error
(b) increase the number of bits at the output
(c) increase the range of input voltage that can be converted
(d) improve the speed of operation

Q.263 A weighted resistor digital to analog converter using n-bits required a total of
(a) $2n$ precision resistors
(b) $n - 1$ precision resistors
(c) n precision resistors
(d) $n + 1$ precision resistors

Q.264 The number of comparators required for 12-bit flash type ADC is
(a) 4000 (b) 4095
(c) 3095 (d) 2512

Q.265 A clock rate of one megahertz operating 12-stage counter of a counter-type ADC would need a maximum conversion time of approximately
(a) 4.1 ms (b) 8 ms
(c) 7.1 ms (d) 3.2 ms

Q.266 Which of the following is not true in the circuit for digital-to-analog voltage converter?
(a) Some logic circuitry
(b) A reference voltage
(c) A resonant circuit
(d) A resistor network

Q.267 In a D/A converter made with binary weighted resistors, it is difficult to maintain high accuracy as the number of bits increases, because
(a) as the resistance values of the most significant bits become larger, the circuit takes a very long time to settle.
(b) obtaining stable, precise resistors with a large spread in their values is very difficult
(c) the LSB resistance consume too much current, and power
(d) the finite gain of the opamp increases inaccuracies

Q.268 Match **List-I** with **List-II** and select the correct answer using the codes given below the lists:

List-I
A. SAR type ADC
B. Flash type ADC
C. Dual slope type ADC
D. Counter type ADC

List-II
1. Simultaneous ADC
2. Integrating type ADC
3. Digital ramp ADC
4. Uses ring counter

Codes:

	A	B	C	D
(a)	3	1	2	4
(b)	4	1	3	2
(c)	2	1	4	3
(d)	4	1	2	3

Q.269 Match **List-I (Types of ADCs)** with **List-II (Characteristics)** and select the correct answer using the codes given below the lists:

List-I
A. Integrating type ADC
B. Ramp type ADC
C. SAR type ADC
D. Flash type ADC

List-II
1. Fastest
2. Most accurate
3. Simplest type of A/D converter
4. Most widely used type of A/D converter

Codes:

	A	B	C	D
(a)	2	1	3	4
(b)	1	2	4	3
(c)	2	3	4	1
(d)	4	3	2	1

Q.270 **Assertion (A):** The flash converter uses no clock signal.
Reason (R): These is no timing or sequencing period in the flash converter.
(a) Both A and R are true and R is the correct explanation of A.
(b) Both A and R are true but R is not the correct explanation of A.
(c) A is true but R is false.
(d) A is false but R is true.

Q.271 An ADC has a total conversion time of 200 μs. What is the highest frequency that its analog input should be allowed to contain?
(a) 5 kHz (b) 2.5 kHz
(c) 7.5 kHz (d) 10 kHz

Q.272 Consider the following statements associated with various analog to digital converters:
1. The counter-type ADC is somewhat slower, but can digitize high resolution signals.
2. The flash type is the most expensive ADC while the dual-slope ADC is cheapest.
3. The successive-approximation type ADC is the most widely used type of ADC.
4. The conversion time is independent of input analog voltage in successive-approximation type ADC.

Which of the statements given above are correct?
(a) 1, 3 and 4 (b) 1, 2 and 3
(c) 1, 2, 3 and 4 (d) 2, 3 and 4

Q.273 Which of the following ADC do not use a DAC?
(a) SAR type ADC
(b) Ramp type ADC
(c) Dual slope type ADC
(d) Both (a) and (c)

Q.274 Which of the following statements is/are correct?
1. The weighted-resistor DAC(digital-to-analog converter) is the most popular DAC.
2. Switched-current DACs are faster than the switched-voltage DACs.
3. The output of a DAC is a true analog quantity.

(a) 2 and 3 (b) 2 only
(c) 1 only (d) 1 and 3

Q.275 A 6-bit DAC has a step size of 50 mV. The percentage resolution of this DAC is around
(a) 1.59% (b) 2.22%
(c) 3.30% (d) 2.85%

Q.276 A certain 6-bit DAC has a full-scale output of 2 mA and a full-scale error of ±0.5%. What is the range of possible outputs for an input of 100000?
(a) 1014 μA to 1024 μA
(b) 1014 μA to 1034 μA
(c) 1004 μA to 1034 μA
(d) 1004 μA to 1024 μA

Q.277 For the 4-bit weighted-resistor DAC shown in figure, what is the full-scale output, if $R_f = R = 1\ k\Omega$ and the inputs are 0 V and 5 V?

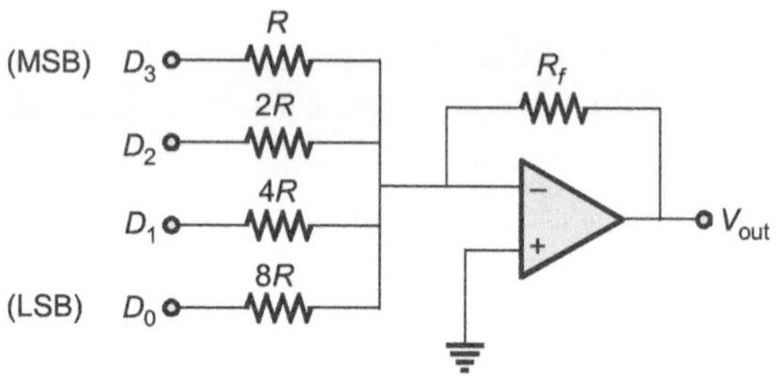

(a) –6.625 V (b) – 8.825 V
(c) – 9.375 V (d) – 5.755 V

Q.278 In a 4-bit DAC, for a digital input of 0100, an output current of 10 mA is produced. What is the output current for a digital input of 1011?
(a) 18.90 mA (b) 27.5 mA
(c) 25.5 mA (d) none of these

Q.279 An 8-bit SAR type ADC has a resolution of 30 mV. What will be its digital output for an analog input of 2.86 V?
Which of the statements given above is/are not correct ?
(a) 01100000_2 (b) 01001011_2
(c) 10000011_2 (d) 01011111_2

Q.280 An ADC is used to convert signals in the range of 1 kHz to 50 kHz. The maximum conversion time would be equal to

(a) 15 μs (b) 5 μs
(c) 20 μs (d) 10 μs

Q.281 **Assertion (A):** In a DAC, the staircase output will have no downward steps as the binary input is incremented from 0 to full-scale value.

Reason (R): A DAC is said to be monotonic, if its output decreases when the binary input is incremented.

(a) Both A and R are true and R is the correct explanation of A.
(b) Both A and R are true but R is not the correct explanation of A.
(c) A is true but R is false.
(d) A is false but R is true.

Q.282 A 4-bit R-2R ladder type D/A converter having resistor values of R = 10 kΩ and $2R$ = 20 kΩ, uses V_R of 10 V. The resolution of the D/A converter is

(a) 62.5 μA (b) 37.8 μA
(c) 58.5 μA (d) 46.2 μA

Q.283 The current I through resistance 'R' in the circuit shown below is _____ . (Assume R = 1 Ω)

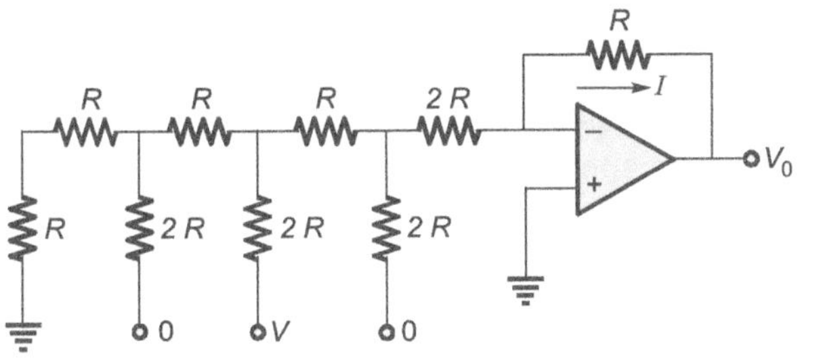

(a) $\frac{V}{4}$ (b) $\frac{V}{12}$
(c) $\frac{V}{6}$ (d) $\frac{V}{3}$

Q.284 A 10-bit ADC having a full scale output voltage of 10.24 V is designed to have a ± LSB/2 accuracy. If the ADC is calibrated at 25°C and the operating temperature ranges from 0°C to 50°C, then the maximum net temperature coefficient of the ADC should not exceed

(a) ± 200 μV/°C (b) ± 400 μV/°C
(c) ± 600 μV/°C (d) ± 800 μV/°C

Q.285 There are two ADCs, 10-bit digital ramp ADC and a 10-bit successive approximation ADC. If both utilize a 500 kHz clock frequency, how much digital ramp slower than successive approximation (with respect to conversion time)?

(a) 10 times (b) 20 times
(c) 50 times (d) 100 times

Q.286 Match **List-I (Type of n-bit ADC)** with **List-II (Characteristic)** and select the correct answer using the codes given below the lists:

List-I
A. Flash converter
B. Successive approximation
C. Counter ramp
D. Dual slope

List-II
1. Integrating type
2. Fastest converter
3. Maximum conversion time = n clock cycle
4. Uses a DAC in its feedback path

Codes:

	A	B	C	D
(a)	1	4	3	2
(b)	1	3	4	2
(c)	2	4	3	1
(d)	2	3	4	1

Q.287 The full scale input voltage of an ADC is 10 V. The resolution required is 5 mV. The minimum number of bits required for ADC is

(a) 12 (b) 11
(c) 10 (d) 6

Q.288 The output voltage range of a 12-bit DAC is 0 to 10 V. The integer number of 516 causes an analog output of

(a) 2.88 V (b) 4.17 V
(c) 0.516 V (d) 1.26 V

10. Semiconductor Memories

Q.289 EPROM contents can be erased by exposing it to

(a) Burst of microwaves
(b) Intense heat radiations
(c) Ultraviolet rays
(d) Infrared rays

Q.290 Which of the following memory is volatile?

(a) ROM (b) EPROM
(c) PROM (d) RAM

Q.291 The access time of ROM using bipolar transistors is about
(a) 1 msec (b) 1 μsec
(c) 1 nsec (d) 1 sec

Q.292 In a RAM, information can be stored
(a) by the user, number of times
(b) by the manufacturer only once
(c) by the user, only once
(d) by the manufacturer, a number of times

Q.293 The process of entering data into a ROM is called
(a) charging the ROM
(b) programming the ROM
(c) burning in the ROM
(d) none of the above

Q.294 How many address bits are required to represent a 32 k memory?
(a) 10 bits (b) 16 bits
(c) 12 bits (d) 14 bits

Q.295 Which of the following memory require refreshing?
(a) ROM (b) EPROM
(c) DRAM (d) SRAM

Q.296 Which of the following memory system can not be accessed randomly?
(a) ROM (b) EPROM
(c) DRAM (d) Magnetic tape

Q.297 The information in ROM is stored
(a) by the manufacturer during fabrication of the device
(b) by the user using the ultraviolet rays
(c) by the user any number of times
(d) by the user once and only once

Q.298 In a virtual memory system, the address used by the programmer belongs to
(a) address space
(b) main memory address
(c) memory space
(d) physical address

Q.299 A circuit in which connections to both AND and OR arrays can be programmed is called
(a) PLA (b) ROM
(c) RAM (d) PAL

Q.300 Dynamic RAM consumes ____ power and is ____ than the static RAM.
(a) more, slower (b) more, faster
(c) less, faster (d) less, slower

Q.301 Cache memory works on the principle of
(a) locality of memory
(b) locality of reference
(c) locality of data
(d) locality of reference and memory

Q.302 Memory interlacing technique is used to address the memory modules in order to have
(a) faster access to a block of data
(b) higher average utilization
(c) reduced complexity in mapping hardware
(d) both (a) and (b)

Q.303 When a ROM is designed to convert a binary word to a Gray code, the number of input bits to the ROM is ____ the number of output bits of the ROM.
(a) greater than (b) equal to
(c) less than (d) not related to

Q.304 Which of the following memories stores the most number of bits?
(a) a 5 M × 4 memory
(b) a 1 M × 12 memory
(c) a 1 M × 16 memory
(d) a 5 M × 8 memory

Q.305 **Assertion (A):** A magnetic tape is called the sequential access memory.
Reason (R): To access a memory location on a magnetic tape, it is necessary to wind or unwind the tape and go through a series of addresses before reaching the address desired.
(a) Both A and R are true and R is the correct explanation of A.
(b) Both A and R are true but R is not the correct explanation of A.
(c) A is true but R is false.
(d) A is false but R is true.

Q.306 **Assertion (A):** Microcomputer programs stored in ROMs are referred to as software.
Reason (R): ROM normally means the mask programmed ROM.
(a) Both A and R are true and R is the correct explanation of A.

(b) Both A and R are true but R is not the correct explanation of A.
(c) A is true but R is false.
(d) A is false but R is true.

Q.307 Match **List-I (Type of memory)** with **List-II (Characteristic)** and select the correct answer using the codes given below the lists:

List-I	List-II
A. RAM	1. Used for mass storage
B. Floppy disk	2. Volatile
C. ROM	3. Portable and is of low cost
D. Magnetic disc	4. Non-volatile

Codes:

	A	B	C	D
(a)	2	1	4	3
(b)	4	3	2	1
(c)	2	3	4	1
(d)	4	2	1	3

Q.308 A single-sided double-density 8-inch floppy disk has tracks numbered 0 to 76 and sectors 0 to 25. What is the total storage capacity of the disc?
(a) 468,512 bytes (b) 512,512 bytes
(c) 468,468 bytes (d) 512,468 bytes

Q.309 Match **List-I** with **List-II** and select the correct answer using the codes given below the lists:

List-I
A. Static RAM
B. Dynamic RAM
C. ECL RAMs
D. NVRAM

List-II
1. Latch storage cell
2. Capacitive storage cell
3. Used for cache memory
4. Contain static RAM array and an EEPROM arran on same chip

Codes:

	A	B	C	D
(a)	2	3	4	1
(b)	1	2	4	3
(c)	2	1	3	4
(d)	1	2	3	4

Q.310 Consider the following statements:
1. Periodic reacharging of the memory cells of a DRAM is called "refreshing" the DRAM.
2. Storing the contents of a ROM is called "programming" the ROM.
3. PROMs are non-programmable ROMs and EPROMs are non-erasable PROMs.
4. Dynamic memory is constructed using both BJT and MOS technology.
5. DRAM controllers are used to refresh large DRAM chips.

Which of the statements given above are correct?
(a) 1, 2 and 5 (b) 2, 3 and 5
(c) 1 and 3 (d) 1, 2 and 4

Q.311 Consider the following statement:
1. Non-volatility means that the data are lost when the power goes off.
2. Volatility means that the memory does not lose its data when the power goes off.
3. Magnetic core memory is a volatile memory.
4. Magnetic memories are faster than semiconductor memories.

Which of the statements given above is/are not correct?
(a) 1 and 2 (b) 1, 2 and 3
(c) 1, 2, 3 and 4 (d) 2 and 3

Q.312 The total capacity of a ROM with 1024 rows and 4 columns is
(a) 4096 (b) 1024
(c) 256 (d) 512

Q.313 The number of bits required to address a 4 k memory is
(a) 11 (b) 12
(c) 13 (d) 14

Answers Digital Electronics

1. (c)	2. (a)	3. (d)	4. (b)	5. (d)	6. (a)	7. (c)	8. (b)
9. (a)	10. (b)	11. (c)	12. (a)	13. (a)	14. (d)	15. (b)	16. (c)
17. (c)	18. (b)	19. (d)	20. (d)	21. (b)	22. (a)	23. (c)	24. (d)
25. (a)	26. (c)	27. (d)	28. (a)	29. (b)	30. (a)	31. (c)	32. (a)
33. (d)	34. (a)	35. (c)	36. (c)	37. (b)	38. (d)	39. (d)	40. (d)
41. (c)	42. (d)	43. (c)	44. (b)	45. (a)	46. (b)	47. (a)	48. (c)
49. (d)	50. (b)	51. (b)	52. (d)	53. (a)	54. (c)	55. (b)	56. (c)
57. (a)	58. (c)	59. (d)	60. (b)	61. (a)	62. (d)	63. (c)	64. (d)
65. (c)	66. (d)	67. (a)	68. (c)	69. (c)	70. (c)	71. (d)	72. (d)
73. (a)	74. (d)	75. (b)	76. (d)	77. (d)	78. (b)	79. (d)	80. (c)
81. (d)	82. (a)	83. (b)	84. (b)	85. (a)	86. (a)	87. (d)	88. (c)
89. (d)	90. (a)	91. (c)	92. (d)	93. (b)	94. (c)	95. (a)	96. (c)
97. (b)	98. (d)	99. (c)	100. (d)	101. (d)	102. (a)	103. (b)	104. (d)
105. (b)	106. (c)	107. (c)	108. (c)	109. (b)	110. (c)	111. (d)	112. (c)
113. (a)	114. (d)	115. (c)	116. (b)	117. (d)	118. (a)	119. (c)	120. (c)
121. (b)	122. (a)	123. (d)	124. (b)	125. (c)	126. (a)	127. (d)	128. (b)
129. (a)	130. (c)	131. (a)	132. (d)	133. (b)	134. (c)	135. (d)	136. (a)
137. (b)	138. (d)	139. (c)	140. (a)	141. (b)	142. (d)	143. (a)	144. (b)
145. (d)	146. (d)	147. (d)	148. (b)	149. (d)	150. (d)	151. (d)	152. (a)
153. (b)	154. (b)	155. (a)	156. (c)	157. (b)	158. (d)	159. (c)	160. (d)
161. (b)	162. (a)	163. (c)	164. (b)	165. (d)	166. (b)	167. (c)	168. (a)
169. (c)	170. (b)	171. (b)	172. (d)	173. (a)	174. (c)	175. (b)	176. (c)
177. (d)	178. (a)	179. (b)	180. (d)	181. (c)	182. (c)	183. (d)	184. (c)
185. (c)	186. (b)	187. (c)	188. (a)	189. (b)	190. (d)	191. (d)	192. (c)
193. (a)	194. (b)	195. (c)	196. (d)	197. (b)	198. (b)	199. (c)	200. (b)
201. (b)	202. (c)	203. (a)	204. (a)	205. (c)	206. (d)	207. (b)	208. (d)
209. (a)	210. (b)	211. (c)	212. (a)	213. (d)	214. (b)	215. (d)	216. (c)
217. (d)	218. (b)	219. (a)	220. (b)	221. (d)	222. (c)	223. (c)	224. (d)
225. (c)	226. (d)	227. (d)	228. (c)	229. (c)	230. (b)	231. (b)	232. (c)
233. (d)	234. (b)	235. (d)	236. (c)	237. (d)	238. (b)	239. (b)	240. (d)

241. (b)	242. (c)	243. (a)	244. (d)	245. (a)	246. (d)	247. (a)	248. (c)
249. (d)	250. (b)	251. (b)	252 (c)	253. (d)	254. (b)	255. (a)	256. (d)
257. (c)	258. (a)	259. (c)	260. (d)	261. (b)	262. (a)	263. (c)	264. (b)
265. (a)	266. (c)	267. (b)	268. (d)	269. (c)	270. (a)	271. (b)	272. (c)
273. (c)	274. (b)	275. (a)	276. (d)	277. (c)	278. (b)	279. (d)	280. (d)
281. (c)	282. (a)	283. (b)	284. (b)	285. (d)	286. (d)	287. (b)	288. (d)
289. (c)	290. (d)	291. (b)	292. (a)	293. (b)	294. (b)	295. (c)	296. (d)
297. (a)	298. (a)	299. (c)	300. (d)	301. (b)	302. (c)	303. (b)	304. (d)
305. (a)	306. (d)	307. (c)	308. (b)	309. (d)	310. (a)	311. (c)	312. (a)
313. (b)							

Explanations

1. (c)

A binary number can be converted into decimal number by multiplying the binary numbers 1 or 0 by their weight and adding the product.

Conversion of 101111 is done as follows:

$$(101111)_2 = (1\times 2^5 + 0\times 2^4 + 1\times 2^3 + 1\times 2^2 + 1\times 2^1 + 1\times 2^0) = 47$$

Conversion of $(0.1101)_2$ is done as follows:

$$(0.1101)_2 = 1\times 2^{-1} + 1\times 2^{-2} + 0\times 2^{-3} + 1\times 2^{-4} = 0.8125$$

Hence, $(101111\cdot 1101)_2 = (47.8125)_{10}$

3. (d)

- "Hamming Code" provides a mathematical way to add one or more parity bits to a data character in order to detect and correct errors.
- "Gray code" is also called a "unit-distance code" because here two adjacent code numbers differ by only one bit.
- "Excess-3 code" is called a "reflective code" because here the code for 9 is the complement of the code for 0, 8 for 1, 7 for 2, 6 for 3, and 5 for 4.
- "BCD code" is as "weighted code" or "8421 code" because it uses the binary number system to specify the decimal number 0 to 9.

4. (b)

$$(7\cdot FD6)_{16} = (0111.111111010110)_2$$

The binary number is converted into octal number as follows:

000	111	.	111	111	010	110	Binary
↓	↓		↓	↓	↓	↓	
0	7		7	7	2	6	Octal

Thus, $(7\cdot FD6)_{16} = (7.7726)_8$

On converting the decimal number $(7864)_{10}$ into hexadecimal number we obtain the result $(EB8)_{16}$ as follows:

16	7864		
16	491	8	→ 8(LSB)
16	30	11 →	B
16	1	14 →	E
	0	1 →	(MSB)

Thus, $(7864)_{10} = (1EB8)_{16}$

5. (d)

Let, $(x)_{10} = (x_3 x_2 x_1 x_0)_2$

$$\begin{array}{r} 1\;0\;0\;1\;1\;0 \\ x_3\;x_2\;x_1\;x_0 \\ \hline x_0\;0\;0\;x_0\;x_0\;0 \\ x_1\;0\;0\;x_1\;x_1\;0\;\; \\ x_2\;0\;0\;x_2\;x_2\;0\;\;\;\; \\ x_3\;0\;0\;x_3\;x_3\;0\;\;\;\;\;\; \\ \hline 1\;0\;1\;0\;1\;0\;1\;1\;0 \end{array}$$

From above multiplication, it is clear that,

$$x_0 = 1$$
$$x_1 = 0$$
$$x_2 = 0$$
$$x_3 = 1$$

Thus, $(x)_{10} = (x_3 x_2 x_1 x_0)_2 = (1001)_2 = (9)_{10}$

6. (a)

In signed binary number representation, the range of number that can be represented by an n-bit number is $(-2^{n-1}-1)$ to $(+2^{n-1}-1)$.

For $n = 6$, the range is –31 to + 31.

7. (c)

- The binary equivalent of $(22)_{10} = (10110)_2$ as follows:

2	22		
2	11	0	↑ LSB
2	5	1	
2	2	1	
2	1	0	
	0	1	MSB

$\therefore \quad (22)_{10} = (10110)_2$

- The GRAY code of $(10110)_2$ is obtained as follows:

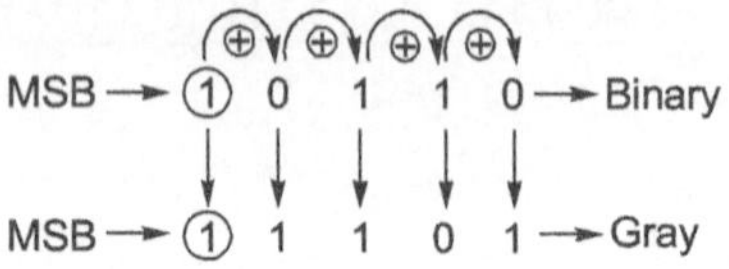

Thus, $(10110)_2 = (11101)_{GRAY}$

- The octal equivalent of $(22)_{10} = (26)_8$ obtained as follows:

$$(22)_{10} = (10110)_2$$
$$= \underbrace{010}_{2}\ \underbrace{110}_{6} = (26)_8$$

- The hexadecimal equivalent of $(22)_{10}$ is obtained as follows:

16	22		↑
16	1	6	LSB
	0	1	MSB

$\therefore \quad (22)_{10} = (10110)_2 = (16)_{16}$

8. (b)

Given,

$$b_3 b_2 b_1 b_0 = (0101)_2$$
$$\therefore \quad h_1 = b_3 \oplus b_2 \oplus b_0$$
$$= 0 \oplus 1 \oplus 1 = 0 \text{ (MSB)}$$
$$h_2 = b_3 \oplus b_1 \oplus b_0 = 0 \oplus 0 \oplus 1 = 1$$
$$h_3 = b_3 = 0$$
$$h_4 = b_2 \oplus b_1 \oplus b_0 = 1 \oplus 0 \oplus 1 = 0$$
$$h_5 = b_2 = 1$$
$$h_6 = b_1 = 0$$
$$h_7 = b_0 = 1 \text{ (LSB)}$$

∴ The required 7-bit even parity Hamming code is (0 1 0 0 1 0 1).

9. (a)

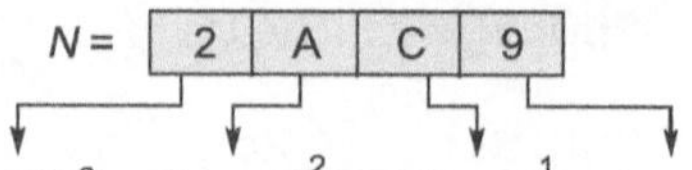

$$(2 \times 16^3) + (10 \times 16^2) + (12 \times 16^1) + (9 \times 16^0) = 10{,}953$$

Thus, $(2AC9)_{16} = (10{,}953)_{10}$

Now, the decimal number is converted into the number of base 7 as follows:

7	10953		
7	1564	5	→ (LSB) ↑
7	223	3	
7	31	6	
7	4	3	
	0	4	→ (MSB)

Thus, $(2AC9)_{16} = (43635)_7$

10. (b)

Digital circuits are often called switching circuits, because the voltages in a digital circuit are assumed to be switching from one value to another instantaneously, that is, the transition time is assumed to be zero.

Reason is also a correct statement because digital circuit is a logic circuit, because each type of digital circuit obeys a certain set of logic rules. Hence, both assertion and reason are true but reason is not the correct explanation of assertion.

11. (c)

The advantages of digital systems over analog systems are as follows:

1. Easier to design.
2. Less affected by noise (unwanted electrical signals).
3. More accuracy and precision.
4. Simple and economical fabrication into ICs.

12. (a)

$$\begin{array}{r} 1\ 0\ 1\ 0\ .\ 0\ 1\ 0 \\ -\quad 1\ 1\ 1\ .\ 1\ 1\ 1 \\ \hline 0\ 0\ 1\ 0\ .\ 0\ 1\ 1 \end{array}$$

In binary system, we take borrow as 2 for subtraction. Thus, $(1010.01)_2 - (111.111)_2 = (0010.011)_2$.

13. (a)

Both assertion and reason are correct and reason is the correct explanation of assertion because when subtraction is performed using complement method, then instead of subtracting one number from other, the complement of the subtrahend is added to the minued. Thus, subtraction is also performed by adders only as in case of addition. Due to this reason the hardware requirement is reduced while performing subtraction by complement method.

14. (d)
On multiplying the given hexadecimal number with the positional weightage, we get:

$(A0F9.0EB)_{16}$
$= (10 \times 16^3) + (0 \times 16^2) + (15 \times 16^1) + (9 \times 16^0)$
$+ (0 \times 16^{-1}) + (14 \times 16^{-2}) + (11 \times 16^{-3})$
$= 40960 + 0 + 240 + 9 + 0 + 0.0546 + 0.0026$
$= (41209.0572)_{10}$

15. (b)

On converting 100_b to decimal, we have:

$1 \times b^2 + 0 \times b^1 + 0 \times b^0 = 16$

or, $b^2 = 16$

or, $b = 4$

16. (c)

- If we take the first number to be in binary (base 2), its equivalent will be $(16)_{10}$ i.e. $(10000)_2 = (16)_{10}$.
- The next number is 121; it is equal to $(16)_{10}$ in base 3 i.e. $(121)_3 = (16)_{10}$.
- 100 is equal to $(16)_{10}$ in base 4 i.e. $(100)_4 = (16)_{10}$.
- $(31)_5 = (16)_{10}$
- $(24)_6 = (16)_{10}$
- $(22)_7 = (16)_{10}$
- $(20)_8 = (16)_{10}$

Hence, the missing number is 31 in base 5.

17. (c)

- $1234 + 5432 = 6666$

i.e.
$$\begin{array}{r} 1\,2\,3\,4 \\ +5\,4\,3\,2 \\ \hline 6\,6\,6\,6 \end{array}$$

Since the largest digit used is 6, this arithmetic operation is valid in any number system with base ≥ 7.

- $\sqrt{41} = 5$

Let the base be 5.

Then, its decimal equivalent operation is,

$\sqrt{4b+1} = 5$ or, $4b + 1 = 25$ or, $b = 6$

Thus, the base is 6.

- $\dfrac{302}{20} = 12.1$

Expressing in decimal, we have:

$$\frac{3b^2+2}{2b} = b+2+\frac{1}{b}$$

or, $3b^2 + 2 = 2b^2 + 4b + 2$

or, $b^2 - 4b = 0$ or $b = 4$

(Neglecting $b = 0$)

Thus, the base is 4.
Hence, only (*ii*) is correct.

18. (b)

- A code is said to be self-complementing, if the code word of the 9's complement of N i.e. of (9–N) can be obtained from the code word of N by interchanging all the 0's and 1s.
Thus, statement–1 is correct.
- There are six illegal combinations in 4-bit BCD codes which are 1010, 1011, 1100, 1101, 1110 and 1111 and hence they are not part of the 8421 BCD code system. Thus, statement-2 is also correct.
- The gray code cannot be used for mathematical operations because it is non-weighted code. Thus, statement-3 is also correct.
- The excess-3 code is also a non-weighted code, but can be used for mathematical operations because it is a sequential code. Thus, statement-4 is also correct.

19. (d)

The ASCII code is a alphanumeric code because it represents alphanumeric information i.e. letters of the alphabet and decimal numbers as a sequence of 0s and 1s.

20. (d)

The 1's complement system has a disadvantage of having two zero representation i.e. a positive 0 (all 0s) and a negative 0 (all 1s). But, the 2's complement system has a unique zero.
Hence, assertion is not a correct statement.

23. (c)

The Gray code is a cyclic code in which successive code words in this code differ in one bit position only i.e. it is a unit distance code. In options (a), (b) and (d) both the 4-bit binary numbers differ in one bit position only while in option (c) both the 4-bit binary numbers differs in two bit positions as shown below.

$$\begin{matrix} 1 & 0 & \boxed{1} & \boxed{0} \\ 1 & 0 & \boxed{0} & \boxed{1} \end{matrix}$$

Hence, option (c) does not forms a gray code pair of two given 4-bit binary numbers.

25. (a)

2's complement representation of given number is $(1101)_2$.

Since MSB is 1, therefore the required number must be negative.

1's complement of $(1101)_2$ is $(0010)_2$.

2's complement of 0010 = $(0010)_2 + (0001)_2 = (0011)_2$.

Thus, the required number = –3 (Decimal equivalent of $(0011)_2 = 3$).

26. (c)

- $(345)_8 = (8^2 \times 3) + (8^1 \times 4) + (8^0 \times 5) = (229)_{10}$
  (Octal to decimal conversion)
- $(11011.011)_2 = \underbrace{0001}_{1}\,\underbrace{1011}_{B}\cdot\underbrace{0110}_{6} = (1B.6)_{16}$
  (Binary to Hexadecimal conversion)
- $(47)_{16} = (0100\,0111)_2 = \underbrace{001}_{1}\,\underbrace{000}_{0}\,\underbrace{111}_{7} = (107)_8$
  (Hexadecimal to Binary and then to Octal Conversion)

28. (a)

Let, $B = (b_3\, b_2\, b_1\, b_0)_2$

Now,

$$\begin{array}{r} b_3\; b_2\; b_1\; b_0 \\ \times \quad 0\;\; 1\;\; 1 \\ \hline b_3\; b_2\; b_1\; b_0 \\ b_3\; b_2\; b_1\; b_0\;\;\; \\ 0\;\; 0\;\; 0\;\; 0\;\;\;\;\;\; \\ \hline 1\;\; 1\;\; 0\;\; 1\;\; 1 \end{array}$$

Thus, $b_0 = 1$

$b_1 + b_0 = 1 \quad \therefore b_1 = 0$

$b_2 + b_1 + 0 = 0 \quad \therefore b_2 = 0$

$b_3 + 0 = 1 \quad \therefore b_3 = 1$

Thus, $B = (b_3 b_2 b_1 b_0)_2 = (1001)_2$

29. (b)

The Gray code is not a BCD code.

31. (c)

Given binary number = 1010101

It's 1's complement = 0101010

$\therefore$ 2's complement of given number 1010101 is

= 1's complement of 1010101 + 0001

= 0101011

32. (a)

In 2's complement representation of data, range with n-bit is

$-(2^{n-1})$ to $+(2^{n-1} - 1)$

For 8-bits, range is $-(2^{8-1})$ to $+(2^{8-1} - 1)$

i.e. -128_{10} to $+127_{10}$.

Hence, the most negative number that can be represented by it is -128_{10}.

33. (d)

ASCII code is a seven bit code.

34. (a)

To represent each digit of a decimal number in BCD, four bits are required.

Hence, to represent an eight digit decimal number in BCD, number of bits required = $8 \times 4 = 32$.

35. (c)

XS-3 code or Excess-3 code is a self complementing code.

37. (b)

Here, $X_1 = Y_1$

$X_2 = Y_1 \oplus Y_2$

and $X_3 = Y_2 \oplus Y_3$

Thus, the binary number $Y_1\, Y_2\, Y_3$ is converted into gray code $X_1\, X_2\, X_3$.

38. (d)

Decimal	BCD	Gray Code	Decimal equivalent of Gray Code
0	0000	0000	0
1	0001	0001	1
2	0010	0011	3
3	0011	0010	2
4	0100	0110	6
5	0101	0111	7
6	0110	0101	5
7	0111	0100	4
8	1000	1100	12
9	1001	1101	13
			Sum = 53

39. (d)

Number of XOR gates required

$$= (n-1) = (6-1) = 5$$

(n = number of binary bits)

40. (d)

The three Boolean operators are NOT, OR, AND. These are also called basic gates for digital electronics.

41. (c)

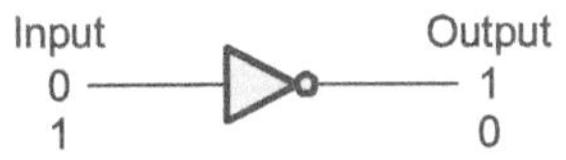

NOT gate is used to invert the input signal.

42. (d)

In negative logic:

- logic 0 = Some positive voltage
  logic 1 = Zero voltage
  or
- logic 0 = Some higher voltage level (i.e. + 5 V)
  logic 1 = Some lower voltage level (i.e. + 2 V)

43. (c)

$$A + 1 = 1 \text{ (Always)}$$

Hence, option (c) is not correct.

44. (b)

Using distribution theorem, we have:

$$\bar{A} + A\bar{B} = (\bar{A} + A)\cdot(\bar{A} + \bar{B})$$
$$= 1\cdot(\bar{A} + \bar{B}) = \bar{A} + \bar{B}$$

45. (a)

Given, $Y = AB + (A+B)(\bar{A}+B)$

$$= AB + A\bar{A} + AB + \bar{A}B + B.B$$
$$= AB + 0 + AB + \bar{A}B + B$$
$$= AB + \bar{A}B + B$$
$$= B(A + \bar{A} + 1) = B\cdot 1 = B$$

46. (b)

Given, $Y = A(B+C)(\bar{C}+\bar{D})$

$$\therefore \quad \bar{Y} = \overline{A(B+C)(\bar{C}+\bar{D})}$$
$$= \bar{A} + \overline{(B+C)} + \overline{(\bar{C}+\bar{D})}$$

(Using De Morgan's theorem)

$$= \bar{A} + \bar{B}\cdot\bar{C} + \bar{\bar{C}}\cdot\bar{\bar{D}}$$
$$= \bar{A} + \bar{B}\,\bar{C} + CD$$

47. (a)

POS form (Product of Sum form) of expression is suitable for circuit using NOR gates.

49. (d)

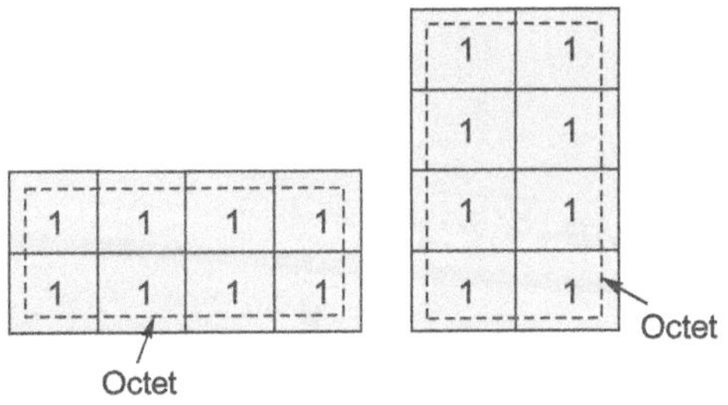

50. (b)

Statement 2 and 5 are not correct because there is no possibility of having negative and fractional numbers is Boolean algebra. Also, there are only two constant 0 and 1 (don't care (x) is not a constant) within the Boolean system.

51. (b)

Boolean algebra and Binary number system are different from each other because in Boolean algebra 1 + 1 = 1 while in the binary number system 1 + 1 = 10. There are lot of other examples to prove this.

Reason is also a correct statement because AND, OR and NOT are the three basic operations which are used in Boolean algebra. However, reason is not the correct explanation of assertion.

52. (d)

- 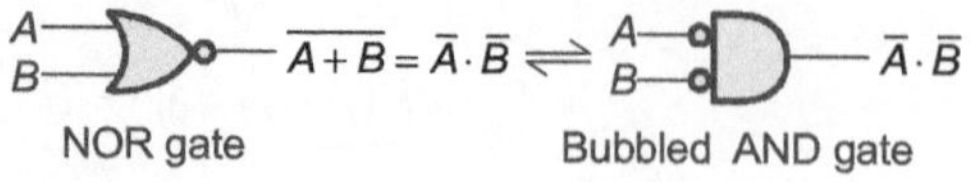

 Hence, statement-1 is not correct.

- $\overline{AB + CD + EFG + ...} = \overline{AB}\cdot\overline{CD}\cdot\overline{EFG}\cdot....$

 (SOP form) (POS form)

 Hence, statement-2 is correct.

- $\overline{A\cdot B\cdot C.......} = \bar{A}+\bar{B}+\bar{C}+........$

 Hence, staement-3 is correct.

- 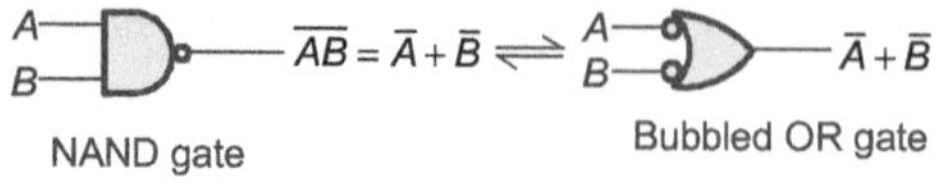

 Hence, statement-4 is not correct.

- $\overline{(AB)\,(CD)\,(EFG).....} = \overline{AB} + \overline{CD} + \overline{EFG} + ...$

 (POS form) (SOP form)

Hence, statement 5 is correct.

53. (a)

$\overline{\overline{AB}+\bar{A}+AB} = \overline{\overline{AB}}\cdot\bar{\bar{A}}\cdot\overline{AB} = AB\cdot A\cdot\overline{AB} = 0$

(Since, $(AB)\cdot(\overline{AB}) = 0$)

54. (c)

- $AB + A\bar{B}C + B\bar{C}$

$= A(B+\bar{B}C)+B\bar{C}$

$= A(B+\bar{B})\,(B+C)+B\bar{C}$

$= AB+AC+B\bar{C}$

$= AB(C+\bar{C})+AC+B\bar{C}$

$= ABC+AB\bar{C}+AC+B\bar{C}$

$= AC(B+1)+B\bar{C}(1+A)$

$= AC+B\bar{C}$

- $\overline{(A+\overline{BC})}\,(A\bar{B}+ABC)$

$= (\bar{A}\cdot\overline{\overline{BC}})\,(A\bar{B}+ABC)$

$= \bar{A}BC\cdot A\bar{B}+\bar{A}BC\cdot ABC$

$= 0 + 0 = 0$

- $A[B+\bar{C}\,\overline{(AB+A\bar{C})}]$

$= A[B+\bar{C}(\overline{AB}\cdot\overline{A\bar{C}})]$

$= A[B+\bar{C}(\bar{A}+\bar{B})(\bar{A}+C)]$

$= A[B+\bar{C}(\bar{A}+\bar{B}C)]$

$= A\left[B+\bar{A}\bar{C}+\bar{B}C\cdot\bar{C}\right]$

$= A[B+\bar{A}\bar{C}] \quad (\because B\cdot C\bar{C}=0)$

$= AB+A\bar{A}\bar{C}$

$= AB + 0 = AB$

- $A\bar{B}C+B+B\bar{D}+AB\bar{D}+\bar{A}C$

$= A\bar{B}C+\bar{A}C+B(1+\bar{D}+A\bar{D})$

$= C(\bar{A}+A\bar{B})+B$

$= C(\bar{A}+A)\,(\bar{A}+\bar{B})+B$

$= C(\bar{A}+\bar{B})+B$

$= \bar{A}C+\bar{B}C+B$

$= \bar{A}C+(B+\bar{B})\,(B+C)$

$= \bar{A}C+B+C$

$= C(\bar{A}+1)+B$

$= C + B$

55. (b)

From given logic diagram,

$f = (A+AB)\,(B+BC)\,(C+AB)$

$= A(1+B)\cdot B(1+C)\,(C+AB)$

$= AB\,(C+AB)$

$= ABC + AB\cdot AB$

$= ABC + AB$

$= AB\,(C+1) = AB$

56. (c)

A K-map map can be used for problems involving any number of variables. But, for more than six variables it becomes tedious to solve the problem sign K-map. Hence, the use of K-map is usually limited to six-variables. Thus, reason is a false statement.

59. (d)

Denoting the non-complemented variable by 1 and the complemented variable by 0, the equation for the output is

$$y = \bar{X}_8\bar{X}_4X_2+\bar{X}_8X_4$$

or, $y = \bar{X}_8\bar{X}_4X_2(X_1+\bar{X}_1)+\bar{X}_8X_4(X_2+\bar{X}_2)(X_1+\bar{X}_1)$

$= 0011 + 0010 + 0100 + 0101 + 0110 + 0111$

$= \Sigma m\,(2, 3, 4, 5, 6, 7)$

The input is a 4-bit BCD, therefore there will be 6 invalid combinations (10, 11, 12, 13, 14, 15) and the corresponding outputs are don't cares.

Hence, the Boolean equation will be

$$y = \Sigma m\,(2, 3, 4, 5, 6, 7) + d(10, 11, 12, 13, 14, 15)$$

The K-map for the abovo Boolean equation is shown below:

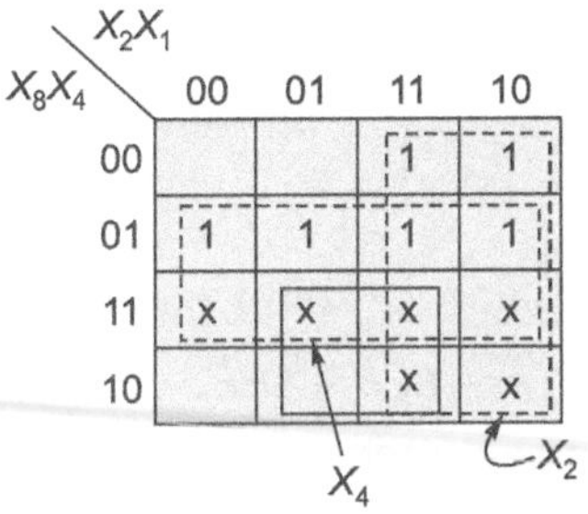

From the above K-map, the simplex output of the given logic circuit will be:

$$y = X_2 + X_4$$

60. (b)

From the given statements, the Boolean expression must be in the POS form given by

$$y = (A_3 + \bar{A}_2 + \bar{A}_1 + \bar{A}_0)\,(\bar{A}_3 + A_2 + \bar{A}_1 + \bar{A}_0)\,(\bar{A}_3 + \bar{A}_2 + \bar{A}_1 + \bar{A}_0)$$

$$= [(A_3 + \bar{A}_2 + \bar{A}_1 + \bar{A}_0)\,(\bar{A}_3 + \bar{A}_2 + \bar{A}_1 + \bar{A}_0)]\,[(\bar{A}_3 + \bar{A}_1 + \bar{A}_0 + A_2)\,(\bar{A}_3 + \bar{A}_1 + \bar{A}_0 + \bar{A}_2)]$$

(Since, $X \cdot X = X$)

$$= [(\bar{A}_2 + \bar{A}_1 + \bar{A}_0)]\,[(\bar{A}_3 + \bar{A}_1 + \bar{A}_0)]$$

$(\because (X+Y)(\bar{X}+Y) = X\bar{X} + Y = Y)$

$$= \bar{A}_2\bar{A}_3 + \bar{A}_1 + \bar{A}_0$$

61. (a)

Given, $f = \Sigma m\,(1, 5, 6, 12, 13, 14) + d\,(2, 4)$ (In SOP form)

$= \pi\, M\,(0, 3, 7, 8, 9, 10, 11, 15) \cdot d(2, 4)$ (In POS form)

The K-map for SOP and POS expressions of output are shown below.

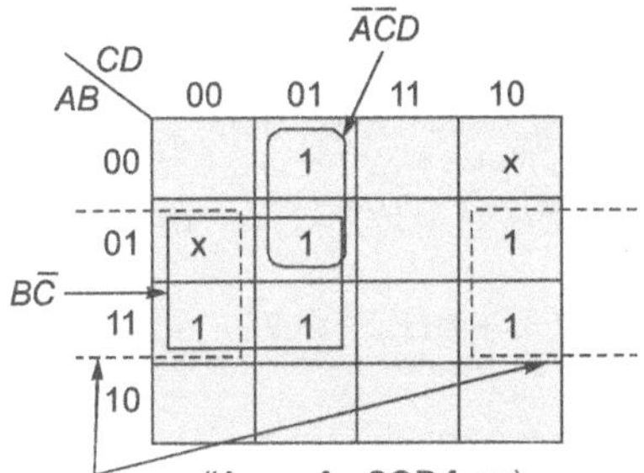

(K-map for SOP form)

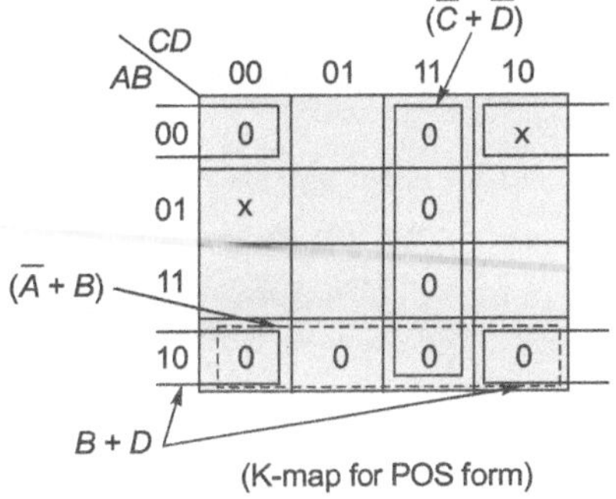

(K-map for POS form)

In SOP form output is $f = B\bar{D} + B\bar{C} + \bar{A}\bar{C}D$

In POS form output is $f = (\bar{A} + B)\,(B + D)\,(\bar{C} + \bar{D})$

The circuit realization for output f using SOP and POS forms respectively are shown below:

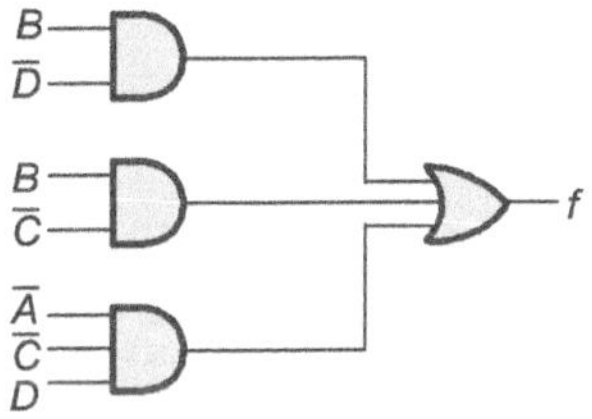

(Using SOP output)

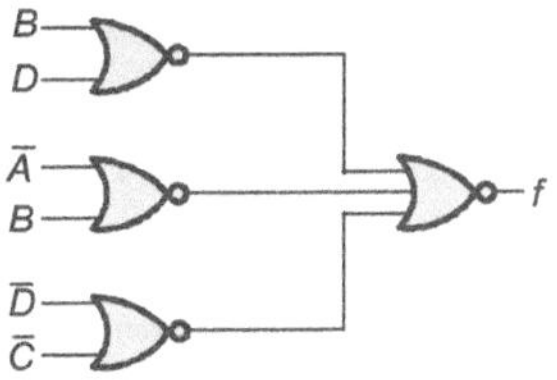

(Using POS output)

We see that total number of gate inputs required using SOP output is 10 while that for POS output is 9. Hence, minimum number of inputs required to implement the minimized expression of output is 9 (using POS form).

62. (d)

Given, $f = (A + B + C)\,(\bar{A} + B + \bar{C})\,(\bar{A} + \bar{B} + \bar{C})\,(A + \bar{B} + \bar{C})\,(\bar{A} + \bar{B} + C)$

The max terms are:

000, 101, 111, 011, 110

$= \pi M\,(0, 3, 5, 6, 7)$

$= \Sigma m\,(1, 2, 4)$

The K-map is shown below.

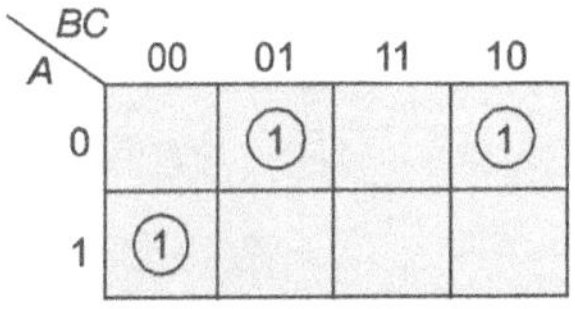

Hence, output $f = A\bar{B}\bar{C} + \bar{A}\bar{B}C + \bar{A}B\bar{C}$

63. (c)

For n number of variables, maximum possible self dual expressions are $2^{2^{n-1}}$.

For $n = 4$, maximum self dual expressions

$$= 2^{2^{4-1}} = 2^{2^3} = 2^8 = 256$$

64. (d)

The shaded area represents

$$Y = A\bar{B}\bar{C} + \bar{A}\bar{B}C + \bar{A}B\bar{C}$$

65. (c)

The truth table for the input-output is shown below.

Inputs			Output
X	Y	Z	W
0	0	0	1
0	0	1	0
0	1	0	1
0	1	1	0
1	0	0	1
1	0	1	0
1	1	0	0
1	1	1	0

Thus, output $W = \bar{X}\bar{Y}\bar{Z} + \bar{X}\bar{Y}Z + \bar{X}Y\bar{Z} + X\bar{Y}\bar{Z}$

66. (d)

For logic circuit-A, output

$$= (\overline{A+B} + AB)$$

$$= \bar{A}\bar{B} + AB = A \odot B$$

$$= (\bar{A}+B)(A+\bar{B})$$

For logic circuit-B, output

$$= (AB + \bar{A}\bar{B}) = A \odot B$$

$$= (\bar{A}+B)(A+\bar{B})$$

For logic circuit-C, output

$$= (A+B)\cdot(\overline{AB})$$

$$= (A+B)(\bar{A}+\bar{B}) = A\bar{B} + \bar{A}B$$

67. (a)

- $A\cdot(A + B) = AA + AB = A + AB$
  $= A(1 + B) = A$
- $AB + BC + \bar{B}C = AB + C(B + \bar{B}) = AB + C$
- $\bar{A}B + AB + \bar{A}\bar{B} = \bar{A}(B + \bar{B}) + AB = \bar{A} + B$
- $A + A\bar{B} + \bar{A}B = A(1 + \bar{B}) + \bar{A}B$
  $= A + \bar{A}B = A + B$

68. (c)

DE \ ABC	000	001	011	010	110	111	101	100
00	0	0	1	1	1	1	1	0
01	0	0	0	0	1	0	1	1
11	1	1	0	0	1	0	0	1
10	0	1	1	1	0	0	0	0

From above K-map, output

$$= B\bar{D}\bar{E} + A\bar{C}E + \bar{A}B\bar{E} + A\bar{B}C\bar{D} + \bar{A}\bar{B}CD + \bar{A}\bar{B}DE$$

Hence, number of minterms in the minimized output are 6.

69. (c)

Given:

$f = \Sigma m\,(0, 2, 3, 6, 7) + \Sigma d\,(8, 10, 11, 15) = \pi M\,(1, 4, 5, 9, 12, 13, 14) + \pi M(8, 10, 11, 15)$

The K-map is shown below:

AB \ CD	00	01	11	10
00	1	0	1	1
01	0	0	1	1
11	0	0	x	0
10	x	0	x	x

Output in SOP form is $f = \bar{A}C + \bar{B}\bar{D}$.

Output in POS form is $f = (C + \bar{D})(\bar{B} + C)\bar{A}$.

Since output can be either in SOP form or in POS form, therefore option (c) is correct.

70. (c)

$A + A = A$. Hence option (c) is not correct.

71. (d)

$A(B + A) = AB + AA = AB + A = A(B + 1) = A$

72. (d)

Given, $\quad Y = ABC + A\bar{B} + AB\bar{C} + \bar{A}$

$$= ABC + A\bar{B}(C + \bar{C}) + AB\bar{C} + \bar{A}(B + \bar{B})(C + \bar{C})$$

$$= \underset{111}{ABC} + \underset{101}{A\bar{B}C} + \underset{100}{A\bar{B}\bar{C}} + \underset{110}{AB\bar{C}} +$$

$$\underset{011}{\bar{A}BC} + \underset{010}{\bar{A}B\bar{C}} + \underset{001}{\bar{A}\bar{B}C} + \underset{000}{\bar{A}\bar{B}\bar{C}}$$

The K-map is shown below:

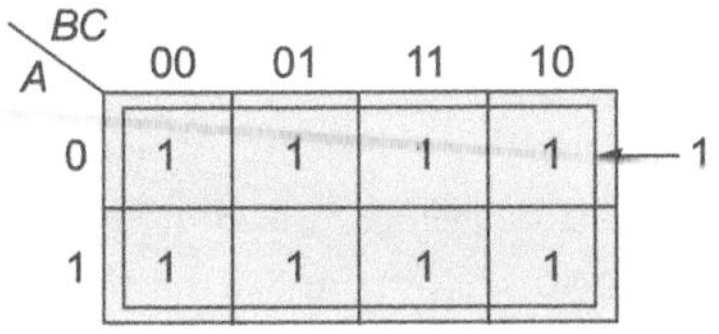

From K-map, it is clear that, $Y = 1$.

73. (a)

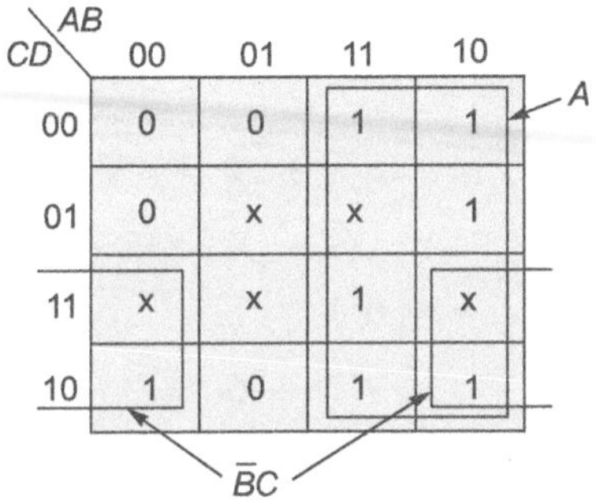

The expression for output is $Y = A + \bar{B}C$.

75. (b)

Given expression is,

$$f = \overline{\overline{AB} + \bar{A} + AB}$$

For finding dual of any expression, all '+' signs are changed to '·' sign, all '·' signs are changed to '+' sign.

Hence, dual of f

$$= \overline{\overline{A+B} \cdot \bar{A}(A+B)} = \overline{0 \cdot \bar{A}} = \bar{0} = 1$$

(Since, $\overline{(A+B)} \cdot (A+B) = 0$)

76. (d)

Given expression:

$$= A(\bar{B} + A)B$$

$$= (A + B\bar{B})(\bar{B} + A)(B + A\bar{A})$$

$$(\because B\bar{B} = A\bar{A} = 0)$$

$$= (A+B)(A+\bar{B})(\bar{B}+A)(A+B)(\bar{A}+B)$$

$$(\because X + Y\bar{Y} = (X+Y)(X+\bar{Y}))$$

$$= (A+B)(A+\bar{B})(\bar{A}+B)$$

$\because$ $(A+B) \cdot (A+B) = (A+B)$

and $(A+\bar{B})(A+\bar{B}) = (A+\bar{B})$

$= (0\,0)\,(0\,1)\,(1\,0)$

$= \pi M\,(0, 1, 2)$

77. (d)

Given, $f = \overline{(A+\bar{B})(C+\bar{D})}$

Complement of $f = (A+\bar{B})(C+\bar{D})$

Changing ANDs to ORs and ORs to ANDs we get the result $= A \cdot \bar{B} + C \cdot \bar{D}$

After complementing the variables, result

$$= \bar{A}B + \bar{C}D$$

Thus, the demorganized result of given function

$$= \bar{A}B + \bar{C}D$$

Note: For demorganization of any expression,

1. Complement the entire given expression.
2. Change all the ANDs to ORs and all the ORs to ANDs.
3. Complement each of the individual variables.

78. (b)

$$\overline{(A+B)}\,\overline{(B+\bar{C})} = (\bar{A}\bar{B})(\bar{B}C) = \bar{A}\bar{B}C$$

$$\overline{\overline{(A+B)}\,\overline{(B+\bar{C})}} = (A+B)+(B+\bar{C}) = A+B+\bar{C}$$

$$\overline{\overline{(\bar{A}+B)}\,\overline{(B+C)}} = (\bar{A}+B)+(B+C) = \bar{A}+B+C$$

From given truth table, $Z = A + B + \bar{C}$.
Thus, option (b) is correct.

79. (d)

From K-map, output

$$= \bar{A}\bar{B}C + \bar{A}B\bar{C} + A\bar{B}\bar{C} + ABC$$

$$= C(\bar{A}\bar{B} + AB) + \bar{C}(\bar{A}B + A\bar{B})$$

$$= C(\overline{\bar{A}B + A\bar{B}}) + \bar{C}(\bar{A}B + A\bar{B})$$

$$= A \oplus B \oplus C$$

80. (c)

- Positive logic AND gate is equivalent to bubbled NOR gate.
- Positive logic OR gate is equivalent to bubbled NAND gate.
- Positive logic NAND gate is equivalent to bubbled OR gate.
- Positive logic NOR gate is equivalent to bubbled AND gate .

81. (d)

$f = A + BC\bar{D}$ can be implemented using 4 NAND gates as shown below:

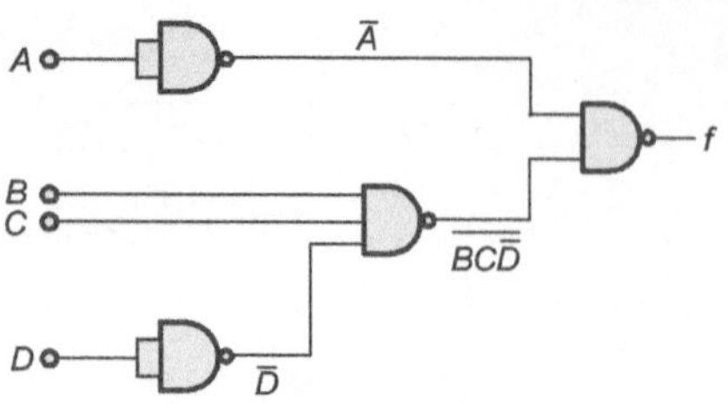

Here, $f = \overline{\bar{A}(\overline{BC\bar{D}})} = A + BC\bar{D}$

82. (a)

Truth table of positive logic AND gate

A	B	Y = AB
0	0	0
0	1	0
1	0	0
1	1	1

Truth table after interchanging all 1's and 0's

A	B	Y = AB
1	1	0
1	0	0
0	1	0
0	0	1

Thus, we see that after interchanging all 1's and 0's of inputs in the truth table of positive logic AND gate, the resulting truth table obtained is the truth table of negative logic OR gate. Similarly, we can prove that if all 0's and 1's inputs in the truth table of NAND gate are interchanged, then the gate becomes negative logic NOR gate. Hence, both assertion and reason are true and reason is the correct explanation of assertion.

83. (b)

- EX-OR gate can be used as a controlled inverter as shown below:

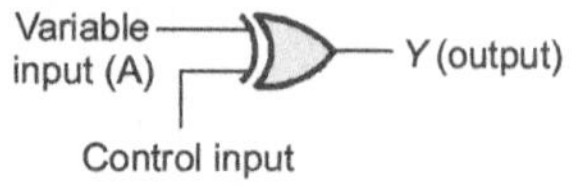

For control input = 1, $Y = \bar{A}$
For control input = 0, $Y = A$

- NOT gate is basically used as an inverter or for complementation purposes.
- EX-NOR gate can be used as even-odd parity checker because the output of EX-NOR gate is 1 if the number of 1's in its inputs is even; if the number of 1's is odd, the output is 0.
- NOR gate is an universal gate.

84. (b)

An important property of the EX-NOR gate is that it can be used for bit comparison. The output of an EX-NOR gate is 1 (high) if both the inputs are similar, i.e. both are 0 or 1 (not necessarily even no. of high inputs); otherwise its output is 0 or low (not necessarily odd number of high inputs). Hence, EX-NOR gate can be used as a "one-bit comparator". Thus, assertion is a correct statement. Reason is also a true statement but not a correct explanation of assertion.

85. (a)

$$X = C\cdot\overline{(B\bar{C}+\bar{B}C)}\cdot(A\bar{B}+\bar{A}B)$$

Here, (0, 1, 1) gives the output $X = 1$.

86. (a)

- $X \oplus 1 = \bar{X}\cdot 1 + X\cdot\bar{1} = \bar{X} + 0 = \bar{X}$
- $(X+Y)\cdot\overline{XY} = (X+Y)(\bar{X}+\bar{Y})$
  $= X\bar{Y} + \bar{X}Y = X \oplus Y$
- $X \odot 1 = \bar{X}\cdot\bar{1} + X\cdot 1 = X$
- $XY + (\overline{X+Y}) = XY + \bar{X}\cdot\bar{Y} = X \odot Y$

87. (d)

The EX-NOR output of a number of variables is HIGH, only when an even number of input variables (including 0) are LOW. The output waveform will be represented by option (d).

88. (c)

The truth table of AND gate is as shown below.

X	Y	Z	Output (f)
0	0	0	0
0	0	1	0
0	1	0	0
0	1	1	0
1	0	0	0
1	0	1	0
1	1	0	0
1	1	1	1

The wave forms of inputs X, Y, Z and output Y are drawn below. Time period of inputs X, Y and Z are respectively 2 ms, 1 ms and 0.5 ms.

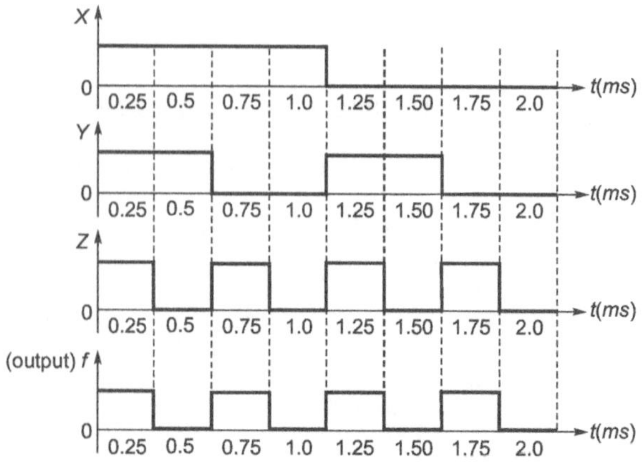

Hence, output frequency

$$= \frac{1}{\text{Output time period}} = \frac{1}{0.25\text{m}} = 4 \text{ kHz}$$

89. (d)

The correct o/p waveform for given NOR gate in option (d) is shown below:

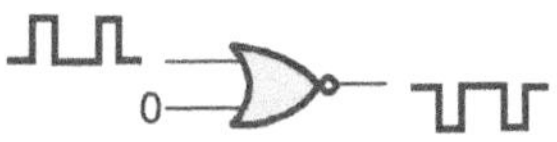

Truth Table for NOR gate with one input fixed at 0 is shown below.

A	B	Output
0	0	1
0	1	0

90. (a)

AND, OR, NOT are three basic gates which follow both commutative and associate law. However, the universal gates NAND and NOR don't follow associate law.

91. (c)

From given logic diagram, output is:

$$f = \overline{(\bar{A} + \overline{A+B})\cdot(\bar{B} + \overline{B+C})}$$

$$= \overline{(\bar{A} + \overline{A+B})} + \overline{(\bar{B} + \overline{B+C})}$$

$$= \bar{\bar{A}}\cdot\overline{\overline{(A+B)}} + \bar{\bar{B}}\cdot\overline{\overline{(B+C)}}$$

$$= A\cdot(A+B) + B\cdot(B+C)$$

$$= A\cdot A + AB + BB + BC$$

$$= A + AB + B + BC$$

$$= A(1+B) + B(1+C)$$

$$= A + B$$

92. (d)

Given expression is $f = \pi M(0, 1, 2, 3, 4, 7)$.
The K-map for this expression is shown below.

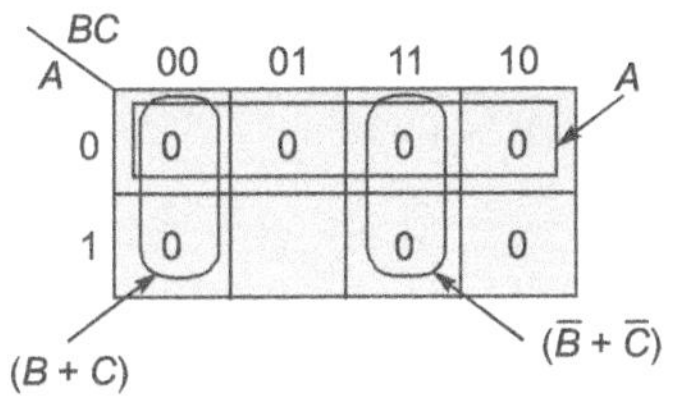

The minimized expression in POS form is

$$f = A\cdot(B+C)\cdot(\bar{B}+\bar{C})$$

The minimized expression '*f*' can be implemented using two input NOR gate as shown below.

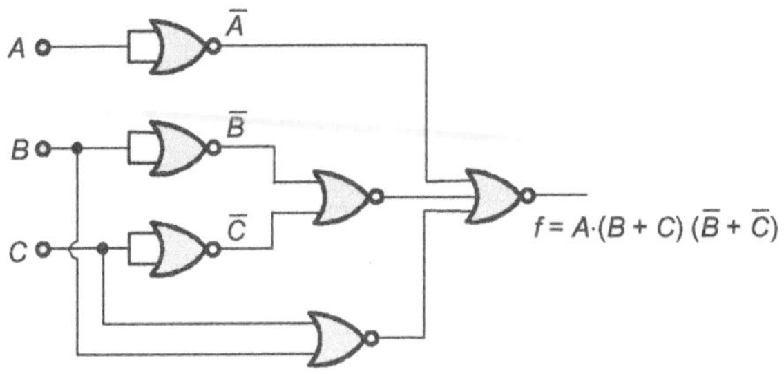

Thus, we require six NOR gates.

93. (b)

- Statement-1 is correct because positive and negative logics give rise to a basic duality in all Boolean identities. Here, an AND gate becomes an OR gate and an OR gate becomes an AND gate and vice-versa.
- For obtaining dual expression, variables are not complemented. Hence, statement-2 is not correct.
- NAND and NOR are the only two universal gates while NOT, OR, AND are three basic gates. Thus, statement-3 is correct.
- Universal gates follow associate law but not commutative law. Hence, statement-4 is correct.
- An inversion bubble on logic diagram at input indicates an active-LOW input. Thus, statement-5 is not correct.

Hence, statements 1 and 3 are correct while statements 2, 4 and 5 are false.

94. (c)

OR gate using NAND gate:

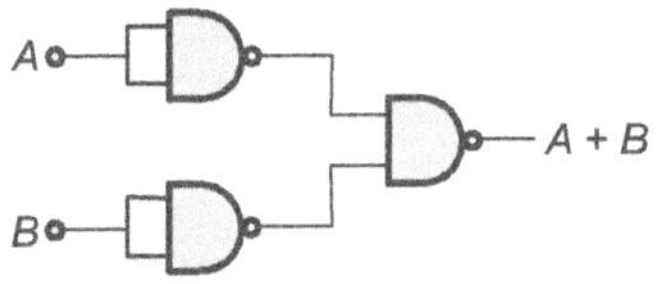

EX-NOR gate using NAND gate:

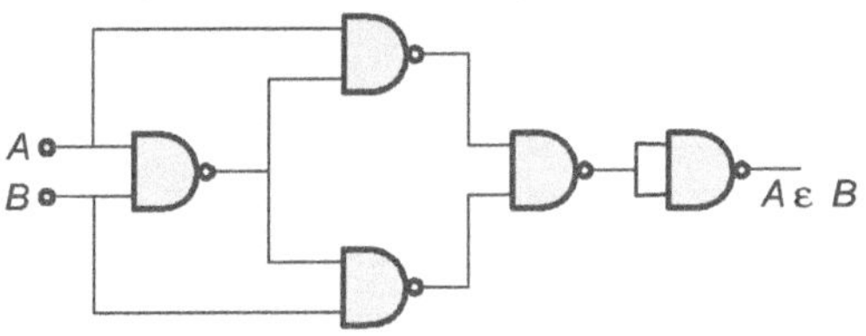

Thus, we require 3 NAND gates for an OR gate while 5 NAND gates for an EX-NOR gate.

95. (a)

Given, $\bar{A}+AB=0$

or, $(\bar{A}+A)(\bar{A}+B) = 0$ or $1\cdot(\bar{A}+B) = 0$

i.e. $\bar{A}$ must be 0 and B must be 0

Hence, $A = 1$ and $B = 0$

Also, $AB = AC$ or $1\cdot B = 1\cdot C$ or $B = C$

Hence, $C = 0$ ($\because B = 0$)

Also, $AB+A\bar{C}+CD = \bar{C}D$

or, $1\cdot 0 + 1\cdot 1 + 0\cdot D = 1\cdot D$ or $D = 1$

Thus, the values of A, B, C and D are given by $A = 1$, $B = 0$, $C = 0$ and $D = 1$.

96. (c)

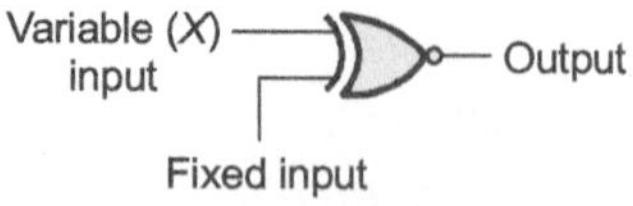

Output = $X\cdot 1+\bar{X}\cdot\bar{1} = X = \text{Input}$

97. (b)

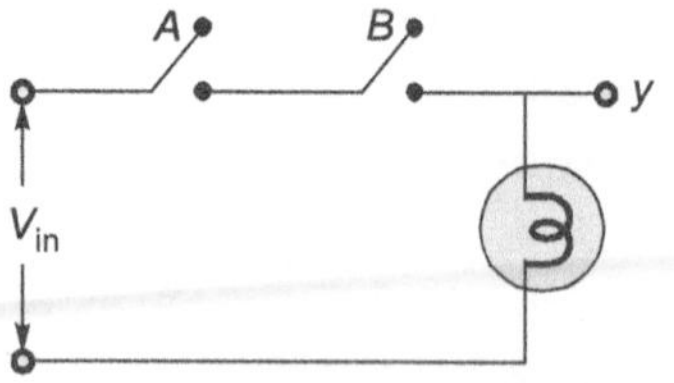

The truth table for above circuit is shown below.

A	B	Y
0	0	0
0	1	0
1	0	0
1	1	1

Hence, the output y is equivalent to output of an AND gate having two inputs A and B. Thus, an AND gate is equivalent to a series switching circuit.

98. (d)

For same inputs, $A\oplus A = 0$

For different inputs, $\bar{A}\oplus A = 1$

99. (c)

For same inputs, output (f) = 0

For different inputs, output (f) = 1

Hence, given truth table represents an EX-OR gate.

101. (d)

Truth tables:

EX-NOR gate

Inputs	Output
0 0	1
0 1	0
1 0	0
1 1	1

NOR gate

Inputs	Output
0 0	1
0 1	0
1 0	0
1 1	0

Thus, the gate can be either an EX-NOR or a NOR gate.

102. (a)

- When $V_1 = 0$ volt and $V_2 = 5$ volt, diode D_1 will be on.

  Using voltage division rule,

$$V_0 = \left(\frac{R_s}{R+R_s}\right)V_R$$

$$= \left(\frac{0.5}{0.5+10}\right)\times 5 = 0.24 \text{ volt}$$

- When $V_1 = V_2 = 5$ volt, then both diodes are off, no current flows anywhere, and $V_0 = 5$ volt.
- When $V_1 = V_2 = 0$ volt, then both diodes D_1 and D_2 will be on and their two R_s's will effectively come in parallel.

$$\text{Thus, } V_0 = \left(\frac{\frac{R_s}{2}}{\frac{R_s}{2}+R}\right)\cdot V_R = \left(\frac{0.25}{0.25+10}\right)\times 5 = 0.12 \text{ V}$$

103. (b)

The truth table for the given condition is shown below:

Input			Output
A	B	C	Y
0	0	0	0
0	0	1	0
0	1	0	0
0	1	1	1
1	0	0	0
1	0	1	1
1	1	0	1
1	1	1	1

The K-map for above truth table is shown below.

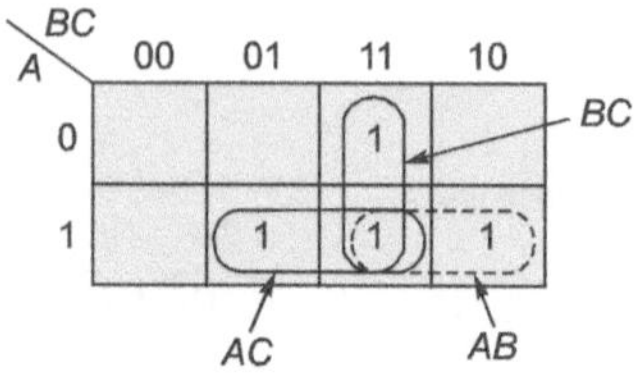

Thus, $Y = AB + BC + CA$ which can be implemented using 3 AND gates and 1 OR gate (total 4 basic gates).

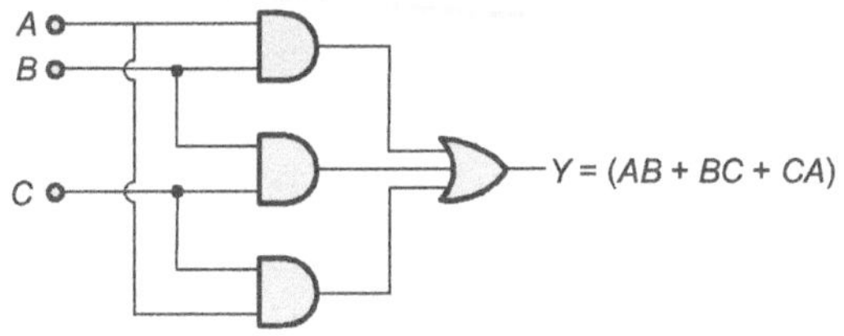

104. (d)

Let the output of second last NAND gate be X.

Then, $$X = \overline{\overline{(\overline{AB}\cdot A)}\cdot\overline{(\overline{AB}\cdot B)}}$$
$$= \overline{AB}\cdot A + \overline{AB}\cdot B$$
$$= (\bar{A}+\bar{B})\cdot A + (\bar{A}+\bar{B})\cdot B$$
$$= A\bar{B} + \bar{A}B = A \oplus B$$

Thus, $$f = \bar{X} = \overline{A \oplus B} = A \odot B$$
$$= AB + \bar{A}\bar{B}$$

Given circuit represents EX-NOR gate using NAND gate.

105. (b)

A two input OR gate truth table is shown below:

Inputs		Output
A	B	Y
0	0	0
0	1	1
1	0	1
1	1	1

From the truth table of OR gate it is clear that output is 0's only when all the inputs are 0's, otherwise for all other input combination output is always 1's. Hence, in the column for the output of the OR gate, the number of zeros is always 1's.

106. (c)

- Statement-1 is correct because any circuit can be realized using AND, OR and NOT gates.
- The two logic gates NAND and NOR can realize any logic circuit single-handedly. Hence, these two gates are called universal building blocks. AND, OR and NOT are called basic gates. Thus, statement-2 is not correct.
- Both NAND and NOR gates can perform all the three basic logic functions (AND, OR and NOT). Therefore, AND/OR/INVERT logic can be converted to NAND logic or NOR logic. Hence, statement-3 is correct.
- 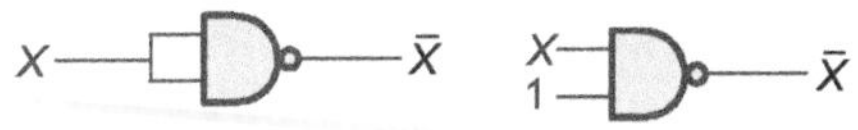

NAND gate as an Inverter — NAND gate as controlled Inverter

Statement-4 is not correct.

107. (c)

Since three or more variable X-NOR gates do not exist, therefore when a number of variables are to be X-NORed, a number of two input X-NOR gates have to be used. Thus, reason is not a correct statement.

108. (c)

When n is even, $X \odot X \odot X \ldots\ldots = 1$

When n is odd, $X \odot X \odot X \ldots\ldots\ldots = X$

109. (b)

From the given circuit, output is

$$F = \overline{\overline{\overline{[(A+\bar{C})\cdot B\bar{D}]}}\cdot[\overline{(A+\bar{C})\cdot B\bar{D}}]}$$
$$= \overline{[(A+\bar{C})+\overline{B\bar{D}}]\cdot[\overline{(A+\bar{C})\cdot B\bar{D}}]}$$
$$= [\overline{\overline{B\bar{D}}}+(A+\bar{C})\overline{(A+\bar{C})}]$$
$$(\because (X+Y)(X+Z) = X+YZ)$$
$$= [\overline{\overline{B\bar{D}}}+0] = \overline{\overline{B\bar{D}}} \qquad (\because X\bar{X} = 0)$$
$$= B\bar{D}$$

Therefore, the given logic circuit can be simplified as shown below.

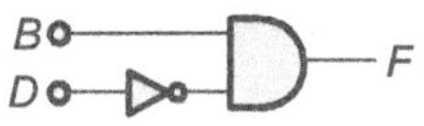

Thus, it require a AND gate and a NOT gate.

110. (c)

In case of ECL circuit, the floating inputs are considered as logic '0'. Hence output

$$Z = \overline{0\cdot(X+Y)} = 1$$

111. (d)

Since one of the input of EX-OR gate is grounded, therefore output is

$$Y = 0 \oplus X = \bar{0}\cdot X + 0\cdot\bar{X} = X$$

112. (c)

$$f = \frac{1}{2n\,t_{pd}} = \frac{1}{2\times 5\times 100\times 10^{-12}} = 1\text{ GHz}$$

113. (a)

When input = 11111

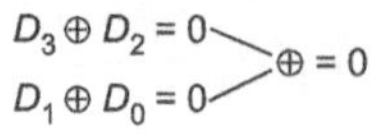

and, $0 \oplus D_2 = 1 \Rightarrow$ Error
Similarly for 10000.

114. (d)

A	B	T_1	T_2	T_3	Y
0	0	OFF	OFF	ON	0
0	1	OFF	ON	ON	0
1	0	ON	OFF	ON	0
1	1	ON	ON	OFF	1

i.e. output of Y represents an AND gate.

115. (c)

- When a MUX is used to implement a logic function, the logic variables are applied to the MUX's select or control input.
  For an example, if we want to implement a NOT gate using a 2×1 MUX, then the circuit will be represented as shown below:

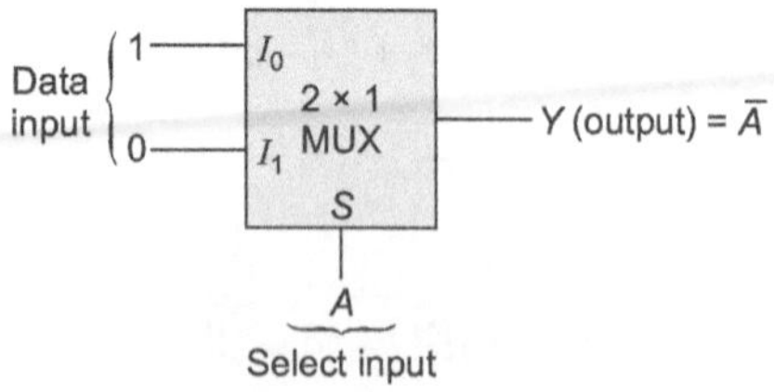

Thus, output

$$Y = I_0\bar{S} + I_1 S$$
$$= 1\cdot\bar{A} + 0\cdot A = \bar{A}$$

Hence, statement-1 is not correct.

- Statement-2 is correct because the internal circuit for a DEMUX (data distributor) and a decoder are basically same.

117. (d)

Given, function $F(A, B, C) = \Sigma(1, 3, 5, 6)$ has three variables. Hence, it can be implemented using a multiplexer (4×1 MUX) with two select inputs and four data inputs.
The implementation table is shown below:

	D_0	D_1	D_2	D_3
$\bar{A}$	0	(1)	2	(3)
A	4	(5)	(6)	7
	0	1	A	$\bar{A}$

Now, the given three variable functions can be implemented using 4-to-1 multiplexer as shown below:

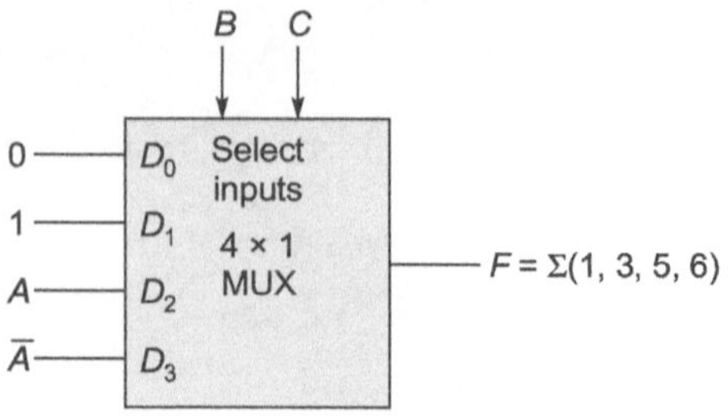

Hence, for implementation of given function we require one 4×1 MUX and a NOT gate.

118. (a)

The truth table for half adder is shown below:

Inputs		Sum	Carry
X	Y	S	C
0	0	0	0
0	1	1	0
1	0	1	0
1	1	0	1

Here, sum output is $S = \bar{X}Y + X\bar{Y}$ and carry output is $C = XY$.

119. (c)

No. of input lines = 2^m (where, m = No. of select/control inputs) or, $32 = 2^m$ or $m = 5$.

120. (c)

- Serial adders are used where circuit minimization is more important than speed because they require one clock pulse for each pair of bits added and hence slower than parallel adders. Thus, statements 2 and 4 are not correct.
- Statements 1 and 3 are correct.

121. (b)

- The multiplexer can be viewed as a function generator because we can easily set or change the logic function it implements.
- The multiplexer acts like a digitally controlled multi-position switch because the digital code applied to the SELECT inputs determine which data inputs will be switched to the output.

Hence, both assertion and reason are correct but reason is not the correct explanation of assertion.

122. (a)

In cases where non-trivial, a multiple-output expressions of the same input variables are required, a decoder could be advantageous than a multiplexer because in such cases, one multiplexer is required for each output, whereas it is likely that only one decoder supported with a few gates would be required.

123. (d)

The speed of operation of the parallel binary adder is limited by the carry propagation delay through all stages. Hence, assertion is not correct.

127. (d)

The truth table for the given problem is shown below:

C	D_3	D_2	D_1	Y
0	X	X	X	0
1	0	0	0	0
1	0	0	1	1
1	0	1	0	1
1	1	0	0	1

The K-map for the above truth table is shown below:

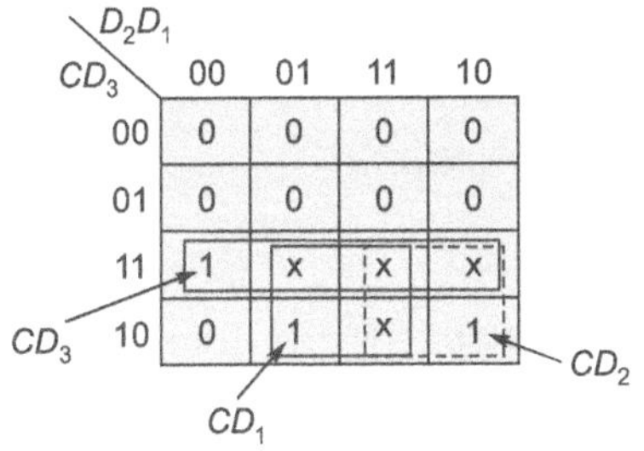

Thus output, $Y = CD_3 + CD_2 + CD_1$

$= C(D_1 + D_2 + D_3)$

128. (b)

Let us consider D for door, I for ignition and L for light. Then, conditions to activate the alarm are:

(i) The headlights are ON while the ignition is OFF.

i.e. $L = 1$, $I = 0$ and D may be anything.

(ii) The door is open while the ignition is ON.

i.e. $D = 1$, $I = 1$, L may be anything.

Also, alarm will sound if logic circuit output is zero.

Therefore, output (Y) for above condition is zero and for rest of the condition it is 1 which is shown in following truth table.

D	I	L	Y
0	0	0	1
x	0	1	0
0	1	0	1
0	1	1	1
1	0	0	1
1	1	x	0

K-map for above truth table is shown below.

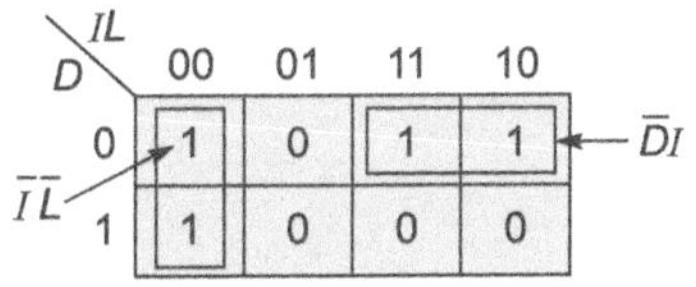

Thus, the output of the given logic circuit is

$$Y = \bar{I}\bar{L} + \bar{D}I$$

129. (a)

Invalid BCD inputs are from decimal 10 to 11.

The K-map is shown below:

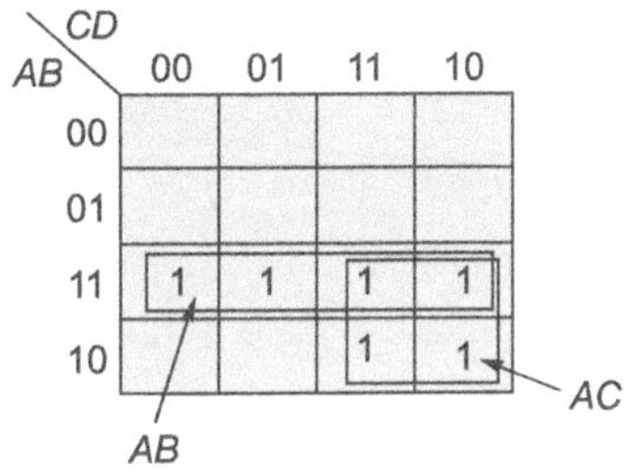

Thus, output $F = AB + AC$, which can be implemented using three 2-input NAND gates as shown below.

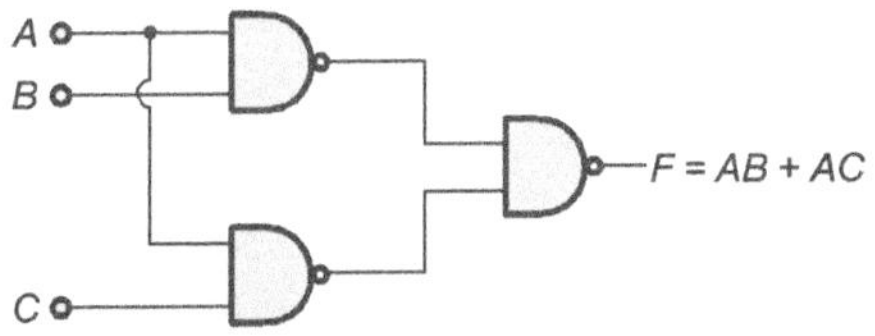

130. (c)

Let the binary inputs to be added be (A_4 A_3 A_2 A_1) and (B_4 B_3 B_2 B_1). A 4-bit adder using cascade connection of four 1-bit full adder is shown below. Here, C represents for carry and S for sum.

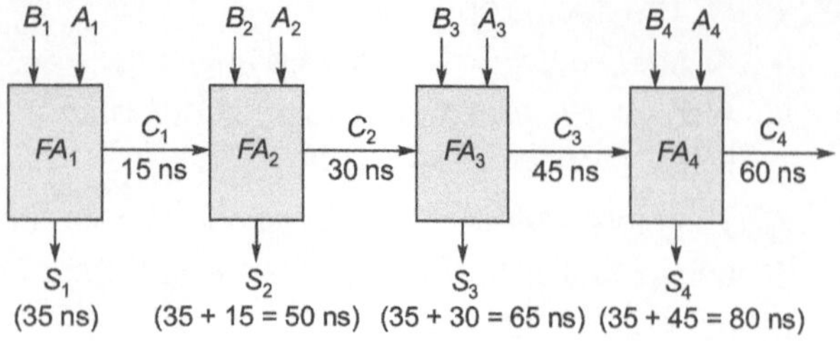

It is clear from above circuit that each addition requires 80 ns.

Therefore, maximum number of additions that can be performed by this 4-bit adder

$$= \frac{1}{80\ ns} = \frac{10^9}{80}/\text{sec} = 125 \times 10^5 \text{ additions/sec}$$

131. (a)

Let the two inputs variations be X and Y.

EX-NOR gate using 2X1 MUX:

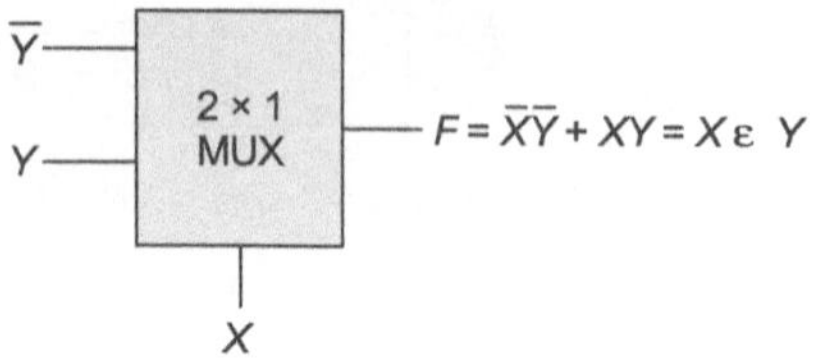

Here, we require two 2 × 1 MUX.

NAND gate using 2 × 1 MUX:

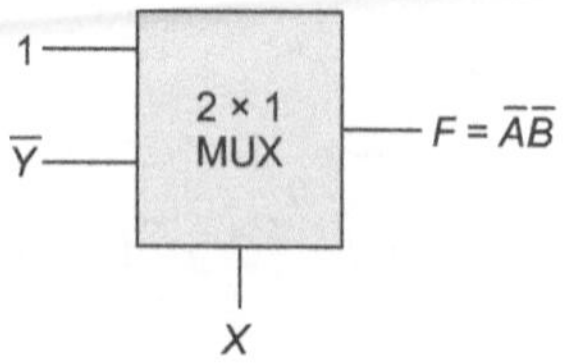

Thus, we require two number of 2 × 1 MUX for NAND gate implementation.

132. (d)

Output, $f = \bar{X}\bar{Y} \cdot 1 + \bar{X}Y \cdot 0 + X\bar{Y} \cdot 0 + XY \cdot 1$

$= \bar{X}\bar{Y} + XY = X \odot Y$

= EX-NOR gate

133. (b)

Given, logic expression is $f(A, B, C) = \pi M(0, 3, 5)$
$= \Sigma m\ (1, 2, 4, 6, 7)$.

Here, B and C has to be control inputs while A as data inputs. We form the table as shown below:

	I_0	I_1	I_2	I_3
$\bar{A}$	0	①	②	3
A	④	5	⑥	⑦
	A	$\bar{A}$	1	A

Thus, $I_0 = A, I_1 = \bar{A}, I_2 = 1, I_3 = A$

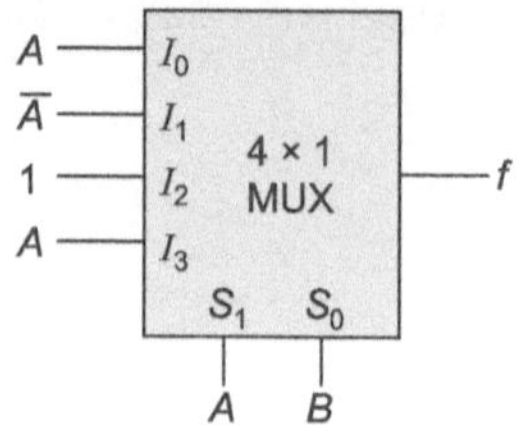

134. (c)

A full adder is used to add three input bits (A, B and C) to give two outputs namely sum and carry as shown below:

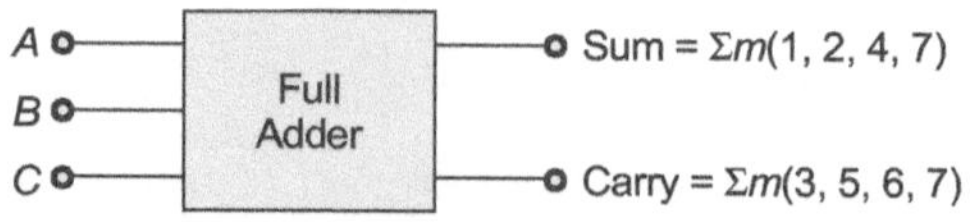

For its implementation we require either of the following:

- One 3 × 8 decoder and two OR gates.
- Two half adders and one OR gate.
- Nine NAND/NOR gates.

135. (d)

To add two n-bit numbers using parallel adder circuit, we require

(n – 1) full adders and 1-half adder

or

n-full adders

or

(2n – 1) half adders and (n – 1) OR gates

Thus, to add two 4-bit numbers using parallel adder circuit, we require

3-full adder and 1-half adder

or

4-full adders

or

7-half adder and 3 OR gates

Hence, option (d) is correct.

136. (a)

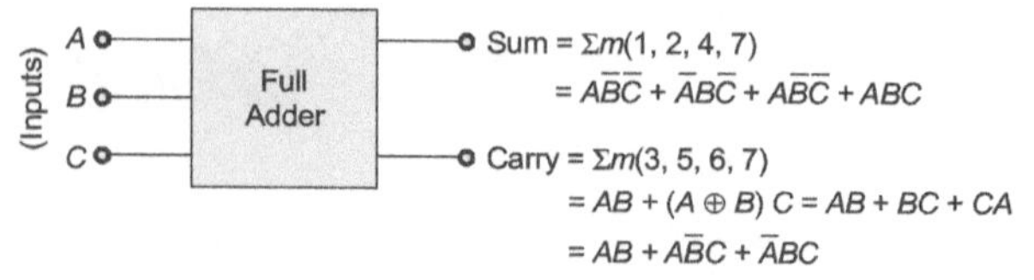

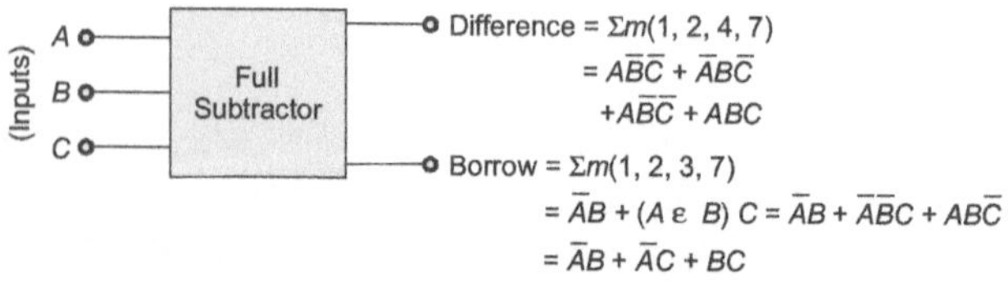

137. (b)

The correction to be applied in decimal adder to the generated sum is 00110. When the four bit sum is more than 9, then the sum Is invalid. In such cases, +6 (i.e. 0110) is added to the four bit sum to skip the six invalid states. If a carry is generated when adding 6, the carry is added to the next four bit group.

138. (d)

The gates required to build a half adder are EX-OR gate and AND gate. Figure below shows the logic diagram of half adder.

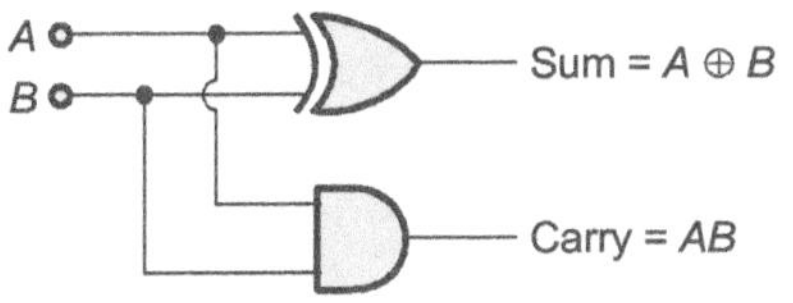

139. (c)

The code where all successive numbers differ from their preceding number by single bit is Gray code. (It is an unweighted code. The most important characteristic of this code is that only a single bit change occurs when going from one code number to next).

140. (a)

The device which changes from serial data to parallel data is demultiplexer because it takes in data from one line and directs it to any of its *N* outputs depending on the status of the select inputs.

141. (b)

A decoder converts binary words into alphanumeric characters i.e. it converts BCD to seven segment.

142. (d)

From given circuit, we have:

$$y_1 = x_1$$
$$y_2 = x_1 \oplus x_2$$
$$y_3 = x_2 \oplus x_3$$

Thus, the given circuit will convert binary code $x_1 x_2 x_3$ into Gray code $y_1\, y_1\, y_3$.

143. (a)

From given circuit, output is

$$Z = \overline{[\overline{(X+\overline{Y})}+\overline{X}]} = (X+\overline{Y})\cdot(\overline{\overline{X}})$$
$$= (X+\overline{Y})\cdot X = X + X\overline{Y}$$
$$= X(1+\overline{Y}) = X$$

144. (b)

In even parity generator if number of '1' is odd then output will be zero. Here, input 0111 has odd number of 1's, therefore output of even parity generator will be zero.

145. (d)

The truth table for given condition is shown below:

A	B	C	Y
0	0	0	1
0	0	1	1
0	1	0	1
0	1	1	1
1	0	0	0
1	0	1	0
1	1	0	0
1	1	1	1

The K-map for above truth table is shown below.

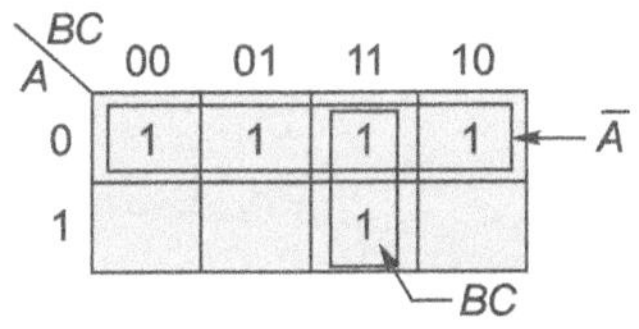

Thus, output $Y = (\overline{A} + BC)$ which can be implemented using two 2-input NAND gate as shown below.

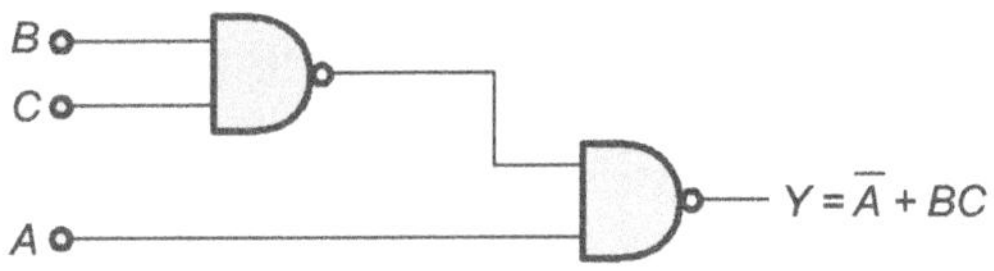

146. (d)

$$Z = D[\overline{A}\overline{B}\overline{C} + \overline{A}\overline{B}C + \overline{A}BC + A\overline{B}\overline{C} + ABC]$$
$$= D[\overline{A}\overline{B}(\overline{C}+C) + BC\,(\overline{A}+A) + A\overline{B}\overline{C}]$$
$$= D[\overline{A}\overline{B} + BC + A\overline{B}\overline{C}] = D[\overline{B}(\overline{A}+A\overline{C}) + BC]$$
$$= D[\overline{B}\,\overline{A} + \overline{B}\overline{C} + BC] = D(B \odot C + \overline{A}\overline{B})$$

147. (d)

Here, $S_2 = C + 0 = C$; $S_1 = B$; $S_0 = A$

Truth table:

$S_2 = C$	$S_1 = B$	$S_0 = A$	Y
0	0	0	1
0	0	1	0
0	1	0	0
0	1	1	1
1	0	0	0
1	0	1	1
1	1	0	1
1	1	1	0

$$Y = \bar{A}\bar{B}\bar{C} + AB\bar{C} + A\bar{B}C + \bar{A}BC$$
$$= \bar{C}(\bar{A}\bar{B} + AB) + C(A\bar{B} + \bar{A}B)$$
$$= C(\overline{A \oplus B}) + C(A \oplus B)$$
$$= C \odot A \oplus B$$

148. (b)

Let the output of MUX with select input C be f_1.

Then, $f_1 = CB + \bar{C} \cdot 0 = CB$

Now, output $f = \bar{f_1} + f_1\bar{A}$

$$= \overline{CB} + CB\bar{A} = \overline{CB} + \bar{A}$$
$$= \bar{C} + \bar{B} + \bar{A} = \overline{ABC}$$

= Output of a NAND gate with inputs A, B and C

150. (d)

A parallel adder is faster than a serial adder but require more hardware for implementation.

151. (d)

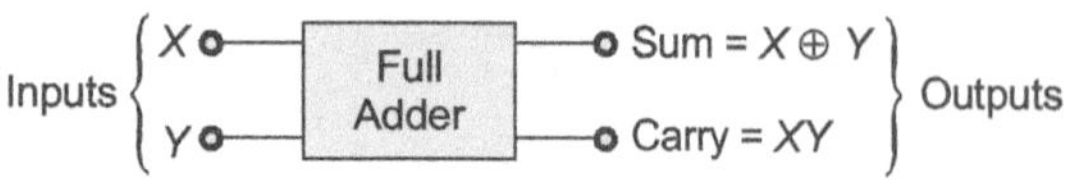

Sum = $X \oplus Y$ (requires EX-OR gate)

Carry = XY (requires AND gate)

152. (a)

Let, $A = 5$ and $B = 4$

Then, $A - B = 5 - 4 = 1$

Using 1's complement, we can find $A - B$ as follows:

+5 = 0101

+4 = 0100

1's complement of +4 = 1011

$$\begin{array}{r} 0101 \\ +1011 \\ \hline 10000 \\ + \\ \hline 0001 \end{array}$$

∴

Thus, result of $A - B = (0001)_2 = (1)_{10}$.

Thus, first of all complement of B is found out and then A is added to it for finding the result of $A - B$.

154. (b)

For J-K flip-flop, the output is clearly defined for all combinations of two inputs. For S-R flip-flop output is not defined when $S = R = 1$. A D flip-flop has only one input.

155. (a)

Truth table for J-K flip-flop is shown below.

Clock	J	K	Q^+
0	x	x	Q
1	0	0	Q
1	0	1	0
1	1	0	1
1	1	1	$\bar{Q}$

When $J = 1$ and $K = 0$, output (Q^+) is always set upon the occurrence of the active clock transition.

156. (c)

In a J-K flip-flop, toggle means change the output to the opposite state. A J-K flip-flop toggles when $J = K = 1$ for which $Q^+ = \bar{Q}$ i.e. $Q = 0$, $\bar{Q} = 1$ or $Q = 1$, $\bar{Q} = 0$.

157. (b)

When, $S = 1, R = 0$

Output, $(Q^+) = 1$ (high)

158. (d)

Maximum time taken for all flip-flops to stabilize is (75 ns × 8) + 50 ns = 650 ns.

Frequency of operation must be less than

$$\frac{1}{650ns} \approx 1.5 \text{ MHz}$$

Thus, option (d) is correct.

159. (c)

When $J = 1$, $K = 1$ and the clock, next state will be complement of the present state. Thus,

$Q^+ = \bar{Q} = \bar{1} = 0$.

161. (b)

From given circuit, $J_0 = \bar{Q}_1$, $K_0 = 1$ and $J_1 = Q_0$, $K_1 = 1$. When both the flip-flop are cleared, their outputs are $Q_0 = Q_1 = 0$. After the clear input goes high, each pulse on the x-input will cause a change in the flip-flop outputs. The truth table for the given circuit will be as shown below.

In consecutive time intervals, the values of Q_1Q_0 are 00, 01, 10, 00, 01,...... Thus, the circuit generates the count sequence : 0, 1, 2, 0, 1,.... Hence, given circuit is a Mod-3 counter.

Time Interval	FF0			FF1		
	J_0	K_0	Q_0	J_1	K_1	Q_1
Clear	1	1	0	0	1	0
t_1	1	1	1	1	1	0
t_2	0	1	0	0	1	1
t_3	1	1	0	0	1	0
t_1	1	1	1	1	1	0

163. (c)

- Combinational circuits are often faster than sequential circuits since the combinational circuits do not require memory whereas the sequential circuits need memory devices to perform their operations in sequence. Hence, statement -1 is not correct.
- In an asynchronous circuit, events are allowed to occur without any synchronisation. In such a case, the system becomes unstable which results in difficulties. Hence, statement-2 is not correct.
- Statement-3 is correct which is the definition of a combinational circuit.
- In a combinational circuit, for a change in the input, the output appears immediately, except for the propagation delay through circuit gates. Thus, statement-4 is not correct.
- In a sequential circuit, an output signal is a function of the present input signals and a sequence of the past input signals i.e. the past output signals since the output signals are fed back to the input side. Hence, statement-5 is correct.

Thus, statements 3 and 5 are only correct.

164. (b)

A latch is memory device with the capability of storing one binary digit of information because the latch output will remain set/reset until the trigger pulse is given to change the state.

A basic latch is made up of cross coupled inverters as shown below.

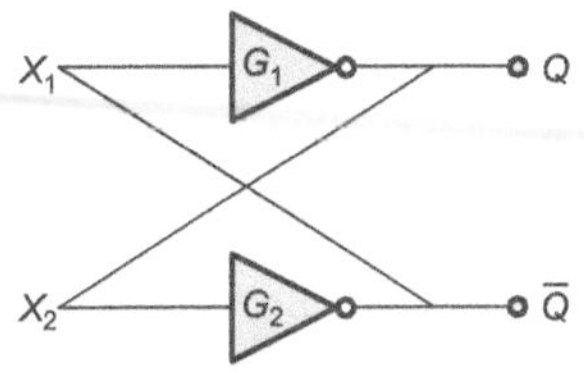

Thus, both assertion and reason are true but reason is not the correct explanation of assertion.

165. (d)

The truth table for *S-R* and *J-K* flip-flop are shown below.

S	R	Q^+
0	0	Q
0	1	0
1	0	1
1	1	Indeterminate

J	K	Q^+
0	0	Q
0	1	0
1	0	1
1	1	$\bar{Q}$

It is clear from the above truth table that when $S = R = 1$, the output of *S-R* flip-flop is indeterminate. However, when $J = K = 1$, the output of *J-K* flip-flop is $\bar{Q}$ (not indeterminate). Therefore, the indeterminate condition of the *S-R* flip-flop is permitted in a *J-K* flip-flop.

A *J-K* flip-flop has a characteristic similar to that of an S-R flip-flop which is evident from the above truth tables.

Hence, assertion is false but reason is true.

166. (b)

Let the flip-flop is initially reset.

The truth table for given circuit is shown below.

Clock	J	K	Q^+
0	1	1	0
1	1	1	0
0	1	1	1
1	1	1	1
0	1	1	0
1	1	1	0
⋮	⋮	⋮	⋮

Thus, the waveform for Q will be as shown below.

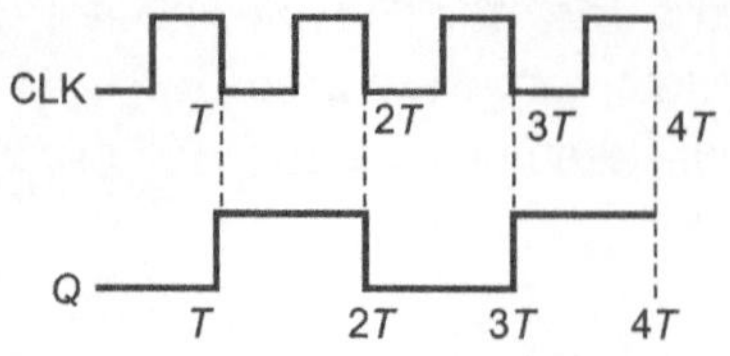

167. (c)

If $J = K = 1$, then the truth table will be as shown below (Initially flip-flop is reset).

Clock	J	K	Q	Q^+
1	1	1	0	1
1	1	1	1	0
1	1	1	0	1
1	1	1	1	0
⋮	⋮	⋮	⋮	⋮

Thus, the given *J-K* flip-flop will work as a divide-by-2 element.

168. (a)

- The *J-K* flip-flop is the most versatile and most widely used of all the flip-flops.
- Transparent latch is associated with D-flip flop since the clocked *D* latch is called a transparent *D* latch,because its output follows the input when the clock is HIGH.
- Master-slave flip-flops are called pulse-triggered flip-flops, because the length of time required for its output to change state equals the width of one (clock) pulse.
- *T* flip-flop is not widely available as commercial items.

171. (b)

An unclocked flip-flop is called a latch, because the output of the flip-flop latches on to '1' or '0' immediately after the input is applied. Hence, assertion is a correct statement. Reason is also a true statement but not the correct explanation of assertion.

172. (d)

- Statement-1 is correct.
- The inputs to a master-slave flip-flops must not change when the clock is HIGH. If they change, the flip-flop gives an output corresponding to the last combination of inputs present before the clock terminals. Hence, statement-2 is not correct.
- In a master-slave flip-flop, master is edge-triggered while slave is level triggered. Thus, statements-3 is not correct.
- Data lock-out flip-flops are nothing but master-slave flip-flops in which the master is an edge triggered flip-flop. Thus, statement-4 is correct.

173. (a)

$Q = 0$ (Initially)

$J = J_1J_2$	$K = K_1K_2$	Q
0	0	0
0	1	0
0	0	0
0	0	0
1	1	1
0	0	1
0	0	1
1	1	0

Thus, serial output data at Q

$= 00001110 = 14$ (Decimal value)

174. (c)

Given flip-flop represents *S-R* latch using NOR gates.

When $A = 1$, $B = 1$ we will get invalid or unusable state.

175. (b)

Here, $D = X + \bar{Y}Q$

If $X = S$ and $Y = R$ then, the given circuit represents S-R flip-flop using D flip-flop.

176. (c)

Race around condition occurs in *J-K* flip-flop when $J = K = 1$. This condition can be avoided if the propagation delay of flip-flop is more than pulse-width of the clock but less than the clock time period.

i.e. $2\, t_p < \Delta t < T$

177. (d)

For conversion of S-R flip-flop to

A. **J-K flip flop:**

$S = JQ'$ and $R = KQ$

B. **T Flip-flop:**

$S = TQ'$ and $R = TQ$

C. **D flip-flop:**

$S = D$ and $R = \bar{D}$

178. (a)

For a T flip-flop, $Q_{n+1} = T \oplus Q_n = T\bar{Q}_n + \bar{T}Q_n$

180. (d)

The truth table will be as shown below.

CLK	$J = \bar{Q}$	$K = 1$	Q_n	$\bar{Q}_n$
0	–	–	0	1
1	1	1	1	0
2	0	1	0	1
3	1	1	1	0
4	0	1	0	1
5	1	1	1	0
6	0	1	0	1

Thus, output Q for 6 clock pulse = 101010.

181. (c)

The truth table of J-K flip-flop is shown below.

Clock	J	K	$Q_n + 1$
0	–	–	Q_n
1	0	0	Q_n
1	0	1	0
1	1	0	1
1	1	1	$\bar{Q}_n$

Given, $Q_n = 0$ initially,
When clock pulse is applied, the state of FF changes to '1' i.e. $Q_{n+1} = 1$, which is possible.

if, $J = 1, K = 0$ ...(i)

or, $J = 1, K = 1$ ...(ii)

Thus, $J_n = 1$ and K_n = 'X' (don't care) satisfy the above two conditions.

182. (c)

A FF is set when $Q = 1$; $\bar{Q} = 0$.

A FF is reset when $Q = 0$; $\bar{Q} = 1$.

183. (d)

The circuit for obtaining a T-flip-flop using a J-K flip-flop is shown below.

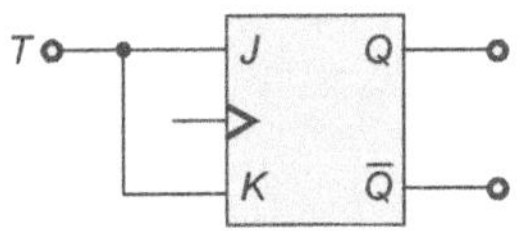

Thus, no any extra gate is required.

184. (c)

When $S = 1$, $R = 1$ the output is not defined i.e. unpredictable which can be either '0' or '1'.

185. (c)

In the S-R flip-flop when $S = 1$ and $R = 1$, the output is invalid (or not defined). This disadvantage of the S-R flip-flop can be eliminated using the J-K flip-flop. But, in a J-K flip-flop when $J = 1$ and $K = 1$ and propagation delay of flip-flop is less than the pulse-width, then race around condition occurs. This undersirable phenomenon is overcome by using the master-slave flip-flop.

186. (b)

A register in which data gets shifted towards left or right when clock pulses are applied is called a shift register. Hence, data can be changed from spacial code to temporal code by using shift registers.

187. (c)

Suppose the content of register be 0101 (i.e. 5). After one bit left shift operation, the content of register becomes 1010 (i.e. 10). Thus, the decimal value of new content of register get double. i.e. multiple of two of the initial decimal value.

188. (a)

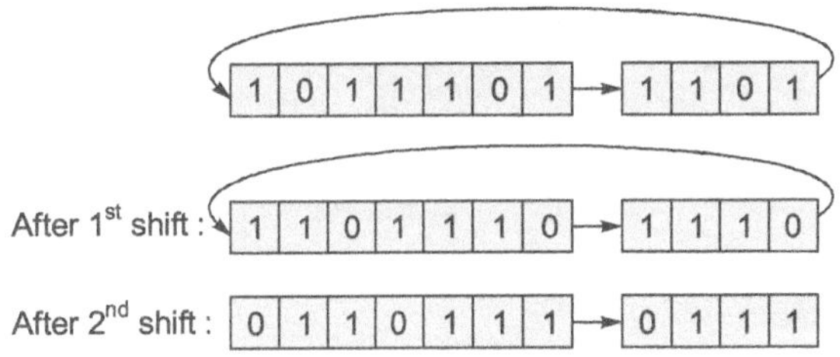

Thus, the contents of register after 1st and 2nd right shift operations are respectively 1110 and 0111.

190. (d)

- Statements-1 and 2 are correct.
- In a bidirectional shift register, data can be shifted from left-to-right or right-to-left. Hence, statement-3 is not correct.
- In a shift register, data is said to be in parallel form if the bits are available simultaneously. Hence, statements-4 is not correct.

191. (d)

A shift register has no specified sequence of states except in certain very specialized applications, whereas a counter has a specified sequence of states. Hence, assertion is a false statement. Shift registers are basically used for the storage and transfer of digital data and, therefore are very much similar to counters. Hence, reason is a correct statement.

192. (c)

A shift register can be constructed using *D*, *S-R* or *J-K* flip-flops. Hence, statement 3 is not correct. Statements 1, 2 and 4 are correct.

193. (a)

PIPO/SISO/SIPO/PISO all these four shift registers are static shift registers because each one of the memory elements (i.e. FFs) used to build these registers can retain the data bit indefinitely since once loaded, the contents of each element of the register remain the same. Hence, both assertion and reason are true and reason is the correct explanation of assertion.

194. (b)

Since the data is entered serially, therefore eight clock pulses will be required to shift an 8-bit number.

Given, $T_{clk} = \dfrac{1}{f_{clk}} = \dfrac{1}{10 \times 10^{-6}} = 100 \text{ ns}$

Hence, the time required $= n\,T_{clk}$

$= 8 \times 100 \text{ ns} = 800 \text{ ns}$

195. (c)

Most of the buffer register use *D* flip-flop because they do nothing more than storing a binary bit. A *D*-flip-flop is used to store a binary bit temporary. Hence, reason is a false statement.

197. (b)

J-K flip-flops are mostly used in shift registers.

199. (c)

CLK	S_{in}	*A*	*B*	*C*	*D*
0	1	0	1	1	0
1	0	1	0	1	1
2	1	0	1	0	1
3	1	1	0	1	0

200. (b)

The output of XOR gate is $z = b_{i+1} \oplus b$ and this output shift the register to left.

Initially, $z = 0$ (z = output of XOR gate)

After 1st clock $z = b_7 \oplus 0 = b_7$

After 2nd clock $z = b_7 \oplus b_6$

After 3rd clock $z = b_6 \oplus b_5$

After 4th clock $z = b_5 \oplus b_4$

This shows that given circuit acts as binary to gray code converter.

202. (c)

Total time required $= 8 \times t_p = \dfrac{8}{f} = \dfrac{8}{10^6} = 8\ \mu s$

204. (a)

The number of flip-flops required for Mod-M counter is n and the relation is given by

$M \leq 2^n$

or, $16 \leq 2^n$

or, $2^4 \leq 2^n$

or, $n \geq 4$

or, $n = 4$

205. (c)

With n – flip-flops, number of states in a ring counter is n.

With 5 – flip flops, number of states in a ring counter is 5.

206. (d)

Given, input frequency to *T*-flip flop $= f_i = 100$ Hz. When three *T*-flip flops are cascaded, then total number of states in this cascade combination $= 2^3 = 8$.

Now, final output frequency $= \dfrac{f_i}{2^n}$

$= \dfrac{100 \text{ Hz}}{2^3} = 12.5 \text{ Hz}$

207. (b)

In synchronous counter all the flip - flops change state at the same time.

208. (d)

A decade counter counts 10 states from 0 to 9 (0000 to 1001). Thus, four flip-flops are required to construct a decade counter.

209. (a)

A mod-6 (Mod-M) and Mod-4 (Mod-N) counter when cascaded will work as a Mod-24 (Mod-MN) counter as shown below.

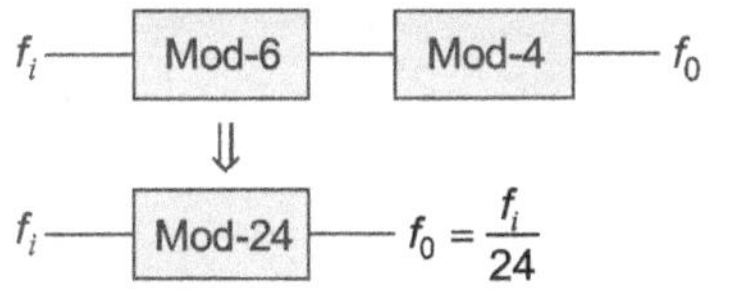

Thus, output frequency

$$f_0 = \frac{f_i}{24} = \frac{48\text{ kHz}}{24} = 2\text{ kHz}$$

211. (c)

For a state change to ripple through all n stages,

$$T_c = nt_{pd}$$

Therefore, the clock frequency is constrained by

$$\frac{1}{T_c} = f_c = \frac{1}{nt_{pd}}$$

$$\therefore \quad t_{pd} = \frac{1}{nf_c}$$

or, $$t_{pd}(\text{min}) = \frac{1}{10 \times 10 \times 10^6} = 10\text{ ns}$$

212. (a)

Since all the FFs change state simultaneously in synchronization with the clock pulse, the propagation delays of FFs do not add together (as in Asynchronous counters) to produce the overall delay. Due to this reason the synchronous coutners can operate at much higher frequencies than those that can be used in asynchronous counters. Hence, both assertion and reason are true and reason is the correct explanation of assertion.

213. (d)

The number of FFs n is to be selected such that the number of states $N \leq 2^n$. With n *FFs*, the largest count possible is $2^n - 1$.

Therefore,

$$2^n - 1 = 16{,}383$$

or, $$n = \log_2 16{,}384 = 14$$

So, the number of FFs required is 14.

∴ Frequency at the output of last stage is

$$f_{14} = \frac{f_c}{2^{14}}$$

$$= \frac{8.192}{16{,}384}\text{ MHz} = 0.5\text{ MHz}$$

$$= 500\text{ kHz}$$

215. (d)

Ring counter and Johnson counter suffers from the problem of lock-out i.e. if the counter finds itself in an unused state, it will persist in moving from one unused state to another and will never find its way to a used state. Thus, assertion is a false statement.

217. (d)

- Statements-1 and 2 are correct.
- CLEAR and PRESET are the two asynchronous inputs being used. Hence, statement-3 is not correct.
- Only asynchronous counter suffers from decoding errors. Thus, statement-4 is not correct.

218. (b)

Given, total number of states in ripple counter required = 25.

With n flip-flops, number of states in a ripple counter = 2^n.

With 4 flip-flops, maximum number of states = $2^4 = 16$.

With 5 flip-flops, maximum number of states = $2^5 = 32$.

Thus, we require minimum 5 flip-flops for a 25 : 1 ripple counter and since there are 4 flip-flops available on a chip, therefore we require at least 2 such chips.

219. (a)

For a BCD counter, number of bits = 4 (equal to number of states). For 4-bits we require 2 flip-flops. (Since $2^n = 2^2 = 4$ states).

220. (b)

It is a down counter because '0' state of previous FFs changes the state of next FF. We may trace the following sequence considering the initial states $(Q_C\, Q_B\, Q_A)_2 = (0\,0\,0)_2$:

FF C			FF B			FF A			Next States		
J_C	K_C	$\bar{C}$	J_B	K_B	$\bar{B}$	J_A	K_A	$\bar{A}$	C^+	B^+	A^+
1	1	1	1	1	1	1	1	1	1	1	1
0	0	0	0	0	0	1	1	0	1	1	0
0	0	0	1	1	0	1	1	1	1	0	1
0	0	0	0	0	1	1	1	0	1	0	0
1	1	1	1	1	1	1	1	1	0	1	1
0	0	1	0	0	0	1	1	0	0	1	0
0	0	1	1	1	0	1	1	1	0	0	1
0	0	0	0	0	1	1	1	0	0	0	0

The above truth table shows that there are total 8-states. Hence, given circuit is a MOD-8 down counter.

221. (d)

- 10-bit ring counter is a Mod-10 counter. So, it divides the 160 kHz input by 10. Therefore, W = 16 kHz.
- The 4-bit parallel counter is a Mod-16 counter. So, it divides the 16 kHz by 16. Therefore, X = 1 kHz.
- The Mod-25 ripple counter produces a frequency at Y = 40 Hz (1 kHz/25 = 40 Hz).
- The 4-bit Johnson counter is a Mod-8 counter. So, it divides the 40 Hz by 8. Therefore, Z = 5 Hz.

223. (c)

The truth table for given circuit is shown below.

$J_A = \bar{Q}_B$	$K_A = 1$	$J_A = Q_B$	$K_B = 1$	Q_A	Q_B	
X	X	X	X	X	0	Mod-3
1	1	0	1	1	0	
1	1	1	1	0	0	
0	1	0	1	0	0	

224. (d)

After 1st pulse : $Y_1 = 0$, $Y_2 = 1$
After 2nd pulse : $Y_1 = 0$, $Y_2 = 0$
After 3rd pulse : $Y_1 = 1$, $Y_2 = 0$
After 4th pulse : $Y_1 = 1$, $Y_2 = 1$

226. (d)

Number of clock pulses that have occurred may be
3 or 3 + 16 or 3 + 16 + 16 or 3 + 16 + 16 + 16
i.e. 3 or 19 or 35 or 51

229. (c)

Since, A = 1, D_1 will be OFF.
With S = 1, the transistor Q will be in saturation and therefore the collector of Q will be at 0.2 V.
Since B = 0, which corresponds to 0 volt, which is less than 0.2 volt, only D_2 will be on.
For D_3 to be on, Y will have to be (0.2 + 0.6) = 0.8 volt, which will make the forward voltage across D_3 to be 0.8 volt, which is not possible. Hence, diode D_3 will be off, and Y will be at 0.6 volt (which is the cut-in voltage of D_2).

$$\text{Therefore, } I_{D_2} = \left(\frac{12-0.6}{15k}\right) = 0.76 \text{ mA}$$

$$\text{and } I_{D_3} = 0 \text{ A}$$

232. (c)

One of the biggest advantage of CMOS is that it has lowest power consumption among all logic family members. Due to which it is widely used for making digital circuits.

233. (d)

Emitter coupled logic (ECL) is a non-saturated logic where the transistors operate only in active and cut-off region resulting in higher speed of operation.

235. (d)

The digital logic family which has minimum power dissipation is CMOS.

236. (c)

ECL is the fastest logic among the logic families. High speeds are possible in ECL because the transistors are used in difference amplifier configuration, in which they are never driven into saturation and thereby the storage time is eliminated.

237. (d)

ECL is the digital logic family having lowest propagation delay time and highest power dissipation. Lowest propagation delay time is possible in ECL because the transistors are used in difference amplifier configuration, in which they are never driven into saturation and thereby the storage time is eliminated.

238. (b)

Schottky transistor reduces the storage time delay by preventing the transistor from going deep into saturation due to which it reduces the propagation delay. Due to this reason Schottky transistors are preferred over normal transistor because of their lower propagation delay.

240. (d)

The CMOS has the greatest complexity and the lowest packing density of the MOS families, but it possesses the important advantages of higher speed and much less power dissipation.

243. (a)

- ECL has low noise-margin of about 250 mV due to which it is unreliable for use in heavy industrial environment.
- Since ECL circuits produce the output and its complement, therefore it eliminates the use of inverters.
- Due to the low output impedance, ECL has larger fan-out of around 25.
- ECL is costly.

Hence, statements 1, 3 and 4 are only the characteristics possessed by ECL.

244. (d)

By connecting a Schottky barrier diode between the base and the collector of each transistor in TTL speed is increased and has no effect on fan-out. Hence, assertion is a false statement.

245. (a)

- CMOS has highest fan-out of around 50.
- ECL is fastest logic family (tpd = 1 ns).
- RTL has lowest noise-margin of around 0.2 V.
- HTL has highest figure of merit of around 5000 pJ.

248. (c)

- Most modern MOSFET circuitry is constructed using NMOS devices because they are three times faster and two times denser than their PMOS counterparts. Hence, statement-1 is not correct.
- The ECL is the fastest of all logic families and is mainly used in supercomputers. Hence, statement-2 is not correct.
- Statement-3 and 4 are also correct.

249. (d)

The "figure of merit" or the "speed-power product" is the product of average propagation delay and the average power dissipation.

i.e. $$\text{FOM} = (tpd) \times P_{diss} = (ns) \times (mW) = \text{Pico-Joule}$$

Thus, the unit of figure of merit is energy (Joule). For the better performance of an IC family figure of merit should be low. Hence, assertion is a false statement.

250. (b)

ECL has highest power dissipation while CMOS has lowest power dissipation. Power dissipation of TTL is little bit less than ECL.

251. (b)

$$(\text{Fan-out})_L = \frac{8\text{ mA}}{0.36\text{ mA}} = 22$$

$$(\text{Fan-out})_H = \left|\frac{400\,\mu A}{20\,\mu A}\right| = 20$$

$$\therefore \quad \text{Fan-out} = (22,20)_{min} = 20$$

252. (c)

The overall voltage is 3.9 V – 0.2 V = 3.7 V

For rise time, the 10% voltage point is at

0.2 V + 10% (3.9 V) = 0.59 V

For fall time, the 90% voltage point is at

0.2 V + 90% (3.9 V) = 3.71 V

253. (d)

The logic levels are $V_{CE(sat)}$ = 0.2 V for the logic '0' state and V_{CC} = 5 V for the logic '1' state. If at least one input is in the '0' state, then the corresponding diode D conducts and V_P = 0.2 + 0.7 = 0.9 V. For the diodes D_1 and D_2 to conduct, a voltage of (2) (0.7) = 1.4 V is required. Hence, these diodes are cut-off and V_{BE} = 0 V. Therefore, Q is OFF and the output voltage is 5V i.e. Y = 1. This confirms the first three rows of the NAND truth table.

If all inputs are HIGH, then all input diodes are OFF. Hence, both D_1 and D_2 conduct, and Q is turned ON. The voltage at P is

$$V_P = V_{D1} + V_{D2} + V_{BE(sat)} = 0.7 + 0.7 + 0.8 = 2.2\text{V}$$

Then, voltage across each input diode is 5 V – 2.2 V = 2.8 V in the opposite direction, thus confirming that the input diode *D* is off. And hence further we can proceed to find the other rows of a NAND gate corresponding to inputs *A*, *B* and *C*.

256. (d)

Output of NAND must be LOW for LED to emit light. So both inputs to NAND must be HIGH. If any or both switches are closed, output of AND will be LOW. If both switches are open, output of XOR will be LOW. So, there can't be both inputs HIGH for NAND and therefore, LED doesn't emit light.

257. (c)

Each state of diode logic perform AND function.

So, $Z = \overline{(AB)C} = \overline{ABC}$

258. (a)

Let, $V_1 = V_2 = 0\ V$, then M_3 will be ON, M_1 and M_2 OFF and M_4 ON.
Hence, $V = -V_{DD}$.
Let $V_1 = 0$ V and $V_2 = -V_{DD}$, then M_3 will be ON, M_1 OFF, M_4 OFF, M_2 ON and hence $V_0 = -V_{DD}$.
Let $V_1 = -V_{DD}$ and $V_2 = 0V$, then M_3 will be OFF, M_4 ON, M_2 OFF and hence, $V_0 = -V_{DD}$.
Finally, if $V_1 = V_2 = -V_{DD}$, then M_2 will be ON, hence $V_0 = 0$ V.
Thus, the given CMOS gate satisfies the function of a negative NAND gate.

260. (d)

Speed of conversion is maximum in flash type (or parallel-comparator A/D) A/D converter because the comparisons of the input voltage are carried out simultaneously.

261. (b)

In flash type A/D converter, the conversion time is independent of the number of bits. In this type of A/D converter, an array of comparators are connected in parallel and which compares the input voltage at a particular ratio of the reference voltage.

264. (b)

For a *n*-bit flash type *ADC*, number of comparators required = $2^n - 1$.
Thus, for $n = 12$, number of comparators required $= 2^{12} - 1 = 4096 - 1 = 4095$

265. (a)

For a counter type ADC,

$$\text{conversion time} = \frac{1}{f} \times (2^n - 1)$$

Given, $n = 12$, $f = 1$ MHz

$$\therefore \text{Conversion time} = \frac{1}{10^6} \times (2^{12} - 1)$$

$$= \frac{4095}{10^6} \text{ sec}$$

$$= 4.095 \text{ ms} \approx 4.1 \text{ ms}$$

266. (c)

There is no any resonant circuit available in a digital-to-analog converter.

269. (c)

- Dual slope or integrating type is the most accurate A/D converter.
- Ramp type or counter type is the simplest type of the A/D converter.
- SAR (Successive approximation register) type is one of the most widely used type of ADCs.
- A flash type is the fastest A/D converter because it requires only one clock pulse for getting the output so that conversion time is equal to the clock time period.

270. (a)

The flash converter utilizes the parallel differential comparators, that compare reference voltages with the analog input voltage. It does not uses any timing or sequencing period. It uses only resistors, comparators and priority encoder. Due to this reason no clock pulse is required in this type of A/D converter.

271. (b)

An ADC must perform at a rate equal to at least twice the frequency of the highest component of the input.

i.e. the average conversion time = $\frac{t_c(\max)}{2}$

or, $t_c(\max) = 2 \times 200\ \mu s$
$= 400\ \mu s$

∴ The highest frequency that the analog signal can contain is

$$\frac{1}{t_c(\max)} = \frac{1}{400\ \mu s} = 2.5 \text{ kHz}$$

273. (c)

Both SAR type and ramp type ADC uses a digital-to-analog converter (DAC). A dual-slope type (or an integrating type) ADC uses an integrator instead of a DAC. By using an integrator with a large time constant, the noise signals at the input are suppressed or attenuated.

274. (b)

- The disadvantage of the weighted-resistor DAC is, that a different -valued precision resistance is needed for each bit position of the digital input. Due to this reason the R-2R ladder type DAC is widely used since the disadvantage of weighted-resistor DAC is avoided in this type of DAC. Hence, statement-1 is not correct.
- Statement-2 is correct.
- The output of a DAC is not a true analog quantity, because it can take on only specific values. In that sense, it is actually digital. Thus, the output of a DAC is a "pseudo-analog" quantity. Hence, statement-3 is not correct.

275. (a)

With 6 bits, there will be $2^6 - 1 = 63$ steps of size 50 mV each.

Therefore, the full-scale output = 63 × 50 mV = 3.15 V.

$$\therefore \text{ \% resolution} = \frac{50 \text{ mV}}{3.15 \text{ V}} \times 100$$

$$= \frac{1}{63} \times 100 = 1.587\% \approx 1.59\%$$

276. (d)

The step size = $\frac{2 \text{ mA}}{63} = 31.7\ \mu A$

Since, $(100000)_2 = (32)_{10}$, the ideal output should be 32 × 31.7 = 1014 μA.

The error can be as much as ± 0.5 % × 2 mA = 10 μA.

Thus, the actual output can deviate by ± 10 μA from the ideal value of 1014 μA, and therefore, the actual output can be anywhere from 1004 μA to 1024 μA i.e. (1014 ± 10) μA.

277. (c)

The MSB passes with a gain of 1, so, its weight = 5 V.

The next bit (D_2) passes with a gain of 1/2, so, its weight = 5/2.

The D_1 bit passes with a gain of 1/4, so its weight = 5/4.

The D_0 bit passes with a gain of 1/8, so its weight = 5/8.

Therefore, the full scale output (when $R_f = R = 1\ k\Omega$) is

$$V_{out} = -\left(1 + \frac{1}{2} + \frac{1}{4} + \frac{1}{8}\right) \times 5 = -9.375 \text{ V}$$

278. (b)

Output current = 10 mA for a digital input of $(0100)_2$ i.e. $(4)_{10}$.

Analog output = k × digital input

$$\therefore \quad k = \frac{10 \text{ mA}}{4} = 2.5 \text{ mA}$$

Hence, the output current for a digital input of 1011 = $(11)_{10}$ × 2.5 = 27.5 mA

279. (d)

Since, $\frac{2.86\ V}{30 \text{ mV}} = 95.3$, the step 95 would produce 2.85 V and step 96 would produce 2.88 V. The SAR type ADC always produces a final output, that is, at the step below the analog input. Therefore, for the case of $V_A = 2.86$ V, the digital result would be $(95)_{10} = (01011111)_2$.

280. (d)

Since, the highest input frequency is 50 kHz, conversions should be performed at the rate of $2 \times 50 \times 10^3 = 100 \times 10^3$ conversions/sec.

(Since $t_c(\max) = 2 \times$ Average conversion time)

$\therefore$ The maximum allowable conversion time

$$= \frac{1}{100 \times 10^3} = 10\ \mu s$$

281. (c)

Reason is a false statement because a DAC is called non-monotonic, if the output decreases when the binary input is incremented while it is called monotonic if its output increases as the binary input is incremented from one value to the next.

282. (a)

$$\text{Resolution of LSB} = \frac{1}{2^n} \times \frac{V_R}{R}$$

$$= \frac{1}{2^4} \times \frac{10\ V}{10 \times 10^3\ \Omega} = 62.5\ \mu A$$

283. (b)

Here, $R_{Th} = 3\ \Omega$ (across input terminals of OP-AMP)

$V_{Th} = \frac{V_R}{4R}$ (across input terminals of OP-AMP).

$$\therefore \quad V_0 = \frac{-R_f}{R_{TH}} \times V_{Th} = \frac{-R}{3} \times \frac{V}{4R} = -\frac{V}{12}$$

$$\therefore \quad I = -\frac{V_0}{R} = \frac{V}{12R} = \frac{V}{12} \quad (R = 1\ \Omega)$$

284. (b)

$$\text{LSB} = \frac{1}{2^{10}} = 0.000977$$

$$\therefore \quad \text{Accuracy} = \pm \frac{\text{LSB}}{2}$$

$$= \pm\ 0.000488 \text{ or } 400\ \mu V/^\circ C$$

285. (d)

Digital ramp conversion time $= (2^n - 1) \times T_{CLK}$

$= 1023 \times 2\ \mu s = 2046\ \mu s$

For SAR type ADC,

conversion time = 10 clock period time

$= nT_{CLK,}\ (n = 10)$

$= 10 \times 2\ \mu s = 20\ \mu s$

Thus, digit ramp is 100 times slower than SAR type ADC .

287. (b)

For an ADC, resolution $= \frac{V_f}{2^n - 1}$

$= 5 \times 10^{-3}$ V (Given)

or, $\frac{10}{2^n - 1} = 5 \times 10^{-3}$ V

or, $\frac{10}{2^n - 1} = \frac{10}{2000}$

or, $2^n - 1 = 2000$ or $n \approx 11$

288. (d)

For 12-bit DAC, resolution

$$= \frac{10}{2^{12} - 1} = \frac{10}{4095} \text{ volt}$$

For an integer number of 516, analog output voltage

$= 516 \times$ Resolution

$$= 516 \times \frac{10}{4095}$$

$$= \frac{5160}{4095} = 1.26 \text{ volt}$$

289. (c)

The ultraviolet rays passes through a window in the IC package to the EPROM chip where it releases stored charges. Thus, the stored contents are erased. Hence, EPROM contents can be erased by passing it to ultraviolet rays.

290. (d)

A volatile memory is one in which contents get erased as soon as the power goes off. RAM falls in the category of volatile memory.

291. (b)

The access time of ROM using bipolar transistors is about 1 μ sec.

292. (a)

Random Access Memory (RAM) is used by the user, number of times.

294. (b)

Given memory size = 32 k = $2^5 \times 2^{10} = 2^{15}$. Thus, 15 address lines are required. Since out of the given options, only 16 bits can address the given memory of 32 k, therefore we require 16 address bits.

296. (d)

Magnetic tape memory system can not be accessed randomly. It can only be accessed sequentially.

298. (a)

An address used by programmers in a system supporting virtual memory concept is called virtual address and the set of such addresses are called address space.

300. (d)

Dynamic RAMs are fabricated using only MOS technology and they are noted for their high capacity, low power requirement and moderate speed (slower than static RAM).

304. (d)

$$5\text{ M} \times 8 = 40\text{ M (Highest size)}$$
$$5\text{ M} \times 4 = 20\text{ M}$$
$$1\text{ M} \times 12 = 12\text{ M}$$
$$1\text{ M} \times 16 = 16\text{ M]}$$

306. (d)

Microcomputer programs stored in ROMs are referred to as "firmware" because they are not subject to change. Hence, assertion is a flase statement. Reason is a correct statement because ROM normally means the mask programmed ROM since it is programmed by the manufactures and hence can't be programmed by the user.

308. (b)

Capacity of disc

$$= (77\text{ tracks}) \times \left(26 \frac{\text{sectors}}{\text{track}}\right) \times \left(256\ \frac{\text{bytes}}{\text{sector}}\right)$$

= 512,512 bytes

310. (a)

- PROMs are programmable ROMs and EPROMs are erasable PROMs.
- Dynamic memory is constructed only using MOS technology. Thus, statements 2 and 3 are not correct.

311. (c)

- Statements-1 and 2 are not correct.
- Magnetic core memory is a non-volatile random access memory. Hence, statement-3 is not correct.
- Magnetic memories are slower and cheaper than semiconductor memories. Hence, statement-4 is not correct.

Thus, none of the given statement is/are correct.

312. (a)

Total capacity of ROM = 1024 × 4 = 4096

313. (b)

Given,

Memory size = 4 k bytes
= 4 × 1024 bytes
= $2^2 \times 2^{10}$ bytes
= 2^{12} bytes = 2^n bytes

(where, n = number of address lines)

∴ n = 12 = required number of bits

■■■■

UNIT 9

Analog Electronics

1. Semiconductor Physics

Q.1 The mobility of free electrons and holes in pure germanium are 3800 and 1800 $cm^2/V\text{-}s$ respectively. The corresponding values for pure silicon are 1300 and 500 $cm^2/V\text{-}s$, respectively. Assuming $n_i = 2.5 \times 10^{13}\ cm^{-3}$ for germanium and $n_i = 1.5 \times 10^{10}\ cm^{-3}$ for silicon at room temperature, the values of intrinsic conductivity for germanium and silicon are respectively given by

(a) 4.32×10^{-6} S/cm and 0.0224 S/cm
(b) 4.32×10^{-6} S/cm and 0.325 S/cm
(c) 0.0224 S/cm and 4.32×10^{-6} S/cm
(d) 0.325 S/cm and 4.32×10^{-6} S/cm

Q.2 What is the concentration of holes and electrons in n-type Silicon at 300°K, if the conductivity is 30 S/cm?
Assume at 300°K, $n_i = 1.5 \times 10^{10}/cm^3$, $\mu_n = 1300\ cm^2/V\text{-}s$ and $\mu_p = 500\ cm^2/V\text{-}s$.

(a) $1.56 \times 10^2\ cm^{-3}$ and $1.44 \times 10^{18}\ cm^{-3}$ respectively
(b) $1.56 \times 10^4\ cm^{-3}$ and $1.44 \times 10^{21}\ cm^{-3}$ respectively
(c) $1.44 \times 10^2\ cm^{-3}$ and $1.56 \times 10^{18}\ cm^{-3}$ respectively
(d) $1.44 \times 10^4\ cm^{-3}$ and $1.56 \times 10^{21}\ cm^{-3}$ respectively

Q.3 When an electric field is applied across a semiconductor, free electrons in it will accelerate due to the applied field, and gain energy. This energy can be lost as heat when the electrons

(a) recombine with holes
(b) collide with atoms in the crystal
(c) radiate energy while being accelerated
(d) collide with other electrons

Q.4 If a current of 1.6 μA is flowing through a conductor, the number of electrons crossing a particular cross-section per second will be

(a) 10^{13}
(b) 1.6
(c) 10^{19}
(d) 1.6×10^{-6}

Q.5 Match **List-I** with **List-II** and select the correct answer using the codes given below the lists:

List-I	**List-II**
A. Semiconductor	1. Zero energy gap
B. Metal	2. Small energy gap
C. Insulator	3. Large energy gap
	4. Non-existent energy gap

Codes:

	A	B	C
(a)	3	1	2
(b)	3	4	2
(c)	2	1	3
(d)	2	4	3

Q.6 If elements in column IV of the periodic table are placed in increasing order of their atomic number, the order will be

(a) *C, Ge, Sn, Si*
(b) *C, Sn, Ge, Si*
(c) *Ge, Si, C, Sn*
(d) *C, Si, Ge, Sn*

Q.7 **Assertion (A):** The drift velocity is in the direction opposite to that of the electric field.
Reason (R): At each inelastic collision with an ion, an electron loses energy, and a steady-state condition is reached where a finite value of drift speed is attained.

(a) Both A and R are true and R is the correct explanation of A.
(b) Both A and R are true but R is not the correct explanation of A.
(c) A is true but R is false.
(d) A is false but R is true.

Q.8 The density and mobility of electrons in a conductor are respectively $10^{20}/cm^3$ and 800 $cm^2/V\text{-}s$. If a uniform electric field of 1 V/cm exists across this conductor, then the electron current density would be approximately

(a) 11 kA/cm^2 (b) 9 kA/cm^2
(c) 13 kA/cm^2 (d) 18 kA/cm^2

Q.9 A semiconductor is doped with a donor density N_D and no acceptors. If the intrinsic concentration is n_i, then the free electron density(n) will be equal to

(a) $\frac{N_D + \sqrt{4n_i^2 + N_D^2}}{2}$ (b) $\frac{n_i + \sqrt{n_i^2 + 4N_D^2}}{2}$

(c) $\frac{N_D + \sqrt{n_i^2 + 4N_D^2}}{2}$ (d) $\frac{n_i + \sqrt{4n_i^2 + N_D^2}}{2}$

Q.10 In a p-type semiconductor, $p = 10^{16}$/cm^2, and μ_p = 400 cm^2/V-s. If a magnetic field (B) of 5×10^{-4} Weber/cm^2 is applied in the x-direction, and an electric field of 2000 V/m is applied in $+y$ direction. The value of electric field caused due to the "Hall effect" is
(a) –400 V/cm in $-z$ direction
(b) –200 V/cm in $+z$ direction
(c) –200 V/cm in $-z$ direction
(d) –400 V/cm in $+z$ direction

Q.11 In a semi-conductor holes exist in
(a) the forbidden energy band
(b) the valence band
(c) no band
(d) the conduction band

Q.12 In silicon, phosphorus acts as impurity that _______ electrons and boron acts as impurity that _______ electrons.
(a) donates, donates (b) accepts, donates
(c) donates, accepts (d) None of these

Q.13 A conductor has an electron density n, and an electron mobility μ_n. When an electric field is applied across it, electrons will move with a drift velocity 'v' in the direction of the applied field. The conductivity of the semiconductor and the current density of the conductor will be respectively given by
(a) $q\mu_n n$ and qnv (b) qnv and nv
(c) qnv and $q\mu_n n$ (d) $q\mu_n n$ and $q\mu_n nv$

Q.14 A particular semiconductor in equilibrium has 10^{16}/cm^3 donor atoms, and 1.1×10^{17}/cm^3 acceptor atoms. If the intrinsic carrier density (n_i) of the semiconductor is 10^{12}/cm^3, then the electron density in it will be
(a) 10^{12}/cm^3 (b) 1.1×10^{17}/cm^3
(c) 10^{16}/cm^3 (d) 10^7/cm^3

Q.15 Consider the following statements:
1. Current due to carrier diffuse exists in both semiconductors and metals.
2. Current due to carrier drift occurs in both semiconductors and metals.
3. The intrinsic concentration of carriers is independent of temperature.
4. Metals are unipolar while semiconductors have bipolar nature.

Which of the above statements is/are correct?
(a) 3 and 4 only (b) 2 and 4 only
(c) 1, 2 and 3 only (d) 4 only

Q.16 Which of the following components are all active components?
(a) a capacitor and an inductor
(b) a resistor and an inductor
(c) a diode, a BJT and an FET
(d) an opamp, a BJT and thermionic triode

Q.17 An intrinsic semiconductor at absolute zero temperature
(a) behaves like an insulator
(b) has few holes and same number of electrons
(c) behaves like a metallic conductor
(d) has a large number of holes

Q.18 Capacitors for integrated circuits
(a) can be made with very high values of capacitance.
(b) cannot be made using diffusion techniques.
(c) can be made using silicon dioxide as the dielectric.
(d) are always discrete components connected externally.

Q.19 Doping materials are called impurities because they
(a) make semiconductors less than 100 percent pure.
(b) alter the crystal structures of the pure semiconductors .
(c) decrease the number of charge carriers.
(d) change the chemical properties of semiconductors.

Q.20 Consider the following statements for a n-type semiconductor:

1. Conduction is due to the movement of electrons in the conduction band.
2. Electrons constitute the majority charge carriers.
3. It is formed by adding an acceptor impurity to germanium.
4. Fermi level lies closer to the valence band between the conduction and valence band.

Which of the above statements are not true?
(a) 1, 2 and 4 (b) 3 and 4
(c) 1 and 2 (d) 2 and 3

Q.21 When an intrinsic semiconductor is doped with an acceptor type impurity
(a) electrons are generated and the material becomes p-type.
(b) electrons are generated and the material is called n-type.
(c) holes are generated and the material becomes p-type.
(d) holes are generated and the material becomes n-type.

Q.22 An *n*-type semiconductor as a whole is
(a) negatively charged.
(b) positively charged.
(c) positively or negatively charged depending on doping.
(d) electrically neutral.

Q.23 Conduction electrons have more mobility than holes because they
(a) need less energy to move.
(b) are lighter.
(c) have negative charge.
(d) experience collision less frequently.

Q.24 An electron in the conduction band
(a) is located near the top of the crystal.
(b) has no charge.
(c) has a higher energy than an electron in the valence band.
(d) is bound to its parent atom.

Q.25 Match **List-I (Temperature dependent parameters)** with **List-II (Variation with temperature)** and select the correct answer using the codes given below the lists:

List-I
A. Diffusion constant
B. Intrinsic concentration
C. Mobility of charge carriers

List-II
1. Increases with temperature
2. Independent of temperature
3. Decreases with temperature

Codes:

	A	B	C
(a)	1	2	3
(b)	2	1	3
(c)	2	3	1
(d)	3	1	2

Q.26 Consider the following statements:
1. The ratio of diffusion constant for electron D_e to the mobility of electrons μ_e is proportional to T.
2. Mass action law is given by: $n = pn_i^2$.
3. Hall's effect can be used to measure electric field intensity.
4. In a semiconductor conductivity mainly depends on mobility of charge carriers.
5. The Hall voltage developed in a metal is in microvolts while in semiconductors it is in millivolts.

Which of these statements are correct?
(a) 2, 4 and 5 only (b) 1, 3 and 4 only
(c) 2, 3 and 5 only (d) 1 and 5 only

Q.27 **Assertion (A):** Gallium arsenide is a direct band semiconductor having faster switching capabilities and high temperature operating capabilities.
Reason (R): A substance for which the width of the forbidden energy region is relatively small is called a semiconductor.
(a) Both A and R are true and R is the correct explanation of A.
(b) Both A and R are true but R is not the correct explanation of A.
(c) A is true but R is false.
(d) A is false but R is true.

Q.28 **Assertion (A):** The transport of charges in a semiconductor may be accounted for by a mechanism called diffusion, not ordinarily encountered in metals.
Reason (R): It is possible to have a non-uniform concentration of particles in a metal.
(a) Both A and R are true and R is the correct explanation of A.

(b) Both A and R are true but R is not the correct explanation of A.
(c) A is true but R is false.
(d) A is false but R is true.

Q.29 **Assertion (A):** The higher the value of mobility of a charge carrier, the smaller will be its tendency to diffuse.
Reason (R): The average distance that an excess charge carrier can diffuse during its life time is called the diffusion length.
(a) Both A and R are true and R is the correct explanation of A.
(b) Both A and R are true but R is not the correct explanation of A.
(c) A is true but R is false.
(d) A is false but R is true.

Q.30 Match **List-I** with **List-II** and select the correct answer using the codes given below the lists:
List-I
A. Silicon
B. Gallium arsenide
C. Silicon dioxide
D. Germanium
List-II
1. Faster switching capability
2. Narrower energy gap
3. Smaller conductivity
4. Natural Insulator
Codes:

	A	B	C	D
(a)	3	1	4	2
(b)	1	2	4	3
(c)	3	4	1	2
(d)	1	2	3	4

Q.31 In a semiconductor it is observed that three quarter of the current is carried by electrons and one quarter by holes. If at this temperature the drift speed of electrons is three times that of the holes, the ratio of electrons to holes in the semiconductor would be
(a) 3 (b) 1
(c) 1/2 (d) 2

Q.32 What is the diffusion current in a piece of germanium having concentration gradient of 1.5×10^{22} electrons/m^3 and $D_n = 0.0012$ m^2/s?
(a) 4.45 A/m^2 (b) 3.33 A/m^2
(c) 1.88 A/m^2 (d) 2.88 A/m^2

2. Semiconductor Diodes

Q.33 The barrier voltage (V_0 or V_r) in a junction diode is the effect of
(a) the voltage needed to make the semiconductor material behave as a conductor.
(b) the emf required to move the holes fast enough to have the mobility equal to that of the electrons.
(c) the p-side and n-side of the junction forming a battery.
(d) the recombination of charge carriers across the junction leaving behind the opposite charged ions.

Q.34 In a zener diode large reverse current is due to
(a) presence of impurity
(b) lower resistance in reverse biased region
(c) rupture of bonds
(d) collision

Q.35 For thermionic emission
(a) a material with low work function is preferable.
(b) a material with high work function is preferable.
(c) the work function of the material has no importance.
(d) none of the above

Q.36 The most commonly used type of electron emission in electron tubes is
(a) secondary emission
(b) photo-electron emission
(c) thermionic emission
(d) field emission

Q.37 When forward bias is applied to a junction diode, it
(a) reduces the majority - carrier current to zero
(b) reduces the minority - carrier current to zero
(c) increases the potential barrier
(d) decreases the potential barrier

Q.38 The emission of electrons in a vaccum diode is achieved by
(a) heating
(b) electron bombardment
(c) magnetic field
(d) electrostatic field

Q.39 The main reason why electrons can tunnel through a *p-n* junction is that
(a) barrier potential is very low
(b) they have high energy
(c) impurity level is low
(d) depletion layer is extremely thin

Q.40 The forbidden energy gap in semiconductors
(a) lies just above the conduction band
(b) lies just below the valence band
(c) is the same as the valence band
(d) lies between the valence band and the conduction band

Q.41 The barrier potential for a Ge p-n junction is
(a) 0.6 V (b) 0.1 V
(c) 0.2 V (d) 0.5 V

Q.42 With the increase in temperature, the resistivity of an intrinsic semiconductor decreases. This is because, with the increase of temperature
(a) the carrier concentration decreases, but the mobility of carriers increases.
(b) the carrier concentration increases, but the mobility of carriers decreases.
(c) the carrier concentration remains the same but the mobility of carriers decreases.
(d) both the carrier concentration and mobility of carriers decreases.

Q.43 Which of the following is a unipolar device?
(a) *p-n* junction diode
(b) Schottky diode
(c) Zener diode
(d) Tunnel diode

Q.44 The breakdown mechanism is a lightly doped *p-n* junction under reverse biased condition is called
(a) avalanche breakdown
(b) high voltage breakdown
(c) zener breakdown
(d) breakdown by tunneling

Q.45 Space charge region around a *p-n* junction
(a) contains both free electrons and holes
(b) contains one type of mobile carriers depending on the level of doping of the *p* or *n* regions
(c) contains electrons only as free carriers
(d) does not contain mobile carriers

Q.46 Match **List-I** with **List-II** and select the correct answer using the codes given below the lists:

List-I / **List-II**

A. Photo diode — 1.
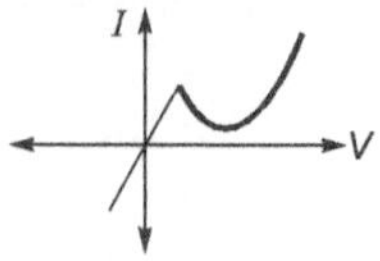

B. Zener diode — 2.
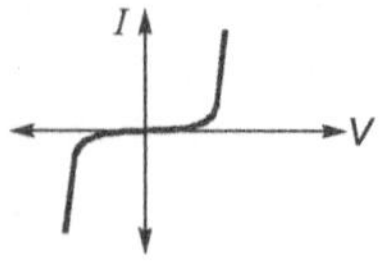

C. P-N Junction diode — 3.
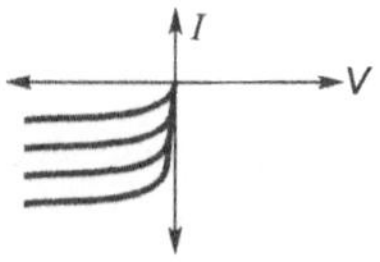

D. Tunnel diode — 4.
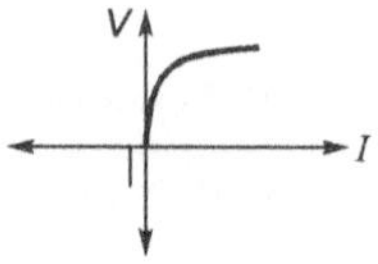

Codes:

	A	B	C	D
(a)	1	2	4	3
(b)	3	4	2	1
(c)	3	2	4	1
(d)	1	4	2	3

Q.47 Match **List-I (Doping level)** with **List-II (Diode type)** and select the correct answer using the codes given below the lists:

List-I	List-II
A. Highest (1 : 10^3)	1. Photo diode
B. Moderate (1 : 10^8)	2. Zener diode
C. Lowest (1 : 10^{11})	3. PN junction diode
D. Slightly high (1:10^5)	4. Tunnel diode

Codes:

	A	B	C	D
(a)	4	3	1	2
(b)	4	2	1	3
(c)	2	1	4	3
(d)	1	2	4	3

Q.48 For the circuit shown below, if $V = 10$ volt, $R = 1\text{ k}\Omega$ and the breakdown voltage of the zener diode be 6 V, then the value of I_z would be equal to

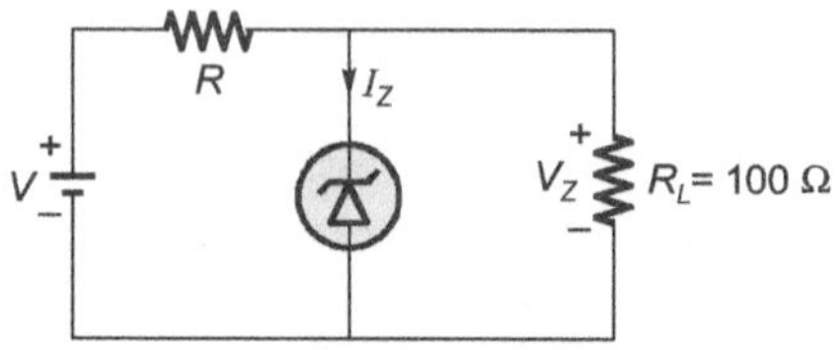

(a) 3.3 mA (b) 0 A
(c) 2 mA (d) 1.2 mA

Q.49 The mean lifetime of holes for a Germanium diode is 10 ns, and a forward current of 0.1 mA flows through it. The diffusion capacitance of the diode at room temperature is
(a) 16.8 nF (b) 23.6 pF
(c) 44.4 nF (d) 38.5 pF

Q.50 **Assertion (A):** The dynamic resistance of a diode is constant.
Reason (R): Dynamic resistance of a diode is used for small-signal operation and is defined as the reciprocal of the slope of the volt-ampere characteristic.
(a) Both A and R are true and R is the correct explanation of A.
(b) Both A and R are true but R is not the correct explanation of A.
(c) A is true but R is false.
(d) A is false but R is true.

Q.51 **Assertion (A):** An electric field must build up across the p-n junction in such a direction that a hole drift current will tend to flow across the junction from the n to the p side in order to counter balance the diffusion current.
Reason (R): Since the concentration of holes in the *p* side is much greater than that in the *n* side, a very large hole diffusion current tends to flow across the junction from the *p* to the *n* material.
(a) Both A and R are true and R is the correct explanation of A.
(b) Both A and R are true but R is not the correct explanation of A.
(c) A is true but R is false.
(d) A is false but R is true.

Q.52 **Assertion (A):** Silicon is preferred for manufacturing zener diodes.
Reason (R): Reverse saturation current is less in case of silicon.
(a) Both A and R are true and R is the correct explanation of A.
(b) Both A and R are true but R is not the correct explanation of A.
(c) A is true but R is false.
(d) A is false but R is true.

Q.53 In many commercial applications, temperatures of electronic circuits can vary from –50°C to 125°C. To maintain a constant current in a diode over this temperature range, by how much will the diode voltage have to change?
(a) 1.1 V (b) 0.89 V
(c) 0.44 V (d) 0.65 V

Q.54 Consider the following statements:
1. In a forward-biased *p-n* junction, the injected hole current in n-region is proportional to the total charge *Q* of the injected minority carrier holes.
2. If a p-n junction is heavily doped, breakdown voltage will reduce.
3. In a Zener diode, *p* and *n* materials are heavily doped.
4. A Zener diode has characteristics similar to that of an ideal current source.
5. Potential barrier decreases at high temperature.

Which of these statements are correct?
(a) 1, 2, 3 and 5 (b) 2, 4 and 5
(c) 1, 2, 4 and 5 (d) 2, 3 and 5

Q.55 In a reverse biased p-n junction, almost no current flows because
(a) electrons and holes recombine before they can cross the junction.
(b) only minority carriers, whose densities are very small, contribute to the current, when they cross the junction.
(c) the electric field in the neutral regions is very small.
(d) all the applied voltage appears across the ohmic contacts.

Q.56 A diode with a forward bias of 0.8 V is carrying 2.6 mA of current at room temperature. If $\eta = 2$ for this diode, the dynamic resistance (*r*) of the diode will be
(a) 308 Ω (b) 10 Ω
(c) 20 Ω (d) 616 Ω

Q.57 A varactor is a reverse biased p-n junction whose capacitance varies with voltage due to the variation in the
(a) conductivity of the space charge region
(b) injected minority carrier density
(c) capacitance of the neutral *p* and *n* regions
(d) space charge layer width

Q.58 Consider the following statements associated with various types of special diodes:

1. The tunnel diode is widely used as a rectifier.
2. Schottky diode is a unipolar device because of absence of minority carriers in reverse direction.
3. A varactor diode is optimised for its variable capacitance.
4. A Zener diode has sharp breakdown voltage at low reverse voltage.
5. PIN diodes are used at high frequencies (more than 300 MHz).

Which of these statements are correct?

(a) 1, 2, 3 and 5 (b) 2, 3, 4 and 5
(c) 1, 3 and 5 (d) 2, 3 and 4

Q.59 What is the forward resistance of a silicon *p-n* junction diode, when the forward current is 5 mA at $T = 300°K$?

(a) 10.34 Ω (b) 22.65 Ω
(c) 8.20 Ω (d) 16.45 Ω

Q.60 A silicon diode has a saturation current of 7.5 μA at room temperature 300°K. The saturation current at 400°K would be

(a) 5.57 mA (b) 8.20 mA
(c) 7.68 mA (d) 8.12 mA

Q.61 For the diode circuit shown below, the cut-in voltage of each diodes are 0.7 V, forward resistance is zero while reverse resistance is infinite. The voltage V_1 is

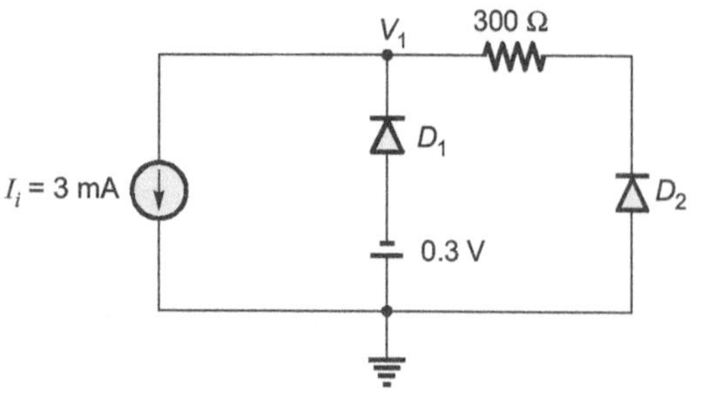

(a) –0.3 V (b) –0.9 V
(c) –1.0 V (d) none of these

Q.62 At large forward biases, tunnel diodes conduct due to

(a) avalanche multiplication in the space charge layer.
(b) the diffusion of carriers across the space charge layer.
(c) displacement current of the diffusion capacitance.
(d) tunneling of carriers across the forbidden energy gap.

Q.63 The electrical power output of a photodiode is maximum when a

(a) small reverse bias exists across it.
(b) large reverse bias exists across it.
(c) small forward bias exists across it.
(d) small forward current flows through it, irrespective of the bias.

Q.64 When light falls on a reverse biased photodiode, the current through it increases because of

(a) optical generation of electron-hole pairs near, and within, the space charge layer.
(b) increased diffusion of minority carriers due to the light.
(c) tunneling.
(d) optical generation of electron-hole pairs far away from the space charge layer.

Q.65 **Assertion (A):** A photodiode is always operated under reverse biased condition.
Reason (R): the forward current in a photodiode is in mA while the reverse current is in the range of μA.

(a) Both A and R are true and R is the correct explanation of A.
(b) Both A and R are true but R is not the correct explanation of A.
(c) A is true but R is false.
(d) A is false but R is true.

Q.66 Match **List-I (Diodes)** with **List-II (Applications)** and select the correct answer using the codes given below the lists:

List-I	List-II
A. Photodiode	1. FM modulators
B. Varactor diode	2. Microwave oscillators
C. Schottky diode	3. Alarm system
D. Tunnel diode	4. Digital computers

Codes:

	A	B	C	D
(a)	4	2	1	3
(b)	3	2	4	1
(c)	4	1	3	2
(d)	3	1	4	2

3. Field Effect Transistors

Q.67 FET is advantageous in comparison with BJT because of
(a) high gain-bandwidth product
(b) high input impedance
(c) its current controlled behaviour
(d) high noise immunity

Q.68 For the operation of a depletion-type *n*-MOSFET, the gate voltage has to be
(a) high positive (b) zero
(c) high negative (d) low positive

Q.69 JFET has main drawback of
(a) having low input impedance
(b) being noisy
(c) having high output impedance
(d) having small gain-bandwidth product

Q.70 The extremely high input impedance of a MOSFET is primarily due to the
(a) depletion of current carriers
(b) silicon-dioxide layer
(c) extremely small leakage current of its gate capacitor
(d) negative gate-source voltage

Q.71 In a JFET, at pinch-off voltage applied on the gate
(a) the drain current begins to decrease.
(b) the drain-to-source voltage is close to zero volts.
(c) the drain current is almost at saturation value.
(d) the drain current becomes almost zero.

Q.72 What is the value of transconductance when the reverse gate voltage of JFET changes from 4.0 to 3.9 V and the drain current changes from 1.3 to 1.6 mA?
(a) 3 mΩ (b) 1 mΩ
(c) 2.5 mΩ (d) 3.5 mΩ

Q.73 A FET has a drain current of 4 mA. If $I_{DSS} = 8$ mA and $V_{GS(off)} = -6$ V, then the value of gate to source voltage would be
(a) –2.62 V (b) –1.25 V
(c) –1.76 V (d) None of these

Q.74 The gate controls
(a) the width of the channel
(b) the gate voltage
(c) the drain current
(d) all of the above

Q.75 Which of the following statements associated with FET is/are correct compared to BJT?
1. It has higher cut-off frequencies and switching speed.
2. The performance of FET is relatively more affected by ambient temperature changes.
3. FETs are cheaper to produce.
4. FETs are suitable for the fabrication of ICs because they occupy less space than BJTs.

(a) 1 only (b) 2 and 4
(c) 1, 3 and 4 (d) 1 and 4

Q.76 **Assertion (A):** FET does not suffer from thermal breakdown.
Reason (R): FET has a positive temperature coefficient for resistance.
(a) Both A and R are true and R is the correct explanation of A.
(b) Both A and R are true but R is not the correct explanation of A.
(c) A is true but R is false.
(d) A is false but R is true.

Q.77 In the saturation region, the JFET transfer characteristics are
(a) exponential (b) linear
(c) parabolic (d) hyperbolic

Q.78 The threshold voltage (V_T) of a MOSFET is defined as
(a) the drain-source voltage at which the transistor goes into saturation.
(b) the gate-source voltage at which the transistor goes into saturation.
(c) the drain-source voltage at which a predefined value of drain current starts flowing.
(d) the gate-source voltage at which a predefined value of drain current starts flowing.

Q.79 In a JFET, the drain current magnitude decreases as the reverse bias across the gate-channel increases because a larger reverse bias causes
(a) more recombination in the transition region.
(b) the transition region to be wider, which narrows the thickness of the conducting channel.

(c) the mobile carriers to travel slower in the channel, by reducing their mobility.
(d) the reverse bias current to increase, thus diverting the channel current away from the drain.

Q.80 The common source JFET amplifier has
(a) a very high input impedance and a relatively low voltage gain.
(b) a high input impedance and a very high voltage gain.
(c) a low input impedance and a voltage gain less than 1.
(d) none of the above.

Q.81 **Assertion (A):** In the JFET, the transverse electric field across the reverse biased *p-n* junction controls the conductivity of the channel.
Reason (R): In enhancement and depletion types of MOSFET, the transverse electric field induced across an insulating layer deposited on the semiconductor material controls the conductivity of the channel.
(a) Both A and R are true and R is the correct explanation of A.
(b) Both A and R are true but R is not the correct explanation of A.
(c) A is true but R is false.
(d) A is false but R is true.

Q.82 Four important parameters of MOSFET and JFET have comparatively larger or smaller values. In comparison to a JFET, a MOSFET will have which one of the following combinations?

	Gate leakage Current	Drain resistance	Switching speed	Input resistance
(a)	Larger	Larger	Smaller	Smaller
(b)	Smaller	Larger	Larger	Larger
(c)	Smaller	Smaller	Larger	Larger
(d)	Larger	Smaller	Smaller	Larger

Q.83 Match **List-I (Device)** with **List-II (Applications)** and select the correct answer using the codes given below the lists:

List-I	List-II
A. CMOS	1. Relaxation oscillator
B. JFET	2. Analog signal switching
C. MOSFET	3. Portable systems
D. UJT	4. Voltage variable resistor

Codes:

	A	B	C	D
(a)	2	4	3	1
(b)	3	1	4	2
(c)	4	3	2	1
(d)	3	4	2	1

Q.84 Consider the following statements:
1. In enhancement *n*-channel MOSFET an induced *n*-type channel can be produced between the source and drain if V_{GS} is negative.
2. A FET is more sensitive to variations in input voltage in comparison to an ordinary transistor.
3. A FET is characterized by transconductance.
4. UJT has only one *p-n* junction and three leads.

Of these statements which are not correct is/are?
(a) 4 only (b) 1 and 2 only
(c) 2 only (d) 1 and 4 only

Q.85 What are the values of V_P and I_D for $V_{GS} = 0.5$ V of an *N*-channel FET that has $I_{DSS} = 10$ mA with drain voltage of 5 V and $g_{mo} = 5$ mA/V?
(a) 2 V and 7.65 mA (b) 4 V and 3.83 mA
(c) 4 V and 7.65 mA (d) 2 V and 3.83 mA

Q.86 An N-channel JFET has $I_{DSS} = 1$ mA and $V_P = -5$ V. The value of its maximum transconductance is
(a) 0.4 mS (b) 1.0 mS
(c) 0.25 mS (d) 0.6 mS

Q.87 N-channel FETs are superior to P-channel FETs, because
(a) they have higher input impedance
(b) mobility of electrons is greater than that of holes
(c) they consume less power
(d) they have high switching speed

Q.88 In a FET as V_{GS} is changed from zero to increasing reverse bias, the value of g_m
(a) increases (b) decreases
(c) remains constant (d) falls suddenly to zero

Q.89 A properly biased JFET will act as a
(a) current controlled current source
(b) voltage controlled voltage source
(c) voltage controlled current source
(d) current controlled voltage source

Q.90 A FET is a better chopper than a BJT because it has
(a) lower off-set voltage
(b) higher series ON resistance
(c) a bitable multivibrator
(d) a schmitt trigger circuit

Q.91 The variation of drain current with gate-to-source voltage ($I_D - V_{GS}$) characteristic of MOSFET is shown in the figure below.
The MOSFET is

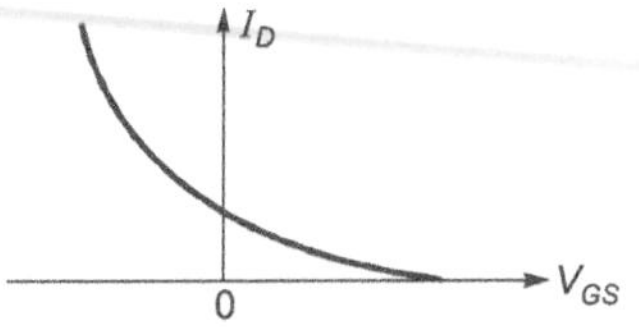

(a) an n-channel depletion mode device
(b) an n-channel enhancement mode device
(c) a p-channel depletion mode device
(d) a p-channel enhancement mode device

Q.92 In a MOSFET, the pinch-off voltage refers to
(a) drain-to-source voltage at which drain-to-source current is zero.
(b) gate-to-source voltage at which gate-to-source current is zero.
(c) drain-to-source voltage at which gate-to-source current is zero.
(d) gate-to-source voltage at which drain-to-source current is zero.

Q.93 FET tuned amplifier with g_m = 5 mA/V, r_d = 20 kΩ has resonant impedance of 20 kΩ. The gain at resonance is given by
(a) 200 (b) 100
(c) 50 (d) 25

4. Diode Circuits

Q.94 The theoretical maximum efficiency of a bridge rectifier circuit is
(a) 48.2% (b) 40.53%
(c) 81.06% (d) 82%

Q.95 Ripple factor of a full-wave rectifier without filter will be
(a) 0.48 (b) 0.20
(c) 1.21 (d) 0.24

Q.96 A full wave rectifier delivers 50 W to a load of 200 Ω. If the ripple factor is 1%, the AC ripple voltage across the load is
(a) $\frac{1}{2}$ V (b) 1 V
(c) $\frac{2}{3}$ V (d) $\frac{3}{2}$ V

Q.97 A half wave rectifier is used to supply 24 V DC to a resistive load of 500 Ω and the diode has a forward resistance of 50 Ω. What is the maximum value of the AC voltage required at the input?
(a) 82.94 V (b) 66.67 V
(c) 94.80 V (d) none of these

Q.98 What is the value of inductance to be used in the inductor filter connected to a full-wave rectifier operation at 60 Hz to provide a DC output with 4% ripple for a 100 Ω load?
(a) 0.88 H (b) 1.25 H
(c) 1.34 H (d) 1.56 H

Q.99 Diodes are used to clip voltages in circuits because they act as
(a) current sources under certain bias conditions
(b) voltage sources under certain bias conditions
(c) inductors that can remove current spikes
(d) dependent current sources whose current is clipped by the load resistor value

Q.100 When a capacitor is connected across the output terminals of a half or full-wave rectifier, the output voltage
(a) essentially becomes a DC voltage.
(b) becomes sinusoidal with a phase lag of 90 degrees.
(c) exhibits sharp spikes.
(d) is shifted up or down by the amplitude of the incoming sinusoid.

Q.101 The equivalent DC output voltage of a full-wave rectifier is _____ the equivalent DC output voltage of a half-wave rectifier.
(a) equal to (b) not related to
(c) half (d) double

Q.102 **Assertion (A):** A high value of resistor is connected across the diode in a clamper circuit for better performance.
Reason (R): If this resistor is not present, the circuit will respond to sudden decrease in the magnitude of the input pulse train, but the response to sudden increase in the amplitude will cause improper response.

(a) Both A and R are true and R is the correct explanation of A.
(b) Both A and R are true but R is not the correct explanation of A.
(c) A is true but R is false.
(d) A is false but R is true.

Q.103 Assertion (A): Bridge arrangement is preferred over centre-tapped transformer arrangement for full-wave rectification.

Reason (R): The bridge arrangement improves the filtering action.

(a) Both A and R are true and R is the correct explanation of A.
(b) Both A and R are true but R is not the correct explanation of A.
(c) A is true but R is false.
(d) A is false but R is true.

Q.104 For the network shown below, the input swing of 30 volts is equal to the

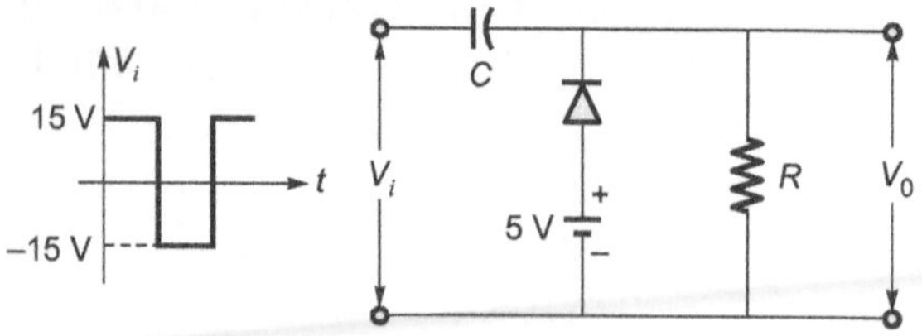

(a) output swing of 20 volts
(b) output swing of 15 volts
(c) output swing of 30 volts
(d) output swing of 25 volts

Q.105 The output voltage V_0 for the circuit shown below for the given sinusoidal input signal will be represented as

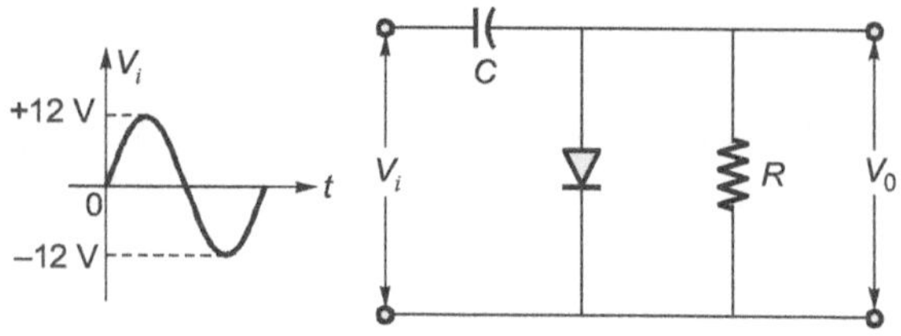

(a) 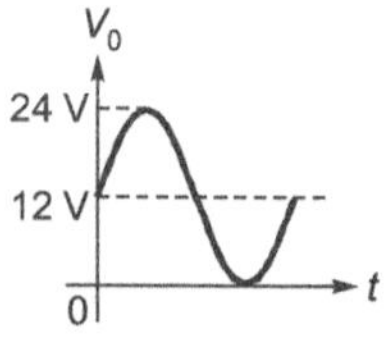

(b) 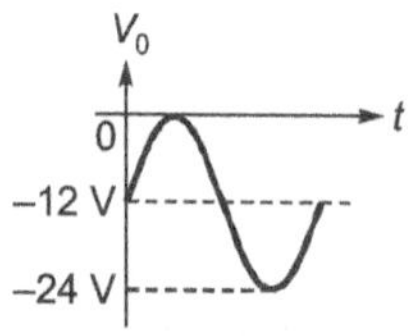

(c) 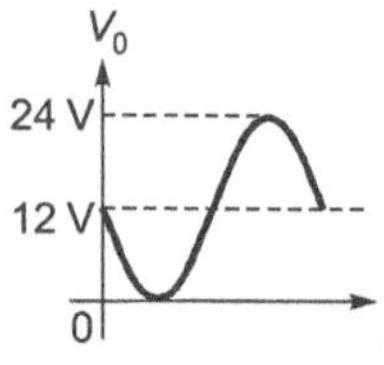

(d) 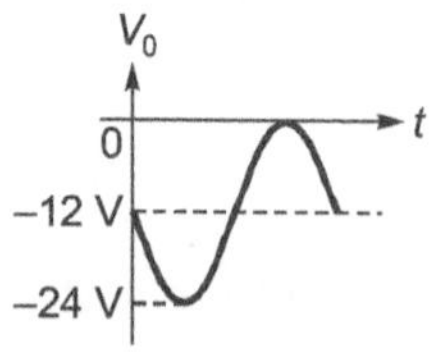

Q.106 Match **List-I (Circuits)** with **List-II (Applications)** and select the correct answer using the codes given below the lists:

List-I
A. Diode comparators
B. Diode clampers
C. Astable multivibrator
D. Diode clippers

List-II
1. Slicer
2. Oscillator
3. DC level to an AC signal
4. Square wave from a sine wave

Codes:

	A	B	C	D
(a)	3	4	2	1
(b)	4	1	2	3
(c)	3	4	1	2
(d)	4	3	2	1

Q.107 The circuit shown below represents a

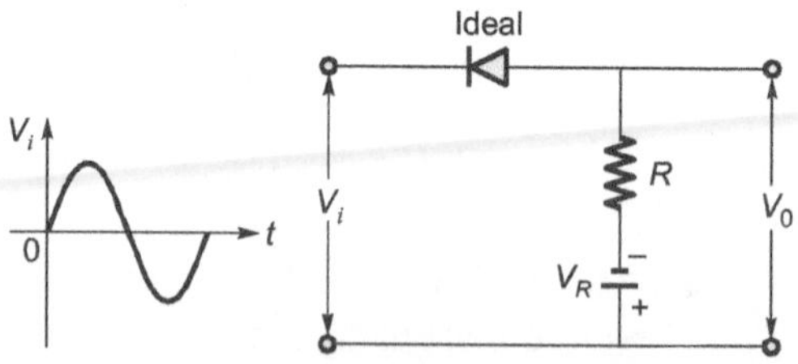

(a) biased positive clipper
(b) biased negative clipper
(c) clamper
(d) none of the above

Q.108 Assertion (A): Clipping circuit controls the shape of the waveform by clipping or removing a portion of the applied wave.

Reason (R): It needs minimum three components namely an ordinary diode, a resistor and a capacitor.

(a) Both A and R are true and R is the correct explanation of A.
(b) Both A and R are true but R is not the correct explanation of A.
(c) A is true but R is false.
(d) A is false but R is true.

Q.109 Consider the following statements associated with the diode rectifier circuits:

1. The maximum reverse voltage, which can be applied before the breakdown point is reached, is called the "peak inverse voltage".

2. In a full-wave rectifier, the current in each of the diode flows for the whole input signal cycle.
3. The ratio of direct or average value of the output to the effective value of AC component present in the rectifier output is called the "ripple factor".
4. Availability of low-cost silicon diodes has made bridge circuit arrangement more economical than centre trapped transformer arrangement despite its requirement of four diodes.

Which of the above statements are correct?

(a) 1, 2 and 3 (b) 1 and 4
(c) 2, 3 and 4 (d) 2 and 4

Q.110 In the Zener regulator shown below, the output requirements are 5 V, 20 mA. If $I_{z(min)}$ and $I_{z(max)}$ are 5 mA and 80 mA, then the value of limiting resistor (R) and the load resistance (R_L) will be respectively

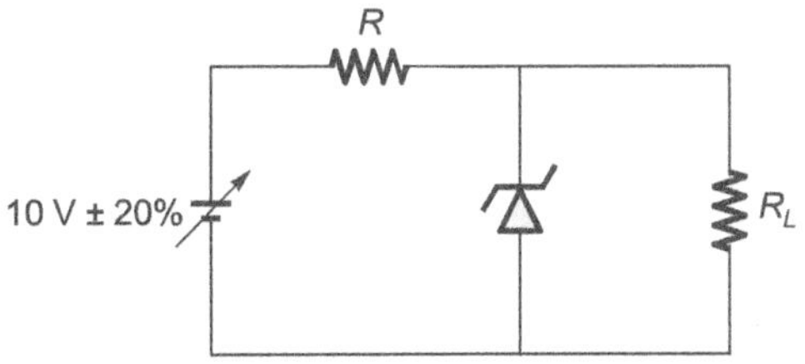

(a) 190 Ω and 250 Ω
(b) 100 Ω and 200 Ω
(c) 100 Ω and 250 Ω
(d) 120 Ω and 250 Ω

Q.111 For the zener voltage regular circuit shown below, the output voltage is required to be maintained at 5 volt with a load current of 10 mA. The zener wattage is 400 mW and input voltage is specified as 10 V ± 2 V.

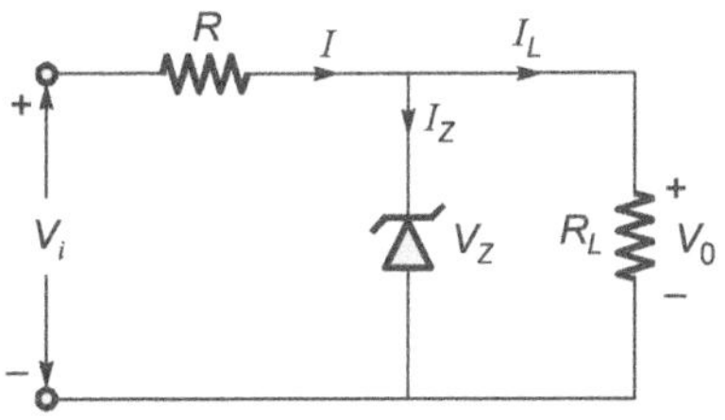

One of the possible value of series resistance R will be

(a) 266 Ω (b) 302 Ω
(c) 68 Ω (d) none of the these

Q.112 The Zener diode V_{Z1} in the figure shown below has the reverse saturation current of 20 mA and reverse breakdown voltage of 100 V whereas the corresponding values for diode V_{Z2} are 40 μA and 40 V.

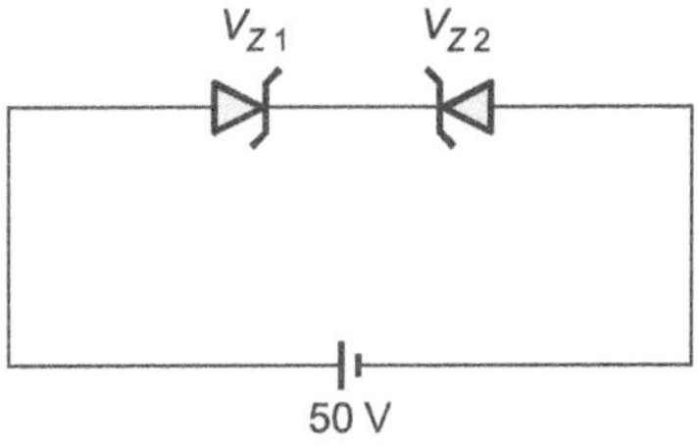

The current through the circuit is

(a) 20 μA anticlockwise
(b) 20 μA clockwise
(c) 40 μA anticlockwise
(d) 40 μA clockwise

Q.113 Match **List-I (Diode Circuits)** with **List-II (Output-Input Characteristics)** and select the correct answer using the codes given below the lists:

List-I

A. 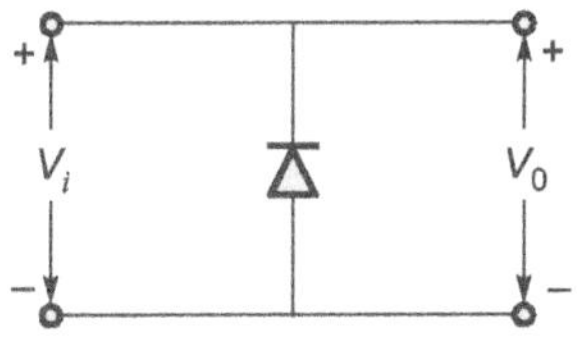

B. 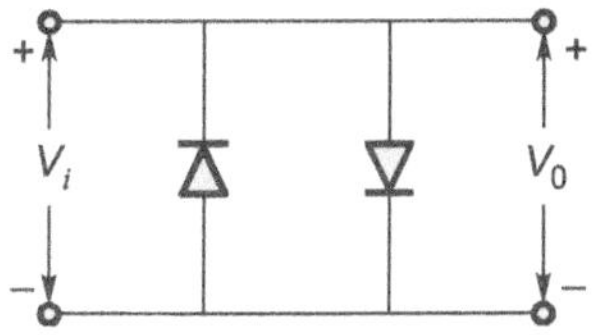

C. 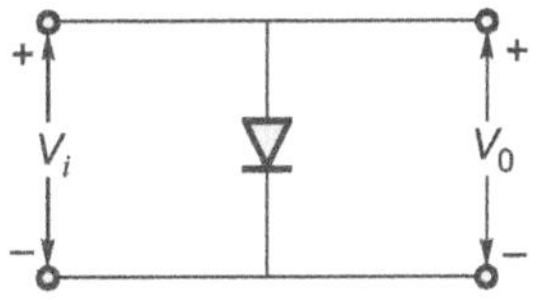

List-II

1. 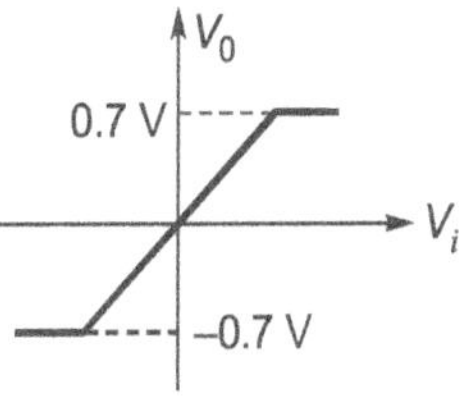

2. 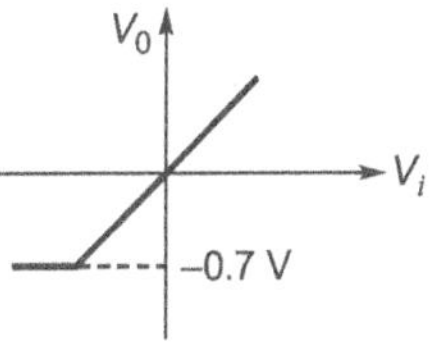

3. 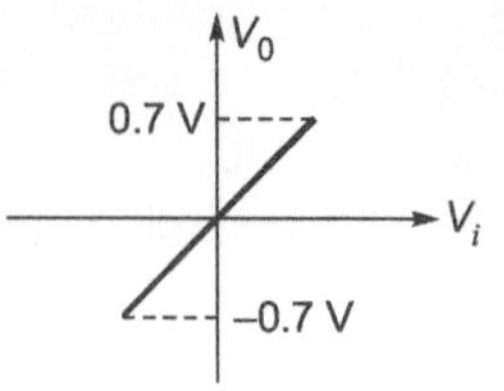

4. 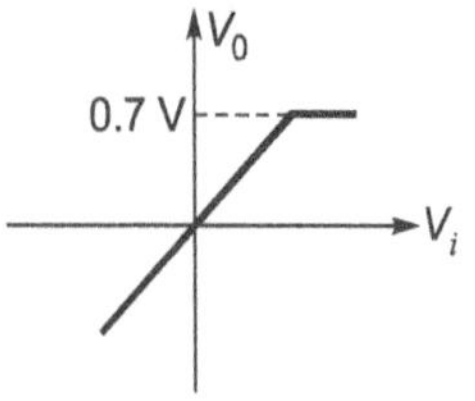

Codes:

	A	B	C
(a)	4	1	2
(b)	2	3	4
(c)	2	1	4
(d)	4	3	2

Q.114 What is the output voltage of clipper circuit shown in figure below for the sinusoidal input voltage with its peak voltage of 100 V, assuming both diodes ideal?

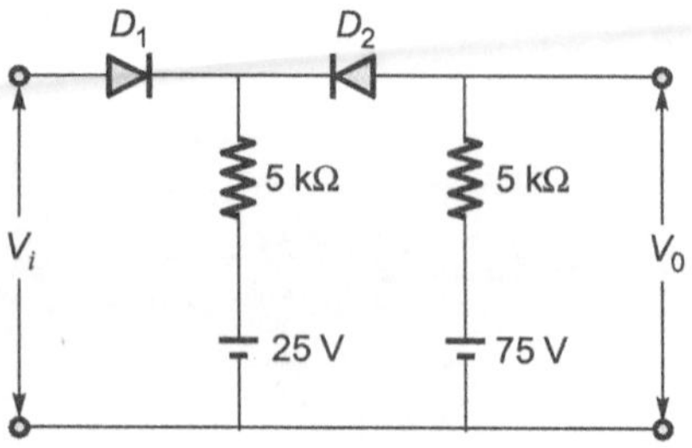

(a) 65 V (b) 50 V
(c) 45 V (d) 30 V

Q.115 For the circuit shown below, assume that the diodes are ideal. The meter reading would be

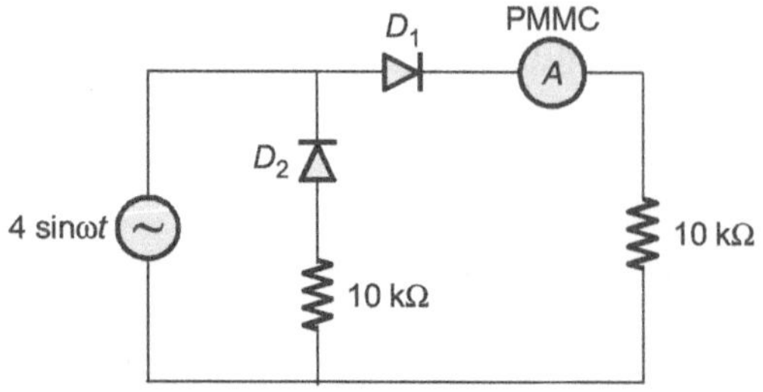

(a) $\frac{0.4}{\pi}$ mA (b) 0.4 mA

(c) $\frac{0.8}{\pi}$ mA (d) $0.4\sqrt{2}$ mA

Q.116 Two identical diodes, D_1 and D_2 are connected back to back as shown in figure below. The reverse saturation current I_0 of each diode is 10^{-8} A and the breakdown voltage is 50 V. The voltage drop across diodes D_1 and D_2 will be respectively (assume kT/q = 25 mV)

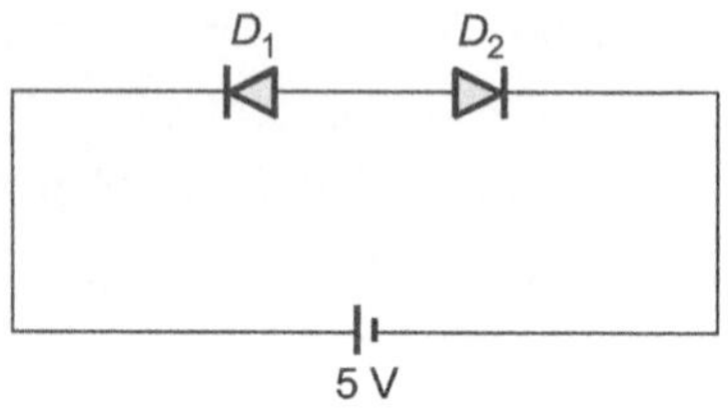

(a) 4.931 V and 0.0693 V
(b) 0.0173 V and 4.983 V
(c) 0.0693 V and 4.931 V
(d) 4.983 V and 0.0173 V

Q.117 The 6 V Zener diode shown in figure has zero zener resistance and a knee current of 5 mA.

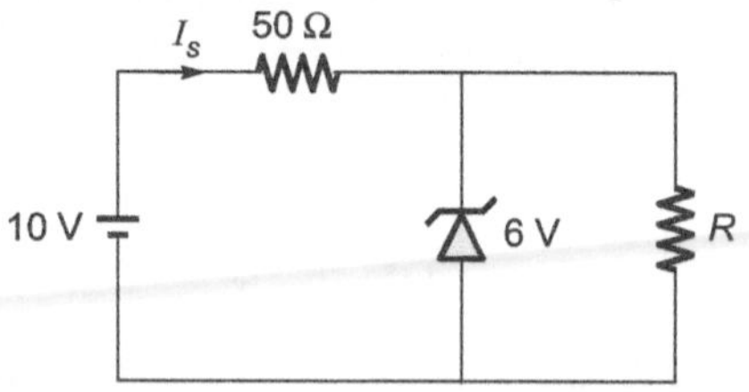

The minimum value of R so that the voltage across it does not fall below 6 V is

(a) 1.2 Ω (b) 80 Ω
(c) 1.2 Ω (d) 0 Ω

Q.118 What are the values of V_0 and I for the diode circuit shown below?
(Assume the drop across the diodes to be 0.7 V)

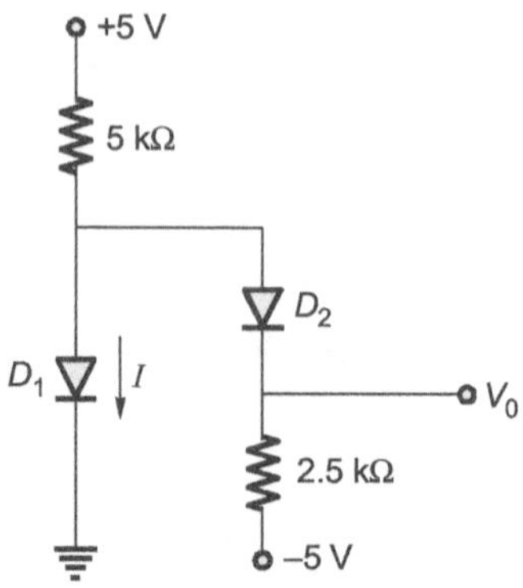

(a) –2.2 V and –0.25 mA respectively
(b) –1.9 V and –0.14 mA respectively
(c) –2.2 V and –0.14 mA respectively
(d) –1.9 V and –0.25 mA respectively

Q.119 A DC power supply has no-load voltage of 30 V, and a full-load voltage of 25 V at a full-load current of 1 A. Its output resistance and load regulation, respectively, are

(a) 5 Ω and 20% (b) 25 Ω and 20%
(c) 5 Ω and 16.7% (d) 25 Ω and 16.7%

Q.120 Higher order active filters are used for variable
(a) bandwidth
(b) gain in the pass-band
(c) impedance
(d) roll-off rate

Q.121 In the figure shown below, the diode of Op-amps are ideal. For an input V_{in} = sin (ωt) the output voltages V_0 is

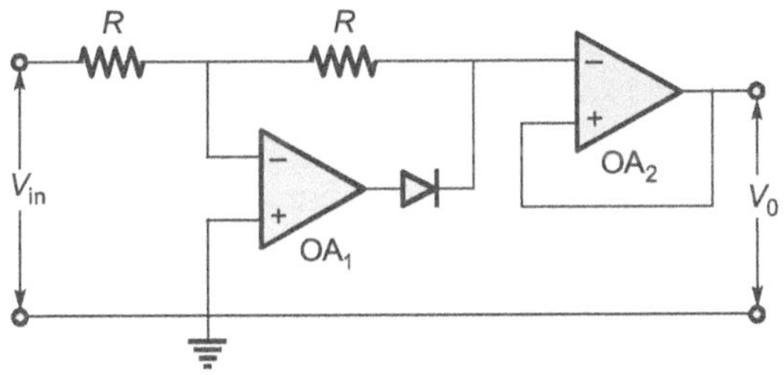

(a) full wave rectifier with a peak value of +1 V
(b) full wave rectifier with a peak value of –1 V
(c) half wave rectifier with a peak value of –1 V
(d) half wave rectifier with a peak value of +1 V

Q.122 The cutin voltage for each diode in figure is V_γ = 0.6 V. Each diode current is 0.5 mA. The value of R_1, R_2 and R_3 will be respectively

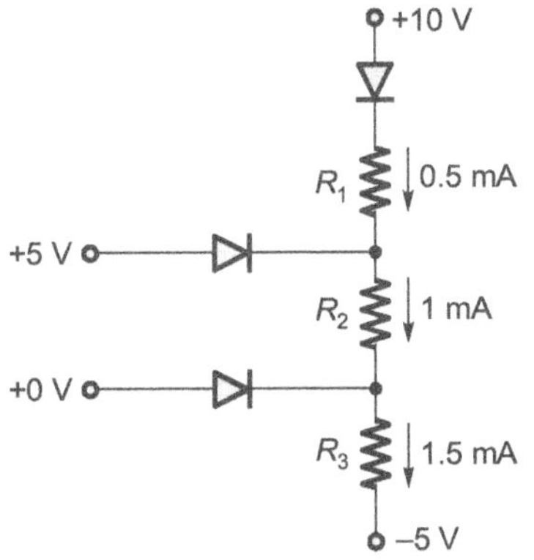

(a) 10 kΩ, 5 kΩ, 2.93 kΩ
(b) 6 kΩ, 3 kΩ, 3.43 kΩ
(c) 5 kΩ, 6 kΩ, 4.93 kΩ
(d) 6 kΩ, 8 kΩ, 6.43 kΩ

Q.123 In the circuit shown below, diodes D_1 and D_2 are ideal. The current i_1 and i_2 are respectively

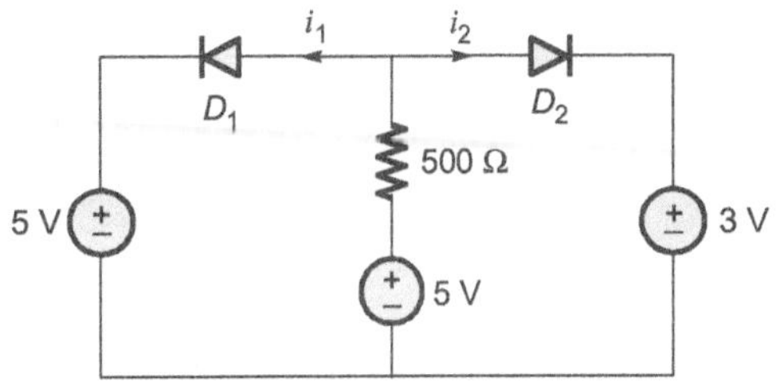

(a) 0 and 4 mA (b) 4 mA and 0
(c) 0 and 8 mA (d) 8 mA and 0

Q.124 In a feedback series regulator circuit, the output voltage is regulated by controlling the
(a) magnitude of the input voltage
(b) gain of the feedback transistor
(c) reference voltage
(d) voltage drop across the series pass transistor

Q.125 In a bridge rectifier, the input AC source has an amplitude of 5 V, and the load resistance is 100 Ω. The diodes have an on resistance of 10 Ω, and a cutin voltage of 0.6 V. The output voltage is

(a) 0.89 volt (b) 1.25 volt
(c) 0.67 volt (d) 1.69 volt

Q.126 The peak inverse voltage across the diode in a full-wave rectifier made with two diodes and a centre-tapped transformer is _____ that of a bridge rectifier.

(a) equal to (b) not related to
(c) half (d) double

5. BJT Characteristics and Biasing

Q.127 In a transitor, the value of α is 0.97. The value of current gain of common emitter configuration is

(a) 22.32 (b) 32.33
(c) 33.32 (d) 32.22

Q.128 A transistor has I_B = 100 μA and I_C = 2 mA. If I_B changes by 25 μA and I_C changes by 0.6 mA, the change in the value of β would be

(a) $\frac{3}{2}$ (b) $\frac{5}{4}$

(c) $\frac{1}{4}$ (d) $\frac{4}{5}$

Q.129 For a transistor having $\alpha = 0.98$, $I_{CBO} = 5\ \mu A$ and $I_B = 100\ \mu A$. The value of I_C is

(a) 5.15 mA (b) 2.25 mA
(c) 3.65 mA (d) 4.45 mA

Q.130 For the circuit shown below, the collector to emitter voltage is

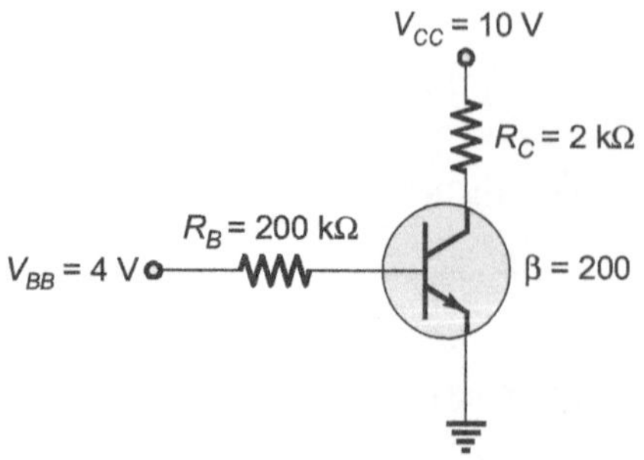

(a) 1.25 V (b) 3.40 V
(c) –2.30 V (d) 0 V

Q.131 The reverse leakage current of the transistor when connected in CB configuration is 0.2 μA and it is 18 μA when the same transistor is connected in CE configuration. The value of α and β of the transistor for a base current of 30 mA will be respectively

(a) 0.988 and 120 (b) 0.966 and 42.5
(c) 0.988 and 89 (d) 0.966 and 120

Q.132 The value of V_C for the transistor shown in figure below is

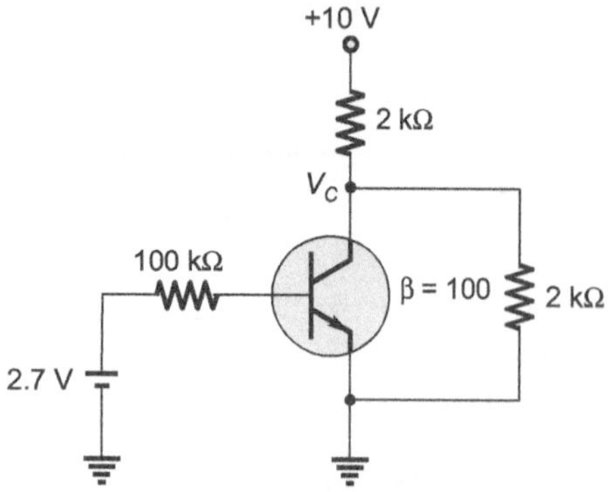

(a) 3 volt (b) $\frac{1}{2}$ volt
(c) 2 volt (d) none of these

Q.133 Match **List-I (Transistor configuration)** with **List-II (Applications)** and select the correct answer using the codes given below the lists:

List-I
A. Common emitter
B. Common base
C. Common collector

List-II
1. For impedance matching
2. For high frequency circuits
3. For audio frequency circuits

Codes:

	A	B	C
(a)	2	1	3
(b)	1	2	3
(c)	2	3	1
(d)	3	2	1

Q.134 **Assertion (A):** The dependency of base width on collector to emitter voltage is known as the "Early effect".

Reason (R): Due to "Early effect", α of the transistor is increased by a larger value while maintaining a constant value of β.

(a) Both A and R are true and R is the correct explanation of A.
(b) Both A and R are true but R is not the correct explanation of A.
(c) A is true but R is false.
(d) A is false but R is true.

Q.135 In the common-emitter configuration, if the transistor is in the saturation region, then

(a) $I_C > I_E$ (b) $I_C < \beta I_B$
(c) $I_E < I_B$ (d) $I_B > \beta I_C$

Q.136 In forward mode *n-p-n* BJT, if we increase the voltage V_{CC}, the collector current increases

(a) because due to base width decrease, less carrier combine in the base region,
(b) as the gradient of the minority carriers in the base region becomes steeper.
(c) due to ohm's law, higher V_{CC} causes higher current.
(d) due to both the reasons (a) and (b).

Q.137 An emitter follower has high input impendance because

(a) there is negative feedback in the base-emitter circuit.
(b) large emitter resistance is used.
(c) the emitter-base junction is highly reverse biased.
(d) large biasing resistance is used.

Q.138 The CE configuration amplifier circuits are preferred over CB configuration amplifier circuits because they have

(a) high input resistance and low output resistance
(b) larger amplification factor
(c) lower amplification factor
(d) none of the above

Q.139 A device whose characteristics are very close to that of an ideal current source is

(a) a BJT in CB mode
(b) a gas diode
(c) a triode
(d) a BJT in CE mode

Q.140 The input resistance of a common-collector configuration will be of the order of

(a) $\simeq$ 90 kΩ (b) $\simeq$ 150 kΩ
(c) 150 kΩ (d) 300 kΩ and above

Q.141 The common collector amplifier is also Known as

(a) Collector follower
(b) Source follower
(c) Emitter follower
(d) Base follower

Q.142 Consider the statements associated with various types of transistor configuration of a BJT:

1. In cut-off region, emitter and collector junctions are both reverse biased.
2. In saturation region, emitter junction is forward biased while collector junction can be either forward biased or reverse biased.
3. In active region, the collector junction is biased in the reverse biased and the emitter junction in the forward biased.
4. The phase shift introduced to the input in CC and CE configurations are each zero degree.

Which of the above statements are correct ?

(a) 1 and 3 (b) 2 and 4
(c) 1, 2 and 3 (d) 1, 3 and 4

Q.143 In an *n-p-n* transistor biased in the active region:

(a) electrons diffuse across the emitter-base junction and drift across the base-collector junction.
(b) electrons drift across the emitter-base junction and diffuse across the base-collector junction.
(c) electrons drift across the emitter-base junction and drift across the base-collector junction.
(d) electrons diffuse across the emitter-base junction and diffuse across the base-collector junction.

Q.144 **Assertion (A):** An emitter follower is widely used in electronic instruments.

Reason (R): The voltage gain of emitter follower is very high.

(a) Both A and R are true and R is the correct explanation of A.
(b) Both A and R are true but R is not the correct explanation of A.
(c) A is true but R is false.
(d) A is false but R is true.

Q.145 **Assertion (A):** Common base amplifier has very low input resistance.

Reason (R): Low emitter ac resistance r_e' shunts all other resistances.

(a) Both A and R are true and R is the correct explanation of A.
(b) Both A and R are true but R is not the correct explanation of A.
(c) A is true but R is false.
(d) A is false but R is true.

Q.146 Consider the following statements:

1. In a *p-n-p* transistor with normal bias, the collector junction is of low resistance.
2. *p-n-p* and *n-p-n* transistors can be interchanged with each other.
3. The circuit arrangement of a transistor used as a buffer is CC (Common Collector).
4. In a CE connected transistor, V_{CE} will be equal to V_{CC} when base current is zero.
5. In a transistor, emitter junction breakdown is due to "Zener effect" while collector junction breakdown is due to "Avalanche effect".

Which of the statements given above are not correct ?

(a) 3 and 5 only (b) 2, 3 and 4 only
(c) 2, 3 and 5 only (d) 1 and 2 only

Q.147 Match **List-I** with **List-II** and select the correct answer using the codes given below the lists:

List-I

A. Common emitter Configuration
B. Common collector Configuration
C. Base width modulation
D. Common base Configuration

List-II

1. Early Effect
2. Highest input resistance
3. Inverting amplifier
4. Lowest power gain

Codes:

	A	B	C	D
(a)	3	2	1	4
(b)	3	4	1	2
(c)	4	3	2	1
(d)	1	2	3	4

Q.148 For the transistor shown in the figure below, $\beta = 25$. The range of V_1 such that, $1.0 \le V_{CE} \le 4.5$ V is

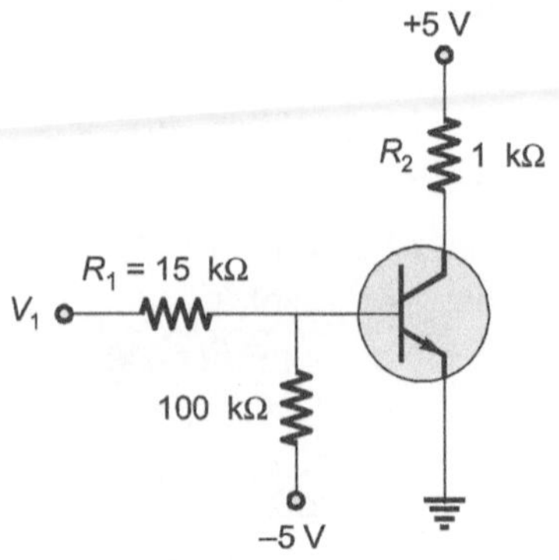

(a) $1.86 \le V_1 \le 3.96$ V (b) $2.81 \le V_1 \le 4.46$ V
(c) $1.43 \le V_1 \le 79$ V (d) $2.18 \le V_1 \le 3.69$ V

Q.149 In the bipolar source of figure the diode voltage and transistor V_{BE} voltages are equal. If base current is neglected then collector current is

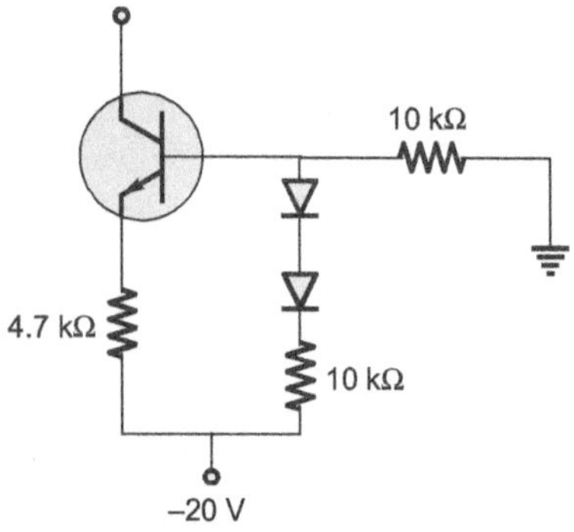

(a) 6.43 mA (b) 2.13 mA
(c) 1.48 mA (d) 9.19 mA

Q.150 The Early voltage of a BJT is $V_A = 75$ volt. The minimum required collector current, such that the output resistance is atleast $r_0 = 200$ kΩ, is
(a) 1.67 mA (b) 5 mA
(c) 0.375 mA (d) 0.75 mA

Q.151 The thinnest region in a BJT is_______ and the most doped region is_______.
(a) emitter, collector (b) base, collector
(c) collector, emitter (d) base, emitter

Q.152 In an *n-p-n* transistor biased in the active region, as the magnitude of the collector-base voltage is increased,
(a) the base current increases because more electrons are injected from the emitter.
(b) the base current increases because more holes are injected from the base to the collector.
(c) the emitter current decreases because the base-emitter junction gets slightly less forward biased.
(d) the collector current increases slightly because the base width reduces.

Q.153 The common-emitter is the most widely used amplifier configuration because it has
(a) a very high input resistance and a low output resistance.
(b) a low voltage gain and a low output resistance.
(c) a low input resistance and a high output resistance.
(d) a high voltage gain and a high current gain.

Q.154 A common-base configuration is useful when
(a) a high current gain is required
(b) a low output resistance is required
(c) a low input resistance is required
(d) a high voltage gain and low output resistance are required

Q.155 The BJT amplifier which offers highest input impendance and least voltage gain is
(a) CB (b) CC
(c) CE (d) cascade amplifier

Q.156 The circuit using a BJT with β = 50 and V_{BE} = 0.7 V is shown in the figure. The base current I_B and collector voltage V_C are respectivoly

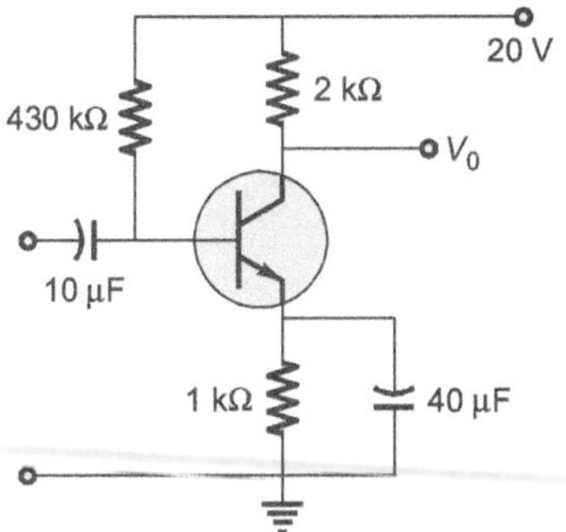

(a) 43 μA and 11.4 V (b) 40 μA and 16 V
(c) 45 μA and 11 V (d) 50 μA and 10 V

6. Transistor Biasing & Thermal Stabilization

Q.157 If a change in base current does not change the collector current, the transistor amplifier is said to be
(a) cut-off (b) complemented
(c) saturated (d) critical

Q.158 In a BJT, if the collector junction is reverse biased and the emitter junction is reverse - biased, it is said to operate in
(a) cut-off region (b) saturation region
(c) active region (d) none of these

Q.159 On applying a positive voltage signal to the base of a normally biased *n-p-n* CE transistor amplifier
(a) emitter current will fall
(b) collector current will fall
(c) base current will fall
(d) collector voltage will become less positive

Q.160 **Assertion (A):** The process of giving proper supply voltages and resistances for obtaining the desired Q-point is called biasing.
Reason (R): In order to produce distortion free output in amplifier circuits, the supply voltages and resistances in the circuit must be suitably chosen.
(a) Both A and R are true and R is the correct explanation of A.
(b) Both A and R are true but R is not the correct explanation of A.
(c) A is true but R is false.
(d) A is false but R is true.

Q.161 Match **List-I (Transistor parameters)** with **List-II (Temperature effect)** and select the correct answer using the codes given below the lists:

List-I
A. Base-emitter voltage (V_{BE})
B. Transistor current gain (β)
C. Reverse saturation current (I_{CO})

List-II
1. Independent of temperature
2. Decreases by 2.5 mV per °C
3. Increases with temperature
4. Increase by 7% per °C

Codes:

	A	B	C
(a)	2	1	4
(b)	2	3	4
(c)	4	3	2
(d)	1	4	2

Q.162 For the transistor circuit shown below, the Q-point values are given as I_{CQ} = 1 mA and V_{CEQ} = 6 V

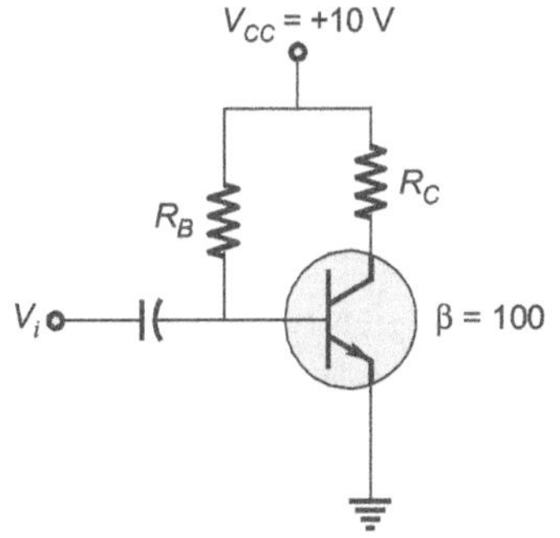

If V_{BE}(on) = 0.7 V, the values of the R_B and R_C will be respectively
(a) 1.28 MΩ and 4 kΩ
(b) 0.93 MΩ and 4 kΩ
(c) 0.93 MΩ and 2.5 kΩ
(d) 1.28 MΩ and 2.5 kΩ

Q.163 For the transistor circuit shown below, β = 100 and V_{BE}(on) = 0.7 V.

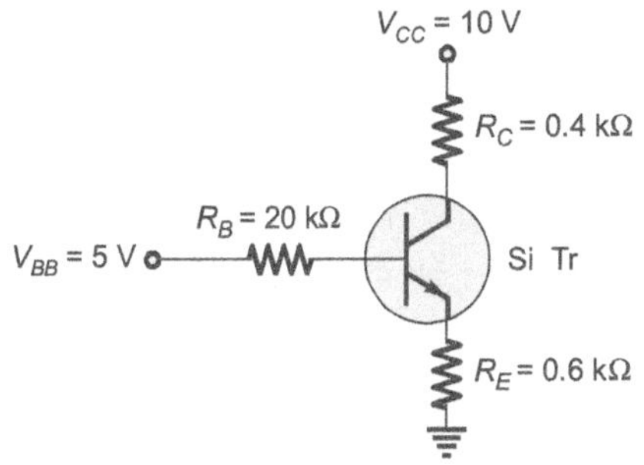

The transistor is working under
(a) active region with $V_{CEQ} = 2.34$ volt
(b) saturation region with $V_{CEQ} = 0.2$ volt
(c) active region with $V_{CEQ} = 4.63$ volt
(d) cut-off region with $V_{CEQ} = 9.9$ volt

Q.164 In a collector-to-base CE amplifier circuit having $V_{CC} = 12$ V, $R_C = 250\ \Omega$, $I_B = 0.25$ mA, $\beta = 100$ and $V_{CEQ} = 8$ V, the stability factor will be approximately given by (neglect V_{BE})
(a) 57 (b) 62
(c) 47 (d) 68

Q.165 A self-bias circuit (with an emitter resistor) stabilizes the ________ of the circuit against variations in temperature and process parameters.
(a) current gain (b) base- emitter voltage
(c) collector current (d) base current

Q.166 In typical circuits, the stabilization factor $S(=\Delta I_C/\Delta I_{CO})$ is
(a) < 1 (b) > 1
(c) = 1 (d) << 1

Q.167 Assertion (A): Current mirrors are widely used in analog integrated circuits for DC biasing.
Reason (R): Current mirrors provide highly stable DC bias points without the need of a lot of biasing resistors.
(a) Both A and R are true and R is the correct explanation of A.
(b) Both A and R are true but R is not the correct explanation of A.
(c) A is true but R is false.
(d) A is false but R is true.

Q.168 Consider the following statements:
1. Self-bias is superior to the fixed bias from the consideration of stability of operating point in the CE amplifier.
2. Bias stabilization is used to prevent thermal runaway.
3. Improper biasing of a transistor results in the distortion of the output signal.
4. A transistor is said to be in quiscent stage when no signal is applied to the input.

Which of the statements given above are correct?
(a) 1, 2 and 3 (b) 3 and 4
(c) 1, 2 and 4 (d) 1, 2, 3 and 4

Q.169 For the circuit shown below, what is the value of R_C needed if $V_C = 2$ V ?

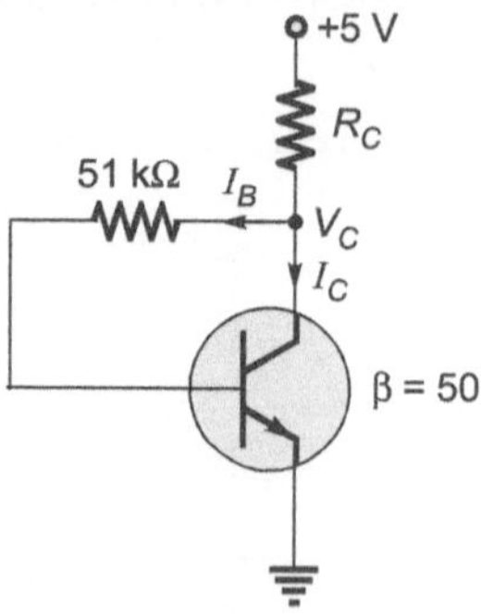

(a) 5.3 kΩ (b) 3.5 kΩ
(c) 4.8 kΩ (d) 2.3 kΩ

Q.170 Assuming the current source to be ideal, all transistors to be identical, and $I_C = \beta I_B$, the stability factor $\frac{\partial I_{C_1}}{\partial \beta}$ for the current mirror circuit shown below would be

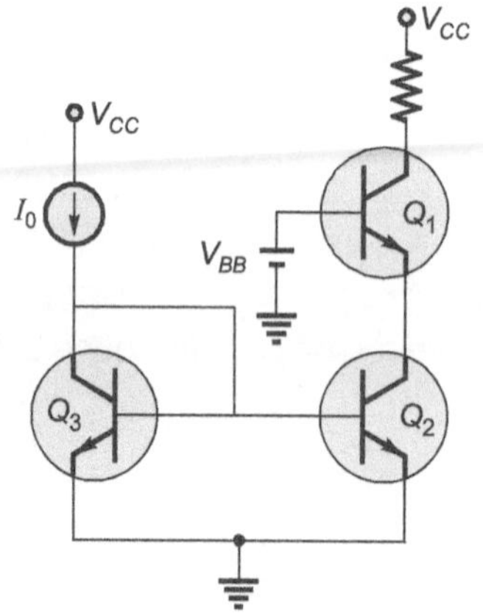

(a) $I_0\left(\frac{(3\beta+4)}{\beta(\beta+1)(\beta+2)}\right)$

(b) $I_0\left(\frac{\beta(\beta+2)}{(\beta+1)^2(3\beta+4)^2}\right)$

(c) $I_0\left(\frac{\beta(3\beta+4)}{(\beta+1)^2(\beta+2)^2}\right)$

(d) $I_0\left(\frac{(\beta+1)^2}{(3\beta+4)^2(\beta+2)^2}\right)$

Q.171 A reverse biased diode placed in parallel with the base-emitter junction makes the circuit insensitive to variations in ______ with changes in temperature.

(a) V_{BE}
(b) I_{CO}
(c) β
(d) all of these parameters

Q.172 The static (DC) and dynamic (AC) load lines in a transistor amplifier are different when
(a) a capacitive load is connected at the output.
(b) the coupling capacitor exist at the input port
(c) the collector resistance varies with temperature.
(d) a resistive load is connected through a coupling capacitor at the output.

Q.173 Thermal runaway is not possible in FET because as the temperature of FET increases,
(a) the mobility decreases
(b) the transconductance increase
(c) the drain current increases
(d) the mobility increases

Q.174 The two transistors in figure are identical. If $\beta = 25$, the current I_{C2} is

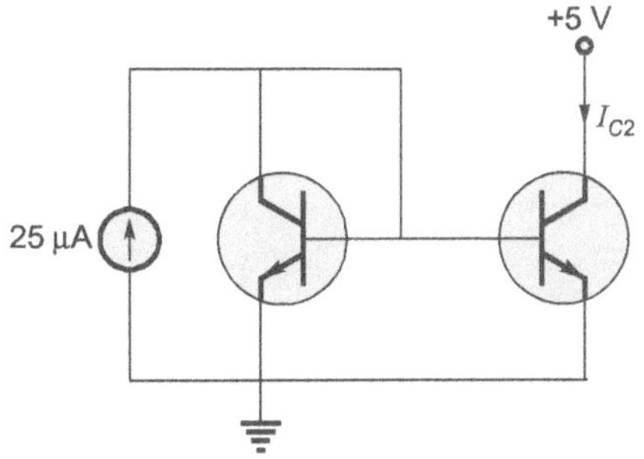

(a) 28 μA (b) 23.2 μA
(c) 26 μA (d) 24 μA

Q.175 The input resistance of a common emitter stage can be increased by
1. unbypassing emitter resistance
2. bootstrapping
3. biasing it at low quiescent current
4. using compounded BJTs

The correct sequence in descending order of the effectiveness of these methods is
(a) 2, 4, 1 and 3 (b) 4, 3, 2 and 1
(c) 2, 4, 3 and 1 (d) 4, 2, 3 and 1

Q.176 For the transistor amplifier shown in figure below, V_{CC} = 12 V, R_1 = 8 kΩ, R_2 = 4 kΩ, R_C = 1 kΩ, R_E = 1 kΩ.
Assuming V_{BE} = 0.7 V, the operating point Q be given as

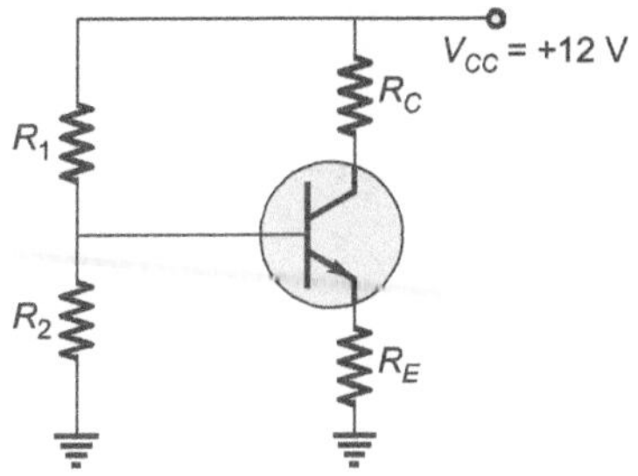

(a) Q (2.2 V, 2.8 mA) (b) Q (6.8 V, 1.6 mA)
(c) Q (5.4 V, 3.3 mA) (d) none of these

Q.177 In a self bias circuit using BJT, if the stability factor is represented as $S = \frac{\partial I_C}{\partial I_{CO}}$, then the rate of change of I_C w.r.t. β, keeping I_{CO} and V_{BE} constant will be given by (I_C = collector current of the transistor)

(a) $\frac{SI_C}{\beta^2}$ (b) $\frac{SI_C}{(1+\beta)^2}$

(c) $\frac{S(1+\beta)I_C}{\beta}$ (d) $\frac{SI_C}{\beta(1+\beta)}$

Q.178 The forward current gain (α) of a bipolar transistor can be increased by
1. reducing the recombination lifetime in the emitter.
2. increasing the emitter doping density.
3. increasing the base doping density.
4. reducing the base width.

Select the correct code from the given options.
(a) 2 and 3 only (b) 2 and 4 only
(c) 1, 2 and 4 only (d) 4 only

Q.179 What is the value of thermal resistance for the 2N338 transistor for which the manufacturer specifies $P_{C,max}$ = 125 mW at 25 °C free-air temperature and maximum junction temperature of 150 °C?
(a) 1.0 °C/mW (b) 1.45 °C/mW
(c) 1.2 °C/mW (d) 0.2 °C/mW

Q.180 The stability factor (S_1) for the circuit shown below is given by

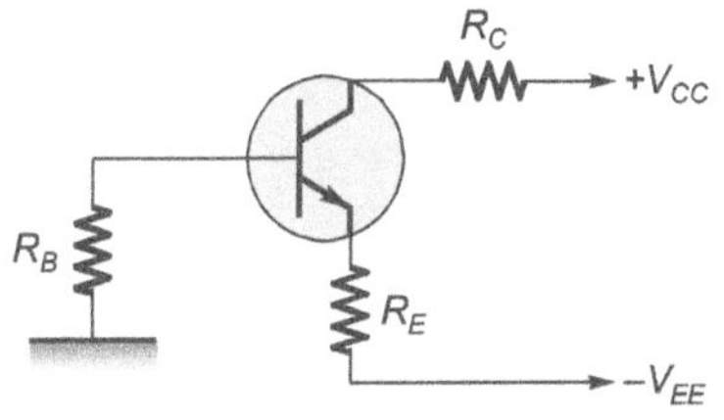

(a) $\dfrac{\beta(R_B + R_E)}{R_B + \beta R_E}$ (b) $\dfrac{(1+\beta)(R_B + R_E)}{R_B + \beta R_E}$

(c) $\dfrac{\beta(R_B + R_E)}{R_B + (1+\beta)R_E}$ (d) $\dfrac{(1+\beta)(R_B + R_E)}{R_B + (1+\beta)R_E}$

Q.181 For the cascade amplifier shown below, the quiescent current through transistors is 1 mA and the collector voltage V_{C2} = 6 V and V_{C1} = 3 V. Assuming V_{BE} = 0.7 V and very high value of β, the values of R_1 and R_2 will be respectively

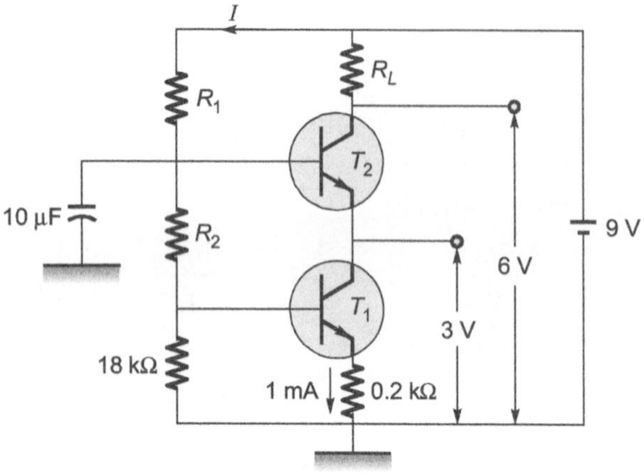

(a) 167 kΩ and 82 kΩ
(b) 167 kΩ and 56 kΩ
(c) 106 kΩ and 56 kΩ
(d) 106 kΩ and 82 kΩ

Q.182 In the common-emitter amplifier shown in figure, the transistor has a forward gain of 100 and V_{BE} = 0.6 V. Assume I_{CO} to be negligible. The value of R_1 and R_C such that the transistor has a collector current of 1 mA and V_{CE} = 2.5 V are respectively given by

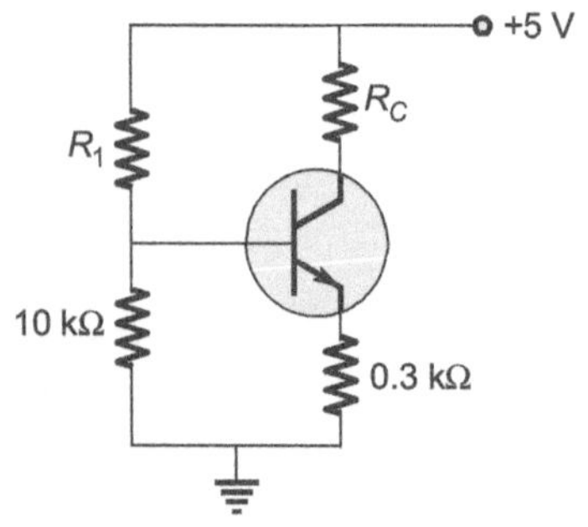

(a) 41 kΩ and 2.2 kΩ
(b) 36 kΩ and 1.1 kΩ
(c) 41 kΩ and 1.1 kΩ
(d) 36 kΩ and 2.2 kΩ

7. BJT as an Amplifier

Q.183 Removing bypass capacitor across the emitter-leg resistor in a CE amplifier causes
(a) decrease in current gain
(b) increase in voltage gain
(c) increase in current gain
(d) decrease in voltage gain

Q.184 The current gain of a common-emitter amplifier ______ as the load resistance is increased.
(a) increases
(b) decreases
(c) in unchanged
(d) increase and then decreases

Q.185 Cascading amplifier stages to obtain a high gain is best done with
(a) common-emitter stages
(b) common-base stages
(c) common-collector stages
(d) a combination of successive common-base and common-emitter stages

Q.186 The h-parameter equivalent circuit is the most common representation of the transistor because
(a) it represents the admittance parameters of the transistor.
(b) it is most useful in the ON and OFF states of transistor.
(c) it is usable over the entire frequency band of interest to the transistor.
(d) the parameters values correspond to input impedance, output admittance and current gain for the most common applications of the transistor.

Q.187 Assertion (A): The voltage gain of the emitter follower is very high.
Reason (R): The emitter follower increases the power level of the signal.
(a) Both A and R are true and R is the correct explanation of A.
(b) Both A and R are true but R is not the correct explanation of A.
(c) A is true but R is false.
(d) A is false but R is true.

Q.188 For a cascade amplifier shown below, the value of voltage gain V_0/V_s is given by
(Use h_{ie} = 1 km, h_{fe} = 100, h_{re} = 0, and h_{oe} = 0)

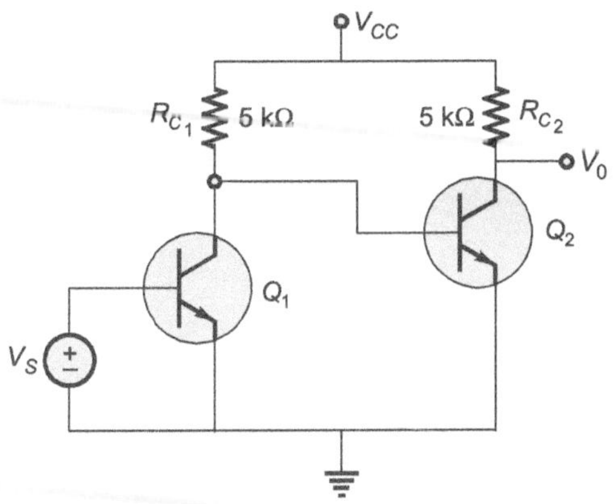

(a) 52415 (b) 41666
(c) 62877 (d) 39624

Q.189 The *h*-parameters for the transistor shown below are given as:
h_{ie} = 10 kΩ, h_{fe} = 100, h_{re} = 0, and h_{oe} = 0

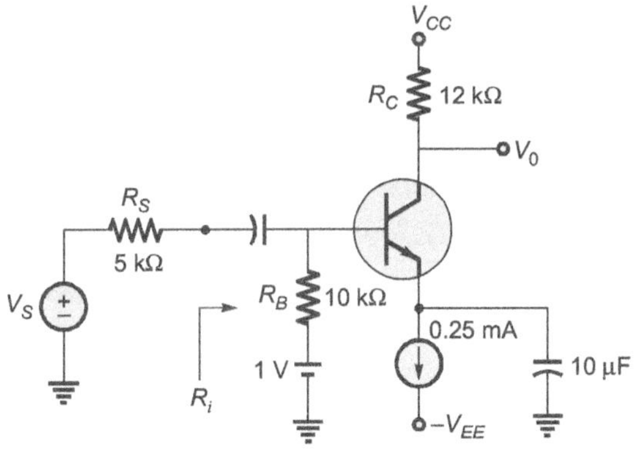

The value of voltage gain $\left(\frac{V_0}{V_s}\right)$ and input resistance (R_i) for the circuit shown above are respectively given by

(a) –60 and 5 kΩ (b) –40 and 8 kΩ
(c) –60 and 8 kΩ (d) –40 and 5 kΩ

Q.190 Assertion (A): The Darlington connection of three or more transistors is usually impractical.
Reason (R): The Darlington emitter follower has a higher input resistance and a voltage gain less close to unity than does a single-stage emitter follower.

(a) Both A and R are true and R is the correct explanation of A.
(b) Both A and R are true but R is not the correct explanation of A.
(c) A is true but R is false.
(d) A is false but R is true.

Q.191 Bootstrapping is used in emitter follower configuration to

(a) stabilize the voltage gain against process variations.
(b) increase the current gain.
(c) reduce the output resistance.
(d) increase the input resistance.

Q.192 Consider the following statements associated with the use of h-parameters to describe a transistor:

1. They are easy to measure.
2. They are real numbers for all frequency range.
3. They are easily convertible from one configuration to other.
4. The smallest value of h-parameter, for a transistor configuration is that of short circuit input impedance (h_i).

Which of the statements given above are correct?

(a) 2 and 4 (b) 2 only
(c) 1 and 3 (d) 1 only

Q.193 The value of input impedance and voltage gain of the common base circuit shown below are respectively (Take V_{BE} = 0.6 V)

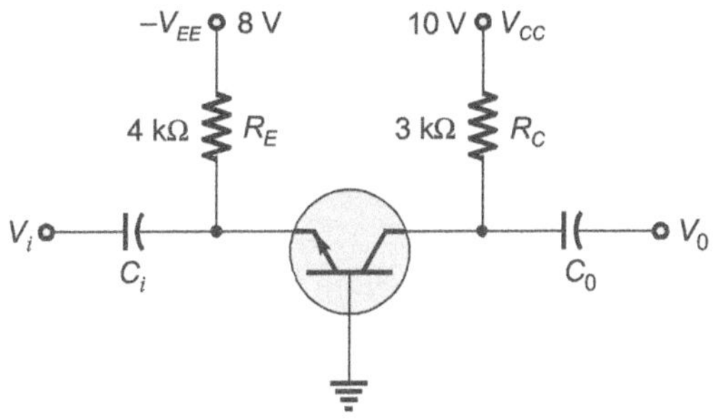

(a) 14 Ω and 213.5
(b) 12 kΩ and 222
(c) 3 kΩ and 213.5
(d) 14 kΩ and 222

Q.194 The circuit shown in figure is biased such that V_{CE2} = 0 when V_S = 0. Neglecting the base current of both transistors and forward bias across V_{BE} for both transistors to be 0.7, the value of R_E would be equal to

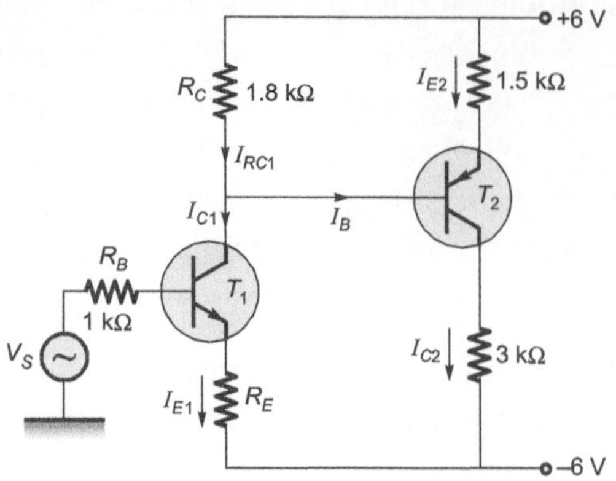

(a) 3 kΩ (b) 1.5 kΩ
(c) 2 kΩ (d) 4 kΩ

Q.195 A cascade stage is shown in figure below. If the transconductance of T_1 and T_2 are $g_{m1} = i_{c1} / V_{be1}$ and $g_{m2} = i_{c2}/V_{be2}$, the overall transconductance $g_m = i_{c2}/V_{be1}$ is

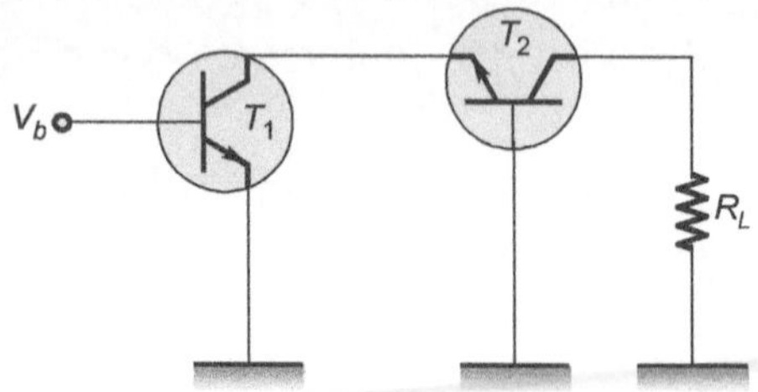

(a) more than g_{m1} (b) less than g_{m1}
(c) equal to g_{m1} (d) none of these

Q.196 An emitter follower with β = 100 is biased at I_C = 0.25 mA. The voltage source connected at its input has its internal resistance of 2 kΩ. What is the value of R_E such that it produces the output resistance of 110 Ω? (Take V_T = 25 mV)
(a) 2 kΩ (b) 1.5 kΩ
(c) 2.5 kΩ (d) 1 kΩ

Q.197 Two stages of BJT amplifiers are cascaded by RC coupling. The voltage gain of the first stage is 20 and that of the second stage is 30. The overall gain of the coupled amplifier is
(a) $(20 + 30)$ (b) (20×30)
(c) $\left(\frac{20 \times 30}{2}\right)$ (d) $(20 + 30)^2$

Q.198 The main characteristics of a Darlington amplifier are
(a) high input impedance, low output impedance and high current gain.
(b) low input impedance, low output impedance and low current gain.
(c) high input impedance, high output impedance and high current gain.
(d) low input impedance, low output impedance and high current gain.

Q.199 Which of the following parameters is used for distinguishing between a small signal and a large-signal amplifier?
(a) Harmonic distortion
(b) Frequency response
(c) Voltage gain
(d) Input/Output impedances

Q.200 Consider the following statements:
1. The coupling capacitor mainly affect the lower cutoff frequency.
2. The phase reversal between output and input takes place only for voltage wave and not for current waves, in a transistor amplifier in CE mode.
3. The DC collector current in a transistor circuit is limited by the junction capacitance.
4. Negative DC feedback through R_E is responsible for the stabilization of the operating point in a potential divider bias circuit.

Of these statement:
(a) only 2 and 3 are correct
(b) only 2, 3 and 4 are correct
(c) only 1, 2 and 4 are correct
(d) only 1 and 3 are correct

Q.201 A transistor has h_{fe} = 100, h_{ie} = 5.2 kΩ and $r_{bb'}$ = 0. At room temperature V_T = 26 mV. The collector current $|I_C|$ will be
(a) 10 mA (b) 5 mA
(c) 1 mA (d) 0.5 mA

Q.202 A transistor is used in common emitter mode in an amplifier circuit. When a signal of 30 mV is added to the base emitter voltage, the base current changes by 30 μA and collector current by 3 mA. The load resistance is 5 kΩ. The value of transconductance and voltage gain is given by

(a) $g_m = 0.10$ A/V and $A_V = 1000$
(b) $g_m = 0.25$ A/V and $A_V = 250$
(c) $g_m = 0.10$ A/V and $A_V = 500$
(d) $g_m = 0.20$ A/V and $A_V = 750$

Q.203 The transistor in the amplifier shown has the following parameters:
$h_{fe} = 100$, $h_{ie} = 2\ k\Omega$, $h_{re} = 0$, $h_{oe} = 0.05$ m Mho.
'C' is very large. The output impedance is

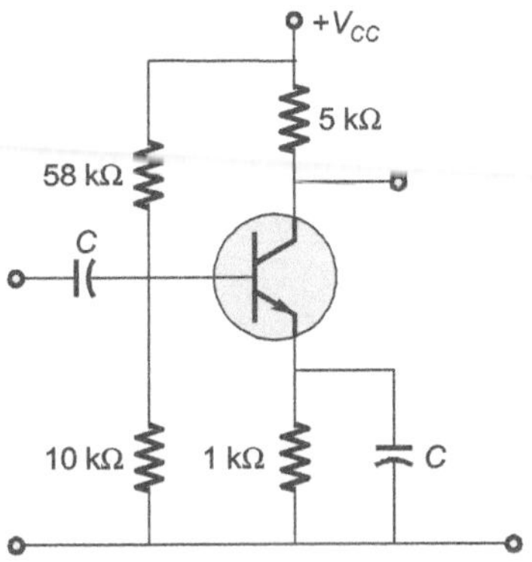

(a) 20 kΩ (b) 16 kΩ
(c) 5 kΩ (d) 4 kΩ

Q.204 If R_S is the source resistance, the output resistance of an emitter-follower using the simplified hybrid model would be

(a) $\frac{h_{ie} + R_S}{1 + h_{fe}}$ (b) $\frac{h_{ie} + R_S}{h_{fe}}$

(c) $R_S + \frac{1}{h_{oe}}$ (d) $\frac{1}{h_{oe}}$

Q.205 In the CE amplifier with $R_L = 4000\ \Omega$, given that $h_{fe} = 100$, $h_{oe} = \frac{1}{36 \times 10^3}$, $h_{ie} = 1000$.
The current gain $|A_i|$ is
(a) 400 (b) 25
(c) 90 (d) 100

Q.206 Assertion (A): On extending the hybrid model for two-port network to a transistor it is assumed that the signal excursion about the Q-point is small.
Reason (R): Small signal operation is that in which the AC input signal voltages and currents are in order of ±10% of Q-point voltage and currents.
(a) Both A and R are true and R is the correct explanation of A.
(b) Both A and R are true but R is not the correct explanation of A.
(c) A is true but R is false.
(d) A is false but R is true.

Q.207 For a common-collector configuration of an n-p-n transistors, the hybrid parameters are given as: $h_{ic} = h_{ie}$, $h_{rc} = 1$, $h_{fc} = -(1 + h_{fe})$ and $h_{oc} = h_{oe}$.
Match **List-I (Hybrid parameters of CB configuration)** with **List-II (Values)** and select the correct answer using the codes given below the lists:

List-I	List-II
A. h_{ib}	1. $\left(\frac{h_{ie}h_{oe}}{1+h_{fe}} - h_{re}\right)$
B. h_{rb}	2. $\left(\frac{-h_{fe}}{1+h_{fe}}\right)$
C. h_{fb}	3. $\left(\frac{h_{ie}}{1+h_{fe}}\right)$
D. h_{ob}	4. $\left(\frac{h_{oe}}{1+h_{fe}}\right)$

Codes:

	A	B	C	D
(a)	1	2	4	3
(b)	3	1	2	4
(c)	2	4	3	1
(d)	3	1	4	2

Q.208 Consider the following statements associated with "r_e" model of transistors:
1. This model employs transistors beta and resistance values only.
2. It gives simple and easy procedure.
3. In this model, results obtained have a fairly good accuracy for the study of amplifier circuits.
4. The required parameters are easily available in this model.

Which of the above statements are correct?
(a) 2, 3 and 4 (b) 1 and 4
(c) 1, 3 and 4 (d) 1, 2, 3 and 4

Q.209 A common-emitter transistor amplifier can be constructed by connecting an external load (R_L) and a signal source as shown in figure and biasing the transistor properly.

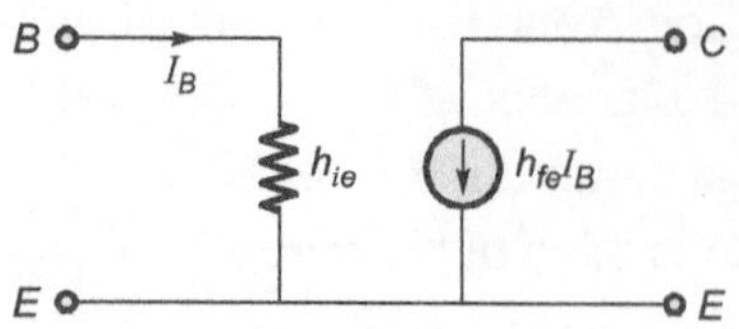

Match **List-I (r_e model CE amplifier)** with **List-II (Corresponding values)** and select the correct answer using the codes given below the lists:

List-I	List-II
A. Voltage gain (A_V)	1. $-\beta$
B. Current gain (A_I)	2. $-\dfrac{R_L}{r_e}$
C. Input impedance (Z_i)	3. ∞
D. Output impedance (Z_0)	4. β
	5. βr_e
	6. 0

Codes:

	A	B	C	D
(a)	4	2	3	6
(b)	4	1	6	3
(c)	2	4	3	5
(d)	2	1	5	3

8. Basic FET Amplifiers

Q.210 When V_{GS} of a JFET changes from –2.4 to –2 V, the drain current changes from 1 mA to 1.4 mA. What is the value of transconductance?

(a) 2 mA/V (b) 0.5 mA/V
(c) 1 mA/V (d) none of these

Q.211 In the CS amplifier shown in figure, if $R_D = 5\ k\Omega$, $R_G = 10\ M\Omega$, $\mu = 50$, and $r_d = 35\ k\Omega$.

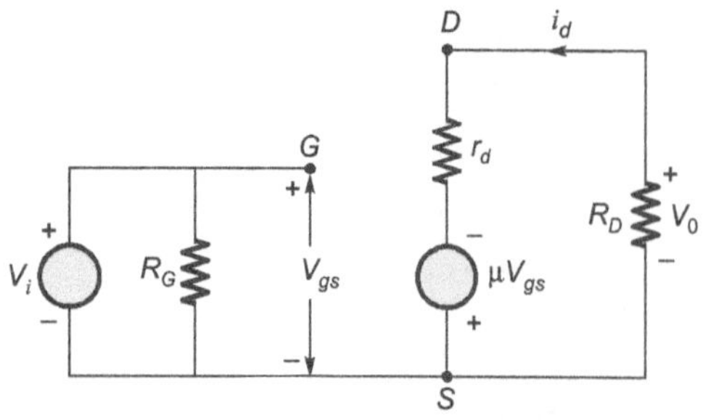

The voltage gain (A_V) of the amplifier is

(a) –4.25 (b) –2.25
(c) –6.25 (d) –5.25

Q.212 Consider the following statements associated with small signal FET model and *h*-parameter model of BJT (in CS configuration and CE configuration respectively):

1. Both FET and BJT have a Norton's output circuit, but the current generated in BJT depends on the input current and not on the input voltage as in FET.
2. There is no feedback from output to input in the FET, whereas a feedback exist in the BJT through the parameter h_{re}.
3. The high input resistance of the FET is replaced by an input resistance of about 1 kΩ for a CE amplifier.

Which of the above statements is/are correct?

(a) 2 only (b) 1, 2 and 3
(c) 3 only (d) 2 and 3 only

Q.213 **Assertion (A):** FET is a much more ideal amplifier than the BJT at low frequencies.

Reason (R): FET has high input impedance and there is no feedback from output to input.

(a) Both A and R are true and R is the correct explanation of A.
(b) Both A and R are true but R is not the correct explanation of A.
(c) A is true but R is false.
(d) A is false but R is true.

Q.214 Match **List-I (Small Signal FET Amplifier Circuits)** with **List-II (Voltage Gain)** and select the correct answer using the codes given below the lists:

List-I	List-II
A. Common drain	1. $\dfrac{-\mu R_D}{R_D + r_d}$
B. Common source	2. $\dfrac{(g_m r_d + 1)R_D}{R_D + r_d}$
C. Common gate	3. $\dfrac{\mu R_S}{(\mu+1)R_S + r_d}$

Codes:

	A	B	C
(a)	1	2	3
(b)	2	1	3
(c)	3	2	1
(d)	3	1	2

Q.215 An N-channel JFET has I_{DSS} = 8 mA and V_P = –5 V. The minimum values of V_{DS} for pinch-off region and the drain current I_{DS}, for V_{GS} = –2 V in the pinch-off region are respectively.

(a) 2 V and 1.8 mA (b) 3 V and 2.9 mA
(c) 2 V and 2.9 mA (d) 3 V and 1.8 mA

Q.216 Assertion (A): The N-channel MOSFET may turn-on prematurely.
Reason (R): The drain resistance of P-channel MOSFET is three times higher than that for an identical N-channel MOSFET.
(a) Both A and R are true and R is the correct explanation of A.
(b) Both A and R are true but R is not the correct explanation of A.
(c) A is true but R is false.
(d) A is false but R is true.

Q.217 Assertion (A): The input noise is less in N-channel JFET than that in P-channel JFET.
Reason (R): The transconductance of N-channel JFET is smaller than that of P-channel JFET.
(a) Both A and R are true and R is the correct explanation of A.
(b) Both A and R are true but R is not the correct explanation of A.
(c) A is true but R is false.
(d) A is false but R is true.

Q.218 A JFET amplifier with a voltage divider biasing circuit, shown in figure has the following parameters:
$V_P = -2$ V, $I_{DSS} = 4$ mA, $R_D = 910\ \Omega$, $R_S = 3$ kΩ, $R_1 = 12$ MΩ, $R_2 = 8.75$ MΩ and $V_{DD} = 24$ V.

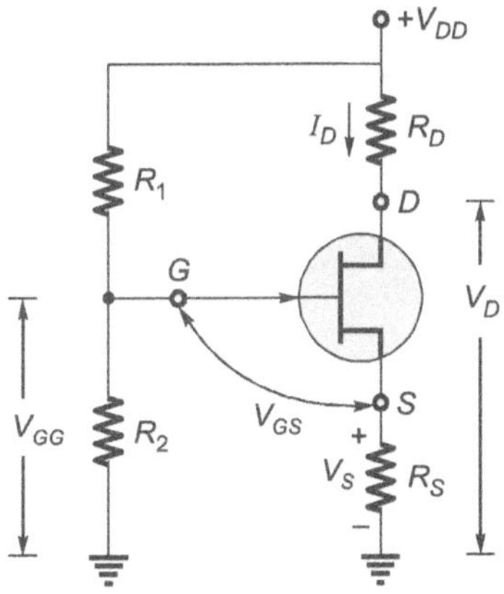

The value of the drain current I_D at the operating point is
(a) 4.72 mA (b) 2.29 mA
(c) 3.20 mA (d) 3.39 mA

Q.219 The value of source resistance required to self bias an N-channel JFET with $I_{DSS} = 40$ mA, $V_P = -10$ V, $V_{GS} = -5$ V and $V_{DSQ} = 5$ V is
(a) 177 Ω (b) 350 Ω
(c) 250 Ω (d) 400 Ω

Q.220 In the common drain amplifier shown below, if $R_S = 4$ kΩ, $R_G = 10$ MΩ, $\mu = 50$ and $r_d = 35$ kΩ, then the voltage gain A_V of the amplifier is approximately equal to

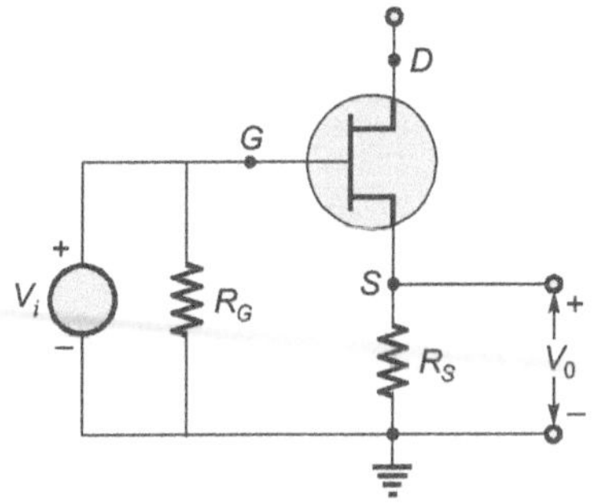

(a) 0.66 (b) 0.84
(c) 0.77 (d) 0.92

Q.221 The value of V_0/V_{in} for the circuit shown below is (Assume M_1 in saturation)

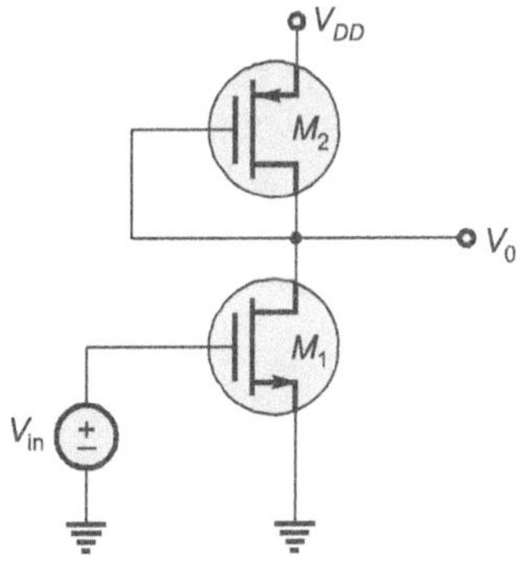

(a) $-\sqrt{\dfrac{\mu_n(W/L)_n}{\mu_P(W/L)_P}}$ (b) $-\dfrac{\mu_n(W/L)_n}{\mu_P(W/L)_P}$

(c) $-\sqrt{\dfrac{\mu_P(W/L)_P}{\mu_n(W/L)_n}}$ (d) $-\dfrac{\mu_P(W/L)_P}{\mu_n(W/L)_n}$

Q.222 Consider the following statements:
1. In most linear applications of field-effect transistors the device is operated in the constant-current portion of its output characteristics.
2. In the region before pinch-off, where V_{DS} is small, the FET is useful as a voltage variable resistors (VVR).
3. The VVR can be used in AGC amplifier to vary the voltage gain of multistage amplifier.

Which of the above statements is/are correct?

(a) 2 only (b) 1 and 2 only
(c) 1, 2 and 3 (d) 2 and 3 only

Q.223 In the saturation region, the $I_D - V_{GS}$ characteristics of a MOSFET are

(a) linear (b) quadratic
(c) exponential (d) hyperbolic

Q.224 The source-bias circuit provides DC biasing for the FET. The resistance from gate to ground (R_S) in that circuit is necessary because, without it,

(a) the input time-constant will become infinite.
(b) the input AC signal will have no path to ground.
(c) there will be no DC potential established at the gate terminal.
(d) the circuit will actually work fine; R_g is not really necessary.

Q.225 In an FET common-source high frequency amplifier, which one of the following is the correct expression for input capacitance C_i?

(a) $C_i = C_{gs} + (1 - A_V)\, C_{gd}$
(b) $C_i = C_{gs} + (1 - 1/A_V)\, C_{gd}$
(c) $C_i = C_{gd} + (1 - A_V)\, C_{gs}$
(d) $C_i = C_{gd} + (1 - 1/A_V)\, C_{gs}$

Q.226 A JFET is set up as a source follower. Given, μ = 200, r_d = 100 kΩ and source load resistance R_L = 1 kΩ. The output resistance R_0 is given approximately by

(a) 1000 Ω (b) 500 Ω
(c) 332 Ω (d) 666 Ω

Q.227 A FET tuned amplifier with g_m = 5 mA/V, r_d = 20 kΩ has a resonant impedance of 20 kΩ. The gain at resonance is given by

(a) 200 (b) 100
(c) 50 (d) 25

Q.228 The unity gain bandwidth f_T of a JFET is given by

(a) $g_m/2\pi C_{sg}$ (b) $C_{sg}/2\pi g_m$
(c) $g_m/2\pi f C_{sg}$ (d) $C_{sg}/2\pi f g_m$

Q.229 The transconductance for a JFET having I_{DSS} = 8 mA, V_P = –5 V and biased to operate at V_{GS} = –1.8 volt will be given by

(a) 4.4 mS (b) 2.2 mS
(c) 3.3 mS (d) 1.1 mS

9. Frequency Response

Q.230 The mid frequency gain of the amplifier circuit shown below with high value of beta of the BJT is

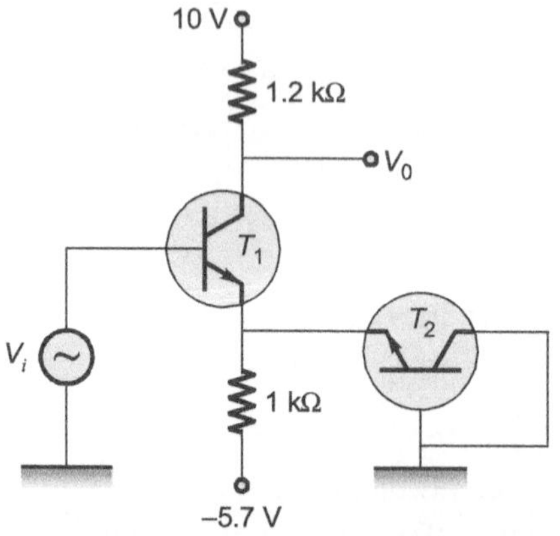

(a) –0.98 (b) –1.2
(c) –0.82 (d) –1.8

Q.231 In a common source MOSFET amplifier circuit operating at 10 kHz, the drain resistance R_d = 200 kΩ. If MOSFET parameters are g_m = 1.5 mA/V, r_d = 50 kΩ, C_{ds} = 1.0 pF and C_{gd} = 2.0 pF, the voltage gain is equal to

(a) 30 (b) –30
(c) 60 (d) –60

Q.232 An npn BJT has g_m = 38 mΩ/V, $C_\mu = 10^{-14}$ F, $C_\pi = 4 \times 10^{-13}$ F and DC current gain, β_0 = 90. For this transistor f_T and f_β are respectively

(a) $f_T = 1.64 \times 10^8$ Hz and $f_\beta = 1.47 \times 10^{10}$ Hz
(b) $f_T = 1.47 \times 10^{10}$ Hz and $f_\beta = 1.64 \times 10^8$ Hz
(c) $f_T = 1.33 \times 10^{12}$ Hz and $f_\beta = 1.47 \times 10^{10}$ Hz
(d) $f_T = 1.47 \times 10^{10}$ Hz and $f_\beta = 1.33 \times 10^{12}$ Hz

Q.233 Given for an FET, g_m = 9.5 mA/volt, total capacitance = 500 pF. For a voltage gain of –30, the bandwidth will be

(a) 100 kHz (b) 630 MHz
(c) 3 MHz (d) 19 MHz

Q.234 The drain gate capacitance of a junction FET is 2 pF. Assuming a common source voltage gain of 20, what is the input capacitance due to miller effect?

(a) 21 pF (b) 40 pF
(c) 42 pF (d) 10 pF

Q.235 Assertion (A): In the high frequency model of FET, the voltage amplification drops drastically with increase in frequency.

Reason (R): The high frequency model is identical with the low frequency model except that the capacitances between various terminals (gate, source and drain) is added.

(a) Both A and R are true and R is the correct explanation of A.
(b) Both A and R are true but R is not the correct explanation of A.
(c) A is true but R is false.
(d) A is false but R is true.

Q.236 For a single stage CE amplifier, g_m = 50 mA/V, C_e = 1 pF and C_c = 0.2 pF. The frequency at which the short circuit common-emitter current gain attains unit magnitude is given by

(a) 2.2 GHz (b) 3.3 GHz
(c) 5.5 GHz (d) 6.6 GHz

Q.237 When the load resistance is zero (short circuit) in a common-emitter configuration, the current flowing from the base to the collector node through the collector capacitor C_c, is small because

(a) the emitter capacitance (C_e) is smaller than C_c
(b) the short circuit permits no current flow from base to collector
(c) the base-collector resistance ($r_{b'c}$) is very high
(d) the emitter capacitance (C_e) is larger than C_c

Q.238 Consider the following statements:

1. For finding the transistor behaviour at high frequency it is necessary to examine the diffusion mechanism.
2. At low frequency it is assumed that the transistor responds instantly to changes of input voltage or current.
3. The hybrid-π model is valid under dynamic conditions when the rate of change of V_{BE} is small enough so that the base incremental current I_b is small compared with the collector incremental current I_c.
4. The parameter f_T is a frequency characteristic of a transistor.

Which of the above statements are correct?

(a) 3 and 4 (b) 1 and 3
(c) 1, 2, 3 and 4 (d) 1 and 2

Q.239 A modern bipolar transistor can have C_e = 1 pF. If g_m = 50 mA/V, the f_T of a common emitter amplifier will be

(a) 20 MHz (b) 80 MHz
(c) 8 GHz (d) 2 GHz

Q.240 The high frequency hybrid-π model of a bipolar transistor has ______ capacitance.

(a) 2 (b) 3
(c) 1 (d) zero

Q.241 In the active region of operation, the collector base capacitance is due to the

(a) diffusion charge in emitter
(b) diffusion charge in base
(c) emitter-base transition region
(d) base-collector transition region

Q.242 An amplifier has a midband gain of 100, a lower 3-dB frequency (f_L) of 0.1 Hz and an upper 3-dB frequency (f_H) of 100 Hz. The frequency range for which the gain is at least 99 is

(a) 3.5 f_L to f_H/3.5 (b) 7 f_L to f_H/7
(c) 0.7 f_L to f_H/0.7 (d) 2 f_L to f_H/2

Q.243 Assertion (A): The individual 3-dB frequencies for each stage in non-interacting cascaded stages of amplifiers can not be defined.

Reason (R): To obtain the overall transfer function of non-interacting stages, the transfer gains of the individual stages are multiplied together.

(a) Both A and R are true and R is the correct explanation of A.
(b) Both A and R are true but R is not the correct explanation of A.
(c) A is true but R is false.
(d) A is false but R is true.

Q.244 Due to cascading of '*n*' non-interacting amplifiers some of its parameters either increase, decrease or remains constant. Match **List-I (Parameters)** with **List-II (Effect of cascading)** and select the correct answer using the codes given below the lists:

List-I	List-II
A. Gain	1. Reduces
B. Bandwidth	2. Unchanged
C. Gain bandwidth product	3. Increases
D. Rise time	

Codes:

	A	B	C	D
(a)	3	2	1	1
(b)	2	3	1	2
(c)	3	1	2	3
(d)	2	1	3	3

Q.245 Emitter bypass capacitors cause a
(a) high frequency pole in the transfer function
(b) high-pass response
(c) low-pass response
(d) high frequency zero in the transfer function.

Q.246 Consider the following statements associated with choice of transistor configuration in cascade amplifiers:
1. The choice of transistor configuration depends on the maximum voltage gain expected from the cascade amplifier.
2. For input stage, CE configuration is popularly used.
3. For input and output stages, the impedance matching of the source and the input circuit of the first stage is the main consideration.
4. For intermediate stages CC or CB configuration is popularly used.

Which of the above statements is/are not correct?
(a) 2 and 4 (b) 4 only
(c) 1, 2 and 4 (d) 1 and 3

Q.247 Which of the following is not true about RC coupling?
(a) Its non- linear distortion is higher.
(b) It has better fidelity over a wide frequency range.
(c) Its frequency response is much better over the audio frequency range.
(d) All the above options are correct.

Q.248 The rise times of certain amplifier is 10 ns. The approximate bandwidth of the amplifier is
(a) 25 MHz (b) 3.5 GHz
(c) 2.5 GHz (d) 35 MHz

Q.249 For a CE amplifier, the h-parameters are:
h_{fe} = 400 and h_{ie} = 10 kΩ
The circuit details are : R_S = 600 Ω, R_L = 5 kΩ, R_E = 1 kΩ, V_{CC} = 12 V, R_1 = 15 kΩ, R_2 = 2.2 kΩ and C_E = 50 μF.
The lower 3-dB point is
(a) 160 Hz (b) 135 Hz
(c) 120 Hz (d) 148 Hz

Q.250 The high frequency-π model for a CE transistor is shown below.

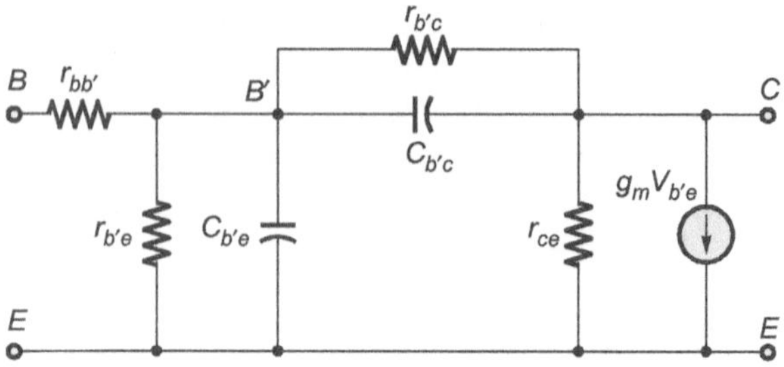

Which of the following relations between low frequency h-parameters and high frequency parameters for a transistor is not correct?

(a) $g_m = \dfrac{|I_C|}{V_T}$ (b) $r_{bb'} = h_{ie} - r_{b'e}$

(c) $g_{ce} = \dfrac{r_{b'e}}{h_{re}}$ (d) $r_{b'e} = \dfrac{h_{fe}}{g_m}$

Q.251 A multistage amplifier has a gain of 70 dB. If the gain of first and second stages are 20 dB and 30 dB respectively, then the gain of third stage will be
(a) 10000 (b) 100
(c) 1000 (d) 10

10. Differential Amplifiers

Q.252 A resitive loaded and biased differential amplifier circuit is shown in figure. Neglecting the base currents and assuming matched transistors with $V_A = \infty$, β = 100, what are the values of R_C and R_E to meet the following specifications?
Differential mode gain (double ended) = –500,
Common mode rejection ratio = 500,
Differential mode input resistance = 2 MΩ.
(Use V_T = 26 mV, $V_{BE(ON)}$ = 0.7 V and $V_{CE\,(SAT)}$ = 0.2 V)

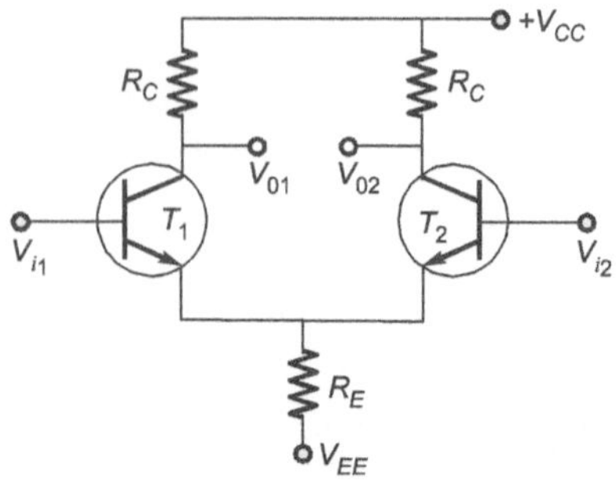

(a) $R_C = 5\ M\Omega$ and $R_E = 1.5\ M\Omega$
(b) $R_C = 2\ M\Omega$ and $R_E = 1.5\ M\Omega$
(c) $R_C = 2\ M\Omega$ and $R_E = 5\ M\Omega$
(d) $R_C = 5\ M\Omega$ and $R_E = 2.5\ M\Omega$

Q.253 In the level shifter circuit shown in figure below, the internal drop across each diode $V_D = V_{BE2} = 0.7$ V and h_{fe} is very large.
The value of R for V_0 to become zero is

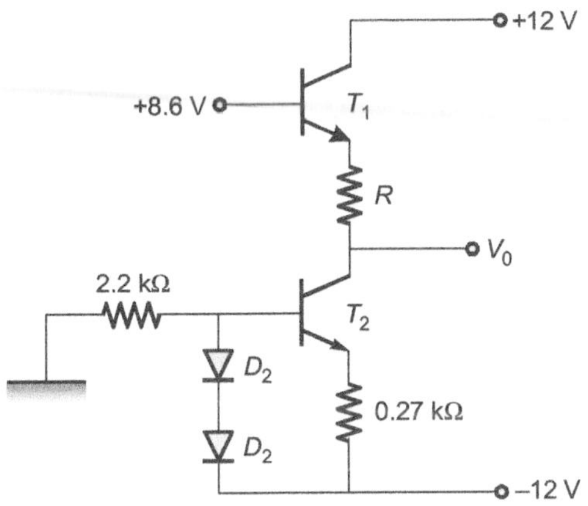

(a) 2.00 kΩ (b) 3.04 kΩ
(c) 2.25 kΩ (d) 2.80 kΩ

Q.254 A differential amplifier, amplifies
(a) and mathematically differentiates the average of the voltages on the two input lines.
(b) and differentiates the sum of the two input waveforms.
(c) the difference of voltages between the two input lines.
(d) and differentiates the input waveform on one line when the other line is grounded.

Q.255 Most of linear *ICs* are based on the two transistor differential amplifier because of its
(a) high voltage gain
(b) high CMMR
(c) high input resistance
(d) input voltage dependent linear transfer characteristic

Q.256 A differential amplifier has $R_L = 10\ k\Omega$ (equal values in both collectors), $h_{ie} = 1\ k\Omega$, $R_e = 50\ k\Omega$, $h_{fe} = 100$. The common mode gain is given by
(a) 1500 (b) 20
(c) 0.2 (d) 0.1

Q.257 In a single-stage differential amplifier, the output offset voltage is basically dependent on the mismatch of
(a) V_{BE}, I_B and β (b) V_{BE} and I_B
(c) I_B and β (d) V_{BE} and β

Q.258 Assertion (A): It is not so easy to design d.c. amplifiers using transistors.
Reason (R): The values of h_{fe}, V_{BE} and I_{CBO} vary with temperature.
(a) Both A and R are true and R is the correct explanation of A.
(b) Both A and R are true but R is not the correct explanation of A
(c) A is true but R is false
(d) A is false but R is true

Q.259 Assertion (A): The ability of a differential amplifier to reject a differential mode signal is called its "figure of merit".
Reason (R): The ideal value of figure of merit of a differential amplifier is infinite.
(a) Both A and R are true and R is the correct explanation of A.
(b) Both A and R are true but R is not the correct explanation of A
(c) A is true but R is false
(d) A is false but R is true

Q.260 A differential amplifier has (*i*) *CMMR* = 1000 and (*ii*) *CMRR* = 10000.
The first set of inputs is $V_1 = 100\ \mu V$ and $V_2 = -100\ \mu V$.
The second set of inputs is $V_1 = 1100\ \mu V$ and $V_2 = 900\ \mu V$.
What is the percentage difference in output voltage obtained for the two sets of input voltages?
(a) 1.0% (b) 0.25%
(c) 0.5% (d) 0.15%

Q.261 The *Q*-point (V_c and I_B) for the differential amplifier shown in figure below is given by:

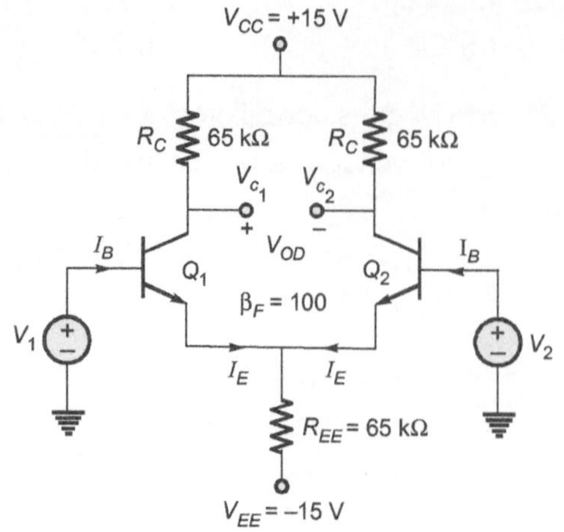

(a) $V_c = 6.6$ V and $I_B = 1.25$ μA
(b) $V_c = 7.9$ V and $I_B = 1.09$ μA
(c) $V_c = 6.6$ V and $I_B = 1.09$ μA
(d) $V_c = 7.9$ V and $I_B = 1.25$ μA

Q.262 The value of V_{DS} and I_{DS} for the MOSFETs in the differential amplifier shown below are respectively
(Given, $I_{SS} = 175$ μA, $R_D = 65$ kΩ, $K_n = 3$ mA/V^2 and $V_{TN} = 1$ V)

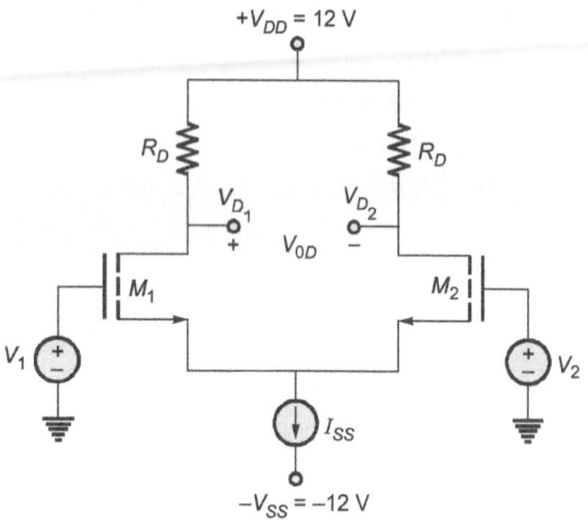

(a) $V_{DS} = 7.55$ V and $I_{DS} = 87.5$ μA
(b) $V_{DS} = 5.25$ V and $I_{DS} = 66$ μA
(c) $V_{DS} = 7.55$ V and $I_{DS} = 66$ μA
(d) $V_{DS} = 5.25$ V and $I_{DS} = 87.5$ μA

Q.263 In an emitter-coupled differential amplifier with an emitter resistance (R_e), the value of R_e cannot be increased to very high values because as R_e is increased
(a) the emitter currents decrease, so h_{ie} decreases, and hence the differential mode gain decreases.
(b) the common-mode gain increases, so the CMMR decreases.
(c) the base-collector voltages increase, so the probability of avalanche breakdown increases.
(d) the base currents increase, so the input resistance decreases.

Q.264 In a differential amplifier, if the emitter resistor is replaced by an ideal current source, then the CMMR becomes infinite because
(a) an ideal current source provides a very high slew rate.
(b) the base-emitter voltages of the differential transistor pair become truly constant independent of the input voltages .
(c) the differential mode gain becomes infinite.
(d) an ideal current source offers an infinite source resistance, which makes the common mode gain zero.

Q.265 For the differential amplifier shown in figure below, let $V_{CC} = V_{EE} = 5$ V, $R_e = 1$ kΩ, $R_c = 1$ kΩ, and $R_s = 0$. For the transistors assume $h_{fe} = 100$, $h_{oe} = 0$. What is the differential gain A_d for dc common-mode voltage equal to –2 V?

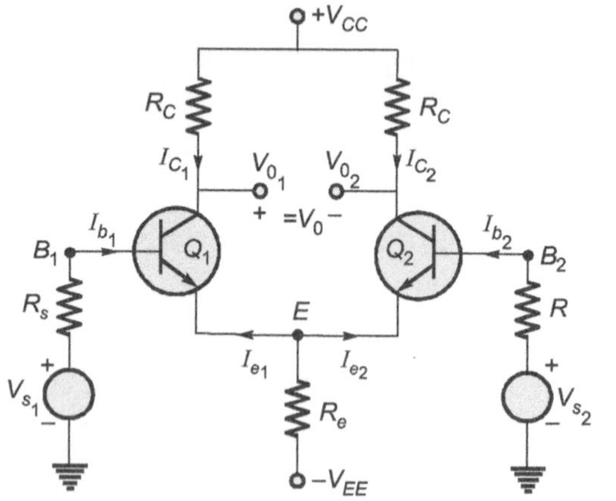

(a) 26
(b) 20
(c) 23
(d) 28

11. Feedback Amplifiers

Q.266 When an amplifier is provided with current-shunt feedback, its
(a) input impedance increases and output impedance decreases.
(b) input impedance decreases and output impedance increases.

(c) input and output impedances both increases.
(d) input and output impedances both decreases.

Q.267 An amplifier has an open-loop gain of 1000 and a feedback ratio of 0.04. If the open-loop gain changes by 10% due to temperature, the percentage change in gain of the amplifier with feedback will be

(a) 0.75% (b) 0.50%
(c) 0.25% (d) 0.45%

Q.268 A voltage-series negative feedback amplifier has a voltage gain without feedback of A = 500, input resistance R_i = 3 kΩ and feedback ratio β = 0.01. The voltage gain A_f and input resistance R_{if} of the amplifier with feedback are respectively

(a) 83.33 and 18 kΩ (b) 55.54 and 22 kΩ
(c) 83.33 and 22 kΩ (d) 55.54 and 18 kΩ

Q.269 **Assertion (A):** The emitter follower inherently exhibits 100% negative feedback.
Reason (R): The emitter follower is an example of voltage-series feedback.

(a) Both A and R are true and R is the correct explanation of A.
(b) Both A and R are true but R is not the correct explanation of A.
(c) A is true but R is false.
(d) A is false but R is true.

Q.270 For the emitter follower circuit shown below, match **List-I** with **List-II** and select the correct answer using the codes given below the lists:

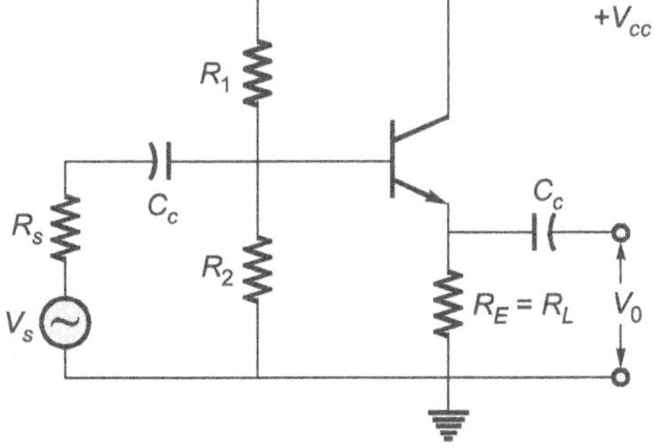

List-I
A. Voltage gain (A_v)
B. Input resistance (R_i)
C. Current gain (A_i)
D. Output resistance (without feedback)

List-II
1. $1 + h_{fe}$
2. $\frac{h_{ie} + R_s}{1 + h_{fe}}$
3. $1 - \frac{h_{ie}}{R_i}$
4. $h_{ie} + (1 + h_{fe}) R_E$

Codes:

	A	B	C	D
(a)	4	3	2	1
(b)	3	4	2	1
(c)	3	4	1	2
(d)	4	3	1	2

Q.271 An amplifier has a mid-band gain of 125 and a bandwidth of 250 kHz. A 4% negative feedback is introduced to obtain a new bandwidth and gain. If the bandwidth is to be restricted to 1 MHz, the feedback ratio will be

(a) 1.2% (b) 0.8%
(c) 1.8% (d) 2.4%

Q.272 An amplifier has voltage gain with feedback of 100. If the gain without feedback changes by 20% and the gain with feedback should not vary more than 2%, the values of open loop gain and feedback ratio will be respectively given by

(a) 1000 and 1.9% (b) 10000 and 0.9%
(c) 1000 and 0.9% (d) 10000 and 1.9%

Q.273 Match **List-I (Type of negative feedback)** with **List-II (Common names)** and select the correct answer using the codes given below the lists:

List-I
A. Voltage-shunt feedback
B. Current-series feedback
C. Voltage-series feedback
D. Current-shunt feedback

List-II
1. Trans-conductance amplifier
2. Current amplifier
3. Trans-resistance amplifier
4. Voltage amplifier

Codes:

	A	B	C	D
(a)	1	3	4	2
(b)	3	1	4	2
(c)	1	3	2	4
(d)	3	1	2	4

Q.274 Negative feedback enhances all performance parameters of an amplifier except its
(a) gain (b) 3-dB noise
(c) noise figure (d) input impedance

Q.275 Shunt mixing is generally easy to identify because it
(a) has a resistor in parallel with the input source.
(b) has a resistor whose one end is connected to the input node, such that it draws current away from that node.
(c) has a shunt capacitance from the output to ground.
(d) has a shunt resistance from the input node to ground.

Q.276 When current sampling is used, the output resistance ______.
(a) increases
(b) decreases
(c) depends on the type of mixing
(d) decreases only for FET circuits

Q.277 A transductance amplifier should have
(a) a high R_i and a low R_o
(b) a low R_i and a low R_o
(c) a low R_i and a high R_o
(d) a high R_i and a high R_o

Q.278 Match **List-I** with **List-II** and select the correct answer using the codes given below the lists:

List-I	List-II
A. Return difference	1. $1/1 + \beta A$
B. Sensitivity factor	2. $1 + \beta A$
C. Return ratio	3. β
D. Feedback factor	4. βA

Codes:

	A	B	C	D
(a)	1	3	2	4
(b)	2	3	4	1
(c)	4	1	2	3
(d)	2	1	4	3

Q.279 The amplifier gain varies with frequency mainly due to
(a) Miller effect
(b) interstate transformers
(c) presence of external and internal capacitances
(d) logarithmic increase in its output power

Q.280 In an amplifier, the coupling capacitors are employed for
(a) limiting the bandwidth
(b) preventing of dc mixing with input or output
(c) controlling the output
(d) matching the impedances

Q.281 Feedback in amplifier always helps in
(a) increasing its gain
(b) stabilize its gain
(c) reducing its input impedance
(d) controlling its output

Q.282 Consider the following statements associated with the type of feedback introduced to an amplifier:
1. Amplifier with positive feedback produces instability but still it is useful.
2. Negative feedback in amplifier reduces the gain but increases stability.
3. Negative feedback reduces the amplifier output impedance but increases the input impedance.
4. Negative feedback always increases the thermal stability as well as frequency stability of the output signal.

Which of the above statements are correct?
(a) 1, 2, 3 and 4 (b) 2, 3 and 4
(c) 1, 2 and 4 (d) 1, 3 and 4

Q.283 For the feedback amplifier shown below, $A\beta$ is very large. What would be the value of R_E, if R_F is set to be 1.2 kΩ to produce a close loop gain of 25?

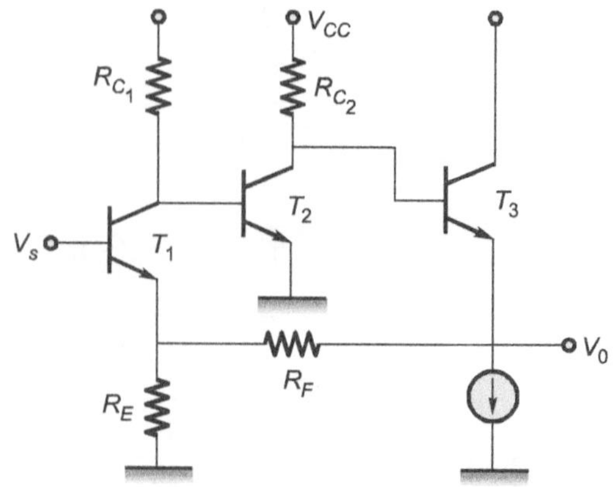

(a) 100 Ω (b) 25 Ω
(c) 50 Ω (d) 75 Ω

Q.284 A series-series type feedback circuit shown in figure below realizes a voltage controlled current source. Biasing circuit has not been shown. If the loop gain $A\beta$ is large, then I_0/V_s is equal to

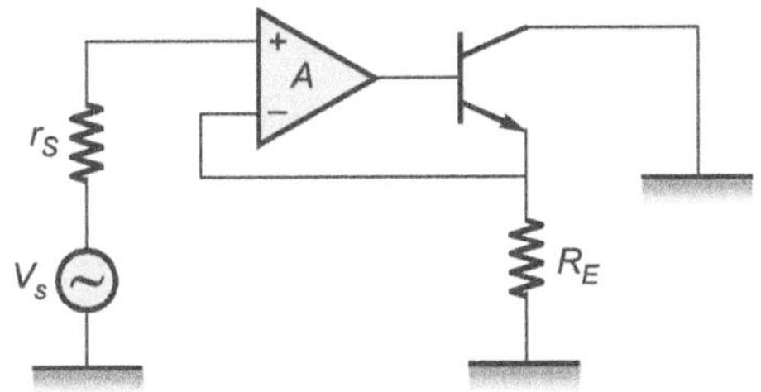

(a) $\frac{1}{R_E}$ (b) $\frac{1}{r_s}$

(c) $\frac{1}{r_E + r_s}$ (d) none of these

Q.285 The voltage gain without negative feedback is 40 dB. What is the new voltage gain if 3% negative feedback is introduced?

(a) 15 (b) 25

(c) 42 (d) 12

Q.286 The open circuit voltage gain 40 dB of an amplifier changes by 5%. How much change in voltage gain will result with 2% negative feedback?

(a) 0.24% (b) 1.66%

(c) 0.60% (d) 0.38%

Q.287 An amplifier without feedback has a voltage gain of 60 dB. The lower and upper cut-off frequencies of this amplifier are respectively 100 Hz and 100 kHz. Inclusion of negative feedback reduces the gain to 20 dB. The new bandwidth with negative feedback is

(a) 1 MHz (b) 100 kHz

(c) 120 kHZ (d) 10 MHz

Q.288 n negative feedback the return ratio is _____.

(a) 0 (b) greater than 0

(c) 1 (d) greater than 1

Q.289 In an amplifier with negative feedback, the bandwidth is

(a) increased by a factor of β

(b) decreased by a factor of β

(c) increased by a factor of $(1 + A\beta)$

(d) not affected at all by the feedback

Q.290 If the feedback signal is returned to the input in series with the applied voltage, the input impedance _______.

(a) increases (b) decreases

(c) becomes infinity (d) does not change

Q.291 Consider the following statements:

1. The decibel gain of a cascaded amplifier is equal to the product of individual gain.
2. The decibel gain is important because it tallies with human ear response.
3. The gain of R-C coupled amplifier may increase or decrease when the signal frequency is increased.
4. Due to degenerative feedback, bandwidth and gain of the amplifier will be very large.
5. Collector to base bias circuit is an example of voltage-shunt feedback amplifier.

Of these statements:

(a) 2 and 5 are not correct

(b) 2, 4 and 5 are not correct

(c) 3, 4 and 5 are not correct

(d) 1 and 4 are not correct

Q.292 The feedback used in the circuit shown in the figure can be classified as

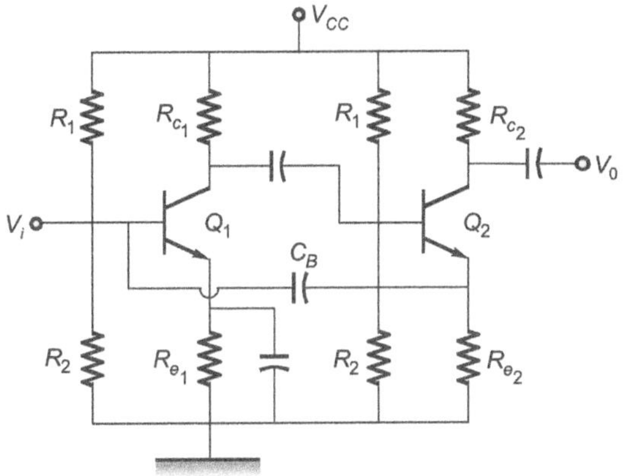

(a) shunt-shunt feedback

(b) shunt-series feedback

(c) series-shunt feedback

(d) series-series feedback

Q.293 A two stage amplifier with negative feedback has an overshoot when a damping factor K is

(a) less than unity (b) greater than unity

(c) zero (d) negative

Q.294 The common emitter current gain-bandwidth product of a transistor (f_T) is defined as the frequency at which

(a) alpha of the transistor falls by 3-dB
(b) beta of the transistor falls by 3-dB
(c) beta of the transistor falls by unity
(d) power gain of the transistor falls to unity

Q.295 In a feedback amplifier the parameter that is stabilized, is transconductance 9 Mho. The type of amplifier is
(a) voltage series feedback
(b) current series feedback
(c) voltage shunt feedback
(d) current shunt feedback

Q.296 Match **List-I** with **List-II** and select the correct answer using the codes given below the lists:

List-I
A. Series-shunt
B. Series-series
C. Shunt-shunt
D. Shunt-series

List-II
1. Current to voltage converter
2. Voltage to current converter
3. Voltage
4. Current

Codes:

	A	B	C	D
(a)	1	2	3	4
(b)	3	2	1	4
(c)	3	2	4	1
(d)	1	4	3	2

Q.297 Assertion (A): Negative feedback is used to improve the performance of electronic amplifiers.
Reason (R): The stabilisation of the d.c. operating point of a transistor amplifier is accomplished by the use of negative feedback as far as the d.c. potential is concerned and the operating point is kept constant in the case of change in temperature or a change in the h_{fe} or β of a transistor.
(a) Both A and R are true and R is the correct explanation of A.
(b) Both A and R are true but R is not the correct explanation of A.
(c) A is true but R is false.
(d) A is false but R is true.

12. Large Signal Amplifiers

Q.298 Assertion (A): The harmonic distortion of an amplifier increases at higher power levels.
Reason (R): For high quality audio amplifiers, the maximum THD is 0.7% at the desired listening level.
(a) Both A and R are true and R is the correct explanation of A.
(b) Both A and R are true but R is not the correct explanation of A.
(c) A is true but R is false.
(d) A is false but R is true.

Q.299 The advantages of class *B* as compared with class *A* operation are
1. efficiency is higher.
2. possible to obtain greater power output.
3. less distortion and noise interference.
4. negligible power loss at no input signal.

Of these the statements:
(a) 2 and 3 are correct
(b) 1, 2, 3 and 4 are correct
(c) 1, 2 and 4 are correct
(d) 2, 3 and 4 are correct

Q.300 For a given transistor, the thermal resistance is 8 °C/W and the ambient temperature T_A is 27°C. If the transistor dissipates 3 W of power, the junction temperature T_j will be
(a) 51 °C (b) 45 °C
(c) 38 °C (d) 3 °C

Q.301 In a class B amplifier, $V_{CE(min)}$ = 2 V and supply voltage V_{CC} = 15 V. The collector circuit efficiency is approximately
(a) 78.0% (b) 56.0%
(c) 68.0% (d) 72.0%

Q.302 In a class A amplifier, $V_{CE(max)}$ = 15 V, $V_{CE(min)}$ = 1 V. The overall efficiency for series fed load and transformer coupled load are respectively
(a) 18.28% and 33.33%
(b) 23.33% and 33.33%
(c) 18.28% and 43.75%
(d) 23.33% and 43.75%

Q.303 For a transistor, T_j = 160°C, T_A = 40°C and $\theta_{J\text{-}A}$ = 80°C/W. What is the power that the transistor can safely dissipate in free air?

(a) 2.5% (b) 0.5%
(c) 1.5% (d) 1.25%

Q.304 The maximum theoretical efficiency of a class A amplifier can be
(a) 50% (b) 78%
(c) 100% (d) 25%

Q.305 Crossover distortion refers to the distortion caused due to
(a) non-linearities in the input-output characteristics of the transistor.
(b) the exponential I_C-V_{BE} characteristics of the bipolar transistor.
(c) the crossing over the collector voltage from negative to positive values.
(d) the finite input voltage required to turn on transistors biased in cutoff in a class B configuration.

Q.306 Assertion (A): In a system where the power supply is limited, such as those operating from solar cells or a battery, the output power is usually delivered through a push-pull class B transistor circuit.
Reason (R): It is possible to obtain greater power output, higher efficiency and negligible power loss at no signal with the help of a push-pull class B transistor circuit.
(a) Both A and R are true and R is the correct explanation of A.
(b) Both A and R are true but R is not the correct explanation of A.
(c) A is true but R is false.
(d) A is false but R is true.

Q.307 A common-emitter amplifier, with a self-bias circuit to bias the transistor in the active mode, is an example of a class _____ amplifier.
(a) A (b) AB
(c) C (d) B

Q.308 In a self-biased common-emitter configuration, non-linear distortion
(a) is negligible.
(b) causes only odd-order harmonics in the output.
(c) increases as the negative feedback is increased.
(d) increases as the amplitude of the ac signal increases.

Q.309 Compared to class B amplifiers, class A amplifiers have
(a) a lower efficiency, and less harmonic distortion.
(b) a higher efficiency, and more power consumption.
(c) a higher efficiency, and less harmonic distortion.
(d) lower power consumption, and more harmonic distortion.

Q.310 In a class C amplifier
(a) efficiency and distortion both are maximum.
(b) efficiency is minimum but distortion is maximum.
(c) efficiency is maximum but distortion is minimum.
(d) efficiency and distortion both are minimum.

Q.311 In push-pull amplifiers there occurs cancellation
(a) of all odd order harmonics
(b) of odd harmonics as small as possible
(c) of even order harmonics
(d) none of the above

Q.312 In a class A amplifier, there will be a worst condition when
(a) signal input is maximum
(b) load resistance is high
(c) there is a transformer coupling
(d) signal input is zero

Q.313 The output power of a power amplifier is several times its input power. This is possible due to the fact that
(a) there is a positive feedback in the circuit.
(b) step us transformer is used in the circuit.
(c) power amplifier converts a part of the input dc power into ac power.
(d) a negative resistance is introduced.

Q.314 Assertion (A): If an extremely large amount of power over a narrow range of frequencies is to be obtained, tuned class B or class C operation is generally preferred.
Reason (R): The transistor power amplifier stage should be so designed that the power conversion efficiency is maximum and the collector dissipation of the transistor is minimum.
(a) Both A and R are true and R is the correct explanation of A.

(b) Both A and R are true but R is not the correct explanation of A.
(c) A is true but R is false.
(d) A is false but R is true.

Q.315 In class A amplifier, conduction extends over 360° because Q-point is
(a) located on load line
(b) located near saturation point
(c) centred on load line
(d) located at or near cutoff point

Q.316 Match **List-I (Power Amplifiers)** with **List-II (Applications)** and select the correct answer using the codes given below the lists:

List-I
A. Class C
B. Class D
C. Class B
D. Class A

List-II
1. Used in transmitters, pulse and digital circuits
2. As a voltage amplifier
3. Used in tuned circuits
4. As an audio frequency power amplifier

Codes:

	A	B	C	D
(a)	4	1	3	2
(b)	1	2	3	4
(c)	1	2	4	3
(d)	3	1	4	2

Q.317 Heat sink is used in the power amplifier circuit
(a) to increase the voltage gain of the power amplifier.
(b) to increase the collector dissipation ratings of the transistor.
(c) to reduce the heat loss in a transistor.
(d) to increase the output power.

Q.318 In a class B push pull amplifier:
1. the effect of ripple voltages contained in the power supply will be balanced out.
2. distortion is less for a given power output.
3. the d.c. components of the collector current oppose each other magnetically in the transformer core, thereby eliminating any tendency towards core saturation leading to non-linear distortion.
4. all odd harmonics are balanced out and the second harmonic acts as the principle source of distortion, provided the two transistors Q_1 and Q_2 are identical.

Which of the statements given above are correct?
(a) 1, 2 and 3 (b) 3 and 4
(c) 1 and 3 (d) 2, 3 and 4

Q.319 For tuned radio frequency we use
(a) class A amplifier (b) class C amplifier
(c) class B amplifier (d) class AB amplifier

Q.320 Match **List-I** with **List-II** and select the correct answer using the codes given below the lists:

List-I
A. Class A amplifier
B. Class B amplifier
C. Class C amplifier
D. Class AB amplifier

List-II
1. Used as harmonic generator
2. Eliminates crossover distortion
3. Excellent thermal stability
4. Suffers from crossover distortion
5. Amplifier acts as a switch

Codes:

	A	B	C	D
(a)	1	2	3	5
(b)	2	1	5	3
(c)	3	4	1	2
(d)	3	2	5	4

Q.321 Consider the following statements:
1. Transformer coupling is mostly used for voltage amplification.
2. Direct coupling is used for amplification of extremely low frequency signals.
3. In class C amplifier, the bias voltage is beyond the cutoff voltage.
4. The collector current in quiescent condition is zero when the amplifier is working in class A operation.
5. In class AB amplifier, the transistor operates between the two extremes defined for class A and class B amplifiers.

Which of the statements given above is/are correct?
(a) 1, 2 and 5 (b) 1 and 4 only
(c) 2, 3 and 5 (d) 2 and 3 only

Q.322 The maximum power developed across a load resistance R_L is 2 W for the input signal $i_i = I_{im} \sin\omega_0 t$. What is the maximum power developed across the load if the signal is expressed as

$$i_i = \frac{I_{im}}{2}\sin\omega_0 t + \frac{I_{im}}{2}\sin 3\omega_0 t?$$

(a) $\frac{1}{2}$ W (b) 2 W
(c) $\frac{3}{2}$ W (d) 1 W

Q.323 For the circuit shown below, $V_{CC} = 15$ V, $I = 10$ mA and $R_L = 1$ kΩ for an output voltage of 5 V peak sinusoid.

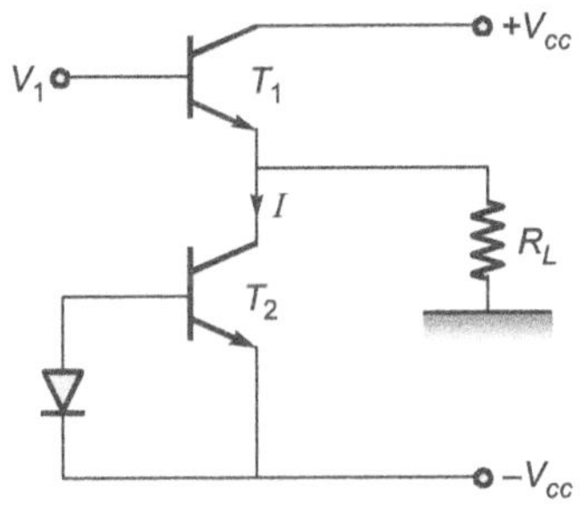

The efficiency of the above circuit is
(a) 4.2% (b) 3.8%
(c) 2.6% (d) 5.5%

Q.324 In a class A amplifier, the output current flows for
(a) half the cycle of the input signal
(b) a part of the cycle of the input signal
(c) 3/4th of the cycle of the input signal
(d) the full cycle of the input signal

Q.325 The power conversion efficiency of an output stage is defined as
(a) (Load power + supply power) / (Load power – supply power)
(b) (Load power + supply power) / Supply power
(c) Load power / Supply power
(d) Supply power / Load power

Q.326 The type of power amplifier which exhibits crossover distortion in its output is
(a) Class C (b) Class B
(c) Class A (d) Class AB

Q.327 A radio frequency signal contains three frequency components, 870 kHz, 875 kHz and 880 kHz. The signal needs to amplified. The amplifier used should be
(a) tuned voltage amplifier
(b) wide band amplifier
(c) push-pull amplifier
(d) audio frequency amplifier

Q.328 Two stages of BJT amplifiers are cascaded by RC coupling. The voltage gain of the first stage is 10 and that of the second stage is 20. The overall gain of the coupled amplifiers is
(a) (10 × 20)/2 (b) (10 + 20)
(c) $(10 + 20)^2$ (d) (10 × 20)

Q.329 Which of the following parameters is used for distinguishing between a small-signal and a large-signal amplifiers?
(a) Overall gain (b) Distortion
(c) Instability (d) Bandwidth

Q.330 If a class *C* power amplifier has an input signal with frequency of 200 kHz and the width of collector current pulses of 0.1 mS, then the duty cycle of the amplifier will be
(a) 1% (b) 2%
(c) 10% (d) 20%

Q.331 The open circuit voltage and resistance of a microphone are 15 mV (rms) and 100 kΩ respectively. An amplifier with an input impedance (resistive) of 50 kΩ is used for amplifying the microphone output so as to deliver 2 watt power to a load of 8 Ω. The voltage gain of amplifier is
(a) 400 (b) 800
(c) 1600 (d) 3200

Q.332 A class B push-pull type amplifier with transformer coupled load uses two transistor rated 10 W each. What is the maximum power output one can obtain at the load from the circuit?
(a) 40 W (b) 50 W
(c) 60 W (d) 70 W

13. Operational Amplifier

Q.333 In a differential amplifier an ideal CMMR is
(a) –1 (b) zero
(c) infinity (d) +1

Q.334 For the circuit shown below, the output voltage is given by

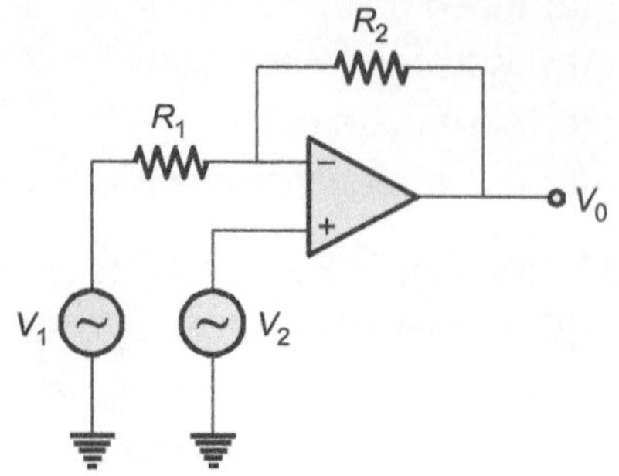

(a) $V_0 = \frac{R_2}{R_1}(V_2 - V_1) - V_2$

(b) $V_0 = \frac{(V_2 - V_1)}{R_1 + R_2}$

(c) $V_0 = \frac{R_2}{R_1}(V_2 - V_1)$

(d) $V_0 = \frac{R_2}{R_1}(V_2 - V_1) + V_2$

Q.335 An ideal operational amplifier has input impedance of

(a) 1 MΩ (b) infinity
(c) 1 Ω (d) zero

Q.336 The order of input resistance in 741 OP-AMP is

(a) 1 to 10^4 Ω (b) 10^5 Ω
(c) 10^3 Ω (d) 10^6 Ω

Q.337 An ideal operational amplifier has

1. non-zero offset voltage.
2. characteristics which do not drift with temperature.
3. an infinite voltage gain.
4. a CMMR more than 120 dB.
5. infinite bandwidth.

Of these statements:

(a) 2, 3 and 5 are correct
(b) 1, 2 and 4 are correct
(c) 2, 3 and 4 are correct
(d) 3, 4 and 5 are correct

Q.338 An inverting amplifier using the 741C must have a flat response upto 40 kHz. The gain of the amplifier is 10 and slew rate is 0.5 V/μS. The maximum peak-to-peak input signal that can be applied without distorting the output is approximately equal to

(a) 0.32 V (b) 0.60 V
(c) 0.40 V (d) 0.28 V

Q.339 For a given op-amp, CMMR = 10^5 and differential gain $A_d = 10^5$. The common-mode gain of the op-amp is

(a) 10 (b) 0.5
(c) 1 (d) 100

Q.340 For the op-amp circuit shown below, R_1 = 1 kΩ and R_F = 10 kΩ. The feedback factor β of this amplifier is

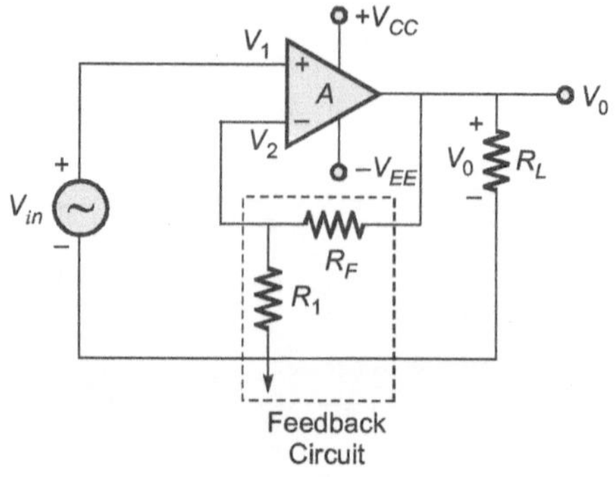

(a) 7.6% (b) 8.8%
(c) 6.5% (d) 9.1%

Q.341 For an astable multivibrator shown below, R_1 = 116 kΩ, R_2 = 100 kΩ and ± V_{sat} = ± 14 V.

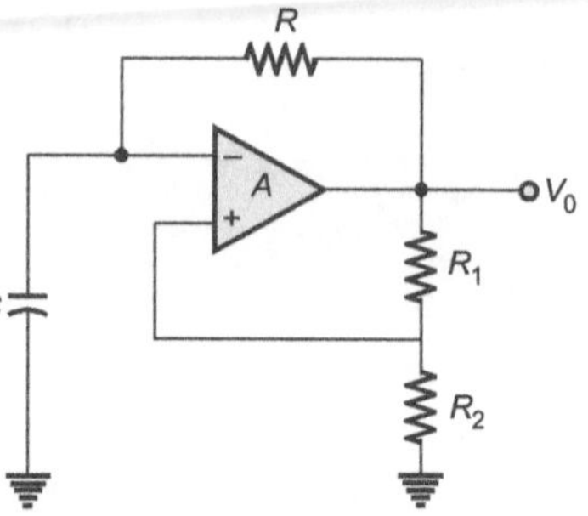

The time constant of the circuit to produce an output of frequency 1 kHz will be

(a) 1 ms (b) 0.5 ms
(c) 1.5 ms (d) 0.75 ms

Q.342 The op-amp circuit shown below is initially nulled having a supply voltages of ± 15 V. The output voltage V_0 is

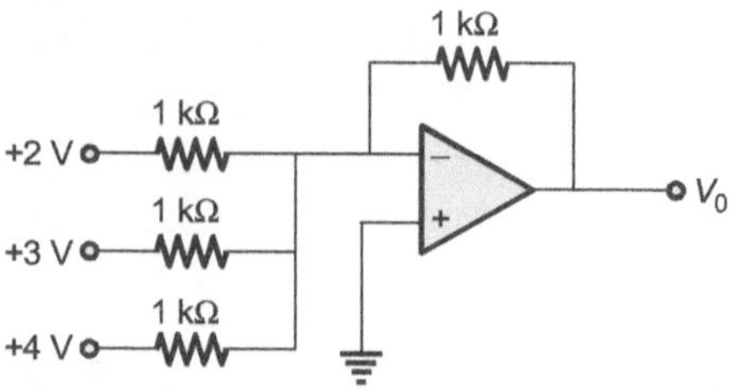

(a) –9 V (b) 3 V
(c) –6 V (d) 9 V

Q.343 An op-amp with a slew rate of 0.5 V/μs is used as an inverting amplifier with a gain of 50. The sinusoidal input signal has a variable frequency and maximum amplitude of 20 mV peak. Assuming the amplifier to be initially nulled, what is the maximum frequency of the input at which the output will be undistorted?

(a) 88.0 kHz (b) 78.4 kHz
(c) 81.0 kHz (d) 79.6 kHz

Q.344 Assertion (A): When connected in open-loop, the op-amp functions as a very high gain amplifier.

Reason (R): The three open-loop configurations of op-amp are namely, inverting, non-inverting and differential amplifiers.

(a) Both A and R are true and R is the correct explanation of A.
(b) Both A and R are true but R is not the correct explanation of A.
(c) A is true but R is false.
(d) A is false but R is true.

Q.345 Assertion (A): An unity gain non-inverting operational amplifier draws a high current from the source.

Reason (R): An unity gain non-inverting operational amplifier is used for impedance matching applications.

(a) Both A and R are true and R is the correct explanation of A.
(b) Both A and R are true but R is not the correct explanation of A.
(c) A is true but R is false.
(d) A is false but R is true.

Q.346 Match **List-I (OP-AMP circuits)** with **List-II (Applications)** and select the correct answer using the codes given below the lists:

List-I
A. Instrumentation amplifiers
B. Differentiator
C. Multivibrators
D. Integrators

List-II
1. High pass filter
2. Measurement and control of temperature, humidity and light intensity
3. Low pass filter
4. Timing networks

Coded:

	A	B	C	D
(a)	4	3	2	1
(b)	4	1	2	3
(c)	2	1	4	3
(d)	2	4	1	3

Q.347 For the op-amp circuit shown below both the diodes are non-ideal (cut-in voltage = 0.7 V).

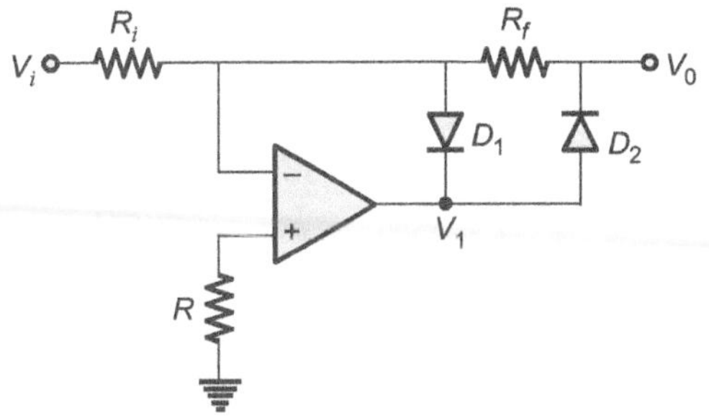

Which of the following output equations are correct for the above op-amp circuit?

1. $V_0 = \frac{-R_f}{R_i} V_i$ for $V_i < 0$ V
2. $V_0 = 0$ V for $V_i > 0$ V
3. $V_1 = -0.7$ V for $V_i > 0$ V
4. $V_1 = \frac{R_f}{R_1} V_i + 0.7$ V for $V_i < 0$ V

(a) 1, 2, 3 and 4 (b) 1, 2 and 4
(c) 2, 3 and 4 (d) 1 and 2

Q.348 Consider an ideal op-amp (infinite voltage gain). Let V_1 and V_2 be the values of independent voltage sources connected to the positive and negative input terminals, respectively, and let V_0 be the output voltage. If $V_1 > V_2$, then, V_0 will be

(a) zero (b) infinite
(c) unpredictable (d) $(V_1 - V_2)$

Q.349 In order to ensure that the output voltage of an op-amp is zero, when both its inputs are grounded,

(a) the current incident at the output node are carefully designed.
(b) internal negative feedback is used.
(c) the totem-pole output transistors are designed to have exactly equal cutin voltages.
(d) an external offset balancing circuit is used at the input terminals.

Q.350 Connecting a capacitor in parallel with the feedback resistor of an inverting amplifier causes
(a) the dc gain to reduce.
(b) an additional low frequency pole in the transfer function.
(c) the circuit to become a high pass filter.
(d) an additional low frequency zero in the transfer fuction.

Q.351 The slew rate of an amplifier is defined as the
(a) highest frequency at which the output can change without distortion.
(b) the highest possible rate of change of the output voltage with respect to time.
(c) the slowest rate of change of the input voltage allowable in an RC-coupled amplifier.
(d) the rate of change of the bias current of the differential amplifier as a function of the input differential voltage.

Q.352 In an ideal differential amplifier, the input signals are $V_1 = +50\ \mu V$ and $V_2 = -50\ \mu V$ respectively however its values are 1,050 μV and 950 μV for the practical amplifier. If the common mode rejection ratio is 100, then what is the percentage difference in output voltage obtained for the above two sets of input signals?
(a) 10% (b) 5%
(c) 20% (d) 2.5%

Q.353 Match **List-I (Op-Amp circuit)** with **List-II (Type of filter)** and select the correct answer using the codes given below the lists:

List-I

A.
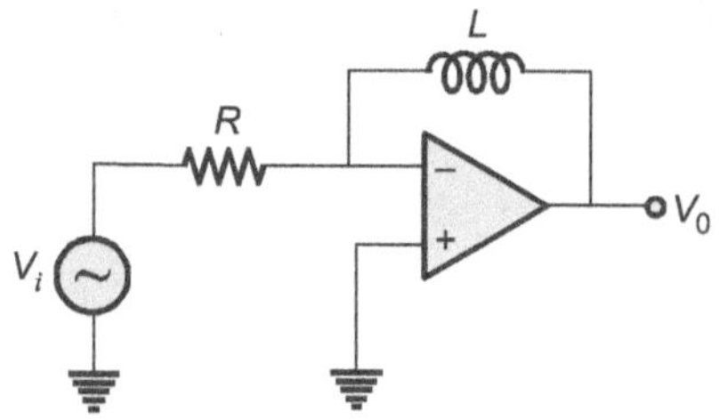

B.
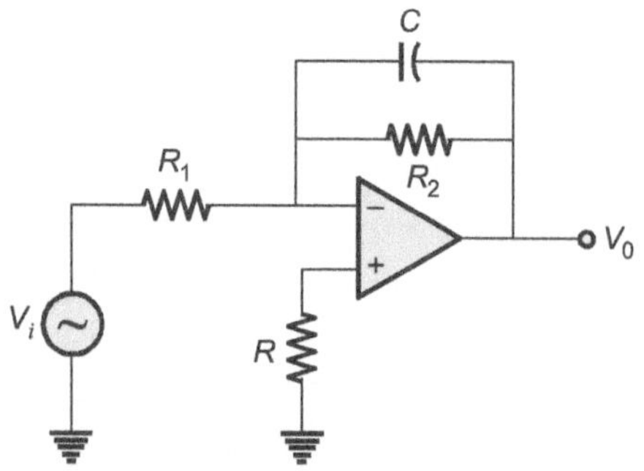

C.
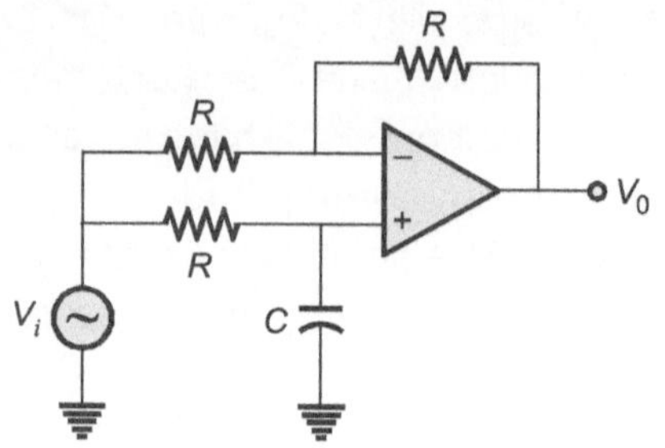

List-II
1. Low pass filter
2. All pass filter
3. High pass filter

Codes:

	A	B	C
(a)	2	1	3
(b)	1	2	3
(c)	3	1	2
(d)	3	2	1

Q.354 The common mode rejection ratio (CMMR) of a differential amplifier (where A_d = differential gain, A_c = common mode gain) is defined as
(a) $\frac{A_d - A_c}{A_d}$ (b) $20 \log_{10} \frac{A_c}{A_d}$
(c) $\frac{A_d}{A_c}$ (d) $20 \log_e \frac{A_c}{A_d}$

Q.355 The voltage follower is commonly used as
(a) Isolator (b) Switch
(c) Regulator (d) None of these

Q.356 The purpose of phase lag compensation is to
(a) make the op-amp stable at very low value of gain.
(b) make the op-amp stable at very high values of gain.
(c) increase the bandwidth.
(d) reduce the unity gain frequency.

Q.357 The voltage gain of an ideal voltage follower is
(a) 0 (b) < 1
(c) infinity (d) 1

Q.358 If slew rate of op-amp is very high, so that triggering pulse reach to $+V_{sat}$ or $-V_{sat}$ is
(a) very slow
(b) very fast
(c) not reach $+V_{sat}$ or $-V_{sat}$
(d) none of the above

Q.359 Noise on the input of the comparator can cause the output to
(a) change erratically between two states
(b) hang-up in one state
(c) produce the amplified noise signal
(d) go to zero

Q.360 The zero level detector is one application of a
(a) summing amplifier
(b) differentiator
(c) comparator
(d) integrator

Q.361 Consider the following statements associated with operational amplifiers:
1. Op-amp is a current controlled voltage device.
2. An ideal op-amp has infinite input resistance and zero gain.
3. Inverting amplifier gain is independent of source resistance.
4. An op-amp is used in IC regulators as an error amplifier.

Which of the statements given above are correct?
(a) 1, 3 and 4 (b) 3 and 4 only
(c) 2, 3 and 4 (d) 1 and 2 only

Q.362 **Assertion (A):** Op-amps, when used as Comparators, do not need frequency compensation.
Reason (R): Op-amp Comparators are always used in open-loop configurations, and hence the Barkhausen criterion is never met.
(a) Both A and R are true and R is the correct explanation of A.
(b) Both A and R are true but R is not the correct explanation of A.
(c) A is true but R is false.
(d) A is false but R is true.

Q.363 For the instrumentation amplifier shown in figure below, assuming the opamps to be ideal, the output voltage is given by

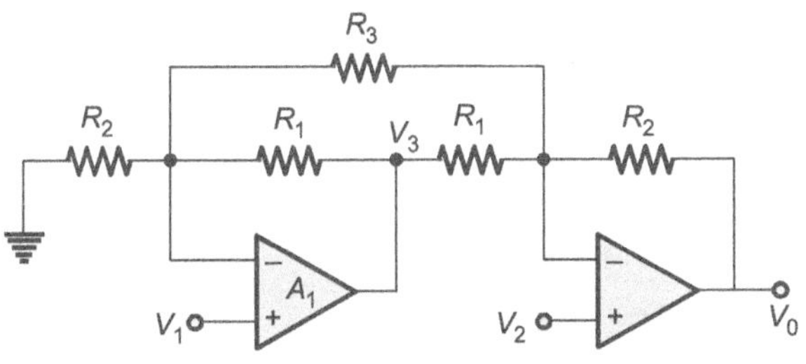

(a) $V_0 = \left(1+\frac{R_2}{R_1}+\frac{2R_3}{R_2}\right)(V_1 - V_2)$
(b) $V_0 = \left(1+\frac{R_2}{R_1}+\frac{R_2}{R_3}\right)(V_1 - V_2)$
(c) $V_0 = \left(1+\frac{R_2}{R_1}+\frac{R_3}{R_2}\right)(V_2 - V_1)$
(d) $V_0 = \left(1+\frac{R_2}{R_1}+\frac{2R_2}{R_3}\right)(V_2 - V_1)$

Q.364 A square wave input of 8 V peak to peak magnitude and frequency 2 MHz is applied to a voltage follower which produces the triangular output as shown in figure below. What is its slew rate?

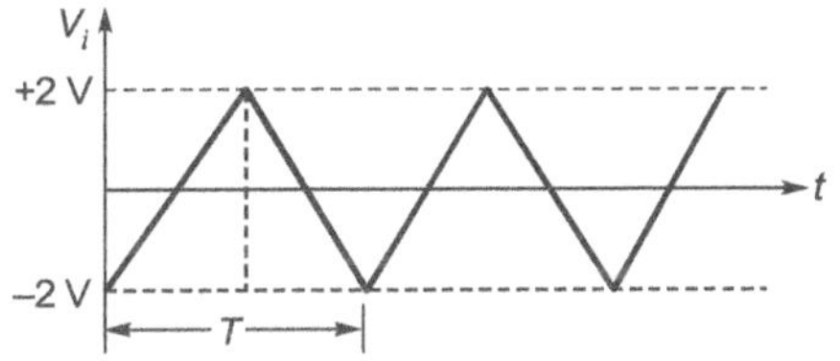

(a) 16 V/μs (b) 8 V/μs
(c) 20 V/μs (d) 4 V/μs

Q.365 The characteristic of a diode in a logarithmic amplifier using op-amp is given by the relationship as $I_D = I_s\,(e^{qV/\eta KT} - 1)$, where V is the forward voltage and η is the ideality factor = 1 (*Ge*). What is the value of input voltage to result in output voltage $V_0 = 0$, if $R = 100\ \text{k}\Omega$, $I_s = 1\ \mu\text{A}$ and $\frac{kT}{q} = 26\ \text{mV}$?
(a) 10 V (b) 0.5 V
(c) 0.1 V (d) 2 V

Q.366 The input impedance of the op-amp circuit shown below is

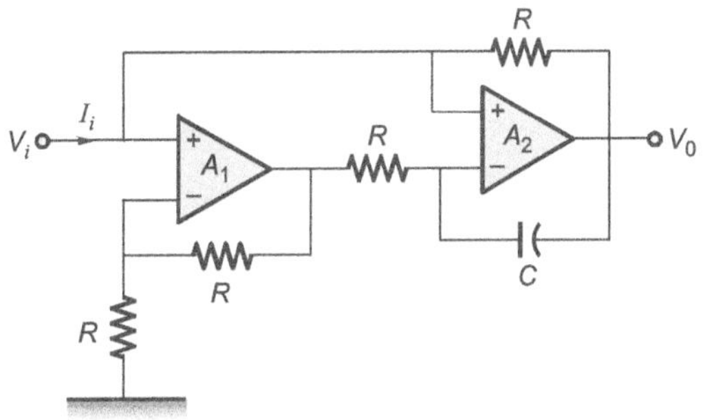

(a) purely capacitive (b) reactive
(c) purely resistive (d) purely inductive

Q.367 The output voltage of the circuit shown below is:

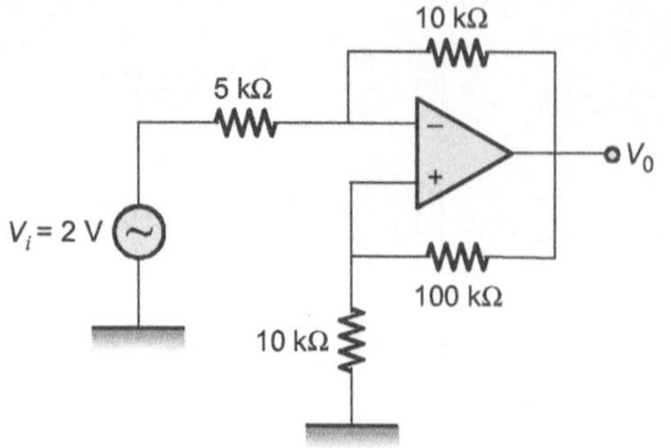

(a) –7.5 V (b) – 6.5 V
(c) –3.5 V (d) –5.5 V

Q.368 For the circuit shown below, what is the value of resistor R for the condition that both inputs V_- and V_+ should be amplified by the same amount?

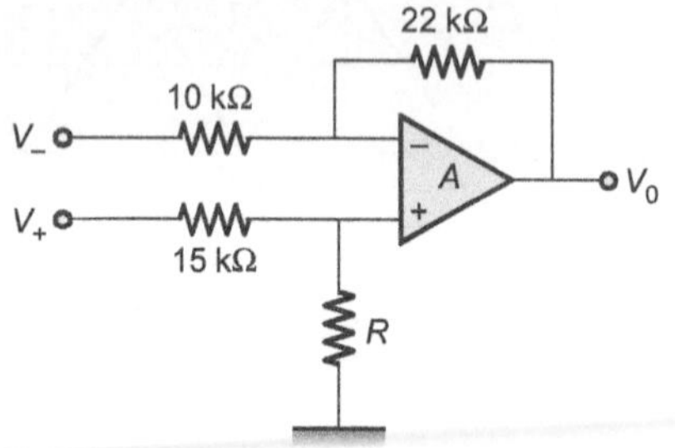

(a) 27 kΩ (b) 33 kΩ
(c) 25 kΩ (d) 52 kΩ

Q.369 For the op-amp circuit shown below, the output get saturated at time t equal to

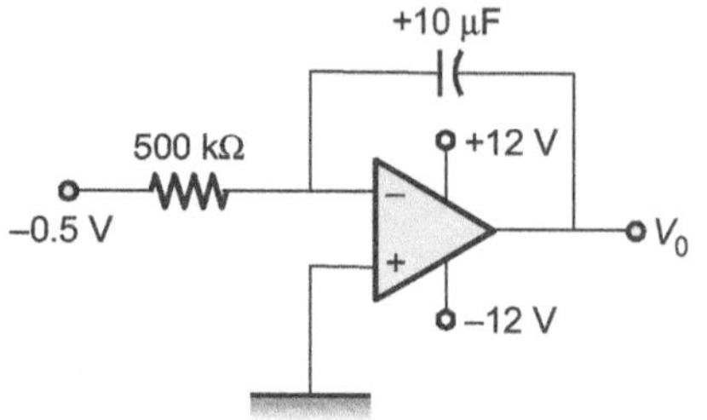

(a) 120 s (b) 90 s
(c) 150 s (d) none of these

Q.370 The circuit using an op-amp is shown in the given figure. It has

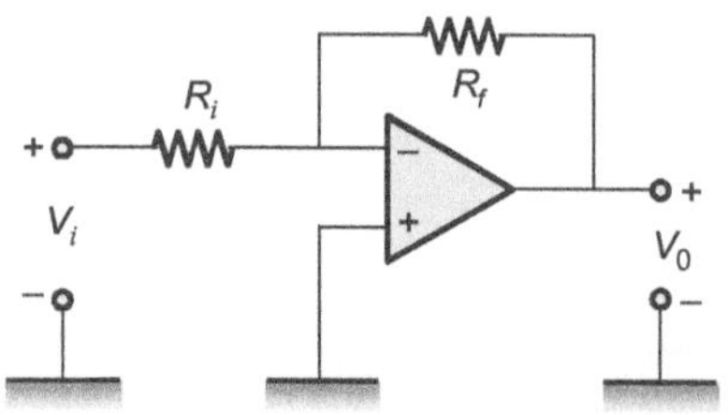

(a) Voltage series feedback
(b) Voltage shunt feedback
(c) Current shunt feedback
(d) Current series feedback

Q.371 Assume that the operational amplifier shown in the figure below is ideal. The current 'I' through the 1 kΩ resistor is

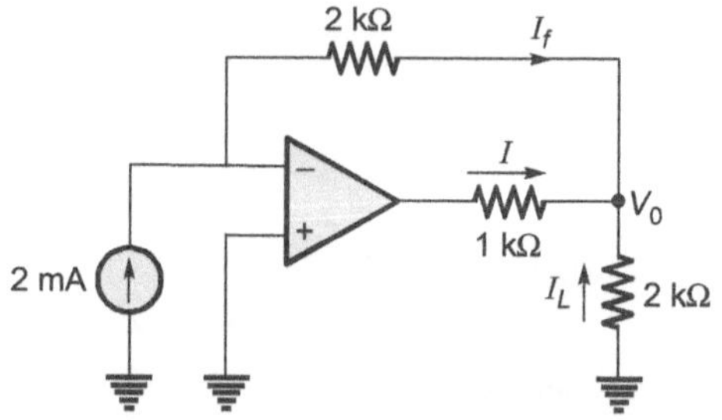

(a) 2 mA (b) 4 mA
(c) –4 mA (d) 0 mA

Q.372 Consider the op-amp circuit shown below.

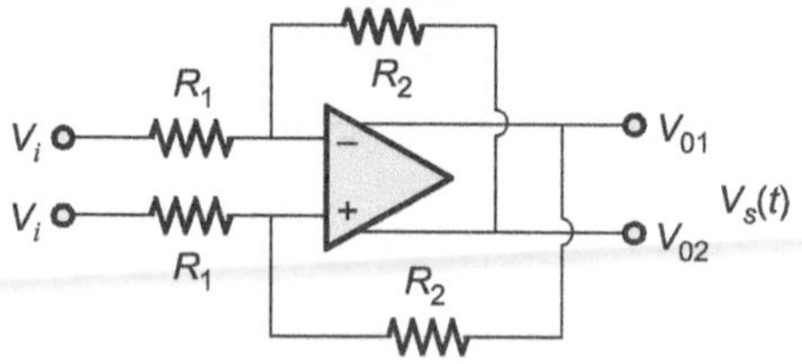

The differential gain $\left(\frac{V_{01}-V_{02}}{V_i}\right)$ is given by

(a) $1+\frac{R_2}{R_1}$ (b) $1+\frac{2R_2}{R_1}$

(c) $\frac{R_2}{R_1}$ (d) $\frac{2R_2}{R_1}$

Q.373 In the inverting op-amp circuit shown below, the resistor R_g is chosen as $R_1 || R_2$ in order to

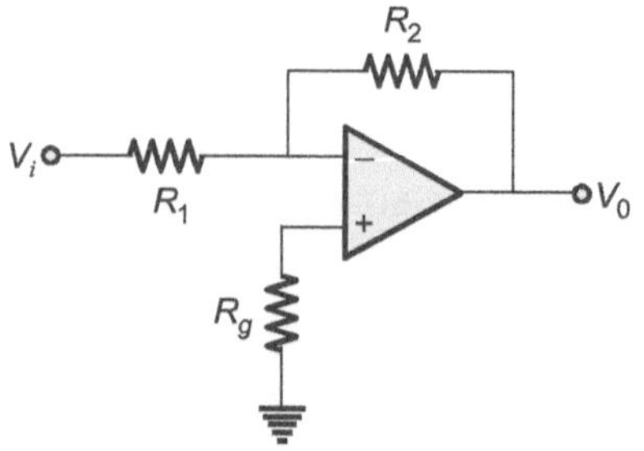

(a) increase gain
(b) reduce offset voltage
(c) reduce offset current
(d) increase CMMR

14. Signal Generators and Waveform Shaping Circuits

Q.374 Hysteresis is desirable in Schmitt-trigger because

(a) it would prevent noise from causing false triggering.

(b) effects of temperature would be compensated.

(c) devices in the circuit should be allowed time for saturation and unsaturation.

(d) energy is to be stored/discharged in parasitic capacitances.

Q.375 Each stage of a three stage amplifier without feedback has identical pole frequencies and its open loop transfer function is expressed as:

$$A_3(s) = \frac{A_0^3}{\left(1+\frac{s}{\omega_H}\right)^3}$$

What is the maximum value of feedback factor β_{max} for which the amplifier remains stable?

(a) $\frac{8}{A_0^3}$ (b) $\frac{16}{A_0^3}$

(c) $\frac{27}{A_0^3}$ (d) None of these

Q.376 Assertion (A): If the magnitude of loop gain of the amplifier is unity then the amplifier gives an a.c. output without any a.c. input signal.

Reason (R): The property of the regenerative feedback is utilised in oscillators.

(a) Both A and R are true and R is the correct explanation of A.

(b) Both A and R are true but R is not the correct explanation of A.

(c) A is true but R is false.

(d) A is false but R is true.

Q.377 A 555 time basically operates in

(a) monostable multivibrator mode

(b) astable multivibrator mode

(c) either as monostable or as an astable multivibrator mode

(d) none of the above

Q.378 In the Hartley oscillator, L_2 = 0.4 mH and C = 0.004 μF. If the frequency of the oscillator is 120 kHz, then the value of L_1 would be (neglect the mutual inductance)

(a) 0.50 mH (b) 2.25 mH

(c) 0.02 mH (d) 0.04 mH

Q.379 A tank circuit contains an inductance of 1 mH. What is the range of tuning capacitor value if the resonant frequency ranges from (540 - 1650) kHz?

(a) 3.0 pF ≤ C ≤ 75 pF

(b) 9.3 pF ≤ C ≤ 86.86 pF

(c) 3.0 pF ≤ C ≤ 86.86 pF

(d) 9.3 pF ≤ C ≤ 75.0 pF

Q.380 For an RC phase-shift oscillator using BJT, R = 10 kΩ, C = 0.01 μF and R_C = 2.2 kΩ. The minimum current gain needed for sustained oscillations is around

(a) 156 (b) 101

(c) 96 (d) 188

Q.381 In a Wien-bridge oscillator, if the value of R is 100 kΩ, and frequency of oscillation is 10 kHz, then the value of capacitance C would be

(a) 108 pF (b) 66 μF

(c) 159 pF (d) 125 μF

Q.382 Assertion (A): In a crystal oscillator, it is necessary to enclose the crystal in a temperature controlled oven to achieve the frequency stability of the order of 1 part in 10^{10}.

Reason (R): A crystal has very high Q as a resonant circuit, which results in good frequency stability for the oscillator.

(a) Both A and R are true and R is the correct explanation of A.

(b) Both A and R are true but R is not the correct explanation of A.

(c) A is true but R is false.

(d) A is false but R is true.

Q.383 Assertion (A): All oscillators require positive feedback for their operation.

Reason (R): Any circuit which is used to generate a periodic voltage without an a.c. input signal is called an oscillator.

(a) Both A and R are true and R is the correct explanation of A.

(b) Both A and R are true but R is not the correct explanation of A.

(c) A is true but R is false.

(d) A is false but R is true.

Q.384 Consider the following statements associated with the astable multivibrator:

1. It is a voltage to frequency converter.
2. It is a source of production of harmonic frequencies of higher order.
3. It can be operated as an oscillator over a wide range of audio and radio frequencies.
4. It is used in the construction of digital voltmeter and SMPS.

Which of the statements given above is/are correct?

(a) 1, 2, 3 and 4 (b) 2 and 4 only
(c) 1, 3 and 4 only (d) 2 and 3 only

Q.385 What is the frequency of oscillation for an asymmetrical astable multivibrator shown below with components values $R_1 = 2\ k\Omega$, $R_2 = 10\ k\Omega$, $C_1 = 0.01\ \mu F$ and $C_2 = 0.05\ \mu F$?

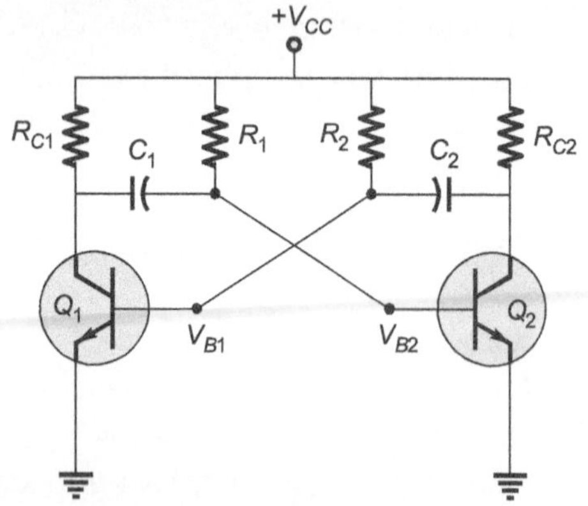

(a) 4.0 kHz (b) 3.3 kHz
(c) 2.25 kHz (d) 2.8 kHz

Q.386 Consider the following statements associated with a monostable multivibrator:

1. The monostable multivibrator is used to function as an adjustable pulse width generator.
2. It is used to generate uniform width pulse from a variable width input pulse train.

Select the correct answer from the given options:

(a) Both statement 1 and 2 are correct.
(b) Both statements 1 and 2 are false.
(c) Statement 1 is correct while 2 is false.
(d) Statement 2 is correct while 1 is false.

Q.387 Match **List-I (Type of Oscillators)** with **List-II (Oscillating frequency)** and select the correct answer using the codes given below the lists:

List-I

A. Radio frequency oscillator
B. Ultra high frequency oscillator
C. Very high frequency oscillator
D. Microwave frequency oscillator

List-II

1. 100 MHz
2. 25 kHz
3. 5 GHz
4. 500 MHz

Codes:

	A	B	C	D
(a)	3	2	1	4
(b)	1	4	2	3
(c)	2	4	1	3
(d)	1	2	3	4

Q.388 Match **List-I (Oscillators)** with **List-II (Operating principle)** and select the correct answer using the codes given below the lists:

List-I

A. Crystal Oscillator
B. Clapp Oscillator
C. Hartley Oscillator
D. Wien bridge Oscillator

List-II

1. Null condition
2. Parallel resonance
3. Series resonance
4. Piezoelectric effect

Codes:

	A	B	C	D
(a)	3	2	4	1
(b)	4	3	2	1
(c)	1	2	3	4
(d)	1	3	2	4

Q.389 Match **List-I** with **List-II** and select the correct answer using the codes given below the lists:

List-I

A. Schmitt Trigger
B. Bistable multivibrator
C. Monostable multivibrator
D. Astable multivibrator

List-II

1. Memory elements in shift registers and counters.

2. In digital voltmeter and SMPS
3. Wave shaper
4. Generation of clean and sharp pulses from the distorted pulses

Codes:

	A	B	C	D
(a)	3	1	4	2
(b)	3	2	4	1
(c)	1	4	3	2
(d)	4	2	1	3

Q.390 For the oscillator circuit shown below the value of gain for which the circuit will give sustained oscillation in the output is

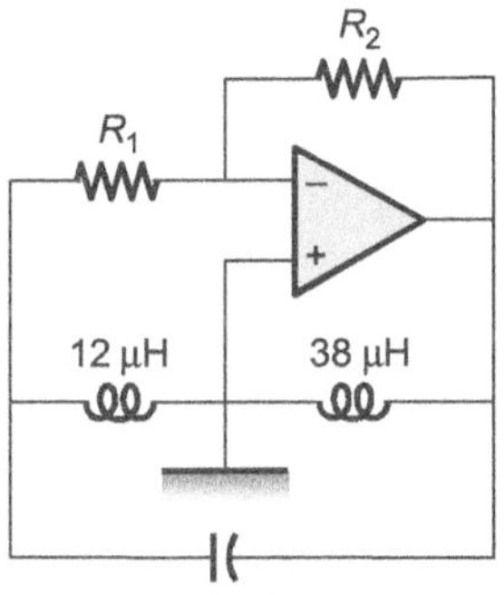

(a) 3.0 (b) 2.6
(c) 4.0 (d) 1.8

Q.391 **Assertion (A):** In every practical oscillator the loop gain is slightly larger than unity, and the amplitude of the oscillations is limited by the onset of nonlinearity.

Reason (R): Oscillations will not be sustained if, at the oscillator frequency, the magnitude of the product of the transfer gain of the amplifier and the magnitude of the feedback factor of the feedback network are less than unity.

(a) Both A and R are true and R is the correct explanation of A.
(b) Both A and R are true but R is not the correct explanation of A.
(c) A is true but R is false.
(d) A is false but R is true.

Q.392 The Barkhausen criterion states that for oscillations to occur in a feedback loop, at a particular frequency

(a) the magnitude of the loop gain must be greater than zero when the phase shift is 180°.
(b) the magnitude of the loop gain must be greater than unity when the phase shift is less than 360°.
(c) the magnitude of the loop gain must be unity and the phase shift must be 360°.
(d) there must be positive feedback.

Q.393 Phase shift oscillators, in which the RC sections in the feedback path provide a phase shift of 180°, are difficult to design with bipolar transistors because

(a) the base current loads the phase shift network.
(b) bipolar transistors have a current gain much greater than unity.
(c) the base-emitter voltage is close to 0.7 V irrespective of the current.
(d) bipolar transistors cause a phase shift of 180° and not 360°.

Q.394 Colpitts and Hartley oscillators belong to a general class of oscillators that use

(a) voltage -shunt feedback
(b) current-series feedback
(c) voltage-series feedback
(d) current-shunt feedback

Q.395 A quartz crystal is widely used in the design of sinusoidal oscillators because

(a) the crystal can produce electrical oscillations when pressure is applied to it.
(b) its piezoelectric properties make it superior to transistor amplifiers.
(c) the crystal provides precise positive feedback.
(d) the crystal is equivalent to a very high Q LC-tuned circuit, whose characteristics are extremely stable with respect to temperature and time.

Q.396 When designing amplifiers, positive feedback is hardly ever used to increase the gain because

(a) it is not advantageous to increase gain at the expense of reduced bandwidth.
(b) positive feedback leads to potential instability.
(c) positive feedback increases power dissipation significantly.
(d) positive feedback makes the amplifier nonlinear.

Q.397 A circuit for pulse generation using IC 555 timer is shown below.

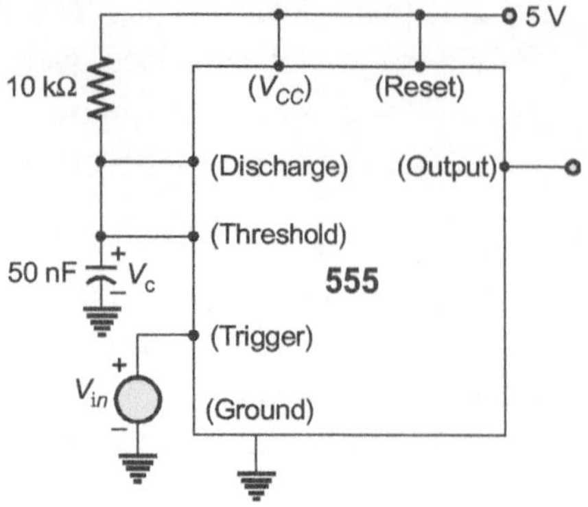

If a short pulse is applied at the input (v_{in}) at t = 0, the width of the pulse generated at the output would be (Assume β = 20 for the transistor)

(a) 550 μs (b) 310 ns
(c) 375 ns (d) 625 μs

Q.398 In the above problem, what is the time taken by the capacitor to discharge back to zero after the output has gone low?

(a) 550 ns (b) 22 μs
(c) 29 μs (d) 325 ns

Q.399 A schmitt trigger circuit achieves hysteresis by utilizing

(a) the magnetic properties of a transformer core
(b) avalanche multiplication in a zener (tunnel diode)
(c) the Barkhausen principle
(d) regenerative positive feedback

Q.400 In sample-and-hold circuits, the op-amp buffer is essential because

(a) it prevents the hold capacitor voltage from changing due to loading by the subsequent circuitry.
(b) without it the switch, when closed, will short the input to the output.
(c) the switch resistance will otherwise load the output.
(d) the hold capacitor has a high impedance.

Q.401 The frequency band from 30 MHz to 300 MHz is of

(a) radio-frequency oscillator
(b) very high-frequency oscillator
(c) audio-frequency oscillator
(d) video-frequency oscillator

Q.402 An R-C phase shift oscillator will not produce any oscillation until and unless the voltage gain of its internal amplifier is

(a) less than unity (b) around 3
(c) unity (d) more than 29

Q.403 In a Wien bridge oscillator, if the resistance in the positive feedback circuit are decreased the frequency

(a) remains the same
(b) fluctuates in an erratic fashion
(c) decreases
(d) increases

Q.404 A relaxation oscillator is one which

(a) produces non-sinusoidal waves
(b) oscillates continuously
(c) relaxes indefinitely
(d) produces sinusoidal waves

Q.405 Schmitt trigger is basically ______ multivibrator and converts slowly varying waveform into ______ .

(a) a monostable, sine wave
(b) an astable, sine wave
(c) a bistable, square wave
(d) a monostable, square wave

Q.406 An R-C shift oscillator is shown in the figure. Which of the following statements is **not** true, if f is the frequency of oscillation?

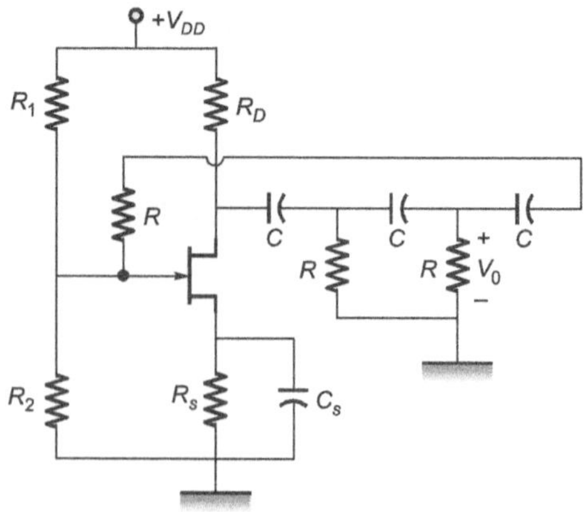

(a) For R = 200 kΩ, $C = \dfrac{1}{20\pi}$ nF, $f = \dfrac{50}{\sqrt{6}}$ kHz

(b) For R = 400 kΩ, $C = \dfrac{1}{40\pi}$ nF, $f = \dfrac{50}{\sqrt{6}}$ kHz

(c) The gain of the FET should be more than three for proper operation.

(d) The gain of the FET should be more than twenty nine for proper operation.

Q.407 Consider the following statements:

1. Positive feedback is essential for the operation of an oscillator.
2. In a transistor Hartley oscillator, inductive feedback is used.
3. Clapp oscillator is a modification of Hartley oscillator.
4. Conventional bistable multivibrators are preferred over Schmitt triggers in applications that require the circuit to be triggered back and forth between the two stable states.

Which of the statements given above are correct?

(a) 2, 3 and 4 (b) 1, 2 and 4
(c) 1, 2 and 3 (d) 1, 3 and 4

Q.408 A triangular-square wave generator uses
(a) a sine wave oscillator and a comparator
(b) an integrator and a comparator
(c) a differentiator and a comparator
(d) a sine wave oscillator and a clipper

Q.409 The most commonly used amplifier in sample and hold circuits is
(a) a unity gain inverting amplifier.
(b) a unity gain non-inverting amplifier.
(c) an inverting amplifier with a gain of 10.
(d) an inverting amplifier with a gain of 100.

Q.410 Consider an *RC* phase shift oscillator with $R = 1$ kΩ and $C = 500$ pF. The frequency at which the phase angle produced is 45° is given by

(a) 1.5×10^5 Hz (b) 3.2×10^5 Hz
(c) 2.5×10^5 Hz (d) 2.8×10^5 Hz

Q.411 FET phase-shift oscillator has
(a) voltage-series feedback
(b) Voltage shunt feedback
(c) Current series feedback
(d) Current shunt feedback

Q.412 A Hartley oscillator is used for generating
(a) very low frequency oscillation
(b) radio frequency oscillation
(c) microwave oscillation
(d) audio frequency oscillation

Q.413 Match **List-I** with **List-II** and select the correct answer using the codes given below the lists:

List-I
A. Colpitt oscillator
B. Hartley oscillator
C. Crystal oscillator
D. Clapp oscillator

List-II
1. Excellent frequency stability
2. Tapped inductance oscillator
3. Designed with one inductor and three capacitors
4. Tapped capacitance oscillator

Codes:

	A	B	C	D
(a)	1	2	3	4
(b)	3	2	4	1
(c)	2	4	1	3
(d)	4	2	1	3

Answers Analog Electronics

1. (c)	2. (a)	3. (b)	4. (a)	5. (d)	6. (d)	7. (b)	8. (c)
9. (a)	10. (d)	11. (b)	12. (c)	13. (a)	14. (d)	15. (b)	16. (d)
17. (a)	18. (c)	19. (c)	20. (b)	21. (c)	22. (d)	23. (a)	24. (c)
25. (b)	26. (d)	27. (b)	28. (c)	29. (d)	30. (a)	31. (b)	32. (d)
33. (d)	34. (c)	35. (a)	36. (b)	37. (d)	38. (a)	39. (d)	40. (d)
41. (c)	42. (b)	43. (b)	44. (a)	45. (d)	46. (c)	47. (a)	48. (b)
49. (d)	50. (d)	51. (a)	52. (b)	53. (c)	54. (a)	55. (b)	56. (c)
57. (d)	58. (b)	59. (a)	60. (c)	61. (c)	62. (b)	63. (c)	64. (a)
65. (a)	66. (d)	67. (b)	68. (c)	69. (d)	70. (b)	71. (c)	72. (a)
73. (c)	74. (d)	75. (d)	76. (a)	77. (c)	78. (d)	79. (b)	80. (a)

81. (b)	82. (c)	83. (d)	84. (b)	85. (c)	86. (a)	87. (b)	88. (b)
89. (c)	90. (a)	91. (d)	92. (d)	93. (c)	94. (c)	95. (a)	96. (b)
97. (a)	98. (d)	99. (b)	100. (a)	101. (d)	102. (a)	103. (c)	104. (c)
105. (b)	106. (d)	107. (a)	108. (c)	109. (b)	110. (d)	111. (a)	112. (d)
113. (c)	114. (b)	115. (a)	116. (d)	117. (b)	118. (b)	119. (b)	120. (d)
121. (d)	122. (a)	123. (a)	124. (b)	125. (d)	126. (d)	127. (b)	128. (d)
129. (a)	130. (b)	131. (c)	132. (a)	133. (d)	134. (c)	135. (b)	136. (d)
137. (a)	138. (b)	139. (a)	140. (d)	141. (c)	142. (a)	143. (a)	144. (c)
145. (a)	146. (d)	147. (b)	148. (a)	149. (b)	150. (c)	151. (d)	152. (d)
153. (d)	154. (c)	155. (b)	156. (b)	157. (c)	158. (a)	159. (d)	160. (a)
161. (b)	162. (b)	163. (c)	164. (a)	165. (c)	166. (d)	167. (a)	168. (d)
169. (d)	170. (c)	171. (b)	172. (d)	173. (a)	174. (b)	175. (a)	176. (c)
177. (d)	178. (b)	179. (a)	180. (d)	181. (c)	182. (a)	183. (d)	184. (b)
185. (a)	186. (d)	187. (d)	188. (b)	189. (a)	190. (b)	191. (d)	192. (c)
193. (a)	194. (c)	195. (c)	196. (b)	197. (b)	198. (a)	199. (d)	200. (c)
201. (d)	202. (c)	203. (a)	204. (a)	205. (c)	206. (b)	207. (b)	208. (d)
209. (d)	210. (c)	211. (c)	212. (b)	213. (a)	214. (d)	215. (b)	216. (b)
217. (c)	218. (d)	219. (a)	220. (b)	221. (a)	222. (c)	223. (b)	224. (c)
225. (a)	226. (c)	227. (c)	228. (d)	229. (b)	230. (b)	231. (d)	232. (b)
233. (a)	234. (c)	235. (a)	236. (d)	237. (d)	238. (c)	239. (c)	240. (a)
241. (d)	242. (b)	243. (d)	244. (c)	245. (b)	246. (a)	247. (a)	248. (d)
249. (c)	250. (c)	251. (d)	252. (d)	253. (b)	254. (c)	255. (b)	256. (d)
257. (c)	258. (a)	259. (d)	260. (c)	261. (b)	262. (a)	263. (a)	264. (d)
265. (c)	266. (b)	267. (c)	268. (a)	269. (b)	270. (c)	271. (d)	272. (c)
273. (b)	274. (a)	275. (b)	276. (a)	277. (d)	278. (d)	279. (c)	280. (b)
281. (d)	282. (a)	283. (c)	284. (a)	285. (b)	286. (b)	287. (d)	288. (b)
289. (c)	290. (a)	291. (b)	292. (b)	293. (a)	294. (c)	295. (b)	296. (b)
297. (a)	298. (b)	299. (c)	300. (a)	301. (c)	302. (d)	303. (c)	304. (a)
305. (d)	306. (a)	307. (a)	308. (d)	309. (a)	310. (a)	311. (c)	312. (d)
313. (c)	314. (a)	315. (c)	316. (d)	317. (b)	318. (a)	319. (b)	320. (c)
321. (c)	322. (d)	323. (a)	324. (d)	325. (c)	326. (b)	327. (a)	328. (d)
329. (d)	330. (b)	331. (b)	332. (b)	333. (c)	334. (d)	335. (b)	336. (d)
337. (a)	338. (c)	339. (c)	340. (d)	341. (b)	342. (a)	343. (d)	344. (b)
345. (d)	346. (c)	347. (a)	348. (b)	349. (d)	350. (b)	351. (b)	352. (a)
353. (c)	354. (c)	355. (a)	356. (d)	357. (d)	358. (b)	359. (a)	360. (c)
361. (b)	362. (a)	363. (d)	364. (a)	365. (c)	366. (d)	367. (d)	368. (b)
369. (a)	370. (b)	371. (c)	372. (b)	373. (c)	374. (a)	375. (a)	376. (b)
377. (c)	378. (d)	379. (b)	380. (a)	381. (c)	382. (b)	383. (d)	384. (a)
385. (d)	386. (a)	387. (c)	388. (b)	389. (a)	390. (c)	391. (b)	392. (c)
393. (a)	394. (c)	395. (d)	396. (b)	397. (a)	398. (c)	399. (d)	400. (a)
401. (b)	402. (d)	403. (d)	404. (a)	405. (c)	406. (c)	407. (b)	408. (c)
409. (b)	410. (b)	411. (a)	412. (b)	413. (d)			

Explanations

1. (c)

The intrinsic conductivity for germanium is

$$\sigma_i = qn_i(\mu_n + \mu_p)$$
$$= (1.6 \times 10^{-19})(2.5 \times 10^{13})(3800 + 1800)$$
$$= 0.0224 \text{ S/cm}$$

The intrinsic conductivity for silicon is

$$\sigma_i = qn_i(\mu_n + \mu_p)$$
$$= (1.6 \times 10^{-19})(1.5 \times 10^{10})(1300 + 500)$$
$$= 4.32 \times 10^{-6} \text{ S/cm}$$

2. (a)

The conductivity of an n-type silicon is $\sigma = q_n \mu_n$.
Concentration of electrons,

$$n = \frac{\sigma}{q\mu_n}$$

$$n = \frac{300}{(1.6 \times 10^{-19})(1300)}$$
$$= 1.44 \times 10^{18} \text{ cm}^{-3}$$

Using "mass-action law",

$$np = n_i^2$$

∴ Concentration of holes,

$$p = \frac{n_i^2}{n} = \frac{(1.5 \times 10^{10})^2}{1.44 \times 10^{18}}$$
$$= 1.56 \times 10^2 \text{ cm}^{-3}$$

3. (b)

When an electron is accelerated by the potential applied to a semiconductor, the energy gained from the field may then be transferred to an atom when the electron collides with the atom.

4. (a)

Given, $I = 1.6 \times 10^{-6}$ A
$= 1.6 \times 10^{-6}$ Coulomb/second

Charge crossing a particular cross-section per second = 1.6×10^{-6} C

Hence, number of electrons crossing a particular cross-section per second

$$= \frac{1.6 \times 10^{-6}}{1.6 \times 10^{-19}} = 10^{13}$$

5. (d)

Energy band diagram for semiconductor, metal and insulator are shown below.

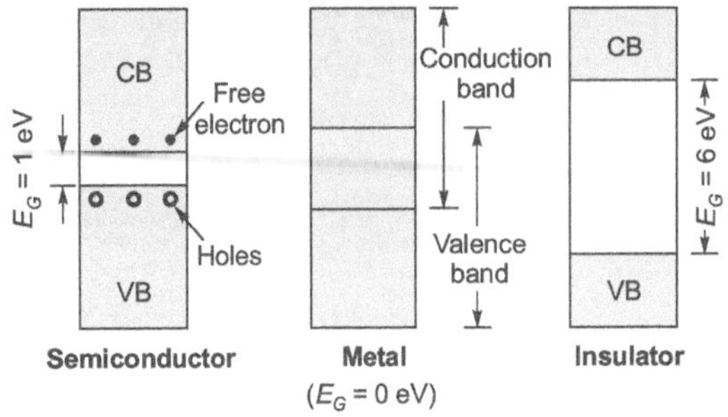

7. (b)

Both assertion and reason are individually correct statements. However, the reason for assertion is that due to the applied electric field, and electrostatic force is developed on the electron and the electrons would be accelerated in a direction opposite to the applied electric field and this motion is called directed motion of electron.

8. (c)

The current density is given by

$$J = \sigma E = nq\mu_n E$$
$$= 10^{20} \times 1.6 \times 10^{-19} \times 800 \times 1$$
$$= 1.28 \times 10^4 \text{ A/cm}^2$$
$$\approx 1.3 \times 10^4 \text{ A/cm}^2 \approx 13 \text{ kA/cm}^2$$

9. (a)

Given, $N_A = 0$

Using "charge neutrality equation", we have:

$$N_D + p = N_A + n$$

or, $$N_D + p = n \quad \text{...(i)}$$

Using "Mass-action law", we have:

$$np = n_i^2 \text{ or } p = \frac{n_i^2}{n} \quad \text{...(ii)}$$

Substituting value of p from equation (ii) in equation (i), we have:

or, $$N_D + \frac{n_i^2}{n} = n$$

or, $$n^2 - N_D n - n_i^2 = 0$$

$$n = \frac{N_D \pm \sqrt{N_D^2 + 4n_i^2}}{2}$$

Neglecting negative sign, we get:

$$n = \frac{N_D + \sqrt{4n_i^2 + N_D^2}}{2}$$

= free electron density or concentration

10. (d)

The force acting on a charge q placed in a magnetic field B and an electric field E is given by

$$F = qv \times B$$

The velocity of the hole placed in an electric field is

$$v = v_d = \mu_p E$$

Given, $B = (5 \times 10^{-4} \hat{i}_x)$ Wb/cm^2

and $E = (2000\ \hat{i}_y)$ V/cm

So, $F = q(\mu_p E) \times B$

$= q(400 \times 2000\ \hat{i}_y) \times (5 \times 10^{-4} \hat{i}_x)$

$= -q \times 400\ \hat{i}_z$ V/cm ...(i)

Now, $F = +qE$ ...(ii)

($+q$ = charge on a hole)

Comparing equations (i) and (ii), the electric field due to the Hall effect will be –400 V/cm in $+z$ direction.

12. (c)

Boron, Aluminium, Galium, Indium are acceptor impurities (valence electrons = 3) while Phosphorous, Arsenic, Antimoney, Bismuth are donor impurities (valence electrons = 5).

13. (a)

Conductivity of semiconductor

= carrier conc. × charge × mobility

$= nq\mu_n$

Current density of conductor,

$J = \sigma E = nq\mu_n E$

$= nqv$ (since drift velocity $v = \mu_n E$)

14. (d)

Given, $N_A = 1.1 \times 10^{17}/\text{cm}^3 \approx p$

$N_D = 10^{16}/\text{cm}^3$

$n_i = 10^{12}/\text{cm}^3$

We know that,

$$np = n_i^2 \text{ or } n = \frac{n_i^2}{p}$$

Here, $p \approx N_A$

So, $n = \dfrac{n_i^2}{N_A} = \dfrac{10^{24}}{1.1 \times 10^{17}}$

$= \dfrac{10}{11} \times 10^7/\text{cm}^3 \approx 10^7/\text{cm}^3$

15. (b)

- Diffusion current due to concentration gradient exists only in semiconductor. Hence, statement-1 is not correct.
- Drift current due to applied electric field occurs both in metals and semiconductors. Hence, statement-2 is correct.
- The intrinsic concentration of carriers is given by $n_i^2 = A_0 T^3 e^{-E_{G0}/KT}$
  (T = temperature and A_0 = constant)
  Hence, statement-3 is not correct.
- The current in semiconductor is due to both hole and electron flow due to which it is bipolar in nature while the current in metals is due to flow of electrons only due to which it is unipolar. Hence, statement-4 is correct.

17. (a)

At 0 kelvin (absolute zero temperature), both valence band and conduction band are empty, therefore intrinsic semiconductor at 0 kelvin will behave like a perfect insulator.

19. (c)

Doping materials are called impurities (acceptor impurities or donor impurities) because they increases the carrier concentration (concentration of holes or electrons) and hence, increases the conductivity.

20. (b)

- In a n-type semiconductor, $n >> p$, hence conduction is due to the movement of electrons in the conduction band.
- Sine $n >> p$, therefore electron constitute the majority charge carriers.
- n-type semiconductor is formed by adding donor impurities to germanium.
- Fermi level lies just below the conduction band between the conduction and valence band.
  Hence, statements 3 and 4 are not correct.

21. (c)

When acceptor type impurity is added (like Boron and Aluminium) to an intrinsic semiconductor $p >> n$ therefore the holes in valence bands becomes far greater than numbers of electrons due to which semiconductor becomes p-type.

22. **(d)**

All semiconductors are electrically neutral i.e. the net charge is zero.

24. **(c)**

Conduction band has higher energy level than a valence band.

25. **(b)**

- Diffusion constant, $D = \mu V_T$.
  As $T\uparrow$, $\mu\downarrow$ and $V_T\uparrow \Rightarrow D$ = constant i.e. independent of temperature.
- Intrinsic concentration, $n_i \propto T^{3/2} \Rightarrow n_i$ increases with temperature.
- Mobility, $\mu \propto \dfrac{1}{\text{(Temperature)}} \Rightarrow \mu$ decreases with temperature.

26. **(d)**

- $$\frac{\text{Diffusion constant}}{\text{Mobility}} = \frac{D_e}{\mu_e} = V_T = \frac{T}{11600}$$

  i.e. $\dfrac{D_e}{\mu_e} \propto T$
- Mass action law: $np = n_i^2$.
- Hall's effect can be used to measure magnetic field intensity (B).
- In a semi conductor, conductivity mainly depends upon carrier concentration.
- Hall voltage, $V_H \propto \dfrac{1}{\sigma}$.

Since metals have high conductivity, therefore V_H is low (in μV).
Since semiconductors have low conductivity, therefore V_H is high (in mV).
Hence, only statements 1 and 5 are correct.

27. **(b)**

- Both assertion and reason are correct statements. However, the reason for GaAs to have faster switching capability is due to its higher electron mobility (μ_n). The operating temperature of GaAs is higher (> 200°C) compared to Ge and *Si*.
- A semiconductor has the small width of the forbidden energy region i.e. around 1 eV.

28. **(c)**

Since it is possible to have a non-uniform concentration of particles in a semiconductor (concentration gradient) but, not in a metal, therefore there is diffusion current in a semiconductor but, not present in a metal.

29. **(d)**

Assertion (A) is not a correct statement.
According to Einstein equation, we have

$$\frac{D}{\mu} = V_T \quad \begin{pmatrix} \text{Here, } D = \text{Diffusion constant;} \\ \mu = \text{ Mobility of charge carriers;} \\ V_T = \text{Thermal voltage} \end{pmatrix}$$

Here, $D \propto \mu$
Hence, higher the value of mobility of a charge carrier, the greater will be its tendency to diffuse.

31. **(b)**

Total current, $I = I_n + I_p$

Given, $I_n = \dfrac{3}{4}I$ and $I_p = \dfrac{I}{4}$

Also, $v_n = 3\,v_p$

Now, $I_n = n_n qvA$ and $I_p = p_n qv_p A$

$$\therefore \quad \frac{I_n}{I_p} = \frac{n_n q v_n A}{q p_n v_p A} = \frac{n_n v_n}{p_n v_p} = \frac{n_n \times 3v_p}{p_n v_p}$$

$$\text{or,} \quad \frac{(3/4)I}{(1/4)I} = \frac{3n_n}{p_n} \text{ or } 3 = \frac{3n_n}{p_n} \text{ or } \frac{n_n}{p_n} = 1$$

Hence, the ratio of electrons to holes present in the semiconductor is unity.

32. **(d)**

$$J = qD_n\left(\frac{dn}{dx}\right)\text{Amp/m}^2$$
$$= 1.6 \times 10^{-19} \times 0.0012 \times 1.5 \times 10^{22}$$
$$= 2.88 \text{ A/m}^2$$

33. **(d)**

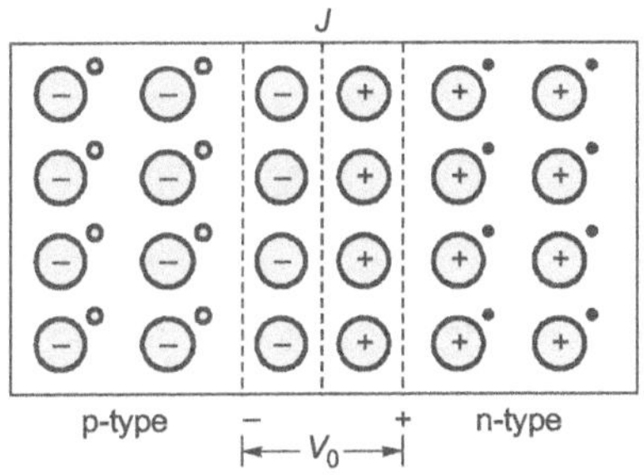

Due to the density gradient across the junction, holes will initially diffuse to the right across the junction, and electrons to the left. The positive holes which neutralizes the acceptor ions near the junction in the p-type silicon disappears as a result of combination with electrons which have diffused across the junction. Similarly, electrons in n-type semiconductor combines with holes.

34. (c)

The large reverse current flow in zener diode is due to rupturing-off or tearing-off of covalent bends within the depletion layer. This phenomena is called zener breakdown phenomena. It is due to large electric field intensity.

35. (a)

The work function of a metal represents the minimum amount of energy that must be given to the fastest-moving electron at the absolute zero of temperature in order for this electron to be able to escape from the metal. Hence, low work function (E_W) is required for thermionic emission.

37. (d)

Width of depletion layer, $W \propto \frac{1}{\sqrt{\text{Forward Bias}}}$

Hence, when forward bias is applied to a *p-n* junction diode, width of the depletion layer is reduced i.e. potential barrier is decreased.

39. (d)

In a p-n junction tunnel diode, the depletion layer width is negligible and therefore, the charge carriers will be penetrating the depletion layer almost at the speed of light and this property is called tunnelling of charge carriers across the junction.

40. (d)

The energy level diagram for a semiconductor is shown below.

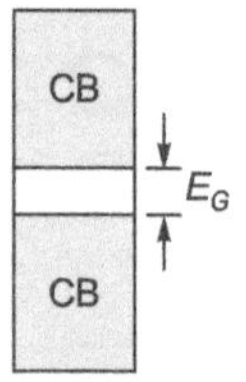

E_G is called "band gap" or "forbidden energy gap" which lies between the valence band and the conduction band.

41. (c)

Typical value of "Barrier voltage" or "Contact potential" for a *p-n* junction is:

V_0 = 0.2 V (for *Ge p-n* junction diode)
= 0.7 V (for *Si p-n* junction diode)

42. (b)

In an intrinsic semiconductor, as the temperature increases mobility is slightly reduced since mobility of charge carrier is inversely proportional to temperature. This will slightly reduce the conductivity. But, due to temperature effect, large number of covalent bonds are broken which creates electrons and holes and this increases the conductivity by a larger value. As a result of this resistivity is decreased $\left(\text{since } \rho \propto \frac{1}{\sigma}\right)$.

43. (b)

Schottky diode is a majority carrier device or diode i.e. unipolar while p-n junction diode, Zener diode and Tunnel diode all have current flow due to both majority and minority carrier i.e. they are bipolar device.

44. (a)

The breakdown voltage of any *p-n* junction diode is given by

$$V_{Br} \propto \frac{1}{\text{Doping}}$$

When doping is less, V_{Br} is more and hence breakdown occurs due to avalanche effect.

45. (d)

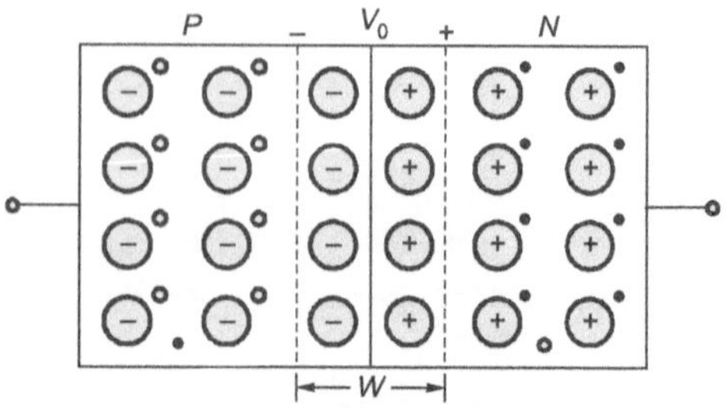

Open circuited P-N Junction

W is called "space charge region" or "depletion region" in which electron and hole concentrations are zero i.e. doesn't contain mobile carriers.

48. **(b)**

Let the Zener diode to be off.

Then, $$V_z = \left(\frac{100}{100+1000}\right)\times 10 = 0.91\text{ V}$$

Since V_z is less than the diode breakdown voltage (6 V), therefore the Zener diode will be in off state so that I_z = 0 A.

49. **(d)**

The diffusion capacitance of the diode is given by

$$C_D = \frac{\tau I_f}{\eta V_T}$$

Given, I_f = 0.1 mA, τ = 10 ns

and V_T = 26 mV (At room temperature)

For Ge diode,

$$\eta = 1$$

$$\therefore \quad C_D = \frac{10\times 10^{-9}\times 0.1\times 10^{-3}}{1\times 26\times 10^{-3}} = 38.5\text{ pF}$$

50. **(d)**

The dynamic resistance of a diode is given by

$$g = \frac{dI}{dV} = \frac{I_0 e^{Vd/\eta V_T}}{\eta V_T}$$

Here, g is a variable quantity depending upon the operating voltage. Thus, assertion is not a correct statement.

53. **(c)**

For either silicon or germanium (at room temperature), $\frac{dV}{dT} \approx -2.5$ mV/°C

Hence, the change in voltage required will be $2.5\times 10^{-3}\times[125-(-50)] \simeq 0.44$ V

54. **(a)**

- According to the "charge-control description of a diode",

$$I = \frac{Q}{\tau_P}$$

where, τ_P = Mean lifetime for holes

Thus, the diode current (which consists of holes crossing the junction from p to the n side) is proportional to the stored charge Q of excess minority carriers. Hence, statement-1 is correct.

- Since, $V_{Br} \propto \frac{1}{\text{Doping}}$, therefore if doping concentration increases, breakdown voltage will reduce and vice-versa. Thus, statement-2 is correct.
- The p and n materials in a Zener diode are heavily doped i.e. doping level is 1 : 10^5.
- Statement-4 is not correct because a Zener diode has characteristics similar to that of an ideal voltage source.
- Since, $\frac{dV}{dT} \approx -2.5$ mV/°C, therefore potential barrier decreases with temperature. Thus, statements 1, 2, 3 and 5 are only correct.

56. **(c)**

Dynamic resistance of the diode is given by

$$r = \frac{\eta V_T}{I_f}$$

Given, η = 2, V_T = 26 mV (at room temperature) and I_f = 2.6 mA.

$$\therefore \quad r = \frac{2\times 26\times 10^{-3}}{2.6\times 10^{-3}} = 20\ \Omega$$

57. **(d)**

The varactor, also called a varicap, tuning or voltage variable capacitor diode, is a junction diode with a small impurity dose at its junction, which has the useful property that its junction or transition capacitance is easily varied electronically. As the capacitance is inversely proportional to the distance between the plates $\left(C_T \propto \frac{1}{W}\right)$, the transition capacitance C_T varies inversely with the reverse voltage as shown in figure below.

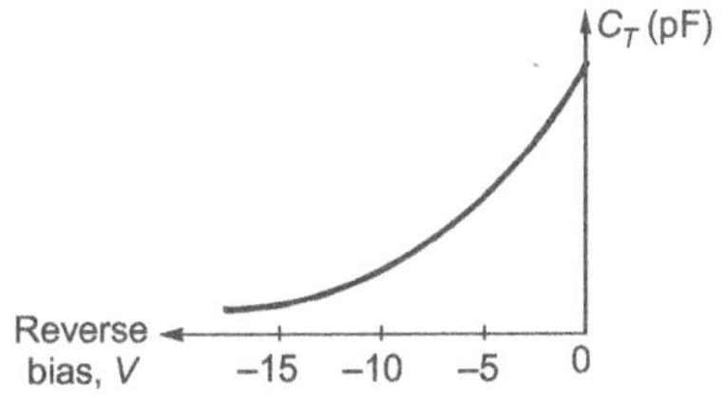

58. **(b)**

- Due to its negative resistance, a Tunnel diode is widely used as an oscillator not as a rectifier. Hence, statement-1 is not correct.

- Statements 2, 3 and 4 are correct.
- Due to the intermediate layer (intrinsic layer) between p and n regions in a PIN diode, the capacitance between these two regions is reduced. As a result it allows a faster response time for a PIN diode. Hence, PIN diodes are used at high frequencies (more than 300 MHz).

59. (a)

Forward resistance of a p-n junction diode is

$$r_f = \frac{\eta V_T}{I_f}$$

where, $$V_T = \frac{T}{11600}$$

$$= \frac{300}{11600} V \quad (\text{at } T = 300°\text{K})$$

$\eta = 2$ and $I_f = 5$ mA

$$\therefore \quad r_f = \frac{2 \times \frac{300}{11600}}{5 \times 10^{-3}} = 10.34\ \Omega$$

60. (c)

Given: $I_{0_1} = 7.5 \times 10^{-6}$ A

at $T_1 = 300°$ K $= 27°$C

and $T_2 = 400°$ K $= 127°$C

The saturation current at 400° K is

$$I_{0_2} = I_{0_1} \times 2^{(T_2 - T_1)/10}$$
$$= 7.5 \times 10^{-6} \times 2^{(127-27)/10}$$
$$= 7.5 \times 10^{-6} \times 2^{10}$$
$$= 7680 \times 10^{-6} = 7.68 \text{ mA}$$

61. (c)

Let both the diodes be ON.

Replacing the diodes with 0.7 V batteries (by cut-in voltage), the circuit is obtained as shown in figure below.

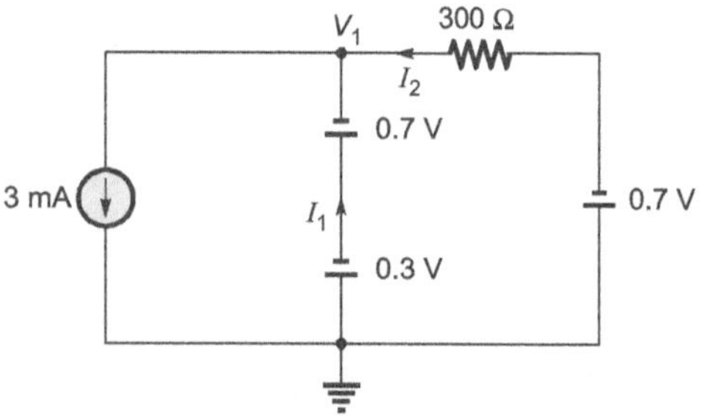

Here, $$I_2 = \frac{(-0.7+1)}{300} = 1 \text{ mA}$$

and, $I_1 = 3 - 1 = 2$ mA

Both these current values are consistent with the assumption of the two diodes being on. Therefore, our solution is correct.

Hence, $V_1 = -1$ volt

62. (b)

For large forward biases, the operating point of tunnel diode falls into exponential region and in exponential region tunnel diodes conduct due to the diffusion of carriers across the space charge layer.

65. (a)

The reverse current without light in a diode is in the range of μA. The change in this current due to the light is also in the range of μA. Thus, such a change can be significantly observed in the reverse current. If the photodiode is forward biased, the current flowing through it is in mA. The applied forward biased voltage takes the control of the current instead of light.

Hence, both assertion and reason are correct and reason is the correct explanation of assertion.

67. (b)

As the input circuit of FET is reversed biased, FET exhibits a much higher input impedance (in the order of 100 MΩ) and lower output impedance. So, FET can act as an excellent buffer amplifier but the BJT has low input impedance because its input circuit is forward biased.

69. (d)

JFET has main drawback of having small gain bandwidth product due to the junction capacitive effects and produce more signal distortion except for small signal operation.

71. (c)

The "drain characteristic" or the "output characteristic" is shown below. Here, in the pinch-off region, as V_{DS} is increased then I_D almost remains constant.

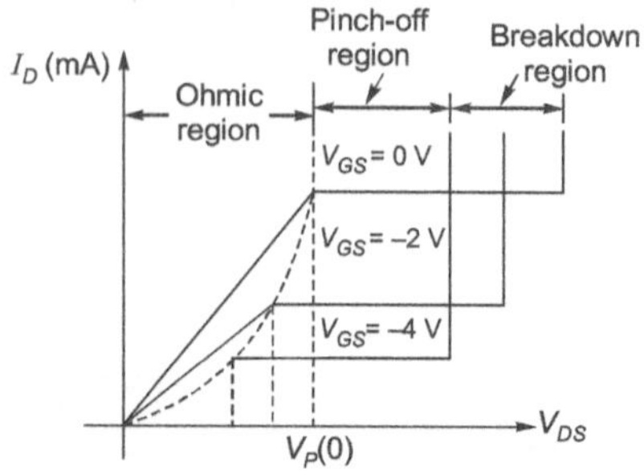

72. (a)

Given, $\Delta V_{GS} = 4 - 3.9 = 0.1$ V

$\Delta I_D = 1.6 - 1.3 = 0.3$ mA

Therefore, transductance,

$$g_m = \frac{\Delta I_D}{\Delta V_{GS}} = \frac{0.3 \times 10^{-3}}{0.1} = 3\text{ m}\Omega$$

73. (c)

Given, $V_{GS(\text{off})} = -6\text{ V} = V_P$

$$I_D = I_{DSS}\left[1 - \frac{V_{GS}}{V_{GS(\text{off})}}\right]^2$$

or, $$4 = 8\left[1 + \frac{V_{GS}}{6}\right]^2$$

or, $$1 + \frac{V_{GS}}{6} = \sqrt{\frac{4}{8}} = \frac{1}{\sqrt{2}} \text{ or } V_{GS} = -1.76\text{ V}$$

75. (d)

- Since FET does not suffer from minority carrier storage effects, it has higher switching speeds and cut-off frequencies.`
- The performance of FET is relatively unaffected by ambient temperature changes compared to BJT.
- BJTs are cheaper to produce.
- FETs are much easier to fabricate and are particularly suitable for ICs because they occupy less space than BJTs.

Hence, statements 1 and 4 are only correct.

77. (c)

The transfer characteristic, giving the relationship between I_D and V_{GS}, can be approximated by the parabola

$$I_D = I_{DSS}\left(1 - \frac{V_{GS}}{V_P}\right)^2$$

78. (d)

The transfer characteristic of an N-MOSFET indicating V_T is shown below.

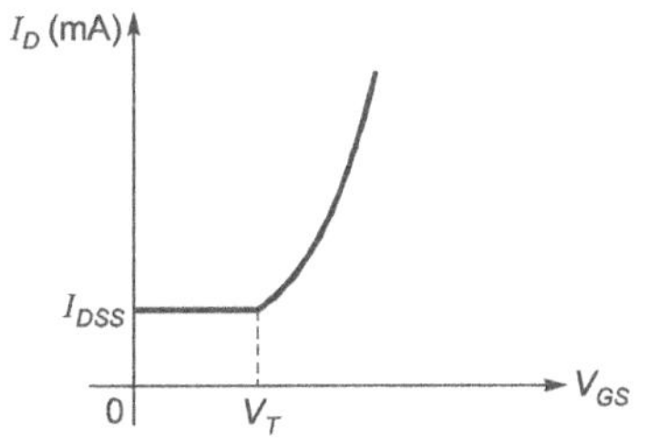

82. (c)

Device	Gate leakage Current	Drain resistance	Switching speed	Input resistance
MOSFET	10^{-12} A	1 to 50 kΩ	Larger	10^{10} to 10^{15} Ω
JFET	10^{-9} A	0.1 to 1 MΩ	Smaller	10^{8} Ω

83. (d)

- Special digital CMOS circuits are available which involve near-zero power dissipation and very low voltage and current requirements. This makes them most suitable for portable systems.
- Since JFET is voltage controlled device, it is used as a voltage variable resistor in operational amplifiers and tone controls.
- Due to zero offset voltage and its symmetrical nature, MOSFET is useful in analog signal switching.
- UJT is used as a relaxation oscillator for generating sawtooth waveforms.

85. (c)

$$V_P = \frac{2I_{DSS}}{g_{mo}} = \frac{2 \times 10}{5} = 4 \text{ volt}$$

and, $$I_D = I_{DSS}\left(1 - \frac{V_{GS}}{V_P}\right)^2 = 10\left(1 - \frac{0.5}{4}\right)^2$$

$$= 10\,(1 - 0.125)^2$$

$$= 10 \times 0.765 = 7.65\text{ mA}$$

86. (a)

$$I_D = I_{DSS}\left(1 - \frac{V_{GS}}{V_P}\right)^2$$

and $$g_m = \frac{\partial I_D}{\partial V_{GS}} = 2I_{DSS}\left(1 - \frac{V_{GS}}{V_P}\right)\left(-\frac{1}{V_P}\right)$$

$$\therefore \quad g_{m(\max)} = g_{mo}\big|_{V_{GS} = 0\text{ V}}$$

$$= \frac{2I_{DSS}}{|V_P|} = \frac{2 \times 1 \times 10^{-3}}{5} = 0.4\text{ mS}$$

87. (b)

n-channel FETs are superior to p-chanel FETs because n-channel FETs are faster than p-channel FETs since $\mu_n > \mu_P$.

88. (b)

$$g_m = \left.\frac{\Delta I_{DS}}{\Delta V_{GS}}\right|_{V_{DS} = \text{const.}}$$

$$\therefore \quad g_m \propto \frac{1}{\Delta V_{GS}}$$

89. (c)

Due to high input impedance of JFET, it acts as a voltage controlled current source i.e. $I_G \approx 0$ A.

93. (c)

$$\text{Gain} = \frac{g_m r_d}{\left(1+\frac{r_d}{R}\right)} = \frac{100}{2} = 50$$

94. (c)

For a bridge rectifier circuit,

$$\%\eta = \frac{P_{dc}}{P_{ac}} \times 100 = \frac{I_{dc}^2 R_L}{I_{ac}^2 (R_L + 2R_f)}$$

For ideal diode,

$R_f = 0$

$$\%\eta = \frac{I_{dc}^2}{I_{ac}^2} \times 100 = \frac{(2I_m/\pi)^2}{(I_m/\sqrt{2})^2} \times 100 = 81.06\%$$

and hence maximum efficiency, $\eta = 81.06\%$.

96. (b)

DC power delivered to the load is

$$P_{DC} = \frac{V_{DC}^2}{R_L}$$

$$\therefore \quad V_{DC} = \sqrt{P_{DC} \times R_L} = \sqrt{50 \times 200} = 100 \text{ V}$$

The ripple factor,

$$\text{r.f.} = \frac{V_{AC}}{V_{DC}}$$

or, $$0.01 = \frac{V_{AC}}{100} \text{ or } V_{AC} = 1 \text{ volt}$$

Therefore, the AC ripple voltage across the load,

$$V_{AC} = 1 \text{ volt}$$

97. (a)

Average value of load current,

$$I_{DC} = \frac{V_{DC}}{R_L} = \frac{24}{500} = 48 \text{ mA}$$

Maximum value of load current,

$$I_m = \pi \times I_{DC} = \pi \times 48 \text{ mA} = 150.8 \text{ mA}$$

Therefore, maximum AC voltage required at the input is

$$V_m = I_m \times (r_f + R_L) = 150.8 \times 10^{-3} \times 550 = 82.94 \text{ V}$$

98. (d)

The ripple factor for an inductor filter is

$$\text{r.f.} = \frac{R_L}{3\sqrt{2}\omega_0 L}$$

$$\therefore \quad 0.04 = \frac{100}{3\sqrt{2}(2\pi \times 60 \times L)} = \frac{0.0625}{L}$$

or, $$L = \frac{0.0625}{0.04} = 1.56 \text{ H}$$

101. (d)

$$V_{DC} = 2\frac{V_m}{\pi} \text{ (for a full-wave rectifier)}$$

$$V_{DC} = \frac{V_m}{\pi} \text{ (for a half-wave rectifier)}$$

103. (c)

No filtering action is provided by bridge arrangement. Hence, reason is a false statement.

104. (c)

- During the negative half cycle of input voltage V_i, diode is forward biased and acts as short-circuit. Hence capacitor changes to V_C volts.

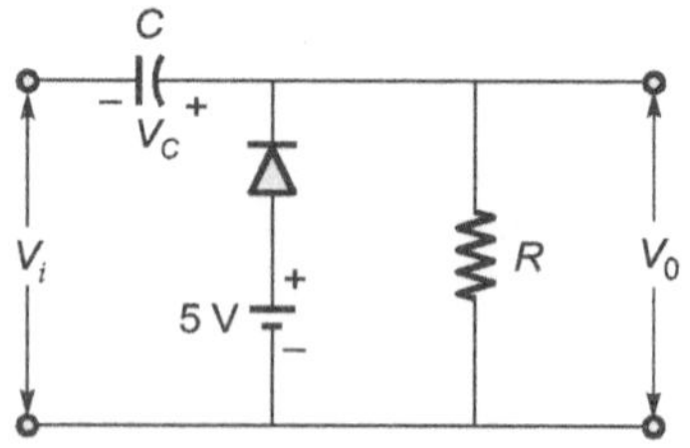

Using KVL,

$$-5 + V_C - 15 = 0$$

or, $V_C = 20$ volts

Hence, $V_0 = 5$ volts

- During the positive half cycle of input voltage V_i, diode is reverse biased and acts as open-circuit.

Applying KVL, we have

$$15 = -20 + V_0$$

or, $V_0 = 35$ volts

The waveforms for V_i and V_0 will be as shown below:

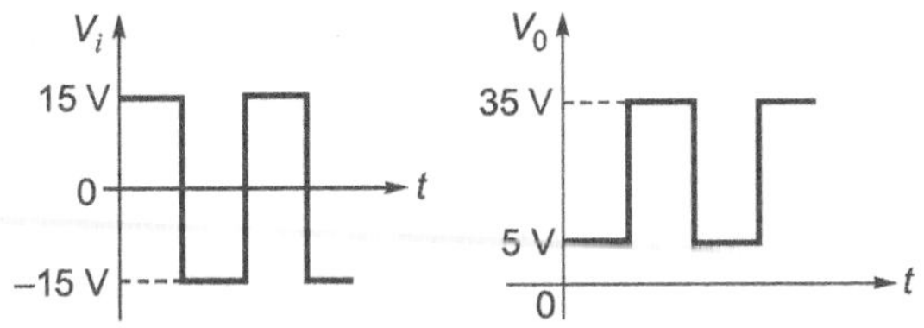

Thus, the input swing of 30 V is equal to the output swing of 30 V.

105. (b)

- **When V_i is positive:**
  Diode is ON and acts as short circuit.
  Output voltage,
  $$V_0 = 0 \text{ V}$$
  The capacitor charges to 12 volts as shown below and acts like a battery.

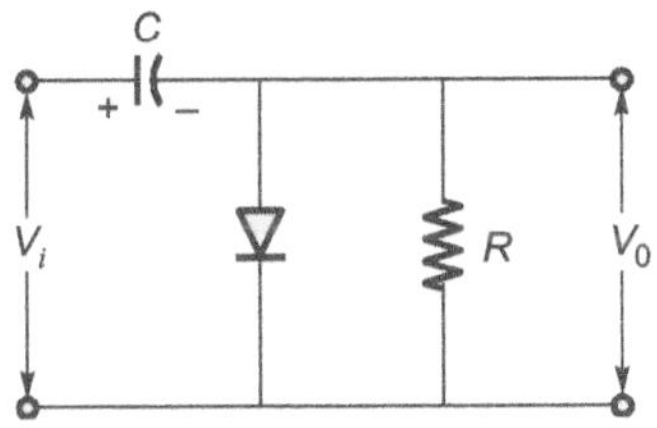

- **When V_i is negative:**
  Diode is OFF and acts as open circuit.
  Output voltage,
  $$V_0 = -12 - 12 = -24 \text{ Volt}$$

This gives negatively clamped voltage and the resultant output waveform will be as shown in figure below.

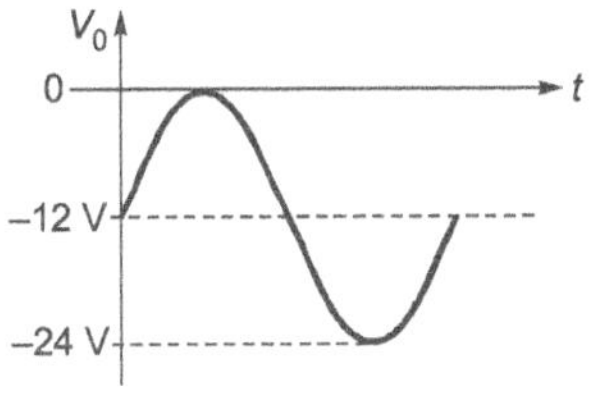

107. (a)

- **When V_i is positive:**
  Diode is OFF so that, $V_0 = -V_R$
- **When V_i is negative:**

$\Rightarrow$ $|V_i| < |V_R|$, Diode is OFF so that, $V_0 = -V_R$.

$\Rightarrow$ $|V_i| \geq |V_R|$, Diode is ON so that, $V_0 = V_i$.

Hence, output waveform will be as shown below.

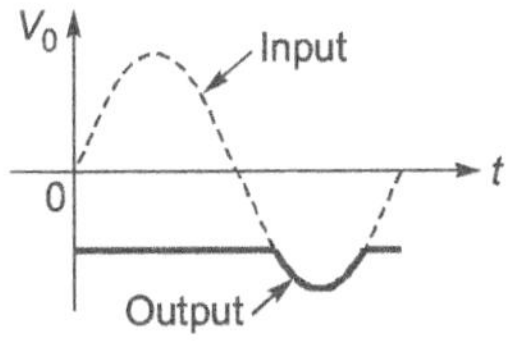

108. (c)

A clamper circuit requires a capacitor not a clipping circuit. Hence, reason is not a correct statement.

109. (b)

- In a full-wave rectifier each diode conducts during one half-cycle. Hence, statement-2 is not correct.
- Ripple factor is the ratio of rms value of AC component to the DC component in the output.

$$\text{r.f.} = \frac{V_{r,\text{rms}}}{V_{DC}}$$

Hence, statement - 3 is also not correct.

110. (d)

Given, minimum Zener current = $I_{Z(\min)}$ = 5 mA when the input voltage is minimum.

Here, the input voltage varies between 8 V to 12 V.

(Since V_i = 10 V ± 20%).

$$\therefore \quad V_{i(\min)} = 8 \text{ V}$$

Given, I_L = 20 mA and voltage across load,

$$V_0 = 5 \text{ volts}$$

$$\therefore \quad R_L = \frac{V_0}{I_L} = \frac{5}{20 \times 10^{-3}} = 250\ \Omega$$

$\therefore$ Limiting resistance,

$$R = \frac{V_{i(\min)} - V_0}{I_{Z(\min)} + I_L} = \frac{8-5}{(5+20) \times 10^{-3}} = 120\ \Omega$$

111. (a)

Given, V_0 = 5 V, I_L = 10 mA

Load resistance,

$$R_L = \frac{V_0}{I_L} = \frac{5}{10 \times 10^{-3}} = 500\ \Omega$$

Maximum Zener current,

$$I_{Z(\max)} = \frac{400 \times 10^{-3}}{5} = 80 \text{ mA}$$

(Since $P_{Z(\max)}$ = 400 mW)

The minimum input voltage required will be when $I_z = 0$.

Under this condition,

$$I = I_L = 10 \text{ mA}$$

Minimum input voltage,

$$V_{i(\min)} = V_0 + IR \qquad \ldots(i)$$

Given, V_i = 10 V ± 2 V

$\therefore \quad V_{i(min)} = 10 - 2 = 8 \text{ V} \quad \text{...}(ii)$

and $\quad V_{i(max)} = 10 + 2 = 12 \text{ V}$

From equations (*i*) and (*ii*), we have

$$8 = 5 + (10 \times 10^{-3})\, R$$

or, $$R_{max} = \frac{3}{10 \times 10^{-3}} = 300\ \Omega$$

Now, maximum input voltage,

$$V_{i(max)} = 5 + [(80 + 10) \times 10^{-3}]\, R$$

or, $$12 = 5 + (90 \times 10^{-3})\, R$$

or, $$R_{min} = \frac{7}{90 \times 10^{-3}} = 77.77\ \Omega$$

Hence, $77.77\ \Omega < R < 300\ \Omega$

Thus, one of the possible value of out of the given options is $R = 266\ \Omega$.

112. (d)

The Zener diode V_{Z2} is reverse biased and V_{Z1} is forward biased. As both Zener diodes V_{Z1} and V_{Z2} are connected in series, the reverse saturation current 40 μA of V_{Z2} will flow clockwise in the circuit as 50 V reverse bias appears across the V_{Z2} diode.

114. (b)

For, $\quad V_i < 25$ volts,

Diode D_1 will be OFF while diode D_2 will be ON.

$$\therefore \quad V_0 = 75 - \left(\frac{75 - 25}{10\ \text{k}\Omega}\right)(5\ \text{k}\Omega)$$

$$= 75 - 25 = 50 \text{ volts}$$

115. (a)

The PMMC ammeter will read average value of current. The current through the PMMC ammeter will flow only for positive half cycle of input voltage while current through it will be zero for the negative half cycle.

Hence, $$I_{avg} = I_{DC} = \frac{I_m}{\pi}$$

$$I_m = \frac{V_m}{10\text{ k}} = \frac{4}{10 \times 10^3} = 0.4 \text{ mA}$$

$$\therefore \quad I_{avg} = I_{DC} = \frac{0.4}{\pi} \text{ mA} = \text{Meter reading}$$

116. (d)

Diode D_1 is reverse biased and diode D_2 is forward biased, but as both the diodes form a close loop, the reverse saturation current flows through both the diodes.

$$I_f = I_0 = I_0(e^{V_2/V_T} - 1)$$

or, $$e^{V_2/V_T} = 2 \text{ or } \frac{V_2}{V_T} = ln\, 2$$

or, $$V_2 = V_T\, ln\, 2 = (25 \text{ mV}) \times ln\, 2$$
$$= 25 \times 10^{-3} \times 0.693$$
$$= 17.33 \text{ mV} = 0.01733 \text{ V}$$

$\therefore$ Voltage drop across D_2 is

$$V_2 \approx 0.0173 \text{ volt}$$

and voltage drop across D_1 is

$$V_1 \approx 5 - 0.0173 \approx 4.983 \text{ volts}$$

117. (b)

$$I_s = \left(\frac{10 - 6}{50}\right) = \frac{4}{50} = 80 \text{ mA}$$

The maximum load current = 80 − 5 = 75 mA

Hence, minimum load resistance

$$= \frac{6}{75 \times 10^{-3}} = 80\ \Omega$$

118. (b)

If the diode D_1 conducts, the current I must flow as indicated in figure below.

Here, $$I = \left(\frac{5 - 0.7}{5\ \text{k}\Omega}\right) - \left(\frac{5 + 0.7 - 0.7}{2.5\ \text{k}\Omega}\right)$$
$$= 0.86 - 2 = -1.14 \text{ mA}$$

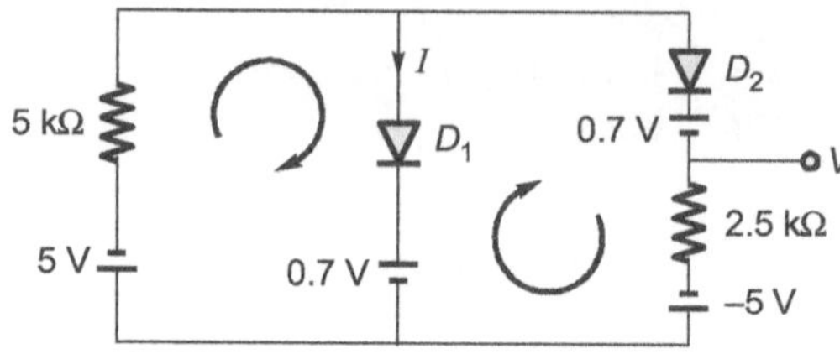

Since, I is negative, therefore D_1 does not conduct.

$$\therefore \quad V_0 = -5 + \frac{(10 - 0.7) \times 2.5\text{ k}}{7.5\text{ k}}$$

$$= -5 + \frac{9.3 \times 2.5\text{ k}}{7.5\text{ k}} = -1.9 \text{ V}$$

119. (b)

$$\%\ \text{regulation} = \left(\frac{V_{NL} - V_{FL}}{V_{FL}}\right) \times 100$$

$$= \left(\frac{30 - 25}{25}\right) \times 100 = 20\%$$

$$\text{Output resistance} = \frac{V_{FL}}{I_{FL}} = \frac{25}{1} = 25\ \Omega$$

120. (d)

Higher order active filters are used for variable roll-off rate.

122. (a)

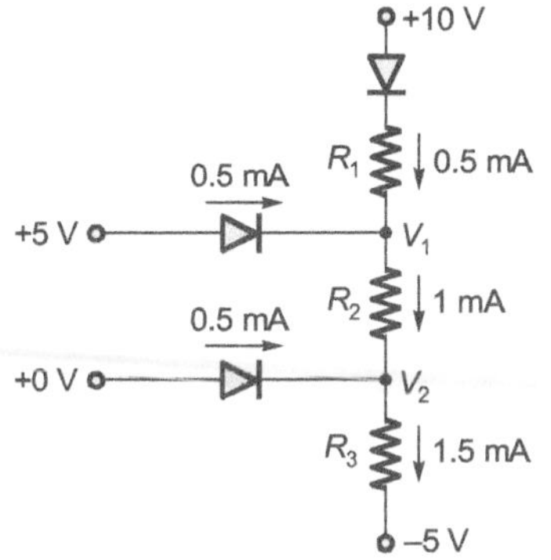

$V_1 = 5 - 0.6 = 4.4$ Volt

$$R_1 = \left(\frac{10 - 0.6 - 4.4}{0.5 \times 10^{-3}}\right) = 10\ \text{k}\Omega$$

$$R_2 = \left(\frac{V_1 - V_2}{1 \times 10^{-3}}\right) = \left(\frac{4.4 - (-0.6)}{1 \times 10^{-3}}\right) = 5\ \text{k}\Omega$$

$$R_3 = \left(\frac{V_2 - (-5)}{1.5 \times 10^{-3}}\right) = \left(\frac{-0.6 + 5}{1.5 \times 10^{-3}}\right) = 2.93\ \text{k}\Omega$$

($\because V_2 = 4.4 - 5 = -0.6$ volt)

123. (a)

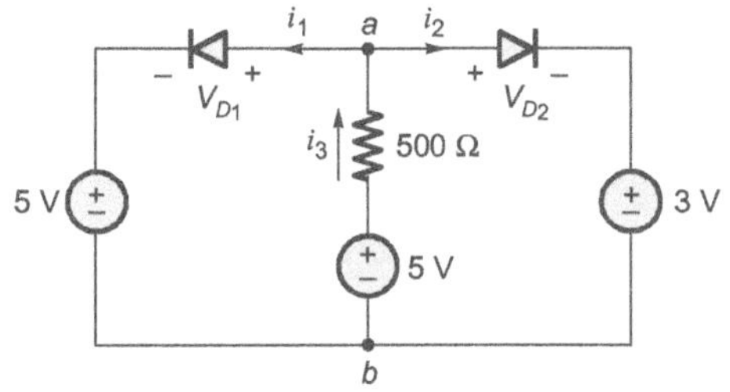

For $i_3 > 0$, $V_{ab} < 5$ volts and $V_{D1} \le 0$

Hence, $i_1 = 0$

and $i_2 = \dfrac{5-3}{500} = 4$ mA

125. (d)

The output voltage (V_{DC}) is

$$V_{DC} = 2\left(\frac{V_m}{\pi} \cdot \frac{R_L}{(R_L + R_f)} - V_\gamma\right)$$

$$= 2\left(\frac{5}{\pi} \cdot \frac{100}{(100+10)} - 0.6\right)$$

$= 1.69$ volt

126. (d)

Rectifier Circuit	PIV
Half wave	V_m
Centre tapped full-wave	$2V_m$
Bridge type full-wave	V_m

127. (b)

Current gain of CE configuration is

$$\beta = \left(\frac{\alpha}{1-\alpha}\right) = \left(\frac{0.97}{1-0.97}\right) = 32.33$$

128. (d)

Given, $I_B = 100$ mA and $I_C = 2$ mA (old values)

$$\therefore \quad \beta = \frac{I_C}{I_B} = \frac{2 \times 10^{-3}}{100 \times 10^{-6}} = 20$$

Change in values of I_B and I_C are

$\Delta I_B = 25\ \mu$A and $\Delta I_C = 0.6$ mA

$\therefore$ New values of the currents are:

$I'_B = 100 + 25 = 125\ \mu$A

and $I'_C = 2 + 0.6 = 2.6$ mA

$\therefore$ New values of β is

$$\beta = \frac{I'_C}{I'_B} = \frac{2.6 \times 10^{-3}}{125 \times 10^{-6}} = 20.8$$

Hence, change in the value of β is

$\Delta\beta = \beta_2 - \beta_1 = 20.8 - 20$

$= 0.8 = \dfrac{4}{5}$

129. (a)

Given, $\alpha = 0.98$,

$I_{CBO} = I_{CO} = 5\ \mu$A

and $I_B = 100$ mA

The collector current,

$$I_C = \beta I_B + (1 + \beta) I_{CO}$$

$$= \left(\frac{\alpha}{1-\alpha}\right) I_B + \frac{I_{CO}}{(1-\alpha)}$$

$$= \frac{0.98 \times 100 \times 10^{-6}}{(1-0.98)} + \frac{5 \times 10^{-6}}{(1-0.98)}$$

$= 5.15$ mA

130. (b)

Let the transistor be operating under active region.

$$I_B = \frac{V_{BB} - V_{BE}}{R_B} = \frac{4 - 0.7}{200 \times 10^3} = 16.5\ \mu A$$

(V_{BE} = 0.7 (Si Transistor) = default value)

The collector current is

$$I_C = \beta I_B = 200 \times 16.5 \times 10^{-6} = 3.3\text{ mA}$$

The emitter current is

$$I_E = I_C + I_B$$
$$= 3.3 \times 10^{-3} + 16.5 \times 10^{-6} = 3.3165\text{ mA}$$
$$\therefore\ V_{CE} = V_{CC} - I_C R_C$$
$$= 10 - 3.3 \times 10^{-3} \times 2 \times 10^3 = 3.4\text{ volt}$$

131. (c)

Given, $I_{CBO} = 0.2\ \mu A$ and $I_{CEO} = 18\ \mu A$

$I_B = 30$ mA

Now, $I_{CEO} = (1 + \beta)\, I_{CBO}$

or, $$\beta = \left(\frac{I_{CEO}}{I_{CBO}} - 1\right)$$

or, $$\beta = \left(\frac{18}{0.2} - 1\right) = 89$$

Now, $$I_{CEO} = (1 + \beta) I_{CBO} = \frac{I_{CBO}}{1 - \alpha}$$

or, $$\alpha = \left(1 - \frac{I_{CBO}}{I_{CEO}}\right)$$

$$= \left(1 - \frac{0.2 \times 10^{-6}}{18 \times 10^{-6}}\right) = 0.988$$

Hence, $\alpha = 0.988$ and $\beta = 89$

132. (a)

Let the transistor be in active mode.

KVL for the base circuit gives,

$$I_B = \left(\frac{2.7 - 0.7}{100}\right) = 0.02\text{ mA}$$

$$\therefore\ I_C = \beta I_B = 100 \times 0.02\text{ mA} = 2\text{ mA}$$

KVL at the collector node yields,

$$\frac{V_C - 10}{2\text{ k}} + \frac{V_C}{2\text{ k}} + I_C = 0$$

or, $2V_C - 10 + (2\text{ k})I_C = 0$

or, $2V_C = 10 - (2\text{ k})(2\text{ m}) = 6$

or, $V_C = 3$ volt

134. (c)

Due to early effect, I_B is reduced while I_C is increased. Hence, $\alpha = \frac{I_C}{I_E}$ is increased by a smaller value while $\beta = \frac{I_C}{I_B}$ is increased by a larger value. Thus, reason is not a correct statement.

135. (b)

In Saturation region,

$$I_C \le I_{C\,max}$$

i.e. $$I_C \le \beta I_B$$

139. (a)

The output characteristic of a CB transistor configuration is shown in figure below.

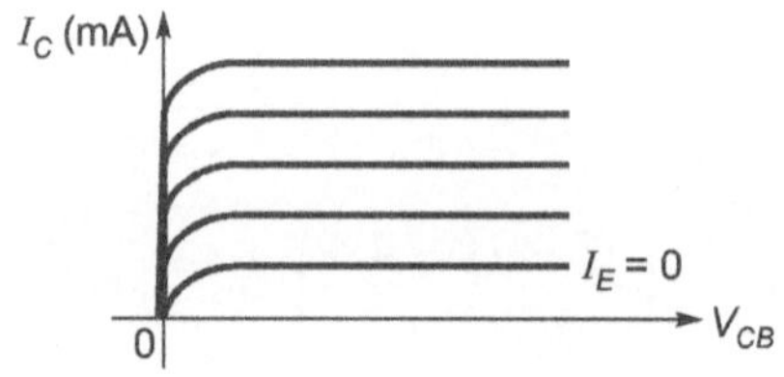

It is clear that for a constant value of I_E, I_C is independent of V_{CB} and the curves are parallel to the axis of V_{CB}.

Hence, a CB transistor acts as an ideal current source.

141. (c)

A common collector amplifier is also known as Emitter follower because the output voltage (emitter voltage) will follow the input voltage.

142. (a)

- In saturation region, both emitter and collector junctions are forward biased.
- The phase shift introduced to the input signal in CC configuration is 0° while in CE configuration is 180°.

Hence, statements 2 and 4 are not correct.

144. (c)

The voltage gain of emitter follower (or common collector configuration) is less than unity i.e. a low value.

Hence, reason is not true.

146. (d)

- For transistor operation, with normal bias, collector junction is reverse biased due to which collector junction is of high resistance.

- *p-n-p* and *n-p-n* transistors cannot be interchanged with each other.
- CC configuration has highest input resistance due to which it is used as buffer amplifier for impedance matching.
- In CE transistor,

$$V_{CE} = V_{CC} - I_C R_C$$

When $I_B = 0$,

I_C is also equal to zero so that

$$V_{CE} = V_{CC}$$

- Breakdown voltage, $V_{Br} \propto \dfrac{1}{\text{(Doping Conc.)}}$

Since emitter is heavily doped while collector is moderately doped, therefore, emitter junction breakdown is due to "Zener effect" while collector junction breakdown is due to "Avalanche effect."

Thus, statements 1 and 2 are not correct.

148. (a)

For $V_{CE} = 4.5$ V,

$$I_C = \frac{5-4.5}{1\text{k}} = 0.5 \text{ mA}$$

$$I_B = \frac{0.5}{25} = 0.02 \text{ mA}$$

$R_1 = 15$ kΩ, $R_2 = 100$ kΩ

$$I_{R2} = \frac{0.7-(-5)}{100\text{ k}} = 0.057 \text{ mA}$$

$$I_{R1} = I_{R2} + I_B = 0.057 + 0.02 = 0.077 \text{ mA}$$

$$V_1 = I_{R1} R_1 + V_{BE} = (0.077) \times 15 + 0.7 = 1.855 \text{ V} \approx 1.86 \text{ Volt}$$

For, $V_{CE} = 1.0$ V

$$I_C = \frac{5-1}{1\text{k}} = 4 \text{ mA}$$

$$I_B = \frac{4}{25} = 0.16 \text{ mA}$$

$$I_{R2} = 0.057 \text{ mA}$$

$$I_{R1} = 0.057 + 0.16 = 0.217 \text{ mA}$$

$$V_1 = (0.217) \times 15 + 0.7 = 3.96 \text{ V}$$

So, $1.86 \le V_1 \le 3.96$ V

149. (b)

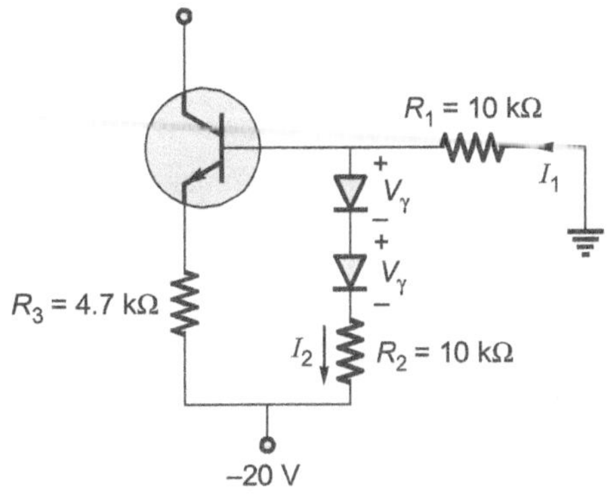

Here, $I_C = I_E,\ I_1 = I_2 = I_2$

$$I_2 = \frac{0 - 2V_\gamma - V^-}{R_1 + R_2}$$

or, $$I_2 = \frac{-2V_\gamma + 20}{R_1 + R_2} \quad ...(i)$$

Applying KVL, we have:

$$2V_\gamma + I_2 R_2 = V_{BE} + I_C R_3 \quad ...(ii)$$

Substituting I_2 from equation (*i*) into equation (*ii*), we get:

$$2V_\gamma + \frac{R_2}{R_1 + R_2}(-2V_\gamma + 20) = V_{BE} + I_C R_3$$

Now, $V_{BE} = V_\gamma$,

$R_1 = R_2 = 10$ kΩ

$$\therefore\ 2V_\gamma + \frac{1}{2}(-2V_\gamma + 20) = V_\gamma + I_C R_3$$

or, $10 = I_C R_3$

or, $$I_C = \frac{10}{R_3} = \frac{10}{4.3\text{ k}} = 2.13 \text{ mA}$$

∴ Collector current, $I_C = 2.13$ mA

150. (c)

$$r_0 = \frac{V_A}{I_{CQ}}$$

or, $$I_{CQ} = \frac{V_A}{r_0} = \frac{75}{200\text{ k}} = 0.375 \text{ mA}$$

151. (d)

Base is provided with smallest area to reduce transit type or journey time for the majority carriers while travelling from emitter to collector. Emitter is heavily doped to inject its majority carrier into base.

152. (d)

I_C increases because base width reduces due to early effect.

153. (d)

A common-emitter amplifier has very high voltage gain.
($A_V = -116.6$) and high current gain ($A_I = -44.5$) due to which it is most widely used.

154. (c)

A common-base configuration has highest voltage gain, lowest current gain, input resistance and highest output resistance compared to a common-emitter and common collector configuration.

156. (b)

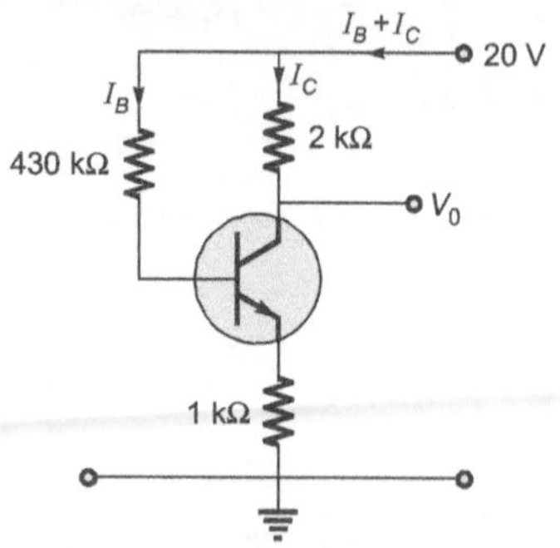

Capacitors will be open circuited.

Now, $20 = (430\text{ k})I_B + 0.7 + (I_B + I_C)(1\text{ k})$

and $I_C = 50\, I_B$

Thus, $19.3 = (430\text{ k})I_B + (51 I_B)(1\text{ k})$

or, $I_B = 40\ \mu\text{A}$

$\therefore$ $I_C = 50 \times 40\ \mu\text{A} = 2\text{ mA}$

Now, $\dfrac{20 - V_0}{2\text{k}} = I_C$

or, $V_0 = 16$ volts

158. (a)

C-B Junction	E-B Junction	Transistor operation
RB	RB	Cut-off
RB	FB	Active
FB	FB	Saturation

162. (b)

The collector resistance is:

$$R_C = \frac{V_{CC} - V_{CEQ}}{I_{CQ}} = \frac{10 - 6}{1 \times 10^{-3}} = 4\text{ k}\Omega$$

Base current is,

$$I_B = \frac{I_{CQ}}{\beta} = \frac{1 \times 10^{-3}}{100} = 10\ \mu\text{A}$$

The base resistance,

$$R_B = \frac{V_{CC} - V_{BE}(\text{on})}{I_{BQ}} = \frac{10 - 0.7}{10 \times 10^{-6}}$$

$$= 0.93\text{ M}\Omega$$

163. (c)

Let the transistor be operating under active region.
Applying KVL, we have:

$$V_{BB} = I_B R_B + V_{BE}(\text{on}) + I_E R_E$$

We know that,

$$I_E = I_B + I_C = (1 + \beta) I_B$$

The base current,

$$I_B = \frac{V_{BB} - V_{BE}(\text{on})}{R_B + (1+\beta)R_E}$$

$$= \frac{5 - 0.7}{20 \times 10^3 + 101 \times 600}$$

$$= 53.34\ \mu\text{A}$$

Therefore, $I_C = \beta I_B = 100 \times 53.34 \times 10^{-6}$

$= 5.334\text{ mA}$

$\therefore$ $I_E = I_C + I_B$

$= (5.334 \times 10^{-3} + 53.34 \times 10^{-6})$

$= 5.38\text{ mA}$

Now, $V_{CE} = V_{CC} - I_C R_C - I_E R_E$

$= 10 - 5.334 \times 10^{-3} \times 400 - 5.38 \times 10^{-3} \times 600$

$= 4.63$ volt

Since $0.1\text{ V} < V_{CE} < V_{CC}$, therefore transistor is working under active region with $V_{CEQ} = 4.63$ volt.

164. (a)

For a collector-to-base CE amplifier, stability factor is given by

$$S = \frac{1+\beta}{1+\beta\left(\dfrac{R_C}{R_C + R_B}\right)}$$

Here, $R_B = \dfrac{V_{CEQ}}{I_B} = \dfrac{8}{0.25 \times 10^{-3}} = 32\text{ k}\Omega$

$$\therefore\ S = \frac{101}{1 + 100\left(\dfrac{0.25}{32 + 0.25}\right)} = 56.9 \approx 57$$

165. (c)

A self-bias circuit is used to establish a stable operating point in the self-biasing configuration. If I_C tends to increase, say, because I_{CO} has risen as a result of an elevated temperature, the current in R_e increases. As a consequence of the increase in voltage drop across R_e, the base current is decreased. Hence, I_C will increase less than it would have been no self-biasing resistor R_e.

166. (d)

For better stability of the circuit, stability factor S must be as small as possible.

169. (d)

Using KVL, we have:

$$-5 + R_C(I_B + I_C) + R_B I_B + 0.7 = 0$$

Since, $V_{CE} = V_C = 2$ V, therefore transistor is in active mode.

$\therefore$ $I_C = \beta I_B$

So, $-5 + I_B[(\beta + 1)R_C + R_B] + 0.7 = 0$

or, $$I_B = \frac{4.3}{51R_C + R_B} = \frac{4.3}{51(R_C + 1)} \quad ...(i)$$

Now, $V_C = 5 - R_C(I_B + I_C)$

$= 5 - (R_C)\,51\, I_B$

or, $$V_C = 2 = 5 - R_C\frac{4.3}{(R_C + 1)}$$

(Using equation (i))

or, $R_C = 2.3$ kΩ

170. (c)

Applying KCL at the collector of Q_3 gives,

$$I_0 = I_{C3} + I_{B3} + I_{B2}$$
$$= I_{C3} + 2I_{B3}$$
$$= I_{C_3}\left(1 + \frac{2}{\beta}\right) = I_{C_3}\left(\frac{\beta + 2}{\beta}\right)$$

or, $$I_{C3} = I_0\left(\frac{\beta}{\beta + 2}\right) = I_{C2} = I_{E1}$$

Also, $$I_{C1} = I_{E_1}\left(\frac{\beta}{\beta + 1}\right)$$

Hence, $$I_{C1} = I_0\left(\frac{\beta}{\beta + 2}\right)\left(\frac{\beta}{\beta + 1}\right)$$

Therefore, $$S'' = \frac{\partial I_{C_1}}{\partial \beta} = \frac{\partial}{\partial \beta}\left[I_0\left(\frac{\beta^2}{(\beta + 1)(\beta + 2)}\right)\right]$$

or, $$S'' = I_0\left(\frac{\beta(3\beta + 4)}{(\beta + 1)^2(\beta + 2)^2}\right)$$

171. (b)

The circuit for the "diode compensation for I''_{CO} is shown below.

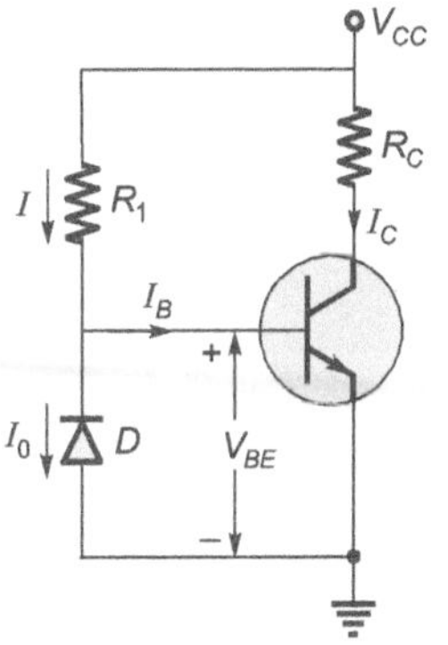

Here, if the diode and the transistor are of the same type and material, the reverse saturation current I_0 of the diode will increase with temperature at the same rate as the transistor current I_{CO}.

Here, $$I = \left(\frac{V_{CC} - V_{BE}}{R_1}\right) \approx \frac{V_{CC}}{R_1} = \text{Constant}$$

Since the diode is reverse biased by $V_{BE} = 0.2$ V (for Ge device), the current through D is I_0 and $I_B = I - I_0$.

Now, $I_C = \beta I_B + (1 + \beta)I_{CO}$

$= \beta I - \beta I_0 + (1 + \beta)I_{CO}$

Thus, if $\beta >> 1$ and if I_0 of D and I_{CO} of transistor track each other over the desired temperature range, then I_C remains essentially constant so that the circuit becomes insensitive toe variations in I_{CO} with changes in temperature.

174. (b)

Both transistor are in the forward active region.

Here, $I_{C1} + I_{B1} + I_{B2} = 25$ μA

Since the transistors are identical and have the same V_{BE}, therefore

$$I_{C2} = I_{C1},\ I_{B1} = I_{B2}$$

$\therefore$ $I_{C1} + 2I_{B1} = (2 + \beta)I_{B1} = 25$ μA

or, $$I_{C2} = \beta I_{B2} = \beta I_{B1} = \left(\frac{\beta}{\beta + 2}\right) \times 25\ \mu\text{A}$$

$$= \frac{25}{27} \times 25 \times 10^{-6}\ \text{A}$$

$= 23.15$ μA » 23.2 μA

176. (c)

The voltage across R_2 is $V_2 = \left(\frac{R_2}{R_1 + R_2}\right)V_{CC}$

Therefore, $V_2 = \frac{4\times10^3}{12\times10^3}\times12 = 4$ V

Now, $V_2 = V_{BE} + I_E R_E$

$\therefore \quad I_E = \frac{V_2 - V_{BE}}{R_E} = \frac{4-0.7}{1\times10^3} = 3.3$ mA

$I_C \approx I_E = 3.3$ mA

$\therefore \quad V_{CE} = V_{CC} - I_C(R_C + R_E)$

$= 12 - 3.3\times10^{-3}\times2\times10^3$

$= 5.4$ volt

$\therefore \quad Q \equiv (5.4\text{ V}, 3.3\text{ mA})$

177. (d)

Stability factor S is given by

$$S = \frac{\partial I_C}{\partial I_{Co}} = \frac{1+\beta}{1+\beta\left(\frac{R_E}{R_E+R_B}\right)}$$

Now,
$$S'' = \left.\frac{\partial I_C}{\partial \beta}\right|_{I_{CO}, V_{BE}=\text{const.}}$$

$$= \frac{I_C}{\beta\left[1+\beta\left(\frac{R_E}{R_E+R_B}\right)\right]}$$

$$= \frac{SI_C}{\beta(1+\beta)}$$

178. (b)

- When emitter is heavily doped, more majority carriers will reach into base due to which I_C will increase and hence α increases.
- Due to early effect, base width is reduced as a result of which I_C is slightly increased. Hence, α of the transistor is increased. Hence, 2 and 4 are correct.

179. (a)

Given, $T_j = 150$ °C, $T_A = 25$ °C,

$P_{C,\max} = 125$ mW

We know that,

$T_j - T_A = \theta P_{C,\max}$

or, $\theta = \left(\frac{T_j - T_A}{P_{C\max}}\right) = \frac{150-25}{125} = 1$°C/mW

180. (d)

Applying KVL, we get:

$V_{EE} = R_B I_B + V_{BE} + R_E I_E$

or, $V_{EE} = (R_B + R_E) I_B + R_E I_C + V_{BE}$ ...(i)

Now, $I_C = \beta I_B + (1+\beta) I_{CO}$

or, $I_B = \left(\frac{I_C - (1+\beta)I_{CO}}{\beta}\right)$ ...(ii)

Putting value of I_B from equation (ii) into (i), we get:

$$V_{EE} - V_{BE} = (R_B + R_E)\left(\frac{I_C - (1+\beta)I_{CO}}{\beta}\right) + R_E I_C$$

or, $$I_C = \frac{\beta(V_{EE} - V_{BE}) + (1+\beta)I_{CO}(R_B + R_E)}{R_B + (1+\beta)R_E}$$

$$\therefore \quad S = S_1 = \frac{\partial I_C}{\partial I_{CO}} = \frac{(1+\beta)(R_B + R_E)}{R_B + (1+\beta)R_E}$$

181. (c)

Since β is very large, we can neglect the base current of the transistors.

Now, $I = \frac{9}{R_1 + R_2 + 18\text{ k}}$,

$V_{B2} = 3 + 0.7 = 3.7$ volt

$V_{B1} = 0.2 + 0.7 = 0.9$ V

Hence, current through 18 kΩ resistor

$= \frac{0.9}{18\text{ k}} = 0.05\text{ mA} = I$

$\therefore \quad R_2 = \frac{3.7-0.9}{0.05\times10^{-3}} = 56$ kΩ

and $R_1 = \frac{9-3.7}{0.05\text{ m}} = 106$ kΩ

182. (a)

The Thevenin's equivalent model is shown below.

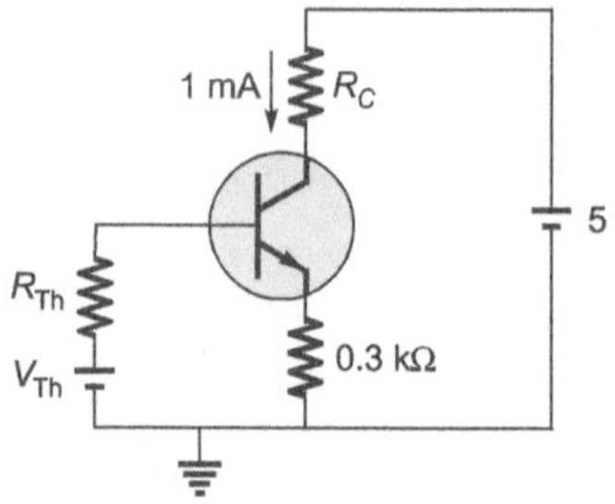

Here, $R_{Th} = \left(\frac{10R_1}{R_1+10}\right)$ kΩ

and $$V_{Th} = \left(\frac{50}{R_1+10}\right) V$$

Applying KVL, we get:

$$\left(\frac{50}{R_1+10}\right) = 0.01\left(\frac{10R_1}{R_1+10}\right) + 0.6 + 0.3$$

$(I_E \approx I_C)$

or, $$\left(\frac{50}{R_1+10}\right) = \left(\frac{0.1R_1}{R_1+10}\right) + 0.9$$

or, $$50 = 0.1\, R_1 + 0.9\, R_1 + 9$$

or, $$R_1 = 50 - 9 = 41\ \text{k}\Omega$$

Now, $R_C \times 1\ \text{mA} = 5 - 2.5 - 0.3 = 2.2$ Volt

Hence, $$R_C = \frac{2.2}{1\,\text{mA}} = 2.2\ \text{k}\Omega$$

Thus, $R_1 = 41\ \text{k}\Omega$ and $R_C = 2.2\ \text{k}\Omega$

184. (b)

As the load resistance is increases, collector current reduces and hence current gain (β) is decreased.

185. (a)

Since the short-circuit current gain h_{fe} of a common-emitter stage is much greater than unity, it is possible to increase the voltage amplification by cascading such stages. Hence, in a cascade, the intermediate transistors should be connected in a common-emitter configuration.

187. (d)

The common-collector transistor configuration is called the emitter follower because its voltage gain is close to unity (low) and hence the emitter follows the input signal.
Hence, assertion is a false statement.

188. (b)

The equivalent circuit of the given cascade amplifier can be drawn as shown below.

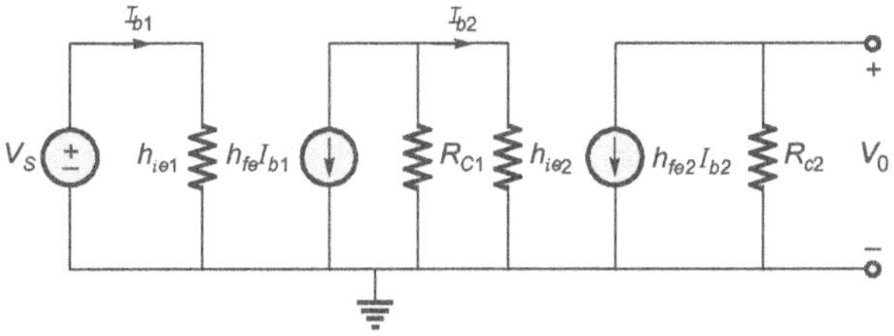

Working forward from the input, we get:

$$I_{b1} = \frac{V_s}{h_{ie1}}$$

and $$I_{b2} = -\left(\frac{R_{C1}}{R_{C1}+h_{ie2}}\right)h_{fe1}I_{b1}$$

$$= -\left(\frac{R_{C1}}{R_{C1}+h_{ie2}}\right)h_{fe1}\frac{V_s}{h_{ie1}}$$

Also, $$V_0 = -R_{C2}\,h_{fe2}\,I_{b2}$$

$$= -R_{C2}h_{fe2}\left(\frac{-R_{C1}}{R_{C1}+h_{ie2}}\right)h_{fe1}\frac{V_s}{h_{ie1}}$$

Substituting values yields,

$$\frac{V_0}{V_s} = 15\cdot 100\left(\frac{5}{5+1}\right)\cdot 100$$

$$= 41666$$

189. (a)

Figure below shows the small-signal circuit (with D.C. voltage source and capacitors short circuited).

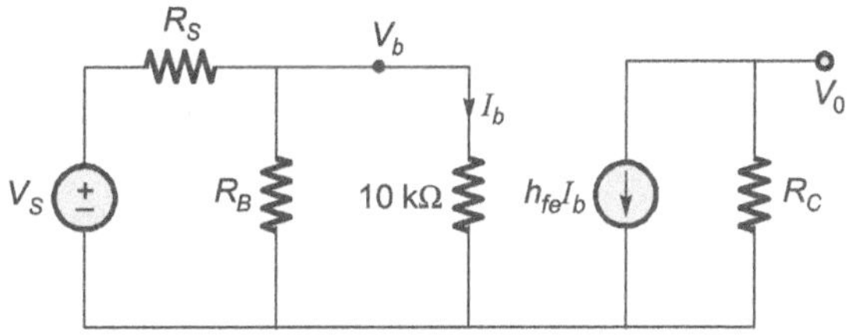

From the above circuit,

$$V_0 = -h_{fe}\, I_b\, R_C$$

$$V_b = \frac{R_B\,||\,h_{ie}}{(R_S + R_B\,||\,h_{ie})} \times V_S$$

$$I_b = \frac{V_b}{h_{ie}}$$

Therefore, $$V_0 = -h_{fe}\cdot\frac{R_C}{h_{ie}}\cdot\frac{R_B\,||\,h_{ie}}{(R_S + R_B\,||\,h_{ie})}\cdot V_S$$

Substituting all values yields,

$$\frac{V_0}{V_S} = -100\cdot\frac{12}{10}\cdot\frac{(10\,||\,10)}{(5+10\,||\,10)} = -60$$

and $$R_i = R_B\,||\,h_{ie} = 5\ \text{k}\Omega$$

190. (b)

The Darlington connection of three or more transistors is usually impractical because leakage current of the first transistor is amplified by the second due to which the overall leakage current may become high.
Reason is individually a correct statement. Hence, both assertion and reason are correct but reason is not the correct explanation of assertion.

191. (d)

The input resistance of emitter follower can be increased by bootstrapping the darlington circuit through the addition of capacitor C_0 between the first collector (C_1) and the second emitter (E_2).

192. (c)

h-parameters are real numbers upto radio frequency only. The smallest value of h-parameter is that of open circuit output admittance (h_0). Hence, statements 2 and 4 are not correct.

193. (a)

Using AC equivalent circuit for the common base configuration, we can find that,

$$Z_i = R_E \,||\, r_e \text{ and } A_V = \frac{R_C}{r_e}$$

Now, $$|I_E| = \frac{V_{EE} - V_{BE}}{R_E} = \frac{8 - 0.6}{4000} = 1.85 \text{ mA}$$

$$r_e = \frac{26 \text{ mV}}{I_E} = \frac{26}{1.85} = 14.05\ \Omega$$

∴ $$Z_i = R_E || r_e = 4000 || 14.05 = 14\ \Omega$$

and $$A_V = \text{voltage gain}$$

$$= \frac{R_C}{r_e} = \frac{3000}{14.05} = 213.5$$

194. (c)

Applying KVL, we have:

$$6 - 1.5\text{ k}\, I_{C2} - V_{EC2} - 3\text{ k}\, I_{C2} + 6 = 0$$

(Since $I_B = 0$, therefore $I_{E2} = I_{C2}$ and $I_{RC1} = I_{C1} = I_{E1}$)

or, $(4.5\text{ k})I_{C2} = 12$ V (Since $V_{CE2} = 0$)

or, $$I_{C2} = \frac{12}{4.5\text{ k}} = 2.67 \text{ mA}$$

Also, $$V_{RE2} = 6 - I_{E2} \times (1.5\text{ k})$$

$$= 6 - 2.67 \times 1.5 = 4 \text{ volt}$$

∴ $$V_{RC1} = V_{RE2} + V_{BE} = 4 + 0.7 = 4.7 \text{ V}$$

∴ $$I_{C1} = \frac{4.7}{1.8\text{ k}} = 2.61 \text{ mA} = I_{E1} = I_{RC1}$$

When, $V_i = 0$,

$$V_{BE1} + R_E I_{E1} = 6$$

or, $$R_E = \frac{6 - V_{BE1}}{I_{E1}} = \frac{6 - 0.7}{2.61 \text{ mA}}$$

$$= \frac{5.3}{2.61 \times 10^{-3}} = 2.03\text{ k}\Omega \approx 2\text{ k}\Omega$$

195. (c)

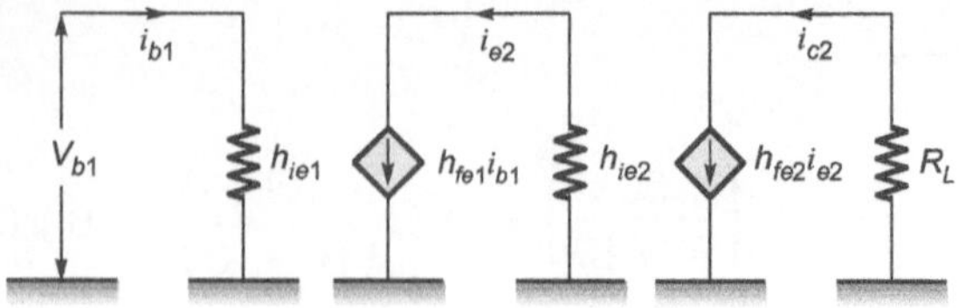

Here, $$i_{c2} = h_{fb2} \cdot i_{e2} \text{ and } h_{ib2} = \frac{h_{ie2}}{1 + h_{fe2}}$$

$$V_{be1} = h_{ie}\, i_{b1}$$

Also, $$g_m = \frac{i_{c2}}{V_{be1}} = \frac{h_{fb2} \cdot i_{e2}}{V_{be1}} = \frac{h_{fb2} \cdot i_{c1}}{V_{be1}}$$

$$= \frac{h_{fb2} \cdot h_{fe1} \cdot i_{b1}}{V_{ie1} i_{b1}} = \frac{h_{fb2} \cdot h_{fe1}}{h_{ie1}}$$

$$= g_{m1} \cdot h_{fb2} \simeq g_{m1} \quad (\because h_{fb1} \approx 2)$$

Hence, $$g_m \approx g_{m1}$$

196. (b)

$$g_m = \frac{I_C \text{ (mA)}}{V_T \text{ (mA)}} = \frac{0.25}{25} = 0.01 \text{ S}$$

$$\beta = 100 = g_m r_\pi$$

or, $$r_\pi = \frac{100}{g_m} = \frac{100}{0.01} = 10\text{ k}\Omega$$

Now, $$r_0 = \frac{r_s + r_\pi}{1 + \beta} = \frac{2\text{ k} + 10\text{ k}}{101} = 1188\ \Omega$$

Desired effective output resistance

$$= 110\ \Omega = R_E || 118.8$$

or, $$110 = \frac{118.8 \times R_E}{118.8 + R_E}$$

or, $$118.8 R_E = 118.8 \times 110 + 110 R_E$$

or, $$(118.8 - 110)R_E = 8.8 R_E = 13068$$

or, $$R_E = 1.485\text{ k}\Omega \approx 1.5\text{ k}\Omega$$

197. (b)

The voltage gain of a multi-stage amplifier is equal to the product of the gains of the individual stages.

201. (d)

$$h_{ie} = r_{bb'} + r_\pi$$

Here, $$r_{bb'} = 0$$

Therefore, $$h_{ie} = r_\pi = 5.2\text{ k}\Omega$$

Now, $$r_\pi = \frac{\beta}{g_m} \text{ and } g_m = \frac{I_C}{V_T}$$

$$\therefore \quad r_\pi = \frac{\beta}{(I_C / V_T)} \text{ or } r_\pi = \frac{\beta V_T}{I_C}$$

$$\therefore \quad I_C = \frac{\beta V_T}{r_\pi} = \frac{h_{fe} V_T}{r_\pi} = \frac{100 \times 26 \times 10^{-3}}{5.2 \times 10^3}$$

$$= 0.5 \text{ mA}$$

202. (c)

$$g_m = \frac{\Delta I_C}{\Delta V} = \frac{3 \text{ mA}}{30 \text{ mV}} = 0.10 \text{ A/V}$$

and $$A_V = \frac{V_0}{V_i} = \frac{(5 \text{ k}) \times (3 \text{ mA})}{30 \times 10^{-3}} = 500$$

203. (a)

$$Y_0 = h_0 - \left(\frac{h_f h_r}{h_i + R_S}\right)$$

$$= 0.05 \times 10^{-3} - \left(\frac{100 \times 0}{2 \times 10^3 + R_S}\right)$$

or, $$Z_0 = \frac{1}{0.05 \times 10^{-3}} = 20 \text{ k}\Omega$$

204. (a)

If R_S is the source resistance, the output resistance of an emitter-follower using the simplified hybrid model would be $\left(\frac{h_{ie} + R_S}{1 + h_{fe}}\right)$.

205. (c)

Current gain,

$$A_i = \frac{h_{fe}}{1 + h_{oe} R_L}$$

$$= \frac{100}{\left(1 + \frac{1}{36 \times 10^3} \times 4 \times 10^3\right)} = \frac{100}{1 + \frac{1}{9}}$$

or, $$A_i = \frac{100 \times 9}{10} = 90$$

206. (b)

Both assertion and reason are correct statements.

The correct reason for assuming small signal excursion about the Q-point is to keep the transistor parameters constant over the signal excursion.

209. (d)

The r_e model for CE amplifier is shown below.

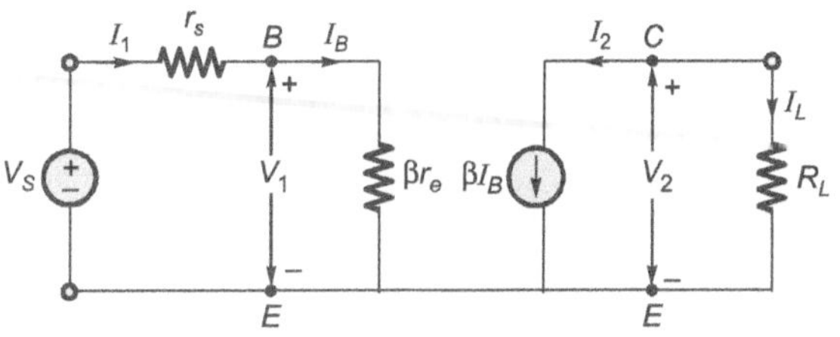

Here, voltage gain,

$$A_V = \frac{V_2}{V_1} = -\frac{\beta I_B R_L}{I_B \beta r_e} = \frac{-R_L}{r_e}$$

Current gain,

$$A_I = \frac{I_L}{I_1} = -\frac{\beta I_B}{I_B} = -\beta$$

Input impedance,

$$Z_i = \frac{V_1}{I_B} = \frac{I_B(\beta r_e)}{I_B} = \beta r_e$$

Output admittance, $Y_0 = \frac{I_2}{V_2}$ with $V_S = 0$

and $R_L = \infty$

With $V_S = 0$, we have:

$I_B = 0$ and $I_2 = 0$;

Hence, $Y_0 = 0$

$\therefore$ Output impedance,

$$Z_0 = \frac{1}{Y_0} = \infty$$

210. (c)

$$10 \; V_{GS1} = 2.4 - 2 = 0.4 \text{ V}$$

$$\Delta I_D = 1.4 - 1 = 0.4 \text{ mA}$$

$$\text{Transconductance} = \frac{\Delta I_D}{\Delta V_{GS}} = \frac{0.4}{0.4} = 1 \text{ mA/V}$$

211. (c)

The output voltage is,

$$V_0 = \left(-\frac{R_D}{R_D + r_d}\right) \mu V_{gs}$$

($V_{gs} = V_i$ = input voltage)

$$\therefore \text{Voltage, gain } A_V = \frac{V_0}{V_i} = \left(-\frac{\mu R_D}{R_D + r_d}\right)$$

$$= \left(\frac{-50 \times 5 \times 10^3}{5 \times 10^3 + 35 \times 10^3}\right)$$

$$= \left(\frac{-250 \times 10^3}{40 \times 10^3}\right) = -6.25$$

212. (b)

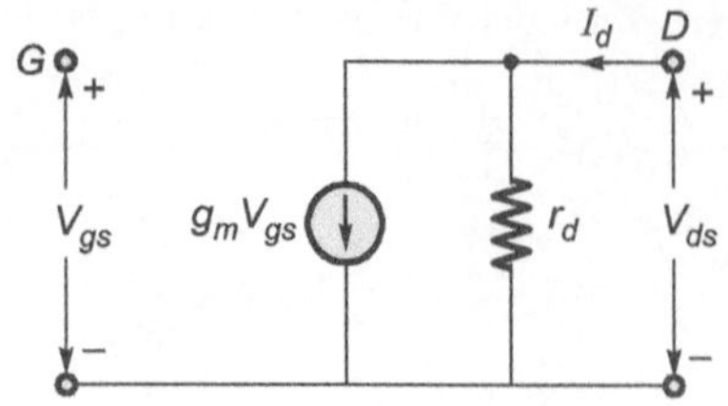

Small signal model of FET in CS configuration

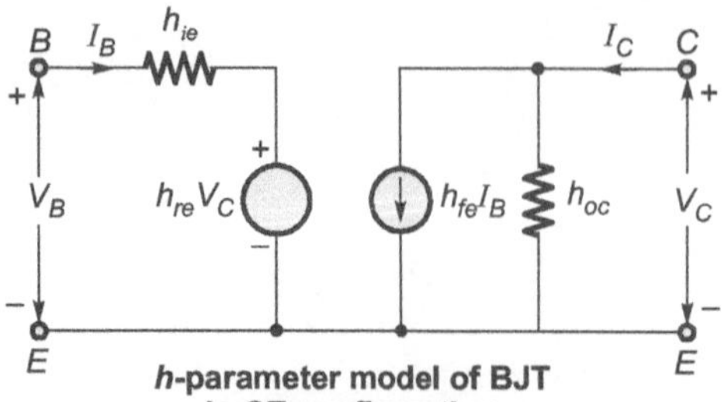

***h*-parameter model of BJT in CE configuration**

By observing the above two modes, we can find that all the statements are correct.

215. (b)

The minimum value of V_{DS} for pinch-off to occur for V_{GS} = –2 V is

$$(V_{DS})_{min} = V_{GS} - V_P = -2 - (-5) = 3\text{ V}$$

$$I_{DS} = I_{DSS}\left[1-\frac{V_{GS}}{V_P}\right]^2$$

$$= 8\times10^{-3}\left[1-\frac{(-2)}{(-5)}\right]$$

$$= 2.88\text{ mA} \approx 2.9\text{ mA}$$

216. (b)

The N-channel MOSFET may turn on prematurely due to the positively charged contaminants. Reason is also a correct statement but not the correct explanation of assertion.

217. (c)

The transconductance of N-channel JFET is more than that of P-channel JFET. Hence, reason is not a correct statement.

218. (d)

We have: $$V_{GG} = \left(\frac{R_2}{R_1+R_2}\right)V_{DD}$$

$$= \frac{8.75\times10^6\times24}{(12+8.75)\times10^6} = 10\text{ V}$$

We know that,

$$I_D = I_{DSS}\left(1-\frac{V_{GS}}{V_P}\right)^2$$

$$= I_{DSS}\left(1-\frac{V_{GG}-I_DR_S}{V_P}\right)^2$$

(where, $V_{GS} = V_{GG} - I_D R_S$)

or, $$I_D = 4\times\left(1-\frac{10-I_D\times3}{-2}\right)^2$$

or, $9I_D^2 - 73I_D + 144 = 0$

On solving we get,

$$I_D = 3.39\text{ mA or }4.72\text{ mA}$$

As I_D = 4.72 mA > 4 mA (i.e. I_{DSS})

therefore, I_{DQ} = 3.39 mA

219. (a)

$$I_D = I_{DSS}\left[1-\frac{V_{GS}}{V_P}\right]^2$$

$$= 40\times10^{-3}\left[1-\frac{(-5)}{(-10)}\right]^2$$

$$= 28.28\text{ mA}$$

Therefore, $$R_S = \left|\frac{V_{DSQ}}{I_D}\right| = \frac{5}{28.28\times10^{-3}}$$

$$= 176.8\ \Omega \simeq 177\ \Omega$$

220. (b)

The small signal equivalent circuit of common-drain amplifier is shown below.

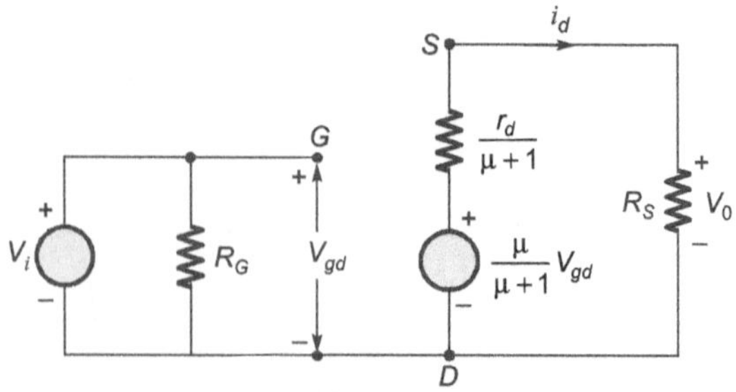

Using Thevenin's theorem, the output voltage is

$$V_0 = \left(\frac{R_S}{R_S+\dfrac{r_d}{\mu+1}}\right)\times\frac{\mu}{\mu+1}V_{gd}$$

$$= \frac{\mu R_S V_{gd}}{(\mu+1)R_S + r_d}$$

(where, $V_{gd} = V_i$ = input voltage)

∴ Voltage gain is

$$A_V = \frac{V_0}{V_i} = \frac{\mu R_S}{(\mu+1)R_S + r_d}$$

$$= \frac{50 \times 4 \times 10^3}{(50+1)\times 4\times 10^3 + 35\times 10^3}$$

$$= 0.836 \approx 0.84$$

221. (a)

M_2 is also saturated because its gate and drain are shorted together. The small signal circuit is shown below:

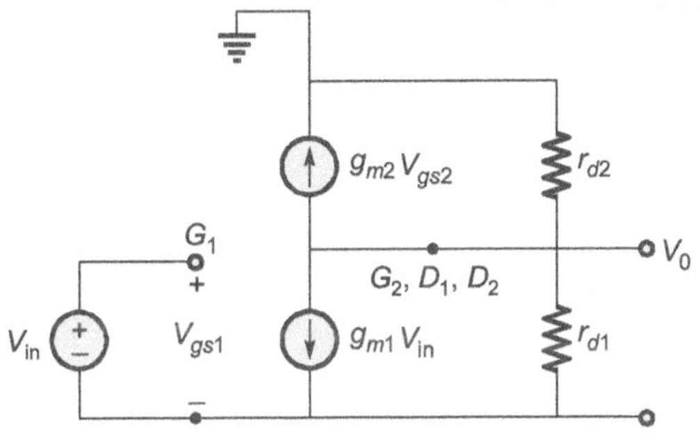

Here, $V_0 = -g_{m1}(r_{d1} || r_{d2} || 1/g_{m2})\, V_{in}$

For most MOSFET transistor, r_{d1} and r_{d2} are much greater than $1/g_{m2}$.

Hence, $$\frac{V_0}{V_{in}} = -\frac{g_{m1}}{g_{m2}}$$

Now, $$g_m = \sqrt{2\frac{\mu_n C_{ox} W}{L}}$$

∴ $$\frac{V_0}{V_{in}} = -\sqrt{\frac{\mu_n (W/L)_n}{\mu_P (W/L)_P}}$$

= Required voltage gain

223. (b)

For amplifier operation, MOSFET is always operated in saturation region.

Since, $I_D = I_{DSS}\left(1 - \frac{V_{GS}}{V_P}\right)^2$ in saturation region,

therefore the $I_D - V_{GS}$ characteristics of a MOSFET are quadratic as evident from the above equation.

226. (c)

We know that,

$$\mu = g_m r_d$$

or, $$g_m = \frac{200}{100\text{ k}} = 2\times 10^{-3}\text{ S}$$

Output resistance,

$$R_0 = \frac{1}{g_m} || r_d || R_L$$

$$= \left(\frac{1}{2\times 10^{-3}}\right) || (100\times 10^3) || (1000) = 332.226\ \Omega$$

227. (c)

$$A_V = \text{Gain at resonance} = g_m R$$

Here, $$R = r_d || R_{max}$$

$$= \frac{20\times 20}{20+20} = \frac{20\times 20}{40} = 10\text{ k}\Omega$$

∴ $$A_V = 5\times 10^{-3} \times 10 \times 10^3 = 50$$

229. (b)

$$g_m = -\frac{2I_{DSS}}{V_P}\left(1 - \frac{V_{GS}}{V_P}\right)$$

$$= -\frac{2\times(8\times 10^{-3})}{-4}\left[1 - \frac{-1.8}{-4}\right] = 2.2\text{ mS}$$

230. (b)

Applying KVL, we get:

$V_{BE} + (1\text{ k}\Omega) I_{E1} = 5.7$

or, $$I_{E1} = \left(\frac{5.7 - 0.7}{1\text{k}\Omega}\right) = 5\text{ mA}$$

or, $$r_\pi = h_{fe} r_e$$

or, $$r_\pi = 5\, h_{fe}$$

Now, $$A_V = \left[\frac{-h_{fe}R_C}{h_{ie} + (1+h_{fe})R_E}\right]$$

$$= \frac{-h_{fe}R_C}{5h_{fe} + (1+h_{fe})R_E}$$

$$\approx \frac{-R_C}{R_E} = -\frac{1.2}{1} = -1.2$$

or, this diagram is CE unbypass amplifier so,

$$A_V = -\frac{R_C}{R_E} = -\frac{1.2}{1} = -1.2$$

232. (b)

$$f_T = \frac{g_m}{2\pi(C_\pi + C_\mu)}$$

$$= \frac{38\times 10^{-3}}{2\pi(4\times 10^{-13} + 10^{-14})} = 1.47\times 10^{10}\text{ Hz}$$

and, $$f_\beta = \frac{f_T}{\beta_0} = \frac{1.47\times 10^{10}}{90} = 1.64\times 10^8\text{ Hz}$$

235. (a)

The high frequency model of FET is shown below.

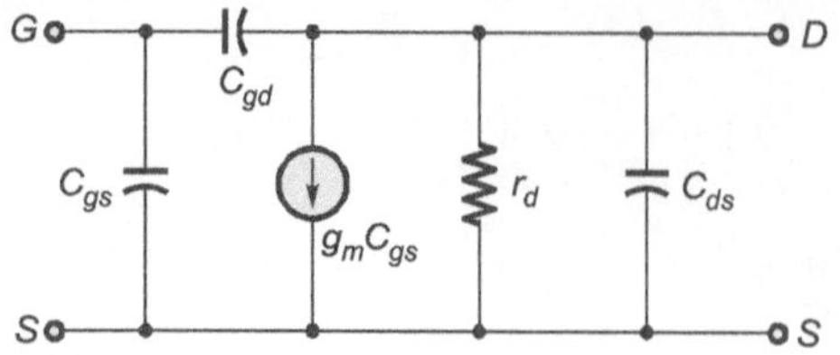

Because of the internal capacitances (C_{gd}, C_{gs} and C_{ds}), feedback exists between the output and input circuits of the FET in its high frequency model and hence the voltage amplification drops drastically with the increase in frequency. Moreover, there is no feedback from output to input in the FET small signal (low frequency) model.

Hence, both assertion and reason are correct and reason is the correct explanation of assertion.

236. (d)

The frequency at which the short circuit common-emitter current gain attains unit magnitude is given by

$$f_T = \frac{g_m}{2\pi(C_c + C_e)} = \frac{50 \text{ mA/V}}{2\pi(1.2 \text{ pF})} = 6.6 \text{ GHz}$$

237. (d)

C_e can be as small as 1 pF, and C_c as small as 0.2 pF.

Thus, C_e is larger than C_c

238. (c)

- The mechanism of the transport of charge carriers from emitter to collector is essentially one of diffusion. Hence, for finding the transistor behaviour at high frequency it is necessary to examine the diffusion mechanism. Hence, statement-1 is correct.
- Statement-2 is also correct.
- Statement-3 is also correct.
- The frequency at which the short-circuit common-emitter current gain attains unit magnitude is represented by f_T. Thus, Statement - 4 is also correct.

239. (c)

$$C_e \approx \frac{g_m}{2\pi f_T} \text{ or } f_T \approx \frac{g_m}{2\pi C_e}$$

or, $$f_T = \frac{50 \times 10^{-3}}{2\pi \times 10^{-12}} = 7.957 \text{ GHz} \approx 8 \text{ GHz}$$

240. (a)

The hybrid-π model for a transistor as shown below includes two capacitances. The collect-junction capacitance $C_c = C_{b'c}$ is the measured CB output capacitance with the input open (I_E = 0), and is usually specified by manufacturers as C_{ab}. The other capacitance C_e represents the sum of the emitter diffusion capacitance C_{De} and the emitter junction capacitance C_{Te}.

i.e. $$C_e = C_{De} + C_{Te} \approx C_{De}$$

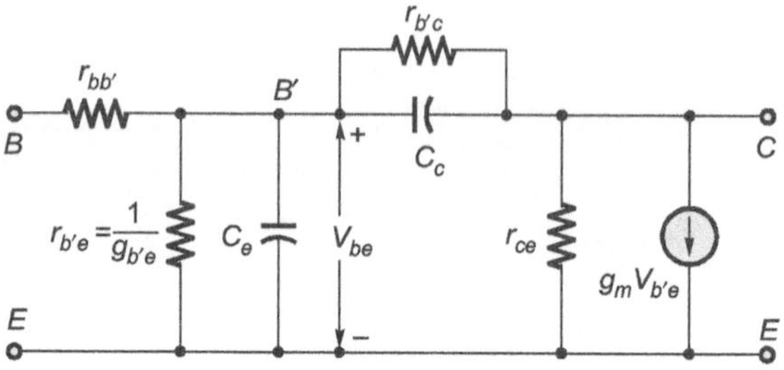

242. (b)

The magnitude of the gain of the amplifier near the upper 3-dB frequency is given by

$$|A_H(jf)| = \frac{100}{\sqrt{1+\left(\frac{f}{f_H}\right)^2}}$$

If the frequency at which the gain becomes 99 is f_1, then

$$99 = \frac{100}{\sqrt{1+\left(\frac{f_1}{100}\right)^2}}$$

or, $$1+\left(\frac{f_1}{100}\right)^2 = \left(\frac{100}{99}\right)^2$$

or, $$f_1 = 14.2 \text{ Hz}$$

Similarly, near the lower 3-dB frequency, we have:

$$|A_L(jf)| = \frac{100}{\sqrt{1+\left(\frac{f_L}{f}\right)^2}}$$

$$99 = \frac{100}{\sqrt{1+\left(\frac{0.1}{f_2}\right)^2}}$$

$$\text{or,} \quad 1+\left(\frac{0.1}{f_2}\right)^2 = \left(\frac{100}{99}\right)^2$$

$$\text{or,} \quad f_2 = 0.7 \text{ Hz}$$

Hence, the frequency range over which the gain reduces by less than 1 percent or gain is 99 is 0.7 Hz to 14.2 Hz i.e. $7f_L$ to $f_H/7$.

243. (d)

In non-interacting stages of cascaded amplifiers, the individual 3-dB frequencies for each stages are defined due to which only the overall transfer function is the product gains of individual stages. Hence, assertion is not a correct statement.

244. (c)

When *'n'* number of non-interacting stages of amplifier are cascaded, then

- Gain is increased.
- Gain bandwidth product remains constant.
- Bandwidth is reduced.
- Rise time is increased.

246. (a)

- Statement-1 is correct.
- For input stage CC or CB configuration is popularly used thereby securing proper impedance match and sacrificing voltage or current gain.
- Statement-3 is correct.
- Common emitter configuration is popularly used in the intermediate stages since the short-circuit current gain h_{fe} is very much greater than unity. It is thus possible to obtain a high voltage gain by cascading CE stages.

Hence, statements 2 and 4 are wrong.

247. (a)

One of the differences in performance of an RC coupled amplifier over single stage is that an RC coupled amplifier has higher non-linear distortion.

248. (d)

We know that,

$$\text{BW} \approx \frac{0.35}{t_r} \quad \text{or} \quad \text{BW} = \frac{0.35}{10\times 10^{-9}} = 35 \text{ MHz}$$

249. (c)

Using the simplified *h*-parameter equivalent circuit for CE amplifier, the lower 3-dB point is given by

$$f_1 = \frac{(1+h_{fe})}{(R_S+h_{ie})2\pi C_E}$$

$$= \frac{1+400}{(600+10\times 10^3)\times 2\pi\times 50\times 10^{-6}}$$

$$= 120 \text{ Hz}$$

250. (c)

$$g_{ce} = \frac{1}{r_{ce}} = h_{oe} - (1+h_{fe})\, g_{b'c}$$

$$\text{and} \quad r_{b'c} = \frac{1}{g_{b'c}} = \frac{r_{b'e}}{h_{re}}$$

251. (d)

Gain of multistage amplifier is

$$A^* = A_1 \cdot A_2 \cdot A_3$$

$$\text{In dB,} \quad (A^*)_{dB} = (A_1)_{dB} + (A_2)_{dB} + (A_3)_{dB}$$

$$\text{or,} \quad (A_3)dB = (A^*)_{dB} - [(A_1)_{dB} + (A_2)_{dB}]$$

$$= 70 - [20 + 30]$$

$$= 70 - 50 = 20 \text{ dB}$$

$$\therefore \quad (A_3)_{dB} = 20 \text{ dB} = 20 \log A_3$$

$$\text{or,} \quad A_3 = 10$$

252. (d)

$R_{i1} = R_{i2} = 2h_{ie} = 2\ \text{M}\Omega$ or $h_{ie} = 1\ \text{M}\Omega$

$$g_m = \frac{h_{fe}}{h_{ie}} = 100\ \mu S$$

$$A_{dm} = -g_m R_C = -500 = -\frac{h_{fe}}{h_{ie}}$$

= Differential mode gain

$$A_{cm} = -\frac{R_C}{2R_E} = \text{Common-mode gain}$$

$$\text{Hence,} \quad CMRR = \frac{A_{dm}}{A_{dc}} = 2g_m R_E = 500$$

$$\text{or,} \quad R_E = \frac{500}{2\times 100\ \mu S} = 2.5\ \text{M}\Omega$$

$$\text{Now,} \quad g_m R_C = 500 \text{ or } R_C = \frac{500}{g_m}$$

$$\text{or,} \quad R_C = \frac{500}{100\times 10^{-6}} = 5\ \text{M}\Omega$$

Thus, $R_C = 5\ \text{M}\Omega$ and $R_E = 2.5\ \text{M}\Omega$

253. (b)

The given circuit is redrawn as shown below.

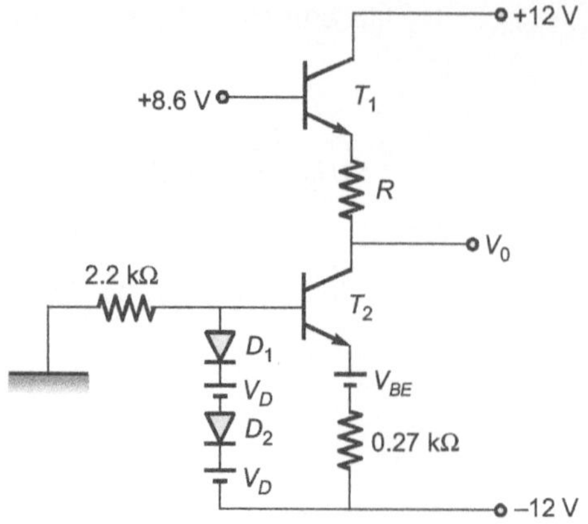

Writing loop equation containing two diodes and V_{BE}; we have:

$$2V_D = V_{BE} + I_E R_E$$

or,
$$I_E = \frac{2V_D - V_{BE}}{270} = \frac{2\times 0.7 - 0.7}{270}$$

$$= \frac{0.70}{270} = 2.6 \text{ mA}$$

$$V_{E1} = 8.6 - V_{BE} = 7.9 \text{ V}$$

Thus,
$$R = \frac{7.9}{2.6} = 3.04 \text{ k}\Omega$$

254. (c)

The goal of designing differential amplifiers is to minimize the effect of common-mode input signal and amplify the difference of the voltages between the two input lines.

256. (d)

Common- mode gain,

$$A_C = \left(\frac{100\times 10\text{k}}{1\text{k} + 100\text{k}\times 100}\right) = 0.1$$

257. (c)

In single stage differential amplifier, the output offset voltage is basically dependent on the mismatch of I_B and β.

258. (a)

The values of h_{FE}, V_{BE} and I_{CBO} vary with temperature and due to the variation in any of these quantities the output voltage may change which may not be distinguished from a change in the input signal voltage. Due to this reason it is not so easy to design d.c. amplifiers using transistors. Hence, both assertion and reason are true and reason is the correct explanation of assertion.

259. (d)

Assertion is not true because "figure of merit" (or CMMR) of a differential amplifier is to reject a common mode signal (V_C).

260. (c)

We know that,

$$V_0 = A_c V_c + A_d V_d$$

$$= A_d V_d\left(1 + \frac{A_c V_c}{A_d V_d}\right) \quad \text{...(i)}$$

Now, $$CMMR = \frac{A_d}{A_c} \text{ or } \frac{A_c}{A_d} = \frac{1}{CMMR} \quad \text{...(ii)}$$

From equation (i) and (ii), we have:

$$V_0 = A_d V_d\left(1 + \frac{1}{CMMR}\cdot\frac{V_c}{V_d}\right)$$

For the first set, we have:

$$V_d = [100 - (-100)]\ \mu\text{V} = 200\ \mu\text{V}$$

$$V_c = \frac{V_1 + V_2}{2} = \frac{100 + (-100)}{2} = 0$$

$$\therefore \quad V_{0_1} = 200 A_d\left(1 + \frac{1}{1000}\cdot\frac{0}{200}\right)$$

$$= 200\ A_d \mu\text{V} \quad \text{...(iii)}$$

For the second set, we have:

$$V_d = (1100 - 900)\ \mu\text{V} = 200\ \mu\text{V}$$

$$V_c = \frac{V_1 + V_2}{2} = \frac{1}{2}(1100 + 900)\ \mu\text{V}$$

$$= 1000\ \mu\text{V}$$

$$\therefore \quad V_{0_2} = 200 A_d\left(1 + \frac{1}{1000}\cdot\frac{1000}{200}\right)$$

$$= 201\ A_d \mu\text{V} \quad \text{...(iv)}$$

From equation (iii) and (iv), the percentage difference in output voltage obtained for the two sets of input voltages

$$= \left(\frac{201 - 200}{200}\right)\times 100\% = 0.50\%$$

261. (b)

Applying KVL, we get:

$$V_{BE} + 2I_E R_{EE} - V_{EE} = 0$$

or, $$I_E = \frac{V_{EE} - V_{BE}}{2R_{EE}}$$

$$= \frac{15 - 0.7}{2 \times 65 \times 10^3} = 110\ \mu A$$

Now, $$I_C = \alpha_F I_E = \frac{100}{101} I_E$$

$$= \frac{100}{101} \times 110 \times 10^{-6}\ A \approx 109\ \mu A$$

$$I_B = \frac{I_C}{\beta_F} = \frac{109 \times 10^{-6}}{100} \approx 1.09\ \mu A$$

Also, $$V_c = V_{cc} - I_c R_c$$
$$= 15 - (109 \times 10^{-6} \times 65 \times 10^3)$$
$$\approx 7.9\ V$$

Hence, both transistors of the differential amplifier are biased at a Q-point having $I_B = 1.09\ \mu A$ and $V_c = 7.9$ V.

262. (a)

$$I_{DS} = \frac{I_{SS}}{2} = \frac{175}{2} = 87.5\ \mu A$$

Now, $$V_{GS} = V_{TN} + \sqrt{\frac{I_{SS}}{K_n}}$$

$$= 1 + \sqrt{\frac{175 \times 10^{-6}}{3 \times 10^{-3}}} = 1.242\ V$$

$\therefore$ $$V_{DS} = V_{DD} - I_{DS} R_D + V_{GS}$$
$= 12 - (87.5 \times 10^{-6} \times 65 \times 10^3) + 1.242$
$= 7.55$ V

Checking for saturation,

$$V_{GS} - V_{TN} = 0.242\ V \text{ and } V_{DS} \geq 0.2$$

Thus, both transistors in the differential amplifier are biased at Q-point of V_{DS} = 7.55 V and $I_{DS} = 87.5\ \mu A$.

265. (c)

With $R_s = 0$,

$$A_d = \frac{1}{2} \frac{h_{fe} R_c}{h_{ie}}$$

Now, $$\frac{h_{fe}}{h_{ie}} = \frac{I_c}{V_T}$$

$\therefore$ $$A_d = \frac{1}{2} \frac{I_c R_c}{V_T}$$

For a common-mode d.c. voltage V_{CM} applied at the bases of Q_1 and Q_2 (because $R_s = 0$), we can write

$$V_E = V_{CM} - 0.7$$

$$I_{Re} = \frac{V_E - (-V_{EE})}{R_e} = \frac{V_{CM} - 0.7 + V_{EE}}{R_e}$$

Due to symmetry,

$$I_{C_1} = I_{C_2} = \frac{I_{R_e}}{2}$$

Putting everything together, we get

$$A_d = \frac{1}{2} \cdot \frac{1}{2} \frac{(V_{CM} - 0.7 + V_{EE})}{R_e} \cdot \frac{R_c}{V_T}$$

Substituting the values in above equation, we get:

$$A_d = \frac{1}{4} \times \frac{(V_{CM} + 4.3)}{1} \times \frac{1}{25\ mV}$$

For, $V_{CM} = -2$ V

$$A_d = \frac{1}{4} \times \frac{(-2 + 4.3)}{1} \times \frac{1}{25 \times 10^{-3}} = 23$$

266. (b)

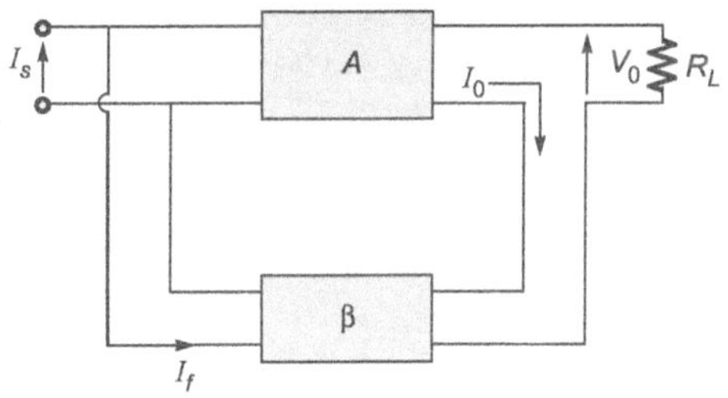

Here, $$R_{if} = \left(\frac{R_i}{1 + \beta A}\right)$$

and $$R_{0f} = R_0(1 + \beta A)$$

267. (c)

Given, $A = 1000$, $\beta = 0.04$ and $\frac{dA}{A} = 10\%$

The percentage change in gain of the amplifier with feedback is

$$\frac{dA_f}{A_f} = \frac{dA}{A}\left(\frac{1}{1 + \beta A}\right)$$

$$= 10 \times \left(\frac{1}{1 + 1000 \times 0.04}\right) = 0.25\%$$

268. (a)

Given, $A = 500$, $R_i = 3\ k\Omega$ and $\beta = 0.01$

Voltage gain with feedback is

$$A_f = \frac{A}{1 + \beta A} = \frac{500}{1 + 500 \times 0.01}$$

$$= \frac{500}{6} = 83.33$$

For a voltage-series negative feedback amplifier, input resistance with negative feedback is

$$R_{if} = (1 + A\beta)R_i$$
$$= (1 + 500 \times 0.01) \times 3 \times 10^3$$
$$= 18\ k\Omega$$

269. (b)

The emitter follower inherently exhibits 100% negative feedback because the voltage at the emitter follows the input voltage. Reason is also a correct statement since the emitter follower simultaneously increases input resistance and decreases output resistance. Hence, both assertion and reason are correct but reason is not the correct explanation of assertion.

271. (d)

Due to negative feedback, bandwidth becomes

$$BW_f = (1 + \beta A)\ BW$$

Given, $BW = 250$ kHz

$A = 125$, $BW_f = 1$ MHz

Let β' be the feedback ratio required to restrict the bandwidth to 1 MHz.

Then, $1 \times 10^6 = (1 + 125\beta') \times 250 \times 10^3$

or, $(1 + 125\beta') = \dfrac{1 \times 10^6}{250 \times 10^3} = 4$

or, $\beta' = \dfrac{3}{125} = 0.024 = 2.4\%$

272. (c)

We know that,

$$\frac{dA_f}{A_f} = \frac{dA}{A}\left(\frac{1}{1+\beta A}\right)$$

or, $0.02 = 0.2 \times \dfrac{1}{1+\beta A}$

or, $1 + \beta A = \dfrac{0.2}{0.02} = 10$

Also, gain with feedback is

$$A_f = \frac{A}{1+\beta A}$$

or, $100 = \dfrac{A}{10}$

or, $A = 1000 =$ Open-loop gain

Now, $1 + \beta A = 10$ or $\beta A = 9$

or, $\beta = \dfrac{9}{1000} = 0.009 = 0.9\%$

274. (a)

Due to negative feedback, gain of the amplifier is reduced

i.e. $A_f = \left(\dfrac{A}{1+\beta A}\right)$

276. (a)

Due to current sampling, the output resistance increases while due to voltage sampling, the output resistance reduces.

277. (d)

A transconductance amplifier uses current-series feedback due to which both input and output resistances are increased.

281. (d)

Depending upon the type of feedback (+ve or –ve), the gain, its input and output impedance may increase or decrease. However, both +ve and –ve feedback can control its output.

282. (a)

- Positive feedback produces instability i.e. reduces the stability of the amplifier but still it is useful (in oscillators).
- Negative feedback in amplifiers reduces the gain and increases stability of the amplifier.
- Due to negative feedback input impedance is increased while output impedance is reduced.
- Statement-4 is also correct.

Hence, all the statements are correct.

283. (c)

The closed loop voltage gain is given by

$$A_{vf} = \frac{V_0}{V_s} = \frac{A_v}{1+A_v\beta} \approx \frac{A_v}{A_v\beta}$$

(since A_{vb} is very large)

$$= \frac{1}{\beta} = \frac{R_E + R_F}{R_E}$$

Given, $A_{vf} = 25$, $R_F = 1.2\ k\Omega$

∴ $25 = \dfrac{R_E + 1.2\ k}{R_E}$

or, $25\ R_E = R_E + 1.2\ k$

or, $R_E = \dfrac{1.2\ k}{24} = \dfrac{1200}{24} = 50\ \Omega$

284. (a)

Voltage across the emitter resistance R_E is

$$V_E = I_0 R_E = V_s$$

(Due to virtual ground concept)

or, $$\frac{I_0}{V_s} = \frac{1}{R_E}$$

285. (b)

Given, $40\text{ dB} = 20\log 10^2 = 20\log\left(\frac{V_0}{V_i}\right)$

$\therefore$ Voltage gain $= \frac{V_0}{V_i} = 100 = A_v$

(open loop gain)

For 3% negative feedback, $\beta = 3\%$

and $$A_{vf} = \frac{A_v}{1+\beta A_v} = \frac{100}{1+100\times 0.03} = 25$$

286. (a)

We know that,

$$\frac{dA_f}{A_f} = \frac{dA}{A}\left[\frac{1}{1+\beta A}\right]$$

Open loop gain = 40 dB or $A = 100$

$$\therefore \quad \frac{dA_f}{A_f} = 5\times\frac{1}{1+100\times 0.02} = \frac{5}{3}\%$$

$$= 1.66\%$$

287. (d)

Given, $60\text{ dB} = 20\log 10^3 = 20\log A_V$

$A_v = 1000$, $f_L = 100$ Hz, $f_H = 100$ kHz,

$20\text{ dB} = 20\log 10 = 20\log A_{vf}$

$A_{vf} = 10$

Now, $$A_{vf} = \frac{1000}{1+1000\,\beta} = 10$$

or, $$\beta = \frac{99}{1000} = 0.099$$

$$f_{LF} = \frac{f_L}{1+\beta A} = \frac{100}{100} = 1\text{ Hz}$$

$$f_{HF} = f_H(1+\beta A)$$
$$= (100\text{ kHz})(100)$$
$$= 10\text{ MHz}$$

$\therefore$ New bandwidth

$$= f_{HF} - f_{LF} = 10\text{ MHz} - 1\text{ Hz}$$
$$= 10\text{ MHz}$$

288. (b)

Return ratio (β) lies between 0 and 1.

290. (a)

This is a case of series mixer due to which input impedance is increased.

291. (b)

- The decibel gain of a cascaded amplifier is equal to sum of the individual gain.
- Due to degenerative feedback (negative feedback), bandwidth is increased while gain is reduced.

  Hence, statements 1 and 4 are not correct.

292. (b)

Given figure represents "a two stage self-bias circuit" in cascade and with a feedback existing from emitter of second transistor to the base of first transistor which is a case of "shunt-series" or "shunt-current" or "current-shunt feedback".

299. (c)

Compared to class A operation, class B operation has

- higher efficiency ($\eta = 78.5\%$)
- greater output power
- more distortion and noise interference.
- negligible power loss at no input signal i.e. no power drain since transistor in class B operation will not consume any power if signal is not applied.

Hence, statements 1, 2 and 4 are correct.

300. (a)

We know that,

$$T_j = T_A + \theta P_D$$

or, $$T_j = 27°\text{C} + (8\text{ °C/W})\times 3\text{ W}$$
$$= 27°\text{C} + 24°\text{C} = 51°\text{C}$$

301. (c)

Collector circuit efficiency is

$$\eta = \frac{\pi}{4}\left(1-\frac{V_{CE(\min)}}{V_{CC}}\right)\times 100\%$$

$$= \frac{\pi}{4}\left(1-\frac{2}{15}\right)\times 100\% = 68.06\% \approx 68\%$$

302. (d)

For series-fed load,
overall efficiency,

$$\eta = 25\left(\frac{V_{max} - V_{min}}{V_{max}}\right)\%$$

$$= \frac{25(15-1)}{15} = 23.33\%$$

For transformer-coupled load,
overall efficiency,

$$\eta = 50\left(\frac{V_{max} - V_{min}}{V_{max} - V_{min}}\right)\%$$

$$= 50\left(\frac{15-1}{15+1}\right) = 43.75\%$$

303. (c)

The power dissipated by the transistor is

$$P_D = \frac{T_J - T_A}{\theta_{JA}} = \frac{160-40}{80} = \frac{120}{80} = 1.5 \text{ W}$$

307. (a)

The Q-point of class-A amplifier is located in active region i.e. at the mid-point of dc load line as shown below.

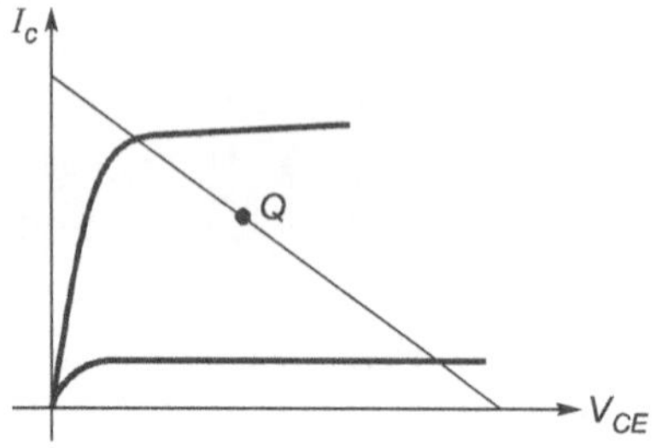

309. (a)

Efficiency of class-A is around 25% while that of class B is 78.5%.

The operating point of class-A amplifier is in active region while that of class-B is in cutt-off region. Hence, class-A has less harmonic distortion than class-B.

310. (a)

The Q-point of class-C amplifier is located just below the cutoff due to which it has maximum distortion.

The efficiency of class-C amplifier is highest (87.5%).

311. (c)

A push-pull circuit will balance out all even harmonics in the output and the third harmonic term acts as the principle source of distortion, provided the two transistors Q_1 and Q_2 are identical.

312. (d)

In a class A amplifier, when signal is not applied i.e. signal input is zero it will suffer from power drain (i.e. it will consume maximum power).

313. (c)

The main aim of a large signal amplifier, otherwise called as a power amplifier is to deliver a substantial amount of power to a load. Power amplifiers convert a part of the input d.c. power of supply into the a.c. power to the load due to which the output power is several times its input power.

317. (b)

The heat sink is a relatively large, finned, usually black metallic heat conducting device placed in close contact with the transistor case or flange. Larger the heat sink, smaller will be its thermal resistance which will improve the dissipation capability of the transistor.

318. (a)

Statement-4 is not correct because all even harmonics are balanced out or cancelled in the expression of output current. Thus, it contain only third harmonic terms.

319. (b)

The output of class C amplifier is a highly distorted version of its input. It could not be used in an application requiring high fidelity, such as an audio amplifier. Class C amplifiers are used primarily in high-power, high frequency applications such as radio-frequency transmitters.

321. (c)

The Q-point of class A amplifier lies in middle of dc load line.

Hence, statement 4 is not correct.

Statement 1 is also not correct.

Hence, statements 2, 3, and 5 are only correct only.

322. (d)

$$P_{L(\max)} = \left[\left\{\frac{I_{im}}{2\sqrt{2}}\right\}^2 + \left\{\frac{I_{im}}{2\sqrt{2}}\right\}^2\right] R_L$$

$$= \frac{I_{im}^2}{4} R_L = \frac{1}{2}\left\{\frac{I_{im}}{\sqrt{2}}\right\}^2 R_L$$

$$= \frac{1}{2} P_{L(\max)} = \frac{2}{2} = 1\text{ W}$$

323. (a)

Power delivered by the supply voltage is

$$P_s = P_{cc(+)} + P_{cc(-)}$$

$$= 15 \times 10 + 15 \times 10 = 300\text{ mW}$$

$$P_{L(\max)} = \left(\frac{V_{L(\max)}}{\sqrt{2}}\right)^2 \Big/ R_L = \left(\frac{5}{\sqrt{2}}\right)^2 \Big/ 1\text{k}$$

$$= 12.5\text{ mW}$$

$$\therefore \text{ Efficiency, } \eta = \left(\frac{P_{L(\max)}}{P_{cc(+)} + P_{cc(-)}}\right) = \frac{12.5\text{ mW}}{300\text{ mW}}$$

$$= 0.0417 \approx 4.2\%$$

325. (c)

The power conversion efficiency is the ability of an active device in converting the d.c. power of the supply into the a.c. power delivered to the load.

$$\%\eta = \frac{\text{signal power delivered to the load}}{\text{d.c. power supplied to input circuit}} \times 100\%$$

326. (b)

In class-B, the devices being biased at cut-off, one device stops conducting before the other device starts conducting leaving to crossover distortion.

327. (a)

We need to amplify 3 signal frequencies i.e. 870 kHz, 875 kHz and 880 kHz. These frequencies lie in a bandwidth of 10 kHz and we should use only tuned voltage amplifiers to amplify them.

328. (d)

The voltage gain of a multistage amplifier is equal to the product of the gains of the individual stages.

330. (b)

Given, $f = 200$ kHz, $T = 0.1\ \mu$s

$$\text{Duty cycle} = 200 \times 10^3 \times 0.1 \times 10^{-6}$$

$$= 20 \times 10^4 \times 1 \times 10^{-7}$$

$$= 20 \times 10^{-3}$$

$$= \frac{20}{1000} \times 100 = 2\%$$

331. (b)

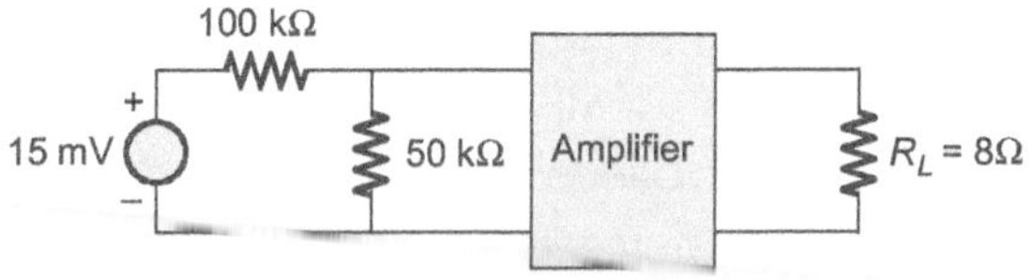

Thevenin equivalent i/p impedance is

$$R_{th} = \frac{100 \times 50}{100 + 50} = 33.33\text{ k}\Omega$$

$$\therefore \quad V_i = 15 \times 10^{-3} \times \frac{50 \times 10^3}{150 \times 10^3}$$

$$= 5 \times 10^{-3}\text{ V}$$

Now, $$P_L = \frac{V_0^2}{R_L} = 2,$$

or, $$V_0^2 = 2 \times 8 = 16$$

or, $$V_0\text{ (rms)} = 4\text{ volt}$$

$$\therefore \quad A_v = \frac{V_0}{V_i} = \frac{4}{5 \times 10^{-3}} = 800$$

332. (b)

In class-B operation, the output signal power is equal to 5 times the power dissipated by single transistor circuit. Hence, maximum power output obtained at the load from the circuit = 5 × 10 = 50 Watts.

334. (d)

Using superposition theorem, the output voltage is given by

$$V_0 = \frac{-R_2}{R_1} V_1 + \left(1 + \frac{R_2}{R_1}\right) V_2$$

$$= \frac{R_2}{R_1}(V_2 - V_1) + V_2$$

335. (b)

For an ideal OPAMP, input impedance is infinite while output impedance is zero.

336. (d)

For a practical OPAMP input impedance is of the order of $10^6\ \Omega$.

337. (a)

An ideal operational amplifier has
- zero offset voltage
- characteristics which do not drift with temperature
- infinite voltage gain
- infinite CMMR
- infinite bandwidth

Hence, statements 2, 3 and 5 are only correct.

338. (c)

We know that,

$$SR = \left.\frac{dV_0}{dt}\right|_{(max)}$$

$$= \frac{dV_0}{dV_i} \times \frac{dV_i}{dt} = |A_{CL}| V_m \omega_m$$

$$0.5\ V/\mu s = 10 \times V_m \times 2\pi \times 40 \times 10^3$$

$$0.5 \times 10^6 = 4 \times 10^5 \times 2\pi \times V_m$$

$$V_m = 0.199$$

$$\text{Peak to peak} = 2V_m = 2 \times 0.199$$

$$= 0.398 \simeq 0.4\ V$$

Note: If $|A_{CL}|$ is not given take $|A_{CL}| = 1$.

339. (c)

$$\text{CMMR} = \frac{A_d}{A_c} = 10^5$$

$\therefore$ The common mode gain,

$$A_c = \frac{A_d}{\text{CMMR}} = \frac{10^5}{10^5} = 1$$

340. (d)

Given op-amp circuit represents a closed-loop non-inverting amplifier. The feedback factor is given by

$$\beta = \frac{R_1}{R_1 + R_F} = \frac{1\text{k}}{1\text{k} + 10\text{k}}$$

$$= \frac{1}{11} = 0.091 = 9.1\%$$

341. (b)

The time period of the given astable multivibrator circuit is given by

$$T = 2RC\,ln\left(\frac{1+\beta}{1-\beta}\right)$$

Here, $$\beta = \frac{R_2}{R_1 + R_2}$$

$$\therefore \quad T = 2RC\,ln\left(\frac{R_1 + 2R_2}{R_1}\right)$$

$$= 2RC\,ln\left(\frac{116 \times 10^3 + 2 \times 100 \times 10^3}{116 \times 10^3}\right)$$

$$= 2RC\,ln\left(\frac{316 \times 10^3}{116 \times 10^3}\right) = 2RC\,ln\,(2.724) \approx 2\,RC$$

(Since $ln\,(2.724) \approx ln\,e = 1$)

Given, $$f = 1\text{ kHz},\ T = \frac{1}{f} = 1\text{ ms}$$

$$\therefore \quad 2\,RC = 10^{-3}\text{ sec}$$

or, $$RC = 0.5 \times 10^{-3}\text{ sec} = 0.5\text{ ms}$$

342. (a)

Given circuit represents an inverting summer using op-amp.

Here, $R = R_1 = R_2 = R_3 = R_f = 1\ \text{k}\Omega$

and $V_1 = +2\text{ V},\ V_2 = +3\text{ V},\ V_3 = +4\text{ V}$

The output voltage is

$$V_0 = \frac{-R_f}{R}(V_1 + V_2 + V_3)$$

$$= \frac{-1 \times 10^3}{1 \times 10^3}(2 + 3 + 4) = -9\text{V}$$

343. (d)

$$SR = \left.\frac{dV_0}{dt}\right|_{max} = \frac{dV_0}{dV_i} \times \frac{dV_i}{dt}$$

$$= |A_{CL}| \times \left.\frac{dV_i}{dt}\right|_{max}$$

$$SR = |A_{CL}| V_m \omega_m$$

$$0.5\ V/\mu s = 50 \times 20 \times 10^{-3} \times 2\pi f_m$$

$$f_m = 79.6\text{ kHz}$$

344. (b)

In open-loop configuration, the op-amp functions as a very high gain amplifier because the term open-loop indicates that no feedback, in any form, is given to the input from the output. The three open-loop configurations of op-amp are inverting, non-inverting and differential amplifiers. Hence, both assertion and reason are correct but reason is not the correct explanation of assertion.

345. (d)

An unity gain non inverting operational amplifier is also called "voltage follower" which offers very high input impedance of the order of MΩ and very low output impedance. Therefore, this circuit draws negligible current from the source. Hence, assertion is not a correct statement. The voltage follower can be used as a buffer between a high impedance source and a low impedance load for impedance matching applications. Thus, reason is a correct statement.

347. (a)

Given op-amp circuit represents a non-saturating half-wave precision rectifier.

- When $V_i < 0$, D_1 is RB while D_2 is FB, therefore $V_0 = \frac{-R_f}{R_1}V_i$.
- When $V_i > 0$, D_1 is FB while D_2 is RB, therefore $V_0 = 0$ V.
- When $V_i > 0$, $V_1 = -0.7$ V.
- When $V_i < 0$, $V_1 \approx \frac{R_f}{R_1}\ V_i + 0.7$ V

349. (d)

A real operational amplifier exhibits an unbalance caused by a mismatch of the input transistors. This mismatch results in unequal bias currents flowing through the input terminals, and also requires that an offset voltage be applied between the two input terminals to balance the amplifier output

350. (b)

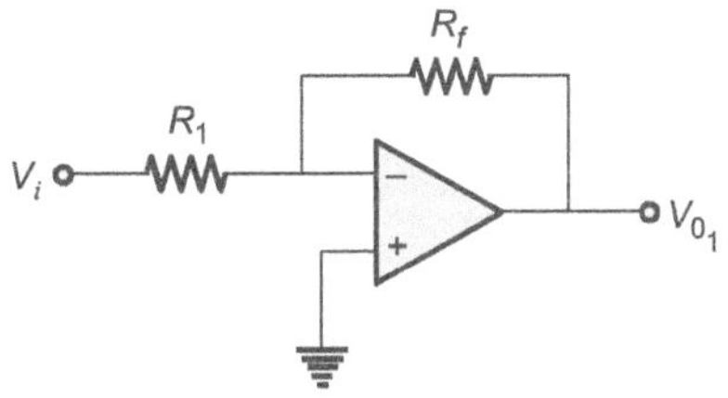

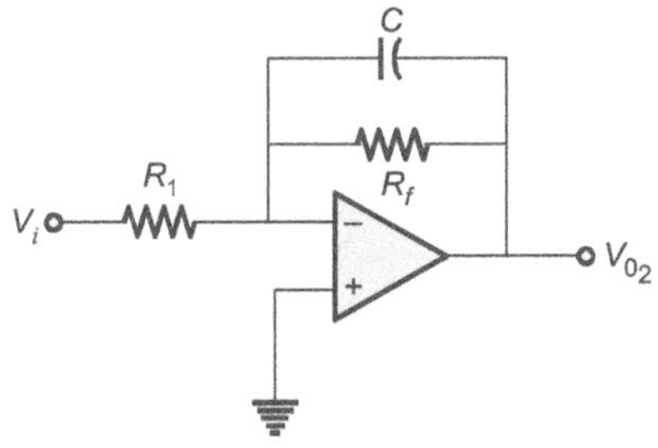

The transfer function of the inverting amplifier is

$$\frac{V_{0_1}(s)}{V_i(s)} = \frac{-R_f}{R_1}$$

When a capacitor (C) is connected across the feedback resistor (R_f), new transfer function becomes

$$\frac{V_{0_2}(s)}{V_i(s)} = -\frac{R_f}{R_1(1+R_f Cs)}$$

$$= -\frac{R_f}{R_1 R_f C\left(s+\frac{1}{R_f C}\right)}$$

$$= \frac{-\frac{1}{R_1 C}}{\left(S+\frac{1}{R_f C}\right)}$$

Hence, when a capacitor is connected across the feedback resistor an additional low frequency pole in the transfer function occurs.

351. (b)

Slew rate is defined as the maximum rate at which the output voltage of an op-amp can change in a time of 1 μs.

$$S_R = \left.\frac{dV_0}{dt}\right|_{t=1\mu s}$$

352. (a)

In the first case differential amplifier is ideal one.

Therefore, $V_{0_1} = (V_1 - V_2)\,A_d$

$= [50-(-50)]\,A_d = 100\,A_d\,\mu V$

In the second case, differential amplifier is practical one.

Therefore, $V_d = (V_1 - V_2)\,A_d = 100\,\mu V$

$$V_c = \frac{1}{2}(1050+950) = 1000\,\mu V$$

$$\therefore \quad V_{0_2} = A_d V_d + A_c V_c$$

$$= A_d V_d\left[1+\frac{A_c V_c}{A_d V_d}\right]$$

$$= A_d V_d\left[1+\frac{V_c}{V_d\,(\text{CMMR})}\right]$$

$$= 100 A_d\left[1+\frac{1000}{100\times 100}\right]\mu V$$

or, $V_{0_2} = (100\, A_d \times 1.1)\,\mu V$
$= 110\, A_d \mu V$

$\therefore$ Percentage difference in output voltage for the given two sets of input signals

$$= \left(\frac{V_{0_2} - V_{0_1}}{V_{0_1}}\right) \times 100\%$$

$$= \left(\frac{110-100}{100}\right) \times 100\% = 10\%$$

354. (c)

$$\text{CMMR} = \frac{A_d}{A_c} = \frac{\text{Differential mode gain}}{\text{Common mode gain}}$$

Also, $\text{CMMR (in dB)} = 20 \log_{10}\left(\frac{A_d}{A_c}\right)$

357. (d)

Voltage follower is a non-inverting amplifier with unit gain (A = 1).

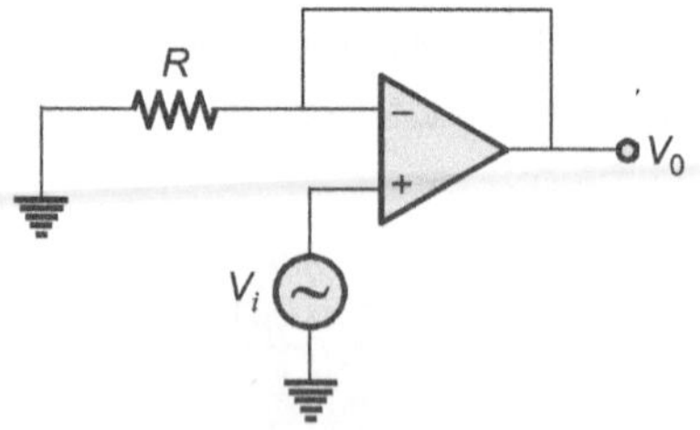

360. (c)

Op-Amp under open-loop configuration can be used as

- Comparator
- Zero crossing/level detector

361. (b)

- Op-amp is a voltage controlled voltage device.
- An ideal op-amp has infinite input resistance as well as voltage gain.
- Gain of an inverting amplifier is $A_v = \frac{-R_f}{R_1}$ which is independent of source resistance.
- An op-amp is used in IC regulators as an error amplifier.

Hence, statements 3 and 4 are only correct.

363. (d)

Applying KCL at node $V_{(-)}$ of A_1, we have:

$$\frac{V_1}{R_2} + \frac{V_1 - V_3}{R_1} + \frac{V_1 - V_2}{R_3} = 0$$

or, $$V_1\left(\frac{1}{R_2} + \frac{1}{R_1} + \frac{1}{R_3}\right) - \frac{V_3}{R_1} - \frac{V_2}{R_3} = 0$$

or, $$V_1\left(\frac{1}{R_2} + \frac{1}{R_1} + \frac{1}{R_3}\right) - \frac{V_2}{R_3} = \frac{V_3}{R_1} \quad ...(i)$$

Again, applying KCL at node $V_{(-)}$ of A_2, we have:

$$\frac{V_2 - V_0}{R_2} + \frac{V_2 - V_3}{R_1} + \frac{V_2 - V_1}{R_3} = 0$$

or, $$V_2\left(\frac{1}{R_2} + \frac{1}{R_1} + \frac{1}{R_3}\right) - \frac{V_0}{R_2} - \frac{V_1}{R_3} = \frac{V_3}{R_1} \quad ...(ii)$$

Subtracting equation (*ii*) from equation (*i*), we have:

$$V_1\left(\frac{1}{R_1} + \frac{1}{R_2} + \frac{1}{R_3} + \frac{1}{R_3}\right)$$
$$+V_2\left(-\frac{1}{R_3} - \frac{1}{R_2} - \frac{1}{R_1} - \frac{1}{R_3}\right) + \frac{V_0}{R_2} = 0$$

or, $$V_1\left(\frac{1}{R_1} + \frac{1}{R_2} + \frac{2}{R_3}\right) - V_2\left(\frac{1}{R_1} + \frac{1}{R_2} + \frac{2}{R_3}\right) + \frac{V_0}{R_2} = 0$$

or, $$\frac{V_0}{R_2} + (V_1 - V_2)\left(\frac{1}{R_1} + \frac{1}{R_2} + \frac{2}{R_3}\right) = 0$$

or, $$\frac{V_0}{R_2} = \left(\frac{1}{R_1} + \frac{1}{R_2} + \frac{2}{R_3}\right)(V_2 - V_1)$$

or, $$V_0 = \left(1 + \frac{R_2}{R_1} + \frac{2R_2}{R_3}\right)(V_2 - V_1)$$

364. (a)

Given, $f = 2\text{ MHz}$

$\therefore$ $T = \frac{1}{f} = 0.5\,\mu\text{sec}$

$\therefore$ $$\text{Slew rate} = \frac{4\text{ V}}{\left(\frac{0.5}{2}\right)\mu s} = 16\text{ V}/\mu s$$

365. (c)

Given, $I_D = I_s(e^{qV/\eta KT} - 1)$

or, $$e^{qV/KT} = \frac{I_D}{I_s} + 1$$

or, $V = \frac{KT}{q} ln\left(\frac{I_D}{I_s}+1\right)$

or, $V = 0.026\, ln\left(\frac{I_D}{I_s}+1\right) \approx 0.026\, ln\frac{I_D}{I_s}$

For the logarithmic amplifier shown below,

$V_0 = V_D = V$

and $I_D = \frac{V_i}{R}$

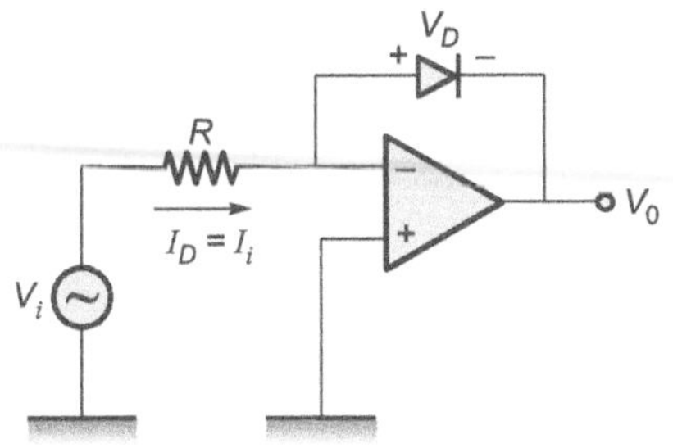

Thus, $V_0 = V_D = 0.026\, ln\left(\frac{V_i}{RI_s}\right) = 0$ (Given)

or, $\frac{V_i}{RI_s} = 1$ or $V_i = RI_s = 10^5 \times 10^{-6} = 0.1$ V

366. (d)

Let the output of op-amp with gain A_1 be V_{01}.

Now, $V_{01} = V_i\left(1+\frac{R}{R}\right) = 2V_i$ ...(i)

(Non-inverting amplifier output)

Also, $V_0 = -\frac{1}{CR}\int V_{01}\,dt$

or, $V_0(s) = -\frac{V_{01}(s)}{sCR} = -\frac{2V_i(s)}{sCR}$ ...(ii)

Now, $V_0(s) = -I_i(s)R = -2\frac{V_i(s)}{sCR}$

(using equation (ii))

$$\frac{V_i(s)}{I_i(s)} = \frac{sCR^2}{2} = sL_{eq} = j\omega L_{eq}$$

where, $L_{eq} = \frac{CR^2}{2}$

Hence, the input impedance of given op-amp circuit is purely inductive.

367. (d)

Applying *KCL* at the inverting input $V_{(-)}$, we get:

$$\frac{V_i - V_{(-)}}{5\text{ K}} = \frac{V_{(-)} - V_0}{10\text{ K}}$$

or, $2V_i - 2V_{(-)} = V_{(-)} - V_0$

or, $V_{(-)} = \left(\frac{2V_i + V_0}{3}\right)$ ...(i)

Also, $V_{(-)} = \left(\frac{10\text{ K}}{10\text{ K}+100\text{ K}}\right)V_0 = \frac{1}{11}V_0$ ...(ii)

Equating equations (i) and (ii), we have:

$$\frac{2V_i + V_0}{3} = \frac{1}{11}V_0$$

or, $22V_i + 11V_0 = 3V_0$

or, $V_0 = -\frac{22}{8}V_i = -\frac{22}{8}\times 2 = -5.5$ V

368. (b)

We have:

$$A_+ = \left(1+\frac{22}{10}\right)\left(\frac{R}{R+15}\right)V_+$$

and $A_- = -\left(\frac{22}{10}\right)V_-$

According to question, $|A_+| = |A_-|$

$\therefore (1+2.2)\left(\frac{R}{R+15}\right) = 2.2$ (Since $V_+ = V_-$)

or, $\frac{R}{R+15} = \frac{2.2}{3.2}$ or $32R = 22R + 330$

or, $R = \frac{330}{10} = 33$ kΩ

369. (a)

For the op-amp circuit output to be in saturation,

$$V_0 = 12\text{ V} = -\frac{1}{RC}\int(-0.5)dt = \frac{0.5}{RC}\int dt$$

or, $12 = \frac{0.5}{500\times10^3\times10^{-5}}\times t$

or, $t = \frac{12\times5}{0.5} = 120$ seconds

370. (b)

Given circuit represents an inverting amplifier. Type of feedback mechanism in an inverting amplifier is voltage-shunt feedback.

371. (c)

$$I_f = 2\text{ mA} = \frac{0-V_0}{2\text{ k}}$$

or, $V_0 = -4$ V

$$I_L = \frac{-V_0}{2\text{ k}} = \frac{+4}{2\text{ k}} = 2\text{ mA}$$

Applying KCL, we have:

$$I_f + I_L + I = 0$$

or, $I = -(I_f + I_L)$

or, $I = -(2 + 2) = -4\text{ mA}$

373. (b)

$$V_{01} = \left(1+\frac{R_2}{R_1}\right)V_i\ ;\ V_{02} = \frac{-R_2}{R_1}V_i$$

$\therefore$ $$V_{01} - V_{02} = \left(1+\frac{2R_2}{R_1}\right)V_i$$

or, $$\left(\frac{V_{01}-V_{02}}{V_i}\right) = \left(1+2\frac{R_2}{R_1}\right)$$

374. (a)

Hysteresis is desirable in a Schmitt trigger because it prevents noise from causing false triggering.

375. (a)

Given, $$A_3(s) = A = \frac{A_0^3}{\left(1+\frac{s}{\omega_H}\right)^3}$$

or, $$1 + A\beta = 1+\frac{A_0^3}{\left(1+\frac{s}{\omega_H}\right)^3}\beta$$

$$= 1+\frac{A_0^3}{\left(1+j\frac{\omega}{\omega_H}\right)^3}\beta$$

The phase change may change $A\beta$ to negative. This can happen only when the total change in the phase angle of $\left(1+j\frac{\omega}{\omega_H}\right)^3$ becomes 180°. The required phase change in one stage i.e. $\left(1+j\frac{\omega}{\omega_H}\right)$ should be 60°. i.e. $\tan^{-1}\frac{\omega}{\omega_H} = 60°$ or,

$$\frac{\omega_C}{\omega_H} = \sqrt{3}.$$

$\therefore$ Here, ω_C is the frequency at which the phase angle changes the sign. The value of $A\beta$ at this frequency is

$$= \frac{A_0^3}{(1+j\sqrt{3})^3}\beta$$

$$= \frac{\beta A_0^3}{(\sqrt{1+3})^3}\angle -180°$$

$$= -\frac{\beta A_0^3}{(2)^3} = -\frac{\beta A_0^3}{8}$$

For stability, $\beta A = -1$ or $-\frac{\beta_{max}A_0^3}{8} = -1$

or, $$\beta_{max} = \frac{8}{A_0^3}$$

376. (b)

The closed-loop gain with positive feedback is

$$A_f = \left(\frac{A}{1-\beta A}\right)$$

If the magnitude of loop gain i.e. $|\beta A| = 1$, then $A_f = \infty$. Hence, the gain of the amplifier with positive feedback is infinite and the amplifier gives an a.c. output without any a.c. input signal. Thus, the amplifier acts as an oscillator.

Regenerative or positive feedback is utilized in oscillators.

Hence, both assertion and reason are correct but reason is not the correct explanation of assertion.

378. (d)

The frequency of Hartley oscillator is given by

$$f_0 = \frac{1}{2\pi\sqrt{(L_1+L_2)C}}$$

or, $$L_1 = \frac{1}{4\pi^2 f_0^2 C} - L_2$$

$$= \frac{1}{4\pi^2(120\times10^3)^2\times0.004\times10^{-6}} - 0.4\times10^{-3}$$

$= 0.44 \times 10^{-3} - 0.4 \times 10^{-3} = 0.04$ mH

379. (b)

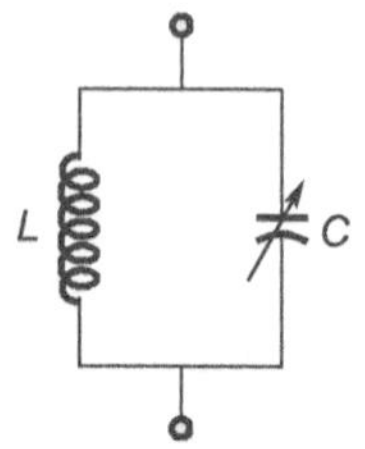

Tank circuit

For the given tank circuit,

$$f_0 = \frac{1}{2\pi\sqrt{LC}} \text{ or } C = \frac{1}{(2\pi f_0)^2}$$

Given, f_0 = (540 – 1650) kHz, L = 1 mH

$$\therefore \quad C_{max} = \frac{1}{4\pi^2 \times (540\times10^3)^2 \times 1\times10^{-3}}$$
$$= 86.86 \text{ pF}$$

$$C_{min} = \frac{1}{4\pi^2 \times (1650\times10^3)^2 \times 1\times10^{-3}} = 9.3 \text{ pF}$$

$\therefore$ 9.3 pF $\le C \le$ 86.86 pF

380. (a)

For sustained oscillations, the minimum value of current gain or forward current gain is

$$\beta = h_{fe} = \left(23 + 29\frac{R}{R_C} + \frac{4R_C}{R}\right)$$
$$= \left(23 + 29\times\frac{10}{2.2} + 4\times\frac{2.2}{10}\right) = 155.6 \approx 156$$

381. (c)

The operating frequency of a Wien-bridge oscillator is given by

$$f_0 = \frac{1}{2\pi RC} \text{ or } C = \frac{1}{2\pi f_0}$$
$$= \frac{1}{2\pi \times 100\times10^3 \times 10\times10^3} = 159 \text{ pF}$$

382. (b)

Both assertion and reason are individually correct statements. The correct reason for the assertion is that the resonant frequency of the crystal is temperature dependent, therefore it is necessary to enclose the crystal in a temperature controlled oven to achieve the required frequeny stability.

383. (d)

All oscillators do not require positive feedback for their operation because if the positive resistance of the LC tank circuit is cancelled by introducing the right amount of negative resistance across the tank circuit, then the steady oscillation can be maintained. Some of the devices which possess negative resistance as mentioned above are dynatron, transitron, UJT and tunnel diode. Hence, assertion is not correct.

385. (d)

The period of oscillation for an asymmetrical astable multivibrator is

$$T = 0.693\,(R_1C_1 + R_2C_2)$$
$$= 0.693\,(2\times10^3\times0.01\times10^{-6} + 10\times10^3\times0.05\times10^{-6})$$
$$= 360.36\,\mu\text{s}$$

$\therefore$ The frequency of oscillation,

$$f = \frac{1}{T} = \frac{1}{360.36\times10^{-6}}$$
$$= 2.775 \text{ kHz} \approx 2.8 \text{ kHz}$$

387. (c)

The frequency ranges for given oscillators are:

Oscillator Type	Frequency Range
Radio frequency Osc. (RFO)	20 kHz – 30 MHZ
Ultra high frequency Osc.	300 MHZ – 3 GHz
Very high frequency Osc.	30 MHZ – 300 MHZ
Microwave frequency Osc.	> 3 GHz

390. (c)

Given circuit represents a Hartley oscillator using an op-amp.

Here, L_1 = 12 μH and L_2 = 38 μH.

The minimum gain required in the circuit to produce sustained oscillation in the output is

$$|A| \ge \frac{L_2}{L_1} \text{ or } |A| \ge \frac{38}{12} \text{ or } |A| \ge 3.166$$

For $|A| = 4$, we will get sustained oscillation in output (All other values of $|A|$ are less than 3.166).

391. (b)

An oscillator in which the loop gain is exactly unity is an abstraction completely unrealizable in practice. It is accordingly necessary, in the adjustment of a practical oscillator, always to arrange to have $|\beta A|$ somewhat larger (say 5%) than unity in order to ensure that, with incidental variations in transistor and circuit parameters, $|\beta A|$ shall not fall below unity.

Reason is also a true statement but not the correct explanation of assertion.

392. (c)

According to Barkhausen criterion, for sustained oscillation in output, $-A\beta = 1$ and phase shift = 360° (or integral multiple of 2π).

396. (b)

Positive feedback increases the gain of the amplifier (only advantage of positive feedback) on the other hand bandwidth of the amplifier is reduced to a greater extent due to positive feedback. Due to this reason for amplifier design, positive feedback is hardly ever used.

397. (a)

Given, $R_A = 10\text{ k}\Omega$, $C = 50$ nF
$V_{CC} = 5$ V and $\beta = 20$

The pulse width (T) is

$$T = 1.1\, R_A C$$
$$= 1.1 \times (10\text{ k}) \times (50\text{ n})$$
$$= 550\ \mu\text{s}$$

398. (c)

For $t > 550$ ns, the capacitor will discharge through the discharge transistor. The base current of the transistor will be

$$I_b = \frac{V_{CC} - 0.7}{10\text{ k}} = 0.43\text{ mA}$$

$\therefore \quad I_c = \beta I_b = 0.86 \times 10^{-2}$ A

Assuming a constant discharge current (I_c),

$$I_c = C\frac{dV_c}{dt} = C\frac{\Delta V_c}{\Delta t}$$

or, $$\Delta t = C\frac{dV_c}{I_c} = 50 \times 10^{-9} \times \frac{5}{0.86 \times 10^{-2}}$$
$$= 29.07\ \mu\text{s}$$

402. (d)

For an RC phase shift oscillator, the minimum voltage gain required to produce sustained oscillations in the output is:

$|A| \geq 29$ (using op-amp); $\mu \geq 29$ (using FET)

403. (d)

The frequency of oscillation for a Wien bridge oscillator is

$$\omega_{osc} = \frac{1}{\sqrt{R_1 R_2 C_1 C_2}}$$

(When $R_1 \neq R_2$ and $C_1 \neq C_2$) $= \dfrac{1}{RC}$

(When $R_1 = R_2$ and $C_1 = C_2$)

Thus, when R (or R_1 and R_2) decreases, then ω_{osc} (or f_{osc}) increases.

406. (c)

For Wien bridge oscillator,

$$f_{osc} = \frac{1}{2\pi RC\sqrt{6}}$$

For $R = 200\text{ k}\Omega$, $C = \dfrac{1}{20\pi}$ nF

$$\therefore \quad f_{osc} = \frac{1}{2\pi \times (200 \times 10^3) \times \frac{1}{20\pi} \times 10^{-9}\sqrt{6}}$$

$$= \frac{10^5}{2\sqrt{6}}\text{ Hz} = \frac{50}{\sqrt{6}}\text{ kHz}$$

For $R = 400\text{ k}\Omega$, $C = \dfrac{1}{40\pi}$ nF

$$\therefore \quad f_{osc} = \frac{1}{2\pi \times (400 \times 10^3) \times \frac{1}{40\pi} \times 10^{-9}\sqrt{6}}$$

$$= \frac{50}{\sqrt{6}}\text{ kHz}$$

For producing sustained oscillation in output, $|\mu| \geq 29$. Hence, option (c) is not correct.

407. (b)

Statement-3 is not correct because Clapp oscillator is a modification of Colpitt oscillator where the variable inductor L is replaced by a variable capacitor C_3 in series with an inductor L and output is inductively provided.

410. (b)

$$\tan\phi = \frac{1}{2\pi f RC} = \tan 45° = 1$$

or, $$f = \frac{1}{2\pi RC} = \frac{1}{2 \times 3.14 \times 1 \times 10^3 \times 500 \times 10^{-12}}$$
$$= 3.2 \times 10^5\text{ Hz}$$

412. (b)

The oscillating frequency of a Hartley Oscillator lies between 20 kHz to 30 MHz which lies in the range of radio frequency.

■■■■

UNIT 10

Microprocessor

1. Introduction to 8085 and Its Functional Organisation

Q.1 A programme written in machine language is called the
(a) hardware program
(b) source program
(c) object program
(d) both (a) and (c)

Q.2 In 8085 microprocessor the number of interrupts maskable and non-maskable are
(a) four and one respectively
(b) one and four respectively
(c) five and zero respectively
(d) zero and five respectively

Q.3 The interrupt pin TRAP of a 8085 microprocessor is
(a) edge triggered
(b) level triggered
(c) both edge and level triggered
(d) neither edge nor level triggered

Q.4 Match **List-I** with **List-II** and select the correct answer using the code given below the lists:

List-I
A. Compiler
B. Assembler
C. Interpreter
D. System Software

List-II
1. High-level language program into machine codes, statementwise.
2. High-level language program into machine language program.
3. Controls the overall operation of a computer.
4. Assembly language program into a machine language program.

Codes:

	A	B	C	D
(a)	1	4	2	3
(b)	2	3	1	4
(c)	1	2	3	4
(d)	2	4	1	3

Q.5 A micro controller differs from a microprocessor in terms of
(a) data bus width and clock speed
(b) I/O interfaces and instruction decoding
(c) memory configuration and I/O interfaces
(d) memory configuration and instruction decoding

Q.6 A microprocessor program written in assembly language is translated into machine language. The number of instructions in the machine language when compared with the number of instructions in assembly language is
(a) same
(b) more only
(c) less only
(d) either more or less

Q.7 **Assertion (A):** The writing of a program in assembly language is much easier and faster as compared to the writing of program in machine language.
Reason (R): An assembly language is a low-level language.
(a) Both A and R are true and R is the correct explanation of A.
(b) Both A and R are true but R is not the correct explanation of A
(c) A is true but R is false
(d) A is false but R is true

Q.8 Consider the following statements:
1. A cross assembler runs on the microcomputer on which it produces machine codes.
2. A resident assembler runs on a computer other than that for which it produces object codes.

3. The program written in hexadecimal system is converted into binary system by a program called Monitor program.
4. Macros are the assembler directive which are commands to the assembler itself.

Which of the above statements are correct?
(a) 1, 2 and 3 only (b) 2, 3 and 4 only
(c) 1 and 3 only (d) 3 and 4 only

Q.9 Match **List-I** with **List-II** and select the correct answer using the codes given below the lists:

List-I
A. Operating system
B. Locator
C. Editor
D. Debugger

List-II
1. Allows user to examine the contents of registers and memory locations after each step of execution.
2. A program or collection of programs which are usually stored on discs.
3. A program which assigns specific memory addresses for the machine codes of the program to be loaded into the memory.
4. A program which allows user to write source program or text into RAM.

Codes:

	A	B	C	D
(a)	3	4	2	1
(b)	2	1	4	3
(c)	2	3	4	1
(d)	1	2	3	4

Q.10 The number of input pins of 8085 microprocessor are
(a) 21 (b) 19
(c) 40 (d) 27

Q.11 Which of the following statements is correct? In 8085 microprocessor, the READY signal is useful when the CPU communicates with
(a) a fast peripheral device
(b) a slow peripheral device
(c) a PPI chip
(d) a DMA controller chip

Q.12 Match **List-I** with **List-II** and select the correct answer using the codes given below the lists:

List-I
A. ALE
B. PSW
C. CMA
D. RLC

List-II
1. Rotate Accumulator left
2. Compare with accumulator
3. Program status word
4. Address latch enable
5. Program stack word
6. Arithmetic logic enabled
7. Complement accumulator
8. Rotate accumulator left through carry

Codes:

	A	B	C	D
(a)	6	5	2	8
(b)	6	5	7	1
(c)	4	3	2	8
(d)	4	3	7	1

Q.13 The highest integer that can be stored in an 8-bit accumulator is
(a) 256 (b) 64
(c) 255 (d) 144

Q.14 The number of general purpose 8-bit registers in 8085 microprocessor are
(a) 6 (b) 9
(c) 3 (d) 12

Q.15 Which of the following statements about program counter (PC) is/are correct?
1. It performs logical operations.
2. It is a cell in ROM.
3. During execution of the current instructions, its contents changes.
4. It is a register.

Select the correct answer using the code given below:
(a) 1 and 2 (b) 3 and 4
(c) 4 only (d) 2 and 3

Q.16 In 8085 pairing of registers B, C, D, E, H, L is:
(a) B–C, D–L, H–E (b) B–D, C–E, H–L
(c) B–C, D–E, H–L (d) either of these

Q.17 In a microprocessor, the program counter points to the address location from where the

(a) current byte is to be added
(b) next byte is to be stored
(c) current byte is to be fetched
(d) next byte is to be fetched

Q.18 Consider the table given below where s_1 and s_0 are status signals:

Microprocessor operation	S_1	S_0
X	0	0
Y	1	1
Z	1	0

Here x, y and z respectively stands for
(a) no operation, opcode fetch and write operation.
(b) no operation, opcode fetch and read operation.
(c) opcode fetch, no operation and write operation.
(d) opcode fetch, no operation and read operation.

Q.19 Microprocessor based instruments can be used to measure
(a) temperature only
(b) pressure only
(c) flow only
(d) all the quantities in (a), (b) and (c)

Q.20 **Assertion (A):** Magnetic cores are generally used in main memory of a digital computer.
Reason (R): Magnetic cores are slow and volatile.
(a) Both A and R are true and R is the correct explanation of A.
(b) Both A and R are true but R is not the correct explanation of A.
(c) A is true but R is false.
(d) A is false but R is true.

Q.21 Consider the following statements:
1. Accumulator is a hardware which stores one of the operand to be maintained by the ALU.
2. Compilers used in microcomputers are classified as application softwares.
3. Mnemonic symbols are used to assist human memory.
4. A PC counts the number of programs run after starting.

Which of the statements given above are correct?
(a) 1 and 3 (b) 2, 3 and 4
(c) 1 and 2 (d) 1, 2 and 4

Q.22 Which interrupt has the lowest priority?
(a) TRAP (b) RST 4.5
(c) INTR (d) RST 7.5

Q.23 What is the vector address of interrupt RST 6?
(a) 0038 H (b) 0034 H
(c) 0024 H (d) 0030 H

Q.24 Which of the following is a software interrupt?
(a) RST 4.5 (b) RST 0
(c) both (a) and (b) (d) none of these

Q.25 Match **List-I (Interrupts)** with **List-II (Characteristics)** and select the correct answer using the codes given below the lists:

List-I	List-II
A. RST 7.5	1. Software interrupt
B. RST 4.5	2. Non-vectored
C. INTR	3. Edge-sensitive
D. RST 7	4. Non-maskable

Codes:

	A	B	C	D
(a)	1	2	4	3
(b)	4	3	2	1
(c)	3	4	2	1
(d)	2	1	3	4

Q.26 The number of control and status signals in 8085 microprocessor are respectively
(a) 2 and 3 (b) 2 each
(c) 3 and 2 (d) 2 and 4

Q.27 **Assertion (A):** The lower order address lines of 8085 microprocessor are shared with the data lines.
Reason (R): The ALE signal is used to distinguish between lower order address lines and data lines.
(a) Both A and R are true and R is the correct explanation of A.
(b) Both A and R are true but R is not the correct explanation of A.
(c) A is true but R is false.
(d) A is false but R is true.

Q.28 If the status lines S_1 and S_0 is HIGH, then 8085 microprocessor is performing:
(a) Halt operation
(b) HOLD operation
(c) Opcode fetch operation
(d) None of the above

Q.29 Which of the following option is not correct about the various flags in 8085 microprocessor?
(a) The sign-flag is set to zero, if the most significant bit of the result of an arithmetic or logic operation is zero.
(b) The zero-flag is set if the ALU operation result is zero.
(c) The parity flag is set if the result has odd parity.
(d) The carry flag is set if the ALU operation result generates a carry.

Q.30 Consider the following statements associated with the stack pointer:
1. It is automatically initialized with an instruction LXI SP, 16 bit address.
2. It points to top of the stack.
3. It is a 16 bit register.
4. It is a UP/DOWN counter.
5. It is incremented by two after execution of each POP instruction and decremented by two after execution of each PUSH instruction.

Which of the statements given above is/are correct?
(a) 2, 4 and 5 (b) 1, 3 and 4
(c) 1, 2, 3, 4 and 5 (d) 2, 3 and 5

Q.31 Pseudo-instructions are basically
(a) assembler directives
(b) instructions that are ignored by the microprocessor
(c) false instructions
(d) instructions that are treated like comments

Q.32 Which microprocessor pins are used to request and acknowledge a DMA transfer?
(a) ready and wait (b) HOLD and HL DA
(c) reset and ready (d) none of these

Q.33 The first modern computer was called _____ .
(a) UNIVAC-I (b) FLOW-MATIC
(c) ENIAC (d) INTEL

Q.34 NMI input is
(a) level sensitive
(b) edge-sensitive
(c) both edge and level sensitive
(d) neither edge nor level sensitive

Q.35 The data bus in 8080 A/8085 microprocessor is a group of
(a) eight bidirectional lines that are used to transfer 8-bits between the microprocessor and its I/O and memory.
(b) eight lines used to transfer data among the registers.
(c) eight unidirectional lines that are used for I/O devices.
(d) sixteen bidirectional lines that are used for data transfer between the microprocessor and memory.

Q.36 What is the direction of control bus?
(a) Unidirectional into μP
(b) Unidirectional out of μP
(c) Bidirectional
(d) Mixed direction i.e. some lines into μP and some others out of μP

Q.37 Every processor must necessarily have
(a) data bus
(b) data bus and address bus
(c) control bus
(d) address bus, data bus and control bus

2. Microprocessor Interfacing

Q.38 The number of address lines required to address 8 K bytes of memory is
(a) 14 (b) 13
(c) 16 (d) 15

Q.39 The number of memory chips of size 1 K × 4-bits required to build a memory bank of size 16 K × 8 bits is
(a) 8 (b) 16
(c) 32 (d) 64

Q.40 The logic circuit used to generate the active low chip select (CS) by an 8085 microprocessor to address a peripheral is shown in figure below.

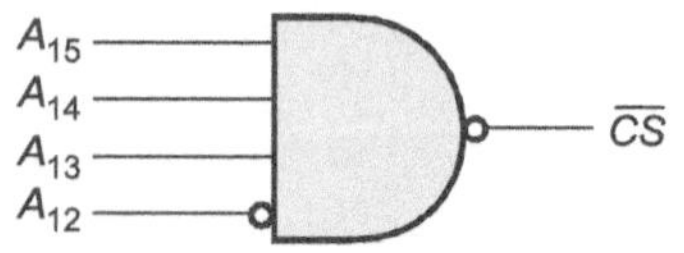

The address range for which the device will respond is:
(a) $E000_H - EFFF_H$ (b) $F000_H - FFFF_H$
(c) $F001_H - FFFF_H$ (d) $E001_H - EFFC_H$

Q.41 Eight memory chips of size 64 × 8 bits have their address buses connected together. What will be the size of the resultant memory?
(a) 256 × 16 bits (b) 128 × 32 bits
(c) 64 × 64 bits (d) 512 × 8 bits

Q.42 What is the address range for 4 K RAM used in 8085 microprocessor if $CS = \bar{A}_{15} A_{14} A_{13}$ is used as the chip select logic?
(a) 7000_H to $7FFF_H$ (b) 6000_H to $6FFF_H$
(c) Either (a) or (b) (d) None of these

Q.43 An 8-bit microcontroller has an external RAM with the memory map from 8000 H to 9FFF H. The number of bytes this RAM can store is:
(a) 8191 (b) 8000
(c) 8193 (d) 8192

Q.44 Consider the following statements associated with memory mapped I/O scheme of communication with microprocessor:
1. It reduces the memory space available.
2. Arithmetic or logical operation can be directly performed with I/O data.
3. The processor cannot manipulate I/O data residing in interface registers with the same instructions that are used to manipulate memory location.
4. The processor on treats an interface register as a part of the memory system.

Which of the statements given above is/are not correct?
(a) 1 and 4 only (b) 3 only
(c) 3 and 4 only (d) 2 only

Q.45 The I/O devices in 8085 microprocessor can be used in:
(a) Memory mapped I/O or I/O mapped I/O
(b) I/O mapped I/O only
(c) memory mapped I/O only
(d) none of the above

Q.46 The maximum number of memory and I/O devices that can be addressed in a 8085 microprocessor based system is:
(a) infinite (b) $2^8 + 2^{16}$
(c) 2^8 (d) 2^{16}

Q.47 The capacity of a memory chip is 8192 bits and it has 2048 rows. Then the organisation of the chip is
(a) word organized (b) byte organized
(c) nibble organized (d) bit organized

Q.48 **Assertion (A):** Semiconductor memories are preferred.
Reason (R): Semiconductor memories are non-volatile and of large size.
(a) Both A and R are true and R is the correct explanation of A.
(b) Both A and R are true but R is not the correct explanation of A
(c) A is true but R is false
(d) A is false but R is true

Q.49 The size of the memory, a microprocessor can access is determined by the number of _____ lines the microprocessor
(a) data (b) control
(c) address (d) status

Q.50 In an 8085 microprocessor system with memory mapped I/O
(a) I/O devices have 8-bit addresses.
(b) I/O devices are accessed using IN and OUT instructions.
(c) there can be a maximum of 256 input devices and 256 output devices.
(d) arithmetic and logic operations can be directly performed with the I/O data.

Q.51 A ROM is used to store the table for multiplication of two 8-bit unsigned integers
(a) 256 × 16 (b) 64 K × 8
(c) 4 K × 16 (d) 64 K × 16

Q.52 An advantage of memory interfacing is that
(a) a larger memory is obtained
(b) effective speed of the memory is increased
(c) the cost of the memory is reduced
(d) a non-volatile memory is obtained

Q.53 The capacity of memory unit is defined by the number of words multiplied by the number of bits/word. How many separate address and data lines are needed for a memory of 4 K × 16?

(a) 10 address, 16 data lines
(b) 11 address, 8 data lines
(c) 12 address, 16 data lines
(d) 12 address, 12 data lines

Q.54 Consider a 4 KB ROM with 4 output bits. If the encoder is square, how many bits are needed for X-address and Y-address?
(a) 32, 32 (b) 128, 128
(c) 64, 64 (d) 256, 256

Q.55 The highest address in a 48 K memory, in hexadecimal system is:
(a) FFFF H (b) FEEE H
(c) CDEF H (d) BFFF H

Q.55 The memory address of the last location of a 1 K byte memory chip is given as 0FBFF H. What will be the address of the first location?
(a) 0F817 H (b) 0F818 H
(c) 0F800 H (d) 0F801 H

Q.57 In an 8-bit microcomputer having 8 K bytes of RAM memory, the length of SP will be:
(a) 5 (b) 8
(c) 11 (d) 13

Q.58 The Intel Pentium processor, address bus consists of two sets of signals: 29 address line (A_{31} – A_3) and 8 byte enable (BE_7 – BE_0) is 0000 0108 H and all byte enable lines ($BE_7 - BE_0$) are assorted, then the processor's intention is to address
(a) one byte at address 0000 0108 H
(b) eight byte at locations 00000101 to 00000108 H
(c) eight bytes at locations 00000108 to 0000010F H
(d) one byte at location 0000 010F H

Q.59 A typical cell, for dynamic RAM can be implemented by using how many MOS transistors?
(a) Six (b) Five
(c) One (d) Two

Q.60 A microprocessor has 24 address lines and 32 data lines. If it uses 10 bits of opcode, the size of its memory buffer register is:
(a) 22 bits (b) 24 bits
(c) 32 bits (d) 14 bits

Q.61 Consider the memory system with the following parameters:
T_c = Cache access time = 100 ns
T_m = Main memory access time = 1200 ns.
If we would like to use the cache per paging and the hit ratio of finding the page table entry in the cache is 98%, the effective of average paged memory access time in seconds is equal to
(a) 200 (b) 340
(c) 1200 (d) 2400

Q.62 A 100 K byte memory is managed using variable partitions but no compaction. It currently has two partitions of sizes 200 K bytes and 260 K bytes respectively. The smallest allocation request in K bytes that could be denied is for
(a) 151 (b) 181
(c) 231 (d) 541

Q.63 The decoding circuit shown in figure below has been used to generate the active low chip select signal for a microprocessor peripheral (the address lines are designated as A_0 to A_7 for input addresses). The peripheral will correspond to input addresses in the range.

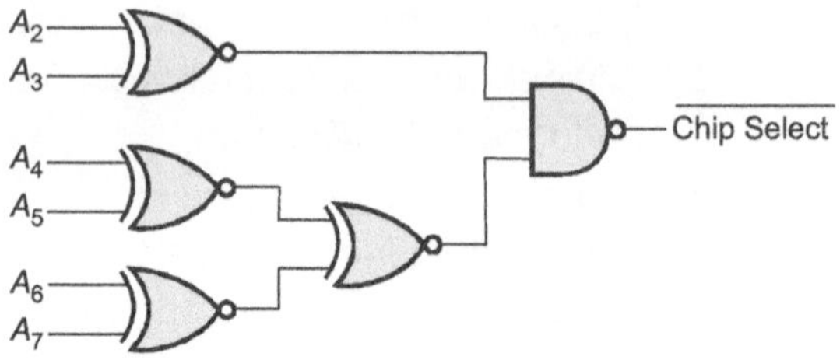

(a) 60 H to 63 H (b) A4 H to A7 H
(c) 30 H to 33 H (d) 70 H to 73 H

Q.64 Match **List-I** with **List-II** and select the correct answer using the codes given below the lists:

List-I	List-II
A. SID, SOD	1. Wait state
B. READY	2. Interrupt
C. TRAP	3. Serial data transfer
D. ALE	4. Memory or I/O read/write
	5. Address Latch control

Codes:

	A	B	C	D
(a)	3	1	5	2
(b)	3	1	2	5
(c)	4	3	2	5
(d)	4	3	1	2

Q.65 Suppose 64 KB, ROM ICs are available in abundance. 1 MB ROM can be obtained from
(a) 16 ICs in a row
(b) 16 ICs in a column
(c) 8 ICs in a column and 2 ICs in a row
(d) None of the above

Q.66 Match **List-I (Type of Memory)** with **List-II (Used As)** and select the correct answer using the code given below the lists:

List-I	List-II
A. DRAM	1. Cache memory
B. SRAM	2. Main memory
C. Parallel Access Registers	3. BIOS
D. ROM	4. CPU register

Codes:

	A	B	C	D
(a)	1	2	3	4
(b)	2	1	4	3
(c)	1	2	4	3
(d)	2	1	3	4

Q.67 **Assertion (A):** The read/write memory consists of registers which are made of group of flip-flops.
Reason (R): A flip-flop is a basic element of memory.
(a) Both A and R are true and R is the correct explanation of A.
(b) Both A and R are true but R is not the correct explanation of A
(c) A is true but R is false
(d) A is false but R is true

Q.68 **Assertion (A):** 14 address lines are required to address a 4096 byte memory.
Reason (R): The number of address lines in a memory chip are determined from the number of registers in the memory chip.
(a) Both A and R are true and R is the correct explanation of A.
(b) Both A and R are true but R is not the correct explanation of A.
(c) A is true but R is false.
(d) A is false but R is true.

Q.69 **Assertion (A):** Interfacing of I/O mapped I/O devices with the microprocessor is easy.
Reason (R): The time required for I/O mapped I/O devices to communicate with microprocessor is less compared to memory mapped I/O devices.
(a) Both A and R are true and R is the correct explanation of A.
(b) Both A and R are true but R is not the correct explanation of A.
(c) A is true but R is false.
(d) A is false but R is true.

3. Instruction Sets and Data Formats

Q.70 A microprocessor using a 3 MHz clock has three '*T*' states in each machine cycle. What is the time taken to complete the execution of an instruction if the instruction cycle of this instruction needs 4 machine cycles?
(a) 1 μs (b) 333 ns
(c) 4 μs (d) 1333 ns

Q.71 The addressing mode in which the machine language instruction itself includes the data is called
(a) direct (b) indexed
(c) immediate (d) indirect

Q.72 The content of TEMP location after the execution of the following program in a 8085 microprocessor is:
MVI B, 00 H
MVI A, 1C H
DCR B
DAA
STA TEMP
HLT
(a) 10 H (b) 22 H
(c) 1F H (d) 46 H

Q.73 What is the addressing mode used in the instruction PUSH D
(a) Immediate (b) Register
(c) Direct (d) Register indirect

Q.74 Which of the following instruction does not use the stack?
(a) JMP (b) CALL
(c) RET (d) None of these

Q.75 An instruction cycle is the time in which, a hard-wired controller completes four functions. The correct sequence of these functions is

(a) Execute-decode-fetch-update
(b) Fetch-update-decode-execute
(c) Fetch-decode-update-execute
(d) Decode-execute-fetch-update

Q.76 What is the effect of a DAD H instruction?
(a) To shift each bit one position to the right with a zero inserted in MSB position
(b) To shift each bit one position to the left with a zero inserted in LSB position
(c) To shift each bit one position to the right
(d) To shift each bit one position to the left

Q.77 The following program is executed in the 8085 microprocessor:

Address	Code	Mnemonics
203A	3E 20	MVIA, 20 H
203C	2A 3A 20	LHLD 203A H
203F	86	ADD M
2040	76	HLT

What are the contents of the accumulator after the execution of above program?
(a) 40 H (b) 7C H
(c) 20 H (d) 5E H

Q.78 In an 8085 microprocessor, the accumulator contents are AAH. After the execution of the instruction 'CPI 99H' in the microprocessor
(a) both carry and zero flags will be set
(b) both carry and zero flags will be reset
(c) carry flag will be set but zero flag will be reset
(d) carry flag will be reset but zero flag will be set

Q.79 Which of the following instruction will alter the contents of accumulator?
(a) CMP C (b) CPI 3A H
(c) ANI 5C H (d) ORA A

Q.80 Match **List-I (Addressing modes of 8085)** with **List-II (Instructions)** and select the correct answer using the codes given below the lists:

List-I	List-II
A. Immediate	1. LDA 3000 H
B. Implied	2. MOV A, B
C. Register	3. LXI H, 2050 H
D. Direct	4. RRC

Codes:

	A	B	C	D
(a)	3	2	1	4
(b)	3	2	4	1
(c)	3	4	2	1
(d)	2	3	1	4

Q.81 A 8085 microprocessor uses a crystal frequency of 5 MHz. The duration of one T-state would be
(a) 0.4 μs (b) 5.0 μs
(c) 0.2 μs (d) 2.5 μs

Q.82 In 8085 with 2 MHz clock frequency, what is the time delay obtained after execution of 4 NOP instructions?
(a) 6 μs (b) 9 μs
(c) 5 μs (d) 8 μs

Q.83 Consider the following program for 8085 microprocessor:

```
      XRA A
      LXI B, 0007 H
LOOP: DCX B
      JNZ LOOP
```

The loop will be executed
(a) infinite times (b) 6 times
(c) once (d) 7 times

Q.84 Consider the following:
The 16-bit register/register pair which cannot be stored in stack using PUSH instruction is
1. HL 2. SP
3. PSW 4. PC
Select the correct answer using the codes given below:
(a) 1 and 2 (b) 1 and 3
(c) 2 and 3 (d) 2 and 4

Q.85 Which of the following 8085 instruction does not affect the flags?
(a) DCR A (b) INX B
(c) INR A (d) ORA C

Q.86 Consider the following 8085 microprocessor program segment:

```
         MVI A, 92 H
         ORA A
         JP PORT 1
         XRA A
PORT 1 : OUT F2 H
         HLT
```

The output at port F2 H after the execution of the above program would be
(a) 50 H (b) 92 H
(c) 00 H (d) 7 F H

Q.87 Match **List-I (Characters)** with **List-II (Interpretations)** and select the correct answer using the codes given below the lists:

List-I
A. Colon (:)
B. Blank ( )
C. Comma (,)
D. Semicolon (;)

List-II
1. Separates operation and operand field
2. Separates operands
3. Beginning of comment field
4. End of label field

Codes:

	A	B	C	D
(a)	2	3	4	1
(b)	4	1	2	3
(c)	2	1	4	3
(d)	4	3	2	1

Q.88 Which of the 8085 instructions uses the stack?
(a) JMP (b) RET
(c) DCR A (d) MOV

Q.89 An intel 8085 microprocessor is executing the following program:

```
        MVI A, 10 H
        MVI B, 10 H
BACK :  NOP
        ADD B
        RLC
        JNC BACK
        HLT
```

How many times the NOP instruction in the above program will be executed?
(a) 3 times (b) once
(c) 9 times (d) None of these

Q.90 The advantage of immediate addressing is
(a) to decode the operand easily
(b) to execute the instruction immediately
(c) not to fetch operand from memory
(d) to allow both positive and negative operands

Q.91 In register indirect addressing mode of 8085 microprocessor, data is stored
(a) in the register pair
(b) in a fixed location of the memory
(c) at the address contained in the register pair
(d) in the accumulator

Q.92 If a 8085 microprocessor program uses all available Jump instructions, each only once, then for this program, the total memory (in bytes) occupied by the Jump instructions would be equal to
(a) 24 (b) 18
(c) 30 (d) 27

Q.93 An arithmetic operation in the 8085 microprocessor sets the sign and parity flags. The contents of the accumulator after the execution of the operation can be
(a) 10101101 (b) 01100111
(c) 10110100 (d) 00101101

Q.94 What is the content of memory location 2070 H after the execution of the following 8085 program?

```
LXI B, 2070 H
MVI A, 8F H
MVI C, 68 H
SUB C
ANI 0F H
STAX B
HLT
```

(a) 09 H (b) OF H
(c) 04 H (d) 07 H

Q.95 An instruction of the 8085 microprocessor that requires both memory read and memory write machine cycles is:
(a) RST 1 (b) LHLD 8088
(c) MVI M, 8F (d) ADD M

Q.96 **Assertion (A):** In 8085 microprocessor, the instruction DAA is used for both BCD addition and BCD subtraction
Reason (R): With DAA instruction, the binary contents of accumular are changed to its two digit BCD equivalent.
(a) Both A and R are true and R is the correct explanation of A.

(b) Both A and R are true but R is not the correct explanation of A.
(c) A is true but R is false.
(d) A is false but R is true.

Q.97 Consider the following statements associated with various instructions of a 8085 microprocessor:
1. CMA performs the logical NOT operation.
2. HLT is a machine control instruction which uses an implicit addressing mode.
3. NOP instruction is used to introduce a delay of 4T states between two successive instructions.
4. All the instructions related to stack have indirect addressing mode.

Which of the statements given above are correct?
(a) 1, 3 and 4 (b) 2, 3 and 4
(c) 1, 2 and 3 (d) 1, 2 and 4

Q.98 **Assertion (A):** The stack pointer works on the principle of last-in-first-out.
Reason (R): Stack pointer is always incremented by 2 after the execution of each PUSH instruction and is decremented by 2 after the execution of each POP instructions.
(a) Both A and R are true and R is the correct explanation of A.
(b) Both A and R are true but R is not the correct explanation of A
(c) A is true but R is false
(d) A is false but R is true

Q.99 **Assertion (A):** In 8085 microprocessor can identity 256 different input or 256 different output peripheral mapped I/O devices, with address ranging from 00H to FF H.
Reason (R): The peripheral mapped I/O devices are identified with 8-bit port address.
(a) Both A and R are true and R is the correct explanation of A.
(b) Both A and R are true but R is not the correct explanation of A
(c) A is true but R is false
(d) A is false but R is true

Q.100 **Assertion (A):** Call instructions are used to call subroutine during execution of a main program.
Reason (R): When microprocessor encounters a return instruction, the execution of the program is returned to the main program.
(a) Both A and R are true and R is the correct explanation of A.
(b) Both A and R are true but R is not the correct explanation of A
(c) A is true but R is false
(d) A is false but R is true

Q.101 Match **List-I (Instructions of 8085)** with **List-II (Addressing Modes)** and select the correct answer using the code given below the lists:

	List-I		List-II
A.	LHLD 3000 H	1.	Register
B.	SPHL	2.	Immediate
C.	DAA	3.	Direct
D.	JMP 2000 H	4.	Implied

Codes:

	A	B	C	D
(a)	4	1	2	3
(b)	2	4	3	1
(c)	1	3	4	2
(d)	3	1	4	2

Q.102 Match **List-I (Instructions)** with **List-II (Execution time)** and select the correct answer using the codes given below the lists:

	List-I		List-II
A.	LXI B, 3000 H	1.	6 T-states
B.	PCHL	2.	10 T-states
C.	PUSH PSW	3.	16 T-states
D.	XTHL	4.	10 T-states

Codes:

	A	B	C	D
(a)	4	3	2	1
(b)	3	1	4	2
(c)	4	1	2	3
(d)	4	2	1	3

Q.103 Consider the following statements:
1. Unconditional CALL instruction in 8085 microprocessor requires five machine cycles and total 18 T-states.
2. No flag is affected by data transfer instructions.
3. The contents of registers are unaffected during execution of HLT instruction.

4. Stack pointer must be initiated before using PUSH, POP or call instructions.
5. In PSW (Program Status Word), the accumulator is high order register and the flag register is low order register.

Which of the statements given above are correct?

(a) 1, 2, 3, 4 and 5 (b) 1, 3 and 5
(c) 3, 4 and 5 (d) 2, 3 and 5

Q.104 During which of the following T-states does the CPU sends the address to memory or I/O and the ALE signal for de-multiplexing?

(a) T_4 (b) T_3
(c) T_2 (d) T_1

Q.105 What will be the contents of register *A* after the execution of the following instructions in a 8085 microprocessor?

MOV B, 8C H
MOV A, 7E H
ADD A, B

(a) 6 A and carry flag is set
(b) 6 A and carry flag is reset
(c) 0 A and carry flag is set
(d) 0 A and carry flag is reset

Q.106 A microprocessor can be brought out of the wait state by

(a) pressing RESET key
(b) giving an interrupt
(c) pressing RESET Key or by giving an interrupt
(d) none of the above

Q.107 Which of the following statements is/are not true?

1. The group of machine cycle is called a state.
2. A machine cycle consists of one or more instruction cycle.
3. An instruction cycle is made up of machine cycles and a machine cycle is made up of number of states.

(a) 1 and 3 (b) 2 only
(c) 3 only (d) 1 only

Q.108 In an 8 bit microcomputer, the fetch cycles required to fetch an 8 byte instruction will be ______.

(a) 8 (b) 32
(c) 16 (d) None of these

Q.109 What do the symbol [ ] indicates?

(a) Direct addressing
(b) Register addressing
(c) Indirect addressing
(d) None of these

Q.110 In a generic microprocessor, instruction cycle time is

(a) shorter than machine cycle time
(b) larger than machine cycle time
(c) exactly double the machine cycle time
(d) exactly the same as the machine cycle time

Q.111 In a multiprocessor configuration, two co-processors are connected to the host 8085 processor. The two co-processor instruction sets

(a) must be the same
(b) must overlap
(c) must be disjoint
(d) must be the same as that of the host

Q.112 The fetching, decoding and executing of an instruction is broken down into several time intervals. Each of these intervals, involving one or more clock periods, is called a

(a) instruction cycle (b) machine cycle
(c) process cycle (d) none of these

Q.113 With 2's complement representation, the range of values that can be represented on the data bus of an 8-bit microprocessor is given by

(a) −128 to +127 (b) −128 to +128
(c) −127 to +128 (d) −256 to +256

Q.114 A sequence of two instructions that multiplies the contents of the DE register pair by 2 and stores the result in the HL register pair (in 8085 assembly language) is

(a) XCHG and DAD B
(b) STHL and DAD H
(c) PCHL and DAD D
(d) XCHG and DAD H

Q.115 Return from a subroutine is affected by

(a) a jump instruction
(b) an RST instruction
(c) a RET instruction
(d) a hardware interrupt signal

Q.116 Consider the following statements:

1. Counters and time delays can be designed using software.

2. In 8085, 8-bit registers can be combined as register pairs (BC, DE and HL) to manipulate 16-bit data.
3. Instruction DCX and INX do not affects flags.
4. Instruction DCR and INR do not affects flags.

Which of the statements given above are correct?

(a) 1, 2 and 3 (b) 3 and 4
(c) 1 and 2 (d) 1 and 3

Q.117 Call in computer system means
(a) to execute a part of main program
(b) to call a sub-program to be executed
(c) to transfer control to specified closed subroutine
(d) none of the above

Q.118 The length of a bus cycle in 8086/8088 is four clock cycles T_1, T_2, T_3, T_4 and an intermediate number of wait state clock cycles denoted by T_W. The wait states are always inserted between
(a) T_1 and T_2 (b) T_2 and T_3
(c) T_3 and T_4 (d) T_4 and T_1

Q.119 If PUSH and POP operation on a stack takes 1 unit time, then how much time would it take to delete an element in n^{th} position from bottom?
(a) 1
(b) n
(c) n^2
(d) cannot be determined from the given data

Q.120 Consider the following statements:
1. Negative operands cannot be used.
2. When immediate operand changes, the programme should be reassembled.
3. The programme is difficult to read.
4. The size of operand is restricted by word length of the computer.

Disadvantages of immediate addressing include.

(a) 1 and 2 (b) 2 and 4
(c) 2 and 3 (d) 1 and 4

4. Peripheral Devices

Q.121 In a microcomputer, wait states are used to
(a) make the processor wait during a DMA operation.
(b) make the processor wait during an interrupt processing.
(c) make the processor wait during a power shutdown.
(d) interface slow peripherals to the processor.

Q.122 A 'DMA' Transfer implies
(a) direct transfer of data between memory and accumulator.
(b) direct transfer of data between memory and I/O devices without the use of μP.
(c) transfer of data exclusively within μP registers.
(d) a fast transfer of data between μP and I/O devices.

Q.123 Direct memory Access channel (DMA) facilitates data to move into and out of the system
(a) one first come first serve basis
(b) without equal time delay
(c) without subroutine
(d) without program intervention

Q.124 Bootstrap is a technique designed
(a) to transfer control to specified location
(b) to suspend the main program for a while
(c) to bring itself into a desired state by means of its own action
(d) none of the above

Q.125 The synchronization between microprocessor and memory is done by
(a) ALE signal (b) HOLD signal
(c) READY signal (d) None of these

Q.126 A USART chip provides
(a) half duplex operation
(b) full duplex operation
(c) duplex operation
(d) full duplex operation but cannot work in asynchronous mode

Q.127 For which of the following devices, is DMA the most suitable?
(a) Keyboard (b) Mouse
(c) Joy stick (d) Hard disc

Q.128 Consider the following statements associated with microprocessors:
1. Debug is synonymous with exponent.
2. Direct-Memory Access Channel (DMA) facilitates to move into and out of the system without program interruption.

3. Double precision employs double signal speed.
4. Dump means copying data from internal storage to external storage.

Which of the statements given above are correct?

(a) 2 and 4 (b) 1, 2 and 4
(c) 2 and 3 (d) 1, 3 and 4

Q.129 Match **List-I (Request)** with **List-II (Device)** and select the correct answer using the codes given below the lists:

List-I	**List-II**
A. Interrupt	**1.** I/O device
B. Wait	**2.** DMA controller
C. Hold	**3.** Memory

Codes:

	A	B	C
(a)	1	2	3
(b)	1	3	2
(c)	2	1	3
(d)	2	3	1

Q.130 What must be the contents of the control word of Intel 8255 for mode 0 (operation) and for the following ports configuration?

Port A - output, Port B - output

Port C_{lower} – Output, Port C_{upper} – Input

(a) 85 H (b) 86 H
(c) 87 H (d) 88 H

Q.131 Consider the following statements:

The advantage of cycle stealing in DMA is that

1. It increases the maximum I/O transfer rate.
2. It reduces the interference by the DMA controller in the CPU; memory access.
3. It is beneficially employed for I/O devices with shorter bursts of data transfer.

(a) 1 and 2 (b) 1 and 3
(c) 2 and 3 (d) 1, 2 and 3

Q.132 An input port is

(a) tri-state buffer with register
(b) select logic
(c) input device
(d) read write memory

Q.133 Which of the following methods could be used for I/O operation in an 8085 microprocessor?

1. Handshaking
2. Interrupting
3. Direct Memory Access

Select the correct answer using the codes given below:

(a) 1 and 2 (b) 1 and 3
(c) 2 and 3 (d) 1, 2 and 3

Q.134 Consider the following statements associated with DMA operation:

1. DMA operation ends with DMA controller sending an interrupt to processor.
2. DMA controller has bidirectional READ and WRITE control lines.
3. DMA operation begins with DMA controller sending a DMA request to processor.
4. DMA controller has a bidirectional address bus.

Which of the statements given above are correct?

(a) 2, 3 and 4 (b) 1, 2 and 3
(c) 1, 2 and 4 (d) 1, 2, 3 and 4

Q.135 When a processor uses handshaking, the following handshaking signals are present:

1. Acknowledge 2. Data transfer
3. Status 4. Strobe

What is the correct sequence in which the signal appear?

(a) Strobe-Status-Data Transfer-Acknowledge
(b) Status-Data Transfer-Acknowledge-Strobe
(c) Status-Strobe-Data Transfer-Acknowledge
(d) Strobe-Data Transfer-Acknowledge-Status

Q.136 8251 is a

(a) UART
(b) USART
(c) Programmable Interrupt Controller
(d) Programmable Interval timer/counter

Q.137 By what factor does the 8284 A clock generator divide the crystal oscillator's output frequency?

(a) One (b) Two
(c) Three (d) Four

Q.138 If the programmable counter timer 8254 is set in mode 1 and is to be used to count six events, the output will remain at logic 0 for _____ number of counts.

(a) 5 (b) 6
(c) 0 (d) All of these

Q.139 Assertion (A): The cycle stealing mode of DMA data transfer is used in hard disk drives.
Reason (R): The cycle stealing mode is suitable when long block of data is to be transferred with minimum interference with microprocessor
(a) Both A and R are true and R is the correct explanation of A.
(b) Both A and R are true but R is not the correct explanation of A.
(c) A is true but R is false.
(d) A is false but R is true.

Q.140 Assertion (A): DMA data transfer is faster than either Interrupt I/O or Polling based I/O for very large data transfers.
Reason (R): 8237/8257 ICs are DMA controllers.
(a) Both A and R are true and R is the correct explanation of A.
(b) Both A and R are true but R is not the correct explanation of A.
(c) A is true but R is false.
(d) A is false but R is true.

Q.141 To configure port A and port B as output ports, port C not being used in the 8155 programmable interface, the control word should have the value
(a) 02 H (b) 03 H
(c) 60 H (d) 01 H

Q.142 The interface Chip used for data transmission between 8085 and a 16-bit ADC is
(a) 8259 (b) 8255
(c) 8253 (d) 8251

5. Introduction to Microprocessor 8086

Q.143 Consider the following statements:
1. A total of about one million bytes can be directly addressed by the 8086 microprocessors.
2. 8086 has thirteen 16 bit- registers.
3. 8086 has eight flags.
4. Compared to 8086, the 80286 provides a higher degree of memory protection.

Which of the statements given above are correct?
(a) 2, 3 and 4 (b) 1, 3 and 4
(c) 1, 2 and 4 (d) 1, 2 and 3

Q.144 Which of the following pairs of 8086 microprocessor's interrupt is not correctly matched?
(a) INT 0 : Divide by zero
(b) INT 1 : Single step
(c) INT 2 : Break point
(d) INT 4 : Overflow

Q.145 Match **List-I (2 pins of 8086)** with **List-II (Status)** and select the correct answer using the codes given below the lists:

List-I

	$\overline{BHE}$	A_0
A.	0	0
B.	0	1
C.	1	0
D.	1	1

List-II
What is read/written?
1. 1-byte from/to odd address
2. 1-byte from/to even address
3. One 16-bit word
4. NOP

Codes:

	A	B	C	D
(a)	4	2	1	3
(b)	3	1	2	4
(c)	4	1	2	3
(d)	3	2	1	4

Q.146 Match **List-I (8085 Register)** with **List-II (8086 Register)** and select the correct answer using the codes given below the lists:

List-I	List-II
A. A	1. CH
B. H	2. AL
C. L	3. BL
D. B	4. BH

Codes:

	A	B	C	D
(a)	4	2	3	1
(b)	2	4	1	3
(c)	4	2	1	3
(d)	2	4	3	1

Q.147 Consider the following instructions executed in 8086;
PUSH AX ; AX has 20 Hex in it
PUSH BX ; BX has 34 Hex in it
POP AX
ADD AX, BX
POP G

The value stored in *G* would be

(a) 20 Hex (b) 34 Hex
(c) 54 Hex (d) 68 Hex

Q.148 When an 8086 microprocessor executes an INT type instruction, it

(a) resets both IF and TF flags
(b) resets all flags
(c) sets both IF and TF flags
(d) resets both CF and TF flags

Q.149 Match **List-I (Addressing modes of 8086)** with **List-II (8086 Instructions)** and select the correct answer using the codes given below the lists:

List-I

A. Register indirect addressing mode
B. Indexed addressing mode
C. Direct addressing mode
D. Based addressing mode

List-II

1. MOV BX, [SI + 1528 H]
2. ADD BL, [0 1 0 3]
3. MOV CL, [BX + 05 H]
4. MOV BX, [BP]

Code:

	A	B	C	D
(a)	4	3	2	1
(b)	4	1	2	3
(c)	2	4	1	3
(d)	3	2	1	4

Q.150 Number of times the instruction sequence below will loop before coming out of loop is:

MOV AL, 00 H
A1 : INC AL
JNZ A1

(a) 00 (b) 01
(c) 255 (d) 256

Q.151 What will be the contents of register AL after the following program has been executed in a 8086 microprocessor?

MOV AL, 7E
MOV BL, 8C
ADD AL, BL

(a) 0A and carry flag is reset
(b) 6A and carry flag is set
(c) 0A and carry flag is set
(d) 6A and carry flag is reset

Q.152 Direction flag is used with

(a) string instructions
(b) stack instructions
(c) arithmetic instructions
(d) branch instructions

Q.153 In the instruction FADD, F stands for

(a) Far (b) Floppy
(c) Floating (d) File

Q.154 The number of condition flags and control flags in a 8086 microprocessor are respectively

(a) 4 and 5 (b) 6 and 3
(c) 7 and 2 (d) 5 and 4

Q.155 The number of addressing modes available in a 8086 microprocessor are

(a) 7 (b) 9
(c) 6 (d) 8

Q.156 **Assertion (A):** Microprocessor 8086 is a general purpose register type processor.

Reason (R): For ALU operation all general purpose register (AX, BX, CX and DX) act as an accumulator in a 8086 microprocessor.

(a) Both A and R are true and R is the correct explanation of A.
(b) Both A and R are true but R is not the correct explanation of A
(c) A is true but R is false
(d) A is false but R is true

Q.157 **Assertion (A):** Intel's 16-bit processors are 8086, 80186 and 80286.

Reason (R): The Intel 8086, 80186 and 80286 microprocessors have the different basic set of register, instruction and addressing modes.

(a) Both A and R are true and R is the correct explanation of A.
(b) Both A and R are true but R is not the correct explanation of A.
(c) A is true but R is false.
(d) A is false but R is true.

■■■■

Answers Microprocessor

1. (c)	2. (a)	3. (c)	4. (d)	5. (c)	6. (a)	7. (b)	8. (d)
9. (c)	10. (a)	11. (b)	12. (d)	13. (c)	14. (a)	15. (b)	16. (c)
17. (d)	18. (b)	19. (d)	20. (c)	21. (a)	22. (c)	23. (d)	24. (b)
25. (c)	26. (a)	27. (b)	28. (d)	29. (c)	30. (c)	31. (a)	32. (b)
33. (c)	34. (c)	35. (d)	36. (d)	37. (d)	38. (b)	39. (c)	40. (a)
41. (c)	42. (c)	43. (d)	44. (b)	45. (a)	46. (d)	47. (b)	48. (c)
49. (c)	50. (d)	51. (d)	52. (b)	53. (c)	54. (b)	55. (d)	56. (c)
57. (d)	58. (a)	59. (a)	60. (d)	61. (b)	62. (b)	63. (a)	64. (b)
65. (b)	66. (c)	67. (a)	68. (d)	69. (b)	70. (c)	71. (c)	72. (b)
73. (d)	74. (a)	75. (b)	76. (b)	77. (a)	78. (b)	79. (c)	80. (c)
81. (a)	82. (d)	83. (c)	84. (d)	85. (b)	86. (c)	87. (b)	88. (b)
89. (a)	90. (a)	91. (a)	92. (d)	93. (c)	94. (d)	95. (c)	96. (d)
97. (a)	98. (c)	99. (a)	100. (b)	101. (d)	102. (c)	103. (a)	104. (d)
105. (c)	106. (c)	107. (b)	108. (a)	109. (c)	110. (b)	111. (d)	112. (b)
113. (a)	114. (d)	115. (c)	116. (a)	117. (c)	118. (b)	119. (d)	120. (b)
121. (d)	122. (b)	123. (d)	124. (c)	125. (c)	126. (b)	127. (d)	128. (a)
129. (b)	130. (d)	131. (d)	132. (a)	133. (d)	134. (a)	135. (d)	136. (b)
137. (c)	138. (b)	139. (d)	140. (b)	141. (b)	142. (b)	143. (c)	144. (c)
145. (d)	146. (d)	147. (a)	148. (c)	149. (b)	150. (d)	151. (c)	152. (a)
153. (c)	154. (b)	155. (d)	156. (c)	157. (c)			

Explanations

2. (a)

There is only one interrupt in 8085 microprocessor which is non-maskable which is TRAP.
The interrupts which are maskable are INTR, RST 7.5, RST 6.5 and RST 5.5 (total four).

3. (c)

TRAP is an interrupt in 8085 microprocessor which is both edge and level triggered.

5. (c)

A micro controller differs from a microprocessor in terms of memory configuration and I/O interfaces. It has both on chip memory and on chip ports.

6. (a)

Since for each mnemonics written assembly language there is only one machine code, therefore the number of instructions in the machine language when compared with the number of instructions in assembly language is same.

7. (b)

- A machine language program is written in the form of 0's and 1's. Since the each bit has to be entered individually the entry of a program is very slow in machine language compared to assembly language.
- Both assembly language and machine language are microprocessor - specific and a microprocessor - specific language is known as a low - level language. Hence, both assertion and reason are correct but reason is not a correct explanation of assertion.

8. (d)

- A cross-assembler is an assembler that runs on a computer other than that for which it produces object codes. Hence, statement-1 is not correct.
- A resident assembler or a self assembler is an assembler which runs on the microcomputer on which it produces object codes (machine codes). Hence, statement-2 is not correct.
- Statements 3 and 4 are correct.

10. (a)

There are total 40 pins in 8085 microprocessor out of which 27 are output pins, 21 are input pins and 8 pins (AD_0 – AD_7) are shared between output and input signals.

11. (b)

The ready signal is an active low signal, input to the 8085 as an external request from a slow peripheral, to indicate that the peripheral is not yet ready for data transfer. Thus, the ready signal is useful when the CPU communicates with a slow peripheral device.

13. (c)

Accumulator is a 8-bit resistor, therefore the highest integer than can be stored = $2^8 - 1$ = 255.

Thus, the lowest integer that can be stored = 0 and the highest integer that can be stored = 255.

14. (a)

There are total six general purpose registers in 8085 namely B, C, D, E, H and L.

15. (b)

The program counter is a 16-bit special purpose register which is used to store the 16-bit address of the next byte to be fetched from the memory (RAM) or address of next instruction to be execute. It is incremented automatically after fetching each byte from the memory. Hence, statements 3 and 4 are only correct.

20. (c)

Magnetic cores are slow and non-volatile. Hence, reason is false.

21. (a)

- Compilers and interpreters are programs not application softwares.
- A program counter holds the address of the next instruction to be executed.

Hence, statements 2 and 4 are not correct.

22. (c)

TRAP (RST 4.5) has highest priority and INTR has the lowest priority.

23. (d)

For RST 6, $n = 6$

$$\therefore \quad 6 \times 8 \xrightarrow{\text{Hex}} (0030)_H$$

Hence, vector address of RST 6 is $(0030)_H$.

24. (b)

RST 4.5 (or TRAP) is a hardware interrupt.

26. (a)

Control signals are: $\overline{RD}$ and $\overline{WR}$

Status signals are: $IO/\bar{M}$, S_0 and S_1.

27. (b)

For reducing the number of pins in microprocessor, lower order address lines are shared or time division multiplexed with data bus. Reason is also a correct statement but not the correct explanation of assertion.

28. (d)

For S_1 = 1 and S_0 = 1, there are two types of operations possible.

If $IO/\bar{M} = 0$, S_1 = 1, S_0 = 1, it is opcode fetch operation.

If $IO/\bar{M} = 1$, S_1 = 1, so S_0 = 1, it is interrupt acknowledge operation.

Hence, none of the option is correct.

29. (c)

The parity flag is set when the result has even parity. Hence option (c) is not correct.

31. (a)

Pseudo-instructions are command to the assembler. All pseudo-operations start with a period. Pseudo-instructions are composed of a pseudo-operation which may be followed by one or more expressions.

Hence, option (a) is correct.

32. (b)

The HOLD pin is an input that is used to request a DMA action and the HLDA pin is an output that acknowledges the DMA action.

33. (c)

ENIAC (Electronic Numerical Integrator and Computer) was the first general purpose electronic computer. It was a complete digital computer capable of being reprogrammed to solve a full range of computing problems. ENIAC was designed to calculate artillery firing for the U.S. Army's Ballistic Research Laboratory.

34. (c)

Non-maskable interrupt (NMI) is both edge and level triggered that requests an interrupt on the positive edge (0 to 1 transition) and on the negative edge (1 to 0 transition).

38. (b)

Size of memory $= 8$ K bytes

$= 8 \times 1024 \times 8$ bits

$= 2^{13} \times 8$ bits

$= 2^n \times m$ bits

Here, n = No. of address lines = 13

and m = No. of data lines = 8

39. (c)

Size of given chip

$= 1\text{ K} \times 4$ bits

$= 1024 \times 4$ bits

Size of required memory bank

$= 16\text{ K} \times 8$ bits

$= 16 \times 1024 \times 8$ bits

$\therefore$ Required number of memory chips

$$= \frac{16 \times 1024 \times 8}{1024 \times 4} = 32$$

40. (a)

For chip select $\overline{CS}$ signal to be low, we have:

$A_{15} = 1$, $A_{14} = 1$, $A_{13} = 1$ and $A_{12} = 0$

The address range with this combination will be as follows:

A_{15}	A_{14}	A_{13}	A_{12}	A_{11}	A_{10}	A_9	A_8	A_7	A_6	A_5	A_4	A_3	A_2	A_1	A_0	
1	1	1	0	0	0	0	0	0	0	0	0	0	0	0	0	$= E000_H$
1	1	1	0	1	1	1	1	1	1	1	1	1	1	1	1	$= EFFF_H$
Chip select				Register select												

Hence, the address range is $E000_H - EFFF_H$.

41. (c)

Size of the given memory chip

$= 64 \times 8$ bits $= 2^6 \times 8$ bits

$= 2^n \times m$ bits

Here, $n = 6$ and $m = 8$

Hence, the chip has 6 address lines and 8 data lines.

Now, when eight memory chips of size $2^6 \times 8$ bits have their address buses connected together, then the address lines will be same but data lines will become $8 \times 8 = 64$. Thus, the size of the resultant memory will be 64×64 bits.

42. (c)

For 4 K RAM ($4 \times 1024 = 2^{12}$), the number of address lines required will be 12. Hence, out of the 16 address lines, 12 addresses lines (i.e. A_{11} to A_0) will be used for addressing the memory location in a chip and the remaining four (i.e. A_{15} to A_{12}) will be used for selecting the chip. Thus, the chip select signal is $CS = \bar{A}_{15} A_{14} A_{13}$. So, the address lines $A_{15} = 0$, $A_{14} = 1$ and $A_{13} = 1$. The line A_{12} may be 0 or 1 and hence we will get separate range corresponding to $A_{12} = 1$ and $A_{12} = 0$.

- **When $A_{12} = 1$:**

$\bar{A}_{15}$	A_{14}	A_{13}	A_{12}	A_{11}	A_{10}	A_9	A_8	A_7	A_6	A_5	A_4	A_3	A_2	A_1	A_0	
0	1	1	1	0	0	0	0	0	0	0	0	0	0	0	0	$= 7000_H$
0	1	1	1	1	1	1	1	1	1	1	1	1	1	1	1	$= 7FFF_H$
Chip select				Register select												

Thus, the address range with $A_{12} = 1$ is 7000_H to $7FFF_H$.

- **When $A_{12} = 0$:**

$\bar{A}_{15}$	A_{14}	A_{13}	A_{12}	A_{11}	A_{10}	A_9	A_8	A_7	A_6	A_5	A_4	A_3	A_2	A_1	A_0	
0	1	1	0	0	0	0	0	0	0	0	0	0	0	0	0	$= 6000_H$
0	1	1	0	1	1	1	1	1	1	1	1	1	1	1	1	$= 6FFF_H$
Chip select				Register select												

Thus, the address range with $A_{12} = 0$ is 6000_H to $6FFF_H$.

43. (d)

One memory location of an 8-bit microcontroller external RAM will store 8 bits of data or 1 byte. The number of memory locations in the memory range from 8000 H to 9FFF H are equal to $(9FFF - 8000)_H + 1 = 2000$ H.

Hence, the number of bytes the memory locations from 8000 H to 9FFF H can store is equal to 2000 H or 8192.

44. **(b)**

Statement-3 is not correct because in memory mapped I/O scheme of communication with microprocessor, the processor manipulates registers with same instruction that are used to manipulate memory locations.

45. **(a)**

The I/O devices in 8085 microprocessor can be used in memory mapped I/O or I/O mapped I/O.

46. **(d)**

An 8085 microprocessor has 16 address lines and 8 data lines. Thus, it can address maximum of 2^{16} I/O devices and memory.

50. **(d)**

Memory mapped I/O has following features:

1. I/O have 16-bit addresses.
2. Devices are accessed like registers
3. Mass devices = $2^{16} = 256 \times 256$
4. Arithmetic/Logic operations can be done

52. **(b)**

An advantage of memory interlacing is that the effective speed of the memory is increased.

53. **(c)**

$$4\text{ K} = 2^{10} \cdot 2^2 = 2^{12}$$

$\therefore$ No. of address lines = 12

$$4\text{ K} \times 16 = 2^{12} \cdot 2^4 = 2^{12} \times 16$$

$\therefore$ No. of data lines = 16

57. **(d)**

Size of RAM = 8 K bytes = $2^3 \cdot 2^{10}$ bytes = 2^{13} bytes

$\therefore$ No. of address lines = 13 = length of SP

61. **(b)**

We may consider the hit ratio of 98% as the probability of finding the page table entry if the associative memory (cache). Accordingly, with 98% probability, the page memory access time will be:

Cache access time for page number memory access time

= 100 ns + 120 ns

= 220 ns

The other 2% of the time, the search in each will fail and the page number will be obtained through and additional normal memory access.

Accordingly:

Cache access time + memory access for page number + memory access time = 100 ns + 120 ns + 120 ns = 340 ns

68. **(d)**

Size of memory is defined as $2^n \times m$ bits, where 'm' is the number of data lines and 'n' is the number of address lines.

Given memory size

$$= 4096 \text{ bytes}$$
$$= 4 \times 1024 \text{ bytes}$$
$$= 2^2 \times 2^{10} \text{ bytes}$$
$$= 2^{12} \times 8 \text{ bits}$$
$$= 2^n \times m \text{ bits}$$

Thus, no. of address lines required = n = 12.

Hence, assertion is not correct.

69. **(b)**

Both assertion and reason are individually correct statements. Interfacing of I/O mapped I/O devices with the microprocessor is easy because hardware requirement is less for I/O mapped I/O device interfacing compared to memory mapped I/O devices.

70. **(c)**

Number of T-states in each machine cycle = 3

So, number of T-states in 4 machine cycles = $3 \times 4 = 12$

Given, f = 3 MHz

$\therefore$ Time of one T-state = $\frac{1}{3}\mu s = T$

Thus, time required for execution of the instruction

$$= 12\,T = 12 \times \frac{1}{3}\mu s = 4\ \mu s$$

71. **(c)**

In immediate addressing mode, an 8 bit or 16 bit data is given in the instruction as operand.

72. **(b)**

MVI B, 00 H ; B ← 00 H

MVI A, 1C H ; A ← 1C H

DCR B

Decrement the content of B by one (initially B = 00 H)

$\therefore$ B = FF H

DAA ; DAA instruction changes the binary values of contents of accumulator to BCD.

Since lower order nibble of A is more than 9 (i.e. C), AC flag is set and upper nibble is less than 9, therefore DAA adds 0110 H to lower order nibble of *A* to adjust binary result to Binary coded decimal.

$$\begin{array}{rr} A: 1C\,H = & 0001\ 1100 \\ & +\ 0110 \\ \hline & 0010\ 0010 \end{array}$$

Hence, A = 22 H

STA TEMP ; TEMP = 22 H

HLT

Thus, the content of TEMP is 22 H.

73. (d)

PUSH D stores the content of register pair DE on two top locations of stack. It is a 1 byte instruction and uses register indirect addressing mode.

74. (a)

CALL and RET instruction uses stack while JMP instruction does not uses stack.

75. (b)

An instruction cycle is the time in which, a hard-wired controller completes the following four functions:

(i) fetching of code from memory
(ii) updating the program counter
(iii) decoding of the instruction code
(iv) execution of the instruction

77. (a)

203A MVI A, 20 H ; $A \leftarrow 20$ H

203C LHLD 203A H ; $L \leftarrow 3EH$ and $H \leftarrow 20H$

203F ADDM ;

$$\begin{array}{rr} A = & 00100000 \\ +B = & 00100000 \\ \hline & 001000000 \end{array}$$

So, A = 40 H and CY = 0

2040 HLT ; Halts the execution of program

Hence, the content of accumulator after the execution of given program are 40 H.

78. (b)

Initially contents of accumulator A = AA H.

The instruction CPI 99 H compares the data 99H with the content of accumulator (A). This comparison is made by subtracting 99 H from contents of A i.e. AA H.

The contents of accumulator will remain unaffected but status of result will be reflected by flags.

$$\begin{array}{rr} A: & 1010\ 1010 \\ 99\,H: & 1001\ 1001 \\ \hline & \boxed{0}0001\ 0001 \end{array}$$

Thus, CY = 0 and Z = 0

Hence, after comparison both carry and zero flags will be reset.

79. (c)

- Instruction 'CMP C' compares the contents of register C with the content of accumulator A but contents of A and C will remain unchanged.
- Instruction 'CPI 3A H' compares the 8-bit data 3A H with the contents of accumulator A but contents of A and C will remain unchanged.
- In instruction 'ANI 5C H', the 8-bit data 5C H is ANDed with contents of A and result is stored in A. So, the content of accumulator changes.
- In instruction 'ORA A', contents of A are ORed with contents of A itself and result is stored in A. But, the result of operation remains same as contents before execution of ORA A.

Hence, the contents of *A* does not change.

81. (a)

$$f_{\text{clock}} = \frac{1}{2} \times (\text{crystal frequency})$$

$$= \frac{1}{2} \times 5 = 2.5\text{ MHz}$$

Since one T-state is precisely equal to one time period of clock, therefore duration of one T-state is

$$T = \frac{1}{f_{\text{clock}}} = \frac{1}{2.5 \times 10^6} = 0.4\ \mu\text{s}$$

82. (d)

A NOP instruction requires only one machine cycle (Opcode fetch) and thus 4T-states.
Therefore, number of T-states required in 4 NOP instructions = 4 × 4 = 16

Now, times of one T-state, $T = \frac{1}{f_{clock}} = \frac{1}{2}\mu s$

∴ Time required for 4 NOP instructions to be executed = 16 T = $16 \times \frac{1}{2}\mu s = 8\ \mu s$

83. (c)

XRA A; — This instruction performs EXOR operation on the content of A with contents of A and the result is stored in A only.
A ⊕ A = 00 H
Hence, the zero flag is set.

LXI B, 007 H; — The content of BC pair becomes 0007 H.

LOOP : DCX B; — It decrements the contents of BC pair by one.

JNZ LOOP; — Jumps the execution of program to LOOP if zero flag is not set.

In the above program, after the execution of XRA A instruction, zero flag is set and after the execution of instruction DCX B, the zero flag remains unaffected. Hence, the execution comes out of LOOP during first instance itself as Z flag is already set. So, loop runs only once.

84. (d)

Push R_P ; stores the contents of register pairs (R_P) on $R_P \rightarrow$ BC, DE, HL and PSW two top location of stack.
Here, PUSH instruction cannot be used for storing the data of SP and PC on top of stack.

85. (b)

Instruction INX B does not affect any of the flags of 8085 microprocessor.

86. (c)

MVI A, 92 H ; A ← 92 H

ORA A ; Here, the contents of accumulator are ORed with the contents of accumulator itself and result is stored in accumulator. The result of accumulator remains same as before execution of ORA A instruction.
Thus, A = 92 H = 10010010
As bit D_7 of A is '1', therefore sign flag is set i.e. S = 1.

JP PORT 1 ; Jumps the execution of program to PORT 1 if sign flag is set.
Since, S = 1, therefore the execution of program will jump to PORT 1.

XRA A ; Content of A are XORed with the contents of A and result is stored in A.

$$\begin{array}{rr} A \oplus A: & 10010010 \\ & \underline{10010010} \\ & 00000000 \end{array}$$

∴ A = 00 H
Since content of A is zero, therefore sign flag S = 0.

OUT F2 H ; Contents of A i.e. 00 H are sent to I/O device connected at port address F2 H.

HLT ; Halts the program execution

Hence, the output at port F2 H after the execution of above program is F2 H.

88. (b)

When RET instruction is executed by microprocessor it takes three machine cycles. During first machine cycle microprocessor fetches the opcode and during second and third machines it reads the contents of top of stack and stores in temporary register pair WZ. The contents of WZ pair are placed on address bus and program counter is loaded with WZ + 1 i.e. address of next byte to be fetched. So, RET instruction uses the stack.

89. (a)

```
        MVI A, 10 H  ;  A ← 10 H
        MVI B, 10 H  ;  B ← 10 H
BACK :  NOP          ;  No operation
        ADD B        ;  A = A + B
```

$$\begin{array}{rr} A = 10\text{ H}: & 00010000 \\ B = 10\text{ H}: & \underline{+00010000} \\ & 00100000 \end{array}$$

Thus, A = 20 H and CY = 0

RLC ; Rotates contents of accumulator to left without carry flag.

Before rotation: CY = 0

0	0	1	0	0	0	0	0

After location: CY = 0

0	1	0	0	0	0	0	0

Thus, A = 40 H and CY = 0

JNC BACK ; Jumps to BACK if CY = 0

HLT ; Halts the execution

After the execution of RLC instruction three times, CY flag will be set. Thus, the instruction NOP will be executed three times.

91. (a)

When the data transfer is with the memory or I/O device and address of memory or I/O device is not included in the instruction rather it is indicated through a memory pointer, the addressing mode is called register indirect addressing mode.

92. (d)

Jump instructions used in 8085 microprocessor are as follows:

1. JMP : Unconditional
2. JP : Jump on plus
3. JM : Jump on minus
4. JPE : Jump on even parity
5. JPO : Jump on odd parity
6. JC : Jump on carry
7. JNC : Jump on carry
8. JZ : Jump on zero
9. JNZ : Jump on no zero

The total number of jump instructions = 9.

Number of bytes required for each jump instructions = 3.

Hence, total memory bytes required by all jump instruction is $9 \times 3 = 27$ bytes.

93. (c)

8085 has five flags which reflect the status of arithmetic and logical operations performed on the content of accumulator with some exceptions. The different flags are shown by the flag register.

D_7	D_6	D_5	D_4	D_3	D_2	D_1	D_0
S	Z	X	AC	X	P	X	CY

Flag register

Given, sign and parity flags are set i.e. $S = 1$ and $P = 1$. Since parity flag is set, therefore the content of accumulator must have even number of one's. Hence, the possible answer will be as in option (c).

94. (d)

LXI B 2070 H : B ← 20 H and C ← 70 H

MVI A, 8F H : A ← 8F H

MVI C, 68 H : C ← 68 H

SUB C : A = A − C

A : 8F H
−C : 68 H
27 H

ANI 0F H : The content of A (i.e. 27 H) are ANDed with 0F H.

A : 0010 0111
0F H : 0000 1111
00000 111

Thus, A = 07 H

STAX B : Stores the content of A i.e. 07 H at the address stored in BC pair.

Thus, 2070 H ← 07 H

HLT : Halts the program execution

Hence, after execution of the above program, the content of memory location 2070 H are 07 H.

95. (c)

For the instruction MVI M, 8F microprocessor first reads the data 8F from memory and then write the same at memory location whose address is stored in HL pair. Thus, this instruction requires both memory read and memory write machine cycles.

96. (d)

In 8085 microprocessor, the instruction DAA (Decimal Adjust Accumulator) is used for BCD addition only not for BCD subtraction. Hence, assertion is not correct.

97. (a)

- HLT is a machine control instruction which has no addressing mode. Hence, statement-2 is not correct.
- The instructions related to stack are PUSH R_P, POP R_P and XTHL. All these instructions uses indirect addressing mode.
- Statements 1 and 3 are also correct.

98. (c)

Reason is a false statement because stack pointer is always decremented by 2 after the execution of each PUSH instruction and is incremented by 2 after the execution of each POP instruction.

101. (d)

LHLD 3000 H : Load HL pair with the contents of given address 3000 H and next address 3001 H. Since address of memory is given in the instruction itself, therefore it uses direct addressing mode.

SPHL : The content of HL pair are copied in stack pointer. Since data transfer is between two registers, therefore SPHL uses register addressing mode.

DAA : Decimal Adjust Accumulator. Here, binary contents of accumulator are changed to its two digit BCD equivalent. This instruction modifies the contents of accumulator without using any operand and hence uses implicit or implied addressing mode.

JMP 2000 H : Jump execution of program immediately at the given 16-bit address i.e. 2000 H. Since the 16 bit data (i.e. 2000 H) is given as operand here, therefore this instruction uses immediate addressing mode.

104. (d)

During the first clocking period in a bus cycle, which is called T_1, the address of the memory or I/O location is sent out and the control signals ALE, $DT/\overline{R}$ and $IO/\overline{M}$ are also output.

105. (c)

MOV B, 8CH ; B← 8C H
MOV A, 7E H ; A ← 7E H
ADD A, B ; Add the content of register B (i.e. 8C H) with the content of register A (i.e. 7E H) and store the result in A.

A : 7E H
B : 8C H
\+
1 0 A

Hence, A = 0 A H and CY = 1 (set)

106. (c)

A microprocessor can be brought out of the wait state either by pressing RESET key or by giving an interrupt.

107. (b)

An instruction cycle consists of several machine cycles. Hence, option (b) is not correct.

108. (a)

Since one byte of memory is required to store one opcode (or one fetch cycle), therefore for fetching an 8 byte instruction, number of fetch cycles required will be 8.

110. (b)

The total time required to execute an instruction is known as an instruction cycle. An instruction cycle may consists of 1 to 5 machine cycle hence it is larger than machine cycle time.

111. (d)

The two co-processor instruction sets must be the same as that of the host.

113. (a)

For n-bit, range = $-(2^{n-1})$ to $+(2^{n-1} - 1)$
(2's complement representation)
∴ For 8-bit, range = −128 to +127

116. (a)

Instruction DCR and INR affects flags. Hence, statement-4 is not correct.

118. (b)

Wait states are always provided between T_2 and T_3 to provide synchronization between slow peripheral devices and microprocessor.

122. (b)

The DMA is a process of high speed direct data transfer between memory and external peripherals (I/O device) without programming intervention.

130. **(d)**

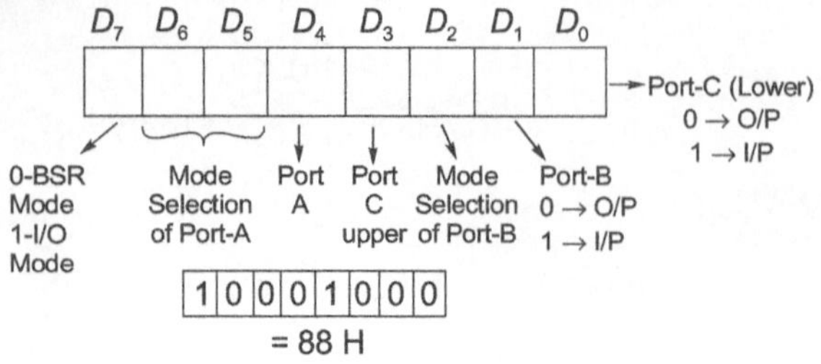

133. **(d)**

- Handshaking is used in asynchronous data transfer. This technique is used when the speed of microprocessor and that of I/O device is not matching.
- Interrupt is used for giving restart instruction from I/O devices.
- DMA is high speed data transfer between memory and external peripherals (I/O device) without programming intervention.

136. **(b)**

The Intel 8251 is a programmable communication interface. It is USART.

137. **(c)**

The Oscillator output is steered through to the divide by-3 counter.

138. **(b)**

Output continues for the total length of the count. Hence, option (b) is correct.

139. **(d)**

Assertion is not correct because in hard disk drives, Burst mode of DMA data transfer is used where data transfer can be stopped or slowed down without loss of data. This mode (Burst mode) transfers block of data in one go.

140. **(b)**

Both assertion and reason are individually correct statements. The correct reason for assertion is that DMA takes control of system buses and needs no processor intervention during the data transfer.

141. **(b)**

To configure port-A and port-B as output ports, port-C not being used in the 8155 programmable interface, the control word should have the value 03 H.

142. **(b)**

It is called PPI (Programming Peripheral Interface) which is used for data transmission between 8085 and a 16-bit ADC.

143. **(c)**

8086 microprocessor has total 9 flags namely– OF, DF, IF, TF, S, Z, AC, P and C. Hence, statement-3 is not correct.

151. **(c)**

MOV, AL, 7E : 7E data is copied in Al
$\therefore$ AL = 7E H

MOV BL, 8C : 8C data is copied in BL.
$\therefore$ BL = 8C H

ADD AL, BL : The content of BL (i.e. 8C H) are added with the content of AL (i.e. 7E H) and result is stored in AL itself.

AL : 7E
BL : 8C
______
[1] 0 A

Thus, AL = 0A H and CY = 1 (set).

152. **(a)**

Direction flag is used only with string instructions.

153. **(c)**

FADD means "add two floating numbers".

154. **(b)**

There are total 9 flags in 8086 microprocessor. Out of which 6 flags are condition flags and remaining 3 are control flags.

155. **(d)**

There are total 8 addressing modes in 8086 microprocessor.

156. **(c)**

Unlike 8085, which is an accumulator based processor, microprocessor 8086 is a general purpose register type processor because for ALU operation all general purpose register (AX, BX, CX and DX) act as an accumulator.
Hence, both assertion and reason are correct and reason is the correct explanation of assertion.

UNIT 11

Power Electronics

1. Power Semiconductor Diode & Transistors

Q.1 In a Schottky diode junction is formed between
(a) two type of metals
(b) two type of semiconductors
(c) metal and semiconductor
(d) None of the above

Q.2 **Assertion (A) :** A Schottky diode has very high reverse voltage rating as compared to alloyed and diffused diodes.
Reason (R) : In a Schottky diode the metallic layer is surrounded by a ring of insulating material such as SiO_2.
(a) Both A and R are true and R is the correct explanation of A.
(b) Both A and R are true but R is not the correct explanation of A.
(c) A is true but R is false.
(d) A is false but R is true.

Q.3 Consider the following statements:
1. By increasing the rate of recombination and removal of minority carrier, reverse recovery period can be reduced.
2. In fast recovery diode, rate of recombination is decreased.
3. MOSFET and transistors are used in high power applications.
4. IGBT is having all the good properties of a transistors and MOSFET.

Which of the statements given above are correct?
(a) 1 and 4 (b) 1, 2, 3 and 4
(c) 2 and 3 (d) 1, 2 and 3

Q.4 A power transistor
1. cannot be easily paralleled.
2. is susceptible to secondary breakdown.
3. has low transconductance than MOSFET.
4. has negative temperature coefficient on collector current.
5. has low switching speed than MOSFET.

Which of the above characteristics are possessed by a power transistor?
(a) 1, 3 and 5 (b) 1, 2 and 5
(c) 2, 3 and 4 (d) 3, 4 and 5

Q.5 Match **List-I (Power Semiconductor Devices)** with **List-II (Characteristics)** and select the correct answer using the codes given below the lists:

List-I
A. Power BJT
B. Power MOSFET
C. IGBT
D. Schottky diode

List-II
1. Exponential I-V characteristics
2. Suitable for high frequency inverters and choppers
3. Square law I-V characteristics
4. Offers ease of control and high current density

Codes:

	A	B	C	D
(a)	3	1	2	4
(b)	1	3	4	2
(c)	1	3	2	4
(d)	3	1	4	2

Q.6 Which of the following are not a unidirectional current devices?
1. BJT 2. Triac
3. MOSFET 4. GTO
5. IGBT

(a) 4 and 5 (b) 2 and 4
(c) 5 only (d) 2 only

Q.7 Match **List-I (Power devices)** with **List-II (Symbols)** and select the correct answer using the codes given below the lists:

List-I

A. IGBT
B. MCT
C. Programmable Unijunction Transistor (PUT)
D. Silicon Controlled Switch (SCS)

List-II

1. 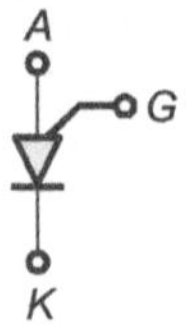

2. 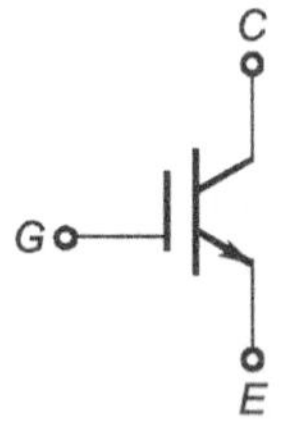

3. 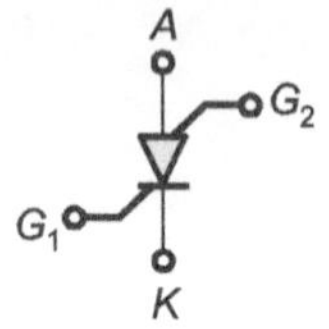

4. 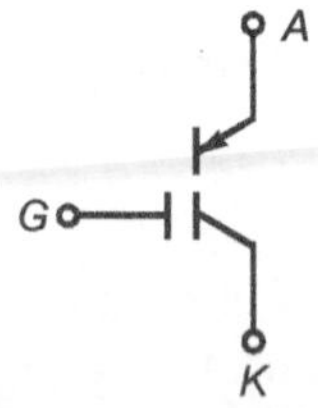

Codes:

	A	B	C	D
(a)	2	4	1	3
(b)	4	2	3	1
(c)	2	4	3	1
(d)	4	2	1	3

Q.8 **Assertion (A):** With increased reverse bias, the reverse blocking safe operating area increases in size.

Reason (R): The reverse blocking safe operating area is a plot of collector current versus collector-emitter voltage.

(a) Both A and R are true and R is the correct explanation of A.
(b) Both A and R are true but R is not the correct explanation of A.
(c) A is true but R is false.
(d) A is false but R is true.

Q.9 A power transistor has its switching waveforms as shown in figure. If the average power loss in the transistor is limited to 300 W, the switching frequency at which the transistor can be operated is approximately equal to

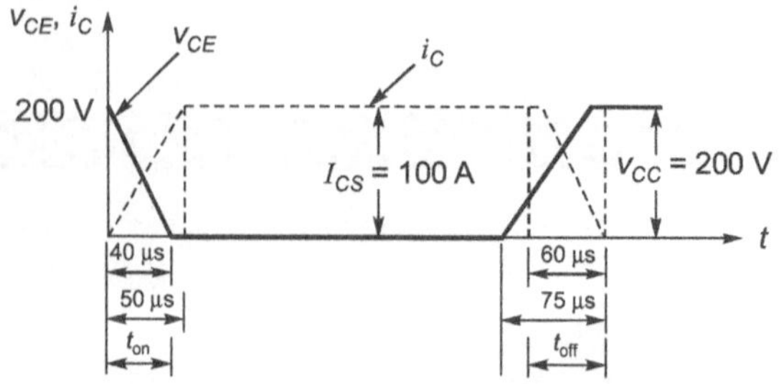

(a) 0.98 kHz
(b) 1.25 kHz
(c) 1.12 kHz
(d) 1.65 kHz

Q.10 Match **List-I (Power devices)** with **List-II (Applications)** and select the correct answer using the codes given below the lists:

List-I

A. Schottky diode
B. BJT
C. IGBT
D. MOSFET

List-II

1. SMPS
2. DC and AC motor drives
3. As a low input impedance device
4. High-frequency instrumentation

Codes:

	A	B	C	D
(a)	4	3	2	1
(b)	2	4	3	1
(c)	4	2	1	3
(d)	1	2	4	3

Q.11 Germanium is not used in the fabrication of thyristors because

(a) it is much costlier than silicon
(b) it has much lower thermal conductivity than silicon
(c) both (a) and (b)
(d) its breakdown voltage is much larger than that of silicon

Q.12 For a diode, reverse recovery time is defined as the time between the diode current becomes zero and the instant reverse recovery current decays to

(a) 10% of reverse peak current I_{RM}
(b) 15% of I_{RM}
(c) zero
(d) 25% of I_{RM}

Q.13 Reverse recovery current in a diode depends upon
(a) storage charge
(b) forward field current
(c) peak inverse voltage
(d) temperature

Q.14 As compared to BJT, a power MOSFET has
(a) higher switching losses and lower conduction loss.
(b) lower switching losses and lower conduction loss.
(c) lower switching losses and higher conduction loss.
(d) higher switching losses and higher conduction loss.

Q.15 Which of the following option is correct?
(a) Both MOSFET has BJT have negative temperature coefficient.
(b) MOSFET has positive temperature coefficient whereas BJT has negative temperature coefficient.
(c) MOSFET has negative temperature coefficient whereas BJT has positive temperature coefficient.
(d) Both MOSFET and BJT have positive temperature coefficient.

Q.16 Secondary breakdown does not occurs in
(a) BJT but occurs in MOSFET
(b) both MOSFET and BJT
(c) MOSFET but occurs in BJT
(d) None of the above

Q.17 The duration of reverse recovery transient affects which of the following aspect in a power diode?
(a) Forward threshold voltage
(b) Average current
(c) Peak inverse voltage
(d) Speed of response

Q.18 The main advantage of IGBT over SCR in power electronics is
(a) very high reliability
(b) self cooling property
(c) self commutating property
(d) reduced weight

Q.19 Conductivity modulation is a phenomenon which occurs in
(a) Power BJT (b) Power MOSFET
(c) GTO thyristor (d) IGBT

Q.20 **Assertion (A) :** A diode with a softness factor more than unity is called snappy-recovery diode.
Reason (R) : Softness factor is a measure of the voltage transients that occur during the time diode recovers.
(a) Both A and R are true and R is the correct explanation of A.
(b) Both A and R are true but R is not the correct explanation of A.
(c) A is true but R is false.
(d) A is false but R is true.

Q.21 Schottky transistors work faster than regular bipolar transistors because
(a) in Schottky transistors the base-collector junction is replaced with a Schottky diode, which is a faster device.
(b) the Schottky diode prevents base-collector junction from becoming forward biased, and thus the transistor does not go into deep saturation.
(c) in Schottky transistors, the Schottky contact to the base region prevents minority carrier storage in the base.
(d) the base resistance in Schottky transistors is negligible small.

Q.22 If Aluminium is deposited upon a lightly-doped *n* region, then
(a) a Schottky diode is obtained
(b) an Ohmic contact is obtained
(c) a constant capacitance is obtained
(d) a high-valued constant resistance can be obtained

Q.23 For the circuit shown below, the power-loss in the on-state is (MOSFET parameters are: $t_r = 2\ \mu s$, $R_{DS(on)} = 0.2\ \Omega$, duty cycle $D = 0.7$ and $f = 30$ kHz)

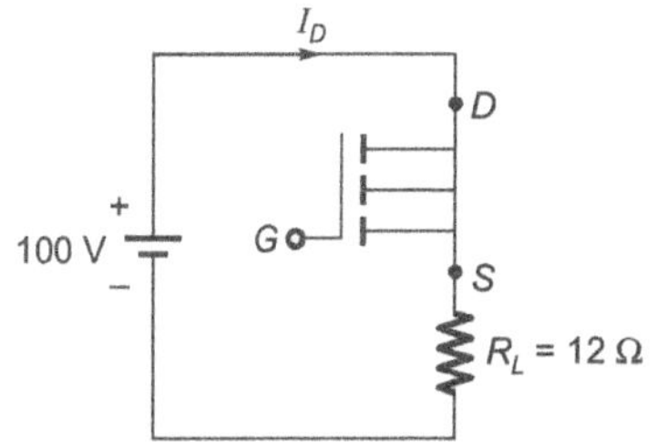

(a) 6.28 W (b) 9.41 W
(c) 5.00 W (d) 2.22 W

Q.24 The IGBT used in the circuit of given figure has the following data:
t_{ON} = 3 μs, t_{off} = 1.2 μs, D (duty cycle) = 0.7, $V_{CE(sat)}$ = 2 V and f_s = 1 kHz.

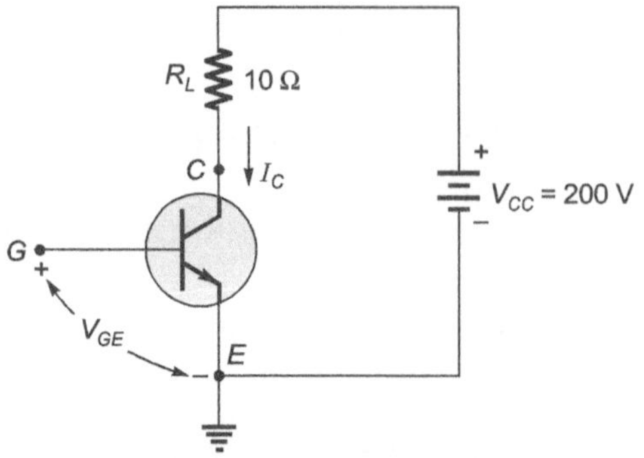

The switching power loss during turn-on is around
(a) 2 W (b) 1.5 W
(c) 1 W (d) 3.5 W

Q.25 The power loss in MOSFET-1 and MOSFET-2 for the circuit shown below are respectively

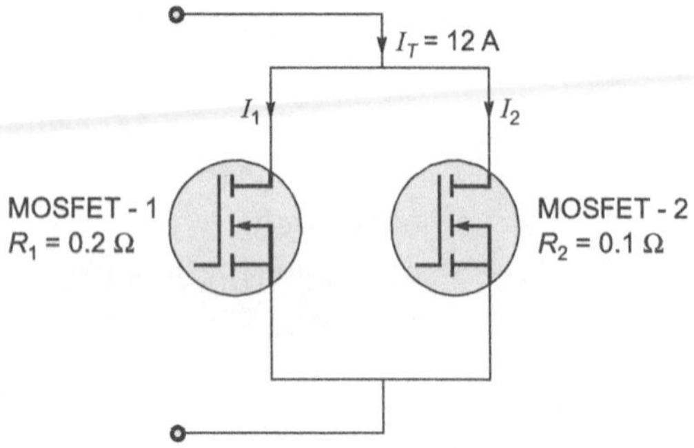

(a) 6.4 W and 3.2 W (b) 2.8 W and 7.2 W
(c) 3.2 W and 6.4 W (d) 7.2 W and 2.8 W

Q.26 IGBT is used for applications in
(a) low power (b) medium power
(c) high power (d) none of these

Q.27 The device which has bipolar current conduction and bipolar voltage conduction are
(a) TRIAC, BJT (b) BJT, TRIAC
(c) SCR, TRIAC (d) TRIAC

Q.28 The correct sequence of the following devices in the increasing order of turn-off times is
(a) MOSFET, IGBT, BJT, SCR
(b) IGBT, MOSFET, BJT, SCR
(c) SCR, BJT, MOSFET, IGBT
(d) MOSFET, BJT, IGBT, SCR

2. Diode Rectifiers

Q.29 The circuit shown below is initially relaxed. With switch closed at t = 0, the conduction time of diode is around

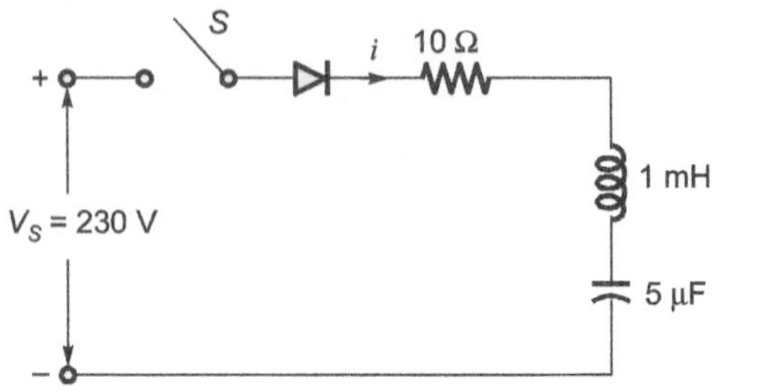

(a) 225 μs (b) 237 μs
(c) 256 μs (d) 202 μs

Q.30 The effect of using freewheeling diode are :
1. the load performance is improved.
2. the system efficiency is improved.
3. it prevents the output voltage from becoming negative.

Choose the correct code from the given options.
(a) 1 and 2 (b) 1 and 3
(c) 1, 2 and 3 (d) 2 and 3

Q.31 A single-phase 230 V, 1 kW heater is connected across single-phase 230 V, 50 Hz supply through a diode as shown in figure. The power delivered to the heater element is

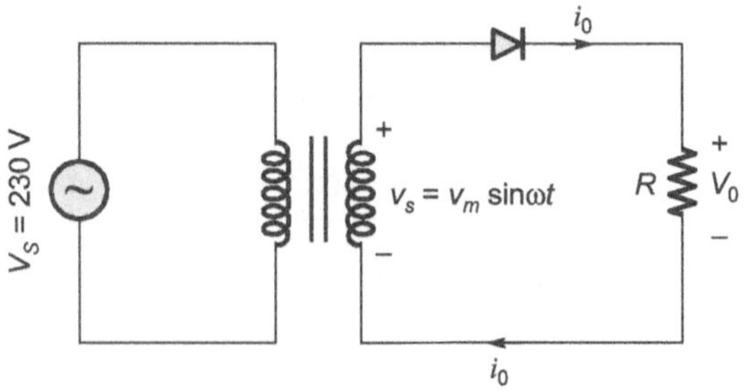

(a) 0.5 kW (b) 1.0 kW
(c) 2.5 kW (d) 0.75 kW

Q.32 A single-phase full-bridge diode rectifier delivers a load current of 10 A, which is ripple free. The rms and average values of diode currents are respectively
(a) 10 A, 5 A (b) 5 A, 7.07 A
(c) 7.07 A, 10 A (d) 7.07 A, 5 A

Q.33 In the circuit shown below, the diode states at the extremely large negative value of the input voltage v_i are

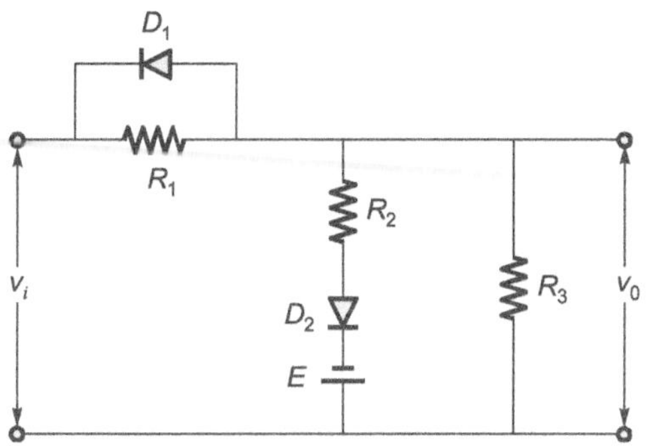

(a) D_1 off, D_2 on (b) D_1 off, D_2 off
(c) D_1 on, D_2 on (d) D_1 on, D_2 off

Q.34 A single-phase full-bridge diode rectifier delivers a constant load current of 20 A. The rms and average values of source current are respectively
(a) 0 A, 20 A (b) 20 A, 20 A
(c) 10 A, 20 A (d) 10 A, 10 A

Q.35 A diode whose internal resistance is 20 Ω is to supply power to a 1 kΩ load from a 230 V (rms) source of supply as shown in figure below. The percentage regulation from no load to the given load is

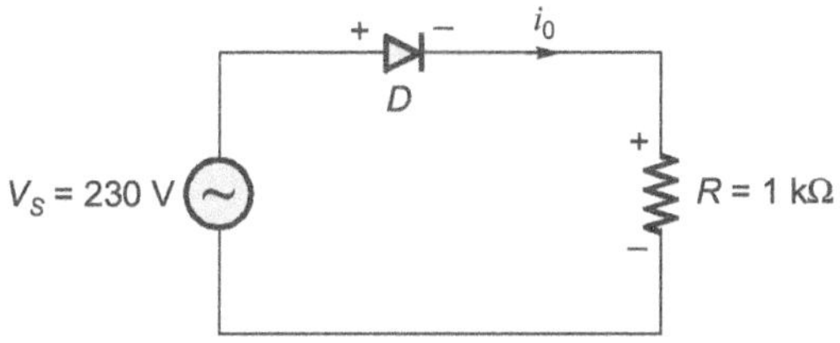

(a) 2.20% (b) 0.78%
(c) 1.96% (d) 1.67%

Q.36 **Assertion (A) :** A single-phase half-wave rectifier circuit is popularly known as one-pulse rectifier.
Reason (R) : A single-phase half-wave rectifier circuit is used to convert ac power to dc.
(a) Both A and R are true and R is the correct explanation of A.
(b) Both A and R are true but R is not the correct explanation of A.
(c) A is true but R is false.
(d) A is false but R is true.

Q.37 **Assertion (A) :** In 3-phase converters, the ripple frequency of the converter output voltage is higher than in single phase converter.
Reason (R) : The load current is mostly discontinuous in 3-phase converters.
(a) Both A and R are true and R is the correct explanation of A.
(b) Both A and R are true but R is not the correct explanation of A.
(c) A is truo but R is false.
(d) A is false but R is true.

Q.38 **Assertion (A) :** In a half-wave rectifier with R-L load and with freewheeling diode, the input Volt-Ampere demand is less.
Reason (R) : With freewheeling diode, the stored energy is in the load itself.
(a) Both A and R are true and R is the correct explanation of A.
(b) Both A and R are true but R is not the correct explanation of A.
(c) A is true but R is false.
(d) A is false but R is true.

Q.39 Consider the following statements associated with a full-wave rectifier with centre-tapped transformer:
1. Its transformer utilization factor is better.
2. Its output voltage has more ripple content.
3. It requires a larger size transformer.
4. The peak inverse voltage of diode is same as in full-wave bridge rectifier.

Which of these statements are correct?
(a) 1 and 3 (b) 2 and 4
(c) 1 and 4 (d) 1, 2 and 3

Q.40 Match **List-I (Three-phase rectifier circuits)** with **List-II (Average output voltage)** and select the correct answer using the codes given below the lists:

List-I	List-II
A. Six-pulse bridge rectifier	1. $\frac{3\sqrt{3}}{2\pi}V_m$
B. Mid-point six-pulse rectifier	2. $\frac{3V_m}{\pi}$
C. Three-pulse rectifier	3. $\frac{3\sqrt{3}}{\pi}V_m$

Codes:

	A	B	C
(a)	3	1	2
(b)	3	2	1
(c)	2	3	1
(d)	1	2	3

Q.41 A full-wave rectifier is used to charge a 24 V battery as shown in figure. A 5 Ω resistance is connected for current limiting.
The ripple voltage (ac component) is (assume ripple-free load current).

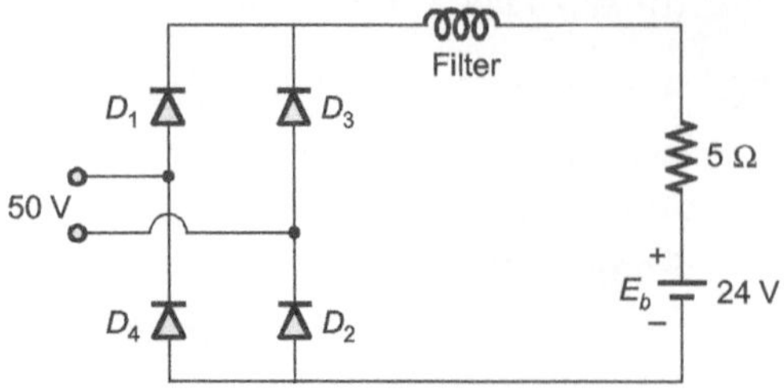

(a) 15.5 V (b) 18.6 V
(c) 14.0 V (d) 21.8 V

Q.42 A 3-ϕ bridge rectifier is used to charge a 48 V battery.
The input votage is 50 V and current limiting resistance is 5 Ω. Assuming ripple-free current, the average output voltage is

(a) $\frac{50\sqrt{3}}{\pi}$ volt (b) $\frac{25\sqrt{2}}{2\pi}$ volt
(c) $\frac{150\sqrt{2}}{\pi}$ volt (d) $\frac{125\sqrt{3}}{2\pi}$ volt

Q.43 The rms value of a half-wave rectified symmetrical square wave current of 2 A is
(a) 1 A (b) $\sqrt{3}$ A
(c) $1/\sqrt{2}$ A (d) $\sqrt{2}$ A

Q.44 A single-phase one-pluse diode rectifier is feeding on *RL* load with freewheeling diode across the load. For conduction angle β, the freewheeling diode and the main diode would, respectively, conduct for
(a) π, β (b) β − π, π
(c) π, β − π (d) π − β, π

Q.45 A 200 Ω resistive load is connected across a full-wave rectifier as shown in figure. What is the value of the capacitor required to be connected across the load so that the ripple factor is less than 5%?

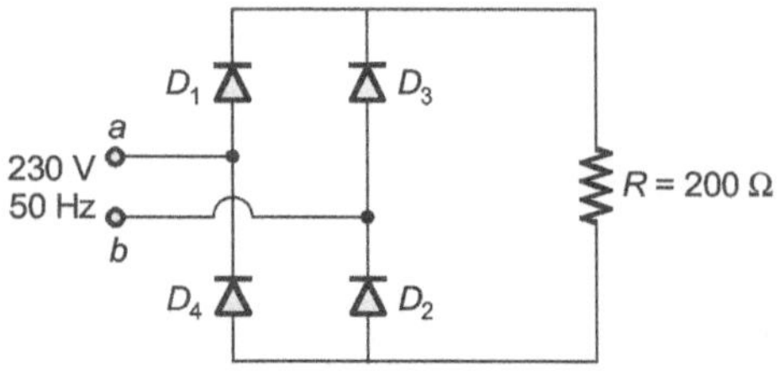

(a) 396.6 μF (b) 325.5 μF
(c) 378.5 μF (d) 302.0 μF

Q.46 Consider the following statements:
1. Bridge rectifiers are better because of TUF and PIV.
2. Peak inverse voltage (PIV) of six-pulse converter is lower.
3. Filter circuits are used to remove the harmonics generated at the input of rectifier.
4. Freewheeling diode is used for capacitive and inductive loads.
5. Half wave rectifier is rarely used because of low TUF, high ripple content and low frequency of rectification.

Which of these statements are not correct?
(a) 2, 3 and 4 (b) 1, 2 and 5
(c) 3 and 4 (d) 2, 4, and 5

Q.47 On increasing the number of pulses in rectification, the form factor, ripple frequency and efficiency
(a) all increases.
(b) decreases, decreases and increases respectively.
(c) decreases, increases and increases respectively.
(d) increases, decreases and increases respectively.

Q.48 What is the total angular period in the circuit shown below for which diode provides half wave rectification?

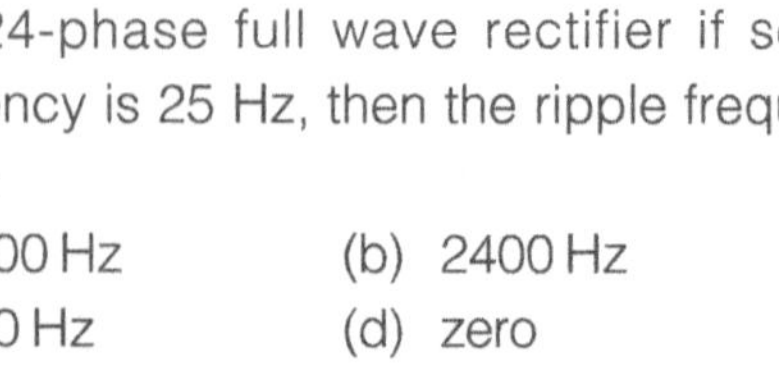

(a) 180° (b) 150°
(c) 120° (d) 90°

Q.49 In a 24-phase full wave rectifier if source frequency is 25 Hz, then the ripple frequency will be
(a) 1200 Hz (b) 2400 Hz
(c) 600 Hz (d) zero

Q.50 The circuit shown in figure is a 3-phase rectifier. The source is a symmetrical 3-phase four-wire system. The line-to-line voltage of the source is 100 V. The supply frequency is 400 Hz. The ripple frequency at the output is

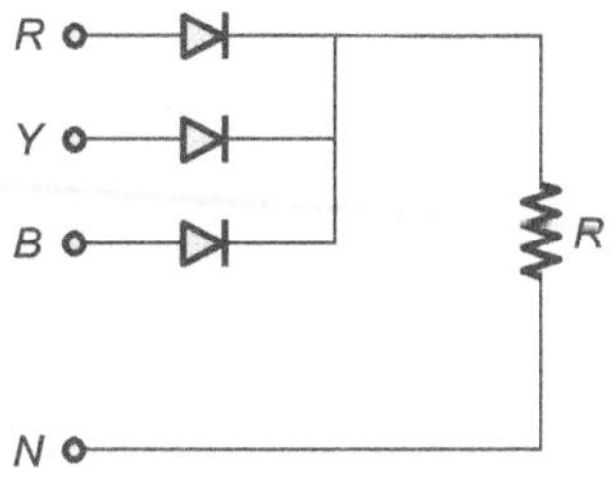

(a) 400 Hz (b) 800 Hz
(c) 1200 Hz (d) 2400 Hz

Q.51 The circuit in figure shows a full-wave rectifier. The input voltage is 230 V(rms) single-phase ac. The peak inverse voltage across the diodes D_1 and D_2 is

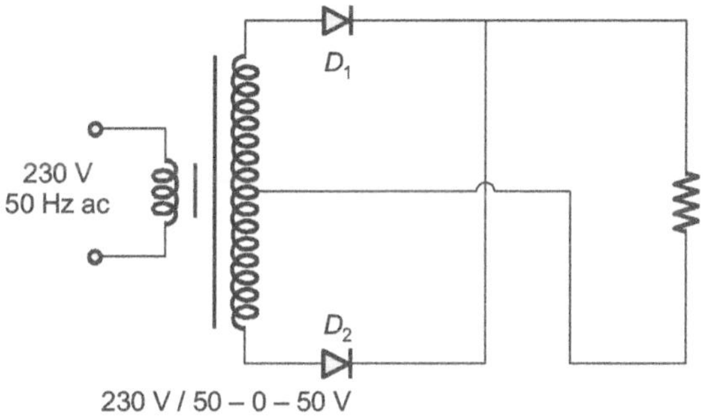

(a) $100\sqrt{2}$ V (b) 100 V
(c) $50\sqrt{2}$ V (d) 50 V

Q.52 If the commutation angle of a diode rectifier (due to source inductance effect) is μ, then the inductive voltage regulation will be

(a) $\frac{1+\cos\mu}{2}$ (b) $1+\frac{\cos\mu}{2}$
(c) $1-\frac{\cos\mu}{2}$ (d) $\frac{1-\cos\mu}{2}$

3. Thyristors

Q.53 In a silicon controlled rectifier (SCR) in its forward cutoff region injecting a current into the cathode gate terminal can
(a) increase the generation of electron-hole pairs in the reverse-biased space charge region.
(b) increase the α's of the two transistors, and cause the SCR to turn on.
(c) increase the displacement current in the junction capacitance.
(d) increase the recombination rate in the neutral regions.

Q.54 By controlling the point in each half cycle when the SCR turns on, the SCR-based control circuit achieves
(a) high efficiency but high cost
(b) low cost but low efficiency
(c) high efficiency but increased distortion
(d) low distortion but high cost

Q.55 **Assertion (A) :** The operation of an SCR can be explained in a very simple way by considering it in terms of two transistors.
Reason (R) : In the two transistor analogy of the SCR, the SCR can be considered as an npn and a pnp transistor, where the collector of one transistor is attached to the emitter of the other and vice-versa.
(a) Both A and R are true and R is the correct explanation of A.
(b) Both A and R are true but R is not the correct explanation of A.
(c) A is true but R is false.
(d) A is false but R is true.

Q.56 After firing an SCR, the gate pulse is removed. The current in the SCR will
(a) rise up.
(b) remain the same.
(c) rise a little and then fall to zero.
(d) immediately fall to zero.

Q.57 Thyristor is not made using germanium, because
(a) its current handling capacity is less than silicon.
(b) it is a stable semiconductor.
(c) it has high leakage current.
(d) group V elements cannot be doped with it.

Q.58 Thyristor is a semiconductor switch which is
(a) unilateral and bistable
(b) bilateral and bistable
(c) unilateral and astable
(d) bistable and astable

Q.59 **Assertion (A) :** If the one-state anode current drops below a minimum level, designated as the holding current I_H, the thyristor reverts to the forward blocking state.
Reason (R) : The loop gain of the equivalent pnp-npn transistors fall below unity and the regenerative hold-on action ceases.

(a) Both A and R are true and R is the correct explanation of A.
(b) Both A and R are true but R is not the correct explanation of A.
(c) A is true but R is false.
(d) A is false but R is true.

Q.60 The main reason for connecting a pulse transformer at the output stage of a thyristor triggering circuit is to
(a) reduce the turn on time of thyristor
(b) amplify the power of the triggering pulse
(c) avoid spurious triggering of the thyristor due to noise
(d) provide electrical isolation

Q.61 Surge current rating of an SCR specifies the maximum
(a) repetitive current with rectangular wave
(b) non-repetitive current with rectangular wave
(c) repetitive current with sine wave
(d) non-repetitive current with sine wave

Q.62 The drawback of *R*-firing circuit for SCR is
(a) firing angle limitation to $\pi/2$
(b) reduced response time
(c) high power loss
(d) requirement of high on-time of SCRs

Q.63 The $\frac{dv}{dt}$ effect in SCR can result in
(a) false triggering
(b) high rate of rise of anode voltage
(c) low capacitive charging current
(d) increased junction capacitance

Q.64 If the latching current in the circuit shown in figure is 2 mA, the minimum width of the gating pulse required to properly turn-on the SCR is

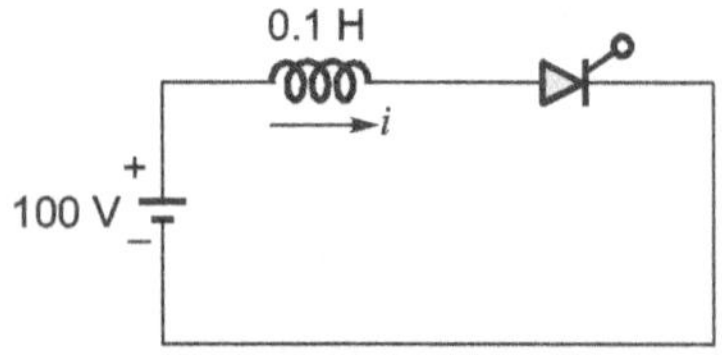

(a) 4 μs (b) 3.5 μs
(c) 1 μs (d) 2 μs

Q.65 A typical V-I characteristic for a thyristor in ON state is shown in figure. What is the average power-loss due to the rectangular current pulses of I_{av} = 100 A and for conduction angle of 180°?

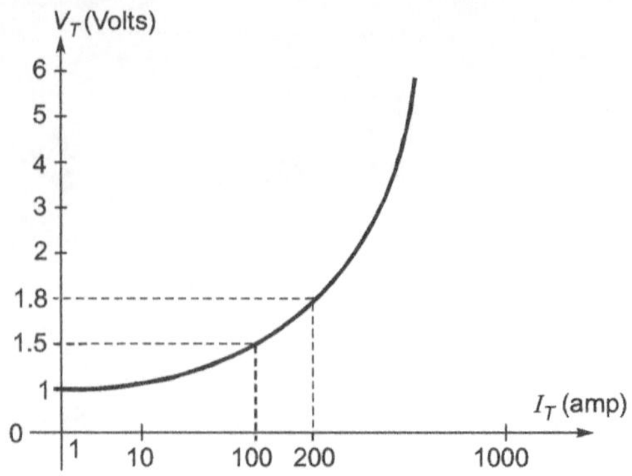

(a) 150 W (b) 180 W
(c) 75 W (d) 120 W

Q.66 Match **List-I (Current rating of SCR)** with **List-II (Features)** and select the correct answer using the codes given below the lists:

List-I
A. $\frac{di}{dt}$ Rating
B. Latching current
C. Gate current
D. Holding current

List-II
1. Controls conduction angle of SCR
2. Associated with turn-off process
3. Associated with junction temperature
4. Associated with turn-on process

Codes:

	A	B	C	D
(a)	1	2	4	3
(b)	2	1	4	3
(c)	4	3	2	1
(d)	3	4	1	2

Q.67 Match **List-I (Power semiconductor devices)** with **List-II (Characteristic)** and select the correct answer using the codes given below the lists:

List-I
A. Diac
B. GTO
C. Triac
D. ASCR

List-II
1. Turns off by a negative pulse
2. Bi-directional diode
3. Reverse blocking capability is unimportant
4. Bi-directional thyristor

Codes:

	A	B	C	D
(a)	2	1	4	3
(b)	1	2	3	4
(c)	3	1	4	2
(d)	2	4	3	1

Q.68 Consider the following statements:
1. Triacs have higher *dv/dt* rating compared to SCRs.
2. Triacs can be triggered with positive or negative polarity voltages.
3. The LASER will stay 'OFF' if the light disappears.
4. The RCT is simply an asymmetric thyristor with a monolithically integrated, antiparallel diode in a single silicon chip.
5. Transistors are generally preferred for their low conduction losses and thyristors for their rugged turn-off capabilities.

Which of the statements given above are correct?
(a) 1, 2, 4 and 5 (b) 2, 3 and 4
(c) 2 and 4 (d) 1 and 2

Q.69 If the $V_g - I_g$ characteristics of an SCR is assumed to be a straight line passing through the origin with a gradient of 3×10^3, the required gate voltage V_g is
(Given E_{gs} = 10 V and allowable P_g = 12 mW)
(a) 4 V (b) 6 V
(c) 3 V (d) 1 V

Q.70 For an SCR the gate cathode characteristic is given by a straight line with a gradient of 16 volts per amp passing through the origin, the maximum turn-on time is 4 μs and the minimum gate current required to obtain this quick turn-on is 500 mA. If the gate source voltage is 15 V, the resistance to be connected in series with the SCR gate is
(a) 10 Ω (b) 7 Ω
(c) 5 Ω (d) 14 Ω

Q.71 A thyristor has an $\int i^2 dt$ rating of 15 amp^2-s and is being used to supply the circuit of figure shown below from a 120 V a.c. supply when a fault occurs, short circuiting the 10 Ω resistors to earth.

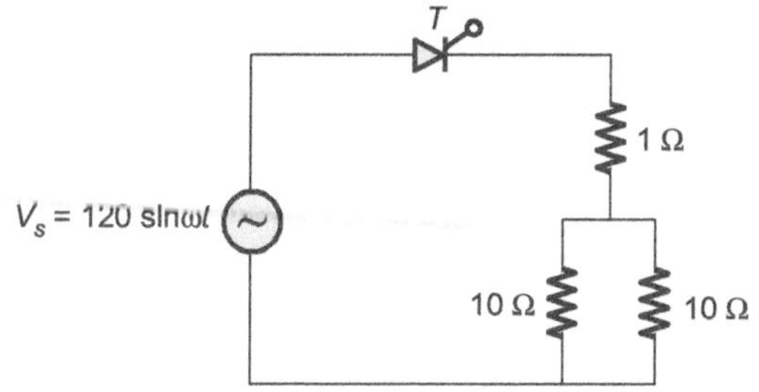

What is the shortest fault clearance time to be achieved if damage to the thyristor is to be prevented?
(a) 0.6 ms (b) 0.79 ms
(c) 1.04 ms (d) 1.20 ms

Q.72 The circuit shown below uses an SCR with $I_{g(min)}$ = 0.1 mA and $V_{g(min)}$ = 0.5 V. The diode is silicon and the peak amplitude of the input is 24 volts. The trigger angle α for R_v = 100 kΩ and R_{min} = 10 kΩ is

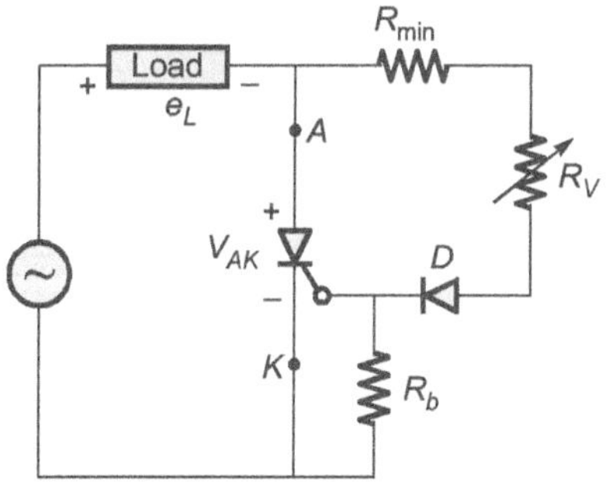

(a) 30.6° (b) 45.2°
(c) 38.5° (d) 60°

Q.73 A UJT relaxation oscillator for triggering an SCR as shown below, the UJT has following characteristics:
η = 0.7, I_P (peak current) = 50 μA,
V_V (Valley voltage) = 2 V
I_V(valley current) = 6 mA, I_{E0} = 2 mA,
R_{BB} = 7 kΩ, $V_{g(min)}$ = 0.2 V and *ln* = 3.33 = 1.2.

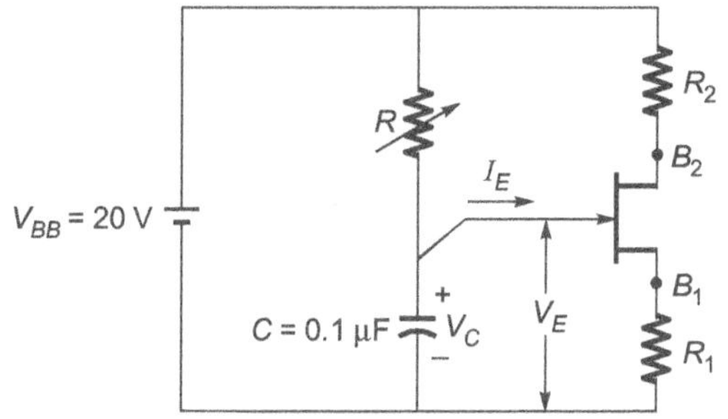

The limits for the output frequency of the oscillator is

(a) 88.8 Hz < f_0 < 2.78 kHz
(b) 69.44 Hz < f_0 < 1.26 kHz
(c) 88.8 Hz < f_0 < 1.26 kHz
(d) 69.44 Hz < f_0 < 2.78 kHz

Q.74 The period time T of the UJT relaxation oscillator is given by

(a) $T = RC\, ln\,(1 - \eta)$ (b) $T = RC\, ln\,(\eta - 1)$

(c) $T = RC\, ln\left(\frac{1}{1-\eta}\right)$ (d) $T = RC\, ln\left(\frac{1}{\eta-1}\right)$

Q.75 A 100 A SCR is to be used in parallel with a 150 A SCR. The on-state voltage drops of the SCRs are 2.1 and 1.75 V, respectively. What is the value of series resistance to be connected with each SCR if the two SCRs have to share the total current of 250 A in proportional to their ratings?

(a) 7 mΩ (b) 5 mΩ
(c) 3 mΩ (d) 1 mΩ

Q.76 What is the number of SCRs, each with rating of 500 V, 75 A required in each branch of a series combination for a circuit with total voltage rating of 7.5 kV and having a derating factor of 14%?

(a) 15 (b) 16
(c) 17 (d) 18

Q.77 In the previous question what is the number of SCRs (rating of each SCR remaining the same) required in each branch of a parallel combination for a circuit with total current rating of 1 kA and having same derating factor?

(a) 15 (b) 16
(c) 17 (d) 18

Q.78 Match **List-I (Firing circuits of SCR)** with **List-II (Characteristics)** and select the correct answer using the codes given below the lists:

List-I
A. R-firing circuit
B. UJT firing circuit
C. R-C firing circuit

List-II
1. Turned-on and Turned-off in controlled manner
2. Limited range of firing angle
3. Higher range of firing angle

Codes:

	A	B	C
(a)	3	1	2
(b)	2	1	3
(c)	2	3	1
(d)	1	2	3

Q.79 Consider the following statements associated with R and RC- firing circuits of SCR:

1. They can be employed in power circuits having only one thyristor.
2. They are capable of close loop control only.
3. Both the circuits are capable for providing high frequency gate signal.
4. There is electrical isolation between control circuit and power circuit in both R and RC- firing circuits.

Which of the statements given above is/are correct?

(a) 1 and 3 (b) 1, 2 and 4
(c) 1 only (d) 4 only

Q.80 The thyristor shown in figure is operated at 5 kHz. What is the turn-on loss if the turn-on time, t_{on}, is 3 μs?

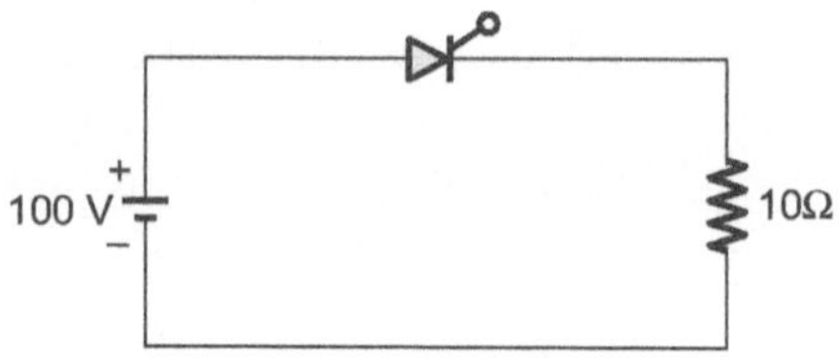

(a) 2.5 Watt (b) 1.0 Watt
(c) 1.5 Watt (d) 2.0 Watt

Q.81 The Thyristor connected in the circuit shown in figure has di/dt rating of 200 A/μsec. The value of series inductance for its protection is

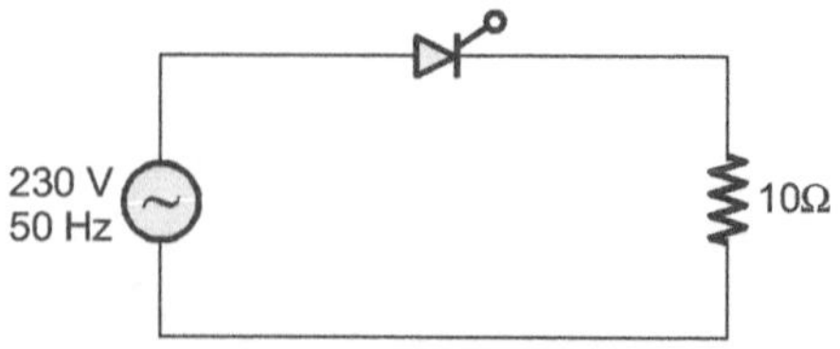

(a) 0.8 μH (b) 1.2 μH
(c) 1.6 μH (d) None of these

Q.82 Ten thyristors are connected in series. Each thyristor has a rating of 4 kV and maximum leakage current difference is 10 mA. If the string efficiency is 82.5%, then the value of each resistance of static equalisation circuit will be:

(a) 87.7 kΩ (b) 88.7 kΩ
(c) 78.8 kΩ (d) 77.8 kΩ

Q.83 **Assertion (A) :** Snubber circuit is connected to protect the thyristor from unnecessary turn-on due to high *di*/*dt*.
Reason (R) : Snubber circuit reduces transient over voltage during reverse recovery period and provides dynamic equalisation in a string of thyristors.
(a) Both A and R are true and R is the correct explanation of A.
(b) Both A and R are true but R is not the correct explanation of A.
(c) A is true but R is false.
(d) A is false but R is true.

Q.84 An SCR has half-cycle surge current rating of 3000 A for 50 Hz supply. Its one-cycle surge current rating and I^2t rating are respectively
(a) 1.6 kA and 45 kA-sec
(b) 2.1 kA and 45 kA-sec
(c) 1.6 kA and 33 kA-sec
(d) 2.1 kA and 33 kA-sec

Q.85 Match **List-I (Undesirable phenomenon for an SCR)** with **List-II (Effects)** and select the correct answer using the codes given below the lists:

List-I
A. Overcurrent
B. High *dv*/*dt*
C. High *di*/*dt*
D. Overvoltage

List-II
1. False triggering of SCR
2. Damage of SCR due to reverse breakdown
3. Breakdown of SCR due to high junction temperature
4. Local hot spots near the gate connection

Codes:

	A	B	C	D
(a)	2	4	3	1
(b)	3	1	2	4
(c)	3	1	4	2
(d)	4	3	2	1

Q.86 An SCR can be turned-on if
1. it is forward biased.
2. gate pulse width is equal to the turn-on time of SCR.
3. anode to cathode voltage is equal to finger voltage.
4. the gate triggering synchronizes with the ac supply.

Choose the correct code from the given options.
(a) 1 and 4 (b) 2, 3 and 4
(c) 1 and 2 (d) 1, 2 and 4

Q.87 An SCR can withstand a maximum junction temperature of 120°C with an ambient temperature of 75°C. If the SCR has thermal resistance from junction to ambient as 1.5 °C/W, the maximum power dissipation allowed is:
(a) 60 W (b) 50 W
(c) 30 W (d) 80 W

Q.88 The reason of connecting resistance and capacitance across gate circuit is to protect the SCR gate against
(a) *dv*/*dt* (b) overvoltages
(c) overcurrent (d) noise signals

Q.89 The function of connecting a zener diode in an UJT circuit, used for the triggering of SCRs, is to
(a) provide a constant voltage to UJT to prevent erratic firing.
(b) expedite the generation of triggering pulses.
(c) provide a variable voltage to UJT as the source voltage changes.
(d) delay the generation of triggering pulses.

Q.90 Thyristors are classified as inverter grade or converter grade based on
(a) turn-on time
(b) forward breakdown voltage
(c) holding current
(d) turn-off time

Q.91 Which of the following option is not true about a GTO compared to a conventional thyristor?
(a) Its magnitude of latching and holding current is more.
(b) On state voltage drop and the associated loss is more.
(c) Triggering gate current is lower.
(d) Its reverse blocking capability is less than its forward blocking capability.

Q.92 A pulse transformer is used for pulse isolation in case of

(a) power MOSFET, IGBT and TRIAC
(b) an SCR, TRIAC and IGBT
(c) an SCR, power transistor and GTO
(d) an SCR, TRIAC and GTO

Q.93 How many SCRs are to be connected in series with 800 V rating to be used for a 3 kV circuit using derating factor of 15%?
(a) 6 (b) 4
(c) 5 (d) 3

Q.94 The TRIAC can be used only in
(a) rectifiers
(b) multi-quadrant chopper
(c) cycloconverter
(d) inverter

Q.95 Which of the following cause permanent damage to an SCR?
1. High *di*/*dt* 2. High temperature rise
3. High current 4. High *dv*/*dt*
Select the correct code from the given options:
(a) 1, 2 and 4 (b) 2, 3 and 4
(c) 1, 3 and 4 (d) 1, 2, 3 and 4

Q.96 Peak inverse current and reverse recovery time depends on
(a) stored charge, *di*/*dt*, forward current
(b) stored charge, *di*/*dt*
(c) *di*/*dt*, forward current
(d) stored charge, forward current

Q.97 Figure shows a thyristor with the standard terminations of anode (*A*), cathode (*K*), gate (*G*) and the different junctions named J_1, J_2 and J_3. When the thyristor is turned on and blocking

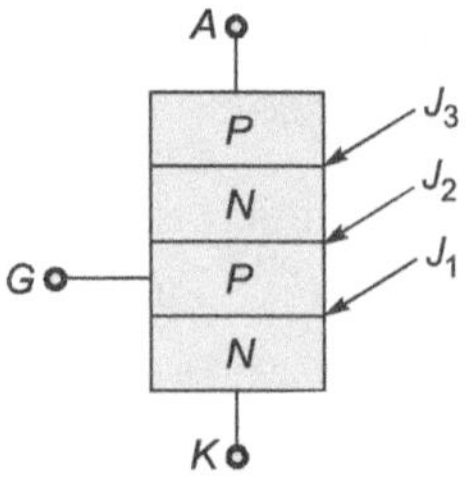

(a) J_1 and J_2 are forward biased and J_3 is reverse biased.
(b) J_1 and J_3 are forward biased and J_2 is reverse biased.
(c) J_1 is forward biased and J_2 and J_3 are reverse biased.
(d) J_1, J_2 and J_3 are all forward biased.

Q.98 In a thyristor, anode current is made up of
(a) electrons only
(b) holes only
(c) electrons and holes
(d) none of the above

Q.99 The turn-off time is longer than the turn-on time in SCR because
(a) the anode and cathode junctions get reverse biased while gate junction is still forward biased
(b) the forward break-over voltage is high
(c) the gate pulse has been removed
(d) none of the above

Q.100 For the SCR circuit in the given figure, which one of the following waveforms represent V_T, when input voltage V is sinusoidal?

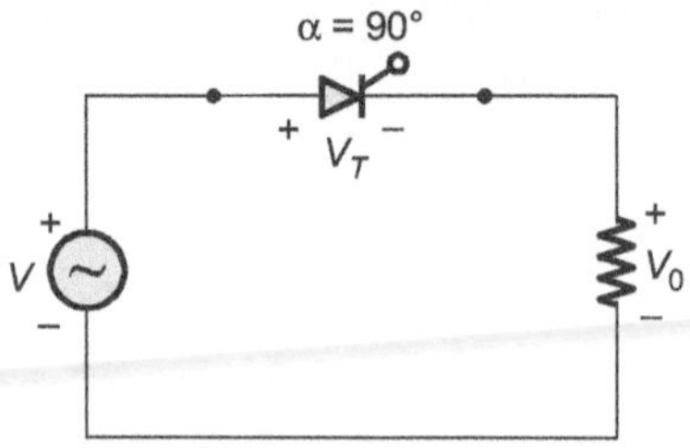

(a)
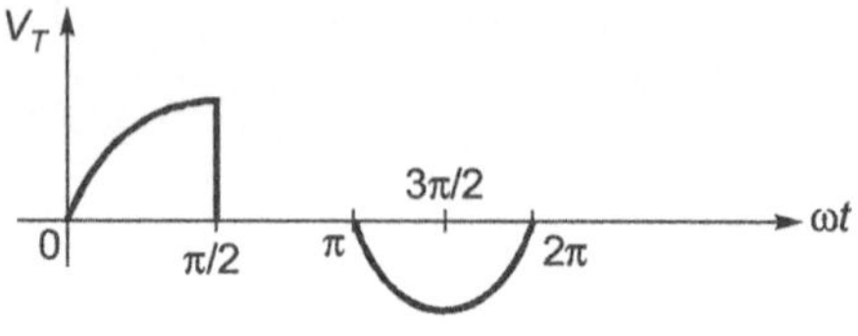

(b)
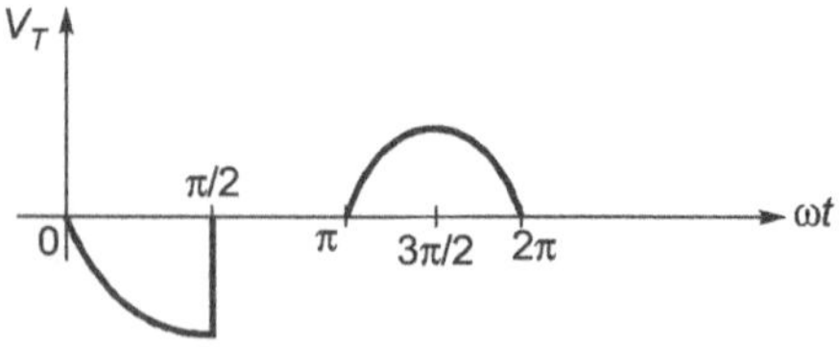

(c)
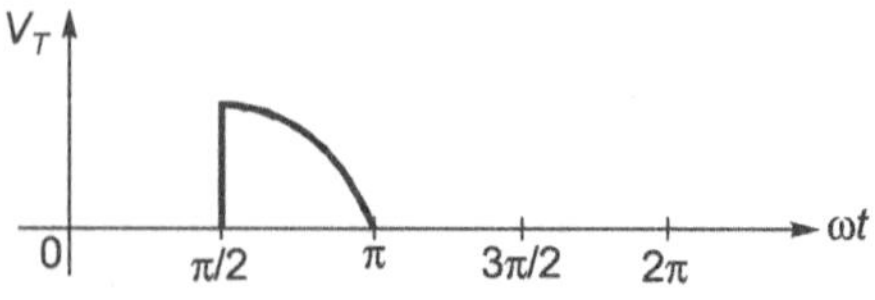

(d)
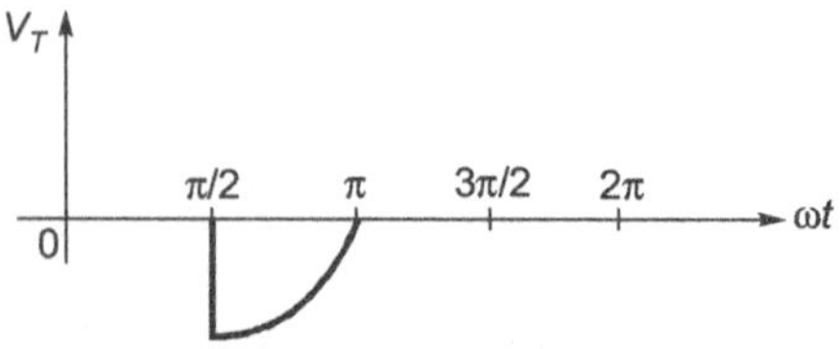

Q.101 The triggering circuit of a thyristor is shown in figure. The thyristor requires a gate current of 10 mA for guaranteed turn-on. The value of R required for the thyristor to turn-on reliably under all conditions of V_b variation is

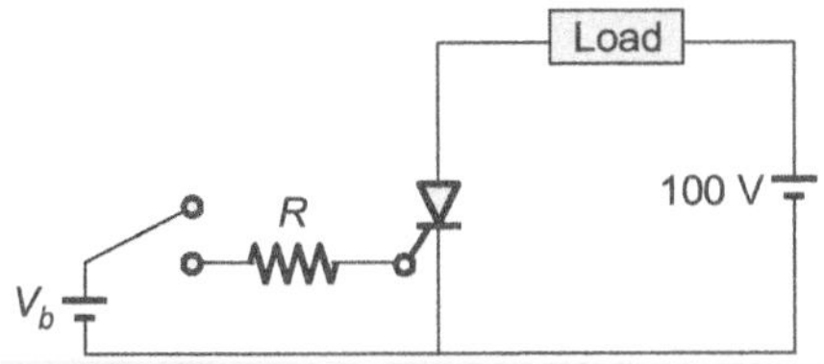

$(V_b = 12 \pm 4\text{V})$

(a) 10 kΩ (b) 1.6 kΩ
(c) 1.2 kΩ (d) 0.8 kΩ

4. Thyristor Commutation Techniques

Q.102 To turn-off or commutate a thyristor
(a) forced commutation is used
(b) reverse blocking voltage is applied
(c) gate current is made zero
(d) none of the above

Q.103 Most SCRs can be turned-off by voltage reversal during negative half cycle of the ac supply for
(a) all frequencies
(b) frequencies upto 300 kHz
(c) frequencies upto 30 kHz
(d) frequencies upto 300 Hz

Q.104 Turn-off time of a thyristor
(a) depends on load
(b) depends upon junction temperature
(c) depends upon forward current
(d) both (b) and (c)

Q.105 Turn-on time (t_{on}) of an SCR is related to its turn-off time (t_{off}) in which of the following way?
(a) $t_{on} < t_{off}$
(b) $t_{on} > t_{off}$
(c) t_{off} increases with t_{on}
(d) t_{on} depends on L/R ratio of load

Q.106 Assertion (A) : It is necessary to keep the thyristor reverse biased for a finite period before a forward anode voltage can be reapplied.
Reason (R) : Once the SCR starts conducting an appreciable forward current, the gate has no control on it and the device can be brought back to the blocking state only by reducing the forward current level below that of the holding current.
(a) Both A and R are true and R is the correct explanation of A.
(b) Both A and R are true but R is not the correct explanation of A
(c) A is true but R is false.
(d) A is false but R is true.

Q.107 Match **List-I (Commutation techniques)** with **List-II (Alternative name)** and select the correct answer using the codes given below the lists:

List-I
A. Class-E
B. Class-F
C. Class-B, C and D
D. Class-A

List-II
1. Forced commutation
2. Load commutation
3. External-pulse commutation
4. Line commutation

Codes:

	A	B	C	D
(a)	3	1	4	2
(b)	4	2	1	3
(c)	3	4	1	2
(d)	2	4	3	1

Q.108 Consider the following statements associated with various commutation techniques of a thyristor:
1. In forced commutation, the commutating components L and C carry load current continuously.
2. Forced commutation is usually employed in dc choppers.
3. In load commutation, L and C are connected in parallel with the load or C in series with the load.
4. Load commutation is employed in series inverters.
5. In class-C commutation, the main thyristor that is to be commutated is connected in parallel with the load.

Which of the statements given above are correct?
(a) 2 and 4 (b) 1, 3 and 5
(c) 2, 4 and 5 (d) 1, 2, 3 and 5

Q.109 For the class-B commutation circuit shown below, the initial voltage across capacitor is 230 volt. For a constant load current of 300 A, the conduction time for the auxiliary thyristor is approximately equal to

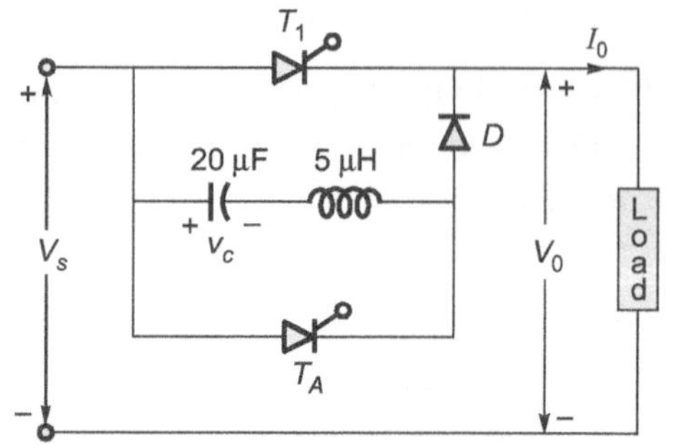

(a) 22.60 μs (b) 16.16 μs
(c) 31.40 μs (d) 28.80 μs

Q.110 For the circuit shown below, the peak value of the current through thyristors T_1 and T_2 are

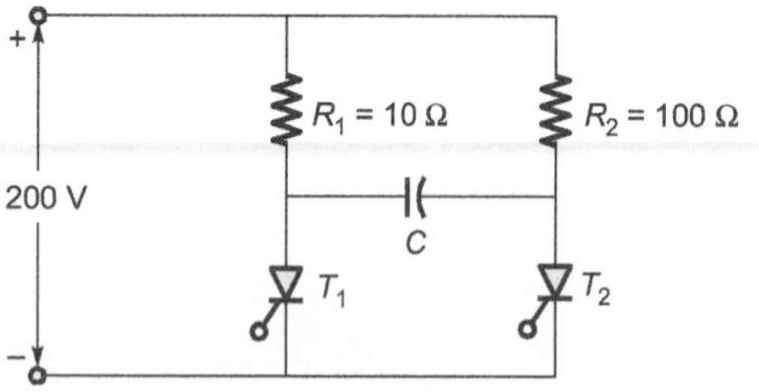

(a) 16 A and 24 A respectively
(b) 24 A and 42 A respectively
(c) 16 A and 42 A respectively
(d) 24 A and 16 A respectively

Q.111 The SCR in the circuit shown below is forced commutated by a circuitry not shown in the figure. The SCR has minimum charging current of 5 mA to turn it on and its junction capacitance is 25 pF.

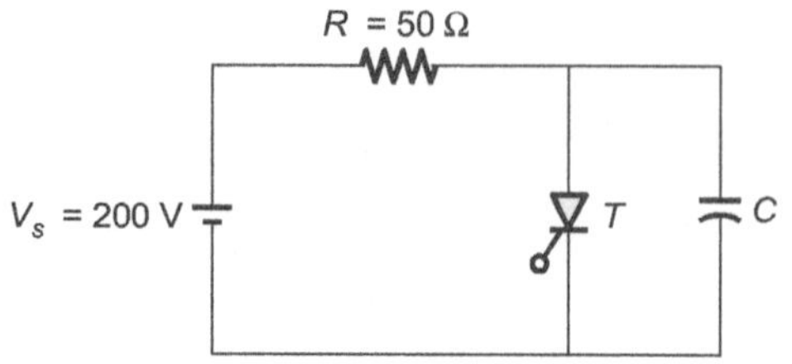

What is the minimum value of C so that SCR does not get turned-on due to re-applied dv/dt?

(a) 0.01 μF (b) 0.02 μF
(c) 0.03 μF (d) 0.04 μF

Q.112 Match **List-I (Commutation techniques)** with **List-II (Applications)** and select the correct answer using the codes given below the lists:

List-I
A. Forced commutation
B. Natural commutation
C. Load commutation

List-II
1. AC voltage controllers
2. Series inverters
3. Choppers

Codes

	A	B	C
(a)	3	2	1
(b)	2	1	3
(c)	1	2	3
(d)	3	1	2

Q.113 For the class-C commutation circuit, the d.c. source voltage V_s is 120 V and current through R_1 and R_2 = 20 A. The turn-off time of both the SCRs is 60 μs. The value of commutating capacitance C for successful commutation is

(a) 14.4 μF (b) 13.8 μF
(c) 12.2 μF (d) 11.0 μF

Q.114 For the class-D commutation circuit shown below,

$I_{L(max)}$ = 50 A, t_{off} of SCR_1 = 30 μs,
Chhoping frequency f = 500 Hz and the load voltage variation required is 10 to 100%.

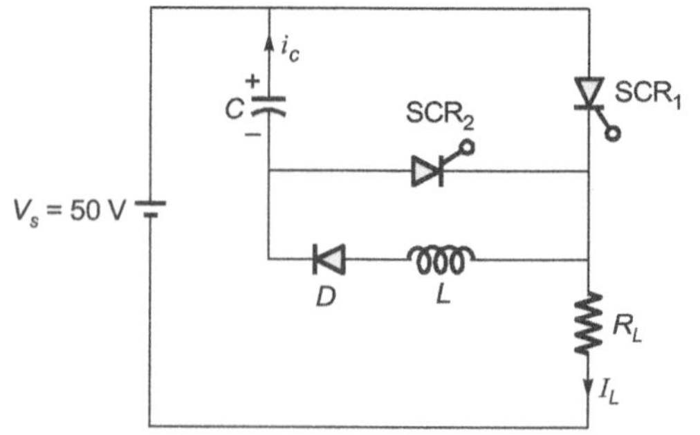

Assuming 50% tolerance on turn-off time of SCR_1 for reliable operation, the value of commutating capacitor C would be equal to

(a) 25 μF (b) 35 μF
(c) 45 μF (d) 55 μF

Q.115 For the circuit shown in figure, the capacitor voltage after SCR gets self-commutated is

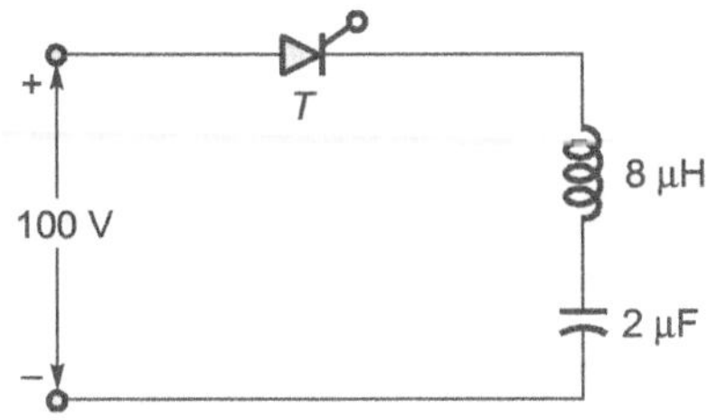

(a) 400 V (b) 100 V
(c) 300 V (d) 200 V

Q.116 In the previous problem, the conduction time for thyristor is _____ μs.
(a) 25.12 (b) 12.56
(c) 6.28 (d) 3.14

Q.117 For the circuit shown below, the maximum value of current through thyristor T_A and T_1 are respectively.

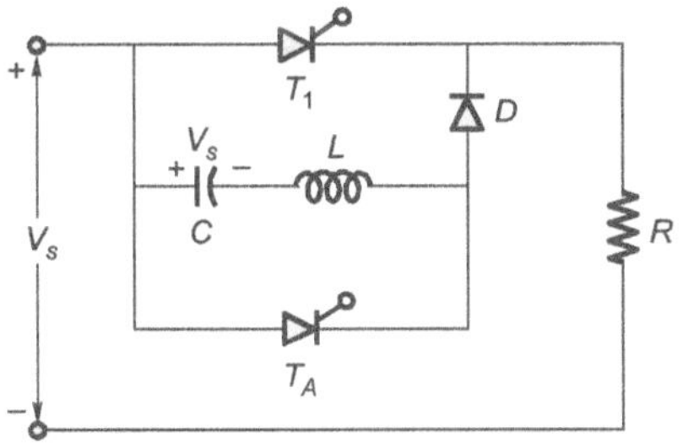

(a) $\frac{V_s}{R} + V_s\sqrt{\frac{C}{L}}$ and $\frac{V_s}{R}$

(b) $V_s\sqrt{\frac{C}{L}}$ and $\frac{V_s}{R} + V_s\sqrt{\frac{C}{L}}$

(c) $\frac{V_s}{R}$ and $V_s\sqrt{\frac{C}{L}}$

(d) $V_s\sqrt{\frac{C}{L}}$ and $\frac{V_s}{R}$

Q.118 Consider the following statements:
1. No additional circuit is required to turn-off the thyristor in line commutation.
2. In some applications thyristor current is forcibly reduced to zero and the load current is diverted to some alternate path and the thyristor is turned-off.
3. In self commutation, the value of commutation elements L and C are so selected that $R > 2\sqrt{L/C}$.
4. Class-C commutation method is called complementary voltage commutation because each thyristor is used to commutate the other thyristor.

Which of the statements given above is/are correct?
(a) 2 only (b) 1, 2 and 4 only
(c) 2 and 4 only (d) 1, 3 and 4 only

Q.119 **Assertion (A) :** Class-E commutation method requires a pulse transformer.
Reason (R) : Whenever, thyristor is required to be turned-off in a class-E commutation circuit, a pulse is applied to the transformer.
(a) Both A and R are true and R is the correct explanation of A.
(b) Both A and R are true but R is not the correct explanation of A.
(c) A is true but R is false.
(d) A is false but R is true.

Q.120 During the turn-off process of a thyristor the current flow does not stop at the instant when the current reaches zero but continues to flow to a peak value in the reverse direction. This is due to
(a) commutation failure
(b) hole storage effect
(c) presence of reverse voltage across the thyristor
(d) protective inductance in series with the thyristor

Q.121 For the circuit shown in figure, the peak SCR current is

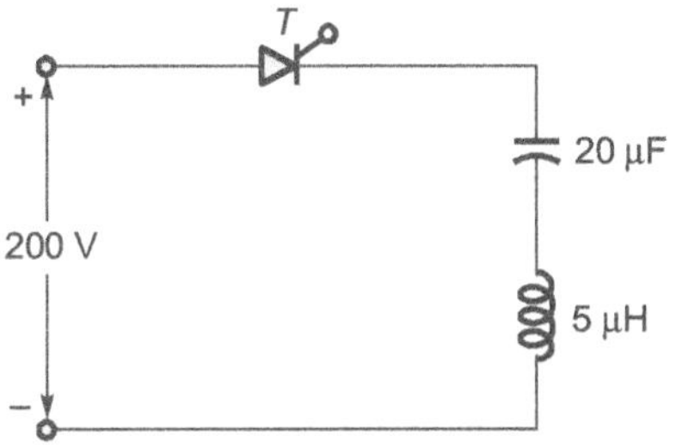

(a) 200 A (b) 400 A
(c) 300 A (d) 100 A

5. Phase Controlled Rectifiers

Q.122 In an SCR half wave rectifier, the forward breakdown voltage of SCR is 110 V for a gate current of 1 mA. If a 50 Hz sinusoidal voltage of 220 V peak is applied, the time during which SCR remains OFF would be equal to (assume load resistance is 100 Ω and the holding current to be zero)

(a) 1.67 ms (b) 2.25 ms
(c) 1.88 ms (d) 2.66 ms

Q.123 When an SCR full wave rectifier is connected across a sinusoidal voltage of 400 sin 314t, the RMS value of the current flowing through the device is 20 A. The power rating of the SCR is
(a) 13.82 kW (b) 16.68 kW
(c) 12.47 kW (d) 11.87 kW

Q.124 Assertion (A) : Single phase half-wave converter introduces a dc component into the supply line.
Reason (R) : The supply current taken from the source is unidirectional and is in the form of dc pulses.
(a) Both A and R are true and R is the correct explanation of A.
(b) Both A and R are true but R is not the correct explanation of A.
(c) A is true but R is false.
(d) A is false but R is true.

Q.125 Assertion (A) : By using freewheeling diode, load performance becomes better.
Reason (R) : Freewheeling diode prevents the load voltage from becoming negative.
(a) Both A and R are true and R is the correct explanation of A.
(b) Both A and R are true but R is not the correct explanation of A.
(c) A is true but R is false.
(d) A is false but R is true.

Q.126 The advantage of using a freewheeling diode with bridge type ac to dc converter is
(a) reliable speed control
(b) reduced cost of system
(c) regenerative braking
(d) improved input power factor

Q.127 A single-phase full converter is connected across 230 V ac. Its average output voltage when firing angle is 30° is approximately equal to
(a) 89.5 V (b) 193 V
(c) 179 V (d) None of these

Q.128 SCRs with peak forward rating of 1 kV and average on-state current rating of 40 A is used in single-phase mid-point converter. If the factor of safety is 2.5, then the power that this converter can handle is

(a) $\frac{16}{\pi}$ kW (b) $\frac{8}{\pi}$ kW
(c) $\frac{4}{\pi}$ kW (d) $\frac{2}{\pi}$ kW

Q.129 The main advantages of the dual converter with circulating current operation is/are
1. the efficiency and power factor are more.
2. it has faster operation than dual converter without circulating current.
3. the thyristors are rated for lower current.
Choose the correct code from the following options:
(a) 1 and 2 (b) 1 and 3
(c) 2 only (d) 3 only

Q.130 The average value of voltage of a single-phase ac voltage converter with resistive load is (α is the firing angle).

(a) $\frac{V_m}{2\pi}(1+\cos\alpha)$ (b) $\frac{V_m}{2\pi}\cos\alpha$
(c) $\frac{V_m}{\pi}(1+\cos\alpha)$ (d) $\frac{V_m}{\pi}\cos\alpha$

Q.131 A single-phase controlled rectifier bridge consists of one SCR and three diodes as shown in figure. Assuming the continuous current conduction, the average output voltage is equal to (α is firing angle of thyristor T_1).

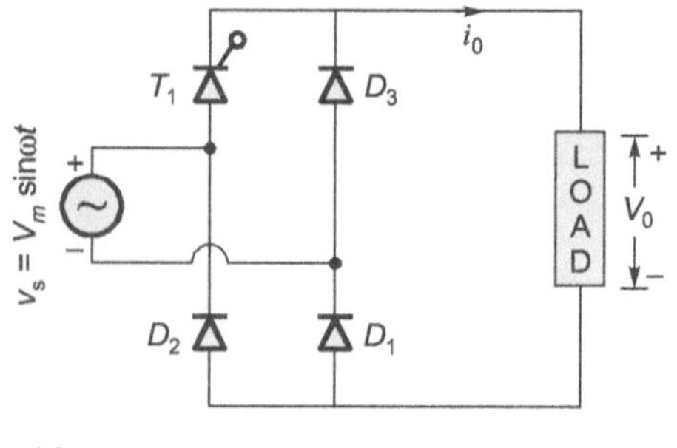

(a) $\frac{V_m}{\pi}(2+\sin\alpha)$ (b) $\frac{V_m}{2\pi}(3+\cos\alpha)$
(c) $\frac{V_m}{\pi}(1+\cos\alpha)$ (d) $\frac{V_m}{2\pi}(1+\sin\alpha)$

Q.132 A four quadrant operation requires
(a) two full converters connected back to back
(b) two full converters is series
(c) two semi-converters connected back to back
(d) two full converters connected in parallel

Q.133 Consider the following statements associated with single phase full converters:

1. Mid-point converter configuration is used in case the terminals on dc side have to be grounded.
2. The transformer rating in mid-point converter is double the load rating.
3. SCRs are subjected to a peak inverse voltage of 2 V_m in single phase mid-point converter.
4. Bridge converter is preferred over mid-point converter.

Which of these statements are correct?

(a) 1, 3 and 4 (b) 2, 3 and 4
(c) 1, 2 and 3 (d) 1, 2, 3 and 4

Q.134 In a single phase full converter with resistive load and firing angle α, the load current is non-zero and zero, for

(a) $\pi + \alpha$ and α respectively
(b) α and $\pi - \alpha$ respectively
(c) π and $\pi - \alpha$ respectively
(d) $\pi - \alpha$ and α respectively

Q.135 **Assertion (A):** The commutation overlap is more predominant is semiconverters than in full converters.

Reason (R): The commutation period in seconds, when outgoing and incoming SCRs are conducting, is known as the overlap period.

(a) Both A and R are true and R is the correct explanation of A.
(b) Both A and R are true but R is not the correct explanation of A.
(c) A is true but R is false.
(d) A is false but R is true.

Q.136 For a single-phase full converter having a full-load current of I_0 match **List-I (Overlap angle)** with **List-II (Output voltage)** and select the correct answer using the codes given below the lists (L_s = source Inductance):

List-I	List-II
A. $\mu = 0°$	1. $\frac{2V_m}{\pi}\cos\alpha - \frac{\omega L_s}{\pi} I_0$
B. $\mu = 180°$	2. 0
C. $\mu < 180°$	3. $\frac{2V_m}{\pi}\cos\alpha$

Codes:

	A	B	C
(a)	1	2	3
(b)	2	3	1
(c)	3	2	1
(d)	2	1	3

Q.137 Match **List-I** with **List-II** and select the correct answer using the codes given below the lists:

List-I

A. 3-phase semiconverter ($\alpha < 60°$)
B. Dual converter without circulating current
C. Dual converter with circulating current
D. 3-phase semiconverter ($\alpha \geq 60°$)

List-II

1. Three-pulse converter
2. Load current may be continuous or discontinuous
3. Faster operation
4. Six-pulse converter

Codes:

	A	B	C	D
(a)	4	3	2	1
(b)	4	2	3	1
(c)	1	3	2	4
(d)	2	1	4	3

Q.138 Which of the following converters can feed power in any of the four quadrants?

(a) Full converter
(b) Dual converter
(c) Semi-converter
(d) Combination of a semi and full converter

Q.139 Consider the following statements:

1. In a full converter, direction of current cannot reverse.
2. Semi-converters are single quadrant converters.
3. A full converter can operate as a two-quadrant converter.
4. A full converter operates as a rectifier in first quadrant and as an inverter in the second quadrant.
5. In a full converter, direction of current cannot reverse but polarity of output voltage can be reversed.

Which of the statements given above are correct?

(a) 1, 2, 3 and 5 (b) 2, 3 and 4
(c) 2, 4 and 5 (d) 1, 3 and 4

Q.140 AC-to-DC circulating current dual converters are operated with which of the following relationship between their triggering angles (α_1 and α_2)?
(a) $\alpha_1 - \alpha_2 = 180°$ (b) $\alpha_1 + \alpha_2 = 90°$
(c) $\alpha_1 + \alpha_2 = 360°$ (d) $\alpha_1 + \alpha_2 = 180°$

Q.141 A single-phase half-wave controlled converter with resistive load is operated from a 120 V, 50 Hz supply. If the average output voltage is 25% of the maximum possible output voltage, the firing angle of the thyristor would be equal to
(a) 30° (b) 45°
(c) 60° (d) None of these

Q.142 A single phase fully controlled thyristor bridge converter supplies a load consisting of *R*, *L* and *E* as shown in figure below. The inductance *L* in the circuit is so large that the output current may be considered to be virtually constant. The firing angle α is

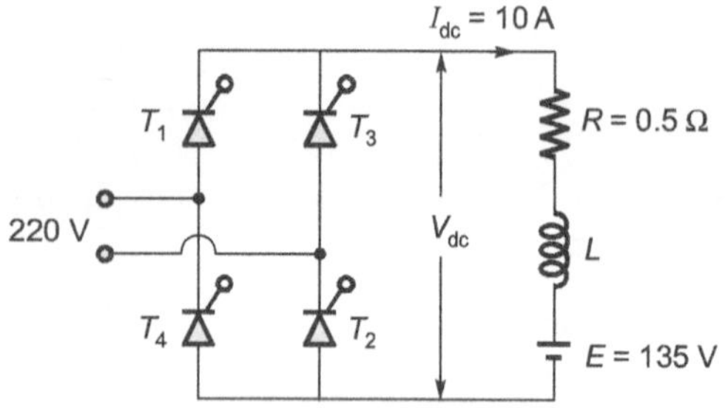

(a) 30° (b) 45°
(c) 60° (d) 120°

Q.143 Consider the following statements:
1. Fully-controlled and half controlled circuits both are capable of regeneration.
2. The higher the pulse number, the lower is the magnitude of ripple voltage and current in the load-circuit and the a.c. supply.
3. Bridge circuits and half-wave circuits both require same number of thyristors to carry the same current.
4. Half-wave circuits tend to be used for low-voltage, high-current use.

Which of the statements given above is/are **not** correct?
(a) 1, 2 and 4 (b) 2 only
(c) 1 and 3 only (d) 3 only

Q.144 Match **List-I** with **List-II** and select the correct answer using the codes given below the lists:

List-I
A. Leading displacement factor
B. Full converter
C. Asymmetrical firing of SCRs
D. Semiconverter

List-II
1. Bidirectional power flow
2. Compensates line voltage drop
3. Unidirectional power flow
4. Improves power factor

Codes:

	A	B	C	D
(a)	2	1	4	3
(b)	2	3	4	1
(c)	4	3	2	1
(d)	2	1	3	4

Q.145 In a 3-ϕ bridge converter the dc load current is 90 A, the commutating inductance per phase is 50 μH and the 3-ϕ power supply is 2.3 kV, 50 Hz. The reduction in output voltage due to overlap is around
(a) 2.65 V (b) 1.35 V
(c) 2.22 V (d) 0.95 V

Q.146 The holding current of SCRs in 3-ϕ full-converter is 200 mA and the delay time is 2.5 μsec. Converter is supplied from a star-connected 208 V, 50 Hz supply, has a load of $R = 2\ \Omega$, $L = 8$ mH and the firing angle is 60°. The minimum width of gate pulse t_p for $L = 0$ H is
(a) 7.9 μsec (b) 6.8 μs
(c) 4.0 μs (d) 2.5 μs

Q.147 The nature of load current, i.e., whether load is continuous or discontinuous in controlled rectifiers
(a) depends only on the type of load
(b) does not depend on type of load and firing angle delay
(c) depends only on the firing angle delay
(d) depends both on the type of load and firing angle delay

Q.148 In a 3-phase full converter, the output voltage during overlap is equal to
(a) source voltage
(b) zero

(c) $\frac{3V_{ml}}{\pi}\cos\alpha - \frac{3\omega L_s}{\pi}I_0$

(d) $\frac{3V_{ml}}{\pi}\cos\alpha$

Q.149 The peak inverse voltage in ac to dc converter system is highest in
(a) 3-phase bridge converter
(b) 3-phase half-wave converter
(c) single-phase full wave mid-point converter
(d) single-phase full-converter

Q.150 The frequency of the ripple in the output voltage of a 3-phase semiconverter depends upon
(a) the load circuit parameters
(b) firing angle and load resistance
(c) firing angle and the supply frequency
(d) firing angle and load inductance

Q.151 Resonant converters are basically used to
(a) reduce the switching losses
(b) eliminate harmonics
(c) generate large peaky voltage
(d) convert a square wave into a sine wave

Q.152 A transformer is having source voltage of 400 volt and turns ratio of 2 : 1. The transformer is centre-tapped. If the secondary is connected to single-phase full-wave converter, PIV per SCR will be

(a) $200\sqrt{2}$ V (b) $100\sqrt{2}$ V
(c) 200 V (d) 100 V

Q.153 If a 1-ϕ half-wave controlled rectifier is fired at an angle of 90°, then ripple factor of the output rectifier wave will be

(a) $\sqrt{\frac{\pi^2}{2}}$ (b) $\sqrt{\frac{\pi^2-2}{2}}$
(c) $\sqrt{\frac{\pi^2-1}{2}}$ (d) $\sqrt{\frac{\pi^2-3}{2}}$

Q.154 In a single-phase full converter, if load current is I and ripple free, then average thyristor current is

(a) $\frac{1}{2}I$ (b) $\frac{1}{3}I$
(c) $\frac{1}{4}I$ (d) I

Q.155 The voltage across an R-C (parallel) load connected through a full-wave rectifier bridge will have which of the following waveform shown below?

(a)
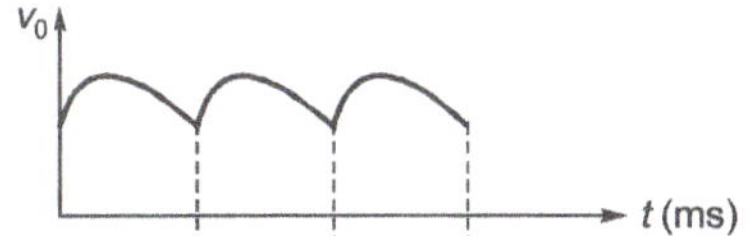

(b)
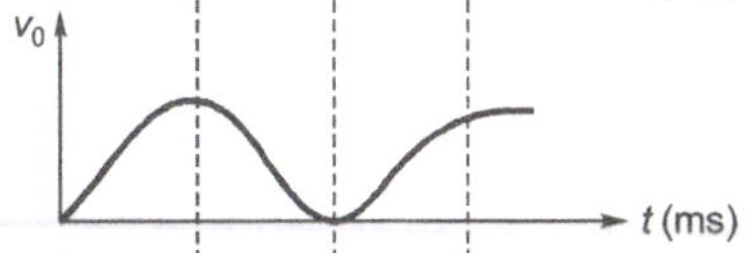

(c)
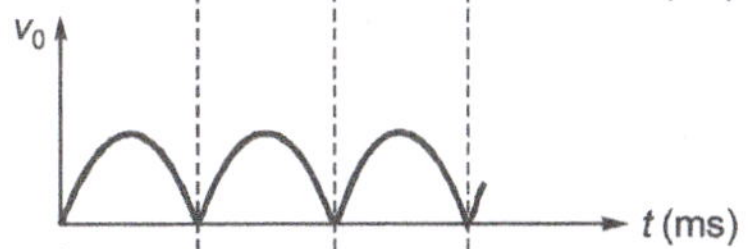

(d)
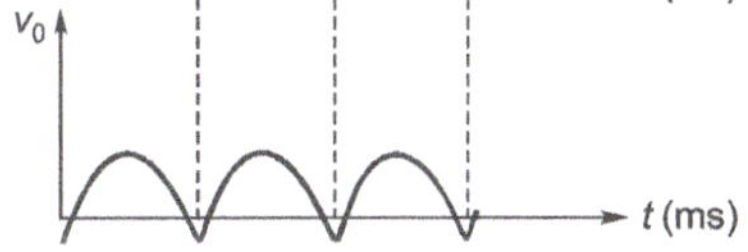

Q.156 A half wave diode rectifier is feeding a pure inductive load as shown in figure. Assuming the diode to be ideal, voltage waveform V_L across the load terminal is approximately similar to

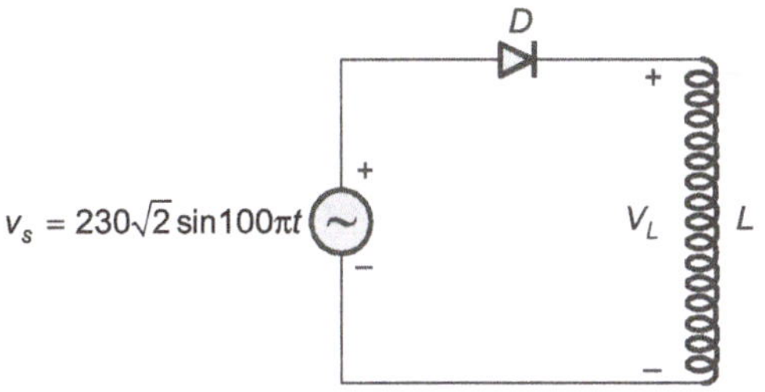

(a)
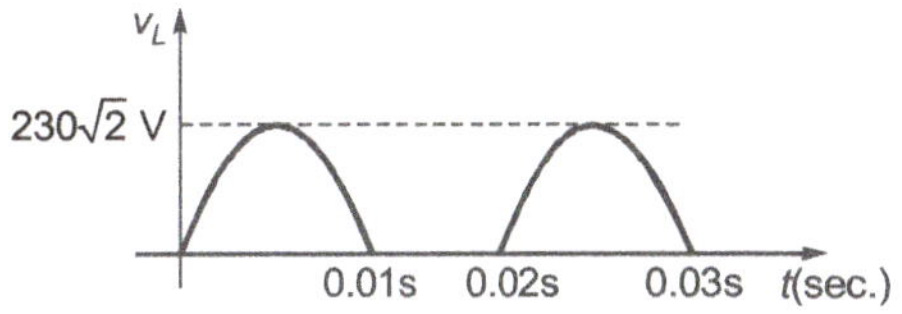

(b)
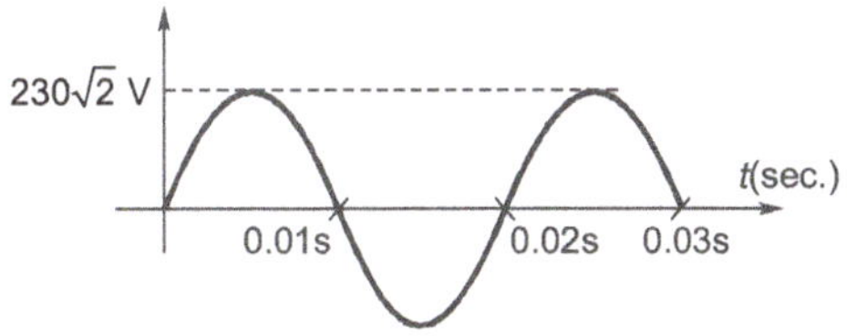

(c)
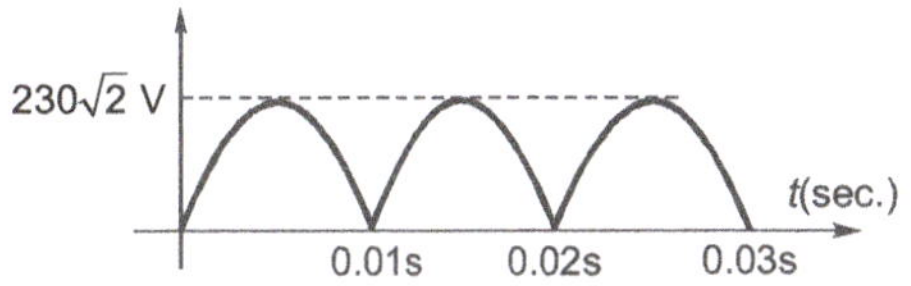

(d)
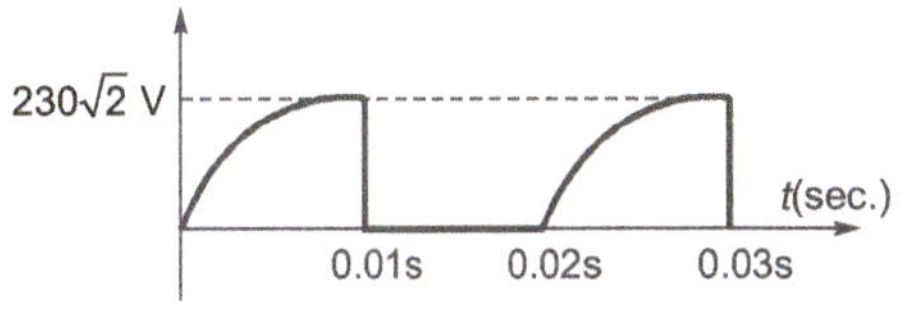

6. Choppers

Q.157 A d.c. chopper circuit connected to a 100 V d.c. source supplies an inductive load having 40 mH in series with a resistance of 5 Ω. A freewheeling diode is placed across the load. The load current varies between the limits of 10 A and 12 A. The time ratio (T_{on}/T_{off}) of the chopper is

(a) 1.65 (b) 0.98
(c) 1.22 (d) 0.75

Q.158 A step-up chopper is used to deliver load voltage of 500 V from a 220 V d.c. source. If the blocking period of the thyristor is 80 μs, the required pulse width is

(a) 101.8 μs (b) 50.8 μs
(c) 152.4 μs (d) 92.4 μs

Q.159 **Assertion (A) :** The principal of step-up chopper can be employed for the regenerative braking of dc motors.

Reason (R) : The step-up chopper can be employed in regenerative braking only for increasing motor speeds.

(a) Both A and R are true and R is the correct explanation of A.
(b) Both A and R are true but R is not the correct explanation of A.
(c) A is true but R is false.
(d) A is false but R is true.

Q.160 A dc chopper has a T_{ON} of 1 ms and its frequency is 500 Hz. What will be its duty cycle?

(a) 25% (b) 50%
(c) 80% (d) 75%

Q.161 An RLE load is operating in a chopper circuit from a 500 volts dc source as shown in figure

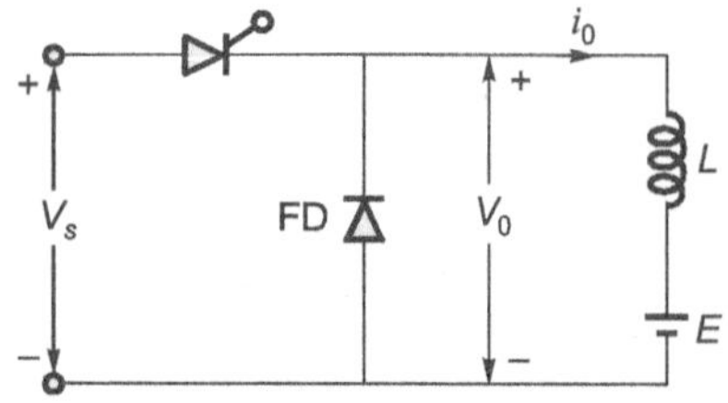

For the load, $L = 0.06$ H, $R = 0$ Ω an duty cycle of 0.2, what is the chopping frequency to limit the amplitude of load current excursion to 10 A?

(a) 202.48 Hz (b) 98.60 Hz
(c) 125.0 Hz (d) 133.33 Hz

Q.162 In a thyristorised dc-dc chopper, the commutation of main SCR can be of

(a) voltage commutation, current commutation or load commutation.
(b) voltage commutation, current commutation or line commutation.
(c) current commutation or line commutation.
(d) natural commutation or load commutation.

Q.163 A 220 V dc shunt motor runs at 1500 rpm at no-load. The motor is fed through a type-A chopper. If the armature resistance is 1 Ω and duty cycle of chopper is 60%, the speed of the motor when it draw a current of 20 A would be approximately equal to

(a) 636 rpm (b) 842 rpm
(c) 764 rpm (d) 535 rpm

Q.164 If the motor in the previous problem is driven by an active load at 1710 rpm, then the type of chopper suitable for speed control and its duty cycle for an input voltage of 250 V will be respectively

(a) first quadrant and 7.7%
(b) second quadrant and 6.5%
(c) first quadrant and 6.5%
(d) second quadrant and 7.7%

Q.165 A four-quadrant chopper is driving a separately excited dc motor load. The motor parameters are $R = 0.1$ Ω, $L = 10$ mH. The supply voltage is 200 V d.c. If the rated current of the motor is 10 A with $E_b = 150$ V and motor is driving the rated torque, then it is operating under

(a) forward motoring mode
(b) reverse braking mode
(c) reverse motoring mode
(d) forward braking mode

Q.166 A single-quadrant type A chopper is operating with the following specifications:
Ideal battery of 220 V;
on-time t_{on} = 1 ms;
off-time t_{off} = 1.5 ms
The ripple and form factors will be respectively given by

(a) 0.98 and 1.20 (b) 1.58 and 1.22
(c) 1.22 and 1.58 (d) 1.58 and 1.20

Q.167 Match **List-I (Chopper circuit)** with **List-II (Characteristic)** and select the correct answer using the codes given below the lists:

List-I	List-II
A. Type-A	1. Step-up chopper
B. Type-B	2. Load voltage can be positive or negative
C. Type-C	3. Step-down chopper
D. Type-D	4. Suitable for motoring and regenerative braking of dc motors

Codes:

	A	B	C	D
(a)	3	4	1	2
(b)	2	1	3	4
(c)	1	2	4	3
(d)	3	1	4	2

Q.168 A step-up chopper has source voltage and duty cycle of V_{dc} and K. The output voltage of this chopper is

(a) $V_{dc}(1 - K)$ (b) $\frac{V_{dc}}{1+K}$

(c) $\frac{V_{dc}}{1-K}$ (d) $V_{dc}(1 + K)$

Q.169 Consider the following statements:

1. DC chopper can be used in both dc and ac drives.
2. For four-quadrant operation dual converter is required.
3. Output voltage from a chopper circuit depends both on load current and duty cycle
4. In a step-down source, current can be discontinuous if duty cycle is low.

Which of the statements given above are correct?

(a) 1, 2 and 3 (b) 2 and 4

(c) 1, 3 and 4 (d) 1 and 4

Q.170 For the type-A chopper shown below, the voltage drop across the chopper when it is on is 2 volt. If the duty cycle is 0.4, then the chopper efficiency would be approximately equal to

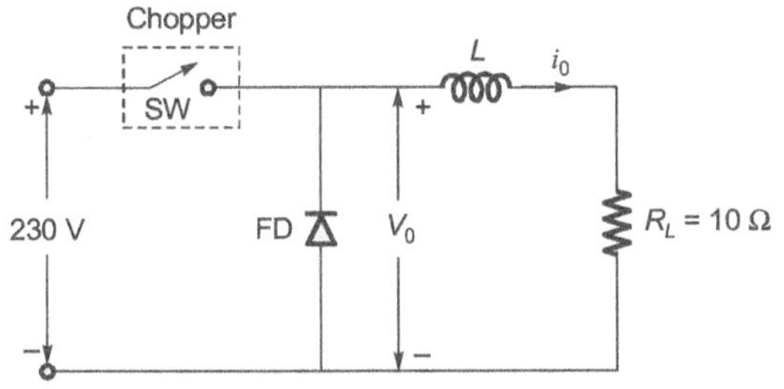

(a) 99% (b) 98%

(c) 97% (d) 96%

Q.171 In a type-A chopper, the maximum value of ripple current is

(a) directly proportional to chopping frequency only.

(b) directly proportional to the circuit inductance only.

(c) directly proportional to chopping frequency and inversely proportional to the circuit inductance.

(d) inversely proportional to chopping frequency and the circuit inductance.

Q.172 Match **List-I (Thyristor chopper circuits)** wtih **List-II (Characteristics)** and select the correct answer using the codes given below the lists:

List-I

A. Current commutated chopper

B. Voltage commutated chopper

C. Load commutated chopper

List-II

1. Diode in antiparallel with main SCR
2. Used in high power circuits
3. No requirement of any commutating inductor

Codes:

	A	B	C
(a)	3	2	1
(b)	1	3	2
(c)	1	2	3
(d)	3	1	2

Q.173 In dc choppers, per unit ripple is maximum when duty cycle α is

(a) 0.9 (b) 0.7

(c) 0.5 (d) 0.2

Q.174 If T is the chopping period, T_{on} the on-time period of chopper and T_{off} the off-time period of the chopper, then consider the following statements associated with the methods of controlling the output voltage of a chopper :

1. In pulse-width modulation (PWM) scheme, T is kept constant while T_{on} is varied.
2. In variable-frequency modulation (VFM) scheme, T is varied while either of T_{on} or T_{off} is kept constant.

3. In PWM scheme, the output voltage can be varied between zero to source voltage.
4. The large T_{off} in VFM scheme may make the load current discontinuous.
5. PWM scheme is better than VFM scheme.

Which of the statements given above are correct?

(a) 1, 2, 3, 4, and 5 (b) 2, 3 and 5
(c) 1, 3 and 4 (d) 2, 3, 4 and 5

Q.175 In dc choppers, the waveforms for input and output voltages are respectively
(a) continuous, discontinuous
(b) both discontinuous
(c) discontinuous, continuous
(d) both continuous

Q.176 A class C chopper is operated from a 220 V battery. The load is a dc motor with $R = 0.1\ \Omega$, $L = 10$ mH and $E_b = 100$ V. The duty cycle of the chopper to achieve regenerative braking at the rated current of 10 Ampere would be equal to
(a) 25% (b) 45%
(c) 66% (d) 72%

Q.177 Assertion (A) : In dc choppers, it is essential to provide a separate commutation circuitry to commutate the main power SCR.
Reason (R) : A chopper is dc equivalent to an ac transformer having continuously variable turns ratio.
(a) Both A and R are true and R is the correct explanation of A.
(b) Both A and R are true but R is not the correct explanation of A.
(c) A is true but R is false.
(d) A is false but R is true.

Q.178 A DC link chopper as compared to AC link is
(a) efficient (b) less costly
(c) bulky (d) both (a) and (b)

Q.179 If an ac chopper is feeding an inductive load, the firing pulse to the SCR
(a) may have a width equal to turn ON time of the SCR.
(b) should be a series of pulses of short duration.
(c) should be a single pulse of long duration.
(d) should be train of pulses of duration equal to the conduction period of the SCR.

Q.180 A boost-regulator has an input voltage of 5 V and the average output voltage of 15 V. The duty cycle is
(a) $\frac{3}{2}$ (b) $\frac{2}{3}$
(c) $\frac{5}{2}$ (d) $\frac{15}{2}$

7. Inverters

Q.181 Assertion (A): For high power applications, inverters are used instead of transistors.
Reason (R): For high power applications, inverter is operated in active region.
(a) Both A and R are true and R is the correct explanation of A.
(b) Both A and R are true but R is not the correct explanation of A.
(c) A is true but R is false.
(d) A is false but R is true.

Q.182 Match **List-I (Type of Inverter circuits)** with **List-II (Characteristics)** and select the correct answer using the codes given below the lists:

List-I
A. Voltage Source Inverter
B. 120° Conduction Inverter
C. Current Source Inverter
D. 180° Conduction Inverter

List-II
1. Suitable for a load of low leakage impedance
2. Each device turns on at regular interval of 60°
3. Suitable for multi machine control
4. Preferable for a delta connected load

Codes:

	A	B	C	D
(a)	1	4	3	2
(b)	3	4	1	2
(c)	2	3	4	1
(d)	1	3	2	4

Q.183 Consider the following statements associated with CSI and VSI:
1. In CSI, commutation is load dependent process.
2. Thyristorised current source fed inverter has inherent four quadrant operation.
3. In VSI dynamic braking is applicable during ac line failure.

4. MOSFET and transistor are more suitable for CSI compared to VSI.

Which of the statements given above is/are correct?

(a) 1, 2 and 3 (b) 3 and 4
(c) 2 only (d) 1 only

Q.184 A single-phase half bridge inverter has supply voltage of 200 V. For a load resistance of 10 Ω, the output power is equal to

(a) 2000 W (b) 500 W
(c) 225 W (d) 1000 W

Q.185 **Assertion (A):** The terminal voltage of a voltage source inverter remains substantially constant with variations in load.

Reason (R): Any short-circuit across the terminals of a voltage source inverter causes current to rise very fast.

(a) Both A and R are true and R is the correct explanation of A.
(b) Both A and R are true but R is not the correct explanation of A.
(c) A is true but R is false.
(d) A is false but R is true.

Q.186 The single-phase half-bridge inverter has a resistive load of 10 Ω and the centre-tap dc input voltage is 96 V. The fundamental power consumed by the load is

(a) 628.6 Watt (b) 525.0 Watt
(c) 746.5 Watt (d) 824.4 Watt

Q.187 A single-phase half-bridge inverter has a resistive load of R = 3 Ω and the dc input voltage V_{dc} = 24 volts. The harmonic factor of the lowest order harmonic would be equal to

(a) 33.33% (b) 25.0%
(c) 66.67% (d) 50.0%

Q.188 A single-phase transistorized bridge inverter has a resistive load of R = 3 Ω and the dc input voltage of 48 volts. The total harmonic distortion is approximately equal to

(a) 28.80% (b) 23.33%
(c) 42.62% (d) 48.43%

Q.189 A full-bridge bipolar PWM inverter is fed from a 240 V battery and is driving an RL load. The fundamental output voltage for a modulation index of 0.8 is

(a) 135 V (b) 215 V
(c) 205 V (d) 165 V

Q.190 A three-phase bridge inverter is fed from a dc source of V_{dc} volts. The inverter is operated in 180° conduction mode and is supplying a purely resistive star-connected load. The rms value of the output line voltage is given by

(a) $\frac{V_{dc}}{\sqrt{2}}$ (b) $\sqrt{\frac{2}{3}}V_{dc}$

(c) $\sqrt{3}\,V_{dc}$ (d) $\sqrt{\frac{3}{2}}\,V_{dc}$

Q.191 A current source inverter can be

(a) load commutated
(b) force commutated
(c) either load or force commutated
(d) neither load nor force commutated

Q.192 Match **List-I (Type of Inverter circuits)** with **List-II (Applications/Limitations)** and select the correct answer using the codes given below the lists:

List-I

A. Adjustable voltage inverter
B. Pulse width modulated inverter
C. Current source inverter
D. Voltage source inverter

List-II

1. Limited speed range of motor
2. Capable of regeneration
3. Good input power factor at all frequencies
4. Possibility of short circuit across the source

Codes:

	A	B	C	D
(a)	2	4	3	1
(b)	2	1	4	3
(c)	1	2	3	4
(d)	1	3	2	4

Q.193 In current source inverters load voltage waveform V_0 and load current waveform i_0 respectively

(a) depends on load impedance Z, does not depend on Z.
(b) depends on Z, depends on Z.
(c) does not depend on Z, does not depend on Z.
(d) does not depend on Z, depends on Z.

Q.194 The three basic types of adjustable frequency controllers are namely PWM, CSI and adjustable voltage inverters (AVI). Match **List-I (Type of controller)** with **List-II (Input devices)** and select the correct answer using the codes given below the lists:

List-I	List-II
A. PWM	1. Diodes
B. CSI	2. MCTs
C. AVI	3. GTOs

Codes:

	A	B	C
(a)	1	2	3
(b)	1	3	2
(c)	2	1	3
(d)	3	1	2

Q.195 A single phase IGBT bridge inverter, compared to a single pulse PWM control, multiple pulse PWM
(a) gives higher harmonic content in output voltage.
(b) gives higher maximum rms output voltage.
(c) requires more number of switching devices.
(d) gives lower harmonic content in the output voltage.

Q.196 Assertion (A): The lower order harmonics are reduced by using some technique while the higher order harmonics are reduced by using filter.
Reason (R): The cost of the filter is reduced and at the same time transient response is also improved to a great extent.
(a) Both A and R are true and R is the correct explanation of A.
(b) Both A and R are true but R is not the correct explanation of A.
(c) A is true but R is false.
(d) A is false but R is true.

Q.197 In a single-pulse modulation of PWM inverters if pulse width is 120° then
(a) 5th harmonic will be eliminated
(b) 3rd harmonic will be eliminated
(c) 7th harmonic will be eliminated
(d) none of the above

Q.198 If for a single-phase half-bridge inverter, the amplitude of output voltage is V_0 and the output power is P, then their corresponding values for a single-phase full-bridge inverter would be respectively
(a) $\frac{V_0}{2}$ and $\frac{P}{2}$ (b) $2V_0$ and $2P$
(c) $2V_0$ and P (d) $2V_0$ and $2P$

Q.199 In a CSI, if the frequency of voltage is f, then frequency of output voltage is
(a) $f/2$ (b) f
(c) $f/4$ (d) $2f$

Q.200 A single-phase full bridge inverter can operate in load-commutation mode in case load consists of
(a) RLC critically damped
(b) RLC underdamped
(c) RLC overdamped
(d) RC

Q.201 Inverters are used in
(a) induction heating (b) UPS
(c) both (a) and (b) (d) None of these

Q.202 Inverters designed from BJT are preferably used in saturation region than in active region because of
(a) high efficiency (b) high power factor
(c) both (a) and (b) (d) none of these

Q.203 The feedback diodes in a d.c. thyristor inverter
(a) freewheel the load current
(b) provide reverse bias effectively to the thyristors for turn-off
(c) improve the switching properties of the inverter
(d) improve the harmonic distortion of the inverter output current

Q.204 An inverter has a periodic output voltage with the output waveform as shown in figure.

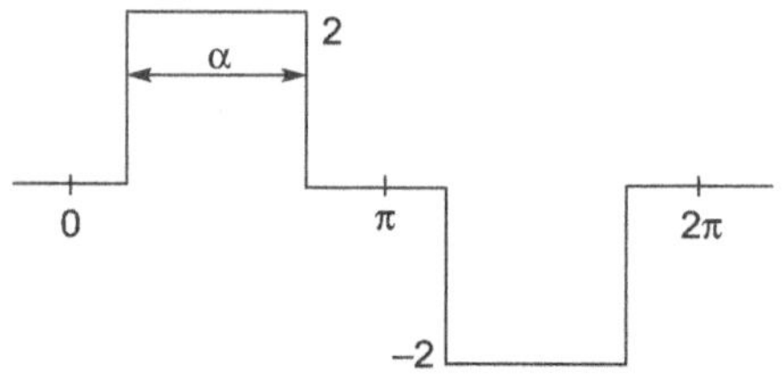

When the conduction angle α = 120°, the rms value of the output voltage is
(a) 1.56 V (b) 1.10 V
(c) 0.90 V (d) None of the these

Q.205 In a three-phase voltage source inverter operating in square wave mode, the output line voltage is free from
(a) 3rd harmonic (b) 7th harmonic
(c) 11th harmonic (d) 13th harmonic

Q.206 A single-phase bridge inverter delivers power to a series connected RLC load with $R = 2\ \Omega$, $\omega L = 8\ \Omega$. For this inverter-load combination, load commutation is possible in case the magnitude of $1/\omega C$ in ohm is
(a) 10 (b) 8
(c) 6 (d) zero

8. AC Voltage Controllers

Q.207 A single-phase a.c. regulator shown in figure feeds power to a resistive load from a 230 V, 50 Hz source. The maximum value of average thyristor current is approximately equal to

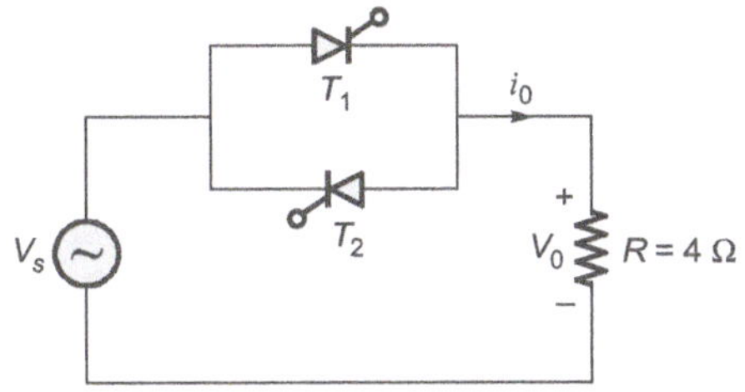

(a) 22.4 A (b) 25.9 A
(c) 28.8 A (d) 18.5 A

Q.208 A single-phase half-wave ac regulator shown in figure feeds power to a resistive load of 6 Ω from a 230 V, 50 Hz source. If the firing angle of SCR is 90°, then the rms value of output voltage would be equal to

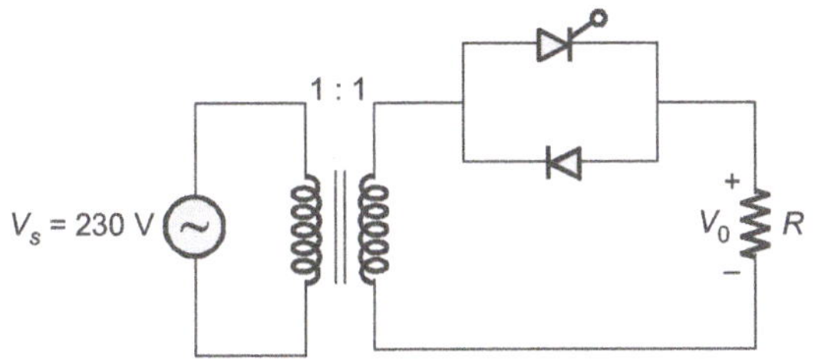

(a) $230\sqrt{2}$ volt (b) $115\sqrt{2}$ volt
(c) $115\sqrt{3}$ volt (d) $\frac{230}{\sqrt{3}}$ volt

Q.209 A single phase a.c. voltage regulator with R-L load has the resistance and inductive reactance of respectively 4 Ω and 3 Ω. The supply voltage is 230 V, 50 Hz. The control range of firing angle is
(a) $36.87° \le \alpha \le 180°$ (b) $\alpha \le 36.87°$
(c) $36.87° \le \alpha \le 150°$ (d) None of these

Q.210 **Assertion (A) :** The a.c. regulators are not suitable for closed-loop control.
Reason (R) : The a.c. regulators converts fixed mains voltage directly to variable alternating voltage without a change in the frequency.
(a) Both A and R are true and R is the correct explanation of A.
(b) Both A and R are true but R is not the correct explanation of A.
(c) A is true but R is false.
(d) A is false but R is true.

Q.211 Consider the following statements:
1. By connecting a reverse parallel pair of thyristors between a.c. supply and load, the voltage applied to the load can be controlled.
2. A.C. regulators require complex commutation circuitry.
3. A.C. regulators are highly affected by harmonics in the supply current and load voltage waveforms.
4. Continuous gating increases the size of the pulse transformer required.

Which of the statements given above are correct?
(a) 2, 3 and 4 (b) 1, 2, 3 and 4
(c) 1, 2 and 3 (d) 1, 3 and 4

Q.212 Match **List-I** with **List-II** and select the correct answer using the codes given below the lists:

List-I
A. RCT
B. Half-wave A.C. voltage regulator
C. Full-wave A.C. voltage regulator
D. DIAC

List-II
1. Thyristor in antiparallel with a diode.
2. Two diodes in antiparallel
3. Two thyristors in antiparallel

Codes:

	A	B	C	D
(a)	2	2	3	2
(b)	1	1	3	2
(c)	2	1	2	3
(d)	3	2	1	2

Q.213 **Assertion (A) :** Integral cycle control is preferred for heating load and for motor-control drives.

Reason (R) : Integral cycle control introduces less harmonics into the supply system.

(a) Both A and R are true and R is the correct explanation of A.
(b) Both A and R are true but R is not the correct explanation of A.
(c) A is true but R is false.
(d) A is false but R is true.

Q.214 In a single-phase a.c. voltage controller with R-load, maximum power is delivered to the load when firing angle α is equal to

(a) 90° (b) 30°
(c) 0° (d) 180°

Q.215 Line regulation is determined by

(a) load current
(b) change in output voltage and input voltage
(c) load current and zener current
(d) change in load resistance and output voltage

Q.216 A single-phase a.c. regulator uses ____ thyristors connected in _____.

(a) 2, parallel (b) 4, anti-parallel
(c) 2, anti-parallel (d) 4, parallel

Q.217 A single-phase voltage controller has input voltage of 230 V, 50 Hz and a load of $R = 20\ \Omega$. For 6 cycles on and 4 cycles off, the input power factor is

(a) $\sqrt{\frac{3}{5}}$ (b) $\sqrt{\frac{3}{2}}$

(c) $\sqrt{\frac{5}{3}}$ (d) $\sqrt{\frac{2}{3}}$

Q.218 A single-phase voltage controller feeds power to a resistance of 5 Ω. The source voltage is 100 V rms. For a firing angle of 90°, the rms value of thyristor current in amperes is

(a) 5 (b) 10
(c) 20 (d) 15

Q.219 A single-phase voltage controller, using two SCRs in antiparallel, is found to be operating as a controlled rectifier because

(a) load is RL and pulse gating is used.
(b) load is R and pulse gating is used.
(c) load is RL and continuous gating is used.
(d) load is R and high-frequency carrier gating is used.

Q.220 In a single phase voltage controller with RL load, α is the firing angle, ϕ is the load phase angle and β is the extinction angle. For this voltage controller, output power can be controlled if

1. $\alpha > \phi$ 2. $\alpha < \phi$
3. $\beta < \pi$ 4. $\beta > \pi$
5. $(\beta - \alpha) > \pi$ 6. $(\beta - \alpha) < \pi$

Choose the correct code from the given options:

(a) 2, 3 and 5 (b) 2, 4 and 5
(c) 1, 3 and 6 (d) 1, 4 and 6

Q.221 A single-phase voltage controller feeds power to a resistive load of 5 Ω from a 230 V, 50 Hz source. The ratio of maximum value of rms thyristor current to maximum average thyristor current is

(a) $\frac{\pi}{\sqrt{2}}$ (b) $\frac{\sqrt{\pi}}{2}$

(c) $\frac{\pi}{2}$ (d) $\sqrt{\frac{\pi}{2}}$

Q.222 Internal cycle control

(a) cannot used on inductive loads.
(b) can be used only for loads with large time constants.
(c) does not introduce harmonics in the supply lines.
(d) is very fast in action.

Q.223 A.C. voltage regulators are widely used in

(a) traction drives
(b) fan drives
(c) synchronous motor drives
(d) slip power recovery scheme of slip-ring inductor motor

Q.224 A single phase ac voltage regulator is fed from a 50 Hz supply system. If it supplies a load comprising a resistance of 2 Ω connected in series with an inductance of 6.36 mH, then the range of firing angle 'α' providing controlled voltage would be

(a) $0° < \alpha < 180°$ (b) $45° < \alpha < 180°$
(c) $90° < \alpha < 180°$ (d) $0° < \alpha < 45°$

9. Cycloconverters

Q.225 Assertion (A) : A cycloconverter is essentially a dual converter but so operated as to produce an alternating output voltage.
Reason (R) : Each SCR in a cycloconverter works as a phase-controlled converter with a varying firing angle.
(a) Both A and R are true and R is the correct explanation of A.
(b) Both A and R are true but R is not the correct explanation of A
(c) A is true but R is false
(d) A is false but R is true

Q.226 A cycloconverter requires
(a) natural commutation in both step-up and step-down mode
(b) forced commutation in both step-up and step-down mode
(c) forced commutation in step-down mode and natural commutation in step-up mode
(d) natural commutation in step-down mode and forced commutation in step-up mode

Q.227 The number of thyristors required for single-phase to single-phase cycloconverter of the mid-point type and for three phase 3-pulse type cycloconverter are respectively
(a) 4 and 6 (b) 8 and 18
(c) 4 and 18 (d) 4 and 36

Q.228 Match **List-I (Power Electronic devices)** with **List-II (Applications)** and select the correct answer using the codes given below the lists:

List-I
A. Controlled rectifier
B. Cycloconverter
C. Inverter
D. Voltage Controller

List-II
1. Solar cells
2. Magnet power supply
3. High-power ac drive
4. Ceiling fan drive

Codes:

	A	B	C	D
(a)	2	3	1	4
(b)	3	1	2	4
(c)	2	4	1	3
(d)	1	2	4	3

Q.229 Consider the following statements associated with cycloconverters:
1. A cycloconverter allows power flow in only one direction.
2. In a cycloconverter, the distortion of the output voltage increases if the ratio of the output and input frequency increases.
3. The cycloconverters are used to provide a variable input frequency from a fixed input frequency but not a fixed frequency from a variable input frequency.
4. The distortion in the output of a cycloconverter is very low at low output frequencies.

Which of the statements given above is/are correct?
(a) 3 only (b) 2 and 4 only
(c) 2 only (d) 1 and 4 only

Q.230 Assertion (A) : The type of three-phase cycloconverter depends on the number of pulses used.
Reason (R) : Higher the number of pulses used more ripple content in the output of three-phase cycloconverters.
(a) Both A and R are true and R is the correct explanation of A.
(b) Both A and R are true but R is not the correct explanation of A.
(c) A is true but R is false.
(d) A is false but R is true.

Q.231 A three-pulse cycloconverter feeds a 1-ϕ load of 190 V, 45 A at a power factor of 0.7 lagging. The required rms supply voltage per phase is approximately equal to
(a) 230 V (b) 224 V
(c) 236 V (d) 240 V

Q.232 Compared to a cycloconverter, a d.c. link converter
(a) is suitable for low frequencies only
(b) requires natural commutation

(c) has high input power factor if diode rectifier is used
(d) all of the above

Q.233 Consider the following statements:
1. Cycloconverter is a frequency changer used to produce ac output at reduced frequency.
2. When cycloconverter operates with R-L load only rectification mode is used.
3. The circuit of a cycloconverter is complex and costly.
4. Three pulse or six pulse cycloconverter can be used for getting three phase output.

Which of the statements given above is/are **not** correct?
(a) 1 and 2 (b) 4 only
(c) 2 and 4 (d) 2 and 3

Q.234 A cycloconverter power electronic equipment is a
(a) device which converts dc to ac
(b) frequency converter which has no intermediate dc state
(c) device which converts ac to dc
(d) none of the above

Q.235 The input frequency of a cycloconverter is more than the output frequency because
(a) the number of switching devices is large compared to that in a cascade converter pair
(b) the cycloconverter employs natural commutation of the SCRs and more number of supply voltage cycles are needed to form one cycle of the output voltage
(c) the cycloconverter is normally used for high voltage and high current applications
(d) there is no intermediate dc link

Q.236 A 3-phase to single-phase conversion device employs a 6-pulse bridge cycloconverter. For an input voltage of 100 V per phase, the fundamental rms value of output voltage is

(a) $\frac{300}{\pi}$ volt (b) $150\frac{\sqrt{3}}{\pi}$ volt
(c) $\frac{150}{\pi}$ volt (d) $\frac{300\sqrt{3}}{\pi}$ volt

Q.237 An inter-group reactor is used in a single-phase cycloconverter circuit to
(a) reduce current ripples
(b) reduce voltage ripples
(c) limit circulating current
(d) limit *di*/*dt* in the semiconductor switch

Q.238 In a single-phase to single-phase cycloconverter, if α_1 and α_2 are the trigger angles of positive converter and negative converter respectively then

(a) $\alpha_1 + \alpha_2 = \frac{\pi}{2}$ (b) $\alpha_1 + \alpha_2 = \pi$
(c) $\alpha_1 + \alpha_2 = \frac{3\pi}{2}$ (d) $\alpha_1 + \alpha_2 = 2\pi$

Q.239 Three-phase to 3-phase cycloconverters employing 18 SCRs and 36 SCRs have the same voltage and current ratings for their component thyristors. The ratio of power handled by 36-SCR device to that handled by 18-SCR device is
(a) 4 (b) 2
(c) 1 (d) $\frac{1}{2}$

10. Power Semiconductor Drives

Q.240 A separately-excited dc motor, operating from a single-phase half controlled bridge at a speed of 1400 rpm, has an input voltage of 330 sin 314*t* and a back emf of 80 V. The SCRs are fired symmetrically at $\alpha = 30°$ in every half cycle and the armature has a resistance of 4 Ω. The average armature current is around
(a) 24 A (b) 32 A
(c) 29 A (d) 16 A

Q.241 A dc series motor is fed from 600 V dc source through a chopper. The dc motor has an armature and series resistances of 40 mΩ and 60 mΩ respectively. The motor torque constant is 4×10^{-3} N-m/A^2. If the average armature current is 300 A (ripple free) then the motor speed for a chopper duty cycle of 60% will be equal to
(a) 225 rad/sec (b) 325 rad/sec
(c) 215 rad/sec (d) 275 rad/sec

Q.242 **Assertion (A)** : Stator-voltage control method for the speed control of three-phase induction motors is used for low-power drives like fans, blowers and centrifugal pumps requiring low starting torque.

Reason (R) : Stator-voltage control method for the speed control of three-phase induction motors uses 3-phase voltage controllers.
(a) Both A and R are true and R is the correct explanation of A.
(b) Both A and R are true but R is not the correct explanation of A.
(c) A is true but R is false.
(d) A is false but R is true.

Q.243 Assertion (A) : Volts/Hertz control is similar to the armature-voltage control method used for the speed control of a dc motor.
Reason (R) : Volts/Hertz control offers speed control above the rated speed of the induction motor.
(a) Both A and R are true and R is the correct explanation of A.
(b) Both A and R are true but R is not the correct explanation of A.
(c) A is true but R is false.
(d) A is false but R is true.

Q.244 At reduced voltage and reduced frequency in ac drives:
1. current gets reduced.
2. pf is improved under maximum torque conditions.
3. the maximum torque is reduced.
4. the maximum torque occurs at higher value of slip.

Choose the correct code from the given options:
(a) 1, 2, 3 and 4 (b) 2 and 4
(c) 1 and 3 (d) 2, 3 and 4

Q.245 Consider the following statements:
1. Static Kramer drive offers speed control of SRIM above synchronous speed only.
2. Static scherbius drive offers speed control below synchronous speed only.
3. In static Kramer drive the slip-power from the rotor circuit is converted to dc voltage which is then converted to line frequency and pumped back to the ac source.
4. The chopper method of speed control for SRIM leads to better efficiency of the drive.

Which of the statements given above is/are **not** correct?
(a) 1, 2 and 3 (b) 4 only
(c) 1 and 3 (d) 1, 2 and 4

Q.246 A constant frequency machine control is used to control dc series motor for a 440 V dc supply as shown in figure. The armature and field resistance are 0.06 Ω and 0.04 Ω respectively. The average armature current is 200 A and the chopper frequency is 500 Hz.

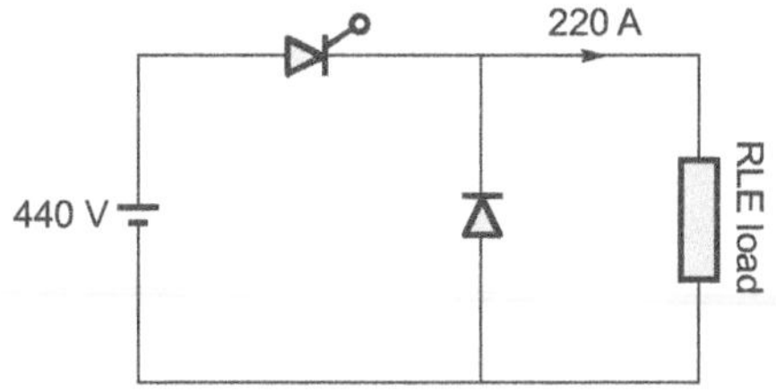

If the back emf is 200 V, then the pulse width would be equal to
(a) 0.5 ms (b) 1 ms
(c) 1.5 ms (d) 2 ms

Q.247 A dc motor can be operated in four quadrants. Match **List-I (Quadrant operation)** with **List-II (Modes of operation)** and select the correct answer using the codes given below the lists:

List-I	List-II
A. I - quadrant	1. Reverse motoring
B. II - quadrant	2. Reverse regeneration
C. III - quadrant	3. Forward motoring
D. IV - quadrant	4. Forward regeneration

Codes:

	A	B	C	D
(a)	1	2	3	4
(b)	3	4	2	1
(c)	3	2	1	4
(d)	3	4	1	2

Q.248 At lower speeds of a motor or at higher firing angle, power factor of the system is
(a) improved
(b) deteriorated
(c) unaffected
(d) may improve or deteriorate

Q.249 Match **List-I (Power Electronic Circuits)** with **List-II (Characteristics)** and select the correct answer using the codes given below the lists:

List-I
A. Three pulse converter
B. Dual converter
C. Half controlled bridges
D. 3-ϕ bridge converter

List-II
1. Each thyristor conducts for 120° or less
2. Better power factor
3. Capable of four quadrant operation
4. Produces saturation in transformer

Codes:

	A	B	C	D
(a)	4	1	2	3
(b)	2	3	4	1
(c)	4	3	2	1
(d)	3	4	1	2

Q.250 A 3-ϕ slip ring induction motor controlled with variable frequency TRC system is operating at a slip of 4%. If the additional resistance introduced during "OFF" period is equal to rotor resistance then the chopper frequency for a pulse width of 2 ms so as to obtain 150% torque at the same slip would be equal to
(a) 250 Hz (b) 500 Hz
(c) 750 Hz (d) 1000 Hz

Q.251 Match **List-I (Methods of speed control of a three-phase induction motor)** with **List-II (Characteristics)** and select the correct answer using the codes given below the lists:

List-I
A. Stator voltage control
B. Pole changing method
C. Rotor resistance control
D. Constant *v/f* control

List-II
1. Inefficient at low value of speed
2. Most suitable for fan-type load
3. Maximum torque is constant
4. Limited number of steps

Codes:

	A	B	C	D
(a)	2	1	4	3
(b)	2	4	1	3
(c)	3	4	1	2
(d)	1	2	3	4

Q.252 For an ac voltage controller fed induction motor drive negotiating a load whose torque requirement does not vary with speed
(a) the fundamental component of current drawn from the supply increases as speed is reduced.
(b) the fundamental component of current drawn from the supply is independent of speed.
(c) the fundamental component of current drawn from the supply decreases as speed is reduced.
(d) none of the above

Q.253 Which of the following option is **not** correct about an ac drive compared to a dc drive?
(a) AC motors are lighter in weight
(b) Power converters for AC drives are simple
(c) Power converters for AC drives generate harmonic in the supply system and load circuit
(d) AC drives are less expensive

Q.254 A delta-connected induction motor being fed by a 3-phase ac to dc inverter and operated in constant *v/f* control mode requires during starting a
(a) star-delta starter
(b) DOL starter
(c) auto-transformer starter
(d) none of the above

Q.255 The armature of a dc motor is fed from a phase controlled rectifier whereas its field is supplied from a constant dc source. To reduce the speed of the motor regeneratively
(a) the polarity of the armature induced voltage should be reversed.
(b) the polarity of the dc voltage should be reversed.
(c) the firing angle of the converter should be increased.
(d) the firing angle should be varied in the range 90° to 180° simultaneously reversing the armature connections.

Q.256 Chopper circuits are used to control the speed of dc motors which are fed from fixed dc voltage sources. Variable voltage output can be obtained by which of the following methods?
1. Adjusting the wave shape.
2. Adjusting the frequency of switching.
3. Varying the 'on' to 'off' time ratio.
4. Using a saturable core reactor in series with the motor.

Choose the correct code from the given options:
(a) 2 and 3 (b) 1 and 2
(c) 1 and 4 (d) 2 and 4

Q.257 As SCR chopper circuit supplies power to a dc motor. The armature current of the motor will be
(a) constant
(b) sinusoidal
(c) exponential rise and decay
(d) none of the above

Q.258 In ac motor control, the ratio of voltage-to-frequency is maintained at constant value
(a) to make minimum use of magnetic circuits
(b) to make maximum use of magnetic circuits
(c) to provide maximum pull-out torque
(d) to maximise the current drawn from the supply to provide torque

Q.259 The phase controlled rectifiers used in speed control of dc motors converts fixed ac supply voltage into _______ output voltage.
(a) variable frequency ac
(b) full rectified dc
(c) variable ac
(d) variable dc

Q.260 The speed of a dc motor can be controlled by using
(a) diodes
(b) transistors
(c) thyristors
(d) either diodes or thyristors

Q.261 In converter control of induction motor's speed control, a d.c. link is used. This is to
(a) provide a constant peak flux density to the air-gap and core insulation
(b) provide necessary d.c. voltage to inverter
(c) vary the slip
(d) trigger the thyristor of the inverter

Q.262 In cycloconverter speed control of an induction motor, cycloconverter frequency
(a) is equal to that of supply frequency.
(b) is higher than the supply frequency.
(c) can be higher or lower than supply frequency.
(d) is always less and about one-third of supply frequency.

Q.263 A fully controlled thyristor bridge drives a d.c. motor. The system is capable of
(a) motoring and breaking in both directions.
(b) only motoring in both directions, no braking.
(c) motoring in forward direction and braking in reverse direction.
(d) only motoring in forward direction, no braking.

11. Switched Mode Power Supply (SMPS)

Q.264 UPS find application in
(a) HVDC
(b) battery powered vehicle
(c) computer power supply
(d) electric traction

Q.265 SMPS are superior to linear power supplies in respect of
(a) regulation and noise
(b) size and efficiency
(c) noise and cost
(d) efficiency and regulation

Q.266 Consider the following statements:
Switched mode power supplies are preferred over the continuous types, because these are:
1. suitable for low power circuits.
2. more efficient.
3. suitable for use in both ac and dc.
4. suitable for high power circuits.

Of these, the correct statements are:
(a) 2 and 4 (b) 1 and 3
(c) 1 and 2 (d) 2 and 3

Q.267 Assertion (A): Switched mode power supply works like a dc chopper.
Reason (R): The output dc voltage of a switched mode power supply is controlled by varying the duty cycle of chopper by PWM or FM techniques.
(a) Both A and R are true and R is the correct explanation of A.
(b) Both A and R are true but R is not the correct explanation of A.
(c) A is true but R is false.
(d) A is false but R is true.

Q.268 Assertion (A): When SMPS is operated in the switching mode, the only energy dissipated occurs during switching on and off
Reason (R): The internal circuit of SMPS consists of resistors of different values.
(a) Both A and R are true and R is the correct explanation of A.

(b) Both A and R are true but R is not the correct explanation of A.
(c) A is true but R is false.
(d) A is false but R is true.

Q.269 Which of the following can be used as switching element in SMPS?
(a) BJT
(b) MOSFET
(c) Both BJT and MOSFET
(d) Diodes

Q.270 The most suitable device for high frequency inversion in SMPS is
(a) BJT (b) IGBT
(c) MOSFET (d) GTO

Q.271 In a switched-mode power supply (SMPS), after conversion of a.c. supply to highly filtered d.c. voltage, a switching transistor is switched ON and OFF at a very high speed by a pulse width modulator (PWM) which generates very-high frequency square pulses. The frequency of the pulses is typically in the range of
(a) 100 Hz - 200 Hz (b) 500 Hz - 1 kHz
(c) 2 kHz - 5 kHz (d) 20 kHz - 50 kHz

12. High Voltage Direct Current (HVDC)

Q.272 Two six-pulse converters, used for bipolar HVDC transmission system, are rated at 1000 MW, ±200 kV. The rms current in each thyristor is:
(a) $2500\sqrt{3}\ A$ (b) $\frac{1250}{\sqrt{3}}$ A
(c) $\frac{2500}{\sqrt{3}}$ A (d) $1250\sqrt{3}$ A

Q.273 HVDC transmission is preferred to EHV-AC because
(a) System stability can be improved.
(b) HVDC terminal equipment are expensive.
(c) harmonic problem is avoided.
(d) VAr compensation is **not** required for HVDC systems.

Q.274 Which one of the following statements in respect to HVDC transmission line is **not** correct?
(a) HVDC link line can operate between two ac systems whose frequencies need not be equal.
(b) Corona loss is much higher in HVDC transmission line.
(c) The power transmission capability of bipolar line is almost the same as that of single circuit ac line.
(d) There is no distance limitation for HVDC transmission by underground cable.

Q.275 Bulk power transmission over long HVDC lines are preferred on account of
(a) simple protection
(b) no harmonic problems
(c) low cost of HVDC terminals
(d) minimum line power losses

Q.276 In a 3-phase converter used in HVDC transmission, the three anode conduct sequentially. Due to overlap caused by the circuit inductances, two anodes conduct simultaneously during the overlap period. The output voltage waveform during this period is the
(a) mean of the two anode voltages, as they conduct together.
(b) sum of the 1st and 2nd anode voltages, because both the anode are conducting.
(c) voltage of the 1st anode, because the 2nd anode has not completely taken over.
(d) voltage of the 2nd anode voltage, because the voltage of this anode is greater than that of the 1st.

Q.277 **Assertion (A) :** In HVDC systems, two six-pulse converters are connected in series on the dc side and in parallel on the ac side.
Reason (R) : The series connection on the dc side helps to meet the high voltage requirement in the HVDC systems.
(a) Both A and R are true and R is the correct explanation of A.
(b) Both A and R are true but R is **not** the correct explanation of A.
(c) A is true but R is false.
(d) A is false but R is true.

Q.278 Consider the following statements associated with HVDC system:
1. Tapping is not provided in HVDC system.
2. HVDC line can not be operated for constant current regulation.

3. There is no limit of the length of HVDC cables.
4. HVDC lines require shunt capacitors.

Which of the statements given above are correct?

(a) 1 and 4 (b) 3 only
(c) 1 and 3 (d) 2 and 4

Q.279 Assertion (A) : The conductors of HVDC lines are loaded upto rated value and they are fully utilized.

Reason (R) : HVDC lines are loaded according to the stability limit.

(a) Both A and R are true and R is the correct explanation of A.
(b) Both A and R are true but R is **not** the correct explanation of A.
(c) A is true but R is false.
(d) A is false but R is true.

■■■■

Answers Power Electronics

1. (c)	2. (d)	3. (a)	4. (b)	5. (b)	6. (d)	7. (a)	8. (d)
9. (c)	10. (a)	11. (c)	12. (d)	13. (b)	14. (c)	15. (b)	16. (c)
17. (c)	18. (a)	19. (d)	20. (d)	21. (b)	22. (a)	23. (b)	24. (a)
25. (c)	26. (b)	27. (d)	28. (a)	29. (b)	30. (c)	31. (a)	32. (d)
33. (d)	34. (a)	35. (c)	36. (b)	37. (c)	38. (a)	39. (a)	40. (b)
41. (d)	42. (c)	43. (d)	44. (b)	45. (c)	46. (a)	47. (c)	48. (c)
49. (a)	50. (c)	51. (a)	52. (d)	53. (b)	54. (c)	55. (c)	56. (b)
57. (c)	58. (a)	59. (a)	60. (d)	61. (b)	62. (a)	63. (a)	64. (d)
65. (b)	66. (d)	67. (a)	68. (c)	69. (b)	70. (d)	71. (c)	72. (a)
73. (d)	74. (c)	75. (a)	76. (d)	77. (b)	78. (b)	79. (c)	80. (a)
81. (c)	82. (d)	83. (d)	84. (b)	85. (c)	86. (a)	87. (c)	88. (d)
89. (a)	90. (b)	91. (c)	92. (d)	93. (c)	94. (c)	95. (a)	96. (a)
97. (b)	98. (c)	99. (a)	100. (a)	101. (d)	102. (a)	103. (b)	104. (d)
105. (d)	106. (b)	107. (c)	108. (a)	109. (c)	110. (b)	111. (b)	112. (d)
113. (a)	114. (c)	115. (d)	116. (b)	117. (d)	118. (b)	119. (a)	120. (b)
121. (b)	122. (a)	123. (c)	124. (a)	125. (b)	126. (d)	127. (c)	128. (a)
129. (c)	130. (c)	131. (b)	132. (c)	133. (a)	134. (d)	135. (d)	136. (c)
137. (b)	138. (b)	139. (a)	140. (d)	141. (d)	142. (b)	143. (c)	144. (a)
145. (b)	146. (d)	147. (d)	148. (c)	149. (c)	150. (c)	151. (a)	152. (a)
153. (b)	154. (a)	155. (a)	156. (b)	157. (c)	158. (a)	159. (c)	160. (b)
161. (d)	162. (a)	163. (c)	164. (d)	165. (a)	166. (c)	167. (d)	168. (c)
169. (b)	170. (a)	171. (d)	172. (c)	173. (c)	174. (a)	175. (a)	176. (b)
177. (b)	178. (d)	179. (d)	180. (b)	181. (c)	182. (b)	183. (a)	184. (d)
185. (b)	186. (c)	187. (a)	188. (d)	189. (a)	190. (b)	191. (c)	192. (d)
193. (a)	194. (b)	195. (a)	196. (a)	197. (b)	198. (d)	199. (d)	200. (b)
201. (c)	202. (c)	203. (a)	204. (d)	205. (a)	206. (a)	207. (b)	208. (c)

209. (a)	210. (d)	211. (d)	212. (b)	213. (a)	214. (c)	215. (b)	216. (c)
217. (a)	218. (b)	219. (a)	220. (d)	221. (c)	222. (b)	223. (b)	224. (b)
225. (a)	226. (d)	227. (c)	228. (a)	229. (b)	230. (c)	231. (a)	232. (c)
233. (d)	234. (b)	235. (d)	236. (a)	237. (c)	238. (b)	239. (a)	240. (b)
241. (d)	242. (b)	243. (c)	244. (a)	245. (d)	246. (b)	247. (d)	248. (b)
249. (c)	250. (a)	251. (b)	252. (b)	253. (b)	254. (b)	255. (c)	256. (a)
257. (c)	258. (b)	259. (d)	260. (c)	261. (a)	262. (d)	263. (d)	264. (c)
265. (a)	266. (c)	267. (a)	268. (c)	269. (c)	270. (c)	271. (d)	272. (c)
273. (d)	274. (b)	275. (d)	276. (a)	277. (a)	278. (c)	279. (c)	

Explanations

1. (c)

In a Schottky diode junction is formed between metal and semiconductor.

2. (d)

A Schottky diode has very low reverse voltage rating as compared to alloyed and diffused diodes. Maximum voltage rating is about 100 V. Thus, assertion is a false statement.

In a schottky diode, a thin epitaxial layer is grown on n-type base (substrate). On the epitaxial layer metal (gold, chromium or platinum) is deposited to form a thin layer. This metallic layer is surrounded by a ring of insulating material such as SiO_2. Thus reason is a true statement.

3. (a)

- In fast recovery diode, reverse recovery time is reduced by addition of gold or platinum due to which rate of recombination is increased. Hence, statements-2 is not correct.
- MOSFET and transistor are used in low and medium power applications. Thus, statements-3 is not correct.

4. (b)

A power transistor

1. can not be easily paralleled because of V_{be} matching problems and local current concentration.
2. is susceptible to secondary breakdown.
3. has high transconductance.
4. has positive temperature coefficient on collector current since I_C increases with temperature.
5. has low switching speed than MOSFET.

6. (d)

Triac is a bidirectional device whereas BJT, MOSFET, GTO and IGBT are unidirectional devices.

8. (d)

The reverse blocking safe operating area (RBSOA) for a power transistor is shown below.

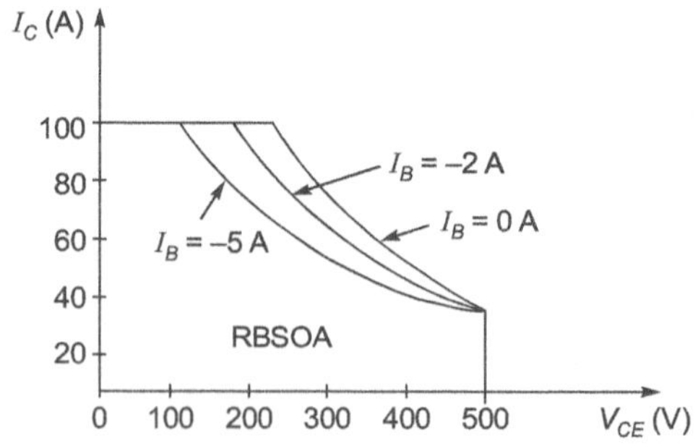

It is clear from above curve that with increased reverse bias, the RBSOA decreases in size.

9. (c)

$$\text{Energy loss during turn-on} = \int_0^{t_{on}} i_c \cdot v_{CE} dt$$

$$= \int_0^{t_{on}} \left(\frac{I_{CS}}{50} \times 10^6 t\right)\left(V_{CC} - \frac{V_{CC}}{40} \times 10^6 t\right) dt$$

$$= \int_0^{t_{on}} (2 \times 10^6 t)(200 - 5 \times 10^6 t) dt$$

$$= 0.1067 \text{ Watt-sec}$$

Energy loss during turn-off

$$= \int_0^{t_{off}} \left(100 - \frac{100}{60} \times 10^6 t\right) \times \left(\frac{200}{75} \times 10^6 t\right) dt$$

$= 0.1603$ Watt-sec

Total energy loss in one cycle

$= 0.1067 + 0.1603$

$= 0.267$ W-sec

Average power loss in transistor

= Switching frequency × energy loss in one cycle

∴ Allowable switching frequency

$$f = \frac{300}{0.267} = 1123.6 \text{ Hz} \approx 1.12 \text{ kHz}$$

11. (c)

Germanium is much costlier than silicon. Germanium has much lower thermal conductivity; its thermal resistance is, therefore, more. As a consequence, germanium thyristors suffer from more losses, more temperature rise and therefore lower operating life. Due to these reasons Germanium is not used in the fabrication of thyristors.

12. (d)

The reverse recovery time t_{rr} is defined as the time between the instant forward diode current becomes zero and the instant reverse recovery current decays to 25%; of its reverse peak value I_{RM}.

13. (b)

$$t_{rr} = \left[\frac{2Q_R}{(di/dt)}\right]^{1/2}$$

Reverse recovery time depends on storage charge and rate of change of current di/dt. The storage charge depends upon the forward diode current I_f. This shows that reverse recovery time depends on forward field current.

15. (b)

MOSFET has positive temperature coefficient for resistance. Due to this reason parallel operation of MOSFETs is feasible. On the other hand BJT has negative temperature coefficient for resistance due to which parallel operation is not feasible for BJT.

16. (c)

In MOSFET, secondary breakdown does not occur, because it has positive temperature coefficient but occurs in BJT as it has negative temperature coefficient.

17. (c)

After the forward diode current in a power diode decays to zero, the diode continues to conduct in the reverse direction because of stored charges in the two layers. Due to this, reverse recovery transient peak inverse voltage of the power diode is affected.

20. (d)

If S-factor = 1 ⇒ Soft-recovery diode

If S-factor < 1 ⇒ Snappy-recovery diode or fast-recovery diode.

Hence, assertion is a false statement.

23. (b)

Given, $V_{DS} = 100$ V ; $R_L = 12\ \Omega$

$R_{DS(on)} = 0.2\ \Omega$; $D = 0.7$

$f = 30$ kHz

Drain current,

$$I_D = \frac{V_{DS}}{R_L + R_{DS(on)}} = \frac{100}{12 + 0.2} = 8.2 \text{ A}$$

Switching period,

$$T = \frac{1}{f} = \frac{1}{30} = 33.33\ \mu\text{sec}$$

On-time is given by

$$t_{on} = D \cdot T = 0.7 \times 33.33 \times 10^{-6}$$

$$= 23.33\ \mu\text{s}$$

Energy loss during on-time,

$$W_{on} = I_D^2 R_{DS(on)} \cdot t_{on}$$

$$= (8.2)^2 \times (0.2) \times (23.33 \times 10^{-6})$$

$$= 313.74\ \mu\text{J}$$

Now, power loss during on-time,

$$P_{on} = W_{on} \cdot f$$

$$= 313.74 \times 10^{-6} \times 30 \times 10^3$$

$$= 9.41 \text{ W}$$

24. (a)

From given figure,

$$I_{C(max)} = \frac{V_{CC} - V_{CE(sat)}}{R_L} = \frac{200 - 2}{10}$$

$$= 19.8 \text{ A}$$

$\therefore \quad I_{C(avg)} = \text{Dury cycle (D)} \times I_{C(max)}$
$= 0.7 \times 19.8 = 13.86$ A

Switching power loss during turn-on

$$= W_{on} \cdot f_s$$

$$= \frac{V_{CE(max)} I_{C(max)} t_{on}}{6} \cdot f$$

$$= \frac{200 \times 19.8 \times 3 \times 10^{-6}}{6} \times 1 \times 10^3$$

$$= 1.98 \text{ W} \approx 2.0 \text{ W}$$

25. (c)

Using current divider rule,

$$I_1 = \left(\frac{R_2}{R_1 + R_2}\right) I_T$$

or $$I_1 = \left(\frac{0.1}{0.3}\right) \times 12 = 4 \text{ A}$$

and $I_2 = I_T - I_1 = 12 - 4 = 8$ A

$\therefore \quad P_1 = I_1^2 \cdot R_1 = 4^2 \times 0.2 = 3.2$ W

and $P_2 = I_2^2 \cdot R_2 = 8^2 \times 0.1 = 6.4$ W

27. (d)

TRIAC is a bidirectional SCR as shown below.

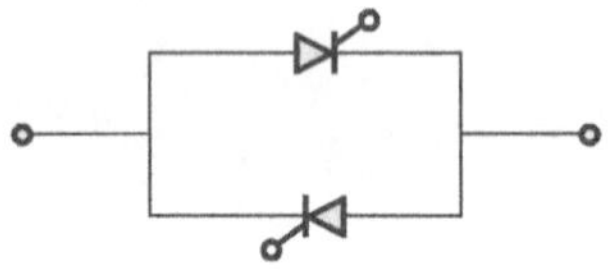

29. (b)

For the series RLC diode circuit, the current $i(t)$ is given by

$$i(t) = \frac{V_s}{\omega_r L} e^{-\xi t} \sin \omega_r t$$

Here, $$\omega_r = \sqrt{\frac{1}{LC} - \left(\frac{R}{2L}\right)^2}$$

$$= \sqrt{\frac{10^{10}}{50} - (5000)^2}$$

$$= 13228.76 \text{ rad/s}$$

The diode will stop conducting when $\omega_r t_1 = \pi$.

$\therefore$ Conduction time of diode,

$$t_1 = \frac{\pi}{\omega_r} = \frac{\pi}{13228.76}$$

$$\approx 237\ \mu\text{s}$$

31. (a)

The rms value of output voltage is:

$$V_{or} = \sqrt{\frac{1}{2\pi}\int_0^{\pi} V_m^2 \sin^2 \omega t \cdot d(\omega t)} = \frac{V_m}{2}$$

or, $$V_{or} = \frac{\sqrt{2} \times 230}{2} \text{ Volt}$$

Heater resistance,

$$R = \frac{230^2}{1000} \Omega$$

$\therefore$ Power absorbed by heater element

$$= \frac{V_{or}^2}{R} = \frac{2 \times 230^2}{4} \times \frac{1000}{230^2}$$

$$= 500 \text{ W} = 0.5 \text{ kW}$$

32. (d)

Given, $I_0 = 10$ A

The rms value of diode current,

$$I_{Dr} = \sqrt{\frac{I_0^2 \pi}{2\pi}} = \sqrt{\frac{I_0^2}{2}} = \sqrt{\frac{10^2}{2}} = \frac{10}{\sqrt{2}} = 7.07 \text{ A}$$

The average value of diode current,

$$(I_D)_{avg} = \frac{I_0 \cdot \pi}{2\pi} = \frac{I_0}{2} = \frac{10}{2} = 5 \text{ A}$$

33. (d)

If v_i is of very large negative value $|v_i| > |E|$, then diode D_1 will be forward biased (ON) while diode D_2 will be reverse biased (OFF).

35. (c)

Given, $R_D = 20\ \Omega$ and $R = 1000\ \Omega$

At no-load, load voltage,

$$V_{on} = \frac{V_m}{\pi} = \frac{\sqrt{2} \times 230}{\pi} = 103.536 \text{ V}$$

At given load, load voltage,

$$V_{01} = \frac{V_m}{\pi} \times \left(\frac{R}{R + R_D}\right)$$

$$= \frac{230\sqrt{2}}{\pi} \times \frac{1000}{1020} = 101.506 \text{ V}$$

$\therefore$ Voltage regulation

$$= \left(\frac{V_{on} - V_{01}}{V_{on}}\right) \times 100$$

$$= \left(\frac{103.536 - 101.506}{103.536}\right)$$

$$= 1.96\%$$

36. (b)

A single-phase half-wave rectifier is popularly known as one-pulse converter because for one cycle of supply voltage, there is one half-cycle of output, or load voltage. Hence, both assertion and reason are true but reason is not the correct explanation of assertion.

37. (c)

The load current is mostly continuous in 3-phase converters. Hence reason is a false statement.

39. (a)

In a full-wave rectifier with centre-tapped transformer:

- TUF is equal to 67.2% which is better.
- Output voltage has less ripple content which is 48.3%.
- Size of transformer required is more.
- PIV is $-2V_m$ while in bridge rectifier PIV is $-V_m$.

Thus, statements 1 and 3 are only correct.

41. (d)

Given, $V_S = 50$ V (rms value)

Since the load current is continuous, therefore D_1 and D_2 conduct in positive half cycle. And D_3 and D_4 conduct in negative half cycle. Due to continuous conduction, average output voltage is

$$V_0 = \frac{2V_m}{\pi} = \frac{2\sqrt{2}\times 50}{\pi} = 45 \text{ volt}$$

Rms value of output voltage is

$$V_{or} = \frac{V_m}{\sqrt{2}} = 50 \text{ Volt}$$

∴ Ripple voltage (ac component) is

$$V_r = \sqrt{V_{or}^r - V_0^2} = \sqrt{50^2 - 45^2} = 21.8 \text{ V}$$

42. (c)

For a 3-ϕ bridge rectifier with a charging battery, the average output voltage is

$$V_0 = \frac{3\sqrt{3}V_m}{\pi}$$

(where, $V_m = \frac{\sqrt{2}V_l}{\sqrt{3}}$ (V_m is the peak value of phase voltage))

Thus, $$V_0 = \frac{3\sqrt{2}V_l}{\pi} = \frac{3\sqrt{2}\times 50}{\pi} = \frac{150\sqrt{2}}{\pi} \text{ volt}$$

43. (d)

The rms value of current = $\frac{2}{\sqrt{2}}$ A = $\sqrt{2}$ A.

45. (c)

When a capacitor '*C*' is connected across the load *R* then, the circuit becomes a full-wave bridge rectifier with capacitive filter. Then, the ripple factor of the circuit with capacitive filter will be given by

$$RF = \frac{V_r}{V_0} = \frac{1}{\sqrt{2}(4fRC - 1)}$$

or, $$0.05 = \frac{1}{\sqrt{2}(4\times 50\times 200\times C - 1)}$$

or, $$(40000\,C - 1) = \frac{20}{\sqrt{2}} = 10\sqrt{2}$$

or, $$40000\,C - 1 = 14.14$$

or, $$C = \frac{15.14}{40000} = 378.5\ \mu\text{F}$$

46. (a)

- The PIV for 6-pulse bridge rectifier is $\sqrt{3}\,V_m$ whereas for mid-point rectifier it is $-2\,V_m$. Thus, peak inverse voltage (PIV) of six-pulse converter is higher.
- Filter circuits are used to remove the harmonics generated at the output of rectifier.
- Freewheeling diode is used only for inductive load for maintaining the instantaneous load voltage always positive.

Thus, statements 2, 3 and 4 are not correct.

48. (c)

Here, $V_m \sin\theta_t = E$

$$\theta_t = \sin^{-1}\left(\frac{E}{V_m}\right)$$

or, $$\theta_t = \sin^{-1}\left(\frac{50}{100}\right) = 30°$$

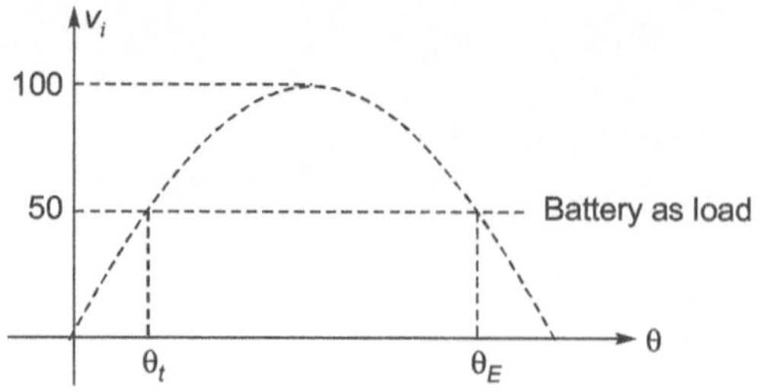

Angular conduction period

$$= 180° - 2\theta_t \quad (\theta_E = \theta_t)$$
$$= 180° - 2 \times 30° = 120°$$

49. (a)

Ripple frequency = $2 \times n \times f$

(Here, n = no. of phase, f = supply frequency)

$\therefore$ Ripple frequency = $2 \times 24 \times 25$ = 1200 Hz

Note: For half wave rectifier, ripple frequency = $n \times f$.

50. (c)

Ripple frequency for 3-ϕ half wave rectifier

$$= 3 \times f = 3 \times 400$$
$$= 1200 \text{ Hz}$$

51. (a)

$$\text{PIV} = 2V_m = 2\sqrt{2}V_{rms}$$
$$= 2\sqrt{2} \times 50$$
$$= 100\sqrt{2} \text{ V}$$

52. (d)

For diode rectifier at full load,

$$(V_0)_{fl} = \frac{2V_m}{\pi} - \frac{\omega L_s I_0}{\pi}$$

At no-load, $I_0 = 0$

$$\therefore \quad (V_0)_{nl} = \frac{2V_m}{\pi}$$

and $$\cos\mu = 1 - \frac{\omega L_s I_0}{V_m} \quad ...(i)$$

$$\text{Voltage regulation} = \frac{V_{nl} - V_{fl}}{V_{fl}} \approx \frac{\omega L_s I_0}{2V_m} \quad ...(ii)$$

From equation (i),

$$\frac{\omega L_s I_0}{2V_m} = 1 + \cos\mu$$

Putting this value in equation (ii), we have

$$\text{voltage regulation} = \frac{1 - \cos\mu}{2}$$

55. (c)

In the two transistor analogy of the SCR, the SCR can be considered as an npn and a pnp transistor, where the collector of one transistor is attached to the base of the other and vice-versa as shown below. Thus, reason is a false statement.

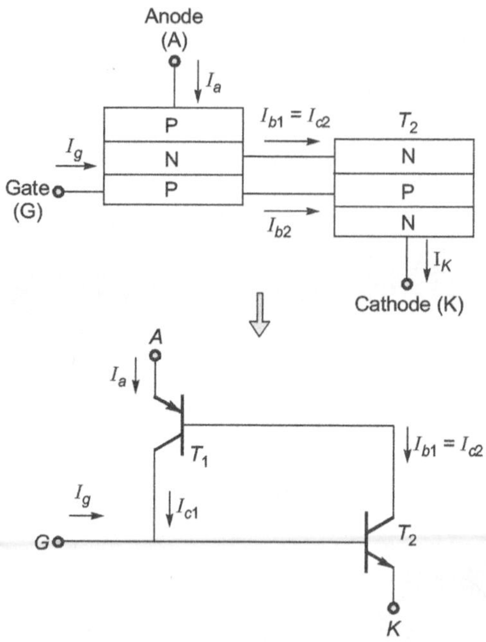

Two transistor analogy of SCR

56. (b)

Once the thyristor (SCR) is turned-on, and the anode current is above latching current level, there is no need of gate current, hence, gate signal can be withdrawn i.e. gate looses control over the anode current.

57. (c)

The leakage current in silicon is of the order of nA while in Germanium it is in μA.

58. (a)

Like the diode, an SCR is an unidirectional device that blocks the current flow from cathode to anode i.e. unilateral. An SCR or a thyristor has three modes of operation namely forward conduction mode, forward blocking mode and reverse blocking mode. Out of these three modes of operation the first two are stable states due to which an SCR is bistable.

60. (d)

Pulse transformers are often used to couple a trigger pulse generator to a thyristor in order to obtain electrical isolation between the two circuits.

61. (b)

The surge current rating indicates the maximum possible non-repetitive or surge current, the device can withstand and this may occur due to non-repetitive faults or short-circuits during the life-span of a thyristor.

62. (a)

The *R*-firing circuit suffers from several disadvantages. One of them is that the trigger angle can be varied only upto an approximate value of 90°.

64. (d)

The circuit equation is,

$$V = L\frac{di}{dt}$$

(i = latching current and t = pulse width)

or, $$dt = \frac{L}{V}di$$

Integrating on both sides, we get:

$$\int dt = \int \frac{L}{V}di \text{ or, } t = \frac{L}{V}i$$

or, $$t_{min} = \frac{0.1}{100}\times 2\times 10^{-3} = 2\ \mu s$$

65. (b)

Let I_T be the current during conduction and zero otherwise.

Therefore, $$I_{av} = \frac{I_T \cdot \beta}{360°}$$

(where, β = Conduction angle)

For $\beta = 180°$,

$$I_T = \frac{I_{av}\times 360°}{\beta} = \frac{100\times 360°}{180°}$$

$$= 200\text{ A}$$

From given figure,

for $I_T = 200$ A, $V_T = 1.8$ volt

∴ Average power loss,

$$P_{av} = (1.8\times 200)\times\frac{180°}{360°} = 180\text{ W}$$

66. (d)

- The *di*/*dt* rating of a thyristor indicates the maximum rate of rise in anode to cathode current and the maximum rate of change of current that the device can withstand during its ON state is called its "critical rate of rise of current." This critical value of *di*/*dt* determines the highest value of junction temperature that the thyristor can safely bear.
- Latching current (I_H) is associated with the turn-on process of SCR.
- Within the safe limit of $I_{g_{min}}$ and $I_{g_{max}}$, the conduction angle of thyristor can be controlled.
- Holding current is associated with turn-off process of SCR.

68. (c)

- Triacs have low *dv*/*dt* rating compared to SCRs.
- Triac is a bidirectional thyristor i.e. two diodes connected in anti-parallel therefore, it can be triggered with positive or negative polarity voltages.
- The LASCR will stay 'ON' even if the light disappears. It will turn 'OFF' only if its anode current is decreased below holding current (I_H).
- Statement-4 is correct.
- Thyristors are generally preferred for their low conduction losses and transistors for their rugged turn-off capabilities.

Thus, statements 2 and 4 are only correct.

69. (b)

The allowable P_g = 12 mW = 0.012 Watt

∴ $$V_g I_g = 0.012 \quad \text{...}(i)$$

Also, gradient $$= \frac{V_g}{I_g} = 3\times 10^3$$

or, $$V_g = 3\times 10^3 I_g$$

Substituting the value of V_g in equation (i), we get:

$$(3\times 10^3 \times I_g \times I_g) = 0.012$$

or, $$I_g = \sqrt{\frac{0.012}{3\times 10^3}} = 2\text{ mA}$$

∴ $$V_g = 3\times 10^3\times 2\times 10^{-3} = 6\text{ V}$$

70. **(d)**

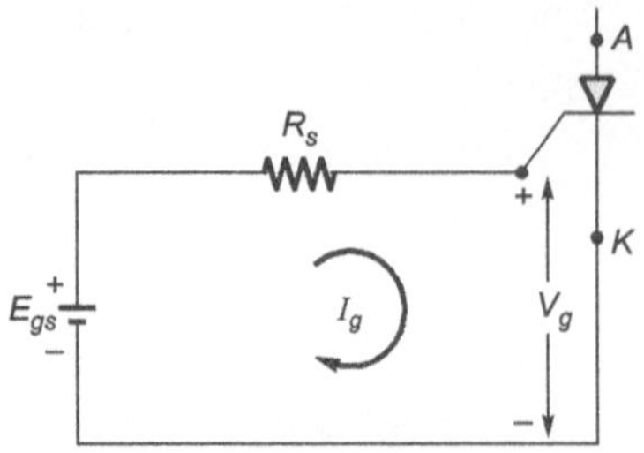

Given, $I_{g(min)} = 500\text{ mA} = 0.5\text{ A}$,

$$\frac{V_g}{I_g} = 16\frac{V}{A} \text{ and } E_{gs} = 15\text{ V}$$

$\therefore$ $V_g = 16\,I_g = 16 \times 0.5 = 8\text{ V}$

From above figure,

$$R_s = \frac{E_{gs} - V_g}{I_g} = \frac{15-8}{0.5} = 14\ \Omega$$

71. **(c)**

The worst case fault occurs when the voltage is at maximum. Assume that the voltage is at the maximum value for the duration of the fault, then,

$$\int_0^{t_c} i^2 dt = \int_0^{t_c} 120^2 dt = 15 \text{ or } 120^2\, t_c = 15$$

$$\therefore \text{ Fault clearance time} = \frac{15}{(120)^2} = 1.04\text{ ms}$$

72. **(a)**

The first step is to determine the instantaneous value of e_s at which triggering will occur. At the SCR triggering point,

$$V_{g(min)} = 0.5\text{ V and } I_{g(min)} = 0.1\text{ mA}$$

Using KVL around the gate circuit, we have,

$$e_s = I_g\,(R_v + R_{min}) + V_D + V_g$$

At the trigger point, we have

$$e_{s(trigger)} = 0.1\text{ mA }(110\text{ k}\Omega) + 0.7 + 0.5 = 12.2\text{ V}$$

Since e_s is a sine-wave, it obeys the expression,

$$e_s = E_{max}\cdot\sin\omega t = E_{max}\cdot\sin(2\pi ft)$$

(where $2\pi ft$ is the phase angle at any instant of time which is equal to α)

Thus, $E_{max} = 24\text{ V}$

$\therefore$ $e_s = 24\sin\alpha$ or $12.2 = 24\sin\alpha$

or, $$\sin\alpha = \frac{12.2}{24} = 0.5083$$

Since sin 30° = 0.5, therefore sin 30.6° will be around 0.5083.

Thus, $\alpha = 30.6°$ = Required trigger angle

73. **(d)**

$$R_{max} = \frac{V_{BB}(1-\eta)}{I_p} = \frac{20(1-0.7)}{50}\times 10^6 = 120\text{ k}\Omega$$

and $$R_{min} = \frac{V_{BB}-V_v}{I_v} = \frac{20-2}{6\times 10^{-3}} = 3\text{ k}\Omega$$

The approximate value of $R_2 = \dfrac{10^4}{\eta V_{BB}}$

$$\therefore \quad R_2 = \frac{10^4}{0.7\times 20} = 714.29\ \Omega$$

Now, $$R_1 = \frac{V_{g(min)}}{I_{E_0}}$$

(where, $V_{g(min)}$ = min. gate voltage required to trigger SCR = 0.2 V)

$$\therefore \quad R_1 = \frac{0.2}{2\times 10^{-3}} = 100\ \Omega$$

Now, the limits on the output frequency are

$$T_{max} = R_{max}C\,ln\left(\frac{1}{1-\eta}\right)$$
$$= 120\text{ k}\times 0.1\times 10^{-6}\,ln\left(\frac{1}{1-0.7}\right)$$
$$= 120\text{ k}\times 0.1\times 10^{-6}\times ln\,3.33$$

or, $T_{max} = 14.44\text{ ms}$ ...(*i*)

and $$T_{min} = R_{min}C\,ln\left(\frac{1}{1-\eta}\right)$$
$$= 3\text{ K}\times 0.1\times 10^{-6}\,ln\,3.33$$
$$= 0.36\text{ ms}$$

Hence, $$f_{min} = \frac{1}{T_{max}} = \frac{1}{14.44\text{m}} = 69.44\text{ Hz}$$

and $$f_{max} = \frac{1}{T_{min}} = \frac{1}{0.36\text{ m}} = 2.78\text{ kHz}$$

75. **(a)**

Dynamic resistance of 100 A SCR1 = $\dfrac{2.1}{100}\ \Omega$

Dynamic resistance of 150 A SCR2 = $\dfrac{1.75}{150}\ \Omega$

Let R be the resistance inserted in series with each SCR. With this, current shared by

$$\text{SCR1} = \frac{250\times\frac{1.75}{150} + R}{\text{Total resistance}} \propto 100 \quad \text{...(i)}$$

and current shared by

$$SCR2 = \frac{\frac{2.1}{100}+R}{\text{Total resistance}} \propto 150 \quad ...(ii)$$

from equation (i) and (ii) we get

$$\frac{\frac{2.1}{100}+R}{\frac{1.75}{150}+R} = \frac{150}{100}$$

$$0.5R = \frac{0.35}{100} \Rightarrow R = 7\ m\Omega$$

76. (d)

Given, E_s = 7.5 kV
= Total applied voltage of thyristor unit.

Derating factor = 14%

Forward voltage rating of each device is

E_D = 500 V

Now, % series derating

$$= \left(1-\frac{E_s}{n_s E_D}\right)\times 100$$

or, $$14 = \left(1-\frac{7500}{n_s \times 500}\right)\times 100$$

or, n_s = 17.44

Thus, $n_s \approx 18$
= Number of SCRs connected in series-string

77. (b)

Given, I_T = 75 A,

Derating factor = 14% and I_m = 1 kA

$$\%\text{ parallel derating} = \left(1-\frac{I_m}{n_p \times I_T}\right)\times 100$$

or, $$14 = \left(1-\frac{1000}{n_p \times 75}\right)\times 100$$

or, $n_p = 15.5 \approx 16$

Thus, number of SCRs connected in parallel = 16.

78. (b)

- One of the disadvantage of *R*-firing circuit is that its firing angle is limited to 90°.
- UJT firing circuit is based on UJT which is a switching device. It can be turned-on and turned-off in a controlled manner.
- In R-C firing circuit, the firing angle is higher i.e. upto 180°.

79. (c)

Both R and R-C firing circuits:

1. can be employed in power circuits having only one thyristor.
2. are capable of open loop control only.
3. are not capable of providing high frequency gate signal.
4. does not provide electrical isolation between control circuit and power circuit.

Hence, statement-1 is only correct.

80. (a)

When the thyristor is turned on, voltage across it falls from 100 V to 0 V and current rises from 0 to 10 A. Figure below shows the various waveforms during turn-on.

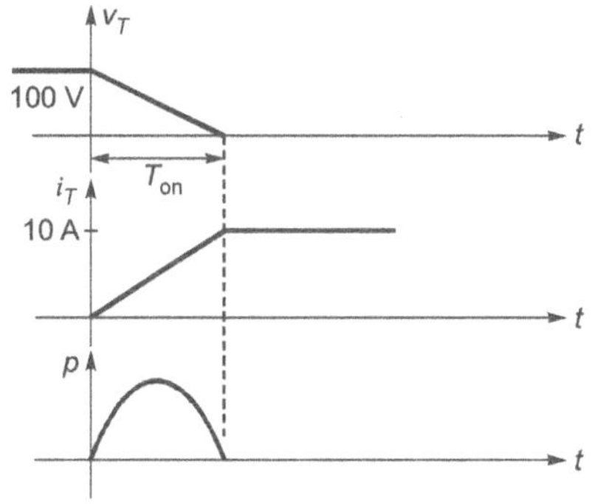

During turn-on,

$$v_T = 100 - \frac{100}{T_{on}}t \quad ...(i)$$

and $$i_T = \frac{10t}{T_{on}} \quad ...(ii)$$

Instantaneous power loss is

$$P = v_T i_T = \left(100 - \frac{100}{T_{on}}t\right)\frac{10}{T_{on}}t$$

$$= \frac{1000}{T_{on}}t - \frac{1000}{T_{on}^2}t^2 \quad ...(iii)$$

Total energy loss during turn-on is

$$W = \int_0^{T_{on}} pdt = \left[\frac{1000}{2}\frac{t^2}{T_{on}} - \frac{1000}{3T_{on}^2}t^3\right]_0^{T_{on}}$$

$$= \frac{500}{3}T_{on} = \frac{500}{3}\times 3\times 10^{-6} = 500\ \mu J$$

Power loss in one second

$= 500 \times 10^{-6} \times 5 \times 10^3 = 2.5$ Watt

81. (c)

Maximum rate of rise of current will take place when the thyristor is turned on at 90° or at peak value of the supply voltage. Since the current at the time of turn-on is negligible, hence,

$$V_P = L\frac{di}{dt}$$

or, $$L = \frac{V_p}{di/dt}$$

$$= \frac{230\sqrt{2}\times 10^{-6}}{200} = 1.6\ \mu H$$

82. (d)

Maximum string voltage is

$$V_{sm} = 4 \times 10 \times 0.825 = 33\text{ kV}$$

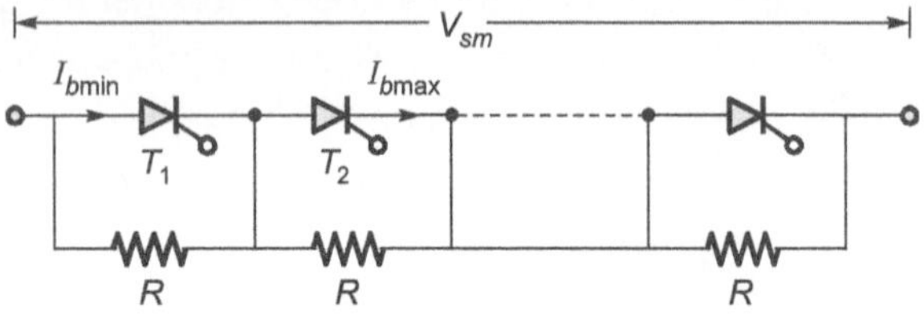

Static equalisation

Here, $n = 10$, $V_p = 4$ kV, $V_{sm} = 33$ kV

$$\Delta I_b = I_{bmax} - I_{bmin} = 10\text{ mA}$$

Hence, the value of each resistance of static equalization circuit is

$$R = \frac{nV_p - V_{sm}}{(n-1)\,\Delta I_b}$$

$$= \frac{(40-33)\times 10^3}{9\times 10\times 10^{-3}} = 77.8\text{ k}\Omega$$

83. (d)

A snubber circuit reduces the rate of rise of voltage, thus prevents *dv/dt* turn on of the thyristor. A snubber circuit is connected across each thyristor and it may be connected on supply side also. Hence, assertion is a false statement.

84. (b)

Let I and I_{sb} be the one-cycle and sub-cycle surge current ratings of the SCR respectively. Then, equating the energies involved in them, we get

$$I^2T = I_{sb}^2 t$$

or, $$I^2 \times \frac{1}{100} = (3000)^2 \times \frac{1}{200}$$

or, $$I = \frac{3000}{\sqrt{2}} = 2121.32\text{ A} \approx 2.1\text{ kA}$$

Now, I^2t rating $= I^2 \times \frac{1}{2f} = \left(\frac{3000}{\sqrt{2}}\right)^2 \times \frac{1}{100}$

$= 45$ kA-sec

85. (c)

- If the thyristor is subjected to overcurrent due to faults, short circuits or surge currents; its junction temperature may exceed the rated valuo and the device may be damaged.
- If the rate of rise of forward voltage *dv/dt* is high, the charging current *i* will be more. This charging current plays the role of gate current and turns on the SCR even when gate signal is zero. Such phenomena of turning-on a thyristor is called false triggering of SCR.
- If the rate of rise of anode current i.e. *di/dt*, is large as compared to the spread velocity of carriers, local hot spot will be formed near the gate connection on account of high current density.
- Transient overvoltages cause permanent damage to the device due to reverse breakdown.

86. (a)

An SCR can be turned-on if:

1. It is forward biased.
2. Gate pulse width is more than turn-on of an SCR.
3. Anode to cathode voltage is more than finger voltage i.e. V_{ak} must be more than a certain voltage above which SCR can be turn-on with the application of gate signal.
4. The gate triggering synchronizes with the ac supply.

All the above conditions must satisfy simultaneously to turn-on the SCR.

87. (c)

We know that,

$$T_j - T_A = \theta P_C$$

Given, $T_j = 120°C$, $T_A = 75°C$

$\theta = 1.5°C/W$

$\therefore \quad 120 - 75 = 1.5\,P_c$

or, $$P_c = \frac{45}{1.5}\text{ W} = 30\text{ W}$$

$\therefore$ Maximum power dissipation $P_c = 30$ W.

88. (d)

An SCR suffers from noise or unwanted trigger pulses which leads to false operation of the main SCR. A capacitor and a resistor are connected across gate to cathode to bypass the noise signal.

89. (a)

With the use of zener diode, small variations in the supply voltage and frequency will not affect the circuit operation so that the UJT will not be fired erratically.

90. (b)

An inverter-grade thyristor has fast turn-off time, generally in the rage of 5 to 50 μs, depending upon voltage rating. For inverter thyristors, the turn-off time is dependent upon the stored charge at the time of current reversal.

91. (c)

Due to the multi cathode structure of GTO, triggering gate current is higher than that required for a conventional SCR.

93. (c)

$$\text{\% of derating factor} = \left(1 - \frac{E_s}{n_s E_D}\right) \times 100$$

or, $$15 = \left(1 - \frac{3000}{n_s \times 800}\right) \times 100$$

or, $$\left(1 - \frac{30}{8n_s}\right) = \frac{15}{100}$$

or, $$\frac{30}{8n_s} = \left(1 - \frac{3}{20}\right) = \frac{17}{20}$$

or, $$n_s = \frac{30}{8} \times \frac{20}{17} = \frac{75}{17} = 4.41$$

Thus, required number of series connected SCRs = 5.

96. (a)

$$I_{rr} = \sqrt{\frac{2Q_{rr}\left(\frac{di_R}{dt}\right)}{S+1}} \text{ and } t_{rr} = \sqrt{\frac{2Q_{rr}(S+1)}{\left(\frac{di_R}{dt}\right)}}$$

(Here, S = Softness factor)

100. (a)

Upto instant π/2, SCR behaves as open circuit. Hence, option (a) is correct.

101. (d)

For reliable turn-on,

$$V_b = 12 - 4 = 8 \text{ V}$$

and $$R = \frac{8}{10 \text{ mA}} = 800\ \Omega$$

$$= 0.8 \text{ k}\Omega$$

102. (a)

To turn-off a thyristor, anode current must be reduced below holding current and a reverse bias must be applied across thyristor for a finite period of time. If all these two conditions are not met simultaneously then forced commutation can be used to turn-off the thyristor.

105. (d)

The turn-on time of an SCR depends on time constant (L/R) of the load.

106. (b)

It is necessary to keep the thyristor reverse biased for a finite period before a forward anode voltage can be reapplied because if a forward voltage is applied immediately after reducing the anode current to zero, it will not block the forward voltage and will start conducting again, although it is not triggered by a gate pulse. Reason is also a true statement but not the correct explanation of assertion.

108. (a)

- In forced commutation, the commutating components L and C do not carry load current continuously.
- Forced commutation is usually employed in dc choppers because natural current zero is not available in the circuits containing dc choppers.
- In load commutation, L and C are connected in series with the load or C in parallel with the load such that overall load circuit is underdamped.
- Load commutation is employed in series inverters.

- In class-C commutation, the main thyristor that is to be commutated is connected in series with the load while the complementary thyristor is connected in parallel with the main thyristor as shown below.

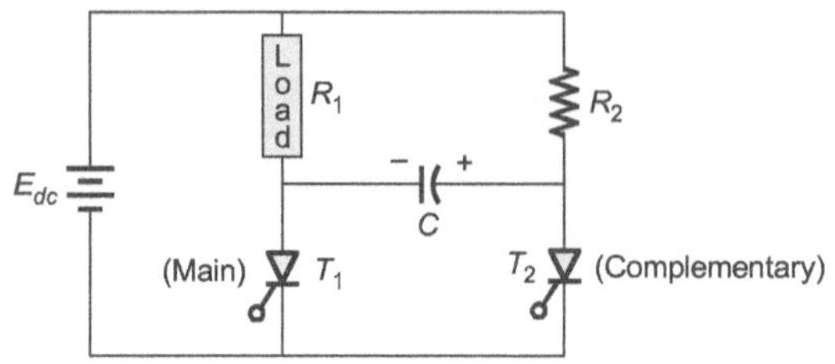

Thus, statements 2 and 4 are only correct.

109. (c)

Given, $L = 5\ \mu H$, $C = 20\ \mu F$

Resonant frequency,

$$\omega_0 = \frac{1}{\sqrt{LC}}$$

$$= \frac{1}{\sqrt{5\times10^{-6}\times20\times10^{-6}}}$$

$$= \frac{10^6}{\sqrt{100}} = 0.1\times10^6 \text{ rad/s}$$

∴ Conduction time for auxiliary thyristor

$$= \frac{\pi}{\omega_0} = \frac{\pi}{0.1\times10^6} = 31.4\ \mu s$$

110. (b)

Given, $V_s = 200$ V, $R_1 = 10\ \Omega$

and $R_2 = 100\ \Omega$

Given circuit represents a class-C commutation circuit.

The peak value of current through thyristor T_1 is

$$(I_P)_{T_1} = V_s\left(\frac{1}{R_1}+\frac{2}{R_2}\right)$$

$$= 200\left(\frac{1}{10}+\frac{2}{100}\right) = 24 \text{ A}$$

The peak value of current through thyristor T_2 is

$$(I_P)_{T_2} = V_s\left(\frac{2}{R_1}+\frac{1}{R_2}\right)$$

$$= 200\left(\frac{2}{10}+\frac{1}{100}\right) = 42 \text{ A}$$

111. (b)

Under steady-state condition, voltage across capacitor is $V_s = 200$ V and SCR conducts having a current through it equal to

$$\frac{V_s}{R} = \frac{200}{50} = 4 \text{ A}$$

When the SCR is forced commutated, the capacitor voltage at time t is

$$v_c = V_s(1 - e^{-t/RC})$$

or, $$\left|\frac{dv_c}{dt}\right| = \frac{V_s}{RC}e^{-t/RC}$$

∴ $$\left.\frac{dv_c}{dt}\right|_{t=0} = \frac{V_s}{RC} \quad ...(i)$$

Let C_j be the junction capacitance of SCR. Now, when SCR get turned-on, then

$$C_j\left(\frac{dv_c}{dt}\right)_{t=0} = 5 \text{ mA}$$

or, $$C_j\cdot\frac{V_s}{RC} = 5\times10^{-3} \quad \text{[Using equation } (i)\text{]}$$

or, $$25\times10^{-12}\times\frac{200}{50C} = 5\times10^{-3}$$

or, $$C = \frac{25\times10^{-12}\times200}{250\times10^{-3}}$$

$$= 0.02\ \mu F$$

Thus, the minimum value of C so that SCR does not get turn-on is 0.02 μF. If $C > 0.02$ μF, SCR will be turn-on.

113. (a)

The circuit diagram for class-C commutation circuit is as shown steady-state current $I = 20$ A (Given)

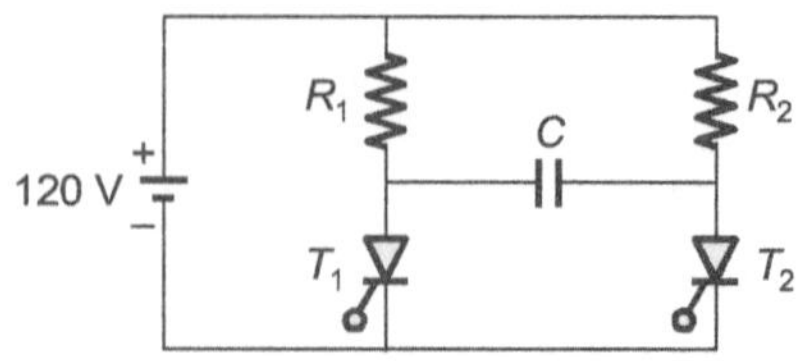

The resistances $R_1 = R_2 = \frac{V_s}{I} = \frac{120}{20} = 6\ \Omega$

Now, for successful commutation, we have

$$C = 1.44\cdot\frac{t_{off}}{R_1} = 1.44\times\frac{60\times10^{-6}}{6} = 14.4\ \mu F$$

114. (c)

Since tolerance on turn-off time is 50% for reliable operation, therefore

$$t_{off} = \left(30 + \frac{50}{100} \times 30\right) = 45\ \mu s$$

Now, the value of the commutating capacitor is

$$C = \frac{I_L t_{off}}{V_s} = \frac{50 \times 45 \times 10^{-6}}{50} = 45\ \mu F$$

115. (d)

When the thyristor is on, it behaves like a diode. The capacitor voltage at time t will be given by

$$v_c(t) = V_s(1 - \cos \omega_0 t)$$

(Here, $V_s = 100$ V)

At time $t = t_0 = \pi/\omega_0$, $i(t) = 0$ and $v_c(t) = -2V_s$.

Thus, after $t = \pi/\omega_0$ sec. or $\pi\sqrt{LC}$ sec., the charging current becomes zero and thyristor is turn-off its own i.e. self commutated.

Hence, voltage across the capacitor after the SCR is self commutated

$$= |2V_s| = 2 \times 100 = 200\ V$$

116. (b)

$$\text{Conduction time of thyristor} = \frac{\pi}{\omega_0} = \pi\sqrt{LC}\ \text{sec}$$

$$= \pi\sqrt{8 \times 10^{-6} \times 2 \times 10^{-6}} = 4\pi \times 10^{-6}\ \text{sec}$$

$= 12.56\ \mu F$

117. (d)

Given figure represents a class-B commutation circuit.

Here, peak value of current through thyristor

$T_A = V_s\sqrt{\frac{C}{L}}$ and peak value of current through thyristor $T_1 = \frac{V_s}{R}$.

118. (b)

- Statements 1 and 2 are correct.
- Statement 3 is not correct because for self commutation (class-A commutation), the circuit should be underdamped i.e. $R < 2\sqrt{L/C}$.
- Settlement 4 is correct.

121. (b)

$$(I_{peak})_{SCR} = V_s\sqrt{\frac{C}{L}}$$

$$= 200\sqrt{\frac{20 \times 10^{-6}}{5 \times 10^{-6}}} = 400\ A$$

122. (a)

We know that,

$$v_1 = V_m \sin\theta$$

or, $110 = 220 \sin\theta$ i.e. $\sin\theta = 0.5$

or, firing angle

$$\theta = 30°$$

Conduction angle

$$= 180° - \theta = 180° - 30° = 150°$$

Now, $\omega t = \theta = 30° = \frac{\pi}{6}$

or, $2\pi \times 50t = \frac{\pi}{6}$

The time during which the SCR remains OFF is

$$t = \frac{1}{2 \times 6 \times 50} = \frac{1}{600} = 1.67\ \text{ms}$$

123. (c)

As the supply voltage is 400 sin 314t,

$$V_m = 400\ V$$

Peak invesre voltage (PIV)

$$\text{PIV} = \sqrt{3}\, V_m = \sqrt{3} \times 400 = 692.8\ V$$

RMS value of current = 20 A

Average value of current,

$$I_{avg} = \frac{\text{RMS value of current}}{\text{Form factor}} = \frac{20}{1.11} = 18\ A$$

∴ Power rating of the SCR

$$= \text{PIV} \times I_{avg}$$

$$= 692.8 \times 18 = 12.47\ kW$$

125. (b)

- By using freewheeling diode, load performance becomes better because load current waveform is improved.
- Whenever load voltage tends to go negative, freewheeling comes into play as a result of which load current is transferred from main thyristor to freewheeling diode, allowing the thyristor to regain its forward blocking capability.

Hence, both assertion and reason are true but reason is not the correct explanation of assertion.

126. (d)

Power delivered to the load, for the same firing angle is more when freewheeling diode is used. As volt-ampere input is almost the same with and without a freewheeling diode, therefore the input power factor with use of freewheeling diode is improved.

127. (c)

The average output voltage for a single-phase full converter is

$$(V_0)_{avg} = \frac{2V_m}{\pi}\cos\alpha$$

Given, $\alpha = 30°$, $V_m = 230\sqrt{2}$ volt

$$\therefore \quad (V_0)_{avg} = \frac{2\times 230\sqrt{2}}{\pi}\cos 30°$$

$$= \frac{230\sqrt{6}}{\pi} = 179.33 \text{ Volt}$$

$$\approx 179 \text{ volt}$$

128. (a)

Maximum voltage across SCR in single-phase mid-point converter is $2V_m$. Therefore, this converter can be design for a maximum voltage of $\frac{1000}{2\times 2.5} = 200$ V.

$\therefore$ Maximum average power that the mid-point converter can handle is

$$(P_{avg})_{max} = \left(\frac{2V_m}{\pi}\cos\alpha\right)(I_T)_{avg}$$

$$= \frac{2\times 200}{\pi}\times 40\times\frac{1}{1000}\text{ kW} = \frac{16}{\pi}\text{ kW}$$

129. (c)

- Circulating current gives rise to more losses in the converters, hence the efficiency and power factor are low. Thus, statement-1 is false.
- In a non-circulating current type dual converter, the normal delay period of 10 to 20 ms is required while no delay period is required in circulating current type dual converter. Therefore, circulating current type dual converter is faster than non-circulating type dual converter. Thus, statement-2 is true.
- As the dual converter with circulating current has to handle load current as well as circulating currents, therefore the thyristors are rated for higher currents. Hence statement-3 is false.

131. (b)

The output voltage waveform v_0 is as shown in figure below.

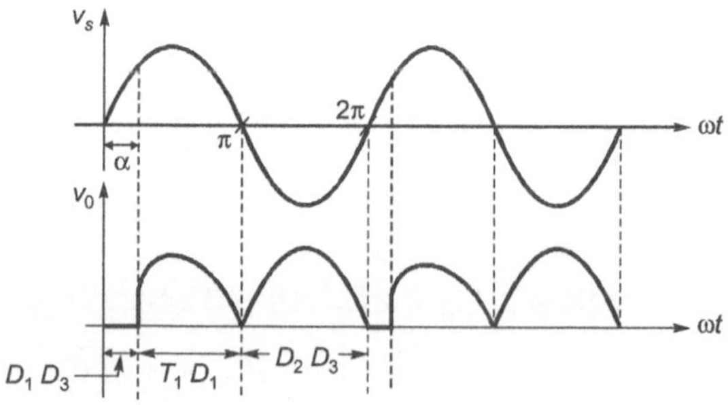

The average output voltage is given by

$$(V_0)_{avg} = \frac{1}{2\pi}\left[\int_\alpha^\pi V_m \sin\omega t\, d(\omega t) - \int_\pi^{2\pi} V_m \sin\omega t \cdot d(\omega t)\right]$$

$$= \frac{V_m}{2\pi}(3+\cos\alpha)$$

132. (c)

In case "four quadrant operation" is required without any mechanical changeover switch, two full converters can be connected back to back to the load circuit. Such an arrangement using two converters in antiparallel and connected to the same dc load is called a "dual converter."

133. (a)

In mid-point converter, each secondary should be able to supply the load power. Due to this reason, the transformer rating in mid-point converter is double the load rating. Thus, statement-2 is not correct.

135. (d)

The commutation overlap is more predominant in full converter than in semiconverter. Hence, assertion is a false statement.

136. (c)

- In a single-phase full converter, as long as $\mu < 180°$, the output voltage will be given by

$$V_0 = \frac{2V_m}{\pi}\cos\alpha - \frac{\omega L_s}{\pi} I_0$$

$$\begin{pmatrix} L_s = \text{Source inductance} \\ I_0 = \text{Full-load current} \end{pmatrix}$$

- When $\mu = 180°$, the load will be permanently short circuited by SCRs and the output voltage will be zero because during the overlap angle, all SCRs will be conducting.
- Since, $V_0 = \frac{V_m}{\pi}[\cos\alpha + \cos(\alpha + \mu)]$,

for $\mu = 0°$, $V_0 = \frac{2V_m}{\pi}\cos\alpha$

138. (b)

A dual converter using two full converters can give four quadrant operation as shown below.

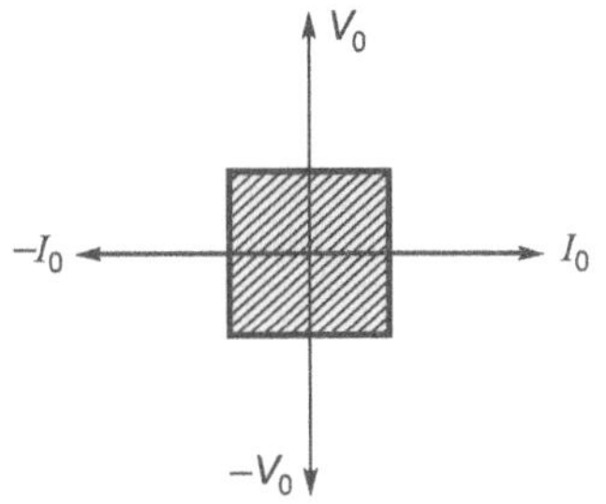

139. (a)

A full converter operates as a rectifier in first quadrant (both V_0, I_0 positive) from $\alpha = 0°$ to 90° and as an inverter (V_0 positive but I_0 negative) from α = 90° to 180° in the fourth quadrant as shown below.

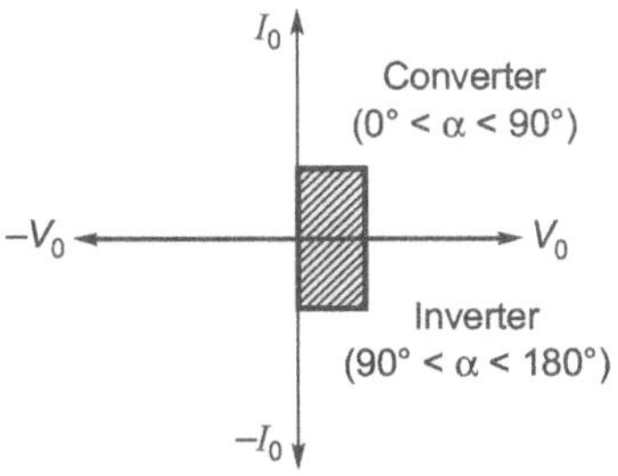

Thus, statement-4 is not correct. Statements 1, 2, 3 and 5 are correct.

140. (d)

In AC-to-DC circulating current dual converters, the firing pulses of the two converters are so adjusted that $\alpha_1 + \alpha_2 = 180°$.

141. (d)

The average output voltage is

$$V_0 = \frac{V_m}{2\pi}(1 + \cos\alpha)$$

The maximum output voltage is obtained when $\alpha = 0°$.

$$\therefore \quad (V_0)_{max} = \frac{V_m}{\pi}$$

According to question,

$$V_0 = 25\% \text{ of } (V_0)_{max}$$

or, $$\frac{V_m}{2\pi}(1 + \cos\alpha) = \frac{1}{4} \times \frac{V_m}{\pi}$$

or, $$1 + \cos\alpha = \frac{1}{2}$$

or, $$\cos\alpha = -\frac{1}{2} \text{ or } \alpha = 120°$$

142. (b)

For a single-phase fully controlled thyristor bridge converter with R, L, E load, the average output voltage is

$$V_{dc} = \frac{2V_m}{\pi}\cos\alpha$$

or, $$V_{dc} = \frac{2 \times \sqrt{2} \times 220}{\pi}\cos\alpha$$

$$= 198 \cos\alpha \qquad \ldots(i)$$

Now, $V_{dc} = I_{dc} R + E = (10 \times 0.5) + 135$

or, $V_{dc} = 140 \text{ V} \qquad \ldots(ii)$

From equations (*i*) and (*ii*), we have:

$$198 \cos\alpha = 140$$

or, $$\cos\alpha = \frac{140}{198} = 0.707$$

or, Firing angle $\alpha = 45°$

143. (c)

- Only fully controlled circuits are capable of inversion and therefore regeneration, while half-controlled circuits are not. Thus, statement-1 is not correct.
- Statement-2 is correct.
- Bridge circuits require twice as many thyristors than half-wave circuits to carry the same current. Thus, statement-3 is not correct.
- Statement-4 is correct.

145. (b)

The reduction in output voltage due to overlap is

$$V_{dc(red)} = \frac{3\omega L_s}{\pi} I_{dc}$$

$$= \frac{3 \times 2\pi \times 50 \times 50 \times 10^{-6}}{\pi} \times 90$$

$$= 1.35 \text{ V}$$

146. (d)

Given, $I_H = 0.2$ A, $t_d = 2.5$ μsec, $\alpha = 60°$
$R = 2\ \Omega$, V_s(line) = 208 V
$f = 50$ Hz, $L = 0$

∴ Phase voltage,

$$(V_s)_P = \frac{208}{\sqrt{3}} = 120 \text{ V}$$

and peak voltage

$$= (V_s)_m = \sqrt{2}(V_s)_P = 169.7 \text{ V}$$

Now, $e = \sqrt{3}(V_s)_m \sin(\omega t + 30°)$

At $\omega t = 60°$, $e = \sqrt{3} \times 169.7 \sin 90° \approx 293.93$ V

Rate of rise of anode current, di/dt at the instant of triggering is

$$\frac{di}{dt} = \frac{e}{L} = \frac{29}{0} = \infty \text{ A/sec}$$

$$\therefore \quad t = \frac{I_H}{\left(\frac{di}{dt}\right)} = \frac{0.2}{\infty} = 0 \text{ μsec}$$

∴ Minimum width of gate-pulse is

$$t_p = t + t_d = 0 + 2.5 = 2.5 \text{ μs}$$

147. (d)

The term continuous means that load current never ceases but continues to flow through SCR or their combination. The term discontinuous is applied to the condition when load current reaches zero during each half cycle before the next SCR in sequence is fired. In practice, the output current may become discontinuous at high values of firing angle or at low values of load current.

148. (c)

Output voltage with no overlap
= Internal voltage of the 3-phase full converter

$$= \frac{3 V_{ml}}{\pi} \cos\alpha$$

Output voltage with overlap

$$= \left(\frac{3 V_{ml}}{\pi} \cos\alpha - \frac{3\omega L_s}{\pi} I_0\right)$$

(where, $\frac{3\omega L_s}{\pi} I_0$ = Average value of fall in output voltage due to overlap)

149. (c)

PIV of 3-ϕ bridge converter = $\sqrt{3}\, V_m = 1.732\, V_m$

PIV of 3-ϕ half wave converter

$$= \sqrt{3}\, V_m = 1.732\, V_m$$

PIV of single-phase full-wave mid-point converter

$$= -2\, V_m \text{ (Highest)}$$

PIV of single-phase full-converter = $-\, V_m$

152. (a)

$$\text{Secondary voltage} = 400 \times \frac{1}{2} = 200 \text{ V}$$

$$\therefore \quad \text{PIV} = 200\sqrt{2} \text{ V}$$

153. (b)

$$\text{Ripple factor} = \frac{V_{ac(rms)}}{V_{dc}} = \sqrt{\frac{V_{or}^2 - V_{dc}^2}{V_{dc}^2}}$$

$$= \sqrt{\frac{V_{or}^2}{V_{dc}^2} - 1} \quad \ldots(i)$$

(V_{or} = Total rms voltage and V_{dc} = d.c. voltage)

For half wave,

$$V_{dc} = \frac{V_m}{2\pi}(1 + \cos\alpha)$$

Since $\alpha = 90°$, therefore $V_{dc} = \frac{V_m}{2\pi}$

Now, $$V_{or}^2 = \frac{1}{2\pi}\int_{\pi/2}^{\pi} V_m^2 \sin^2\theta\, d\theta = \frac{V_m^2}{8}$$

Putting the values of V_{dc} and V_{or}^2 in equation (i), we have

$$\text{Ripple factor} = \sqrt{\frac{V_m^2 \times 4\pi^2}{8 \times V_m^2} - 1} = \sqrt{\frac{\pi^2 - 2}{2}}$$

155. (a)

Since capacitor doesn't allow sudden change of voltage therefore, output voltage will be represented by option (a).

157. (c)

The average value of load current

$$= \frac{10+12}{2} = 11 \text{ A}$$

The maximum value of load current

$$= \frac{V_s}{R} = \frac{100}{5} = 20 \text{ A}$$

Now, the average value of the voltage is

$$(V_0)_{avg} = 100 \times \frac{11}{20} = 55 \text{ V}$$

Also, $$(V_0)_{avg} = \left(\frac{T_{on}}{T_{on}+T_{off}}\right) V_s$$

or, $$\frac{T_{on}}{T_{on}+T_{off}} = \frac{(V_0)_{avg}}{V_s} = \frac{55}{100} = 0.55$$

or, $$T_{on} = 0.55\,(T_{on} + T_{off})$$

or, $$\frac{T_{on}}{T_{off}} = \frac{0.55}{0.45} = 1.22$$

158. (a)

Given, $V_s = 220$ V, $V_0 = 500$ V

and $T_{off} = 80\ \mu s$

For a step-up chopper, output voltage is

$$V_0 = \frac{V_s}{1-\alpha} \quad (\alpha = \text{Duty cycle})$$

or, $$1-\alpha = \frac{V_s}{V_0} = \frac{220}{500} = \frac{11}{25}$$

or, $$\alpha = 1 - \frac{11}{25} = \frac{14}{25}$$

or, $$\frac{T_{on}}{T_{on}+T_{off}} = \frac{14}{25}$$

or, $$25\,T_{on} = 14\,T_{on} + 14\,T_{off}$$

or, $$11\,T_{on} = 14\,T_{off}$$

or, $$T_{on} = \frac{14}{11} \times 80\ \mu s = 101.8\ \mu s$$

159. (c)

The step-up chopper can be employed in regenerative braking both for increasing and decreasing motor speed provided duty cycle α is so adjusted that $V_s/(1-\alpha)$ exceeds the fixed source voltage V_0. Hence, reason is a false statement.

160. (b)

On time of chopper,

$$T_{ON} = 1 \text{ ms}$$

Time period of chopper,

$$T = \frac{1}{f} = \frac{1}{500} \text{ sec} = 2 \text{ ms}$$

$\therefore$ Duty cycle of chopper,

$$\alpha = \frac{T_{ON}}{T} = \frac{1\text{ms}}{2\text{ ms}} = 0.5 = 50\%$$

161. (d)

The average output voltage,

$$V_0 = \alpha V_s$$

Since average voltage drop across L is zero, therefore,

$$E = V_0 = \alpha\, V_s = 0.2 \times 500 = 100 \text{ volts}$$

During T_{ON}, the voltage drop across inductance $L = (V_s - E)$ as shown by waveforms shown below.

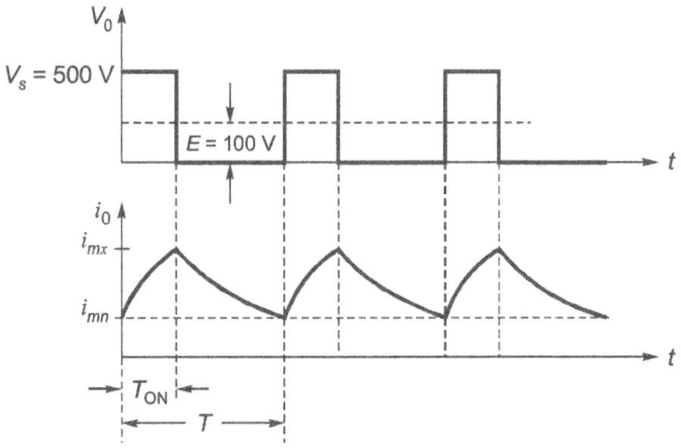

During T_{ON}, volt-time area applied to inductance

$$= (500 - 100)\, T_{ON}$$
$$= 400\, T_{ON} \text{ volt-sec} \quad ...(i)$$

During T_{ON}, the current through L rises from I_{mn} to I_{mx}. Thus, the volt-time area across L during this current change is

$$= \int_0^{T_{ON}} v_L\, dt = \int_0^{T_{ON}} L\frac{di}{dt} dt$$

$$= \int_{I_{mn}}^{I_{mx}} L \cdot di = (I_{mx} - I_{mn})$$

$$= L \cdot \Delta I \quad ...(ii)$$

Equating equations (*i*) and (*ii*), we have

$$400\, T_{ON} = L \cdot \Delta I$$

or, $$T_{ON} = \frac{L \cdot \Delta I}{400} = \frac{0.06 \times 10}{400} = 1.5 \text{ ms}$$

$\therefore$ Chopping frequency,

$$f = \frac{1}{T} = \frac{\alpha}{T_{ON}} = \frac{0.2}{1.5 \times 10^{-3}} = 133.33 \text{ Hz}$$

163. (c)

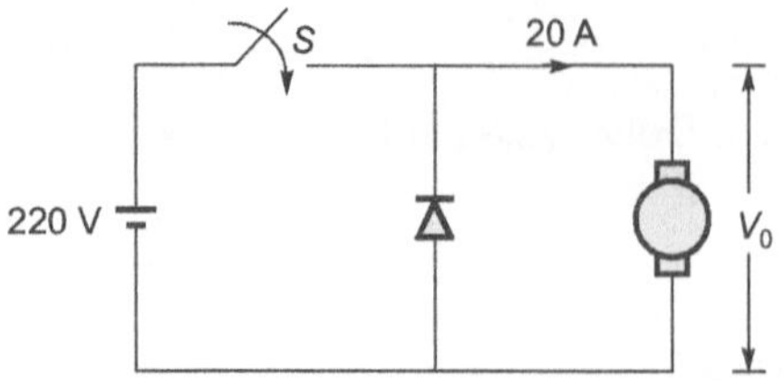

The output voltage,

$$V_0 = \alpha V_s = 0.6 \times 220 = 132 \text{ V}$$

∴ Back emf

$$E_{b_2} = 132 - 20 \times 1 = 112 \text{ V}$$

At no-load, N_1 = 1500 rpm and E_{b_1} = 220 V

Now, $E_b \propto N$

$$\therefore \quad \frac{E_{b_1}}{E_{b_2}} = \frac{N_1}{N_2}$$

or, $$N_2 = \frac{E_{b_2}}{E_{b_1}} \times N_1$$

or, $$N_2 = \frac{112}{220} \times 1500 = 763.6 \text{ rpm}$$

$$\approx 764 \text{ rpm}$$

164. (d)

As the motor is running at 1710 rpm its back emf is

$$E_{b_2} = \frac{N_2}{N_1} \times E_{b_1} = \frac{1710}{1500} \times 220 = 250.8 \text{ V}$$

Since $E_{b_2} > V$, therefore current flows in the reverse direction and hence second quadrant chopper will be used with the motor.

Now, $$V_0 = E_{b_2} - I_a R_a = 250.8 - 20 \times 1$$

$$= 230.8 \text{ volt}$$

Also, $$V_0 = (1 - \alpha) V_s$$

or, $$230.8 = (1 - \alpha) \times 250$$

or, $$1 - \alpha = 0.9232 \quad \text{or} \quad \alpha = 0.0768$$

or, $$\alpha = 7.68\% \approx 7.7\%$$

165. (a)

For a four-quadrant chopper, the average voltage in all the four-modes is given by

$$V_0 = 2 V_s (\alpha - 0.5)$$

The average current,

$$i_0 = \frac{V_0 - E_b}{R} = \frac{2V_s(\alpha - 0.5) - E_b}{R}$$

or, $$10 = \frac{2 \times 200(\alpha - 0.5) - 150}{0.1}$$

or, $$\alpha = 0.877$$

Since $\alpha > 0.5$, the motor is operating under forward motoring mode.

166. (c)

For a type-A chopper, average output voltage

$$(V_0)_{avg} = \sqrt{\alpha} V_s$$

$$\therefore \text{ Form factor} = \frac{(V_0)_{rms}}{(V_0)_{avg}} = \frac{\sqrt{\alpha} V_s}{\alpha V_s} = \frac{1}{\sqrt{\alpha}}$$

$$= \frac{1}{\sqrt{0.4}} = 1.58$$

$$\text{Ripple factor} = \sqrt{(FF)^2 - 1} = \sqrt{\frac{1}{\alpha} - 1}$$

$$= \sqrt{\frac{1-\alpha}{\alpha}} = \sqrt{\frac{1-0.4}{0.4}} = \sqrt{1.5}$$

$$= 1.22$$

167. (d)

- Type-A chopper is also called "step-down chopper" as average output voltage V_0 is always less than the input dc voltage V_s.
- Type-B chopper is also called "step-up chopper" as load voltage $V_0 = \left(E + L\frac{di}{dt}\right)$ which is more than source voltage V_s.
- Type-C chopper configuration is suitable for motoring and dregenerative braking of dc motors.
- Type-D chopper characteristic is shown below

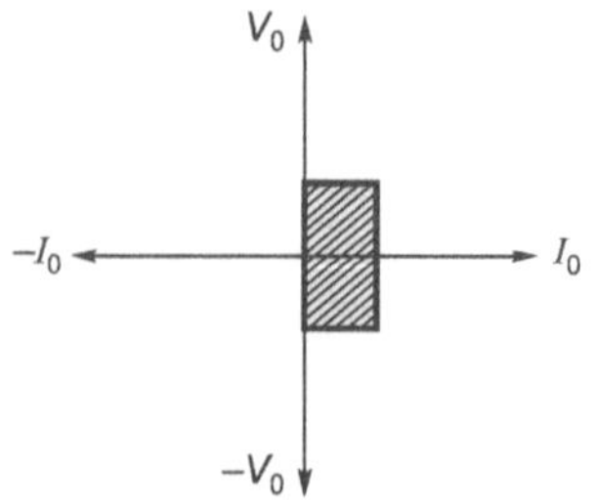

169. (b)

- DC choppers can be used in only dc drives.
- Output voltage from a chopper circuit depends only on duty cycle and is independent of load current.

Hence, statements 1 and 3 are not correct.

170. (a)

Given, $V_s = 230$ V

Load resistance,

$$R_L = 10\,\Omega, \quad \alpha = 0.4$$

When chopper is ON,

output voltage = $(V_s - 2)$ volt

When chopper is OFF,

output voltage = 0 volt

Average output voltage

$$= (V_s - 2)\,\alpha = 0.4\,(230 - 2)$$

$$= 91.2 \text{ volt}$$

Rms output voltage,

$$V_{or} = \sqrt{\alpha}\,(V_s - 2)$$

$$= \sqrt{0.4}\,(230 - 2) = 144.2 \text{ volt}$$

Power output delivered to load is

$$P_0 = \frac{V_{or}^2}{R_L} = \frac{(144.2)^2}{10}$$

$$= 2079.364 \text{ Watt}$$

Power input to chopper;

$$P_i = V_s \cdot I_0 = 230 \times \frac{91.2}{10}$$

$$= 2097.6 \text{ Watt}$$

Chopper efficiency

$$= \frac{P_0}{P_i} = \frac{2079.364}{2097.6} \times 100$$

$$= 99.13\% \approx 99\%$$

171. (d)

The maximum value of ripple current in a type-A chopper is given by

$$\Delta I_{max} = \frac{V_s}{4fL}$$

Thus, $$\Delta I_{max} \propto \frac{1}{f}$$

$$\propto \frac{1}{L}$$

172. (c)

- In a current-commutated chopper a diode is connected in antiparallel with the main thyristor so that voltage drop across the diode reverse biases the main SCR.
- A voltage-commutated chopper is generally used in high-power circuits where load fluctuation is not very large.
- In a load-commutated chopper no commutating inductor is required that is normally costly, bulky and noisy.

176. (b)

Rated load current,

$$i_0 = \frac{\alpha V_s - E_b}{R}$$

For regeneration this current should be negative.

$$\therefore \quad -10 = \left(\frac{220\alpha - 100}{0.1}\right)$$

or, $$-1 = 220\,\alpha - 100$$

or, $$\alpha = \frac{99}{220} = 0.45$$

$\therefore$ Duty cycle of chopper,

$$\alpha = 45\%$$

177. (b)

Since the input supply to a dc chopper is fixed dc voltage therefore, natural commutation is not possible in a dc chopper. Due to this reason a separate commutation circuit is required to commutate the main power SCR in a dc chopper. Reason is also a true statement but not the correct explanation of assertion.

180. (b)

Boost-regulator is a step-up chopper.

Output voltage,

$$V_0 = \frac{V_s}{1 - \alpha}$$

or, $$1 - \alpha = \frac{V_s}{V_0} = \frac{5}{15}$$

or, $$1 - \alpha = \frac{1}{3} \text{ or } \alpha = \frac{2}{3}$$

181. (c)

In low-power electronic circuits oscillators are used for converting dc power into ac power. These oscillator use transistors for converting dc voltage into sinusoidal ac voltage. Since transistor is used in active region, therefore there is substantial loss of power which decreases efficiency. In high power applications inverters are used instead of transistors and the inverters operate in saturation region or cut-off region. Thus, assertion is true but reason is false.

183. (a)

- In CSI at light load, commutation time is considerably increased, which can restrict the highest frequency. Hence, commutation is load dependent process in CSI. Thus, statement-1 is correct.
- A VSI requires an additional line commutated converter for reverse power flow while it is not so for a CSI i.e. a CSI has inherent four quadrant operation. Thus, statement-2 is correct.
- In VSI, dynamic braking is applicable during ac line failure. Thus, statement-3 is correct.
- MOSFET and transistor are more suitable for VSI not for CSI because in CSI large transient voltage is produced during commutation. Hence, statement-4 is not correct.

184. (d)

Output is a square wave with an amplitude of 100 V.
RMS value of a square wave is equal to its peak value.
So, rms value of output voltage is

$$V_0 = 100\text{ V}$$

∴ RMS value of load current is

$$I_0 = \frac{V_0}{R} = \frac{100}{10} = 10\text{ A}$$

Power delivered to load is

$$P_0 = V_0 I_0$$
$$= 100 \times 10 = 1000\text{ W}$$

185. (b)

A voltage source inverter is one in which the d.c. source has small or negligible impedance. Due to low internal impedance, the terminal voltage of a VSI remains substantially constant with variations in load. Reason is also a correct statement because due to low time constant of internal impedance any short-circuit across the terminals of a VSI causes current to rise very fast. Hence, both assertion and reason are true but reason is not the correct explanation of assertion.

186. (c)

The nth harmonic-component of output voltage is

$$V_{on} = \frac{\sqrt{2}V_{dc}}{n\pi}$$

and $\quad V_{0,\,rms} = \frac{V_{dc}}{2} = 96\text{ V or } V_{dc} = 192\text{ volt}$

∴ RMS value of fundamental component of output voltage is

$$V_{01} = \sqrt{2}V_{dc}/\pi$$

$$\therefore \quad V_{01} = \frac{\sqrt{2}\times 192}{\pi} = 0.9 \times 96\text{ volt}$$
$$= 86.40\text{ volt}$$

∴ Fundamental power,

$$P_{01} = \frac{V_{01}^2}{R} = \frac{(86.40)^2}{10} = 746.5\text{ Watt}$$

187. (a)

The lowest harmonic is third harmonic.

∴ HF for the 3rd harmonic = $HF_3 = \frac{V_{3,\,rms}}{V_{1,\,rms}}$

Now, $\quad V_{0,n} = \sum_{n=1,3,5}^{\infty} \frac{2V_{dc}}{n\pi}$

$$\therefore \quad V_{1,\,rms} = \frac{2V_{dc}}{\pi} \text{ and } V_{3,\,rms} = \frac{2V_{dc}}{3\pi}$$

$$\therefore \quad HF_3 = \frac{2V_{dc}}{3\pi} \times \frac{\pi}{2V_{dc}} = \frac{1}{3}$$

$$\therefore \quad \%\,HF_3 = \frac{1}{3}\times 100 = 33.33\%$$

188. (d)

$$V_{0,\,rms} = \frac{V_{dc}}{2} = 48\text{ V or, } V_{dc} = 96\text{ V}$$

$$V_{0,\,rms} = 48\text{ volt}$$

$$V_{1,\,rms} = \frac{\sqrt{2}V_{dc}}{\pi} = 0.45\,V_{dc}$$
$$= 0.45 \times 96 = 43.2\text{ V}$$

The rms harmonic voltage is

$$V_n = \sqrt{V_{0,\,rms}^2 - V_{1,\,rms}^2}$$
$$= \sqrt{48^2 - 43.2^2} = 20.92\text{ V}$$

$$\therefore \quad THD = \frac{\sqrt{V_{0,\,rms}^2 - V_{1,\,rms}^2}}{V_{1,\,rms}}$$
$$= \frac{20.92}{43.2} = 0.48425$$

$$\therefore \quad \%\,THD = 48.43\%$$

189. (a)

The fundamental output voltage for a modulation index of M is

$$V_{0\,(fund)} = 0.707 \times M \times V_{dc} = 0.707 \times 0.8 \times 240 = 135 \text{ V}$$

191. (c)

A CSI can be either load commutated or force commutated.

192. (d)

- By using adjustable voltage inverter, speed range of motor is limited because of motor cogging at 6 Hz and below.
- In pulse width modulated inverter, input p.f. is good at all frequencies.
- A CSI has the capability of regeneration back to the a.c. line because d.c. link polarity can be reversed.
- In VSI, the misfiring of switching devices may cause short circuit across the source and create serious problems.

193. (a)

In a CSI, load current rather than load voltage is controlled, and the inverter output voltage is dependent upon the load impedance and the output voltage waveform. The load current and hence its waveform is independent of load impedance due to which a CSI has inherent protection against short-circuit across its terminals.

195. (a)

In case of single-pulse width modulation (PWM), the width of the pulse is adjusted to reduce the harmonic. However, a single phase IGBT bridge inverter produces a square wave. This square wave contains $33\frac{1}{3}\%$ 3rd harmonic, 20% 5th harmonic and $14\frac{1}{2}\%$ 7th harmonic.

197. (b)

The rms value of amplitude of harmonic voltage of a single pulse modulated wave is given by

$$V_{ln} = \frac{4\,V_{dc}}{\sqrt{2}\,n\pi} \sin\frac{np}{2}$$

(where, p = width of pulse an V_{dc} = supply dc voltage)

If the 3rd harmonic is to be eliminated, then

$$E_{L3} = 0$$

i.e. $$\frac{4V_{dc}}{3\sqrt{2}\,\pi} \sin\frac{3p}{2} = 0$$

or, $$\sin\frac{3p}{2} = \sin\pi$$

or, $$p = \frac{2\pi}{3} = 120°$$

= Required pulse width

198. (d)

Inverter circuit	Amplitude of output voltage	Output power
1-ϕ half-bridge	$\frac{V_{dc}}{2} = V_0$	P
1-ϕ full-bridge	$V_{dc} = 2V_0$	$2P$

200. (b)

In a 1-ϕ full bridge inverter if RLC load is underdamped, then the two thyristors (namely T_1 and T_2) shown in figure will get commutated naturally and therefore no commutation circuitry will be needed. Thus, load commutation will be possible.

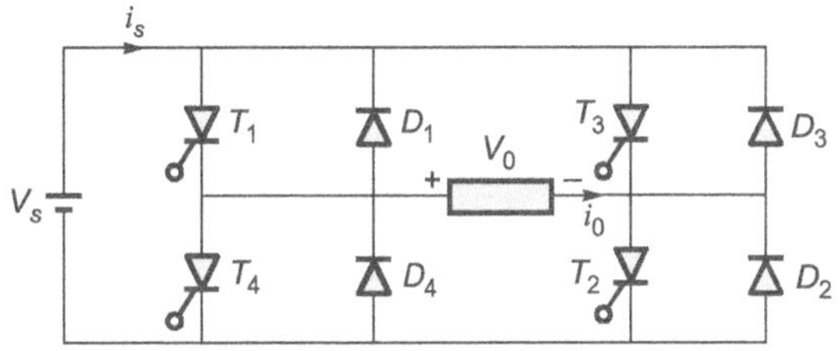

204. (d)

$$V_{or} = V_s\sqrt{\frac{2d}{\pi}}$$

(Here, $2d$ = pulse width = 120°)

∴ $$V_{or} = 2\sqrt{\frac{120}{180}} = 1.63 \text{ V}$$

206. (a)

Load commutation is possible when load is capacitive.

i.e. $X_C > X_L$

or, $\tan\phi = \dfrac{X_C - X_L}{R} = +\text{ve}$

So, $X_C > 8\ \Omega$

Thus, option (a) is correct.

207. (b)

Let the firing angle be α.

The current through the thyristor flows from α to π for the first cycle of 2 π radians.

∴ Average thyristor current is

$$(I_T)_{avg} = \frac{1}{2\pi}\int_0^{\pi}\frac{V_m}{R}\sin\omega t\, d\omega t$$

$$= \frac{V_m}{2\pi R}(1+\cos\alpha)$$

$(I_T)_{avg}$ is maximum when $\alpha = 0°$

$$\therefore\ (I_T)_{avg} = \frac{V_m}{\pi R} = \frac{230\sqrt{2}}{\pi\times 4} = 25.88\text{ A} \approx 25.9\text{ A}$$

208. (c)

Given, $V_s = 230$ volt

The secondary voltage of the transformer is

$$e_s = \sqrt{2}\,V_s\sin\omega t = V_m\sin\omega t$$

Let the firing angle be α, then the output voltage and current waveform will be as shown below:

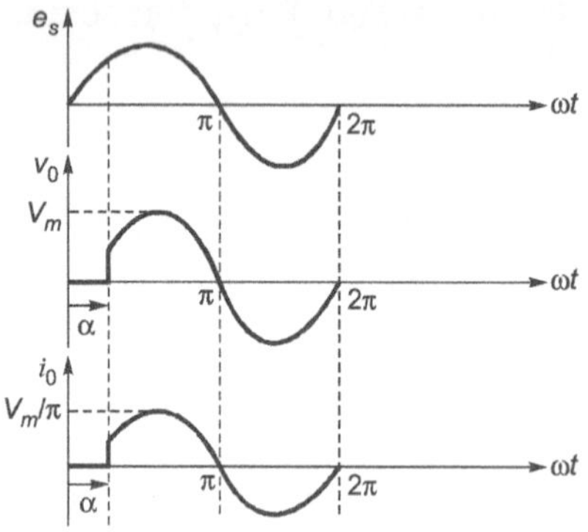

∴ RMS output voltage is

$$(V_0)_{rms} = \left\{\frac{1}{2\pi}\left[\int_{\alpha}^{\pi}2V_s^2\sin^2\omega t\, d(\omega t) + \int_{\pi}^{2\pi}2V_s^2\sin^2\omega t\, d(\omega t)\right]\right\}^{1/2}$$

$$= \left\{\frac{2V_s^2}{4\pi}\left[\int_{\alpha}^{\pi}2V_s^2(1-\cos 2\omega t)d(\omega t) + \int_{\pi}^{2\pi}(1-\cos 2\omega t)\, d(\omega t)\right]\right\}^{1/2}$$

$$= V_s\left[\frac{1}{2\pi}\left(\pi - \alpha + \frac{\sin 2\alpha}{2} + 2\pi - \pi\right)\right]^{1/2}$$

$$\left(\because \alpha = \frac{\pi}{2}\right)$$

$$\therefore\quad (V_0)_{rms} = 230\sqrt{\frac{3}{4}} = 115\sqrt{3}\text{ volt}$$

209. (a)

For controlling the load, the minimum firing angle α = load-phase angle ϕ.

$$\therefore\ \alpha = \phi = \tan^{-1}\left[\frac{X_L}{R}\right] = \tan^{-1}\left[\frac{3}{4}\right] = 36.87°$$

Control range of firing angle is $36.87° \le \alpha \le 180°$.

210. (d)

The a.c. regulators can be used for closed loop control because they require low control power and have faster response. Hence, assertion is a false statement.

211. (d)

Statement-2 is not correct because since the a.c. regulators are phase-controlled controllers, thyristors and Triacs are line commutated and as such no complex commutation circuitry is required in these controllers.

214. (c)

The power delivered to the load resistance R is

$$P = \frac{V_s^2}{\pi R}\left[(\pi - \alpha) + \frac{1}{2}\sin 2\alpha\right]$$

Maximum power is delivered when α = 0 and has a value of

$$P_{max} = \frac{V_s^2}{R}$$

215. (b)

$$\text{Voltage regulation} = \text{Line regulation} = \frac{V_1 - V_2}{V_2}$$

(V_1 = Input voltage and V_2 = Output voltage at full load).

216. (c)

The circuit diagram for a single-phase a.c. regulator is shown in figure below. Here two thyristor are connected in anti-parallel.

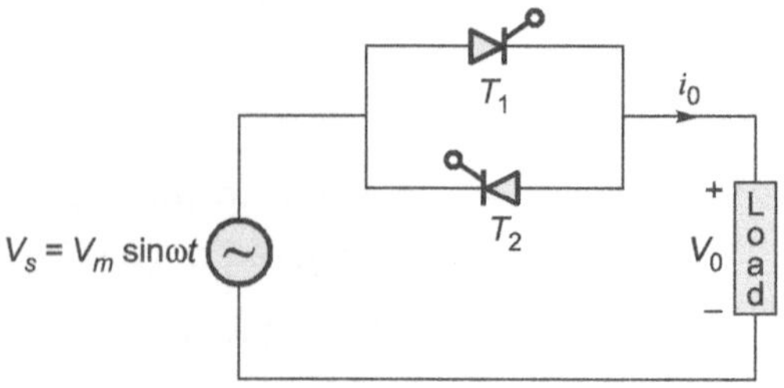

217. (a)

For integral cycle control, the input power factor of a single-phase voltage controller is

$$\cos\phi = \sqrt{K} = \sqrt{\frac{n}{m+n}}$$

(where, m = number of on cycle and n = number of off cycle).

218. (b)

The rms value of thyristor current is

$$(I_T)_{rms} = \frac{V_m}{2R\sqrt{\pi}}\left[(\pi-\alpha)+\frac{1}{2}\sin 2\alpha\right]^{1/2}$$

Given, $\alpha = \frac{\pi}{2}$, $R = 5\ \Omega$

and $V_s = 100$ volt i.e. $V_m = 100\sqrt{2}$ volt

$$\therefore\quad (I_T)_{rms} = \frac{100\sqrt{2}}{2\times 5\sqrt{\pi}}\left[(\pi-\pi/2)+\frac{1}{2}\sin\pi\right]^{1/2}$$

$$= \frac{100\sqrt{2}}{10\sqrt{\pi}}\left[\frac{\pi}{2}\right]^{1/2} = \frac{10\sqrt{2}}{\sqrt{\pi}}\times\frac{\sqrt{\pi}}{\sqrt{2}} = 10\text{ A}$$

220. (d)

The ac output power can be controlled only for $\alpha > \phi$, $\beta > \pi$ and $(\beta - \alpha) < \pi$.

221. (c)

$$(I_T)_{avg} = \frac{V_m}{2\pi R}(1+\cos\alpha)$$

$$\therefore\quad (I_T)_{avg}\Big|_{max} = \frac{V_m}{\pi R} \qquad (\text{For } \alpha = 0°)$$

Also, $$(I_T)_{rms} = \frac{V_m}{2R\sqrt{\pi}}\left[(\pi-\alpha)+\frac{1}{2}\sin 2\alpha\right]^{1/2}$$

$$\therefore\quad (I_T)_{rms}\Big|_{max} = \frac{V_m}{2R} \qquad (\text{For } \alpha = 0°)$$

$$\therefore\quad \frac{(I_T)_{rms}\big|_{max}}{(I_T)_{avg}\big|_{max}} = \frac{V_m}{2R}\times\frac{\pi R}{V_m} = \frac{\pi}{2}$$

224. (b)

$$\tan^{-1}\left(\frac{\omega L}{R}\right) < \alpha < 180°$$

or, $$\tan^{-1}\left(\frac{2\pi\times 50\times 6.36\times 10^{-3}}{2}\right) < \alpha < 180°$$

or, $\tan^{-1}(1) < \alpha < 180°$ or $45° < \alpha < 180°$

226. (d)

A step-down cycloconverter does not require forced commutation. It requires phase-controlled converters which need only line, or natural commutation which is provided by ac supply. On the other hand a step-up cycloconverter requires forced commutation.

227. (c)

A 3-phase to 3-phase cycloconverter require three sets of three-phase half-wave circuit employing a total of 18 thyristors. On the other hand a single-phase to single-phase cycloconverter of mid-point type require four thyristors.

229. (b)

- The cycloconverter can handle load of any power factor and allows power flow in both the directions.
- The output voltage wave shape of a cycloconverter inevitably contains harmonic distortion components in addition to the required sinusoidal component. The distortion of the output voltage increases if the ratio of the output and input frequency increases.
- The cycloconverters are normally used to provide either a variable input frequency from a fixed input frequency or a fixed frequency from a variable input frequency.
- The distortion in the output of a cycloconverter is very low at low output frequencies.

Thus, statements 2 and 4 are only correct.

230. (c)

The amount of ripple content in the output of a 3-ϕ cycloconverter can be reduced by increasing the number of pulses used. Thus, higher the number of pulses used less ripple content in the output of the three-phase cycloconverter. Hence, reason is a false statement.

231. (a)

For an m-pulse cycloconverter, the rms output voltage per phase is

$$V_{or} = V_{ph}\left(\frac{m}{\pi}\right)\sin\left(\frac{\pi}{m}\right)$$

Here, $m = 3$ and $V_{or} = 190$ V

$\therefore$ $$190 = V_{ph}\left(\frac{3}{\pi}\right)\sin\left(\frac{\pi}{3}\right)$$

or, $$V_{ph} = \frac{190\times\pi}{3\sin 60^\circ} = \frac{190\times 2\pi}{3\sqrt{3}}$$
$$= 229.74\text{ V} \approx 230\text{ V}$$

232. (c)

Compared to a cycloconverter, a d.c. link converter:

- is extremely suitable for high frequencies.
- requires forced commutation for the inverter even though the rectifier operates on phase control principle.
- has high input p.f. if diode rectifier is used.

233. (d)

- Statement 1 is correct.
- When cycloconverter operates with R-L load, both rectification and inversion modes are used. Thus, statement-2 is not correct.
- Since the operation of cycloconverter is based on the line commutation, therefore the circuit is simple and cheap. Hence, statement-3 is not correct.
- Statement-4 is correct.

234. (b)

Cycloconverter is a frequency changer, which converts ac voltage at supply frequency into an ac voltage of desired voltage and frequency without an intermediate dc stage.

236. (a)

For an m-pulse 3-ϕ to 1-ϕ cycloconverter, rms output voltage is

$$V_{or} = V_{ph}\left(\frac{m}{\pi}\right)\sin\left(\frac{\pi}{m}\right)$$

Given, $m = 6$ and $V_{ph} = 100$ V

$\therefore$ $$V_{or} = 100\times\left(\frac{6}{\pi}\right)\sin\left(\frac{\pi}{6}\right)$$
$$= 100\times\frac{6}{\pi}\times\frac{1}{2} = \frac{300}{\pi}\text{ volt}$$

240. (b)

Given, $\alpha = 30^\circ$, $R_a = 4\ \Omega$, $V_m = 330$ V
and $E_a = 80$ V

For a single-phase semiconverter feeding a separately-excited motor, average output voltage is

$$V_0 = \frac{V_m}{\pi}(1+\cos\alpha) \quad ...(i)$$

Also, $$V_0 = V_t = E_a + I_aR_a \quad ...(ii)$$

From equation (*i*) and (*ii*), we have:

$$\frac{V_m}{\pi}(1+\cos\alpha) = E_a + I_aR_a$$

or, $$\frac{330}{\pi}(1+\cos 30^\circ) = 80 + 4I_a$$

or, $$196.01 = 80 + 4I_a$$

or, $$I_a = \frac{196.01-80}{4} = 29\text{ A}$$

241. (d)

Given, $\alpha = 0.6$, $I_a = 300$ A,
$K = 4\times 10^{-3}$ N-m/A^2, $r_a = 0.04\ \Omega$
$r_{se} = 0.06\ \Omega$ and $V_s = 600$ V

For a dc series motor,

$$\alpha V_s = E_a + I_aR$$
$$= KI_a\omega_m + I_a(r_a + r_{se})$$

or, $$0.6\times 600 = 4\times 10^{-3}\times 300\times\omega_m + 300\,(0.04+0.06)$$

or, $$360 = 1.2\,\omega_m + 30$$

or, $$\omega_m = \frac{330}{1.2} = 275\text{ rad/sec}$$

242. (b)

Stator-voltage control method for the speed control of three-phase induction motors is used for low-power drives only because this method introduces pronounced harmonic contents and input supply power factor for the voltage controller is quite low, which is usually being used in this method of speed control.

Thus, both assertion and reason are true but reason is not the correct explanation of assertion.

243. (c)

In both volts/hertz control for controlling the speed of IM and armature-voltage control method for controlling the speed of dc motor, the range of speed control lies from standstill upto rated speed. Thus, both these methods of speed control are similar to each other. Hence, assertion is true but reason is false.

245. (d)

- In static Kramer drive, speed of SRIM can be controlled below synchronous speed only. Hence, statement-1 is not correct.
- Static scherbius drive scheme offers speed control both below and above synchronous speed. Hence, statement-2 is not correct.
- Statement-3 is correct.
- In chopper method of speed control for SRIM, the slip power is dissipated in the external resistance and it leads to poor efficiency of the drive. Thus, statement-4 is not correct.

246. (b)

Given, $V_{dc} = V_t = 440$ V, $f = 500$ Hz,

$R = R_a + R_{se} = 0.1\ \Omega$, $E = 200$ V

$I_a = 200$ A

Output voltage of the chopper is

$$V_0 = \alpha V_{dc} \quad ...(i)$$

Also, $$V_0 = I_a R + E \quad ...(ii)$$

From equations (*i*) and (*ii*), we have:

$$\alpha V_{dc} = I_a R + E$$

or, $$440\,\alpha = 200 \times 0.1 + 200 = 220$$

or, $$\alpha = \frac{220}{440} = 0.5$$

Now, $$T = \frac{1}{f} = \frac{1}{500} = 2 \text{ ms}$$

$\therefore$ $$T_{on} = \alpha T = 0.5 \times 2 \text{ ms} = 1 \text{ ms}$$

$\therefore$ On period of chopper

= Required pulse width = 1 ms

250. (a)

Given, slip $s = 0.04$, $T_{ON} = 2$ ms

We know that torque in a slip ring induction motor, for constant slip is $T \propto R$

(R = Rotor resistance referred to stator)

Let R_1 and R_2 be the values of rotor resistance before and after introduction of control. Then,

$$T_2 = 150\%\ T_1 \text{ or } \frac{T_2}{T_1} = 1.5$$

or, $$\frac{R_2}{R_1} = 1.5 \text{ or } R_2 = 1.5\,R_1$$

Now, additional resistance introduced during OFF period is equal to rotor resistance.

$\therefore$ $$R_2 = \frac{R_1 T_{ON} + (R_1 + R_1)T_{OFF}}{T_{ON} + T_{OFF}}$$

$$= \frac{R_1 T_{ON} + 2R_1 T_{OFF}}{T_{ON} + T_{OFF}}$$

or, $$1.5R_1 = \frac{R_1(T_{ON} + 2T_{OFF})}{T_{ON} + T_{OFF}}$$

or, $1.5\,(T_{ON} + T_{OFF}) = T_{ON} + 2\,T_{OFF}$

or, $0.5\,T_{ON} = 0.5\,T_{OFF}$ or $T_{ON} = T_{OFF}$

Given, $T_{ON} = 2$ ms

$\therefore$ $T_{ON} = T_{OFF} = 2$ ms

$\therefore$ $T = T_{ON} + T_{OFF} = 2 + 2 = 4$ ms

Hence, chopper frequency

$$= \frac{1}{4 \times 10^{-3}} = 250 \text{ Hz}$$

253. (b)

Power converters for AC drives or AC motors control are more complex compared to DC motors. Hence, option (b) is not correct.

255. (c)

For decreasing the speed of the motor regeneratively, the power must flow from motor to the ac source and this is feasible only if motor counter emf is reversed because then $E_a I_a$ would be negative. Since the direction of current can't be reversed as SCRs are unidirectional devices, therefore for regenerative braking, the polarity of E_a must be reversed which is possible by reversing the direction of motor field current by increasing delay angle (or firing angle) by more than 90°.

257. (c)

The circuit with the waveform is shown below:

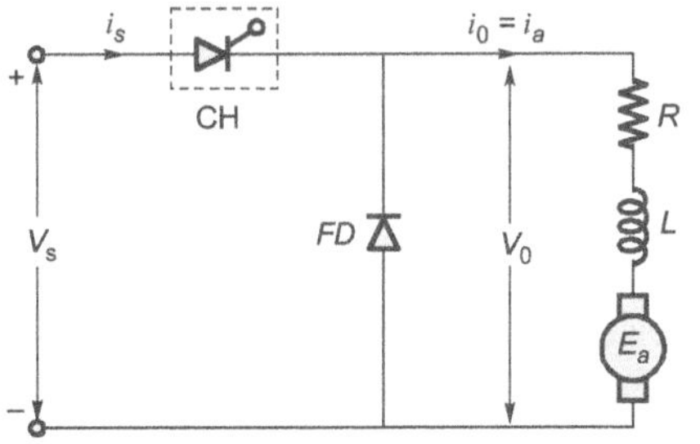

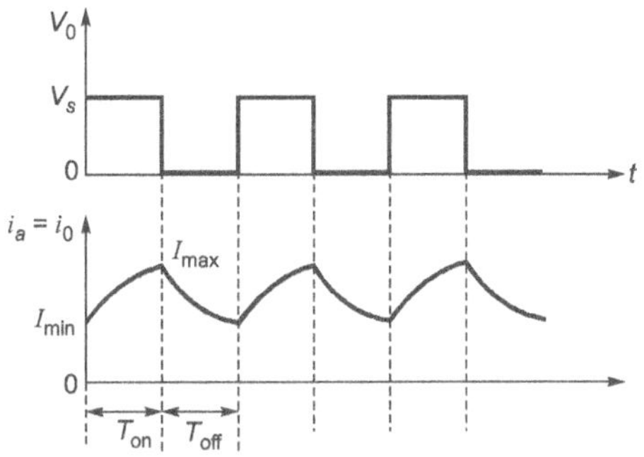

As shown in the above waveform, the motor armature current will rise during chopper on period and fall during off period. Hence, the armature current will be exponential rise and decay.

258. (b)

In constant *v/f* method of speed control of 3-ϕ IM, the starting torque is inversely proportional to supply frequency even if air-gap flux is kept constant. Thus, when *v/f* is kept constant there is maximum use of magnetic circuit.

259. (d)

Phase controlled rectifiers are used to convert fixed ac supply voltage into variable dc output voltage.

260. (c)

By using thyristors, the output voltage can be controlled to a greater extent by varying the firing angle and hence wide range of speed control of a dc motor is possible. Diode is a unilateral semiconductor device (uncontrolled device) and hence can't be used for speed control of a dc motor which requires variable output voltage for speed control.

265. (a)

In a SMPS, by operating the on/off switch very rapidly, ac ripple frequency rises which can be easily filtered by *L* and *C* filter circuits and hence noise is reduced to a great extent. Also, SMPS is less sensitive to input voltage variation and hence regulation is excellent.

268. (c)

There are no resistors present in SMPS internal circuit. There are inductor and capacitors present which are of physically small size.

269. (c)

MOSFETs are widely used as switching elements in SMPS. However, BJT can also be used.

270. (c)

MOSFETs have the lowest switching losses. So they are the most suitable devices for the SMPS.

272. (c)

The dc transmission voltage

$$= 200 + 200 = 400 \text{ kV}$$

Direct current in the transmission lines is

$$I_d = \frac{1000 \times 10^6}{400 \times 10^3} = 2500 \text{ A}$$

In a 3-phase full converter each thyristor conducts for 120° for a periodicity of 360°.

∴ Rms current rating of thyristor

$$= I_d\sqrt{\frac{120}{360}} = \frac{2500}{\sqrt{3}} \text{A}$$

273. (d)

Reactive power compensation or VAr compensation of any kind is not required in HVDC line.

274. (b)

Corona loss ∝ $(f + 25)$.

In HVDC system, $f = 0$ Hz, therefore corona loss is reduced compared to EHV-AC system, Hence, option (b) is not correct.

278. (c)

- Unlike EHV-AC lines, tappings cannot be provided in HVDC system. Suitable and effective DC circuit breakers are not available, hence, it is not possible to isolate the faulty line if fault occurs at the point. Hence, statement-1 is correct.
- HVDC line can be operated for constant current regulation or constant voltage regulation. It is easily accomplished with the help of appropriate control of thyristors of the converters. Hence, statement-2 is not correct.
- There is no limit of the length of HVDC cables. Hence, statement-3 is correct.
- Unlike EHV-AC which require shunt capacitors for maintaining the voltage stability there is no requirement of shunt capacitors in HVDC systems. Hence, stateemnt-4 is not correct.

279. (c)

Reason is a false statement because HVDC lines are loaded upto thermal limit.

■■■■

UNIT 12

Communication Systems

1. Introduction to Communication Systems

Q.1 The very high frequency (VHF) range extends from
(a) 3-30 MHz (b) 30-300 MHz
(c) 300-3000 MHz (d) 3000-30,000 MHz

Q.2 Match **List-I** with **List-II** and select the correct answer using the codes given below the lists:

List-I
A. Channel
B. Carrier
C. Modulation
D. Distortion

List-II
1. The process by which some characteristics of a carrier is varied by an information signal.
2. Any undesirable change in an information signal.
3. A signal that can be modulated by an information signal.
4. A path for transmission of signal.

Codes:

	A	B	C	D
(a)	4	3	2	1
(b)	2	1	4	3
(c)	4	1	3	2
(d)	4	3	1	2

Q.3 The function of the input transducer in a communication system is
(a) to transmit the message signal.
(b) to modulate the message signal.
(c) to convert message sound signal into electrical signal.
(d) none of the above

Q.4 Communication is the process of
(a) keeping in touch
(b) broadcasting
(c) exchanging information
(d) entertainment by electronics

Q.5 Which of the following is **not** a major communication medium?
(a) Free space (b) Water
(c) Wires (d) Fiber-optic cable

Q.6 The process of transmitting two or more information signals simultaneously over the same channel is called
(a) multiplexing (b) telemetry
(c) detection (d) modulation

Q.7 Radio signals are made up of
(a) voltage and currents
(b) electric and magnetic fields
(c) electrons and protons
(d) noise and data

Q.8 Two key barriers to human communication are
(a) distance and ignorance
(b) language and cost
(c) distance and language
(d) distance and cost

Q.9 The communication medium causes the signal to be
(a) amplified (b) modulated
(c) attenuated (d) interfered with

Q.10 Match **List-I (Frequencies spectrum)** with **List-II (Frequency range)** and select the correct answer using the codes given below the lists:

List-I
A. Extremely low frequency
B. Low frequencies
C. Voice frequencies
D. Very low frequencies

List-II
1. 30-300 kHz
2. 300-3000 Hz
3. 3-30 kHz
4. 30-300 Hz

Codes:

	A	B	C	D
(a)	4	1	2	3
(b)	4	3	2	1
(c)	3	4	1	2
(d)	2	3	1	4

Q.11 **Assertion (A):** For the transmission of baseband signals, modulation is required.
Reason (R): A baseband signal cannot be transmitted without modulation.
(a) Both A and R are true and R is the correct explanation of A.
(b) Both A and R are true but R is **not** the correct explanation of A.
(c) A is true but R is false.
(d) A is false but R is true.

Q.12 In a communication system, noise is most likely to affect the signal
(a) at the transmitter
(b) in the channel
(c) in the information source
(d) at the desination

2. Fourier Analysis of Signals, Energy and Power Signals

Q.13 The signals $x_1(t)$ and $x_2(t)$ are given by

$$x_1(t) = \sin 15\pi t \text{ and } x_2(t) = \sin\sqrt{2}\,\pi t$$

If the periods of $x_1(t)$ and $x_2(t)$ be T_1 and T_2 respectively, then T_1/T_2 is equal to
(a) $15\sqrt{2}$
(b) $\frac{15}{\sqrt{2}}$
(c) $\frac{\sqrt{2}}{15}$
(d) None of these

Q.14 **Assertion (A):** A frequency domain representation of a signal shows the frequency content and indicates the shape of the waveform or the combined amplitude of all the input components at any specific time.
Reason (R): In frequency domain, a signal is represented by its frequency spectrum.
(a) Both A and R are true and R is the correct explanation of A.
(b) Both A and R are true but R is **not** the correct explanation of A.
(c) A is true but R is false.
(d) A is false but R is true.

Q.15 Match **List-I (Operations on $x(t)$)** with **List-II ($X(\omega)$/Fourier transform)** and select the correct answer using the codes given below the lists:

List-I	List-II
A. Time shift	1. $\frac{1}{j\omega}X(\omega) + \pi X(0)\,\delta(\omega)$
B. Time differentiation	2. $e^{-j\omega t_0}X(\omega)$
C. Time integration	3. $X(\omega - \omega_0)$
D. Frequency shift	4. $(j\omega)^n X(\omega)$

Codes:

	A	B	C	D
(a)	2	1	4	3
(b)	2	4	1	3
(c)	3	4	1	2
(d)	3	1	4	2

Q.16 Fourier series analysis is a tool used to
1. analyse any periodic and non-periodic signal.
2. find how many frequency components are present in the signal.
3. find the relative phase difference between various frequency components.
4. represents a waveform in the form of sum of infinite number of exponential terms.
5. find the amplitude of various terms in the series.

Which of the statements given above are correct?
(a) 2, 3 and 5
(b) 1, 3 and 4
(c) 1, 2 and 3
(d) 2, 3 and 5

Q.17 The Fourier transform of the decaying exponential pulse shown in figure is

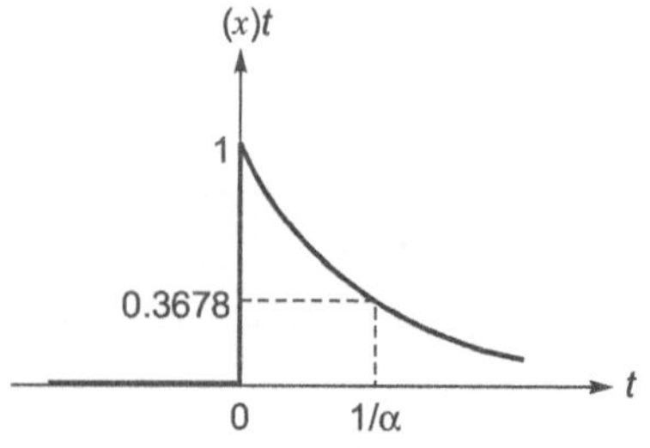

(a) $\frac{1}{(\alpha + j2\pi f)}$
(b) $\frac{1}{(\alpha - j\pi f)}$
(c) $\frac{1}{(\alpha - j2\pi f)}$
(d) $\frac{1}{(\alpha + j\pi f)}$

Q.18 The energy of a causal exponential pulse shown in figure is

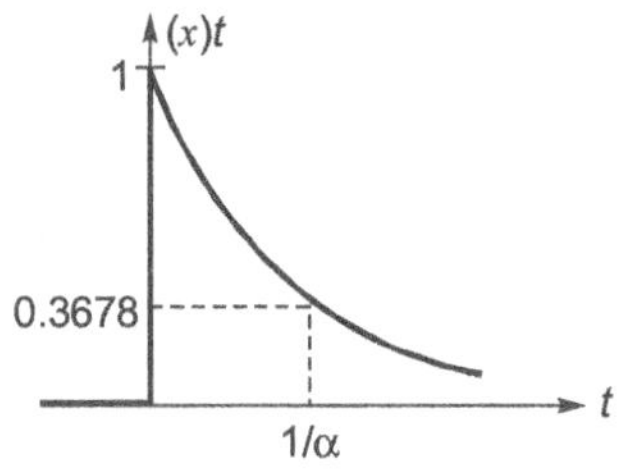

(a) $\frac{1}{2\alpha}$ (b) $-\alpha$

(c) $-\frac{1}{\alpha}$ (d) 2α

Q.19 The power spectral density of a power signal $x(t)$ is shown in figure below.

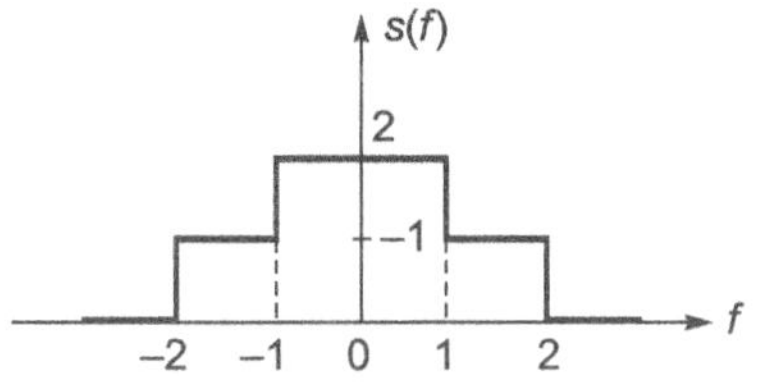

The average power of the signal is

(a) 4 (b) 1

(c) 6 (d) 3

Q.20 **Assertion (A):** The Parseval's theorem implies superposition of the average powers.

Reason (R): The interpretation of the Parseval's theorem is that the total average power of the signal $x(t)$ can be found by squaring and adding the heights of the amplitude lines in the spectrum of the periodic signal $x(t)$.

(a) Both A and R are true and R is the correct explanation of A.

(b) Both A and R are true but R is **not** the correct explanation of A.

(c) A is true but R is false.

(d) A is false but R is true.

Q.21 Which of the following option is **not** correct for obtaining the Fourier transform of a signal $x(t)$?

(a) $x(t)$ should be single valued in any finite time interval T.

(b) $x(t)$ should have infinite number of discontinuities in any finite interval T.

(c) $x(t)$ should have finite number of maxima and minima in any time interval of time T.

(d) $x(t)$ should be absolutely integrable function.

Q.22 The Fourier transform of a unit step function is

(a) $\frac{-j}{\pi f}$ (b) $\frac{1}{j2\pi f}$

(c) 1 (d) $\frac{1}{j\pi f}$

Q.23 Consider the following statements:

1. With the increase in signaling rate, the width of each pulse is reduced.
2. A signal can be band limited or time limited or both band limited and time limited simultaneously.
3. Differentiating the signal in time domain is equivalent to multiplying its Fourier transform by $(j2\pi f)$.
4. Compression in time domain results in expansion of frequency spectrum and vice-versa.

Which of the above statements is/are **not** correct?

(a) 1 and 3 (b) 2 and 4

(c) 3 only (d) 2 only

Q.24 Match **List-I** with **List-II** and select the correct answer using the codes given below the lists:

List-I

A. Transmitter
B. Fourier series
C. Spectrum analyzer
D. Compression

List-II

1. For displaying signals in frequency domain.
2. Provides more gain for low-level than for higher-level signals.
3. Converts an information signal into a form suitable for propagation along a channel.
4. Represents periodic functions as a series of sinusoids.

Codes:

	A	B	C	D
(a)	1	2	4	3
(b)	3	4	1	2
(c)	3	1	4	2
(d)	1	4	2	3

Q.25 Addition of two periodic signals will always be

(a) periodic

(b) aperiodic
(c) may or may **not** be periodic
(d) insufficient data

Q.26 The trigonometric Fourier series of a periodic time function can have only
(a) cosine terms
(b) sine terms
(c) cosine and sine terms
(d) d.c. and cosine terms

Q.27 Which of the following **cannot** be the Fourier series expansion of a periodic signal?
(a) $x(t) = 2\cos t + 3\cos 3t$
(b) $x(t) = 2\cos \pi t + 7\cos t$
(c) $x(t) = \cos t + 0.5$
(d) $x(t) = 2\cos 1.5\pi t + \sin 3.5\pi t$

Q.28 The power of the signal,

$$s(t) = 8\cos\left(20\pi t - \frac{\pi}{2}\right) + 4\sin(15\pi t) \text{ is}$$

(a) 40 (b) 41
(c) 42 (d) 82

3. Theory of Random Variable and Noise

Q.29 A certain random variable has the CDF (cumulative distribution function) given by:

$$F_x(x) = \begin{cases} 0 & \text{for} \quad x \le 0 \\ kx^2 & \text{for} \quad 0 < x \le 10 \\ 100K & \text{for} \quad x > 10 \end{cases}$$

Which of the following options is/are **not** correct?
(a) $P(x \le 5) = 0.25$
(b) $K = 0.01$
(c) $P(5 < x \le 7) = 0.30$
(d) both (a) and (b)

Q.30 The probability density function of a random variable is given by

$$f_x(x) = \begin{cases} K, & a < x \le b \\ 0, & \text{otherwise} \end{cases}$$

The value of constant K is

(a) $(b - a)$ (b) $\frac{1}{(b-a)}$

(c) $(a - b)$ (d) $\frac{1}{(a-b)}$

Q.31 Over a certain binary communication channel (BCC), symbol '0' is transmitted with a probability of 0.4 and symbol '1' is transmitted with a probability of 0.6. The probability of error when '1' is transmitted is 10^{-4} and that when a '0' is transmitted is 10^{-6}.
What is the probability of error in the channel?
(a) 0.604×10^{-4} (b) 0.302×10^{-4}
(c) 0.106×10^{-4} (d) 0.204×10^{-4}

Q.32 The joint PDF of the random variable x and y is given by,

$$f_{xy}(x, y) = C \cdot e^{-(ax + by)} u(x) \cdot u(y)$$

where, a and b are constants. The value of C is

(a) $\frac{a}{b}$ (b) 1

(c) ab (d) $\frac{b}{a}$

Q.33 Consider the following statements associated with conditional probability density function (PDF):
1. The conditional PDF is basically a probability density function (PDF).
2. The conditional PDF is a non-negative function.
3. The area under a conditional PDF is always equal to 1.
4. The conditional density functions reduce to the marginal density functions.

Which of the statements given above are correct?
(a) 1, 3 and 4 (b) 1, 2, and 3
(c) 2, 3, and 4 (d) 1, 2, 3 and 4

Q.34 **Assertion (A):** If the random process $x(t)$ changes very rapidly with time, the auto-correlation function will decrease rapidly to zero.
Reason (R): The autocorrelation function $R_x(\tau)$ is a measure of interdependence of two random variables obtained by observing the random process $x(t)$ at time τ seconds apart.
(a) Both A and R are true and R is the correct explanation of A.
(b) Both A and R are true but R is **not** the correct explanation of A.
(c) A is true but R is false.
(d) A is false but R is true.

Q.35 Match **List-I** with **List-II** and select the correct answer using the codes given below the lists:

List-I

A. Gaussian distribution

B. Probability density function

C. Spectral densities

D. Autocorrelation function

List-II

1. The derivative of cumulative distribution function w.r.t,. some dummy variable.
2. A measure of similarity between a signal or process and its replica by a variable amount.
3. Used for continuous random variables.
4. used to represent random process in frequency-domain.

Codes:

	A	B	C	D
(a)	3	4	1	2
(b)	4	3	1	2
(c)	2	1	4	3
(d)	3	1	4	2

Q.36 Thermal noise is also known as

(a) shot noise (b) johnson noise
(c) flicker noise (d) none of these

Q.37 Consider the following statements associated with different types of noise occurring in nature:

1. The flicker noise occurs at high operating frequencies (few MHz).
2. Shot noise is produced due to random variations in the arrival of electrons at the output electrode of an amplifying device.
3. The thermal noise is proportional to temperature and the value of the resistance and is independent of bandwidth.
4. Noise can also be represented by fictitious temperature called noise temperature.

Which of the statements given above is/are **not** correct?

(a) 1 and 4 (b) 2 only
(c) 1 and 3 (d) 3 only

Q.38 A radio receiver with 10 kHz bandwidth has a noise figure of 30 dB. The noise power at input is (take K = 1.38×10^{-23} J/Kelvin and assume room temperature of 300 K)

(a) 2.22×10^{-14} W (b) 5.65×10^{-12} W
(c) 1.68×10^{-12} W (d) 4.14×10^{-14} W

Q.39 **Assertion (A):** A white noise is sometimes called gaussian noise.

Reason (R): A white noise is analogous with white light which is a superposition of all visible spectral components.

(a) Both A and R are true and R is the correct explanation of A.
(b) Both A and R are true but R is **not** the correct explanation of A.
(c) A is true but R is false.
(d) A is false but R is true.

Q.40 Match **List-I** with **List-II** and select the correct answer using the codes given below the lists:

List-I

A. Shot noise

B. Partition noise

C. Flicker noise

D. Transit time noise

List-II

1. Occurs at very high frequencies
2. Popularly called 1/f noise
3. Higher in a transistor than that in a diode
4. Appears as a randomly varying noise current superimposed on the output

Codes:

	A	B	C	D
(a)	2	1	3	4
(b)	4	3	1	2
(c)	4	3	2	1
(d)	4	1	3	2

Q.41 The signal power and noise power measured at the input of an amplifier are 150 μW and 1.5 μW respectively. If the signal power at the output is 1.5 W and noise power is 40 mW, the amplifier noise factor would be equal to

(a) 2.66 (b) 1.39
(c) 3.47 (d) 4.40

Q.42 The signal to noise ratio at the input of an amplifier is 40 dB. If the noise figure of an amplifier is 20 dB, the signal to noise ratio (in dB) at the amplifier output is

(a) 10 dB (b) 60 dB
(c) 20 dB (d) none of these

Q.43 An amplifier has a noise figure of 6 dB. It's equivalent noise temperature is (Assume T_0 = 300 °K)

(a) 900 °K (b) 200 °K
(c) 600 °K (d) None of these

Q.44 In a radio receiver an RF amplifier and a mixer are connected in cascade as shown in figure below:

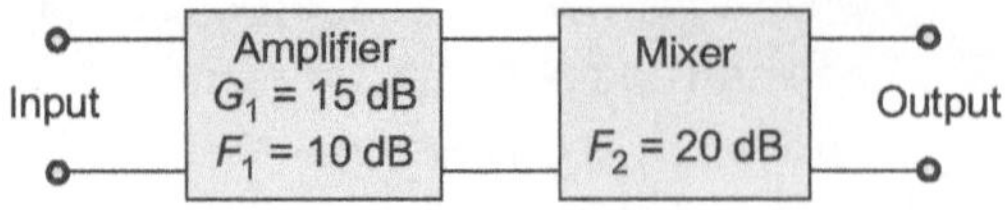

The amplifier has a noise figure of 10 dB and the power gain of 15 dB. The noise figure of the mixer stage is 20 dB. The overall noise figure referred to the input is (Use $\log_{10} 15 = 31.62$)

(a) 11.11 (b) 22.22
(c) 5.50 (d) 13.13

Q.45 A radio receiver with equivalent noise bandwidth of 10 kHz has a noise figure of 20 dB. If input SNR to receiver is 40 dB, the output SNR is

(a) 100 dB (b) 20 dB
(c) 60 dB (d) 40 dB

Q.46 The ideal value of noise figure in dB is

(a) infinite (b) zero
(c) between –1 to +1 (d) unity

Q.47 The figure shown below (power spectral density vs frequency curve) represents one of the characteristics of

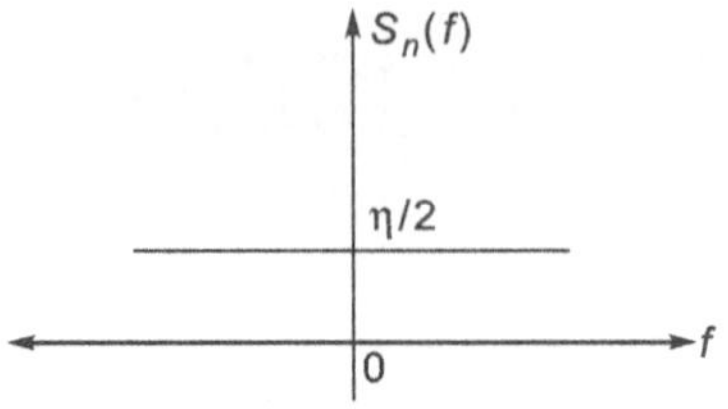

(a) Shot noise (b) Johnson noise
(c) White noise (d) Flicker noise

Q.48 The power spectral density of a real valued random process is ______ function of frequency.

(a) an even
(b) an odd
(c) neither an even nor an odd
(d) none of the above

Q.49 If the value of bandwidth of the noise spectrum is doubled and the value of resistor creating thermal noise is halved, then the rms noise voltage

(a) increases by 4 times
(b) remains unchanged
(c) reduces by half
(d) increases by two times

Q.50 The value of resistor creating thermal noise is doubled. The noise power generated is therefore

(a) halved (b) quadrupled
(c) doubled (d) unchanged

Q.51 Which one of the following is not a useful quantity for comparing the noise performance of receivers?

(a) Input noise voltage
(b) Equivalent noise resistance
(c) Noise temperature
(d) Noise figure

Q.52 Indicate the noise whose source is in a category different from that of the other three

(a) Solar noise
(b) Cosmic noise
(c) Atmospheric noise
(d) Galactic noise

Q.53 Consider a white noise with mean 10 V and a power spectrum of its non-dc component equal to 100 V^2/Hz. The power spectrum of non-zero mean white noise is given by

(a) $10 + 100\,\delta(\omega)$ (b) $100 + 10\,\delta(\omega)$
(c) $100 + 200\,\pi\delta(\omega)$ (d) $100 + 20\,\pi\delta(\omega)$

Q.54 If $x(t)$ is random process with mean value 2 and autocorrelation function $R_x(\tau) = 4\,[2e^{-0.3|t|} + \sin t]$. Let '$x$' be the gaussian random variable obtained by sampling the process at $t = t_i$ and

$$Q(K) = \int_K^{\infty} \frac{1}{\sqrt{2\pi}} e^{-t^2/2}\, dt$$. The probability that

$[x \le 1]$ is

(a) $1 - Q(0.5)$ (b) $Q(0.5)$
(c) $Q\left(\frac{1}{\sqrt{2\pi}}\right)$ (d) $1 - Q\left(\frac{1}{\sqrt{2\pi}}\right)$

Q.55 A random signal $x(t)$ has an autocorrelation function of

$$R_x(\tau) = \begin{cases} 5 - 3|\tau|; & |\tau| \le 1 \\ 2 & ; |\tau| > 1 \end{cases}$$

How much the signal average power is embedded in the d.c. component?

(a) 5 W (b) 3 W
(c) 8 W (d) 2 W

Q.56 Consider the system given below:

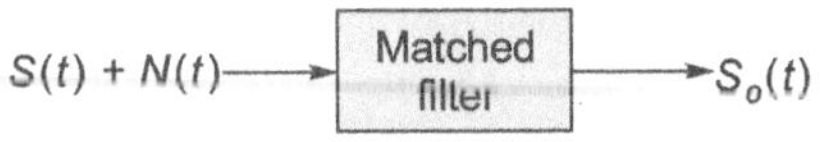

Here $s(t) = A$; $0 < t \leq T$ and $N(t)$ is white noise with PSD $= \frac{N_0}{2}$, the output SNR is equal to

(a) $\frac{A^2T}{N_0}$ (b) $\frac{A^2T}{2T_0}$

(c) $\frac{2A^2T}{N_0}$ (d) $\frac{A^2T}{4N_0}$

Q.57 A 1 mW video signal having a bandwidth of 1 GHz is transmitted to a receiver through a cable that has 20 dB loss. If the effective one sided noise spectrum density at the receiver is 10^{-15} Watt/Hz, the SNR at receiver is

(a) 10 dB (b) 20 dB

(c) 40 dB (d) 50 dB

Q.58 In the given figure, the noise figure and power gain of individual amplifiers are shown. The overall noise figure is

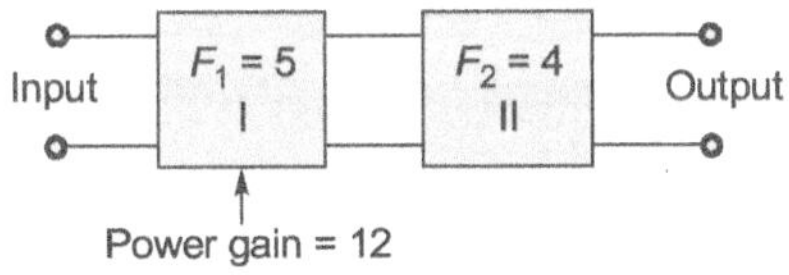

(a) $\frac{21}{4}$ (b) $\frac{25}{4}$

(c) $\frac{15}{4}$ (d) $\frac{18}{4}$

4. Amplitude Modulation

Q.59 The carrier amplitude after AM varies between 4 volts and 1 volt. The depth of modulation is

(a) 72% (b) 55%

(c) 60% (d) 80%

Q.60 A sinusoidal carrier has an amplitude of 10 V and frequency 30 kHz. It is amplitude modulated by a sinusoidal voltage of amplitude 3 V and frequency 1 kHz. The spectrum of AM wave will be represented as

(a)

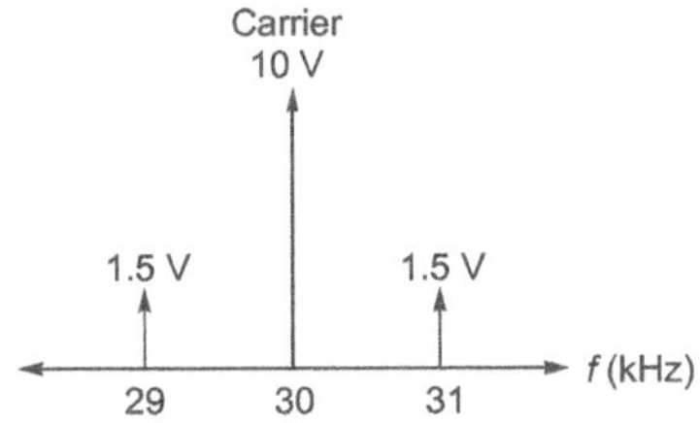

(b)

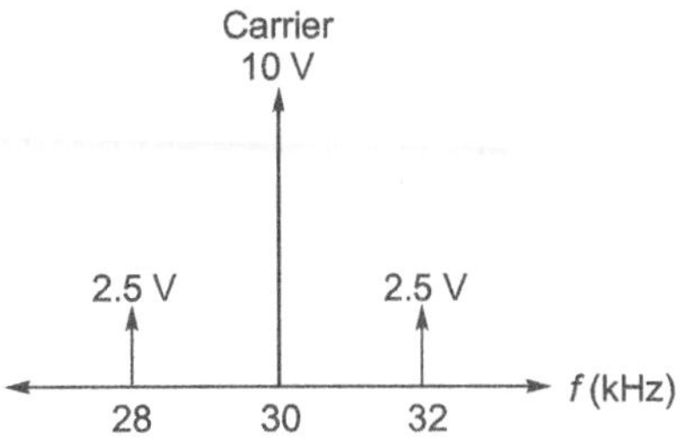

(c)

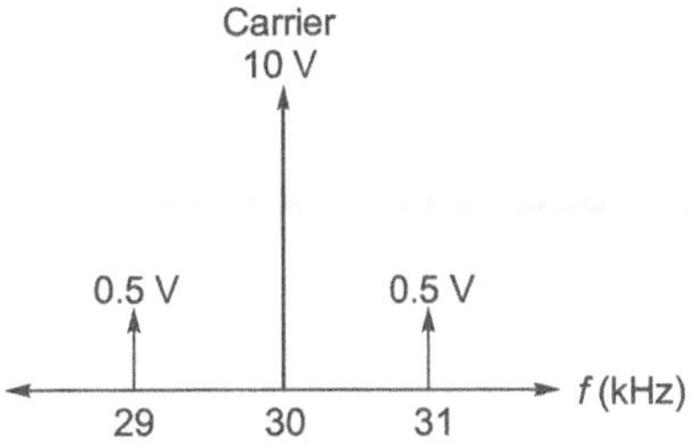

(d)

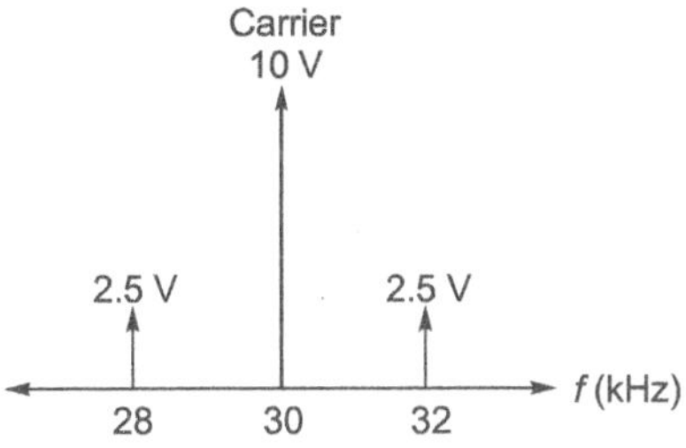

Q.61 **Assertion (A):** In amplitude modulation, the higher percentage of modulation is preferred.

Reason (R): The modulation index of more than 100% can be achieved without compromising the envelope distortion.

(a) Both A and R are true and R is the correct explanation of A.

(b) Both A and R are true but R is **not** the correct explanation of A.

(c) A is true but R is false.

(d) A is false but R is true.

Q.62 A 400 Watts carrier is modulated to a depth of 75 percent. Assuming the modulating signal to a sinusoidal one, the total power in the amplitude-modulated wave is

(a) 625 Watt (b) 426.8 Watt

(c) 634.3 Watt (d) 512.5 Watt

Q.63 The antenna current of an AM transmitter is 8 A if only the carrier is sent, but it increases to 8.93 A if the carrier is modulated by a single sinusoidal wave. What is the antenna current when the percent of modulation is 80%?

(a) 8.6 A (b) 9.2 A

(c) 6.6 A (d) 7.0 A

Q.64 The antenna current of an AM transmitter is 10 A when it is modulated to depth of 30% by an audio signal. It increases to 11 A when another signal modulates the carrier signal. The modulation index due to second signal is

(a) 62% (b) 66%
(c) 67% (d) 64%

Q.65 The modulation index of the low level emitter modulator is 0.7, voltage gain without modulating signal is $A_Q = 100$, carrier frequency $f_c = 1$ MHz. The amplitude of carrier is 10 mV. If the modulating frequency is 1 kHz, the range of voltage gain is

(a) 40 to 150 (b) 30 to 150
(c) 40 to 170 (d) 30 to 170

Q.66 The types of distortions which can occur in the envelope detector output is/are

(a) only diagonal clipping.
(b) only negative peak clipping.
(c) both diagonal clipping and negative peak clipping.
(d) neither diagonal clipping nor negative peak clipping.

Q.67 Consider the following statements associated with the synchronous detection method for detection/demodulation of AM signals:

1. The circuit complexity is very less.
2. It introduces frequency and phase errors only.
3. It does **not** require any locally generated synchronous carrier.
4. It can be used for DSB-SC, SSB and VSB but not for DSB-FC.

Which of the statements given above is/are correct?

(a) 1, 3 and 4 (b) 2 and 4 only
(c) 1, 2 and 3 (d) 3 and 4 only

Q.68 **Assertion (A):** The principle of envelope detection of an AM signal is based on multiplying the incoming signal with the locally generated synchronous carrier and pass the product through a LPF.

Reason (R): An envelope detector produces an output signal that follows the envelope of the input AM signal exactly.

(a) Both A and R are true and R is the correct explanation of A.
(b) Both A and R are true but R is **not** the correct explanation of A.
(c) A is true but R is false.
(d) A is false but R is true.

Q.69 The bandwidth of AM wave is (ω_c is the carrier frequency and ω_m the frequency of modulating signal)

(a) $2\omega_m$ (b) $\omega_c + \omega_m$
(c) ω_m (d) $\omega_c - \omega_m$

Q.70 **Assertion (A):** Double sideband-full carrier (DSB-FC) system is a power inefficient system.

Reason (R): In double sideband-full carrier (DSB-FC) system a much higher power is transmitted than what is actually required?

(a) Both A and R are true and R is the correct explanation of A.
(b) Both A and R are true but R is **not** the correct explanation of A.
(c) A is true but R is false.
(d) A is false but R is true.

Q.71 Match **List-I** with **List-II** and select the correct answer using the codes given below the lists:

List-I
A. Product modulator
B. Balanced modulator
C. Frequency discrimination
D. Vestigial sideband transmission

List-II
1. Generation of SSB modulated wave
2. As efficient as the SSB
3. Generation of DSB-SC wave
4. Suppression of the carrier

Codes:

	A	B	C	D
(a)	1	2	3	4
(b)	4	3	1	2
(c)	3	4	1	2
(d)	2	1	3	4

Q.72 Match **List-I** with **List-II** and select the correct answer using the codes given below the lists:

List-I	List-II
A. AM	1. Suitable for digital signals
B. VSB	2. Radio broadcasting
C. TDM	3. Transmission of video signals

Code:

	A	B	C
(a)	1	2	3
(b)	2	3	1
(c)	2	1	3
(d)	1	3	2

Q.73 What is the percent power saving for the SSB signal if the AM wave is modulated to a depth of 100%?
(a) 33.0% (b) 66.66%
(c) 50% (d) 83.33%

Q.74 What is the percent power saving for a DSB-SC signal for the percent modulation of 50%?
(a) 88.88% (b) 66.66%
(c) 33.33% (d) 60%

Q.75 Match **List-I (Types of AM signal)** with **List-II (Bandwidth)** and select the correct answer using the codes given below the lists:

List-I	List-II
A. DSB-FC	1. $2f_m$
B. DSB-SC	2. Between f_m and $2f_m$
C. SSB	3. f_m
D. VSB	

Codes:

	A	B	C	D
(a)	2	1	3	3
(b)	1	2	2	3
(c)	1	1	3	2
(d)	1	3	2	2

Q.76 In an AM system, the modulating signal is sinusoidal with frequency f_m Hz. If 80% modulation is used, then the ratio of the total side-band power in the modulation signal to the total power is approximately equal to
(a) 0.24 (b) 0.32
(c) 0.44 (d) 0.38

Q.77 Consider the following statements associated with various types of AM signals:
1. The transmission efficiency of vestigial sideband transmission is the highest among all the AM techniques available.
2. The bandwidth requirement of a standard AM technique is the highest.
3. Single sideband technique is popularly used for TV broadcasting.
4. Double sideband suppressed carrier technique is used for radio broadcasting.

Which of the statements given above is/are **not** correct?
(a) 2 only (b) 1 and 3 only
(c) 1, 2 and 3 (d) 3 only

Q.78 The number of AM broadcast stations that can be accommodated in a 100 kHz bandwidth for the highest modulating frequency of 5 kHz with no guard band is
(a) 10 (b) 20
(c) 5 (d) 50

Q.79 The value of modulation index (m_a) for distortion free transmission is
(a) 0 (b) 1
(c) between 0 and 1 (d) infinite

Q.80 The frequency components which contain the message information in AM transmission is/are:
(a) carrier, upper sideband and lower sideband
(b) carrier
(c) upper sideband and lower sideband
(d) upper sideband

Q.81 The maximum value of the total sideband power in AM is
(a) P_c (b) $\frac{P_c}{2}$
(c) $\frac{3}{2}P_c$ (d) None of these

Q.82 If m_a is the modulation index, then the transmission efficiency of an AM signal is given by
(a) $\left(\frac{m_a^2}{1+m_a^2}\right)$ (b) $\left(\frac{m_a}{2+m_a^2}\right)$
(c) $\left(\frac{1}{1+m_a^2}\right)$ (d) $\left(\frac{m_a^2}{2+m_a^2}\right)$

Q.83 For an AM signal represented by

$e_{AM} = 10\,(1 + 0.4 \cos 6280\, t) \cos (6.28 \times 10^6 t)$

The value of modulation index and carrier frequency are respectively.
(a) 0.2 and 1 MHz (b) 0.4 and 2 MHz
(c) 0.2 and 2 MHz (d) 0.4 and 1 MHz

Q.84 Which of the following statements is most correct?

(a) E_m should be equal to E_c
(b) E_m should be greater than E_c
(c) E_m should be less than or equal to E_c
(d) Any of the above

Q.85 Which of the following options related to AM is correct?

(a) Sideband power is always constant.
(b) Carrier power is constant.
(c) Bandwidth is infinite.
(d) Total transmitted power is constant.

Q.86 Consider the vestigial signal sideband modulated system used to modulate a TV signal having bandwidth equal to 4 MHz. If roll-off factor is '1', then the bandwidth required to transmit the signal is given by

(a) 5 MHz (b) 4 MHz
(c) 6 MHz (d) 7 MHz

Q.87 For a message signal $m(t) = 10 \cos 100\, t$ with 60% modulation, the maximum envelope time will be

(a) 10.3 ms (b) 13.3 ms
(c) 33.3 ms (d) 10 ms

Q.88 If the carrier of a 100 percent modulated AM wave is suppressed, the percentage power saving will be

(a) 50 (b) 100
(c) 150 (d) 66.66

Q.89 Which of the following circuits can **not** demodulate SSB?

(a) Balance modulator
(b) Product modulator
(c) BFO
(d) Phase discriminator

Q.90 The amplitude modulated waveform $s(t) = A_c\,[1 + k_i m(t)] \cos\omega_c t$ is fed to an ideal envelope detector. The maximum magnitude of $K_a m(t)$ is greater than 1. Which of the following could be the detector output?

(a) $A_c\, m(t)$ (b) $A_c^2[1+K_a m(t)]^2$
(c) $|A_c[1+K_a m(t)]|$ (d) $A_c[1+K_a m(t)]^2$

5. Angle Modulation

Q.91 Match **List-I** with **List-II** and select the correct answer using the codes given below the lists:

List-I
A. Pre-emphasis
B. Varactor
C. Voltage-controlled Oscillator
D. De-emphasis

List-II
1. Voltage-variable capacitor
2. Uses a low-pass filter in a receiver
3. Frequency can be controlled by changing an external control voltage
4. Uses a high-pass filter in an FM transmitter.

Codes:

	A	B	C	D
(a)	2	3	1	4
(b)	4	3	1	2
(c)	2	1	3	4
(d)	4	1	3	2

Q.92 The amount of frequency shift in FM is

(a) directly proportional to amplitude of the modulating signal.
(b) inversely proportional to amplitude of the modulating signal.
(c) independent of amplitude of the modulating signal.
(d) proportional to frequency deviation.

Q.93 According to Carson's rule, the bandwidth of FM wave is given by ($\Delta\omega$ is the frequency deviation and ω_m the modulating frequency)

(a) $(\Delta\omega + \omega_m)$ (b) $(\Delta\omega + 2\omega_m)$
(c) $2(\Delta\omega + \omega_m)$ (d) $(2\Delta\omega + \omega_m)$

Q.94 We can obtain PM from frequency modulator by passing the modulating signal through a

(a) differentiator (b) integrator
(c) buffer (d) none of these

Q.95 A FM wave is represented by the following equation:

$$v = 10 \sin [5 \times 10^8\, t + 4 \sin 1250 t]$$

Its maximum deviation is around.

(a) 995 Hz (b) 796 Hz
(c) 1989 Hz (d) 199 Hz

Q.96 In a FM system, the modulating frequency f_m = 1 kHz, the modulating voltage V_m = 2 volt and the deviation is 6 kHz. If the modulating voltage is raised to 4 volt then what is the new deviation?

(a) 4 kHz (b) 6 kHz
(c) 8 kHz (d) 12 kHz

Q.97 Match **List-I (Type of modulation)** with **List-II (Noise immunity)** and select the correct answer using the codes given below the lists:

List-I	List-II
A. Frequency modulation	1. Moderate
B. Amplitude modulation	2. Lowest
C. Phase modulation	3. Highest

Codes:

	A	B	C
(a)	1	3	2
(b)	1	2	3
(c)	3	2	1
(d)	3	1	2

Q.98 Consider the following statements associated with angle modulations:

1. In FM, the frequency deviation is proportional to modulating frequency.
2. In PM, the frequency deviation is independent of modulating signal.
3. Amplitude of both FM and PM are constant.
4. It is possible to receive FM on a PM receiver and vice-versa.
5. Signal to noise ratio of PM is better than that of FM.

Which of the statements given above are correct?

(a) 2 and 5 only (b) 3 and 4 only
(c) 1, 4 and 5 (d) 2, 3 and 4

Q.99 Match **List-I (Type of transmission)** with **List-II (Applications)** and select the correct answer using the codes given below the lists:

List-I

A. Wideband FM
B. AM
C. Narrowband FM
D. Vestigial sideband transmission

List-II

1. Radio broadcasting
2. Used for transmission of television signals
3. Used for high quality music transmission
4. Used for speech transmission (Police wireless, ambulance etc.)

Codes:

	A	B	C	D
(a)	3	1	4	2
(b)	4	1	3	2
(c)	3	2	4	1
(d)	4	2	3	1

Q.100 **Assertion (A):** There is less adjacent-channel interference in FM than in AM.

Reason (R): Standard frequency allocation provide a guard band between commercial FM stations.

(a) Both A and R are true and R is the correct explanation of A.
(b) Both A and R are true but R is **not** the correct explanation of A.
(c) A is true but R is false.
(d) A is false but R is true.

Q.101 **Assertion (A):** The modulation index of FM changes in the process of mixing.

Reason (R): Mixing of the FM wave with a local oscillator frequency will produce sum and difference frequency components at the output of the mixer.

(a) Both A and R are true and R is the correct explanation of A.
(b) Both A and R are true but R is **not** the correct explanation of A.
(c) A is true but R is false.
(d) A is false but R is true.

Q.102 What is the bandwidth (in kHz) of commercial FM transmission assuming frequency deviation Δf = 75 kHz and bandwidth of modulating signal f_m = 15 kHz?

(a) 160 kHz (b) 180 kHz
(c) 200 kHz (d) 210 kHz

Q.103 An angle-modulated signal is given as:

$$x_c(t) = 10 \cos [(10^8)\pi t + 5 \sin 2\pi (10^3)t]$$

The maximum phase deviation and the maximum frequency deviation are respectively

(a) 10 rad and 5 kHz (b) 10 rad and 10 kHz
(c) 5 rad and 10 kHz (d) 5 rad and 5 kHz

Q.104 Maximum frequency deviation of FM is
(a) ±200 kHz (b) ±15 kHz
(c) ±75 kHz (d) ±125 kHz

Q.105 For an FM signal, the maximum deviation occurs at
(a) the positive and negative peaks of the modulating signal.
(b) the point deviation occurs near the zero crossing points of the modulating signal.
(c) different points of the modulating signal.
(d) none of the above

Q.106 The deviation in a narrowband FM system is
(a) > 10 kHz (b) < 10 kHz
(c) < 5 kHz (d) < 1 kHz

Q.107 For the broadcasting applications
(a) PM is used
(b) Narrowband FM is used
(c) AM is used
(d) Wideband FM is used

Q.108 Which of the following range of modulating frequencies is suitable for the wideband FM system?
(a) 0 – 6 kHz (b) 30 Hz to 15 kHz
(c) 30 Hz to 3 kHz (d) 0 – 3 kHz

Q.109 Consider the following statements:
1. In direct method of FM generation, the carrier signal is directly affected by the baseband or modulating signal.
2. The frequency stability is very high in the Armstrong method of FM generation.
3. The Armstrong method uses the phase modulator to generate a frequency modulated wave.
4. In direct method, the baseband or modulating signal directly modulates the carrier.

Which of the statements given above is/are correct?
(a) 1 and 4 only (b) 3 and 4 only
(c) 1, 2, 3 and 4 (d) 1, 2 and 3

Q.110 Which of the following is **not** a basic component of phase locked loop (PLL)?
(a) Diode
(b) Multiplier
(c) Loop filter
(d) Voltage controlled Oscillator

Q.111 With the increase in the modulation index of an FM wave, the number of sidebands having significant amplitude will
(a) decrease
(b) increase
(c) remain constant
(d) either increase or decrease

Q.112 FM can be generated using PM by
(a) passing the modulating signal through a low pass filter
(b) passing the modulating signal through a high pass filter
(c) using the pre-emphasis
(d) using the de-emphasis

Q.113 An angle modulated signal with carrier frequency $\omega_c = 2 \times 10^6$ rad/sec is given by
$s(t) = \cos 2\pi (2 \times 10^6 t + 30 \sin 150t + 40 \cos 150t)$
The maximum frequency deviation is
(a) $15 \times 10^3 \pi$ rad/s (b) $30 \times 10^3 \pi$ rad/s
(c) $10^3 \pi$ rad/s (d) $60 \times 10^3 \pi$ rad/s

Q.114 If the transmission bandwidth in FM is doubled the SNR will
(a) also double
(b) improve four times
(c) decrease by one fourth
(d) be unaffected

Q.115 In the spectrum of a frequency-modulated wave
(a) the carrier frequency disappears when the modulation index is large.
(b) the amplitude of any sideband depends on the modulation index.
(c) the total number of sidebands depends on the modulation index.
(d) the carrier frequency cannot disappear.

Q.116 The difference between phase and frequency modulation
(a) is purely theoretical because they are the same in practice.
(b) is too great to make the two systems compatible.
(c) lies in the poorer audio response of phase modulation.
(d) lies in the different definitions of the modulation index.

Q.117 In an FM stereo multiplex transmission, the
(a) sum signal modulates the 19 kHz sub-carrier.
(b) difference signal modulates the 19 kHz sub-carrier.
(c) difference signal modulates the 38 kHz sub-carrier.
(d) difference signal modulates the 67 kHz sub-carrier.

6. Sound Broadcast Transmitting and Superhetrodyne Receivers

Q.118 In order to minimize the noise power a filter is introduced
(a) before modulator
(b) after demodulator
(c) before demodulator
(d) after modulator

Q.119 The output signal to noise ratio depends upon
(a) the type of modulation used at the transmitter only.
(b) the type of demodulation used at the receiver only.
(c) the type of modulation used at the transmitter and the type of demodulation used at the receiver.
(d) none of the above

Q.120 **Assertion (A):** The figure of merit of a receiver is always more than unity.
Reason (R): The figure of merit must be as high as possible because higher value of figure of merit indicates better noise performance of the receiver.
(a) Both A and R are true and R is the correct explanation of A.
(b) Both A and R are true but R is **not** the correct explanation of A.
(c) A is true but R is false.
(d) A is false but R is true.

Q.121 Consider the following statements:
1. The figure of merit of DSB-SC and SSB receivers with coherent detection is 1.
2. The figure of merit of an AM receiver with envelope detection is more than 1.
3. If the average modulated signal power and the average noise power in the message bandwidth is same as that of DSB-SC receiver then the output SNR and figure of merit of SSB coherent receiver is same as that of the DSB SC receiver.
4. A square-law demodulator operates above threshold on a weaker signal than with an envelope demodulator.

Which of the above statements are correct?
(a) 1, 3 and 4 (b) 2 and 4
(c) 1, 2 and 4 (d) 2 and 3

Q.122 Which of the following is **not** an uncoded system?
(a) PPM (b) DM
(c) FM (d) PAM

Q.123 The main function of a radio receiver is to
1. amplify the modulating frequency signal.
2. amplify the selected R.F. (radio frequency) signal.
3. select the desired signal and reject the unwanted signals.
4. intercept the incoming modulated signal by the receiving antenna.
5. detect the modulated signal to get back the original modulating or baseband signal.

Of these, the correct statements are:
(a) 2, 4 and 5 (b) 1, 3, 4 and 5
(c) 1, 2, 3 and 4 (d) all of these

Q.124 Match **List-I (Classification of radio receivers)** with **List-II (Applications)** and select the correct answer using the codes given below the lists:

List-I
A. A.M. Broadcast receivers
B. F.M. Broadcast receivers
C. Communication receivers
D. Radar receivers

List-II
1. Operate in VHF or UHF bands.
2. Used to receive radio detection and ranging
3. Operate on long, medium and short wave bands.
4. Used for reception of telegraph and short wave telephone signals.

Codes:

	A	B	C	D
(a)	3	2	4	1
(b)	3	1	4	2
(c)	1	3	2	4
(d)	1	4	3	2

Q.125 The sensitivity of a superheterodyne receiver does **not** depend on
(a) the gain of the RF amplifier
(b) the noise figure of the receiver
(c) the receiving frequency
(d) the gain of the IF amplifier

Q.126 Match **List-I (Superheterodyne receiver characteristics)** with **List-II (Features/Cause)** and select the correct answer using the codes given below the lists:

List-I
A. Double spotting
B. Sensitivity
C. Fidelity
D. Selectivity

List-II
1. Ability to reproduce all the modulating frequencies equally.
2. Occurs due to inadequate image-frequency rejection.
3. Ability to rejected unwanted signal.
4. Ability to amplify weak signals.

Codes:

	A	B	C	D
(a)	2	4	1	3
(b)	2	1	4	3
(c)	1	3	2	4
(d)	1	2	3	4

Q.127 The last and the first stages in a superheterodyne receiver are respectively
(a) IF amplifier and AF amplifier
(b) RF amplifier and AF amplifier
(c) Mixer and R.F. amplifier
(d) A.F. amplifier and R.F. amplifier

Q.128 For a broadcast superheterodyne AM receiver having no RF amplifier, the loaded Qualify factor Q of the antenna coupling circuit is 100. If the intermediate frequency is 455 kHz and signal frequency is 1 kHz, the image frequency would be equal to
(a) 1455 kHz (b) 1910 kHz
(c) 2455 kHz (d) 1545 kHz

Q.129 For a receiver with IF and RF frequencies of 455 kHz and 900 kHz respectively, the image frequency rejection ratio for a pre-selector Q of 80 is
(a) 120.8 (b) 80.06
(c) 80.04 (d) 98.08

Q.130 If the IF and RF frequencies for a superheterodyne receiver are respectively 455 kHz and 900 kHz, then the local oscillator frequency will be
(a) 2265 kHz (b) 1810 kHz
(c) 910 kHz (d) 1355 kHz

Q.131 Which of the following is **not** an effect caused due to superheterodyning?
(a) Variation in bandwidth
(b) High sensitivity
(c) High selectivity
(d) High adjacent channel rejection

Q.132 **Assertion (A):** The gain provided by the RF amplifier will result in improved signal to noise ratio.
Reason (R): The RF amplifier provides image frequency rejection.
(a) Both A and R are true and R is the correct explanation of A.
(b) Both A and R are true but R is **not** the correct explanation of A.
(c) A is true but R is false.
(d) A is false but R is true.

Q.133 The resonant frequency of RF amplifier of a receiver is 1 MHz and its bandwidth is 10 kHz. The quality factor is
(a) 10 (b) 0.1
(c) 1000 (d) none of these

Q.134 Automatic gain control (AGC) is used to
(a) increase the selectivity
(b) increase the fidelity
(c) balance the fading
(d) increase the sensitivity

Q.135 The range of superheterodyne receiver tuning when $f_{LO} > f_c$ with broadcast frequency ranges 540 to 1600 kHz is
(a) 85 – 1145 kHz (b) 540 – 1600 kHz
(c) 995 – 2055 kHz (d) 1450 – 2510 kHz

Q.136 The noise temperature is about 4306 K. The noise figure is (If T_0 = 17°C)
(a) 3 dB (b) 6 dB
(c) 12 dB (d) 24 dB

Q.137 To prevent overloading of the last IF amplifier in a receiver, one should use
(a) squelch
(b) variable sensitivity
(c) variable selectivity
(d) double conversion

Q.138 If an FET is used as the first AF amplifier in a transistor receiver, this will have the effect of
(a) improving the effectiveness of the AGC
(b) reducing the effect of negative-peak clipping
(c) reducing the effect of noise at low modulation depths
(d) improving the selectivity of the receiver

Q.139 One of the main function of the RF amplifier in a superheterodyne receiver is to
(a) provide improved tracking
(b) permit better adjacent-channel rejection
(c) increase the tuning range of the receiver
(d) improve the rejection of the image frequency

Q.140 A receiver has poor IF selectivity. It will therefore also have poor
(a) blocking
(b) double-spotting
(c) diversity reception
(d) sensitivity

Q.141 Match **List-I** with **List-II** and select the correct answer using the codes given below the lists:

List-I	List-II
A. IF Noise	1. Ganged tuning
B. Image frequency	2. Spurious response
C. Station selector	3. Leaky capacitor filter
D. Excessive hum	4. Mixer stage

Codes:

	A	B	C	D
(a)	3	2	1	4
(b)	1	2	3	4
(c)	4	3	1	2
(d)	4	2	1	3

7. Pulse Modulation

Q.142 The Nyquist rate for a continuous-time signal (x) is
$x(t) = 6\cos 50\pi t + 20\sin 300\pi t - 10\cos 100\pi t$
(a) 300 Hz (b) 150 Hz
(c) 100 Hz (d) 50 Hz

Q.143 The Nyquist interval for the signal

$$x(t) = \frac{1}{2\pi}\cos(4000\,\pi t)\cos(1000\,\pi t) \text{ is}$$

(a) 0.1 ms (b) 0.2 ms
(c) 0.3 ms (d) 0.4 ms

Q.144 A sine wave of frequency f_m and amplitude A_m is applied to a delta modulator having step size Δ. The slope overload distortion will **not** occur if

(a) $A_m < \dfrac{\Delta}{2\pi f_m T_s}$ (b) $A_m = \dfrac{\Delta}{2\pi f_m T_s}$

(c) both (a) and (b) (d) $A_m > \dfrac{\Delta}{2\pi f_m T_s}$

Q.145 The biggest disadvantages of PCM is
(a) its inability to handle analog signals.
(b) the high error rate which its quantizing noise introduces.
(c) its incompatibility with TDM.
(d) the large bandwidths that are required for it.

Q.146 **Assertion (A)** : PCM has much better noise immunity as compared to PAM, PWM and PPM systems.
Reason (R) : PCM does not contain any information in the width or the position of the pulses.
(a) Both A and R are true and R is the correct explanation of A.
(b) Both A and R are true but R is **not** the correct explanation of A.
(c) A is true but R is false.
(d) A is false but R is true.

Q.147 **Assertion (A)** : The signaling rate and transmission channel bandwidth is quite large for delta modulation compared to PCM.
Reason (R) : There is no analog to digital converter required in delta modulation unlike PCM.
(a) Both A and R are true and R is the correct explanation of A.
(b) Both A and R are true but R is **not** the correct explanation of A.
(c) A is true but R is false.
(d) A is false but R is true.

Q.148 The bandwidth of TV video plus audio signal is 4.5 MHz. If this signal is converted into PCM bit stream with 1024 quantization levels and the signal is sampled at the rate 20% above Nyquist rate, then the number of bits/sec of the resulting signal is
(a) 196 M bits/sec (b) 108 M bits/sec
(c) 88 M bits/sec (d) 216 M bits/sec

Q.149 The output signal to noise ratio (SNR) of a 10-bit PCM was found to be 30 dB. The desired SNR is 42 dB. It was decided to increase the SNR to the desired value by increasing the number of quantization levels. The percentage increase in the transmission bandwidth required for this SNR is
(a) 20% (b) 15%
(c) 10% (d) 5%

Q.150 A telephone signal with cut-off frequency of 4 kHz is digitized into 8 bit PCM, sampled at Nyquist rate. The quantization S/N ratio is
(a) 45.2 dB (b) 46.6 dB
(c) 49.8 dB (d) 51.5 dB

Q.151 Which of the following pulse modulation systems is analog?
(a) PCM (b) Differential PCM
(c) PWM (d) Delta modulation

Q.152 Match **List-I** with **List-II** and select the correct answer using the codes given below the lists:

List-I
A. Adaptive Delta Modulation
B. Pulse Code modulation
C. Delta Modulation
D. Differential PCM

List-II
1. Introduces quantizing noise
2. Reduces slope overload error
3. Uses a predictor to predict the value of next sample
4. Introduces slope overload error

Codse:

	A	B	C	D
(a)	1	3	2	4
(b)	1	4	3	2
(c)	2	4	1	3
(d)	2	1	4	3

Q.153 Granular noise is associated with
(a) PCM (b) DPCM
(c) DM (d) QAM

Q.154 In PCM, the biggest disadvantage compared to analog modulation is
(a) large bandwidth
(b) large noise
(c) inability to handle analog signals
(d) incompatibility with TDM system

Q.155 The signal to quantization noise ratio in a PCM system depends upon:
1. sampling rate
2. number of quantization levels
3. message signal bandwidth

Select the correct code from the given options.
(a) 1, 2 and 3 (b) 2 and 3 only
(c) 2 only (d) 3 only

Q.156 Match **List-I** with **List-II** and select the correct answer using the codes given below the lists:

List-I	**List-II**
A. PAM	1. Introduces idle noise
B. DM	2. Highest noise interference
C. PWM	3. Highest signal to noise ratio
D. PPM	4. Similar to frequency modulation

Codes:

	A	B	C	D
(a)	4	1	2	3
(b)	2	1	4	3
(c)	4	3	2	1
(d)	2	3	1	4

Q.157 In pulse code modulation (PCM), the information is transmitted in the form of
(a) variation in the width of pulses
(b) variation in the position of pulses
(c) code words each of n-bit length
(d) variation in the amplitude of pulses

Q.158 According to Shannon-Hartley theorem, the channel capacity depends upon
(a) bandwidth only
(b) S/N ratio only
(c) bandwidth and S/N ratio
(d) none of the above

Q.159 Consider the following statements:
1. PCM allows the transmission of analog signals in the form of digital signals.

2. PCM is popularly used for the radio and TV broadcasting applications.
3. A large step size is required to reduce slope overload noise.
4. The signal-to-noise ratio for either PCM or delta modulation signals can often be improved by using companding.

Which of the statements given above are correct?

(a) 1, 3 and 4 (b) 2 and 3 only
(c) 1, 2 and 3 (d) 3 and 4 only

Q.160 PCM systems require regenerative repeaters over long distances. The correct sequence of the operations in which such a repeater performs is
(a) timing, equalisation and decision making.
(b) equalization, timing and decision making.
(c) timing, thresholding and equalisation.
(d) thresholding, timing and equalization.

Q.161 In order to reduce quantizing noise, one must
(a) increase the number of standard amplitudes.
(b) send pulses whose sides are more nearly vertical.
(c) use an R.F. amplifier in the receiver.
(d) increase the number of samples per second.

Q.162 Consider the following statements:
1. PWM needs synchronization between the transmitter and receiver.
2. The PWM pulse have variable power contents.
3. The bandwidth requirement of PWM is higher than that of a PAM signal.
4. The power contents of PPM signal does **not** remain constant.
5. PPM pulses need higher bandwidth as compared to PAM.

Which of the statements given above are **not** correct?

(a) 2 and 5 only (b) 1, 3, 4 and 5
(c) 1 and 3 only (d) 1 and 4 only

Q.163 If the sampling takes place at a rate which is lower than the Nyquist rate then
(a) reconstruction of the signal is not possible.
(b) an error called aliasing takes place.
(c) no effect on the reconstruction signal.
(d) none of the above

Q.164 With increase in the sampling frequency
(a) the adjacent spectrum will overlap
(b) the adjacent spectrum will touch each other
(c) a guard band is created between the adjacent spectrums
(d) all of the above

Q.165 The circuit required to generate the flat topped PAM signal is
(a) a simple integrator circuit
(b) a sample and hold circuit
(c) a differentiator circuit
(d) none of the above

Q.166 In an ADM system, the output signal amplitudes for 1's and 0's are
(a) fixed and the repetition rate is also fixed.
(b) fixed but the repetition rate is variable.
(c) variable and the repetition rate is also variable.
(d) variable but the repetition rate is fixed.

Q.167 A sinusoidal signal with peak to peak amplitude of 1.536 V in quantized into 128 levels using a mid-rise uniform quantizer. The quantization noise power is
(a) 48×10^{-6} W (b) 12×10^{-6} W
(c) 0.768 (d) 3.072

Q.168 Which of the following PCM schemes is depicted in the below figure?

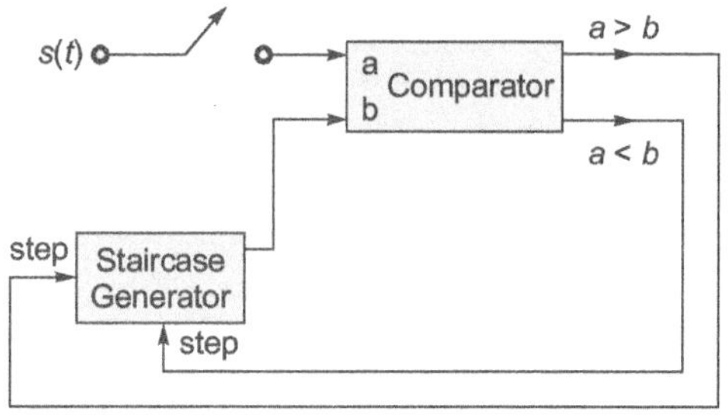

(a) Adaptive DM (b) Differential PCM
(c) Companding (d) Delta Modulation

Q.169 In PCM, the conditional probability of error is proportional to
(a) noise power (N_0)
(b) $\sqrt{N_0}$
(c) Peak signal energy (E_{max})
(d) $\sqrt{E_{max}}$

Q.170 Which of the following signals can be applied to delta modulator whose size is 0.1 V and sampling frequency is 20 π kHz so that no slope-overload occurs?
(a) 2 sin (1200 πt) (b) 1 sin (2600 πt)
(c) 3 sin (1000 πt) (d) 5 sin (400 πt)

Q.171 Time-division multiplex
(a) can be used with PCM only.
(b) combines five groups into a supergroup.
(c) stacks 24 channels in adjacent frequency slots.
(d) interleaves pulses belonging to different transmissions.

Q.172 If $x(t) = 5t + 1$; the sampler operates at 10 samples/sec and the step size is 1 V. The output of delta modulator is
(a) 1 1 1 1 0 1 1 1 1 0 1 1 1 1 0.....
(b) 1 1 0 1 1 0 1 1 0 1 1 0 1 1 0.....
(c) 1 1 1 0 1 1 1 0 1 1 1 0 1 1 1 0......
(d) 1 1 1 0 0 1 1 1 0 0 1 1 1 0 0 1 1 1......

8. Modern Digital Modulation and Detection Techniques

Q.173 Assertion (A): Binary phase shift keying (BPSK) is the most efficient of the three digital modulation, i.e., ASK, FSK and PSK.
Reason (R): In BPSK, phase of sinusoidal carrier is changed according to the data bit to be transmitted.
(a) Both A and R are true and R is the correct explanation of A.
(b) Both A and R are true but R is **not** the correct explanation of A.
(c) A is true but R is false.
(d) A is false but R is true.

Q.174 Match **List-I** with **List-II** and select the correct answer using the codes given below the lists:
List-I
A. BPSK
B. AM
C. QPSK
D. ASK
List-II
1. A group of two binary bits is represented by one phase state.
2. A binary bit is represented by one phase state.
3. Carrier ON or OFF depending on whether '1' or '0' is to be transmitted.
4. Detected using envelope detector.

Codes:

	A	B	C	D
(a)	1	3	2	4
(b)	2	3	1	4
(c)	1	4	2	3
(d)	2	4	1	3

Q.175 Consider the following statements:
1. BPSK has a bandwidth which is lower than that of a BFSK signal.
2. BPSK yields the maximum value of probability of error compared to all the three digital modulation techniques i.e. ASK, FSK and PSK.
3. Binary FSK has the highest system complexity.
4. Binary ASK is demodulated using coherent detection while binary FSK and PSK are demodulated using envelope detection.

Which of the statements given above are correct?
(a) 1 and 3 (b) 1, 2 and 4
(c) 2 and 3 (d) 2, 3 and 4

Q.176 In a digital continuous-time communication system, the bit rate of NRZ data stream is 1 Mbps and carrier frequency of transmission is 100 MHz. What is the bandwidth requirement of the channel in BPSK and QPSK systems respectively?
(a) 2 MHz and 4 MHz (b) 1 MHz and 2 MHz
(c) 4 MHz and 2 MHz (d) 2 MHz and 1 MHz

Q.177 It is required to transmit data in an authorised bandwidth of 20 MHz at a bit rate of 90 Mb/s. The number of bits/symbol is
(a) 8 (b) 7
(c) 9 (d) 6

Q.178 Bandwidth of BPSK is ____ than that of BFSK.
(a) higher (b) equal
(c) lower (d) either lower or higher

Q.179 The error probability of QPSK is
(a) better than BPSK (b) inferior to BPSK
(c) same as BPSK (d) none of these

Q.180 Comparison of MSK and QPSK scheme shows that
(a) MSK requires less power
(b) QPSK requires less power
(c) filtering is simple in MSK
(d) there is no comparison between them

Q.181 The bit rate of a digital communication system is 34 Mbps. The modulation scheme is QPSK. The baud rate of the system is
(a) 68 Mbps (b) 34 Mbps
(c) 17 Mbps (d) 85 Mbps

Q.182 Which of the following modulation scheme is most affected by noise?
(a) PSK (b) ASK
(c) FSK (d) DPSK

Q.183 MSK is basically
(a) QPSK (b) FSK
(c) ASK (d) none of these

Q.184 **Assertion (A):** Differential shift keying (DPSK) has higher probability of error than binary phase shift keying (BPSK).
Reason (R): DPSK need carrier at the receiving end.
(a) Both A and R are true and R is the correct explanation of A.
(b) Both A and R are true but R is not the correct explanation of A.
(c) A is true but R is false.
(d) A is false but R is true.

Q.185 Consider the following statements:
1. Binary phase shift keying (BPSK) has a very good noise immunity.
2. The major drawback of BFSK is its high bandwidth requirement.
3. Noise interference in differential phase shift keying (DPSK) is more.
4. In M-ary PSK bandwidth reduces with increase in the number of bits per symbol, (N).
5. A number of correctly tuned filters are required for the reception of M-ary FSK signal.

Which of these statements given above are correct?
(a) 1, 2, 3, 4 and 5 (b) 2, 4 and 5
(c) 1, 3 and 4 (d) 2, 3, 4 and 5

Q.186 Which of the following is **not** an advantage of QPSK system?
(a) BW is small
(b) Noise immunity is high
(c) Generation and detection is quite simple
(d) Bit rate doubles for same baud rate

Q.187 If in a digital communication, the bit rate is f_b bits/sec, the BW for QPSK and BPSK signals are
(a) f_b and $2f_b$ respectively
(b) $2f_b$ and f_b respectively
(c) $2f_b$ and $4f_b$ respectively
(d) $4f_b$ and $2f_b$ respectively

Q.188 In a BPSK signal detector the local oscillator has a fixed phase error of 20°. This phase error deteriorates the SNR at the output by a factor of
(a) $\cos 20°$ (b) $\cos^2 20°$
(c) $\cos 70°$ (d) $\cos^2 70°$

Q.189 The bits stream 01001 is differentially encoded using 'Delay and EX-OR' scheme for DPSK transmission. Assuming the reference bit as '1' and assigning phases of '0' and π for 1's and 0's respectively in the encoded sequence, the transmitted phase sequence becomes
(a) $\pi 0 \pi \pi 0$ (b) $0 \pi \pi 0 0$
(c) $0 \pi \pi \pi 0$ (d) $\pi \pi \pi 0 \pi$

Q.190 For a given data rate, the bandwidth B_p of a BPSK signal and the bandwidth B_o of the OOK signal are related as
(a) $B_p = \frac{B_o}{4}$ (b) $B_p = \frac{B_o}{2}$
(c) $B_p = B_o$ (d) $B_p = 2B_o$

Q.191 A source delivers symbols X_1, X_2, X_3 and X_4 with probabilities $\frac{1}{2}, \frac{1}{4}, \frac{1}{8}$ and $\frac{1}{8}$ respectively. The entropy of the system is
(a) 1.75 bits per second
(b) 1.75 bits per symbol
(c) 1.75 symbols per second
(d) 1.75 symbols per bit

Q.192 Match **List-I (Digital systems)** with **List-II (Minimum bandwidth required)** and select the correct answer using the codes given below the lists:

List-I	List-II
A. BPSK	1. f_b
B. DPSK	2. $2f_b$
C. QPSK	3. $4f_b$
D. M-ary PSK	4. $3f_b/N$

Codes:

	A	B	C	D
(a)	2	1	3	2
(b)	1	2	3	4
(c)	2	1	1	4
(d)	4	1	2	3

Q.193 A message signal of $m(t) = 10 \cos 2\pi \times 10^4$ in given to 10-bit PCM system resulting binary sequence is transmitted through space by using binary signalling scheme. If modulation scheme used is PSK then transmission bandwidth is

(a) 3125, 128 kHz (b) 312.5, 20 MHz
(c) 125,64 kHz (d) 62.5, 40 MHz

Q.194 A communication channel of bandwidth 75 kHz is required to transmit binary data at a rate of 0.1 Mb/s using raised cosine pulses. The roll-off factor a is

(a) 0.4 (b) 0.5
(c) 0.8 (d) zero

9. Information Theory and Coding

Q.195 The probabilities of the five possible outcomes of an experiment are given as:

$$P(x_1) = \frac{1}{2}, P(x_2) = \frac{1}{4}, P(x_3) = \frac{1}{8}, P(x_4) = P(x_5) = \frac{1}{16}$$

If there are 16 outcomes per second then the rate of information would be equal to

(a) 25 bits/sec (b) 30 bits/sec
(c) 35 bits/sec (d) 40 bits/sec

Q.196 Assertion (A): The Shannon-Hartley law shows that we can exchange increased bandwidth for decreased signal power for a system with given capacity *C*.

Reason (R): The bandwidth and the signal power place a restriction upon the rate of information that can be transmitted by a channel.

(a) Both A and R are true and R is the correct explanation of A.
(b) Both A and R are true but R is **not** the correct explanation of A.
(c) A is true but R is false.
(d) A is false but R is true.

Q.197 The differential entropy $H(x)$ of the uniformly distributed random variable x with the following probability density function for a = 1 is

$$f_x(x) = \begin{cases} \frac{1}{a} & \text{for } 0 \le x \le a \\ 0 & \text{elsewhere} \end{cases}$$

(a) –1 (b) 0.5
(c) 1 (d) 0

Q.198 A channel has a bandwidth of 8 kHz and signal to noise ratio of 31. For same channel capacity, if the signal to noise ratio is increased to 61, then, the new channel bandwidth would be equal to

(a) $16\frac{\log 62}{\log 31}$ (b) $8 \log\frac{16}{31}$ kHz
(c) $16 \log\frac{31}{16}$ (d) $8\frac{\log 32}{\log 62}$ kHz

Q.199 A source delivers symbols m_1, m_2, m_3 and m_4 with probabilities $\frac{1}{2}, \frac{1}{4}, \frac{1}{8}$ and $\frac{1}{8}$ respectively. The entropy of the system is

(a) 1.7 bits/sec (b) 1.75 bits/symbol
(c) 1.75 bits/symbol (d) 1.75 symbol/bit

Q.200 A communication channel with AWGN has a BW of 4 kHz and an SNR of 15. Its channel capacity is

(a) 1.6 kbps (b) 16 kbps
(c) 32 kbps (d) 456 kbps

Q.201 A source generates 4 messages. The entropy of the source will be maximum when

(a) all probabilities are equal
(b) one of the probabilities equal to 1 and two others are zero
(c) the probabilities are $\frac{1}{2}, \frac{1}{4}$ and $\frac{1}{2}$
(d) the two of the probabilities are $\frac{1}{2}$ each and other zero

Q.202 During transmission over a communication channel bit errors occurs independently with probability $\frac{1}{2}$. If a block of 3 bits are transmitted the probability of at least one bit error is equal to

(a) $\frac{1}{2}$ (b) $\frac{3}{2}$

(c) $\frac{7}{8}$ (d) $\frac{5}{8}$

Q.203 A 1200 band data stream is to be sent over a non-redundant frequency hopping system. The maximum bandwidth for the spread spectrum signal is 10 MHz. If no overlap occurs, the number of channels are equal to

(a) 2175 (b) 3125

(c) 4167 (d) 5286

Q.204 Consider the binary Hamming code of block length 31 and rate equal to (26/31). Its minimum distance is

(a) 3 (b) 5

(c) 26 (d) 31

Q.205 Equalizing pulses in TV are sent during

(a) horizontal blanking

(b) vertical blanking

(c) the serrations

(d) the horizontal retrace

Q.206 Four signals $g_1(t)$, $g_2(t)$, $g_3(t)$ and $g_4(t)$ are to be multiplexed and transmitted. $g_1(t)$ and $g_4(t)$ have a bandwidth of 4 kHz and the remaining two signals have bandwidth of 8 kHz. Each sample requires 8 bit for encoding. What is the minimum transmission bit rate of the system?

(a) 512 kbps (b) 16 kbps

(c) 192 kbps (d) 384 kbps

Q.207 In a communication system, each message (1 or 0) is transmitted three times in order to reduce the probability of error. The detection is based on the majority rule at the receiver. If P_e is the probability of bit error, the probability of error for this communication system will be:

(a) $3p_e^2 - 2p_e^3$ (b) $1 - p_e^2 - p_e^3$

(c) p_e^3 (d) $p_e^2(1 - p_e)$

■■■■

Answers Communication Systems

1. (b)	2. (d)	3. (c)	4. (c)	5. (c)	6. (a)	7. (b)	8. (d)
9. (c)	10. (a)	11. (c)	12. (b)	13. (c)	14. (d)	15. (b)	16. (d)
17. (a)	18. (a)	19. (c)	20. (a)	21. (b)	22. (b)	23. (d)	24. (b)
25. (c)	26. (d)	27. (b)	28. (a)	29. (c)	30. (b)	31. (a)	32. (c)
33. (d)	34. (a)	35. (c)	36. (b)	37. (c)	38. (d)	39. (b)	40. (c)
41. (a)	42. (c)	43. (a)	44. (d)	45. (b)	46. (b)	47. (c)	48. (a)
49. (b)	50. (d)	51. (c)	52. (c)	53. (c)	54. (b)	55. (d)	56. (c)
57. (a)	58. (a)	59. (c)	60. (a)	61. (c)	62. (d)	63. (a)	64. (b)
65. (d)	66. (c)	67. (b)	68. (d)	69. (a)	70. (a)	71. (c)	72. (b)
73. (d)	74. (a)	75. (c)	76. (a)	77. (b)	78. (a)	79. (b)	80. (c)
81. (b)	82. (d)	83. (d)	84. (c)	85. (b)	86. (c)	87. (b)	88. (d)
89. (d)	90. (c)	91. (d)	92. (a)	93. (c)	94. (a)	95. (b)	96. (d)
97. (c)	98. (b)	99. (a)	100. (a)	101. (d)	102. (b)	103. (d)	104. (c)
105. (a)	106. (c)	107. (d)	108. (b)	109. (c)	110. (a)	111. (b)	112. (a)
113. (a)	114. (b)	115. (b)	116. (d)	117. (c)	118. (c)	119. (c)	120. (d)
121. (a)	122. (b)	123. (d)	124. (b)	125. (c)	126. (a)	127. (d)	128. (b)
129. (a)	130. (d)	131. (a)	132. (b)	133. (d)	134. (c)	135. (c)	136. (c)
137. (b)	138. (b)	139. (d)	140. (a)	141. (d)	142. (a)	143. (b)	144. (c)

145. (d)	146. (a)	147. (d)	148. (b)	149. (a)	150. (c)	151. (c)	152. (d)
153. (c)	154. (a)	155. (c)	156. (b)	157. (c)	158. (c)	159. (a)	160. (b)
161. (a)	162. (d)	163. (b)	164. (c)	165. (b)	166. (a)	167. (b)	168. (a)
169. (d)	170. (d)	171. (d)	172. (c)	173. (b)	174. (d)	175. (a)	176. (d)
177. (c)	178. (c)	179. (c)	180. (a)	181. (c)	182. (b)	183. (b)	184. (c)
185. (a)	186. (c)	187. (a)	188. (b)	189. (c)	190. (c)	191. (b)	192. (c)
193. (b)	194. (b)	195. (b)	196. (b)	197. (d)	198. (d)	199. (b)	200. (b)
201. (a)	202. (a)	203. (c)	204. (a)	205. (b)	206. (d)	207. (a)	

Explanations

1. (b)

Very high frequency (VHF) has a frequency range of 30-300 MHz and wavelength range of 10 to 1 m.

3. (c)

The message from the information source may or may not be electrical in nature. In a case when the message produced by the information source is not electrical in nature, an input transducer is used to convert it into a time-varying electrical signal.

4. (c)

Communication is simply the basic process of exchanging information.

6. (a)

Multiplexing may be defined as the process of combining several message signals together and send them over the same communication channel. i.e. FDM, TDM and CDM.

9. (c)

The communication medium is the channel through which the message travels from the transmitter to the receiver. During the process of transmission and reception the signal gets distorted due to noise introduced in the system.

11. (c)

Though the baseband signals can be transmitted directly we require modulation because the baseband transmission has many limitations which can be overcome using modulation. Hence, assertion is true but reason is false.

13. (c)

$x_1(t) = \sin 15\pi t$ is a periodic signal having period T_1 equal to $\frac{2\pi}{15\pi} = \frac{2}{15}$.

$x_2(t) = \sin\sqrt{2}\pi t$ is a periodic signal having period T_2 equal to $\frac{2\pi}{\sqrt{2}\pi} = \sqrt{2}$.

$$\therefore \quad \frac{T_1}{T_2} = \frac{2}{15\sqrt{2}} = \frac{\sqrt{2}}{15}$$

14. (d)

A frequency domain representation of a signal shows the frequency content but not necessarily indicate the shape of the waveform or the combined amplitude of all the input components at any specific time. Hence, assertion is a false statement. Reason is a true statement because in frequency domain, a signal is represented by its frequency spectrum and to obtain frequency spectrum of a signal, Fourier series and Fourier transform are used.

15. (b)

Operation	$x(t)$	$X(\omega)$/Fourier transform
• Time shift	$x(t-t_0)$	$e^{-j\omega t_0} X(\omega)$
• Time differerentiation	$\frac{d^n x(t)}{dt^n}$	$(j\omega)^n X(\omega)$
• Time integration	$\int_{-\infty}^{t} x(\tau)d\tau$	$\frac{1}{j\omega}X(\omega) + \pi X(0)\delta(\omega)$
• Frequency shift	$x(t)e^{j\omega_0 t}$	$X(\omega-\omega_0)$

16. (d)

Fourier series analysis is a tool used to

1. analyse any periodic signal only.

2. find how many frequency components are present in the signal.
3. find the relative phase difference between various frequency components.
4. represents a periodic waveform in the form of sum of infinite number of sine and cosine terms.
5. find the amplitude of various frequency components of the series.

Hence, statements 2, 3 and 5 are only correct.

17. (a)

The given signal can be represented as

$$x(t) = e^{-\alpha t}u(t) \qquad (\text{Time constant, } \tau = \frac{1}{\alpha})$$

$\therefore$ Fourier transform of $x(t)$ is

$$F[x(t)] = \int_{-\infty}^{\infty} e^{-\alpha t}u(t)\cdot e^{-j2\pi ft}dt$$

$$= \int_{0}^{\infty} e^{-\alpha t}\cdot e^{-j2\pi ft}dt = \int_{0}^{\infty} e^{-(\alpha+j2\pi f)t}dt$$

$$= \frac{-1}{(\alpha+j2\pi f)}\left[e^{-(\alpha+j2\pi f)t}\right]_0^{\infty}$$

$$= \frac{-1}{(\alpha+j2\pi f)}\left[e^{-\infty}-e^{0}\right]_0^{\infty} = \frac{1}{(\alpha+j2\pi f)}$$

18. (a)

Given, $x(t) = e^{-\alpha t}u(t)$

$\therefore$ Normalized energy of given signal is

$$E = \int_{-\infty}^{\infty} |x(t)|^2 dt = \int_{0}^{\infty} (e^{-\alpha t})^2 dt$$

$$= \int_{0}^{\infty} e^{-2\alpha t}dt = \frac{-1}{2\alpha}[e^{-\infty}-e^{0}] = \frac{1}{2\alpha}$$

19. (c)

The total average power of a signal is the area of curve under $s(f)$.

Hence, $$P = \int_{-\infty}^{\infty} S(f)\,df$$

$$= \int_{-2}^{-1} 1\,df + \int_{-1}^{1} 2\,df + \int_{1}^{2} 1\,df$$

$= [1(-1+2) + 2(1+1) + 1(2-1)]$

$= 6$

21. (b)

For obtaining the Fourier transform of a signal $x(t)$, it must have Infinite number of discontinuities in any finite interval T.

22. (b)

$$u(t) = \begin{cases} 1, & \text{for } t \geq 0 \\ 0 & t > 0 \end{cases}$$

$$\therefore \quad F[u(t)] = \int_{-\infty}^{\infty} u(t)\cdot e^{-j2\pi ft}\cdot dt$$

$$= \int_{0}^{\infty} 1\cdot e^{-j2\pi ft}\cdot dt = \frac{1}{j2\pi ft}$$

23. (d)

- With the increase in signaling rate, the width of each pulse is reduced i.e. compression in time domain takes place. Thus, statement-1 is correct.
- A signal cannot be band limited and time limited simultaneously. Hence, statement-2 is not correct.
- Since, $F\left[\frac{d}{dt}x(t)\right] = j2\pi f\,X(f/\alpha)$

  i.e. $\frac{d}{dt}x(t) \xleftrightarrow{F} j2\pi f\,X(f)$,

  hence, differentiating the signal in time domain is equivalent to multiplying its fourier transform by $(j2\pi f)$. Thus, statement-3 is correct.
- $x(\alpha t) \xleftrightarrow{F} \frac{1}{|\alpha|}X(f/\alpha)$

  Here, $x(\alpha t)$ is a time scaled signal and $X(f/\alpha)$ is the frequency scaled signal.

$\Rightarrow$ If $\alpha < 1$, $x(\alpha t)$ will be compressed but $X(f/\alpha)$ will be an expanded one.

$\Rightarrow$ If $\alpha > 1$, $x(\alpha t)$ will be an expanded signal while $X(f/\alpha)$ will represent a compressed version of $X(f)$.

Thus, statement-4 is correct.

26. (d)

The trigonometric Fourier series of a periodic time function can have d.c. and cosine terms.

27. (b)

$x(t) = (2\cos\pi t + 7\cos t)$ does not satisfy the Drichlet condition, therefore, it can't be expanded as Fourier series.

28. (a)

Power is given by

$$P = \frac{A^2}{2}$$

$$\therefore \quad P = \frac{(8)^2}{2} + \frac{(4)^2}{2} = 32 + 8 = 40$$

29. (c)

$$F_x(x) = 100\text{ K for } x > 10$$

Since, $F_x(x) =$ for $x > x_n$ will be 1, therefore

$$100\text{ K} = 1 \text{ or } K = 0.01$$

Now, $P(x \le 5) = P(0 < x \le 5)$

$$= F_x(5) - F_x(0)$$

$$= kx^2\Big|_{x=5} = 0.01 \times 5^2 = 0.25$$

Also,

$$P(5 < x \le 7) = F_x(7) - F_x(5)$$

$$= kx^2\Big|_{x=7} - kx^2\Big|_{x=5}$$

$$= (0.01 \times 49) - (0.01 \times 25) = 0.24$$

Hence, option (c) is not correct.

30. (b)

Since the area under the probability density function is equal to 1, therefore.

$$\int_{-\infty}^{\infty} f_x(x)\, dx = 1$$

or, $$\int_a^b K\, dx = 1 \text{ or } K(b-a) = 1$$

or, $$K = \frac{1}{(b-a)}$$

31. (a)

BCC with different probabilities:

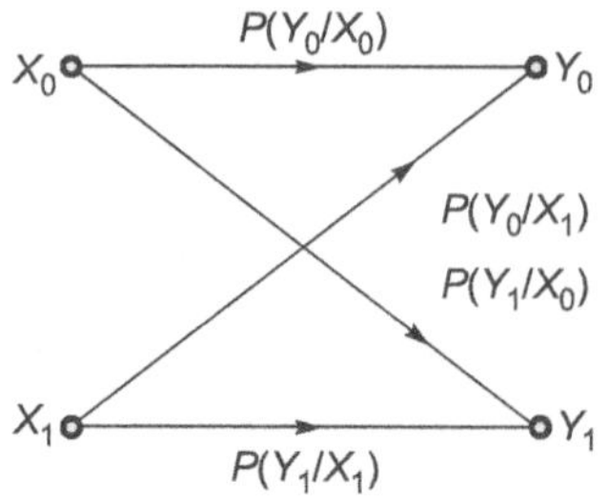

According to question,

$$P(X_0) = 0.4,\ P(X_1) = 0.6,$$

$$P(Y_1/X_0) = 10^{-6} \text{ and } P(Y_0/X_1) = 10^{-4}$$

Here, $P(Y_0/X_0) =$ Probability of error when '1' is transmitted

$P(Y_1/X_0) =$ Probability of error when '0' is transmitted

$\therefore$ Probability of error in channel

$$P(E) = P(Y_0/X_1)P(X_1) + P(Y_1/X_0)P(X_0)$$

$$= (10^{-4} \times 0.6) + (10^{-6} \times 0.4)$$

or, $$P(E) = 0.604 \times 10^{-4}$$

32. (c)

The joint CDF (cumulative distribution function) is

$F_{xy}(\infty, \infty) = P(X \le \infty, Y \le \infty)$, which will be equal to 1 because the probability of $X \le \infty$ and $Y \le \infty$ covers all the possible values of random variables X and Y.

$$\therefore \quad F_{xy}(\infty, \infty) = 1$$

According to the definition of CDF, we have

$$F_{xy}(\infty, \infty) = \int_{-\infty}^{\infty}\int_{-\infty}^{\infty} f_{xy}(x \cdot y)\, dx dy = 1$$

Substituting the value of PDF, we have:

$$F_{xy}(\infty, \infty) = C\int_{-\infty}^{\infty}\int_{-\infty}^{\infty} e^{-(ax+by)}u(x)\, u(y) dx dy$$

or, $$1 = C\int_0^{\infty} e^{-(ax)}\, dx \cdot \int_0^{\infty} e^{-by}\, dy$$

or, $$1 = \frac{C}{ab}[e^{-ax}]_0^{\infty} \cdot [e^{-by}]_0^{\infty}$$

or, $$1 = \frac{C}{ab}\left[e^{-\infty} - e^{-0}\right]\left[e^{-\infty} - e^{0}\right]$$

or, $$1 = \frac{C}{ab}[-1][-1] \text{ or } C = ab$$

33. (d)

- The conditional PDF is basically a PDF. Since $f_y(y/x) \ge 0$ i.e. the conditional PDF is a non-negative function.
  Hence, statements 1 and 2 are correct.
- $\int_{-\infty}^{\infty} f_y(y/x)\, dy$ and $\int_{-\infty}^{\infty} f_x(x/y)\, dx = 1$ i.e. the area under a conditional PDF is always equal to 1. Hence, statement-3 is also correct.

- If the random variables x and y are statistically independent, then
$$f_y(y/x) = f_y(y) \text{ and } f_x(x/y) = f_x(x)$$
This means that the conditional density functions reduce to the marginal density functions. Hence, statement-4 is also correct.

34. (a)

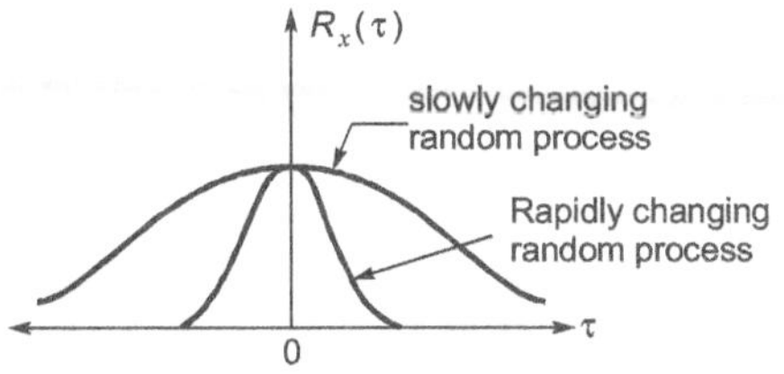

From above figure we can conclude that both assertion and reason are true and reason is the correct explanation of assertion.

36. (b)

The electronic components generate thermal, shot noise. The thermal noise is also called Johnson noise.

37. (c)

- The flicker noise occurs at low operating frequencies (few KHz). Hence, statement-1 is not correct.
- The thermal noise voltage is given by,

  $E_n = \sqrt{4KTBR}$.

  Hence, it is proportional to temperature, bandwidth and the value of resistance. Thus, statement-3 is not correct.
- Statements 1 and 4 are correct.

38. (d)

Noise figure,

$$F_{dB} = 10 \log F$$

or, $30 = 10 \log F$ or $F = 1000$

Noise power at input,

$$P_{ni} = FKTB$$
$$= 1000 \times 1.38 \times 10^{-23} \times (273 + 27) \times 10 \times 10^3$$
$$= 4.14 \times 10^{-14} \text{ W}$$

39. (b)

A white noise is sometimes called a gaussian noise because it has a gaussian distribution i.e. the PDF of white noise has the shape of gaussian PDF. Reason is also a true statement because the white noise contains all the frequency components in equal proportion due to which it is analogous with light which is a superposition of all visible spectral components.

Hence, both assertion and reason are true but reason is not the correct explanation of assertion.

40. (c)

- The shot noise is produced because of the random variations in the arrival of electrons (or holes) at the output electrode of an amplifying device. Hence, it appears as a randomly varying noise current superimposed on the output.
- Partition noise is generated when the current gets divided between two or more paths. It is generated due to the random fluctuations in the division. Therefore, the partition noise in a transistor will be higher than that in a diode.
- The flicker noise will appear at frequencies below a few kilohertz. Due to this reason it is sometimes called 1/f noise.
- Transit time noise occurs at a very high frequency.

41. (a)

Given, $P_{si} = 150\ \mu\text{W}, P_{ni} = 1.5\ \mu\text{W},$
$P_{so} = 1.5\text{ W}, P_{no} = 40\text{ mW}$

$$\text{Noise factor} = \frac{P_{si}}{P_{ni}} \times \frac{P_{no}}{P_{so}}$$
$$= \frac{150 \times 10^{-6}}{1.5 \times 10^{-6}} \times \frac{40 \times 10^{-3}}{1.5} = 2.66$$

42. (c)

Noise figure (dB)

$$= \left(\frac{S}{N}\right)_i (\text{dB}) - \left(\frac{S}{N}\right)_0 (\text{dB})$$

$$\therefore \left(\frac{S}{N}\right)_0 (\text{dB}) = \left(\frac{S}{N}\right)_i (\text{dB}) - \text{Noise figure (dB)}$$
$$= 40 - 20 = 20 \text{ dB}$$

43. (a)

The noise figure is $F_{dB} = 10 \log_{10} (F_{ratio})$

$\therefore$ Noise factor F

$$= F_{ratio} = \text{Antilog } (F_{dB}/10)$$
$$= \text{Antilog } (0.6) = 3.98 \simeq 4$$

Equivalent noise temperature is given by

$$T_{eq} = (F-1)T_0$$
$$= (4-1) \times 300 = 600°K$$

44. (d)

Given,

$F_2 = 20$ dB, $F_1 = 10$ dB and $G_1 = 15$ dB

$F_2 = 20$ dB = 100 : 1 power ratio

[dB = $10 \log_{10} F_2$]

$F_1 = 10$ dB = 10:1 power ratio [dB = $10 \log_{10} F_1$]

and $G_1 = 15$ dB = 31.62 : power ratio

[dB = $10 \log_{10} G_1$]

The overall noise figure referred to the input is

$$F = F_1 + \frac{(F_2-1)}{G_1}$$
$$= 10 + \frac{(100-1)}{31.62} = 13.13$$

45. (b)

Given,

SNR at input = 40 dB

or, $40 = 10 \log(\text{SNR})_i$

or, $(\text{SNR})_i = \text{Antilog}(4) = 10^4$

Now,

Noise figure = 20 dB = 10 log F

or, F = Antilog (2) = 100

Now,

$$\text{Noise figure, } F = \frac{\text{SNR at input}}{\text{SNR at output}}$$

or, SNR at output

$$= \frac{\text{SNR at input}}{\text{F}}$$
$$= \frac{10000}{100} = 100 = 20 \text{ dB}$$

46. (b)

The ideal value of noise figure is 0 dB.

47. (c)

White noise is the noise whose power spectral density is uniform over the entire frequency range of interest. Hence, given figure will represent a white noise.

48. (a)

The psd of a real valued random process is an even function of frequency

i.e. $S_x(-f) = S_x(f)$

49. (b)

The rms noise voltage due to thermal noise is

$$E_n = \sqrt{4KTBR}$$

Here, B = Bandwidth and R = resistance

Since T is constant, therefore,

$$E_n \propto \sqrt{BR}$$

When B is increased two times and R is halved then, E_n will remain unchanged.

50. (d)

Thermal noise power depends only on absolute temperature. It is independent of resistor values since thermal noise power is given by

$$P_n = \text{kTB Watts}$$

where, k = Boltzmann's constant (in J/K)

T = Temperature of conductor in (K)

B = Bandwidth of the noise spectrum (Hz)

53. (c)

$$m = 0,\ S_0 = 100$$

So, $C_n(\tau) = S_0\delta(\tau) = 100\,\delta(\tau)$

So, autocorrelation

$$= C_n(\tau) + (m)^2$$
$$= 100\delta(\tau) + 100$$

Taking Fourier transform,

$$S_n(\omega) = S_0 + 2\pi[m]^2\,\delta(\omega)$$
$$= 100 + 200\,\pi\delta(\omega)$$

54. (b)

$$\sigma^2 = R_x(0) - m^2 = 4[2] - 4 = 4$$

$\therefore$ $\sigma = 2$

$$P[x \le 1] = \int_{-\infty}^{1} \frac{1}{\sqrt{8\pi}} e^{\frac{-(x-2)^2}{8}} dx$$

Let, $\frac{(x-2)^2}{8} = \frac{y^2}{2}$

$\therefore$ $x - 2 = 2y$; $\therefore$ $dx = 2y$

Also, when $x = -\infty$, $y = -\infty$

$$\therefore \int_{-\infty}^{0.5} \frac{1}{\sqrt{2\pi}} e^{\frac{-y^2}{2}} dy$$

Let, $y = -z$, $dy = -dz$

$$\therefore \quad P[x \le 1] = \int_{0.5}^{-\infty} \frac{1}{\sqrt{2\pi}} e^{\frac{-z^2}{2}} dz = Q(0.5)$$

55. **(d)**

We can write

$$R_x(\tau) = [2 + 3(1 - |\tau|)] \; ; \; |\tau| \leq 1$$
$$2 \; ; \; |\tau| > 1$$

$\therefore \quad R_x(t) = 2 + 3\,\Delta_1(t)$

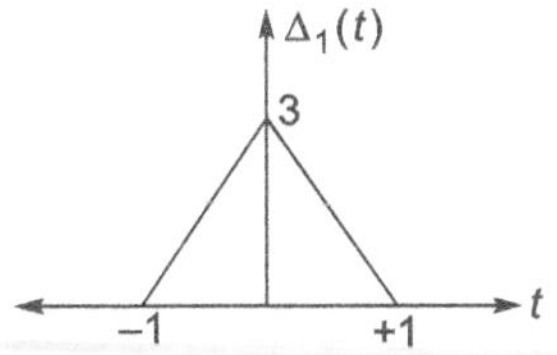

$$\text{PSD} = FT[R_x(\tau)] = 2\delta(f) + 3\sin c^2(f)$$

Area of delta function in PSD is due to the D.C. component.

So, average power in D.C. component = 2 W.

57. **(a)**

We have:

$$-20 = 10\log\left(\frac{P_0}{P_i}\right)$$

or, $\quad P_0 = 10^{-5}$ Watt

Now, $\quad N_0 = \frac{\eta}{2}f_i = 10^{-15} \times 10^9 = 10^{-6}$ Watt

or, $\quad \text{SNR} = \frac{10^{-5}}{10^{-6}} = 10$

$\therefore \quad (\text{SNR})_{dB} = 10$ dB

58. **(a)**

Overall noise figure is

$$F = F_1 + \left(\frac{F_2 - 1}{G_1}\right)$$

$$= 5 + \left(\frac{4-1}{12}\right) = 5 + \frac{1}{4} = \frac{21}{4}$$

59. **(c)**

We know that,

Modulation index;

$$m_a = \frac{V_{max} - V_{min}}{V_{max} + V_{min}}$$

Given, $\quad V_{max} = 4$ volt and $V_{min} = 1$ volt

$$\therefore \quad m_a = \frac{4-1}{4+1} = \frac{3}{5} = 0.6 \text{ or } 60\%$$

60. **(a)**

Given, $\quad V_c = 10$ V, $V_m = 3$ V,
$f_c = 30$ kHz, $f_m = 1$ kHz

$\therefore$ Modulation index,

$$m_a = \frac{V_m}{V_c} = \frac{3}{10} = 0.3$$

The sideband frequencies are as under:

$$f_{USB} = f_c + f_m = 30 + 1 = 31 \text{ kHz}$$
$$f_{LSB} = f_c - f_m = 30 - 1 = 29 \text{ kHz}$$

Amplitude of each sideband

$$= \frac{m_a}{2} \times V_c = \frac{0.3 \times 10}{2} = 1.5 \text{ volts}$$

Hence, the spectrum of AM wave will be as shown in figure below.

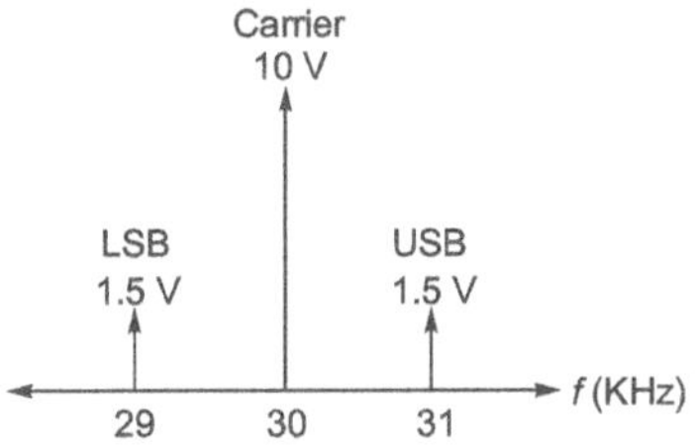

61. **(c)**

In amplitude modulation, the higher percentage of modulation is preferred because the higher value of percentage of modulation results into a strong and more intelligible received signal which is less likely to get contaminated or lost in the noise. A modulation index of more than 100% results into over modulation which introduces envelope distortion. Hence, assertion is true but reason is false.

62. **(d)**

Given, $\quad P_c = 400$ Watt, $m_a = 0.75$

The total power in the amplitude modulated wave is:

$$P_t = P_c\left(1 + \frac{m_a^2}{2}\right)$$

$$= 400\left[1 + \frac{0.75^2}{2}\right] = 512.5 \text{ Watt}$$

63. **(a)**

Given, $\quad I_c = 8$ a, $m_a = 0.8$

We know that, antenna current is

$$I_t = I_c\sqrt{1 + \frac{m_a^2}{2}}$$

$$= 8 \times \sqrt{1 + \frac{0.8^2}{2}} = 8\sqrt{1.32}$$

$$= 8 \times 1.149 \approx 9.2 \text{ A}$$

64. (b)

Given, $I_{t1} = 10$ A, $m_1 = 0.3$, $I_{t2} = 11$ A

We know that, the overall modulation index is:

$$m_t = \sqrt{m_1^2 + m_2^2}$$

or, $$m_2 = \sqrt{m_t^2 - m_1^2}$$

Now, $$I_{t1} = I_C\sqrt{1+\frac{m_1^2}{2}}$$

or, $$I_c = \frac{I_{t1}}{\sqrt{1+\frac{m_1^2}{2}}} = \frac{10}{\sqrt{1+\frac{(0.3)^2}{2}}}$$

$$= \frac{10}{\sqrt{1.045}} = 9.78\text{A}$$

After modulating with the second signal, we have

$$I_{t2} = I_C\sqrt{1+\frac{m_t^2}{2}}$$

or, $$m_t = \sqrt{2\left[\left(\frac{I_{t2}}{I_c}\right)^2 - 1\right]} = \sqrt{2\left[\left(\frac{11}{9.78}\right)^2 - 1\right]}$$

or, $m_t = 0.73$

$\therefore$ $$m_2 = \sqrt{m_t^2 - m_1^2}$$

$$= \sqrt{(0.73)^2 - (0.3)^2} = 0.66 = 66\%$$

65. (d)

We have:

$$A = A_Q[1 + m\sin(2\pi f_m t)]$$

or, $$A_{min} = A_Q[1 + m(-1)]$$
$$= 100[1 - 0.7] = 30$$

and, $$A_{max} = A_Q[1 + m(1)] = 100[1 + 0.7]$$
$$= 170$$

Hence, the range of A = 30 to 170

67. (b)

- The circuit complexity of synchronous detection is very high. Hence, statement-1 is not correct.
- Statement-2 is correct.
- Synchronous detection require a locally generated synchronous carrier. Hence, statement-3 is not correct.
- Statement-4 is correct because synchronous detection is used for DSB-SC, SSB and VSB while envelope detection is used for DSB-FC only.

68. (d)

The principle of envelope detection of an AM signal is based on rectifying the incoming signal and passing it through a LPF. Hence, assertion is false. Assertion is the principle of synchronous detection.

69. (a)

The difference between the two extreme frequencies is equal to the bandwidth of the AM wave.

$\therefore$Bandwidth, $B = (\omega_c + \omega_m) - (\omega_c - \omega_m) = 2\omega_m$

70. (a)

The total power transmitted by an AM wave is given by

$$P_t = P_c + \frac{m_a^2}{4}P_c + \frac{m_a^2}{4}P_c$$

Out of the above total power, the carrier does not contain any information and one sideband is redundant. Hence, out of the total power

$P_t = \left[1+\frac{m_a^2}{2}\right]P_c$, the wasted power is given by:

$$\text{Power wastage} = P_c + \frac{m_a^2}{4}P_c = \left[1+\frac{m_a^2}{4}\right]P_c$$

Hence, both assertion and reason are true and reason is the correct explanation of assertion.

72. (b)

AM $\rightarrow$ used for radio broadcasting

VSB $\rightarrow$ used for transmission of video signals

FDM $\rightarrow$ used for analog signal communication like AM and FM system

73. (d)

$$\text{Power saving} = \frac{\left(P_c + \frac{m_a^2}{4}P_c\right)}{P_c\left(1+\frac{m_a^2}{2}\right)} = \frac{\left(1+\frac{m_a^2}{4}\right)}{\left(1+\frac{m_a^2}{2}\right)}$$

For $m = 1$,

$$\%\text{ Power saving} = \frac{\left(1+\frac{1}{4}\right)}{\left(1+\frac{1}{2}\right)}\times 100$$

$$= \frac{1.25}{1.5}\times 100 = 83.33\%$$

74. (a)

Given, $m_a = 0.5$

$$P_t = P_c\left[1+\frac{m_a^2}{2}\right] = 1.125\,P_c$$

(for $m_a = 0.5$)

For $m_a = 0.5$, the % power saving for DSB-SC signal is

$$= \frac{P_c}{1.125P_c}\times 100 = 88.88\%$$

76. (a)

The AM signal is expressed as

$$x_{AM}(t) = A\cos\omega_c t + \frac{m_a}{2}[\cos(\omega_c - \omega_m)t + \cos(\omega_c + \omega_m)t]$$

$$P_c = \text{Carrier power} = \frac{A^2}{2}$$

$$P_s = \text{Sideband power} = \frac{1}{4}m_a^2 A^2$$

Total power is given by

$$P_t = P_c + P_s = \left(1+\frac{m_a^2}{2}\right)\frac{A^2}{2}$$

$$\therefore \frac{\text{Sideband power}}{\text{Total power}} = \frac{\left[\frac{m_a^2 A^2}{4}\right]}{\left[\frac{(2+m_a^2)A^2}{4}\right]} = \frac{m_a^2}{2+m_a^2}$$

For, $m = 0.8$,

$$\text{required ratio} = \frac{0.8^2}{2+0.8^2} = 0.24$$

77. (b)

- VSB transmission has moderate transmission efficiency. SSB technique has the highest transmission efficiency among all the available AM techniques (i.e. DSB FC, DSB SC, SSB and VSB). Hence, statement-1 is not correct.
- The bandwidth requirement of a standard AM technique i.e. DSB-FC is the highest (i.e. $2f_m$). Hence, statement-2 is correct.
- For T.V. broadcasting VSB is used while for point-to-point communication SSB is used. Hence, statement-3 is not correct.
- Statement-4 is correct.

78. (a)

BW of each AM station = $2\,\omega_m$ = 10 kHz

$$\text{Number of stations} = \frac{100\times 10^3}{10\times 10^3} = 10$$

79. (b)

For a distortion free transmission, the maximum value of the modulation index (m_a) is unity.

80. (c)

Carrier does not contain any useful information.

81. (b)

$$P_t = P_c\left(1+\frac{m_a^2}{2}\right)$$

Total sideband power

$$= P_{SB} = P_t - P_c$$

$$= \frac{m_a^2}{2}P_c$$

$$P_{SB}\big|_{max} = \frac{P_c}{2} \qquad (\text{when, } m_a = 1)$$

83. (d)

$$e_{AM} = A_C\cos\omega_c t\,[1 + m_a\cos\omega_m t] \quad \ldots(i)$$

Given,

$$e_{AM} = 10\cos(6.28\times 10^6 t)\,[1 + 0.4\cos 6280t] \quad \ldots(ii)$$

Comparing equations (*i*) and (*ii*), we have:

$m_a = 0.4$

and $\omega_c = 6.28\times 10^6 = 2\pi f_c$

$$\therefore \quad f_c = \frac{6.28\times 10^6}{2\pi} = 1\times 10^6 \text{ Hz}$$

$$= 1 \text{ MHz}$$

85. (b)

Carrier power is always constant in AM which is equal to P_c (always).

86. (c)

$$(BW)_{VSSB} = \left(1+\frac{r}{2}\right)f_m \text{ and } r = 1 \quad (\text{Given})$$

$$\therefore \quad (BW)_{VSSB} = \left(1+\frac{1}{2}\right)\cdot 4 = 6 \text{ MHz}$$

87. **(b)**

$$RC \le \frac{1}{\omega_m} \cdot \frac{\sqrt{1-m_a^2}}{m_a}$$

Max. envelope time is

$$t_{envelope} = RC = \frac{1}{\omega_m} \cdot \frac{\sqrt{1-m_a^2}}{m_a}$$

$$= \frac{1}{100} \cdot \frac{\sqrt{1-(0.6)^2}}{0.6} = 13.3 \text{ ms}$$

88. **(d)**

$$P_t' = P_c + \frac{P_c m_a^2}{2}$$

$\therefore$ $$P_t' = \frac{3}{2}P_c \quad (\text{If } m_a = 1)$$

If carrier is suppressed, then

$$P_t = \frac{P_c m_a^2}{2}$$

or, $$P_t = \frac{1}{2}P_c \quad (\text{If } m_a = 1)$$

$\therefore$ Power saving

$$P_s = P_c$$

$$\therefore \ \%\text{power saving} = \frac{P_c}{\frac{3}{2}P_c} \times 100\% = 66.66\%$$

92. **(a)**

For FM, the frequency changes with the change in the amplitude of modulating signal value.

93. **(c)**

According to Carson's rule, the bandwidth of a single-tone wideband FM is

$$BW = 2(\Delta\omega + \omega_m)$$

94. **(a)**

A PM wave can be generated using a frequency modulator as shown in figure below:

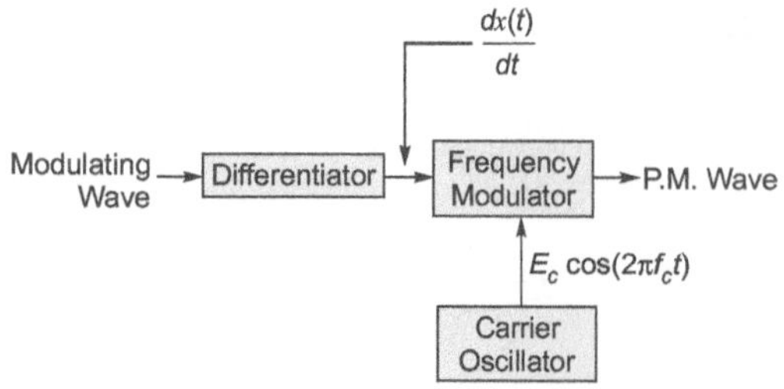

95. **(b)**

Given, $v = 10 \sin [5 \times 10^8 t + 4 \sin 1250 t]$

Comparing the above equation with the standard for FM given by

$$e_{FM} = s(t)$$
$$= E_c \sin [\omega_c t + m_f \sin \omega_m t],$$

we have modulating frequency,

$$f_m = \frac{1250}{2\pi} \text{Hz}$$

and modulation index,

$$m_f = 4$$

$\therefore$ Maximum deviation,

$$\Delta f = m_f \times f_m$$

$$= 4 \times \frac{1250}{2\pi} \cong 796 \text{ Hz}$$

96. **(d)**

Given, $f_m = 1$ kHz, $V_m = 2$ V, $\Delta f = 6$ kHz

We know that,

$$\Delta f = K_f V_m$$

or, $$K_f = \frac{\Delta f}{V_m} = \frac{6 \text{ kHz}}{2 \text{ V}} = 3 \text{ kHz/V}$$

When, $V_m' = 4$ V, the new deviation in frequency is

$$\Delta f' = K_f V_m' = (3 \text{ kHz/V}) \times 4 \text{ V}$$
$$= 12 \text{ kHz}$$

97. **(c)**

The noise immunity of FM is better than AM and PM while the noise immunity of PM is better than AM but worse than FM.

98. **(b)**

- In FM, the frequency deviation is proportional only to the amplitude variation of the modulating signal and it is independent of modulating frequency f_m. Hence, statement-1 is not correct.
- In PM, the frequency deviation increases with increase in modulating frequency f_m and decreases with decrease in modulating frequency. Thus, frequency deviation in PM is proportional to modulating frequency. Hence, statement-2 is false.
- Statement-3 is correct.
- It is possible to receive FM on a PM receiver and PM on a FM receiver. Hence, statement-4 is correct.

- Signal to noise ratio of PM is inferior to that in FM. Hence, statement-5 is not correct.

101. (d)

The centre frequency f_c will change due to mixing. It will increase to $(f_c + f_0)$ if the sum component is selected and it will reduce to $(f_c - f_0)$, if the difference component is selected. Hence, reason is a true statement. Assertion is a false statement because due to mixing, the frequency deviation remains constant and hence the modulation index of FM does not changes.

102. (b)

The deviation ratio for commercial FM transmission is

$$D = \frac{\Delta f}{f_m} = \frac{75 \text{ kHz}}{15 \text{ kHz}} = 5$$

Bandwidth of FM signal using Carson's rule is

$$BW = 2\Delta f\left(1+\frac{f_m}{\Delta f}\right) = 2\Delta f\left(1+\frac{1}{D}\right)$$

$$= 2 \times 75 \times 10^3 \left(1+\frac{1}{5}\right) = 180 \text{ kHz}$$

103. (d)

Given, $x_c(t) = 10 \cos[(10^8)\pi t + 5 \sin 2\pi(10^3)t]$

Comparing the above equation with standard FM wave equation, we have:

$$\theta(t) = \omega_c t + \phi(t)$$
$$= (10^8)\pi t + 5 \sin 2\pi(10^3)t$$

Now, $\phi'(t) = 5(2\pi)(10^3)\cos 2\pi(10^3)t$

$\therefore$ Maximum phase deviation is

$$|\phi(t)|_{max} = 5 \text{ rad}$$

and the maximum frequency deviation will be

$$\Delta\omega = |\phi'(t)|_{max} = 5(2\pi)(10^3) \text{ rad/s}$$

or, $\Delta f = 5$ kHz

105. (a)

The maximum frequency deviation in FM occurs at the point where the rate of change of modulating signal is maximum i.e. at the positive and negative peaks of the modulating signal.

106. (c)

Practically, the narrowband FM systems have m_f less than 1. The maximum permissible frequency deviation is restricted to about 5 kHz.

107. (d)

Wideband FM is used in the entertainment broadcasting applications such as FM radio, TV etc.

108. (b)

The range of modulating frequency for wideband FM is 30 Hz to 15 kHz while for narrowband FM is 30 Hz to 3 kHz.

109. (c)

- In direct method of FM generation, it is not easy to get a high order stability in carrier frequency because the carrier signal is directly affected by the baseband or modulating signal. Hence, statement-1 is correct.
- The frequency stability is very high in Armstrong method of FM generation due to the use of crystal oscillator. Hence, statement-2 is correct.
- Statement-3 is correct.
- Statement-4 is also correct.

110. (a)

PLL is basically a negative feedback system which may be used to extract a baseband signal from a frequency modulated carrier. It consists of 3 major components namely multiplier, loop filter and a voltage controlled oscillator (VCO) as shown below.

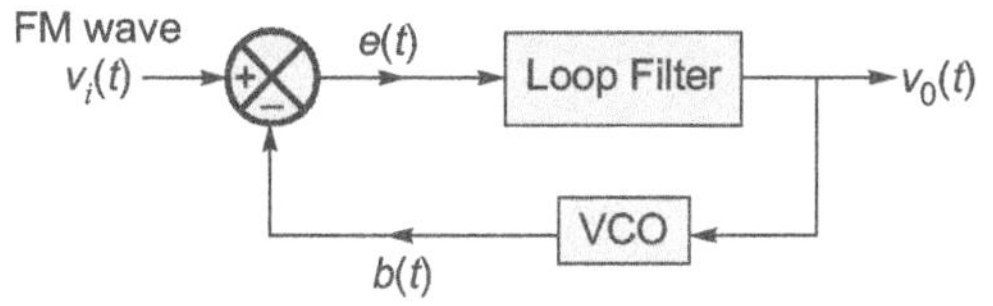

111. (b)

The modulation index of an FM controls the number of sidebands in it. Thus, we can have infinite no. of sidebands in FM.

112. (a)

FM can be generated using PM by passing the modulating signal through a low pass filter (or an integrator).

113. (a)

Instantaneous frequency is given by

$$\omega_t = \frac{d\theta(t)}{dt}$$
$$= 2\pi \times 10^6 + 60\pi \times 150 \cos 150t$$
$$- 80\pi \times 150 \sin 150t$$
$$\Delta\omega = \omega_1 - \omega_c = 15000\,\pi \cos(150\,t + \alpha)$$
$$\therefore \quad \Delta\omega_{max} = 15 \times 10^3 \pi \text{ rad/s}$$

114. (b)

$$\text{SNR} \propto (\text{BW})^2$$

118. (c)

In order to minimize the noise power that is presented to the demodulator of a receiver, a filter is introduced before the demodulator as shown below.

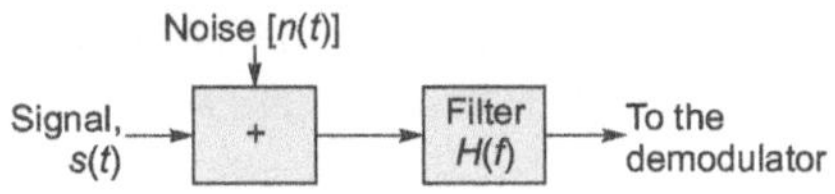

119. (c)

Output signal to noise ratio (SNR) is more useful measure of noise performance.

$$\text{SNR} = \frac{\text{Average power of demodulated message signal}}{\text{Average power of the noise measured at the output of the demodulator}}$$

SNR depends both on the type of modulation used at the transmitter and the type of demodulation used at the receiver.

120. (d)

Assertion is a false statement because figure of merit of a radio receiver can be less than, greater than or equal to 1 depending on the type of modulation.

121. (a)

- Statement-1 is correct.
- The figure of merit of an AM receiver with envelope detection is less than 1.

$$\text{i.e. FOM} = \frac{\left(\frac{m}{V_m}\right)^2 P}{1+\left(\frac{m}{V_m}\right)^2 P} \text{ which is} < 1.$$

Hence, statement-2 is false.

- Statement-3 and 4 are correct.

122. (b)

DM (Delta modulated system) is a coded system.

125. (c)

The sensitivity of a superheterodyne receiver does not depend on the receiving frequency. The selectivity of a superheterodyne receiver depends on the receiving frequency. It becomes somewhat worse when the receiving frequency is raised.

128. (b)

Given, $f_i = 455$ kHz, $f_s = 1000$ kHz
$Q = 100$

The image-frequency is

$$f_{si} = f_s + 2f_i$$
$$= 1000 + (2 \times 455)$$
$$= 1000 + 910 = 1910 \text{ kHz}$$

129. (a)

Given:

$IF = 455$ kHz $= f_i$, $f_s = 900$ kHz and $Q = 80$

Image frequency,

$$f_{si} = f_s + 2f_i$$
$$= 900 + (2 \times 455)$$
$$= 1810 \text{ kHz}$$

Image frequency rejection ratio is

$$\alpha = \sqrt{1+Q^2\rho^2}$$

Here, $$\rho = \left(\frac{f_{si}}{f_s} - \frac{f_s}{f_{si}}\right) = \left(\frac{1810}{900} - \frac{900}{1810}\right)$$
$$= 1.51$$

$$\therefore \quad \alpha = \sqrt{1+Q^2\rho^2}$$
$$= \sqrt{1+(80)^2 \times (1.51)^2} = 120.8$$

130. (d)

The local oscillator frequency is

$$f_0 = f_s + f_i$$
$$= (900 + 455) \text{ kHz}$$
$$= 1355 \text{ kHz}$$

131. (a)

Due to superheterodyning, the BW remains constant over the entire operating range.

132. (b)

The gain provided by the R.F. amplifier will result in improved signal to noise ratio because the incoming weak signal is raised to a higher level with the help of R.F. amplifier before it is fed at the input of the mixer stage which contributes most of the noise generated in the receiver. Reason is also a true statement but not the correct explanation of assertion.

133. (d)

Given, $f_0 = 1$ MHz, BW = 10 KHz

$$\text{Quality factor} = \frac{f_0}{\text{BW}} = \frac{1\times 10^6}{10\times 10^3} = 100$$

134. (c)

Automatic gain control (AGC) is the simplest and the most effective means adopted to balance the fading.

135. (c)

$$540 < f_c < 1600 \text{ kHz}$$
$$f_{LO} = f_c + 455 \text{ kHz}$$

For $f_c = 540$ kHz and 1600 kHz

$$995 < f_{LO} < 2055 \text{ kHz}$$

136. (c)

$$F = \left(1+\frac{T_{eq}}{T_0}\right) = \left(1+\frac{4306}{290}\right) = 15.85$$

$$(F)_{\text{dB}} = 12 \text{ dB}$$

142. (a)

Given, $x(t) = 6\cos 50\pi t + 20\sin 300\pi t - 10\cos 100\pi t$ ...(*i*)

Also, $x(t) = A_1\cos\omega_1 t + A_2\cos\omega_2 t + A_3\cos\omega_3 t$ ...(*ii*)

Comparing equations (*i*) and (*ii*), we have:

$$f_1 = \frac{\omega_1}{2\pi} = \frac{50\pi}{2\pi} = 25 \text{ Hz}$$

$$f_2 = \frac{\omega_2}{2\pi} = \frac{300\pi}{2\pi} = 150 \text{ Hz};$$

$$f_3 = \frac{\omega_3}{2\pi} = \frac{100\pi}{2\pi} = 50 \text{ Hz}$$

Thus, the highest frequency component of the given message signal will be

$$f_{max} = 150 \text{ Hz}$$

$\therefore$ Nyquist rate $= 2f_{max} = 2 \times 150 = 300$ Hz

143. (b)

Given signal is

$$x(t) = \frac{1}{2\pi}\cos(4000\pi t)\cos(1000\pi t)$$
$$= \frac{1}{4\pi}[\cos(4000\pi t + 1000\pi t) + \cos(4000\pi t - 1000\pi t)]$$
$$= \frac{1}{4\pi}[\cos 5000\pi t + \cos 3000\pi t]$$

or, $$x(t) = \frac{1}{4\pi}[\cos\omega_1 t + \cos\omega_2 t]$$

Here, $\omega_1 t = 5000\pi t$

$\therefore$ $\omega_1 = 5000\pi$ rad/s

$\therefore$ $f_1 = 2500$ Hz

Similarly, $\omega_2 = 3000\pi$ rad/s

$\therefore$ $f_2 = 1500$ Hz

Hence, the maximum frequency present in $x(t)$ is $f_1 = 2500$ Hz

$\therefore$ Nyquist rate is given as

$$f_s = 2f_m$$

(where, f_m = Maximum frequency present in the signal)

Here, $f_m = f_1 = 2500$ Hz

$\therefore$ Nyquist rate is

$$f_s = 2f_m = 2 \times 2500 = 5000 \text{ Hz} = 5 \text{ kHz}$$

Nyquist interval is given as

$$T_s = \frac{1}{2f_m} = \frac{1}{2\times 2500}\,s = \frac{1}{5000}\,s = 0.2 \text{ ms}$$

144. (c)

Let, $x(t) = A_m\sin(2\pi f_m t)$

The slope of $x(t)$ will be maximum when derivative of $x(t)$ w.r.t. 't' will be maximum.

$$\text{Maximum slope} = \frac{\Delta}{T_s}$$

Hence, for slope overload distortion not to occur,

$$\max\left|\frac{d}{dt}x(t)\right| > \frac{\Delta}{T_s}$$

or, $$\max\left|\frac{d}{dt}A_m\sin(2\pi f_m t)\right| > \frac{\Delta}{T_s}$$

or, $$\max\left|A_m 2\pi f_m\cos(2\pi f_m t)\right| > \frac{\Delta}{T_s}$$

or, $A_m 2\pi f_m > \frac{\Delta}{T_s}$ or $A_m > \frac{\Delta}{2\pi f_m T_s}$

Hence, slope overload distortion doesn't occur

if $A_m \le \frac{\Delta}{2\pi f_m T_s}$

146. (a)

In practice, the transmitted pulses usually have slightly sloping sides (edges). As the noise is superimposed on them, the width and the position of the regenerated pulses is changed. This distorts the information contents in the PWM and PPM signals. Since PCM does not contain any information in the width or the position of the pulses, therefore PCM has much better noise immunity as compared to PAM, PWM and PPM systems. Hence, both assertion and reason are true and reason is the correct explanation of assertion.

147. (d)

Since, the delta modulation transmits only one bit for one sample, therefore the signaling rate and transmission channel bandwidth is quite small for delta modulation compared to PCM. Hence, assertion is a false statement.

The transmission and receiver implementation is very simple for delta modulation since there is no A/D converter required in delta modulation unlike PCM. Hence, reason is a true statement.

148. (b)

Given, $f_m = 4.5$ MHz, $q = 1024$

f_s = 20% above the Nyquist rate

$= 1.2 \times 2 f_m$

$= 1.2 \times 2 \times 4.5$ MHz

$= 10.8$ MHz

Now, quantization level, $q = 2^n$

(n = No. of bits/sec)

$\therefore \quad 2^n = 1024 = 2^{10}$

or, $\quad n = 10$

$\therefore$ bits/sec is $r_b = nf_s = 10 \times 10.8$ MHz

$= 108$ M bits/sec

149. (a)

Given, SNR = 30 dB, $n = 10$,

Desired value of SNR = 42 dB

With the increase in n by 1 bit, the value of SNR increases by 6 dB. Therefore, to increase the value of SNR by 12 dB, it is necessary to increase n by 2,

$\therefore \quad n = 10 + 2 = 12$

Now, BW of PCM system $= \frac{1}{2} nf_s$

Therefore, BW with $n = 10$ is

$$(BW)_{10} = \frac{1}{2} \times 10 f_s = 5 f_s$$

and BW for $n = 12$ is

$$(BW)_{12} = \frac{1}{2} \times 12 f_s = 6 f_s$$

$\therefore$ change in BW; $\Delta W = 6f_s - 5 f_s = f_s$

Hence, % change in BW $= \frac{f_s}{5f_s} \times 100\% = 20\%$

150. (c)

$$(SNR)_q = (1.8 + 6\,n) \text{ dB}$$
$$= 1.8 + (6 \times 8)$$
$$= 49.8 \text{ dB}$$

151. (c)

PAM, PWM and PPM are examples of analog pulse modulation techniques while PCM is a digital pulse modulation.

153. (c)

Granular noise occurs in delta modulation when the step size is too large compared to small variations in the input signal.

154. (a)

PCM compared to analog modulation require large transmission bandwidth. This is the main drawback of PCM compared to analog modulation.

155. (c)

Signal to quantization noise ratio is given by

$$SN_qR = \frac{3}{2} 2^{2n} = \frac{3}{2}(4)^{BW/f_m} = \frac{3}{2} Q^2$$

Here, BW = Bandwidth of message signal

$Q = 2^n$ = No. of quantization levels

156. (b)

- In PAM, noise interference is high.
- Delta modulation introduces idle noise (also called Granular noise).
- PWM is simple to implement and is similar to frequency modulation.
- PPM has highest SNR.

157. (c)

The information in PCM is transmitted in the form of code words. It is in the form of digital pulses of constant amplitude, width and position.

158. (c)

According to Shannon-Hartely theorem, the channel capacity is given by

$$C = B\log_2\left[1+\frac{S}{N}\right]$$

Here, B = Bandwidth and S/N = signal to noise ratio.

159. (a)

- In PCM, the analog signals are first coded into digital form, and then transmitted. It is in the form of digital pulses of constant amplitude, width and position. Hence, statement-1 is correct.
- Due to higher bandwidth requirement and complicated circuitry of PCM, it is not used for the radio and TV broadcasting applications. Hence, statement-2 is false.
- Slope overload noise occurs in delta modulation. To reduce the error or slope overload noise, the step size must be increased when slope of signal is high. Hence, statement-3 is true.
- Companding is a non-uniform quantization. It is used to improve the signal to quantization noise ratio of weak signal and can be used for PCM and DM. Hence, statement-4 is correct.

160. (b)

The most important feature of PCM system lies in its ability to control the effects of distortion and noise when the PCM wave travels on the channel. PCM accomplishes this capacity by means of using a chain of regenerative repeaters. Such repeaters are spaced close enough to each other on the transmission path. The regenerative performs three basic operations namely equalization, timing and decision making.

163. (b)

When a continuous time band-limited signal is sampled at a rate lower than Nyquist rate $f_s < 2f_m$, the signal is under-sampled and some amount of aliasing is produced in this under-sampling process. Aliasing is the phenomenon in which a high frequency component in the frequency-spectrum of the signal takes identity of a lower-frequency component in the spectrum of the sampled signal.

165. (b)

In flat-top sampling or rectangular pulse sampling, the top of the samples remains constant and is equal to the instantaneous value of the baseband signal $x(t)$ at the start of sampling. A sample and hold circuit is used to generate the flat top samples.

167. (b)

$$\text{Step size, } D = \frac{V_{p-p}}{Q} = \frac{1.536}{128}$$

$$\text{QNP} = \frac{\Delta^2}{12} = \frac{\left(\frac{1.536}{128}\right)^2}{12} = 12\times10^{-6}\text{ W}$$

170. (d)

The condition for no slop occur overload is

$$|m(t)|_{max} = \omega A \le \Delta f_s$$

or, $A \le \dfrac{\Delta f_s}{\omega}$

where, $\Delta = 0.1$ V,

$f_s = 20\,\pi$ kHz,

$\omega = 400\,\pi$ rad/sec

and $A = 5$ V

173. (b)

BPSK is the most efficient of the three digital modulation, i.e. ASK, FSK and PSK because BPSK is used for high bit rates. Reason is also a true statement but not the correct explanation of assertion.

175. (a)

- Statement-1 is correct.
- BPSK has the best performance of all the three digital modulation techniques in presence of noise. It yields the minimum value of probability of error compared to ASK, FSK and PSK. Hence, statement-2 is false.
- Statement-3 is correct since binary FSK has highest system complexity while binary ASK has least system complexity.
- Binary ASK is demodulated using envelope detection while binary FSK and PSK are demodulated using coherent detection. Hence, statement-4 is false.

176. (d)

The bit period of NRZ data stream is

$$T_b = \frac{1}{f_b} = \frac{1}{1\,\text{MHz}} = 1\,\mu\text{s}$$

In BPSK, each binary bit is a symbol, therefore symbol duration $T_s = T_b = 1\,\mu\text{s}$.

∴ The bandwidth required for BPSK system is

$$(\text{BW})_{\text{BPSK}} = 2f_b = \frac{2}{T_b} = \frac{2}{1\times 10^{-6}} = 2\,\text{MHz}$$

In QPSK, we group two successive bits to form one symbol, therefore symbol duration,

$$T_s = 2T_b = 2\,\mu\text{s}.$$

$$(\text{BW})_{\text{QPSK}} = f_b = \frac{1}{T_b} = \frac{2}{2\,\mu\text{s}} = 1\,\text{MHz}$$

177. (c)

Given, $f_b = 90$ Mb/s and BW = 20 MHz

Hence, the technique that can be used here is M-ary PSK. The bandwidth of an M-ary PSK system is given by

$$\text{BW} = \frac{2f_b}{N}$$

(Here, N = No. of bits/symbol)

$$\therefore \quad 20\,\text{Mb/s} = \frac{2\times 90\,\text{MHz}}{N}$$

or, $$N = \frac{180}{20} = 9$$

178. (c)

Minimum bandwidth required in BPSK = $2f_b$.
Minimum bandwidth required in BFSK = $4f_b$.
Hence, bandwidth of BPSK is lower than that of BFSK.

179. (c)

The error probability of QPSK is better (or lower) than that of BPSK due to which QPSK is used for very high bit rate data transmission.

180. (a)

More energy (or power) is contained in the main lobe of MSK (about 99% of the total energy or power) as compared to the QPSK (about 90% of the total energy or power).

181. (c)

In BPSK, baud rate is half the bit rate due to which there is more effective utilization of the available bandwidth of the transmission channel.

Given, bit rate = 34 Mbps

$$\therefore \quad \text{Baud rate} = \frac{1}{2}\times \text{Bit rate} = \frac{1}{2}\times 34\,\text{Mbps} = 17\,\text{Mbps}$$

183. (b)

Minimum shift keying (MSK) is basically frequency shift keying (FSK).

184. (c)

- Since DPSK uses two successive bits for its reception, error in the first bit creates error in the second bit therefore, error propagation in DPSK is more. On the other hand, in BPSK single bit can go in error since detection of each bit is independent. Hence, DPSK has higher probability of error than BPSK.
  Thus, assertion is true.
- Reason is a false statement because DPSK does not require carrier at the receiving end unlike BPSK due to which the complicated circuitry for generation of local carrier is not required in DPSK. Thus, reason is a false statement.

186. (c)

The generation and detection of QPSK is quite complex. Hence, option (c) is not correct or not an advantage of QPSK.

187. (a)

Bandwidth for QPSK = f_b and
Bandwidth for BPSK = $2f_b$
(where, f_b = bit rate in bits/sec)

188. (b)

The probability of error for BPSK signal is:

$$P_c = \frac{1}{2}\text{erfc}\left[\frac{E_s}{\eta}\cos^2\phi\left(1-2\frac{|\tau|}{T}\right)\right]^{\frac{1}{2}}$$

Here, $\tau = 0$, and $\phi = 20°$ (Given)

Then, $$P_c = \frac{1}{2}\text{erfc}\left[\frac{E_s}{\eta}\cos^2 20\right]^{\frac{1}{2}} \quad ...(i)$$

Minimum probability of error is given by

$$P_c = \frac{1}{2}\text{erfc}\left[\frac{E_s}{\eta}\right]^{\frac{1}{2}} \quad ...(ii)$$

Comparing equations (*i*) and (*ii*), the phase error comes out to be:

Phase error = $\cos^2 20°$

189. (c)

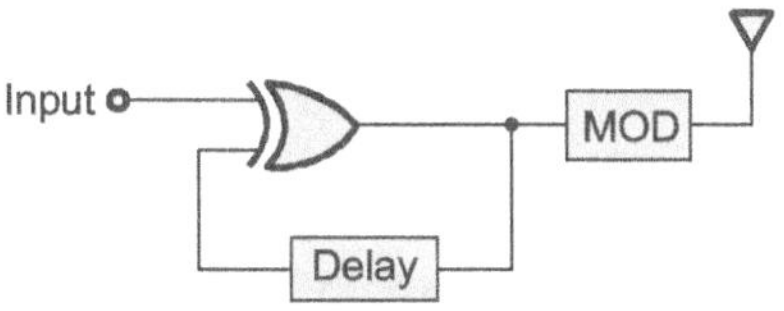

A	B	C
0	0	0
0	1	1
1	0	1
1	1	0

$Y = A \oplus B$
$= A\bar{B} + \bar{A}B$

Given that,
Reference bit = 1
logic 0 $\rightarrow \pi$
logic 1 $\rightarrow 0$

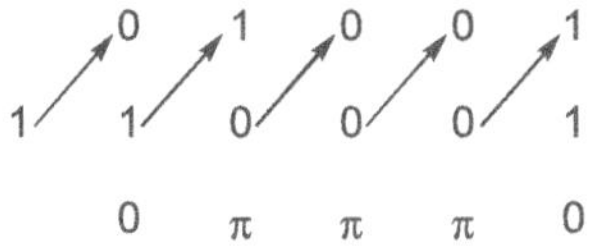

190. (c)

Bandwidth of BPSK signal is:

$$B_p = 2f_b$$

and bandwidth of on-off keying is:

$$B_o = 2f_b$$

$$\therefore \quad B_p = B_o$$

191. (b)

Entropy is

$$H = \frac{1}{2}\log_2 2 + \frac{1}{4}\log_2 4 + \frac{1}{8}\log_2 8 + \frac{1}{8}\log_2 8$$

$= 1.75$ bits/symbol

193. (b)

PSK BW $= 2R_b$

Bit rate, $R_b = nf_s$
$= 10 f_s = 10(2f_m)$
$= 10 \times 20 \times 10^3 = 200$ kHz

PSK BW $= 2R_b = 2 \times 200 = 400$ kHz

194. (b)

$$T_b = \frac{1}{0.1\times 10^6} = 10^{-5}s$$

Now, $1 + \alpha = 2\times f_b T_b$
$= 2 \times 75 \times 10^3 \times 10^{-5} = 1.5$

or, $\alpha = 0.5$

195. (b)

The entropy of the system is

$$H(x) = \sum_{i=1}^{5} P(x_i)\log_2\frac{1}{P(x_i)} \text{ bits/symbol}$$

or, $$H(x) = \frac{1}{2}\log_2 2 + \frac{1}{4}\log_2 4 + \frac{1}{8}\log_2 8 + \frac{1}{16}\log_2 16 + \frac{1}{16}\log_2 16$$

$$= \frac{1}{2} + \frac{2}{4} + \frac{3}{8} + \frac{4}{16} = \frac{15}{8}$$

$= 1.875$ bits/outcome

Now, rate of outcome $r = 16$ outcomes/sec (Given)

$\therefore$ The rate of information R is

$$R = rH(x) = 16 \times \frac{15}{8}$$

$= 30$ bits/sec

196. (b)

According to Shannon-Hartley law, the channel capacity is expressed as

$$C = B\log_2\left(1+\frac{S}{N}\right)$$

Thus, if signal power is more, bandwidth will be less and vice-versa. Thus, assertion is a true statement. Reason is also a true statement because the rate of information that can be transmitted depends on bandwidth and signal to noise power. Thus, both assertion and reason are true but reason is not the correct explanation of assertion.

197. **(d)**

We know that the differential entropy of x is given by

$$H(x) = -\int_{-\infty}^{\infty} f_x(x) \log_2 f_x(x)\, dx \text{ bits/sample}$$

Using the given probability density function, we have:

$$H(x) = -\int_0^a \frac{1}{a} \log_2 \frac{1}{a} dx = \log_2 a$$

For $a = 1$, $H(x) = \log_2 1 = 0$

198. **(d)**

Given, $B_1 = 8$ kHz, $\left(\frac{S}{N}\right)_1 = 31$

and $\left(\frac{S}{N}\right)_2 = 61$

We know that channel capacity is

$$C = B \log_2 \left(1 + \frac{S}{N}\right)$$

Since channel capacity remains constant, therefore

$$B_1 \log_2 \left(1 + \left(\frac{S}{N}\right)_1\right) = B_2 \log_2 \left(1 + \left(\frac{S}{N}\right)_2\right)$$

or, $8 \log_2 (1 + 31) = B_2 \log_2 (1 + 61)$

or, $B_2 = \dfrac{8 \log_2 32}{\log_2 62}$

$\therefore$ New bandwidth $= 8 \dfrac{\log 32}{\log 62}$ kHz

199. **(b)**

The entropy of the system is

$$H(x) = \sum_{i=1}^{4} P(m_i) \log_2 \frac{1}{P(m_i)} \text{ bits/symbol}$$

$$= \frac{1}{2}\log_2 2 + \frac{1}{4}\log_2 4 + \frac{1}{8}\log_2 8 + \frac{1}{8}\log_2 8$$

$$= \frac{1}{2} + \frac{1}{2} + \frac{3}{8} + \frac{3}{8} = \frac{7}{4} = 1.75 \text{ bit/symbol}$$

200. **(b)**

Channel capacity is

$$C = B \log_2 \left(1 + \frac{S}{N}\right)$$

$= 4 \times 10^3 \log_2 (1 + 15)$

$= 4 \times 10^3 \log_2 16$ bps

$= 16$ kbps

202. **(a)**

$(1 - p)^n + np\,(1 - p)^{n-1}$ = Required probability

(Here, n = no. of bits and p = probability)

$\therefore$ Required probability

$$= \left(1 - \frac{1}{2}\right)^3 + \frac{3}{2}\left(1 - \frac{1}{2}\right)^2 = \frac{1}{2}$$

203. **(c)**

The hopped bandwidth

= 2 × Band rate

= 2 × 1200 = 2400 Hz

So, no. of channels

$$= \frac{10 \times 10^6}{2400} = 4167$$

204. **(a)**

Minimum distance in hamming code = 3.

206. **(d)**

Bit rate = n.m.f_s

(n = No. of bits and m = No. of signals)

and $f_s = 2f_m$

$\therefore$ Bit rate $= (8 \times 2 \times 2 \times 4) + (8 \times 2 \times 2 \times 8)$

$= 384$ kbps

Here, 2 signals have $f_m = 4$ kHz and other 2 signals have $f_m = 8$ kHz. That's why we have to take both separately.

207. **(a)**

$$\text{Probability error} = {}^3C_2\, p_e^2 (1 - p_e) + {}^3C_3\, p_e^3$$

$$= 3p_e^2 - 2p_e^3$$

■■■■

UNIT 13

Signals & Systems

1. Introduction to Signals

Q.1 What is the total energy of the rectangular pulse shown in figure below?

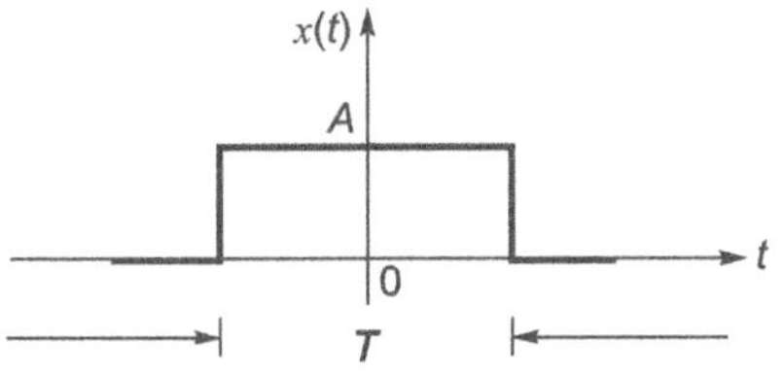

(a) AT (b) $A^2 T$
(c) $A^2 T^2$ (d) AT^2

Q.2 The signal $x(t) = A\cos(\omega t + \phi)$ is
(a) an energy signal
(b) a power signal
(c) an energy as well as a power signal
(d) neither an energy nor a power signal

Q.3 The auto-correlation function $R_x(\tau)$ satisfies which one of the following properties?
(a) $R_x(\tau) = -R_x(-\tau)$ (b) $R_x(\tau) = R_x(-\tau)$
(c) $R_x(\tau) \geq -R_x(0)$ (d) $R_x(\tau) \geq 1$

Q.4 The auto-correlation function $R_x(\tau)$ of the signal $x(t)$ = V sinωt is given by

(a) $\frac{1}{2}V^2\cos\omega\tau$ (b) $V^2\cos\omega\tau$
(c) $V^2\cos^2\omega\tau$ (d) $2V^2\cos^2\omega\tau$

Q.5 The sampling of a function $f(t) = \sin(2\pi f_0 t)$ starts from zero-crossing. The signal can be detected, if sampling time T is

(a) $T = \frac{1}{2f_0}$ (b) $T > \frac{1}{2f_0}$

(c) $T < \frac{1}{2f_0}$ (d) $T \leq \frac{1}{2f_0}$

Q.6 What is the power and energy of the unit step sequence $u(n)$ respectively?
(a) $\infty, 0$ (b) $0, \infty$

(c) $\frac{1}{2}, 0$ (d) $\frac{1}{2}, \infty$

Q.7 Let $\delta(t)$ is the delta function the value of integral

$$\int_{-\infty}^{+\infty} \delta(t)\cos\left(\frac{3t}{2}\right)dt \text{ is}$$

(a) 1 (b) – 1
(c) 0 (d) $\pi/2$

Q.8 If a signal $f(t)$ has energy 'E' the energy of the signal $f(2t)$ is equal to
(a) E (b) $E/2$
(c) $2E$ (d) $4E$

Q.9 Consider the sequence

$$x[n] = [-4 - j5 \quad 1 + j2 \quad 4]$$
$$\uparrow$$

the conjugate anti-symmetric part of the sequence is
(a) $[-4 - j2.5 \quad j2 \quad 4 - j2.5]$
(b) $[-j2.5 \quad 1 \quad j2.5]$
(c) $[-j5 \quad j2 \quad 0]$
(d) $[-4 \quad 1 \quad 4]$

Q.10 The function $x(t)$ is shown in the figure. Even and odd parts of a unit step function $u(t)$ are respectively

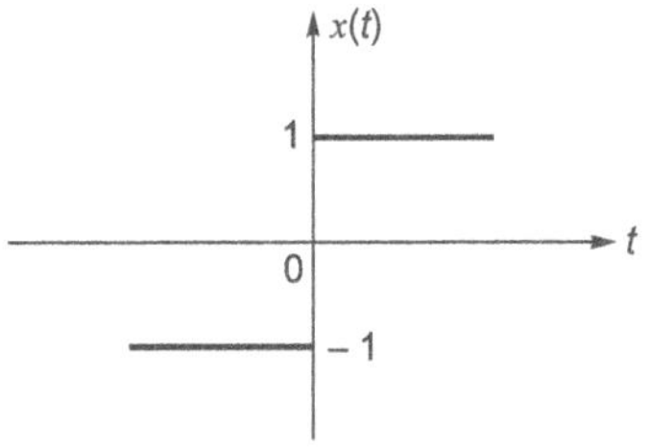

(a) $\frac{1}{2}, \frac{1}{2}x(t)$ (b) $-\frac{1}{2}, \frac{1}{2}x(t)$

(c) $\frac{1}{2}, -\frac{1}{2}x(t)$ (d) $-\frac{1}{2}, -\frac{1}{2}x(t)$

Q.11 The power in the signal,

$$S(t) = 8\cos\left(20\pi t - \frac{\pi}{2}\right) + 4\sin(15\pi t) \text{ is}$$

(a) 40 (b) 41
(c) 42 (d) 82

Q.12 The dirac delta function $\delta(t)$ is defined as

(a) $\delta(t) = \begin{cases} 1 \ ; & t = 0 \\ 0 \ ; & \text{otherwise} \end{cases}$

(b) $\delta(t) = \begin{cases} \infty \ ; & t = 0 \\ 0 \ ; & \text{otherwise} \end{cases}$

(c) $\delta(t) = \begin{cases} 1 \ ; & t = 0 \\ 0 \ ; & \text{otherwise} \end{cases}$ and $\int_{-\infty}^{\infty} \delta(t)\, dt = 1$

(d) $\delta(t) = \begin{cases} \infty \ ; & t = 0 \\ 0 \ ; & \text{otherwise} \end{cases}$ and $\int_{-\infty}^{\infty} \delta(t)\, dt = 1$

Q.13 A signal $m(t)$ with bandwidth 500 Hz is first multiplied by a signal $g(t)$ where

$$g(t) = \sum_{k=\infty}^{\infty} (-1)^k \, \delta(t - 0.5 \times 10^{-4} k)$$

The resulting signal is then passed through an ideal low pass filter with bandwidth 1 kHz. The output of the low pass filter would be

(a) δ (b) $m(t)$
(c) 0 (d) $m(t)\,\delta(t)$

Q.14 Which of the following is false statement?

(a) $t\,\delta(t) = 0$
(b) $\cos t \cdot \delta(t - \pi) = -\delta(t - \pi)$
(c) $\delta(t) = \int u(t)\, dt$
(d) $t\,\delta'(t) = \delta(t)$

Q.15 Consider the voltage wave form shown in below figure the equation for $V(t)$ is

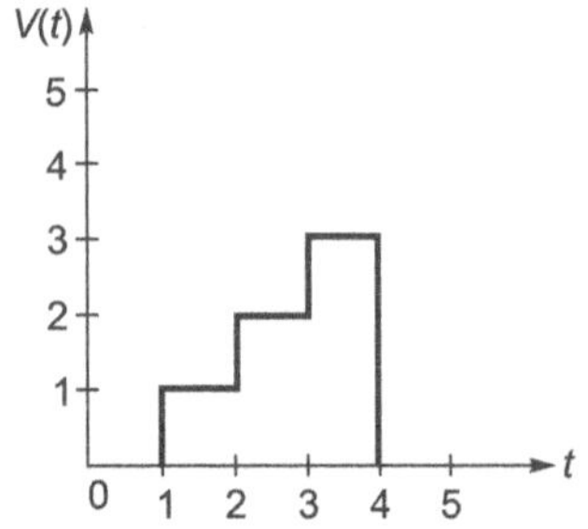

(a) $u(t-1) + u(t-2) + u(t-3)$
(b) $u(t-1) + 2u(t-2) + 3u(t-3)$
(c) $u(t-1) + u(t-2) + u(t-3) + u(t-4)$
(d) $u(t-1) + u(t-2) + u(t-3) - 3u(t-4)$

Q.16 A signal is described by $x(t) = r(t-4) - r(t-6)$. Where $r(t)$ is a unit ramp function starting at $t = 0$. The signal $x(t)$ is represented as

(a)
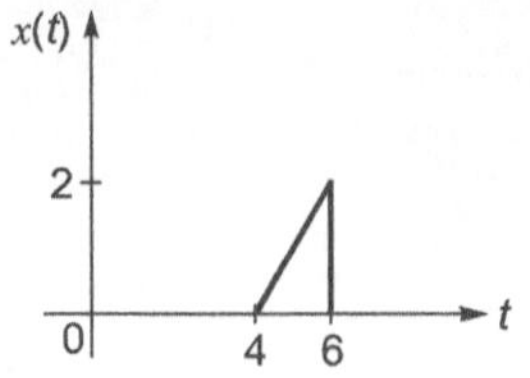

(b)
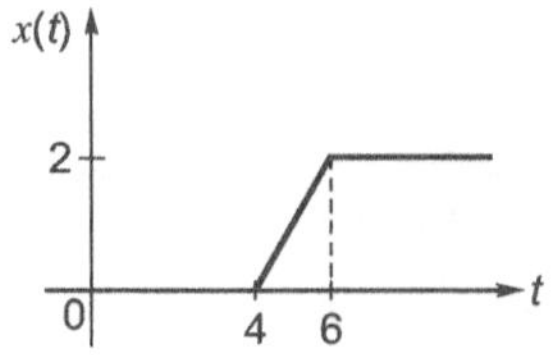

(c)
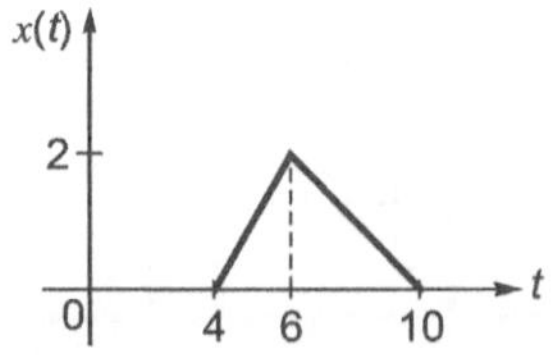

(d)
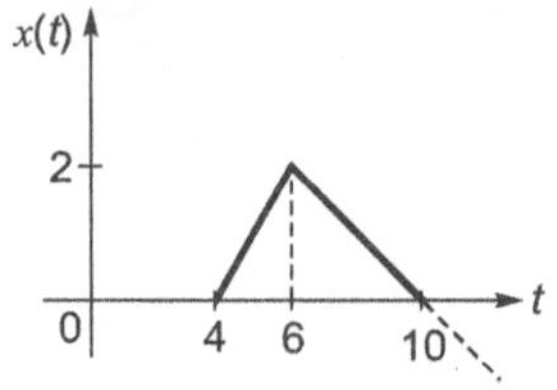

Q.17 Consider the signal $x(t) = \delta(t+2) - \delta(t-2)$. The value of E_∞ for the signal

$$y(t) = \int_{-\infty}^{t} x(\tau)\, d\tau \text{ is}$$

(a) 4 (b) 2
(c) 1 (d) ∞

Q.18 A discrete time signal is given below,

$$x[n] = \cos\frac{\pi n}{9} + \sin\left(\frac{\pi n}{7} + \frac{1}{2}\right)$$

the period of the signal is

(a) Periodic with period $N = 126$
(b) Periodic with period $N = 32$
(c) Periodic with period $N = 252$
(d) Not periodic

Q.19 $x[n] = (-0.5)^n\, u[n]$

Which of the following is correct?

(a) $x[n]$ is an energy signal.
(b) $x[n]$ is an power signal.
(c) $x[n]$ is an energy signal as well as power signal.
(d) $x[n]$ is neither energy nor power signal.

Q.20 If, (i) $x_1(t) = 2\sin\pi t + \cos 4\pi t$
(ii) $x_2(t) = \sin 5\pi t + 3\sin 13\pi t$

Which of the above two are periodic?

(a) (i) only (b) (ii) only

(c) both (i) and (ii) (d) none of the above

Q.21 Consider the continuous time signal $x(t) = \delta(t+2) - \delta(t-2)$. The value of E_∞ for the signal $y(t) = \int_{-\infty}^{t} x(\tau)\, d\tau$ is

(a) 2 (b) 0

(c) 4 (d) 1

Q.22 The fundamental period of the signal $x[n] = 1 + e^{j4\pi n/7} - e^{j2\pi n/5}$ is

(a) 35 (b) 70

(c) 140 (d) 25

Q.23 PDF of random variable is given below. Its CDF can be plotted as,

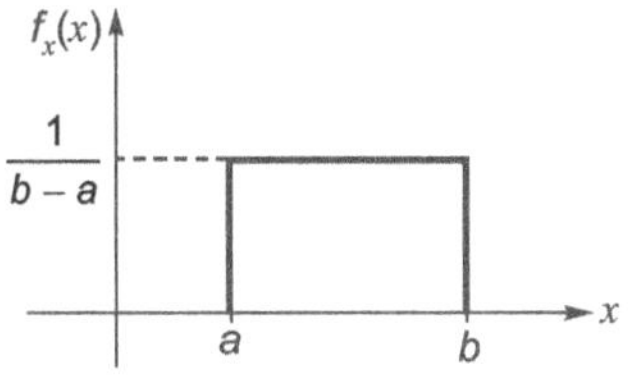

(a)

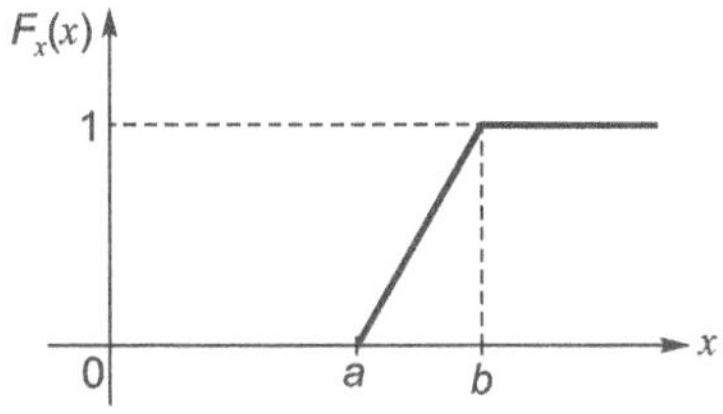

(b)

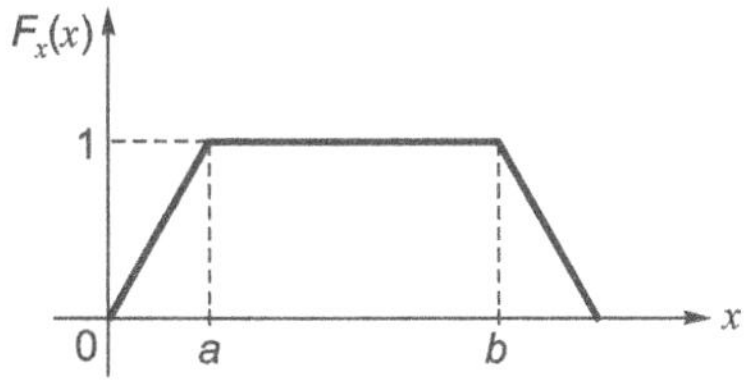

(c)

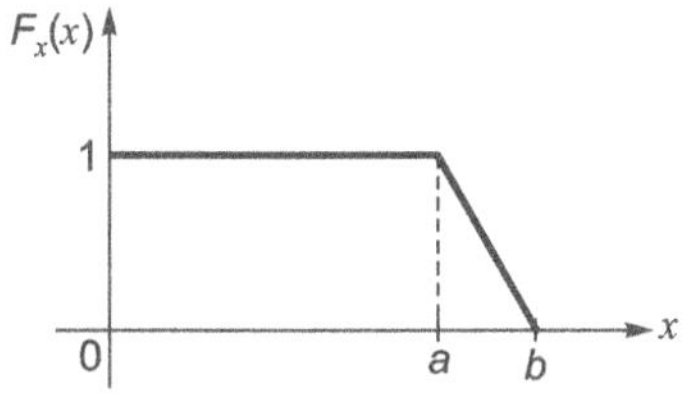

(d) None of the above

Q.24 Statement-1: $x[n] = \cos\left[\frac{1}{4}n\right]$ is non periodic.

Statement-2: $x(t) = e^{j\left[\left(\frac{\pi}{2}\right)t - 1\right]}$ is periodic.

Choose correct option:

(a) Both statement are true.

(b) Statement 1 is false, but statement 2 is true.

(c) Statement 1 is true, but statement 2 is false.

(d) Both statements are false.

Q.25 Consider a continuous time signal given by $x(t) = \delta(t+2) - \delta(t-2)$. The value of energy for $y(t) = \int_{-\infty}^{t} x(\tau)\, d\tau$ is

(a) $\frac{1}{4}$ (b) 1

(c) 2 (d) 4

Q.26 Consider signal x(t) given by

$x(t) = \sin\omega_0 t \times \cos\sqrt{2}\,\omega_0 t + \cos\omega_0 t \times \sin\sqrt{2}\,\omega_0 t;$

the signal is

(a) Periodic, $T = \frac{\sqrt{2}\pi}{1+\omega_0}$

(b) Periodic, $T = \frac{2\pi}{\sqrt{2}+\omega_0}$

(c) Periodic, $T = \frac{2\pi}{(1+\sqrt{2})\cdot\omega_0}$

(d) Not periodic

Q.27 The auto correlation of the signal $x(t) = e^{-t}u(t)$ is

(a) $\frac{1}{2}e^{t}u(-t) + \frac{1}{2}e^{-t}u(t)$

(b) $\frac{e^{t}}{2} + \frac{1}{2}(e^{-t} - e^{t})\,u(t)$

(c) $\frac{1}{2}e^{-t}u(-t) + \frac{1}{2}e^{-t}u(t)$

(d) $\frac{1}{2}e^{t}u(-t) - \frac{1}{2}e^{-t}u(t)$

Q.28 A sequence $x(n)$ has non-zero values as shown in figure. The sequence

$$y(n) = \left\{x\left(\frac{n}{2} - 1\right) \text{ for '}n\text{' even}\right.$$
$$= 0, \text{ for '}n\text{' odd}$$

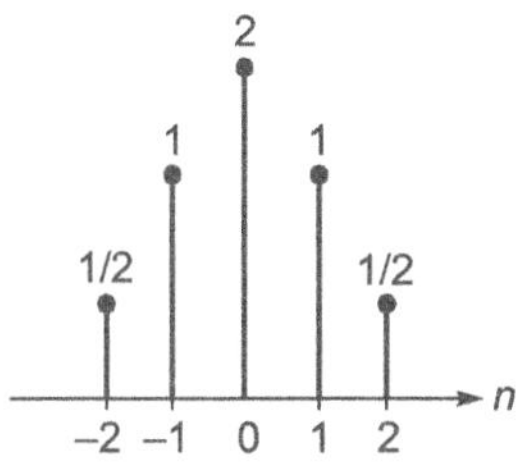

(a)
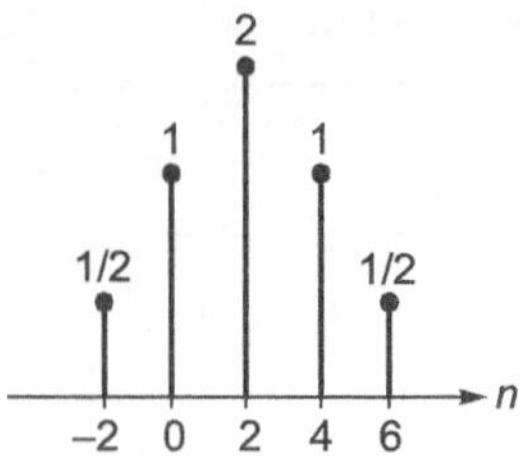

(b)
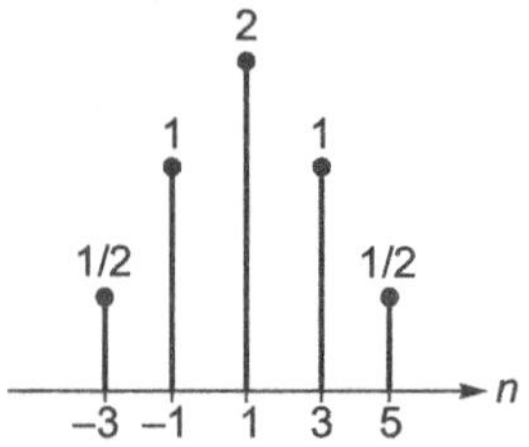

(c)
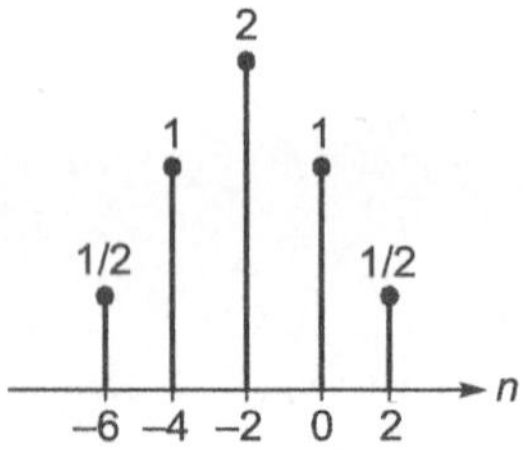

(d)
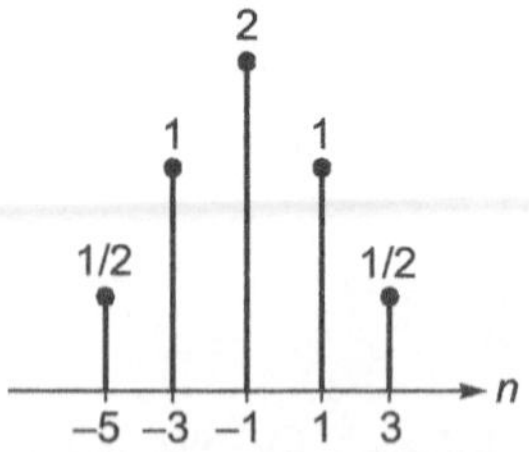

Q.29 The signal $f(t) = (t-1)\,u(t-1) - (t-2)\,u(t-2) - u(t-4)$ shows the figure below.

(a)
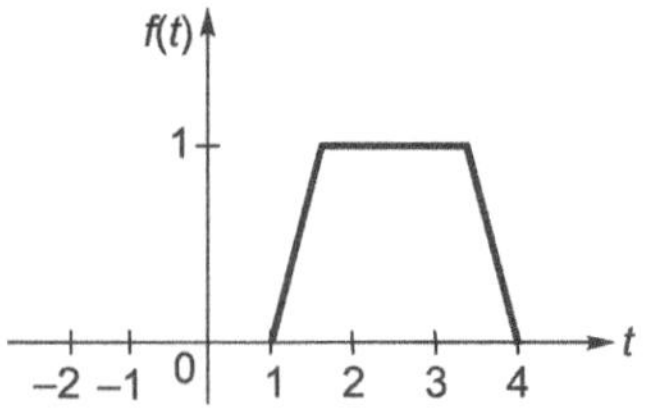

(b)
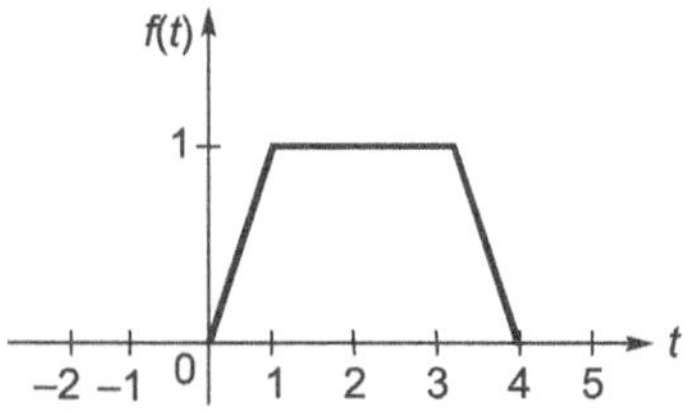

(c)
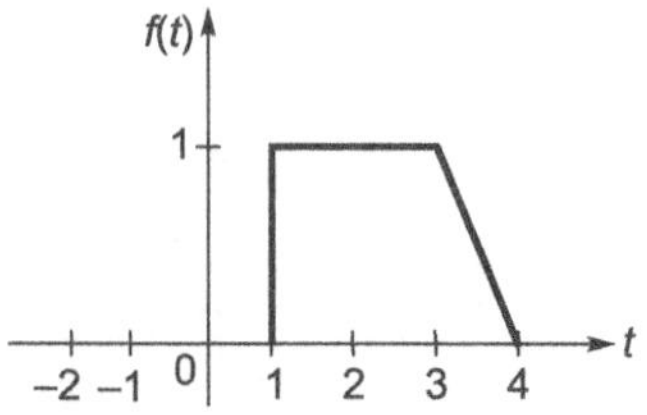

(d)
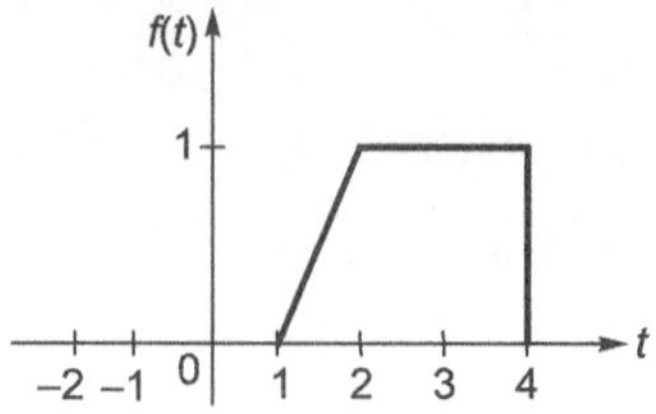

Q.30 $x[n]$ and $y[n]$ are given below figure.

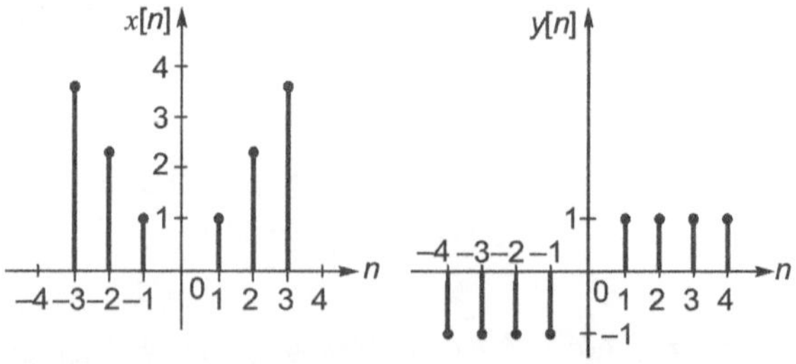

The signal,

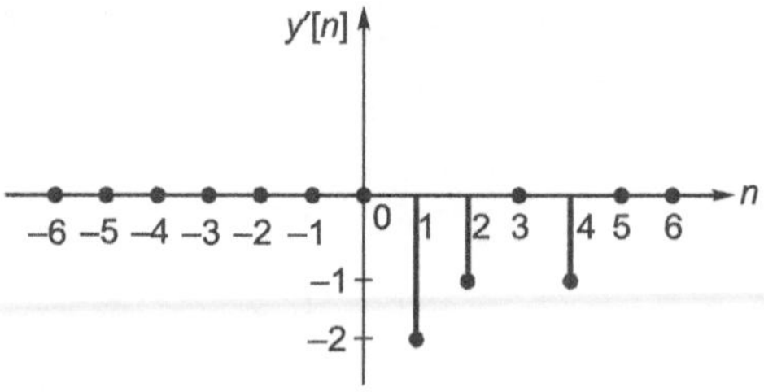

represents

(a) $y'[n] = x[n-3] \cdot y[-n]$
(b) $y'[n] = x[3-n] \cdot y[n]$
(c) $y'[n] = x[-n-3] \cdot y[-n]$
(d) $y'[n] = x[3-n] \cdot y[-n]$

2. Introduction to Systems

Q.31 The convolution of $f(t)$ with itself is given to be $\int_0^t f(\tau)\,d\tau$. Then what is $f(t)$?

(a) The unit ramp function
(b) Equal to '1'
(c) The unit step function
(d) The unit impulse function

Q.32 Which one of the following is the impulse response of the system whose step response is given as $c(t) = 0.5(1 - e^{-2t})\,u(t)$?

(a) $e^{-2t}u(t)$
(b) $0.5\,\delta(t) + e^{-2t}u(t)$
(c) $0.5\,\delta(t) - 0.5\,e^{-2t}u(t)$
(d) $0.5\,e^{-2t}\,u(t)$

Q.33 The unit impulse response of a linear time invariant system is the unit step function $u(t)$ for $t > 0$, the response of the system to an excitation $e^{-at}\,u(t)$, $a > 0$ will be

(a) ae^{-at} (b) $\left(\frac{1}{a}\right)(1 - e^{-at})$
(c) $a(1 - e^{-at})$ (d) $1 - e^{-at}$

Q.34 A system with an input $x(t)$ and output $y(t)$ is described by the relation: $y(t) = t\,x(t)$. This system is
(a) Linear and time-invariant.
(b) Linear and time varying.
(c) None-linear and time invariant.
(d) Non-linear and time varying.

Q.35 If a function $f(t)\,u(t)$ is shifted to the right side by t_0, then the function can be expressed as
(a) $f(t - t_0)\,u(t)$ (b) $f(t)\,u(t - t_0)$
(c) $f(t - t_0)\,u(t - t_0)$ (d) $f(t + t_0)\,u(t + t_0)$

Q.36 The impulse response of a causal, linear, time-invariant, continuous time system is $h(t)$. The output $y(t)$ of the same system to an input $x(t)$. Where $x(t) = 0$ for $t < -2$ is

(a) $\int_0^t h(\tau) \times (t - \tau)d\tau$ (b) $\int_{-2}^{t} h(\tau) \times (t - \tau)\,d\tau$

(c) $\int_{-2}^{t-2} h(\tau) \times (t - \tau)\,d\tau$ (d) $\int_0^{t+2} h(\tau) \times (t - \tau)\,d\tau$

Q.37 The unit step response of a system is given by $(1 - e^{\alpha t})\,u(t)$, the impulse response is given by
(a) $e^{-\alpha t}\,u(t)$ (b) $\alpha\,e^{-\alpha t}\,u(t)$
(c) $\frac{1}{\alpha}\,e^{-\alpha t}\,u(t)$ (d) $-\alpha\,e^{\alpha t}\,u(t)$

Q.38 Figure I and Figure II, shows the input $x(t)$ to a linear time invariant system and the impulse response $h(t)$ of the system

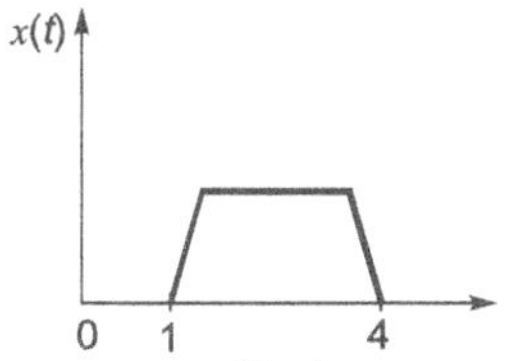

Fig. I

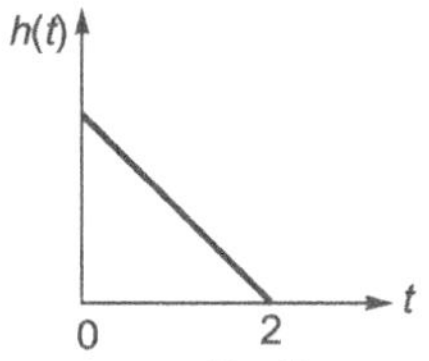

Fig. II

the output of the system is zero everywhere except for the time interval.
(a) $0 < t < 4$ (b) $1 < t < 6$
(c) $1 < t < 5$ (d) $0 < t < 5$

Q.39 A signal $f(t) = \cos 8\pi t + 0.5 \cos 4\pi t$ is instantaneously sampled. The maximum allowable value of sampling interval T_s in sec is

(a) $\frac{1}{4}$ sec (b) $\frac{1}{8}$ sec
(c) $\frac{1}{4\pi}$ sec (d) $\frac{1}{8\pi}$ sec

Q.40 The impulse response $h[n]$ of a linear time invariant system is given by
$$h[n] = u[n + 3] + u[n - 3] - 2u[n - 7],$$
the above system is
(a) stable but not causal.
(b) stable and causal.
(c) causal but unstable.
(d) unstable and not causal.

Q.41 The system $y(t) = tx(t) + 4$ is
(a) non-linear, time-varying and unstable.
(b) linear, time-varying and unstable.
(c) non-linear, time invariant and unstable.
(d) non-linear, time-varying and stable.

Q.42 The system described by the difference equation $y(n) - 2y(n - 1) + y(n - 2) = x(n) - x(n - 1)$ has $y(n) = 0$ and $n < 0$. If $x(n) = \delta(n)$, then $y(2)$ will be
(a) 2 (b) 1
(c) zero (d) –1

Q.43 The function $y(n) = x(n) - x(n - 1)$ is a
(a) Time variant and static system.
(b) Time variant and dynamic system.
(c) Time invariant and static system.
(d) Time invariant and dynamic system.

Q.44 The discrete time system described by $y(n) = x(n^2)$ is
(a) Causal, linear and time varying
(b) Causal, linear and time invariant
(c) Non causal, linear and time varying
(d) Non causal, linear and time varying

Q.45 The convolution $x(n)$ of two signals.

$$x_1(n) = \{1, -2, 1\} \text{ and } x_2(n) = \begin{cases} 1 : 0 \le n \le 5 \\ 0 : \text{otherwise} \end{cases}$$

(a) $x(n) = \{1, -1, 0, 0, 0, 0, -1, 1\}$
$\uparrow$

(b) $x(n) = \{1, -1, 0, 0, 0, 0, -1, 1\}$
$\uparrow$

(c) $x(n) = \{1, -1, 0, 0, 0, 0, -1, 1\}$
$\uparrow$

(d) None of these

Q.46 A system has the input-outuput relation given by,

$$y[n] = T\{x[n]\} = n \times x[n]$$

Which of the following is not correct?

(a) System is memoryless.
(b) System is causal.
(c) System is linear.
(d) System is unstable.

Q.47 Given:

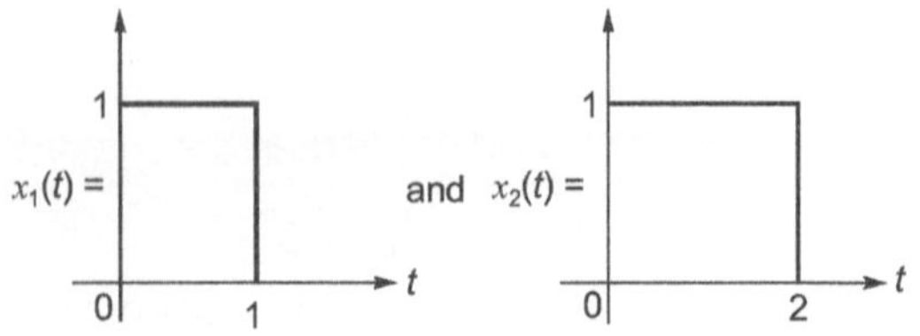

then $x_1(t) * x_2(t)$ is

(a) $r(t) - r(t-1) - r(t-2) + r(t-3)$
(b) $r(t) - r(t-1) + r(t-3)$
(c) $r(t) - 2r(t-1) + r(t-3)$
(d) none of the above

Q.48 The discrete time signal is defined by

$$x(n) = \begin{cases} 1, & n = 1 \\ -1, & n = -1 \\ 0, & n = 0 \text{ and } |n| > | \end{cases}$$

If, $y(n)$ defined by $y(n) = x(n) + x(-n)$, then the value of $y(n)$ is

(a) 0 (b) 1
(c) 2 (d) ∞

Q.49 In memoryless system

(a) zero state response is zero.
(b) zero input response is zero.
(c) both response are zero.
(d) both response are finite.

Q.50 The impulse response $h[n]$ of a linear time invariant system given by

$$h[n] = u[n+3] + u[n-2] - 2u[n-7]$$

Where u[n] is the unit step sequence. The above systems is

(a) stable but not causal.
(b) stable and causal.
(c) causal but unstable.
(d) unstable and not causal.

Q.51 Which of the following can be impulse response of causal system?

(a)
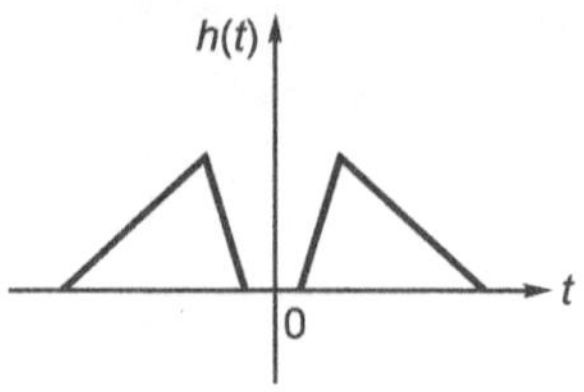

(b)
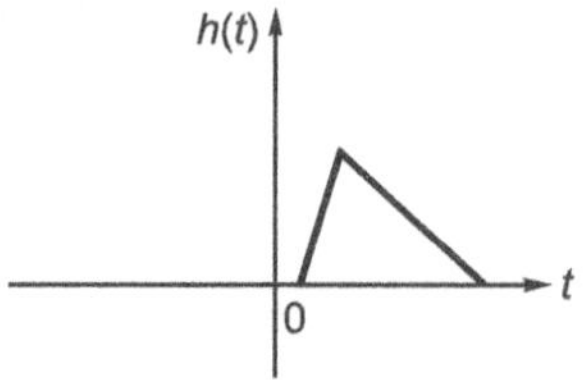

(c)
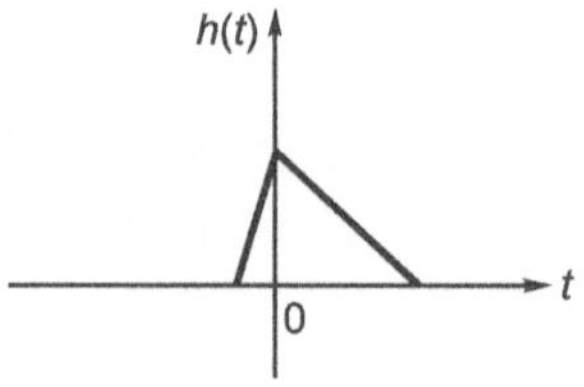

(d)
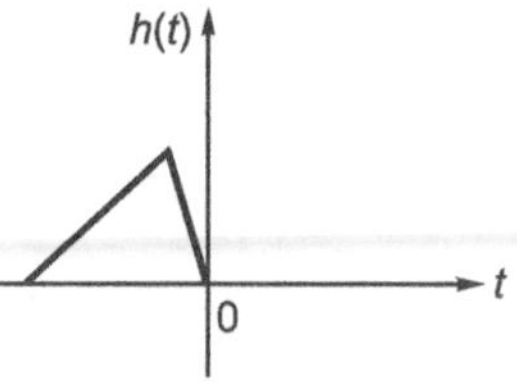

Q.52 A system with input $x[n]$ and output $y[n]$ is given as $y[n] = \left(\sin\frac{5}{6}\pi n\right) \cdot x(n)$. The system is

(a) linear, stable and invertible.
(b) non-linear, stable and non-invertible.
(c) linear, stable and non-invertible.
(d) linear, unstable and invertible.

Q.53 The given options non-invertible system is

(a) $y(t) = x(t-4)$ (b) $y(t) = \int_{-\infty}^{t} x(\tau)\, d\tau$

(c) $y(t) = \frac{dx(t)}{dt}$ (d) None of these

Q.54 $x[n]$ and $h[n]$ are $\{\underset{\uparrow}{1}, -2, 3\}$ and $\{\underset{\uparrow}{0}, 0, 1, -1, 2, 2, 1, 3\}$ respectively.

The convolution $y[n] = x[n] * h[n]$ is

(a) $\{1, -2, 4, 1, 1, 1\}$
$\uparrow$

(b) $\{0, 0, 3\}$
$\uparrow$

(c) {0, 0, 3, 1, 1, 1, 1}
↑

(d) {0, 0, 1, –1, 2, 2, 1, 3}
↑

Q.55 Consider two sequences $x[n]$ and $h[n]$ of length '4' is given by

$$x[n] = \cos\left(\frac{\pi}{2}n\right);\ n = 0, 1, 2, 3$$

and $$h[n] = \left(\frac{1}{2}\right)^n;\ n = 0, 1, 2, 3$$

If $y[n] = x[n] * h[n]$, then $y[3]$ is
(By definition of circular convolution)

(a) $\frac{-3}{4}$ (b) $\frac{+3}{4}$

(c) $\frac{-3}{8}$ (d) $\frac{+3}{8}$

Q.56 $x[n] * u[n - n_0]$ is equal to

(a) $\sum_{k=-\infty}^{n} x[k]$ (b) $\sum_{k=-\infty}^{n_0} x[k]$

(c) $\sum_{k=-\infty}^{n-n_0} x[k]$ (d) $\sum_{k=n_0}^{\infty} x[k]$

Q.57 Impulse response h[n] for the causal LTI discrete time system satisfying the difference equation

$$y[n] - \frac{1}{2}y[n-2] = 2x[n] - x[n-2] \text{ is}$$

(a) $\delta[n] + (-1)^n 2^{n-1} u[n-1]$
(b) $(-1)^n 2^{n-1} u[n-1]$
(c) $2\delta[n]$
(d) None of the above

Q.58 Which of the below is false?
(i) $y[n] = x[2 - n]$ is non-causal.
(ii) $y[n] = x[n]$ cos non is causal.
(iii) $y[n] = \text{sgn}[x[n]]$ is non-causal.

(a) (i) (b) (ii)
(c) (iii) (d) None of these

Q.59 If, $h_1(n) = \left(\frac{1}{2}\right)^n;\ n \geq 0$

$$h_2(n) = \left(\frac{1}{3}\right)^n;\ n \geq 0$$

represented impulse response of two systems. Find I.R. of the series cascaded network.

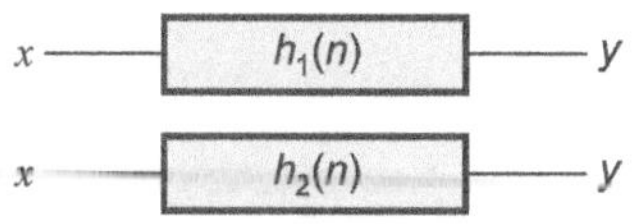

(a) $3(2)^n u(n) + 2(3)^n u(n)$
(b) $3(2)^n u(n) - 2(3)^n u(n)$
(c) $3\left(\frac{1}{2}\right)^n u(n) - 2\left(\frac{1}{3}\right)^n u(n)$
(d) $3\left(\frac{1}{2}\right)^n u(n) + 2\left(\frac{1}{3}\right)^n u(n)$

Q.60 Step response of discrete time LTI system is given as $S(n) = \left(\frac{1}{2}\right)^n u(n)$. Then the impulse response of the system will be

(a) $1, \frac{1}{2}, \frac{1}{4}, \ldots$ (b) $-1 - \frac{1}{2}, -\frac{1}{4}, \ldots$

(c) $1, -\frac{1}{2}, -\frac{1}{4}, \ldots$ (d) 1, 1, 1, 1,...

Q.61 Consider the three statements given below:
Statement-1 : Response $(-2)^n u(n)$ may or may not be stable.
Statement-2 : Response $(-2)^n u(n)$ is unstable.
Statement-3 : Response $(-2)^n u(-n-1)$ is unstable.

	1	2	3
(a)	True	True	False
(b)	False	False	True
(c)	True	True	True
(d)	False	True	False

Q.62 Input causal signal $t^2 + t$ is convolved with 't' (causal) to produce output $y(t)$. If the output for similar type signal is $\frac{1}{3}y(3t)$. Then the corresponding impulse response and input signal will be

(a) $3t$, $3t^2 + 3t$ (b) $3t$, $9t^2 + 3t$

(c) $\frac{t}{3}, \frac{(9t^2+3t)}{3}$ (d) $\frac{t}{3}$, $t^2 + t$

Q.63 Which of the following system is linear?
(i) $y(t) = t\,x(t)$
(ii) $y(t) = t\,x^2(t)$
(iii) $y(t) = x(2t)$

(a) only (i) (b) (i) and (ii)
(c) only (iii) (d) None of these

Q.64 Consider a discrete time LTI system with impulse response h[n] is given by

$$h[n] = \left(-\frac{1}{2}\right)^n u[n-1]$$

Statement-1: The system is causal.
Statement-2: The system is also stable
(a) Both statements are true.
(b) Statement 1 true, but statement 2 is false.
(c) Statement 1 is false, but statement 2 is true.
(d) Both statements are false.

Q.65 **Statement-A:** system described by

$$\frac{dy(t)}{dt} + 5y(t) + 2 = x(t) \text{ is non linear.}$$

Statement-B: A discrete time LTI system with impulse response $h[n] = \begin{cases} 1\ ; & n = 0, 1 \\ 0\ ; & \text{otherwise} \end{cases}$.

The input-output relationship of the system is $y[n] = x[n-1]$.
Choose the correct option:
(a) Both statements are true.
(b) Statement A is true, but statement B is false.
(c) Statement A is false, but statement B is true.
(d) Both statements are false.

Q.66 A continuous time LTI is described by

$$y(t) = T\{x(t)\} = \frac{1}{T}\int_{t-T/2}^{t+T/2} x(\tau)\, d\tau$$

The impulse response is

(a)
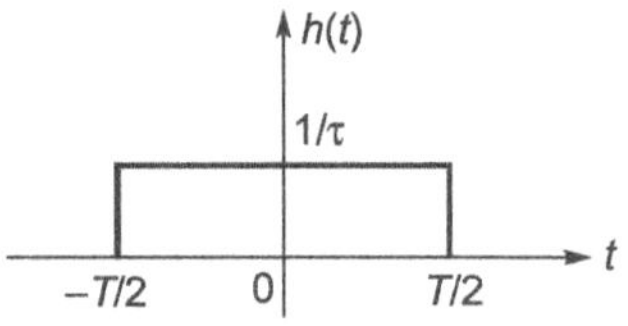

(b)
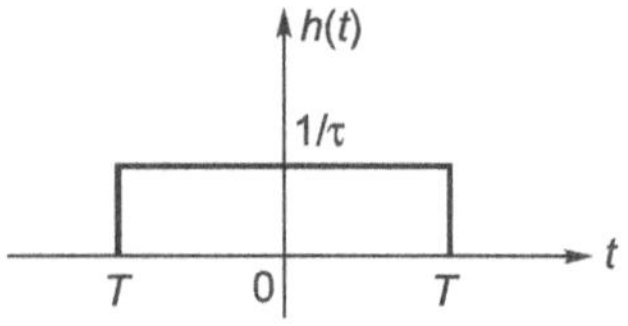

(c)
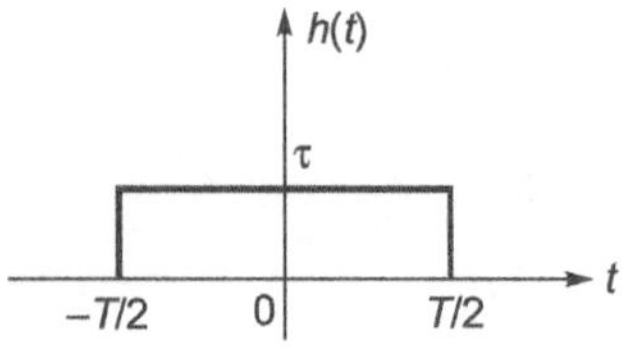

(d)
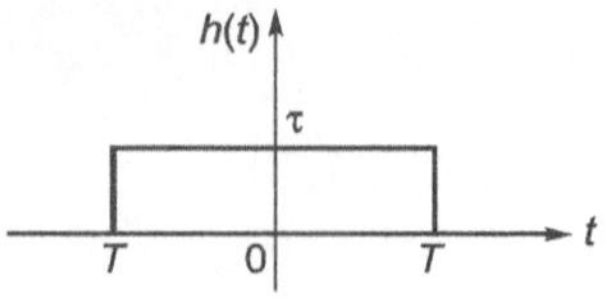

Q.67 Consider a signal $x(t)$ given by

$x(t) = u(t+0.5) - u(t-0.5)$ and $h(t) = e^{j\omega_0 t}$
The output $y(t) = x(t) * h(t)$
Value of ω_0 which ensure that $y(0) = 0$ is
(a) π (b) 2π
(c) $\frac{\pi}{2}$ (d) $\frac{\pi}{4}$

Q.68 Consider three system with following input-output relationship:

System-1: $y_1(n) = \begin{cases} x(n/2) & : n \text{ even} \\ 0 & : n \text{ odd} \end{cases}$

System-2: $y_2(n) = x(n) + \frac{1}{2}x(n-1) + \frac{1}{4}x(n-2)$

System-3: $y_3(n) = x(2n)$

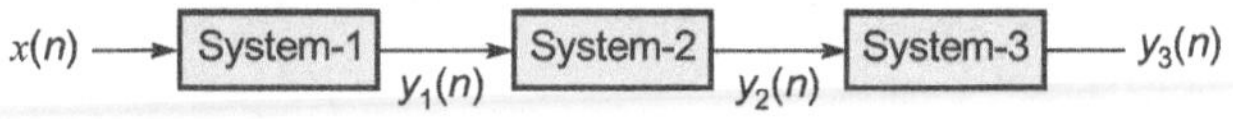

Output $y_3(n)$ is equal to

(a) $x(n) + \frac{1}{2}x(n-1)$

(b) $x(n) + \frac{1}{2}x(n-1) - x(n-2)$

(c) $x(n) + \frac{1}{2}x(n-1) + \frac{1}{4}x(n-2)$

(d) $x(n) - \frac{1}{2}x(n-1) + \frac{1}{4}x(n-2)$

Q.69 Consider:
(i) $y[n] = x[2-n]$ is non-causal
(ii) $y[n] = x[n] \cos \omega_0 n$ is causal
(iii) $y[n] = \text{sgn}[x[n]]$ is non-causal
Which of the above is false?
(a) only 1 (b) only 2
(c) only 3 (d) None

Q.70 Consider for a LTI the impulse response and output are given by

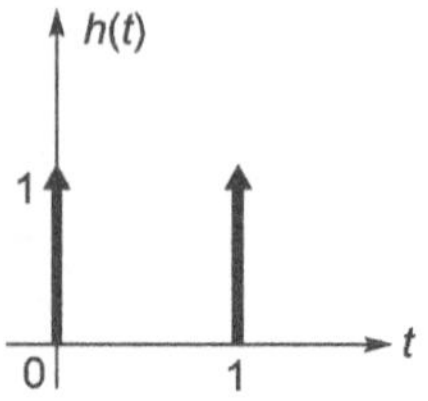

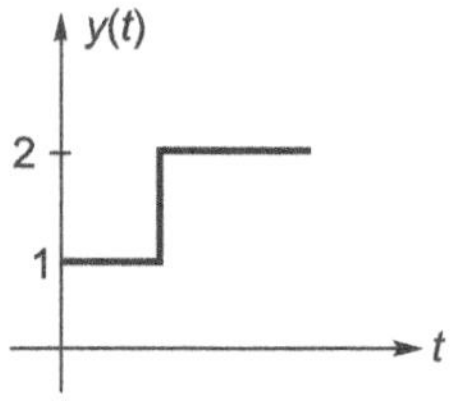

The input signal $x(t)$ should be equal to

(a)

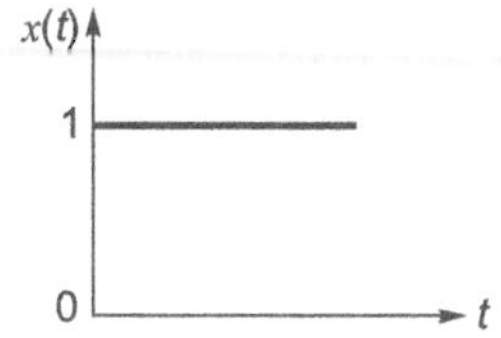

(b)

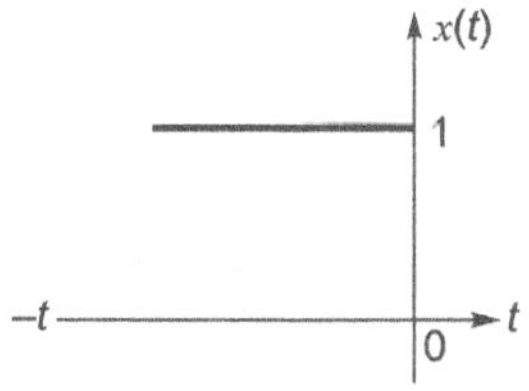

(c)

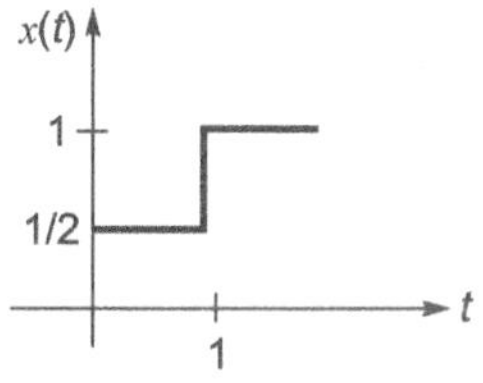

(d)

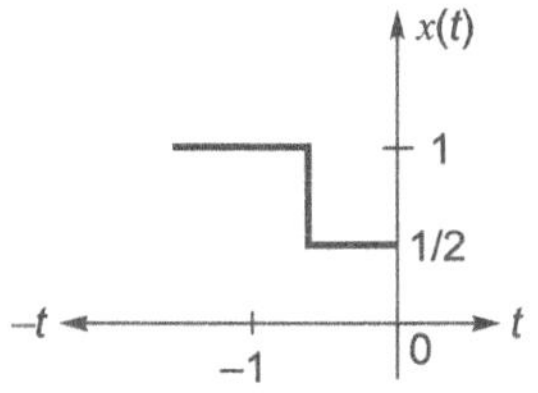

Q.71 Consider a continuous time LTI system with

$$x(t) = \begin{cases} 1 \ ; & 0 \le t \le 1 \\ 0 \ ; & \text{elsewhere} \end{cases}$$

and $\qquad h(t) = x(t/\alpha)$ where, $0 < \alpha \le 1$

then output $y(t) = x(t) * h(t)$; $\alpha \le t \le 1$ is given by

(a) $\frac{1}{\alpha}$ (b) α

(c) $\frac{2}{\alpha}$ (d) 2α

Q.72 The convolution of $x[n] = \cos(\pi/_2 n)\ u[n]$ and $h[n] = u[n-1]$ is $f[n]\ u[n-1]$. The function $f[n]$ is

(a) $\begin{cases} 1, & n = 4m+1, \quad 4m+2 \\ 0, & n = 4m, \quad 4m+3 \end{cases}$

(b) $\begin{cases} 0, & n = 4m+1, \quad 4m+2 \\ 1, & n = 4m, \quad 4m+3 \end{cases}$

(c) $\begin{cases} 1, & n = 4m+1, \quad 4m+3 \\ 0, & n = 4m, \quad 4m+2 \end{cases}$

(d) $\begin{cases} 0, & n = 4m+1, \quad 4m+3 \\ 1, & n = 4m, \quad 4m+2 \end{cases}$

Q.73 Consider a causal LTI system with frequency response $H(j\omega) = \dfrac{1}{j\omega + 3}$. The system produces the output to input $x(t)$ as follows:

$$y(t) = e^{-2t}\,u(t) - e^{-4t}\,u(t)$$

The input $x(t)$ is

(a) $(2e^{-4t} - 3e^{-3t})\ u(t)$ (b) $e^{-4t}\ u(t)$

(c) $(3e^{-4t} - 2e^{-3t})\ u(t)$ (d) $-e^{-4t}\ u(t)$

Q.74 The impulse response of $h[n]$ of linear time invariant system is given as

$$h[n] = \begin{cases} -2\sqrt{2} & ; n = 1, -1 \\ 4\sqrt{2} & ; n = 2, -2 \\ 0 & ; \text{otherwise} \end{cases}$$

If the input to the above system is the sequence $e^{j\pi n/4}$, then output is

(a) $4\sqrt{2}\ e^{j\pi n/4}$ (b) $4\sqrt{2}\ e^{-j\pi n/4}$

(c) $4e^{j\pi n/4}$ (d) $-4e^{j\pi n/4}$

Q.75 A discrete time LTI system with impulse response

$$h[n] = \left(\frac{1}{2}\right)^n u[n] + \left(-\frac{1}{2}\right)^n u[n-1]$$

The system is

(a) Causal and stable.

(b) Non-Causal and stable.

(c) Causal and unstable.

(d) Non-causal and unstable.

Q.76 The $y(n) = x[n] * h[n]$ is in given figure.

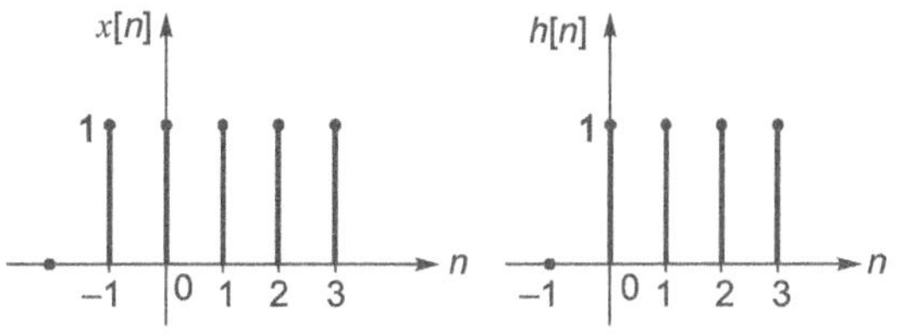

(a) {1, 2, 3, 3, 4, 3, 2, 1}

(b) {1, 2, 3, 4, 4, 3, 2, 1}

(c) {1, 2, 3, 4, 3, 3, 2, 1}

(d) {1, 2, 3, 0, 0, 3, 2, 1}

Q.77 Given sinusoidal signal $x(t) = \cos(21t)$, with sampling interval T_s such that $x[n] = x(\eta T_s)$ is a periodic sequence. The fundamental period is if, $T_s = 0.3\pi$ sec.

(a) 5 (b) 10

(c) 15 (d) 20

3. Continuous-time Fourier Series

Q.78 If, $f(t) = -f(-t)$ and $f(t)$ satisfy the dirichlet conditions then f(t) can be expanded in a fourier series containing
(a) only sine terms
(b) only cosine terms
(c) cosine terms and constant terms
(d) sine terms and constant terms

Q.79 The trigonometric Fourier series expansion of an odd function shall have
(a) only sine terms
(b) only cosine terms
(c) odd harmonics of both sine and cosine terms
(d) none of the these

Q.80 A periodic triangular wave is shown in figure its fourier components will consists only of

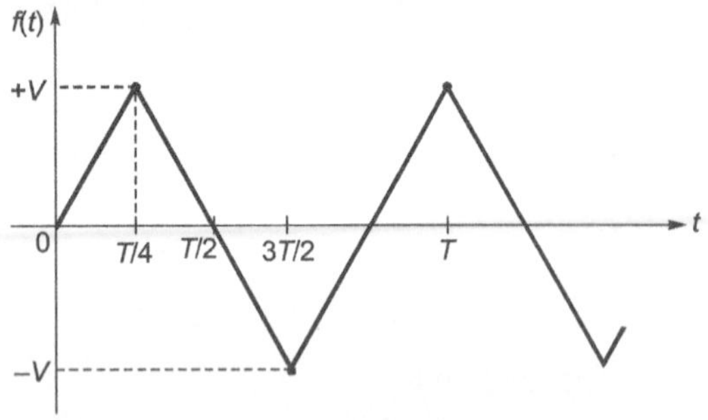

(a) all cosine terms (b) all sine terms
(c) odd cosine terms (d) odd sine terms

Q.81 Determine the Fourier series coefficient for given periodic signal $x(t)$ is

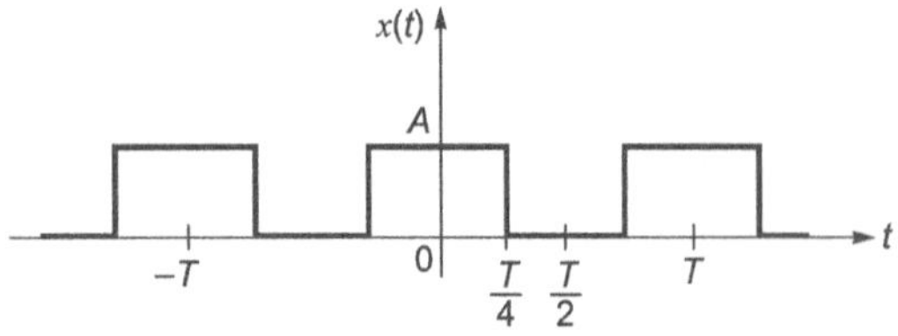

(a) $\frac{A}{j\pi k}\sin\left(\frac{\pi}{2}k\right)$ (b) $\frac{A}{\pi jk}\cos\left(\frac{\pi}{2}k\right)$
(c) $\frac{A}{\pi k}\sin\left(\frac{\pi}{2}k\right)$ (d) $\frac{A}{\pi k}\cos\left(\frac{\pi}{2}k\right)$

Q.82 Consider the three continuous time signals with fundamental period of $T = 1/2$

$x(t) = \cos 4\pi t$
$y(t) = \sin 4\pi t$
$z(t) = x(t) \cdot y(t)$

The Fourier co-efficient of $z(t)$ are given by
(a) $\frac{1}{4}, -\frac{1}{4}$ (b) $\frac{1}{2j}, -\frac{1}{2j}$
(c) $\frac{1}{2}, -\frac{1}{2}$ (d) $\frac{1}{4j}, -\frac{1}{4j}$

Q.83 Consider a continuous time periodic signal $x(t)$ given by

$$x(t) = \sum_{k=-100}^{100} \cos(k\pi)\, e^{jk\frac{2\pi}{50}t}$$

the signal is
(a) Real and Even
(b) Real and odd
(c) Complex and even
(d) Complex and odd

Q.84 The Fourier series coefficient of time domain signal have been given. Determine the corresponding time domain signal and choose correct option.

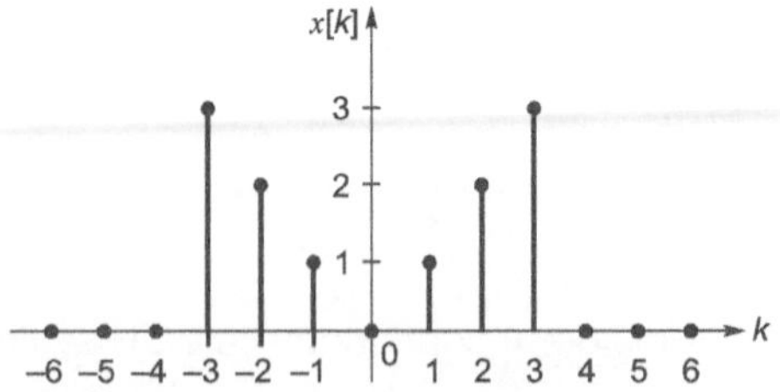

(a) $3\cos 3\pi t + 2\cos 2\pi t + \cos\pi t$
(b) $3\sin 3\pi t + 2\sin 2\pi t + \sin\pi t$
(c) $6\sin 3\pi t + 4\sin 2\pi t + 2\sin\pi t$
(d) $6\cos 3\pi t + 4\cos 2\pi t + 2\cos\pi t$

Q.85 If $f(x) = |\cos x|$; $(-\pi < x < \pi)$, then the fourier series coefficient a_0 and b_2 is
(a) $\frac{4}{3\pi}, 0$ (b) $\frac{4}{\pi}, 0$
(c) $\frac{1}{3\pi}, \frac{2}{3\pi}$ (d) $\frac{5}{3\pi}, \frac{4}{3\pi}$

4. Continuous Time Fourier Transformer (CTFT)

Q.86 If $f(t)$ is an even function, then what is its fourier transform $F(j\omega)$?
(a) $\int_0^\infty f(t)\cos(2\omega t)\,dt$ (b) $2\int_0^\infty f(t)\cos(\omega t)\,dt$
(c) $2\int_0^\infty f(t)\sin(\omega t)\,dt$ (d) $\int_0^\infty f(t)\sin(2\omega t)\,dt$

Q.87 If the fourier transform of $f(t)$ is $f(j\omega)$, then what is the fourier transform of $f(-t)$?
(a) $f(j\omega)$
(b) $f(-j\omega)$
(c) $-F(j\omega)$
(d) complete conjugate of $f(j\omega)$

Q.88 The fourier transform of real value time signal has
(a) Odd symmetry (b) Even symmetry
(c) No symmetry (d) Conjugate symmetry

Q.89 The signal $x(t)$ is a real and odd function of 't' the $x(\omega)$ is
(a) a real and even function of ω
(b) an imaginary and odd function of ω
(c) an imaginary and even function of ω
(d) a real and odd function of ω

Q.90 A linear system has the transfer function $H(j\omega) = \dfrac{1}{(1+1j\omega)}$. When it is subjected to an input white noise process with a constant spectral dencity 'A' the spectral density of the output will be
(a) $\dfrac{A}{(1+j\omega)}$ (b) $\dfrac{A}{(1+j\omega)^2}$
(c) $\dfrac{A}{(1+\omega^2)}$ (d) $\dfrac{A}{\sqrt{(1+\omega^2)}}$

Q.91 The auto-correlation of a wide sense stationary random process is given by $e^{-2|\tau|}$. The peak valve of the spectral density is
(a) 2 (b) 1
(c) $e^{-1/2}$ (d) e

Q.92 Which one of the following is the correct fourier transform of the unit step signal?
$u(t) = 1$ for $t \geq 0$
$= 0$ for $t \leq 0$
(a) $\pi\delta(\omega)$ (b) $\dfrac{1}{j\omega}$
(c) $\dfrac{1}{j\omega} + \pi\delta(\omega)$ (d) $\dfrac{1}{j\omega} + 2\pi\delta(\omega)$

Q.93 The fourier transform of a conjugate symmetric function is always
(a) Imaginary
(b) Conjugate anti-symmetric
(c) Conjugate symmetric
(d) Real

Q.94 The inverse fourier transform of sgn(f)
(a) $-j\pi t$ (b) $\dfrac{j}{\pi t}$
(c) $\dfrac{1}{j\pi t}$ (d) None of these

Q.95 Fourier transform $f(j\omega)$ of an arbitrary signal has the property
(a) $f(j\omega) = f(-j\omega)$ (b) $f(j\omega) = f^*(-j\omega)$
(c) $f(j\omega) = -f(-j\omega)$ (d) $f(j\omega) = -f^*(-j\omega)$

Q.96 If $x(n) = \left(\dfrac{1}{2}\right)^n u(n)$, $y(n) = x^2(n)$ and $y(e^{j\omega})$ be the fourier transform of $y(n)$, then $y(e^{j0})$ is
(a) $\dfrac{1}{4}$ (b) 2
(c) 4 (d) $\dfrac{4}{3}$

Q.97 For a signal $x(t)$, the fourier transform is $X(f)$, then inverse fourier transform of $x(3f+2)$ is given by
(a) $\dfrac{1}{2}X\left(\dfrac{t}{2}\right)e^{j3\pi t}$ (b) $\dfrac{1}{3}X\left(\dfrac{t}{3}\right)e^{-j4\pi t/3}$
(c) $3X(3t)\,e^{-j4\pi t}$ (d) $X(3t+2)$

Q.98 The Fourier transform of given signal $x(t)$

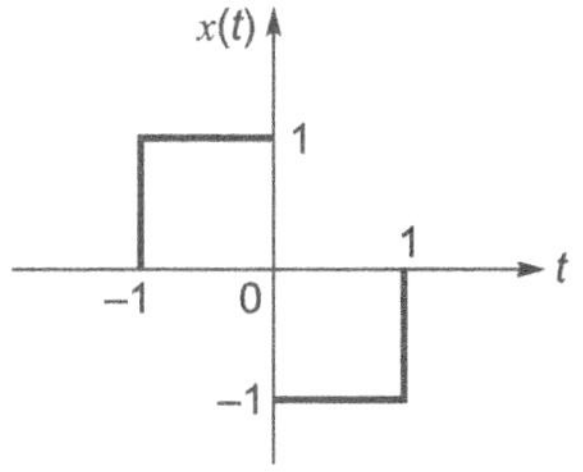

(a) $\dfrac{2\sin\omega - 2}{\omega}$ (b) $\dfrac{2\cos\omega - 2}{j\omega}$
(c) $2j\omega\cos\omega$ (d) $2j\omega\sin\omega$

Q.99 Fourier transform of $x(t) = e^{at}u(-t)$, $a > 0$ is
(a) $\dfrac{1}{a+j\omega}$ (b) $\dfrac{a}{j\omega}$
(c) $\dfrac{-a}{j\omega}$ (d) $\dfrac{1}{a-j\omega}$

Q.100 Determine the fourier transform of the signal $x(t)$ shown in figure.

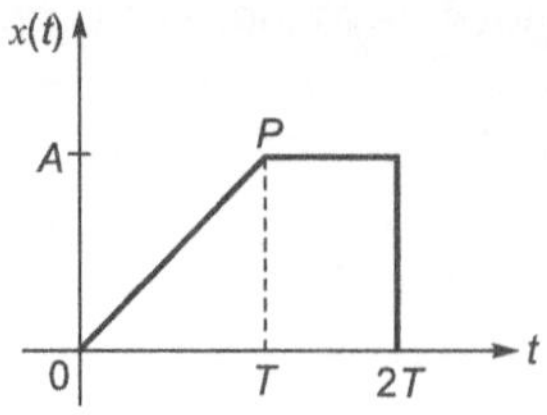

(a) $\frac{A}{j\omega}\left[e^{-j\omega}\,\text{sinc}\left(\frac{\omega T}{2}\right)-e^{-j2\omega T}\right]$

(b) $\frac{A}{j\omega}\left[e^{-j\omega T/2}\,\text{sinc}\left(\frac{\omega T}{2}\right)+e^{-j2\omega T}\right]$

(c) $\frac{A}{j\omega}\left[e^{-j\omega T/2}\,\text{sinc}\left(\frac{\omega T}{2}\right)-e^{-j2\omega T}\right]$

(d) None of these

Q.101 Match **List-I [Function $f(t)$]** with **List-II [Fourier transform $F(\omega)$]** and select the correct answer using the codes given below the lists:

List-I	List-II
A. $f(t-t_0)$	1. $f(\omega-\omega_0)$
B. $f(t)\,e^{j\omega_0 t}$	2. $\frac{1}{2\pi}[F_1(\omega)*F_2(\omega)]$
C. $f_1(t)\cdot f_2(t)$	3. $F(\omega)\,e^{-j\omega t_0}$
D. $\int_{-\infty}^{2} f(t)\,dt$	4. $\frac{f(\omega)}{j\omega}+\pi x(0)\,\delta(\omega)$
	5. $\frac{f(\omega)}{j\omega}$

Codes:

	A	B	C	D
(a)	1	3	2	4
(b)	3	1	2	5
(c)	1	3	4	5
(d)	3	1	2	4

Q.102 Consider a continuous time low pass filter whose impulse response $h(t)$ is known to be real and whouse frequency response magnitude is given by

$$|H(j\omega)| = \begin{cases} 1;\ \omega \le 200 \\ 0;\ \text{otherwise} \end{cases}$$

Determine the value of $h(t)$ if the group delay function is specified as $t(\omega) = 5$.

(a) $\frac{\sin[200\pi(t-5)]}{\pi(t-5)}$ (b) $\frac{\sin[200\pi t]}{\pi t}$

(c) $\frac{5\sin[200\pi t]}{\pi t}$ (d) $\frac{5\sin[200\pi(t-5)]}{\pi t}$

Q.103 Consider a signal $x_1(t)$ having a fourier transform $x_1(j\omega)$. An another signal $x_2(t)$ having fourier transform $x_2(j\omega)$ is related to $x_1(t)$ by

$$x_2(j\omega) = [1+\text{sgn}(\omega)]\,x_1(j\omega)$$

$x_2(t)$ in terms of $x_1(t)$ is equal to

(a) $x_2(t) = x_1(t) + x_1(t) * \frac{1}{\pi t}$

(b) $x_2(t) = x_1(t) - x_1(t) * \frac{1}{\pi t}$

(c) $x_2(t) = x_1(t) + jx_1(t) * \frac{1}{\pi t}$

(d) $x_2(t) = x_1(t) + jx_1(t) * \frac{1}{\pi t}$

Q.104 A Fourier transform pair is as follows:

$$x(t) \xleftrightarrow{\text{F.T}} x(j\omega) = \frac{2\sin\omega}{\omega}$$

where, $x(t) = \begin{cases} 1, & |t|<| \\ 0, & \text{otherwise} \end{cases}$

The Fourier transform of given signal $y(t)$ is

(a) $4pj\left(\frac{\cos\omega}{\omega}-\frac{\sin\omega}{\omega}\right)$

(b) $2n\left(\frac{\cos\omega}{\omega}-\frac{\sin\omega}{\omega^2}\right)$

(c) $4pj\left(\frac{\cos\omega}{\omega^2}-\frac{\sin\omega}{\omega}\right)$

(d) $2j\left(\frac{\cos\omega}{\omega^2}-\frac{\sin\omega}{\omega}\right)$

Q.105 Suppose: $y(t) = x(t)\cos t$

and $y(j\omega) = \begin{cases} 2, & |\omega| \le 2 \\ 0, & \text{otherwise} \end{cases}$

then $x(t)$ will be

(a) $\frac{4\sin t}{\pi t}$ (b) $\frac{2\sin t}{t}$

(c) $\frac{4\sin t}{t}$ (d) $\frac{2\pi\sin t}{t}$

Q.106 The output $y(t)$ of a linear time invariant system is related to its input $x(t)$ by the following equations:

$y(t) = 0.5x(t-t_d+T) + x(t-t_d) + 0.5x(t-t_d-T)$

The filter transfer function $H(\omega)$ of a such system is given by

(a) $(1+\cos\omega t)\,e^{-j\omega t_d}$ (b) $(1+0.5\cos\omega t)\,e^{-j\omega t_d}$

(c) $(1+\cos\omega t)\,e^{+j\omega t_d}$ (d) $(1+0.5\cos\omega t)\,e^{j\omega t_d}$

Q.107 A signal $x(n) = \sin(\omega_0 n + \phi)$ is the input to a linear time invariant system having a frequency response $H(e^{j\omega})$. If the output of the system $A\,x(n - n_0)$, then the most general form of $\angle H(e^{j\omega})$ will be

(a) $-n_0\omega_0 + \beta$ for any arbitrary real β.
(b) $-n_0\omega_0 + 2\pi k$ for any arbitrary integer k.
(c) $n_0\omega_0 + 2\pi k$ for any arbitrary integer k.
(d) $-n_0\omega_0\,\phi$

Q.108 Auto-correlation function of input process $x(t)$ is $R_x(t) = 4\delta(t)$, then find average power of output random process.

Given that, $H(f) = \dfrac{1}{2 + j2\pi f}$

(a) 1.25 W (b) 1 W
(c) 0.1 W (d) zero

Q.109 The fourier transform of given signal,

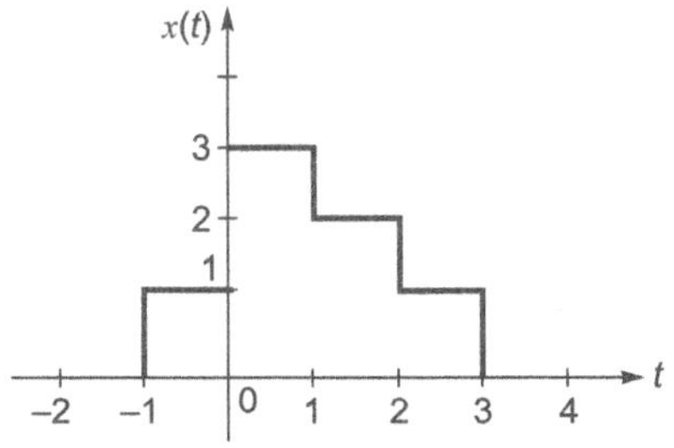

(a) $\dfrac{2 + e^{j\omega} - e^{-j\omega} - e^{-j2\omega} - e^{-j3\omega}}{j\omega}$

(b) $\dfrac{2 - e^{-j\omega} + e^{j\omega} + e^{-j2\omega} + e^{-j3\omega}}{j\omega}$

(c) $\dfrac{2 + 2(e^{j\omega} - e^{-j\omega}) - e^{-j2\omega} - e^{-j3\omega}}{j\omega}$

(d) $\dfrac{2 - e^{+j\omega} - e^{-j\omega} - e^{-j2\omega} - e^{-j3\omega}}{j\omega}$

5. Laplace Transform

Q.110 A voltage having the laplace transform $\dfrac{4S^2 + 3S + 2}{7S^2 + 6S + 5}$ is applied across a 2H inductor having zero initial current. What is the current in the inductor at $t = \infty$?

(a) zero (b) (1/5) A
(c) (2/7) A (d) (2/5) A

Q.111 The function $f(t)$ shown in figure will have laplace transform as:

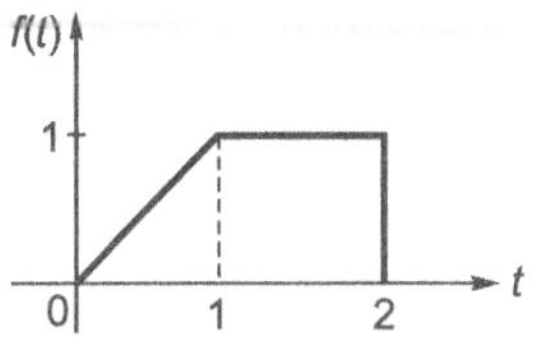

(a) $\dfrac{1}{S^2} - \dfrac{1}{S}e^{-S} - \dfrac{1}{S^2}e^{-2S}$

(b) $\dfrac{1}{S^2}(1 - e^{-S} - e^{-2S})$

(c) $\dfrac{1}{S}(1 - e^{-S} - e^{-2S})$

(d) $\dfrac{1}{S^2}(1 - e^{-S} - Se^{-2S})$

Q.112 Which of the following is the correct Laplace transform of the signal in the given figure?

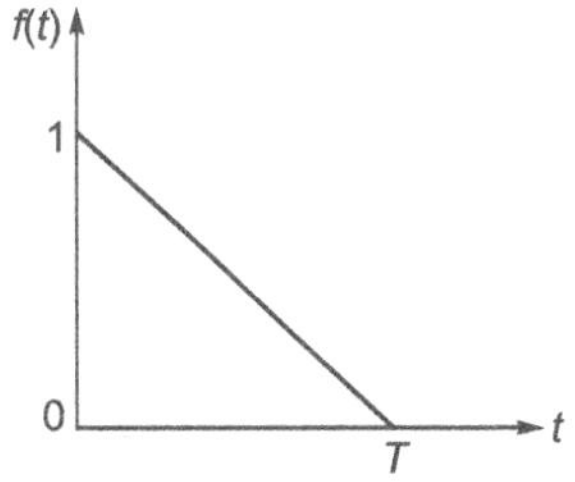

(a) $\dfrac{1}{TS^2}[1 - e^{-TS}(1 + TS)]$

(b) $\dfrac{1}{TS^2}[e^{-TS} - 1 + TS]$

(c) $\dfrac{1}{TS^2}[e^{-TS} + 1 - TS]$

(d) $\dfrac{1}{TS^2}[1 - e^{-TS} + TS]$

Q.113 Given: $x(t) = \delta(t) - \dfrac{4}{3}e^{-t}\,u(t) + \dfrac{1}{3}e^{2t}\,u(t)$ then find $x(s)$ and its ROC:

(a) $x(S) = \dfrac{(S-1)^2}{(S+1)(S-2)}$; $R_e[S] > 2$

(b) $x(S) = \dfrac{(S-1)}{(S+1)(S-2)}$; $R_e[S] > 2$

(c) $x(S) = \dfrac{(S-1)}{(S+1)(S-2)}$; $R_e[S] < 2$

(d) None of these

Q.114 Given: $x(S) = \dfrac{2S^2 + 5S + 5}{(S+1)^2(S+2)}$ and $R_e[S] > -1$ then $x(t)$:

(a) $[2te^{-t} + e^{-t} + 3e^{-2t}]\, u(t)$
(b) $[2te^{-t} - e^{-t} - 3e^{-2t}]\, u(t)$
(c) $[2te^{-t} - e^{-t} + 3e^{-2t}]\, u(t)$
(d) none of these

Q.115 If the input to LTI system is $x(t) = e^{-3t}\, u(t)$. Then the output is $y(t) = [e^{-t} - e^{-2t}]\, u(t)$ the transfer function of the system is

(a) $H(S) = \dfrac{S+3}{S^2 + 3S + 2}$
(b) $H(S) = \dfrac{S-3}{S^2 - 3S - 2}$
(c) $H(S) = \dfrac{S+3}{S^2 - 3S + 2}$
(d) None of these

Q.116 If, $f(t) = \begin{cases} (t-2)^2 & ; t > 1 \\ 0 & ; 0 < t < 1 \end{cases}$ then the value of $L\{t^n f(t)\}$ is equal to

(a) $(-1)\dfrac{d^n}{ds^n}F(s)$ (b) $(-1)^n \dfrac{d}{ds^n}F(s)$
(c) $\dfrac{d^n}{ds^n}F(s)$ (d) $(-1)^n F^n(ds)$

Q.117 The transfer functions of the sytem is given by

$$H(S) = \frac{1}{S^2(S-2)}$$

The impulse response of the system is

(a) $(t^2 * e^{-2t})\, u(t)$ (b) $(t * e^{2t})\, u(t)$
(c) $(te^{-2t})\, u(t)$ (d) $(te^{-t})\, u(t)$

Q.118 The Laplace transform of a continuous signal $x(t)$ is $x(s) = \dfrac{5-s}{S^2 - S - 2}$. If the Fourier transform of the signal exists, then $x(t)$ is

(a) $e^{2t}\, u(t) - 2e^{-t}\, u(t)$
(b) $e^{2t}\, u(-t) + 2e^{-t}\, u(t)$
(c) $-e^{2t}\, u(-t) - 2e^{-t}\, u(t)$
(d) $e^{2t}\, u(-t) - 2e^{-t}\, u(t)$

Q.119 The laplace transform of $i(t)$ is given by $I(S) = \dfrac{2}{S(1+S)}$; As $t \to \infty$, the value of i(t) tends to

(a) 0 (b) 1
(c) 2 (d) ∞

Q.120 Consider the function $f(t)$ having Laplace transform, $f(S) = \dfrac{\omega_0}{S^2 + \omega_0^2}$, $R_e[S] > 0$

the final value of f(t) would be

(a) 0 (b) 1
(c) $-1 \le f(\infty) \le 1$ (d) ∞

Q.121 The transfer function $H(S)$ of a stable system is

$$H(S) = \frac{2S - 1}{S^2 + 2S + 1}$$

the impulse response is

(a) $2u(-t+1) - 3tu(-t+1)$
(b) $(3te^{-t} - 2e^{-t})\, u(t)$
(c) $2u(t+1) - 3tu(t+1)$
(d) $(2e^{-t} - 3t\, e^{-t})\, u(t)$

Q.122 If $L[f(t)] = \dfrac{2(S+1)}{S^2 + 2S + 5}$, then $f(0^+)$ and $f(\infty)$ are given by

(a) 0, 2 respectively (b) 2, 0 respectively
(c) 0, 1 respectively (d) 2/5, 0 respectively

Q.123 Find the laplace transform of time function shown in figure.

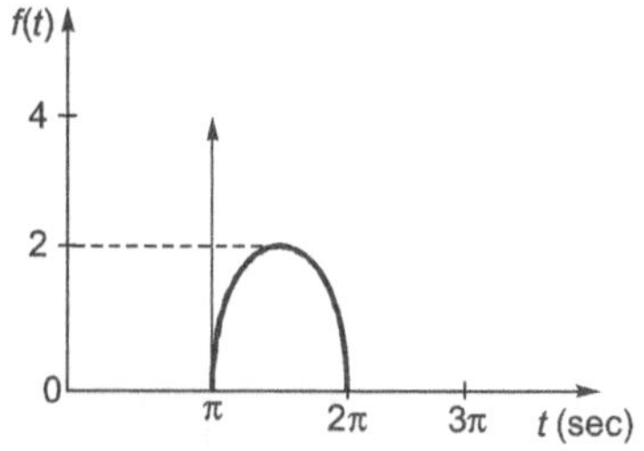

(a) $4e^{-ps} + \dfrac{2[e^{-\pi s} + e^{-2\pi s}]}{[S^2 + 1]}$
(b) $4e^{-ps} + \dfrac{2[e^{-\pi s} - e^{-2\pi s}]}{[S^2 + 1]}$
(c) $2e^{-ps} + \dfrac{2[e^{-\pi s} + e^{-2\pi s}]}{[S^2 + 1]}$
(d) $2e^{-ps} + \dfrac{2[e^{-\pi s} - e^{-2\pi s}]}{[S^2 + 1]}$

Q.124 Consider the following signal:

$$x(t) = e^{-2t}\, u(t) + e^{-t}(\cos 3t)\, u(t)$$

the laplace transform of above system is

(a) $\frac{2S^2+4S+11}{(S^2+2S+10)(S+1)}$

(b) $\frac{2S^2+5S+12}{(S^2+2S+10)(S+2)}$

(c) $\frac{2S^2+5S+12}{(S^2+2S+10)}$

(d) $\frac{2S^2+5S+12}{(S^2+2S+10)S}$

Q.125 Consider a signal $x(t)$ having laplace transform given by,

$$x(S) = \log\left(\frac{S+5}{S+6}\right)$$

The time domain signal $x(t)$ is equal to

(a) $\frac{1}{t}[e^{-6t}-e^{-5t}]$ (b) $\frac{1}{t}[e^{-6t}+e^{-5t}]$

(c) $t[e^{-6t}-e^{-5t}]$ (d) $\frac{1}{t}[e^{-5t}-e^{-6t}]$

Q.126 Given that:

$$L[f(t)] = \frac{S+2}{S^2+1},$$

$$L[g(t)] = \frac{S^2+1}{(S+3)(S+2)},$$

$$h(t) = \int_0^t f(\tau)\,g(t-\tau)\,d\tau$$

then, $L[h(t)]$ is

(a) $\frac{S^2+1}{S+3}$ (b) $\frac{1}{S+3}$

(c) $\frac{S^2+1}{(S+3)(S+2)}$ (d) None of these

Q.127 A causal LTI system is described by the difference equation, $2y[n] = \alpha y[n-2] - 2x[n] + \beta x[n-1]$. The system is stable only if,

(a) $|\alpha| = 2$, $|\beta| < 2$
(b) $|\alpha| > 2$, $|\beta| > 2$
(c) $|\alpha| < 2$, any value of β
(d) $|\beta| < 2$, any value of α

Q.128 Match **List-I (Function in time domain)** with **List-II (Corresponding Laplace Transform F(S))** and select the correct answer using the codes given below the lists:

List-I
A. $\sin\omega_0 t \cdot u(t-t_0)$
B. $\sin\omega_0(t-t_0)\cdot u(t-t_0)$
C. $\sin\omega_0(t-t_0)\cdot u(t)$
D. $\sin\omega_0 t\cdot u(t)$

List-II
1. $\frac{\omega_0}{s^2+\omega_0 2}$
2. $\left\{\frac{\omega_0}{s^2+\omega_0^2}\right\}e^{-t_0 S}$
3. $\frac{e^{-t_0 S}}{\sqrt{S^2+\omega_0^2}}\sin\left(\omega_0 t_0+\tan^{-1}\frac{\omega_0}{s}\right)$
4. $-\frac{1}{\sqrt{S^2+\omega_0^2}}\sin\left(\omega_0 t_0-\tan^{-1}\frac{\omega_0}{s}\right)$

Codes:

	A	B	C	D
(a)	3	1	4	2
(b)	4	2	3	1
(c)	3	2	4	1
(d)	4	1	3	2

Q.129 The laplace transform of $\left(\frac{1-e^t}{t}\right)$

(a) $\log\left(\frac{s}{s-1}\right)$ (b) $\log\left(\frac{s-1}{s}\right)$

(c) $\log\left(\frac{s-1}{s+1}\right)$ (d) $\log\left(\frac{s+1}{s-1}\right)$

Q.130 Inverse Laplace transform of

$$x(S) = \frac{S^2+2S+5}{(S+3)(S+5)^2};\ R_e(S) > -3 \text{ is}$$

(a) $[e^{-3t}+3e^{-5t}-2te^{-5t}]\,u(t)$
(b) $[2e^{-3t}-4e^{-5t}-5te^{-5t}]\,u(t)$
(c) $[2e^{-3t}-e^{-5t}-10te^{-5t}]\,u(t)$
(d) $[e^{-3t}-4e^{-5t}-5te^{-5t}]\,u(t)$

Q.131 The given system, $F(s) = \log\frac{S^2+1}{S(S+1)}$;

Find $f(t)$ at $t = 0$

(a) 1 (b) 2
(c) –1 (d) –2

6. Fourier Analysis of Discrete Time Signals

Q.132 The discrete time fourier transform for the given signal $x[n] = u[n]$ is

(a) $\pi\delta(\Omega) - \dfrac{1}{1+e^{-j\Omega}}$ (b) $\dfrac{1}{1-e^{-j\Omega}}$

(c) $\pi\delta(\Omega) + \dfrac{1}{1-e^{-j\Omega}}$ (d) $\dfrac{1}{1+e^{-j\Omega}}$

Q.133 The discrete time Fourier coefficients $x[k]$ of the signal $x[n] = \sum_{m=-\infty}^{\infty} \delta[n-4m]$

(a) $-\frac{1}{4}$, for all k (b) $\frac{1}{4}$, for all k

(c) $\frac{1}{4}|k|$, for all k (d) $e^{j\frac{\pi}{2}k}$, for all k

Q.134 The zero state respone $y(k)$ for input

$$f(k) = (0.8)^k\, u(k) \text{ is}$$

$$\left[\text{Given}: H(\Omega) = \frac{e^{j\Omega}}{e^{j\Omega} - 0.5}\right]$$

(a) 0

(b) ∞

(c) $\left[-\frac{5}{3}(0.5)^k + \frac{8}{3}(0.8)^k\right]u(k)$

(d) $\left[\frac{5}{3}(0.5)^k - \frac{8}{3}(0.8)^k\right]u(k)$

Q.135 Consider discrete time sequence

$$x(n) = \{-1, 2, -3, 2, -1\}$$
$$\uparrow$$

with DTFT $x(e^{j\omega})$

Value of $\int_{-\pi}^{\pi} x(e^{j\omega})\, d\omega$ is equal to

(a) 3π (b) 6π

(c) -3π (d) -6π

Q.136 Consider a signal $x(n)$ with following factors:

1. $x(n)$ is real and even signal
2. The period of $x(n)$ is $N = 10$
3. $x(11) = 5$
4. $\frac{1}{10}\sum_{n=0}^{9} |x(k)|^2 = 50$

The signal $x(n)$ is

(a) $10\cos\left(\frac{\pi}{5}n\right)$ (b) $5\cos\left(\frac{\pi}{5}n\right)$

(c) $5\cos\left(\frac{\pi}{10}n\right)$ (d) $10\cos\left(\frac{\pi}{10}n\right)$

Q.137 Consider a discrete time signal

$$x(n) = \{-1, 2, -3, 2, -1\}$$

value of $\angle x(e^{j\omega})$ is equal to

(a) $\frac{\pi}{2}$ (b) $\frac{\pi}{4}$

(c) $\frac{3\pi}{4}$ (d) π

Q.138 A low pass filter with impulse response $h_1(n)$ has spectrum $H_1(e^{j\omega})$ shown below.

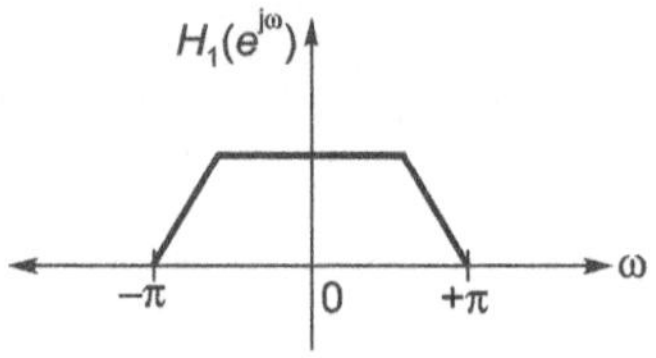

Here only one period has been shown by reversing every second sign of $h_1(n)$ a new filter having impulse response $h_2(t)$ is created. The spectrum of $H_2(e^{j\omega})$ is given by

(a) $H_1(e^{j(\omega-\pi/2)})$ (b) $H_1(e^{j(\omega-3\pi/2)})$

(c) $H_1(e^{j(\omega-\pi)})$ (d) $H_1(e^{j(\omega-\pi/4)})$

Q.139 A red signal $x[n]$ with Fourier transform $x(e^{j\Omega})$ has following property:

1. $x[n] = 0$ for, $n > 0$
2. $x[0] > 0$
3. $I_m\{x(e^{j\Omega})\} = (\sin\Omega - \sin 2\Omega)$
4. $\frac{1}{2}\int_{-\pi}^{\pi} \left|x(e^{j\Omega})\right|^2 d\Omega = 3$

The signal $x[n]$ is

(a) $2\delta[n] - \delta[n+1] - \delta[n+3]$

(b) $\delta[n] + \delta[n+1] - \delta[n+3]$

(c) $\delta[n] + \delta[n+1] - \delta[n+2]$

(d) $2\delta[n] - \delta[n+1] - \delta[n+2]$

Q.140 A causal and stable LTI system has the property that,

$$\left(\frac{2}{3}\right)^n \cdot u[n] \Rightarrow n\left(\frac{2}{3}\right)^n \cdot u[n]$$

The frequency response $H(e^{j\Omega})$ for this system is

(a) $\dfrac{2e^{j\Omega}}{2-3e^{j\Omega}}$ (b) $\dfrac{2e^{-j\Omega}}{2-3e^{-j\Omega}}$

(c) $\dfrac{2e^{j\Omega}}{2-3e^{-j\Omega}}$ (d) $\dfrac{2e^{-j\Omega}}{2-3e^{-j\Omega}}$

Q.141 A 5-point sequence $x[n]$ is given as
$x[-3] = 1, x[-2] = 1, x[-1] = 0, x[0] = 5, x[1] = 1$
Let $x(e^{j\omega})$ denote the discrete time fourier transform of $x[n]$.

The value of $\int_{-\pi}^{\pi} x(e^{j\omega})\, d\omega$ is

(a) 5 (b) 10π
(c) 16π (d) $5 + j10\pi$

Q.142 Fourier coefficient C_1 and C_3 for periodic signal $x[n]$ is

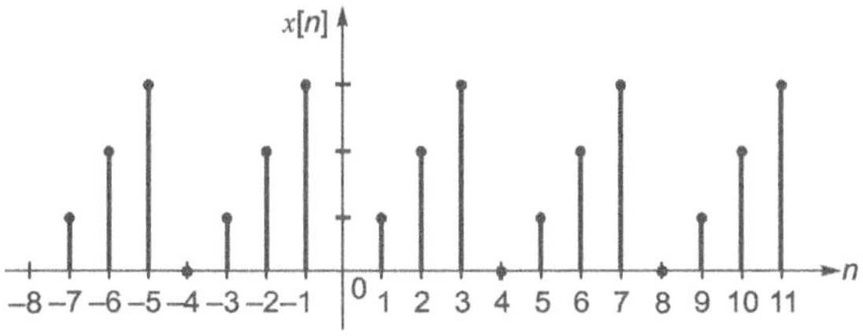

(a) $-\frac{1}{2}+j\frac{1}{2}$ and $-\frac{1}{2}-j\frac{1}{2}$

(b) $-\frac{1}{2}$ and $-\frac{1}{2}-j\frac{1}{2}$

(c) $-\frac{1}{2}$ and $-\frac{1}{2}+j\frac{1}{2}$

(d) $-\frac{j}{2}$ and $\frac{1}{2}-j\frac{1}{2}$

Q.143 The duality of DFT expressed as
(a) $x[n] \leftrightarrow x[-k]$
(b) $x[n] \leftrightarrow x[-k]_{\text{Mod N}}$
(c) $x[n] \leftrightarrow \frac{1}{N}x[-k]$
(d) $x[n] \leftrightarrow \frac{1}{N}x[-k]_{\text{Mod N}}$

7. Z-Transform

Q.144 If, $x(z)$ is $\dfrac{1}{1-z^{-1}}$ with $|z| > 1$, then what is the corresponding x(n)?
(a) e^{-n} (b) e^{n}
(c) $u(n)$ (d) $\delta(n)$

Q.145 For which of the following $x[n]$ cannot find z-transform $x(z)$.

(a) $x[n] = \left(\frac{1}{2}\right)^n u[n] + \left(\frac{1}{3}\right)^n u[n]$

(b) $x[n] = \left(\frac{1}{3}\right)^n u[n] + \left(\frac{1}{2}\right)^n u[-n-1]$

(c) $x[n] = \left(\frac{1}{2}\right)^n u[n] + \left(\frac{1}{3}\right)^n u[-n-1]$

(d) None of the these

Q.146 The z-transform of a signal is given by

$$c(z) = \frac{z^{-1}(1-z^{-4})}{4(1-z^{-1})^2}$$

Its final valve is

(a) $\frac{1}{4}$ (b) zero
(c) 1.0 (d) ∞

Q.147 Which one of the following represents the impulse response of a system defined by $H(z) = z^{-m}$
(a) $u[n-m]$ (b) $\delta[n-m]$
(c) $\delta[m]$ (d) $\delta[m-n]$

Q.148 The z-transform of the time function $\sum_{k=0}^{\infty} \delta(n-k)$ is

(a) $\dfrac{(z-1)}{z}$ (b) $\dfrac{z}{(z-1)^2}$

(c) $\dfrac{z}{(z-1)}$ (d) $\dfrac{(z-1)^2}{z}$

Q.149 The z-transform of the system is $H(z) = \dfrac{z}{z-0.2}$; if the ROC is $|z| < 0.2$, then the impulse response of the system is
(a) $(0.2)^n u[n]$ (b) $(0.2)^n \cdot u[-n-1]$
(c) $-(0.2)^n \cdot u[n]$ (d) $-(0.2)^n \cdot u[-n-1]$

Q.150 If $x(z) = \dfrac{1}{1-az^{-1}}$, the two discrete time signal is
(a) $x(n) = a^n u(n)$ and $x(n) = a^n u(-n)$
(b) $x(n) = a^n u(n)$ and $x(n) = a^n u(-n-1)$
(c) $x(n) = a^n u(n)$ and $x(n) = a^n u(-n-1)$
(d) none of these

Q.151 If $x(n) = \left[7\left(\frac{1}{3}\right)^n u(n) - 6\left(\frac{1}{2}\right)^n u(n)\right]$.

The ROC for the above discrete time signal is

(a) $|z| < \frac{1}{3}$ (b) $|z| < \frac{1}{2}$

(c) $|z| > \frac{1}{2}$ (d) $|z| > \frac{1}{3}$

Q.152 Consider the τ-transform,

$x(z) = 4z^2 + 2 + 3z^{-1}; 0 < |z| < \infty.$

The $x(n)$ is

(a) $x(n) = 4\delta[n-2] + 2\delta[n] + 3\delta[n+1]$

(b) $x(n) = 4\delta[n+2] + 2 + 3\delta[n-1]$

(c) $x(n) = 4\delta[n+2] + 2\delta[n] + 3\delta[n-1]$

(d) none of these

Q.153 A stable and causal system is described by the difference equation,

$$y(n) + \frac{1}{4}y(n-1) - \frac{1}{8}y(n-2) = -2x(n) + \frac{5}{4}x(n-1)$$

The impulses response of the system is

(a) $h(n) = \left(\frac{1}{4}\right)^n u(n) - 3\left(-\frac{1}{2}\right)^n \delta(n)$

(b) $h(n) = \left(\frac{1}{4}\right)^n u(n) - 3u(n)$

(c) $h(n) = \left(\frac{1}{4}\right)^n u(n-1) - 3u(n-1)$

(d) $h(n) = \left(\frac{1}{4}\right)^n u(n) + 3\left(\frac{1}{2}\right)^n u(n)$

Q.154 A linear discrete time system has the characteristics equation, $z^3 - 0.81z = 0$, the system is

(a) is stable

(b) is marginally stable

(c) is unstable

(d) stability can not be assessed from the given information

Q.155 The z-transform of $F(z)$ of the function $f(nT) = a^{nT}$ is

(a) $\frac{z}{z-a^T}$ (b) $\frac{z}{z+a^T}$

(c) $\frac{z}{z-a^{-T}}$ (d) $\frac{z}{z+a^{-T}}$

Q.156 The ROC of z-transform of the sequence

$\left(\frac{5}{6}\right)^n u(n) - \left(\frac{6}{5}\right)^n u(-n-1)$ must be

(a) $|z| < \frac{5}{6}$ (b) $|z| > \frac{5}{6}$

(c) $\frac{5}{6} < |z| < \frac{6}{5}$ (d) $\frac{6}{5} < |z| < \infty$

Q.157 If, the ROC of $x_1[n] + x_2[n]$ is $\frac{1}{3} < |z| < \frac{2}{3}$ then the ROC of $x_1[n] - x_2[n]$ includes

(a) $\frac{1}{3} < |z| < 3$ (b) $\frac{2}{3} < |z| < 3$

(c) $\frac{3}{2} < |z| < 3$ (d) $\frac{1}{3} < |z| < \frac{2}{3}$

Q.158 The z-transform of discrete time signal $x[n] = \{1, 0, -1, 0, 1, -1\}$ is

(a) $1 + 2z^{-2}\, 4z^{-4} + 5z^{-5}$

(b) $1 - z^{-2} + z^{-4} - z^{-5}$

(c) $1 - 2z^2 + 4z^4 - 5z^5$

(d) $1 - z^2 + z^4 - z^5$

Q.159 A all pass causal LTI system is described by the difference equation:

$y[n] - ay[n-1] = b\,x[n] + x[n-1]$

The value of 'b' will be

(a) $-a$ (b) a

(c) $-2a$ (d) $2a$

Q.160 Consider a discrete time signal $x(n)$ having z-transform given by

$x(z) = \frac{1}{z^3} - 1$, if $x(n)$ is right sided then

z-transform $x(z)$ belongs to

(a) Finite series with ROC : $|z| > 1$

(b) Infinite sereis with ROC : $|z| > 1$

(c) Finite sereis with ROC : $|z| < \infty$

(d) Finite sereis with ROC : $|z| < 1$

Q.161 Consider a system with system function:

$$H(z) = \frac{z^3 - 2z^2 + z}{z^2 + \frac{1}{4}z + \frac{1}{8}}$$

(a) System is non-causal.

(b) System is causal.

(c) Data is not sufficient.

(d) None of these

Q.162 Let $x[n] = \left(\frac{1}{2}\right)^n u[n] - \left(-\frac{1}{3}\right)^n u[-n-1]$

The ROC for the z-transform of above sequence is

(a) $|z| > \frac{1}{2}$

(b) $|z| < \frac{1}{3}$

(c) $\frac{1}{3} < |z| < \frac{1}{2}$

(d) z-transform does not exist

Q.163 The sequence $x[n]$ corresponding to $x(z)$ defined as

$$x(z) = \frac{z}{2z^2 - 3z + 1}\ ;\ |z| < \frac{1}{2}$$

will have the value at $n = -2$

(a) 0 (b) 1

(c) 2 (d) 3

Q.164 The direct form-II realisation of IIR filter is shown below.

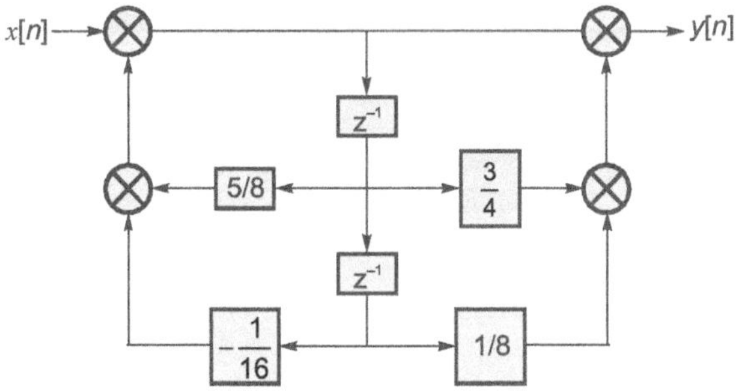

The transfer function is

(a) $H(z) = \dfrac{1 + \frac{5}{8}z^{-1} - \frac{1}{16}z^{-2}}{1 + \frac{3}{4}z^{-1} + \frac{1}{8}z^{-2}}$

(b) $H(z) = \dfrac{1 - \frac{5}{8}z^{-1} + \frac{1}{16}z^{-2}}{1 - \frac{3}{4}z^{-1} - \frac{1}{8}z^{-2}}$

(c) $H(z) = \dfrac{1 + \frac{3}{4}z^{-1} + \frac{1}{8}z^{-2}}{1 - \frac{5}{8}z^{-1} + \frac{1}{16}z^{-2}}$

(d) $H(z) = \dfrac{1 - \frac{5}{8}z^{-1} + \frac{1}{16}z^{-2}}{1 + \frac{3}{4}z^{-1} + \frac{1}{8}z^{-2}}$

Q.165 A sequence of $x(n)$ with the z-transform $x(z) = z^4 + z^2 - 2z + 2 - 3z^{-4}$ is applied as an input to a L.T.I system with the impulse response $h(n) = 2\delta(n-3)$ where,

$$\delta(n) = \begin{cases} 1, & n = 0 \\ 0, & \text{otherwise} \end{cases}$$

the output at $n = 2$ is

(a) −6 (b) 0

(c) 2 (d) −4

Q.166 Find $x[n]$ for the case, when $x(z) = \dfrac{z^2 + z}{(z-1)^2}$

(a) n^2 (b) $2n$

(c) $2n + 1$ (d) $(n + 1)^2$

Q.167 Match **List-I {$x(n)$}** with **List-II {$x(z)$}** and select the correct answer using the codes given below the lists:

List-I

A. $\alpha^n u(n)$

B. $-\alpha^n u(-n-1)$

C. $-n\alpha^n u(-n-1)$

D. $n\alpha^n u(n)$

List-II

1. $\dfrac{az^{-1}}{(1-\alpha z^{-1})^2}$ ROC : $|z| > |\alpha|$

2. $\dfrac{1}{(1-\alpha z^{-1})}$ ROC: $|z| > |\alpha|$

3. $\dfrac{1}{(1-\alpha z^{-1})}$ ROC: $|z| < |\alpha|$

4. $\dfrac{\alpha z^{-1}}{(1-\alpha z^{-1})^2}$ ROC: $|z| < |\alpha|$

Codes

	A	B	C	D
(a)	2	4	3	1
(b)	1	3	4	2
(c)	1	4	3	2
(d)	2	3	4	1

Q.168 Choose the correct option for given causal systems, stable or not

i. $H(z) = \dfrac{z}{(z-1/2)(z-1/4)}$

ii. $H(z) = \dfrac{z(z-2)}{(z-1/2)(z-2)}$

iii. $H(z) = \dfrac{z(z-1)}{(z-1/2)(z-0.6-j0.6)(z-0.6+j0.6)}$

	i	ii	iii
(a)	stable	stable	stable
(b)	stable	unstable	unstable
(c)	stable	unstable	stable
(d)	stable	stable	unstable

Q.169 Find the inverse z-transform of

$$F(z)=\frac{z(2z^2-11z+12)}{(z-1)(z-2)^3}$$

(a) $\left[3+\frac{1}{4}(n^2+n-12)\,2^n\right]u(n)$

(b) $-\left[3+\frac{1}{4}(n^2+n-12)\,2^n\right]u(n)$

(c) $-\left[3+\frac{1}{4}(n^2-12)\,2^n\right]u(n)$

(d) $\left[3+\frac{1}{4}(n^2+n+12)\,2^n\right]u(n)$

Q.170 Consider discrete time signal $x(n)$ is given by

$$x(n)=4^n\,u(n)$$

Also, $x(n)\xleftrightarrow{z.T}x(z)$

Consider another signal $y(n)\xleftrightarrow{z.T}y(z)$

Also, $y(z)=x^2(z)$

then, $y(n)$ is

(a) $n\cdot 4^n\,u(n)$ (b) $(n+1)\,4^n\,u(n+1)$
(c) $(n+1)4^n\,u(n)$ (d) $n\,4^n\,u(n+1)$

Q.171 Consider a z-transform of a signal $x(n)$

$$x(z)=\frac{2+6z^{-1}}{4-2z^{-2}+13z^{-3}}$$

value of $x(1)$ is equal to

(a) 1 (b) $\frac{1}{2}$

(c) $\frac{3}{2}$ (d) $\frac{5}{7}$

Q.172 Consider a discrete time signal $x(n)$, who use z-transform $x(z)$ is given by,

$$x(z)=\frac{z(z-2)}{(z-1)(z-0.5)}$$

value of $x(n)$ for $n=0$ is

(a) 0 (b) 1

(c) $\frac{1}{2}$ (d) $\frac{1}{4}$

Q.173 If z-transform is given by

$$x(z)=\cos(z^{-3}),\ |z|>0$$

the value of $x[12]$ is

(a) $-\frac{1}{24}$ (b) $\frac{1}{24}$

(c) $-\frac{1}{6}$ (d) $\frac{1}{6}$

Q.174 The impulse response of the system in below figure.

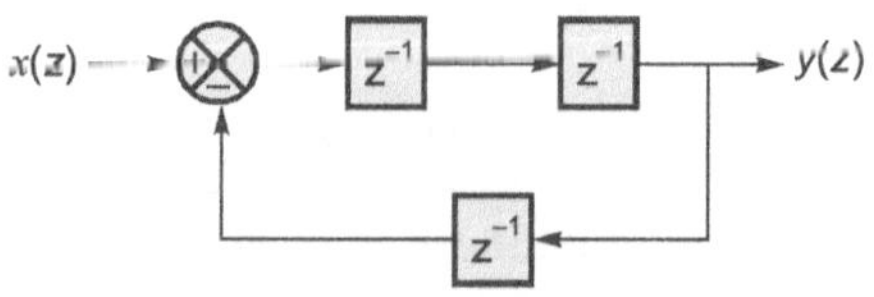

(a) $2^{\left(\frac{n}{2}-2\right)}(1+(-1)^n)\,u[n]+\frac{1}{2}\delta[n]$

(b) $\frac{2^n}{2}(1+(-1)^n)\,u[n]+\frac{1}{2}\delta[n]$

(c) $2^{\left(\frac{n}{2}-2\right)}(1+(-1)^n)\,u[n]-\frac{1}{2}\delta[n]$

(d) $\frac{2^n}{2}(1+(-1)^n)\,u[n]-\frac{1}{2}\delta[n]$

Q.175 The z-transform $x[z]$ of a sequence $x[n]$ is given by $x[z]=\frac{0.5}{1-2z^{-1}}$. It is given that the ROC of $x[z]$ includes the unit circle. The value of $x[0]$ is

(a) – 0.5 (b) 0
(c) 0.25 (d) 0.5

Q.176 The difference equation representation for a system is

$$y[n]-2y[n-1]+y[n-2]=x[n]-x[n-1]$$

If, $y[n]=0$ for $n<0$ and $x[n]=\delta[n]$ then $y[100]$ will be

(a) 1 (b) –1
(c) 0 (d) 100

Q.177 If $U(z)=\frac{2z^2+5z+14}{(z-1)^4}$; the u_2 and u_3 is

(a) 1, 13 (b) 2, 9
(c) 1, 9 (d) 2, 13

Q.178 The system $H(z)=\frac{2z^2-\frac{5}{2}z}{z^2-\frac{5}{2}z+1}$ is

(a) ROC: $|z| > 2$, stable, causal

(b) ROC: $|z| > \frac{1}{2}$, unstable, non- causal

(c) ROC: $\frac{1}{2} > |z| > 2$, stable, non- causal

(d) ROC: $\frac{1}{2} < |z| < 2$, non-stable, non- causal

Q.179 The discrete time LTI system the output $y[n]$ and input $x[n]$ is $u[n]$. Given below figure.

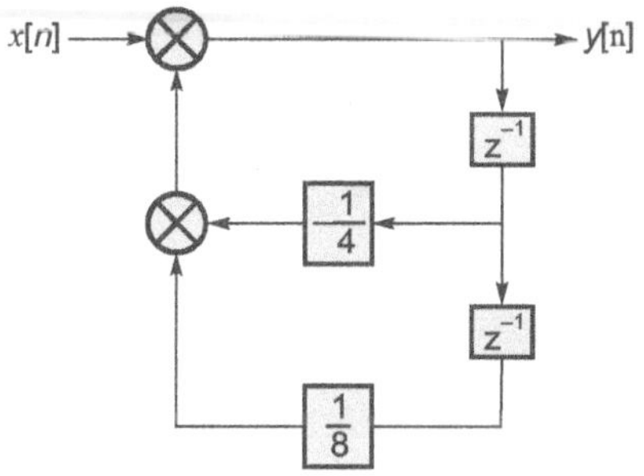

The impulse response of the system is

(a) $2\left(\frac{1}{2}\right)^n u(t) - \left(\frac{1}{4}\right)^n u(t)$

(b) $\left[\frac{1}{2}\left(\frac{1}{4}\right)^n - 2\left(\frac{1}{2}\right)^n\right] u(t)$

(c) $\frac{1}{4}\left(\frac{1}{2}\right)^n u(t) - \frac{1}{2}\left(\frac{1}{4}\right)^n u(t)$

(d) $\left(\frac{1}{2}\right)^n u(t) - 2\left(\frac{1}{4}\right)^n u(t)$

■■■■

Answers Signals and Systems

1. (b)	2. (b)	3. (b)	4. (a)	5. (c)	6. (d)	7. (a)	8. (b)
9. (a)	10. (a)	11. (a)	12. (d)	13. (b)	14. (c)	15. (d)	16. (b)
17. (a)	18. (a)	19. (a)	20. (c)	21. (c)	22. (a)	23. (a)	24. (a)
25. (d)	26. (c)	27. (b)	28. (a)	29. (d)	30. (d)	31. (c)	32. (a)
33. (b)	34. (b)	35. (c)	36. (b)	37. (b)	38. (b)	39. (b)	40. (a)
41. (a)	42. (b)	43. (d)	44. (c)	45. (a)	46. (d)	47. (a)	48. (a)
49. (b)	50. (a)	51. (b)	52. (c)	53. (c)	54. (d)	55. (c)	56. (c)
57. (c)	58. (c)	59. (c)	60. (c)	61. (d)	62. (b)	63. (b)	64. (a)
65. (b)	66. (a)	67. (b)	68. (c)	69. (c)	70. (a)	71. (b)	72. (a)
73. (b)	74. (d)	75. (a)	76. (b)	77. (d)	78. (a)	79. (a)	80. (d)
81. (c)	82. (d)	83. (a)	84. (d)	85. (b)	86. (b)	87. (b)	88. (d)
89. (b)	90. (c)	91. (b)	92. (c)	93. (c)	94. (b)	95. (b)	96. (d)
97. (b)	98. (b)	99. (d)	100. (b)	101. (d)	102. (a)	103. (c)	104. (b)
105. (a)	106. (a)	107. (b)	108. (b)	109. (b)	110. (b)	111. (d)	112. (b)
113. (a)	114. (c)	115. (a)	116. (d)	117. (b)	118. (d)	119. (c)	120. (c)
121. (d)	122. (b)	123. (a)	124. (b)	125. (a)	126. (b)	127. (c)	128. (c)
129. (b)	130. (c)	131. (b)	132. (c)	133. (b)	134. (c)	135. (d)	136. (a)
137. (d)	138. (c)	139. (c)	140. (d)	141. (b)	142. (a)	143. (b)	144. (c)
145. (c)	146. (c)	147. (b)	148. (c)	149. (d)	150. (c)	151. (c)	152. (b)
153. (b)	154. (a)	155. (a)	156. (c)	157. (d)	158. (b)	159. (a)	160. (c)
161. (a)	162. (d)	163. (d)	164. (a)	165. (d)	166. (b)	167. (d)	168. (a)
169. (b)	170. (c)	171. (c)	172. (b)	173. (b)	174. (c)	175. (b)	176. (a)
177. (d)	178. (d)	179. (a)					

Explanations

8. (b)

$$E = \int_{-\infty}^{\infty} f(t)^2 dt$$

and
$$E^1 = \int_{-\infty}^{\infty} f(2t)^2 dt$$

$$= \int_{-\infty}^{\infty} f(p)^2 dp \cdot \frac{1}{2} \quad \begin{bmatrix} \therefore 2t = P \\ dt = dp \cdot \dfrac{1}{2} \end{bmatrix}$$

$$\Rightarrow \quad \boxed{E^1 = \frac{E}{2}}$$

9. (a)

$$x(n) = [-4 - j5, 1 + 2j, 4]$$
$$x^*(-n) = [4, 1 - 2j, -4 + j5]$$
$$x_{CAS}(n) = \frac{x(n) - x^*(-n)}{2}$$
$$= [-4 - 2.5j, 2j, 4 - 2.5j]$$

11. (a)

$$S(t) = 8\cos\left(\frac{\pi}{2} - 20\pi t\right) + 4\sin 15\pi t$$
$$= 8 \sin 20\pi t + 4 \sin 15\pi t$$
$$\Rightarrow \quad P = \frac{8^2}{2} + \frac{4^2}{2} = 32 + 8 = 40$$

16. (b)

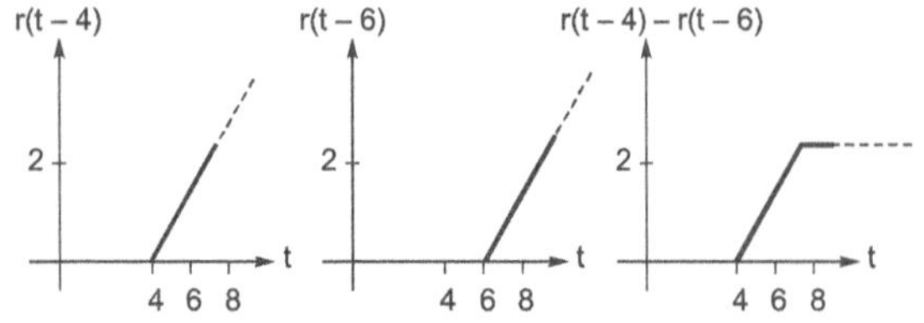

17. (a)

$$y(t) = \int_{-\infty}^{t} x(\tau)\, d\tau$$
$$= \int_{-\infty}^{t} [(\delta(x + 2) - \delta(\tau - 2))]$$
$$= \begin{cases} 0, t < -2 \\ 1, -2 \geq t \leq 2 = E_{\infty} = \int_{-2}^{2} dt = 4 \\ 0, t > 2 \end{cases}$$

18. (a)

Both signal are periodic,

$$N_1 = 18, N_2 = 14$$
$$N = \text{LCM}(18, 14) = 126$$

19. (a)

$$E = \sum_{k=-\infty}^{\infty} \left|x[n]^2\right| = \sum_{k=0}^{\infty} (0.25)^n$$
$$= \frac{1}{1 - 0.25} = \frac{4}{3} < \infty$$

$\therefore$ $x[n]$ is an energy signal.

25. (d)

Since, $\quad y(t) = \int_{-\infty}^{t} [\delta(t + 2) - \delta(t - 2)]dt$

So, $\quad y(t) = u(t + 2) - u(t - 2)$

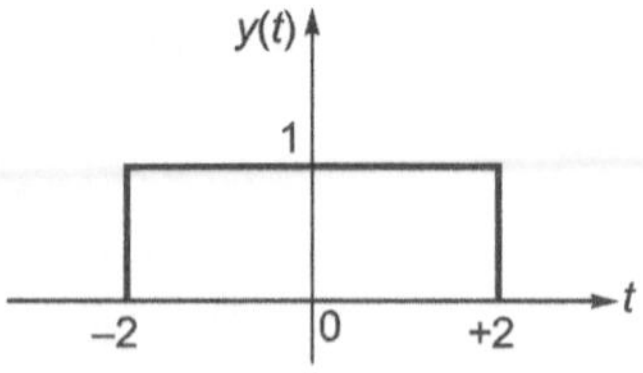

So, $\quad E_y = \int_{-2}^{2} 1 \cdot dt = 4$

26. (c)

Since, $\sin(A + B) = \sin A \cdot \cos B + \cos A \cdot \sin B$

So, $\quad x(t) = \sin(\omega_0 + \sqrt{2}\omega_0)t$

$$T = \frac{2\pi}{\omega} = \frac{2\pi}{(1 + \sqrt{2})\omega_0}$$

27. (b)

$$r_{xx}(\tau) = x(\tau) * x(\tau)$$

For, $t < 0$, $\quad r_{xx}(\tau) = \int_{0}^{\infty} e^t e^{-2\tau} d\tau = \frac{1}{2} e^t$

For, $t \geq 0$, $\quad r_{xx}(\tau) = \int_{t}^{\infty} e^t \cdot e^{-2\tau} d\tau = \frac{1}{2} e^{-t}$

$$r_{xx}(\tau) = \frac{1}{2} e^t (1 - u(t)) + \frac{1}{2} e^{-t} u(t)$$
$$= \frac{1}{2} e^t + \frac{1}{2} (e^{-t} - e^t)\, u(t)$$

28. (a)

$$y(n) = x\left(\frac{n}{2}-1\right), \; n \text{ is even}$$
$$= 0 \text{ ; for, '}n\text{' is odd.}$$
$$n = 0, y(n) = x(-1) = 1$$
$$n = 2, y(n) = x(0) = 2$$
$$n = 4, y(n) = x(1) = 1$$
$$n = 6, y(n) = x(2) = 1/2$$

45. (a)

$$x_1(z) = 1 - 2z^{-1} + z^{-2}$$
and $$x_2(z) = 1 + z^{-1} + z^{-2} + z^{-3} + z^{-4} + z^{-5}$$
$$\Rightarrow x_1(n) \cdot x_2(n) = x_1(z) \cdot x_2(z)$$
$$= [1 - 2z^{-1} + z^{-2}] \cdot [1 + z^{-1} + z^{-2} + z^{-3} + z^{-4} + z^{-5}]$$
$$= [1, 0, 0, 0, 0, 0, -1, 1]$$

47. (a)

Convolution of two signal is

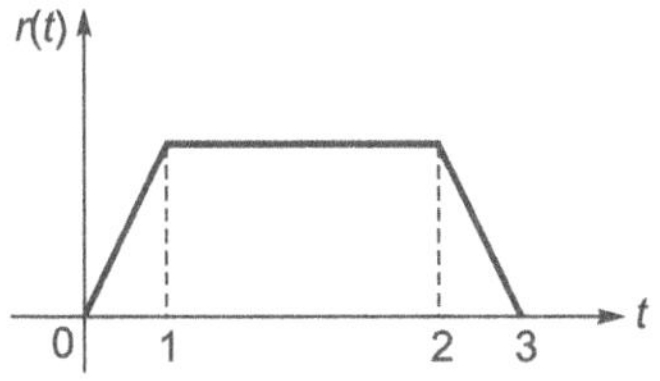

Which is given by,

$$x_1(t) + x_2(t) = r(t) - r(t-1) - r(t-2) + r(t-3)$$

50. (a)

$$\sum_{k=-\infty}^{\infty} h(k) = \sum_{k=-3}^{\infty} u(k+3) + \sum_{k=2}^{\infty} u(k-2)$$
$$= -2\sum_{k=7}^{\infty} u(k-7) = \sum_{k=3}^{6} 1 + \sum_{k=2}^{6} 1$$
$$= 10 + 5 = 15 < \infty$$

for, bounded I/P, and bounded output, system is stable response depends on future value of input signal i.e. $u(n + 3)$. So, system is not causal.

51. (b)

For causal system output depends on the past and present value

$$h(t) = tu(t - t_1) - tu(t - t_2)$$

52. (c)

$$y(n) = \left(\sin\frac{5}{6}\pi n\right) \cdot x(n)$$

Let, $x(n) = \delta(n)$

$\therefore$ $y(n) = \sin 0 = 0$(bounded)

This is BIBO stable.

53. (c)

This system is not invertible, because if, $x(t)$ is any constant, then $y(t) = 0$.

55. (c)

$$y[n] = x[n] * h[n] = \sum_{i=0}^{3} x[i] \cdot h[n-i]_{\text{mod}\,4}$$
$$n = 3, y[3] = 1\left(\frac{1}{8}\right) + (-1)\left(\frac{1}{2}\right) = \frac{-3}{8};$$

It is clear that,

$$x[n] = \{1, 0, -1, 0\},$$
$$h[n] = \left\{1, \frac{1}{2}, \frac{1}{4}, \frac{1}{8}\right\}$$

56. (c)

$$x[n] * u[n-n_0] = \sum_{k=-\infty}^{\infty} x[k] \cdot u[n-k-n_0]$$
$$= \sum_{k=-\infty}^{n-n_0} x[k]$$

62. (b)

We need to use scalling property of the convolution integral.

The property is

if, $y(t) = x(t) * h(t)$

then, $y(at) = |a| x(at) * h(at)$

In given problem,

$$\frac{1}{3}y(3t) = x(3t) * 3t; \qquad (\because 9:3)$$
$$\therefore \quad x(3t) = (3t)^2 + 3t = 9t^2 + 3t$$
$$h(3t) = 3t$$

67. (b)

$$y(t) = x(t) * h(t) = \int_{-0.5}^{0.5} e^{j\omega_0(t-\tau)} d\tau$$
$$\Rightarrow \quad y(t) = \frac{1}{j\omega(t-\tau)} \left| e^{j\omega_0(t-2)} \right|_{-0.5}^{+0.5}$$
$$\Rightarrow \quad y(t) = \frac{-2j}{\omega_0} \sin\left(\frac{\omega_0}{2}\right) e^{j\omega_0 t}$$

Since, $\sin(\pi) = 0$

So, $\omega_0 = 2\pi$

68. (c)

Let output of the system 1 is $\omega(n)$ and output of the system 2 is $z(n)$ then,

$$y_3(n) = z(2n)$$

$$= \omega(2n) + \frac{1}{2}\omega(2n-1) + \frac{1}{4}\omega(2n-2)$$

$$= x(n) + \frac{1}{2}x(n-1) + \frac{1}{4}x(n-2)$$

70. (a)

From the figure,

$$H(t) = \delta(t) + \delta(t-1)$$

So, $H(S) = (1 + e^{-S})$

Also, $y(t) = u(t) + u(t-1)$

So, $y(S) = \frac{1}{S} + \frac{e^{-S}}{S} = \frac{1+e^{-S}}{S}$

$\therefore$ $x(S) = \frac{y(S)}{H(S)}$

$$= \frac{1+e^{-S}}{S} \times \frac{1}{1+e^{-S}} = \frac{1}{S}$$

So, $x(t) = u(t)$

71. (b)

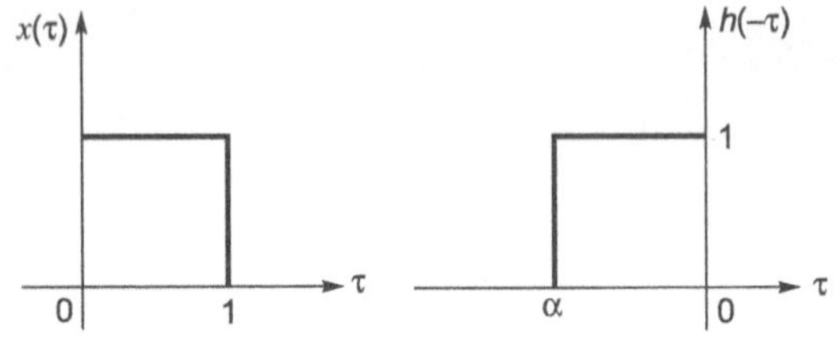

So, for $\alpha \le t \le 1$

$$y(t) = \int_{t}^{t+\alpha} d\tau = t + \alpha - t = \alpha$$

72. (a)

For, $n - 1 < 0$ or $n < 1$, $y[n] = 0$

For $n - 1 > 0$ or $n \ge 1$, $y[n] = \sum_{k=-3}^{n-3} 1 = n + 1$

$$y[n] = (n+1)\,u[n]$$

73. (b)

$$x(j\omega) = \frac{y(j\omega)}{H(j\omega)} = \frac{1}{3+j\omega} - \frac{1}{4+j\omega}$$

$$= \frac{1}{(3+j\omega)(4+j\omega)}$$

$$x(j\omega) = \frac{y(j\omega)}{H(j\omega)} = \frac{1}{(4+j\omega)}$$

$\Rightarrow$ $x(t) = e^{-4t}u(t)$

75. (a)

Both of term follows:

$h[n] = 0$, $n < 0$ and $\sum_{n=-\infty}^{\infty} |h[n]| < \infty$

So, system is causal and stable.

77. (d)

$$T_0 = \frac{2\pi}{\omega_0} = \frac{2\pi}{21}$$

and $\frac{T_s}{T_0} = \frac{T_s}{(2\pi/21)} = \frac{m}{N_0}$

$\Rightarrow$ $T_s = \frac{m}{N_0}\cdot\frac{2\pi}{21} \Rightarrow 0.3\pi = \frac{m}{N_0}\cdot\frac{2\pi}{21}$

$\Rightarrow$ $N_0 = \frac{20}{63}m$

For min, $m = 63$, then $N_0 = 20$ (fundamental period).

81. (c)

$$x[k] = \frac{1}{T}\int_{-T/2}^{+T/2} x(t)\,e^{-jk\omega_0 t}\,dt$$

$$= \frac{1}{T}\int_{-T/4}^{+T/4} Ae^{-jk\omega_0 t}dt$$

$$= \frac{A}{T}\left[\frac{e^{-jk\omega_0 t}}{-jk\omega_0}\right]_{-T/4}^{+T/4} = \frac{A}{\pi k}\sin\left(\frac{\pi k}{2}\right)$$

82. (d)

$$z(t) = x(t)\cdot y(t)$$

So, $z(t) = \cos 4\pi t \cdot \sin 4\pi t = \frac{1}{2}\sin \delta\pi t$

Now, $T = \frac{1}{2}$

So, $\omega_0 = 4\pi$

Let, $z(t) \xleftrightarrow{F.S} a_k$

then, $z(t) = \frac{1}{4j}$

So, $a_2 = \frac{1}{4j}$ and $a_{-2} = \frac{1}{4j}$

83. (a)

Here, $a_k = \begin{cases} \cos(k\pi); -100 \le k \le 100 \\ 0, \quad \text{otherwise} \end{cases}$

Since, a_k is real and even so, also $x(t)$.

84. (d)

$$x[k] = |k|; -3 \le k \le 3$$

$$x(t) = \sum_{k=-\infty}^{\infty} x[k]\, e^{j\pi kt}$$
$$= 3e^{j(-3)\pi t} + 2e^{j(-2)\pi t} + e^{j(-1)\pi t} + e^{j(1)\pi t} + e^{j(2)\pi t} + e^{j(3)\pi t}$$
$$= 6\cos 3\pi t + 4\cos 2\pi t + 2\cos \pi t$$

85. (b)

$f(x) = |\cos x|$, even function.

$$\therefore \quad f(x) = \frac{a_0}{2} + \sum a_n \cos nx$$

Here, $b_n = 0$, So, $b_2 = 0$

$$a_0 = \frac{2}{\pi}\int_0^{\pi} |\cos x|\, dx = \frac{4}{\pi}$$

94. (b)

$$\text{Sgn}(t) \Leftrightarrow \frac{1}{j\pi f}$$

then, $\frac{1}{j\pi t} \Leftrightarrow \text{Sgn}(-f)$

$\therefore \quad \frac{1}{j\pi t} \Leftrightarrow -\text{Sgn}(f)$

$\therefore \quad \boxed{\frac{j}{\pi t} \Leftrightarrow \text{sgn}(f)}$

96. (d)

$$x(n) = \left(\frac{1}{2}\right)^n \cdot u(n)$$

and $y(n) = x^2(n)\left(\frac{1}{2}\right)^{2n} \cdot u^2(n)$

$$\Rightarrow \quad y(n) = \left[\left(\frac{1}{2}\right)^2\right]^n \cdot u(n) = \left(\frac{1}{4}\right)^n u(n)$$

then, $y(z) = \dfrac{1}{1 - \frac{1}{4}z^{-1}}$; put $z = e^{j\omega}$

$$y(j\omega) = \frac{1}{1 - \frac{1}{4}e^{-j\omega}} \Rightarrow y(e^{j0}) = \frac{4}{3}$$

97. (b)

$$x\left[3\left(f + \frac{2}{3}\right)\right] = \frac{1}{3} \times \left(\frac{t}{3}\right) e^{-\frac{j\pi 4t}{3}}$$

Apply scaling and shifting property.

98. (b)

$$x[j\omega] = \int_{-1}^{0} e^{-j\omega t} dt - \int_{0}^{1} e^{-j\omega t} dt = \frac{2\cos\omega - 2}{j\omega}$$

100. (b)

Here, $x(t) = \begin{cases} (A/t)t & ; \text{for } 0 < t < T \\ A & ; \text{for, } T < t < 2T \end{cases}$

$$x(\omega) = F(x(t))$$

$$\Rightarrow \quad x(t) = \int_0^T \frac{A}{T} t e^{-j\omega t} dt + \int_T^{2T} A e^{-j\omega t} dt$$

$$\Rightarrow \quad x(t) = \frac{A}{j\omega}\left[e^{-j\omega/2}\,\text{sinc}\left(\frac{\omega t}{2}\right) - e^{-j2\omega t}\right]$$

102. (a)

Since, $|H(j\omega)| = 1; \quad |\omega| \le 200\pi$
$\qquad 0; \quad$ otherwise

using duality property,

$$h(t) = \frac{\sin 200\pi t}{\pi t}$$

Since group delay is constant and $\tau(\omega) = 5$

So, $h(t) = \dfrac{\sin[200\pi(t-5)]}{\pi(t-5)}$

103. (c)

Since, $x_2(j\omega) = [1 + \text{sgn}(\omega)]\, x_1(j\omega)$

or, $x_2(j\omega) = [1 + j\{-j\text{sgn}(\omega)\}] \cdot x_1(j\omega)$

$x_2(j\omega) = x_1(j\omega) + j[\{-j\text{sgn}(\omega) \cdot x_1(j\omega)\}]$

or, $x_2(t) = x_1(t) + jx_1(t) * \dfrac{1}{\pi t}$

because,

$$x_1(t) * \frac{1}{\pi t} = j\text{sgn}(\omega) \cdot x_1(j\omega)$$

using, Hilbert transform

104. (b)

$$y(t) = t\,x(t) \Rightarrow y(j\omega) = j\frac{d}{d\omega}x(j\omega)$$

$$\Rightarrow \quad y(j\omega) = j\left(\frac{2\cos\omega}{\omega} - \frac{2\sin\omega}{\omega^2}\right)$$

$$= 2j\left(\frac{\cos\omega}{\omega} - \frac{\sin\omega}{\omega^2}\right)$$

107. (b)

$$y(n) = A\,x(n - n_0)$$
$$= A\sin[\omega_0(n - n_0) + \phi]$$
$$\frac{-d\theta(\omega)}{d\omega} = n_0 (= t_g)$$
$$\Rightarrow \quad \theta(\omega) = -n_0 \int d\omega$$
$$= -n_0\omega + k$$

To avoid phase change k should be an integral multiple of 2π

$$\therefore \quad \theta(\omega) = n_0\omega_0 + 2\pi k$$

108. (b)

Average power $= P_y = E\{y^2(t)\} = R_y(0)$.

Also, $R_y(z) = \text{Fourier}^{-1}[S_y(f)]$

$$= \text{Fourier}^{-1}\left[|1 + (f)|^2 \cdot S_x(f)\right]$$

$$R_y(z) = \text{Fourier}^{-1}\left[\frac{1}{2^2 + (2\pi f)^2} \cdot 4\right]$$

$$= 1 \cdot e^{-2|z|}$$

$$\Rightarrow \quad R_y(0) = 1\,\omega$$

$\because$ Here, $S_x(f) = \text{Fourier}^{-1}[4\delta(t)] = 4$

115. (a)

$$\therefore \quad x(S) = \frac{1}{S+3}$$

$$y(S) = \frac{1}{S+3} - \frac{1}{S+2} = \frac{1}{(S+1)(S+2)}$$

$$\Rightarrow \quad H(S) = \frac{y(S)}{x(S)} \Rightarrow H(S) = \frac{S+2}{S^2 + 3S + 2}$$

117. (b)

Impulse response of system is $L^{-1}[H(S)]$

$$\frac{1}{S^2(S-2)} = \frac{1}{S^2} \times \frac{1}{S-2} = (t * e^{+2t})\,u(t)$$

118. (d)

$$x(t) = \frac{5-S}{S^2 - S - 2} = \frac{5-S}{(S+1)(S-2)}$$

$$= \frac{-2}{S+1} + \frac{1}{S-1}$$

$$\therefore \quad x(t) = [-2e^{-t}\,u(t) + e^{2t}\,u(-t)]$$

119. (c)

$$\underset{t\to\infty}{Lt}\; i(t) = \underset{s\to 0}{Lt}\; S \cdot I(s)$$

$$= \underset{s\to 0}{Lt}\; S \cdot \frac{2}{S\,(1+S)} = 2$$

120. (c)

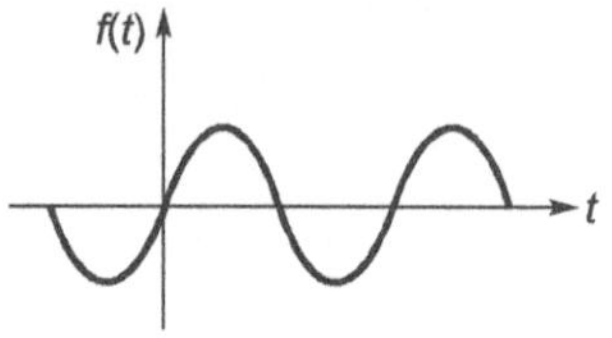

$$L^{-1}[F(S)] = \sin\omega_0 t$$
$$f(t) = \sin\omega_0 t$$

So, $\quad -1 \le f(\infty) \le 1$

121. (d)

$$H(s) = \frac{2}{s+1} - \frac{3}{(s+1)^2};$$

System is stable

$$\Rightarrow \quad h(t) = (2e^{-t} - 3te^{-t})\,u(t)$$

123. (a)

$$F(t) = 4\delta(t - \pi) + \sin(t - \pi) + 2\sin(t - 2\pi)$$

Taking laplace,

$$4e^{-\pi S} + \frac{2\left[e^{-\pi s} + e^{-2\pi s}\right]}{S^2 + 1}$$

126. (b)

Convolution in time domain is multiplication in s-domain.

$$\therefore \quad L[h(t)] = L[f(t)]\,x\,L[g(t)] = \frac{1}{S+3}$$

127. (c)

$$2y(n) = \alpha y(n-2) - 2x(n) + \beta x(n-1)$$

Taking, z-transform,

$$2y(z) = \alpha y(z)\,z^{-2} - 2x(z) + \beta\,x(z)\,z^{-1}$$

$$\Rightarrow y(z)[2-\alpha z^{-2}] = x(z)\,[\beta z^{-1}-2]$$

$$\frac{y(z)}{x(z)} = \left(\frac{\beta z^{-1}-2}{2-\alpha z^{-2}}\right)$$

For system to be stable, β can be of any value.

$$2 - az^{-2} > 0$$

$$2z^2 - \alpha = 0$$

$$\Rightarrow \quad z = \sqrt{\frac{\alpha}{2}} < 1 \Rightarrow |\alpha| < 2$$

For system to be stable all poles should be inside unity circle.

129. (b)

$$L\left(\frac{1-e^t}{t}\right) = \int_s^\infty\left(\frac{1}{s}-\frac{1}{s-1}\right)ds = \log\left(\frac{S-1}{S}\right)$$

130. (c)

$$x(S) = \frac{2}{S+3} - \frac{1}{S+5} - \frac{10}{(S+5)^2}$$

then, $\quad x(t) = [2e^{-3t} - e^{-5t} - 10t\,e^{-5t}]\,u(t)$

131. (b)

$$f(t) = L^{-1}\left[\log\cdot\frac{(S^2+1)}{S(S+1)}\right]$$

$$\Rightarrow \quad t\,f(t) = L^{-1}\left[-\frac{d}{ds}\log\left(\frac{S^2+1}{S(S+1)}\right)\right]$$

$$= -\left[L^{-1}\left\{\frac{d}{ds}\log(S^2+1)\right\} + L^{-1}\left\{\frac{d}{ds}\log S\right\}\right.$$

$$\left. + L^{-1}\left\{\frac{d}{ds}(\log(S+1))\right\}\right]$$

$$= -2\cos t + 1 + e^{-t}$$

$$\Rightarrow \quad f(t) = \frac{(1-2\cos t + e^{-t})}{t}$$

then, $f''(0) = 2$

132. (c)

$$u[n] \xleftrightarrow{\text{DTFT}} x(e^{j\Omega})$$

$$\delta[n] = u[n] - u[n-1]$$

$$\Rightarrow \quad 1 = (1-e^{-j\Omega})\,x(e^{j\omega})$$

Since, $(1 - e^{-j\Omega}) = 0$ for $\omega = 0$, therefore

$$x(e^{j\omega}) = A\delta\,(\Omega) + \frac{1}{1-e^{-j\Omega}},$$

where A is constant

$$u_o[n] = \frac{1}{2} + \frac{1}{2}\delta[n],\ u_0[n]$$

$$= u[n] - \left(\frac{1}{2} + \frac{1}{2}\delta\,[n]\right)$$

$$u_0[n] \xleftrightarrow{\text{DTFT}} A\delta(\Omega) + \frac{1}{1-e^{-j\Omega}} - \pi\delta(\Omega) - \frac{1}{2}$$

The Fourier transform of odd real sequences must be purely imagenary, thus $A = \pi$,

$$u\,[n] \xleftrightarrow{DTFT} \pi\delta(\Omega) + \frac{1}{1-e^{-j\Omega}};\ |\,\Omega\,| \le \pi$$

133. (b)

$$N = 4\ \Omega_0 = \frac{2\pi}{4} = \frac{\pi}{2}$$

$$x[k] = \frac{1}{4}\sum_{n=4}^{3} x[n]e^{-j(\pi/2)nk}$$

$$= \frac{1}{4}\times[0] = \frac{1}{4},\ \text{for all } k$$

134. (c)

$$y(\Omega) = H(\Omega)\cdot F(\Omega)$$

$$\Rightarrow \quad H(\Omega) = \frac{e^{j\Omega}}{e^{j\Omega}-0.5},\quad F(\Omega) = \frac{e^{j\Omega}}{e^{j\Omega}-0.8}$$

$$\Rightarrow \quad y(\Omega) = \frac{e^{2j\Omega}}{(e^{j\Omega}-0.5)\,(e^{j\Omega}-0.8)}$$

The inverse DTFT,

$$y(k) = \left[-\frac{5}{3}(0.5)^k + \frac{8}{3}(0.8)^k\right]u[k]$$

135. (d)

Since, $\quad x(0) = \frac{1}{2\pi}\int_{-\pi}^{\pi} x(e^{j\omega})\,d\omega$

So, $\quad \int_{-\pi}^{\pi} x(e^{j\omega})\,d\omega = 2\pi\,x(e^{j0}) = -6\pi$

136. (a)

Since, $x(n)$ is even with period 10

So, $\quad x(n) = A\cos\left(\frac{\pi}{5}n\right)$

Also, $\frac{1}{10}\sum_{n=0}^{9}|x(k)|^2 = 50$

which gives, $A = 10$

So, $x(n) = 10\cos\left(\frac{\pi}{5}n\right)$

137. (d)

Since, $x(e^{j\omega}) = -e^{j2\omega} + 2e^{j2\omega} - 3 + 2e^{-j\omega} - e^{-j2\omega}$

So, $\angle x(e^{j\omega}) = \pi$ for $u\omega$

138. (c)

Here, $h_2(n) = (-1)^k \cdot h_1(n)$

or, $h_2(n) = e^{j\pi k} h_1(n)$

So, $h_2(j\omega) = h_1[e^{j(\omega-\pi)}]$

140. (d)

$$H(e^{j\Omega}) = \frac{y(e^{j\Omega})}{x(e^{j\Omega})}$$

$$\left(\frac{2}{3}\right)^n u(n) \xleftrightarrow{\text{DTFT}} \frac{1}{1-\frac{2}{3}e^{-j\Omega}}$$

$$n\left(\frac{2}{3}\right)^n u[n] \xleftrightarrow{\text{DTFT}} j\frac{d}{d\Omega}\left(\frac{1}{1-\frac{2}{3}e^{-j\Omega}}\right)$$

$$= \frac{\frac{2}{3}e^{-j\Omega}}{1-\frac{2}{3}e^{-j\Omega}}$$

$$\Rightarrow \quad H(e^{j\Omega}) = \frac{2e^{-j\Omega}}{3-2e^{-j\Omega}}$$

141. (b)

$$x(e^{j\omega}) = e^{3j\omega} + e^{2j\omega} + 0 + 5 + e^{-j\omega}$$

$$\therefore \int_{-\pi}^{\pi} e^{aj\omega}\, d\omega = 0; \text{ if } a \neq 0$$

$$\int_{-\pi}^{\pi} x(e^{j\omega})\, d\omega = \left|\frac{e^{3j\omega}}{3j} + \frac{e^{2j\omega}}{2j} + 5\omega + \frac{e^{-j\omega}}{-j}\right|_{-\pi}^{\pi}$$

$$= 5\pi + 5\pi = 10\pi$$

144. (c)

By def. of z-transform,

$x(z) = \sum_{n=0}^{\infty} x[n]\, z^{-n}$; Here take $x[n] = u[n]$

145. (c)

$$\left(\frac{1}{2}\right)^n u[n] \leftrightarrow \left(\frac{z}{z-\frac{1}{2}}\right)^n \quad |z| > \frac{1}{2} \qquad \text{...(i)}$$

and $$\left(\frac{1}{3}\right)^n u[-n-1] \leftrightarrow \left(\frac{-z}{z-\frac{1}{3}}\right), \quad |z| < \frac{1}{3} \qquad \text{...(ii)}$$

We see that, the ROCs of above two equation (i) and (ii) do not overlap and that there is no common ROC and thus $x[n]$ will not have $x(z)$.

146. (c)

By final value theorem,

$$\underset{n\to\infty}{Lt}\, x[n] = \underset{z\to 1}{Lt}(1-z^{-1})\cdot x(z)$$

$$\Rightarrow \underset{z\to 1}{Lt}\, \frac{1}{4}(1-z^{-1})\frac{z^{-1}(1-z^{-4})}{(1-z^{-1})^2}$$

$$\Rightarrow \underset{z\to 1}{Lt}\, \frac{1}{4}\frac{(z^2-1)(z^2+1)}{z^4(z-1)} = \frac{1}{4}\times 4 = 1$$

154. (a)

$$z^3 - 0.81z = 0$$

then, $z = 0, \pm 0.9$, Here, roots are lie inside the unit circle.

So, system is stable.

155. (a)

z-transform is given by

$$z[f(nT)] = \sum_{n=-\infty}^{\infty} f(nT)z^{-n} = \sum_{n=-\infty}^{\infty}(a^T z^{-1})^n$$

$$\Rightarrow \quad F(z) = \frac{1}{1-a^T z^{-1}} = \frac{z}{z-a^T}$$

156. (c)

z-transform of of $\left(\frac{5}{6}\right)^n u(n) - \left(\frac{6}{5}\right)^n u(-n-1)$

$$= \frac{1}{1-\frac{5}{6}z^{-1}} + \left[1 - \frac{1}{1-\left(\frac{6}{5}\right)^{-1} z}\right]$$

then, overall ROC, is

$$\frac{5}{6} < |z| < \frac{6}{5}$$

158. (b)

$$x[n] = \delta[n] - \delta[n-2] + \delta[n-4] - \delta[n-5]$$

$$\Rightarrow \quad x(z) = 1 - z^{-2} + z^{-4} - z^{-5}, \ z \neq 0$$

159. (a)

For all pass system $|H(e^{j\omega})| = 1$ for all Ω

$$H(e^{j\Omega}) = \frac{b + e^{-j\Omega}}{1 - ae^{-j\Omega}} ; \ |b + e^{-j\Omega}| = 1 - ae^{-j\Omega}|$$

$$\Rightarrow 1 + b^2 + 2b\cos\Omega = 1 + a^2 - 2a\cos\Omega$$

This is possible only if, $\boxed{b = -a}$

163. (d)

$x[n]$ is a left sided sequence

$\therefore$ Carry out division as

$$\begin{array}{r|l} & z + 3z^2 + 7z^3 + 15z^4 + --- \\ 1 - 3z + 2z^2 & z \\ & z - 3z^2 - 2z^3 \\ & \quad 3z^2 - 2z^3 \\ & \quad 3z^2 - 2z^3 \\ & \quad 3z^2 - 9z^3 + 6z^4 \\ & \qquad 7z^3 - 6z^4 \\ & \qquad 7z^3 - 21z^4 + 16z^4 \\ & \qquad 15z^4 - -- \end{array}$$

$\therefore \ x(z) = --- + 15z^4 + 7z^3 + 3z^2 + z$

$\therefore \ x[n] = \{--- 15, 7, 3, 1, 0\}$

165. (d)

$$H(z) = 2z^{-3}$$
$$x(z) = z^4 + z^2 - 2z + 2 - 3z^{-4}$$
$$y(z) = 2z + 2z^{-1} + 4z^{-2} + 4z^{-3} - 6z^{-7}$$
$$y(n) = 2\delta(n+1) + 2\delta(n-1) - 4\delta(n-2) + 4\delta(n-3) - 6\delta(n-7)$$

$\therefore \quad n = 2, \ y(n) = -4$

166. (b)

$$G(z) = z^{n-1}x[z] = \frac{z^{n+1} + z^n}{[z-1]^2}$$

From residue theorem, we have

$$x[n] = R_z = 1$$

Since $x[z]$ is second order pole,

Corresponding residue is given by

$$R_{z=1} = \frac{d}{dz}\,[[z-1]^2\, G[z]]_{z=1}$$
$$= \frac{d}{dz}\left[(z-1)^2 \frac{(z^{n+1} + z^n)}{[z-1]^2}\right]$$
$$= \frac{d}{dz}[z^{n+1} + z^n]$$
$$= [(n+1)z^n + n]\big|_{z=1}$$
$$= [n+1] + n = 2n + 1$$
$$= x[n] = 2n + 1$$

169. (b)

By partial traction,

$$\frac{F(z)}{z} = \frac{-3}{z-1} - \frac{2}{(z-2)^3} - \frac{1}{(z-2)^2} + \frac{3}{(z-2)}$$

Taking inverse z-transform

$$f(n) = -\left[3 + \frac{1}{4}(n^2 + n - 12)\,2^n\right]u(n)$$

170. (c)

Since, $\quad x(n) * x(n) \xleftrightarrow{z.T} x^2(z)$

So, $\quad y(n) = x(n) * x(n)$

$$y(n) = 4^n u(n) * 4^n u(n)$$
$$= (n+1)\,4^n u(n)$$

171. (c)

$$x(1) = \underset{z\to\infty}{Lt}\,[z\{x(z) - x(0)\}]$$

$$x(0) = \underset{z\to\infty}{Lt}\,x(z) = \frac{1}{2}$$

So, $\quad x(1) = \underset{z\to\infty}{Lt}\left[z\left\{\left(\frac{2 + 6z^{-1}}{4 - 2z^{-2} + 13z^{-3}}\right) - \frac{1}{2}\right\}\right]$

$$= \frac{3}{2}$$

172. (b)

$$x(0) = \underset{z\to\infty}{Lt}\,x(z)$$

$$= \underset{z\to\infty}{Lt}\frac{1 - 2z^{-1}}{(1 - z^{-1})(1 - 0.5z^{-1})}$$

$$x(0) = \frac{1}{1} = 1$$

173. (b)

$$\cos\alpha = \sum_{k=0}^{\infty} \frac{(-1)^k}{(2k)!}\alpha^{2k}$$

$$x(z) = \sum_{k=0}^{\infty} \frac{(-1)^k}{(2k)!}(z^{-3k})^{2k}$$

$$\Rightarrow \quad x[n] = \sum_{k=0}^{\infty} \frac{(-1)^k}{(2k)!}\delta[n-6k]$$

$$n = 12 \Rightarrow 12 - 16 = 0, k = 2$$

$$x[12] = \frac{(-1)^2}{4!} = \frac{1}{24}$$

174. (c)

$$[2y(z) + x(z)]\, z^{-2} = y(z)$$

$$H(z) = \frac{z^{-2}}{1-2z^{-2}}$$

$$\Rightarrow \quad h[n] = \frac{-1}{2} + \frac{\frac{1}{4}}{1-\sqrt{2}z^{-1}} + \frac{\frac{1}{4}}{1+\sqrt{2}z^{-1}}$$

$$= \frac{-1}{2}\delta[n] + \frac{1}{4}\left\{(\sqrt{2})^n + (-\sqrt{2})^n\right\}u[n]$$

175. (b)

$$x(z) = \frac{0.5}{1.2z^{-1}};$$

$\because$ ROC includes unit circle

$$\Rightarrow \quad x(n) = -(0.5)\,(2)^{-n}u(-n-1)$$

$$x(0) = 0$$

176. (a)

Put, $n = 0, 1$

then $y[0] = -1$

$$y[1] - 2y[0] = -1$$

$$y[1] = 1$$

Also, put, $n = 2, 3, 4, \ldots, 100$

Always gives the constant value 1.

So, $y[100] = 1$

177. (d)

$$U(z) = \frac{1}{z^2}\cdot\frac{2+5z^{-1}+14z^2}{(1-z^{-1})^4}$$

By initial value theorem,

$$u_0 = \underset{z\to\infty}{Lt}\; U(z) = 0$$

$$u_1 = \underset{z\to\infty}{Lt}\; \{z[U(z) - u_0]\} = 0$$

$$u_2 = \underset{z\to\infty}{Lt}\; \{z^2[U(z) - u_0 - u_1 z^{-1}]\}$$

$$= 2 - 0 - 0 = 2$$

$$u_3 = \underset{z\to\infty}{Lt}\; z^3\{U(z) - u_0 - u_1 z^{-1} - u_2 z^{-2}\}$$

$$= 13$$

179. (a)

LTI system of $H(z)$ is

$$H(z) = \frac{1}{1+\frac{1}{4}z^{-1} - \frac{1}{8}z^{-2}}$$

The impulse response is

$$H(z) = \frac{2}{\left(1+\frac{1}{2}z^{-1}\right)} - \frac{1}{\left(1+\frac{1}{4}z^{-1}\right)}$$

$$\Rightarrow \quad h(n) = \left(\frac{1}{2}\right)^n u(t) - \left(\frac{1}{4}\right)^n u(t)$$

■■■■